William F. Maag Library
Youngstown State University

ANNUAL REVIEW OF PHYSICAL CHEMISTRY

EDITORIAL COMMITTEE (2001)

PAUL ALIVISATOS
LOUIS E. BRUS
JEAN-PIERRE HANSEN
FRANCES HOULE
STEPHEN R. LEONE
ANN E. MCDERMOTT
MARK A. RATNER

RESPONSIBLE FOR THE ORGANIZATION OF VOLUME 52 (EDITORIAL COMMITTEE, 1999)

GERALD T. BABCOCK
LOUIS E. BRUS
ROBERT W. FIELD
MICHAEL L. KLEIN
STEPHEN R. LEONE
MALCOLM F. NICOL
MARK A. RATNER
HERBERT L. STRAUSS
BRENT KOPLITZ (GUEST)

Production Editor: ANNE E. SHELDON
Quality Control Coordinator: MARY A. GLASS
Color Graphics Coordinator: EMÉ O. AKPABIO
Online Quality Control: NOËL J. THOMAS
Subject Indexer: BRUCE TRACY

ANNUAL REVIEW OF PHYSICAL CHEMISTRY

VOLUME 52, 2001

STEPHEN R. LEONE, *Editor*
University of Colorado at Boulder

PAUL ALIVISATOS, *Associate Editor*
University of California, Berkeley

ANN E. MCDERMOTT, *Associate Editor*
Columbia University

www.AnnualReviews.org   science@AnnualReviews.org   650-493-4400

ANNUAL REVIEWS
4139 El Camino Way • P.O. BOX 10139 • Palo Alto, California 94303-0139

ANNUAL REVIEWS
Palo Alto, California, USA

COPYRIGHT © 2001 BY ANNUAL REVIEWS, PALO ALTO, CALIFORNIA, USA. ALL RIGHTS RESERVED. The appearance of the code at the bottom of the first page of an article in this serial indicates the copyright owner's consent that copies of the article may be made for personal or internal use, or for the personal or internal use of specific clients. This consent is given on the condition that the copier pay the stated per-copy fee of $14.00 per article through the Copyright Clearance Center, Inc. (222 Rosewood Drive, Danvers, MA 01923) for copying beyond that permitted by Section 107 or 108 of the US Copyright Law. The per-copy fee of $14.00 per article also applies to the copying, under the stated conditions, of articles published in any *Annual Review* serial before January 1, 1978. Individual readers, and nonprofit libraries acting for them, are permitted to make a single copy of an article without charge for use in research or teaching. This consent does not extend to other kinds of copying, such as copying for general distribution, for advertising or promotional purposes, for creating new collective works, or for resale. For such uses, written permission is required. Write to Permissions Dept., Annual Reviews, 4139 El Camino Way, P.O. Box 10139, Palo Alto, CA 94303-0139 USA.

International Standard Serial Number: 0066-426X
International Standard Book Number: 0-8243-1052-7
Library of Congress Catalog Card Number: A-51-1658

Annual Review and publication titles are registered trademarks of Annual Reviews.
⊗ The paper used in this publication meets the minimum requirements of American National Standards for Information Sciences—Permanence of Paper for Printed Library Materials, ANSI Z39.48-1992.

Annual Reviews and the Editors of its publications assume no responsibility for the statements expressed by the contributors to this *Annual Review*.

Typeset by TechBooks, Fairfax, VA
Printed and Bound in the United States of America

A MEMORIAL TO GERALD T. BABCOCK

This 52nd volume of the *Annual Review of Physical Chemistry* is dedicated to the memory of Gerald T. Babcock. It is now 15 years since Jerry became an Associate Editor of the *Annual Review of Physical Chemistry*. In these years, all of the participants in the *Review* came to love Jerry; his energy, his enthusiasm, his giving unstintingly of himself. He tended to burst into meetings at the last minute, but was remarkably prepared. He brought wisdom to the array of lists of topics and authors, together with excitement. His knowledge was broad-ranging and his excitement about each and every one of the proposed topics was infectious. He was about to become Co-Editor of the *Annual Review of Physical Chemistry*, a task he looked forward to with energy and optimism even while struggling with his increasingly serious medical problems.

Jerry Babcock was born on February 9, 1946. He was a scholarship student at Creighton University during his undergraduate years, and his Ph.D. was obtained at the University of California at Berkeley under the direction of Ken Sauer in 1973. At Berkeley, Herb Strauss instructed Jerry as a student in thermodynamics, and the following year Jerry taught Steve Leone, then himself a new student at Berkeley, while Jerry was a teaching assistant in graduate quantum chemistry. Those days were special for all of us! Who would have thought that our paths would cross again while crafting the *Annual Review of Physical Chemistry* for more than a decade.

Following an NIH postdoctoral fellowship with Graham Palmer at Rice University, Jerry joined the faculty in the Department of Chemistry at Michigan State University in 1976. He was a tremendous leader at Michigan State, beloved by his colleagues in the Department, and eventually became the Department Chair, a University Distinguished Professor, and won the Charles F. Kettering Award of the American Society of Plant Physiologists in 2000. In his scientific career he had over 250 publication, nearly 300 invited talks, 30 students who completed their Ph.D. degrees, and more than 30 postdoctoral associates. His speciality in the application of EPR spectroscopy to biophysical problems, oxygen chemistry in biology, and photosynthesis led to an amazing number of seminal scientific contributions. He kept up his tremendously busy schedule of lecturing, organizing symposia, panel participation, and various editorial board member duties, even in light of his illness. To Jerry, life was to be lived to its fullest, and his enthusiasm, energy, and brilliance was unstoppable, until the very end, on December 22, 2000.

He fulfilled a crucial role at the *Annual Review of Physical Chemistry*, helping to sort out the wide range of topics in biophysical chemistry. Jerry was remarkable in his breadth, so widely read and perceptive that his comments about the importance of nearly any topic or potential author were gemstones of insight.

During our years on the *Annual Review of Physical Chemistry* team, we heard about the rest of Jerry's life in glimpses: his excellent research, his leadership of the Michigan State Department, and his family. We especially appreciated his sons who he brought to our dinners and who are growing to be handsome and energetic in Jerry's image. It is hard to imagine someone so vital not being with us any longer. He will indeed be missed.

Herbert Strauss
University of California, Berkeley
Editor, *Annual Review of Physical Chemistry* 1985–2000

Stephen Leone
Associate Editor, *Annual Review of Physical Chemistry* 1990–2000,
Editor 2001–present

PREFACE

This volume of the *Annual Review of Physical Chemistry* demonstrates the marked breadth of the field of physical chemistry in the twenty-first century, covering such diverse areas as highly detailed gas-phase spectroscopic and collision studies, protein folding and dynamics, coherence effects, and self-assembly. The selection of reviews provides in-depth forays into the exciting scientific areas of interfacial chemistry, resonance phenomena in collision dynamics, single molecule studies, and electron dynamics in quantum dots. The volume also clearly exhibits technological themes such as the relentless quest for higher-resolution microscopy, spectroscopy, and collision studies, the development of innovative ultrafast laser techniques, and the use of coincidence and coherence preparation and detection. The advancement of new techniques continues to revolutionize the field. Powerful theoretical research also plays a prominent role in this volume, providing ways to systematize and understand complex topics and to challenge experimental efforts. Larger and more complex systems and living systems are an inescapable trend in physical chemistry, attested to in this volume.

The volume begins with a Prefatory chapter by Alan Carrington, covering his distinguished career in gas-phase molecular spectroscopy. State-resolved chemistry in neutral gas-phase molecules is covered in detail by the review of Valentini on polyatomic molecule reactions. Wodtke et al discuss the use of stimulated emission pumping to probe unimolecular and collision processes of highly stretched, vibrationally excited molecules. Liu considers dynamical studies in crossed beams, including new insights into reactive resonances. Armentrout considers the reactivity and thermochemistry of size-selected ion clusters by the technique of ion-guided beam mass spectroscopy. Taatjes & Hershberger review the broad field of chemical kinetics related to combustion processes, as studied by infrared laser absorption/gain probing. Continetti reviews ion photofragmentation coincidence methods and the detailed structural information obtained by energy and angular correlations. Neumark reviews elegant ultrafast, time-resolved photoelectron spectroscopy of neutrals and ions. Shore et al detail the results of laser population transfer by adiabatic passage techniques to produce well-defined high-lying states in atoms and molecules. Oudejans & Miller consider the highly nonstatistical vibrational predissociation dynamics of weakly bound cluster species, also with final-state correlations.

Tycko reviews solid state NMR and its application to peptides, protein fibrils, and membrane-bound peptides and proteins. Frydman also reviews the advances in NMR that have made possible the study of polymers, inorganic materials, and biological molecules, considering new challenges available by spin- and quandrupole-nuclei spectroscopies. Fayer reviews the subject of infrared vibrational echoes applied to the dephasing of the stretching mode of CO bound to the active site in proteins, and Dantus considers coherent nonlinear spectroscopies by transient grating and photon echo methods on small molecules in the condensed phase. Nozik reviews the rapidly advancing field of semiconductor nanocrystals and hot electron and electron cooling dynamics with quantum size effects. Nitzan considers the theoretical advances of electron transport through molecules and interfaces, especially related to scanning tunneling microscopy effects and electron transport through molecular bridges. Richmond reviews the use of surface vibrational sum frequency spectroscopy to study the dynamics of water at interfaces. Self-assembly is considered in two reviews, one by Schwartz on the kinetics and mechanisms and one by Golumbfskie & Chakraborty on the control of nanomaterials fabrication and biomimetic assembly. Electroluminescence of organic thin film materials and its optimization is the subject of the review by Gross et al.

Protein and RNA folding is the subject of two reviews, one by Brooks & Shea on the simulation and theory of protein folding and one by Thirumalai et al on the early events in RNA folding. Hansma explores the study of DNA mapping and sizing by atomic force microscopy. The theory of electrophoretic mobility of oligomeric double-stranded DNA is reviewed by McLaughlin & Mohanty. Biological applications of EPR spectroscopy are reviewed by Prisner et al., who explore the local structure and dynamics of paramagnetic biological samples. The study of single biomolecules by fluorescence resonance energy transfer is the novel subject of the review by Deniz et al.

Volume 52 also marks significant changes in personnel. Herb Strauss is stepping down as Editor of the *Annual Review of Physical Chemistry*, after 15 years of outstanding work as Editor and 9 years of contributions before that as an Associate Editor. His wisdom and counsel is irreplaceable and will be greatly missed. Stephen Leone will become the Editor. The untimely loss of Jerry Babcock cut short his anticipated role as co-Editor. Paul Alivisatos and Ann McDermott will be newly appointed Associate Editors. We are delighted to have Anne Sheldon and her remarkable expertise as the Production Editor in Palo Alto.

The Editorial Committee

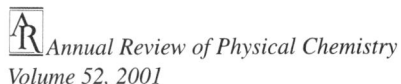
Annual Review of Physical Chemistry
Volume 52, 2001

CONTENTS

Frontispiece—Alan Carrington	xvi
A FREE RADICAL, *Alan Carrington*	1
STATE-TO-STATE CHEMICAL REACTION DYNAMICS IN POLYATOMIC SYSTEMS: CASE STUDIES, *James J. Valentini*	15
RECENT PROGRESS IN INFRARED ABSORPTION TECHNIQUES FOR ELEMENTARY GAS-PHASE REACTION KINETICS, *Craig A. Taatjes and John F. Hershberger*	41
SURFACE BIOLOGY OF DNA BY ATOMIC FORCE MICROSCOPY, *Helen G. Hansma*	71
ON THE CHARACTERISTICS OF MIGRATION OF OLIGOMERIC DNA IN POLYACRYLAMIDE GELS AND IN FREE SOLUTION, *Udayan Mohanty and Larry McLaughlin*	93
MECHANISMS AND KINETICS OF SELF-ASSEMBLED MONOLAYER FORMATION, *Daniel K. Schwartz*	107
CROSSED-BEAM STUDIES OF NEUTRAL REACTIONS: STATE-SPECIFIC DIFFERENTIAL CROSS SECTIONS, *Kopin Liu*	139
COINCIDENCE SPECTROSCOPY, *Robert E. Continetti*	165
SPECTROSCOPY AND HOT ELECTRON RELAXATION DYNAMICS IN SEMICONDUCTOR QUANTUM WELLS AND QUANTUM DOTS, *Arthur J. Nozik*	193
RATIOMETRIC SINGLE-MOLECULE STUDIES OF FREELY DIFFUSING BIOMOLECULES, *Ashok A. Deniz, Ted A. Laurence, Maxime Dahan, Daniel S. Chemla, Peter G. Schultz, and Shimon Weiss*	233
TIME-RESOLVED PHOTOELECTRON SPECTROSCOPY OF MOLECULES AND CLUSTERS, *Daniel M. Neumark*	255
PULSED EPR SPECTROSCOPY: BIOLOGICAL APPLICATIONS, *Thomas Prisner, Martin Rohrer, and Fraser MacMillan*	279
FAST PROTEIN DYNAMICS PROBED WITH INFRARED VIBRATIONAL ECHO EXPERIMENTS, *Michael D. Fayer*	315
STRUCTURE AND BONDING OF MOLECULES AT AQUEOUS SURFACES, *G. L. Richmond*	357

LIGHT-EMITTING ELECTROCHEMICAL PROCESSES, *Neal R. Armstrong, R. Mark Wightman, and Erin M. Gross*	391
REACTIONS AND THERMOCHEMISTRY OF SMALL TRANSITION METAL CLUSTER IONS, *P. B. Armentrout*	423
SPIN-1/2 AND BEYOND: A PERSPECTIVE IN SOLID STATE NMR SPECTROSCOPY, *Lucio Frydman*	463
FROM FOLDING THEORIES TO FOLDING PROTEINS: A REVIEW AND ASSESSMENT OF SIMULATION STUDIES OF PROTEIN FOLDING AND UNFOLDING, *Joan-Emma Shea and Charles L. Brooks III*	499
POLYMER ADSORPTION-DRIVEN SELF-ASSEMBLY OF NANOSTRUCTURES, *Arup K. Chakraborty and Aaron J. Golumbfskie*	537
BIOMOLECULAR SOLID STATE NMR: ADVANCES IN STRUCTURAL METHODOLOGY AND APPLICATIONS TO PEPTIDE AND PROTEIN FIBRILS, *Robert Tycko*	575
PHOTOFRAGMENT TRANSLATIONAL SPECTROSCOPY OF WEAKLY BOUND COMPLEXES: PROBING THE INTERFRAGMENT CORRELATED FINAL STATE DISTRIBUTIONS, *L. Oudejans and R. E. Miller*	607
COHERENT NONLINEAR SPECTROSCOPY: FROM FEMTOSECOND DYNAMICS TO CONTROL, *Marcos Dantus*	639
ELECTRON TRANSMISSION THROUGH MOLECULES AND MOLECULAR INTERFACES, *Abraham Nitzan*	681
EARLY EVENTS IN RNA FOLDING, *D. Thirumalai, Namkyung Lee, Sarah A. Woodson, and D. K. Klimov*	751
LASER-INDUCED POPULATION TRANSFER BY ADIABATIC PASSAGE TECHNIQUES, *Nikolay V. Vitanov, Thomas Halfmann, Bruce W. Shore, and Klaas Bergmann*	763
THE DYNAMICS OF "STRETCHED MOLECULES": EXPERIMENTAL STUDIES OF HIGHLY VIBRATIONALLY EXCITED MOLECULES WITH STIMULATED EMISSION PUMPING, *Michelle Silva, Rienk Jongma, Robert W. Field, and Alec M. Wodtke*	811

INDEXES
Author Index	853
Subject Index	901
Cumulative Index of Contributing Authors, Volumes 48–52	931
Cumulative Index of Chapter Titles, Volumes 48–52	933

ERRATA

An online log of corrections to *Annual Review of Physical Chemistry* chapters (1997 to the present) may be found at http://physchem.AnnualReviews.org/errata.shtml.

Related Articles

From the *Annual Review of Astronomy and Astrophysics*, Volume 39 (2001)

Optical Interferometry, Andreas Quirrenbach

The Development of High Resolution Imaging in Radio Astronomy, Kenneth Irwin Kellermann and James M. Moran

Dusty Circumstellar Disks, Ben Zuckerman

The Cosmic Infrared Background: Measurements and Implications, Michael G. Hauser and Eli Dwek

From the *Annual Review of Biomedical Engineering*, Volume 2 (2000)

Physicochemical Foundations and Structural Design of Hydrogels in Medicine and Biology, N. A. Peppas, Y. Huang, M. Torres-Lugo, J. H. Ward, and J. Zhang

Injury and Repair of Ligaments and Tendons, Savio L-Y. Woo, Richard E. Debski, Jennifer Zeminski, Steven D. Abramowitch, Serena S. Chan Saw, and James A. Fenwick

Cell Mechanics: Mechanical Response, Cell Adhesion, and Molecular Deformation, Cheng Zhu, Gang Bao, and Ning Wang

Microengineering of Cellular Interactions, Albert Folch and Mehmet Toner

Two-Photon Excitation Fluorescence Microscopy, Peter T. C. So, Chen Y. Dong, Barry R. Masters, and Keith M. Berland

In Vivo Near-Infrared Spectroscopy, Peter Rolfe

From the *Annual Review of Biophysics and Biomolecular Structure*, Volume 30 (2001)

Hydrogen Bonding, Base Stacking, and Steric Effects in DNA Replication, Eric T. Kool

Structures and Proton-Pumping Strategies of Mitochondrial Respiratory Enzymes, Brian E. Schultz and Sunney I. Chan

Mass Spectrometry as a Tool for Protein Crystallography, Steven L. Cohen and Brian T. Chait

Probing The Relation Between Force—Lifetime—And Chemistry In Single Molecular Bonds, Evan Evans

NMR Probes of Molecular Dynamics: Overview and Comparison with Other Techniques, Arthur G. Palmer III

Ab Initio Protein Structure Prediction: Progress and Prospects, Richard Bonneau and David Baker

Biomolecular Simulations: Recent Developments in Force Fields, Simulations of Enzyme Catalysis, Protein-Ligand, Protein-Protein, and Protein-Nucleic Acid Noncovalent Interactions, Wei Wang, Oreola Donini, Carolina M. Reyes, and Peter A. Kollman

Chaperonin-Mediated Protein Folding, D. Thirumalai and George H. Lorimer

Protein Folding Theory: From Lattice to All-Atoms Models, Leonid Mirny and Eugene Shakhnovich

From the **Annual Review of Earth and Planetary Sciences**, Volume 29 (2001)

Human Impacts on Atmospheric Chemistry, P. J. Crutzen and J. Lelieveld

Rheological Properties of Water Ice—Applications to Satellites of the Outer Planets, W. B. Durham and L. A. Stern

The Carbon Budget in Soils, Ronald Amundson

From the **Annual Review of Materials Research**, Volume 31 (2001)

Synthesis and Design of Superhard Materials, J. Haines, J. M. Leger, and G. Bocquillon

Photoinitiated Polymerization of Biomaterials, John P. Fisher, David Dean, Paul S. Engel, and Antonios G. Mikos

Functional Biomaterials: Design of Novel Biomaterials, S. E. Sakiyama-Elbert and J. A. Hubbell

Design and Synthesis of Energetic Materials, Laurence E. Fried, M. Riad Manaa, Philip F. Pagoria, and Randall L. Simpson

Block Copolymer Thin Films: Physics and Applications, Michael J. Fasolka and Anne M. Mayes

Synthetic Cells—Self-Assembling Polymer Membranes and Bioadhesive Colloids, Daniel A. Hammer and Dennis E. Discher

ANNUAL REVIEWS is a nonprofit scientific publisher established to promote the advancement of the sciences. Beginning in 1932 with the *Annual Review of Biochemistry*, the Company has pursued as its principal function the publication of high-quality, reasonably priced *Annual Review* volumes. The volumes are organized by Editors and Editorial Committees who invite qualified authors to contribute critical articles reviewing significant developments within each major discipline. The Editor-in-Chief invites those interested in serving as future Editorial Committee members to communicate directly with him. Annual Reviews is administered by a Board of Directors, whose members serve without compensation.

2001 Board of Directors, Annual Reviews
Richard N. Zare, Chairman of Annual Reviews
 Marguerite Blake Wilbur, Professor of Chemistry, Stanford University
John I. Brauman, *J. G. Jackson–C. J. Wood Professor of Chemistry, Stanford University*
Peter F. Carpenter, *Founder, Mission and Values Institute*
W. Maxwell Cowan, *Vice President and Chief Scientific Officer, Howard Hughes*
 Medical Institute, Bethesda
Sandra M. Faber, *Professor of Astronomy and Astronomer at Lick Observatory,*
 University of California at Santa Cruz
Eugene Garfield, *Publisher*, The Scientist
Samuel Gubins, *President and Editor-in-Chief, Annual Reviews*
Daniel E. Koshland, Jr., *Professor of Biochemistry, University of California at Berkeley*
Joshua Lederberg, *University Professor, The Rockefeller University*
Gardner Lindzey, *Director Emeritus, Center for Advanced Study in the Behavioral*
 Sciences, Stanford University
Sharon R. Long, *Professor of Biological Sciences, Stanford University*
J. Boyce Nute, *President and CEO, Mayfield Publishing Co.*
Michael Peskin, *Professor of Theoretical Physics, Stanford Linear Accelerator Ctr.*
Harriet A. Zuckerman, *Vice President, The Andrew W. Mellon Foundation*

Management of Annual Reviews
Samuel Gubins, President and Editor-in-Chief
Richard L. Burke, Director for Production
Paul J. Calvi, Jr., Director of Information Technology
Steven J. Castro, Chief Financial Officer
John W. Harpster, Director of Sales and Marketing

Annual Reviews of
Anthropology
Astronomy and Astrophysics
Biochemistry
Biomedical Engineering
Biophysics and Biomolecular
 Structure
Cell and Developmental
 Biology
Earth and Planetary Sciences
Ecology and Systematics
Energy and the Environment
Entomology
Fluid Mechanics
Genetics
Genomics and Human Genetics
Immunology
Materials Research
Medicine
Microbiology
Neuroscience
Nuclear and Particle Science
Nutrition
Pharmacology and Toxicology
Physical Chemistry
Physiology
Phytopathology
Plant Physiology and Plant
 Molecular Biology
Political Science
Psychology
Public Health
Sociology

SPECIAL PUBLICATIONS
Excitement and Fascination of
 Science, Vols. 1, 2, 3, and 4

A FREE RADICAL

Alan Carrington
Department of Chemistry, University of Southampton, Hampshire SO17 1BJ, England;
e-mail: ac@soton.ac.uk

Key Words spectroscopy, resonance, ions, structure, microwave

■ **Abstract**   This chapter describes my research career, spanning the period from 1955 to 2000. My initial PhD work at the University of Southampton was concerned with the electronic structure and spectra of transition metal complexes and included studies of the electronic spin resonance (ESR) spectra of magnetically dilute single crystals. After a year at the University of Minnesota, I went to Cambridge University and for the next six years studied the ESR spectra of liquid phase organic free radicals. I commenced work on the microwave magnetic resonance (MMR) spectra of gaseous free radicals in 1965, and this work continued until 1975. I moved from Cambridge to Southampton in 1967. In 1975 I turned to the study of gas phase molecular ions, using ion beam methods. In the earlier years of this period I concentrated on simple fundamental species like H_2^+, HD^+, and H_3^+. In the later years until my retirement in 1999, I concentrated on the observation and analysis of microwave spectra involving energy levels lying very close to a dissociation asymptote.

DEDICATION

This chapter is dedicated to the memory of Harry E. Radford, who died while it was being written. Harry was a quiet and shy man, who often worked alone and never indulged in self-promotion. So far as I know, he was never awarded any medals or prizes, nor elected to any academies or learned societies. Nevertheless he was an experimentalist of the highest originality and quality, a theorist of true intellectual depth, and a remarkable pioneer in many of the techniques of studying free radicals that are now commonplace.

SOUTHAMPTON 1955–1959

My research career commenced in 1955 after the completion of my first degree at Southampton University. I am not sure why I went to Southampton in the first place, or why I chose to read for an Honors degree in chemistry. I think that chemistry was the only thing left after I had dismally failed my examinations in Latin and Greek, despite the enthusiasm of a young and persuasive classics teacher. My chemistry teacher predicted only a moderately successful career for me. Deciding on my research field and who was to be my research supervisor was the first problem, but not a difficult one because Martyn Symons, then a young lecturer in the Southampton chemistry department, was carrying out pioneering work on free radicals that interested me. He had an active research collaboration with DJE Ingram, a leading physicist in the relatively new field of electron spin resonance (ESR), and the opportunity to be involved in this subject was also appealing. So I chose him as my supervisor but did not, in fact, work on free radicals. Instead I chose to study the structure and spectra of transition metal oxyanions, particularly the permanganate and manganate ions (1). There must be many people who have been fascinated by the deep purple color of potassium permanganate and wondered how it originates. The permanganate ion, MnO_4^-, is readily reduced in alkaline solution to the less well-known manganate ion, MnO_4^{2-}, which is a deep emerald green. Part of my PhD project was to develop a theoretical understanding of these colors and thereby to construct a theory of their electronic structure. I also managed to grow single crystals of potassium chromate, doped with a low concentration of potassium manganate, and to study the paramagnetic resonance spectrum of these crystals (2). This was challenging work and also somewhat frightening when I look back. The manganate ion has one unpaired electron, and because of fast spin-lattice relaxation, the crystals had to be cooled to very low temperatures before a magnetic resonance spectrum could be observed. Every two months a Dewar flask of liquid hydrogen was delivered to Ingram's laboratory. Its primary purpose was to facilitate Ingram's beautiful work on hemoglobin crystals, but at the end of each liquid hydrogen week, I had a couple of days to study my magnetically dilute manganate crystals. By this time the liquid hydrogen had acquired a layer of liquid air, and given the numerous high-voltage sparks that were part of life in the laboratory, it is surprising that my embryonic research career was not terminated prematurely by a spectacular explosion. However, I survived and mastered the art of measuring and analyzing a highly anisotropic single-crystal spectrum. I carried out similar work on the hypomanganate ion, MnO_4^{3-}, diluted in single crystals of potassium vanadate, and also on the ferrate ion, FeO_4^{2-}, again diluted in single crystals of potassium chromate (2). Both of these ions have triplet ground states. The ultimate theoretical analysis owed much to David Schonland, a lecturer in the mathematics department, and my first feeble steps in understanding quantum mechanics owed much to him. Again with his help I developed a theory of the electronic structure and was able to show, at least to my own satisfaction, that the purple color of potassium permanganate is due to an intense charge-transfer transition in which an electron moves from a nonbonding t_1 orbital delocalized over the four oxygen atoms into degenerate $3d$-type orbitals of t_2 symmetry localized on the manganese ion.

MINNESOTA

The middle of my second year as a research student brought a defining moment for me. Professor John E. Wertz of the University of Minnesota was planning to take a sabbatical year at Oxford, and asked David Ingram if he had anyone finishing their PhD who would take a postdoctoral fellowship at Minnesota and oversee the Wertz laboratory. I was the nearest approximation to this requirement, so after two years as a research student, and therefore a year short of completion, I was appointed a postdoctoral fellow at the University of Minnesota for 1957–1958. At the tender age of 23 I left England for the first time, embarked on my first-ever flight (17 hours in a TWA Super-Constellation which only made it as far as Boston), and finally arrived in the American midwest. One month later I was giving my first-ever lecture, at a US Air Force conference in Washington, DC, reporting on the Wertz laboratory work over the past year, with which I had been familiar for just four weeks. I experienced for the first time the warmth and generosity of American scientists and will forever be grateful to Sam Weissman and George Fraenkel particularly, for their understanding and support at that time. I also attended the second International Free Radicals Conference, a biennial series which has continued to the present time, and with which I have been closely associated in the past.

The John Wertz laboratory contained what was probably the world's best nuclear magnetic resonance (NMR) and ESR spectrometers, home designed and home built, and I was let loose with these machines to follow my nose. I chose to work on the ESR spectra of aromatic ions in the liquid phase and, thanks mainly to the magnificent instrument now at my disposal, was able to improve on the best resolution available at the time (3). I got hooked on proton hyperfine structure in organic free radicals, an addiction that has lasted to this day. Another coincidence, however, also shaped my future. Andrew McLachlan was completing his PhD with Christopher Longuet-Higgins at Cambridge, working on the theory of the spectra, including ESR, of alternant aromatic hydrocarbon ions. Andrew went to spend three months with Martyn Symons in Southampton, while I was at Minnesota, and so we became aware of each other's existence and of our mutual scientific interests. Andrew had some nice new spectra to contemplate, and I became increasingly aware of my relative incompetence in matters theoretical.

Toward the end of my year in Minnesota I was offered an assistant professorship there but eventually declined for four reasons. First and most importantly, I had proposed marriage to Hilary Taylor, a student of English at Southampton, but had not been immediately accepted or rejected. (We have subsequently been married for almost 41 years.) Second, I had my PhD thesis to complete, although that was not a major problem. Third, I had come to realize, particularly through my correspondence with Andrew McLachlan, that my education was by no means complete. I had to learn and understand more theory. Fourth, although I am very fond of the United States, have many friends there, and have subsequently paid many visits lasting anywhere from one week to five months, my English roots are deep, and I always knew that England was where I wanted to live.

RETURN TO SOUTHAMPTON

So I returned to Southampton University for the final year needed to complete my PhD degree. During my three years there I published papers on electronic spectroscopy (4), solid state magnetic resonance, liquid phase reaction kinetics (5), and even electrochemistry. I owe a great deal to Martyn Symons, who taught me much, broadened my education, and above all, communicated his infectious enthusiasm for science. I wrote somewhat nervously to Christopher Longuet-Higgins, enquiring whether I could possibly spend a year in the theoretical chemistry department at Cambridge, and to my surprise he produced money for a one-year postdoctoral fellowship, obtained from General Electric, USA.

CAMBRIDGE 1959–1967

I moved to Cambridge in August 1959, got married in November (with no money and no security), and commenced what turned out to be an eight-year stay, during which I was eventually appointed to a university position, and elected to a fellowship at Downing College. I was a member of a small but remarkable department whose members included Christopher Longuet-Higgins, Frank Boys, John Griffith, Leslie Orgel and Andrew McLachlan, as well as many distinguished overseas sabbatical visitors. Christopher Longuet-Higgins had already decided that some experimental work would not be out of place in his department, and together we obtained a grant to purchase a Varian ESR spectrometer. While waiting for both the money and the equipment to arrive, I continued to work on transition metal complexes, developing a theory to account for the weak temperature-independent paramagnetism of the permanganate ion (6) and also writing a *Quarterly Review* article on the ESR spectra of transition metal ions (7). This was another defining event. I got stuck trying to work out the theory of the anisotropic g-tensor for a d^1 ion (Ti^{3+}) in a tetrahedral complex with a tetragonal distortion, and I went to Christopher for help. The subject was new to him, but in one afternoon, spent alone with me, he worked out the theory from first principles. During the course of this memorable afternoon I learned about operators, Dirac notation, perturbation theory, and effective Hamiltonians. All sorts of pennies dropped for me, and although I do not have the mental equipment to be an original theorist, I did begin to understand what theory was about and, more importantly, became able to analyze and interpret my own spectra!

Once my Varian spectrometer arrived, I commenced work on aromatic radicals and radical ions; it was a continuation of the work I had started at Minnesota. In this work I had some wonderful research students, the first of whom was Jorge dos Santos Veiga, who later became Vice-Rector of Coimbra University. Equally important were Jim Bolton, Peter Todd, and Ian (ICM) Smith, who is now Director of the NRC Molecular Biophysics laboratory in Winnipeg. Looking back at that period I recognize that Sam Weissman was the trailblazer in the field, and many of my spectra, particularly those of substituted benzene anions, were simply resolution-enhanced versions of his. Most of the important theory was due to Harden McConnell. We did, however, make some novel discoveries, one of which was hyperfine line-width alternation in the durosemiquinone cation (8). I am rather pleased with the fact that I was eventually able to prove that the effect was due to intramolecular motion, a dynamic isomerization between *cis* and *trans* forms (9); more distinguished members of the department came up with a variety of ingenious but incorrect explanations. This was the first example of such an effect in ESR spectra, but many subsequent examples have since been observed and valuable kinetic information derived. An interesting additional comment on this work was that George Fraenkel, paying a visit to Cambridge, was particularly puzzled because he had looked at the same free radical and had not observed this novel line-width effect. The reason was that the isomerization rate for the radical in concentrated sulphuric acid was found to be strongly temperature dependent. Although both studies were conducted in solution at room temperature, room temperature at Columbia in New York was about 30°C, whereas in my Cambridge laboratory it was 16°C on a warm day! Another novel result was the first observation of free radical partial alignment in a nematic liquid crystal (10). This work, carried out with Geoffrey Luckhurst, was Geoffrey's idea, stimulated by his work as an undergraduate student at Hull University where George Gray was carrying out pioneering work on the preparation of liquid crystal materials, for which he much later received the Kyoto Prize. We extended the application of liquid phase ESR to seven- (11) and eight-membered (12) ring systems, and also to much larger organic free radicals (13), including heterocyclic systems (14). I also indulged in some theory, in an effort to understand line-width variations in proton hyperfine patterns (15). I scarcely understand my own papers in this field when I reread them now; did I really master the density matrix theory of relaxation?

MID-1960s AND ESR STUDIES

In 1964 or thereabouts I began to feel that most of the fundamental phenomena in liquid phase ESR had been discovered, and that the future of the technique was in the hands of organic chemists and biochemists. For several years I had been attracted to the idea of using ESR methods to study the spectra of gas phase free radicals. Pioneering work on NO and OH had been described by Harry Radford, but the subject had then rather stalled. The arrival of Don Levy from Berkeley, who joined the department as a postdoctoral fellow, and Terry Miller from Kansas, who joined us as a graduate student, provided the impetus for gas phase studies. Initially we used our ESR spectrometer without any modifications; we simply inserted a quartz flow tube through the resonance cavity and set up a microwave discharge a few centimeters upstream of the cavity. Our choice for the first system was either inspired or lucky; we chose to look at mixtures of chlorine and oxygen and were immediately rewarded with a beautiful spectrum of the ClO radical (16) so strong that we could display it on an oscilloscope. This was followed by the spectra of other halogen monoxides (17). As the work developed we introduced new techniques, particularly a microwave cavity suitable for studying Stark effects (18). I was also fortunate to have the opportunity of spending a few months with Jim Hyde at Varian Associates in Palo Alto, California, during which time we developed a new microwave cavity that had particularly large entrance and exit holes. I was allowed to bring the prototype back to Cambridge, the only time I have ever smuggled anything through customs! This new cavity greatly enhanced our gas phase work, but for others it proved to be valuable for studying much larger solid state samples and was also the basis of Varian's later ENDOR cavity. My one and only venture into the industrial market was scientifically and commercially successful; but alas, no one else has ever offered me an industrial consultancy.

This period of my scientific life, with Don and Terry, was exciting and productive. Together with other graduate students we developed both the techniques and the theory (19), although in matters theoretical I was, and always will be, behind Don and Terry. We also had some memorable laughs. I remember well arriving at the daily coffee meeting, feeling somewhat low because my attempts that morning at quartz blowing had been less than admirable. (Don Levy has publicly described my glass blowing as a random-walk process which eventually, and by chance, arrives at the correct destination.) Christopher Longuet-Higgins noted my low spirits, remarked that he had once been quite skilled at glass blowing, and offered to give me a quick lesson. This offer I accepted eagerly, and word went like a brush fire around the University Chemical Laboratory that the Professor of Theoretical Chemistry was about to give a glass blowing demonstration. By the time Christopher arrived in the laboratory, there was an audience of over two hundred graduate students, postdocs, and staff. What followed was a classic. The molten glass sagged, and the blowing resulted in a series of huge glass balloons which exploded spectacularly. As a grand finale, Christopher picked up a piece of red hot glass and rapidly dropped it again. All of this was greeted with thunderous cheers from the audience. Christopher finally left the stage, to tumultuous applause, with the remark that one could get out of practice.

I have noted that inside most theoreticians, there is an experimentalist trying to get out. As the only experimentalist in the department, I had to get used to eminent theoreticians who would lean over my shoulder when I was sitting at the ESR instrument and even adjust the controls for me. Eventually we mounted a special theoretician's panel in the instrument. This consisted of a switch, a light, a potentiometer, and a linked galvanometer. This panel represented the limits of my ability in electronics, but was accepted with good humor by famous colleagues; Leslie Orgel particularly enjoyed it. Don Levy eventually returned to a position at the University of Chicago, and a new opportunity also arose for me. My former department at Southampton had established a new Chair, of Theoretical Chemistry, and they were interested in me. I attended the formal interview with no clear idea as to whether I wanted the job. A short conversation with Kenneth Mather, then the Vice-Chancellor, convinced me I did want it, and when I was offered the Chair, after the interview, I accepted on the spot. I am a poor politician, but I did insist, successfully, that the title of the Chair should not include the word "Theoretical." So I left Cambridge in the summer of 1967, as did Christopher Longuet-Higgins. Andrew McLachlan also left the department at the same time, going to the famous Laboratory of Molecular Biology. I loved Cambridge, and still do. My three children, Sarah, Rebecca and Simon were all born there, and my wife Hilary established a huge reputation for her singing. Above all, however, Christopher Longuet-Higgins was and still is the finest scientist I have ever met. I have tried to live up to the standards he established for all of us in his department.

BACK TO SOUTHAMPTON

Going back to Southampton, I was able to take my equipment with me, and with the help of Terry Miller and Brian Howard, particularly, got going quite quickly. My new laboratory had served as a general university assembly hall in its earlier life, and my largest electromagnet was now situated on the same piece of flooring that had once supported a piano; it was here, over the piano, that I had first met my future wife. John Brown joined my group as a postdoctoral fellow two years later; he ultimately remained in the department for 15 years. We were able to extend our studies of the microwave spectra of free radicals to polyatomic species, obtaining the first such spectrum of a linear triatomic molecule, NCO. Brian Howard developed the theory of the Zeeman effect in a molecule showing Renner-Teller coupling (20). Shortly after that we observed the first spectrum of a nonlinear triatomic, HCO, and John Brown was primarily responsible for developing the theory of the fine and hyperfine structure in such a system (21, 22). He guided my always faltering steps in the use of irreducible spherical tensor methods. It was an exciting time, for my own group, for the chemistry department, and for me personally. In 1971 I was elected a Fellow of the Royal Society, at the tender age of 37. It had been my great good fortune to be independent, with my own research group, at the age of 25. This would be impossible today; most people now appointed to university posts are over 30, and obtaining research grants for equipment takes a lot of effort, and a lot of time.

I took my turn as chairman of the department, but found it difficult to combine this with active participation in my research program. I have never been much good at multitasking. Moreover it became clear that subsequent appointments as a Dean, and then probably as a deputy Vice-Chancellor, would be hard to resist indefinitely. This was not the direction in which I wanted to go, but if I was to avoid it, I had either to move elsewhere or obtain non-university support for my salary. Word circulated that I was open to proposals to move, and I received several offers from US universities; without exception, they were all for chairmanships of departments, which was not what I wanted. It was depressing that so many people saw me as a potential administrator. Finally the Science Research Council (SRC) (as it was then) came to my rescue. George Porter and Ron Mason had, for some time, wished to institute the post of senior research fellow (SRC), to give a few individuals the opportunity of five years devoted solely to research. I became the first recipient of such a fellowship, in 1976, which solved my immediate problem. The problem would have resurfaced in 1981, but I was elected to a Royal Society research professorship in 1979, a position I held for the subsequent 20 years until retirement in 1999. These professorships are the most desirable positions in a university anywhere in the world; the recipients are actually required to devote their time solely to scientific research. Incidentally, the period from 1967 to 1976 was the only time I had the security of a tenured position; the rest of my life has been spent in temporary jobs.

THE EARLY SEVENTIES

Microwave Magnetic Resonance Methods

My research was going well in the early seventies, but I was again becoming conscious that it was time for a change. My microwave magnetic resonance methods owed their success to the small sample volume to be filled with short-lived free radicals, but they were being overtaken, first by pure microwave spectroscopy in the hands particularly of Claude Woods at Wisconsin and Eizi Hirota at Okazaki in Japan. Field-free studies are desirable if one wishes for ultimate accuracy in the determination of molecular constants. At the same time, far-infrared laser magnetic resonance was being developed, principally by Ken Evenson at Boulder. Although the principles underlying this technique are essentially the same as I had used with microwave methods, the much higher resonance frequencies led to huge gains in sensitivity, with which I could not compete effectively. For example, Evenson and his colleagues obtained beautiful spectra of short-lived radicals like CH, a species for which I had searched many times without success.

Gas Phase Molecular Ions

Whereas the high-resolution spectroscopy of neutral free radicals was advancing rapidly, gas phase molecular ions continued to pose problems, despite the pioneering work of Claude Woods. I had long been fascinated by the beautiful quadrupole trap/radiofrequency studies of H_2^+, described some ten years earlier by Dehmelt and Jefferts. They recorded radiofrequency hyperfine transitions, and obtained what were for twenty years the only resolved spectra of this fundamental molecule. I decided to attempt to develop their methods toward the study of slightly more complex molecules. I obtained a research grant and built a radiofrequency quadrupole trap. Peter Sarre joined me as a research student, and I proudly presented him with my quadrupole trap. He shared my enthusiasm for molecular ions, but not my quadrupole trap, and suggested that it might be better to exploit ion beam methods, which were highly developed and commercially available. With some initial reluctance I agreed with his suggestion, and the vacuum system that had been designed and built to house a quadrupole trap was converted into a rather strange and unlikely-looking tandem mass spectrometer system. My quadrupole trap has never been inside a vacuum chamber, and I will always wonder if my ambitions for it would have been realized. It was not the only piece of apparatus I built that was never used. I once had a crazy idea of simulating a radiotelescope/universe combination in the laboratory, and to this end built a beautiful parabolic aluminium reflector to collect microwave radiation from a flame or a molecular discharge. This reflector has also never been inside a vacuum system. For many years it has sat in John Brown's living room as a unique objet d'art and occasional ashtray.

At about the time Peter Sarre arrived, Bill Wing and his colleagues described their beautiful ion beam studies of the vibration-rotation spectrum of HD^+, the first such spectrum for any gas phase molecular ion. They used a carbon monoxide laser to drive the transitions and an indirect method of detecting population transfer. Ion beams were clearly going to be productive and in the following years we had many successes. Apart from Peter Sarre, I was fortunate to be joined by Juliet Buttenshaw, Richard Kennedy and Tim Softley, who all brought their special talents to enhance the program. We obtained both electronic (23) and vibrational spectra, the most remarkable example of the latter being an enormously rich spectrum of the H_3^+ ion (24, 25). This spectrum continues to challenge theoreticians, although I believe the most important aspects are qualitatively understood. An important feature of ion beam experiments is that the ions are often produced and maintained with extensive vibrational excitation, up to and beyond the dissociation limit. We have made use of this fact, and much of our later work has been devoted to studies of the energy levels lying close to a dissociation asymptote. With Richard Kennedy and Juliet Buttenshaw, I obtained some fascinating vibration-rotation spectra of the HD^+ ion in levels near to dissociation (26), and I was able to show that the electron migrates toward the deuterium atom as the molecule approaches dissociation (27). I never anticipated this result, but the reasons became obvious once the observations had been made. How often that is true; it is but one good reason for doing experiments, even on a system one might suppose to be well understood.

OXFORD

In the early eighties I was beginning to feel restless again, and after talking to friends in the Physical Chemistry Laboratory at Oxford, I requested the transfer of my Royal Society Professorship to Oxford, which was granted. I moved in 1984 and was warmly received and supported by all, especially John Rowlinson, the department head. I enjoyed both the department and my College, Jesus College, which elected me as a Professorial Fellow. I found that I was much better at the after-dinner High Table talk than I had been during my younger days at Cambridge. Also during this Oxford period I was elected a Foreign Honorary Member of the American Association of Arts and Sciences. But my career timing could have been better. My wife Hilary, after bringing up our three children, became a Chartered Librarian and obtained a good position in Fareham Public Library. There are many highly qualified women chasing rather few jobs in Oxford, and it gradually became clear that Hilary's budding career would probably be terminated by our move. We had failed to sell our house in Hampshire, and after much agonizing, I decided that the right thing to do was to return to Southampton, who were happy to have me back. The whole process was catalyzed by my meeting the new Southampton

Vice-Chancellor, Gordon Higginson, at a Jesus College feast. For the second time in my life I was putty in the hands of a Vice-Chancellor. The Royal Society, no doubt with a collective sigh, agreed to my return, which occurred in 1987. I have always regretted the disappointment I caused my many friends at Oxford, but I think they understood and respected my reasons. The subsequent progress of my wife's career has more than justified the difficult decision I made.

During my three years at Oxford I had given much thought to the possibility of obtaining electronic spectra of the H_2^+ ion, the simplest of all molecules but highly elusive in the laboratory. Together with Iain McNab and Christine Leach, I designed a new ion beam machine, which was built at Oxford. Our efforts to obtain spectra largely failed but, miraculously, after my return to Southampton, spectra appeared and have kept coming ever since. We were using a two-photon infrared method, which depended on photodissociation as a detection probe, and we continued to study the HD^+ ion. However I was visited by a young Danish physicist, Niels Bjerre, who suggested that I might investigate the possibilities of using electric field dissociation to state-select the very weakly bound near-dissociation levels. This proved to be one of the best suggestions ever made to me, and I have been using the technique ever since (28). We were able to detect and study the first resolved electronic spectra of H_2^+(29), D_2^+(30), and He_2^+(31), involving transitions between the ground state and the long-range van der Waals state in these molecules. Apart from using electric field dissociation as the detection probe, we had also replaced our infrared lasers by microwave sources operating over the range 4 to 170 GHz. There were several reasons for this, one of the most important being that frequency scanning of the resonances using a carbon dioxide laser could be achieved only by doppler tuning the ion beam. In contrast the microwave sources were fully frequency tunable. A second reason was that the doppler line-width in the microwave region was essentially negligible, whereas it was large in our infrared experiments, despite kinematic compression effects.

In succeeding years the microwave experiments were extended to other molecular ions, like He...Ar^+ (32), He...Kr^+ (33), Ne_2^+(34), and HeH_2^+(35), where we could study levels bound by as little as a few MHz. Microwave/microwave double resonance experiments, together with observations of Zeeman splittings, were crucial in the analysis and assignment of the spectra. Several outstanding research students were involved in this work, and Andrew Shaw, as a postdoctoral fellow, made many important contributions. The spectra have raised new problems quite different from those now well understood for molecules at or near their equilibrium configurations (36). (Our spectroscopy overlaps with scattering studies in a most direct manner.) The fact that many of our spectra are now understood with great quantitative accuracy owes much to Jeremy Hutson (Durham). I believe our techniques could be applied to many other molecular systems, and I should have started the work 20 years earlier. I must have done something right, however, and my election as a foreign associate of the US National Academy of Sciences in 1994 means a great deal to me.

RETIREMENT

Retirement, whatever that means apart from a huge reduction in income, occurs at age 65 in the United Kingdom, and it means what it says. I had been considering my options after retirement for at least two years before the event, and at first thought I would attempt to continue my life in research. That is not so easy, however; one requires financial resources to support students and to purchase equipment. It is not easy to obtain the necessary resources, even in mid-career, and it gradually dawned on me that the discontinuity in life brought about artificially by "retirement" should be regarded as an opportunity to develop in other directions. For many years I had wanted to write a book on the spectroscopy of diatomic molecules; indeed John Brown and I had started such a book, dealing with the fundamentals of the theory, some 25 years ago. We had, however, been distracted by research. I decided that retirement provided the opportunity to return to the book. It was less convenient for John, but nevertheless he shared my enthusiasm. We have now written almost a thousand pages, and the first draft is almost finished. The scope of the book has contracted; it is now confined to the rotational spectroscopy of diatomic molecules, but that is still a huge and intellectually demanding subject. I am enjoying the challenge. I also hope to become a better pianist now that I have no excuses for not practicing.

FREE RADICALS

Throughout this chapter I have avoided defining the term "free radical"; indeed I have been avoiding it all my scientific life because there is, to my mind, no satisfactory all-inclusive definition. Some, who adopt the structure-based definition, would say that a free radical possesses one or more unpaired electrons. I think it does not, for example, make scientific sense to include the CH_3 molecule in a discussion of free radicals, but not CH_3^+. So I suppose I am adopting a kinetic definition; my free radicals are transient species, with open or closed electronic shells, that do not come in bottles or cylinders. That definition does not, therefore, include oxygen or NO, but these are very useful prototypes. It would not make sense to discuss the spectroscopic properties of OH, which is a free radical by anyone's definition, without also referring to NO. Similarly the structural and spectroscopic properties of SO and NF, for example, should always be discussed with reference to oxygen. There is, however, no doubt that it is the open shell properties of most free radicals that make them important molecules to study. The links between spectroscopy and electronic structure are never more significant than in open shell molecules because the presence of electron and nuclear spin interactions displays details that uniquely illuminate our understanding of molecules in general. That is one good reason why the spectroscopy of free radicals has captured my interest for a lifetime. The other is that they can be so devilishly difficult to work with; they stretch your imagination, your technical skills, and your patience. Rather like fishing in fact, which is another of my loves.

I have been immensely fortunate in my life. The most important thing in my professional life, from which all else flows, is the science itself. It can provide intense excitement and depressing frustration, in equal measures. It brings one into close contact with young people, undergraduates and postgraduates, at the most imaginative and uninhibited stages of their lives. And it provides an international circle of true friends. Science is international, and I have no patience with nationalistic sentiments of the type politicians, and some scientists who should know better, like to propound. None of us ever starts from square one, and we are always, in part, dependent on the achievements of others, wherever they live and work.

Visit the Annual Reviews home page at www.AnnualReviews.org

LITERATURE CITED

1. Carrington A, Symons MCR. 1956. *J. Chem. Soc.* 1956:3373–80
2. Carrington A, Ingram DJE, Lott KAK, Schonland DS, Symons MCR. 1960. *Proc. R. Soc.* A254:101–10
3. Carrington A, Dravnieks F, Symons MCR. 1959. *J. Chem. Soc.* 1959:947–52
4. Carrington A, Schonland DS. 1960. *Mol. Phys.* 3:331–38
5. Carrington A, Symons MCR. 1960. *J. Chem. Soc.* 1960:284–89
6. Carrington A. 1960. *Mol. Phys.* 3:271–75
7. Carrington A, Longuet-Higgins HC. 1960. *Q. Rev.* 14:427–52
8. Bolton JR, Carrington A. 1962. *Mol. Phys.* 5:161–67
9. Carrington A. 1962. *Mol. Phys.* 5:425–31
10. Carrington A, Luckhurst GR. 1964. *Mol. Phys.* 8:401–2
11. Carrington A, Smith ICP. 1963. *Mol. Phys.* 7:99–100
12. Carrington A, Todd PF. 1964. *Mol. Phys.* 7:533–40
13. Carrington A, Longuet-Higgins HC, Todd PF. 1964. *Mol. Phys.* 8:45–48
14. Carrington A, Longuet-Higgins HC, Todd PF. 1965. *Mol. Phys.* 9:211–15
15. Carrington A, Longuet-Higgins HC. 1962. *Mol. Phys.* 5:447–54
16. Carrington A, Levy DH. 1966. *J. Chem. Phys.* 44:1298–99
17. Carrington A, Dyer PN, Levy DH. 1970. *J. Chem. Phys.* 52:309–14
18. Carrington A, Levy DH, Miller TA. 1967. *Rev. Sci. Instr.* 38:1183–84
19. Carrington A, Levy DH, Miller TA. 1970. *Adv. Chem. Phys.* 18:149–248
20. Carrington A, Fabris AR, Howard BJ, Lucas NJD. 1971. *Mol. Phys.* 20:961–80
21. Bowater IC, Brown JM, Carrington A. 1973. *Proc. R. Soc.* A333:265–88
22. Bolman PHS, Brown JM, Carrington A, Lycett GJ. 1973. *Proc. R. Soc.* A335:113–26
23. Carrington A, Milverton DRJ, Sarre PJ. 1978. *Mol. Phys.* 35:1505–21
24. Carrington A, Kennedy RA. 1984. *J. Chem. Phys.* 81:91–112
25. Carrington A, McNab IR, West YD. 1993. *J. Chem. Phys.* 98:1073–92
26. Carrington A, Buttenshaw JA. 1981. *Mol. Phys.* 44:267–85
27. Carrington A, Kennedy RA. 1985. *Mol. Phys.* 56:935–75
28. Carrington A, McNab IR, Montgomerie CA. 1988. *Chem. Phys. Lett.* 151:258–62
29. Carrington A, McNab IR. 1989. *Chem. Phys. Lett.* 160:237–42

30. Carrington A, McNab IR, Montgomerie CA, Kennedy RA. 1989. *Mol. Phys.* 67:711–38
31. Carrington A, Knowles PJ, Pyne CH. 1995. *J. Chem. Phys.* 102:5979–88
32. Carrington A, Hutson JM, Law MM, Leach CA, Marr AJ, et al. 1995. *J. Chem. Phys.* 102:2379–403
33. Carrington A, Hutson JM, Law MM, Pyne CH, Shaw AM, Taylor SM. 1996. *J. Chem. Phys.* 105:8602–14
34. Carrington A, Gammie DI, Page JC, Shaw AM, Taylor SM. 1999. *Phys. Chem. Chem. Phys.* 1:29–36
35. Carrington A, Gammie DI, Shaw AM, Taylor SM, Hutson JM, 1996. *Chem. Phys. Lett.* 260:395–405
36. Carrington A, Shaw AM, Taylor SM. 1995. *J. Chem. Soc. Faraday Trans.* 91:3725–40

STATE-TO-STATE CHEMICAL REACTION DYNAMICS IN POLYATOMIC SYSTEMS: Case Studies

James J Valentini
Department of Chemistry, Columbia University, New York, New York 10027; e-mail: jjv1@chem.columbia.edu

Key Words chemical kinetics, hot atoms, hydrogen abstraction, collision complex, reaction cross section, energy disposal

■ **Abstract** This review illustrates the experimental study of chemical reaction dynamics using methods that select the quantum states and energy of the reactants and determine the quantum states and energy of the products. The focus is reaction dynamics in systems in which at least one of the reactants or products is a polyatomic molecule. The approach taken is to select four prototype reaction systems as case studies to demonstrate the detail of information and insight that can come from such experiments. Thus, the review is selective and neither claims nor attempts to be comprehensive. Reference to and discussion of theoretical reaction dynamics are included where computational results directly connect with the experiments.

INTRODUCTION

State-to-state chemical reaction dynamics aims to provide mechanistic understanding of reactions by indirect probing of the transition state. What is probed are the connections made there between the quantum states of the reactants and the quantum states of the products. These connections reveal how in the transition state a chemical bond is broken and a new one is formed. The understanding aimed for is in effect a molecular motion picture of the dynamics, a picture of the motion of all the individual atoms as they move in response to the forces generated by the potential energy surface that describes the interactions among the atoms of the system.

The domain in which state-to-state dynamics developed was that of reactions of atoms with diatoms, $A + BC \rightarrow AB + C$. Such reactions are the minimum basis set of the process of chemical transformation—the single bond present in the reactants breaks and the single bond in the products forms. Many of the fundamental concepts of chemical reaction dynamics have been developed or tested in such systems. Only three variables are needed to determine the energy of interaction of the A, B, and C atoms; most commonly these are chosen as the distance between B and C, r_{BC}, the distance between A and B, r_{AB}, and the ABC angle. For this low-dimensionality reaction, extensive experimental measurements, both state-to-state and others, combined with theoretical treatments of the highest level of sophistication, have led to a nearly complete picture of the dynamics of many prototype reactions. From this specific knowledge, understanding of atom + diatom reactions in general has followed.

Reactions in which the reactants have many bonds, or even just two or three, require a motion picture of higher dimensionality and greater complexity. The relatively simple $H + CH_4 \rightarrow H_2 + CH_3$ reaction requires specification of 12 atom-atom distances and angles to describe the motion of its constituent atoms. These polyatomic systems pose fundamental questions that do not appear in atom + diatom reactions. Is the reaction localized at a particular reaction site? Are some bonds in the reactant polyatomic molecule spectators to the reaction? Can the reaction be accelerated by excitation of particular bonds in the reactants? Can the reaction be driven to select the reaction site by such excitation? When there are multiple, identical reaction sites in the polyatom, what is the competition among these sites for reaction? How are the multiple quantum numbers of the products correlated? How does the orientation of the reactants affect reaction? It is these and related questions that the experiments described here aim to answer.

$H + H_2O \rightarrow H_2 + OH$

The $H + H_2O \rightarrow H_2 + OH$ reaction is endoergic, $\Delta H° = 0.64$ eV, and has a reaction barrier height of 0.93 eV (1–5). The reaction has been studied in both the forward and reverse directions. However, the available state-to-state data are primarily for the forward reaction, and that is the only way in which this reaction is discussed here. As a consequence of its high barrier, the dynamics of this reaction can only be studied at high energy. It was one of the first reactions studied (6) using pulsed laser photolytic generation of translationally fast (hot) H atoms (7) to access collision energies above the reaction threshold.

Determination of the absolute total reaction cross section is not a state-to-state measurement, but the results are important to understanding the state-to-state dynamics, particularly the product energy disposal and the dependence of the total cross section on the vibrational state of the reactant H_2O. Wolfrum, Kleinermanns, and co-workers (6, 8–11) have made many measurements of the absolute total reaction cross section for both $H + H_2O \rightarrow H_2 + OH$ and $D + D_2O \rightarrow D_2 + OD$ over a wide range of collision energies. The relative cross sections that were directly measured were made absolute by calibration via comparison with OH yields from photodissociation. The cross sections for $H + H_2O \rightarrow H_2 + OH$ rise from 0.03 Å^2 at 1.0 eV to 0.24 Å^2 at 2.5 eV, whereas the $D + D_2O \rightarrow D_2 + OD$ cross sections are only about 40% of these throughout this range of collision energies.

The fact that the cross section for both isotopomers remains small even at energies far above the energetic barrier can be made comprehensible by a simple model. This is a substantially endoergic reaction that should have, and does have, a late barrier (1–5), one that is more like the products than the reactants. It is well understood that such reactions will not be driven effectively by reactant translational energy (12–14). Exact quantum calculations (15), reduced dimensionality quantum calculations (16–18), and QCT calculations (19, 20) carried out on the Walch-Dunning-Schatz-Elgersma potential (1, 2), or the more recent Isaacson 5 potential (4), give reasonably good agreement with the experimental results, with the exact agreement between theory and experiment depending on the potential energy surface used and the approximations made.

Many measurements of the product state distributions have been made for this reaction. All but one of the measurements have focused on characterization of the OH product. This is so because laser spectroscopic detection of the OH product is much easier than detection of the H_2 product. At a collision energy of 2.5 eV, Kleinermanns & Wolfrum (6) found only v = 0 OH product with a rotational distribution, peaking at N = 2, that showed only 3% of the 1.9 eV available energy in that degree of freedom. This is a result repeatedly obtained by many experimental studies. At collision energies of 1.8 and 1.5 eV, the rotational distribution peaked at N = 3, and only 5% (at 1.8 eV) or 8% (at 1.5 eV) of the available energy was in rotation, and at most 10% of the OH product was in v = 1 (8). Another measurement by Honda et al (21) at 1.5 eV showed 5%–8% of the energy in rotation. At collision energies of 1.0 and 2.2 eV, Jacobs et al (10) found 12% and 5% of the energy in OH rotation. For the H + D_2O → OD + HD reaction the same behavior was observed: No OD (v = 1) product could be detected. The fraction of the energy in rotation was only 2% at a collision energy of 2.5 eV, 4% at 2.2 eV, 5% at 1.8 eV, and 7% at 1.5 eV (9, 11).

These results show that the relative motion of the H and H_2O reactants is not effectively coupled to the OH product rotational and vibrational degrees of freedom. This can be most clearly seen from examination of the actual energy, not the fraction of that available, in rotation. For all these measurements, the energy in rotation is almost constant at 0.10 eV, independent of collision energy. QCT calculations on the H + H_2O/D_2O reactions give results that are in quantitative agreement with this (19, 22). The QCT results also show even more starkly the propensity against deposition of available energy in OH vibration. At a relative energy of 2.5 eV only 0.2% of the available energy appears as OH vibration. Together, these experimental and theoretical results present fairly convincing evidence that the OH bond that is preserved in the reaction behaves as a spectator to the reaction.

All of these measurements were for reaction from the vibrationless ground state of the H_2O/D_2O reactant. To be a spectator means not only that any energy not put in the non-reacting OH mode of water does not appear in the OH product but also that energy put in the nonreacting OH bond of the water stays in that bond and thus appears as OH product motion. This adiabaticity has to be tested by preparing vibrationally excited states of the H_2O reactant. Sinha et al and Hsiao et al (23, 24) have done just that. They measured relative rate constants for reaction of thermal H atoms with vibrationally state-selected reagents. Through vibrational overtone pumping they prepared excited local modes of H_2O. From a reactant state having one quantum in one OH stretch, and three quanta in the other, they observed OH product predominantly in v = 1. From reaction with four quanta of vibration in one of the OH local modes in H_2O, or the OH local mode in HOD, they observed predominantly v = 0 OH/OD product, with only a very small average energy in vibration or rotation of the OH/OD product. (As discussed below, the reaction selectively occurs at the OH that is most highly excited vibrationally.)

These results showing OH adiabaticity can be interpreted as due to two effects. First, the equilibrium OH bond length in water is, at 0.958 Å (25), very close to the bond length in the OH radical itself, 0.970 Å (26). So, from a Franck-Condon-like view, vibrational excitation of the "old" OH bond in the reactive collision seems unlikely. This is, of course, not sufficient to make the OH bond a spectator to the reaction. Energy transfer to the nonreacting OH bond in the entrance channel, and at the transition state, would result in rotational or vibrational excitation of the OH product. Similarly, energy exchange as the products separate could do the same. However, at the high energies of the hot atom experiments, the energy disposal in the reaction could reasonably be expected to be impulsive, with rotational and vibrational excitation of the products being due to repulsion between the H and O in the reactive bond. Because this impulse is delivered very close to the center of mass of the departing OH, excitation of OH rotation or vibration will be ineffective.

The OH radical is a $^2\Pi$ open-shell species that has both rotational and electronic orbital angular momentum, and spin-orbit-interaction-split states. Thus it is characterized by four separate rotational manifolds, and the $H + H_2O$ reaction may populate these manifolds differentially. Since the electronic spectroscopy of the OH shows separate transitions for each rotational quantum number N in each of the four spin-orbit/Λ-doublet states, the distribution of the OH product over these manifolds can be readily measured. And that indeed has been done. All the measurements (6, 10, 11, 21) of these distributions agree. The F_1 and F_2 spin-orbit states are produced in a statistical ratio. The Λ-doublet states on the other hand are very non-statistical. The A'/A'' ratio rises from 1 at N = 1 to 3–6 at the highest N observed. A rotational state distribution that extends to higher N is correlated with this preference. Several possible interpretations of the preference for the A' state have been offered and are summarized by Jacobs et al (10). These include a planar transition state with both the product repulsion and the OH unpaired orbital in that plane, symmetry conservation of the electronic wavefunction at the transition state, interference between the two adiabatic potential energy surfaces, and inelastic scattering in the exit channel.

The single measurement of the quantum state distribution of the molecular hydrogen product of the reaction is that of Adelman et al (27) for HD from the $H + D_2O$ reaction. In contrast to the very small energy disposal in the old OD bond, that in the new HD bond is substantial, about 35% of the available energy on average, and rovibrational states close to the energetic limit are populated. The yields of v = 1 and v = 0 were about the same and some v = 2 was observed, again in contrast to the energy disposal previously characterized for OH/OD products of the H + H_2O/D_2O reaction. Putting these HD measurements together with those on the OD product (9, 11) reveals that most ($\approx$60%) of the available energy ends up in product translation. This is a feature of most H + HA → H_2 + A reactions, where HA is H_2, HD, a hydrogen halide, or an alkane.

Using a polarized Doppler-resolved method (described in the next section) for detection of the OH product of the H + H_2O reaction, Brouard et al (28) concluded that the fraction of the energy in product translation was a bit higher than this, approximately 87%. This method allows extraction of the product angular distribution in favorable circumstances, which are present for this reaction. The results show that the reaction yields OH product scattered broadly over all center-of-mass angles, at collision energies from 1.0 eV to 2.2 eV. These angular distributions are accurately reproduced by QCT calculations (20).

A major issue in reactions of polyatomic molecules is the possibility of mode-selective or bond-selective reaction. Can energy be put in a particular degree of freedom of the reactant to select the bond at which reaction will occur? For the H + H_2O → OH + H_2 and H + HOD → OD + H_2/OH + HD reactions, that answer is an emphatic yes. In a series of elegant experiments, Crim and co-workers (23, 24, 29–31) showed that vibrational overtone excitation of the H_2O or HOD reactant could be used to select the bond that will react with thermal H atoms. A schematic of the energetics of such an experiment, and an example of the results, is depicted in Figure 1. For the HOD reaction, excitation of the $4\nu_{OH}$ mode resulted in a selectivity for reaction at the OH compared with reaction at the OD by a factor of more than 100, and excitation of $5\nu_{OD}$ was similarly effective for selecting reaction at OD rather than OH. For the H_2O reaction, excitation of nearly isoenergetic vibrational modes showed a clear mode-specificity. Depositing three quanta in one OH stretch local mode was far more effective in promoting reaction than two quanta in one OH stretch and one in the other, or two quanta in an OH stretch and two quanta in the bend.

The mode selectivity in this reaction is not restricted to highly excited vibrational states. The Zare group (32, 33) studied the H + HOD → OD + H_2/OH + HD reaction and found that excitation of either the OH or OD stretch to the first excited vibrational state leads to quite selective breaking of the excited bond. Here the reaction had to be studied with hot H atom reactant generated by photolysis of HI, since the vibrational excitation energy is smaller than the reaction endoergicity. For the H + D_2O → HD + OD reaction, the Zare group prepared vibrational states with one quanta in the asymmetric stretch or one quanta in this stretch plus one quanta in the bend. They found that excitation of the bend did not enhance the reaction rate.

There have been quite a few theoretical calculations of this reaction with vibrationally excited water. Many have been reviewed by Bowman & Schatz (34). More recent papers also summarize the theoretical treatments of the reaction (35, 36). For the most part there is at least qualitative agreement between theory and experiment.

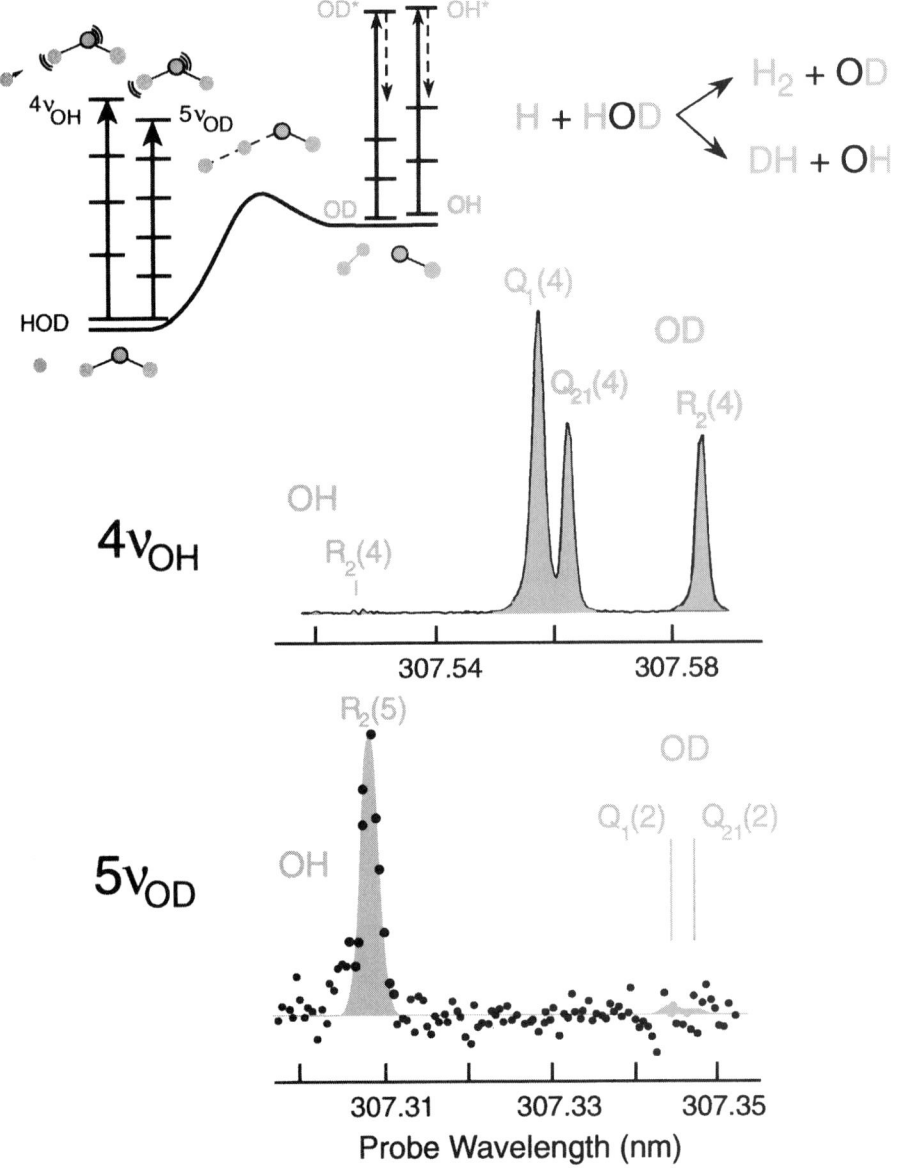

Figure 1 Schematic diagram describing bond selective chemistry in the H + HOD → H_2/HD + OD/OH reaction. The energy level diagram (*top left*) shows the potential energy along the reaction coordinate and, superimposed on it, the vibrational energy levels of the HOD reactant and OH and OD products. The spectra (*bottom*) are laser-induced fluorescence scans over OH and OD transitions. Excitation of $4\nu_{OH}$ yields almost exclusively OD product, excitation of $5\nu_{OD}$ almost exclusively OH. Figure provided by FF Crim and used by permission.

$H + CO_2 \rightarrow OH + CO$

The $H + H_2O \rightarrow H_2 + OH$ reaction is the four-atom prototype of a reaction with a substantial activation energy that proceeds by direct abstraction. The $H + CO_2 \rightarrow OH + CO$ reaction in turn is the four-atom prototype of a reaction with a substantial well that proceeds by complex formation. The energetics of the reaction, showing the energies of various stationary points, are given in Figure 2. The reaction does have a deep well, or rather wells, but is a substantially endoergic reaction, $\Delta H° = 1.11$ eV, so like the $H + H_2O \rightarrow H_2 + OH$ reaction it requires high reactant energies and has been studied using hot H atom reactant.

The $H + CO_2$ reaction, like $H + H_2O$, yields two diatomic products, both of which are accessible to quantum-state-resolved spectroscopic detection. From such measurements the complete distribution of the available energy can be determined, allowing an unusually detailed characterization of the energy disposal in this prototype reaction. The Wolfrum group (37–40) and the Wittig group (41–46) carried out a series of experiments in which they determined the OH rotational state distributions at collision energies from 1.4 to 2.6 eV. They found distributions that populated a wide range of N extending up to near the energetic limit. Surprisal analysis (47) showed that these distributions were characterized by rotational surprisal parameters of 1.7 at the lowest collision energy to 2.8–3.6 at the highest. Overall, 15% (highest energies) to 20% (lowest energies) of the available energy appeared in OH rotation.

At collision energies between 1.9 and 2.6 eV, the Wolfrum group reported no preference for one or the other of the Λ doublet states and only a slight propensity (30% greater yield) for the lower energy spin-orbit state (38–40). The Wittig group reported just the opposite, a small preference (20% greater yield) for the A' Λ-doublet state, and no preference for either of the spin-orbit states (41–43). The reason for these small differences is not obvious. Jacobs et al (40) found about 10% of the available energy in OH vibration. Doppler profiles in the OH LIF spectra showed 60% of the energy in product translation, implying that 15% of the energy ended up in CO rotation and/or vibration.

The energy disposal in the CO fragment was directly measured by Rice & Baronavski (48, 49) and Nickolaisen et al (50). The two measurements do not agree completely. At a collision energy of 2.4 eV, the former showed vibrational distributions that depended on the CO_2 temperature and had $P(v = 0)/P(v = 1) = 1.0$ ($T_{CO_2} = 300$ K) and 0.6, i.e. inverted, ($T_{CO_2} = 70$ K). The latter experimental report gave $P(v = 0)/P(v = 1) = 0.4$ at 300 K. The rotational distributions were similarly not in complete agreement, with the Rice & Baronavski measurement giving more energy in rotation than found in the Nickolaisen experiment. The average energy in rotation/vibration of the CO was 20% in the former but only 15% in the latter. The source of the difference in the results of these two experiments is not clear. Rice & Baronavski also found CO rotational state distributions that became colder as the temperature of the CO_2 reactant was reduced.

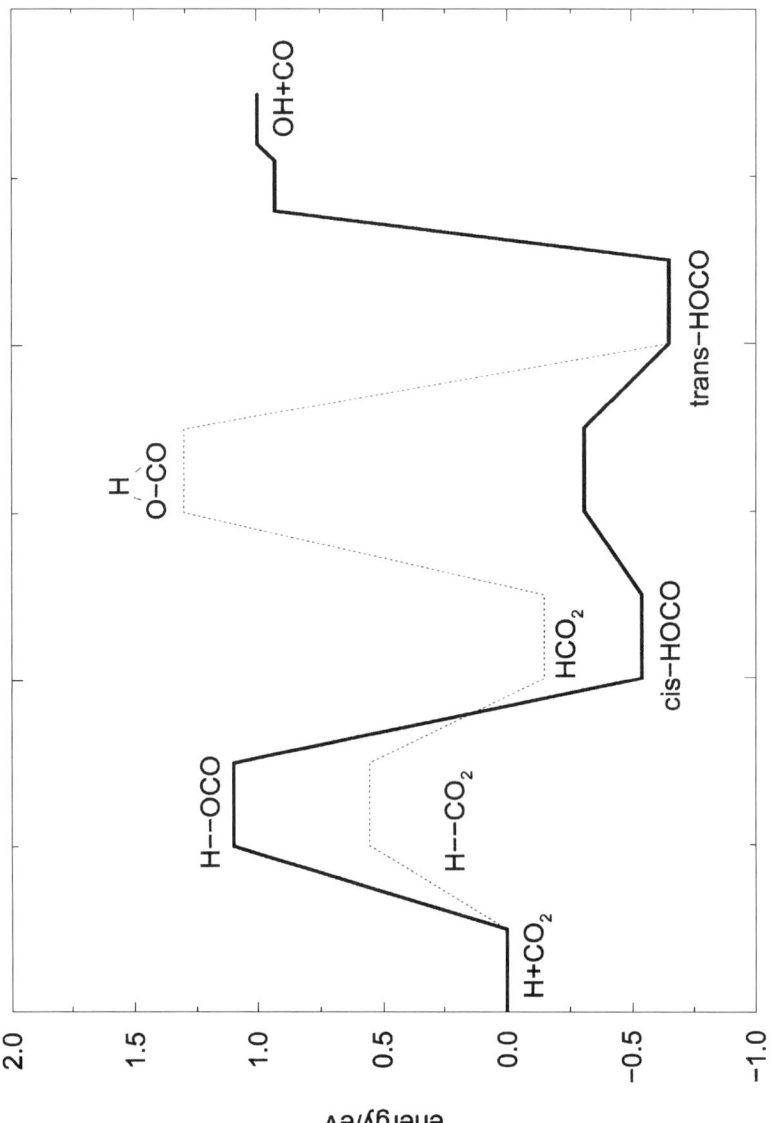

Figure 2 Energy level diagram for the H + CO_2 → OH + CO reaction showing the minima and maxima along the reaction coordinate. Reprinted with permission from Ref. 116. Copyright 1997 *American Chemical Society*.

The distribution of the available energy over the several degrees of freedom of the OH and CO products is clearly non-statistical. Surprisal analysis of both the OH and CO rotational distributions yields large positive surprisal parameters, indicating less than statistical energy in rotation, and the translational energy of the products is greater than statistical. This behavior is not surprising. At the high energies of the hot atom experiments, the lifetime of the HOCO complex is probably not long enough to ensure energy randomization. The measured distributions are mostly recovered by QCT calculations on an ab initio potential energy surface, though there are some significant differences (51, 52).

Wittig and co-workers extended measurements of OH product state distributions to reactions in van der Waals complexes of HX•CO_2, where HX, a hydrogen halide, is the hot atom precursor (41–46, 53). The idea in these studies was to use the structure of the complex to orient the reactants and reduce the range of geometries and impact parameters contributing to the reaction. They found that the OH rotational distribution shifts to lower N, with $\Delta N \approx 4$, compared to the binary collision distribution. Interpretation of this result is complicated by the fact that the geometries of the complexes are not all known and are in any case floppy. Further, the H + CO_2 collision energy in the complex is not well defined as the H atom recoil is confined by the geometry of the complex in what is known as the "squeezed atom" effect (46).

Probably the most interesting use of the HX•CO_2 van der Waals complexes was real-time measurement of product formation. Since the distance between the reactants in the complex is of atomic dimensions and the velocity of the H atom is very large, the time of initiation of the reaction can be well specified, if the laser pulse that photolytically prepares the hot H atom is of short enough duration. And the reaction can be clocked in real time if a similarly short laser pulse is used to detect the OH product.

Zewail and co-workers carried out such an experiment on HI•CO_2, photolyzing HI at three different wavelengths and probing the OH product in two different $v = 0$ rotational states, with a convoluted pump-probe time response of 5 ps (54, 55). They found that their data were best fit with two characteristic times for the H + CO_2 reaction, an induction time representing the time needed to form the HOCO intermediate and a decay time representing the time evolution of products from this complex. They found induction times of 0.7 to 1.2 ps and decay times (HOCO lifetimes) of 1.1 to 4.4 ps; in both cases the times decreased as the H atom energy was increased. The longest decay times are comparable to the classical rotational period of CO_2/HOCO in low J states.

The Wittig group performed the same measurements, though using subpicosecond laser pulses for pump and probe (53, 56, 57). They found appearance times of the OH product of 280 to 610 fs at the same H atom energies as Zewail and co-workers. The appearance times give rates of reaction that match RRKM calculations. The source of the discrepancy between these real-time measurements is not known. The difference in the pump and probe pulse widths does make the characteristics of the OH product detection different—a broad range of

N states centered at N = 5 in the Wittig experiments versus specific N, N = 1 or N = 6, in those of Zewail—but it is not clear how this should effect the results.

In a remarkable series of experiments, Brouard et al have measured the completely state-resolved differential cross sections for the H + CO_2 reaction, using polarized Doppler-resolved laser-induced fluorescence detection of the OH (58–61). The method, developed by a number of groups (62–69), extracts angle-speed distributions of the reaction products through analysis of composite Doppler profiles taken with different propagation/polarization combinations of the photolysis and probe lasers, measured on different branches of the LIF transitions. Analysis of such profiles extracts not only scalar properties of the scattered products but also correlations among linear momentum and angular momentum vectors.

The complicated dynamics of the H + CO_2 reaction provide a richness of results from these measurements that is difficult to capture completely here. Among the findings are a quite sensitive dependence of the angular distributions on rotational quantum number N and on spin-orbit state. At a collision energy of 2.5 eV, the angular distribution is backward peaked for OH(v = 0, N = 1, Ω = 1/2) but nearly isotropic for OH(v = 0, N = 1, Ω = 3/2). For OH(v = 0, N = 9), the peak of the N distribution, the product is slightly forward peaked with no dependence on Ω. The recoil speed distributions are similarly quite sensitive to OH quantum state. Both behaviors are illustrated in Figure 3 for OH(v = 1, N = 1, Ω = 1/2 or 3/2), which shows the dramatic dependence on quantum state.

The OH product distributions are also sensitive to collision energy, some shift from forward peaked to backward peaked for some N, remaining nearly unchanged for others, with corresponding changes in the recoil speed distributions. The OH rotational angular momentum is found to be preferentially aligned in the scattering plane. QCT calculations (70) agree with these observations in some, but not all, respects. This might be expected, as at this high a level of experimental resolution, approximate calculations are not likely to be sufficient to describe all the observables.

Brouard et al interpret their results as revealing reaction dynamics occurring on two different potential energy surfaces, one an excited state surface leading to Ω = 1/2 via a direct reaction with small impact parameters. The other products are believed to form on the ground state surface, either via direct reaction or through the HOCO complex, the routes depending on collision energy and rotational state. The OH rotational polarization suggests that the reaction is not confined to a plane, and that out-of-plane OH-CO torsional forces are also important.

The collection of these results led Brouard et al to identify a reason for the discrepancy in the real-time measurements of the reaction. They suggest that the sub-picosecond experiments of Wittig et al were more sensitive to OH in low N with Ω = 1/2, for which their results imply direct reaction (short reaction time), as opposed to the Zewail et al experiments that monitored OH(N = 1, Ω = 3/2), for which their results imply slower reaction through a HOCO complex.

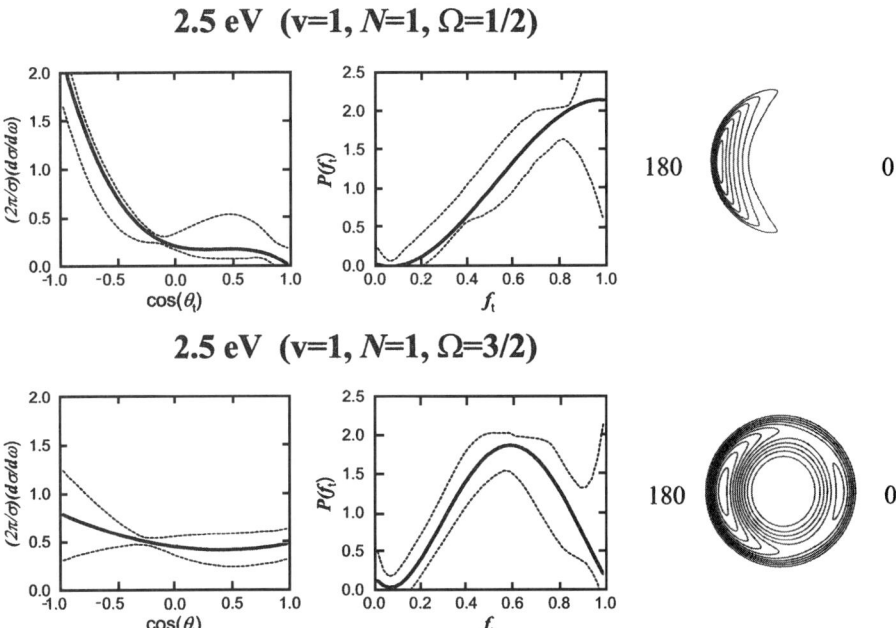

Figure 3 Angular and speed distributions for the OH product of the H + CO_2 → OH + CO reaction at a collision energy of 2.5 eV, for OH(v = 1, N = 1, Ω = 1/2) (*top*) and for OH(v = 1, N = 1, Ω = 3/2) (*bottom*). Cos (θ_t) is the cosine of the scattering angle, while f_t is the fraction of the available energy in translation. The combined angle-speed distributions are plotted as contour maps on the right of the figure, with 180° corresponding to back scattering and 0° to forward scattering. Figure provided by M Brouard and used by permission.

Cl + RH → HCl + R

The abstraction of a hydrogen atom from an alkane to form HCl differs sharply from the reactions of H with H_2O and CO_2. These Cl atom reactions are nearly thermoneutral to mildly exothermic depending on the identity of the alkane, with ΔH^0 ranging from +0.075 eV for CH_4 to approximately −0.30 eV for some sites of larger linear and branched alkanes. The barrier heights are consequently smaller, much smaller, than for the H + H_2O and H + CO_2 reactions. The largest is that for the reaction with CH_4, for which a recent review of all the calculations by Corchado et al recommends a value of 0.36 eV for the classical barrier height and 0.20 eV for the vibrationally adiabatic barrier height (71).

This class of reactions also differs from the four-atom reactions already described here in that it is a truly polyatomic reaction. There is a polyatomic reactant and a polyatomic product. There are multiple reaction sites, sometimes equivalent, sometimes not, for any alkane, and reaction at these different sites can be seen as competitive and possibly displaying different dynamics. More than one, and perhaps several, of the degrees of freedom of the polyatomic reactant and product could be active in the reaction coordinate.

These reactions are quite accessible for state-to-state dynamics experiments. Cl atoms of well-defined energy near and above the reaction barriers can be easily generated through laser photolysis of Cl_2. Sensitive, state-selective detection of the HCl product can be readily effected by resonant multi-photon ionization. Not surprisingly, many experiments have been carried out on this reaction system in the past few years.

With any reaction involving a polyatomic reactant with more than one reactive site, the question of site propensities of reaction immediately comes to mind. For the Cl + RH reaction, this has been investigated by the Koplitz and Dagdigian groups using selectively deuterated alkanes (72–74). With $CH_2CD_2CH_3$ and $CD_3CH_2CD_3$ as reactants at a mean collision energy of approximately 0.30 eV, Yen et al found that the yield of DCl was the same for either alkane, whereas the yield of HCl was about 1.5 times greater for $CH_2CD_2CH_3$ than for $CD_3CH_2CD_3$ (72). Varley & Dagdigian got a similar result at essentially the same collision energy with $CD_3CH_2CD_3$ the production of HCl was 1.1 times greater than DCl (73). With $(CH_3)_3CD$, they observed 3.3 times more HCl than DCl (74). These results calibrate the expectation that abstraction of either a secondary or tertiary H is more favored than abstraction of a primary H atom, quantifying that selectivity as, on average, a factor of 2.8.

Varley & Dagdigian have argued that a larger cone of acceptance around the expected linear Cl-H-C energy minimum for reaction of the secondary and tertiary H can account for these results (73). This seems reasonable, since reaction of secondary and tertiary H is more exoergic than reaction at a primary H, and therefore the secondary and tertiary reactions would be expected to have a smaller reaction barrier height. However, measurements of the thermal rate constants for reactions of Cl with ethane, propane, and n-butane (75) show near zero activation energy, so the importance of differential barrier heights may not be so significant.

All the Cl + RH reactions that have been studied show very little energy released in HCl vibration or rotation. The results are so similar, for a wide range of alkane reactants, that a summary of the results seems repetitive. Simpson et al found that the HCl and DCl products from the Cl + CH_4 and CD_4 reaction both had only about 0.005 eV of rotational energy, 5% of the total available, at a collision energy of 0.16 eV (CH_4) or 0.18 eV (CD_4) (76). Kandel et al found 2% for the Cl + C_2H_6 reaction at 0.24 eV collision energy (77). For the reactions of methane, propane, and isobutane, Varley & Dagdigian found the average HCl rotational energy varied from 0.007 eV for the methane reaction to 0.02 eV for isobutane, consistently about 3% of the total (78). For reaction with $CD_3CH_2CD_3$ at a collision energy of 0.32 eV, the average rotational energy of both the HCl and DCl products was 3% of the total (73). For reaction with $(CH_3)_3CD$ at a comparable energy, the results were the same (74). Park et al found 2% of the available energy in DCl rotation for reaction of Cl with deuterated cyclohexane at relative energies of 0.23 eV and 0.62 eV, with the same fraction in DCl vibration (79). All these results have been cited as evidence of a linear Cl-H-C transition state geometry, though that conclusion rests on an assumption of impulsive energy release. In the cases where the HCl/DCl vibrational distribution was measured, production of HCl (v = 1) was a minor channel (77, 79).

The most dramatic results obtained for the Cl + RH reactions are the quantum-state-resolved product angular distributions. The measurements are based on the same formalism as those of OH angle-speed distributions used by Brouard et al and described in the previous section (62–69). Here, however, the measured quantity is not a composite Doppler profile of a spectroscopic transition. Rather, it is a one-dimensional projection of the product velocity distribution, from the time-of-flight of the product molecules along the extraction axis of a time-of-flight mass spectrometer. Resonant multiphoton ionization effected with the ionizer electrodes set up in a space-focusing condition produces ions (HCl^+ or Cl^+) whose flight times to the detector depend on the velocity of the HCl product before ionization. Temporal resolution of these ion flight times yields the neutral product speed distribution. In favorable cases, this one-dimensional measurement can be inverted to yield the product angular distribution. The Cl + RH → HCl + R reactions fall into one of these favorable cases.

The Zare group first used this approach to investigate the reaction of Cl atoms with vibrationally and rotationally state-selected methane (80). IR absorption prepared the CH_4 in the asymmetric C-H stretch excited state, $v_3 = 1, j = 1$. The v = 1 vibrationally excited HCl product molecules, which can come only from reaction of the vibrationally excited methane reactant, were detected in various low-j states, and the one-dimensional projection of the product recoil velocity distribution was measured for these states. Predominantly forward scattered HCl (v = 1, j = 1) was observed, whereas HCl(v = 1, j = 3) had a backward scattered component as well. With $(CH_3)_3CD$ as alkane reactant Varley & Dagdigian found backward scattered DCl(v = 0, j = 0) but sideways scattered HCl(v = 0, j = 0) (74). With $CD_3CH_2CD_3$ both the DCl and HCl products were scattered over a wide range of angles, with the HCl slightly favoring backward and DCl more sideways.

These results have been analyzed by Wang et al using QCT calculations on the $Cl + CH_4$ reaction (81). They identified two mechanisms for the reaction. One is the expected direct reaction at small impact parameters producing backward scattered and rotationally hot (relatively) HCl. The other is a large impact parameter reaction leading to forward scattered and rotationally cold HCl. This path is what Ben-Nun et al refer to as peripheral reaction (82), and Aker & Valentini (83) describe in their shell model, large impact parameter, grazing collisions.

These QCT calculations were done both in full dimensionality and freezing various degrees of freedom of the incipient CH_3 radical product. The results were substantially invariant, indicating that the methyl radical modes are not particularly active in the reaction. Reduced dimensionality (CH_4 as pseudodiatom) quantum calculations on this reaction also recover the essential features of the experimental results: forward scattered for HCl(v = 1), backward-to-sideways for HCl(v = 0) (84).

The Zare group developed an approach, which they call "core extraction," that improves the determination of the differential cross section from measurement of one-dimensional velocity distributions (85). In this method a spatial mask is placed in the ion flight path, restricting detection to product molecules that have zero velocity component perpendicular to the extraction direction. This turns the one-dimensional measurement into a measurement of a three-dimensional slice of the full velocity distribution and greatly improves the sensitivity of the measured velocity distributions to the product angular distribution. They first demonstrated this method by a new measurement of their previously characterized reaction of Cl with vibrationally excited ($\nu_3 = 1$) CH_4 (85), as well as new measurements on this vibrationally excited CH_4, C-H stretch excited CHD_3, and ground vibrational state CH_4 (76, 86, 87).

They derived differential cross sections that are strongly dependent on the vibrational state of the methane reactant and the rovibrational state of the HCl product. The dependence on j for the v = 1 product is shown in Figures 4 and 5. The former plots the one-dimensional velocity distributions that are the measured quantities. The latter shows the differential cross sections derived from them. For reaction with the vibrationally excited methane isotopomers, the HCl(v = 1) is scattered strongly forward for low j, shifting to forward-backward for higher j. The HCl (v = 0) is scattered backwards to sideways. For reaction with ground vibrational state methane, the HCl(v = 0) product is strongly backscattered. By varying the spatial alignment of the vibrationally excited methane relative to the axis along which the velocity is measured (by rotating the plane of polarization of the IR excitation laser), they discovered a pronounced steric effect in the reaction. Cl atoms making an approach perpendicular to the C-H reactive bond preferentially produced forward scattered HCl(v = 1).

The Zare group interpreted the behavior they observed in terms of open (for vibrationally excited methane) or restricted (ground vibrational state methane) cones of acceptance. Using a hard-sphere model of the scattering, they developed a map of the impact parameters that led to forward and backward scattering. They arrived at the same result that Wang et al did from their QCT calculations: direct, rebound dynamics for the ground state reactions, and also for the v = 0 product of the reaction with vibrationally excited reactant, but peripheral, forward scattering for the v = 1 product from the vibrationally enhanced reaction.

In these experiments, Kandel et al (87) also performed REMPI detection of the methyl radical product from reaction of Cl with CH_4 and CD_4 in the vibrational ground state. They studied the reaction over a range of energies and found velocity distributions that indicated formation of products that were not apparently energetically accessible. They concluded that the products were coming from reaction of CH_4/CD_4 reactants in thermally populated excited bending and torsion vibrational states.

Kandel et al extended these measurements to reaction of Cl with ground state and vibrationally excited ($\nu_5 = 1$) ethane (77, 88). Here too, vibrational excitation produced forward scattered HCl(v = 1), but the HCl from the ground state reaction

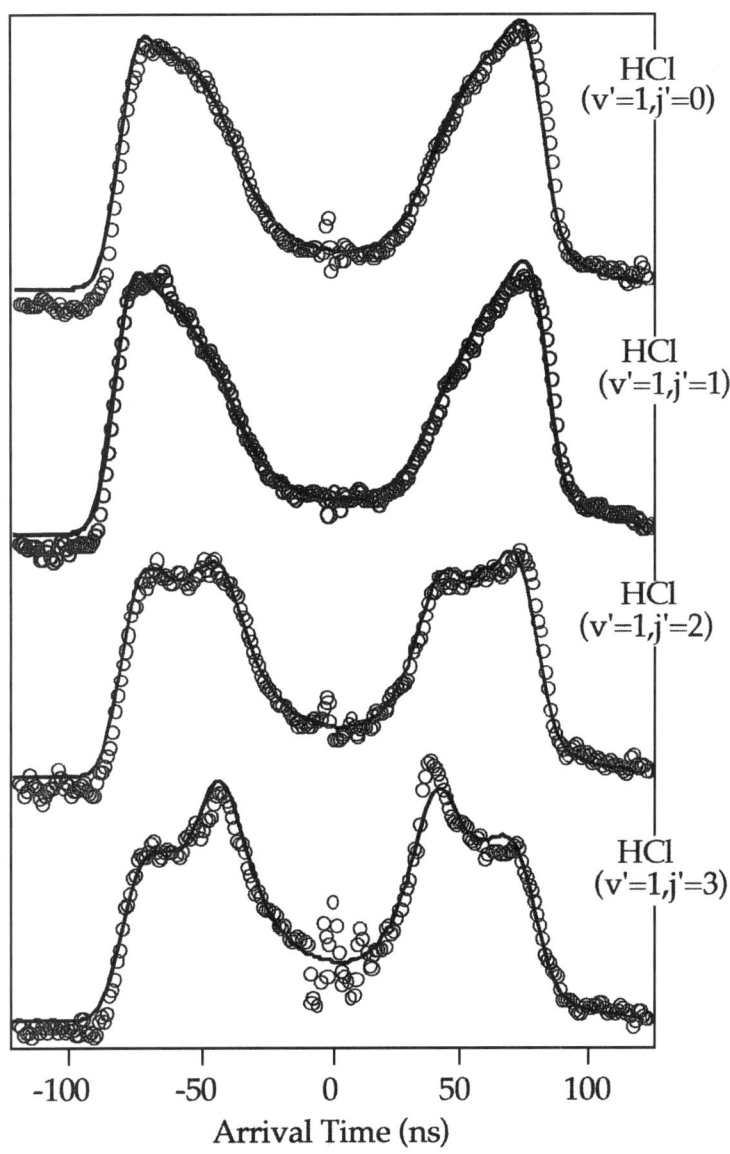

Figure 4 Time-of-flight profiles for the HCl(v = 1, j) product of the Cl + CH$_4$(ν_3 = 1) reaction using the core extraction method. From Reference 86, and used by permission.

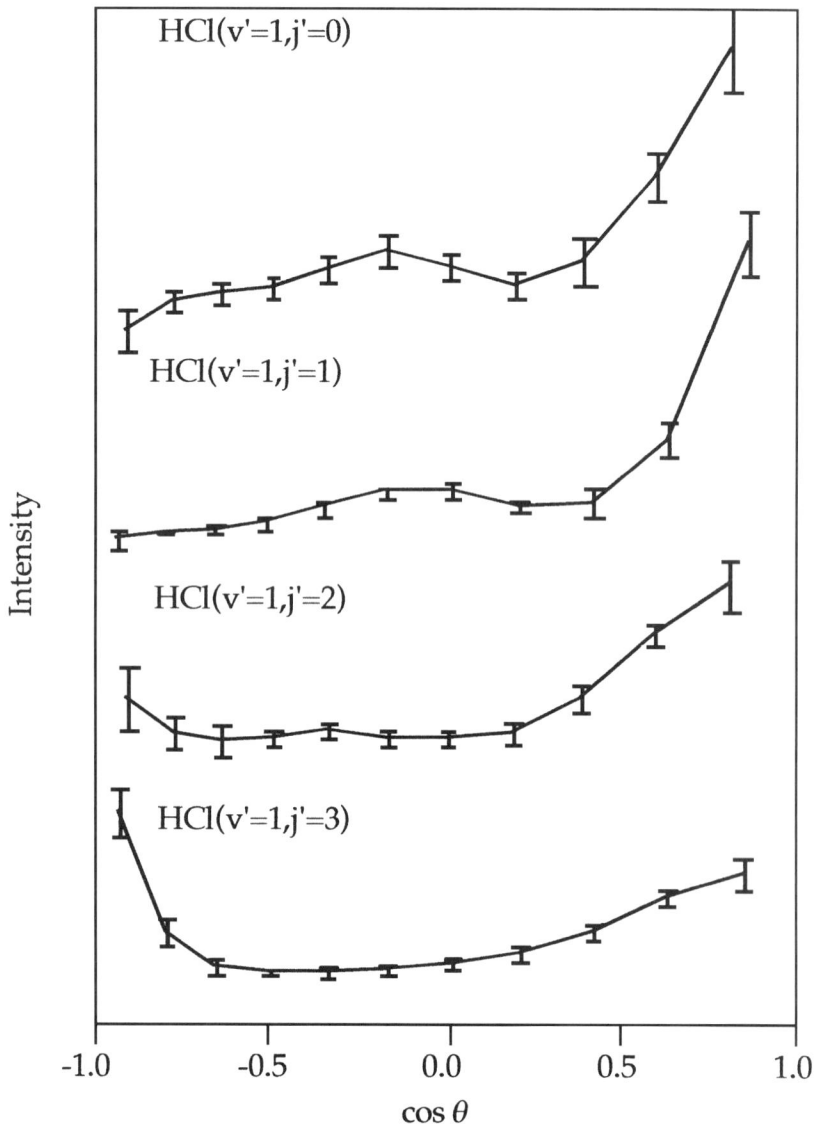

Figure 5 Rotational and vibrational state-resolved differential cross sections derived from the experimental measurements shown in Figure 4. Cos θ is the cosine of the scattering angle, with a value of 1.0 indicating forward scattered product. From Reference 86, and used by permission.

showed a nearly isotropic angular distribution. Using the same line-of-centers model as for the methane reactions, they were able to account for their observations, concluding that the ethane reaction proceeds through a less constrained transition state than the methane reaction. They also were able to show that there is little internal energy in the ethyl radical product.

The Zare group also developed and applied methods for determining differential cross section polarization moments and product rotational alignment (89–91). They found different DCl alignment for the Cl + C_2D_6 and Cl + CD_4 reactions, from which they inferred that the D atom transfer occurs earlier in the reaction of the former than in the latter.

H + RH → H_2 + R

The abstraction of an H atom from an alkane by an H atom is the simplest reaction at a carbon center. Thermochemically these reactions share much in common with the analogous reactions of Cl atoms. They range from nearly thermoneutral for H + CH_4, for which ΔH^0 is −0.04 eV, to mildly exoergic for abstraction of a tertiary hydrogen atom, $\Delta H^0 = -0.52$ eV for isobutane. However, they have substantially higher barriers to reaction and, unlike the Cl reactions, have significant barriers even for the most exoergic of the reactions. The reported barrier heights for these reactions are around 0.5 eV: 0.58 eV for CH_4 (92), 0.48 eV and 0.31 eV for abstraction of the primary and secondary H in propane (93), and 0.40(4) for cyclohexane (94). For the substituted alkane $CDCl_3$, for which experiments are also described below, the barrier is 0.35 eV (95). High-quality potential energy surface calculations for H + CH_4 confirm these substantial barriers (96–99).

Much of the mechanistic structure of chemical reaction dynamics is based on atom + diatom reactions, or reactions that are effectively atom plus diatom. How well do these models describe the dynamics of reactions of polyatomic molecules? The H + alkane reactions are a natural class on which to impose this question. They have multiple reaction sites that are likely to be competitive with one another. But they also are likely to have reaction coordinates with large projections on many modes of the reactants and products. Both the reactant atom and all the reaction sites of the polyatom have the same mass, so kinematic coupling can be large. The time scale of the reaction is comparable to the vibrational periods of the hydrogenic motions in the reactant molecule, and this is another source of dynamical coupling of several modes to the reaction coordinate.

The Valentini group has carried out an extensive series of experiments on the reactions of hydrogen atoms with alkanes, aimed at addressing these questions. To do so requires a reference set of atom + diatom reactions. For these H + alkane reactions, the reference set is composed of the reactions H + HX → H_2 + X, where X is Cl, Br, and I. The hydrogen halide reactions have the same kinematics as the alkane reactions, light + light-heavy → light-light + heavy. They bracket the energetics of the alkane reactions, with ΔH of -0.05 eV to -1.42 eV. This is important in the context of how the Valentini group has approached the questions of the dynamics of polyatomic reactions, classifying reactions in a three-dimensional conceptual space where the dimensions are energetics, kinematics, and structure. For this reference set purpose, Aker et al have carried out detailed characterization of the dynamics, both in state-to-state experiments and QCT simulations (83, 100–102).

Because the H + RH reactions have substantial energetic barriers they have to be studied at high collision energy, and hot atom techniques have been employed here as in the H + H_2O and H + CO_2 reactions already described. Energetic H atoms were generated by photolysis of HI to give H + RH collision energies of 1.5 to 1.6 eV. The first of these reactions studied was the simplest, H + CD_4 → HD + CD_3, using the deuterium isotopomer because the small cross section (measured to be 0.14 Å^2) made interference from the reaction of the H with residual H atom precursor HI appreciable (103). These experiments showed an H_2 vibrational distribution that is common to all the thermoneutral or nearly thermoneutral H + HA → H_2 + A atom + diatom reactions, where HA is HCl, D_2, H_2: Very little conversion of initial relative energy to H_2/HD vibration occurs, in this case only 7% of the 1.5 eV available energy, with no population in v > 1. And, like those atom + diatom analogues, there was little energy in product rotation as well, 9%. However, the methane reaction was quite distinct in one regard. The fraction of the available energy (collision energy−ΔH−vibrational energy) that appears in rotation for the v = 1 product was greater than for the v = 0. This positive correlation of vibrational and rotational excitation was unanticipated, as it is directly opposite the negative correlation (product rotational energy decreasing with increasing vibrational energy) observed for all other bimolecular reactions for which state-to-state dynamics measurements have been made.

This behavior is a defining characteristic of all the H + RH reactions. Germann et al found it for the ethane and propane reactions (104). In recent experiments Picconatto (105) and Srivastava (106) have found it also in reactions of n-pentane, n-hexane, cyclopentane, and cyclohexane. This behavior is illustrated in Figures 6 and 7, where plots of the rotational state distributions are given for some of the alkanes as well as for the HCl and HBr reactions from the reference set. The measurements are shown as solid symbols; the solid curves are linear surprisal fits to the data.

The energy axis in these figures is an unusual one that requires explanation. It is a dimensionless reduced energy, rotational energy divided by an effective total available energy. This effective energy is given by a fraction of the collision energy minus the ΔH of reaction minus the vibrational energy. Except for the fractional collision energy, this is conventional. The fractional collision energy is the nominal collision energy times the square of the sine of the skew angle, γ. This multiplicative term is, for these reactions with a skew angle very close to 45°, 0.50 to 0.54. This collision energy scaling comes from an observation that for all the H + RH and H + HX reactions, the maximum part of the collision energy that

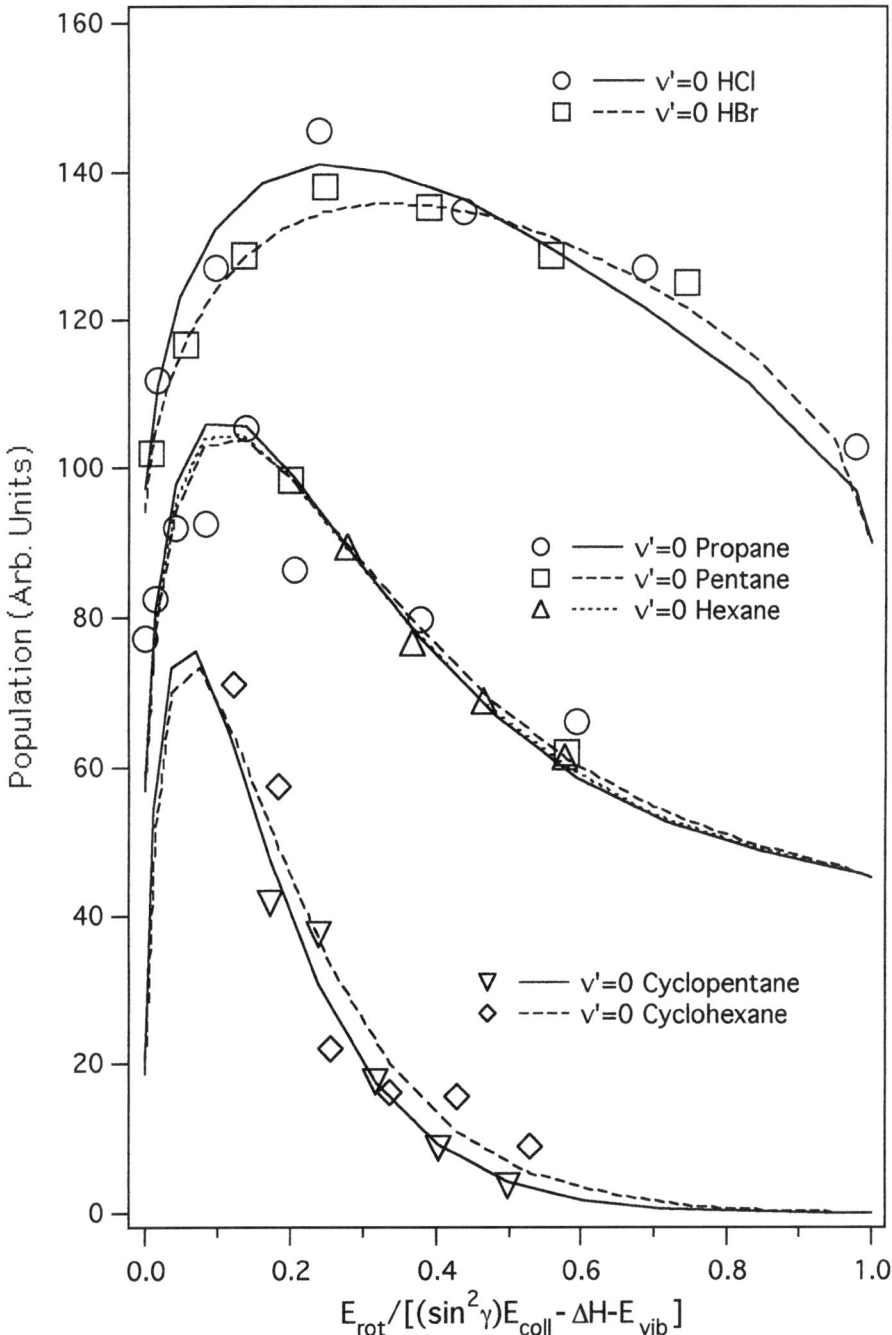

Figure 6 H_2 (v = 0, j) rotational state distributions plotted as a function of reduced rotational energy for some of the H + HX and H + RH reactions. The symbols are the measurements; the solid curves are linear surprisal fits to the data. See text for details. From Picconatto (105) and Srivastava (106) with permission.

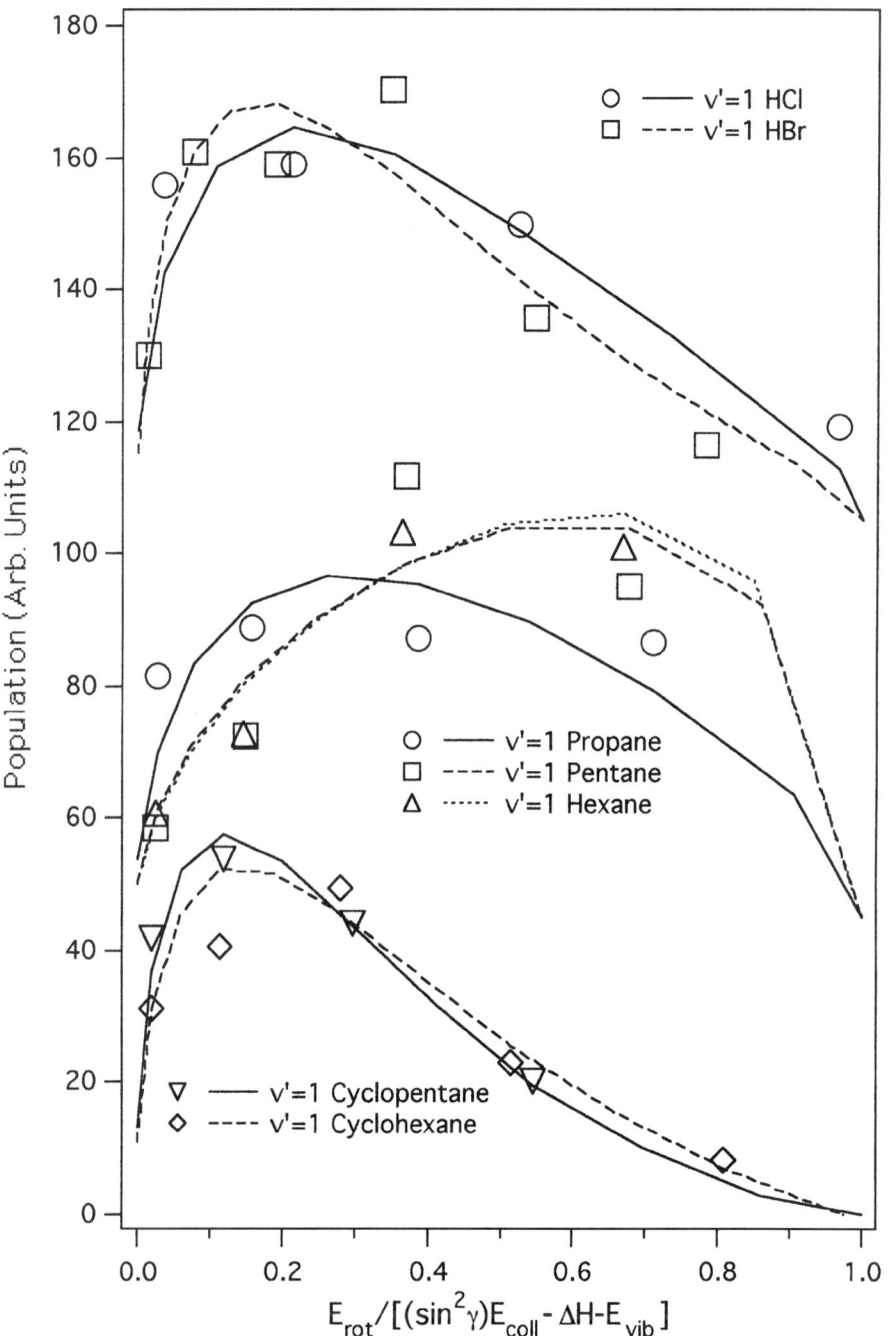

Figure 7 H_2 (v = 1, j) rotational state distributions plotted as a function of reduced rotational energy for some of the H + HX and H + RH reactions. The symbols are the measurements; the solid curves are linear surprisal fits to the data. See text for details. From Picconatto (105) and Srivastava (106) with permission.

ever appears in internal energy of the H_2 product is $\sin^2 \gamma \times E_{coll}$. This has been shown to be a kinematic effect of reactions at high relative energies, and analysis of other reactions with other mass combinations and other skew angles follow it as well (105). It really is a specification of the minimum energy that must appear in product translation, which is $\cos^2 \gamma \times E_{coll}$. If the translational energy is less than that, the trajectory will reflect back into the entrance channel and reaction is precluded.

Although common to all alkane reactions, the quantitative degree of positive rotation-vibration energy correlation depends on the identity of the alkane or, more correctly, the class of the alkane. The effect is greatest in the straight chain alkanes and weakest in the primary alkanes methane and ethane. The cyclic alkanes are intermediate. These trends are clearly evident in Figures 6 and 7. The Valentini group (105, 106) has developed a picture of this behavior in terms of a local collision model, an extension of the shell model of Aker & Valentini (83).

This local collision model, like the shell model, characterizes the reactive collision in terms of a local impact parameter, defined as the distance between the relative velocity vector and a line parallel to it that passes not through the center of mass of the alkane but rather through the nearest H atom. Using this model, Valentini et al interpret the rotational energy distributions of the H_2 product as dependent on two factors. One is the truncation of the local impact parameter opacity function by overlapping cones of acceptance on adjacent H atoms. The second is the effective uncoupling of the simultaneous constraints of energy and angular momentum conservation, made possible by the heavy alkyl radical being a sink for angular momentum at very little cost in energy. Using this model they can explain the variation of the rotational energy disposal for all the reactions that they have studied.

It is noteworthy that the one H + alkane reaction for which the unusual positive correlation of vibrational and rotational energy disposal behavior is not observed is the reaction with cyclopropane. For this reaction H_2 product rotational and vibrational excitation are not positively correlated; they are negatively correlated as in other atom transfer reactions (106). It is also observed for this reaction that the rotational and vibrational energies of the H_2 product extend to energies larger than the total energy available to products for the reaction that yields a cyclopropyl radical as the co-product. The energy disposal is accountable if it is assumed that the co-product is instead an allyl radical.

The H + c-C_3H_6 reaction cannot be direct if this happens, as the time scale of direct abstraction at these collision energies is far too short, of order 10^{-14} s (107), to allow the isomerization from cyclopropyl to allyl, which involves slower heavy atom motion. This implies that the H + c-C_3H_6 reaction is not a direct reaction but proceeds by addition followed by H_2 elimination. Although such reactions are well known for cyclopropane, that a complex-forming reaction should be operative at these energies is not obvious.

The H_2 product state distributions in this cyclopropane reaction are not statistical, but they are also not like those for reactions with typical alkanes. For this reaction the fraction of the available energy that appears as H_2 product rotation is less for v = 1 product than for v = 0. This reaffirms the belief that the positive correlation of H_2 product rotational and vibrational excitation is characteristic of the class of H + RH direct abstraction reactions.

Lanzisera & Valentini have studied the reaction of H atoms with the alkyl halide $CDCl_3$ (108–110). This reactant was selected to test whether vibrational modes of the alkyl radical product in the H + RH reactions might be responsible for some part of the characteristic rovibrational energy disposal in the H_2 product. The motion of the heavy Cl atoms is too slow to follow that of the light and fast H atoms, and the geometry of the CCl_3 radical is pyramidal with a structure essentially the same as that which it has in $CDCl_3$ (111–115). So the CCl_3 vibrations should be inactive in the reaction. However, the CCl_3 co-product is still heavy and structured, with very large moments of inertia, and is therefore a very low energy cost sink for angular momentum. And there is still a restricted cone of acceptance around the reacting D atom due to the presence of the bulky Cl atoms. So, based on the local collision model (105, 106), it would be expected to behave like the alkane reactions, and it does. It shows the same positively correlated rotational and vibrational HD energies that the reactions of the alkanes do.

Interestingly, vibrational excitation of the C-D stretch in the $CDCl_3$ enhances the degree of rotation-vibration coupling in the HD product (110). It also enhances the absolute reaction cross section, even though the added vibrational energy, 0.28 eV, is small compared with the collision energy, 1.6 eV, which is itself well above the 0.35 eV barrier (95) to reaction. Surprisingly, none of the reactant vibrational energy appears as HD product rotation or vibration.

CONCLUSION

These four case studies illustrate the level of detail of both measurement and understanding that can be obtained in state-to-state experiments on polyatomic reactions. The examples are illustrative, and not exhaustive, even at present. Although these four reactions will remain prototypes, it is certain that new prototypes will be added as experimental work in this area grows and expands, as seems likely.

ACKNOWLEDGMENTS

This work is supported by Grant No. DE-FG02-91ER14225 from the Division of Chemical Sciences, Office of Basic Energy Sciences, US Department of Energy. The author thanks Carl Picconatto and Abneesh Srivastava for assistance in the preparation of this review. The author also thanks M Brouard, FF Crim, GC Schatz, and WR Simpson for supplying figures used here.

Visit the Annual Reviews home page at www.AnnualReviews.org

LITERATURE CITED

1. Walch SP, Dunning TH Jr. 1980. *J. Chem. Phys.* 72:1303–11
2. Schatz GC, Elgersma H. 1980. *Chem. Phys. Lett.* 73:21–25
3. Dunning TH, Jr., Harding LB, Kraka E. 1989. In *Supercomputer Algorithms for Reactivity, Dynamics, and Kinetics of Small Molecules*, ed. A Lagana, pp. 57–66. Dordrecht, The Netherlands:Kluwer
4. Isaacson AD. 1997. *J. Chem. Phys.* 107:3832–39
5. Ochoa de Aspuru G, Clary DC. 1998. *J. Phys. Chem. A* 102:9631–37
6. Kleinermanns K, Wolfrum J. 1984. *J. App. Phys. B* 34:5–9
7. Flynn GW, Weston RE, Jr. 1986. *Annu. Rev. Phys. Chem.* 37:551–85
8. Kessler K, Kleinermanns K. 1992. *Chem. Phys. Lett.* 190:145–48
9. Jacobs A, Volpp HR, Wolfrum J. 1992. *Chem. Phys. Lett.* 196:249–54
10. Jacobs A, Volpp HR, Wolfrum J. 1994. *J. Chem. Phys.* 100:1936–45
11. Koppe S, Laurent T, Naik PD, Volpp HR, Wolfrum J. 1994. *Can. J. Chem.* 72:615–24
12. Mok MH, Polanyi JC. 1969. *J. Chem. Phys.* 51:1451–69
13. Polanyi JC, Wong WH. 1969. *J. Chem. Phys.* 51:1439–40
14. Polanyi JC. 1972. *Acc. Chem. Res.* 5:161–68
15. Zhang DH, Light JC. 1996. *J. Chem. Phys.* 104:4544–53
16. Szichman H, Baer M. 1995. *Chem. Phys. Lett.* 242:285–90
17. Szichman H, Baer M, Volpp HR, Wolfrum J. 1999. *J. Chem. Phys.* 111:567–71
18. Nyman G, Clary DC. 1993. *J. Chem. Phys.* 99:7774–86
19. Kudla K, Schatz GC. 1993. *J. Chem. Phys.* 98:4644–51
20. Bradley KS, Schatz GC. 1998. *J. Chem. Phys.* 108:7994–8003
21. Honda K, Takayanagi M, Nishiya T, Ohoyama H, Hanazaki I. 1991. *Chem. Phys. Lett.* 180:321–26
22. Schatz GC, Colton MC, Grant JL. 1984. *J. Phys. Chem.* 88:2971–77
23. Sinha A, Hsiao MC, Crim FF. 1991. *J. Chem. Phys.* 94:4928–35
24. Hsiao MC, Sinha A, Crim FF. 1991. *J. Phys. Chem.* 95:8263–67
25. Hoy AR, Bunker PR. 1979. *J. Mol. Spectrosc.* 74:1–8
26. Huber KP, Herzberg G. 1979. *Molecular Spectra and Molecular Structure*. New York:Van Nostrand Reinhold
27. Adelman DE, Filseth SV, Zare RN. 1993. *J. Chem. Phys.* 98:4636–43
28. Brouard M, Burak I, Markillie GAJ, McGrath K, Vallance C. 1997. *Chem. Phys. Lett.* 281:97–104
29. Sinha A, Hsiao MC, Crim FF. 1990. *J. Chem. Phys.* 92:6333–35
30. Sinha A. 1990. *J. Phys. Chem.* 94:4391–93
31. Metz RB, Thoemke JD, Pfeiffer JM, Crim FF. 1993. *J. Chem. Phys.* 99:1744–51
32. Bronikowski MJ, Simpson WR, Zare RN. 1993. *J. Phys. Chem.* 97:2204–8
33. Bronikowski MJ, Simpson WR, Zare RN. 1993. *J. Phys. Chem.* 97:2194–203
34. Bowman JM, Schatz GC. 1995. *Annu. Rev. Phys. Chem.* 46:169–95
35. Lendvay G, Bradley KS, Schatz GC. 1999. *J. Chem. Phys.* 110:2963–70
36. Zhang DH, Light JC. 1997. *J. Chem. Soc. Faraday Trans.* 93:691–97
37. Kleinermanns K, Wolfrum J. 1984. *Chem. Phys. Lett.* 104:157–59
38. Kleinermanns K, Linnebach E, Wolfrum J. 1985. *J. Phys. Chem.* 89:2525–27
39. Jacobs A, Wahl M, Weller R, Wolfrum J. 1989. *Chem. Phys. Lett.* 158:161–66
40. Jacobs A, Volpp HR, Wolfrum J. 1994. *Chem. Phys. Lett.* 216:51–59

41. Buelow S, Radhakrishnan G, Catanzarite J, Wittig C. 1985. *J. Chem. Phys.* 83:444–45
42. Radhakrishnan G, Buelow S, Wittig C. 1986. *J. Chem. Phys.* 84:727–38
43. Buelow S, Radhakrishnan G, Wittig C. 1987. *J. Phys. Chem.* 91:5409–12
44. Rice J, Hoffmann G, Wittig C. 1988. *J. Chem. Phys.* 88:2841–43
45. Chen Y, Hoffmann G, Oh D, Wittig C. 1989. *Chem. Phys. Lett.* 159:426–34
46. Hoffmann G, Oh D, Chen Y, Engel YM, Wittig C. 1990. *Isr. J. Chem.* 30:115–29
47. Levine RD, Bernstein RB. 1987. *Molecular Reaction Dynamics and Chemical Reactivity*. New York: Oxford Univ. Press
48. Rice JK, Chung C, Baronavski AP. 1990. *Chem. Phys. Lett.* 167:151–54
49. Rice JK, Baronavski AP. 1991. *J. Chem. Phys.* 94:1006–19
50. Nickolaisen SL, Cartland HE, Wittig C. 1992. *J. Chem. Phys.* 96:4378–86
51. Schatz GC, Fitzcharles MS, Harding LB. 1987. *Faraday Discuss. Chem. Soc.* 84:359–69
52. Kudla K, Schatz GC. 1991. *J. Phys. Chem.* 95:8267–73
53. Jaques C, Valachovic L, Ionov S, Bohmer E, Wen Y, et al. 1993. *J. Chem. Soc. Faraday Trans.* 89:1419–25
54. Scherer NF, Khundkar LR, Bernstein RB, Zewail AH. 1987. *J. Chem. Phys.* 87:1451–53
55. Scherer NF, Sipes C, Bernstein RB, Zewail AH. 1990. *J. Chem. Phys.* 92:5239–59
56. Ionov SI, Brucker GA, Jaques C, Valachovic L, Wittig C. 1992. *J. Chem. Phys.* 97:9486–89
57. Ionov SI, Brucker GA, Jaques C, Valachovic L, Wittig C. 1993. *J. Chem. Phys.* 99:6553–61
58. Brouard M, Lambert HM, Payner SP, Simons JP. 1996. *Mol. Phys.* 89:403–23
59. Brouard M, Hughes DW, Kalogerakis KS, Simons JP. 1998. *J. Phys. Chem.* 102:9559–64
60. Brouard M, Hughes DW, Kalogerakis KS, Simons JP. 2000. *J. Chem. Phys.* 112:4557–71
61. Brouard M, Burak I, Hughes DW, Kalogerakis KS, Simons JP, et al. 2000. *J. Chem. Phys.* 113:3173–80
62. Green F, Hancock G, Orr-Ewing AJ. 1991. *Chem. Phys. Lett.* 182:568–74
63. Brouard M, Duxon SP, Enriquez PA, Simons JP. 1992. *J. Chem. Phys.* 97:7414–22
64. Shafer NE, Orr-Ewing AJ, Simpson WR, Xu H, Zare RN. 1993. *Chem. Phys. Lett.* 212:155–62
65. Aoiz FJ, Brouard M, Enriquez PA, Sayos R. 1993. *J. Chem. Soc. Faraday Trans.* 89:1427–34
66. Costen ML, Hancock G, Orr-Ewing AJ, Summerfield D. 1994. *J. Chem. Phys.* 100:2754–64
67. Kim HL, Wickramaaratchi MA, Zheng XN, Hall GE. 1994. *J. Chem. Phys.* 101:2033–50
68. Shafer-Ray NE, Orr-Ewing AJ, Zare RN. 1995. *J. Phys. Chem.* 99:7591–603
69. Aoiz FJ, Brouard M, Enriquez PA. 1996. *J. Chem. Phys.* 105:4964–82
70. Bradley KS, Schatz GC. 1997. *J. Chem. Phys.* 106:8464–72
71. Corchado JC, Truhlar DG, Espinosa-Garcia J. 2000. *J. Chem. Phys.* 112:9375–89
72. Yen YF, Wang Z, Xue B, Koplitz B. 1994. *J. Phys. Chem.* 98:4–7
73. Varley DF, Dagdigian PJ. 1996. *Chem. Phys. Lett.* 255:393–400
74. Varley DF, Dagdigian PJ. 1996. *J. Phys. Chem.* 100:4365–74
75. Lewis RS, Sander SP, Wagner S, Watson RT. 1980. *J. Phys. Chem.* 84:2009–15
76. Simpson WR, Rakitzis TP, Kandel SA, Lev-On T, Zare RN. 1996. *J. Phys. Chem.* 100:7938–47
77. Kandel SA, Rakitzis TP, Lev-On T, Zare RN. 1996. *J. Chem. Phys.* 105:7550–59
78. Varley DF, Dagdigian PJ. 1995. *J. Phys. Chem.* 99:9843–53
79. Park J, Lee Y, Hershberger JF, Hossenlopp

JM, Flynn GW. 1992. *J. Am. Chem. Soc.* 114:58–63
80. Simpson WR, Orr-Ewing AJ, Zare RN. 1993. *Chem. Phys. Lett.* 212:163–71
81. Wang X, Ben-Nun M, Levine RD. 1995. *Chem. Phys.* 197:1–17
82. Ben-Nun M, Brouard M, Simons JP, Levine RD. 1993. *Chem. Phys. Lett.* 210:423–31
83. Aker PM, Valentini JJ. 1990. *Isr. J. Chem.* 30:157–78
84. Nyman G, Yu HG, Walker RB. 1998. *J. Chem. Phys.* 109:5896–904
85. Simpson WR, Orr-Ewing AJ, Rakitzis TP, Kandel SA, Zare RN. 1995. *J. Chem. Phys.* 103:7299–312
86. Simpson WR, Rakitzis TP, Kandel SA, Orr-Ewing AJ, Zare RN. 1995. *J. Chem. Phys.* 103:7313–35
87. Kandel SA, Zare RN. 1998. *J. Chem. Phys.* 109:9719–27
88. Kandel SA, Rakitzis TP, Lev-On T, Zare RN. 1997. *Chem. Phys. Lett.* 265:121–28
89. Orr-Ewing AJ, Simpson WR, Rakitzis TP, Kandel SA, Zare RN. 1997. *J. Chem. Phys.* 106:5961–71
90. Rakitzis TP, Kandel SA, Lev-On T, Zare RN. 1997. *J. Chem. Phys.* 107:9392–405
91. Rakitzis TP, Kandal SA, Zare RN. 1997. *J. Chem. Phys.* 107:9382–91
92. Furue H, Pacey PD. 1990. *J. Phys. Chem.* 94:1419–25
93. Nicholas JE, Vaghjiani G. 1987. *J. Chem. Soc. Faraday Trans. 2* 83:607–18
94. Grief D, Oldershaw GA. 1982. *J. Chem. Soc. Faraday Trans. 1* 78:1189–98
95. Oldershaw GA, Robinson EA. 1978. *Chem. Phys. Lett.* 54:527–29
96. Jordan MJT, Gilbert RG. 1995. *J. Chem. Phys.* 102:5669–82
97. Espinosa-Garcia J, Corchado JC. 1996. *J. Phys. Chem.* 100:16,561–67
98. Takata T, Taketsugu T, Hirao K, Gordon MS. 1998. *J. Chem. Phys.* 109:4281–89
99. Thompson KC, Jordan MJT, Collins MA. 1998. *J. Chem. Phys.* 108:8302–16
100. Aker PM, Germann GJ, Valentini JJ. 1989. *J. Chem. Phys.* 90:4795–808
101. Aker PM, Germann GJ, Valentini JJ. 1992. *J. Chem. Phys.* 96:2756–61
102. Aker PM, Valentini JJ. 1993. *Int. Rev. Phys. Chem.* 12:363–90
103. Germann GJ, Huh YD, Valentini JJ. 1992. *J. Chem. Phys.* 96:1957–66
104. Germann GJ, Huh YD, Valentini JJ. 1992. *J. Chem. Phys.* 98:5746–57
105. Picconatto CA. 2000. PhD dissertation. *The Dynamics of the Photodissociation of HCl dimmer and State-to-State Reactions of Hydrogen Atoms with Linear Alkanes.* New York: Columbia Univ.
106. Srivastava A. 2000. PhD dissertation. *State-to-State Dynamics of the Reaction of Atomic Hydrogen with Cyclic Alkanes.* New York: Columbia Univ.
107. Huang J. 1995. PhD dissertation. *Quasiclassical Trajectory Studies of Polyatomic Reaction Dynamics.* New York: Columbia Univ.
108. Lanzisera DV, Valentini JJ. 1993. *Chem. Phys. Lett.* 216:122–25
109. Lanzisera DV, Valentini JJ. 1994. *J. Chem. Phys.* 101:1165–71
110. Lanzisera DV, Valentini JJ. 1997. *J. Phys. Chem. A* 101:6496–503
111. Andrews L. 1968. *J. Chem. Phys.* 48:972–82
112. Luke BT, Loew GH, McLean AD. 1987. *J. Am. Chem. Soc.* 109:1307–17
113. Hudgens JW, III, Johnson RD, III, Tsai BP, Kafali SA. 1990. *J. Am. Chem. Soc.* 112:5763–72
114. Gustev GL. 1991. *J. Phys. Chem.* 95:5773–83
115. Rodriguez CF, Sirois S, Hopkinson AC. 1992. *J. Org. Chem.* 57:4869–76
116. Setzler JV, Guo H, Schatz GC. 1997. *J. Phys. Chem. B* 101:5352–61

Recent Progress in Infrared Absorption Techniques for Elementary Gas-Phase Reaction Kinetics

Craig A Taatjes[1] and John F Hershberger[2]

[1]Combustion Research Facility, Mail Stop 9055, Sandia National Laboratories, Livermore, California 94551-0969; e-mail: cataatj@ca.sandia.gov
[2]Department of Chemistry, North Dakota State University, Fargo, North Dakota 58105; e-mail: John_Hershberger@ndsu.nodak.edu

Key Words branching fraction, vibrational energy distribution, kinetic isotope effect, linestrengths, photolysis

■ **Abstract** Sensitive and precise measurements of rate coefficients, branching fractions, and energy disposal from gas-phase radical reactions provide information about the mechanism of elementary reactions as well as furnish modelers of complicated chemical systems with rate data. This chapter describes the use of time-resolved infrared laser absorption as a tool for investigating gas-phase radical reactions, emphasizing the exploitation of the particular advantages of the technique. The reaction of Cl atoms with HD illustrates the complementarity of thermal kinetic measurements with molecular beam data. Measurements of second-order reactions, such as the self-reactions of SiH_3 and C_3H_3 radicals, and determinations of product branching fractions in reactions such as $CN + O_2$ rely on the wide applicability of infrared absorption and on the straightforward relationship of absorption to absolute concentration. Finally, investigations of product vibrational distributions, as in the $CN + H_2$ reaction, provide additional insight into the details of reaction mechanisms.

INTRODUCTION

The study of elementary gas-phase reaction kinetics, a venerable area of chemical investigation, continues to play a prominent role in our understanding of fundamental chemistry and of large chemical systems. High-precision laboratory measurements of gas-phase radical reactions are responsible for much of the kinetic data on which fields such as combustion modeling (1) and atmospheric chemistry (2, 3) rely. The investigation of thermal rate constants and branching fractions, coupled with isotopic substitution, also provides an important window into the mechanisms underlying elementary chemical reactions. Reactions under thermal conditions access large areas of the global potential energy surface for the reaction, and collisional energy transfer affects rate coefficients and branching fractions in complex-mediated reactions. Precise and accurate thermal kinetics measurements provide data complementary to that from state-resolved dynamics. Optical detection of the disappearance of reactants, or the appearance of products following pulsed photolytic initiation, have become standard techniques in gas-phase radical reaction kinetics over the past several decades.

Time-resolved absorption spectroscopy has a number of key advantages for the study of chemical reactions. Absorption does not rely on the fate of the excited-state molecule and can thus be applied in quenching environments. It can detect non-fluorescing species and can be used in transitions where the upper state is dissociative. Additionally, the integrated absorption is proportional to the concentration of the molecule. If the lineshape can be measured (or is known), measurements of absolute concentrations are in principle straightforward (4–6). Absorption of a continuous-wave (CW) probe also collects the entire time profile on each photolysis laser pulse, providing excellent time resolution and allowing signal-averaging to be used to increase signal-to-noise ratios.

Absorption in the infrared region has additional advantages. First, infrared absorption is nearly universal: almost every molecule, except homonuclear diatomics, displays some absorption in the infrared. Further, infrared transitions often arise from absorption features of functional groups clustered in frequency regions, making it convenient to probe many similar molecules with the same apparatus. Doppler broadening is smaller in the infrared than in the visible or ultraviolet, providing better resolution of closely-spaced features. In addition, infrared transitions are generally not dissociative, so significant lifetime broadening is uncommon.

The disadvantages of absorption techniques are also familiar. First, absorption is a line-of-sight measurement, and any absorber in the probe path contributes to the signal. Such interferences cause difficulty if atmospheric absorption occurs at the probe wavelength, and can make temperature-dependent measurements more difficult because of the contributions of signals from areas of the reactor that are not effectively temperature-controlled (usually near windows). The most famous disadvantage of absorption probing is low sensitivity. Infrared absorption strengths are often relatively small. By its nature, the absorption signal appears as a small change in transmitted intensity, making it sensitive to amplitude noise of the probe laser. An added feature of any state-resolved absorption that can complicate kinetic measurements is the inherent sensitivity of the absorption signal to the vibrational distribution of the probe molecules.

The present pulsed photolytic methods for studying radical reaction kinetics and dynamics trace back to the development of flash kinetic spectroscopy by Norrish and Porter (7). Spectrally-resolved photography of absorption, or emission from transient species formed by flashlamp pulses, has produced a vast body of spectroscopic data on radicals. Measurement of photochemical processes and reaction kinetics using ultraviolet or visible probes of the radical species also followed rapidly. However, extension to the infrared region was hampered by less efficient detection and less convenient sources; not until the development of reliable tunable infrared laser sources did laser flash kinetic spectroscopy in the infrared become viable (8–15).

A review of high-resolution infrared studies of molecular dynamics in 1985 described the investigation of photochemistry by infrared laser methods as being in an "immature stage," with only a few photolytically initiated reactions having been studied by infrared laser absorption at that date (16). However, since then infrared kinetic spectroscopy has gained wide use for studies of reaction kinetics (8, 17–30). New technical developments in semiconductor lasers and advances in nonlinear optical methods as well as methods to improve absorption sensitivity, like frequency modulation and cavity ringdown, promise still wider application of time-resolved infrared absorption to the study of gas-phase radical reactions.

This chapter describes recent developments in time-resolved infrared absorption techniques and their applications to measurements of gas-phase chemical kinetics. First, a short introduction to the technical basis of the method is given; then several examples are discussed, illustrating the utility of infrared absorption techniques to radical reactions. Much of the development of infrared laser kinetic spectroscopy has been reviewed by others (14–16, 31); this chapter concentrates on applications of time-resolved infrared absorption to measurements of gas-phase radical kinetics. We describe measurements of rate coefficients for reactions of gas-phase radicals with stable molecules, and with other radicals; the use of infrared absorption for determination of branching fractions in radical-molecule reactions; and the measurement of vibrational energy disposal in thermal reactions.

TIME-RESOLVED INFRARED LASER ABSORPTION FOR GAS-PHASE RADICAL KINETICS

General Methodology

The principle of a laser photolysis/CW infrared laser absorption experiment is simple: a reaction is started by photolysis of some precursor molecule, typically by an excimer laser or a harmonic of a Nd:YAG laser, and the course of the subsequent reaction is monitored by absorption of a CW laser probe. Transient digitization of a CW laser signal yields detailed time resolution that can be valuable in discerning a subtle departure from the anticipated time behavior that can be evidence of an experimental artifact or unexpected chemistry. A diagram of a modern experimental apparatus is shown in Figure 1.

Photolysis can easily produce nearly uniform radical concentrations of 10^{14} cm^{-3} in a few nanoseconds. The ideal photolysis would be of a stable, nonreactive precursor, at a wavelength at which the other reactants do not absorb, and which produces only the radical of interest at thermal energies. Often none of these conditions are obtained in a real experiment. Undesired photochemistry from absorption of the reagents can usually be minimized by the choice of

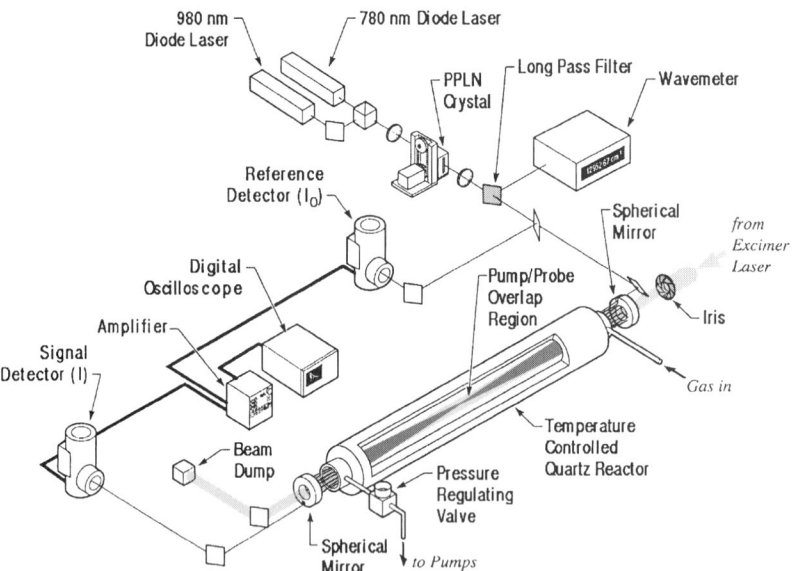

Figure 1 Representative apparatus for investigating reaction kinetics by laser photolysis/time-resolved infrared laser absorption. In the figure the probe beam is generated by difference-frequency mixing of two diode lasers in periodically-poled $LiNbO_3$ and is passed multiple times through the reaction cell using a Herriott cell. The transient absorption is obtained as the difference between balanced signal and reference detectors (I-I_0).

precursors and photolysis wavelength. However, the production of an unwanted counterradical resulting from the photolysis, and the deposition of large amounts of excess energy into the radical fragments, are often unavoidable. Unless the photolyte fragments into two of the desired radicals, e.g. Cl_2 photolysis to produce Cl atoms or $(CN)_2$ photolysis to produce CN radicals, one fragment in the photolysis is always unwanted. Ultraviolet photolysis commonly leaves fragments with tens of kcal mol^{-1} in internal energy that can significantly alter reactivity; care must be taken to ensure that the timescale for reaction greatly exceeds that for collisional relaxation. Investigations of the collisional relaxation of radicals therefore provide important auxiliary information for reactive kinetics studies.

The effect of collisional vibrational relaxation can be directly monitored in some infrared absorption experiments. Nonthermal vibrational energy distributions from photolytic or reactive initiation can be straightforwardly monitored by infrared laser absorption to ensure thermalization (32). The sensitivity of infrared absorption to the vibrational distribution can also be exploited to allow measurement of vibrational energy disposal in thermal reactions, which can provide clues about the underlying reaction mechanism (29, 33–35). The wide applicability of infrared absorption permits the production of several photolysis products to be measured in the same apparatus that can be used to calibrate absorption strengths, as discussed in the section titled "Radical-Radical Kinetics" below.

Determination of rate coefficients or branching fractions, using infrared absorption probing, usually requires time-resolved monitoring of an isolated transition of a reactant or product molecule. Most often this is accomplished by fixing the probe laser at the peak of the absorption profile and monitoring the change in absorbance following the photolysis pulse. The time resolution of a transient absorption technique is usually limited at short times by the detector response, and at long times by the removal of reactants by diffusion. Infrared detectors are slower than their counterparts in the visible and ultraviolet, but sub-microsecond rise times are easily obtainable. Diffusion times at pressure of tens of torr are typically on the order of $0.01-0.1$ s at room temperature, making the range of easily accessible first-order rate constants between 10^2 and 10^6 s^{-1}.

The most common measurement in gas-phase radical kinetics is that of a second-order rate constant that is deduced from the linear dependence of a pseudo–first-order rate coefficient (where one reactant is present in great excess) on the concentration of the excess reagent. In a time-resolved experiment under these conditions, the decay of the concentration of the more dilute reactant (usually a radical formed in the initial photolysis), or the formation of products from a simple bimolecular reaction, is most often governed by a simple exponential. The slope of a plot of the exponential time constant versus the concentration of the excess reactant gives the desired second-order rate coefficient. For more complex systems involving multiple reactions or equilibration processes, the time behavior of reactant disappearance and product appearance will not be simply exponential, and these cases require more detailed analysis. An example from the Nesbitt group (36) is shown in Figure 2, which depicts the biexponential decay of OH in the O_3/OH/HO$_2$ chain reaction. Under the conditions of these experiments, the fast initial decay of the OH represents the induction period of the chain reaction and is related to the sum of the two rate constants; the ratio of the two rate constants is reflected in the ratio of the amplitudes of the slow and fast decays. Because the infrared probing of OH is free of interfering absorption from the other reagents, these studies were able to measure chain kinetics at ten times the ozone concentrations available to experiments using ultraviolet detection.

In the pseudo–first-order limit, only the relative concentration of the minor reagent as a function of time is necessary to determine the rate constant. However, in cases where the absolute concentration of the reactant must be determined, as for second-order rate coefficients and branching fraction measurements, it is important to unambiguously determine the integrated absorption. Often the frequency stability of the probe laser is sufficient to simply to set the probe at the peak of the profile and calculate the integrated absorption from the known line profile. Applications requiring longer-term stability may use line-locking techniques (37, 38). In some cases the integrated absorption is measured by scanning over the entire lineshape to obtain line profiles as a function of time delay. This measurement can be made using a standard gated integrator/boxcar averager combination (39) or

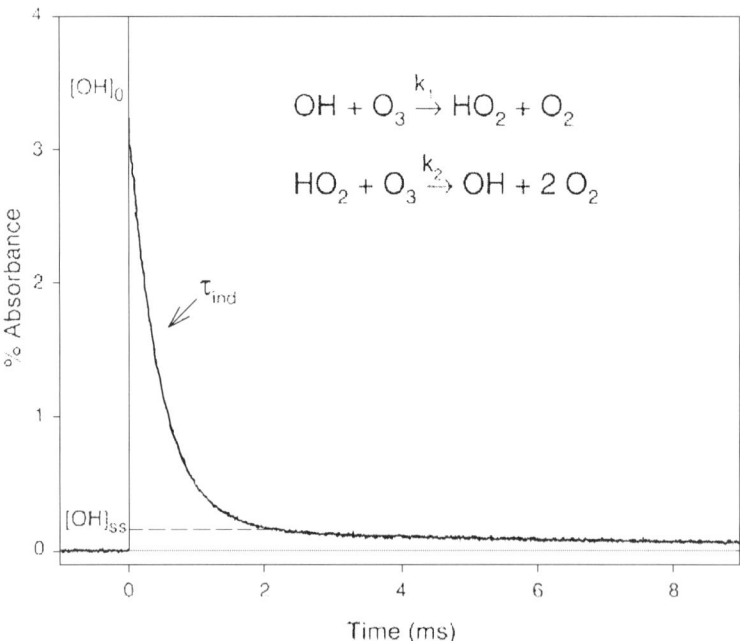

Figure 2 Transient infrared absorption signal corresponding to OH in the OH/HO$_2$/O$_3$ chain reaction system, initiated by photolysis of O$_3$ in the presence of water. Reproduced with permission from Reference (36). Copyright 1998 American Chemical Society.

by rapidly and repeatedly scanning a diode laser in frequency over the transition during the acquisition of the transient absorption trace (19, 29, 40).

Calibration of Infrared Absorption Signals An absorption technique is extremely powerful for determining absolute concentrations of reactants and/or products. These situations occur in the measurement of the rate of a bimolecular reaction that cannot be performed under pseudo–first-order conditions, such as a radical self-reaction. Also, if a reaction has several product channels, quantitative determination of the branching ratios requires the measurement of absolute concentrations. The conversion of an infrared absorption signal into absolute concentrations is a straightforward application of Beer's absorption law that relates the transmitted light intensity I to the incident intensity I_0 by:

$$I = I_0 e^{-\alpha l p}, \qquad 1.$$

where l is the path length, p is the partial pressure of the absorbing species, and α is the absorption coefficient (base e), often expressed in units of cm^{-1} torr^{-1}. In many kinetics experiments, a transient absorption signal $\Delta I = I_0 - I$ is the directly measured quantity. Furthermore, if the probe light is a high-resolution laser source tuned to an individual rotation-vibration line of the probed molecule, the transient signal is sensitive not to the total number density N but to N_{vJ}, the number density of molecules in a particular rotational and vibrational state. In fact, the absorption strength is proportional to the difference in population between the upper and lower states involved in the transition. The effect of nonequilibrium vibrational distributions on time-resolved absorption signals is discussed in the section titled "Vibrational Energy Disposal in Thermal Gas-Phase Radical Reactions" below; the present discussion assumes negligible population in the upper state of the probed transition. A modified form of Beer's law is then:

$$N_{vJ} = \ln\left[\frac{I_0}{I_0 - \Delta I}\right] \frac{f_v^0 f_J^0 N_0}{\alpha l}, \qquad 2.$$

where the Boltzmann factors f_v^0 and f_J^0 are the fraction of molecules in the probed v and J state respectively, under the conditions at which α was measured (for example, a $T_0 = 296$ K Boltzmann distribution). N_0 is Loschmidt's number; $N_0 = 3.26 \times 10^{16}$ molecules cm^{-3} torr^{-1} at 296 K.

For some molecules, tabulated values of linestrengths S_{vJ} are available. For example, the popular HITRAN database (41) provides tabulated values of precise line positions and linestrengths for pure rotation and rotation-vibration lines of many small molecules of interest in atmospheric chemistry. The linestrength may roughly be described as the integral of the absorption coefficient α over the entire lineshape. S_{vJ} has units of cm molecule^{-1}. If the absorption line is dominated by Doppler broadening (as is generally the case for pressures of a few torr or less) the absorption coefficient at line center may be easily obtained from the linestrength:

$$\alpha = \left(\frac{1}{\Delta v_D}\right)\left(\frac{\ln 2}{\pi}\right)^{\frac{1}{2}} N_0 S_{vJ}, \qquad 3.$$

where Δv_D is the Doppler half-width at half maximum:

$$\Delta v_D = \left(\frac{2 k_B T \ln 2}{mc^2}\right)^{\frac{1}{2}} v_0. \qquad 4.$$

Here v_0 is the wavenumber (cm^{-1}) of the line center, k_B is Boltzmann's constant, m is the mass of the probed molecule, and c is the speed of light. At pressures greater than $\sim$5 torr, pressure broadening becomes significant, and the absorption lineshape is a Voigt profile. Although pressure-broadening data are available for some molecules, the simplest procedure for a stable molecule is usually to measure the linestrength independently under pressure conditions that closely mimic those used for the kinetics experiments.

Note that the above equations give N_{vJ}, the number density in a specific v and J state. However, the total number density N of the probed species is usually required for branching ratio measurements. In that case one must either explicitly measure N_{vj} for each significantly populated quantum state or arrange the experimental conditions to ensure a Boltzmann distribution of vibrational and rotational states. If the latter procedure is used, then one obtains:

$$N = \frac{N_{vJ}}{f_v f_J}, \qquad 5.$$

where f_v and f_J are Boltzmann factors determined at the temperature of the experiment, which is not necessarily the same as $T_0 = 296$ K, the temperature for which linestrength data are routinely tabulated. (Author's note: Equations 6 and 7 have been deleted.)

As mentioned previously, pressure broadening will decrease the absorption coefficient compared to that calculated above. It must also be emphasized that measurement on a single rotation-vibration line can determine the total species concentration only if one ensures that rotational and vibrational relaxation processes are completed on the timescale of the experiment. In general, rotational energy transfer is an efficient process, occurring at a roughly gas-kinetic rate. Vibrational energy transfer, however, is often less efficient and depends greatly on both the species probed and the collision partners in the reaction mixture. A substantial body of literature exists on vibrational energy transfer rates, which can be helpful in suggesting a choice of buffer gas for a given experiment. For example, efficient relaxation of vibrationally excited states of CO_2 and N_2O may be accomplished by use of SF_6 buffer gas (42, 43). Relaxation of excited CO, however, is slow in SF_6 (44) but relatively fast in CF_4 (44, 45).

An example of relaxation effects is shown in Figure 3, taken from a study of CO_2 and N_2O product ratios formed in the NCO + NO reaction (46, 47), which is of interest in modeling of NO_x reduction schemes in combustion chemistry (48–50). The traces shown are transient infrared absorption signals obtained while monitoring a single rotation-vibration line of CO_2 products in the ground vibrational state. Under the experimental conditions used, removal of NCO reactants by NO is expected to occur within a few microseconds, based on an NCO + NO rate constant of $(3.2–3.4) \times 10^{-11}$ cm^3 molecule^{-1} s^{-1} (51–53). As shown, the risetimes of the signals are much longer than predicted by the reaction rate and also display significant induction periods. Both the risetimes and induction periods depend markedly on the partial pressure of SF_6 buffer gas, providing evidence that the risetimes are governed by vibrational relaxation of excited vibrational states of CO_2 produced in the NCO + NO reaction. Although much slower than the reaction rate, the relaxation is still fast enough to allow one to discriminate against even longer (∼several ms) timescale processes such as slow secondary chemistry and diffusion of product molecules out of the probed volume.

An additional consideration is the temperature sensitivity of Equation 2. Many time-resolved kinetics experiments use laser photolysis to initiate the chemical reaction by photodissociation of a precursor molecule. Depending on the photophysics of the precursor and the heat capacities of the gases used in the reaction mixture, the photolysis pulse typically induces a small transient local temperature rise on the order of a few Kelvin. This temperature perturbation affects the choice of which rotation-vibration lines to probe. Generally, one should use spectral lines

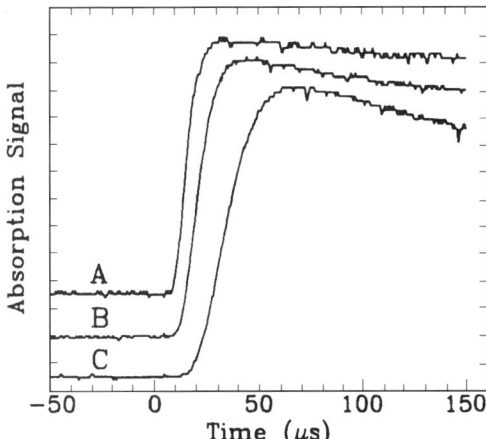

Figure 3 Transient infrared absorption signals for CO_2 (00^00; J = $8 \to 00^01$; J = 9 line at 2355.890 cm^{-1}) products of the NCO + NO reaction, illustrating the effect of SF_6 buffer gas pressure. A: 3.0 torr SF_6; B: 1.5 torr SF_6; C: 0.5 torr SF_6. Reproduced with permission from Reference (47). Copyright 1992 American Chemical Society.

that are fairly close to the maximum of the Boltzmann distribution because these linestrengths are the least sensitive to slight changes in temperature.

Experimental Techniques

The early examples of infrared laser kinetic spectroscopy were driven by the emergence of convenient tunable infrared laser sources such as lead-salt diode lasers (8, 9), color center lasers (13, 54), and difference-frequency generation (10). Similarly, the continued development of more versatile, more stable, smaller, more robust, and cheaper laser sources throughout the infrared region is fueling increased activity in infrared absorption studies of gas-phase chemical kinetics. The combination of new solid-state laser sources with the rapid progress in nonlinear optical methods promises improvements across the entire infrared region.

The first reports of time-resolved infrared laser spectroscopy following photolysis used cooled lead-salt diode lasers, and they still account for the bulk of infrared diode laser absorption measurements of gas-phase kinetics. Diode laser detection is extraordinarily versatile; since diodes can cover most of the infrared out to 30 μm, diode absorption experiments are commonly able to detect most of the products in chemical reactions. The advantages of the lead-salt diode lasers are partially offset by their cryogenic operation, uneven diode quality, and unpredictable tunability (55, 56). Recent developments in semiconductor laser technology have made small, efficient, and reliable infrared lasers available for a myriad of sensing and spectroscopic applications (57). The most rapid progress has been in room-temperature diodes operating in the near-infrared, which are important for communications applications, and in visible diodes. Conversion of room temperature diode laser light using fiber amplifiers and nonlinear optical methods will enhance their application to the chemically useful mid-infrared region (58). Room temperature near-infrared diode lasers have been used directly in kinetics experiments to detect HO_2 radicals in the first overtone of the $H-O_2$ stretch near 1.5 μm (59, 60) and to detect CN radicals in the (A ← X) band near 790 nm (17, 39, 61). The quantum cascade laser is another promising mid-infrared source (62, 63). Pulsed operation at room temperature and CW operation at cryogenic temperatures are possible over a wide range, and distributed feedback versions provide single-mode tunability (64).

Difference-frequency generation (DFG) in bulk $LiNbO_3$ (10, 65) was also used in early transient infrared laser absorption experiments. Advances in the production of nonlinear optical crystals have enabled DFG to be expanded to longer wavelength (66). Materials such as GaSe, $AgGaS_2$, and $AgGaSe_2$ can be used to generate tunable infrared light from difference-frequency mixing of two tunable visible or near-infrared lasers out to 20 μm (67–70). The convenience of difference-frequency generation has been most increased by the recent development of quasi–phase-matched (QPM) materials, especially periodically poled $LiNbO_3$ (PPLN), which allows production of IR difference frequency radiation from 1 μm to 5 μm (71). Other nonlinear crystals can also be periodically poled or otherwise grown with periodic changes in the nonlinear optical indices. Since the bulk phase-matching condition need not be met, laser sources can be chosen for ease of operation. Compact, rugged sources are feasible using solid state laser sources (72), and PPLN has begun to be used in infrared kinetic spectroscopy (73, 74).

Because of the relatively low sensitivity of infrared absorption, many experiments employ some method to increase the sensitivity of the infrared laser probe. The fundamental limit of direct absorption detection is that given by the shot noise (i.e., the square root of the number of photons) of the transmitted laser beam. For a 10 μW beam at 3 μm, this is an absorption of 8×10^{-8} in a 1 Hz bandwidth, or 8×10^{-5} in a 1 MHz bandwidth more typical of a kinetics experiment. Simple subtraction of two detectors to cancel laser amplitude noise can typically reach within a factor of 100 of the shot noise limit, and careful balancing of the detectors can reach within a factor of ten. Beyond the improved characterization of time profiles that higher signal-to-noise allows, improving the sensitivity of the probe permits lower concentrations of radicals to be used, thus minimizing the competition of radical-radical reactions.

Frequency-modulation (FM) or wavelength-modulation (WM) spectroscopy is one means for improving the sensitivity of absorption measurements to near the shot noise limit (75–77). The technique of transient FM spectroscopy and its application to the study of dynamics and kinetics has been recently reviewed (78). Because FM detection measures a differential absorption between carrier and sideband frequencies, the contributions of thermal lensing noise, caused by transient changes in the refractive index after the photolysis pulse, and of changes in broadband absorption because of the photolysis, are greatly reduced. Already in some of the earliest reports of infrared kinetic spectroscopy it was noted that such effects were a dominant source of noise (13). The amplitude of the FM or WM signal is proportional to the absorption strength; if absolute concentrations are needed, some form of calibration must be performed to extract linestrengths.

Taatjes & Oh (60) compared direct absorption signals to wavelength modulated signals in an investigation of the HO_2 self-reaction at room temperature. This simple calibration method relies on the ability to detect the species of interest in direct absorption, which may not always be possible. In the measurement of pseudo–first-order rate coefficients, however, the proportionality of the FM signal to absorption is sufficient.

While subtraction of signal and reference beams and frequency modulation reduce the detection noise, other techniques are available to increase the absorption signal. The simplest method is the use of a multipass arrangement to increase the pathlength through the reaction zone. A White cell (79) is commonly used, which permits the probe beam to fill a large volume. A multipass cell based on the Herriott design (80) has been applied to kinetics measurements by Pilgrim et al (81). The Herriott cell is built on a spherical resonator, where the probe beam follows an off-axis path, tracing out a circle of spots on each mirror and a smaller circle in the center of the cell. The photolysis laser is then sent down the center of the resonator, on-axis, and the overlap between pump and probe beams is thereby confined to the center of the cell, where the temperature can be well controlled. This design permits efficient and adjustable overlap between the photolysis and probe beams. Overlap in the White cell is accomplished by dichroic mirrors or by crossing the photolysis beam at an angle. Multipass cells, such as the White cell or an astigmatic Herriott cell (82), that can nearly fill the volume between the mirrors can have an advantage over the spherical Herriott design when reducing diffusion out of the probe volume is desired.

In the past several years cavity ringdown (CRD) spectroscopy has been widely investigated as a means for ultra-sensitive absorption detection (83–85). As with many spectroscopic innovations, the application to time-resolved kinetics measurements follows closely. The Lin group (86, 87) pioneered the use of visible pulsed CRD for time-resolved kinetics, first investigating reactions of the phenyl radical; visible and UV CRD is now gaining broad application in chemical kinetics studies (85, 88–91). The use of CRD methods in the infrared for measurements of reaction kinetics is an inevitable development. Quantum cascade lasers have already been employed in CRD experiments near 8.5 μm (92). CRD using CW lasers can operate at high enough acquisition rates to obtain entire transient absorption traces on a single photolysis laser pulse (93, 94), enabling signal averaging of repeated photolysis pulses.

MEASUREMENTS OF RATE COEFFICIENTS FOR GAS-PHASE RADICAL REACTIONS

Radical-Molecule Reactions

The measurement of rate coefficients for reactions of radical species with stable molecules, usually carried out in the pseudo–first-order limit with the stable species in excess, is the most common type of kinetic measurement studied by photolytic techniques. Infrared absorption measurements have concentrated on the reactions of species that are not easily accessible by other methods. For example, Moore and coworkers studied the reactions of triplet CH_2 using diode-laser probing (28, 95, 96), and *trans*-HOCO reactions using difference-frequency probing (25). Reactions of HCCO (97) and C_2H (98) were also early targets.

The reactions of Cl atoms with hydrocarbon species produces HCl, a product ideally suited for infrared absorption spectroscopy because of its large rotational constant, high vibrational frequency, and large infrared absorption cross section. Rate constants for many hydrogen abstraction reactions of Cl atoms have been measured by monitoring the time-resolved absorption of the HCl product (29, 35, 38). The reaction of Cl with HD (99) provides a simple but illustrative example of an absolute rate coefficient measured by time-resolved infrared absorption detection of a reaction product, where the complementary nature of thermal kinetics and state-resolved dynamics is clear.

The reaction of Cl atoms with H_2 is the simplest chlorine atom reaction, and investigations of this reaction have played a fundamental role in the development of transition state theory and the description of chain chemistry. In recent years there has been a renewed interest in $Cl + H_2$ and its isotopomers as precise kinetics measurements, new and more powerful theoretical methods and capabilities, and a series of extraordinarily elegant and detailed reaction dynamics experiments have been brought to bear. The experimental thermal rate coefficients of $Cl + H_2$ and D_2 between 296 and 3000 K were evaluated and summarized in 1994 (100). High quality ab initio (101) and semi-empirical potential energy surfaces (102) are now available for this reaction.

Absolute rate coefficients for the reaction of Cl with HD were measured using time-resolved infrared absorption probing of the HCl product following 193 nm photolysis of CCl_2F_2 (99). Chain chemistry, which severely complicates the use of Cl_2 as a precursor, is not important when using chlorofluoromethane photolysis as a Cl source. The production of HCl was monitored using a difference-frequency laser tuned to the R(2) line of the HCl fundamental. Two-tone frequency modulation was used to decrease the sensitivity to thermal lensing noise in the experiments. A simple exponential growth of HCl was observed, and the second-order rate coefficient was measured as the slope of a plot of the time constant versus HD concentration in the standard way.

The rate coefficients measured for Cl + HD as a function of temperature are shown in Figure 4*a*. Because the infrared experiments probe the HCl product, the branching into HCl was also measured by comparing the signal amplitude from the Cl + HD reaction with that from a reference reaction producing only HCl (e.g. $Cl + C_3H_8$) under the same photolysis conditions. This intramolecular kinetic isotope effect, as well as kinetic isotope effects calculated from ratios of the absolute rate coefficients are in relatively good agreement with earlier direct kinetic isotope effect measurements (103, 104). The rate constant for Cl + HD is approximately 30% smaller than the mean of the H_2 and D_2 reactions.

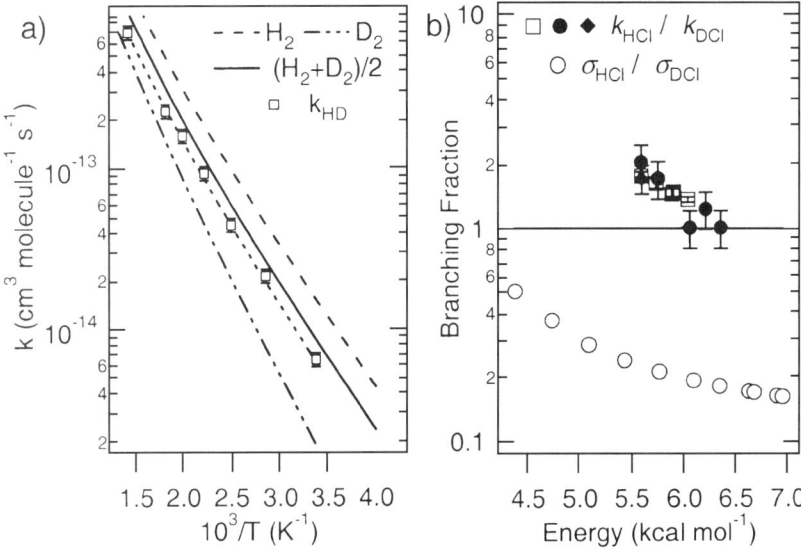

Figure 4 (*a*) Rate coefficients for the reactions of Cl atom with hydrogen molecule isotopomers. The dashed line is the rate coefficient for Cl + H_2, and the dot-dashed line that for Cl + D_2, both taken from the evaluation of reference (100). The mean of the evaluated Cl + H_2 and Cl + D_2 rate coefficients is shown as the thin solid line. The Cl + HD rate coefficient, shown as the individual points with ±2σ error bars, falls ∼30% below the mean of the Cl + H_2 and Cl + D_2 rate coefficients. The fitted temperature dependence of the Cl + HD reaction is shown as the dotted line. Adapted from *Chemical Physics Letters* 306, CA Taatjes, "Infrared frequency-modulation measurements of absolute rate coefficients for Cl + HD → HCl(DCl) + D(H) between 295 and 700 K," copyright (1999), p. 37, with permission from Elsevier Science. (*b*): Comparison of relative reactive cross sections for HCl and DCl production in the molecular beam experiments of Reference (110) (open circles) with thermal branching measurements from References (103) (open squares), (99) (closed circles), and (167) (filled diamond). The average energy for the thermal measurements was calculated using the activation energy of 4.7 kcal mol^{-1} from Reference (99).

Greater insight comes from the comparison of the thermal rate coefficients with reaction dynamics measurements using molecular beams. Recently the reaction of Cl with H_2 has been the subject of exquisite crossed molecular beam measurements by the Casavecchia group (105, 106) and the Liu group (107–110). The Liu group has made measurements of excitation functions (i.e. reactive cross sections as a function of collision energy) and differential cross sections, and they have shown compelling evidence for the reactivity of spin-orbit excited Cl atoms. Excitation functions for Cl reacting with H_2 (109), D_2 (111), and HD (110) have been measured at collision energies ranging from 2.5 to 9 kcal mol^{-1}; the HD and D_2 cross sections are very similar, and both are much less than the cross section for Cl + H_2 (111). The molecular beam cross-section results contrast significantly with the relative thermal rate coefficients, where k_{Cl+HD} is approximately midway between k_{Cl+D_2} and k_{Cl+H_2}. An even more dramatic discrepancy occurs for the intramolecular kinetic isotope effect, where the beam experiments favor H + DCl products by a factor of 3 and thermal measurements favor D + HCl at room temperature by nearly a factor of 2 (108, 110).

Thermal kinetics measurements act as a sum-check on molecular beam experiments, since the weighted sum over state-resolved excitation functions must give the thermal rate constant. Comparing thermal kinetics measurements with molecular beam measurements is not always straightforward because of the elevated collision energies in most molecular beam experiments. The Cl + HD reaction proceeds over a barrier, and it has an activation energy of 4.7 kcal mol^{-1} near room temperature. The activation energy of a reaction is the difference between the average energy of reactive collisions and the average thermal energy (112). Hence, the average energy of a reactive encounter in the room temperature reaction of Cl + HD is more than 5 kcal mol^{-1}, similar to the collision energies in the molecular beam experiments. Figure 4b compares the thermal measurements with the molecular beam measurements as a function of average energy, taking the average energy for the thermal reaction as the activation energy plus the average thermal energy (3/2 RT). However, in a thermal experiment higher vibrations and rotations of the reactants are also populated. The differences between Cl + HD under thermal conditions and in a molecular beam therefore directly suggest differences between high kinetic energy collisions of cold molecules and collisions with significant internal energy (especially rotation) of the HD molecule. The rotational dependence of the intramolecular kinetic isotope effect can be qualitatively explained in a simple classical model (113, 114). The cone of acceptance for D end attack is increased in the non-rotating molecule simply by the displacement of the center of mass away from the center of charge; however, as the molecule rotates, this advantage is lost. Quantum mechanical calculations on the newest ab initio surface are able to reproduce the change in isotopic selectivity between beam and thermal measurements (110).

Radical-Radical Kinetics

The measurement of accurate rate constants of reactions between two transient species represents a difficult problem in chemical kinetics. For a reaction of two different radicals one generally attempts to arrange the experimental conditions such that one species is present in excess, resulting in pseudo–first-order kinetics. In order to obtain a bimolecular rate constant, the absolute concentration of the species in excess must be measured. For a reaction of a transient radical with itself, pseudo–first-order conditions cannot be obtained, and the data must be analyzed using second-order kinetics, which requires knowledge of the absolute concentration versus time. The relative simplicity of calibration procedures in absorption spectroscopy and the wide applicability of infrared absorption are advantages in such experiments. Here, we focus on two examples of the use of infrared spectroscopic measurements of radical-radical kinetics: the $SiH_3 + SiH_3$ reaction and the $C_3H_3 + C_3H_3$ reaction. Other examples of this technique include $HO_2 + HO_2$ (8), $Cl + C_2H_5$ (115), $Cl + CH_3CO$ (116), and $NH_2 + O$ (117) reactions.

The $SiH_3 + SiH_3$ Reaction Silicon hydride radicals are intermediates in chemical vapor deposition (CVD) processes involving silane (SiH_4). Applications of silane CVD include the deposition of thin films of amorphous hydrogenated silicon that are used in the manufacture of low-cost photovoltaic devices. The recombination of silyl (SiH_3) radicals,

$$SiH_3 + SiH_3 \rightarrow Si_2H_6 \rightarrow SiH_2 + SiH_4, \qquad 8.$$

is an extremely important reaction in plasma-enhanced CVD because SiH_3 is one of the primary species involved in film formation. Several reports of this rate constant have appeared and been reviewed recently (118–120). In a careful time-resolved infrared absorption study (120), SiH_3 was generated by photolysis of CCl_4 with 193 nm excimer light, followed by reaction of photolytically generated chlorine atoms with SiH_4:

$$CCl_4 + h\nu \text{ (193 nm)} \rightarrow CCl_3 + Cl \qquad 9.$$

$$Cl + SiH_4 \rightarrow HCl + SiH_3. \qquad 10.$$

SiH_3 radicals were then detected in a time-resolved infrared absorption experiment using a diode laser tuned to 726.901 cm^{-1}, which corresponds to the Q(6,6) transition of the $\nu_2(1^- \leftarrow 0^+)$ band. The approach used to form SiH_3 produces several other transient species, which in principle (few of the relevant rate coefficients are known) could contribute to the SiH_3 decay signal by secondary reactions that produce or destroy SiH_3. However, the authors argue by comparison with analogous reactions that secondary chemistry is too slow to affect their measurements. The key advantage of the two-step approach involving CCl_4 photolysis is that it provides a reliable calibration procedure by forming HCl and SiH_3 in equal yields. Because HCl is a stable species, its infrared linestrengths are well known, and absolute HCl concentrations may be readily measured. Using this technique, the authors were able to obtain an absolute measurement of $[SiH_3]$ versus time over a 600 μs timescale. After small corrections for diffusion loss, they obtained a recombination rate constant of $(7.9 \pm 2.9) \times 10^{-11}$ cm^3 molecule^{-1} s^{-1}.

In spite of extremely careful data analysis, significant uncertainties still exist, attributed primarily to uncertainties in the effective path length of overlap between the photolysis and probe beams. This disadvantage is inherent to multipass cells in which the photolysis and probe beams cannot be exactly copropagated. Often a simple single-pass cell with exactly copropagated laser beams may be advantageous despite the resultant reduction in sensitivity. The optical layout of reference (120) was partly motivated by unavailability of dichroic mirrors that would combine the UV and far-IR beams. More recent spectroscopic work on the SiH_3 radical, however, has been reported. In particular, the v_3 Si-H stretching band in the 2100–2240 cm^{-1} region has been spectroscopically characterized (121). One of the lines in this region was used in a study of SiH$_3$ + NO$_2$ kinetics (122). This spectral region is experimentally more convenient than $\sim$700 cm^{-1} for several reasons, including improved laser sources, improved detectors, and availability of dichroic mirrors.

Propargyl Radical Recombination The propargyl radical, C$_3$H$_3$, is an extremely important intermediate in the chemistry of hydrocarbon combustion. C$_3$H$_3$ may be formed by several reactions, including CH$_2$ + C$_2$H$_2$ → C$_3$H$_3$ + H and HCCO + C$_2$H$_2$ → C$_3$H$_3$ + CO. The recombination reaction is believed to play a role in soot formation. The rate of this recombination was studied by Curl and coworkers (123) using propargyl halides as photolytic precursors:

$$C_3H_3Cl \text{ (or } C_3H_3Br) + h\nu \text{ (193 nm)} \rightarrow C_3H_3 + Cl \text{ (or Br)} \qquad 11.$$

$$\rightarrow C_3H_2 + HCl \text{ (or HBr)}. \qquad 12.$$

The second channel was estimated to have a small quantum yield of only about 0.07 ± 0.01. C$_3$H$_3$ radicals were probed using a color center laser tuned to a rotation-vibration line in the v_1 band near 3300 cm^{-1} (124). HCl and Br photolysis products were also probed with the color center laser (in the case of Br, a hyperfine line $F = 2 \leftarrow 3$ of the $^2P_{1/2} \leftarrow {}^2P_{3/2}$ spin-orbit transition was used). Absolute [Br]$_0$ concentrations were determined from the Br transient signal and a calculated absorption coefficient. Since [Br]$_0$ = [C$_3$H$_3$]$_0$ in the photolysis of C$_3$H$_3$Br, this calibration procedure yielded absolute C$_3$H$_3$ concentrations as well.

As is unavoidable in these types of experiments, several secondary reactions could contribute to the observed [C$_3$H$_3$] decay rates. The authors extensively discuss the reactions

$$Cl + C_3H_3Cl \rightarrow C_3H_3Cl_2 \qquad 13.$$
$$C_3H_3Cl_2 + C_3H_3 \rightarrow \text{products}. \qquad 14.$$

By a series of Cl atom quenching experiments, they conclude that neither Cl atoms nor products of Cl atom reactions contribute significantly to the C$_3$H$_3$ decay rates. Upon fitting the [C$_3$H$_3$] vs time signals to a second-order rate equation as shown in Figure 5, they obtain the rate constant (1.2 ± 0.2) × 10^{-10} cm^3 molecule^{-1} s^{-1}.

A possible source of systematic error in second-order kinetics is inhomogeneous species concentrations in the probed reaction volume (125). For example, hot spots in the photolysis laser beam profile in principle would lead to concentration gradients normal to the propagation direction of the laser. One could characterize the beam profile using a suitable array detector, although the errors that result from such effects are generally minor. Another kind of inhomogeneity is due to a gradient in photolysis pulse energy parallel to the propagation direction, which can arise

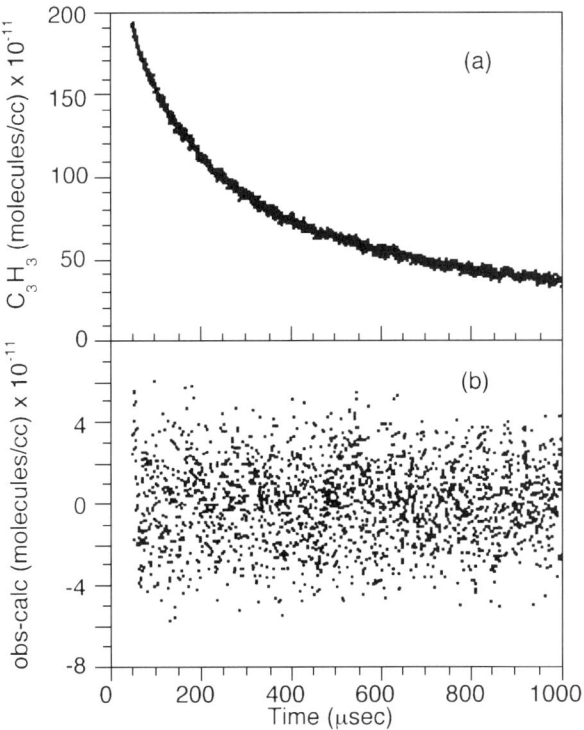

Figure 5 (*a*): C_3H_3 decay following the photolysis of C_3H_3Cl in 1 torr of CO_2 and 15 torr of He. The data were fit to a second-order rate equation. The absolute concentration is calculated from comparison of infrared signal strengths for C_3H_3 and Br obtained from C_3H_3Br photolysis. (*b*) Residuals of the second-order fit. Reproduced with permission from Reference (123). Copyright 1994 American Chemical Society.

either from divergence of the laser beam or, if the gas sample is not optically thin, from absorption of the photolysis light as it travels down the reaction cell. This issue was explicitly addressed in reference (123), and the effects of non-uniform radical concentrations were found to be minor under the experimental conditions used.

PRODUCT BRANCHING MEASUREMENTS USING INFRARED ABSORPTION: $CN + O_2$

The product branching fraction of a reaction is a key indication of the underlying mechanism. As stated earlier, an advantage of absorption spectroscopy in kinetic measurements is the relative ease with which absolute concentrations may be measured. This issue is important for chemical reactions that have more than one possible product channel:

$$A + B \xrightarrow{k_1} P_1 \qquad 15.$$
$$\xrightarrow{k_2} P_2.$$

Straightforward integration of the kinetic equations shows that, under pseudo–first-order conditions of $[B]_0 \gg [A]_0$, the time dependence of the product concentrations are given by

$$[P_1]_t = [A]_0 \frac{k_1}{k_1 + k_2} (1 - e^{-(k_1+k_2)[B]_0 t}), \qquad 16.$$

$$[P_2]_t = [A]_0 \frac{k_2}{k_1 + k_2} (1 - e^{-(k_1+k_2)[B]_0 t}). \qquad 17.$$

The rate of exponential rise of the product concentration is therefore governed by the total rate constant $(k_1 + k_2)$. In order to measure the individual rate constants and therefore the branching ratios $\phi_1 = k_1/(k_1 + k_2)$ and $\phi_2 = k_2/(k_1 + k_2)$, one must measure the absolute concentrations $[P_1]$ and $[P_2]$.

Numerous examples of the use of transient infrared laser spectroscopy to measure product branching ratios of chemical reactions exist in the literature. Some of the reactions that have been studied in this way include $NH_2 + NO$ (54, 126), $NH_2 + NO_2$ (22, 127, 128), $NCO + NO$ (46, 47), $NCO + NO_2$ (129), $NH + NO_x$ (130), $CD + NO$ (131), $CH + N_2O$ (132), $CH + NO_2$ (133), $HCO + NO_2$ (134, 135), $CH_2 + O_2$ (28), $C_2H + O_2$ (98), and $OD + CH_3SCH_3$ (136, 137).

As an example of the use of transient infrared spectroscopy to measure product yields, we consider the $CN + O_2$ reaction in some detail. This reaction has been described as a prototypical example of a barrierless chemical reaction (138). The CN radical plays an important role in NO_x formation in combustion chemistry, as well as interstellar and cometary chemistry. This reaction has the distinction of having the widest temperature range over which experimental kinetic data are available. Using varied techniques including molecular beam methods at low temperature, laser photolysis and time-resolved laser-induced fluorescence at intermediate temperatures, and shock tube experiments at high temperature, total rate constant data spanning the range 13 to 3800 K have been obtained (139–143). Both the experimental data and calculations on this reaction have been recently reviewed (1, 138). Among other applications, this reaction is of interest in efforts to develop empirical rate expressions for barrierless reactions at low temperatures (144).

Compared to the total rate constant data, less attention has been devoted to the branching ratio of this reaction. Three product channels are possible:

$$CN + O_2 \rightarrow NCO + O, \qquad 18a.$$
$$\rightarrow CO + NO, \qquad 18b.$$
$$\rightarrow CO_2 + N. \qquad 18c.$$

Conventional wisdom has long assumed that NCO + O is the major product channel. Several workers, however, have suggested that CO + NO is also a significant product (46, 145, 146). In an early study, absorption spectroscopy using a carbon monoxide laser was used to detect CO with an estimate of the branching fraction $\phi_{18b} \sim 0.06$ at 298 K (145). Experiments using infrared diode laser absorption indicated a much higher but highly uncertain branching fraction $\phi_{18b} = 0.23 \pm 0.10$ at 298 K (46). Infrared emission studies suggested an even higher fraction $\phi_{18b} = 0.29 \pm 0.02$ (146). In all three of these investigations, measurement of this branching ratio was a secondary objective, and possible contamination from competing reactions was not addressed in detail. In particular, the detected CO products could be formed by the secondary reaction

$$NCO + O \rightarrow CO + NO, \qquad 19.$$

even if $\phi_{18b} = 0$. Other secondary reactions (CN + NO, NCO + O_2) are less important. In view of the rather small yield of CO products and uncertainties in the kinetic modeling, further experiments were needed to confirm the presence of channel 18b.

In more recent experiments, the existence of channel 18b has been more convincingly demonstrated and quantified (147). The approach used was to photolyze ICN precursor molecules as before in the presence of O_2, but to also include in the reaction mixture a reagent that reacts quickly with NCO without producing the detected CO product. Because the radical concentrations in these experiments are fairly small ($\sim 10^{13}$), even a small amount of this reagent will effectively shut off the NCO + O secondary reaction. The reagent that was ultimately used was silane, SiH_4. The reaction

$$NCO + SiH_4 \rightarrow HNCO + SiH_3 \qquad 20.$$

is relatively fast, with $k_{20} = 7.0 \times 10^{-12}$ cm^3 molecule^{-1} s^{-1}. The reaction

$$CN + SiH_4 \rightarrow HCN + SiH_3 \qquad 21.$$

is still faster, with $k_{21} = 2.2 \times 10^{-10}$ cm^3 molecule^{-1} s^{-1}, but the ratio between these rate constants is not so great as to prevent modeling simulations from distinguishing their effects, because the NCO + SiH_4 reaction competes with NCO + O, while CN + SiH_4 competes with CN + O_2. Since $[O_2] \gg [O]$, a small amount of SiH_4 will still shut off the NCO + O reaction more effectively than the CN + O_2 reaction.

The result of this experiment is shown in Figure 6. As the SiH_4 pressure is increased, the CO yield decreases, as expected. Kinetic modeling of the important reactions in this system was performed under two contrasting assumptions: 1. that the CO was produced exclusively by secondary chemistry (primarily NCO + O but also NCO + NO), and 2. that the CO was produced exclusively as a minor channel in the CN + O_2 reaction. These modeling predictions are shown in the figure as triangles (model 1) and circles (model 2), respectively. Because of considerations

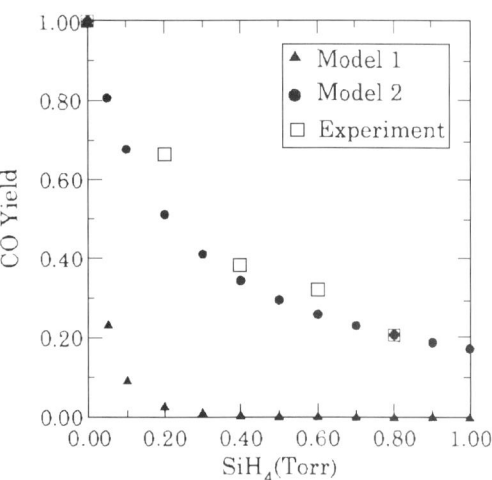

Figure 6 Experimental and modeled dependence of CO yield from the CN + O_2 reaction system on SiH_4 reagent. Model 1 (triangles) assumes that all CO originates from secondary chemistry. Model 2 (circles) assumes that all CO originates from the CN + O_2 reaction. Reproduced with permission from Reference (147). Copyright 1999 American Chemical Society.

described in the preceding paragraph, small amounts of SiH_4 quench the CO production far more effectively in model 1 than in model 2. The experimental data fit model 2 almost perfectly, clearly indicating that most of the observed CO is indeed produced by the CN + O_2 reaction and not by secondary chemistry. The determination of the branching ratio then becomes a straightforward procedure of comparing the CO and NCO yields. As discussed in Reference (147), NCO yields were determined by titration of NCO with NO into N_2O, which is more easily detected and quantified using infrared spectroscopy.

The results of this experiment are that the branching into channel 18b (CN + $O_2 \rightarrow$ CO + NO) was found to be 0.22 ± 0.02 at 296 K. More surprisingly, this branching fraction has a remarkably steep temperature dependence, as shown in Figure 7. At 240 K, the lowest temperature studied, $\phi_{18b} = 0.28$. As T increases, ϕ_{18b} decreases, reaching 0.08 at 650 K. If this trend can be extrapolated to lower and higher temperatures (an open question at this time), it is likely that the CO + NO channel is unimportant at combustion chemistry conditions, but it is very significant and possibly the major product channel under the ultra-low temperature conditions of the molecular beam experiments of Smith et al (139, 140).

It is interesting to speculate on the cause of the temperature dependence of ϕ_{18b}. Although several computational efforts on this reaction have been published, most of the ab initio work has focused on characterization of the entrance channel to a low-energy NCOO intermediate, in an effort to calculate the barrierless radical-radical capture rate. Less attention has been paid to transition states between this intermediate and possible products. One interpretation of the experimental data is that the barriers to the two product channels must be close in energy to the CN + O_2 reactant energy; otherwise, it is difficult to imagine how small changes in temperature in a thermal experiment would have a significant effect. Since NCO + O is favored at higher temperatures, it presumably has the higher barrier.

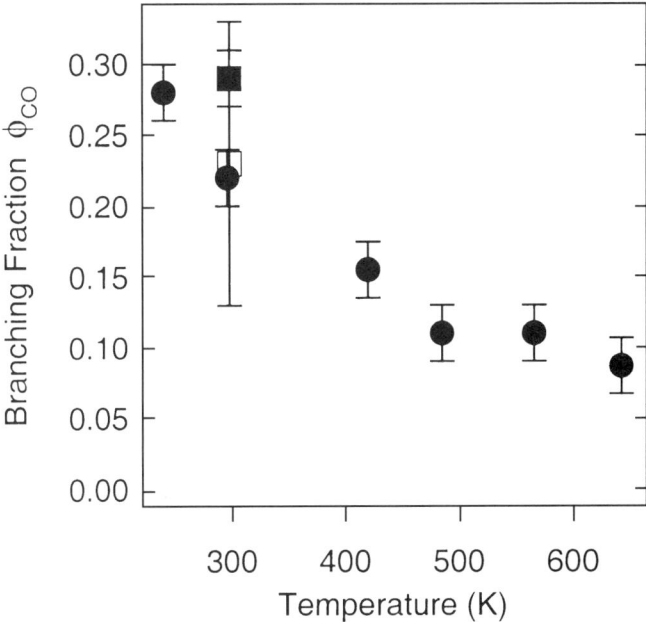

Figure 7 Temperature dependence of the branching ratio into channel 18b. Determinations from Reference (147) are shown as the filled circles. Room temperature measurements of Reference (46) and (146) are shown as the open and filled squares, respectively.

A four-centered transition state to CO + NO, although possibly lower in energy, is entropically disfavored, accounting for the fact that this is the minor channel, except possibly at very low temperatures.

Unfortunately, the little available ab initio work appears to contradict the above interpretation. Mohammad et al (146) reported MP2 calculations on this system, and claimed a very high barrier to CO + NO formation via a 4-centered cyclic transition state. Based on this observation, they proposed that any CO + NO formed results from a dynamic mechanism in which NCO + O is initially formed from an NCOO complex, with only a small amount of product translational energy but a large amount of NCO rotation. Some evidence for significant internal NCO excitation exists in energy disposal studies (148, 149). The suggestion is that rapid rotation of NCO as the O atom departs will present an OCN–O arrangement followed by N-atom abstraction to yield CO + NO before the NCO and O fragments fully separate. Although this is an interesting hypothesis, some caution is in order. The calculations were performed at what is now considered a modest level of theory, especially for radical systems. Obviously, new calculations at much higher levels of theory are needed. A dynamical mechanism such as that proposed would probably need trajectory calculations on a reasonable potential surface to be verified. It would be extremely interesting if such a calculation could predict the observed temperature dependence. It is quite possible that transition states in addition to those reported exist in this system. A further complication is that multiple potential energy surfaces may be involved in this reaction.

VIBRATIONAL ENERGY DISPOSAL IN THERMAL GAS-PHASE RADICAL REACTIONS

The formation of vibrationally excited products in chemical reactions is a marker of the reaction mechanism. For example, study of the differences in vibrational energy distributions from abstraction and from elimination dates back to the first infrared chemiluminescence investigations of chemical reactions (150). Since many infrared absorptions are vibrational transitions, the use of infrared lasers to probe reactions is inherently sensitive to the vibrational energy distribution of the probed molecule. As discussed above, in the usual case where the probe absorption is taken to be proportional to the total population, vibrational disequilibrium is a possible source of experimental error. However, many experiments have utilized time-resolved infrared laser absorption to probe the vibrational distribution of product molecules.

Several investigators have measured vibrationally excited HCl, produced from thermal reactions of Cl atoms, using time-resolved infrared absorption. The convolution of collisional vibrational relaxation and the reactive production of the HCl produces a departure from a simple exponential rise of the absorption signal; however, this departure can be subtle, as seen in Figure 8, where the effects of $\sim$13% HCl (v = 1) production in the Cl + CD_3CH_2OH reaction are shown. For reactions that produce only HCl (v = 1) and HCl (v = 0), Pilgrim & Taatjes (151) developed a method for extracting the reaction rate coefficient and the vibrational branching from measurements of v = 1 $\leftarrow$ 0 absorption profiles under different vibrational quenching conditions that was used in studies of Cl + C_3H_6 (151), Cl + C_3H_4 (152), and Cl + CH_3CH_2OH (73) reactions. Kegley-Owen et al (29) directly determined the vibrational branching in Cl + CH_3CHO by rapidly scanning a diode laser probe, on the timescale of the reaction, over closely-spaced 1 $\leftarrow$ 0 and 2 $\leftarrow$ 1 lines of HCl.

Vibrational energy disposal in polyatomic products is more challenging for time-resolved laser absorption measurements, simply because of the larger number of available vibrational states. Flynn (153, 154) and Mullin (155, 156) and coworkers have made extensive use of infrared absorption to measure vibrational energy deposition into CO_2 and H_2O from highly excited molecules. Since the conditions needed to enforce thermalization of photolytically generated reactants also increase vibrational relaxation of products, many investigations of product vibrational distributions are performed at low pressure, with superthermal translational energy (i.e. "hot" reactants). However, some reactions have been studied under thermal conditions. Vibrational distributions of methyl radicals produced by

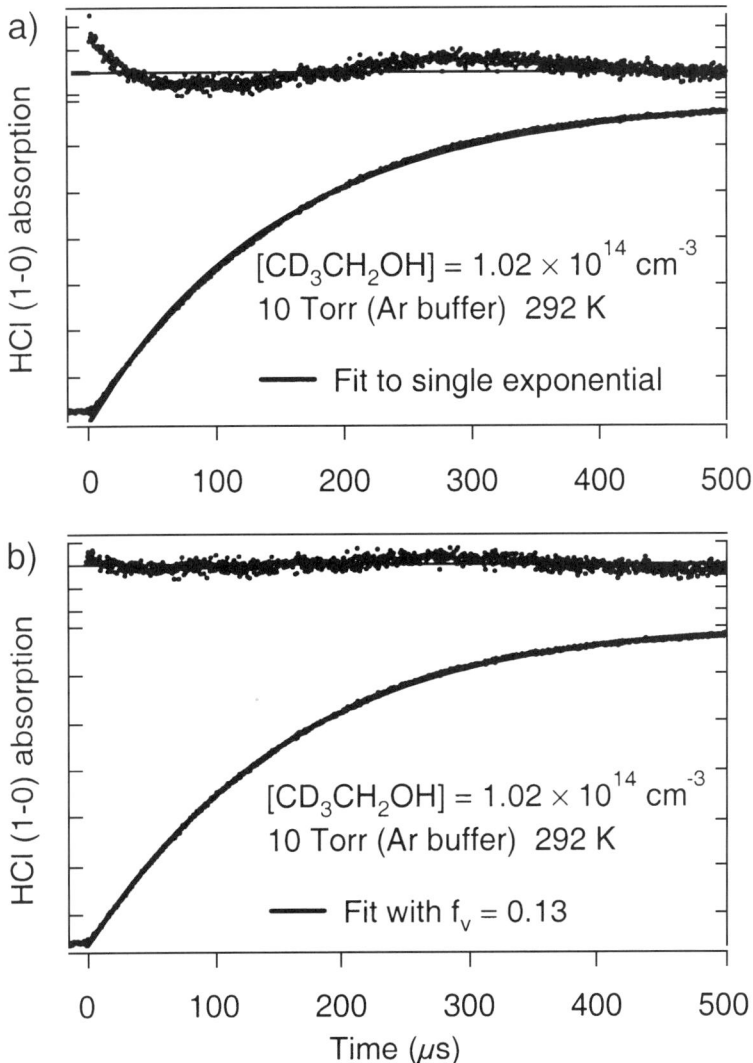

Figure 8 Time-resolved absorption on the P(2) line of the HCl (1 ← 0) transition following 193 nm photolysis of $CF_2Cl_2/CD_3CH_2OH/Ar$ mixture at 295 K. (*a*) Fit to a simple exponential form that ignores possible contributions of HCl ($v = 1$). Residuals (×5) are displayed above. There is a systematic deviation of the signal from single exponential behavior that is particularly visible in the residuals at early times. (*b*) Fit taking HCl ($v = 1$) into account. Residuals (×5) are displayed above. The fit is considerably improved. Reproduced with permission from Reference (73). Copyright 1999 American Chemical Society.

H atom abstraction from methane by F and O (^{1}D) have been probed using diode laser absorption (157, 158). The initial vibrational distribution of CO_2 produced from the reaction OH + CO was measured by Frost et al (34, 159), who inferred from the low vibrational excitation that the O-C-O geometry at the transition state was similar to that of the CO_2 product. Macdonald and coworkers (33, 160, 161) have undertaken measurements of HCN vibrational distributions from CN reactions with H_2 and hydrocarbons. The reaction of CN with H_2 is particularly informative because of the high level of theoretical and dynamical characterization that exists for that system.

In the experiments of Macdonald et al (33), CN radicals are produced by the photolysis of $(CN)_2$, and the HCN product is measured by direct infrared absorption of an F-center laser. In addition, a diode laser can be used to probe the CN radical directly by absorption in the A ← X band near 790 nm, the linestrength of which has been measured in Macdonald's group (61). This enables, in principle, absolute concentrations of both reactants and products to be monitored, a powerful capability that this group has applied to measurements of HCN branching fractions in the CN + HD isotopic reaction (39). In the measurements of the vibrational branching, however, only populations relative to the total HCN (measured as the absorption from the ground state after vibrational relaxation is complete) are needed. Figure 9 shows time-resolved infrared absorption measurements for two vibrational states of HCN produced in the reaction. Note the vibrational inversion on the (101) ← (100) transition, visible as gain in the lower trace in Figure 9. Using a combination of modeling of the vibrational relaxation process and exponential extrapolation on a set of traces for various HCN (v_1, 0, v_3) levels, relative populations could be inferred.

A large amount of C-H stretching excitation is produced in the HCN product, which is expected from a direct-abstraction mechanism. Somewhat surprisingly, significant excitation of the C-N stretch (v_1) also occurs in the reaction, suggesting that the C-N bond strength changes as the reaction proceeds. Only the bendless ($v_2 = 0$) levels were observed, which corresponded to only 27% of the total HCN, indicating a large amount of bending excitation in the reaction. Since the abstraction occurs via a linear HHCN transition state, the amount of bending excitation may provide a clue to the possible role of addition-elimination channels in the reaction. The product vibrational energy distributions in the CN + H_2 can be compared to the differential cross section data from Liu's group on the isotopic variant CN + D_2 (162). The observed backward-peaked angular distributions near the threshold energy are consistent with a dominance of the abstraction mechanism. However, results with a slightly warmer CN beam (163) showed larger forward and side scattering, suggesting that reagent rotation may subtly alter the mechanism.

High-quality ab initio calculations exist on the CN + H_2 reaction; quasiclassical trajectory calculations and quantum dynamics simulations have been performed to generate theoretical vibrational energy distributions (164–166). The calculated distributions also show enhancement of v_1, arising because the C-N stretch

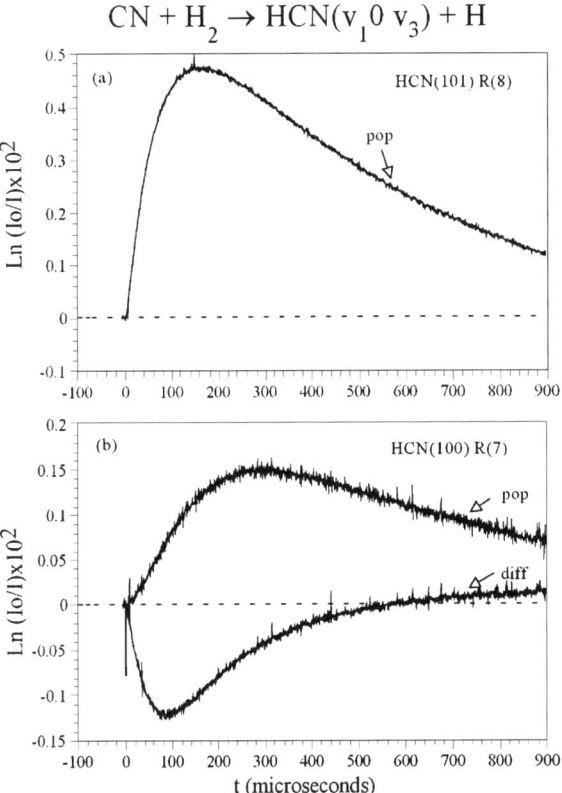

Figure 9 Absorbance (labeled "diff") and extracted populations (labeled "pop") as a function of time for HCN (v_1, 0, v_3) from the CN + H_2 reaction. The large population inversion between HCN (101) and HCN (100) is clearly evident from the gain on the HCN (100) absorption signal in the lower trace. Initial populations of the HCN vibrational states are obtained from modeling vibrational relaxation and from extrapolation back to zero time. Reproduced with permission from Reference (33). Copyright 1998 American Institute of Physics.

cannot be separated from the reaction coordinate. Since the infrared absorption experiments are able to probe the ground state population (unlike emission measurements), the fractional vibrational populations relative to the total population can be ascertained. The experimental and theoretical results are in relatively good agreement. The classical trajectory calculations indicate that significant bend excitation occurs even though abstraction (with a linear transition state) dominates, a result that appears to imply a low frequency for the bend at the transition state. The conjunction of molecular beam data, thermal measurements, and theoretical calculations serves to confirm the picture of CN + H_2 as a direct abstraction reaction, with significant coupling of the CN stretch to the reaction coordinate.

SUMMARY

Time-resolved infrared laser absorption has become a staple technique for the investigation of elementary reaction kinetics. Infrared absorption provides a selective and broadly applicable probe with a straightforward means for absolute concentration measurements. Determinations of thermal rate constants and branching fractions, as can be obtained using infrared absorption, provide key information for understanding global reaction mechanisms. Recent technical progress in infrared laser technology, and in methods for increasing absorption sensitivity, provide the means for greater exploitation of the advantages of this method for study of a broad range of reactions.

ACKNOWLEDGMENTS

The authors acknowledge Prof Kopin Liu, Dr John DeSain, Dr Hope Michelsen, and Dr David Osborn for helpful discussions during the course of preparing this chapter, and thank all who contributed figures. This work is supported by the Division of Chemical Sciences, the Office of Basic Energy Sciences, the US Department of Energy.

Visit the Annual Reviews home page at www.AnnualReviews.org

LITERATURE CITED

1. Baulch DL, Cobos CJ, Cox RA, Frank P, Hayman G, et al. 1994. *J. Phys. Chem. Ref. Data* 23:847–1033
2. Atkinson R, Baulch DL, Cox RA, Hampson RF Jr, Kerr JA, et al. 1997. *J. Phys. Chem. Ref. Data* 26:521–1011
3. DeMore WB, Sander SP, Golden DM, Hampson RF, Kurylo MJ, et al. 1997. *Chemical Kinetics and Photochemical Data for Use in Stratospheric Modeling*. Pasadena, CA: Jet Propulsion Lab. 266 pp.
4. Webster CR, Menzies RT, Hinkley ED. 1988. In *Laser Remote Chemical Analysis*, ed. RM Measures, pp. 163–272. New York: Wiley
5. Smith MAH, Rinsland CP, Fridovich B, Rao KN. 1985. In *Molecular Spectroscopy: Modern Research*, ed. KN Rao, 3:111–248. Orlando, FL: Academic
6. Pugh LA, Rao KN. 1976. In *Molecular Spectroscopy: Modern Research*, ed. KN Rao, 2:165–227. New York: Academic
7. Porter G, West MA. 1974. In *Investigation of Rates and Mechanisms of Reactions: Part II*, Vol. VI, ed. GG Hammes, pp. 367–462. New York: Wiley
8. Thrush BA, Tyndall GS. 1982. *Chem. Phys. Lett.* 92:232–35
9. Laguna GA, Baughcum SL. 1982. *Chem. Phys. Lett.* 88:568–71
10. Petek H, Nesbitt DJ, Ogilby PR, Moore CB. 1983. *J. Phys. Chem.* 87:5367–71
11. Dane CB, Lander DR, Curl RF, Tittel FK, Guo Y, et al. 1988. *J. Chem. Phys.* 88:2121–28
12. Hall JL, Adams H, Kasper JVV, Curl RF, Tittel FK. 1985. *J. Opt. Soc. Am. B* 2:781–85
13. Adams H, Hall JL, Russell LA, Kasper JVV, Tittel FK, Curl RF. 1985. *J. Opt. Soc. Am. B* 2:776–80

14. Hirota E. 1989. *Int. Rev. Phys. Chem.* 8:171–205
15. Hirota E. 1991. *Annu. Rev. Phys. Chem.* 42:1–22
16. Hirota E, Kawaguchi K. 1985. *Annu. Rev. Phys. Chem.* 36:53–76
17. He G, Tokue I, Macdonald RG. 1998. *J. Phys. Chem. A* 102:4585–91
18. Crawford MA, Wallington TJ, Szente JJ, Maricq MM, Francisco JS. 1999. *J. Phys. Chem. A* 103:365–78
19. Cronkhite JM, Stickel RE, Nicovich JM, Wine PH. 1998. *J. Phys. Chem. A* 102:6651–58
20. House PG, Weitz E. 1997. *J. Phys. Chem. A* 101:2988–95
21. Maricq MM, Shi J, Szente JJ, Rimai L, Kaiser EW. 1993. *J. Phys. Chem.* 97:9686–94
22. Meunier H, Pagsberg P, Sillesen A. 1996. *Chem. Phys. Lett.* 261:277–82
23. Opansky BJ, Seakins PW, Pedersen JOP, Leone SR. 1993. *J. Phys. Chem.* 97:8583–89
24. Pagsberg P, Jodkowski JT, Ratajczak E, Sillesen A. 1998. *Chem. Phys. Lett.* 286:138–44
25. Petty JT, Harrison JA, Moore CB. 1993. *J. Phys. Chem.* 97:11194–98
26. Schiffman A, Nelson DD, Robinson MS, Nesbitt DJ. 1991. *J. Phys. Chem.* 95:2629–36
27. Adamson JD, Morter CL, DeSain JD, Glass GP, Curl RF. 1996. *J. Phys. Chem.* 100:2125–28
28. Alvarez RA, Moore CB. 1994. *J. Phys. Chem.* 98:174–83
29. Kegley-Owen CS, Tyndall GS, Orlando JJ, Fried A. 1999. *Int. J. Chem. Kinet.* 31:766–75
30. Li Q, Osborne MC, Smith IWM. 2000. *Int. J. Chem. Kinet.* 32:85–91
31. Bernath PF. 1990. *Annu. Rev. Phys. Chem.* 41:91–122
32. Nizkorodov SA, Harper WW, Blackmon BW, Nesbitt DJ. 2000. *J. Phys. Chem. A* 104:3964–73
33. Bethardy GA, Northrup FJ, He G, Tokue I, Macdonald RG. 1998. *J. Chem. Phys.* 109:4224–36
34. Frost MJ, Sharkey P, Smith IWM. 1991. *Faraday Discuss. Chem. Soc.* 91:305–17
35. Taatjes CA. 1999. *Int. Rev. Phys. Chem.* 18:419–58
36. Kulcke A, Blackmon B, Chapman WB, Kim IK, Nesbitt DJ. 1998. *J. Phys. Chem. A* 102:1965–72
37. Bomse DS. 1991. *Appl. Opt.* 30:2922–24
38. Qian H-b, Turton D, Seakins PW, Pilling MJ. 2000. *Chem. Phys. Lett.* 322:57–64
39. He G, Tokue I, Harding LB, Macdonald RG. 1998. *J. Phys. Chem. A* 102:7653–61
40. Cronkhite JM, Stickel RE, Nicovich JM, Wine PH. 1999. *J. Phys. Chem. A* 103:3228–36
41. Rothman LS, Rinsland CP, Goldman A, Massie ST, Edwards DP, et al. 1998. *J. Quant. Spectrosc. Radiat. Trans.* 60:665–710
42. Fakhr A, Bates RD. 1980. *Chem. Phys. Lett.* 71:381–86
43. Stephenson JC, Moore CB. 1970. *J. Chem. Phys.* 52:2333–40
44. Richman DC, Millikan RC. 1975. *J. Chem. Phys.* 63:2242–44
45. Green WH, Hancock JK. 1973. *J. Chem. Phys.* 59:4326–35
46. Cooper WF, Park J, Hershberger JF. 1993. *J. Phys. Chem.* 97:3283–90
47. Cooper WF, Hershberger JF. 1992. *J. Phys. Chem.* 96:771–75
48. Miller JA, Bowman CT. 1989. *Prog. Energy Combust. Sci.* 15:287–338
49. Miller JA, Bowman CT. 1991. *Int. J. Chem. Kinet.* 23:289–313
50. Perry RA, Siebers DL. 1986. *Nature* 324:657–58
51. Atakan B, Wolfrum J. 1991. *Chem. Phys. Lett.* 178:157–62
52. Cookson JL, Hancock G, McKendrick KG. 1985. *Ber. Bunsenges. Phys. Chem.* 89:335–36
53. Perry RA. 1985. *J. Chem. Phys.* 82:5485–88

54. Hall JL, Zeitz D, Stephens JW, Kasper JVV, Glass GP, et al. 1986. *J. Phys. Chem.* 90:2501–505
55. Tacke M. 1994. *Infrared Phys. Technol.* 36:447–63
56. Mantz AW. 1995. *Spectrochim. Acta A* 51:2211–36
57. Werle P. 1998. *Spectrochim. Acta A* 54:197–236
58. Lancaster DG, Richter D, Curl RF, Tittel FK, Goldberg L, Koplow J. 1999. *Opt. Lett.* 24:1744–46
59. Clifford EP, Farrell JT, DeSain JD, Taatjes CA. 2000. *J. Phys. Chem. A* 104:11549–60
60. Taatjes CA, Oh DB. 1997. *Appl. Opt.* 36:5817–21
61. He G, Tokue I, Macdonald RG. 1998. *J. Chem. Phys.* 109:6312–19
62. Sirtori C, Faist J, Capasso F, Cho AY. 1998. *Pure Appl. Opt.* 7:373–81
63. Capasso F, Gmachl C, Sivco DL, Cho AY. 1999. *Phys. World.* 12:27–33
64. Köhler R, Gmachl C, Tredicucci A, Capasso F, Sivco DL, et al. 2000. *Appl. Phys. Lett.* 76:1092–94
65. Pine AS. 1976. *J. Opt. Soc. Am.* 66:97–108
66. Bordui PF, Fejer MM. 1993. *Annu. Rev. Mater. Sci.* 23:321–79
67. Sumpf B, Rehle D, Kelz T, Kronfeldt HD. 1998. *Appl. Phys. B* 67:369–73
68. Simon U, Waltman S, Loa I, Tittel FK, Hollberg L. 1995. *J. Opt. Soc. Am. B* 12:323–27
69. Chen W, Burie J, Boucher D. 1999. *Spectrochim. Acta A* 55:2057–75
70. Eckhoff WC, Putnam RS, Wang S, Curl RF, Tittel FK. 1996. *Appl. Phys. B* 63:437–41
71. Goldberg L, Burns WK, McElhannon RW. 1995. *Opt. Lett.* 20:1280–82
72. Petrov KP, Curl RF, Tittel FK. 1998. *Appl. Phys. B* 66:531–38
73. Taatjes CA, Christensen L, Hurley MD, Wallington TJ. 1999. *J. Phys. Chem. A* 103:9805–14
74. Sun F, Kosterev A, Scott G, Litosh V, Curl RF. 1998. *J. Chem. Phys.* 109:8851–56
75. Bjorklund GC, Lenth W, Levenson MD, Ortiz C. 1981. In *Laser Spectroscopy for Sensitive Detection,* ed. JA Gelbwachs, 286:153–59. Washington, DC: SPIE
76. Cooper DE, Watjen JP. 1986. *Opt. Lett.* 11:606–8
77. Supplee JM, Whittaker EA, Lenth W. 1994. *Appl. Opt.* 33:6294–6302
78. Hall GE, North S. 2000. *Annu. Rev. Phys. Chem.* 51:243–74
79. White JU. 1942. *J. Opt. Soc. Am.* 32:285–88
80. Herriott D, Kogelnik H, Kompfner R. 1964. *Appl. Opt.* 3:523–26
81. Pilgrim JS, Jennings RT, Taatjes CA. 1997. *Rev. Sci. Instrum.* 68:1875–78
82. McManus JB, Kebabian PL, Zahniser MS. 1995. *Appl. Opt.* 34:3336–48
83. O'Keefe A, Deacon DAG. 1988. *Rev. Sci. Instrum.* 59:2544–51
84. Scherer JJ, Paul JB, O'Keefe A, Saykally RJ. 1997. *Chem. Rev.* 97:5–52
85. Wheeler MD, Newman SM, Orr-Ewing AJ, Ashfold MNR. 1998. *J. Chem. Soc. Faraday Trans.* 94:337–51
86. Yu T, Lin MC. 1993. *J. Am. Chem. Soc.* 115:4371–72
87. Lin MC, Yu T. 1993. *Int. J. Chem. Kinet.* 25:875–80
88. Atkinson DB, Hudgens JW. 2000. *J. Phys. Chem. A* 104:811–18
89. Brown SS, Ravishankara AR, Stark H. 2000. *J. Phys. Chem. A* 104:7044–52
90. Ninomiya Y, Hashimoto S, Kawasaki M, Wallington TJ. 2000. *Int. J. Chem. Kinet.* 32:125–30
91. Tonokura K, Marui S, Koshi M. 1999. *Chem. Phys. Lett.* 313:771–76
92. Paldus BA, Harb CC, Spence TG, Zare RN, Gmachl C, et al. 2000. *Opt. Lett.* 25:666–68
93. Spence TG, Harb CC, Paldus BA, Zare RN, Willke B, Byer RL. 2000. *Rev. Sci. Instrum.* 71:347–53
94. Levenson MD, Paldus BA, Spence TG, Harb CC, Zare RN, et al. 2000. *Opt. Lett.* 25:920–22
95. Darwin DC, Young AT, Johnston HS,

Moore CB. 1989. *J. Phys. Chem.* 93: 1074–78
96. Darwin DC, Moore CB. 1995. *J. Phys. Chem.* 99:13467–70
97. Murray KK, Unfried KG, Glass GP, Curl RF. 1992. *Chem. Phys. Lett.* 192:512–16
98. Lander DR, Unfried KG, Stephens JW, Glass GP, Curl RF. 1989. *J. Phys. Chem.* 93:4109–16
99. Taatjes CA. 1999. *Chem. Phys. Lett.* 306:33–40
100. Kumaran SS, Lim KP, Michael JV. 1994. *J. Chem. Phys.* 101:9487–98
101. Bian WS, Werner HJ. 2000. *J. Chem. Phys.* 112:220–29
102. Allison TC, Lynch GC, Truhlar DG, Gordon MS. 1996. *J. Phys. Chem.* 100: 13,575–87
103. Yaakov YB, Persky A, Klein FS. 1973. *J. Chem. Phys.* 44:3617–26
104. Persky A, Klein FS. 1966. *J. Chem. Phys.* 44:3617–26
105. Alagia M, Balucani N, Cartechini L, Casavecchia P, Volpi GG, et al. 2000. *Phys. Chem. Chem. Phys.* 2:599–612
106. Alagia M, Balucani N, Cartechini L, Casavecchia P, vanKleef EH, et al. 1996. *Science* 273:1519–22
107. Lee S-h, Liu K. 1999. *J. Chem. Phys.* 111:6253–59
108. Lee S-h, Liu K. 1998. *Chem. Phys. Lett.* 290:323–28
109. Lee S-h, Lai LH, Liu K, Chang H. 1999. *J. Chem. Phys.* 110:8229–32
110. Skouteris D, Manolopoulos DE, Bian WS, Werner HJ, Lai LH, Liu K. 1999. *Science* 286:1713–16
111. Liu K. 1999. In press
112. Moore JW, Pearson RG. 1981. *Kinetics and Mechanism.* New York: Wiley
113. Levine RD. 1999. *Faraday Discuss.* 113:77
114. Johnston GW, Kornweitz H, Schechter I, Persky A, Katz B, et al. 1991. *J. Chem. Phys.* 94:2749–57
115. Maricq MM, Szente JJ, Kaiser EW. 1993. *J. Phys. Chem.* 97:7970–77
116. Maricq MM, Ball JC, Straccia AM, Szente JJ. 1997. *Int. J. Chem. Kinet.* 29:421–29
117. Adamson JD, Farhat SK, Morter CL, Glass GP, Curl RF, Phillips LF. 1994. *J. Phys. Chem.* 98:5665–69
118. Matsumoto K, Koshi M, Okawa K, Matsui H. 1996. *J. Phys. Chem.* 100:8796–8801
119. Jasinski JM, Becerra R, Walsh R. 1995. *Chem. Rev.* 95:1203–28
120. Loh SK, Jasinski JM. 1991. *J. Chem. Phys.* 95:4914–26
121. Sumiyoshi Y, Tanaka K, Tanaka T. 1994. *Appl. Surf. Sci.* 79:471–75
122. Quandt RW, Hershberger JF. 1993. *Chem. Phys. Lett.* 206:355–60
123. Morter CL, Farhat SK, Adamson JD, Glass GP, Curl RF. 1994. *J. Phys. Chem.* 98:7029–35
124. Morter CL, Domingo C, Farhat SK, Cartwright E, Glass GP, Curl RF. 1992. *Chem. Phys. Lett.* 195:316–21
125. Fahr A, Laufer AH. 1993. *Int. J. Chem. Kinet.* 25:1029–35
126. Stephens JW, Morter CL, Farhat SK, Glass GP, Curl RF. 1993. *J. Phys. Chem.* 97:8944–51
127. Quandt RW, Hershberger JF. 1996. *J. Phys. Chem.* 100:9407–11
128. Lindholm N, Hershberger JF. 1997. *J. Phys. Chem. A* 101:4991–95
129. Park J, Hershberger JF. 1993. *J. Phys. Chem.* 97:13,647–52
130. Quandt RW, Hershberger JF. 1995. *J. Phys. Chem.* 99:16,939–44
131. Lambrecht RK, Hershberger JF. 1994. *J. Phys. Chem.* 98:8406–10
132. Hovda N, Hershberger JF. 1997. *Chem. Phys. Lett.* 280:145–50
133. Rim KT, Hershberger JF. 1998. *J. Phys. Chem. A* 102:4592–95
134. Guo Y, Smith SC, Moore CB, Melius CF. 1995. *J. Phys. Chem.* 99:7473–81
135. Rim KT, Hershberger JF. 1998. *J. Phys. Chem. A* 102:5898–5902

136. Stickel RE, Zhao Z, Wine PH. 1993. *Chem. Phys. Lett.* 212:312–18
137. Zhao Z, Stickel RE, Wine PH. 1996. *Chem. Phys. Lett.* 251:59–66
138. Smith IWM. 1995. In *Chemical Dynamics and Kinetics of Small Radicals, Part I*, ed. K Liu, A Wagner pp. 214–49. River Edge, NJ: World Sci.
139. Sims IR, Smith IWM. 1988. *J. Chem. Soc. Faraday Trans.* 84:527–39
140. Sims IR, Smith IWM. 1988. *Chem. Phys. Lett.* 151:481–84
141. Atakan B, Wolfrum J. 1991. *Chem. Phys. Lett.* 186:547–52
142. Durant JL, Tully FP. 1989. *Chem. Phys. Lett.* 154:568–72
143. Balla RJ, Castleton KH. 1991. *J. Phys. Chem.* 95:2344–51
144. Hessler JP. 1999. *J. Chem. Phys.* 111:4068–76
145. Schmatjko KJ, Wolfrum J. 1978. *Ber. Bunsenges. Phys. Chem.* 82:419–28
146. Mohammad F, Morris VR, Fink WH, Jackson WM. 1993. *J. Phys. Chem.* 97:11,590–98
147. Rim KT, Hershberger JF. 1999. *J. Phys. Chem. A* 103:3721–25
148. Phillips LF, Smith IWM, Tuckett RP, Whitham CJ. 1991. *Chem. Phys. Lett.* 183:254–63
149. Sauder DG, Patel-Misra D, Dagdigian PJ. 1991. *J. Chem. Phys.* 95:1696–1707
150. Sloan JJ. 1988. *J. Phys. Chem.* 92:18–27
151. Pilgrim JS, Taatjes CA. 1997. *J. Phys. Chem. A* 101:5776–82
152. Farrell JT, Taatjes CA. 1998. *J. Phys. Chem. A* 102:4846–56
153. Flynn GW, Michaels CA, Tapalian HC, Lin Z, Sevy ET, Muyskens MA. 1997. *ACS Symp. Ser.* 678:134–49
154. Weston RE, Flynn GW. 1992. *Annu. Rev. Phys. Chem.* 43:559–89
155. Elioff MS, Fraelich M, Sansom RL, Mullin AS. 1999. *J. Chem. Phys.* 111:3517–25
156. Mullin AS, Schatz GC. 1997. *ACS Symp. Ser.* 678:2–24
157. Sugawara KI, Ito F, Nakanaga T, Takeo H, Matsumura C. 1990. *J. Chem. Phys.* 92:5328–37
158. Suzuki T, Hirota E. 1993. *J. Chem. Phys.* 98:2387–98
159. Frost MJ, Salh JS, Smith IWM. 1991. *J. Chem. Soc. Faraday Trans.* 87:1037–38
160. Bethardy GA, Northrup FJ, Macdonald RG. 1995. *J. Chem. Phys.* 102:7966–82
161. Bethardy GA, Northrup FJ, Macdonald RG. 1996. *J. Chem. Phys.* 105:4533–49
162. Wang J-H, Liu K, Schatz GC, ter Horst M. 1997. *J. Chem. Phys.* 107:7869–75
163. Lai L-H, Wang J-H, Che DC, Liu K. 1996. *J. Chem. Phys.* 105:3332–35
164. Takayanagi T, Schatz GC. 1997. *J. Chem. Phys.* 106:3227–36
165. ter Horst MA, Schatz GC, Harding LB. 1996. *J. Chem. Phys.* 105:558–71
166. Bethardy GA, Wagner AF, Schatz GC, ter Horst MA. 1997. *J. Chem. Phys.* 106:6001–15
167. Xing G, Huang X, Bersohn R, Tsukiyama K, Katz B. 1995. *J. Chem. Phys.* 102:3169–71

SURFACE BIOLOGY OF DNA BY ATOMIC FORCE MICROSCOPY

Helen G Hansma

Department of Physics, University of California, Santa Barbara, California 93106; e-mail: hhansma@physics.ucsb.edu

Key Words biomaterials, nanotechnology, scanning probe microscopy, SFM, SPM

■ **Abstract** The atomic force microscope operates on surfaces. Since surfaces occupy much of the space in living organisms, surface biology is a valid and valuable form of biology that has been difficult to investigate in the past owing to a lack of good technology. Atomic force microscopy (AFM) of DNA has been used to investigate DNA condensation for gene therapy, DNA mapping and sizing, and a few applications to cancer research and to nanotechnology. Some of the most exciting new applications for atomic force microscopy of DNA involve pulling on single DNA molecules to obtain measurements of single-molecule mechanics and thermodynamics.

INTRODUCTION

In the early 1990s, it was not too much of a stretch to write review articles about the first several years of the entire field of biological AFM with ~130 references (1, 2). More recent reviews of biological AFM for only one year periods cited half as many references (3, 4), and now this review, specific for AFM studies of DNA since the mid-1990s, needs >150 references.

For an old-timer in AFM of DNA, one of the most exciting examples of growth in the field is the emergence of leading-edge research by new groups, as in Figure 1 (5). These people often know the biology of their DNA systems in a way that we old-timers do not, since we are often from physics laboratories.

The author of the present review was struck by the high quality and great beauty of many images presented in the literature. Since DNA has become an easy-to-use biomaterial for AFM imaging, many new developments in AFM technology and sample preparation are tried first on DNA. The inclusion of these new developments gives considerable breadth to the topics covered in this review. The figures in the review come mostly from the author's own laboratory, owing to the author's laziness rather than to any lack of beautiful results from other laboratories. The atomic force microscope (Figure 2; see color insert) is now quite familiar to many

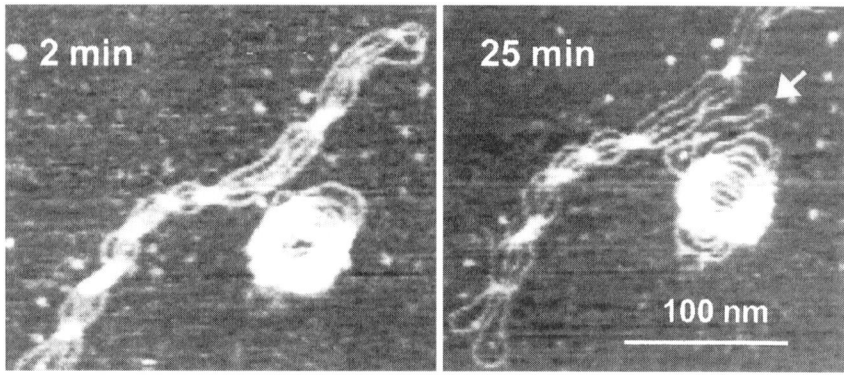

Figure 1 Movie of DNA condensation. The first and last frames from a time-lapse movie of dynamic interactions between toroidal and rodlike DNA condensates show a DNA loop from the toroid extending toward the rod and interacting with it by forming a structure with parallel DNA strands (arrow). DNA condensates were produced by depositing plasmid DNA in Tris-EDTA buffer on AP-mica (5). Images are courtesy of E Spain.

readers. Some of the many recent reviews and news articles about biological AFM are those by Hansma et al (6), Fisher et al (7), Müller et al (8), Vinckier & Semenza (9), Czajkowsky & Shao (10), Miles (11), Shao & Zhang (12), Amato (13), and Stokstad (14).

Surface Biology

AFM has been called unnatural because it looks at biomaterials on surfaces instead of in test tubes. However, living systems are filled with surfaces, especially membranes, so surfaces are arguably more relevant biologically than test tubes. In fact, AFM may be a leading technique in a new field, surface biology, which will grow in this century.

In this context, a valuable direction for biological AFM is the development of flat biologically relevant surfaces. Lipid films are a good example of such a surface. These films have been formed by Langmuir-Blodgett depositions and vesicle fusion onto mica for observing proteins (15–17) and DNA (18, 19) by AFM.

Physical Chemistry

Biological AFM is contributing to classical physical chemistry most directly by providing thermodynamic information about ligand-receptor binding based on measurements of rupture forces for single ligand-receptor pairs. An excellent example of this work is from the Gaub laboratory, whose members have measured rupture forces for single biotin-avidin molecules and derivatives of these molecules (20, 21). These rupture forces correlate with enthalpy changes but not with free-energy changes, indicating that the rupture is adiabatic. The entropic changes attributable to rearrangements of water molecules must occur after the rapid rupture event, because the free-energy changes do not correlate with the rupture forces. Similarly, experiments involving pulling on single DNA molecules are producing data that are being compared with thermodynamic data on DNA base pairing (22–25), as reviewed below. Another interface between physical chemistry and AFM is that predictions involving thermodynamics of DNA curvature and condensation (26) are being tested by AFM (J Sitko, manuscript in preparation).

DNA IMAGING BY ATOMIC FORCE MICROSCOPY

DNA Mapping, Sizing, and Sequence Recognition

Several AFM methods are being developed for mapping or sizing DNA molecules. Members of the Hoh laboratory (27, 28) are imaging DNA restriction fragments and using an automated software package to measure their lengths. This approach has been used for DNAs of up to 2000 base pairs (bp) (700 nm) and has successfully resolved DNAs differing in length by only 17 nm (50 bp). Results obtained with this AFM method are comparable to those from densitometric scans of DNA separated in gels, but the AFM method requires less DNA—approximately the DNA output from only 15 cycles of the polymerase chain reaction (PCR). DNA sizing is useful not only for DNA mapping but also for other applications, such as DNA fingerprinting and screening. Automated DNA sizing is essential for high-throughput analyses.

In a variant of AFM called chemical force microscopy, DNA arrays are scanned with a chemically modified AFM tip containing DNA oligonucleotides complementary to one of the DNAs in the array (29). Friction forces were larger in sample regions containing the complementary DNA than in sample regions containing a control DNA with the same base composition but a noncomplementary sequence. This is a new method for scanning DNA arrays to detect a target sequence. It has the advantage that small arrays can be used, because the bands of each DNA sequence are only ~ 10 μm wide and can be 10- to 100-fold smaller.

Sequence-Specific DNA Structures Other methods for DNA mapping are based on the sequence-specific binding of particles or enzymes (30, 31) or on sequence-specific DNA structures. One sequence-specific DNA structure uses 10- to 27-nucleotide (nt)–long triple-helix-forming pyrimidine (cytosine and thymidine) oligonucleotides end-labeled with biotin and psoralen to target DNA sites. After photo-cross-linking is achieved via psoralen to produce a covalent adduct, streptavidin labeling is performed to increase the size and, hence, the visibility of the adduct (32).

Another sequence-specific DNA structure was created by hybridizing 300-nt RNAs to double-stranded DNA (dsDNA), forming loops or occasionally knots at hybridization sites owing to the presence of the double-stranded hybrid and the remaining single strand of DNA at hybridization sites (33). This method detected target sequences separated by 200 bp on DNA molecules 4–40 kbp long. The presence of loops at target sequences makes this method less sensitive to sample debris than methods that produce lumps at the target sequences.

Height information from AFM gives it an advantage over electron microscopy in two new ways of tagging DNA sites through the formation of 20-nt triple helices (34). Only AFM could detect the triplex-forming 20-nt DNA by the height increase at its DNA binding site. AFM and electron microscopy could both locate the triple-helical complex when the 20-nt DNA was attached to a 74-nt DNA that formed a 33-bp DNA hairpin. The hairpins were perpendicular to the plasmid DNA and were located either above or beside the DNA. AFM also detected the height increase with another 20-mer triple-helix-forming PNA. PNAs are peptide analogues of DNAs. In these PNAs, two PNA 10-mers were joined by a flexible linker that induced a bend or kink in the dsDNA. Predictably, these bends or kinks were identifiable by either electron microscopy or AFM.

G-wires—four-stranded guanosine-rich DNA such as that found in telomeres—formed structures that depended on the metal ions present during self-assembly (35). Single-stranded DNA (ssDNA), dsDNA, and triple-stranded DNA were differentiated by AFM according to their different persistence lengths (ssDNA versus dsDNA) and their different heights (dsDNA versus triple-stranded DNA) (36).

DNA sticky ends could be analyzed by AFM, because linear DNA containing sticky ends became circular in the presence of Mg(II) or Ca(II) when stabilized by uranyl salts (37). Four-nt sticky ends were, predictably, better at circularizing linear DNA than 2-nt sticky ends, and guanine/cytosine (GC) content correlated with circularization.

A Molecular Switch   AFM indicates that the structural transition between cruciform conformations can act as a molecular switch to facilitate or prevent communication between distant regions in DNA (38). DNA cruciforms can form in negatively supercoiled plasmid DNA when palindromic DNA sequences extrude as hairpins from the dsDNA. Cruciforms are important for Holliday junctions, which occur in the process of DNA recombination. AFM demonstrated that cruciforms in DNA plasmids adopt either an X-shaped compact conformation or a plus sign (+)–shaped extended conformation (38, 39). In +-shaped cruciforms, the angle between the short DNA arms is $\sim 180°$; the two side arms of the + are short and the top and bottom arms are long strands of plasmid dsDNA. In X-shaped cruciforms, the two short arms are at the top of the X.

X-shaped cruciforms are the predominant cruciform conformation observed by AFM (39). In X-shaped cruciforms, the short DNA arms sometimes move such that they are parallel to the main DNA. The X-shaped cruciforms are always found at apices or bends in circular DNA molecules. In fluid, the short arms of X-shaped cruciforms are dynamic, with considerable change occurring in the angle between the arms. In contrast, the short arms of the +-shaped cruciforms are fairly static in fluid. The +-shaped cruciforms are found on relatively linear regions of the supercoiled DNA circles.

The recombination protein RuvA stabilizes DNA in a +-shaped cruciform, which is proposed by Shlyakhtenko et al (38) to be the "on" state of the switch. Slithering of supercoiled DNA past the +-shaped cruciform allows one region of the DNA to slide along other regions of the DNA and hence to explore remote regions of the DNA. Such local structures and their conformational transitions during gene expression could have dramatic structural effects on chromatin architecture and function.

Related to this elegant work on molecular switches are the following three classes of palindromic DNA sequences whose conformations have been determined by AFM. Three-way DNA junctions are pyrimidal, not planar. This conclusion, originally arrived at by the use of other techniques (40, 41), has been explored in more detail by the use of AFM (42). Three-way DNA junctions occur when a loop of self-complementary DNA projects from one strand of dsDNA. The angles and conformations that are observed for three-way junctions with short loops (e.g. 7 bp) are different from those observed for such junctions with long loops (e.g. 50 bp). With 7-bp loops, there is a smaller angle and a smaller variation in angle between the two arms of dsDNA. With longer loops, the angle between the two dsDNA arms is larger and more variable and the loop can lie either outside the V of the dsDNA, forming a Y-shaped structure, or within.

AFM and ligation assays both show the same structure for one of the types of lambda DNA insertion complexes that are mediated by the enzyme integrase (43). Insertion complexes insert lambda DNA into the bacterial chromosome. By both AFM and ligation assays, this complex is roughly tetrahedral, with both DNA molecules bent.

Two hypothetical forms for the stem-loop structure of a *Staphylococcus* binding site for a heat shock protein have been proposed by Ohta et al from AFM images that show either planar or nonplanar structures (44). A palindromic DNA sequence that could branch into an X shows instead, by AFM, a Y with heights consistent with a stem-loop-loop-stem in which the two loops bind to each other. This paper presents a useful method for dealing with the problem of variable tip widths by plotting measured diameters for features of interest versus measured widths of DNA.

DNA-Protein Interactions: Conformations, Binding Sites, and Stoichiometries

The analysis of DNA-protein interactions is perhaps the single largest application for AFM of DNA. Most of this research has been performed on dried DNA-protein samples and could be done instead by electron microscopy, but it is done much more easily by AFM. Examples of the uses for DNA-protein AFM are the following: observing the location of protein binding sites—at the ends of DNA (45, 46) or on Z-DNA (47)—and the DNA conformations at these sites, such as looping (33, 48–51), bending (52, 53), supercoiling (54), and determining the protein:DNA stoichiometry at the binding site (55, 56). Several of these papers (45, 46, 48–50) have been reviewed previously (4).

Protein binding to DNA causes an apparent shortening of the DNA molecules if the DNA wraps around part or all of the protein. These data provide information about the length of the DNA that interacts directly with the protein.

In a centromeric DNA-protein complex from the yeast *Saccharomyces cerevisiae*, the DNA is "shortened" by 23 nm (70 bp) by the CBF3 (centromere binding factor 3) protein complex. These 70 bp are only long enough to wrap $\sim$30% of the way around the CBF3 proteins. The more exciting observation was that many of the complexes had four DNA "arms" instead of two (Figure 3; see color insert), suggesting that the CBF3 proteins may be involved in holding together the sister chromatids (152).

Determining the locations of protein binding sites on DNA and measuring approximate protein molecular weights are other uses for AFM of DNA-protein complexes, as reviewed in the following two examples. ATM is the defective gene product in ataxia-telangiectasia, which predisposes affected individuals to neurodegeneration and cancer. The ATM protein is a kinase that binds to DNA ends. The ATM protein exists both as a monomer and as a tetramer, as determined by AFM volume analyses (57). The transcriptional activator protein med8 binds to one or the other, but not both, of two *Saccharomyces* HXK2 DNA consensus sequences. The size of the bound protein is consistent with a monomer, not a dimer (58).

DNA Bending and Bend Angles DNA bends are commonly measured as bend angles—the degree of bending relative to no bending (Figure 4, right panel; see color insert). If the bend generates DNA arms of unequal length, the bend can be characterized as either left-handed or right-handed (Figure 4, left panel; see color insert). Such an analysis of histidine-tagged RNA polymerase (RNAP) elongation complexes showed that over 90% were left-handed on mica (59).

Some DNA methyltransferases bend DNA upon binding, whereas others do not (53). DNA bending by *Eco*RI DNA methyltransferase accelerates base flipping but compromises specificity (60). This important conclusion comes from AFM and other analyses of a wild-type and a mutant form of *Eco*RI DNA methyltransferase. The wild-type enzyme bends DNA, but the mutant in which histidine 235 is replaced with asparagine does not bend DNA. This asparagine mutant has an improved sequence recognition ability.

Photolyase is another enzyme involved in DNA bending. Photolyase binding at damaged DNA sites produces a bend in the DNA of $\sim$36°. In contrast, DNA does not bend when photolyase binds at undamaged sites (61, 62).

RNA Polymerase Active complexes of *Escherichia coli* RNAP with DNA have been imaged in time-lapse movies, with the addition of nucleoside triphosphates initiating the movement of DNA through the RNAP as transcription occurs (63, 64). DNA without a promoter slides on RNAP with a one-dimensional (1-D) diffusion rate of $\sim 10^{-13}$ cm^2/s and an unexpectedly long lifetime of ~ 600 s (65). Occasionally the RNAP appears to hop along the DNA or move by intersegment transfer from one segment of DNA to another via an intermediate stage in which the RNAP simultaneously binds to both DNA sites (65). These experiments continue to be extremely difficult to perform (66).

It is much easier to image RNAP-DNA complexes either in air or fluid, where the RNAP is not in action during the experiment, and these experiments have also generated useful results. RNAP was usually positioned on the side of the DNA, with the DNA arms pointing away from the RNAP (59, 67). DNA was shortened by 6 nm (18 bp) in closed RNAP complexes with the protein cofactor σ^{54}, which is consistent with a short DNA path through the RNAP (67). Complexes imaged in air were similar to those imaged in fluid if air samples were washed with 1-ml volumes of water but were significantly disrupted if samples were washed with 50-ml volumes of water. In another study, using open promoter complexes of DNA with RNAP and the protein cofactor σ^{70}, the apparent DNA length was shortened by 90 bp (30 nm), long enough to wrap $\sim 80\%$ of the way around the RNAP and requiring a bending energy of 60 kJ/mol (68).

Cancer Research AFM has made a few contributions to cancer research. Briefly, the tumor suppressor protein p53 binds preferentially to supercoiled DNA (54). The guanine (G)-binding carcinogen benzo[a]pyrene diol epoxide bends DNA (69). DNA shortening, compaction, and aggregation were observed when various Pd(II) and Pt(II) *cis*-platins and their analogues were complexed to DNA (70).

Neutron-induced DNA damage produces many short DNA fragments, indicating that there may be clusters of double-stranded breaks in the DNA (71). Electron irradiation of plasmid DNA in an aqueous buffer, analyzed by AFM, gives a dose-response curve for DNA length after different levels of radiation (72, 73).

Condensed DNA

DNA Condensation in Solution In several investigations, DNA was condensed in test tubes and the morphologies of the condensates were analyzed by AFM. The condensing agents ranged in size from ~ 50 Da to 80 kDa. The question of whether there is one or more than one pathway of DNA condensation (i.e. whether flowers and toroids are formed by the same or different pathways of DNA condensation) has been raised (74; Figure 5); the answer is not yet clear.

Spermidine condensed DNA into flowerlike shapes (Figure 5, left panel) (74, 75) or toroids of different sizes (76) under quite similar conditions in different laboratories. Thus, the mechanism of DNA condensation must be strongly affected by small changes in the experimental protocol.

A polycation much larger than spermidine that condenses DNA into toroids is asialo-orosomucoid covalently attached to poly(L-lysine) (AsOR-PLL) (75, 77; Figure 5, right panel). The toroids formed with plasmid DNA and AsOR-PLL usually contain one plasmid molecule and five or six lysine residues per DNA

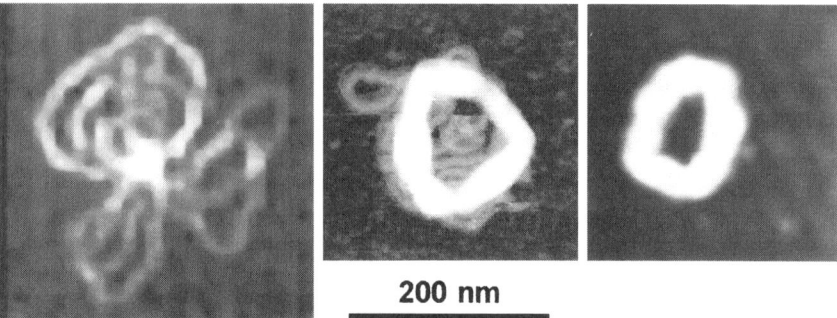

Figure 5 Varieties of DNA condensates: Are they generated by different pathways for condensation? (*Left*) Loopy condensates with one or more dense foci are seen with 30 μM spermidine (see 74). (*Middle*) Loopy toroids and loopy rodlike or branched condensates are seen with Ni(II)-induced condensation of poly(dG-dC)•poly(dG-dC) (78; J Sitko, manuscript in preparation). Toroid height is ~8 nm. (*Right*) Clean toroids and rods are seen with the large (10-kDa) polycation AsOR-PLL, as diagrammed in Figure 6 (see 75, 77).

nucleotide. Long PLL strands on the AsOR-PLL cause DNA to condense into toroids, whereas short PLL strands on the AsOR-PLL result in much less DNA condensation (77; Figure 6). Toroids also form with double-stranded poly(dG-dC) DNA (GC-DNA) and Ni(II) (78; J Sitko, manuscript in preparation); in both cases, the contribution of the ligand to the total toroid volume is much less than with AsOR-PLL. The toroids and other condensates of GC-DNA and Ni(II) have loops around them (Figure 5, center panel). Therefore, DNA forms toroids under a variety of different conditions, and there appears to be a continuum of condensed DNA structures ranging from loopy structures to toroids and rods (77).

Surface-Directed DNA Condensation vs DNA Condensation in Solution In the following two studies, with protamine and with silanes, DNA forms larger condensates when condensation occurs on the sample surface than when condensation occurs in solution. DNA on mica condenses into distinct toroids when protamine is added (79). When DNA and protamine are preincubated in a test tube and applied to mica, a network of DNA, with a few thin toroids in it, occurs. Similarly, mobile cationic silanes on a silicon surface condense DNA into distinct toroids whose size depends on the length of the DNA (80, 81). When silanes in solution and DNA are preincubated in a test tube, looped structures that include flower and sausage shapes form. Finally, ethanol at 20%–35% produces DNA condensates of flower and toroidal shapes when the DNA is on mica in the presence of Mg(II) (82).

Some of the most beautiful AFM images of surface-directed DNA condensation have been acquired in aqueous fluid and show parallel arrays of DNA strands held together at intervals in dense clumps. Some of these structures formed on polyornithine-coated mica in the presence of polyethylimine (83). Imaging of DNA

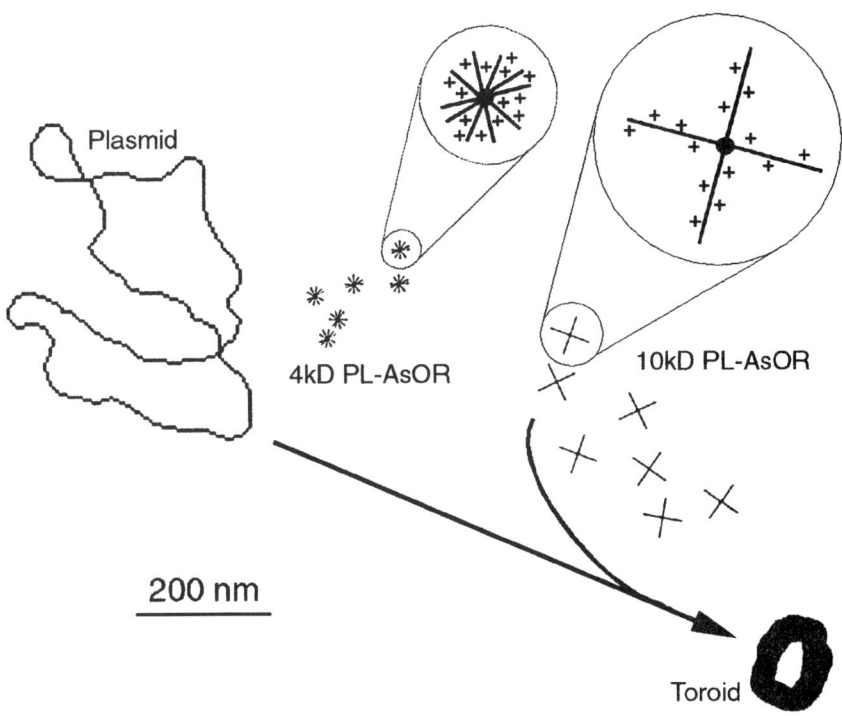

Figure 6 A scale model of DNA condensation. The toroid contains only a single plasmid DNA molecule, as determined on the basis of calculations of the toroid volume by AFM; this is unexpectedly little DNA, according to current theories (77). Scale diagrams of a plasmid DNA molecule, several molecules each of 10-kDa PLL-AsOR and 4-kDa PLL-AsOR, and a typical toroid are shown. The 10-kDa PLL molecule has ~80 lysine residues, giving a mean length of 27 nm; 4-kDa PLL has ~30 lysine residues per molecule, giving a mean length of 11 nm. Magnified views of 10-kDa PLL-AsOR and 4-kDa PLL-AsOR show the charge and arrangement of polylysine chains on AsOR. AsOR is a 38-kDa glycoprotein with 30 acidic amino acid residues per molecule and a net charge of 5 carboxylic acid groups per molecule (154). The diameter of a globular protein of this size is ~5 nm. At a 1:1 (wt/wt) ratio of PLL to AsOR, there is an average of 4 molecules of 10-kDa PLL per AsOR molecule or 10 molecules of 4-kDa PLL per AsOR molecule. This corresponds to ~300 lysine residues per AsOR molecule. Since toroids form with 10-kDa PLL-AsOR at a lysine:nucleotide ratio of 5:1 or 6:1, a single circular plasmid DNA molecule condenses with 300–400 10-kDa PLL-AsOR molecules into a toroid. The 4-kDa PLL-AsOR molecule is much less effective at condensing DNA and reduces its contour length only two- to fourfold.

condensation on an amino-silylated mica surface (AP-mica) under an aqueous buffer shows how DNA condensates can grow in size by means of a DNA loop from one condensate reaching into the adjacent condensate (5; Figure 1).

DNA Condensation for Gene Therapy Nonviral methods of condensing DNA such as those described here are of special interest now that a human death has occurred during viral gene therapy (84). Some of the investigations of DNA condensation have used both AFM and transfection in animal systems (75, 85–87). The variety of design in these DNA condensates is evident from the several examples below.

Compact condensates of DNA with a block copolymer gave less efficient transfection than a more extended condensate in one study (88). In contrast, transfection and condensation correlated in an integrin-targeted DNA condensate with Lipofectin (86). Minimization of endosomal degradation is always one of the objectives of nonviral gene therapies. In yet a third study, it was determined that mid-sized liposomes transfected best (89). Additionally, a low-density lipoprotein vector transfected better than Lipofectin; this condensate was ~ 100 nm in size (90, 91).

Two types of DNA condensate are carefully designed nanostructures. Particles of 15-nm diameter that are a type of biomimetic "protein" are formed by covalently cross-linking diblock amphiphilic micelles. DNA binds to these particles, which provide partial protection of the DNA from nuclease degradation (92). Another condensed DNA nanostructure was obtained with DNA and excess polycation condensed to form positively charged particles, to which successive positive and negative layers were added using PLL and succinylated PLL, respectively (93).

Nucleosomes and Chromosomes "Despite more than 20 years of research, the structure of the chromatin fiber and its molecular determinants remain enigmatic" (94). This provides a rich opportunity for AFM analysis, which is being undertaken by several groups.

Reconstituted nucleosomes were first imaged with the atomic force microscope by Allen et al (95), who showed a correlation between the number of nucleosomes formed on DNA and the shortening of the DNA. Nucleosomes appear to be nonrandomly arranged on DNA, in that there are typically bare stretches of DNA interspersed with dense arrays of nucleosomes in functionally active chromatin and in reconstituted subsaturated nucleosomes (96). Salt and the linker histone H1 both decrease the internucleosome distance in quantifiable ways (94, 97). Reconstituted chromatin does not need histone H3 or the tails of the linker histones for normal morphology, and mild trypsinization affects the structure of reconstituted chromatin (98, 99). Both chromatin and reconstituted nucleosomes are presently analyzed only in air, with or without glutaraldehyde fixation.

In a new advance (100) on an old idea for AFM (101), DNA has been cut from a region of a chromosome and amplified by PCR. The fluorescently labeled PCR products bound to the region of the chromosome at which the cut had been made.

Individual chromatin fibers were detected by AFM of human chromosomes after the chromosomes were enzymatically treated to remove RNA and protein (102). The positions of these structures were compared with stained bands on chromosomes (103).

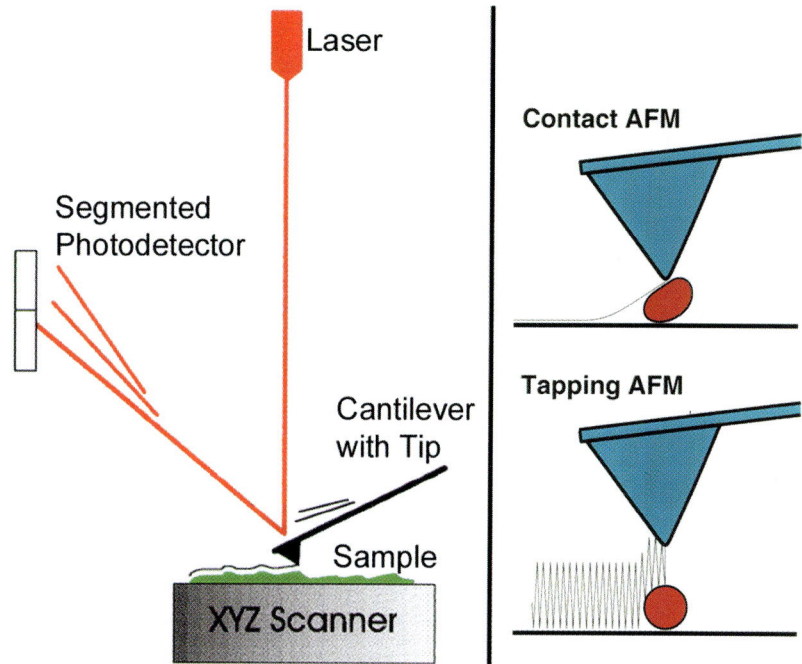

Figure 2 How does the AFM work? The AFM (148) feels the surface of a sample with such a discriminating touch that it can sometimes even detect the individual atoms on the surface of crystal such as gold or silicon (149 151). Resolution is much lower usually a few nanometers on soft biological macromolecules such as DNA and proteins and is even lower on biological cells. (*Left*) The AFM feels surfaces by raster-scanning a small tip back and forth over the sample surface. The tip is on the end of a cantilever, which deflects when the tip encounters features on the sample surface. This deflection is sensed with an optical lever (*red line*): a laser beam reflecting off the end of the cantilever onto a segmented photodiode magnifies small cantilever deflections into large changes in the relative intensity of the laser light on the two segments of the photodiode. In this way, the AFM makes a topographic map of the sample surface. (*Right*) The tip oscillates in Tapping Mode AFM but not in Contact Mode AFM. Tapping Mode is good for minimizing sample damage and movement caused by lateral forces. Contact Mode is good for high resolution imaging of flat samples such as 2-dimensional arrays of proteins or DNA, where lateral forces are small because of the small variability in height.

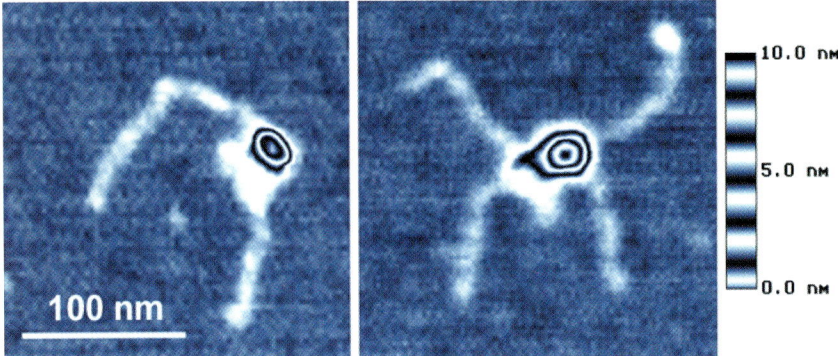

Figure 3 DNA-protein complexes from a yeast centromere. These CBF3-centromere DNA complexes from *Saccharomyces cerevisiae* were imaged by AFM in air. (*Left*) A typical two-armed complex of centromere DNA with CBF3 proteins bound. (*Right*) The bound proteins in four-armed complexes were typically almost twice as tall as and slightly wider than those in two-armed complexes. The presence of these four-armed complexes suggests that CBF3 may be involved in holding sister chromatids together during cell division. Rings show the relative heights of the two protein complexes. The banded height scale is on the far right. (See 78, 152.)

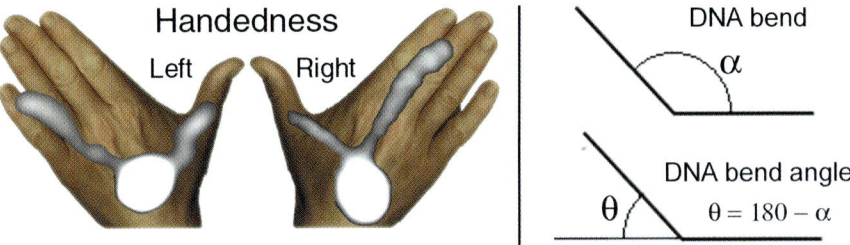

Figure 4 DNA bends. (*Left*) Bends in DNA and DNA-protein complexes can be either left- or right-handed if the two DNA arms are of uneven lengths (59). (*Right*) DNA bend angles are commonly expressed as the deviation from linearity, i.e. 180° minus the angle between the two DNA arms (153).

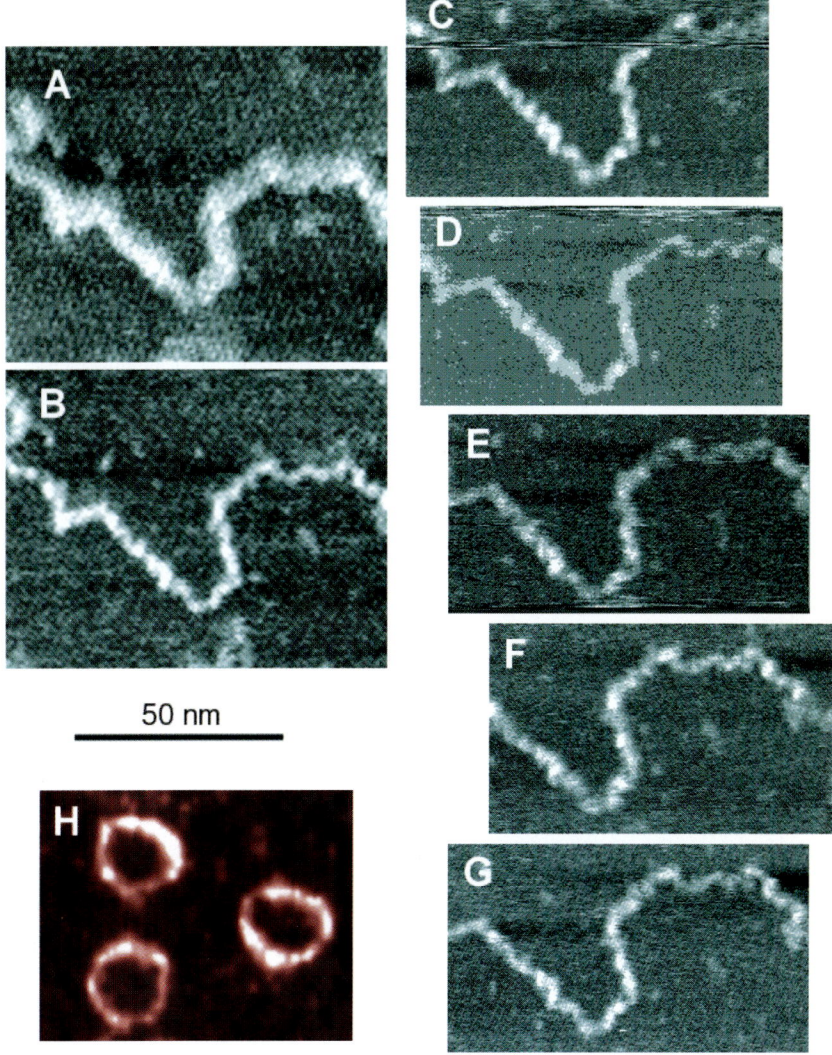

Figure 8 Nanometer resolution of DNA. (*A–G*) dsDNA on mica showing a right-handed double helix (see 155, 156) with an extremely good tip; (*H*) dsDNA minicircles on mica imaged by magnetic alternating-current (MAC mode) AFM (107). (*A, B*) Spontaneous evolution of a tip from bad to good in successive images. (*C–E*) images from successive 86- by 86-nm scans showing a helical periodicity of ~3.4 nm with helical deformations at bends. This unique set of images was taken under propanol; DNA was adsorbed to mica under an aqueous fluid, which resulted in B-DNA lengths (157). (*H*) 168-bp DNA minicircles (42.5 ± 1.1 nm long) in 1 mM $MgCl_2$. Even DNA minicircles as short as 126 bp show holes in their centers (107). Image courtesy of S. Lindsay.

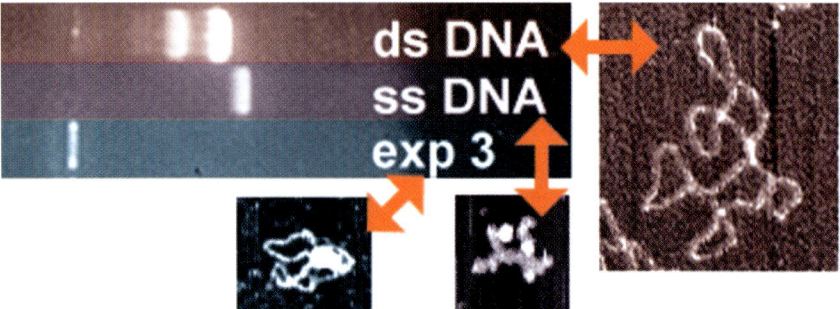

Figure 9 A successful AFM assay for DNA replication in which the electrophoretic assay failed. DNA replication by Taq polymerase was assayed by AFM in aqueous buffers and by agarose gel electrophoresis. The Taq polymerase in experiment (*exp*) 3 formed aggregates with the DNA that prevented it from entering the gel. *ss DNA* (unreplicated; φX-174 virion) and *ds DNA* (replicated; double-stranded φX-174) are shown in the lanes above the gel and in the corresponding AFM images indicated with arrows. In experiment 3, an aliquot of the Taq polymerase reaction mixture was diluted and placed on the cantilever in the fluid cell before engagement. The AFM image shows that a small amount of replication occurred. The corresponding lane on the gel (*exp 3*) shows that most of the DNA has not migrated into the gel, making it impossible to determine the extent of replication by agarose gel electrophoresis. (See 118.)

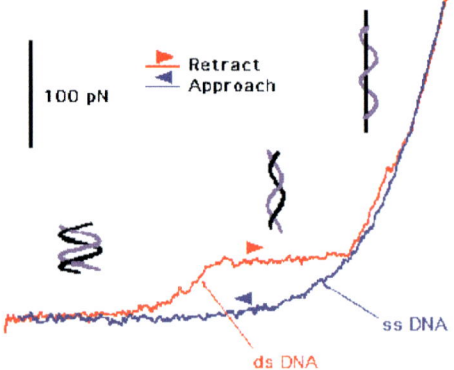

Figure 10 A single molecule of over-stretched DNA. This graph shows a force measurement of a single tethered molecule of Lambda Digest DNA showing the B-S and the melting transition. During the extension of the molecule (*red trace*), the DNA first goes through the B-S transition (*the plateau*), and then melts to single-stranded DNA (ssDNA) at a higher force. During relaxation of the molecule (*blue trace*), the DNA doesn't renneal so the curve is a simple freely-jointed-chain, indicative of ssDNA. The traces were made at a pulling speed of 1mm/second. Data courtesy of H. Clausen-Schaumann and R. Krautbauer, Gaub Laboratory, Ludwig-Maximilians-Universität München. Data were obtained with a cantilever from Park Scientific Microlevers in a Molecular Force Probe from Asylum Research (http://www.asylumresearch.com).

DNA In Fluid

Approximately 15% of published articles on AFM of DNA describe studies involving imaging under aqueous solutions. This seems to be a fair representation of the relative effort and utility of fluid imaging versus air imaging. Some of this research, such as the work on active RNAP complexes, has been reviewed above.

DNA Condensation and Enzymology An exciting example of new science from AFM in fluid comes from time-lapse AFM movies of the endonuclease *Eco*KI (104). As the AFM images show, *Eco*KI dimerizes when it binds to a plasmid with two *Eco*KI binding sites and it is monomeric when bound to a plasmid with one *Eco*KI binding site. The AFM movie shows that a loop of the DNA plasmid becomes shorter as it translocates through the *Eco*KI molecule upon addition of ATP to the DNA-enzyme complex on mica. As *Eco*KI cleaved the DNA, free DNA ends appeared, after which DNA fragments moved away from the *Eco*KI. The measured volume of the *Eco*KI was 75% of the volume expected based on its molecular weight.

The lack of speed has been a major problem with AFM imaging of enzymatic processes. Early images of DNA being successively degraded by DNase I, for example, were captured at a rate of 30–60 s/image (105). The imaging rate is an order of magnitude faster with the new prototype small-cantilever atomic force microscope described below. With this prototype, the degradation of DNA by DNase I has been visualized at scan rates up to 2 s/image (Figure 7). In two fast DNase movies of samples with a 10-fold difference in DNase concentration, the relative reaction rates show a $\sim$10-fold difference.

Resolution and Helix Turns Although an early AFM goal was to sequence single DNA molecules, rarely has the resolution actually been good enough to resolve the major groove of dsDNA. There are, however, a few examples of unusually good resolution on dsDNA.

AFM of DNA packed in a monolayer on a positively charged lipid film has resolved features with the spacing and right-handedness of the major groove of the DNA double helix (18). A series of DNA images produced by using an exceptionally sharp tip under propanol shows the major groove of a well-defined right-handed double helix of 3- to 4-nm periodicity, revealing deformations of the helix where the DNA strand was bent (Figure 8*a–g*; see color insert). Ultra-high vacuum (UHV) noncontact AFM of DNA shows a single image with approximately two turns of an exceptionally fine double helix. This image is not totally convincing, however, because the spacing seems to be 3.3 nm for only one half of a helix turn instead of a full helix turn and the remainder of the DNA in the image looks somewhat like an artifact from a circular tip (106).

The other example of exceptionally fine-resolution AFM on DNA is the finding of minicircles of 126 and 168 bp with holes in their centers (107, 108; Figure 8*h*; see color insert). These DNA minicircles formed bends or kinks of $\sim$78°. There were

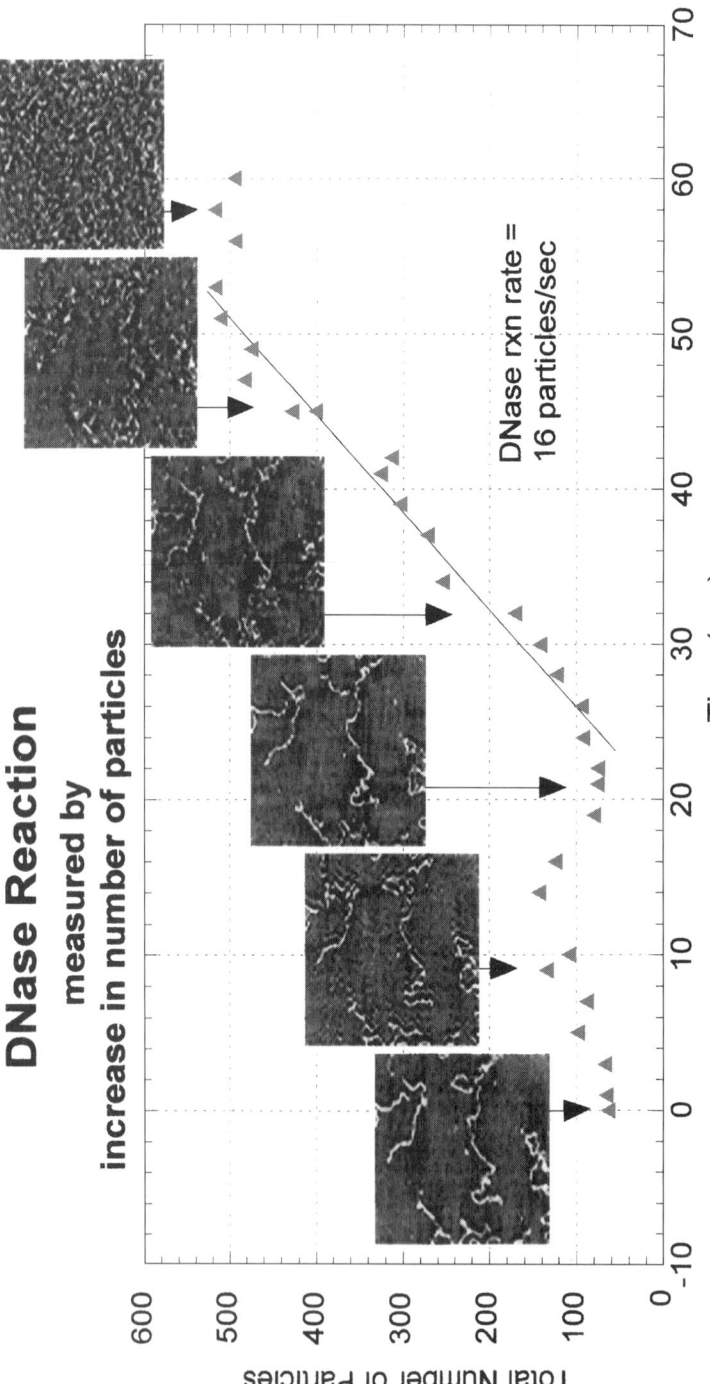

Figure 7 DNase I enzyme kinetics from fast imaging with a small-cantilever atomic force microscope. Each *triangle* represents data from one AFM image. The number of particles per image was determined with the Scion Image software (NIH Image). Six representative images (2 μm by 2 μm) are also shown. DNase molecules were all bound to DNA and did not contribute to the particle count. DNase molecules could be differentiated from DNA molecules by height. The DNase concentration was ~100 DNase molecules per μm^3 of solution, or ~40 DNase molecules in a 100-nm-thick fluid layer above the surface in these image. Small-cantilever AFM images are courtesy of Mario Viani.

four times more kinks in the presence of 1 mM Zn(II) as in 1 mM Mg(II). The kinks were not sequence dependent. Free energies for kink formation were estimated from the frequency of cyclization of these DNA sequences; these free energies were surprisingly small. These minicircles were imaged with an oscillating magnetic field (MAC mode) atomic force microscope (109), which gave an overall resolution comparable to that provided by acoustically driven tapping-mode (TM) AFM (110); therefore the tip(s) may again be the secret of these good images.

Biotechnology and Other Novel Applications for Atomic Force Microscopy of DNA

Two-dimensional DNA crystals have been assembled from designer oligonucleotides that anneal to form interlocking double-crossover structures. Self-assembled nanostructures such as these are envisioned as being "either direct targets in nanofabrication or aids to the construction of such targets" (111). These 2-D DNA nanostructures follow the construction principles of Seeman et al's DNA cubes (112) and linear DNA constructs (113). AFM has become a good tool for imaging 1-D and 2-D DNA constructs. A linear DNA construct, for example, was created from triangular DNA constructs ligated with themselves and with rodlike DNA constructs. AFM analyses of the products revealed zigzag DNA molecules in which the orientation and spacing of the zigzags corresponded to the structures anticipated from the ligations (113).

Other DNA constructs have already aided in the fabrication of nanodevices. Nanowires have been formed by annealing free DNA molecules to complimentary oligonucleotides on a surface and then "growing" silver on the DNA to form conducting wires (114). These functional self-assembled nanodevices may be prototypes for the integrated circuits of the future.

Stretched DNA Stretched DNA up to 80% longer and half as high as normal plasmid DNA in AFM images (115) was perhaps the first overstretched DNA to be observed. Such overstretched DNA is now being analyzed extensively by a number of single-molecule techniques—including AFM—as reviewed below (24, 116, 117).

Polymerase Assays AFM has been used for polymerase assays (118). RNAPs are assayed using a rolling-circle DNA template (118a) that generates a long strand of readily visible RNA as the polymerase polymerizes around and around the DNA minicircle. DNA polymerases are assayed by visualizing the extent of replication of single-stranded ϕX-174 DNA into dsDNA (Figure 9; see color insert).

Michaelis-Menten Kinetics Binding affinities and other kinetic constants for ligand-DNA interactions have been determined by measuring DNA lengths before and after ligand binding. The method is direct, simple, and unambiguous for answering the question of whether the ligand intercalates between the DNA bases, because intercalating ligands lengthen the DNA (119).

Old and New Ways of Preparing Samples

Surfaces for DNA Binding   Two systems are especially good for observing DNA in near-physiological solutions and have been used extensively. One utilizes an unmodified mica surface with a buffer containing one of the following metal ions: Ni(II), Co(II), Zn(II), Mn(II) (118, 120), or Mg(II) (65). The other utilizes an amino-silylated mica surface (AP-mica) (121); DNA in a variety of buffers can be imaged on AP-mica (122).

Unmodified mica is flatter (36) and easier to prepare than AP-mica. AP-mica is preferable for experiments in which metal ions needed for binding to mica cannot be added without perturbing the DNA structures and/or the enzymatic functions that are required for the experiment. Persistence lengths are shorter on AP-mica, approximately half as long as on bare mica (123). A new surface for AFM of DNA is a sapphire surface made hydrophilic with an Na_3PO_4 aqueous solution (124).

In a recent variation on the AP-mica technique, DNA has been covalently attached to a surface, such that it remains on the surface even after being washed with a 1% sodium dodecyl sulfate solution (125). AP-mica was bonded covalently with a psoralen derivative. DNA was applied to this surface. The psoralen derivative intercalates between the DNA bases. Irradiation with 360-nm light cross-links the psoralen derivative to DNA, especially at Ts and TT sequences.

A surface for binding oriented active RNAP molecules was useful in air but not in fluid for observing nucleic acids that associated with the RNAP. Histidine-tagged RNAP molecules were oriented by binding them to an Ni(II)-chelating ligand in an alkane-thiol monolayer on gold. The alkane thiols in the monolayer were functionalized such that 10% were Ni(II) chelating and 90% contained ethylene glycol. The ethylene glycol prevented nonspecific adsorption of proteins and nucleic acids to the surface (126).

Orienting and Elongating DNA Molecules on Surfaces   DNA molecules the length of lambda DNA (16 μm) and even longer have been elongated by fluid evaporation on untreated mica (127), by spin stretching on Mg(II)-mica (128, 129), and by compressed nitrogen flow at an angle of 45° on untreated mica (130). Electrospraying (131) produced many globules of DNA as well as linear DNA, which was not oriented.

Equilibrium versus Kinetic Trapping   Kinetic measurements indicate that DNA diffuses on mica and equilibrates in a 2-D solution before binding irreversibly to the mica. Such a 2-D solution gives DNA conformations that are energetically comparable to the 3-D conformations of the DNA in solution (132–134).

DNA does not equilibrate on glow-discharged mica or H^+-exchanged mica but is instead kinetically trapped on the mica. Kinetic trapping induces 2-D conformations of DNA that do not reflect the 3-D conformations. Ideally the DNA molecules equilibrate on the mica in a "2-D solution," in which case the images reflect the 3-D conformations of the DNA. A statistical polymer chain analysis of DNA images indicated which sample preparation conditions led to equilibrium binding and which led to kinetic trapping (132–134).

Short DNA on Gold Voltage changes induce changes in the orientation of short dsDNAs (15 bp) on thiol-derivatized gold. The 15-bp DNAs are ~2 nm wide and 5 nm long. At more positive voltages, the DNA layer is as thick as the dsDNA. At more negative voltages, the DNA layer is as thick as rodlike 15-bp DNA molecules standing on end on the surface. These changes are reversible (135).

NONIMAGING ATOMIC FORCE MICROSCOPY OF DNA

Stretching DNA Molecules

The forces between the two strands of dsDNA were first measured by Lee and coworkers (136) in a tensile pulling experiment with complementary DNA oligonucleotides on the tip and the sample. The two DNA strands separated at a force of 0.13 nN and an extension of 2.6 times the length of B-DNA both in AFM experiments and in simulations using molecular dynamics (23). In related experiments performed in the Guntherodt laboratory, the cooperative unbinding forces scaled with the number of base pairs ruptured; the data fit the model of a single energy barrier for unbinding and also fit thermodynamic data (25).

Single DNA Molecules Pulling on single long DNA molecules generates force-distance curves such as the one in Figure 10 (see color insert). B-DNA at the left is stretched cooperatively through the B-S transition at ~65 pN to an overstretched form at ~1.7 times its B-DNA contour length (116, 117). At forces above the 65-pN plateau of overstretched DNA, the base pairs melt (24). When both strands of DNA are attached firmly to the tip and the sample surface, the unstretching curve is the same as the stretching curve (22). In Figure 10 (see color insert), however, the DNA molecule did not reanneal when relaxed because one of the DNA strands was broken.

The melting force for dsDNA depends on the pulling velocity, the temperature, the ionic strength, and the DNA sequence. The mechanical energies required to melt single DNA molecules agree well with the enthalpies measured in DNA-melting experiments (22).

Sequence-Dependent Unbinding Forces Entire 16-μm-long lambda DNA molecules have been unzipped not by AFM but by a related single-molecule force-sensing technology (137). The pattern of rupture forces along the length of the DNA mapped with the GC content, with a resolution in the 100–500 base range. The strengths of the rupture forces were 10–15 pN.

Sequence-dependent differences in rupture forces were measured directly for GC versus adenine/thymine (AT) sequences by AFM pulling on poly(dG-dC) and poly(dA-dT) (24). Stretching each of these molecules produced an ssDNA molecule that folded into a hairpin when the pulling force was relaxed. Pulling on these hairpins produced unzipping plateaus of ∼9 pN for AT sequences and ∼20 pN for GC sequences.

The Nanomechanics of Recognition

Intermolecular forces induce cantilever deflections in a new system that transduces molecular recognition into nanomechanical responses. Parallel cantilevers functionalized on one surface with different DNA oligonucleotides showed different deflections upon introduction of a complementary nucleotide into the solution. Single-nucleotide differences were detected with differential cantilever deflections (138). "This method, which was also demonstrated for protein-protein recognition, does not require labeling, optical excitation, or external probes, is compatible with silicon technology, and may easily be run in parallel" (139).

NEW ATOMIC FORCE MICROSCOPY TECHNOLOGY

Single-Walled Nanotube Tips

The new development in AFM tips is that single-walled nanotube (SWNT) tips are finally being produced in larger numbers and by new techniques. SWNT tips have not produced apparent DNA widths as narrow as those in Figure 8, but they have decreased the apparent DNA widths from a typical value of 12–15 nm obtained with most commercial AFM tips to 7–8 nm (140) or 5.6 nm (141). This is already a significant improvement in resolution, having produced beautiful images of Y-shaped antibody molecules (142). As users acquire more experience using SWNT tips, it may become possible to image DNA helix turns on isolated DNA molecules more easily than can be achieved at present. It will be wonderful if we can eventually observe sequence-dependent variability in the major groove of the DNA double helix.

SWNT tips have been made by gluing SWNTs onto AFM tips (141), by electrically transferring SWNTs to tips (140), and by "growing" SWNTs on AFM tips. One method for growing SWNTs on AFM tips involves deposition of metal catalysts on AFM tips and then growth of SWNTs by chemical vapor deposition (143). Another method for growing SWNTs on AFM tips is to flatten the end of a silicon AFM tip by scanning at high force, etch pores in the flat surface, electrochemically deposit an iron catalyst in the pores, and grow SWNTs in the pores by chemical vapor deposition (142). These SWNT tips are too long and must be electrochemically etched in the atomic force microscope before use.

Fast Atomic Force Microscopy Imaging

Because DNA is one of the easier biomaterials to identify in AFM images, much of the work on new AFM technologies involves testing the new atomic force microscopes on DNA. Fast AFM imaging is a major technical objective for new AFM technology because scan rates are typically 0.5–1 min per image with commercial atomic force microscopes and even much slower for many delicate samples or samples with large height variations across their surfaces.

Fast AFM of 4 s/image has been achieved with relatively stiff cantilevers with a 0.5-N/m spring constant (k) by optimizing the atomic force microscope feedback system to keep imaging forces stable and by reducing the scan size and pixel density. As van Noort et al point out, if a 2- by 2-μm image at 512 pixels/line is scanned in 65 s, then a 0.5- by 0.5-μm image at 128 pixels/line has the same resolution and scans in 4 s at the same scan speed of 32 μm/s (144).

A small-cantilever atomic force microscope that images samples faster—<2 s/image for the experiment in Figure 7—and with less thermal noise than commercial atomic force microscopes has been developed (145). Small cantilevers have higher resonance frequencies than commercial cantilevers with the same spring constant, which allows for faster scanning. Small cantilevers are also able to detect smaller forces than conventional cantilevers (146). These small cantilevers are as short as 9 μm in length and as narrow as 3 μm in width (k = 0.06 N/m); commercial cantilevers are typically 100–200 μm long. A major technological development in the small-cantilever atomic force microscope is the reduction of the laser spot size on the cantilever in (Figure 2, see color insert), from $\sim$20 μm in commercial atomic force microscopes to $\sim$3 μm. In addition to exhibiting the fast DNase degradation evident in Figure 7, a small-cantilever atomic force microscope has imaged the GroES caps of the bacterial chaperonin system as they go on and off GroEL (147).

ACKNOWLEDGMENTS

Thanks go to Lee Stone for providing a good writing environment; to Eileen Spain, Jan Hoh, and Stuart Lindsay for permission to use the images in Figures 1, 5 (left panel), and 8h; to Yuri Lyubchenko for helpful discussions about his research; and to Mario Viani, John Sitko, and Lia Pietrasanta for sharing their unpublished data. This work was supported by National Science Foundation grant MCB 9982743.

Visit the Annual Reviews home page at www.AnnualReviews.org

LITERATURE CITED

1. Engel A. 1991. *Annu. Rev. Biophys. Biomol. Struct.* 20:79–108
2. Hansma HG, Hoh J. 1994. *Annu. Rev. Biophys. Biomol. Struct.* 23:115–39
3. Colton RJ, Baselt DR, Dufrene YF, Green JBD, Lee GU. 1997. *Curr. Opin. Chem. Biol.* 1:370–77
4. Hansma HG, Pietrasanta L. 1998.

Curr. Opin. Chem. Biol. 2:579–84
5. Ono MY, Spain EM. 1999. *J. Am. Chem. Soc.* 121:7330–34
6. Hansma HG, Pietrasanta LI, Auerbach ID, Sorenson C, Golan R, Holden PA. 2000. *J. Biomater. Sci. Polym. Ed.* 11:675–84
7. Fisher TE, Oberhauser AF, Carrion-Vazquez M, Marszalek PE, Fernandez JM. 1999. *Trends Biochem. Sci.* 24:379–84
8. Müller DJ, Fotiadis D, Engel A. 1998. *FEBS Lett.* 430:105–11
9. Vinckier A, Semenza G. 1998. *FEBS Lett.* 430:12–16
10. Czajkowsky DM, Shao Z. 1998. *FEBS Lett.* 430:51–54
11. Miles M. 1997. *Science* 277:1845–47
12. Shao Z, Zhang Y. 1996. *Ultramicroscopy* 66:141–52
13. Amato I. 1997. *Science* 276:1982–85
14. Stokstad E. 1997. *Science* 275:1882
15. Weisenhorn AL, Drake B, Prater CB, Gould SA, Hansma PK, et al. 1990. *Biophys. J.* 58:1251–58
16. Brisson A, Bergsma-Schutter W, Oling F, Lambert O, Reviakine I. 1999. *J. Crys. Growth.* 196:456–70
17. Reviakine I, Simon A, Brisson A. 2000. *Langmuir* 16:1473–77
18. Mou J, Czajkowsky DM, Zhang Y, Shao Z. 1995. *FEBS Lett.* 371:279–82
19. Fang Y, Yang J. 1997. *J. Phys. Chem. Ser. B* 101:441–49
20. Florin E-L, Moy VT, Gaub HE. 1994. *Science* 264:415–17
21. Moy VT, Florin E-L, Gaub HE. 1994. *Science* 266:257–59
22. Clausen-Schaumann H, Rief M, Tolksdorf C, Gaub HE. 2000. *Biophys. J.* 78:1997–2007
23. MacKerell AD Jr, Lee GU. 1999. *Eur. Biophys. J.* 28:415–26
24. Rief M, Clausen-Schaumann H, Gaub HE. 1999. *Nat. Struct. Biol.* 6:346–49
25. Strunz T, Oroszlan K, Schafer R, Guntherodt HJ. 1999. *Proc. Natl. Acad. Sci. USA* 96:11277–82
26. Rouzina I, Bloomfield VA. 1998. *Biophys. J.* 74:3152–64
27. Fang Y, Spisz TS, Wiltshire T, D'Costa NP, Bankman IN, et al. 1998. *Anal. Chem.* 70:2123–29
28. Spisz TS, Fang Y, Reeves RH, Seymour CK, Bankman IN, Hoh JH. 1998. *Med. Biol. Eng. Comput.* 36:667–72
29. Mazzola LT, Frank CW, Fodor SP, Mosher C, Lartius R, Henderson E. 1999. *Biophys. J.* 76:2922–33
30. Allison DP, Kerper PS, Doktycz MJ, Spain JA, Modrich P, et al. 1996. *Proc. Natl. Acad. Sci. USA* 93:8826–29
31. Allison DP, Kerper PS, Doktycz MJ, Thundat T, Modrich P, et al. 1997. *Genomics* 41:379–84
32. Pfannschmidt C, Schaper A, Heim G, Jovin TM, Langowski J. 1996. *Nucleic Acids Res.* 24:1702–9
33. Klinov DV, Lagutina IV, Prokhorov VV, Neretina T, Khil PP, et al. 1998. *Nucleic Acids Res.* 26:4603–10
34. Cherny DI, Fourcade A, Svinarchuk F, Nielsen PE, Malvy C, Delain E. 1998. *Biophys. J.* 74:1015–23
35. Marsh TC, Vesenka J, Henderson E. 1995. *Nucleic Acids Res.* 23:696–700
36. Hansma HG, Revenko I, Kim K, Laney DE. 1996. *Nucleic Acids Res.* 24:713–20
37. Revet B, Fourcade A. 1998. *Nucleic Acids Res.* 26:2092–97
38. Shlyakhtenko LS, Hsieh P, Grigoriev M, Potaman VN, Sinden RR, Lyubchenko YL. 2000. *J. Mol. Biol.* 296:1169–73
39. Shlyakhtenko LS, Potaman VN, Sinden RR, Lyubchenko YL. 1998. *J. Mol. Biol.* 280:61–72
40. Shlyakhtenko LS, Appella E, Harrington RE, Kutyavin I, Lyubchenko YL. 1994. *J. Biomol. Struct. Dyn.* 12:131–43
41. Shlyakhtenko LS, Rekesh D, Lindsay SM, Kutyavin I, Appella E, et al. 1994. *J. Biomol. Struct. Dyn.* 11:1175–89
42. Oussatcheva EA, Shlyakhtenko LS, Glass R, Sinden RR, Lyubchenko YL, Potaman VN. 1999. *J. Mol. Biol.* 292:75–86

43. Cassell G, Moision R, Rabani E, Segall A. 1999. *Nucleic Acids Res.* 27:1145–51
44. Ohta T, Nettikadan S, Tokumasu F, Ideno H, Abe Y, et al. 1996. *Biochem. Biophys. Res. Commun.* 226:730–34
45. Yaneva M, Kowalewski T, Lieber MR. 1997. *EMBO J.* 16:5098–112
46. Pang D, Yoo S, Dynan WS, Jung M, Dritschilo A. 1997. *Cancer Res.* 57:1412–15
47. Herbert A, Schade M, Lowenhaupt K, Alfken J, Schwartz T, et al. 1998. *Nucleic Acids Res.* 26:3486–93
48. Cary RB, Peterson SR, Wang J, Bear DG, Bradbury EM, Chen DJ. 1997. *Proc. Natl. Acad. Sci. USA* 94:4267–72
49. Wyman C, Rombel I, North AK, Bustamante C, Kustu S. 1997. *Science* 275:1658–61
50. Rippe K, Guthold M, von Hippel PH, Bustamante C. 1997. *J. Mol. Biol.* 270:125–38
51. Becker JC, Nikroo A, Brabletz T, Reisfeld RA. 1995. *Proc. Natl. Acad. Sci. USA* 92:9727–31
52. Valle M, Valpuesta JM, Carrascosa JL, Tamayo J, Garcia R. 1996. *J. Struct. Biol.* 116:390–98
53. Garcia RA, Bustamante CJ, Reich NO. 1996. *Proc. Natl. Acad. Sci. USA* 93:7618–22
54. Palecek E, Vlk D, Stankova V, Brazda V, Vojtesek B, et al. 1997. *Oncogene* 15:2201–9
55. Wyman C, Grotkopp E, Bustamante C, Nelson HC. 1995. *EMBO J.* 14:117–23
56. Niemeyer CM, Adler M, Pignataro B, Lenhert S, Gao S, et al. 1999. *Nucleic Acids Res.* 27:4553–61
57. Smith GC, Cary RB, Lakin ND, Hann BC, Teo SH, et al. 1999. *Proc. Natl. Acad. Sci. USA* 96:11134–39
58. Moreno-Herrero F, Herrero P, Colchero J, Baro AM, Moreno F. 1999. *FEBS Lett.* 459:427–32
59. Hansma HG, Bezanilla M, Nudler E, Hansma PK, Hoh J, et al. 1998. *Probe Microsc.* 1:117–25
60. Allan BW, Garcia R, Maegley K, Mort J, Wong D, et al. 1999. *J. Biol. Chem.* 274:19269–75
61. van Noort SJT, van der Werf KO, Eker AP, Wyman C, de Grooth BG, et al. 1998. *Biophys. J.* 74:2840–49
62. van Noort J, Orsini F, Eker A, Wyman C, de Grooth B, Greve J. 1999. *Nucleic Acids Res.* 27:3875–80
63. Kasas S, Thomson NH, Smith BL, Hansma HG, Zhu X, et al. 1997. *Biochemistry* 36:461–68
64. Guthold M, Zhu X, Rivetti C, Yang GL, Thomson NH, et al. 1999. *Biophys. J.* 77:2284–94
65. Bustamante C, Guthold M, Zhu X, Yang GL. 1999. *J. Biol. Chem.* 274:16665–68
66. Hansma HG. 1999. *Proc. Natl. Acad. Sci. USA* 96:14678–80
67. Schulz A, Mucke N, Langowski J, Rippe K. 1998. *J. Mol. Biol.* 283:821–36
68. Rivetti C, Guthold M, Bustamante C. 1999. *EMBO J.* 18:4464–75
69. Pietrasanta LI, Smith BL, MacLeod MC. 2000. *Chem. Res. Toxicol.* 13:351–55
70. Onoa GB, Cervantes G, Moreno V, Prieto MJ. 1998. *Nucleic Acids Res.* 26:1473–80
71. Pang D, Berman BL, Chasovskikh S, Rodgers JE, Dritschilo A. 1998. *Radiat. Res.* 150:612–18
72. Pang D, Popescu G, Rodgers J, Berman BL, Dritschilo A. 1996. *Scanning Microsc.* 10:1105–9
73. Pang D, Vidic B, Rodgers J, Berman BL, Dritschilo A. 1997. *Radiat. Oncol. Invest.* 5:163–69
74. Fang Y, Hoh JH. 1998. *J. Am. Chem. Soc.* 120:8903–9
75. Hansma HG, Golan R, Hsieh W, Lollo CP, Mullen-Ley P, Kwoh D. 1998. *Nucleic Acids Res.* 26:2481–87
76. Lin Z, Wang C, Feng X, Liu M, Li J, Bai C. 1998. *Nucleic Acids Res.* 26:3228–34
77. Golan R, Pietrasanta LI, Hsieh W, Hansma HG. 1999. *Biochemistry* 38:14069–76

78. Hansma HG, Pietrasanta LI, Golan R, Sitko JC, Viani M, et al. 2000. *Biol. Struct. Dyn. Conversation* 11:271–76
79. Allen MJ, Bradbury EM, Balhorn R. 1997. *Nucleic Acids Res.* 25:2221–26
80. Fang Y, Hoh JH. 1998. *Nucleic Acids Res.* 26:588–93
81. Fang Y, Hoh JH. 1999. *FEBS Lett.* 459:173–76
82. Fang Y, Spisz TS, Hoh JH. 1999. *Nucleic Acids Res.* 27:1943–49
83. Dunlap DD, Maggi A, Soria MR, Monaco L. 1997. *Nucleic Acids Res.* 25:3095–101
84. Marshall E. 2000. *Science* 287:565, 7
85. Ziady AG, Ferkol T, Gerken T, Dawson DV, Perlmutter DH, Davis PB. 1998. *Gene Ther.* 5:1685–97
86. Hart SL, Arancibia-Carcamo CV, Wolfert MA, Mailhos C, O'Reilly NJ, et al. 1998. *Hum. Gene Ther.* 9:575–85
87. Choi YH, Liu F, Choi JS, Kim SW, Park JS. 1999. *Hum. Gene Ther.* 10:2657–65
88. Wolfert MA, Schacht EH, Toncheva V, Ulbrich K, Nazarova O, Seymour LW. 1996. *Hum. Gene Ther.* 7:2123–33
89. Kawaura C, Noguchi A, Furuno T, Nakanishi M. 1998. *FEBS Lett.* 421:69–72
90. Kim JS, Kim BI, Maruyama A, Akaike T, Kim SW. 1998. *J. Controlled Release* 53:175–82
91. Kim JS, Maruyama A, Akaike T, Kim SW. 1998. *Pharm. Res.* 15:116–21
92. Thurmond KB II, Remsen EE, Kowalewski T, Wooley KL. 1999. *Nucleic Acids Res.* 27:2966–71
93. Trubetskoy VS, Loomis A, Hagstrom JE, Budker VG, Wolff JA. 1999. *Nucleic Acids Res.* 27:3090–95
94. Zlatanova J, Leuba SH, van Holde K. 1998. *Biophys. J.* 74:2554–66
95. Allen MJ, Dong X-F, O'Neill TE, Yau P, Kowalczykowski SC, et al. 1993. *Biochemistry* 32:8390–96
96. Yodh JG, Lyubchenko YL, Shlyakhtenko LS, Woodbury N, Lohr D. 1999. *Biochemistry* 38:15756–63
97. Sato MH, Ura K, Hohmura KI, Tokumasu F, Yoshimura SH, et al. 1999. *FEBS Lett.* 452:267–71
98. Leuba SH, Bustamante C, Zlatanova J, van Holde K. 1998. *Biophys. J.* 74:2823–29
99. Leuba SH, Bustamante C, van Holde K, Zlatanova J. 1998. *Biophys. J.* 74:2830–39
100. Thalhammer S, Stark RW, Muller S, Wienberg J, Heckl WM. 1997. *J. Struct. Biol.* 119:232–37
101. Henderson E. 1992. *Nucleic Acids Res.* 20:445–47
102. Tamayo J, Miles M. 2000. *Ultramicroscopy* 82:245–51
103. Rasch P, Wiedemann U, Wienberg J, Heckl WM. 1993. *Proc. Natl. Acad. Sci. USA* 90:2509–11
104. Ellis DJ, Dryden DT, Berge T, Edwardson JM, Henderson RM. 1999. *Nat. Struct. Biol.* 6:15–17
105. Bezanilla M, Drake B, Nudler E, Kashlev M, Hansma PK, Hansma HG. 1994. *Biophys. J.* 67:2454–59
106. Uchihashi T, Tanigawa M, Ashino M, Sugawara Y, Yokoyama K, et al. 2000. *Langmuir* 16:1349–53
107. Han W, Dlakic M, Zhu YJ, Lindsay SM, Harrington RE. 1997. *Proc. Natl. Acad. Sci. USA* 94:10565–70
108. Han W, Lindsay SM, Dlakic M, Harrington RE. 1997. *Nature* 386:563
109. Wenhai H, Lindsay SM, Tianwei J. 1996. *Appl. Phys. Lett.* 69:4111–13
110. Revenko I, Proksch R. 2000. *J. Appl. Phys.* 87:526–33
111. Winfree E, Liu F, Wenzler LA, Seeman NC. 1998. *Nature* 394:539–44
112. Seeman NC, Kallenbach NR. 1994. *Annu. Rev. Biophys. Biomol. Struct.* 23:53–86
113. Yang XP, Wenzler LA, Qi J, Li XJ, Seeman NC. 1998. *J. Am. Chem. Soc.* 120:9779–86
114. Braun E, Eichen Y, Sivan U, Ben-Yoseph G. 1998. *Nature* 391:775–78
115. Thundat T, Allison DP, Warmack RJ. 1994. *Nucleic Acids Res.* 22:4224–28

116. Smith SB, Cui Y, Bustamante C. 1996. *Science* 271:795–98
117. Cluzel P, Lebrun A, Heller C, Lavery R, Viovy J-L, et al. 1996. *Science* 271:792–94
118. Hansma HG, Golan R, Hsieh W, Daubendiek SL, Kool ET. 1999. *J. Struct. Biol.* 127:240–47
118a. Liu D, Daubendiek SL, Zillman MA, Ryan K, Kool ET. 1996. *J. Am. Chem. Soc.* 118:1587–94
119. Coury JE, McFail-Isom L, Williams LD, Bottomley LA. 1996. *Proc. Natl. Acad. Sci. USA* 93:12283–86
120. Hansma HG, Laney DE. 1996. *Biophys. J.* 70:1933–39
121. Lyubchenko YL, Shlyakhtenko LS, Harrington RE, Oden PI, Lindsay SM. 1993. *Proc. Natl. Acad. Sci. USA* 90:2137–40
122. Bezanilla M, Manne S, Laney DE, Lyubchenko YL, Hansma HG. 1995. *Langmuir* 11:655–59
123. Hansma HG, Kim KJ, Laney DE, Garcia RA, Argaman M, et al. 1997. *J. Struct. Biol.* 119:99–108
124. Yoshida K, Yoshimoto M, Sasaki K, Ohnishi T, Ushiki T, et al. 1998. *Biophys. J.* 74:1654–57
125. Shlyakhtenko LS, Gall AA, Weimer JJ, Hawn DD, Lyubchenko YL. 1999. *Biophys. J.* 77:568–76
126. Thomson NH, Smith BL, Almqvist N, Schmitt L, Kashlev M, et al. 1999. *Biophys. J.* 76:1024–33
127. Wang W, Lin J, Schwartz DC. 1998. *Biophys. J.* 75:513–20
128. Yokota H, Nickerson DA, Trask BJ, van den Engh G, Hirst M, et al. 1998. *Anal. Biochem.* 264:158–64
129. Yokota H, Sunwoo J, Sarikaya M, van den Engh G, Aebersold R. 1999. *Anal. Chem.* 71:4418–22
130. Li J, Bai C, Wang C, Zhu C, Lin Z, et al. 1998. *Nucleic Acids Res.* 26:4785–86
131. Morozov VN, Morozova TY, Kallenbach NR. 1998. *Int. J. Mass Spectrom.* 178:143–59
132. Rivetti C, Guthold M, Bustamante C. 1996. *J. Mol. Biol.* 264:919–32
133. Rivetti C, Walker C, Bustamante C. 1998. *J. Mol. Biol.* 280:41–59
134. Bustamante C, Rivetti C. 1996. *Annu. Rev. Biophys. Biomol. Struct.* 25:395–429
135. Kelley SO, Barton JK, Jackson NM, McPherson LD, Potter AB, et al. 1998. *Langmuir* 14:6781–84
136. Lee GU, Chrisey LA, Colton RJ. 1994. *Science* 266:771–73
137. Essevaz-Roulet B, Bockelmann U, Heslot F. 1997. *Proc. Natl. Acad. Sci. USA* 94:11935–40
138. Fritz J, Baller MK, Lang HP, Rothuizen H, Vettiger P, et al. 2000. *Science* 288:316–18
139. Szuromi P. 2000. *Science* 288:225
140. Nishijima H, Kamo S, Akita S, Nakayama Y, Hohmura KI, et al. 1999. *Appl. Phys. Lett.* 74:4061–63
141. Wong SS, Woolley AT, Wang Odom TW, Huang JL, Kim P, et al. 1998. *Appl. Phys. Lett.* 73:3465–67
142. Cheung CL, Hafner JH, Lieber CM. 2000. *Proc. Natl. Acad. Sci. USA* 97:3809–13
143. Hafner JH, Cheung CL, Lieber CM. 1999. *Nature* 398:761–62
144. van Noort SJ, van Der Werf KO, de Grooth BG, Greve J. 1999. *Biophys. J.* 77:2295–303
145. Viani MB, Schäffer TE, Paloczi GT, Pietrasanta LI, Smith BL, et al. 1999. *Rev. Sci. Instrum.* 70:4300–3
146. Viani MB, Schäffer TE, Chand A, Rief M, Gaub HE, Hansma PK. 1999. *J. Appl. Phys.* 86:2258–62
147. Viani MB, Pietrasanta LI, Thompson JB, Chand A, Gebeshuber IC, et al. 2000. *Nat. Struct. Biol.* 7:644–47
148. Binnig G, Quate CF, Gerber C. 1986. *Phys. Rev. Lett.* 56:930–33
149. Binnig G, Gerber C, Stoll E, Albrecht RT, Quate CF. 1987. *Europhys. Lett.* 3:1281–86

150. Manne S, Hansma PK, Massie J, Elings VB, Gewirth AA. 1991. *Science* 251:183–86
151. Giessibl FJ, Hembacher S, Bielefeldt H, Mannhart J. 2000. *Science* 289:422–25
152. Pietrasanta LI, Thrower D, Hsieh W, Rao S, Stemmann O, et al. 1999. *Proc. Natl. Acad. Sci. USA* 96:3757–62
153. Rees WA, Keller RW, Vesenka JP, Yang GL, Bustamante C. 1993. *Science* 260:1646–49
154. Baumann P, Eap CB, Mueller WE, Tillement J-P. 1989. α_1-*Acid glycoprotein: genetics, biochemistry, physiological functions, and pharmacology: Proceedings of a Meeting held in Prilly-Lausanne, Switzerland, September 1–2, 1988.* New York: Liss. 470 pp.
155. Hansma HG, Laney DE, Bezanilla M, Sinsheimer RL, Hansma PK. 1995. *Biophys. J.* 68:1672–77
156. Hansma HG. 1996. *J. Vac. Sci. Technol. Ser. B* 14:1390–94
157. Hansma HG, Bezanilla M, Zenhausern F, Adrian M, Sinsheimer RL. 1993. *Nucleic Acids Res.* 21:505–12

ON THE CHARACTERISTICS OF MIGRATION OF OLIGOMERIC DNA IN POLYACRYLAMIDE GELS AND IN FREE SOLUTION

Udayan Mohanty and Larry McLaughlin

Eugene F. Merkert Chemistry Center, Boston College, Chestnut Hill, Massachusetts 02467; e-mail: mohanty@bc.edu; larry.mclaughlin@bc.edu

Key Words polyelectrolytes, counterion condensation, anomalous migration, electrophoretic mobility

■ **Abstract** We review a model for the free-solution electrophoretic mobility of oligomeric double-stranded (ds) DNA. We have found that the free-solution mobility of ds DNA increases as the molecular weight of the fragment increases, up to a few hundred base pairs. This insight is combined with recent advances in the nature of counterion condensation theory of very short DNA fragments to describe quantitatively the electrophoretic mobility of oligomeric single-stranded DNA in polyacrylamide gels. The model predicts, in agreement with recent experiments, that significant anomalous migration exists with short DNA sequences, the onset of which is dependent on the size of polyacrylamide gel pores. For terminal phosphate-labeled DNA fragments, the free-solution mobility is no longer proportional to the ratio of the total effective charge and the friction coefficient. These changes in properties affect the characteristics of migration of end-labeled DNA fragments in polyacrylamide gels.

INTRODUCTION

Polyacrylamide gel electrophoresis is a broadly used technique in the investigation of structural and conformational changes in DNA and protein-DNA complexes (1–3). Despite considerable advances in the past two decades, a quantitative understanding of how various-size DNA fragments migrate in polyacrylamide and agarose gels remains elusive (3–36), in part due to the absence of structural information about the nature of the gel matrix and in part because of difficulty in accounting for the ionic effects of the buffer and understanding the structure of the macromolecule (26–31, 35, 36). Qualitative descriptions of the mobility of high-molecular-weight DNA fragments in polyacrylamide gels are based on the reptation model (5–9, 13–25). In contrast, for low-molecular-weight fragments, the electrophoretic mobility of the macromolecule is described by the Ogstron pore size distribution model (26–30, 32–36).

In the reptation model, the gel matrix imposes restraint on the migrating DNA such that its motion is restricted to tubelike passages (5–9, 13–25). The electrophoretic mobility is inversely proportional to the length of the macromolecule (5–9, 13–25). In the Ogstron pore size distribution model, the retardation of a macromolecule during electrophoretic migration is governed by the portion of available volume in the gel (32–36). Weaknesses of the Ogstron model include the facts that it does not account for the ionic effects of the buffer, the structure of the macromolecule, or the interaction of the molecule with the gel matrix (35, 36). In the generalized Ogstron model, attention is focused on the retardation coefficient K_R—a quantity that is evaluated from a semilogarithmic plot of the relative mobility versus the square root of the gel concentration (35, 36). This coefficient is then used to obtain parameters related to the gel fibers, supposing a plausible hypothesis about the structure of the gel (35, 36). If the radius R of the macromolecule is smaller than that of the gel fibers, then the retardation coefficient is given by $K_R = \pi l (r + R)^2$, where l is the gel fiber length per gram of the matrix and r is the radius of the fiber (35, 36).

Stellwagen and coworkers have analyzed in considerable detail the electrophoretic mobility of two 147-bp restriction fragments to elucidate the mechanism of migration of DNA in polyacrylamide gel with tris-borate-EDTA (TBE) buffer of pH 8.3 (35, 36). These fragments were obtained from an *Msp*I digest of plasmid pBR322, and corresponding molecular-weight ladders were constructed (35, 36). Both fragments are GC rich; one contains a single bend at its center due to the $(AT)_5$ tract, while the other fragment does not (35, 36). Both absolute and relative mobilities of the DNA fragments were measured, and Ferguson plots were constructed (35, 36).

Several noteworthy results were obtained (35, 36). The absolute-mobility data indicate that the mobility for the low-molecular-weight fragments goes as $M^{-1/3}$, whereas the reptation mode of transport is valid for higher-molecular-weight DNA (36). The exponent obtained in the relationship for low-molecular-weight fragments, however, depends on the polyacrylamide concentration ($\%T$) in the gel (35, 36). The mobilities obtained for small DNA fragments are in agreement with free-solution mobility values if the concentration of the polyacrylamide gel is assumed to be vanishingly small (35, 36). If the macromolecule could be described by the geometric mean radius $\overline{R}$, then the pore size of the gel could be determined to vary from 5 to 0.6 nm as the monomer concentration ($\%T$) increased from 3.5 to 10.5 (36). On the other hand, if the radius of gyration R_G was used as a gauge of the macromolecule radius, then the pore size varied from 30 to 0.3 nm for the same range of gel concentrations. Furthermore, the effective pore radius is in the range 130 to 70 nm if the data are analyzed based on relative mobilities and the macromolecular size estimate is based on the radius of gyration (35, 36).

The Ogstron model for electrophoretic mobility is obeyed provided that relative mobilities are used and the size of the DNA is based on the radius of gyration (35, 36). Upon decreasing $\%T$—the polyacrylamide concentration, at a constant concentration of the cross-linking agent bisacrylamide ($\%C$), the mobility anomaly of the anomalous migrating 147-bp fragment becomes less pronounced (36). This decrease in anomalous migration does not depend on whether the molecular size is smaller or larger than the radius of gyration R_G. This result suggests that, instead of the pore size, the electrophoretic mobility of the anomalous migrating 147-bp fragments is governed primarily by the acrylamide concentration (36). If the cross-linking agent %C is decreased at constant %T, one observes a decrease in anomalous migration until the size of the pore reaches the same order of magnitude as the radius of gyration (36). With further decreases in %C at constant %T, the anomalous electrophoretic mobility was found to be independent of the pore size (36).

In this account, we summarize a new approach that has been developed in our laboratory for describing quantitatively the electrophoretic mobility of oligomeric DNA fragments in free solution and in polyacrylamide gels. Our work was stimulated by the reported results of Holmes & Stellwagen (35) and Stellwagen (36) described above. Space restrictions prevent us from covering recent advances in understanding the mechanisms by which DNA fragments migrate in agarose gels (11, 12), as well as the gel retardation of intrinsically curved DNA sequences (4) or protein-DNA complexes during gel-shift assays (2, 3).

This review is divided into three parts. In the first part, we describe a recent advance in understanding of the nature of counterion condensation in ionic oligomers. In the second part, we review a model that describes the free-solution mobility of ionic oligomers. Equipped with the findings from the first two parts, we review our work on the mechanisms by which oligomeric DNA fragments migrate in polyacrylamide gel.

ELECTROPHORETIC MOBILITY OF OLIGOMERIC DNA

Extended Debye-Huckel Theory: Ionic Oligomers

The Debye-Huckel potential of the mean force between two ions of charges q_1 and q_2 that are separated by a distance r is given by $(q_1 q_2/\varepsilon)\exp(-\kappa r)/r$, where the Debye screening parameter and the dielectric constant of the solvent are denoted by κ and ε, respectively (37, 38). The work w_P required, in units of $k_B T$, to assemble P ions, each with renormalized charge $q(1-\theta)$, from an infinite distance to a linear configuration is (38)

$$w_P = (1-\theta)^2 \xi \sum_{n=1}^{P}(P-n)\frac{e^{-\kappa nb}}{n}. \qquad 1.$$

Here b is the charge spacing, and ξ is the dimensionless linear charge density of the polymer, defined by the ratio of the Bjerrum length l_B to the charge spacing b (37, 38). The factor $q(1-\theta)$ reduces the bare charge q by θ. It assesses the fact that counterions along the backbone of the DNA are not necessarily dissociated (37–39).

By considering the limit of small Debye screening parameters and the constraint that $L/b \gg 1$ such that κL is finite, an asymptotic expansion of the work w_P leads to (38)

$$w_P \approx -(1-\theta)^2 \xi P \left\{ \ln(\kappa b) + \frac{1}{\kappa L} + e^{-\kappa L} \sum_{n=1}^{\infty} (-1)^n n! \frac{1}{(\kappa L)^{n+1}} \right\}. \quad 2.$$

The logarithmic term in Equation 2 signifies the case for which the length L of the polyion is large. The second and the third terms take into account the finite size of the polyion. Since some of the counterions are adsorbed to the polyion, there is an additional term that must be included into the total work w. This term is the work required to move the counterions from bulk solution to the adsorbed layer around the macromolecule and is expressed in terms of the salt concentration, the fraction of condensed counterions per polyion charge θ, and the partition function of the adsorbed layer (37, 38). On minimizing the total work with respect to θ, one obtains (38)

$$\frac{\partial w}{\partial \theta} \approx -P\xi \left\{ \theta - \left(1 - \frac{1}{\xi}\right) \right\} \ln c_s + B(\theta, \kappa L), \quad 3.$$

where c_s is the salt concentration and the function $B(\theta, \kappa L)$ denotes all terms that are in powers of $(\kappa L)^{-1}$.

For linear charge densities of less than unity, total work is at a minimum when $\theta = 0$. However, if the linear charge density is greater than unity ($\xi \geq 1$), then in dilute solutions ($\kappa \to 0, L \to \infty$; κL held fixed), an equilibrium state is obtained for $\theta = 1 - \xi^{-1}$ (37, 38). In this case, the condensed layer is stable provided $B(\theta = 1 - \xi^{-1}, \kappa L)$, which vanishes. This imposes an explicit constraint on the partition function of the condensed layer—a constraint that forces the characteristics of the condensed layer to be compatible with the ionic interactions in the Debye cloud (37, 38). A significant conclusion from the analysis is that for polyion length $L \approx \kappa^{-1}$, the number of condensed counterions as well as the critical charge density that signals counterion condensation remains unaffected (38).

What happens to counterion condensation for polymer lengths $L \ll \kappa^{-1}$ [i.e. when $\kappa \to 0, L \to \infty, L = o(\kappa^{-1})$]? The work w_P to assemble the P ions in a linear configuration scales as $w_P \approx (1-\theta)^2 \xi P \ln(\frac{L}{b})$. The equilibrium value of θ is obtained by setting the derivative $\partial w / \partial \theta$ to zero (38):

$$\frac{\partial w}{\partial \theta} = -P \ln c_s - 2P\xi(1-\theta)\ln\frac{L}{b} + PA(\theta). \quad 4.$$

The term $A(\theta)$ carries information about the partition function of the condensed layer. In the limit $\kappa \to 0, L \to \infty, L = o(\kappa^{-1})$, the logarithmic terms vanish, and the value of θ is given by (38)

$$\theta \approx 1 - \frac{\xi_{\text{crit}}}{\xi}$$
$$\xi_{\text{crit}} = -\frac{\ln \kappa b}{\ln (L/b)}. \quad 5.$$

Three comments are in order. First, when considering an oligomer shorter than the Debye length, one finds that counterions do in fact condense on the oligomer (38). However, the critical linear charge density ξ_{crit} increases as the Debye length approaches infinity. Second, the fractional extent of the condensed counterions for a polymer chain of length equal to the Debye length is the same as that of a long polyelectrolyte chain (38). The partition functions of the condensed layer in the two cases are, however, different. Third, density functional treatment offers a novel interpretation of the nature of counterion condensation in ionic oligomers (38, 40). Near the middle sections of the polyion, the condensed fraction is $1-1/\xi$. For distances $s \approx \kappa^{-1}$ from the ends of the polymer chain, the condensed fraction begins to decrease and vanishes at the characteristic distance of $\xi_{crit}(s) = \xi$ (38). These predictions are in agreement with computer simulations and experiments on Coulombic end effects (41–46).

The number of counterions per phosphate charge is intimately related to the adsorption excess per monomer (47, 48). The adsorption excess is defined by the slope of the free energy of charging via $\gamma = -(\beta\kappa/2)(\partial g_{elec}/\partial\kappa)$, where $\beta = 1/k_B T$, T is the temperature, and k_B is the Boltzmann constant (47–50). The surface charge density of the polyion, modeled as a cylinder of radius R and length L, within the framework of the nonlinear Poisson-Boltzmann equation, is (47, 50)

$$\sigma^2 = (2/x_o)[1 + \cosh(y_o/2) - 1/\cosh(y_o/2)]^{1/2}\{1 + (1+w^2)^{1/2}\}, \qquad 6.$$

where $w = x_o[1 + \cosh(y_o/2)]$, $x_o = \kappa R$, $y_o = y(x_o)$, $y = e\Psi/k_B T$, e is the proton charge, and Ψ is the electrostatic potential. The adsorption excess per monomer follows from Equation 6 (47, 50):

$$\gamma = (2/x_o\sigma)\{(1+w^2)^{1/2} - (1-x_o)^{1/2}\}. \qquad 7.$$

In the limit $\kappa L \gg 1$, and for large value for $z = \cosh(y_o/2)$, the adsorption excess is precisely that given by the counterion condensation model (37, 47).

FREE-SOLUTION MOBILITY

Free-solution mobility experiments of oligomeric DNA via capillary experiments have been hampered by electroendosmosis flows (EOFs). The attraction of the counterions in the buffer to the silanol residues on the capillary surface is believed to give rise to EOFs (51). In a noteworthy experimental study, Stellwagen and coworkers have recently succeeded in significantly reducing the EOF by coating the interior of the capillary walls with N-substituted acrylamides. This approach enabled accurate measurement of the free-solution electrophoretic mobility of oligomers in tris-acaetate-EDTA (TAE) and TBE buffers (51).

Three important conclusions were reached by Stellwagen and coworkers. First, the quantitative features of the free-solution mobility depend on the choice of the buffer. For an oligomer of a fixed length, mobility in the TBE buffer was greater than that observed in the TAE buffer (51). Second, as expected for a free-draining molecule, the free-solution mobility of DNA is independent of molecular weight for molecular weights larger than ~400 bp to as much as 48.5 kb (51). Third, the free-solution mobility increases with increasing length for oligomeric DNA of less than a few hundred base pairs (51).

We have developed a model that explicitly describes the observed experimental characteristics (52). The forces acting on monomer i of the macromolecule include the random forces as a result of Brownian motions, $\mathbf{F}_{i;\text{rand}}$, the external forces $\mathbf{F}_{\text{ext}}$, the polarization forces $\mathbf{F}_{i;\text{pol}}$, the relaxation field force $q\mathbf{E}_{\text{rel}}$ and frictional forces $\zeta(\mathbf{v}_i - \mathbf{u}(\mathbf{r}_i))$, where $\mathbf{r}_i$ is the coordinate of the ith monomer, v_i is the velocity if the monomer, ζ is the frictional coefficient of the monomer in the solvent, and q is the charge on a monomer (52, 53a). The velocity field of the solvent $\mathbf{u}(\mathbf{r}_i)$ satisfies (52, 53a):

$$\mathbf{u}(\mathbf{r}_i) = \sum_m \overleftrightarrow{O}(\mathbf{r}_i - \mathbf{r}_m)\exp(-\kappa(\mathbf{r}_i - \mathbf{r}_m)) \cdot \mathbf{F}_{\text{ext}} + \overleftrightarrow{O}(\mathbf{r}_i - \mathbf{r}_m)$$
$$\cdot [\mathbf{F}_{m;\text{rand}} + \mathbf{F}_{m;\text{pol}}] + \sum_m \overleftrightarrow{T}_{\text{sc}}(\mathbf{r}_i - \mathbf{r}_m) \cdot q\mathbf{E}_{\text{rel}}(\mathbf{r}_m), \qquad 8.$$

where $\overleftrightarrow{O}$ and $\overleftrightarrow{T}_{\text{sc}}$ are the unscreened and the screened Oseen tensors, respectively.

A charge that is in motion in an ionic solution produces an electric field. The total electric field $E_T(r)$ is in part due to the ionic-charge densities and in part due to the electric field of the moving charges (52, 53a):

$$\mathbf{E}_T(\mathbf{r}) = -\overleftrightarrow{T}(\mathbf{r} - \mathbf{r}') \cdot \mathbf{U}(\mathbf{r}')$$
$$T_{\chi\sigma}(\mathbf{r}) = -\int d^3k \, \exp(i\mathbf{k} \cdot \mathbf{r}) \frac{k_\chi k_\sigma}{(2\pi)^3} \frac{\beta q \lambda_o \kappa^2}{(k^2 + \kappa^2)}. \qquad 9.$$

Here, λ_o is the mobility of the ion, $\mathbf{k}$ is the Fourier transform variable, and the unit vector $\mathbf{k}_\sigma$ is along the σ direction.

Upon balancing the various forces acting on a given monomer i, summing over i, making use of definition of the electrophoretic mobility, substituting Equation 8 for the velocity field in the force balance expression, and ignoring terms higher than quadratic in the velocity field, one obtains the desired expression for the free-solution mobility (52):

$$U = \frac{q O_{\text{eff}}}{1 - qT_o/\zeta}, \qquad 10.$$

where T_o and O_{eff} are equilibrium averages of $\overleftrightarrow{T}$ and the Oseen tensor, respectively. We have evaluated Equation 10 for a double-stranded (ds) DNA based on several reasonable assumptions (52): (*a*) The oligomer is stiff on the distance scale of the Debye screening length; (*b*) effective charge on each oligomer is deduced from the counterion condensation formalism; (*c*) the charge spacing of B-DNA is 1.7 Å; and (*d*) at pH 8.3 in the TBE buffer, EDTA^{3-} is the most dominant species. With the further assumption that the salt concentration is larger than the molarity of the oligomer, the ionic strength of the buffer is found to be 0.0256 M; (*e*), we have ignored the contribution to the electrophoretic mobility from the asymmetric field effect. This is so because the magnitude of the asymmetric field effects at room temperature is small if one uses the limiting value of the molar conductivity of the tris and the acetate ions; (*f*), the monomer-solvent frictional coefficient is approximated by the mobility of the ions λ_o (52, 54).

A plot of the electrophoretic mobility of oligomeric B-DNA in free solution vs the number of phosphates is shown in Figure 1. The mobility increases upon an increase in the molecular weight of the fragment and approaches a plateau for higher-molecular-weight fragments (52). The free-solution mobility of ds B-DNA is larger than that of single-stranded (ss) B-DNA (52). Furthermore, upon decreasing the ionic strength of the TBE buffer, the electrophoretic mobility of the DNA fragments increases (52). These results are in good agreement with experimental data (51).

ANOMALOUS MIGRATION OF DNA

Figure 2 illustrates the experimental data for the gel electrophoretic mobility of DNA fragments of the form $(Np)_n N$, containing residues *n* ranging from 2 to 20, where *N* denotes the number of phosphate residues in TBE buffer (55, 56). An important feature of these data is the existence of anomalous migration of the DNA fragments (55, 56). The electrophoretic mobility of the trimer is approximately the same as that of the DNA sequence of length greater than 14 residues (55, 56). The anomalous migration reaches a maximum of around 5–6 nucleotide residues (55, 56). These features defy rationalization based on the Ogstron and the reptation models (13–22, 32, 33).

We have developed a model that quantitatively characterizes the polyacrylamide gel retardation data of oligomeric DNA fragments. In this model, the electrophoretic mobility of oligomeric ss DNA in polyacrylamide gel is described by the following equation (34, 55, 56):

$$\mu = \mu_o(\kappa, N)\exp(-R/\zeta_{\text{mesh}}). \qquad 11.$$

Here, $\mu_o(\kappa, N)$ is the free-solution mobility, which, as we have seen above, explicitly depends on the Debye screening parameter and the length of the oligomer (see Figure 1). The mesh size of the polyacrylamide gel is denoted by ζ_{mesh}. The coil size *R* of the oligomer is approximately $Nb/2$, where the charge spacing *b* is ≈ 4 Å (55, 56).

The approximate free-solution mobility term appearing in the right-hand side of Equation 11 is given by the ratio of the effective charge of the DNA divided by the friction it experiences as can be seen from Equation 10. The effective charge is calculated within the Manning counterion condensation framework but with the inclusion of Coulombic end effects along the line discussed earlier (55–57). The

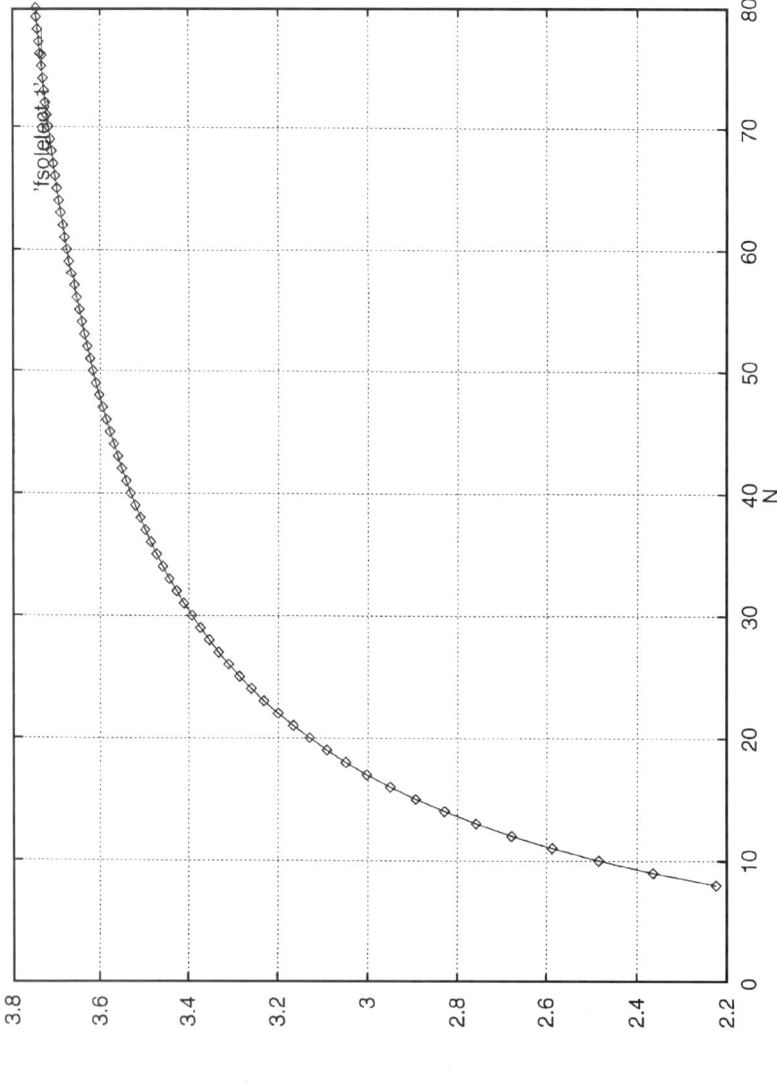

Figure 1 The free-solution mobility U of oligomeric B-DNA as a function of the number of phosphates (N) as predicted by the counterion condensation-screened hydrodynamic model. The ionic strength of the TBE buffer was 0.0256 M. (Reprinted with permission from Reference 52.)

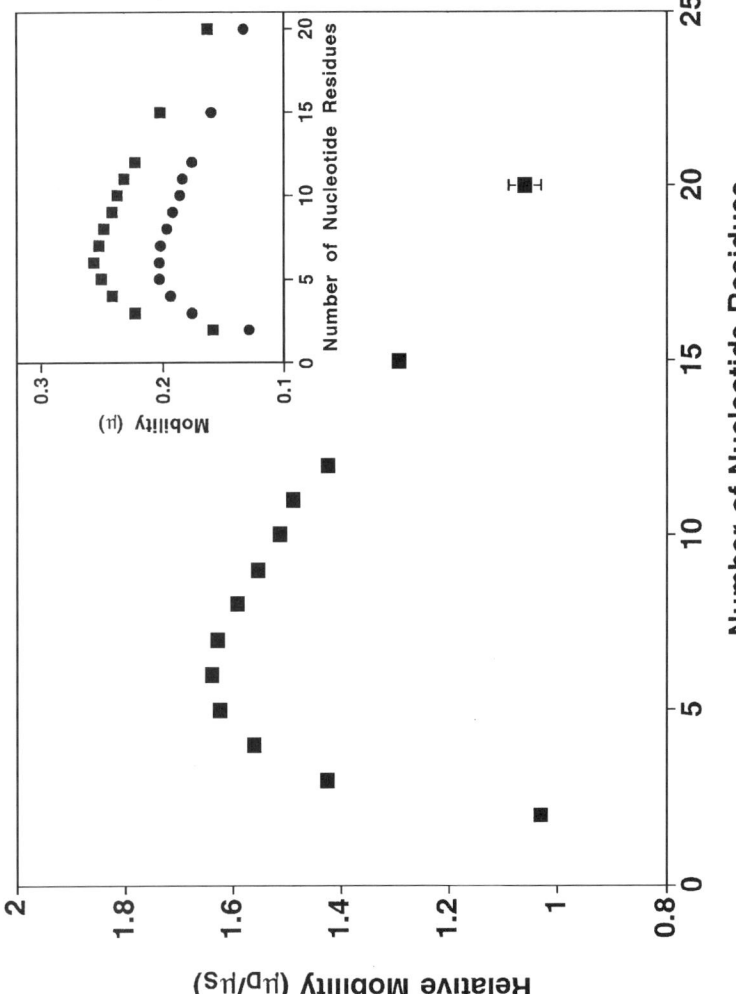

Figure 2 The experimental gel electrophoretic mobility of DNA fragments $(Np)_n$, N relative to the mobility of bromphenol blue marker. μ_D = mobility of DNA fragment and μ_S = mobility of bromphenol. The inset depicts the raw data for the electrophoretic mobility (in units of 10^{-4} cm^2 V^{-1} s^{-1}) from two different experiments (reprinted with permission from Reference 55).

frictional coefficient within the screened hydrodynamic formalism is (54–57)

$$N \frac{6\pi \eta}{\sum_{i=1}^{N} \sum_{j=1; i \neq j}^{N} \langle r_{ij}^{-1} \exp(-\kappa r_{ij}) \rangle}, \qquad 12.$$

where η is the viscosity of the solvent.

The theoretical predictions for the electrophoretic mobilities are illustrated in Figure 3. The model predicts, in agreement with experiments, that the mobility increases with the number of nucleotide residues, reaches a maximum, and then continues to decrease (55–57). A three-residue fragment, for example, migrates with the same electrophoretic mobility as a fragment containing 14 residues (55–57). The location of the maximum mobility depends on the pore size of the gel (55–57). These features have been observed for RNA and DNA fragments of very small size during denaturing polyacrylamide gel electrophoresis (58–61).

What happens to the above results if one considers a composite molecule in which an additional charge is attached to one terminus? This situation occurs, for example, with a terminal phosphomonoester (62). Figure 3 illustrates the gel retardation versus fragment length for ss DNA in TBE buffer. The fragments $(pN)_n$ were synthesized based on a particular 20-mer sequence as described in reference 62. At a 6% (wt/vol) gel concentration, the electrophoretic mobility is independent of fragment length $(pN)_n$ (62). For a 9.5% (wt/vol) gel concentration, there is slight anomalous migration, which disappears at higher gel concentrations (62).

To explain the gel electrophoretic experimental data on end-labeled ss DNA, a key observation is that the free-solution electrophoretic mobility for the composite molecule is not given by the ratio of the total charge to the frictional coefficient (62–64); instead, it is given by (62–64)

$$\mu_o = \frac{\xi_d \mu_d + \xi_p \mu_p}{\xi_d + \xi_p}. \qquad 13.$$

We have denoted by ζ_p and ζ_d the frictional coefficients of the polyion and the oligodeoxynucleotide, respectively. Similarly, μ_d and μ_p are the free-solution mobilities of the oligodeoxynucleotide and the polyion, respectively (62). The gel electrophoretic mobility in the model is (62)

$$\mu = \mu_o / \lfloor 1 + (R/\zeta_{\text{mesh}})^\alpha \rfloor, \qquad 14.$$

where α is an exponent that is a function of the gel concentration. The predictions of the model are shown in Figure 4. Anomalous migration is observed for pore sizes in the range 25–80 Å and is suppressed at higher gel concentrations (62). At very low gel concentrations, the mobility is almost independent of fragment length $(pN)_n$ (62). These results are in agreement with experimental data shown in Figure 5 (62).

SUMMARY

In summary, we have reviewed a model for the free-solution electrophoretic mobility of oligomeric ds DNA. We found that the free-solution mobility of ds DNA increases as the molecular weight of the fragment increases, up to a few hundred

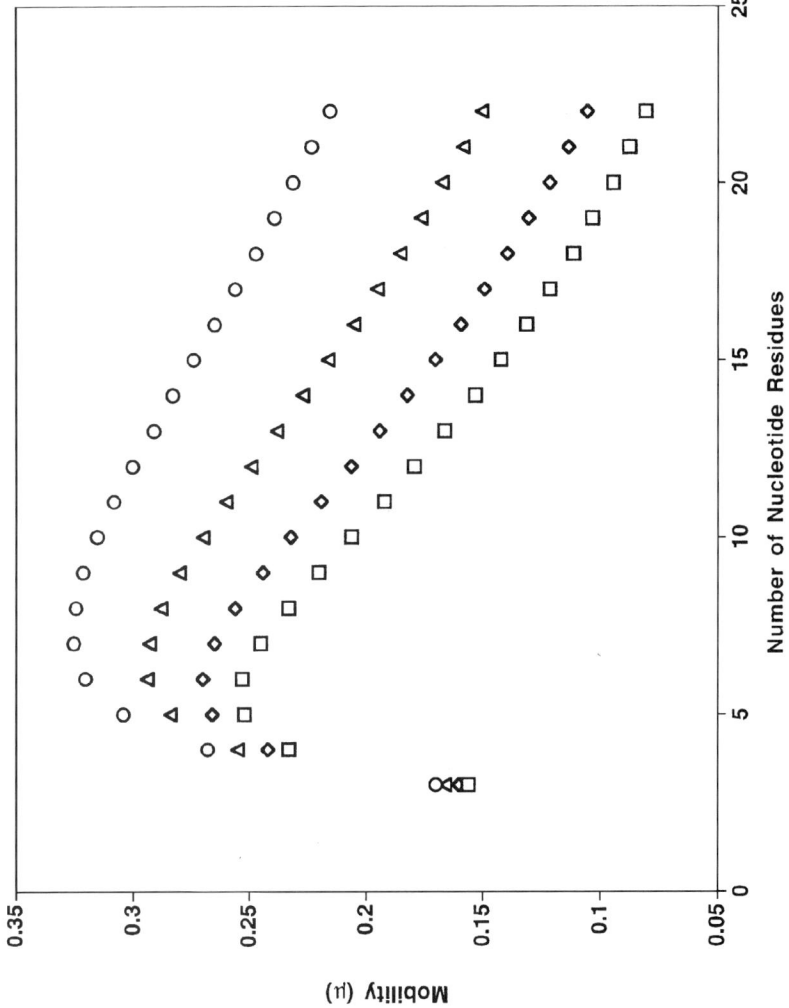

Figure 3 The predicted gel electrophoretic mobility [μ (in units of 10^{-4} cm^2 V^{-1} s^{-1})] as a function of the number of nucleotide residues at the following pore sizes as predicted by the model: 60 Å (*circles*), 40 Å (*triangles*), 30 Å (*diamonds*), and 25 Å (*squares*) (reprinted with permission from Reference 55).

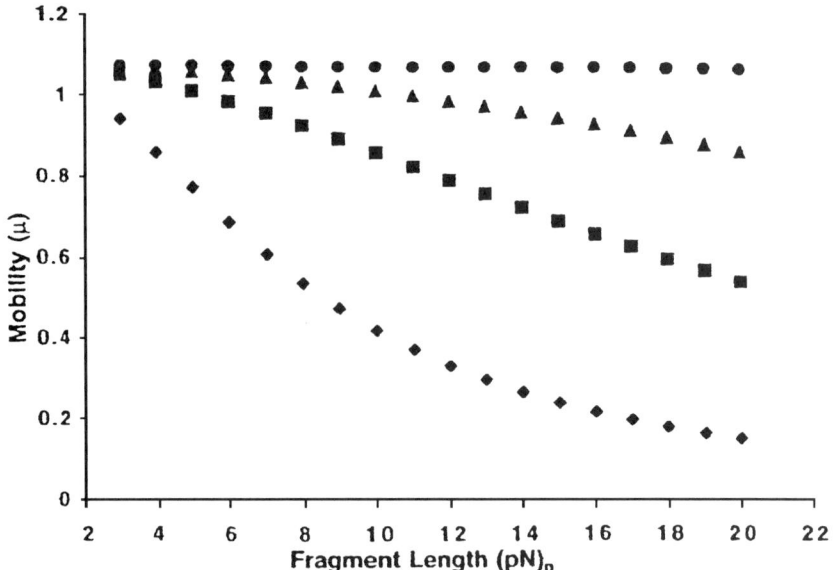

Figure 4 The predicted gel electrophoretic mobility (in units of 10^{-4} cm^2 V^{-1} s^{-1}) as a function of length of the fragments at the following pore sizes as predicted by the model: 16 Å (*diamonds*), 40 Å (*squares*), 80 Å (*triangles*), and 180 Å (*circles*) (reprinted with permission from Reference 62).

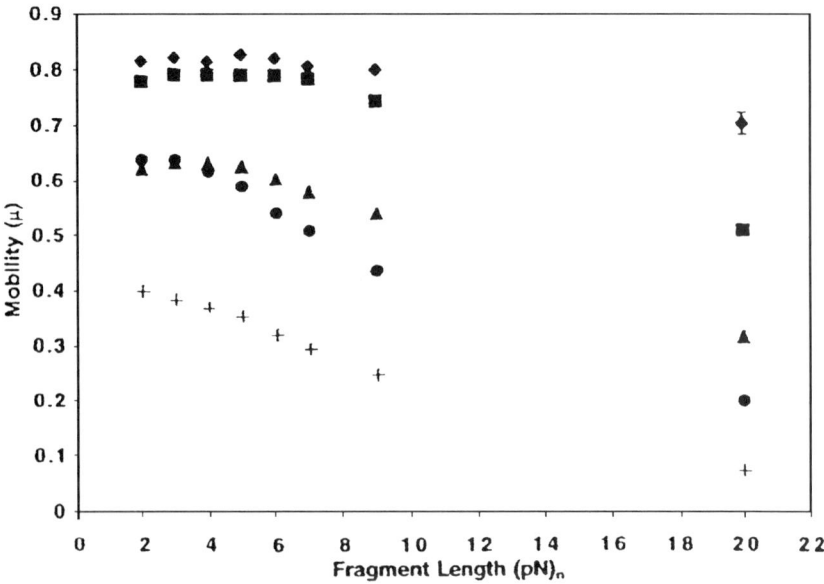

Figure 5 The experimental gel electrophoretic mobility data (in units of 10^{-4} cm^2 V^{-1} s^{-1}) as a function of length of the fragments at the following gel concentrations: 6% (*diamonds*), 9.5% (*squares*), 13% (*triangles*), 16.5% (*circles*), and 20% (*plus signs*) (reprinted with permission from Reference 62).

base pairs. This insight has been used along with recent advances in the nature of counterion condensation theory of very short DNA fragments to describe quantitatively the electrophoretic mobility of oligomeric ss DNA in polyacrylamide gels. The model predicts, in agreement with recent experiments, that significant anomalous migration exists with short DNA sequences, the onset of which is dependent on the size of polyacrylamide gel pores. For terminal phosphate-labeled DNA fragments, the free-solution mobility is no longer proportional to the ratio of the total effective charge and the friction coefficient. These changes in properties affect the characteristics of migration of end-labeled DNA fragments in polyacrylamide gels.

Visit the Annual Reviews home page at www.AnnualReviews.org

LITERATURE CITED

1. Hudson JH, et al. 1995. *Science* 270:1945–54
1a. Wickelgren I. 1995. *Science* 270:1587–88
1b. Drmanac R, et al. 1993. *Science* 260:1649–52
2. Fried M, Crothers DM. 1981. *Nucleic Acids Res.* 23:6505–25
3. Senear DF, Brenowitz M. 1991. *J. Biol. Chem.* 266:13,661–71
4. Mohanty U, McLaughlin LW. 2000. *Physical Chemistry of Polyelectrolytes*, ed. TS Radeva, pp. 665–85. New York: Marcel Dekker.
5. Zimm BH, Lumpkin OJ. 1993. *Macromolecules* 26:226–34
6. Lumpkin OJ, Levene SD, Zimm BH. 1989. *Phys. Rev. A* 39:6557–67
7. Zimm BH. 1991. *J. Chem. Phys.* 94:2187–206
8. Lumpkin OJ, Dejardin P, Zimm BH. 1985. *Biopolymers* 24:1573–93
9. Perkins TT, Smith DE, Chu S. 1994. *Science* 264:819–22
10. Holmes DL, Stellwagen NC. 1989. *J. Biomol. Struct. Dyn.* 7:311–27
11. Stellwagen J, Stellwagen NC. 1994. *Biopolymers* 34:187–201
12. Stellwagen J, Stellwagen NC. 1995. *J. Phys. Chem.* 99:4247–51
13. Doi M, Edwards SF. 1989. *The Theory of Polymer Dynamics*. Oxford, UK: Oxford Univ. Press
14. de Gennes PG. 1971. *J. Chem. Phys.* 55:572–79
15. Duke TA, Semenov AN, Viovy JL. 1992. *Phys. Rev. Lett.* 69:3260–63
16. Doi M, Graessley WW, Helfand E, Pearson DS. 1987. *Macromolecules* 20:1900–1906
17. Viovy JL, Rubinstein M, Colby RH. 1991. *Macromolecules* 24:3587–96
18. Widom B, Viovy JL, Defontaines AD. 1991. *J. Phys. I* 1:1759–84
19. Barkema GT, Caron C, Marko JF. 1995. *Biopolymers* 38:665–67
20. Deutsch JM. 1980. *Electrophoresis of Large DNA Molecules: Theory and Applications*, pp. 81–99. Cold Spring Harbor, NY: Cold Spring Harbor Labor.
21. Deutsch JM. 1988. *Science* 240:922–24
22. Deutsch JM. 1987. *Phys. Rev. Lett.* 59:1255–58
23. Welp KA, Wool RP, Satija SK, Pispas S, Mays J. 1998. *Macromolecules* 31:4915–25
24. Lodge TP, Rotsein NA, Prager S. 1990. *Adv. Chem. Phys.* 129:1–132
25. Herman MF. 1992. *Macromolecules* 25:4931
26. Hino T, Prausnitz JM. 1996. *Appl. Poly. Sci.* 62:1635–40

27. Hu Z, Li C, Li Y. 1993. *J. Chem. Phys.* 99:7108–14
28. Baselga J, Fuentes JH, Pierola IF, Liorente MA. 1987. *Macromolecules* 20:3060–65
29. Haggerty L, Sugarman JH, Prud'homme RK. 1988. *Polymer* 29:1058–63
30. Tong J, Anderson J. 1996. *Biophys. J.* 70:1505–13
31. Oppermann W, Rose S, Rehage G. 1985. *Br. Polym. J.* 17:175–80
32. Ogstron AG. 1958. *Trans. Faraday Soc.* 54:1754–57
33. Chrambach A, Rodbard D. 1971. *Science* 172:44–45
34. Cukier RI. 1984. *Macromolecules* 17:252–55
35. Holmes DL, Stellwagen NC. 1991. *Electrophoresis* 12:253–63
36. Stellwagen NC. 1997. *Electrophoresis* 18:34–44
37. Manning GS. 1978. *Q. Rev. Biophys.* 11:179–246
38. Manning GS, Mohanty U. 1997. *Physica A* 247:196–204
39. Young MA, Jayaram B, Beveridge DL. 1997. *J. Am. Chem. Soc.* 119:59–69
40. Odijk T. 1991. *Physica A* 176:201–5
41. Woodbury CP Jr, Ramanathan G. 1982. *Macromolecules* 15:82–86
42. Anderson CF, Record MT Jr. 1982. *Annu. Rev. Phys. Chem.* 33:191–222
43. Olmsted MC, Bond JP, Anderson CF, Record MT Jr. 1995. *Biophys. J.* 68:634–37
44. Olmsted MC, Anderson CF, Record MT Jr. 1989. *Proc. Natl. Acad. Sci. USA* 86:7766–70
45. Stein VM, Bond JP, Capp MW, Anderson CF, Record MT Jr. 1995. *Biophys. J.* 68:1063–72
46. Mills PA, Rashid A, James TL. 1992. *Biopolymers* 32:1491–501
47. Mohanty U, Ninham BW, Oppenheim I. 1996. *Proc. Natl. Acad. Sci. USA* 93:4342–44
48. Rouzina I, Bloomfield VA. 1996. *J. Phys. Chem.* 100:4292–304
49. Evans DF, Mitchell DJ, Ninham BW. 1984. *J. Phys. Chem.* 88:6344–48
50. Hayter JB. 1992. *Langmuir* 8:2873–76
51. Stellwagen NC, Gelfi C, Righetti GP. 1997. *Biopolymers* 42:687–703
52. Mohanty U, Stellwagen NC. 1998. *Biopolymers* 49:209–14
53. Barrat JL, Joanny JF. 1996. *Adv. Chem. Phys.* 94:1–66
53a. Allison SA, Stigter D. 2000. *Biophys. J.* 78:121–24
54. Manning GS. 1984. *J. Phys. Chem.* 88:6654–61
55. Mohanty U, Searls T, McLaughlin LW. 1998. *J. Am. Chem. Soc.* 120:8275–76
56. Mohanty U, Searls T, McLaughlin LW. 2000. *J. Biomol. Struct. Dyn.* 1:371–75
57. Mohanty JF. 2000. *Chem. Phys. Lett.* 316:558–65
58. Zinkel SS, Crothers DM. 1991. *J. Mol. Biol.* 219:201–5
59. Wu HM, Crothers DM. 1984. *Nature* 308:509–13
60. Mendelman LV, Notarnicola SM, Richardson CC. 1993. *J. Biol. Chem.* 268:27,208–13
61. Mendelman LV, Kuimelis RG, McLaughlin LW, Richardson CC. 1995. *Biochemistry* 34:10,187–93
62. Mohanty U, Searls T, McLaughlin LW. 2000. *J. Am. Chem. Soc.* 122:1225–26
63. Long D, Viovy JL, Adjari A. 1996. *Phys. Rev. Lett.* 76:3858–61
64. Long D, Ajdari A. 1996. *Electrophoresis* 17:1161

MECHANISMS AND KINETICS OF SELF-ASSEMBLED MONOLAYER FORMATION

Daniel K Schwartz

Department of Chemical Engineering, University of Colorado, Boulder, Colorado 80309; e-mail: daniel.schwartz@colorado.edu

Key Words thin film, coatings, SAM, monolayer growth

■ **Abstract** Recent applications of various in situ techniques have dramatically improved our understanding of the self-organization process of adsorbed molecular monolayers on solid surfaces. The process involves several steps, starting with bulk solution transport and surface adsorption and continuing with the two-dimensional organization on the substrate of interest. This later process can involve passage through one or more intermediate surface phases and can often be described using two-dimensional nucleation and growth models. A rich picture has emerged that combines elements of surfactant adsorption at interfaces and epitaxial growth with the additional complication of long-chain molecules with many degrees of freedom.

INTRODUCTION

The adsorption of amphiphilic surfactant molecules at interfaces is a well-known phenomenon that is at the heart of all detergency applications. A single molecular layer (monolayer) of surfactant stabilizes oil droplets and gas bubbles in an aqueous environment, enhancing the stability of emulsions and foams. In addition to adsorption at liquid-liquid and liquid-vapor interfaces, amphiphilic molecules also adsorb at the solid-liquid interface. Self-assembled monolayers (SAMs) are distinguished from ordinary surfactant monolayers by the fact that one end of the molecule (generally the hydrophilic one) is designed to have a favorable and specific interaction with the solid surface of interest (the substrate). This results in the formation of a stable monolayer film that remains intact even after the substrate is removed from solution.

Due to the specific interaction between molecule and substrate, the adsorption can often be carried out in a variety of solvents, polar and nonpolar, allowing greater flexibility in molecular design and, therefore, in the types of surface properties that can be modified and controlled. Since the monolayer films are thin and homogeneous, they have found frequent use as model surfaces in research applications. Much of the interest in these films lies in their potential as inexpensive and versatile surface coatings for applications including control of wetting and adhesion, chemical resistance, biocompatibility, sensitization for photon harvesting, molecular recognition for sensor applications, and many others.

Zisman is often credited with originating the SAM concept in his 1946 paper (1). Work in the early 1980s by Nuzzo & Allara (thiols on gold) (2) and Maoz & Sagiv (trichlorosilanes on silicon oxide) (3) introduced what were to become the two most popular SAM systems and brought SAMs into the popular scientific consciousness. Interest in these monolayer films has continuously increased since that time, and the development and application of surface-sensitive experimental techniques (e.g. scanning probe microscopy, vibrational spectroscopy, and synchrotron X-ray sources) has resulted in an improved understanding of the film structure and growth process. Poirier recently reviewed scanning tunneling microscopy (STM) measurements of thiol-based SAMs (4), and a more general review of SAM structure (with some information about film growth) was previously published by Ulman (5). Ulman's book (6) serves as a useful introduction to SAMs and thin organic films in general. The current review is more narrowly focused on the growth process of a variety of SAM systems.

THE BIG PICTURE: General Growth Mechanisms

Bulk Transport and Adsorption

Many processes are involved in SAM growth. A first step is clearly the solution-phase transport of adsorbate molecules to the solid-liquid interface, which can involve some combination of diffusive and convective transport. This is followed by adsorption on the substrate with some adsorption rate (related to a "sticking" probability). The overall adsorption dynamics may be diffusion-controlled, adsorption-rate controlled, or in an intermediate mixed-kinetic regime. This part of the self-assembly process is closely related to the adsorption of surface-active molecules at the liquid-vapor interface, an area that has been thoroughly studied. Although the typical quantity of interest at the liquid-vapor interface is surface tension rather than surface concentration (or coverage), the two quantities are related by the surface equation of state. In fact, most dynamic adsorption models are actually written in terms of surface concentration and translated into dynamic surface tension predictions, using an equation of state determined by applying the Gibbs equation (7) to equilibrium surface tension data. The dynamics of surfactant adsorption were thoroughly reviewed by Chang & Franses (8), and most of the mathematical development presented by them is directly relevant to the initial adsorption stage of SAM formation. Quantitative aspects of this process are discussed later in this review.

Self-Organization on the Surface

Two-dimensional (2D) molecular organization is a key ingredient for SAM stability and function. This differs from surfactant monolayers at the air-water interface, for example, because their primary function is simply to reduce surface tension. In SAM formation, therefore, there must be an evolution of the molecular order as adsorption progresses and the surface coverage increases. For example, the very early stages of adsorption can be pictured as isolated adsorbed molecules, conformationally disordered and randomly distributed on the substrate. The final film involves close-packed adsorbate molecules with relatively uniform molecular orientation and conformation. Although one might imagine a continuous path from the former structure to the latter, experimental evidence points to a stepwise process that can be thought of as an isothermal path through a quasiequilibrium 2D-phase diagram like the one schematically illustrated in Figure 1. Possible states alluded to in this phase diagram include (*a*) a low-density "vapor" phase in which isolated, mobile adsorbate molecules are randomly deposited on the surface, (*b*) an intermediate-density phase that could involve conformationally disordered molecules or ones lying flat on the surface, and (*c*) a final, high-density "solid" phase in which the molecules are conformationally ordered, close packed, and

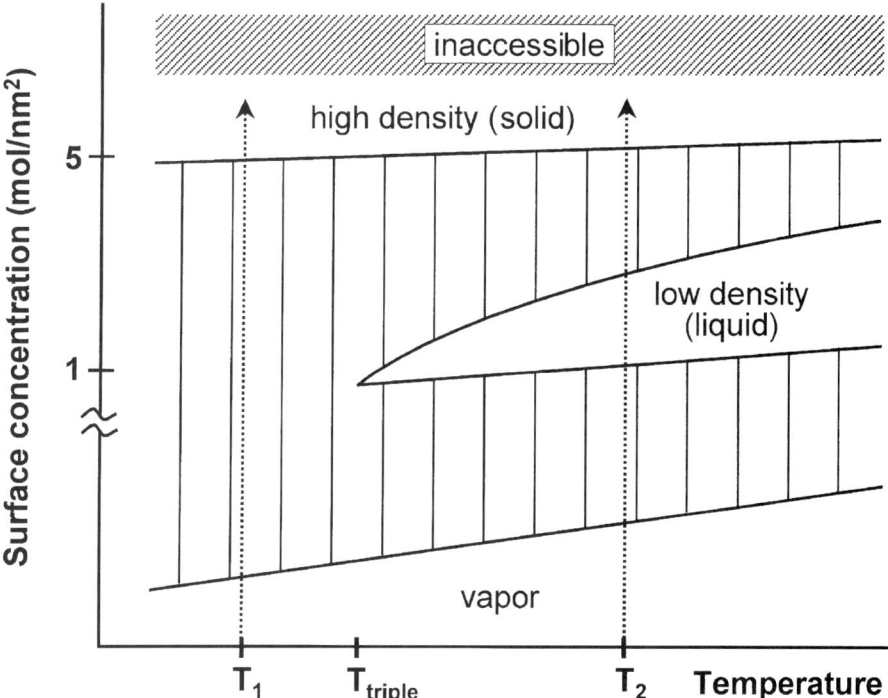

Figure 1 Schematic quasi-equilibrium 2D-phase diagram for a generic SAM system. The *dotted lines* represent hypothetical isothermal paths of SAM growth at temperatures below (T_1) and above (T_2), the triple point (T_{triple}).

standing approximately normal to the surface plane with a possible polar tilt angle of about $\leq 30°$. As discussed below, other states are, of course, possible.

In a hypothetical situation in which the adsorption rate is much slower than any other process, the monolayer system would follow the equilibrium phase diagram. There are two qualitatively different growth processes suggested by the lines at temperatures T_1 and T_2 in Figure 1. If the temperature is lower than the triple point (e.g. temperature T_1), the growth sequence will be similar to the one shown in Figure 2a. Initially, adsorbed molecules will form a dilute 2D-vapor phase. At a relatively low surface concentration, the monolayer will enter a coexistence region between the vapor and the high-density condensed (solid) phase. Domains (islands) of solid phase will nucleate and grow, surrounded by isolated adsorbate molecules in the vapor phase. Eventually, these domains will grow to cover the entire substrate. This mechanism is analogous to the three dimensional (3D) process of crystal nucleation and growth from a vapor phase precursor, and the 2D scenario is typical for epitaxial film growth from the vapor phase (e.g. molecular-beam epitaxy) (9). At a temperature above the triple point (e.g. T_2 in Figure 1), a more complicated progression will occur as illustrated in Figure 2b. When the vapor phase reaches a certain surface concentration, islands of an intermediate, low-density condensed phase will nucleate and grow. This phase may be a disordered 2D-liquid phase or an ordered phase with lower density than the solid phase (e.g. a "lying-down" phase where the molecular axis is parallel to the surface plane). Eventually the vapor phase is completely converted to the low-density condensed phase. As adsorption continues, a second transition occurs involving nucleation, growth, coalescence, etc, of solid-phase islands surrounded by the low-density condensed phase. Note that, at any temperature, a snapshot of an incomplete film during growth will often involve islands of one phase surrounded by another, in particular, islands of solid phase surrounded by either liquid or vapor phase.

It is important to recognize that the picture painted in the previous paragraph is somewhat oversimplified. For example, the adsorption rate will not always be much slower than other surface processes, and, therefore, partial monolayers may be quite far from equilibrium. If the nucleation and growth of condensed-phase domains do not keep up with the deposition rate, the less condensed phase will become "super concentrated" (i.e. it will have a density greater than the equilibrium coexistence concentration), and, thus, its density may vary considerably during the growth of the more condensed phase. This behavior is well known in vapor phase thin-film deposition, where the surface concentration of free adsorbate atoms is understood to vary during island nucleation and growth, and is likely to occur during SAM growth as well. However, the surface concentration in the vapor phase will always be small and amount to a negligible fraction of the molecules on the surface. In the case of a 2D-liquid phase, however, the surface density is not negligible, and, in fact, the film thickness is directly related to the surface concentration. Therefore, variation of the surface density in a 2D liquid in coexistence with solid-phase islands will have a significant effect on the appearance of the partial-monolayer film.

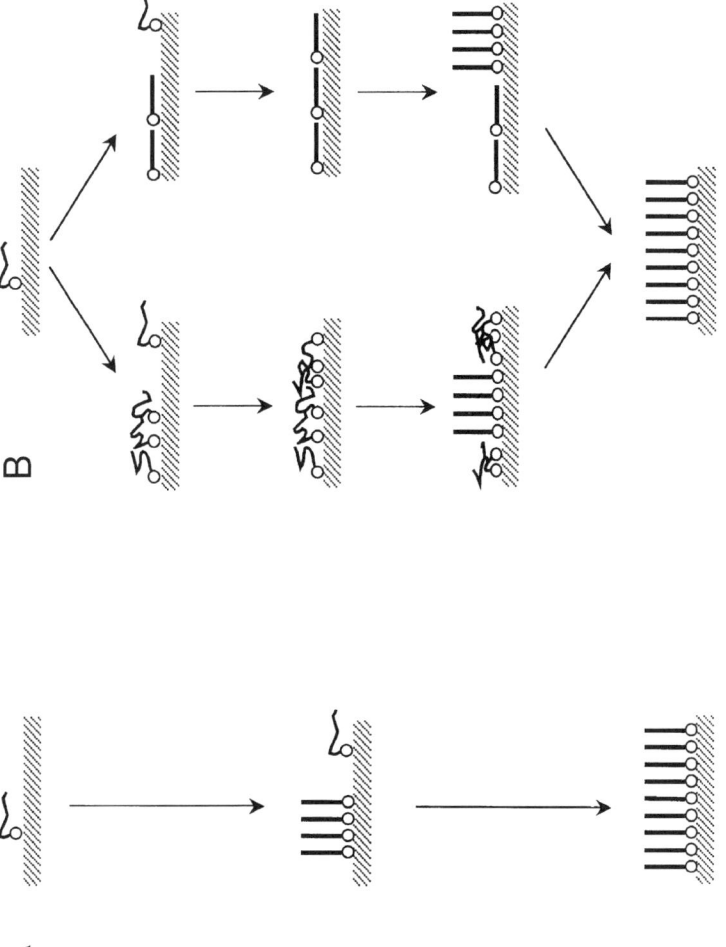

Figure 2 Cartoons depicting typical sequences of a self-assembled monolayer structure during growth below (A) and above (B) a triple point like that shown in Figure 1. (A) Below the triple point, growth proceeds from a 2D-vapor phase, through a solid-vapor coexistence region, to the solid phase. (B) Above the triple point, the SAM must pass through three phases and two coexistence regions. The intermediate low-density phase may be a disordered (liquid) phase, a "lying-down" phase, etc.

The molecules used to create SAMs have numerous degrees of freedom, and, therefore, it is quite possible that the equilibrium phase diagram could be more complicated and involve a greater number of condensed phases than implied in Figure 1. It could include a lying-down phase and a disordered-liquid phase, for example. However, there are numerous other possibilities. Langmuir monolayers of long-chain fatty acids, for example, are known to display a variety of liquid crystalline and crystalline phases (10) that differ in the polar tilt angle, the azimuthal direction of molecular tilt (i.e. nearest-neighbor vs next-nearest-neighbor direction), and rotational freedom (herringbone vs rotator). To date there is no firm experimental evidence for liquid-crystalline phases, transient or equilibrium, in SAMs. Given their ubiquity in Langmuir monolayers and Langmuir-Blodgett films, however, one would not be surprised if they were observed in SAMs with the appropriate experimental studies.

Therefore, although it is certainly overly simplistic, the phase diagram of Figure 1 will be used as a conceptual framework to describe the experimentally observed growth mechanisms of various SAM systems discussed in the following sections. Two general experimental strategies have been used to study monolayer growth: (a) in situ studies under actual deposition conditions in real time and (b) studies on quenched partial monolayers removed from solution and possibly rinsed to remove loosely attached adsorbate molecules. Although in situ experiments have become increasingly important in recent years, many publications in the literature report experiments that used quenched films. The clear advantage of in situ experiments is that one avoids the issue of whether the film structure is altered by the quenching process. This is not a trivial matter, since there is clear evidence that quenching can alter the film coverage and morphology in some molecular systems. On the other hand, working with quenched films permits the use of certain techniques not applicable in situ, such as contact angle and X-ray photoelectron spectroscopy. Furthermore, one can work over a longer range of time scales (i.e. concentrations). Although experiments on quenched films often report reliable and useful information (particularly on qualitative issues), one must view subtle quantitative conclusions based on quenched films with appropriate skepticism until they are confirmed by more direct experiments.

Vapor Phase-Deposited Thiol Films

Although not strictly considered SAMs, films created by vapor phase (molecular beam) deposition of alkylthiols on gold share many structural characteristics with solution-deposited films. Furthermore, studies on these films have the advantages of ultra-high-vacuum substrate cleanliness and the availability of traditional in situ surface characterization techniques. Although it is clear that solvent interactions are potentially important for SAMs (perhaps even more so during the growth process than for complete films), many of the results obtained on vapor-deposited films are relevant to SAM structure and growth, and they are briefly reviewed here.

Poirier & Pylant's (11) STM observations of vapor-deposited thiols on single-crystal Au(111) were the first report of the general mechanism of thiol monolayer growth. Studying the formation process of C_6–C_{10} alkylthiols, both methyl and hydroxy terminated, Poirier & Pylant reported a two-step process starting with the nucleation and growth of islands of "striped" phases from a lower-density lattice-gas phase. Based on the observed periodicity of these striped phases, it was proposed that they consisted of molecules lying flat on the gold surface, in either a head-to-head or a head-to-tail arrangement. The growth of the stripe phase islands was accompanied by the appearance of gold atom vacancies ("pits"). After the surface was covered by the stripe phase, continued deposition resulted in islands of molecules arranged in a way consistent with an epitaxial overlayer $(\sqrt{3} \times \sqrt{3})R30°$ on the Au(111) surface. The lateral density necessary to form this structure implied thiol molecules with an orientation nearly perpendicular to the substrate. This sequence is similar to the one suggested in Figure 2b.

Schreiber and coworkers (12) presented a multitechnique study (X-ray diffraction, atom diffraction, and X-ray photoelectron spectroscopy) of the vapor phase growth process of C_{10} thiol on Au(111). Their results were qualitatively consistent with Poirier & Pylant's observations (11) at low temperatures with a few added details. The atom diffraction suggested that the striped phase became disordered prior to nucleation of the upright $[(\sqrt{3} \times \sqrt{3})R30°]$ phase. Also, experiments at temperatures above 15°C found an additional intermediate 2D-liquid phase between the striped phase and the standing-up solid phase. To incorporate this into a phase diagram like Figure 1, one would have to add another low-density condensed phase. It is interesting that Schreiber and coworkers found that the size of correlated $[(\sqrt{3} \times \sqrt{3})R30°]$ domains that grew from the liquid phase were significantly larger than those that nucleated from the striped phase. This suggested that the defect structure of the final film may be intimately related to deposition conditions and mechanisms. In two subsequent papers (13, 14), the growth process of the $[(\sqrt{3} \times \sqrt{3})R30°]$ phase was studied in greater detail. It was found that at >15°C, the growth rate of the $[(\sqrt{3} \times \sqrt{3})R30°]$ phase was approximately proportional to the adsorbate pressure in the gas phase. However, at lower temperatures, the growth rate was proportional to the square of the pressure, suggesting that a bimolecular process may be rate limiting.

Thiol on Gold Self-Assembled Monolayers

Contact angle and ellipsometry experiments on quenched, incomplete alkanethiol SAMs on gold by Bain et al (15) revealed at least two time scales in the growth process. For a typical solution of 1 mM in ethanol, the contact angle and film thickness were observed to reach ~90% of their final values within the first minute of growth. A slower second step was observed in which the final film properties were reached only after several hours. This suggested multiple processes at work, with time scales differing by ~2 orders of magnitude. They also noted that the monolayer properties of longer-chain n-alkyl thiols ($n > 8$) were consistent but that shorter-chain thiol monolayers were qualitatively different in a way that suggested greater disorder. Sun & Crooks (16) decorated defects in quenched partial SAMs by underpotential electrochemical deposition of Cu. They monitored the decrease in SAM defect density as a function of exposure time using STM and found that the number of defects disappeared on a time scale of several hours, consistent with Bain's (15) results. This study also provided evidence, albeit indirect, that an islanding mechanism was involved in alkylthiol SAM growth. They also observed "pits" ~0.5 nm deep that were characteristic of the SAM-covered gold surface. These pits did not appear to be holes in the SAM layer, however, because Cu islands did not nucleate in these locations. These intriguing results inspired a multitude of measurements designed to confirm and later explain the existence of multiple time scales.

Numerous studies involving measurements of film mass and average thickness were conducted to explore the overall coverage kinetics of thiol SAMs on gold. The results of many of these were qualitatively consistent with the observations of Bain et al (15), finding fast and slow time scales. Shimazu and coworkers (17) performed in situ quartz crystal microbalance (QCM) experiments on ferrocene-substituted thiols in hexane solution. They observed a fast adsorption step (a few seconds in 0.5 mM solution) followed by a process with a time scale ~2 orders of magnitude slower. Their results were consistent with a single molecular layer after 800 s. QCM and STM studies on quenched monolayers by Kim et al (18) detected a slow build-up of multilayers (over a period of days) during C_{18} thiol SAM growth from ethanol solution. Schneider & Buttry's in situ QCM experiments (19) also considered multilayer formation. However, their results suggested gradual conversion of physisorbed multilayers to a chemisorbed monolayer. Schneider & Buttry also observed a significant solvent effect. In dimethylformamide solution, adsorption was rapid; however, a complete monolayer was never formed. In acetonitrile solution, on the other hand, adsorption was slower, but the physisorbed film was slowly converted to a densely packed monolayer. Schneider & Buttry suggested that the final monolayer quality had an inverse relationship with the solubility of the thiol in the solvent. In situ QCM experiments by Fruböse & Doblhofer (20) revealed two distinct time scales in adsorption from 0.1 mM thiol solution—initial adsorption in ~2 min followed by a much slower process taking >1 h. Their measurements of gradually decreasing electrochemical impedance during that latter process suggested that the slow time scale corresponded to "healing" of the SAM. In situ surface plasmon resonance (SPR) experiments by DeBono and coworkers (21) found two adsorption time scales differing by a factor of ~100. For C_{12} or C_{16} thiols from ethanolic solution, the initial fast step resulted in 80% of monolayer coverage. The C_{16} rate constant was faster than the C_{12} for both steps. For the C_6 thiol, only 50% coverage was reached in the initial step. Peterlinz & Georgiadis (22) also performed in situ SPR experiments on n-alkyl thiol SAMs grown from ethanol and heptane solutions (see Figure 3). For ethanol solution, they observed

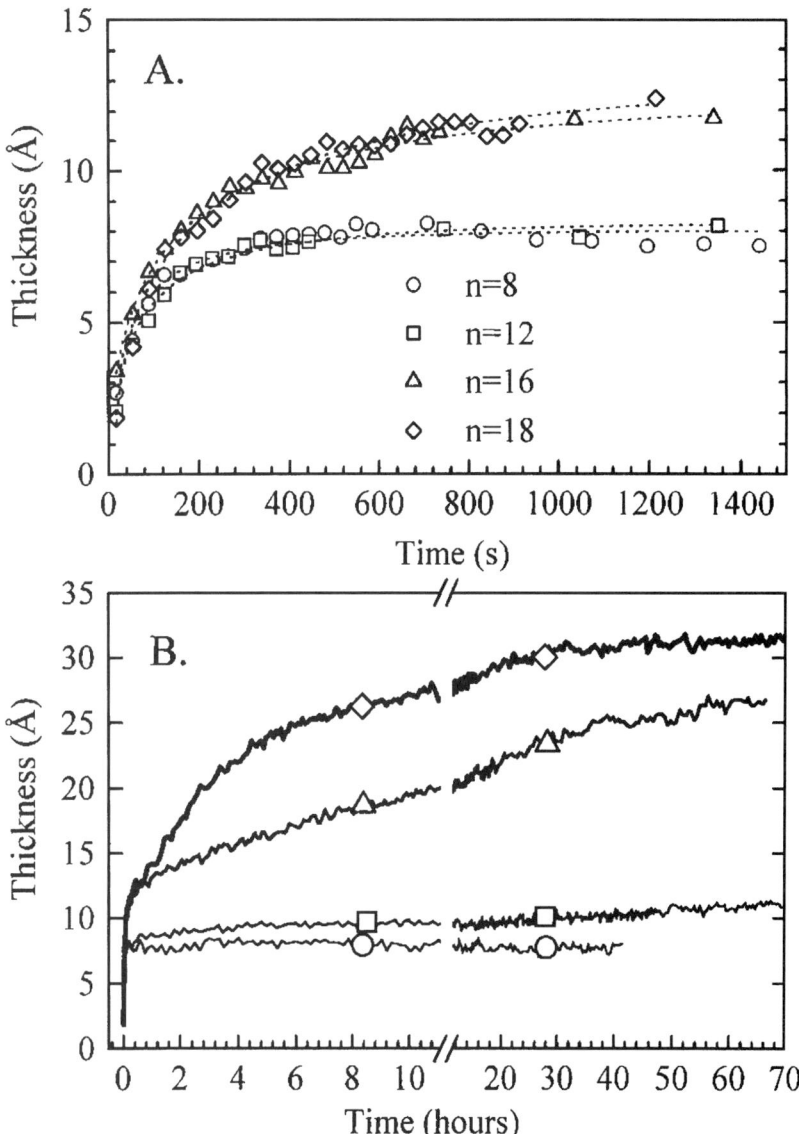

Figure 3 Chain length dependence of formation kinetics for C_8 (*circles*), C_{12} (*squares*), C_{16} (*triangles*), and C_{18} (*diamonds*) thiols from 1.0 mM ethanolic solutions. The film thicknesses were calculated from in situ surface plasmon resonance measurements. Up to three distinct kinetic regimes were observed, depending on chain length. (a) Details of kinetics at short times. (b) Overview of kinetics for 48–72 h. (Figure adapted with permission from Reference 22.)

multiple time scales. For the initial fast adsorption step (25 min for a 1 mM solution), the rate decreased with increasing chain length, whereas the opposite trend was observed in the second step which lasted many hours. Monolayers composed of shorter-chain molecules (C_8 and C_{12}) reached $\geq 80\%$ of the final thickness in the first step while the longer-chain films (C_{16} and C_{18}) obtained only 40%–50%. In the longer-chain thiol systems (C_{16} and C_{18}), a third step with an even longer time scale was observed. Growth from heptane solution for the C_{16} thiol had a first step with a rate somewhat faster than that of ethanol solution. In addition, the film reached $>80\%$ of its final thickness in this first step. This was followed by a gradual evolution to the final film thickness over a period of ~ 2 days. Shao and coworkers (23) used in situ electrochemical methods to show that the growth of an azobenzenethiol also involved a two-step process.

Buck et al (24) noted only a single, relatively fast time scale (e.g. <1 min for 0.045 mM solution) from their in situ second-harmonic-generation experiments on n-alkyl thiols in ethanolic solution. The in situ QCM measurements by Karpovich & Blanchard (25) [later followed up by Schessler et al (26)] also reported only a single time scale for SAMs grown over a wide range of concentrations, using hexane or cyclohexane as the solvent. They noted, however, that the kinetic data for concentrations above ~ 0.1 mM deviated systematically from the predictions of the Langmuir model, which involves only a single time scale. In both of these studies, the measurements may not have extended to long enough times to observe very slow later steps of growth. The in situ SHG experiments of Dannenberger et al (27) also probed only the initial stages of growth. X-ray photoelectron spectroscopy experiments by Kawasaki and coworkers (28) on quenched partial films probed the first 2 h of growth of C_8 thiol adsorbed from 0.001 to 0.1 mM ethanolic solutions on "atomically flat" sputter-grown gold surfaces. In this time period Kawasaki and coworkers observed two-step growth kinetics for the 0.001 mM solution but a one-step growth process for the 0.01 and 0.1 mM solutions.

Spectroscopic techniques have been used to observe the evolution of molecular order within thiol SAMs in the various stages of growth that were previously identified. Near-edge X-ray adsorption fine structure spectroscopy experiments by Hähner et al (29) on quenched films probed the molecular conformation of C_{22} thiol SAMs after 10-min and 43-h immersions in 3 μM ethanolic solution. They reported that the angular dependence of the C-H resonance was highly anisotropic for the 43-h sample, suggesting that the molecules were conformationally ordered and uniformly tilted $\sim 35°$ from the surface normal. Although the data from the 10-min sample were more difficult to analyze, the anisotropy present at the long exposure time was not present, implying that the alkyl chains were disordered. They suggested that the slow time scale observed in kinetics experiments corresponded to an ordering process in which alkyl chains straighten as they disentangle. Truong & Rowntree (30) followed the infrared spectra (in the C-H stretch region) of quenched C_4 thiol SAMs as a function of immersion time in 5 μM methanolic solution. The significant changes in the spectra occurred over the first 10–15 min, so these experiments probed the structural evolution in the initial period of SAM formation. The relative band intensities suggested that the alkyl chains were, on average, lying close to the surface at first and closer to the surface normal after 15 min of immersion. The invariance of the band frequencies suggested that the local molecular environments were insensitive to coverage. These data were consistent with a picture in which islands of vertically oriented molecules form and grow to cover the surface. Terrill and coworkers (31) obtained IR spectra from quenched C_{16} thiol SAMs immersed in ethanolic solutions (10^{-6}–10^{-2} M) for periods of ≤11 days. Using the position of the antisymmetric methylene stretch as a signature of chain disorder, they found that long times (from several hours to several days depending on concentration) were necessary to reach the most conformationally ordered state. They also observed that the ordering was faster on smoother substrates. On the other hand, Bensebaa et al (32) reported that this same peak position reached its ultimate value, representative of well-ordered alkyl chains, after only a 45-s immersion in 5 μM ethanol solution (for quenched films of a C_{22} thiol).

Himmelhaus et al (33) performed sum frequency generation spectroscopy studies on quenched C_{22} thiol SAMs adsorbed from 3 μM ethanolic solution, monitoring the various C-H stretch bands over 2 days of immersion time. They found three distinct regimes of growth. The first stage (initial 5 min of immersion) involved formation of Au-S bonds. The coverage reached 80%–90% after this stage. The second stage (5–15 min of immersion) was characterized by a transition of the hydrocarbon chains from a highly kinked to an all-*trans* conformation. The final stage (20 min to 2 h of immersion) involved reorientation of the terminal methyl groups from a state in which methyl groups were disordered relative to one another to one in which they were aligned. The authors pointed out that this sequence implies that the ordering process can be viewed as consecutive steps originating at the gold interface and moving toward the film surface. Humbert and coworkers (34) performed SFG studies on quenched para-nitroanilino C_{12} thiol SAMs deposited from 2 μM ethanolic solution. They observed a marked change in molecular orientation over the first 30 min, followed by a slower change over the next 90 min, after which their observations ended.

In recent years, several scanning-probe-microscopy experiments have shed light on the thiol growth process. In a sequence of two papers (35, 36), Yamada & Uosaki performed in situ STM experiments monitoring alkylthiol growth on Au(111) from micromolar heptane solutions. They observed three basic steps. Initially, patches of adsorbed molecules were observed, but no periodic structures were detected on molecular-length scales. The authors suggested that these patches might correspond to a disordered phase. In this stage of growth, pits (or vacancy islands) were formed in the gold. The second step involved the appearance of patches in which "striped" patterns were observed. Although several periodic length scales were found, all were greater than the molecular length (and increased systematically with alkyl chain length). These observations and the structural similarities to the features observed on vapor-phase-deposited thiol films suggested that these stripe phase domains were composed of thiol molecules lying down on the surface in various ordered epitaxial arrangements. In the third and final stage of growth, islands of apparently greater film thickness formed and grew to cover the surface. A hexagonal pattern was observed on these islands, consistent with the $(\sqrt{3} \times \sqrt{3})R30°$ epitaxial arrangement of thiol molecules well-known from STM and scattering experiments (as discussed above). This growth sequence is reminiscent of the path discussed in an earlier section of this review through a quasi-equilibrium phase diagram at temperatures above the triple point (Figure 2b). Xu and coworkers (37) performed detailed and quantitative in situ atomic-force microscopy (AFM) experiments to follow the growth of C_{18} and C_{22} thiol molecules from 2-butanol solution (see Figure 4). Their observations were consistent with those of Yamada & Uosaki but provided direct height information. They first observed the formation of patches that were 0.5 nm high, consistent with molecules lying on the surface. At longer exposure times, islands that were 1.8 nm higher than the lying-down phase were observed to nucleate and grow, consistent with a structure in which molecules were approximately vertically oriented. The transition from lying down to standing up was faster for the C_{22} than the C_{18} thiol film. For 0.2 mM C_{18} thiol solution, the time elapsed between the initial appearance of lying-down

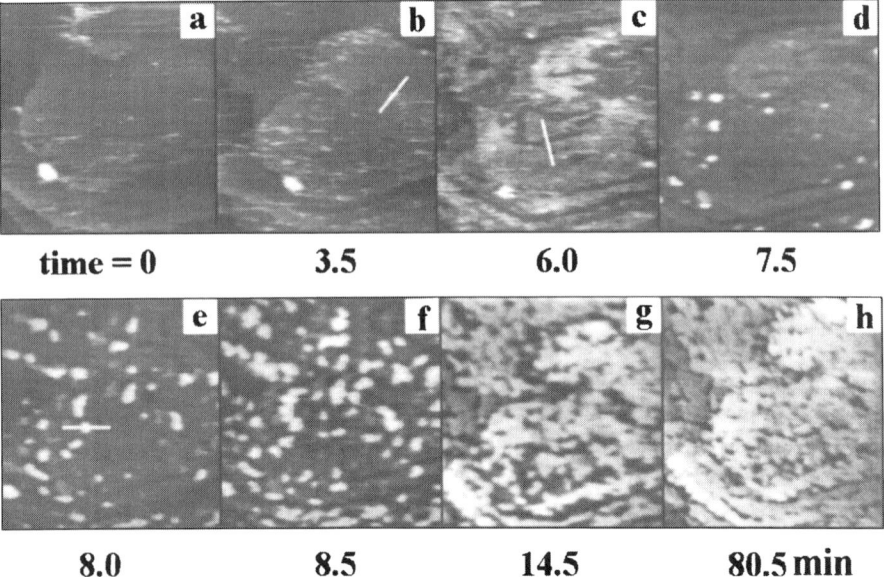

Figure 4 In situ topographic atomic-force microscopy images of Au(111) obtained at various times after injection of a solution of C_{18} thiol (0.2 mM in 2-butanol). The area of each frame is 150 nm × 150 nm^2, and the bright spot in the lower left corner and the Au(111) steps at the bottom of each frame provide landmarks for comparison of the images. (Figure adapted with permission from Reference 37.)

patches, and essentially complete coverage of the standing-up phase was 10–15 min. Tamada et al (38) also observed island growth with AFM on quenched partial monolayers.

Silane Self-Assembled Monolayers

The growth of trichlorosilane (similarly trimethoxy- or triethoxysilane)-based SAMs is unique among SAM systems in that it involves an irreversible covalent cross-linking step. This is critical to the desirable properties of this class of SAMs, including their chemical and mechanical robustness on a variety of substrates. There is also the potential for hydrolytic bond formation to −OH surface groups that would immobilize adsorbate molecules. Again, this is critical to the stability of the final monolayer. The kinetics of this step relative to the self-assembly process can clearly have dramatic implications on the growth mechanisms and final film structure. This adds a number of complications, because the rate of hydrolysis is sensitive to water content, pH, and temperature. It is interesting that, since the molecular packing is ultimately determined by the covalent siloxane network, one does not find long-range molecular order in these SAMs as one does in thiol SAMs, where the epitaxial arrangement on the Au lattice dictates the molecular arrangement.

In two X-ray reflectivity studies, Wasserman et al (39) and Tidswell et al (40) determined the electron density profiles of quenched partial and complete monolayers formed from C_{10}–C_{18} trichlorosilanes on silicon oxide substrates. They found that the structure of partial monolayers was inconsistent with molecular islands. An AFM study by Schwartz and coworkers (41) of quenched C_{18} silane SAMs on mica, on the other hand, explicitly observed 2-nm-high islands that grew to cover the surface with increasing immersion time. They found that the islands were fractal in shape and that the scaling exponent (fractal dimension) of ∼1.7 in the early stages of growth was consistent with 2D-diffusion–limited aggregation. This suggested an island growth mechanism involving collisions between adsorbate molecules moving randomly on the surface and immobile islands. The assumption of irreversible attachment to islands led to the fractal shape. This was essentially a view of the sequence shown in Figure 2*a* from a kinetic (rather than a thermodynamic) perspective and was clearly inconsistent with the conclusions of the prior X-ray studies. However, the mica surfaces were known to have only isolated-exposed −OH sites appropriate for anchoring the monolayer, whereas such sites were ubiquitous on silica substrates. A later AFM study by Bierbaum & Grunze (42) on quenched C_{18} (and longer) silane SAMs on silicon oxide observed similarly shaped islands. They did not observe islands on partial C_3 silane monolayers. Interestingly, second harmonic generation experiments by Zhao & Kopelman (43) showed that only a small minority of the surface −OH groups on the silica surface are actually bound to the SAM. Therefore, the picture of large, interconnected monolayer domains anchored to the substrate only in isolated locations may be appropriate for a variety of substrates. Using AFM,

Kropman and coworkers (44) observed dendritic C_{18} silane islands in quenched partial monolayers prepared even on $SrTiO_3$ substrates.

As mentioned above, the competition between various time scales in silane SAM growth makes the process quite sensitive to variations in preparation conditions. For example, contact angle (45) and IR spectroscopy (46) studies showed that quenched C_{18} silane SAMs prepared on silica substrates at a temperature below $\sim 30°C$ contained well-ordered alkyl chains, while those prepared at higher temperatures contained increasing chain disorder. This led to a quasi-equilibrium picture of silane SAM growth similar to those steps illustrated in Figure 1 and Figure 2 above. However, the nomenclature used was borrowed from the Langmuir monolayer literature; therefore the liquid phase was labeled LE and the solid phase LC.

Carraro et al (47) obtained AFM images of quenched partial C_{18} silane SAMs on silicon oxide over a range of temperatures (see Figure 5). At a low temperature (10°C), dendritic islands were observed to grow and coalesce to cover the surface, while at a high temperature (40°C) only a homogeneous uniform film was

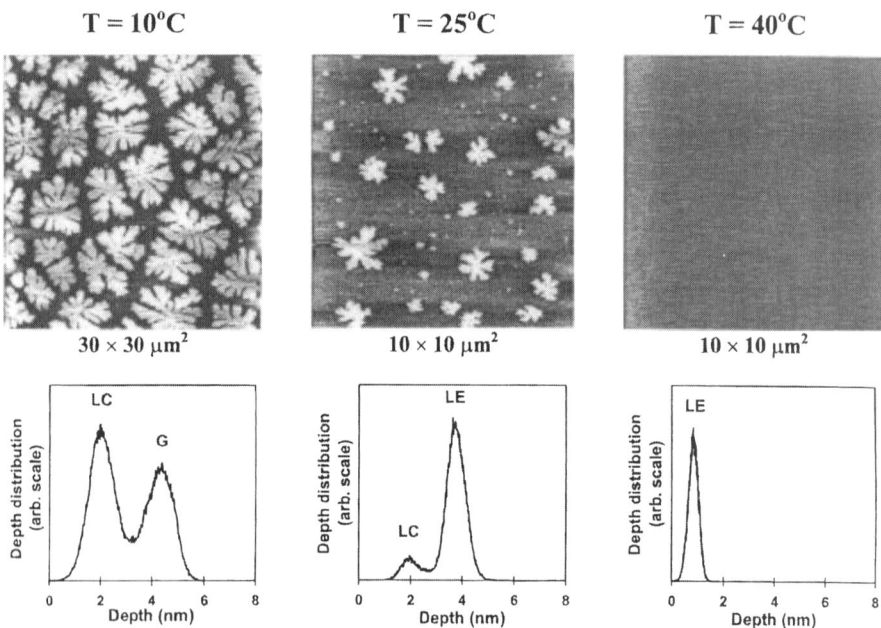

Figure 5 Atomic-force microscopic images of partial octadecyltrichlorosilane self-assembled monolayers on silicon oxide removed from 2 mM solution (using hexadecane-carbon tetrachloride as a solvent) after 30 s of immersion at 10°C, 25°C, and 40°C (left to right). The height distribution histograms are also shown. The differences in structure as a function of temperature are consistent with a quasi-equilibrium-phase diagram as discussed in the text. (Figure adapted with permission from Reference 47.)

observed. At an intermediate temperature (25°C), some dendritic islands were observed to nucleate and grow. However, before they could coalesce and cover the entire surface, the continuous phase between the islands gradually increased its thickness to that of the islands, ending the film formation. A similar AFM study by Goldmann and coworkers (48) observed quenched partial C_{18} silane SAMs on silicon oxide prepared at 12, 21.5, 26.5, 35, and 43°C. They observed regions of three heights that they considered to be vapor, liquid, and solid, respectively (although their nomenclature was G, LE, and LC). At 26.5°C and below, they observed sequential transitions from vapor to liquid to solid involving domain nucleation and growth. The vapor-liquid coexistence region was characterized by a foamlike morphology. At $\geq 35°C$, only the evolution from vapor to liquid was observed. The results of both papers can be interpreted via a phase diagram like Figure 1, where the triple point is between 10 and 12°C if one assumes that the growth is not under quasi-equilibrium conditions. Below the triple point, the growth process is typical of 2D vapor-solid coexistence. Above the triple point (but below 30°C), the vapor-liquid transition is observed followed by the liquid-solid transition. Although solid-phase islands nucleate from the liquid phase, they do not grow quickly enough to maintain quasi-equilibrium conditions, and the surrounding liquid phase becomes more and more super concentrated, therefore thicker. Above 30°C, it is possible that concentrations necessary to nucleate the solid phase are not reached before the SAM growth is quenched by cross-linking or that the surface density necessary for nucleation of the solid phase is not accessible via spontaneous adsorption from solution. A recent experiment by Sung et al (49) on a quenched C_{18} silane SAM explicitly demonstrated a phase transition consistent with the results of these temperature-dependent adsorption studies. A partial film was prepared in the vapor-solid coexistence region at 10°C, removed from solution, and heated to 30 or 60°C. The sample heated to 30°C had a lower area fraction of solid-phase islands than the unheated film and the islands had disappeared completely on the film heated to 60°C. If the film was heated to 60°C and then cooled to 30°C, islands were observed to form; if cooled to 10°C, the islands were larger and covered more of the surface. This study explicitly verified the phase diagram paradigm as well as the mobility of the molecules even after quenching. The mobile state was found to last for several minutes, after which cross-linking and grafting apparently "froze" the film morphology.

In a series of several papers (50–52), the Hoffmann group explored the effect of deposition conditions (water content and solution age) and substrate on the structure of quenched alkylsiloxane SAMs formed at room temperature. Their AFM images suggested simultaneous growth by island formation and a continuous disordered-liquid phase. The relative contributions of the two mechanisms were sensitive to growth conditions. Island growth was favored in deposition solutions with higher water content (in toluene solution) or solutions that had been aged longer. In addition, they found that adsorption was faster on mica than on silicon oxide. They also prepared mica substrates coated with a specific number of silicon oxide layers and found that the growth rate decreased exponentially with increasing oxide coating thickness up to about six layers. In situ IR spectroscopy experiments found that the silane molecules adsorbed initially in a disordered conformation and gradually aligned and stood up as the coverage increased. Furthermore, they observed an enhancement in the adsorption kinetics with increasing water content of the deposition solution. In a later in situ AFM study (53) conducted at room temperature, the same group observed only islands ∼2.5 nm high during C_{18} silane growth. The discrepancy between these observations and the AFM images of quenched films (which showed both islands and continuous phase) casts some doubt on the relevance of the quenched-film studies.

Richter and coworkers (54, 55) performed X-ray reflectivity studies of C_{18} silane film growth on silicon oxide at room temperature from micromolar-concentration heptane solution. In in situ experiments, they found density profiles that suggested that the maximum film thickness did not change during growth but that the average density gradually evolved to that of a complete monolayer. This was consistent with island growth of approximately vertically oriented molecules. Richter and coworkers compared the structure of partially formed monolayers during these in situ experiments with quenched partial monolayers and found systematic differences. The quenching process apparently introduced free area into the film that was not restored by reintroducing the quenched film into solvent. This suggested that some adsorbate molecules were removed during quenching. Although both the average film thickness and density were affected by the quenching process, the density decreased more dramatically, which suggested that regions of relatively densely packed molecules were not significantly affected by quenching, while other, less dense regions lost most of their molecules, with the remaining molecules tilting over dramatically. This experiment again sounds a warning regarding overinterpretation of experiments based on quenched partial monolayers.

Other Self-Assembled-Monolayer Systems: Organic Acids and Ions

Although thiol- and silane-based systems represent the bulk of the SAM literature, there are a number of reports of monolayers based on organic acids or ions. For example, alkyl carboxylic, sulfonic, and phosphonic acids have been demonstrated to form organized monolayers on several metal or metal oxide surfaces. Also, organic ions, such as quaternary ammonium salt detergents, form stable-monolayer films on substrates like mica that have a nonzero net charge at accessible pH values. Aside from the practical significance of expanding the range of substrates that may be coated with SAMs, these systems offer the opportunity to explore how the adsorbate-substrate interaction affects the assembly process, because the type of interactions (acid-base or ionic) are in stark contrast with those in thiol or silane SAMs.

The growth process of octadecylphosphonic acid (OPA) on mica from tetrahydrofuran solution was explored by our group in a series of papers (56–60). AFM observations of quenched samples showed evidence of island growth (56) and suggested that the regions between islands were virtually bare, since the mica atomic lattice could be imaged. Complementary IR spectroscopy and contact angle studies (58) confirmed that the islands were surrounded by a 2D vapor. In situ AFM experiments (57, 59, 60) confirmed these qualitative observations, putting OPA on mica within the class of SAMs that forms via the sequence shown in Figure 2a.

Looking closely at the coverage kinetics of quenched OPA-mica films (58), it was determined that a significant amount of adsorbate molecules (5%–25% coverage over the concentration range of 0.02–2 mM) was deposited during sample removal from solution via a Langmuir-Blodgett–like transfer process. This process also resulted in a dramatically altered island size distribution, as new islands nucleated and existing islands grew during sample removal. After an attempt was made to account for the effects of quenching, the island coverage kinetics were explored as a function of concentration in the range 0.02–2 mM (58). The data were collapsed onto a single curve with the introduction of a concentration-dependent time scale. However, the curve was not in good agreement with Langmuir kinetics. Although the growth kinetics could be explained using a diffusion-limited functional form, the diffusive time scales extracted from the fits were much slower than expected for solution-phase molecular diffusion and did not display the expected concentration dependence. It is interesting that Peterlinz & Georgiadis (22) made similar observations regarding thiol SAM growth. For OPA, however, it is not clear that the apparent form of the coverage kinetics may not simply be an artifact of studying quenched films, since later in situ AFM experiments found that the early stages of growth were consistent with Langmuir kinetics and became complicated in the later stages (which is addressed in more detail in a later section).

Hayes and coworkers (61) explored the monolayer growth of the organic cation octadecyltrimethylammonium bromide (OTAB) on mica from aqueous solution. The room temperature experiments were performed below the Krafft point (the temperature above which micelles are observed). Their AFM, contact angle, and IR spectroscopy observations on quenched partial monolayers found extremely slow growth of islands, with complete monolayers forming only after 2 weeks of immersion time. The atomic lattice of mica could not be imaged with AFM on the areas surrounding the islands, and IR and contact angle data suggested that these regions were covered by a thin film of disordered molecules (a 2D liquid). Thus the OTAB monolayer appeared to grow via a sequence consistent with the one shown in Figure 2b.

Although AFM data may suggest which phase is in coexistence with solid islands by means of island heights, relative friction, or stiffness (using phase contrast), these observations are rather indirect. In general, it is preferable to use complementary techniques that are sensitive to molecular conformation to clarify this issue. In this regard, it is instructive to compare the contact angle (Figure 6) and IR (Figure 7) signatures of 2D vapor-solid coexistence [as exemplified by OPA SAMs on mica (58)] with those of 2D liquid-solid coexistence [using OTAB SAMs on mica as an example (61)].

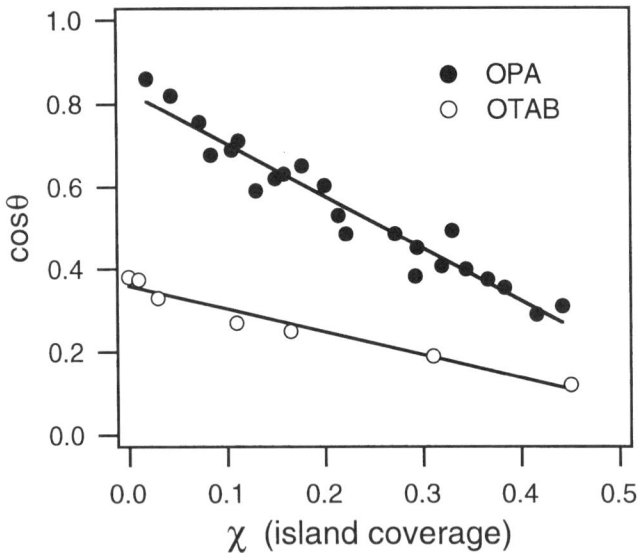

Figure 6 Comparison of the cosine of the contact angle of water to the fractional coverage of partially formed films of octadecyltrimethylammonium bromide (OTAB) on mica (*open circles*) and octadecylphosphonic acid (OPA) on mica (*filled circles*). The different behavior at low coverage suggests that solid-phase islands on partial OPA self-assembled monolayers (SAMs) are surrounded by a two-dimensional (2D) vapor, while those on OTAB SAMs are surrounded by a 2D-liquid phase. (Figure adapted with permission from Reference 61.)

Applying the Cassie equation (62) to the case of solid-phase islands in coexistence with a dilute phase yields the following predictions for the cosine of the contact angle θ:

$$\cos\theta = \cos\theta_{\text{dilute}} + \chi_{\text{island}}(\cos\theta_{\text{island}} - \cos\theta_{\text{dilute}})$$

where χ_{island} is the fractional surface coverage of the island phase and θ_{dilute} and θ_{island} are the contact angles on a surface composed purely of the respective phase. This equation predicts that the extrapolated value of $\cos\theta$ at zero island coverage will be equal to the cosine of the contact angle on a surface composed purely of the dilute phase, which surrounds the islands. As shown in Figure 6, the extrapolated value for OPA is close to unity, implying that water would wet the dilute phase. This is consistent with the dilute phase being bare mica or mica with a very low coverage of adsorbed surfactant molecules—a 2D-vapor phase of OPA molecules. On the other hand, for OTAB, $\cos\theta$ extrapolates to ~0.4 at zero coverage, implying that the dilute phase is fairly hydrophobic, with a contact angle of ~65°. This is consistent with a disordered layer (a 2D liquid), for example.

IR studies of partial OPA and OTAB monolayers were consistent with these conclusions. Figure 7 shows the methylene stretch region of partial monolayers

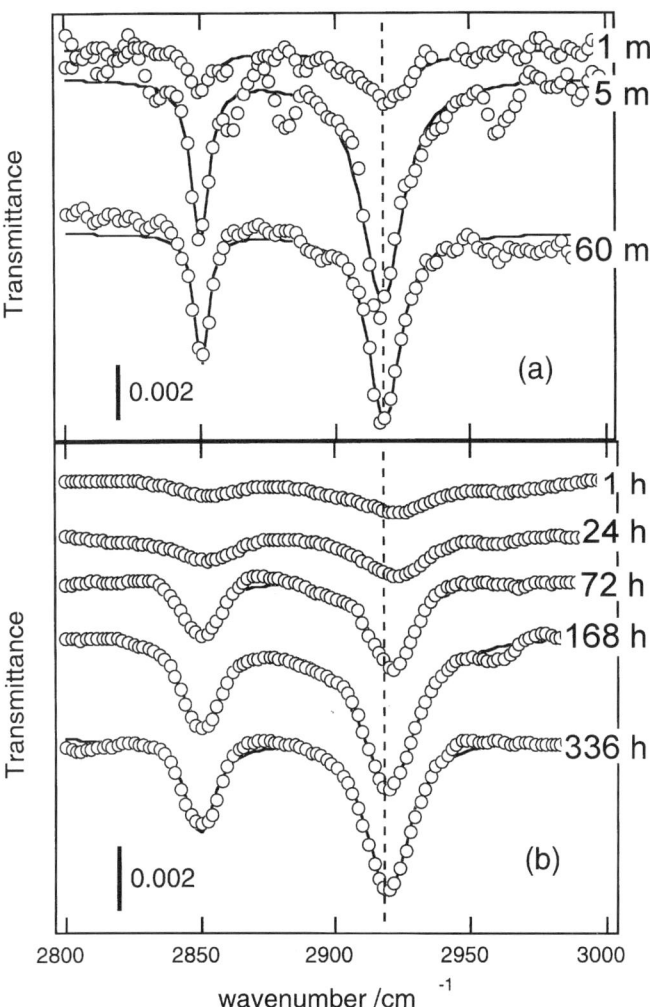

Figure 7 Transmission IR spectra of quenched (*a*) octadecylphosphonic acid (OPA) and (*b*) octadecyltrimethylammonium bromide (OTAB) self-assembled monolayers on mica taken after increasing immersion times in 0.2 mM and 0.1 mM solution, respectively. (Figure adapted with permission from References 58 and 61.)

of OPA (Figure 7*a*) and OTAB (Figure 7*b*) after increasing immersion times (top to bottom in each figure). For OPA, although the peaks were observed to grow as immersion time and island coverage increased, the peak positions remained at the same wavelength. For example, the antisymmetric methylene stretch was found at 2919 cm^{-1}, consistent with well-ordered all-*trans* alkyl chains (63), even at a fractional island coverage of only 0.1. This suggested that a negligible fraction of the molecules on the surface were disordered. On the other hand, the peak positions for OTAB gradually shifted to lower wave numbers with increasing immersion time and island coverage. The antisymmetric peak position moved from 2924 cm^{-1}, consistent with disordered alkyl chains, to 2919 cm^{-1}. For samples partially covered by islands, this peak is presumably the convolution of two peaks, one for the molecules within islands and one for the molecules in the dilute phase between islands. The relative weighting of the two peaks changes with the island coverage, resulting in the apparent peak position shift, which is consistent with a significant coverage of disordered molecules in the region between islands. For both systems, the contact angles and IR spectra tell the same story—the dilute phase surrounding the islands may be either a 2D vapor (as for OPA) or a 2D liquid (as for OTAB), depending on the system chemistry and thermodynamic conditions.

QUANTITATIVE ASPECTS OF GROWTH PROCESSES

Rate Constants/Time Scales

Many reported rate constants or time scales for SAM growth have been collected in Figures 8*a* and 8*b*, cast in terms of time constants. For cases in which rates constants were reported, the time constants were calculated simply as the inverse of the rate constants. Although rate and time constants clearly contain the same information, time constants are presented in the hope that the plotted values will have more intuitive value to the reader. Figure 8*a* displays data for thiol SAMs, and Figure 8*b* includes data for silanes and acid-based monolayers. The large scatter in the thiol data is particularly noticeable. For any concentration in the range 10^{-6}–10^{-3} M, the spread in the measured values of time constants is typically 2 orders of magnitude. Data have been included for a range of chain lengths (C_{12}–C_{22}) and a variety of solvents (mostly alcohols and alkanes). Also, several different theoretical models were used to extract rate or time constants. However, these variables typically introduce variations in rates only of order unity. Data included in Figure 8*a* were determined using a variety of techniques, both in situ and on quenched partial films; however, there is no real pattern or consistency even when considering only individual techniques or methods. Thus, one is left with the impression that there may be real differences in the growth kinetics of thiol SAMs in different laboratories. It is unclear which parameters are not controlled; one possibility that has been suggested is the substrate roughness or microcrystallinity. In contrast, the data in Figure 8*b* is surprisingly consistent even though time scales are included for silane SAMs on a variety of substrates (*open symbols*) in addition to carboxylic and phosphonic acid SAMs (*filled symbols*). Again, the data represent a variety of techniques, some in situ and some on quenched partial films. A casual inspection also reveals that these SAMs grew consistently more slowly than the thiol SAMs.

The *dotted lines* on the graphs in Figure 8 represent the inverse dependence between concentration and time constant (equivalent to a linear dependence of

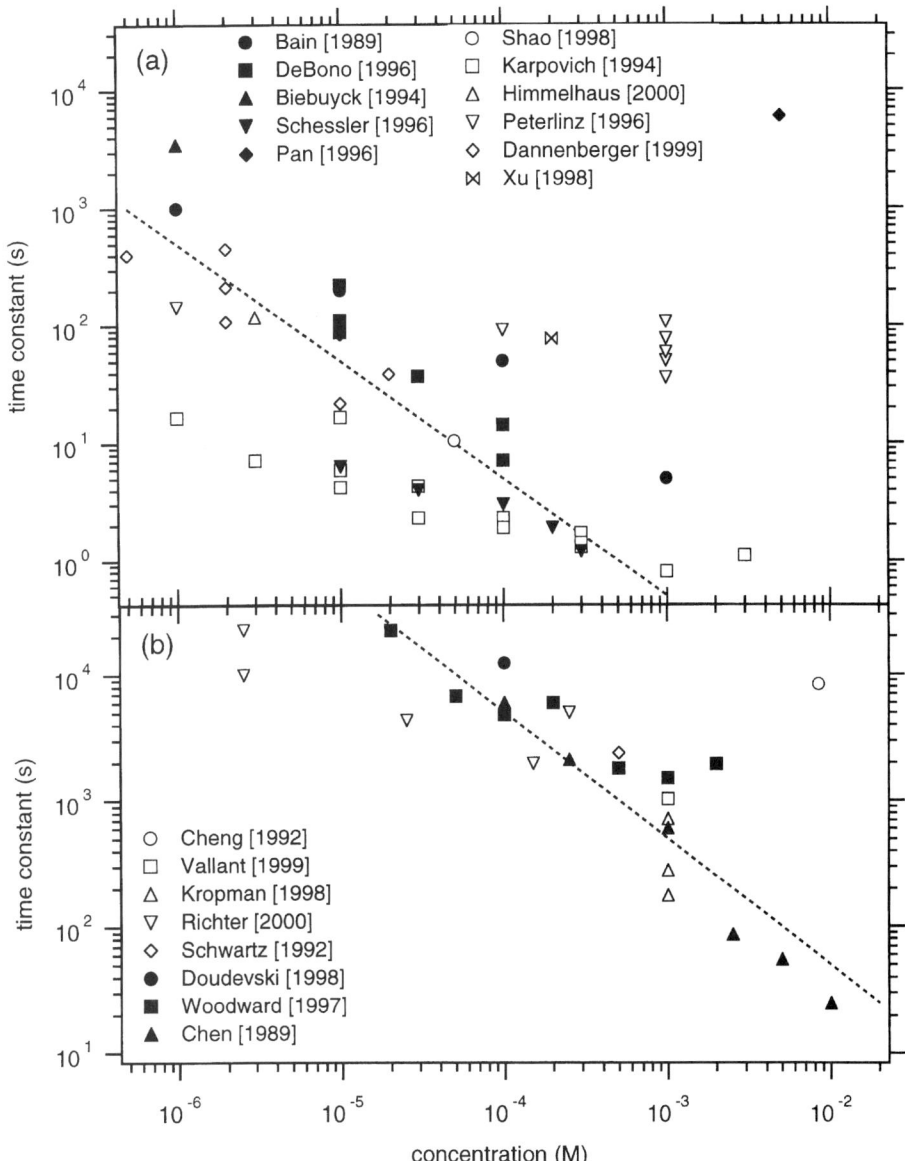

Figure 8 Time scales of self-assembled monolayer (SAM) growth versus solution concentration summarized from a variety of reports. (*a*) All symbols represent Alkanethiol SAMs. (*b*) Silane (*open symbols*) and acid (*filled symbols*) SAMs.

rate on concentration) that one would expect for any model involving adsorption-limited growth. There appears to be a reasonable agreement for silanes and acids (Figure 8b) above 10^{-5} M. Some of the individual thiol data sets (15, 21, 27) in Figure 8a are approximately consistent with the inverse dependence of time constant and concentration. However, others, such as those from the Blanchard group (25, 26) and the Georgiadis group (22), report time constants that decrease much more slowly. It has been suggested (25, 26) that this can be explained by a significant desorption rate, which is not concentration dependent. The discrepancy regarding the concentration dependence of growth kinetics between different laboratories remains unresolved.

Functional Form of Coverage Kinetics

In most cases, the coverage kinetics are compared to the simple Langmuir or Avrami kinetic model, which assumes that the rate of deposition is proportional to the free space on the surface, that is, $d\theta/dt = k(1 - \theta)$, where k is a rate constant. This leads to the integrated expression $\theta = 1 - \exp(-kt)$, which has frequently been used to fit kinetic data for SAM growth, sometimes because it agreed quantitatively with the data and on other occasions simply because it was considered the simplest model to use when the low precision of the data did not justify using a more complicated model. The time constants reported in the previous section are simply the inverse of the rate constant, that is, time constant = $1/k$. In situ nonlinear optical experiments on thiol SAM growth (24, 27, 33, 64) have typically found that Langmuir kinetics [or a variant involving multiple adsorption sites (27)] described the early stages of adsorption reasonably well, as did several in situ SPR (21, 65), QCM (25, 26), and AFM (37) studies. In a careful SPR study that divided the growth process into three regimes, Peterlinz & Georgiadis (22), however, found that the early stages were equally well described by second-order Langmuir kinetics or a diffusion-limited form. A second step was found to obey zero-order kinetics. In electrochemical studies of azobenzene-containing thiol monolayers, Shao et al (23) found that kinetics based on a Frumkin isotherm approach (which includes adsorbate-adsorbate interactions) described the data better than Langmuir kinetics. In situ IR (51, 66) and X-ray reflectivity (55) studies of alkylsilane SAM growth also found reasonable agreement with Langmuir kinetics, except for early times (55, 66) or for solutions with high water content (51). IR spectra of quenched partial monolayers of alkyl carboxylic acids on aluminum oxide (67) were consistent with Langmuir kinetics, as were in situ AFM studies (59) of the early stages of phosphonic acid monolayer growth on mica.

Given what is known about the details of the growth processes for the various SAM systems discussed above, it is somewhat surprising that the simple Langmuir form effectively describes SAM formation that involves island nucleation and growth, multiple 2D-phase transitions, etc. Certainly in some cases the data may simply be too imprecise to justify using other models with greater numbers of free parameters. Also, although it often describes the coverage kinetics well over a large range of the growth, there are clear limitations to the Langmuir model at very early times and for late times. At early times, there is often a discrepancy between observed coverage and the prediction from the Langmuir form. Sometimes an induction period is observed before growth starts. Also, the last 10%–20% of monolayer growth is generally found to evolve with a slower time scale than the earlier regimes. However, the simple Langmuir model is remarkably robust, and thus it is worthwhile to consider what this tells us about the growth process.

For systems in which SAM growth involves a 2D-vapor-to-solid transition, the signal observed by any of the typical techniques is dominated by the molecules in the solid islands. Thus Langmuir kinetics are consistent with a growth rate that is proportional to the area uncovered by islands. If adsorption occurs primarily on these uncovered regions and the overall SAM growth kinetics is adsorption-limited, this would explain why Langmuir kinetics work well in such cases. For cases in which a low-density phase (2D-liquid or lying-down phase) forms before the solid phase, however, it is somewhat more difficult to justify a simple Langmuir model. With thiols, for example, one wonders which part of the growth process is being observed in SPR, QCM, or spectroscopic measurements. Is it the formation of the lying-down phase, the conversion to the standing-up phase, or some combination of the two? It is certainly reasonable that the formation of the original lying-down phase would follow Langmuir kinetics. However, this would account for only $\sim$20% of the final film coverage, whereas many experiments assert that Langmuir kinetics describes the initial 50%–80% of film growth. In the only experiment that could directly separate the two processes, an in situ AFM study (37), growth of the two phases were observed to occur sequentially but with approximately the same time constant. It is not clear, however, that the kinetics of each of the processes have the same concentration dependence. Another possibility is that some of the techniques used might not be particularly sensitive to molecules in the lying-down phase, and/or the signal caused by this phase is part of the experimental baseline. In such a situation, the experiment would report essentially the growth of the solid-phase islands, and Langmuir kinetics would imply that the kinetics are limited by adsorption in the regions covered by the lying-down phase. These assumptions are not particularly satisfying, however, and the connection between the detailed growth mechanism and the macroscopically averaged coverage kinetics remains an open question for thiol SAM growth. Particularly useful in resolving these issues will be in situ techniques that are capable of discriminating between surface-bound molecules that are in different 2D phases.

Solvent Effects

Several studies have addressed the effect of different solvents on the growth kinetics of thiol SAMs; solvents typically used were n-alkanes and ethanol. Peterlinz &

Georgiadis (22) found that the initial stage of thiol SAM growth from heptane solution was ~35% faster than that from ethanolic solution. Karpovich & Blanchard (25) did not observe significant differences using hexane or cyclohexane as solvents. Dannenberger et al (27) found that solvents affected the thiol SAM growth kinetics in the order (fastest to slowest) hexane > ethanol > dodecane > hexadecane. Although this order coincides with the solvent viscosity (which would affect the molecular diffusivity in solution), there is ample evidence to suggest that SAM growth is typically not limited by bulk diffusion at micromolar concentrations or above. It was suggested that a limiting step might involve the displacement of solvent molecules by adsorbate molecules at the surface, so that solvents with stronger surface interactions would result in slower adsorption and SAM growth.

Chain Length Effects

The literature is full of dramatically conflicting reports regarding the effects of chain length on thiol SAM growth kinetics. Regarding the initial fast stage of growth, Bain et al (15) found that C_{18} grew faster than C_{10} from ethanolic solution, Xu and coworkers (37) reported that C_{22} formed more quickly than C_{18} from 2-butanol solution, and Jung & Campbell (65) performed a systematic SPR study and found that the growth rate increased with chain lengths in the range C_2–C_{18} from ethanolic solution. Thus, these studies consistently found that adsorption rate increased with chain length. Other studies reported exactly the opposite trend, however. Peterlinz & Georgiadis (22) reported growth rates for the initial step in the order $C_8 > C_{12} > C_{16} > C_{18}$ from ethanolic solution, and Dannenberger et al (27) found that growth rates obeyed the trend $C_4 > C_{12} > C_{22}$ for both ethanolic and hexane solution. Complicating the matter even further, two additional reports were inconsistent with all of these results. DeBono and coworkers (21) found that the initial stages of growth for C_{16} thiol occurred at about the same rate as C_6 and that both were faster than C_{12} from ethanolic solution. Karpovich & Blanchard (25) found that the early stages of growth for C_8 and C_{18} thiols (from hexane solution) were approximately equal in overall rate. Analyzing the concentration dependence of the growth kinetics, they reported that the adsorption rate for C_{18} was greater than that for C_8, but that the desorption rates had the opposite behavior. There is also some confusion regarding the chain length dependence of the later slow-growth regime. Peterlinz & Georgiadis (22) reported that the rate of this process increased with chain length from C_{12} to C_{16} to C_{18}. DeBono et al (21) also found that C_{16} was faster than C_{12}, but they observed that the trend was reversed for C_6, which was equally as fast as C_{16}.

There is, unfortunately, little basis on which to critically analyze these results. The discrepancies do not divide along the lines of experimental technique, solvent, concentration range, or any other obvious parameter. In addition, it is not intuitively clear which trend should be correct for a given regime. In considering a hypothetical activated process for adsorption, one might think that the enhanced interactions between a longer chain and the surface would lower the energy barrier and increase the adsorption rate (65). On the other hand, if mobility is an issue, longer chains might move more slowly. It is clear that none of the results summarized above are dominated by bulk solution-phase molecular diffusion because of the absolute rates, the details of the time dependence, and the concentration dependence of the rate constants. However, one cannot rule out the importance of molecular mobility in moving through a hypothetical physisorbed layer (22, 27), etc.

Adsorption Energetics

Several approaches have been used to get at the energetics of thiol SAMs. Bain et al (15) determined desorption rates of alkanethiol SAMs into hexadecane at 83°C. Assuming an Arrhenius-type expression, they found that the activation energy for desorption increased by ~0.2 kcal/mol for each methylene group. Their estimate for the absolute activation energy for C_{22} thiol was 28 kcal/mol. Jung & Campbell (65) determined the "sticking probabilities" of various-chain-length thiols by analyzing the observed SAM growth kinetics with a model incorporating molecular diffusion in solution and adsorption from the subsurface layer. Again assuming an activated energy process for adsorption, they reported that the activation free energy for adsorption decreased by ~0.16 kcal/mol per methylene group and that the absolute activation extrapolated to ~11 kcal/mol for zero chain length. The Blanchard group (25, 26) determined the free energy of adsorption, ΔG_{ads}, of C_8 and C_{18} thiols by analyzing the concentration dependence of the observed growth rate constant. They found that $\Delta G_{ads} = -5.5$ kcal/mol for C_{18} and -4.4 kcal/mol for C_8 thiol SAMs. By measuring the temperature dependence of ΔG_{ads} for C_{18}, they found the molar enthalpy of adsorption, $\Delta H_{ads} = 48$ kcal/mol, and the entropy of adsorption, $\Delta S_{ads} = -48$ cal mol^{-1} K^{-1}.

It should be noted that these measurements are not completely consistent. For example, one would expect that ΔG_{ads} should be approximately the difference between the activation energies for desorption and adsorption. Using the values from the above references, this would give approximately $\Delta G_{ads} \approx 8-27 = -19$ kcal/mol for C_{18} thiol compared with the -5.5 kcal/mol quoted by Karpovich et al (25). However, these absolute free energies involve an approximate value of the pre-exponential frequency factor in the Arrhenius expression and are, therefore, somewhat arbitrary. Considering the change with chain length, one finds that the activation energy measurements predict that longer chains will be stabilized by approximately $0.2 + 0.16 = 0.36$ kcal/mol per methylene group. This predicts that ΔG_{ads} for C_{18} should be 3.5 kcal/mol lower than that for C_8, whereas the value quoted is only 1.1 kcal/mol lower. Of course, there were numerous simplifications in the analyses, not least of which was the assumption of Arrhenius-like behavior in experiments performed at only one temperature (15, 65). SAM energetics should be a fruitful area for future work.

Submonolayer Island Nucleation, Growth, and Size Distributions

Doudevski et al (59) analyzed their in situ AFM images of OPA SAM growth on mica to determine the time dependence of the island number density as well as the growth rates of individual islands. The island density (and more generally the full island size distribution) is frequently used in the literature of vapor-phase epitaxial growth to characterize the submonolayer film morphology (9, 68–72). Figure 9 shows the island number density per "site," (a site is calculated as the approximate cross-sectional molecular area) N, as a function of time. Three regimes of growth were observed. For short times (growth regime; <600 s), the number of islands increased, indicating that nucleation of new islands occurred. At intermediate times (aggregation regime; 600–3000 s), the island density was approximately constant—nucleation of new islands virtually stopped, and existing

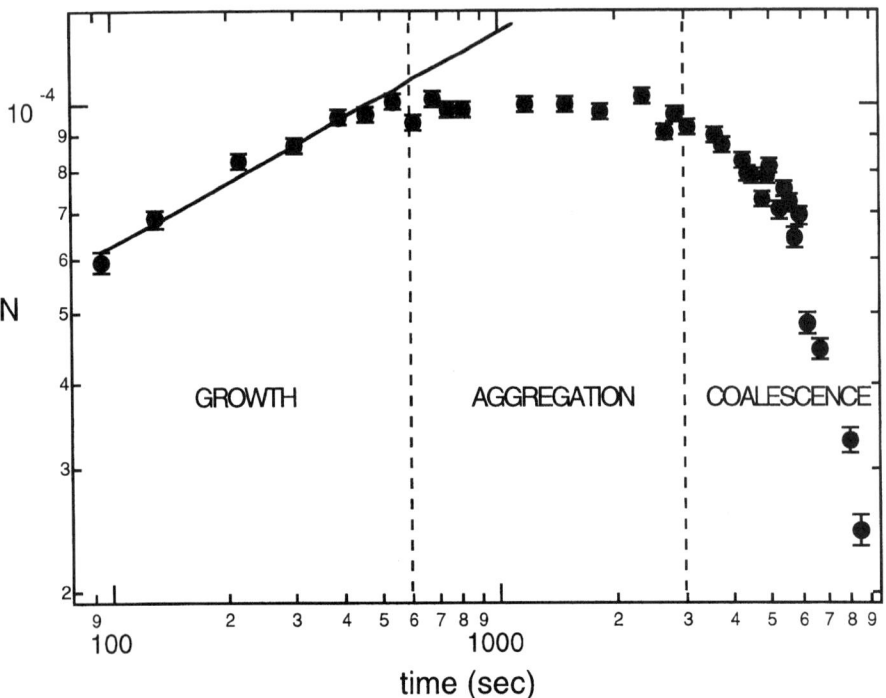

Figure 9 A log-log plot of the number density of islands per site (estimated at 0.25 nm^2) vs time, extracted from in situ atomic-force microscopy images during the growth of octadecylphosphonic acid self-assembled monolayers on mica from 0.15 mM tetrahydrofuran solution. The various regimes of growth are described in the text. The line represents the best fit to a power law time dependence at early times. The best-fit exponent is 0.31 ± 0.5, consistent with the point island model value of one-third. (Figure adapted with permission from Reference 59.)

islands gradually grew. At later times (coalescence regime; >3000 s), the island density decreased rapidly due to the merging of individual islands. At short times, the number density was found to have a power law dependence on deposition time with an exponent of 0.31 ± 0.05, consistent with the point island model prediction of one-third (68, 69). This time dependence also implied that the critical nucleus consisted of two molecules. The growth kinetics of individual islands also had a power law form, with an exponent of 0.70 ± 0.08, again consistent with the point island model prediction of two-thirds (68, 69). By comparing the rates of island nucleation and growth, Doudevski et al (59) inferred a value of the surface diffusivity for adsorbate molecules of $D = 1.1(\pm 0.1) \times 10^6$ sites/s $= 2.9(\pm 0.3) \times 10^{-9}$ cm^2/s.

In other work, Doudevski & Schwartz (60) measured the island size distribution in the aggregation regime (in which the island number density was approximately constant) of OPA SAM growth. As expected qualitatively, with increasing coverage the peak position of the distribution gradually moved to larger island size, and the distribution broadened considerably. They used these distributions to test one of the fundamental assumptions of cluster growth, the dynamic-scaling assumption. The essential concept of the dynamic-scaling assumption is that, at a given stage of growth, there is only a single characteristic length scale. This length can be taken to be S, the average island size, which is a function of the fractional island coverage θ. If this assumption is correct, then the island size distribution function can be written as $N_s(\theta) = \theta S^{-2} f(s/S)$ (68), where $N_s(\theta)$ is the number density of islands containing s molecules at coverage θ. That is, the island size distribution can be factored into two parts—one that contains all dependence on coverage and length scale and another that is a scale-invariant fundamental distribution function, f. Upon applying this scaling form, the island size distributions obtained at various stages of the aggregation regime were found to collapse onto a single function $f(s/S) = S^2 \theta^{-1} N_s(\theta)$, consistent with the dynamic-scaling-assumption prediction. The shape of this fundamental size distribution was different than expected from kinetic Monte Carlo simulations of epitaxial growth (68, 72), in that the peak was shifted to smaller island sizes and the distribution did not extrapolate to zero for small island sizes. This suggested the importance of additional processes not included in these simulations, such as desorption from island edges or long-range interactions. The shape of the distribution did rule out the possible influence of Ostwald ripening, however (73).

MANIPULATING GROWTH WITH EXPERIMENTAL PROBES

Owing to the current interest in nanotechnology, there has been recent interest in using localized experimental probes (typically AFM tips) to fashion small features in SAMs. Although not strictly within the topic of this review, such scanning probe lithography represents a promising technological area that relies in many ways on the SAM growth process. Two approaches are particularly relevant. The Liu group has developed a technique they call nanoshaving/nanografting (74, 75). The concept involves applying a large enough load while scanning with an AFM tip so that molecules are mechanically removed from a SAM. If this procedure is performed while the SAM is in contact with a solution of adsorbate molecules, those molecules adsorb and assemble to fill in the region previously shaved by the AFM tip. Using this method, the group has demonstrated the ability to create features with typical length scales of ≥ 10 nm. The features can be raised or recessed, or they can have a different chemical surface functionality relative to the surrounding surface, depending on the choice of adsorbate molecules used to create the original "resist" SAM and those present in solution during the shaving.

For thiol SAMs, they found that, if the dimension of the "shaved" region was smaller than the extended length of the adsorbate molecule, the regrowth of SAM in the shaved region was approximately an order of magnitude faster than SAM growth on a bare surface (76). In practice, these conditions were obtained by scanning the AFM probe slowly so that the regrowth was faster than the shaving and the dimension of the shaved patch was related to the size of the AFM probe tip. Under these conditions, there were no observable defects or "scars" at the boundary of the grafted features. The authors suggested that, in the spatially confined environment, the usual sequence of growth was altered. Instead of progressing through a lying-down phase to a standing-up phase, the molecules adsorbed directly in a standing-up conformation.

The Mirkin group (77, 78) reported a technique they labeled "dip-pen nanolithography" that involves casting a layer of thiol adsorbate molecules onto an AFM tip and subsequently transferring these molecules to a gold substrate by scanning the tip under ambient conditions. The authors presumed that the thiols are transported to the surface via a water capillary bridge formed between the tip and surface. Although originally used to write SAM features on a bare substrate, this technique has recently been combined with the nanoshaving technique to create features within existing SAMs (79).

CONCLUSIONS

Although kinetic data on SAM growth have been collected for the past decade, we are only beginning to develop a reliable qualitative picture of the process, and we are far from a complete quantitative understanding. Given the current awareness regarding the complexity of the growth process (largely due to the application of scanning probe microscopy), that is, multiple 2D phases forming sequentially via nucleation and growth processes, it is now clear why techniques that do not discriminate between different surface species are insufficient to characterize the growth mechanisms and kinetics. Additionally, clear evidence now exists that quenching the SAM growth process during growth often results in modification of the SAM structure, coverage, and morphology. This reduces our confidence in results obtained by techniques that cannot be used in situ, under actual growth conditions.

The paradigm that is emerging for SAM growth is a fascinating one, combining aspects of surfactant science, epitaxial growth, and nonequilibrium thermodynamics in two dimensions. Qualitative pictures of the growth process are now reasonably well established for several important systems. However, few of the mechanisms are understood at a quantitative level. The temperature dependence of a proposed 2D-phase diagram has been explored for alkylsilane SAMs. The details of island nucleation and growth have been touched on for alkylphosphonic acid SAMs. The basic issues of solvent and chain length effects, etc, have been addressed in numerous studies for thiol SAMs; however, the discrepancies in the literature are dramatic on these issues, both quantitative and qualitative, and essentially unexplainable. A few studies have concerned themselves with the energetics of adsorption and desorption, but, as yet, no consistency has emerged. These are complicated and rich systems with a variety of interactions of the same order in strength, that is, adsorbate-solvent, adsorbate-adsorbate, adsorbate-substrate, etc. Although variable solvent and chain length studies have the ability to address the first and second types of interaction (at least in a limited way), there has been little effort to probe the ways in which the adsorbate-substrate interaction affects the growth process or the effects that qualitatively different types of intermolecular interactions might have. Because SAMs are being proposed as a route for surface modification in increasing numbers of applications, involving a greater variety of substrates and adsorbate chemical functionality, these issues will become increasingly important.

ACKNOWLEDGMENTS

Thanks go to Gang-Yu Liu, Rosina Georgiadis, and Roya Maboudian (and their coworkers) for contributing figures to this manuscript and to Chad Taylor for his careful and critical reading of this manuscript. The author is grateful for support from the National Science Foundation (award 9980250).

Visit the Annual Reviews home page at www.AnnualReviews.org

LITERATURE CITED

1. Bigelow WC, Pickett DL, Zisman WA. 1946. *J. Colloid Interface Sci.* 1:513
2. Nuzzo RG, Allara DL. 1983. *J. Am. Chem. Soc.* 105:105
3. Maoz R, Sagiv J. 1984. *J. Colloid Interface Sci.* 100:465
4. Poirier GE. 1997. *Chem. Rev.* 97:1117–27
5. Ulman A. 1996. *Chem. Rev.* 96:1533–54
6. Ulman A. 1991. *An Introduction to Ultrathin Organic Films*. Boston, MA: Academic
7. Adamson AW, Gast AP. 1997. *Physical Chemistry of Surfaces*. New York: Wiley Interscience, Wiley & Sons
8. Chang C-H, Franses EI. 1995. *Colloids Surf.* 100:1–45

9. Zhang ZY, Lagally MG. 1997. *Science* 276:377–83
10. Kaganer V, Mohwald H, Dutta P. 1999. *Rev. Mod. Phys.* 71(3):779–819
11. Poirier GE, Pylant ED. 1996. *Science* 272:1145–48
12. Schreiber F, Eberhardt A, Schwartz P, Wetterer SM, Lavrich DJ, et al. 1998. *Phys. Rev. B* 57:12476–81
13. Eberhardt A, Fenter P, Eisenberger P. 1998. *Surf. Sci.* 397:L285–90
14. Schwartz P, Schreiber F, Eisenberger P, Scoles G. 1999. *Surf. Sci.* 423:208–24
15. Bain CD, Troughton EB, Tao Y, Evall J, Whitesides GM, Nuzzo RG. 1989. *J. Am. Chem. Soc.* 111:321–25
16. Sun L, Crooks RM. 1991. *J. Electrochem. Soc.* 138:L23–25
17. Shimazu K, Yagi I, Sato Y, Uosaki K. 1992. *Langmuir* 8:1385–87
18. Kim Y-T, McCarley RL, Bard AJ. 1993. *Langmuir* 8:1941–44
19. Schneider TW, Buttry DA. 1993. *J. Am. Chem. Soc.* 115:12391–97
20. Fruböse C, Doblhofer K. 1995. *J. Chem. Soc. Faraday Trans.* 91:1949–53
21. DeBono RF, Loucks GD, Dellamanna D, Krull UJ. 1996. *Can. J. Chem.* 74:677–88
22. Peterlinz KA, Georgiadis R. 1996. *Langmuir* 12:4731–40
23. Shao HB, Yu HZ, Cheng GJ, Zhang HL, Liu ZF. 1998. *Ber. Bunsenges. Phys. Chem.* 102:111–17
24. Buck M, Grunze M, Eisert F, Fischer J, Trager F. 1992. *J. Vac. Sci. Technol. A* 10:926–29
25. Karpovich DS, Blanchard GJ. 1994. *Langmuir* 10:3315–22
26. Schessler HM, Karpovich DS, Blanchard GJ. 1996. *J. Am. Chem. Soc.* 118:9645–51
27. Dannenberger O, Buck M, Grunze M. 1999. *J. Phys. Chem. B* 103:2202–13
28. Kawasaki M, Sato T, Tanaka T, Takao K. 2000. *Langmuir* 16:1719–28
29. Hähner G, Woll C, Buck M, Grunze M. 1993. *Langmuir* 9:1955–58
30. Truong KD, Rowntree PA. 1996. *J. Phys. Chem.* 100:19917–26
31. Terrill RH, Tanzer TA, Bohn PW. 1998. *Langmuir* 14:845–54
32. Bensebaa F, Voicu R, Huron L, Ellis TH, Kruus E. 1997. *Langmuir* 13:5335–40
33. Himmelhaus M, Eisert F, Buck M, Grunze M. 2000. *J. Phys. Chem. B* 104:576–84
34. Humbert C, Buck M, Calderone A, Vigneron JP, Meunier V, et al. 1999. *Phys. Status Solidi A.* 175:129–36
35. Yamada R, Uosaki K. 1997. *Langmuir* 13:5218–21
36. Yamada R, Uosaki K. 1998. *Langmuir* 14:855–61
37. Xu S, Cruchon-Dupeyrat SJN, Garno JC, Liu GY, Jennings GK, et al. 1998. *J. Chem. Phys.* 108:5002–12
38. Tamada K, Hara M, Sasabe H, Knoll W. 1997. *Langmuir* 13:1558–66
39. Wasserman SR, Whitesides GM, Tidswell IM, Ocko BM, Pershan PS, Axe JD. 1989. *J. Am. Chem. Soc.* 111:5852–61
40. Tidswell IM, Ocko BM, Pershan PS, Wasserman SR, Whitesides GM, Axe JD. 1990. *Phys. Rev. B* 41:1111–28
41. Schwartz DK, Steinberg S, Israelachvili J, Zasadzinski JAN. 1992. *Phys. Rev. Lett.* 69:3354–57
42. Bierbaum K, Grunze M, Baski AA, Chi LF, Schrepp W, Fuchs H. 1995. *Langmuir* 11:2143–50
43. Zhao XL, Kopelman R. 1996. *J. Phys. Chem.* 100:11014–18
44. Kropman BL, Blank DHA, Rogalla H. 1998. *Thin Solid Films* 329:185–90
45. Brzoska JB, Shahidzadeh N, Rondelez F. 1992. *Nature* 360:719–21
46. Parikh AN, Allara DL, Azouz IB, Rondelez F. 1994. *J. Phys. Chem.* 98:7577–7590
47. Carraro C, Yauw OW, Sung MM, Maboudian R. 1998. *J. Phys. Chem. B* 102:4441–45
48. Goldmann M, Davidovits JV, Silberzan P. 1998. *Thin Solid Films* 329:166–71
49. Sung MM, Carraro C, Yauw OW, Kim

Y, Maboudian R. 2000. *J. Phys. Chem. B* 104:1556–59
50. Vallant T, Brunner H, Mayer U, Hoffmann H, Leitner T, et al. 1998. *J. Phys. Chem. B* 102:7190–97
51. Vallant T, Kattner J, Brunner H, Mayer U, Hoffmann H. 1999. *Langmuir* 15:5339–46
52. Brunner H, Vallant T, Mayer U, Hoffmann H, Basnar B, et al. 1999. *Langmuir* 15:1899–901
53. Resch R, Grasserbauer M, Friedbacher G, Vallant T, Brunner H, et al. 1999. *Appl. Surf. Sci.* 140:168–75
54. Richter AG, Durbin MK, Yu C-J, Dutta P. 1998. *Langmuir* 14:5980–83
55. Richter AG, Yu CJ, Datta A, Kmetko J, Dutta P. 2000. *Phys. Rev. E* 61:607–15
56. Woodward JT, Ulman A, Schwartz DK. 1996. *Langmuir* 12:3626–29
57. Woodward JT, Schwartz DK. 1996. *J. Am. Chem. Soc.* 118:7861–62
58. Woodward JT, Doudevski I, Sikes HD, Schwartz DK. 1997. *J. Phys. Chem. B* 101:7535–41
59. Doudevski I, Hayes WA, Schwartz DK. 1998. *Phys. Rev. Lett.* 81:4927–30
60. Doudevski I, Schwartz DK. 1999. *Phys. Rev. B* 60:14–17
61. Hayes WA, Schwartz DK. 1998. *Langmuir* 14:5913–17
62. Cassie AB. 1952. *Discuss. Faraday Soc.* 75:5041
63. Porter MD, Bright TB, Allara DL, Chidsey CED. 1987. *J. Am. Chem. Soc.* 109:3559–68
64. Dannenberger O, Wolff JJ, Buck M. 1998. *Langmuir* 14:4679–82
65. Jung LS, Campbell CT. 2000. *Phys. Rev. Lett.* 84:5164–67
66. Cheng SS, Scherson DA, Sukenik CN. 1992. *J. Am. Chem. Soc.* 1114:5436–5437
67. Chen SH, Frank CW. 1989. *Langmuir* 5:978–87
68. Amar JG, Family F, Lam PM. 1994. *Phys. Rev. B* 50:8781–97
69. Tang LH. 1993. *J. Phys. I* 3:935–50
70. Evans JW, Bartz JA, Sanders DE. 1986. *Phys. Rev. A* 34:1434–48
71. Amar JG, Family F. 1996. *Thin Solid Films* 272:208–22
72. Amar JG, Family F. 1995. *Phys. Rev. Lett.* 74:2066–69
73. Zinke-Allmang M, Feldman LC, Grabow MH. 1992. *Surf. Sci. Rep.* 16:377–463
74. Xu S, Liu GY. 1997. *Langmuir* 13:127–29
75. Xu S, Miller S, Laibinis PE, Liu GY. 1999. *Langmuir* 15:7244–51
76. Xu S, Laibinis PE, Liu GY. 1998. *J. Am. Chem. Soc.* 120:9356–61
77. Piner RD, Zhu J, Xu F, Hong S, Mirkin CA. 1999. *Science* 283:661–63
78. Hong SH, Zhu J, Mirkin CA. 1999. *Langmuir* 15:7897–900
79. Amro NA, Xu S, Liu GY. 2000. *Langmuir* 16:3006–9

CROSSED-BEAM STUDIES OF NEUTRAL REACTIONS: State-Specific Differential Cross Sections

Kopin Liu

Institute of Atomic and Molecular Sciences, Academia Sinica, Taipei, Taiwan 10764, Republic of China; e-mail: kpliu@gate.sinica.edu.tw

Key Words reaction dynamics, geometric phase, dynamical resonance, spin-orbit reactivity, mode-specific chemistry

■ **Abstract** Crossed-molecular-beam and laser techniques have enabled experimentalists to measure the state-resolved differential cross sections of elementary chemical reactions. This article reviews recent progress in this area. Particular emphasis is placed on some intriguing physical phenomena associated with a few benchmark reactions and how these measurements help in answering fundamental questions about reaction dynamics. We examine specifically the geometric phase effects in the reaction $H + D_2$, the dynamical resonance phenomenon in $F + HD$, the unusually large spin-orbit reactivity in $Cl(^2P) + H_2$, the insertion reaction $O(^1D) + H_2$, and the mode-specific reactivity in $Cl + CH_4(\nu)$. The give-and-take between experiment and theory in unraveling the physical picture of the dynamics is illustrated throughout this review.

INTRODUCTION

Chemical reaction dynamics is essentially the modern-day approach to the study of reaction kinetics (1). The fundamental aim of a reaction dynamicist is to understand the intimate mechanisms of chemical reactivity at a microscopic level of detail, rather than just measure the rate of chemical reactions under different temperatures and pressures. Crossed-molecular-beam and laser spectroscopic techniques have played a central role in the experimental developments and advances in the field. Over the past few decades, we have witnessed a great expansion and remarkable progress in this field. The activity perhaps peaked in the 1980s, and more recently as the field has matured, the output has leveled off or arguably declined. Interestingly, with the implementation and refinement of several innovative experimental techniques, it was during this leveling-off period that the "Holy Grail"—the state-to-state differential cross section (DCS)—was finally achieved for a few simple chemical reactions. At the same time, the recent development of femtochemistry (2) offers a very appealing new direction in this field and provides a complementary view by allowing the experimentalist to take snapshots of the chemical transformation from reagents to products.

Equally impressive progress has been made on the theoretical front. Chemically accurate (i.e. within 0.2 to $\sim$0.3 kcal mol^{-1} of accuracy) ab initio potential energy surfaces (PESs) are now available for several atom-diatom benchmark systems. The exact quantum mechanical (QM) calculation of reactive scattering dynamics on a single PES has also become possible, allowing direct and detailed comparison with experiments. The comparison between the QM results and those from a quasi-classical trajectory (QCT) calculation on the same PES can reveal the underlying quantum effects in a reaction. Such a close interplay between experiment and theory has been pivotal in deepening our understanding of how a chemical reaction occurs.

Because the field of neutral reaction dynamics over the past decade was exhaustively reviewed by Casavecchia et al in 1999 (3) and in 2000 (4), this article is mainly concerned with more recent experiments and discusses them from a different perspective. This review is "phenomena oriented" in that we devote little space to summarizing all the results for a given reaction, instead giving a critical assessment of a particular aspect of the reaction that illustrates a certain physical phenomenon. The discussion is from an experimentalist's perspective; thus, the description in some cases may not be rigorously accurate from a theoretical point of view. We focus only on a few selected reactive systems for which the state-specific DCS measurements have just become available. It turns out, fortuitously or not, that each of these reactions is accompanied by one or two outstanding physical phenomena that could have profound effects on our fundamental understanding of chemical reactivity in general. From the viewpoint of microscopic reversibility (1), the initial state selection in a reaction is as important as the state-resolved measurement of the product. Hence, the state-specific DCS in this article refers loosely to the state specificity in the reagent and/or product state-resolved DCS. The questions to be addressed here are the following: Is the state-specific DCS essential in resolving a particular problem, and what do we learn from it?

EXPERIMENTAL METHODOLOGIES

Currently, laser-based and non-laser-based techniques are used to measure state-specific DCSs. Non-laser-based techniques involve the universal crossed-beam machine (5). In this approach, the recoil velocity of the reaction product is determined by the time-of-flight (TOF) measurement over a well-defined flight path, L. A rotatable electron impact mass spectrometer is used for detection. Because any neutral species can be ionized by 70- to 200-eV electron bombardment, this detection scheme is, in principle, universal. The TOF resolution is determined largely by $\Delta L/L$, where ΔL is the convolution of the two effective lengths, namely the two-beam-crossing region and the ionizer region. In a state-of-the-art apparatus, $\Delta L/L$ of $\sim$0.015 can be accomplished, which implies an achievable translational energy resolution, $\Delta E/E$, of ~3%. However, because of the speed spreads and the finite angular divergences of the two molecular beams, the actual overall resolution in a typical crossed-beam experiment becomes somewhat worse. Nevertheless, for a reaction with favorable kinematics, this resolution is sufficient to resolve the product vibrational state distribution through energy balance. A notable example is the celebrated experiment of F + H_2 by Neumark et al (6, 7). Very recently, Rusin & Toennies (8), Faubel et al (9), and Baer et al (10, 11) improved the resolution further and obtained a rotational state-resolved TOF spectrum at a particular laboratory angle for the reaction F + $H_2(j = 0) \rightarrow$ HF ($v' = 2, j' = 7$–10) + H at a collision energy of 1.86 kcal mol^{-1} (8). That is quite a heroic effort, and the resolution thus achieved is probably as good as one can hope for using this approach. With the availability of the third-generation synchrotron radiation source and the powerful F_2 excimer laser (photon energy of 7.9 eV), photoionization detection (12–14) has been used to replace the electron impact ionizer. This variation is conceivably capable of somewhat better resolution.

Combination of the crossed-beam technique and laser spectroscopic detection should permit, in principle, the most detailed investigation of the dynamics of a chemical reaction. However, because of limitations of sensitivity, two general strategies were only recently developed to realize this goal. The first strategy was demonstrated in the study of the reaction F + H_2 by Dharmasena et al (15, 16), in which the scattered HF(v', j') rotational-vibrational state is interrogated directly by an infrared laser, and then its angular distribution is measured by a rotatable detector (a bolometer in this case). The advantage of this method is its high state specificity if given sufficient spectroscopic data for the probed species (usually a diatomic molecule). Its drawback is the number of states that need to be measured to get a global picture of the reaction dynamics. Of course, the state-resolved angular distribution can be obtained in other ways after a specific state is tagged by laser, such as by Doppler profile measurement (17) or ion imaging (18). These two methods have been applied in state-specific studies of inelastic scattering processes (17, 19, 20) and in a few earlier reactive scattering studies (21–24) in which angular resolution was quite limited.

The second strategy has three variants, all of which use translational spectroscopy with species and state specificity of laser spectroscopic techniques: ion imaging, H atom Rydberg tagging, and Doppler-selected TOF techniques. In general, the first method is ideal for detecting slowly moving species, whereas the latter two are better suited for fast recoiling products. Kitsopoulos et al (25) used ion imaging to study the H + D_2 reaction. The D atom reaction product was ionized by laser, and its three-dimensional velocity distribution was projected onto a position-sensitive detector as a two-dimensional image. A numerical procedure, inverse Abel transformation, was then used to recover information about the speed and angular distributions of the product from its two-dimensional projection. The resolution was such that the vibrationally excited coproduct, HD(v'), which has different recoil velocities by energy balance, appeared as separated rings in the image. The intensity distribution of each ring in the polar direction gave the vibrational state-resolved angular distribution. More recently, Eppink & Parker (26) introduced the velocity mapping technique, which significantly improves the ion-imaging resolution. It is now almost routine to achieve $\Delta E/E$ of $\sim 2\%$ in a typical photodissociation study. Because of its multiplexing advantage, we expect to see more state-specific DCSs of chemical reactions measured by this method in the near future.

Schnieder et al (27, 32), Wrede et al (28, 30, 31), Wrede & Schnieder (29), and Banares et al (33) developed a novel Rydberg tagging technique for detecting H atoms. The method has high sensitivity and extremely high $\Delta E/E$ ($\sim 0.3\%$). It has been applied to studies of the reactions $H + D_2 \rightarrow HD + D$ and, very recently, $O(^1D) + H_2 \rightarrow OH + H$. In this technique, the H or D products are "tagged" by using two laser photons through a double-resonance excitation to form long-lived high-n Rydberg states. The translational and angular distributions of the nascent H (D) products are monitored via the Rydberg atoms, which are field-ionized at the end of their TOF. The elegance of this method is to eliminate the space charge and stray field effects, which are often the limiting factors in an ion-TOF measurement. However, to achieve the ultra-high $\Delta E/E$, the probe lasers are necessarily focused, making ΔL significantly smaller than the crossed-beam-overlap volume. As a result, extensive modeling (27) is required to obtain the desired DCSs. Moreover, such ultra-high resolution can be fully realized only when this Rydberg H atom-tagging detection scheme is combined with very high Mach number molecular beams. In the studies of the $H + D_2$ and $O(^1D) + H_2$ reactions, nearly complete ro-vibrational state-resolved DCSs were obtained. In the studies, the H or $O(^1D)$ atomic beam was generated by photolyzing a diatomic precursor (HI or O_2, respectively) with a pulsed dye laser or an F_2 excimer laser, respectively. The resulting speed distribution was very narrow ($\Delta v/v < 2\%$). In addition, the target beams were converted to ortho-D_2 and para-H_2, respectively, and were cooled to 77 K before supersonic expansion. Thus, the target beams contained only the ground rotational state with a narrow speed spread ($\Delta v/v \sim 6\%$).

The last method was developed in this laboratory. It combines the Doppler-shift and ion TOF techniques in an orthogonal manner such that the three-dimensional velocity distribution of the reaction product is directly mapped out in a center-of-mass (c.m.) Cartesian coordinate. To take advantage of the large recoil velocity of a light species, this method, which uses a typical commercial dye laser, has been applied to detect the H or D atom products of a number of reactions (34–39). The resolution of this method at present is $\sim 1\%$ (40). Although the resolution is constant in a Cartesian velocity coordinate (i.e. $\Delta v_x \Delta v_y \Delta v_z =$ constant), it yields different speed resolutions along different c.m. scattered angles when cast into the conventional reactive flux velocity contour map (1) in a polar coordinate system. Because of the limited laser bandwidth in Doppler selection, the energy resolution is usually the worst along the probe laser propagation axis, which corresponds to the c.m. forward-backward direction in this approach.

H + H$_2$: Geometric Phase Effects

The hydrogen exchange reaction H + H$_2$ → H$_2$ + H is the simplest of all neutral reactions. The first accurate, fully ab initio PES was published in 1979, hereafter denoted Liu-Siegbahn-Truhlar-Horowitz (LSTH) PES (41, 42). Successively, more elaborate representations of the PES, denoted the double many-body expansion (DMBE) (43), Boothroyd-Keogh-Martin-Peterson (BKMP2) (44), and the exact quantum Monte Carlo (EQMC) [for which an unprecedented absolute accuracy of about 0.6 meV (0.014 kcal mol^{-1} or 5 cm^{-1}) was achieved] (45), have confirmed the essential features and refined the details. Both QCT and exact QM dynamical calculations have also been performed using those PESs, except the latest EQMC PES. On the experimental front, various product state and angular distributions have become available over the past decade. Only during the past few years, however, has the experiment caught up with the theoretical developments, allowing quantitative comparison for this simplest chemical reaction.

One of the most interesting aspects of this reaction concerns the possible effects of the geometric phase (GP) (46–51). This reaction proceeds mainly on the repulsive, strongly collinear, ground state surface. However, a conical intersection occurs between the ground and first excited electronic states at the D_{3h} nuclear configuration. In simple terms, the two electronic states are degenerate for geometries in which the nuclei are at the vertices of an equilateral triangle, and the electronic state energies split in an approximately linear manner for deviations from this geometry, giving rise to a locally cone-shaped appearance of the PESs. The minimum energy of this intersection is around 2.72 eV above the bottom of the H$_2$ well; nevertheless, the low potential barrier (0.4 eV) allows the dynamics of the nuclei to encircle the conical intersection for reaction at much lower energies. The dynamical consequences of this phenomenon were first pointed out by Mead & Truhlar (50). In a classical paper, Longuet-Higgins (51) showed that the real ground state adiabatic electronic wave function becomes double valued (i.e. changes sign) for any closed path that encircles the conical intersection. Because the total wave function must be a single-valued and continuous function of the nuclear coordinates, the nuclear wave function must also change sign. This sign change is the so-called GP or Berry phase, and it can lead to significant interference effects in the nuclear dynamics.

The first indication of an experimental manifestation of GP effects in a chemical reaction was pointed out in a theoretical work by Kuppermann & Wu (52). They found that the product rotational state distribution for the reaction D + H$_2$($v = 1$, $j = 1$) → HD($v' = 1, j'$) + H, which was measured by Zare and coworkers (53) at a collision energy of 1.0 eV, could be well reproduced by calculations including the GP effect on the LSTH PES, whereas calculations without it were noticeably hotter. Two years later, the same authors (54) predicted that for the following reaction,

$$H + D_2(v = 0, j = 0) \rightarrow HD(v', j') + D, \qquad 1.$$

near the collision energy 1.29 eV, the product state-resolved DCSs calculated with or without inclusion of the GP effect differ significantly. Specifically, their GP results showed a pronounced resonance-like backward peak for $v' = 0, j' = 4$ and 5, whereas the same calculations without the GP effect did not show this peak. A collision lifetime analysis of the scattering matrix furnished a resonance lifetime of 164 fs. Moreover, when the same calculations were repeated on the BKMP2 PES (55), whose barrier height is 0.417 eV (only 8.2 meV lower than the LSTH one), the corresponding resonance energy shifted downward by 55 meV and the lifetime dropped to 52 fs. Clearly, the resonance characteristics show extreme sensitivity to the fine details of the PES.

In a series of elegant experiments by Schnieder et al (27, 32), Wrede et al (28, 30, 31), Wrede & Schnieder (29), and Banares et al (33), fully ro-vibrational state-resolved DCSs for Reaction 1 were measured for a wide but discrete range of collision energies of 0.52–0.54 eV (27, 33), 1.27–1.30 eV (27, 29, 32), 2.20 eV (31), and 2.67 eV (30), using the novel Rydberg H atom TOF spectroscopy method. Figure 1 shows some typical kinetic energy spectra of the D atom at $E_c = 1.28$ eV, together with detailed simulation of the experimental data using both QCT and QM calculations on the LSTH PES (32). As can be seen, the unprecedented resolution of the experiment ($\Delta E/E \sim 0.4\%$) is such that the D atom kinetic energy distribution mirrors the fully resolved ro-vibrational distribution of the HD coproducts. These results are the most comprehensive and accurate experimental data ever measured for a chemical reaction and have enabled a very detailed assessment of the fully ab initio QM treatment of elementary chemical reactions. The comparison between the experiment and theories, both QM and QCT, shows excellent concordance. More significantly, because neither the QCT nor QM calculation presented in Figure 1 included the GP effect, the remarkable agreement between theory and experiment led to the conclusion that the GP effect does not play an important role at this energy. This conclusion is the complete opposite of the aforementioned theoretical prediction (54). Further experiments (29) covering the energy range from 1.27 to 1.30 eV and using the same technique also did not find any evidence of the predicted resonance. Unfortunately, these experiments did not cover the forward hemisphere in product angular distributions. As is discussed in the next section, in order to see the resonance "signature", if any, it is essential to measure the distribution over the full angular range.

As mentioned above, the theoretical calculations indicated that the resonance is extremely sensitive to the fine details of the PES. One could argue that the LSTH PES, which was used in the dynamical calculations shown in Figure 1, is not accurate enough to guide the experimentalists in searching for the GP effects. However, Kendrick (56) used an alternative approach to incorporate the GP effect into an accurate QM calculation for the reaction $H + D_2(v, j) \rightarrow HD(v', j') + D$ on both the BKMP2 and LSTH PESs at 126 values of total energy in the range from 0.4 to 2.4 eV, and he showed convincingly that the GP effects cancel out in both the integral and DCSs when the contributions of even and odd partial waves, Js,

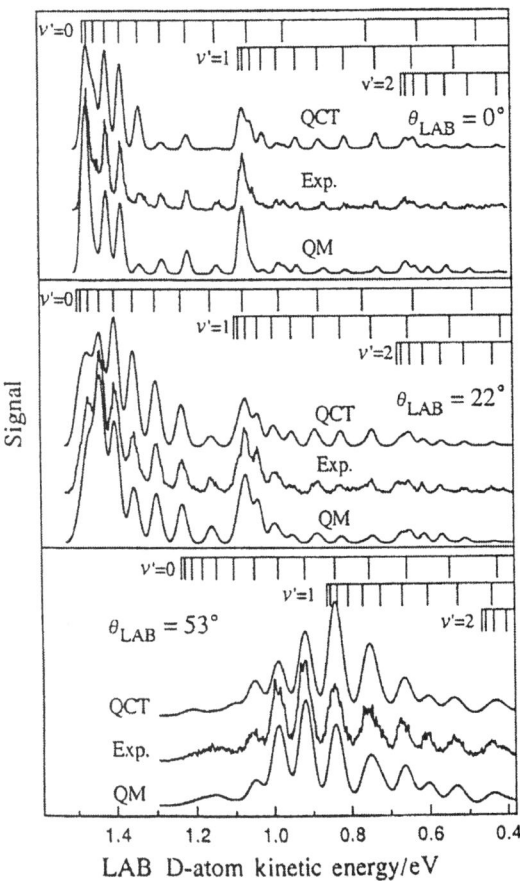

Figure 1 Laboratory D atom kinetic energy spectra at the indicated laboratory scattering angles (θ_{lab}), for the H + D$_2$($v = 0$, $j = 0$) reaction at a collision energy of 1.28 eV, obtained in a high-resolution molecular beam experiment using the technique of Rydberg H atom time-of-flight spectroscopy. The experimental data are shown along with the simulations based on quantum mechanical and quasi-classical trajectory calculations performed on the LSTH potential energy surface (adapted from Reference 32).

are added together. In simple physical terms, the GP effect is to alter only the phase of the scattering amplitude. The resulting out-of-phase behavior from the GP effect alternates phase with even and odd J. However, the Wigner D-functions (56a) exhibit an alternating nuclear exchange symmetry with even and odd J. This alternating symmetry also gives rise to an alternating phase shift and thus offsets the GP effects when all partial waves are included. Because this cancellation is the direct consequence of symmetry, it should be independent of the PES and is expected to hold for all collision energies. In other words, there should be no

GP effects in the reaction H + D_2. This conclusion is in sharp contrast to the original proposition by Kuppermann & Wu (54), but it is in line with the experimental results illustrated in Figure 1 and with those at higher collision energies (29–31).

Do these results mean that there are no experimentally measurable GP effects in the hydrogen exchange reaction? The answer is "not necessarily." In 1980, Mead (48) pointed out that "the interference between reactive and nonreactive scattering amplitudes for identical nuclei is exactly the opposite to what it would be without consideration of the conical intersection." This conclusion is based on the symmetry considerations under permutation of identical nuclei; thus, it should be applicable to the reaction H + H_2. Because the DCS involves a coherent sum over the scattering amplitudes (i.e. one sums and then squares), the GP effect in a chemical reaction should manifest itself as the "correct" phase of the oscillations in the state-resolved angular distributions for the reaction H + H_2. It would be particularly rewarding to have the experimental confirmation of this phenomenon in the near future. One should be cautious to jump to definite conclusions, however, because the angular distributions tend to exhibit oscillations (but with the "wrong" phase in scattered angles), even in the absence of the GP.

Although the cancellation of the GP effect between the even and odd partial waves could be a general result and applicable to all A + B_2 reactions, it does not imply that the conical intersection cannot yield measurable effects in reaction attributes. In 1992, an intriguing reactive excitation function for the reaction H + O_2 → OH + O was reported in a bulb experiment (57). After a typical featureless reaction cross section at low collision energies, a pronounced peak showed up near $E_c = 1.8$ eV, which then declined with further increases in energy. Using the Doppler-resolved laser-induced fluorescence technique, Fei et al (58) later measured the rotationally resolved DCSs for OH($v' = 0, N = 4, 14$, and 17) product at a few collision energies. An anomalous observation was made for the appearance and disappearance of a sideways scattering component to the DCS at and above 1.8 eV, which was ascribed to a transition from adiabatic to diabatic collisions at and above the energy of the C_{2v} conical intersection of this system.

As alluded to earlier, a quantal effect greatly sought for this reaction has been reactive resonance. Two very recent reports (59, 60) presented experimental evidence for resonances in the H + D_2 reaction. Both experiments were photoinitiated reaction studies in a single molecular beam, that is, under coexpansion conditions. A prominent peak at $E_c = 0.94$ eV was observed in the state-resolved integral cross section (ICS) measurement of the reaction H + D_2 → HD($v' = 0$, $j' = 7$) + D, which verifies the QM prediction for a resonance (59). We defer the discussion of the nature of this resonance to the next section. The signature for resonances was also reported in the state-resolved DCS for the reaction H + D_2 → HD($v' = 3, j'$) + D at $E_c = 1.64$ eV (60). Although the QCT simulation agreed with the experimental results rather well for $j' \geq 1$, it failed to reproduce a sharply forward-scattered peak for $j' = 0$, indicative of a resonant mechanism.

$F + H_2$: Quantum Dynamical Resonance at Last

The reactions $F + H_2 \to HF + H$ and its isotopic variants have long played a crucial role in the development of gas phase reaction dynamics. These reactions have been extensively studied in various kinetics and dynamics experiments, as well as with theoretical calculations. In the early 1970s, quantum-scattering calculations of this reaction predicted the existence of isolated resonances (61, 62). However, an empirical PES was used, and the scattering calculation was restricted to one-dimensional dynamics. Thus, it was not totally clear if the predicted resonance was a real physical phenomenon or just a theoretical artifact from the approximations used. Nevertheless, the importance of these resonances cannot be overemphasized because they can, if they indeed exist, provide a very sensitive and direct probe of the interaction potential in the transition state region. The term resonance in this context refers to a transiently formed metastable species produced while the reaction is occurring. Transient intermediates, which can live from just a few vibrational periods to thousands of vibrational periods before dissociating, are well known in many kinds of atomic and molecular processes. The central questions concerning any resonance phenomenon focus on the formation and decay of a quasi-bound state. Deeper insights into the role of resonances in chemical reactions can be gained by investigating the mode of the internal excitation or the quantum assignment of the metastable species, the nature of the attractive interaction that underlies the trapping mechanism, and the coupling of the internal excitation and external (translational) motions for decay. Figure 2 depicts three classes of transition state resonances in chemical reactions. From the onset, it should be pointed out that the distinction among them is not always unambiguous, and a single resonance may exhibit characteristics of all three types.

For a complex-forming reaction, numerous bound and quasi-bound (predissociative) states are generally built upon the deep intermediate well. It is natural to view resonances or quasi-bound states in this case as the continuation of the bound-state spectrum into the continuum. This type of resonance has been extensively studied by spectroscopic means, and its role in chemical kinetics and dynamics was recently documented by Reid & Reisler (63) and by Bowman (64). The second kind of transition state resonance, as illustrated in Figure 2b, can be classified as the vibrational threshold resonance. This resonance typically corresponds to the energetic threshold for a quantized dynamical bottleneck in the transition state region, which gates the flow of reactive flux from reactants to products. In addition to the total energy, total angular momentum, and parity, this quasi-bound state can be characterized by two vibrational quantum numbers (for a three-atom system) corresponding to the modes of motion orthogonal to the

(a) (Quasi-)bound states in a potential well

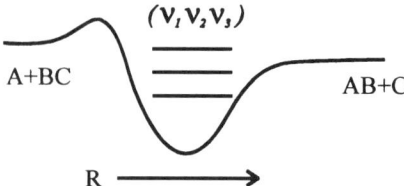

(b) Threshold resonance near the saddle point

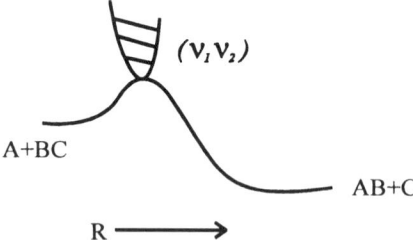

(c) Trapped-state resonance on a repulsive PES

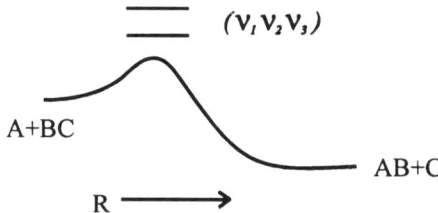

Figure 2 Three types of transition state resonances in chemical reactions. The one shown in panel (*a*) is normally associated with a deep potential well. The resulting bound and predissociative quasi-bound states can be characterized, for a three-atom system, by three vibrational modes. (*b*) Threshold resonance for which only the two motions orthogonal to the unbound reaction coordinate are quantized and thus assignable by vibrational quantum numbers. The dynamical trapped-state resonance is schematically shown in panel (*c*). Despite the repulsive potential energy surface along the reaction coordinate, this metastable state can be assigned by three vibrational quantum numbers.

unbound reaction coordinate. This type of resonance was experimentally confirmed in several unimolecular reactions and was reviewed by Green et al (65) several years ago. Its role in state-selective chemistry and significance in understanding bimolecular reactions were amply emphasized by Truhlar and coworkers (66). It is possible that the recently observed resonance in the H + D$_2$ reaction at $E_c = 0.94$ eV (59), as alluded to in the previous section, belongs to this category. On the other hand, the persistence of the striking oscillations in the QM simulation of state-resolved ICSs over a wide energy range (see Figures 9 and 10 in reference 56) seems to suggest some sort of interference effect as an alternative interpretation. A similar pattern of oscillating reaction probabilities was observed in a theoretical study (67) of the reaction I + HI. Its physical origin was ascribed to the interfering rotational dynamics rather than to the exchange dynamics. Thus, the exact nature of those oscillatory features in H + D_2 remains unknown.

The last type of transition state resonance is the trapped state (compound state) or Feshbach resonance. What makes this resonance special is that, unlike the previous two, it is quasi-bound even along the reaction coordinate on a totally repulsive Born-Oppenheimer PES. In many ways, it behaves like a stable molecule with all three vibrational modes assignable for a three-atom system. Its very existence, no matter how fleeting it is, or the trapping mechanism that requires attractive forces, is dynamical in origin.

Numerous studies have attempted to experimentally realize the dynamical resonance phenomenon that was theoretically predicted for this reaction. In the groundbreaking molecular beam studies of Neumark et al (6, 7), anomalous forward-scattering peaks for HF($v' = 3$) from F + H_2 were observed and attributed to quantum dynamical resonance. In another study, Neumark and coworkers (68–70) used an innovative photodetachment spectroscopic approach to directly probe the transition state. Unfortunately, the experimental search for these resonances has shown that they are particularly elusive to direct observation. The reported "sightings" of dynamical resonance in this reaction have since been shown theoretically (71–73) to be inconclusive because other possible dynamical explanations could not be ruled out. Very recently, we reported a combined experimental and theoretical effort (74, 75) that proves unequivocally the observation of the long-sought quantum dynamical resonance in the reaction F + HD → HF + D. This is the first solid experimental evidence, not just an indication, of this type of resonance in a full collision experiment for any chemical reaction. The only other previous experimental evidence for a trapped-state resonance was the photodetachment spectroscopic study of I + HI (76). The experiment was performed with a rotating-sources, crossed-beam machine and used the Doppler-selected TOF detection scheme. Identifiable signatures in both ICS and DCS were sought experimentally, and the accompanying theoretical simulations and analyses not only confirmed them but also provided deeper insights into the nature of this resonance.

Figure 3 illustrates the resonance fingerprints in ICS (74). Although the total reactive excitation function for the DF + H channel displays a typical over-the-barrier behavior and is in excellent agreement with both QCT (77) and QM (74) simulations using the Stark-Werner (SW) PES (78), the same cannot be said for the HF + D channel. For this channel, a peculiar steplike feature at low energies is quite evident, which is entirely absent in the QCT simulation and, hence, is suspected to be of QM origin. The QM simulation reproduces the steplike peak

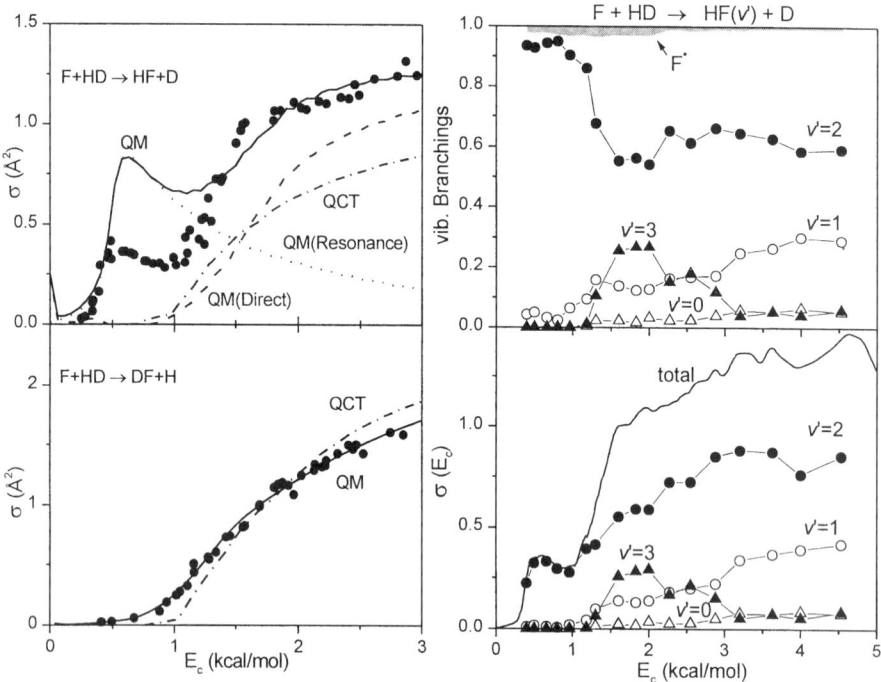

Figure 3 The *left panel* shows the total excitation functions of the two isotope product channels for the F + HD reaction (modified from Reference 74). Both quasi-classical trajectory and quantum mechanical simulations were performed on the Stark-Werner potential energy surface. The quantum mechanical results can be partitioned into the resonance (dotted line) and direct-scattering (dashed line) contributions. The *right panel* shows the product vibrational branching (upper curve) and the vibrational-specific excitation functions (lower curve) of the HF(v') + D isotope channel. The total contribution (not relative cross section) from the spin-orbit excited F*($^2P_{1/2}$) reagent is indicated by the shaded area (adapted from Reference 79).

quite well in position and shape, but the absolute magnitude appears to be too high by a factor of two. The excellent agreement in thresholds for the DF + H channel and for the F + D$_2$ reaction (79) suggests that the barrier height on the SW PES (about 1.1 kcal mol^{-1}) is probably quite accurate. We therefore suspect that the barrier width may be too thin, which results in excess tunneling and thus too large of a steplike feature.

Theoretical analyses (74) proved that the origin of this striking step feature is a dynamical resonance state localized behind the barrier at the collinear F–H–D configuration in the transition state region. It corresponds to a vibrational-excited state of the temporarily formed collision complex and can be assigned as a *(003)* local-mode state with three quanta in the H–F stretch and zeros in the H–D stretch and the bend. In simple terms, the wells in the vibrationally adiabatic potentials provide the necessary attractive forces for dynamically trapping the resonance state (80, 81). It is a fortunate situation that, for this particular reaction, the direct-scattering component is nearly absent at low energies, which allows the resonance to survive the angular momentum or partial-wave averaging and to manifest itself as a distinctive feature, even in the total ICS, for easy identification.

In the vibrational branching of the product, an unusual dependence on collision energies also occurs (79), as shown in Figure 3. At low energies ($E_c \lesssim 1$ kcal mol^{-1}) where only the resonant tunneling mechanism (74, 75) is operative, more than 90% of reactive flux is channeled into HF($v' = 2$) + D. Intuitively, this high specificity in the state distribution of the product follows the well-known Δv propensity rule in a vibrational predissociation process (82), which says that the larger the difference between the initial and final vibrational states, the much less probable is the vibrational predissociation. However, as soon as the formation of HF($v' = 3$) + D becomes energetically allowed, one sees a strong competition in the decay paths of this resonance state. This result accounts for the abrupt rise for HF($v' = 3$) and the concurrent drop for HF($v' = 2$) in the vibrational branching of the reaction products.

The previous controversy about resonance in this reaction has to do with the origin of the anomalous forward peaks in the vibration-resolved angular distributions of the landmark 1985 crossed-beam experiments of Lee and coworkers (6, 7). This anomaly, albeit less marked, also shows up in the QCT simulation (71) using the SW PES, which obviously does not possess quantized resonance states. An alternative interpretation of tunneling through the centrifugal barrier for large-impact-parameter collisions has been proposed (72). A very recent theoretical reinvestigation of the F + H$_2$ reactions, however, tilts the balance back in favor of the original resonance interpretation (83). Figure 4 (see color insert) shows the total (i.e. sum of all v', j' states of the HF or DF product) angular distributions of the F + HD reaction over a wide range of collision energies (75). Experimentally, an oscillatory forward-backward peaking in the HF + D distribution is quite apparent. For the DF + H channel, however, the distribution is rather smooth and localized in a broad swath in the backward hemisphere. It was conjectured that the analogous broad swath in the HF + D channel is characteristic of the direct reaction, and the remaining ridge structure at low energy and the highly oscillatory forward-backward peaking at higher energy are the imprints of the resonant scattering. The QM investigations, shown on the right in Figure 4 (see color insert), appear to confirm it. The quantitative discrepancies between theory and experiment presumably trace back to deficiencies of the SW PES.

The full analysis of the completely state-resolved DCS data revealed far richer information. Nevertheless, it appears that the results elucidated in Figure 3 and Figure 4 (see color insert) already capture the dominant and the most apparent features of the resonant scattering in this reaction. Those fingerprints should provide useful guidance for future searches of resonance phenomena in other reaction systems. From an experimental point of view, a resonance manifests itself most clearly in the evolution of reaction attributes with the collision energy at fine grids. The appearance of collision energy as (at least) one of the variables in both Figure 3 and Figure 4 (see color insert) underscores its necessity in the identification and elucidation of a resonance phenomenon.

As indicated in the vibrational branching plot (Figure 3), the contribution (not reactive cross section) of the spin-orbit-excited $F^*(^2P_{1/2})$ to the total reactivity of this reaction is quite small. This result is in line with the conclusion inferred from a most recent, full QM multisurface calculation (84) and with the previous crossed-beam conclusions (6–11). This finding is currently of considerable interest (79, 84–86) and leads naturally to the next subject.

$Cl(^2P) + H_2$: Spin-Orbit Reactivity

The interest in spin-orbit reactivity is long-standing. Two comprehensive reviews summarize what we currently know about this subject from the kinetics (87) and dynamics (88). A general rule has emerged from those studies by correlating the individual fine-structure state of the reagent and product. It was found that the adiabatically correlated reaction path is generally the preferred one. In other words, the electronically nonadiabatic transition usually makes only a small contribution to total reactivity. The small nonadiabatic contribution from $F^*(^2P_{1/2})$ to the reaction $F(^2P) + H_2$, as just mentioned, is thus consistent with this expectation. In this sense, the recent discovery (89) of a larger reactivity of the spin-orbit excited $Cl^*(^2P_{1/2})$ to H_2 than to the ground state $Cl(^2P_{3/2})$ atom at $E_c = 5.2$ kcal mol^{-1} is truly startling.

The reaction of $Cl(^2P) + H_2 \rightarrow HCl + H$ is slightly endothermic ($\Delta H_0^0 = 1.03$ kcal mol^{-1}). It has played a seminal role in the development of transition state theory and also serves as a textbook example of the kinetic isotope effect (90). The reaction is the abstractive type with a collinear barrier of about 5 kcal mol^{-1}. Extensive kinetics data from 200 to 3000 K (91) and the previously reported crossed-beam dynamical results (92) can all be explained without invoking the nonadiabatic contribution from the spin-orbit excited Cl^* reagent. The surprising result of a large reactivity for $Cl^*(^2P_{1/2})$ was brought about by comparing the doubly DCSs (in angles and speeds) of this reaction from the two different Cl atom beam sources under otherwise identical conditions in a recent crossed-beam investigation (89). Photolysis of Cl_2 (1% seeded in H_2) by a 355-nm laser was used for a nearly pure $Cl(^2P_{3/2})$ beam. A discharge source (5% Cl_2 seeded in He) was adapted to generate a $Cl(^2P)$ beam containing both spin-orbit states. The relative concentrations of the two spin-orbit states from the discharge source were not known at that time. Nevertheless, on the basis of the requirement that the doubly DCS from either spin-orbit reagent must be nonnegative, a lower-bound analysis was performed. Figure 5 summarizes the main findings.

As shown in Figure 5, the reactions from $Cl(^2P_{3/2})$ and $Cl^*(^2P_{1/2})$ make about equal contributions to the observed dynamical attribute. Because it is very unlikely

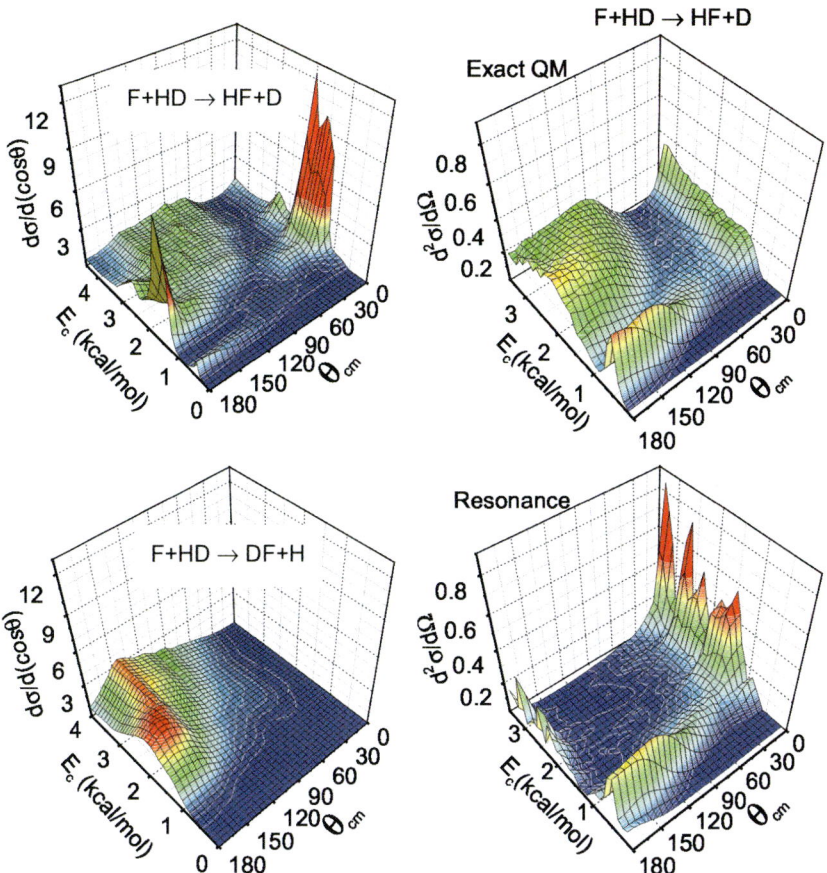

Figure 4 The total angular distributions of the F + HD reaction as a function of collision energies. Both isotope channels are shown experimentally on the left. The QM simulation (*the right upper panel*) and the resonance contribution (*the right lower panel*) for the HF + D isotope channel are also displayed for comparisons. (Adapted from Reference 75).

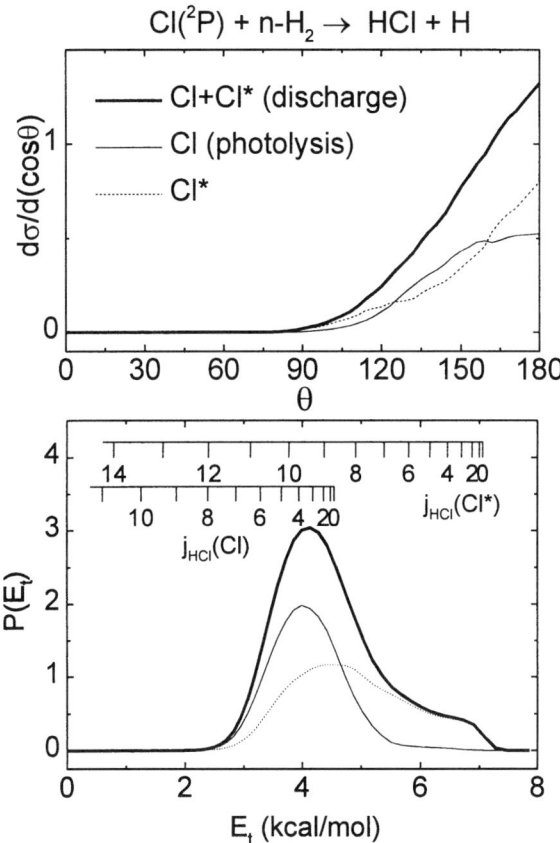

Figure 5 The product center-of-mass angular distributions [$d\sigma/d(\cos\theta)$] and the product translational energy distributions ($d\sigma/dE_t$) from the two different Cl beam sources for the reaction $Cl(^2P) + n\text{-}H_2 \rightarrow HCl + H$ at $E_c = 5.2$ kcal mol^{-1}. The partitionings of the contributions from $Cl(^2P_{3/2})$ (light solid lines) and $Cl^*(^2P_{1/2})$ (dotted lines) are based on a lower-bound analysis (adapted from Reference 89).

that the discharged Cl beam contains more $Cl^*(^2P_{1/2})$ than $Cl(^2P_{3/2})$, this finding implies that the reaction cross section for $Cl^*(^2P_{1/2}) + H_2$ has to be larger than that for $Cl(^2P_{3/2}) + H_2$ at $E_c = 5.2$ kcal mol^{-1}. This conclusion clearly violates the conventional adiabatic correlation rule and is in sharp contrast to what is generally accepted for this reaction. Very recently, a direct determination of the relative cross sections for Cl and Cl* was made in this laboratory. In that experiment, the relative concentration of $Cl^*(^2P_{1/2})$ in the discharge beam was found to be unexpectedly high, about half of the ground state Cl atom. The resulting ratio of $\sigma^*/\sigma = 1.6 \pm 0.4$ for $Cl^*/Cl + H_2$ at 5.2 kcal mol^{-1} supports nearly quantitatively the previous results inferred from a lower-bound analysis (89).

In a related hot-atom experiment (94), the absolute (total) reactive cross sections for the reverse reaction, H + HCl → H_2 + Cl($^2P_{3/2}$)/Cl*($^2P_{1/2}$), were determined at $E_c = 1.0$ eV. It was found that the partial ICS for the Cl*($^2P_{1/2}$) channel is only 6% of that for Cl($^2P_{3/2}$). Despite the large difference in collision energies of the two experiments, this hot-atom result, at first sight, seems contradictory to the conclusion drawn from the above crossed-beam experiment. The results shown in Figure 5 provide a hint in resolving this discrepancy. As shown in the figure, the HCl products from the Cl($^2P_{3/2}$) + H_2 reaction are formed with little rotational energy, whereas those from the Cl*($^2P_{1/2}$) + H_2 reaction are mostly in the higher j states. The reaction cell in the hot-atom experiment is at room temperature; that is, only the low j states of HCl are present. According to the principle of microscopic reversibility (1) and also as shown by the crossed-beam results (89), it is not surprising to see more Cl($^2P_{3/2}$) products than Cl*($^2P_{1/2}$) products in the reverse reaction for a HCl reagent with little rotational energy. An obvious experimental test is to perform the same hot-atom experiment with highly rotationally excited HCl reagents. If the above scenario is correct, then one should observe the opposite, that is, more products in the spin-orbit excited Cl*($^2P_{1/2}$) state than in the ground state Cl($^2P_{3/2}$).

To our best knowledge, this is the first neutral reaction in which the nonadiabatic spin-orbit reactivity supersedes the adiabatic one. Interestingly, the only other previously known bimolecular reaction that shows a higher nonadiabatic spin-orbit reactivity is the isoelectronic ion-molecule reaction Ar$^+$($^2P_{3/2}$, $^2P_{1/2}$) + H_2 → ArH$^+$ + H (95). Its reaction mechanism was thought to involve the crossings and transitions of the Ar$^+$($^2P_{3/2}$) + H_2 and Ar$^+$($^2P_{1/2}$) + H_2 surfaces with the Ar + H_2^+ PES, followed by a rearrangement process. This charge transfer mechanism is unique to an ion-molecule or neutral reaction via the harpooning mechanism and is not applicable to the present case.

A highly accurate ab initio BW2 PES recently became available for the ground state surface of this reaction (96). Dynamical calculations have also been performed with this PES and compared with available experimental results (97, 98). Excellent agreement has been achieved in all cases. It is extremely desirable to have the coupled PESs with explicit inclusion of spin-orbit effects for this reaction. A better understanding of the spin-orbit reactivity of this benchmark system should lead to a deeper appreciation of the role of the spin-orbit interaction and to further insights into nonadiabatic reactivity in general.

O(^1D) + H_2: A Prototypical Insertion Reaction

The interaction of O(^1D) with H_2 involves five singlet PESs, for which the lowest ($1^1A'$) PES correlates directly with the $^1A_1$ ground state of the water molecule. From a fundamental point of view, it is one of the best known complex-forming reactions. As a result, there were numerous experimental and theoretical studies of this reaction in the 1970s and 1980s. Until recently, it was widely accepted that over the energy or temperature range of chemical interest, the reaction proceeds solely via an insertion mechanism on the ground state PES. Thus, this reaction is often regarded as a benchmark for studies of insertion reactions.

The past few years have witnessed intense, renewed studies of this reaction (34–37, 99–105). This resurgence of interest was sparked in part by the surprising and somewhat controversial proposition (106) that an abstraction pathway, which proceeds on excited PESs, also contributes to the total reactivity for $E_c \gtrsim 1.8$ kcal mol^{-1}. The experimental signatures of this elusive, microscopic abstraction pathway are the following: (*a*) an abnormal shape of the reactive excitation function (106), (*b*) a change in product angular distributions from a nearly forward-backward symmetric distribution at low energies to a backward-bias one at higher energies (34–37, 101, 104), and (*c*) a dramatic variation of the angle-specific product translational energy distributions with a slight change in total available energy (36, 37). Theoretically, two accurate ab initio PESs have become available since the abstraction pathway was proposed and are denoted K PES (107, 108) and DK PES (109). In particular, the more recent DK PES includes the couplings of all five adiabatic PESs. The two ab initio PESs are nearly identical, with some subtle differences, and both exhibit a collinear barrier height of 2.3 kcal mol^{-1} on the first excited surfaces. Extensive dynamical calculations, both QCT (105, 110) and approximate QM wave packet (111–114) on either the adiabatic surfaces or the coupled multisurfaces, have also been carried out. At low energies, the theoretical results were in general accordance with experimental findings in terms of product angular and translational energy distributions and the more detailed angle-specific translational energy distributions (35). At higher collision energies, however, significant discrepancies were noted (36). In addition, although the theoretical results support the notion of the participation of the abstraction from excited surfaces, the calculated contribution appears to be smaller than the experimentally deduced value by a factor of 1.5–2.

Because the crossed-beam results obtained so far are of partially state-resolved DCSs at best, it is difficult to pinpoint the cause of discrepancies between theory and experiment at the state-resolved level of detail. On the other hand, using a state-specific Doppler probe under room temperature bulb conditions, Simons and coworkers (99, 100) were able to obtain the angular distributions for a few selected ro-vibrational states of OH products from $O(^1D) + H_2$. However, because only a few individual product states were examined and the collision energy for that bulb experiment was quite broad, it is difficult to compare their results directly to the crossed-beam experiments and to provide a global view of the dynamics of this reaction.

To this end, it is particularly encouraging to see the recent crossed-beam scattering experiment by Liu et al (115), in which a nearly complete ro-vibrational state-resolved DCS was finally achieved. As in the studies of the $H + D_2$ reaction by Wrede et al (27), this remarkable accomplishment was made possible by the

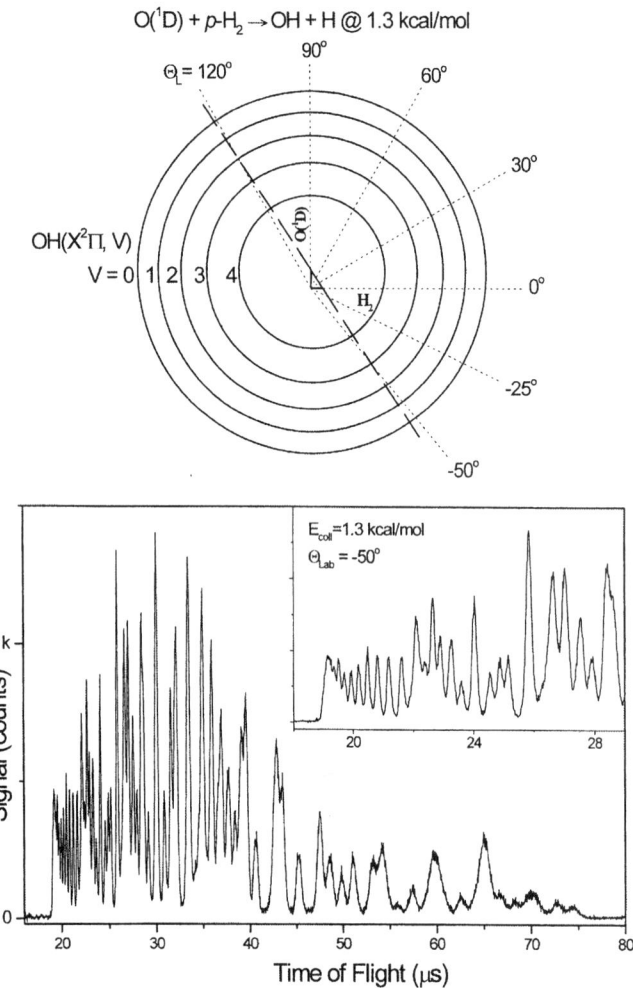

Figure 6 The *top panel* is the Newton diagram for the reaction $O(^1D) + H_2(j = 0) \rightarrow OH(v') + H$ at $E_c = 1.3$ kcal mol^{-1}. The *lower panel* shows the product H atom kinetic energy distribution measured at the laboratory scattering angle $\theta_{lab} = -50°$ using the Rydberg H atom time-of-flight technique.

combination of the Rydberg H atom detection scheme with molecular beams of very high Mach number. As mentioned earlier, the use of both a photolytical source of O_2 for the $O(^1D)$ beam and a precooled para-H_2 beam greatly improved the overall resolution of the experiment. Figure 6 illustrates the result at one of the laboratory angles, which corresponds to the c.m. backward direction, for $O(^1D) + H_2(j = 0)$ at $E_c = 1.3$ kcal mol^{-1}. As shown in the figure, the experimental resolution is truly remarkable ($\Delta E/E \sim 1\%$ at 40 kcal mol^{-1}). Although the electronic fine-structure states of the OH($X^2\Pi$) product cannot be resolved, every peak in the spectrum can be unambiguously assigned to a specific ro-vibrational energy level(s). Preliminary comparison with theoretical results showed excellent agreement. This finding is in accord with the general conclusion drawn from the previous theory-experiment comparisons, albeit with lower resolution, at low collision energies.

The central issue of the O(^1D) + H$_2$ reaction is whether it is a single-surface or multisurface reaction. Mechanistically, does abstraction, which occurs on the excited surfaces, also contribute to the experimental observables in a significant manner? At present, both experiments and theory indicate a participant role of the excited surfaces, but a quantitative agreement is lacking. It would be extremely interesting if the experiment illustrated in Figure 6 could be carried out at higher collision energies (>2 kcal mol^{-1}). The nearly complete state-resolved DCSs would provide the most stringent test of the theory and also shed light on the current discrepancies between theory and experiment.

Cl + CH$_4$(ν_i): Mode-Specific Reactivity

Mode- or bond-specific reactivity is a central theme for reactions more general than atom-diatom systems. This area is largely unexplored territory, particularly in terms of mode-specific DCSs. The past few years have witnessed a series of fascinating bond-specific ICS measurements. The most notable examples are the reactions of an H or Cl atom with HOD (ν_{OH}, ν_{OD}), where ν_{OH}, ν_{OD} denote the bond being excited in the HOD reagent, which were studied in Crim's (116–120) and Zare's (121, 122) laboratories, and those with HCN (ν_1, ν_2, ν_3), where ν_1, ν_2, ν_3 denote the symmetric stretch, bend, and antisymmetric stretch vibrational modes, respectively, which were studied by Crim's (123, 124) and Gericke's (125, 126) groups. These results are truly exciting and have significantly improved our conceptual understanding of how a complicated chemical reaction occurs. There is every reason to believe that many unexpected phenomena in this area remain to be discovered from this type of measurement, and much more can be learned when the mode-specific DCS data become available.

At present, only a few state-specific DCS measurements in the literature provide deeper insights into mode specificity. These measurements involve reactions of Cl and saturated hydrocarbons studied by Zare's (127–136) and Dagdigian's (137–139) groups. We focus on the simplest one, Cl + CH$_4$(ν_i) (127–132). This is a photoinitiated reaction study using a single molecular beam. The angular distribution of the products was determined through the product speed distribution measurements using the core extraction technique in the ion-TOF measurement. The selection of the initial vibrational modes of the polyatomic reagent was achieved by laser excitation. The product rotational alignment or the spatial anisotropy has also been exploited in these experiments. Because this particular aspect was recently reviewed (140–142), it is omitted here.

For the reaction with methane in the ground vibrational state, the product HCl($v' = 0$) was found to have a cold rotational state distribution and was strongly back scattered. Both of these behaviors are consistent with the line-of-center model (1) with a narrow cone of acceptance to reaction. One quantum of C–H asymmetric-stretch vibrational excitation enhanced the reaction rate at $E_c = 0.16$ eV by a factor of ~30. Approximately one third of the product was HCl($v' = 1, j'$) with a cold j' distribution, and the remaining product was HCl($v' = 0, j'$), which is more rotationally excited. A large disparity in state-resolved angular distributions was also seen. Although the HCl($v' = 0, j'$) product remained backward and a bit more side scattered, the HCl($v' = 1$) product was sharply forward scattered for low j' and became more backward scattered for higher j'. These observations are consistent with the opening of the cone of acceptance and the shift in reaction mechanism from a rebound type to a peripheral type (143, 144) as the C–H bond is stretched.

This remarkable state-to-state experiment demonstrates the typical multifaceted nature of a chemical reaction involving a small polyatomic reagent and/or product. It also elucidates the complexity one needs to unravel from such an investigation. It is particularly enlightening to compare the DCSs with different modes of excitation of $CH_4(v_i)$. In this regard, by detecting the CH_3/CD_3 products from the $Cl + CH_4/CD_4$ reactions, Kandel & Zare (127) reported an anomalously fast-moving component in product speed distributions. It was attributed to the reaction of the ground state Cl atom with trace amounts of bending and torsionally excited methane in the beam. Some hints of mode-specific DCSs were also inferred. Further experiment is warranted.

BULB STUDIES

In the context of state-specific DCSs, no review is complete without mentioning some very exciting bulb studies. Crossed-beam techniques are not the only experimental approach to achieve state-specific DCSs; similar information can be obtained in a bulb using the polarized Doppler technique in a pump-and-probe experiment. The experimental strategy was illustrated in one of the pioneering experiments of Park et al (145). They used the polarized photodissociation of the H_2S precursor to generate a velocity-aligned and translationally hot H atom "beam" in a cell. The D atom recoiling from the subsequent collision with the target molecule SiD_4, the $H + SiD_4 \rightarrow SiHD_3 + D$ reaction, was then interrogated by a Doppler detection scheme. The remarkable observation was that the D atom product tended to recoil parallel to the direction of the incident H atom, that is, *k*//*k'*, and about half of the incident H atom energy was retained in the exiting D atom. These results can readily be rationalized if the reaction proceeds through a collinear inversion mechanism, which is reminiscent of an S_N2 reaction.

Several groups subsequently refined this experimental approach in a number of dynamics investigations and derived invaluable information from them. We just commented on the study of $Cl + CH_4$ in Zare's laboratory (127–132), in which the use of a supersonic beam, instead of a room-temperature cell, greatly reduced the spread in initial collision energies. The same approach led to the recent observation of transition state resonances in the H + D_2 reaction (59, 60). We also touched on the study of the H + O_2 reaction by Hall and coworkers (58), in which the change of the reaction mechanism from an adiabatic to a diabatic type due to the presence of a conical intersection was inferred. Simons, Brouard, and coworkers (99, 100, 140) exploited the stereodynamical aspects of several important bimolecular reactions involving OH products, including $O(^1D) + H_2$ and CH_4, as well as H + CO_2 and N_2O. The concept of a variable rotational clock rate in a complex-forming reaction was introduced to rationalize the results. A similar concept, which emphasizes different types of complexes formed from a distribution of impact parameters, was invoked in the crossed-beam studies of the $O(^1D)$ and $S(^1D) + H_2$ reactions (35, 39).

Despite the limited angular resolution of the bulb approach, the product state selection is high, and the dominant state-specific angular features should still be captured. In addition, the use of polarized lasers allows the study of vector correlation in chemical reactions (99, 100, 140–142), an area largely untouched by the crossed-beam technique and yet to be explored. We expect to see more bulb experiments conducted in this manner in the future.

CONCLUDING REMARKS

Reaction dynamics is now a relatively mature field. However, as evident from this review, it continues to offer exciting new perspectives on the fundamental nature of chemical transformation. The state-specific DCS measurement provides the most stringent test of fully ab initio theory, which builds our confidence in the ability of our theoretical treatments. It also often leads to a significant breakthrough in our conceptual understanding of reaction dynamics. In a sense, the close comparison between experiment and theory marks just the beginning of physical understanding. A few examples have been illustrated in this review. With ever-increasing computing capability and more reliable theoretical methodologies, one might well wonder if there is any need for such detailed experiments in the future, or can one simply do a computer experiment to find out what one wants to know without the actual laboratory measurement?

We believe that, at least in the near future, there is great value in additional scattering experiments and the related gas phase chemical dynamics studies in general. We certainly do not need to investigate every elementary reaction in such great detail. However, a number of significant barriers to our understanding still need to be overcome. Theoretically, we are just at the verge of being able to electronically calculate nonadiabatic effects "exactly," in terms of both PESs and quantum dynamics. We need to comprehend better the role of each of these effects, as well as derivative coupling, spin-orbit and Coriolis coupling effects, and their interplay in a chemical reaction. The GP effect in a chemical reaction is intriguing, yet the experimental confirmation is not yet clearly established.

Compared with a direct abstraction or exchange reaction, our mechanistic view of an insertion reaction is far less detailed, even for a three-atom system. We know some, but not much, about mode-, bond-, or site-selective chemistry. These areas will play a bigger role as we move toward larger reactive systems. At the same time, the role of intramolecular vibrational energy redistribution (146) in dictating the feasibility of bond-selective chemistry also needs to be addressed. The current intense activities in the study of four-atom reactions (147–152, 152a), such as OH + H_2, which is relevant to mode-specific reactivity, and OH + CO, which involves a deep well and exhibits unusual kinetic behavior, will prove particularly rewarding. We have seen some fascinating coherence and correlation experiments involving photodissociative events in recent years (153, 154). How will these phenomena manifest themselves in a full collision (155)? All of these questions and several more remain to be answered. Before that happens, the era of "computer experiment only" will remain a distant goal. The answers to these questions are also where the detailed experimental study of a-few-bodies (say, $\lesssim 6$ atoms) reaction systems can continue making the most contribution.

ACKNOWLEDGMENTS

Financial support for this research was provided by the National Science Council of Taiwan. The theoretical collaborations and stimulating discussions with George C Schatz, David E Manolopoulos, and Rex T Skodje are gratefully acknowledged. I am indebted to Yuan T Lee for his continuing encouragement, to Xueming Yang for providing Figure 6 of his unpublished work, and to RTS for his critical reading of this manuscript. The results from this laboratory are due to the hard work of current and former members, particularly Y-T Hsu, S-H Lee, and F Dong.

Visit the Annual Reviews home page at www.AnnualReviews.org

LITERATURE CITED

1. Levine RD, Beinstein RB. 1987. *Molecular Reaction Dynamics and Chemical Reactivity*. New York: Oxford Univ. Press. 535 pp.
2. Zewail AH. 2000. *J. Phys. Chem. A* 104:5660–94
3. Casavecchia P, Balucani N, Volpi GG. 1999. *Annu. Rev. Phys. Chem.* 50:347–76
4. Cassavecchia P. 2000. *Rep. Prog. Phys.* 63:355–414
5. Lee YT. 1987. *Science* 236:793–98
6. Neumark DM, Wodtke AM, Robinson GN, Hayden CC, Lee YT. 1985. *J. Chem. Phys.* 82:3045–66
7. Neumark DM, Wodtke AM, Robinson GN, Hayden CC, Shobatake K, et al. 1985. *J. Chem. Phys.* 82:3067–77
8. Rusin LY, Toennies JP. 2000. *Phys. Chem. Chem. Phys.* 2:501–5
9. Faubel M, Martinez-Haya B, Rusin LY, Tappe U, Toennies JP, et al. 1998. *J. Phys. Chem. A* 102:8695–707
10. Baer M, Faubel M, Martinez-Haya B, Rusin LY, Tappe U, Toennies JP. 1998. *J. Chem. Phys.* 108:9694–710
11. Baer M, Faubel M, Martinez-Haya B, Rusin L, Tappe U, Toennies JP. 1999. *J. Chem. Phys.* 110:10231–34

12. Blank DA, Hemmi N, Suits AG, Lee YT. 1998. *Chem. Phys.* 231:261–78
13. Ahmed M, Peterka DS, Suits AG. 2000. *Phys. Chem. Chem. Phys.* 2:861–68
14. Willis PA, Stauffer HU, Hinrichs RZ, Davis HF. 1999. *Rev. Sci. Instrum.* 70:2606–14
15. Dharmasena G, Copeland K, Young JH, Lasell RA, Phillips TR, et al. 1997. *J. Phys. Chem. A* 101:6429–40
16. Dharmasena G, Phillips TR, Shokhirev KN, Parker GA, Keil M. 1997. *J. Chem. Phys.* 106:9950–53
17. Mestdagh JM, Visticot JP, Suits AG. 1995. In *The Chemical Dynamics and Kinetics of Small Radicals*, part II, ed. K Liu, AF Wagner, pp. 668–729. Singapore: World Sci.
18. Heck AJR, Chandler DW. 1995. *Annu. Rev. Phys. Chem.* 46:335–72
19. Lorenz KT, Westley MS, Chandler DW. 2000. *Phys. Chem. Chem. Phys.* 2:481–94
20. Yonekura N, Gebauer C, Kohguchi H, Suzuki T. 1999. *Rev. Sci. Instrum.* 70:3265–70
21. Murphy EJ, Brophy JH, Kinsey JL. 1981. *J. Chem. Phys.* 74:331–36
22. Girard B, Gouedard G, Vigue J. 1991. *J. Chem. Phys.* 95:4056–69
23. L'Hermite JM, Rahmat G, Vetter R. 1990. *J. Chem. Phys.* 93:434–44
24. Che DC, Liu K. 1996. *Chem. Phys.* 207:367–78
25. Kitsopoulos TN, Buntine MA, Baldwin DP, Zare RN, Chandler DW. 1993. *Science* 260:1605–10
26. Eppink ATJB, Parker DH. 1997. *Rev. Sci. Instrum.* 68:3477–84
27. Schnieder L, Seekamp-Rahn K, Wrede E, Welge KH. 1997. *J. Chem. Phys.* 107:6175–95
28. Wrede E, Schnieder L, Welge KH, Aoiz FJ, Banares L, Herrero VJ. 1997. *Chem. Phys. Lett.* 265:129–36
29. Wrede E, Schnieder L. 1997. *J. Chem. Phys.* 107:786–90
30. Wrede E, Schnieder L, Welge KH, Aoiz FJ, Banares L, et al. 1997. *J. Chem. Phys.* 106:7862–64
31. Wrede E, Schnieder L, Welge KH, Aoiz FJ, Banares L, et al. 1999. *J. Chem. Phys.* 110:9971–81
32. Schnieder L, Seekamp-Rahn K, Borkowski J, Wrede E, Welge KH, et al. 1995. *Science* 269:207–10
33. Banares L, Aoiz FJ, Herrero VJ, D'Mello MJ, Niederjohann B, et al. 1998. *J. Chem. Phys.* 108:6160–69
34. Hsu YT, Liu K. 1997. *J. Chem. Phys.* 107:1664–67
35. Hsu YT, Liu K, Pederson LA, Schatz GC. 1999. *J. Chem. Phys.* 111:7921–30
36. Hsu YT, Liu K, Pederson LA, Schatz GC. 1999. *J. Chem. Phys.* 111:7931–44
37. Hermine P, Hsu YT, Liu K. 2000. *Phys. Chem. Chem. Phys.* 2:581–87
38. Wang JH, Liu K, Schatz GC, ter Horst M. 1997. *J. Chem. Phys.* 107:7869–75
39. Lee SH, Liu K. 1998. *J. Phys. Chem. A* 102:8637–40
40. Wang JH, Hsu YT, Liu K. 1997. *J. Phys. Chem. A* 101:6593–602
41. Siegbahn P, Liu B. 1978. *J. Chem. Phys.* 68:2457–65
42. Horowitz CJ, Truhlar DG. 1978. *J. Chem. Phys.* 68:2466–76. Erratum. 1979. *J. Chem. Phys.* 71:1514
43. Varandas AJC, Brown FB, Mead CA, Truhlar DG, Blais NC. 1987. *J. Chem. Phys.* 86:6258–69
44. Boothroyd AI, Keogh WJ, Martin PG, Peterson MR. 1996. *J. Chem. Phys.* 104:7139–52
45. Wu YS, Kuppermann A, Anderson JB. 1999. *Phys. Chem. Chem. Phys.* 1:929–37
46. Zwanziger JW, Koenig M, Pines A. 1990. *Annu. Rev. Phys. Chem.* 41:601–46
47. Mead CA. 1992. *Rev. Mod. Phys.* 64:51–85
48. Mead CA. 1980. *J. Chem. Phys.* 76:3839–40
49. Yarkony DR. 1998. *Acc. Chem. Res.* 31:511–18
50. Mead CA, Truhlar DG. 1979. *J. Chem. Phys.* 70:2284–96

51. Longuet-Higgins HC. 1975. *Proc. R. Soc. London Ser. A* 344:147–56
52. Kuppermann A, Wu YS. 1993. *Chem. Phys. Lett.* 205:577–87
53. Kliner DAV, Adelman DE, Zare RN. 1991. *J. Chem. Phys.* 95:1648–62
54. Kuppermann A, Wu YS. 1995. *Chem. Phys. Lett.* 241:229–40
55. Kuppermann A. 1996. In *Dynamics of Molecules and Chemical Reactions*, ed. RE Wyatt, JZH Zhang, pp. 411–72. New York: Dekker
56. Kendrick BK. 2000. *J. Chem. Phys.* 112:5679–704
56a. Pack RT, Parker GA. 1987. *J. Chem. Phys.* 87:3888–3921
57. Kebler K, Kleinermanns K. 1992. *J. Chem. Phys.* 97:374–77
58. Fei R, Zheng XS, Hall GE. 1997. *J. Phys. Chem. A* 101:2541–45
59. Kendrick BK, Jayasinghe L, Moser S, Auzinsh M, Shafer-Ray N. 2000. *Phys. Rev. Lett.* 84:4325–28
60. Fernandez-Aloso F, Bean BD, Ayers JD, Pomerantz AE, Zare RN, et al. 2000. *Angew. Chem. Int. Ed. Engl.* 39:2748–52
61. Wu SF, Johnson BR, Levine RD. 1973. *Mol. Phys.* 25:839–56
62. Schatz GC, Bowman JM, Kuppermann A. 1975. *J. Chem. Phys.* 63:674–84
63. Reid SA, Reisler H. 1996. *Annu. Rev. Phys. Chem.* 47:495–525
64. Bowman JM. 1998. *J. Phys. Chem. A* 102:3006–17
65. Green WH, Moore CB, Polik WF. 1992. *Annu. Rev. Phys. Chem.* 43:591–626
66. Chatfield DC, Friedman RS, Schwenke DW, Truhlar DG. 1992. *J. Phys. Chem.* 96:2414–21
67. Grayce BB, Skodje RT, Hutson JM. 1993. *J. Chem. Phys.* 98:3929–44
68. Weaver A, Neumark DM. 1991. *Faraday Discuss. Chem. Soc.* 91:5–16
69. Manolopoulos DE, Stark K, Werner HJ, Arnold DW, Bradforth SE, Neumark DM. 1993. *Science* 262:1852–55
70. Bradforth SE, Arnold DW, Neumark DM, Manolopoulos DE. 1993. *J. Chem. Phys.* 99:6345–58
71. Aoiz FJ, Banares L, Herrero VJ, Rabanos VS, Stark K, Werner HJ. 1994. *Chem. Phys. Lett.* 223:215–26
72. Castillo JF, Manolopoulos DE, Stark K, Werner HJ. 1996. *J. Chem. Phys.* 104:6531–46
73. Manolopoulos DE. 1997. *J. Chem. Soc. Faraday Trans.* 93:673–83
74. Skodje RT, Skouteris D, Manolopoulos DE, Lee SH, Dong F, Liu K. 2000. *J. Chem. Phys.* 112:4536–52
75. Skodje RT, Skouteris D, Manolopoulos DE, Lee SH, Dong F, Liu K. 2000. *Phys. Rev. Lett.* 85:1206–9
76. Waller IM, Kitsopoulos TN, Neumark DM. 1990. *J. Phys. Chem.* 94:2240–42
77. Aoiz FJ, Banares L, Herrero VJ, Rabanos VS, Stark K, et al. 1996. *Chem. Phys. Lett.* 262:175–82
78. Stark K, Werner HJ. 1996. *J. Chem. Phys.* 104:6515–30
79. Dong F, Lee SH, Liu K. 2000. *J. Chem. Phys.* 113:3633–40
80. Pollak E, Child MS. 1981. *Chem. Phys.* 60:23–32
81. Launay JM, LeDourneuf M. 1982. *J. Phys. B* 15:L455–61
82. Ewing GE. 1987. *J. Phys. Chem.* 91:4662–71
83. Chao SD, Skodje RT. 2000. *J. Chem. Phys.* 113:3487–91
84. Alexander MH, Werner HJ, Manolopoulos DE. 1998. *J. Chem. Phys.* 109:5710–13
85. Nizkorodov SA, Harper WW, Chapman WB, Blackmon BW, Nesbitt DJ. 1999. *J. Chem. Phys.* 111:8404–16
86. Chapman WB, Blackmon BW, Nizkorodov S, Nesbitt DJ. 1998. *J. Chem. Phys.* 109:9306–17
87. Donovan RJ, Husain D. 1970. *Chem. Rev.* 70:489–516
88. Dagdigian PJ, Campbell ML. 1987. *Chem. Rev.* 87:1–18

89. Lee SH, Liu K. 1999. *J. Chem. Phys.* 111:6253–59
90. Johnson HS. 1966. *Gas Phase Reaction Rate Theory*, Chap. 13. pp. 229–52. New York: Ronald
91. Kumaran SS, Lim KP, Michael JV. 1994. *J. Chem. Phys.* 101:9487–98
92. Alagia M, Balucani N, Cartechini L, Casavecchia P, van Kleef EH, et al. 1996. *Science* 273:1519–22
93. Deleted in proof
94. Brownsword RA, Kappel C, Schmiechen P, Upadhyaya HP, Volpp HR. 1998. *Chem. Phys. Lett.* 289:241–46
95. Tanaka K, Durup J, Kato T, Koyano I. 1981. *J. Chem. Phys.* 74:5561–71
96. Bian W, Werner HJ. 2000. *J. Chem. Phys.* 112:220–29
97. Manthe U, Bian W, Werner HJ. 1999. *Chem. Phys. Lett.* 313:647–54.
98. Skouteris D, Manolopoulos DE, Bian W, Werner HJ, Lai LH, Liu K. 1999. *Science* 286:1713–16
99. Simons JP. 1997. *J. Chem. Soc. Faraday Trans.* 93:4095–105
100. Alexander AJ, Brouard M, Kalogerakis KS, Simons JP. 1998. *Chem. Soc. Rev.* 27:405–15
101. Alagia M, Balucani N, Casavecchia P, Stranges D, Volpi GG. 1995. *J. Chem. Soc. Faraday Trans.* 91:575–96
102. Aoiz FJ, Banares L, Herrero VJ. 1998. *J. Chem. Soc. Faraday Trans.* 94:2483–500
103. Ahmed M, Peterka DS, Suits AG. 1999. *Chem. Phys. Lett.* 301:372–78
104. Casavecchia P, Balucani N, Alagia M, Cartechini L, Volpi GG. 1999. *Acc. Chem. Res.* 32:503–11
105. Aoiz FJ, Banares L, Herrero VJ. 1999. *Chem. Phys. Lett.* 310:277–86
106. Hsu YT, Wang JH, Liu K. 1997. *J. Chem. Phys.* 107:2351–56
107. Ho TS, Hollebeck T, Rabitz H, Harding LB, Schatz GC. 1996. *J. Chem. Phys.* 105:10472–86
108. Schatz GC, Papaioannou A, Pederson LA, Harding LB, Hollebeck T, et al. 1997. *J. Chem. Phys.* 107:2340–50
109. Dobbyn AJ, Knowles PJ. 1997. *Mol. Phys.* 91:1107–23
110. Schatz GC, Pederson LA, Kuntz PJ. 1997. *Faraday Discuss.* 108:357–74
111. Balint-Kurti GG, Gonzalez AI, Goldfield EM, Gray SK. 1998. *Faraday Discuss.* 110:169–83
112. Gray SK, Goldfield EM, Schatz GC, Balint-Kurti GG. 1999. *Phys. Chem. Chem. Phys.* 1:1141–48
113. Gray SK, Petrongolo C, Drukker K, Schatz GC. 1999. *J. Phys. Chem. A* 103:9448–59
114. Drukker K, Schatz GC. 1999. *J. Chem. Phys.* 111:2451–63
115. Liu X, Lin JJ, Harich SA, Yang X. 2000. *J. Chem. Phys.* 113:1325–28
116. Hsiao MC, Sinha A, Crim FF. 1991. *J. Phys. Chem.* 95:8263–67
117. Sinha A, Hsiao MC, Crim FF. 1991. *J. Chem. Phys.* 94:4928–35
118. Crim FF. 1999. *Acc. Chem. Res.* 32:877–84
119. Sinha A, Thoemke JD, Crim FF. 1992. *J. Chem. Phys.* 96:372–76
120. Thoemke JD, Pfeiffer JM, Metz RB, Crim FF. 1995. *J. Phys. Chem.* 99:13748–54
121. Bronikowski MJ, Simpson WR, Girard B, Zare RN. 1991. *J. Chem. Phys.* 95:8647–48
122. Bronikowski MJ, Simpson WR, Zare RN. 1993. *J. Phys. Chem.* 97:2194–203
123. Pfeiffer JM, Metz RB, Thoemke JD, Woods E III, Crim FF. 1996. *J. Chem. Phys.* 104:4490–501
124. Metz RB, Thoemke JD, Pfeiffer JM, Crim FF. 1994. *Chem. Phys. Lett.* 221:347–52
125. Kreher C, Rinnenthal JL, Gericke KH. 1998. *J. Chem. Phys.* 108:3154–67
126. Kreher C, Theinl R, Gericke KH. 1996. *J. Chem. Phys.* 104:4481–89
127. Kandel SA, Zare RN. 1998. *J. Chem. Phys.* 109:9719–27
128. Orr-Ewing AJ, Simpson WR, Rakitzis TP,

Kandel SA, Zare RN. 1997. *J. Chem. Phys.* 106:5961–71
129. Simpson WR, Rakitzis TP, Kandel SA, Lev-On T, Zare RN. 1996. *J. Phys. Chem. A* 100:7938–47
130. Simpson WR, Rakitzis TP, Kandel SA, Orr-Ewing AJ, Zare RN. 1995. *J. Chem. Phys.* 103:7313–35
131. Simpson WR, Orr-Ewing AJ, Rakitzis TP, Kandel SA, Zare RN. 1995. *J. Chem. Phys.* 103:7299–312
132. Simpson WR, Orr-Ewing AJ, Zare RN. 1993. *Chem. Phys. Lett.* 212:163–71
133. Kandel SA, Rakitzis TP, Lev-On T, Zare RN. 1998. *J. Phys. Chem. A* 102:2270–73
134. Rakitzis TP, Kandel SA, Lev-On T, Zare RN. 1997. *J. Chem. Phys.* 107:9392–405
135. Kandel SA, Rakitzis TP, Lev-On T, Zare RN. 1997. *Chem. Phys. Lett.* 265:121–28
136. Kandel SA, Rakitzis TP, Lev-On T, Zare RN. 1996. *J. Chem. Phys.* 105:7550–59
137. Varley DF, Dagdigian PJ. 1996. *J. Phys. Chem.* 100:4365–74
138. Varley DF, Dagdigian PJ. 1996. *Chem. Phys. Lett.* 255:393–400
139. Varley DF, Dagdigian PJ. 1995. *J. Phys. Chem.* 99:9843–53
140. Brouard M, Simons JP. 1995. In *The Chemical Dynamics and Kinetics of Small Radicals*, part II, ed. K Liu, AF Wagner, pp. 795–841. Singapore: World Sci.
141. Orr-Ewing AJ, Zare RN. 1994. *Annu. Rev. Phys. Chem.* 45:315–66
142. Orr-Ewing AJ, Zare RN. 1995. In *The Chemical Dynamics and Kinetics of Small Radicals*, part II, ed. K Liu, AF Wagner, pp. 936–1063. Singapore: World Sci.
143. Wang X, Ben-Nun M, Levine RD. 1995. *Chem. Phys.* 197:1–17
144. Nyman G, Clary DC, Levine RD. 1995. *Chem. Phys.* 191:223–33
145. Park J, Satyapal S, Tasaki S, Chattopadhyay A, Yi W, Bersohn R. 1991. *Faraday Discuss. Chem. Soc.* 91:73–78
146. Nesbitt DJ, Field RW. 1996. *J. Phys. Chem.* 100:12735–56
147. Clary DC. 1994. *J. Phys. Chem.* 98:10678–88
148. Bowman JM, Schatz GC. 1995. *Annu. Rev. Phys. Chem.* 46:169–95
149. Loomis RA, Lester MI. 1997. *Annu. Rev. Phys. Chem.* 48:637–67
150. Chen Y, Heaven MC. 1998. *J. Chem. Phys.* 109:5171–74
151. Clary DC. 1998. *Science* 279:1879–82
152. Strazisar BR, Lin C, Davis HF. 2000. *Science* 290:958–61
152a. Zhang DH, Collins MA, Lee S-Y. 2000. *Science* 290:961–63
153. Hall GE, Houston PL. 1989. *Annu. Rev. Phys. Chem.* 40:375–405
154. Alexander AJ, Zare RN. 2000. *Acc. Chem. Res.* 33:199–205
155. Grosser J, Hoffmann O, Rakete C. 1997. *J. Phys. Chem. A* 101:7627–33

COINCIDENCE SPECTROSCOPY

Robert E Continetti
Department of Chemistry and Biochemistry, University of California at San Diego, La Jolla, California 92093-0314; e-mail: rcontinetti@ucsd.edu

Key Words reaction dynamics, dissociative photodetachment, dissociative photoionization, molecular-frame photoelectron angular distributions, three-body dissociation

■ **Abstract** The application of coincidence techniques to the study of the reaction dynamics of isolated molecules is reviewed. Coincidence spectroscopy is a powerful approach for carrying out a number of measurements. At its most basic level, coincidence techniques can identify the source of a specific signal, as in the well-known photoelectron-photoion coincidence approach used for several years. By carrying out coincidence experiments in an increasingly differential manner, correlated energy and angular distributions of reaction products may be recorded. Completely energy- and angle-resolved measurements of photoelectrons and ionic or neutral products can reveal molecular-frame photoelectron and photofragment angular distributions and aid in the characterization of dissociative states of molecules and ions. Recent work in this area is reviewed, including examples from studies of dissociative photodetachment, dissociative photoionization, time-resolved studies of dissociative photoionization, and three-body dissociation processes.

INTRODUCTION

The reaction dynamics of transient molecules and clusters is an area of gas-phase physical chemistry with significant fundamental questions that remain to be answered. The dissociative and metastable electronic states involved in photoexcitation processes, as well as the transition state region of the potential energy surface that governs the outcome of bimolecular reactions, have not been characterized as well as the bound states of molecular systems because of a lack of appropriate experimental tools. Studies of the dynamics of unimolecular dissociations and bimolecular reactions are one approach to probing these states. Over the last 30 years, photodissociation and bimolecular reactions have been examined in increasing detail using laser and molecular beam techniques (1, 2). However, the need for further experimental information has continued to drive the development of new experimental techniques for the study of transient species.

Even though the theoretical machinery of quantum chemistry is well developed, there are still important limitations in our ability to calculate the properties of electronically excited and dissociative states of stable molecules and the ground-state properties of transient species such as free radicals, anions, and weakly bound clusters (3, 4). Quantum mechanical methods in reaction dynamics are even more limited; accurate quantum mechanical calculations without dynamical approximations remain limited to systems of four or fewer atoms (5). To continue to drive these fields forward, new and refined experimental observables on dissociative electronic states, ionization processes, and the reaction dynamics of small molecules and clusters are required. The development of coincidence spectroscopies has arisen in part because of this need.

Coincidence spectroscopies can provide important information, unavailable from conventional spectroscopic techniques, regarding the identity and energetics of reactive species, the correlation of product states, the nature of repulsive electronic states, and insights into product angular distributions in the molecular frame. Coincidence techniques are also providing important insights into three-body dissociation processes, an area in which little information beyond rate coefficients has been available in the past. This review focuses on some of the recent applications of coincidence spectroscopies to the energetics and reaction dynamics of isolated molecules and clusters.

Coincidence techniques were originally developed to aid in the identification of the source of specific signals, as in the well-known photoelectron-photoion coincidence (PEPICO) approach used for several years (6). PEPICO experiments have been continually improved in many ways, ranging from the measurement of threshold photoelectrons (7) to recent fully energy- and angle-resolved experiments. This review focuses on the latter experiments in which individual correlated events are accumulated one at a time with subsequent analysis of the correlations contained therein. Thus, this review does not touch on the important contributions made by imaging techniques, wherein correlated properties of reaction products can be measured by spectroscopic state selection (e.g. multiphoton ionization) followed by measurement of product translational energy and angular distributions using charge-coupled-device (CCD)-based detection schemes. These applications were recently reviewed (8–11).

Energetic Correlations

Two types of observables from coincidence experiments are highlighted here: (*a*) the use of energetic correlations to characterize dissociative molecular electronic states and the transition state region of bimolecular reactions and (*b*) the use of angular correlations between photoelectrons and photofragments—and also among multiple photofragments—to characterize the ionization and dissociation dynamics. The energetic correlations involve examination of the photoelectron kinetic energy distribution and how it correlates with different product channels in a molecular dissociation. For either photoionization of neutral molecules or photodetachment of negative ions, the interpretation of these energetic correlations is very similar: The photoelectron kinetic energy distribution contains information about the Franck-Condon overlap between the initial bound state and the metastable or repulsive state that is reached by photoexcitation. The photofragment translational energy distribution, on the other hand, contains information about how the system prepared in the Franck-Condon region couples to the dissociation continuum. In the simplest case of no energy transfer among the products during the dissociation, then, the product translational energy distribution reveals the energetic repulsion between the products relative to the dissociation asymptote (12). Examples of the application of these energetic correlations to understanding dissociative photodetachment (DPD) and dissociative photoionization (DPI) processes and the potential energy surface near the transition state for bimolecular reactions are discussed below.

Molecular-Frame Photoelectron Angular Distributions

An important quantity that can be measured in a coincidence experiment is the molecular-frame photoelectron angular distribution (MF-PAD). It is well known that the laboratory photoelectron angular distribution (PAD) recorded in photoionization or photodetachment of a randomly oriented sample with linearly polarized light is as follows (13):

$$\frac{\partial \sigma}{\partial \Omega_{LAB}} = \frac{\sigma_{total}}{4\pi} [1 + \beta(E) P_2(\cos\theta)], \qquad 1.$$

where σ_{total} is the total photodetachment cross section, θ is the polar angle between the recoil direction and the electric vector of the laser, and $P_2(\cos\theta)$ is the second-order Legendre polynomial in $\cos\theta$. The energy-dependent asymmetry parameter $\beta(E)$ is a sensitive function of both the symmetry of the orbital from which the electron is removed and the photodetachment dynamics. However, owing to the averaging over molecular orientations in the measurement of a laboratory angular distribution, detailed information on the partial-wave composition of the ionization continuum is lost. Measurement of MF-PADs, however, provides considerably more information. As Dill (14) showed, the MF-PAD is given by

$$\frac{\partial \sigma(E)}{\partial \Omega_{MF}} = \sum_{l=0}^{2l^*} \sum_m A_{lm}(E) Y_{lm}(\vartheta, \varphi), \qquad 2.$$

where l^* is the maximum value for the orbital angular momentum of the photodetached electron and m corresponds to the azimuthal quantum number for the photodetached electron. The magnitudes of the energy-dependent coefficients $A_{lm}(E)$ are determined by electric-dipole selection rules and interference between the degenerate photoelectron continuum channels. These coefficients thus contain detailed information on both the photodetachment dynamics and the orbital from which the electron is ejected, determining the relative contributions of the spherical harmonics Y_{lm}, which are referenced to the molecular frame by the polar and azimuthal angles ϑ and φ.

A number of studies of MF-PADs have focused on the use of aligned molecular ensembles produced by photoexcitation (15). However, DPI or DPD processes provide an alternative approach to studying fixed-in-space molecules. When dissociation is rapid relative to molecular rotation (axial recoil) (16), measurement of the recoil angle of the photoelectron relative to the axis of the broken bond in the molecule provides a direct measure of the MF-PAD in a coincidence experiment. Experiments on both DPD and DPI that are sensitive to the MF-PAD are reviewed below.

Photoelectron spectroscopy has also been applied to time-resolved studies of reaction dynamics (17, 18). Because photoelectron spectra are sensitive to changes in the electronic structure as a chemical reaction, isomerization, or nonradiative transition proceeds, femtosecond time-resolved photoelectron kinetic energy spectra have revealed important insights into nonadiabatic processes (19–21) and solvation phenomena (22). As mentioned in the last paragraph, although, PADs and in particular MF-PADs are the most sensitive probes of the electronic structure and ionization dynamics, so there has been an increasing interest in time-resolved studies of these observables. Theoretical efforts include the development of methods for the prediction and analysis of time-resolved PADs and MF-PADs in molecular photoionization (23–28). Experimentally, time-resolved laboratory PADs have been measured (20, 29), and recently studies of MF-PADs have been extended into the femtosecond time domain in the fully energy- and angle-resolved PEPICO experiments on the DPI of NO_2 by Davies et al (30, 31), as reviewed below.

Three-Body Dissociation Dynamics

Coincidence experiments are also well suited to the study of three-body dissociation dynamics. Important insights into both concerted and sequential dissociation reactions have been gained from noncoincident studies of three-body photodissociation processes (32, 33). However, general techniques to perform the coincidence measurements required to directly measure product angular correlations in neutral molecules were not available until the recent DPD experiments in our group (34) and charge-exchange/laser-excitation experiments by Helm and coworkers (35). These experiments join earlier studies of the three-body dissociation dynamics of multiply charged ions (36–38) in revealing three-dimensional product momenta. Such measurements allow the investigation of not only the partitioning of energy in three-body dissociation processes, but also a second important molecular-frame quantity: the angular correlation of the momenta of the three atomic or molecular fragments. We refer to this quantity as the molecular-frame differential cross section (MF-DCS), which gives immediate insights into the dynamics, including the dissociation lifetime and molecular geometry at the instant of dissociation. In particular, if the three-body dissociation is rapid (the axial recoil approximation mentioned above, although the choice of axis in the three-body dissociation is system-dependent), this is a direct molecular-frame measurement, sensitive to both the molecular structure and dissociation dynamics of the system in question.

Recent examples of coincidence studies of DPD, DPI, and three-body dissociation dynamics are discussed below. First, a brief review of experimental techniques is presented.

EXPERIMENTAL TECHNIQUES

Recent progress in the area of coincidence spectroscopy has been enabled by developments in both detection technology and the increased availability of new light sources, including synchrotrons and laboratory laser systems with kilohertz repetition rates. The experiments discussed in this review are all coincidence measurements of the three-dimensional velocity distributions of the atomic, molecular, or photoelectron products of a dissociation event. To carry out a coincidence measurement successfully, several factors must be considered. The product collection efficiency must be sufficiently high and the sources of background sufficiently small that a valid correlation between observed events can be made. In addition, these experiments must be carried out under conditions of relatively low signal rates and high duty cycles to overcome the problems associated with finite detection efficiencies (12, 39). In this review, two applications of coincidence spectroscopy are discussed: photoionization of neutral molecules in a molecular beam and fast ion-beam experiments involving photodetachment of negative ions or charge-exchange neutralization of cations.

The first priority in setting up a coincidence experiment is to ensure that the products can be detected with a high efficiency. If the products are charged, as in a photoionization process that will produce a free electron and a photoion, extraction with an appropriate field allows collection of all the products. In a DPI process, for example, producing a free electron, a photoion, and a photoneutral, measurement of the time- and position-of-arrival of the charged products (electron and ion) at a detector provides the information required for a complete kinematic description of the dynamics. This is the approach taken in the energy- and angle-resolved PEPICO experiments reviewed here. Takahashi and coworkers (40), for example, recently described an imaging PEPICO spectrometer that uses opposed time- and position-sensitive detectors to record the two-dimensional projection of the three-dimensional velocity distribution. With short-pulse laser or synchrotron radiation and time- and position-sensitive detectors with 0.1-ns timing resolution, a similar opposed detector geometry has been used to directly record the three-dimensional velocity distributions of both photoelectrons and photoions in coincidence. This technique has been used in DPI studies (41–43), including the recent time-resolved experiments of Hayden and coworkers (30).

If the products are neutral, the detection problem is more challenging. The simplest approach to carrying out coincidence studies involving neutral products is the use of fast (keV) ion beams followed by photodetachment (in the case of anions), charge exchange neutralization (in the case of cations), or collision-induced dissociation. Fast ion beams have the important property of allowing efficient (~50%) detection of neutral products with laboratory energies in excess of 1 keV upon impact with a microchannel-plate-based detector (44). In addition, in a fast beam, the products are constrained by momentum conservation to a limited angular range in the laboratory frame because of the high beam velocity. This situation allows the full solid angle in the center-of-mass (CM) frame to be covered with a single detector, similar to the techniques for charged particles discussed above.

The fast-beam method coupled with time- and position-sensitive particle detection was pioneered by DeBruijn & Los (45) in studies of the dissociative charge exchange of diatomic molecules. Measurement of the time and position of particle arrival, coupled with the knowledge of the parent mass and velocity, allows determination of the product mass ratio, scattering angles, and CM translational energy release (E_T) in the dissociation. In contrast to studies of ions, where the product mass can be determined with high accuracy by time-of-flight mass spectrometry, in a fast-beam experiment, the product mass ratio is determined by momentum conservation in the CM frame. The finite size of the parent beam typically limits the mass resolution to $m/\Delta m \approx 15$ (12). The low mass resolution is a challenge when the branching ratio between several open channels must be determined in the dissociation of a polyatomic molecule. In three-body dissociation experiments where all the products are detected, the time and positions of impact give the product velocities and recoil angles in the CM frame. Identification of the products requires some chemical intuition concerning the possible dissociation channels. Then, given the conservation of mass and momentum in the CM frame, the mass of each of the products can be determined (34). The error in determining product masses in three-body dissociation is more significant, with $m/\Delta m \approx 2.5$.

Fast-beam coincidence spectroscopies have several applications, including the dissociative charge exchange of diatomic (46) and polyatomic molecules (47, 48), photodissociation processes of cations (49), and neutral species generated by photodetachment (50, 51). It is beyond the scope of this review to cover these two-body dissociation processes in detail. The restriction here will be to processes that produce three or more products, including DPD: $ABC^- + h\nu \rightarrow A + BC + e^-$ and higher-order processes.

Coincidence studies of DPD require detection of the energy and angle of recoil of each photoelectron produced. In a fast ion beam, detection has occurred in two ways. The first approach involves large, solid-angle, short-flight-path time-of-flight photoelectron detectors with detection efficiencies from 2 to 10% of all photoelectrons (34, 52). These spectrometers require subnanosecond lasers since the photoelectron flight time for electron-volt-range photoelectrons may be only 50–100 ns. The time and position of electron impact is used to determine the actual flight distance for the photoelectron, angle of recoil, and laboratory energy. By knowing the velocity of the negative ion beam, the laboratory energy of the photoelectron can be corrected to give the CM photoelectron kinetic energy, eKE. With this approach, an energy resolution of $\Delta E_{FWHM}/E \sim 5\%$ (where FWHM is full width at half maximum) can be achieved routinely.

A second approach is to extract the photoelectrons with a small electric field in a space-focusing time-of-flight configuration, as was previously applied to photoionization processes (30). In this configuration, timing resolution becomes critical, as the range of photoelectron flight times for forward and backward scattered electrons (relative to the detector) may be <10 ns. In fast-beam experiments in our laboratory, we have achieved an energy resolution of $\Delta E_{FWHM}/E \sim 12\%$ with this approach. We recently reported on photodetachment imaging in a fast-negative-ion beam study of the electron affinity of CF_3 (53).

Studies of three-body dissociation processes producing neutral products in fast beams face the special challenge that all three products strike the particle detector typically within a few hundred nanoseconds. Thus, the detection technique must be carefully chosen. In fast-beam processes producing two fragments, for example, two individual detectors can be used, such as the coincidence wedge-and-strip anode configuration used by Continetti et al (50), which uses charge division on a patterned anode (54), an approach that is limited for multiparticle applications by the several microseconds of dead time per particle. In 1992, Brenot & Durup-Ferguson (55) discussed the various experimental techniques available for multiparticle time- and position-sensitive detection, and several technological approaches to this problem have resulted from studies of molecular structure by Coulomb explosion techniques (56, 57).

There is an increasing use of hybrid techniques involving CCD-based detectors for imaging, which are coupled with less dense discrete-anode photomultiplier tubes for acquiring timing information. This type of detector has been used by Amitay et al (58) in dissociative recombination studies and will likely see further application in coincidence experiments, in particular as the readout rate for CCD cameras continues to increase. In addition, several groups have begun to use crossed-delay line anodes for recording time- and position-of-impact data (34, 59, 60). Delay line methods have an inherent advantage over charge division techniques because they use fast timing signals to determine the position of impact. With fast timing signals, fast logic and switching techniques can be used to set up multihit encoding electronics. This approach has been taken in developing the quadrant crossed-delay-line anode in our laboratory for the study of three-body dynamics (34). This device can record the time and position of arrival of up to eight particles arriving within 10 ns if they are spread out over the detector correctly. This redundancy in detection provides a nearly ideal three-body detector, although some kinematic configurations remain difficult to study in a single experiment.

COINCIDENCE STUDIES OF DISSOCIATIVE PHOTODETACHMENT

Energy and Angular Correlations: O_4^-

Coincidence studies of DPD processes provide a novel way to probe both photodetachment dynamics and the properties of the dissociative states of transient neutral molecules. As discussed above, these experiments are based on the production of mass-selected negative-ion beams, photodetachment, and recording the kinetic energies and recoil angles of the photoelectron and any photofragments produced after dissociation of the nascent neutral species.

Coincidence studies of DPD dynamics build on the important studies of photodetachment of stable negative ions that correspond to unstable neutral species (61–65). The photoelectron spectra in these cases can be analyzed to gain insights into a neutral potential energy surface that is unstable with respect to isomerization or dissociation. Analysis of these spectra constitutes a spectroscopy of the transition state. The photoelectron spectra provide information on the Franck-Condon overlap between the bound anion and the dissociative or unstable region of the neutral potential energy surface. When the nascent neutral species undergoes dissociation, insights into the potential energy surface for bimolecular reactions can be gained, as shown for example by the studies of Neumark and coworkers (62) on the hydrogen exchange reactions X + HY, where X and Y are F, Cl, Br, or I (62), and by studies of the fundamental F + H_2 reaction (63). This approach has also been used by Lineberger and coworkers (61) to study isomerization processes such as conformational changes in cyclooctatetraene. These photodetachment studies have stimulated theoretical studies interpreting the observed photoelectron spectra in terms of the potential energy surfaces and the scattering dynamics for these fundamental reactions (66, 67).

For DPD, however, a more complete coincidence measurement is required to fully characterize both the Franck-Condon region (from the photoelectron spectrum) and the ensuing dissociation dynamics (from the product translational energy distribution). In the photoelectron-photofragment coincidence (PPC) experiments developed in our laboratory, complete kinematic characterization of two- and three-body DPD processes is achieved by detection of the photoelectron in coincidence with the atomic or molecular products (34, 68, 69). Determination of the energies and recoil angles of all products allows determination of the correlated photoelectron-photofragment kinetic energy release and MF-PADs for photodetachment, as shown in studies of O_4^- (70). These experiments have recently been extended to three-body dissociation dynamics, as discussed below. Here, the DPD dynamics of O_4^- and our recent study of the transition-state dynamics of the OH + $H_2O \rightarrow H_2O$ + OH reaction (71) are reviewed.

As an example of the application of PPC spectroscopy to the study of DPD, consider the case of O_4^-:

$$O_4^- + h\nu \rightarrow O_2\left(^3\Sigma_g^-\right) + O_2\left(^3\Sigma_g^-\right) + e^-. \qquad 3.$$

O_4^- is essentially an oxygen dimer bound together by a delocalized electron, with a dissociation energy of 0.46 eV (72). Photodetachment of the anion produces two O_2 molecules in close proximity, with approximately 0.4 eV of repulsive energy. Dissociation occurs on the time scale of molecular vibration. This is an example of a direct DPD process. The experimental signature of these dynamics can be seen most clearly in the work we carried out on this system at 532 nm (2.33 eV) (69, 70). The energetic correlations in a system like this can be displayed in the form of a two-dimensional photoelectron-photofragment correlation spectrum, $N(E_T, eKE)$. This spectrum shows the number of events with specific E_T and eKE as a contour map,

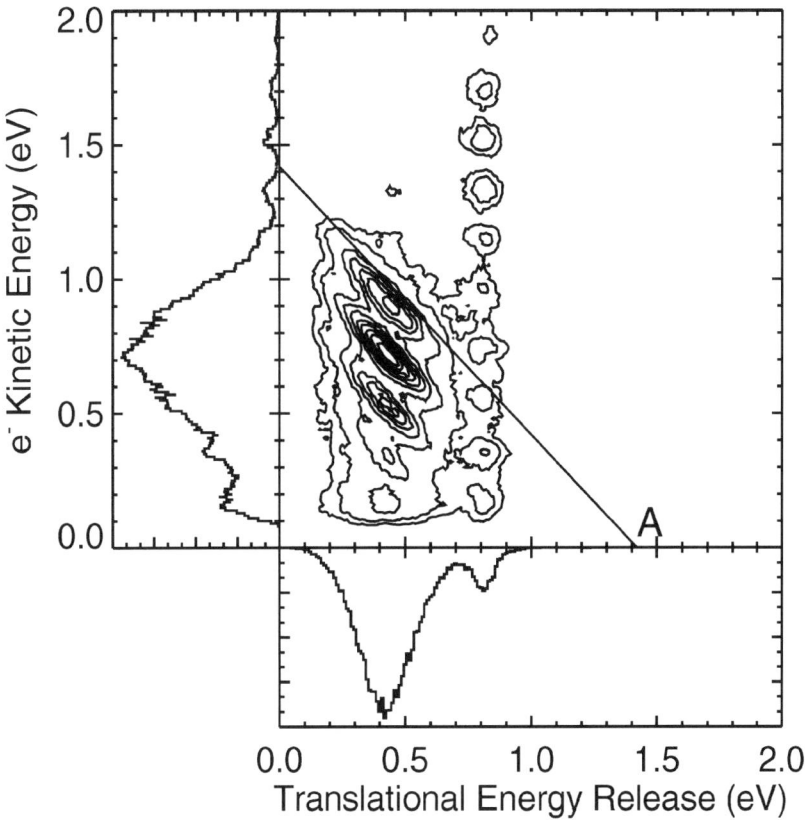

Figure 1 $N(E_T, eKE)$ correlation spectrum for O_4^- recorded at 532 nm shown as a contour plot of the energy partitioning in dissociative photodetachment. The one-dimensional $N(E_T)$ and $N(eKE)$ spectra are shown along the x and y axes, respectively. The maximum kinetic energy for photodetachment to $O_2(X, v = 0) + O_2(X, v = 0) + e^-$ is shown as limit A. The diagonal features correspond to DPD to different correlated O_2 product vibrational states, for example, $(v = 0, v = 0)$, $(v = 0, v = 1)$, etc (adapted from Reference 70 with permission from Elsevier Science).

as shown in Figure 1. The photoelectron and photofragment translational energy spectra obtainable in noncoincidence experiments are shown alongside the appropriate axes in this figure and represent integration of the coincidence spectrum over the complementary variable. The photoelectron spectrum, dominated by a broad peak at 0.7 eV, reproduces that originally obtained by Johnson and coworkers (73). Fine structure observed at higher kinetic energies is a result of the photodissociation of O_4^-: $O_4^- + h\nu \rightarrow O_2(^1\Delta_g) + O_2^-(^2\Pi_g)$, followed by photodetachment of the O_2^- product by a second photon. The translational energy spectrum is dominated by a broad peak at 0.4 eV, with a smaller narrow peak at 0.8 eV.

The correlation observed between these one-dimensional distributions is striking: A series of diagonal lines is observed despite the lack of structure in either the photoelectron or translational-energy spectra. This is an example of a system exhibiting vibrationally adiabatic dissociation dynamics on a repulsive surface under conditions in which little rotational excitation is possible in the dissociation. Because of the steepness of the repulsive neutral surface, photodetachment to the various vibrationally adiabatic final states overlaps considerably in the photoelectron spectrum. Nonetheless, due to constrained exit channel dynamics, with no exchange of vibrational energy and limited rotational excitation, clearly resolved diagonal lines are observed in the correlation spectrum. Another way to view this is in terms of a spectrum of the total translational energy release, $N(E_{TOT})$, where $E_{TOT} = eKE + E_T$ summed for each event. This spectrum is shown in Figure 2, which reveals a well-resolved spectrum with peaks spaced by O_2 product vibrations (70). The spacings of the first five vibrational levels of O_2 are shown as combs on the spectrum for reference. Structural information on gas-phase O_4^- can be obtained from this spectrum.

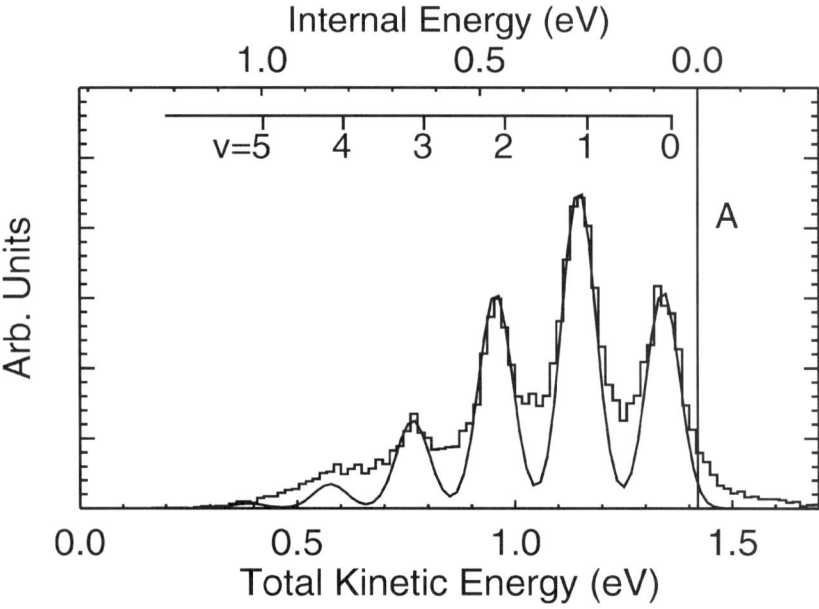

Figure 2 $N(E_{TOT})$ spectrum for the dissociative photodetachment of O_4^- at 532 nm. The energetic limit is the same as that shown in Figure 1. The top axis indicates the internal energy, and the bottom axis shows the total translational energy ($eKE + E_T$). The data are represented by points, and the solid-line fit to the data was generated by a Franck-Condon simulation assuming two O_2 subunits with equal O–O bond lengths of 1.272 Å (adapted from Reference 69 with permission from the American Physical Society).

A simple simulation was performed by calculating the Franck-Condon overlap of the vibrational wave functions of two perturbed O_2 molecules with two free ground-state O_2 molecules. The results of this calculation are plotted as a solid line in Figure 2, and it was found that assuming equal bond lengths of 1.272 Å gave excellent agreement with the data. This result indicates that the excess electron is delocalized over the two O_2 moieties in a symmetric structure, consistent with recent ab initio calculations by Aquino et al (74).

The dynamics observed in the DPD of O_4^- at 532 nm provide an example of what will occur in a direct DPD with a constrained product state distribution. Photodetachment occurs to a repulsive state of the neutral surface, and the time scale of the dissociation of the neutral fragments is a function of the steepness of the repulsive surface accessed by photodetachment. In Reference 70, application of a bound-free Franck–Condon formalism to analysis of the repulsive states of O_4 was made.

Other limiting cases of DPD dynamics as observed by PPC spectroscopy can be established by considering the time scale for photodetachment and the subsequent dissociation of the neutral complex. A second limiting case is sequential DPD, where the electron departs and then later the neutral system dissociates, either by predissociation or unimolecular decomposition on the electronic ground state. In this case, resolved vibrational structure may be observed in the photoelectron spectrum, and the dissociation dynamics of specific vibronic states of the neutral complex may be determined in the coincidence experiment. The best example of this type of process, with sharp structure in the photoelectron spectrum and no diagonal structure observed in the correlation spectrum, is the DPD of O_3^- (75). In this case, predissociation of the initial relatively long-lived vibronic states produced by photodetachment gives rise to horizontal structure along the E_T axis in the correlation spectrum. As these techniques are extended to polyatomic systems, there is no guarantee that these limiting dynamics can be identified, owing to the larger number of internal degrees of freedom. An example of this situation is the DPD process $N_3O_2^- + h\nu \rightarrow NO + N_2O + e^-$ (76). The correlation spectrum for this system shows no structure other than that dictated by energy conservation. This system may not behave vibrationally adiabatically in DPD, or rotational excitation in the exit channel, and the limited energy resolution of the experiment may prevent observation of structure in the correlation spectrum.

Finally, in PPC experiments, the angular correlations between the photoelectrons and photofragments can also be recorded. For a direct DPD, such as O_4^-, the direction of recoil of the O_2 products provides a record of the distribution of O_4^- anions in the laboratory that absorbed photons. Because the O_2 photoproducts were found to have a highly anisotropic angular distribution, and the photoelectrons from photodetachment of O_4^- are also anisotropically distributed in space (77), the angular correlations between the photoelectrons and photofragments were examined for evidence of an anisotropic MF-PAD. In these experiments, the photoelectron detector subtended only 4% of the total solid angle, so the presence of an anisotropic MF-PAD can be identified by examining events in which an electron and two O_2 products are detected and comparing the observed O_2 photofragment angular distribution with that found from events with only two O_2 products detected. Detection of the photoelectron in coincidence, in a direction nearly orthogonal to the electric vector of the laser, breaks the cylindrical symmetry around the electric vector of the laser in the laboratory frame. This observation was reported in References 69 and 70. In these experiments, the O_2 photofragment angular distribution is nearly $\sin^2 \theta$ at 532 nm. Examining only those events coincident with detection of a photoelectron showed a pronounced reduction of $O_2 + O_2$ photofragment recoils detected parallel to the direction of photoelectron detection. The existence of this angular correlation shows that photodetachment and dissociation of the nuclear framework of O_4 both occur on a time scale that is fast relative to molecular rotation (~1–10 ps) and that the photoelectron preferentially recoils perpendicularly to the dissociation coordinate in O_4.

The evidence for a strong anisotropy of the MF-PAD implies that a limited number of electron partial waves contribute in the detachment process. Electronic structure calculations on O_4^- by Aquino et al (74) have shown that O_4^- has D_{2h} symmetry with a ^2A_u electronic ground state. Application of symmetry considerations, as discussed by Brauman and coworkers (78), indicates that d-wave photodetachment is expected near the threshold for this system (70). This conclusion also can be reached by qualitative consideration of the a_u highest occupied molecular orbital (HOMO) for O_4^-. This HOMO has three angular nodes and is thus analogous to an atomic f orbital, resulting in both d-wave ($l = 2$) and g-wave ($l = 4$) photodetachment following the atomic selection rules of $\Delta l = \pm 1$ (79). Above threshold, however, the higher-angular-momentum states of the photoelectron continuum are accessible and will lead to an energy-dependent PAD, which has been observed in experiments at 355 and 266 nm (70). Now that detailed electronic structure calculations on O_4^- have become available (74), a more detailed interpretation of the photodetachment dynamics in this interesting anion, using electron-molecule scattering calculations, should be possible.

Transition-State Dynamics: $OH + H_2O \rightarrow H_2O + OH$

The thermoneutral identity reaction $OH + H_2O \rightarrow H_2O + OH$ is one of the simplest hydrogen abstraction reactions involving the hydroxyl radical and provides an excellent case for studying transition-state dynamics by DPD. The groundwork for this study was laid by the photoelectron studies of Arnold et al (65) of the $H_3O_2^-$ anion. They observed a broad photoelectron spectrum with at least four identifiable peaks at 266 nm (4.66 eV). The spectrum was observed to undergo significant changes, in both intensities and positions, upon deuteration, implying that the spectral features were primarily related to motion of the hydrogen atoms in the complex. They also carried out ab initio calculations and one-dimensional Franck-Condon simulations of the spectrum, confirming the role played by motion of the transferred H atom in the neutral $OH(H_2O)$ complex produced by photodetachment.

The transition-state dynamics of this reaction have been studied using DPD of $OH^-(H_2O)$ and $OD^-(D_2O)$ at 258 nm (4.80 eV) using PPC spectroscopy (71).

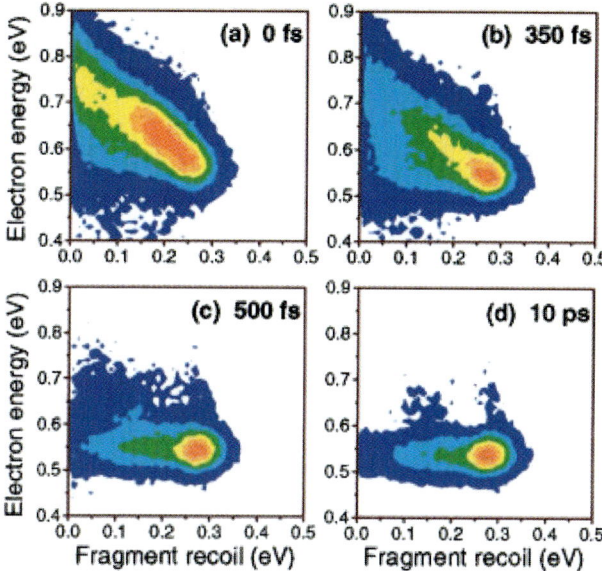

Figure 7 Time-resolved photoelectron-photoion kinetic energy correlation spectra for the DMI of NO_2^- at 375.3 nm. The number of events are plotted on a linear false-color scale with red representing the peak of the distribution. Reproduced from [Davies, 1999 (30)] with permission from the American Institute of Physics.

DPD is the only process that occurs at this photon energy:

$$OH^-(H_2O) + h\nu \rightarrow OH + H_2O + e^-. \qquad 4.$$

The quantum yield for dissociation of the neutral complex into two fragments is unity. No stable $OH(H_2O)$ clusters are produced because the geometry of the anion leads to no Franck-Condon overlap with the $OH(H_2O)$ van der Waals geometry. The photoelectron spectra are consistent with the previous results of Neumark and coworkers (65). The PPC spectrum for the $OH^-(H_2O)$ anion, however, reveals considerably more information.

The correlation of the broad peaks in the electron kinetic energy spectrum (eKE) and photofragment translational energy release (E_T) for $OH^-(H_2O)$ reveals a series of four diagonal ridges (shown in Figure 3). As shown in Reference 71, the $OD^-(D_2O)$ anion gives analogous results, with five features observed owing to the lower stretching frequencies in the deuterated neutral complex. The previous observations on O_4^- suggest that DPD onto vibrationally adiabatic curves in the repulsive region near the transition state for the neutral bimolecular reaction, correlating with the different asymmetric-stretch vibrational states of the H_2O products, is responsible for these features. This assignment cannot be made on energetic grounds alone, however, because the asymmetric stretch of H_2O and the stretching frequency of OH are nearly degenerate. Quantum chemical calculations on the anion and the neutral complex indicate that in the H–O–H–O–H structure the exterior O–H bonds are nearly at the equilibrium value, so it is expected that the product OH will be vibrationally cold, with excitation of the asymmetric stretch of the water that received the shared H atom (71).

As discussed above for O_4^-, examination of the total translational energy spectra $N(E_{TOT})$ is another informative way to view the $N(E_T, eKE)$ correlation spectra. In the $N(E_{TOT})$ spectra for the $H_3O_2^-$ and $D_3O_2^-$ systems shown in Figure 4, the diagonal features in the correlation spectra appear as a resolved spectrum of the correlated product vibrational distribution. Examination of the offset of the vibrational peaks from the internal energy origin (dissociation asymptote) shows that rotational and bending excitation in the products is small. Compared with the photoelectron spectra, the $N(E_{TOT})$ spectra show more structure and a nicely resolved progression in both the nondeuterated and deuterated case. The observed peak spacing of $\sim$0.42 eV (3390 cm^{-1}) in $H_3O_2^-$ and $\sim$0.33 eV (2660 cm^{-1}) in $D_3O_2^-$ is consistent with the interpretation of excitation of the antisymmetric stretch vibration in the water product.

Some further comments on the nature of the $N(E_{TOT})$ spectra and how they differ from the photoelectron spectra should be made. The spacing and width of the peaks in a photoelectron spectrum are determined by the Franck-Condon overlap between the stable anion and the dissociative neutral surface. The lifetime broadening effects of a steeply sloped repulsive surface yield broad features in the photoelectron spectrum. The translational energy release, E_T, between the atomic or molecular products in DPD is governed by both the region of Franck-Condon overlap with the neutral surface and any subsequent transfer of energy from internal

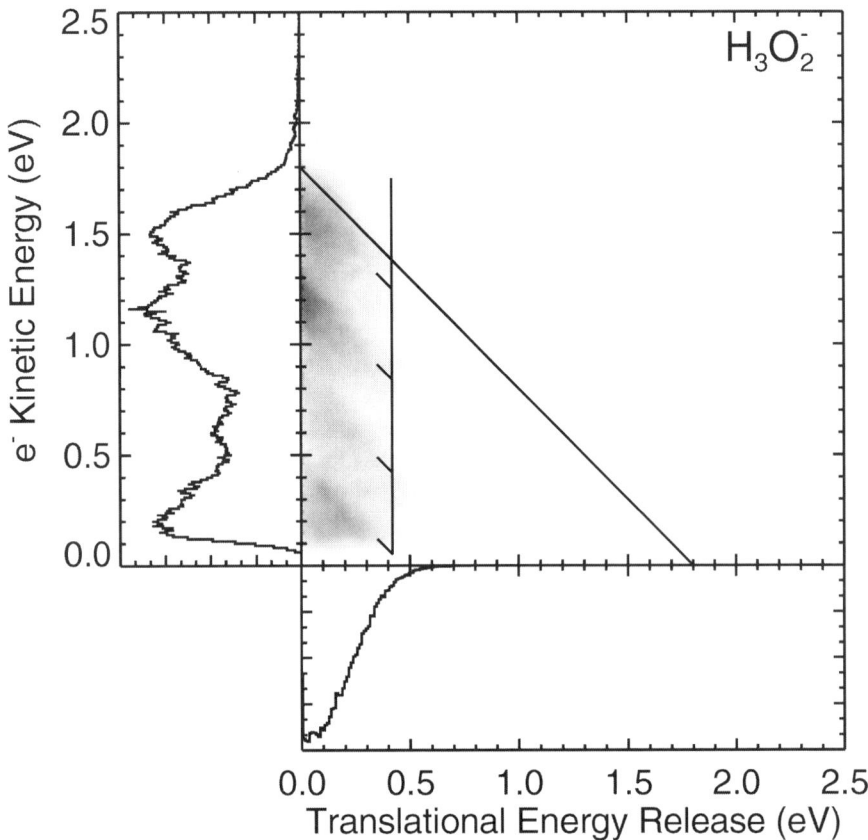

Figure 3 Photoelectron-photofragment energy correlation spectrum [$N(E_T, eKE)$] represented as a two-dimensional gray-scale histogram for $H_3O_2^- + h\nu \to H_2O + OH + e^-$ at 258 nm. The comb shows the asymmetric-stretch spacing of the H_2O product, and the maximum energy available is shown by the diagonal line (reproduced by permission of the Royal Society of Chemistry from Reference 71).

to external degrees of freedom in the dissociation, as discussed in the introduction. The total translational energy release, $E_{TOT} = eKE + E_T$, however, is determined solely by energy conservation. Thus, in the case of $H_3O_2^-$, $E_{TOT} = E_{h\nu} - D^0(OH^- - H_2O) - EA(OH) - E_{int}[OH(i) + H_2O(j)]$. Here, $D^0(OH^- - H_2O)$ is the bond dissociation energy of $H_3O_2^- \to OH^- + H_2O$, $EA(OH)$ is the adiabatic electron affinity of OH, and E_{int} is the sum of the internal energy of states i and j in the products. Irrespective of the exit channel dynamics, the partitioning of E_{TOT} into eKE and the sum $E_T + E_{int}$ is determined by the bound-free Franck-Condon overlap between the anion and neutral surface. The $N(E_{TOT})$ spectrum is a direct, resolved measure of the product internal state distribution

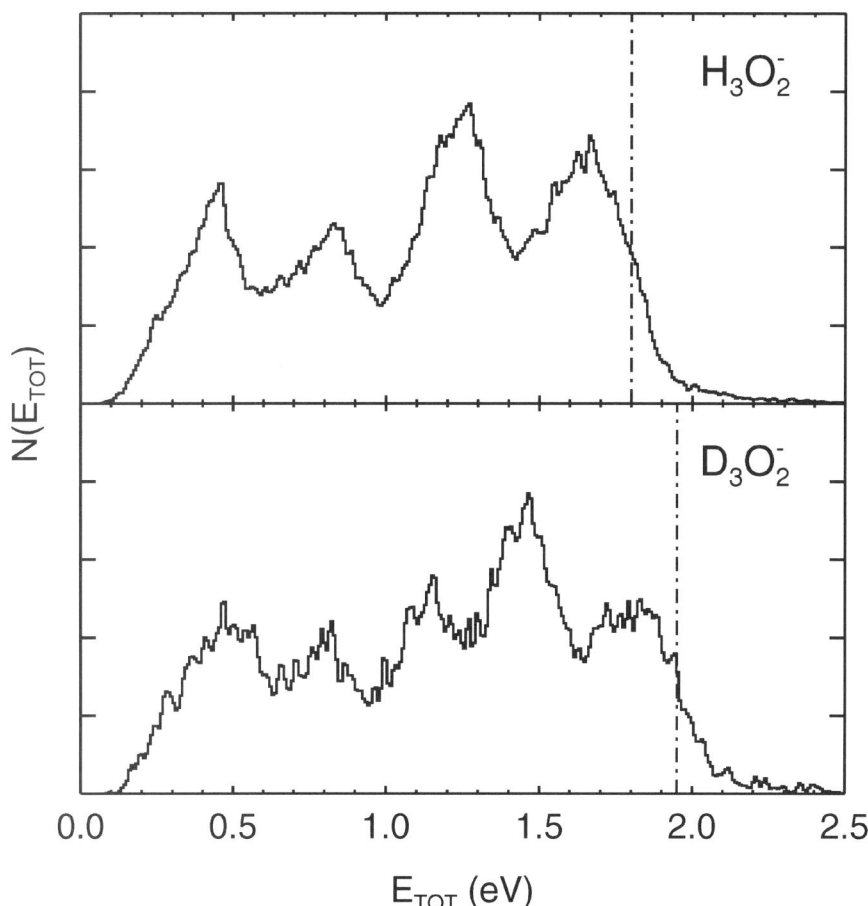

Figure 4 Vibrationally resolved product translational energy [$N(E_{TOT})$] spectra for the dissociative photodetachment (DPD) of $H_3O_2^-$ and $D_3O_2^-$ at 258 nm. The dashed lines represent the maximum energy available for DPD of these isotopic systems (reproduced by permission of the Royal Society of Chemistry from Reference 71).

for this system because it exhibits constrained, vibrationally adiabatic dissociation dynamics.

These results should provide an important test of both ab initio potential energy surfaces and reaction dynamics calculations for this system. Recent theoretical efforts on simulating the photoelectron spectra and extracting information on the reactive potential energy surfaces in related systems (67, 80–82) will need to be extended to take into account the measurement undertaken here—the coincident measurement of the photoelectron kinetic energy and the photofragment translational energy distributions.

COINCIDENCE STUDIES OF DISSOCIATIVE PHOTOIONIZATION

Highly differential coincidence measurements of DPI processes are also becoming more routine in recent years. In this review, the primary focus is on DPI of neutral molecules via direct excitation to a repulsive ionic state. These processes lead to one neutral species, one ion, and a free electron. There have also been extensive studies of the dissociation of multiply ionized molecules. In fact, many of the important coincidence techniques used today, such as PEPICO and threshold PEPICO (TPEPICO), were developed and applied to the study of the dissociation dynamics of multiply ionized molecules (36). The current generation of highly differential, energy- and angle-resolved PEPICO experiments use full-solid-angle imaging detectors for high sensitivity and good counting statistics. Similar to the DPD experiments discussed above, important insights can be gained from examining both photoelectron-photoion energy and angular correlations. DPI processes have received considerably more attention as a source of MF-PADs than studies of DPD in anions, as discussed below.

An example of the newest experiments on DPI processes is provided by the study of Lafosse et al (43) on vector correlations in the DPI of NO in the vacuum UV. This experiment, carried out with 22- to 25-eV photons, made use of time- and position-sensitive detection of the full solid angle for both photoelectrons and photoions. In this arrangement, the kinetic energy and recoil angle correlations for the photoproducts (electron and cation) can be recorded, providing characterization of the dissociative electronic states and a direct measure of the MF-PAD. In this study, $NO(X\ ^2\Pi)$ was ionized to the dissociative $NO^+(c\ ^3\Pi)$ or $NO^+(B\ ^1\Pi)$ states, with rapid dissociation to $N^+(^3P) + O(^3P) + e^-$. Measuring the N^+ and e^- in coincidence thus gives a kinematically complete characterization of this DPI process.

An example of the energetic correlations seen in such an experiment is shown in Figure 5, where three processes are identified in the two-dimensional histogram of the correlation of the photoelectron and photoion kinetic energy. Processes I and II both correspond to ground state products $N^+(^3P) + O(^3P) + e^-$, as shown by their position along the energetic limit for this process marked L_1. The different photoelectron kinetic energies, however, allow assignment of process I to DPI via the $NO^+(c^3\Pi)$ state, whereas process II is DPI via the higher lying $NO^+(B\ ^1\Pi)$ state. The photoelectron kinetic energy near 0.3 eV identifies process III as also occurring via the $NO^+(B\ ^1\Pi)$ state. The limits L_2 and L_3 in this case, however, correspond to excited-state dissociation channels, producing $N^+(^1D)$ or $O(^1D)$ excited products, respectively. This coincidence experiment allows direct determination of the branching ratios between these different channels for the first time. In addition to the energetic correlations, the photoelectron–photoion angular correlations yielded the MF-PAD for feature I in the correlation spectrum, providing further insights into the DPI dynamics.

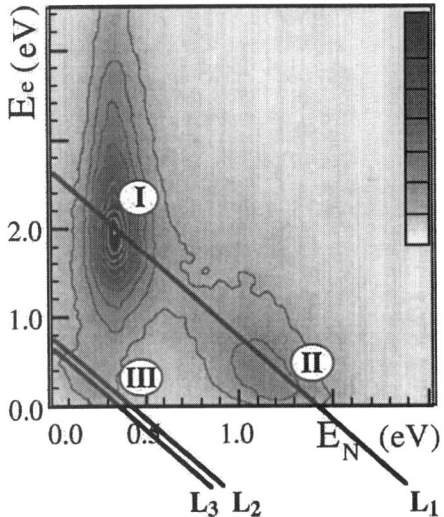

Figure 5 Photoelectron-photoion kinetic energy correlation diagram for NO + hν → N^+ + O + e^-, where hν is 23.6 eV. The limits marked L_1, L_2, and L_3 correspond to the dissociation limits $N^+(^3P)$ + $O(^3P)$ + e^-, $N^+(^1D)$ + $O(^3P)$ + e^-, and $N^+(^3P)$ + $O(^1D)$ + e^-, respectively (reproduced from Reference 43 with permission from the American Physical Society).

When dissociation is rapid relative to molecular rotation, the axial recoil approximation holds, and fully angle-resolved PEPICO experiments directly yield the MF-PAD. Several experiments have probed the MF-PAD in recent years, but now, with high-sensitivity energy- and angle-resolved PEPICO experiments becoming possible, the quality of the measured MF-PADs has improved considerably. In addition to the studies of NO by Lafosse et al (43), recent reports have appeared on the H_2 (83), O_2 (40, 84), CO (85), N_2 (86, 87), and CO_2 (86, 87) molecules.

For larger polyatomic molecules, Downie & Powis (42, 88, 89) have reported fully angle-resolved PEPICO studies of CF_3I. They examined the photoelectron–photoion recoil vector correlations in the DPI of CF_3I by photoionization of the $5a_1$ valence orbital with 21.2-eV radiation from a He discharge lamp. At this photoionization wavelength, DPI pathways leading to both CF_3^+ + $I(^2P_{1/2}, ^2P_{3/2})$ and I^+ + CF_3 are open. An example of the electron–CF_3^+ recoil vector correlations is shown in Figure 6. For valence photoionization, it is expected that examination of the electron-ion vector correlations will give the same results for these two channels if the axial recoil approximation holds, in agreement with the results of Downie & Powis over a wide range of energies. At lower available energies, however, the more endothermic I^+ channel shows differences consistent with dynamical effects in this dissociation channel and the breakdown of the axial recoil approximation. Downie & Powis (42) also carried out calculations on the electron-molecule half collision that occurs on photoionization, recovering the major features of the observed MF-PAD for the stable polyatomic molecule, CF_3I.

A number of collisional and photoinduced processes involving atoms and, more recently, molecules have been carried out using the cold target recoil ion

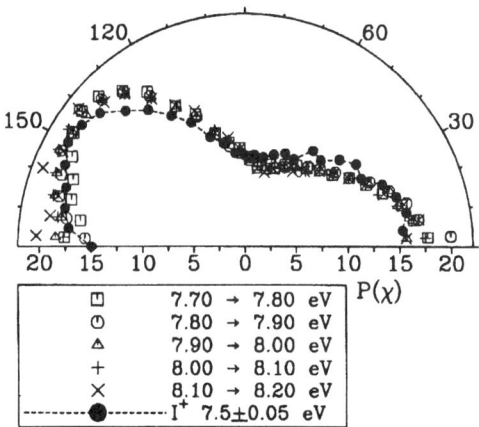

Figure 6 Polar plot of photoelectron-cation recoil vector correlations over a range of photoelectron kinetic energies. The probability $P(\chi)$ as a function of the included angle χ between the photoelectron and ion recoil direction in the He(I) DPI of CF_3I is plotted here. The data for I^+ (solid circles) is plotted against the conjugate angle $\chi' = 180° - \chi$, so that the direction $\chi = 180°$ always corresponds to ejection of the electron from the I atom end of the molecule (reproduced from Reference 42 with permission from the American Institute of Physics).

momentum spectroscopy (COLTRIMS) technique (41, 90). These experiments are entirely analogous to those already discussed in that the recoil momenta of the products are measured directly to provide a kinematically complete measure of the process under study. An example of the recent extension of these techniques to molecular problems is given by the study of the double ionization of D_2 (91). In this experiment, the momenta of both deuterons and one of the two photoelectrons were measured, yielding kinematically complete data on the dynamics of this DPI process. This process provides an important test of electron correlation phenomena because both electrons are removed by a single photon. COLTRIMS experiments have been recently reviewed in detail (90).

TIME-RESOLVED DYNAMICS—PHOTOELECTRON-PHOTOION ENERGY AND ANGULAR DISTRIBUTIONS: NO_2

Advances in ultrafast laser technology coupled with the increasing use of sophisticated time- and position-sensitive detectors have paved the way for carrying out time-resolved coincidence studies on dissociating systems. Hayden and coworkers (30) recently constructed a time-resolved PEPICO spectrometer that allows complete three-dimensional energy and angular distributions to be collected in coincidence for photoelectrons and photoions. More conventional time-resolved PEPICO experiments by Radloff (18) looked at the time evolution of the photoionization of ammonia clusters, benzene dimers, and other systems in which photoelectron spectra for many different species can be acquired as a function of time using the coincident photoion to sort the results. Gas-phase ultrafast studies can often be challenging because of the propensity for the occurrence of multiphoton processes. Kinematically complete coincidence experiments, like those carried out by Hayden and coworkers (30), can play an important role in ultrafast studies by helping to verify what photoexcitation pathways are operative. As seen in the studies of DPD and DPI reviewed above, a coincidence measurement of the kinetic energies of the photoelectron and the photofragments helps deduce the final states of the products and, when multiphoton processes are possible, determine how many photons are involved.

The first time-resolved coincidence study of photoelectron-photoion kinetic energy partitioning is that of the dissociative multiphoton ionization (DMI) of NO_2 at 375 nm reported by Davies et al (30). Previous studies of the DMI of NO_2 at this wavelength were interpreted in terms of a one-photon excitation to the lowest dissociative state of NO_2, followed by three-photon ionization of the NO products (92). By carrying out a fully energy- and angle-resolved PEPICO experiment, Davies et al (30) were able to show that DMI with 150-fs laser pulses at this wavelength occurs via three-photon excitation to a state of NO_2 correlating to the $NO(C^2\Pi) + O$ dissociation limit, with ionization of the nascent NO by a fourth photon. Figure 7 (see color insert) shows the time evolution of the photoelectron-photoion kinetic energy correlations observed in this experiment. The photoion kinetic energy axis shows the CM E_T in the dissociation of NO + O that occurs on the excited state of NO_2. At short times, a diagonal structure is observed, similar to the DPD of O_4^- and the DPI of NO discussed above. This diagonal structure is a measure of the distribution of the original molecular wave packet over the repulsive state of NO_2 that is accessed by three-photon excitation. As the bond breaks, the correlation spectra evolve at longer times to show along the y-axis the photoelectron spectrum of $NO(C^2\Pi)$ and along the x-axis the translational energy distribution of the $NO(C^2\Pi)$ products. The photoelectron–photoion kinetic energy correlations can thus be used to examine the breaking of the O–NO bond over a time scale of a few hundred femtoseconds.

Since the dissociation is so rapid and the experiment records the recoil angles of the coincident photoelectrons and photoions, the DMI of NO_2 provides another example of how coincidence experiments can be used to determine the MF-PAD. The results of Davies et al (31) shown in Figure 8 illustrate the time evolution of the MF-PAD during the dissociation. The data shown here represent the angular distribution of the photoelectrons for only the ions that recoil along the electric vector of the laser, and one can see how this evolves from a noncentrosymmetric angular distribution at short times to a centrosymmetric one at long times, as the now free NO rotates away from the recoiling O atom. At short times, as shown

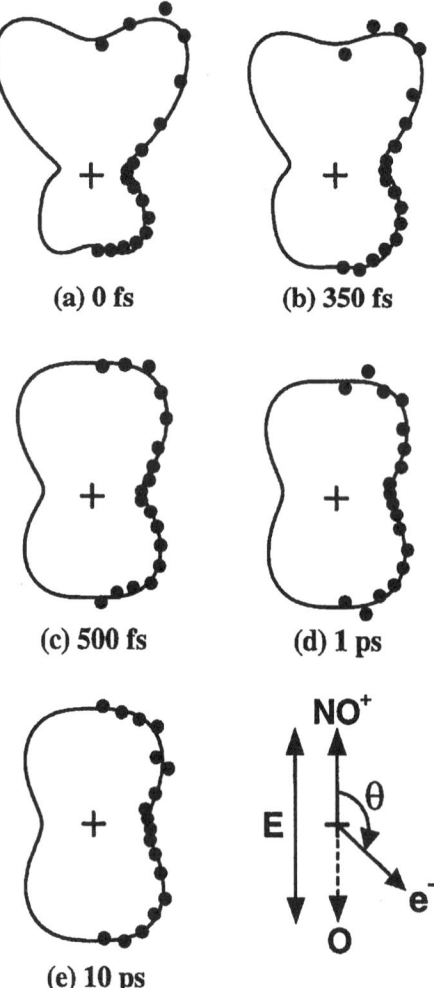

Figure 8 Time-resolved molecular-frame photoelectron angular distributions for the dissociative multiphoton ionization of NO_2^- at 375.3 nm. The angular distribution (in number of events per unit solid angle) is shown by the distance from the origin as a function of the angle θ shown in the lower right-hand corner of the figure. The data were recorded with both pump and probe laser polarization parallel and show the angular correlations for ions recoiling along that axis. The solid-line fits represent the best-fit sum of Legendre polynomials (reproduced from Reference 31, with permission from the American Physical Society).

in Figure 8, there is a marked propensity for both the photoelectron and NO^+ to recoil in the same direction in the molecular frame, defined by the dissociating bond in NO_2.

Qualitatively, the evolution of the angular distributions can be understood in terms of the effect that the close proximity of the recoiling O atom has on the photoelectron. The presence of the neutral O atom causes the interaction potential seen by the departing photoelectron to be highly nonspherical, with a strong inversion asymmetry. The ejected electron scatters from this nonspherical potential, leading to a larger contribution of odd and higher-order partial waves than observed in free $NO(C^2\Pi)$. Quantum interference between these odd and even partial waves can lead to the lobed structure and forward-backward asymmetry seen in these MF-PADs at short time delays. This is similar to previously observed forward-backward asymmetric angular distributions seen in MF-PADs for inner-shell ionization of CO (85) and dissociative autoionization of excited O_2 molecules (84).

Quantitative interpretation of these time-dependent results poses a significant challenge for both quantum chemistry and reaction dynamics theory. The dissociation occurs on a highly excited dissociative electronic state of NO_2, and detailed prediction of the PADs requires solving the electron-molecule scattering problem for the dissociating NO_2. The appearance of experimental observables like this, however, should motivate further studies on the time-resolved properties of PADs.

THREE-BODY DISSOCIATION DYNAMICS

An increasing number of studies are applying coincidence techniques to the characterization of three-body dissociation dynamics. Important insights into three-body photodissociation processes of stable molecules have been made previously by noncoincidence methods, using measurements of the uncorrelated asymptotic properties of the photofragments, such as momentum and quantum state, to infer the three-body dissociation dynamics (32, 33). Coincidence measurements have the benefit of not requiring significant assumptions about the nature of the dissociation process (concerted/sequential) or geometry at the transition state. The coincidence studies of three-body dissociation reactions allow a direct determination of the product angular correlations and partitioning of momenta, yielding a detailed picture of the dissociation dynamics and even structural insights. Recent experiments on DPD, as well as charge-exchange and collision-induced dissociation studies of three-body dynamics, are reviewed here.

Three-body dissociation processes in molecular physics were first studied using coincidence techniques in Coulomb explosion processes involving multiply ionized molecules. These experiments focused on either multiple photoionization (38, 93) or collisional ionization of a target molecule (37). Such processes have an important advantage for coincidence experiments in that they always produce at least two positively charged fragments that can be detected relatively easily.

Hsieh & Eland (38) and Lavollée (60), for example, have studied the three-body dissociation dynamics of species such as OCS^{2+} and SO_2^{3+}, respectively. Because two out of the three momenta of the products are all that are required to fully characterize a three-body process, this measurement is sufficient. These measurements, however, are not truly kinematically complete because they do not record the energy of the photoelectrons in the case of photoionization.

PPC experiments provide a new approach to studying three-body dissociation in the ground and low-lying excited states of neutral species. After the development of a new multiparticle detector (34), we applied this technique to the study of the three-body dynamics of O_6 (94, 95) and $O_3(D_2O)$ (96) complexes produced upon photodetachment of the corresponding anions. In these studies, multihit time- and position-sensitive particle detectors were used to record the three-dimensional momentum distributions of the photoelectron and three photofragments produced in the DPD of these anions.

PPC studies of three-body dissociation dynamics can be used to study energy partitioning, photoelectron-photofragment angular correlations, and laboratory angular distributions similar to the two-body processes discussed above. For the purpose of this review, the focus is on the vector correlations of the three heavy particles produced in three-body DPD. To directly view the product angular correlations, we follow an approach similar to that used by Hsieh & Eland (38) and Lavollée (60) for the three-body dissociation dynamics of doubly charged cations. Since the three particle recoil velocities are directly measured, once the product masses are determined by evaluating the conservation of momentum, the partitioning of product momenta can be examined.

One of the first systems examined for three-body DPD was the O_6^- cluster formed on the addition of an O_2 molecule to the O_4^- system discussed above. Energy partitioning in the three-body DPD of O_6^-:

$$O_6^- + h\nu \rightarrow O_2 + O_2 + O_2 + e^- \qquad 5.$$

is very similar to that observed for O_4^-. In particular, vibrationally resolved $N(E_T, eKE)$ correlation spectra are observed (94). A more detailed view into the three-body half-collision dynamics can be seen in the MF-DCS spectra shown in Figure 9a for the breakup of O_6^- at 388 nm (3.20 eV). In this spectrum, the CM momentum of the fastest particle is constrained to lie along the x-axis. On an event-by-event basis, the direction and magnitude of the momentum vectors of the other products are plotted in the plane of the figure. The partitioning of momentum in the DPD of O_6^- reveals an anisotropic distribution in which two of the O_2 products carry away the majority of the momentum, and the third O_2 plays the role of a spectator. These dynamics are consistent with an O_4^- core weakly interacting with the third O_2. Because of the weakly bound open-shell character of O_6^-, it is difficult to carry out reliable ab initio calculations to confirm the structure of this system. However, given that O_6^- is bound by only ~0.11 eV relative to O_4^-, a weak electrostatic interaction with the third O_2 is very plausible (94).

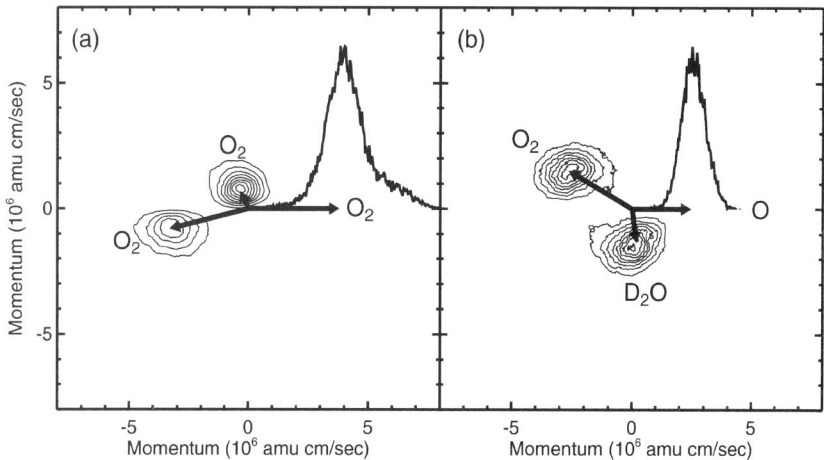

Figure 9 Momentum-space molecular-frame differential cross section spectra recorded in the three-body dissociative photodetachment of O_6^- (*panel A*) and $O_3^-(D_2O)$ (*panel B*). In *panel A*, the fastest O_2 product is chosen as the reference particle along the x-axis, whereas in *panel B*, the fastest of the two lighter particles is assigned to be the O atom and chosen as the reference particle as discussed in the text. The vectors show the peaks of the correlated distributions in both frames (adapted from References 95 and 96, by permission. Copyright (2001) American Chemical Society).

The three-body recombination reaction $O + O_2 + M \rightarrow O_3 + M$ is responsible for the formation of stable ground-state O_3 in the atmosphere. Therefore, this system is an important prototypical three-body system to study. Photodetachment of $O_3^-(D_2O)$ clusters at 258 nm (4.80 eV) was used to prepare the dissociative low-lying triplet and singlet excited states of $O_3(D_2O)$, following an earlier study of the dissociation dynamics of these states in free O_3 (75). Photodetachment to the triplet states of O_3 in the complex leads to three-body dissociation as follows:

$$O_3^-(D_2O) + h\nu \rightarrow O + O_2 + D_2O + e^-, \qquad 6.$$

with no evidence observed for intracluster reaction or quenching of the excited states in the complex at the level of excitation used in this experiment (96). Once again, the detailed dynamics of the three-body dissociation of the excited $O_3(D_2O)$ complex can be studied by examining the angular correlations of the product recoil vectors. In this case, the MF-DCS is generated by transforming the data from the laboratory frame to a molecular breakup frame such that for each event, the velocity vector of the O atom is chosen as the principal axis, and the recoil velocities of the other two particles are then plotted in this CM reference frame as shown in Figure 9*b*.

The mass resolution in these experiments is currently limited: O_2 can be fully resolved from O and D_2O, but the lighter fragments cannot be fully resolved from one another. Because of this limitation, the assumption is made in generation of the MF-DCS that the O atom is the fastest particle. The MF-DCS shows that the centroid of the O_2 distribution recoils backward relative to the O atom, whereas the centroid of the D_2O feature is actually observed to be slightly forward scattered relative to the recoiling O atom. The striking result from the MF-DCS is that the product momenta are partitioned such that the heaviest fragment, O_2, carries away most of the momentum, with the lighter O and D_2O fragments recoiling in the other direction in the CM. This observation is not affected by the assumption that the O atom is the fastest product.

To gain further insights into the three-body dissociation dynamics indicated by the MF-DCS, density functional theory calculations of the structure of the parent $O_3^-(D_2O)$ complex were performed. Two nearly isoenergetic local minima were found: an asymmetric C_s complex with D_2O hydrogen bonded to one end of O_3^- and a double-hydrogen-bonded C_{2v} complex. The calculated C_s geometry showed that the O–D bond in the water moiety and the $O–O_2$ bond interacting with the D_2O are both longer than their respective equilibrium bond lengths. This breaking of the C_{2v} symmetry of O_3 in the cluster is expected to promote the antisymmetric stretch dissociation of the lengthened $O–O_2$ bond, which has been shown to be the reaction coordinate for dissociation of the low-lying excited states of O_3 in previous studies (75).

The asymmetric structure of the cluster is consistent with the observations if the lengthened $O–O_2$ bond breaks, causing the recoiling O atom to interact with D_2O, transferring some momentum to the nearly equal mass D_2O. The heavier O_2 product then carries away the largest fraction of the momentum as it recoils in the other direction. Dissociation of the C_{2v} symmetry complex would not be expected to lead to such a striking partitioning of momentum between the light and heavy fragments. The three-body DPD of O_6^- and $O_3^-(D_2O)$ provides examples of the insights into three-body dissociation dynamics and even the structures of transient neutral species that can be obtained in photoelectron–multiple-photofragment coincidence experiments.

A second type of kinematically complete three-body dissociation experiment on neutral species has been carried out by Müller and coworkers (35), who used the charge-exchange technique to study the three-body breakup of neutral triatomic hydrogen. In this study, specific excited ro-vibronic states of H_3 were prepared by laser excitation, taking advantage of the single long-lived rotational state of neutral H_3 formed by charge exchange of a 3-keV H_3^+ beam with cesium. Using a time- and position-sensitive multiparticle detector to detect the three H atoms in coincidence, they found that the three-body dissociation dynamics exhibit a strong dependence on the initially excited Rydberg states of H_3. They also adapted the three-body analysis procedures developed by Dalitz (97) for nuclear scattering studies and applied increasingly to problems in atomic and molecular physics (98). The Dalitz plots show the partitioning of the square of product momenta and provide fundamental insights into the three-body dynamics, similar to the

MF-DCS discussed above. These representations of the three-body dynamics are being used increasingly, as discussed recently by Maul & Gericke (33), and can also be applied to the DPD experiments discussed above.

A final current topic to be reviewed here is the three-body collision-induced dissociation of Na_n^+ clusters in collisions with He at energies from 160 to 260 eV. These experiments, carried out by Barat et al (99), use multicoincidence detection of neutral and ionic products to study the three-body dissociation reactions $Na_n^+ \rightarrow Na_{n-2}^+ + Na + Na$, where n is 3, 4, and 5. In these collision-induced dissociation processes, in addition to the individual velocity vectors of the products, it is useful to examine the deflection of the centroid of the three products from the collision with He. This deflection provides a measure of the momentum transfer in the collisions. The dynamics observed for the Na_3^+ system are consistent with electronic excitation to a repulsive ionic state with prompt dissociation. In the larger clusters, however, the dominant mechanism involves momentum transfer to the Na atoms in the collision with He with ejection of a fast neutral Na atom, followed by dissociation of the excited ionic fragment. These types of collisions are characterized by one fast neutral product, one slow neutral product, and an ionic product with a momentum matching that of the slow neutral product of the subsequent dissociation of the excited ionic fragment. Kinematically complete coincidence measurements like those reported by Barat et al (99) provide the necessary information to unravel the complicated three-body dissociation dynamics observed in full collisions at high energies.

CONCLUDING REMARKS

Coincidence spectroscopies will likely play an increasing role in the elucidation of the energetics and reaction dynamics of transient species. Further advances in detection technology and laser and synchrotron excitation sources will provide the required tools. One promising development is the potential for new pixellated detectors that should become available as a spin-off of technology originally developed for particle physics experiments. Research in this area is currently being driven by the need for real-time high-throughput X-ray crystallography applications in biophysics (100). These devices will share many of the characteristics of the current CCDs used in imaging studies in chemical dynamics, but they will have the added benefit of full three-dimensional information: x,y position and particle time-of-flight. Of course, the application of detection schemes like this will need to be coupled with improved, higher-speed data acquisition interfaces, which will likely occur.

In conclusion, we have discussed some of the diverse recent applications of coincidence spectroscopic techniques to the study of the energetics, dissociation, and ionization dynamics of isolated molecules and clusters. We have focused on the unique types of information that can be derived from coincidence studies of these fundamental processes in chemical dynamics, including characterization of dissociative states, regions of potential energy surfaces near the transition state for bimolecular reactions, MF-PADs, and three-body dissociation dynamics.

ACKNOWLEDGMENTS

I thank the members of my group, past and present, who have contributed to our coincidence studies of DPD processes. Research support from the Air Force Office of Scientific Research, the National Science Foundation, and the U.S. Department of Energy is gratefully acknowledged. REC is a Camille Dreyfus Teacher Scholar and a David and Lucile Packard Fellow in Science and Engineering.

Visit the Annual Reviews home page at www.AnnualReviews.org

LITERATURE CITED

1. Butler LJ, Neumark DM. 1996. *J. Phys. Chem.* 100:12801–16
2. Casavecchia P. 2000. *Rep. Prog. Phys.* 63:355–414
3. Mayer PM, Parkinson CJ, Smith DM, Radom L. 1998. *J. Chem. Phys.* 108:604–15
4. Roos BO. 1999. *Acc. Chem. Res.* 32:137–44
5. Palma J, Clary DC. 2000. *J. Chem. Phys.* 112:1859–67
5a. Ng C, ed. 2000. *Photoionization and Photodetachment*, Vol. 10A, B. Singapore: World Scientific
6. Weitzel K-M. 2000. See Ref. 5a, Vol. 10A, pp. 539–600
7. Morioka Y. 2000. See Ref. 5a, Vol. 10A, pp. 347–93
8. Heck AJR, Chandler DW. 1995. *Annu. Rev. Phys. Chem.* 46:335–72
9. Houston PL. 1996. *J. Phys. Chem.* 100:12757–70
10. Parker DH. 2000. See Ref. 5a, Vol. 10A, pp. 3–46
11. Suits AG, Continetti RE. 2001. In *Imaging in Chemical Dynamics*, ed. AG Suits, RE Continetti, *ACS Symp. Ser. 770*. Washington, DC: Am. Chem. Soc.
12. Continetti RE. 2000. See Ref. 5a, Vol. 10B, pp. 748–808
13. Cooper J, Zare RN. 1968. *J. Chem. Phys.* 48:942–45
14. Dill D. 1976. *J. Chem. Phys.* 65:1130–33
15. Park H, Zare RN. 1996. *J. Chem. Phys.* 104:4554–67
16. Zare RN. 1972. *Mol. Photochem.* 4:1–37
17. Hayden CC, Stolow A. 2000. See Ref. 5a, Vol. 10A, pp. 91–126
18. Radloff W. 2000. See Ref. 5a, Vol. 10A: pp. 127–81
19. Cyr DR, Hayden CC. 1996. *J. Chem. Phys.* 104:771–74
20. Wang L, Kohguchi H, Suzuki T. 1999. *Faraday Discuss.* 113:37–46
21. Blanchet V, Zgierski M, Seideman T, Stolow A. 1999. *Nature* 401:52–54
22. Lehr L, Zanni MT, Frischkorn C, Weinkauf R, Neumark DM. 1999. *Science* 284:635–38
23. Seideman T. 1997. *J. Chem. Phys.* 107:7859–68
24. Althorpe SC, Seideman T. 1999. *J. Chem. Phys.* 110:147–55
25. Seideman T. 2000. *J. Chem. Phys.* 113:1677–80
26. Reid KL, Underwood JG. 2000. *J. Chem. Phys.* 112:3643–49
27. Underwood JG, Reid KL. 2000. *J. Chem. Phys.* 113:1067–74
28. Arasaki Y, Takatsuka K, Wang K, McKoy V. 2000. *J. Chem. Phys.* 112:8871–84
29. Reid KL, Field TA, Towrie M, Matousek P. 1999. *J. Chem. Phys.* 111:1438–45

30. Davies JA, LeClaire JE, Continetti RE, Hayden CC. 1999. *J. Chem. Phys.* 111:1–4
31. Davies JA, Continetti RE, Chandler DW, Hayden CC. 2000. *Phys. Rev. Lett.* 84:5983–86
32. Maul C, Gericke KH. 1997. *Int. Rev. Phys. Chem.* 16:1–79
33. Maul C, Gericke KH. 2000. *J. Phys. Chem. A* 104:2531–41
34. Hanold KA, Luong AK, Clements TG, Continetti RE. 1999. *Rev. Sci. Instrum.* 70:2268–76
35. Müller U, Eckert T, Braun M, Helm H. 1999. *Phys. Rev. Lett.* 83:2718–21
36. Eland JHD. 1991. In *VUV Photoionization and Photodissociation of Molecules and Clusters*, pp. 297–343. Singapore: World Sci.
37. Werner U, Beckord K, Becker J, Lutz HO. 1995. *Phys. Rev. Lett.* 74:1962–65
38. Hsieh S, Eland JHD. 1997. *J. Phys. B* 30:4515–34
39. Stert V, Radloff W, Schulz CP, Hertel IV. 1999. *Eur. Phys. J. D* 5:97–106
40. Takahashi M, Cave JP, Eland JHD. 2000. *Rev. Sci. Instrum.* 71:1337–44
41. Doerner R, Mergel V, Spielberger L, Achler M, Khayyat K, et al. 1997. *Nucl. Instrum. Methods B* 24:225–31
42. Downie P, Powis I. 1999. *J. Chem. Phys.* 111:4535–47
43. Lafosse A, Lebech M, Brenot JC, Guyon PM, Jagutzki O, et al. 2000. *Phys. Rev. Lett.* 84:5987–90
44. Wiza JL. 1979. *Nucl. Instrum. Methods* 162:587–601
45. DeBruijn DP, Los J. 1982. *Rev. Sci. Instrum.* 53:1020–26
46. van der Zande WJ, Koot W, Los J. 1989. *J. Chem. Phys.* 91:4597–602
47. Beijersbergen JHM, van der Zande WJ, Kistemaker PG, Los J, Drewello T, Nibbering NMM. 1992. *J. Phys. Chem.* 96:9288–93
48. Helm H, Walter CW. 1993. *J. Chem. Phys.* 98:5444–49
49. Helm H, Cosby PC. 1987. *J. Chem. Phys.* 86:6813–22
50. Continetti RE, Cyr DR, Osborn DL, Leahy DJ, Neumark DM. 1993. *J. Chem. Phys.* 99:2616–31
51. Bise RT, Choi H, Neumark DM. 1999. *J. Chem. Phys.* 111:4923–32
52. Hanold KA, Sherwood CR, Garner MC, Continetti RE. 1995. *Rev. Sci. Instrum.* 66:5507–11
53. Deyerl H-J, Alconcel LS, Continetti RE. 2001. *J. Phys. Chem.* In press
54. Martin C, Jelinsky P, Lampton M, Malina RF, Anger HO. 1981. *Rev. Sci. Instrum.* 52:1067–74
55. Brenot J-C, Durup-Ferguson M. 1992. *Adv. Chem. Phys.* 82:309–99
56. Belkacem A, Faibis A, Kanter EP, Koenig W, Mitchell RE, et al. 1990. *Rev. Sci. Instrum.* 61:945–52
57. Levin J, Feldman H, Baer A, Ben-Hamu D, Heber O, et al. 1998. *Phys. Rev. Lett.* 81:3347–50
58. Amitay Z, Zajfman D. 1997. *Rev. Sci. Instrum.* 68:1387–92
59. Ali I, Dorner R, Jagutzki O, Nuttgens S, Mergel V, et al. 1999. *Nucl. Instrum. Methods B* 149:490–500
60. Lavollée M. 1999. *Rev. Sci. Instrum.* 70:2968–74
61. Wenthold P, Hrovat DA, Borden WT, Lineberger WC. 1996. *Science* 272:1456–59
62. Metz RB, Bradforth SE, Neumark DM. 1992. *Adv. Chem. Phys.* 81:1–61
63. Manolopoulos DE, Stark K, Werner H-J, Arnold DW, Bradforth SE, Neumark DM. 1993. *Science* 262:1852–55
64. de Beer E, Kim EH, Neumark DM, Gunion RF, Lineberger WC. 1995. *J. Chem. Phys.* 99:13627–36
65. Arnold DW, Cangshan X, Neumark DM. 1995. *J. Chem. Phys.* 102:6088–99
66. Thompson WH, Miller WH. 1994. *J. Chem. Phys.* 101:8620–27
67. Clary DC, Gregory JK, Jordan MJT, Kauppi E. 1997. *J. Chem. Soc. Faraday Trans.* 93:747–53

68. Hanold KA, Sherwood CR, Continetti RE. 1995. *J. Chem. Phys.* 103:9876–79
69. Hanold KA, Garner MC, Continetti RE. 1996. *Phys. Rev. Lett.* 77:3335–38
70. Hanold KA, Continetti RE. 1998. *Chem. Phys.* 239:493–509
71. Deyerl H-J, Luong AK, Clements TG, Continetti RE. 2000. *Faraday Discuss. Chem. Soc.* 115:147–60
72. Sherwood CR, Hanold KA, Garner MC, Strong KM, Continetti RE. 1996. *J. Chem. Phys.* 105:10803–11
73. Posey LA, Deluca MJ, Johnson MA. 1986. *Chem. Phys. Lett.* 131:170–74
74. Aquino A, Walch S, Taylor PR. 2000. *J. Chem. Phys.* 114:3010–17
75. Garner MC, Hanold KA, Resat MS, Continetti RE. 1997. *J. Phys. Chem. A* 101:6577–82
76. Resat MS, Zengin V, Garner MC, Continetti RE. 1998. *J. Phys. Chem.* 102:1719–24
77. Sherwood CR, Garner MC, Hanold KA, Strong KM, Continetti RE. 1995. *J. Chem. Phys.* 102:6949–52
78. Reed KJ, Zimmerman AH, Andersen HC, Brauman JI. 1976. *J. Chem. Phys.* 64:1368–75
79. Gygax R, McPeters HL, Brauman JI. 1979. *J. Am. Chem. Soc.* 101:2567–70
80. Thompson WH. 1999. *J. Phys. Chem. A* 103:9500–5
81. Thompson WH. 1999. *J. Phys. Chem. A* 103:9506–11
82. Lavender HB, McCoy AB. 2000. *J. Phys. Chem. A* 104:644–51
83. Ito K, Adachi J, Hall R, Motoki SES, Soejima K, Yagishita A. 2000. *J. Phys. B* 33:527–33
84. Golovin AV, Heiser F, Quayle CJK, Morin P, Simon M, et al. 1997. *Phys. Rev. Lett.* 79:4554–57
85. Heiser F, Gessner O, Viefhaus J, Wieliczek K, Hentges R, Becker U. 1997. *Phys. Rev. Lett.* 79:2435–37
86. Shigemasa E, Adachi J, Oura M, Yagishita A. 1995. *Phys. Rev. Lett.* 74:359–62
87. Pavlychev AA, Fominykh NG, Watanabe N, Soejima K, Shigemasa E, Yagishita A. 1998. *Phys. Rev. Lett.* 81:3623–26
88. Downie P, Powis I. 1999. *Phys. Rev. Lett.* 82:2864–67
89. Downie P, Powis I. 2000. *Faraday Discuss. Chem. Soc.* 115:103–17
90. Doerner R, Mergel V, Jagutzki O, Spielberger L, Ullrich J, et al. 2000. *Phys. Rep.* 330:95–192
91. Doerner R, Brauning H, Jagutzki O, Mergel V, Achler M, et al. 1998. *Phys. Rev. Lett.* 81:5776–79
92. Singhal RP, Kilic HS, Ledingham KWD, Kosmidis C, McCanny T, et al. 1996. *Chem. Phys. Lett.* 253:81–86
93. Lavollée M, Brems V. 1999. *J. Chem. Phys.* 110:918–26
94. Hanold KA, Luong AK, Continetti RE. 1998. *J. Chem. Phys.* 109:9215–18
95. Luong AK, Clements TG, Continetti RE. 2000. See Ref. 11, pp. 313–25
96. Luong AK, Clements TG, Continetti RE. 1999. *J. Phys. Chem. A* 103:10237–43
97. Dalitz RH. 1953. *Philos. Mag.* 44:1068–80
98. Wiese LM, Yenen O, Thaden B, Jaecks DH. 1997. *Phys. Rev. Lett.* 79:4982–85
99. Barat M, Brenot JC, Dunet H, Fayeton JA, Picard YJ. 2000. *J. Chem. Phys.* 113:1061–66
100. Datte P, Beuville E, Beche JF, Cork C, Earnest T, et al. 1997. *Nucl. Instrum. Methods A* 391:471–80

SPECTROSCOPY AND HOT ELECTRON RELAXATION DYNAMICS IN SEMICONDUCTOR QUANTUM WELLS AND QUANTUM DOTS

Arthur J Nozik

The National Renewable Energy Laboratory, Center for Basic Sciences, 1617 Cole Boulevard, Golden, Colorado 80401, and Department of Chemistry, University of Colorado, Boulder, Colorado 80309; e-mail: anozik@nrel.nrel.gov

Key Words hot-carrier cooling, high efficiency photovoltaics, semiconductor quantum dot arrays, superlattices, high efficiency photo electrochemistry

■ **Abstract** Photoexcitation of a semiconductor with photons above the semiconductor band gap creates electrons and holes that are out of equilibrium. The rates at which the photogenerated charge carriers return to equilibrium via thermalization through carrier scattering, cooling by phonon emission, and radiative and nonradiative recombination are important issues. The relaxation processes can be greatly affected by quantization effects that arise when the carriers are confined to regions of space that are small compared with their deBroglie wavelength or the Bohr radius of bulk excitons. The effects of size quantization in semiconductor quantum wells (carrier confinement in one dimension) and quantum dots (carrier confinement in three dimensions) on the respective carrier relaxation processes are reviewed, with emphasis on electron cooling dynamics. The implications of these effects for applications involving radiant energy conversion are also discussed.

INTRODUCTION

Semiconductors have a fundamental threshold energy for the absorption of photons; this critical energy is termed the band gap (E_g); photons with energies less than the band gap are not absorbed. The absorption of a photon creates an electron in the conduction band and leaves behind a positive charge, termed a hole, in the valence band. When the absorbed photon energy is greater than the band gap, the distribution of the excess energy between the electron and hole is given by

$$\Delta E_e = (h\nu - E_g)\left[1 + m_e^*/m_h^*\right]^{-1} \qquad 1.$$

$$\Delta E_h = (h\nu - E_g) - \Delta E_e, \qquad 2.$$

where m_e^* and m_h^* are the effective masses of electrons and holes, respectively, ΔE_e is the energy difference between the conduction band and the initial energy of the

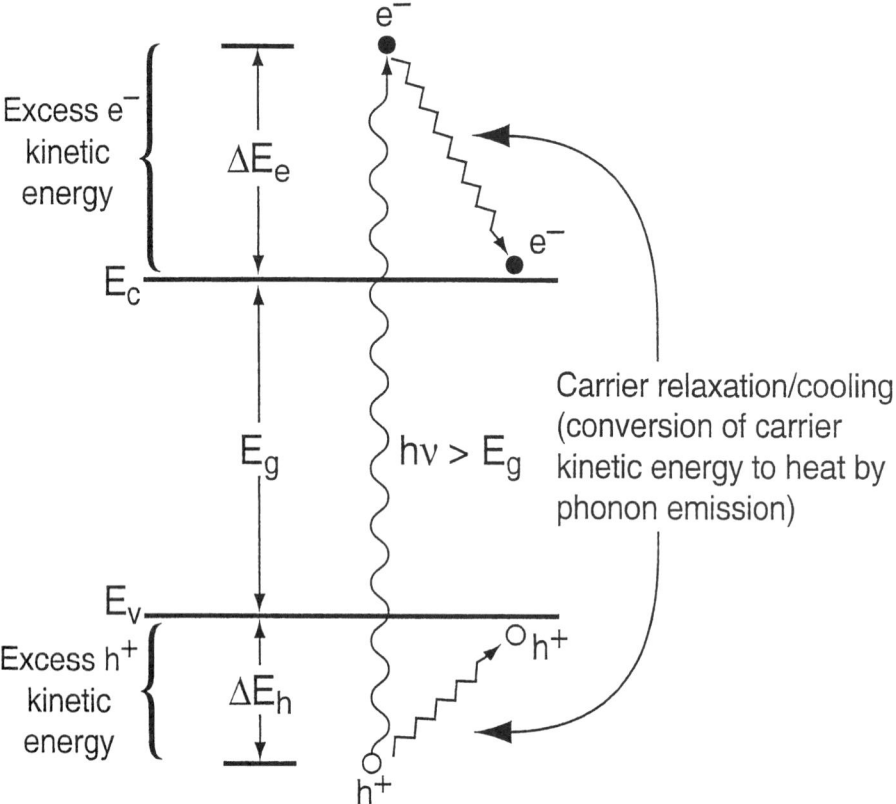

Figure 1 Hot carrier relaxation/cooling dynamics in semiconductors.

photogenerated electron, and ΔE_h is the energy difference between the valence band and the photogenerated hole (see Figure 1) (1, 2). These excess energies appear in the respective charge carriers (electrons and holes) as kinetic energy; such photogenerated carriers produced by the absorption of photons bigger than the bandgap are far from equilibrium. At thermal equilibrium the carriers would have a Boltzmann distribution determined by the temperature of the system, with electrons being within kT of the bottom of the conduction band and holes being within kT of the top of the valence band.

The photogenerated carriers can be produced by either a light pulse or steady-state illumination. In the former case, the carriers seek and establish equilibrium over a time period that depends upon the specific rates of the various equilibration pathways. In the latter case, with continuous optical pumping, a steady-state condition is reached in which a steady-state population of excited, nonequilibrated carriers is produced; the characteristics of this steady-state population will also depend upon the balance between the kinetics of the various equilibration pathways and the rate and energetics of carrier generation. In the discussion immediately below we consider pulsed excitation for reasons of simplicity.

Pulsed excitation with a single wavelength produces initial carrier populations of electrons and holes that are nearly mono-energetic; they may not be perfectly mono-energetic because of possible multiplicities of hole states that are available for the optical transition. This possibility is strongest for semiconductors exhibiting quantum confinement (3). In either case, the initial carrier distributions are not Boltzmann-like, and the first step toward establishing equilibrium is for the electrons and holes to interact separately among themselves through their respective carrier-carrier collisions and intervalley scattering to form separate Boltzmann distributions of electrons and holes. The two Boltzmann distributions of electrons and holes can then be separately assigned an electron and hole temperature that reflects the distributions of kinetic energy in the respective charge carrier populations. If photon absorption produces electrons and holes, each with initial excess kinetic energies at least kT above the conduction and valence bands, then both initial carrier temperatures are always above the lattice temperature; these carriers are called hot carriers (i.e. hot electrons and hot holes). This first stage of relaxation or equilibration occurs very rapidly (<100 fs) (4, 5), and this process is often referred to as carrier thermalization (i.e. formation of a thermal distribution described by Boltzmann statistics).

After the separate electron and hole populations come to equilibrium among themselves in less than 100 fs, they are still not yet in equilibrium with the lattice. The next step of equilibration is for the hot electrons and hot holes to equilibrate with the semiconductor lattice. The lattice temperature of the semiconductor is determined by equilibration of its quantized lattice vibrations (phonons) with the surroundings. The lattice temperature is the ambient temperature and is lower than the initial hot-electron and hot-hole temperatures. Equilibration of the hot carriers with the lattice is achieved through carrier-phonon interactions (phonon emission through electron- and hole-scattering) whereby the excess kinetic energy of the carriers is transferred from the carriers to the phonons; the phonons involved in this process are the longitudinal optical (LO) phonons. This may occur by each carrier undergoing separate interactions with the phonons or in an Auger process in which the excess energy of one carrier type is transferred to the other type, which then undergoes the phonon interaction. The phonon emission results in cooling of the carriers and heating of the lattice until the carrier and lattice temperatures become equal. This process is termed carrier cooling, but some researchers also refer to it as thermalization; however, this latter terminology can cause confusion with the first stage of equilibration that just establishes the Boltzmann distribution among the carriers. Here, we restrict the term thermalization to the first stage of carrier relaxation and we refer to the second stage as carrier cooling (or carrier relaxation) through carrier-phonon interactions.

The final stage of equilibration results in complete relaxation of the system. The electrons and holes can recombine, either radiatively or nonradiatively, to produce the final electron and hole populations that existed in equilibrium in the dark before photoexcitation. Another important possible pathway following photoexcitation of semiconductors is for the photogenerated electrons and holes to undergo spatial separation. Separated photogenerated carriers can subsequently produce a photovoltage and a photocurrent (photovoltaic effect) (1, 2, 6); alternatively, the separated carriers can drive electrochemical oxidation and reduction reactions (generally labeled redox reactions) at the semiconductor surface (photoelectrochemical energy conversion) (7). These two processes form the basis for devices/cells that convert radiant energy (solar energy) into electrical (1, 2, 6) or chemical free energy (photovoltaic cells and photoelectrochemical cells, respectively) (7).

The dynamics of carrier cooling in semiconductors is an extremely important factor for many practical device applications. For example, high-efficiency negative electron affinity photocathodes depend upon hot-electron injection from p-type semiconductors into a vacuum (1). Hot-electron transistors depend upon hot-electron transport (2). For the application of quantum dots in novel quantum dot lasers, it is critical that hot photogenerated carriers relax before photon emission (8). For applications involving photon (radiant) energy conversion, the ultimate thermodynamic limit on conversion efficiency can be increased from about 32% for carriers fully equilibrated with the lattice in a single threshold (single band gap) semiconductor photoconverter, to about 66% for conversion utilizing hot carriers that have not undergone any interaction with phonons and are thus converted at their initial carrier temperatures (9, 10).

There are two fundamental ways to utilize hot carriers for enhancing the efficiency of photon conversion. One way produces an enhanced photovoltage, and the other way produces an enhanced photocurrent. The former requires that the carriers be extracted from the photoconverter before they cool (9, 10), whereas the latter requires the energetic hot carriers to produce a second (or more) electron-hole pair through impact ionization (11, 12). This latter process is the inverse of an Auger process whereby two electron-hole pairs recombine to produce a single highly energetic electron-hole pair. In order to achieve the former, the rates of photogenerated carrier separation, transport, and interfacial transfer across the semiconductor-molecule interface must all be fast compared with the rate of carrier cooling (10, 13–15). The latter requires that the rate of impact ionization (i.e. inverse Auger effect) is greater than the rate of carrier cooling. These processes and their potential applications in novel quantum dot solar cells are discussed later in "Quantum Dot Solar Cells."

Hot electrons and hot holes generally cool at different rates because they generally have different effective masses; for most inorganic semiconductors electrons have effective masses that are significantly lighter than holes and consequently cool more slowly. Another important factor is that hot-carrier cooling rates are dependent upon the density of the photogenerated hot carriers (i.e. the absorbed light intensity) (4, 5, 16); this effect is discussed in "Hot-Electron Cooling Dynamics in Quantum Wells and Superlattices." Here, most of the dynamical effects we discuss are dominated by electrons rather than holes; therefore, we restrict our subsequent discussion primarily to the relaxation dynamics of photogenerated electrons.

Finally, in recent years it has been proposed (10, 13, 14, 17–20), and experimentally verified in some cases (see "Hot-Electron Cooling Dynamics in Quantum Wells and Superlattices" and "Experimental Determination of Relaxation/Cooling Dynamics and a Phonon Bottleneck in Quantum Dots"), that the relaxation dynamics of photogenerated carriers may be markedly affected by quantization effects in the semiconductor (i.e. in semiconductor quantum wells, quantum wires, quantum dots, superlattices, and nanostructures). That is, when the carriers in the semiconductor are confined by potential barriers to regions of space that are smaller than or comparable to their deBroglie wavelength or to the Bohr radius of excitons in the semiconductor bulk, the relaxation dynamics can be dramatically altered; specifically, the hot-carrier cooling rates may be dramatically reduced, and the rate of impact ionization could be enhanced and become competitive with the rate of carrier cooling (21). These effects are discussed extensively in this review.

RELAXATION DYNAMICS OF HOT ELECTRONS IN QUANTUM WELLS AND SUPERLATTICES

Semiconductors show dramatic quantization effects when charge carriers are confined by potential barriers to small regions of space where the dimensions of the confinement are less than their deBroglie wavelength; the length scale at which these effects begin to occur ranges from about 10 to 50 nm for typical semiconductors (groups IV, III-V, II-VI). In general, charge carriers in semiconductors can be confined by potential barriers in one, two, or three spatial dimensions. These regimes are termed quantum films, quantum wires, and quantum dots, respectively (see Figure 2). Quantum films are also more commonly referred to simply as quantum wells (QWs). This is because quantum films were the first quantized semiconductor nanostructures to be studied extensively beginning in the 1970s (22). These structures were produced by molecular beam epitaxy (MBE) and metallo-organic chemical vapor deposition (MOCVD) and represented textbook examples of a one-dimensional quantum well (23–27).

One-dimensional quantum wells, hereafter called quantum films or just quantum wells, are usually formed through epitaxial growth of alternating layers of semiconductor materials with different band gaps. A single QW is formed from one semiconductor sandwiched between two layers of a second semiconductor having a larger band gap; the center layer with the smaller band gap forms the QW while the two layers sandwiching the center layer create the potential barriers (28) (see Figure 6.6 of Reference 28). Two potential wells are actually formed in the QW structure; one is for conduction band electrons, the other for valence-band holes. The well depth for electrons is the difference (i.e. the offset) between the conduction band edges of the well and barrier semiconductors, whereas the well depth for holes is the corresponding valence band offset (see Figure 6.6 of Reference 28). If the offset for either the conduction or valence bands is zero, then only one carrier will be confined in a well.

Quantization Configurations

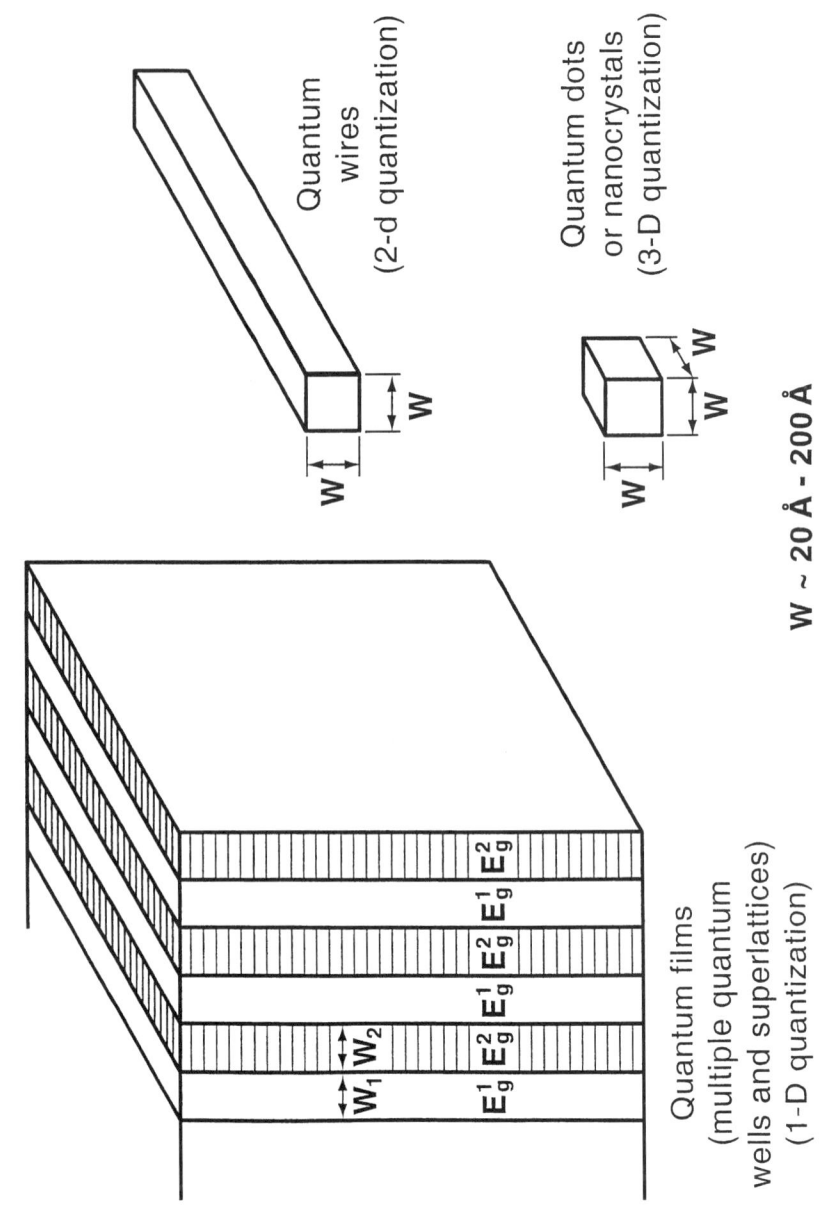

Multiple quantum well structures consist of a series of QWs (i.e. a series of alternating layers of wells and barriers). If the barrier thickness between adjacent wells prevents significant electronic coupling between the wells, then each well is electronically isolated; this type of structure is termed a multiple quantum well (MQW). Alternatively, if the barrier thickness is sufficiently thin to allow electronic coupling between wells (i.e. there is significant overlap of the electronic wavefunctions between wells), then the electronic charge distribution can become delocalized along the direction normal to the well layers. This coupling also leads to a broadening of the quantized electronic states of the wells; the new broadened and delocalized quantized states are termed minibands (see Figure 3). A multiple QW structure that exhibits strong electronic coupling between the wells is termed a superlattice. The critical thickness at which miniband formation just begins to occur is about 40 Å (27, 29); the electronic coupling increases rapidly with decreasing thickness, and miniband formation is very strong below 20 Å (27) (see Figure 18, Chapter 1 of Reference 27). Superlattice structures yield efficient charge transport normal to the layers because the charge carriers can move through the minibands; the narrower the barrier, the wider the miniband and the higher the carrier mobility. Normal transport in MQW structures (thick barriers) requires thermionics emission of carriers over the barriers or, if electric fields are applied, field-assisted tunneling through the barriers (30).

Energy Levels and Density of States in Quantum Wells and Superlattices

The confinement of electrons or holes in potential wells leads to the creation of discrete energy levels in the wells, compared to the continuum of states in bulk material; quantization also leads to a major change in the density of states. The energy levels can be calculated by solving the Schrödinger equation for the well-known "particle in a box" problem. Using the effective mass envelope function approximation (27, 31–34), the electron wave function, Ψ_n, is

$$\Psi_n = \sum_{W,B} e^{i\mathbf{k}\cdot\mathbf{r}} U_\mathbf{k}^{W,B}(\mathbf{r}) \phi_n(z), \qquad 3.$$

where W and B refer to well and barrier, respectively; k is the transverse electron wave vector; z is the growth direction, $U_k^{W,B}(\mathbf{r})$ is the Bloch wave function in W or B; and $\phi_n(z)$ is the envelope wave function. The envelope wave function is determined by solving the Schrödinger equation:

$$\left(\frac{\hbar^2}{2m^*} \frac{\partial^2}{\partial z^2} + V_c(z) \right) \phi_n(z) = E_n \phi_n(z), \qquad 4.$$

Figure 2 Three types of quantization configurations. The type depends upon the dimensionality of carrier confinement.

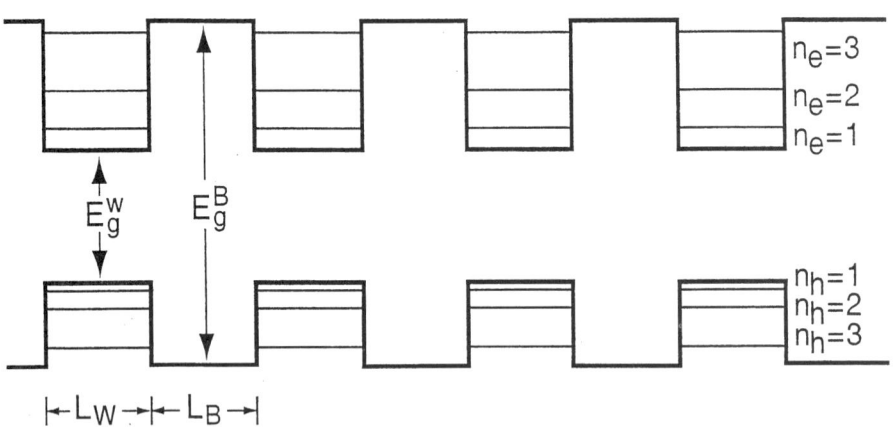

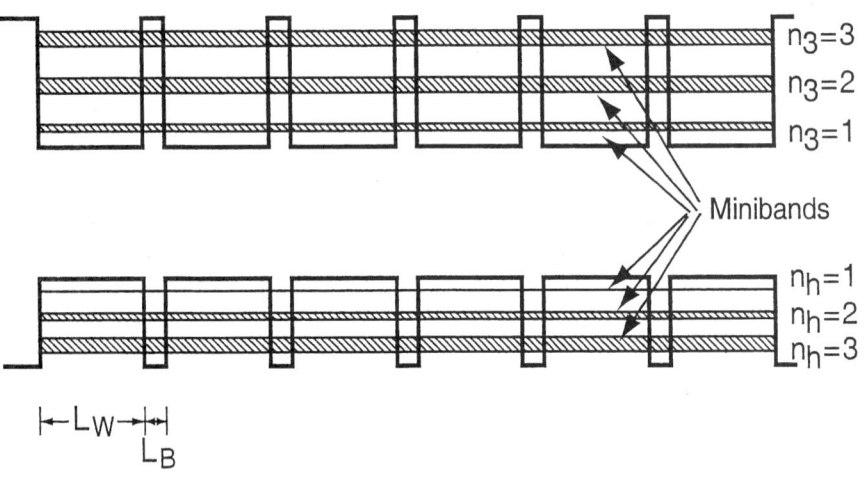

Figure 3 Difference in electronic states between multiple quantum well structures (barriers >40 Å) and superlattices (barriers <40 Å); miniband formation occurs in the superlattice structure, which permits carrier delocalization.

where $V_c(z)$ is the potential barrier function, and E_n is the quantized energy level in the well.

Considering first isolated QWs that are infinitely deep, the solution to Equation 4 becomes very simple because the wave functions must be zero at the well-barrier interfaces. This leads to the well-known solutions:

$$\psi_n = A \sin \frac{n\pi z}{L_w} \qquad 5.$$

$$E_n = \frac{\hbar^2}{2m^*} \left(\frac{n\pi}{L_w}\right)^2, n = 1, 2, 3, \ldots, \qquad 6.$$

where L_w is the well width.

For the more realistic case of finite barrier heights, where the effective masses in the well and barrier are taken to be equal,

$$\phi_n(z) = A \cos kz \quad \text{(inside well)} \qquad 7.$$

$$= B \exp[-k(z - L_w/2)] \quad \text{(outside well)}, \qquad 8.$$

where

$$E_n = \frac{\hbar^2 k^2}{2m^*} - V_o \quad -V_o < E_n < 0 \qquad 9.$$

and

$$\cos(kL_W/2) = k/k_o \quad \text{for} \quad \tan kL_W/2 > 0 \qquad 10.$$

$$\sin(kL_W/2) = k/k_o \quad \text{for} \quad \tan kL_W/2 > 0, \qquad 11.$$

where

$$k_o = 2m^* V_o/\hbar^2, \quad \text{and} \quad V_o \text{ is the barrier height.} \qquad 12.$$

There is always at least one state in the well no matter how thin it is; the number of bound states is given by:

$$1 + Int\left[\frac{(2m^* V_o L_W^2)^{1/2}}{\pi^2 \hbar^2}\right], \qquad 13.$$

where Int(x) indicates the integer part of the argument.

The calculation of the hole levels is much more complicated because the band structure of many important semiconductors has hole bands with fourfold degeneracy at $k = 0$. This leads to heavy and light holes with different effective masses. Consequently, in the QW a double set of hole energy levels is formed with different spacings between levels—one set for the light holes, the second set for the heavy holes (see Figure 6.6 of Reference 28). Solutions to the problem have been reported for both infinite (31, 33) and finite (34) potential barriers.

For superlattices several approaches have been used to calculate the energy level structure of the minibands (31, 33, 34). One approach is to use a tight-binding model for the multiple wells leading to a Bloch-like envelope function of the form (27, 35).

$$\psi_q^i(z) = \frac{1}{\sqrt{N_w}} \sum_l e^{iqlx} \phi^i(z - lx), \qquad 14.$$

where N_w is the number of wells, $\phi(z - lx)$ is the ith wave function of the QW centered at $z = lx$ and q is the Bloch wave vector.

This approach leads to results that are summarized in Figure 18, Chapter 1 of Reference 27, which shows the energy levels as a function of barrier thickness. That figure shows the formation of minibands with energy band widths that increase rapidly with decreasing barrier widths below 40 Å.

The density of states (DOS or N(E)) also shows profound changes with the dimensionality of quantization. For ideal bulk semiconductors with simple parabolic energy bands, the DOS has a square root dependence on electron energy

$$N(E)_{Bulk} = \sqrt{2}(m^*)^{3/2}\pi^{-2}\hbar^{-3}E^{1/2}. \qquad 15.$$

For an ideal QW the DOS shows a step-like function with each plateau having a DOS of:

$$N(E)_{QW} = nm^*/\pi\hbar^2, \qquad 16.$$

where n is the quantum number of the state.

For a superlattice the steepness of the steps is destroyed by the dispersion of the N states in a miniband (23, 27):

$$N(E)_{SL} = N\frac{m^*}{\pi\hbar^2} \arccos\left(\frac{\varepsilon_i - E_i - s_i}{2t_i}\right) \qquad 17.$$

where ε_i, s_i, and t_i are defined by Equations 29 and 30, Chapter 1 of Reference 27. These DOS functions for bulk semiconductors, QWs, superlattices, quantum wires, and quantum dots are shown in Figure 6.2 of Reference 28 and in Figure 17, Chapter 1 of Reference 27; the DOS for quantum dots exhibits discrete values at the discrete quantized energy levels.

Fabrication of Quantum Wells and Superlattices

QWs are produced through epitaxial growth of crystalline films via either molecular beam epitaxy (MBE) or metallo-organic chemical vapor deposition (MOCVD). Both of these techniques are capable of creating epitaxial layers of sufficient quality to produce quantization effects. These qualities include film thickness uniformity and interfacial abruptness (both within a few atomic layers), crystalline perfection, and compositional uniformity.

Molecular Beam Epitaxy In this system an ultrahigh–vacuum chamber is outfitted with a number of evaporation (i.e. effusion) cells, each controlled by a separate shutter, which supply fluxes of molecular beams of the desired atomic species. The beams can be turned on and off within 0.1 sec. Growth rates are typically 5 Å/sec. An ion-beam sputtering gun is used first to ion etch the surfaces to remove impurities and imperfections. The substrate is maintained at about 500–700°C during growth. Various spectroscopic capabilities, such as mass spectrometry, Auger spectroscopy, and reflection high-energy electron diffraction, are included in the chamber to control the process and provide analytical data on the quality of the films. The atomic and molecular species that can be produced by MBE include Ga, Al, In, As, Sb, Sn, Be, Ge, Se, Te, Cd, Hg, Zn, Mn, Pb, and Si. The most common QW materials produced by MBE are the III-V semiconductor binary and ternary compounds, such as $GaAs/Al_xGa_{1-x}As$, $GaSb/Al_xGa_{1-x}Sb$, $InAs/GaAs$, $Ga_xIn_{1-x}As/Al_xIn_{1-x}As$, $GaAs/Ga_xIn_{1-x}P_2$), and $InP/Ga_xIn_{1-x}As$. Some II-VI semiconductor QWs have also been prepared, such as CdTe/HgTe and $ZnSe/Zn_{1-x}Mn_xSe$.

Metallo-Organic Chemical Vapor Deposition The MOCVD process is generally only used to prepare III-V semiconductor QWs. In this process the group III metals are introduced into a reaction chamber in the form of metallo-organic vapors that react at high temperatures with gaseous precursors of the nonmetalloid group V component of the desired semiconductor compound to form a crystalline, epitaxial film on a heated substrate. Ga, Al, and In are commonly introduced as trimethyl gallium (TMG), trimethyl aluminum (TMA), and trimethyl indium (TMI). As and P are usually introduced as arsine and phosphine, although less toxic compounds such as tertiary butyl arsine and tertiary butyl phosphine are being increasingly utilized. The organometallic compounds are kept in liquid form at a sufficiently low, but constant, temperature and are swept into the reaction chamber at a controlled composition by sparging hydrogen gas at a controlled flow rate through the liquid metallo-organics. Gaseous arsine or phosphine is fed directly into the system from gas cylinders through flow and pressure controllers; ultrapure hydrogen is used as the carrier gas for all reactant flows. The single crystal substrate is placed on a graphite block that is heated either by radio frequency (RF) or resistance heaters to 650 to 750°C. Growth rates are typically 5–10 Å/sec.

The reaction chamber is operated either at atmospheric pressure or at reduced pressure (50–100 torr). The low pressure system can yield very sharp interfaces between the semiconductor heterojunctions and very uniform epilayers (see Figure 4).

Optical Spectroscopy of Quantum Wells and Superlattices

Optical Absorption Optical transitions between quantum levels can occur upon excitation with light. The interband transition probability is the product of an optical matrix element (M) times a DOS. M involves the electric dipole operator

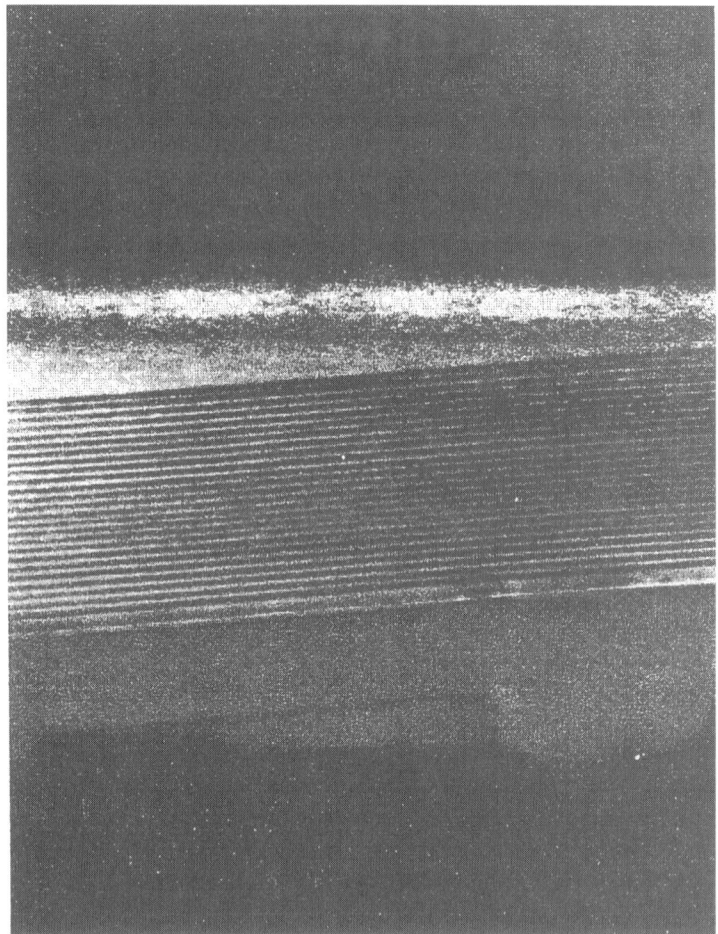

Figure 4 TEM of GaAs/Al$_{0.32}$Ga$_{0.68}$As superlattice structure. GaAs wells are 52 Å thick, and the Al$_{0.32}$Ga$_{0.68}$As barriers are 17 Å thick.

and can be expressed as (23)

$$M = \langle \phi_e(z)e^{ik_e \bullet r}U_{cke}(r)|\eta|\phi_h(z)e^{ik_h \bullet r}U_{vkh}(r)\rangle, \qquad 18.$$

where $\phi_e(z)$ and $\phi_h(z)$ are the electron and hole envelope wave functions, k_e and k_h are electron and hole wave vectors, η is the polarization vector of the light, and $U_{cke}(r)$ and $U_{vkh}(r)$ are the Bloch functions. The integration of Equation 18 produces a factor $[\phi_e(z)\,\phi_h(z)]$, which becomes unity for transitions between electron and hole states with the same quantum number. Thus, the selection rule for optical interband transitions is that $\Delta n = 0$, where n is the quantum number of the energy level in the well. Some additional factors affecting optical transitions are: (*a*) Heavy hole–electron transitions are expected to be three times stronger than light hole–electron transitions; (*b*) transitions involving different *n* values become possible in the presence of an electric field and also when finite wells are considered; and (*c*) parity is a good quantum number, and transitions involving the same parity are allowed.

In light of the above discussion, the optical absorption spectrum of a quantum film is expected to consist of a series of steps, with the position of these steps corresponding to the transitions between heavy- or light hole–quantum states and electron quantum states following the selection rule $\Delta n = 0$. Furthermore, because the widths of the wells are commonly smaller than the calculated diameter of an exciton, the exciton binding energy is greatly increased in QWs, and the excitons can be stable even at room temperature. Thus, the absorption spectra of QWs can be expected to show exciton peaks, even at room temperature, which occur at energies below the step. Such spectra have indeed been observed by many workers (22, 35).

Photoreflectance Spectroscopy Photoreflectance (PR) is one of the powerful spectroscopic techniques for characterizing the electronic energy level structure of QWs. PR is particularly sensitive to the optical transitions in the well because the observed signal reflects the third derivative of the optical constants (36, 37). The experiment is performed by modulating the electric field in the space charge region of the sample with a pump beam having a photon energy above the band gap, and scanning the sample with an overlapping probe beam that is swept through the desired photon energy range. The pump beam is typically from a laser, whereas the probe beam is at low intensity and produced by passing white light from a lamp through a monochromator. The reflectivity of the probe beam is measured as a function of photon energy (E), and the fractional change in reflectivity is given by

$$\Delta R/R = \text{Re} \left| \sum_j \left[C_j e^{i\theta_j} (E - E_{gj} + i\Gamma_j)^{-p_j} \right] \right|, \qquad 19.$$

where subscript *j* refers to the *j*th transition, *C* is the amplitude, E_g is the transition energy, Γ is the broadening parameter, $\mathit{2}$ is the phase, and *p* denotes the type of critical point for the transition and the order of its derivative (38). The value of p depends upon whether the transitions involve localized (e.g. excitonic) or delocalized states (e.g. minibands). The sample can be run in air or as an electrode in a photoelectrochemical cell; in the former case the space charge is produced through equilibration with intrinsic surface states.

PR spectroscopy is very sensitive (29), as illustrated in Figure 1 of Reference 29, in which the PR spectrum of a superlattice containing 20 periods of 250 Å GaAs wells and 17 Å $Al_{0.28}Ga_{0.72}As$ barriers is shown; 32 optical transitions can be clearly seen in the spectrum (29). Because of the thin barriers in this superlattice sample, strong miniband formation occurs with appreciable miniband widths. The calculated energy levels for this sample showing the miniband structure is presented in Figure 6.13 of Reference 28. Transitions to the top and bottom of a given miniband can be observed in PR spectroscopy; the predicted 32 transitions satisfying the selection rule $\Delta n = 0$ are compared with the observed PR transitions in Figure 6.12c of Reference 28. The agreement is excellent. It is quite impressive that PR can resolve the transitions between minibands even when the minibands overlap each other in energy.

Hot Electron Cooling Dynamics in Quantum Wells and Superlattices

Hot-electron cooling times can be determined from several types of time-resolved photoluminescence (PL) experiments. One technique involves hot luminescence nonlinear correlation (39–41), which is a symmetrized pump-probe type of experiment. Figure 2 of Reference 39 compares the hot-electron relaxation times as a function of the electron energy level in the well for bulk GaAs and a 20-period MQW of GaAs/Al$_{0.38}$Ga$_{0.62}$As containing 250 Å GaAs wells and 250 Å Al$_{0.38}$Ga$_{0.62}$As barriers. For bulk GaAs the hot-electron relaxation time varies from about 5 ps near the top of the well to 35 ps near the bottom of the well. For the MQW the corresponding hot-electron relaxation times are 40 ps and 350 ps.

Another method uses time-correlated single photon counting to measure PL lifetimes of hot electrons. Figure 5 shows 3-dimensional (3-D) plots of PL intensity as a function of energy and time for bulk GaAs and a 250 Å GaAs/ 250 Å Al$_{0.38}$Ga$_{0.62}$As MQW (16). It is clear from these plots that the MQW sample exhibits much longer-lived hot luminescence (i.e. luminescence above the lowest n = 1 electron to heavy-hole transition at 1.565 eV) than bulk GaAs. Depending upon the emitted photon energy, the hot PL for the MQW is seen to exist beyond times ranging from hundreds to several thousand ps. On the other hand, the hot PL intensity above the band gap (1.514 eV) for bulk GaAs is negligible over most of the plot; it is only seen at the very earliest times and at relatively low photon energies.

Calculations were performed (16) on the PL intensity versus time and energy results to determine the time dependence of the quasi-Fermi-level, electron temperature, electronic specific heat, and ultimately the dependence of the characteristic hot-electron cooling time on electron temperature.

The cooling, or energy–loss, rate for hot electrons is determined by longitudinal optical (LO) phonon emission through electron-LO-phonon interactions. The time constant characterizing this process can be described by the following expression (42–44):

$$P_e = -\frac{dE}{dt} = \frac{\hbar\omega_{LO}}{\tau_{avg}} \exp(-\hbar\omega_{LO}/kT_e), \qquad 20.$$

where P_e is the power loss of electrons (i.e. the energy-loss rate), $\hbar\omega_{LO}$ is the LO phonon energy (36 meV in GaAs), T_e is the electron temperature, and τ_{avg} is the time constant characterizing the energy-loss rate.

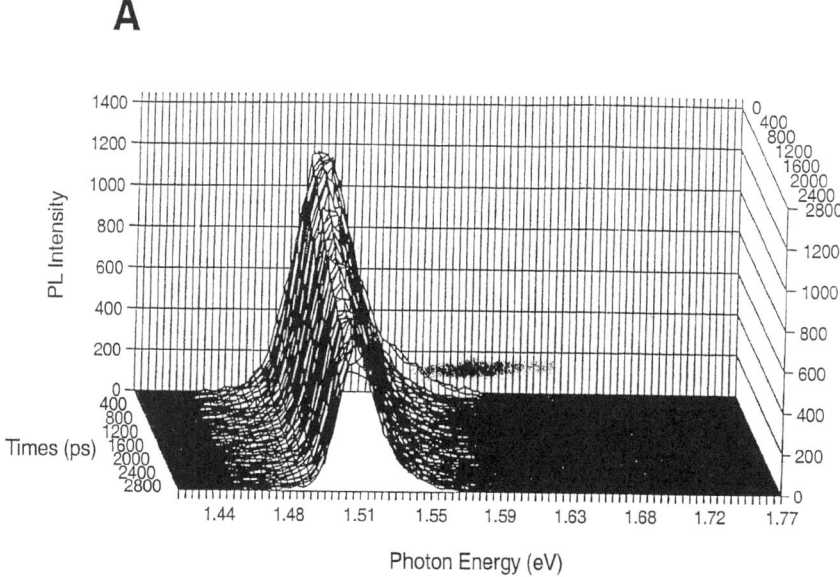
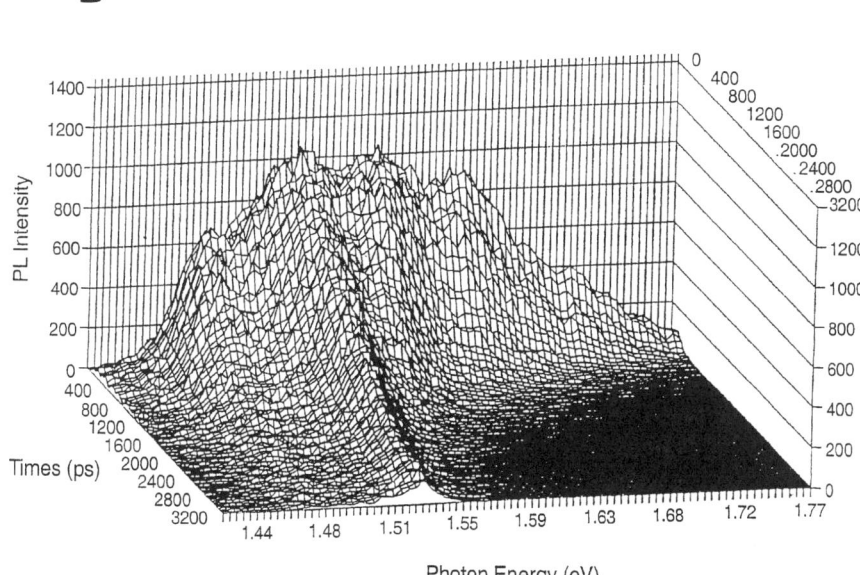

Figure 5 Three-dimensional plots of photoluminescence intensity versus time and photon energy for (*A*) bulk GaAs and (*B*) 250 Å GaAs/250 Å Al$_{0.38}$Ga$_{0.62}$As MQW (16).

The electron energy–loss rate is related to the electron temperature–decay rate through the electronic specific heat. Because at high light intensity the electron distribution becomes degenerate, the classical specific heat is no longer valid. Hence, the temperature- and density-dependent specific heat for both the QW and bulk samples need to be calculated as a function of time in each experiment so that τ_{avg} can be determined.

The results of such calculations (presented in Figure 2 of Reference 16) show a plot of τ_{avg} versus electron temperature for bulk and MQW GaAs at high and low carrier densities. These results show that at a high carrier density [$n_c \sim (2-4) \times 10^{18}$ cm^{-3}] the τ_{avg} values for the MQW are much higher (τ_{avg} = 350–550 ps for T_e between 440 and 400 K) compared to bulk GaAs (τ_{avg} = 10–15 ps over the same T_e interval). Alternatively, at a low carrier density [$n_c \sim (3-5) \times 10^{17}$ cm^{-3}] the differences between the τ_{avg} values for bulk and MQW GaAs are much smaller.

A third technique used to measure cooling dynamics is PL upconversion (16). Time-resolved luminescence spectra were recorded at room temperature for a 4000-Å bulk GaAs sample at the incident pump powers of 25, 12.5, and 5 mW. The electron temperatures were determined by fitting the high-energy tails of the spectra; only the region that is linear on a semilogarithmic plot was chosen for the fit. The carrier densities for the sample were 1×10^{19}, 5×10^{18}, and 2×10^{18} cm^{-3}, corresponding to the incident excitation powers of 25, 12.5, and 5 mV, respectively. Similarly, spectra for the MQW sample were recorded at the same pump powers as the bulk. Figure 6 shows τ_{avg} for bulk and MQW GaAs at the three light intensities, again showing the much slower cooling in MQWs (by up to two orders of magnitude).

The difference in hot-electron relaxation rates between bulk and quantized GaAs structures is also reflected in time-integrated PL spectra. Typical results are shown in Figure 7 for single photon counting data taken with 13 ps pulses of 600 nm light at 800 kHz focused to about 100 μm with an average power of 25 mW (45). The time-averaged electron temperatures obtained from fitting the tails of these PL spectra to the Boltzman function show that the electron temperature varies from 860 K for the 250 Å/250-Å MQW to 650 K for the 250 Å/17-Å superlattice, whereas bulk GaAs has an electron temperature of 94 K, which is close to the lattice temperature (77 K). The variation in the electron temperatures between the quantized structures can be attributed to differences in electron delocalization between MQWs and superlattices (SLs), and the associated nonradiative quenching of hot-electron emission.

As shown above, the hot-carrier cooling rates depend upon photogenerated carrier density; the higher the electron density, the slower the cooling rate. This effect is also found for bulk GaAs, but it is much weaker compared with quantized

Figure 6 Time constant for hot-electron cooling (τ_{avg}) versus electron temperature for bulk GaAs and GaAs multiple quantum wells at three excitation intensities (16).

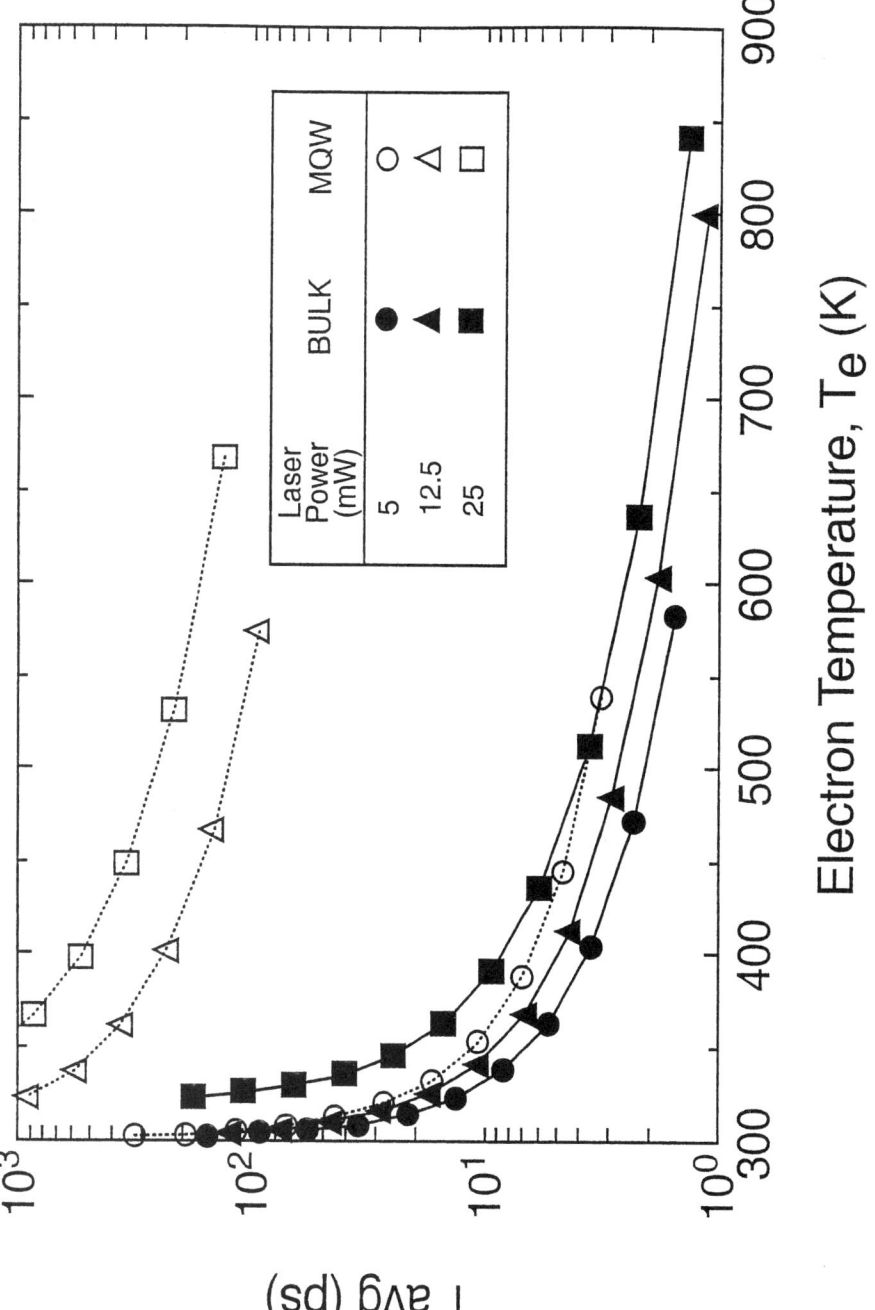

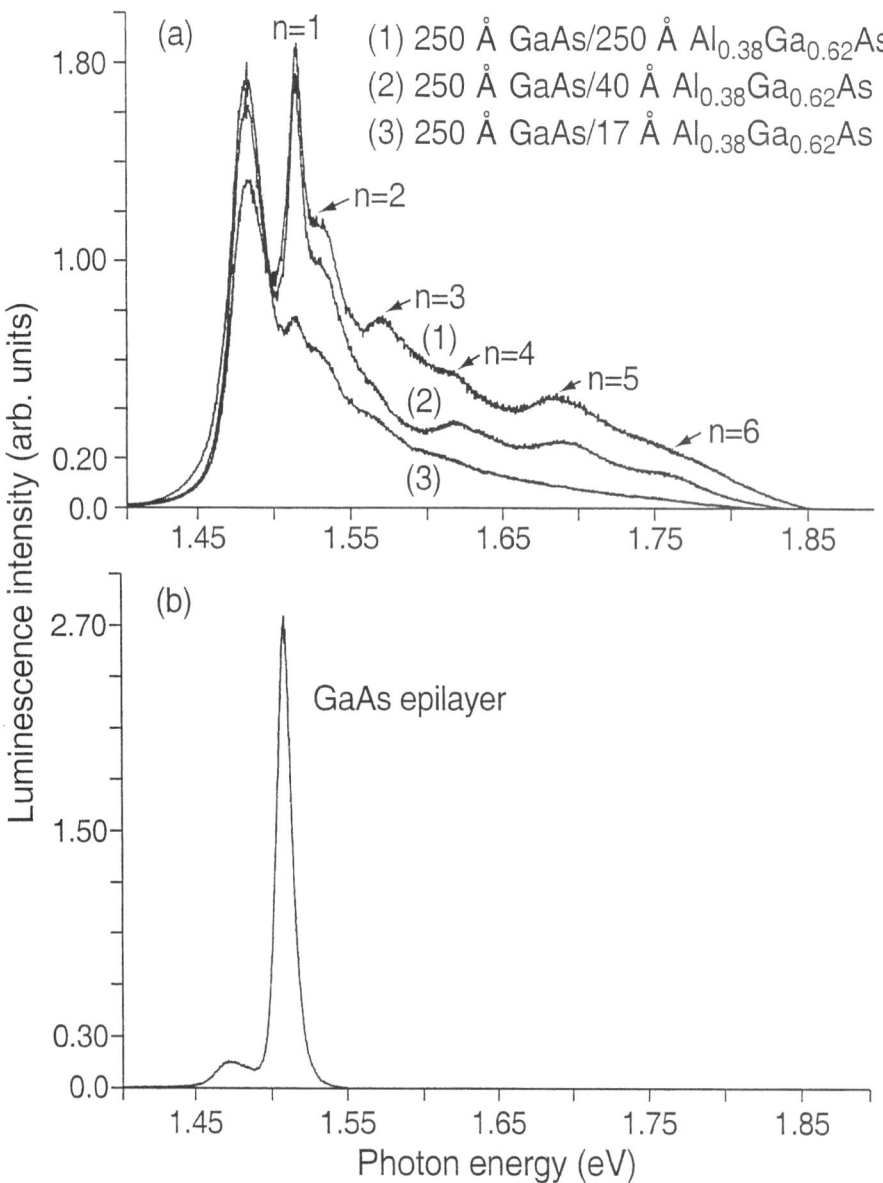

Figure 7 (*a*) Time-integrated photoluminescence spectra for multiple quantum wells and SLs showing hot luminescence tails and high energy peaks arising from hot-electron radiative recombination form upper quantum levels. (*b*) Equivalent spectrum for bulk GaAs showing no hot luminescence (45).

GaAs. The most generally accepted mechanism for the decreased cooling rates in GaAs QWs is an enhanced "hot-phonon bottleneck" (46–48). In this mechanism a large population of hot carriers produces a nonequilibrium distribution of phonons (in particular, LO phonons that are the type involved in the electron-phonon interactions at high carrier energies) because the LO phonons cannot equilibrate fast enough with the crystal bath; these hot LO phonons can be reabsorbed by the electron plasma to keep it hot. In QWs the phonons are confined in the well and they exhibit slab modes (47), which enhance the hot-phonon bottleneck effect.

RELAXATION DYNAMICS OF HOT ELECTRONS IN QUANTUM DOTS

As discussed above in "Hot-Electron Cooling Dynamics in Quantum Wells and Superlattices," slowed hot-electron cooling in QWs and superlattices that is produced by a hot phonon bottleneck requires very high light intensities to create the required photogenerated carrier density of greater than about 1×10^{18} cm^{-3}. This required intensity, possible with laser excitation, is many orders of magnitude greater than that provided by solar radiation at the Earth's surface (maximum solar photon flux is about 10^{18} cm^{-2} s^{-1}; assuming a carrier lifetime of 1 ns and an absorption coefficient of 1×10^5 cm^{-1}, this translates into a photoinduced electron density of about 10^{14} cm^{-3} at steady state). Hence, it is not possible to obtain slowed hot-carrier cooling in semiconductor QWs and superlattices with solar irradiation via a hot-phonon bottleneck effect; solar concentration ratios greater than 10^4 would be required, resulting in severe practical problems.

However, the situation with 3-D confinement in quantum dots (QDs) is potentially more favorable. In the quantum dot case, slowed hot-electron cooling is theoretically possible even at arbitrarily low light intensity; this effect is simply called a phonon bottleneck, without the qualification of requiring hot phonons (i.e. a nonequilibrium distribution of phonons). Furthermore, there is a possibility that the slowed cooling could make the rate of impact ionization (inverse Auger effect) an important process in QDs (21). PL blinking in QDs (intermittent PL as a function of time) has been explained (49, 50) by an Auger process whereby if two electron-holes pairs are photogenerated in a QD, one pair recombines and transfers its recombination energy to one of the remaining charge carriers, ionizing it over the potential barrier at the surface into the surface region. This creates an electric dipole in the QD that quenches subsequent radiative emission after subsequent photon absorption; after some time the ionized electron can return to the QD core, and the PL is turned on again. Because this Auger process can occur in QDs, the inverse Auger process, whereby one high energy electron-hole pair (created from a photon with $h\nu > E_g$) can generate two electron-hole pairs, can also occur in QDs (21). The following discussion first presents general properties of quantum dots followed by a discussion of the hot-carrier cooling dynamics.

Energy Levels and Density of States in Quantum Dots

The earliest and simplest treatment of the electronic states of a QD is based on the effective mass approximation (EMA); the simple EMA treatment was subsequently improved by incorporating the **k** • **p** approach that has been commonly used to calculate the electronic structure of bulk semiconductor and QW structures.

The EMA rests on the assumption that if the QD is larger than the lattice constants of the crystal structure, then it will retain the lattice properties of the infinite crystal and the same values of the carrier effective masses; the electronic properties of the QD can then be determined by simply considering the modification of the energy of the charge carriers produced by the quantum confinement. Thus, the electronic properties are determined by solving the Schrödinger equation for a particle in a 3-D box. The zeroth order approximation is a perfectly spherical QD with infinite potential walls at the surface. Strong confinement is defined as the case where the QD size is small compared with the deBroglie wavelength of electrons in the box or compared with the Bohr radius of the electrons; this is the case for II-VI and III-V semiconductors. Taking into account the Coulomb interaction between electrons and holes that is enhanced owing to confinement in the QD, the Hamiltonian can then be written as

$$\widehat{H} = -\frac{\hbar^2 \nabla_e^2}{2m_e^*} - \frac{\hbar^2 \nabla_h}{2m_h^*} - \frac{e^2}{\varepsilon |r_e - r_h|} + V_e(r_e) + V_h(r_h) \qquad 21.$$

and

$$\widehat{H}\Psi(r) = E\Psi(r), \qquad 22.$$

where V_e and V_h are the confining potentials, r_e and r_h are the distances of the electron and hole from the center of the QD, and ε is the dielectric constant of the semiconductor.

Analytical solutions of Equations 21 and 22 are difficult because center-of-mass motion and reduced mass motion cannot be separated as independent coordinates. Various approaches to solving this problem have been used. These include perturbation theory (51, 52), variational calculations (53–59), matrix diagonalization (60, 61), and Monte Carlo methods (62). Perturbation theory (51, 52) and variational calculations (57) lead to a solution of the form

$$E_{\min} = \frac{\hbar^2 \pi^2}{2R^2}\left[\frac{1}{m_e^*} + \frac{1}{m_h^*}\right] - \frac{1.8 e^2}{\varepsilon R} - 0.25 E_{Ryd}^*, \qquad 23.$$

where $E_{\min}$ is the lowest energy separation between hole and electron states in the QD, E_{Ryd}^* is the bulk exciton binding energy in meV, and R is the QD radius. $E_{\min}$ is often referred to as the band gap of the QD because it represents the threshold energy for photon absorption, blue-shifted from the bulk band gap, E_g.

Improvements in the EMA model involve accounting for band nonparabolicity and hole-state mixing (63–68), finite barrier heights (56), and surface polarization (69).

In the effective mass **k** • **p** approach, the QD wavefunctions are expanded as a linear combination of N_b bulk Bloch functions at the **k** = 0 Brillouin zone center (Γ point) (64, 67); for these calculations values of N_b up to 8 (termed the 8-band **k** • **p** model) have been used (70). The effective mass **k** • **p** approach has been used very extensively for QDs, but recently its validity has been attacked (71–73) and vigorously defended (74, 75). An alternative approach based on direct diagonalization and pseudopotentials has been proposed as a better theoretical approach to calculate the electronic states in QDs (71, 72); the debate continues. A comparison of the calculated electron and hole states for a 28-Å InP QD using a 6-band **k** • **p** model (76) and the direct diagonalization pseudopotential model (76) indicates that there are more electronic states for the latter than the former, and that in the latter the highest lying hole state is s-like rather than p-like. However, improvements in the **k** • **p** parameters can reverse the character of the lowest level hole states and increase the total number of electronic states as well (74).

Because of the 3-D spatial confinement in QDs, the solution of the Schrödinger equation results in describing the electronic states in the QD by three quantum numbers plus spin. A commonly used notation (77, 78) is for the electron states to be labeled as nL_e and the hole states as nL_F, where n is the principal quantum number (1, 2, 3, etc), L is the orbital angular momentum (S, P, D, etc), and F is the total angular momentum (F = L + J, and J = L + S), where S is the spin, and the projection of F along a magnetic axis is m_F = −F to +F. Thus, electron states become $1S_e$, $2S_e$, $1P_e$, etc, and hole states become $1S_{1/2}$, $1S_{3/2}$, and $1P_{1/2}$, etc. For optical transitions in ideal spherical QDs, the selection rules are Δn = 0, ΔL = 0, ±2; and ΔF = 0, ±1. These ideal selection rules can be broken by nonspherical QDs and strong hole-state mixing.

Synthesis of Quantum Dots

Colloidal Quantum Dots   The most common approach to the synthesis of colloidal QDs is the controlled nucleation and growth of particles in a solution of chemical precursors containing the metal and the anion sources (controlled arrested precipitation) (79–81). The technique of forming monodisperse colloids is very old, and can be traced back to the synthesis of gold colloids by Michael Faraday in 1857 (80, 81). A common method for II-VI colloidal QD formation is to rapidly inject a solution of chemical reagents containing the group II and group VI species into a hot and vigorously stirred solvent containing molecules that can coordinate with the surface of the precipitated QD particles (79, 81, 82). Consequently, a large number of nucleation centers are initially formed, and the coordinating ligands in the hot solvent prevent or limit particle growth via Ostwald ripening (i.e. the growth of larger particles at the expense of smaller particles to minimize the higher surface free energy associated with smaller particles). Further improvement of the resulting size distribution of the QD particles can be achieved through size-selective precipitation (81, 82), whereby slow addition of a nonsolvent to the colloidal solution of particles causes precipitation of the larger-sized particles (the solubility of molecules with the same type of chemical structure decreases with increasing size). This process can be repeated several times to narrow the size distribution of II-VI colloidal QDs to a few percent of the mean diameter.

The synthesis of colloidal III-V QDs is more difficult than for II-VI QDs because (a) the synthesis must be conducted in rigorously air-free and water-free atmospheres, (b) it generally requires higher reaction temperatures and much longer reaction times, and (c) it involves more complicated organometallic chemistry. The best results to date have been obtained for InP QDs (83–88). In this synthesis, an indium salt [In(C_2O_4)$_3$, InF$_3$, or InCl$_3$] is reacted with trimethylsilylphosphine [P(Si(CH$_3$)$_3$)$_3$] in a solution of trioctylphosphine oxide (TOPO) and trioctylphosphine (TOP) to form a soluble InP organometallic precursor species that contains In and P in a 1:1 ratio. The InP precursor species in the TOPO/TOP solution is then heated for several days at a temperature ranging from 270 to 290°C, depending upon desired QD properties. A transmission electron microscope (TEM) picture of InP QDs showing lattice imaging is presented in Figure 8. One difference with the synthesis for II-VI materials is that more than 1 day of heating at reaction temperature is required to form crystalline III-V QDs, whereas II-VI QDs form immediately upon injection of the reactants into the hot TOPO/TOP solution.

The resulting InP QDs contain a capping layer of TOPO, which can be readily exchanged for several other types of capping agents, such as thiols, pyridines, amines, and polymers. The size distribution of the InP QDs can be further narrowed to less than 10% through selective precipitation techniques. Finally, they can be studied in the form of colloidal solutions or powders or dispersed in transparent polymers or organic (for low temperature studies) glasses; capped InP QDs recovered as powders can also be redissolved to form transparent colloidal solutions.

In addition to InP QDs, similar methods have been used to produce QDs of GaP (83, 89), GaInP$_2$ (83, 89), GaAs (83, 89–92), and InAs (93); higher temperatures (400°C) are required for GaP and GaInP$_2$ QDs. However, the quality (especially narrow size distributions) of these other III-V QDs is not yet as high as that obtained for InP QDs; this is primarily because less effort has been spent thus far on these other QD preparations.

Quantum Dots Produced by Stranski-Krastinow Epitaxial Growth QDs can also be produced via epitaxial growth from the vapor phase on appropriate substrates; the growth can be done using either MBE (see *"Molecular Beam Epitaxy"*) or MOCVD (see *"Metallo-Organic Chemical Vapor Deposition"*) systems. In the Stranski-Krastinow (SK) mode of QD formation, a thin semiconductor film is deposited by MBE or MOCVD onto another semiconductor substrate material that has a mismatch in lattice constant with the material being deposited. If the lattice mismatch is sufficiently high and the film has a higher surface energy than the substrate, then after the initial growth of a few monolayers (called the wetting layer), subsequent deposition forms small islands that are in the QD size regime. The size, shape, perfection, and density of these SK QDs can be controlled by controlling the conditions in the MBE or MOCVD reactor. The bulk of work on these SK QDs

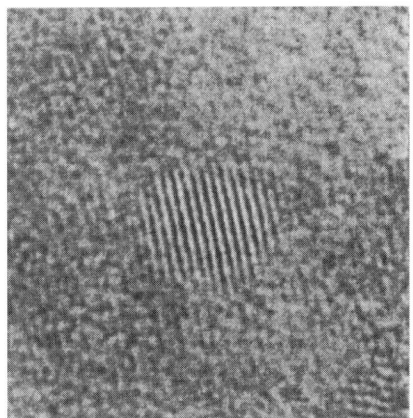

Figure 8 High-resolution TEM of a 60-Å diameter InP QD oriented with the +111, axis in the plane of the micrograph. The bottom plate shows a rare dislocation defect.

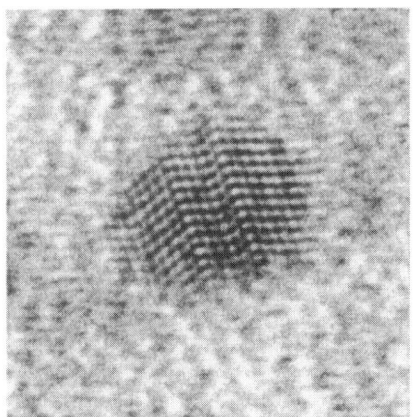

has been on $InGa_xAs_{1-x}$ QDs grown on GaAs substrates (94); InP QDs grown on GaAs or $Al_xGa_{1-x}As$ substrates has also been reported (95), as has work on GeQDs on Si (95a). The size of SK QDs is generally larger than that of colloidal QDs; they typically have a lateral dimension of about 1000 to 2000 Å and a height of 100 to 500 Å. For III-V semiconductors these sizes produce quantization effects because their Bohr radius is large, but the degree of quantization is not as strong as that in colloidal QDs in which the sizes range from 15 to 100 Å. The shape of the InAs/GaAs SK QDs has been reported to be a square-based pyramid (96); however, recent results suggest a parallelogram base and C_{2v} symmetry (97, 98).

Another configuration of SK-type QDs is possible. In this case a quantum well (QW) is first formed that has a thin outer barrier (about 100 Å). SK islands of another semiconductor material having a lattice mismatch with the barrier are then deposited on top of the barrier. These SK islands produce a parabolic strain field that propagates down through the barrier into the QW. In the case of InP SK islands (called stressor islands) formed on an AlGaAs/GaAs/AlGaAs QW, the strain field expands the GaAs lattice and reduces the GaAs band gap in the QW just below the InP islands (95). Thus, the GaAs QW beneath the InP stressor islands is converted into a QD because the 1-D confinement of carriers is transformed into 3-D confinement. An important feature of this type of QD is that both the well and barrier region of the QD are made of the same material. This eliminates interface defects and interface states that complicate the relaxation behavior of photogenerated carriers.

Optical Properties of Quantum Dots

In this section the optical properties of InP QDs are discussed. The optical properties of other III-V and II-VI QDs exhibit the same general behavior as InP QDs, and the latter is used to exemplify some of the important features of this behavior.

Optical Absorption and Global Photoluminescence The absorption and emission spectra of initially prepared InP QDs with a mean diameter of 32 Å are shown in Figure 1 of Reference 87. Optical absorption and PL spectra were collected at room temperature; samples were prepared by dispersing washed QDs in toluene.

The absorption spectrum shows an excitonic peak at about 590 nm. The PL spectrum (excitation at 500 nm) shows two emission bands: a weaker one near the band edge with a peak at 655 nm, and a second, stronger, broader band that peaks above 850 nm. The PL band with deep red-shifted subgap emission peaking above 850 nm is attributed to radiative recombination in QD surface states produced by phosphorous vacancies (87). These surface defects and the accompanying red-shifted emission can be completely removed by etching the InP QDs with hydrogen fluoride (HF); intense band-edge emission from the InP QDs can then be observed. After etching, the PL intensity increases by a factor of 10, and the near-infrared emission produced by the deep surface traps is completely removed (87). The near-band-edge quantum yield increases from 30% to 60% as the temperature decreases from 300 to 10 K (87).

The room temperature absorption spectra as a function of QD size ranging from 26 to 60 Å (measured by TEM) are shown in Figure 9 (88). The absorption spectra show one or more broad excitonic peaks that reflect substantial inhomogeneous line broadening arising from the QD size distribution; as expected, the spectra shift to higher energy as the QD size decreases (88). The color of the InP QD samples changes from deep red (1.7 eV) to green (2.4 eV) as the diameter decreases from 60 to 26 Å. Bulk InP is black with a room-temperature band gap of 1.35 eV and an absorption onset at 918 nm. Higher energy transitions above the first excitonic peak in the absorption spectra can also be easily seen in QD samples with mean diameters equal to or greater than 30 Å. The spread in QD diameters is generally about 10% and is somewhat narrower in samples with larger mean diameters; this is why higher energy transitions can be resolved for the larger-sized QD ensembles. All the prepared QD nanocrystallites are in the strong confinement regime because the Bohr radius of bulk InP is about 100 Å. Figure 9 shows typical room-temperature absorption and global emission spectra of the InP colloids, with mean particle diameters ranging from 26 to 60 Å as measured by TEM (88); the excitation energy for all QD sample ensembles in Figure 9 was 2.48 eV, well above their absorption onset in each case.

The global emission peaks in Figure 9 show an increasing red shift from the first excitonic absorption peak as the mean QD size decreases from 60 to 26 Å. The global (nonresonant) red shift is as large as 300 meV for samples with the smallest mean diameter (88). Global PL is defined as that observed when the excitation energy is much higher than the energy of the absorption threshold exhibited in the absorption spectrum produced by the ensemble of QDs in the sample; thus, a large fraction of all the QDs in the sample are excited. The resulting global PL shows a very broad peak (line width of 175–225 meV) that is red-shifted by 100–300 meV from the first absorption peak. The broad PL line width is caused by inhomogeneous line broadening arising from the ~10% size distribution sampled in the global PL experiment. The large global red shift is caused by the volume dominance of the larger particles in the size distribution; the latter will absorb most of the incident photons and will show large red shifts because the PL excitation energy is well above their lowest transition energy.

Size-Selected Photoluminescence If the excitation energy is restricted to the on-set region of the absorption spectrum of the QD ensemble, then a much narrower range of QD sizes is excited; these QDs will have the larger particle sizes in the ensemble. Consequently, the PL spectra from this type of excitation show narrower line widths and smaller red shifts with respect to the excitation energy. This technique is termed fluorescence line narrowing (FLN), the resulting PL spectra being considerably narrowed.

FLN spectra at 10 K are shown in Figure 3 in Reference 88 for InP QDs with a mean diameter of 32 Å (88). FLN/PL spectra are shown for a series of excitation energies (1.895 eV to 2.07 eV) spanning the absorption tail near the onset of absorption for this sample (88). The global PL spectrum produced when the excitation energy (2.41 eV) is deep into the high-energy region of the absorption spectrum is also shown.

FLN spectra can be combined with photoluminescence excitation spectra to determine the resonant red shift (88). The resultant resonant red shift at T = 11 K as a function of PL excitation energy ranges from 4 to 10 meV.

Origin of Resonant Red Shift The origin of the resonant red shift in InP has been recently analyzed theoretically (99, 100). Four possible models were considered, and the experimental results were consistent with a model in which the emission

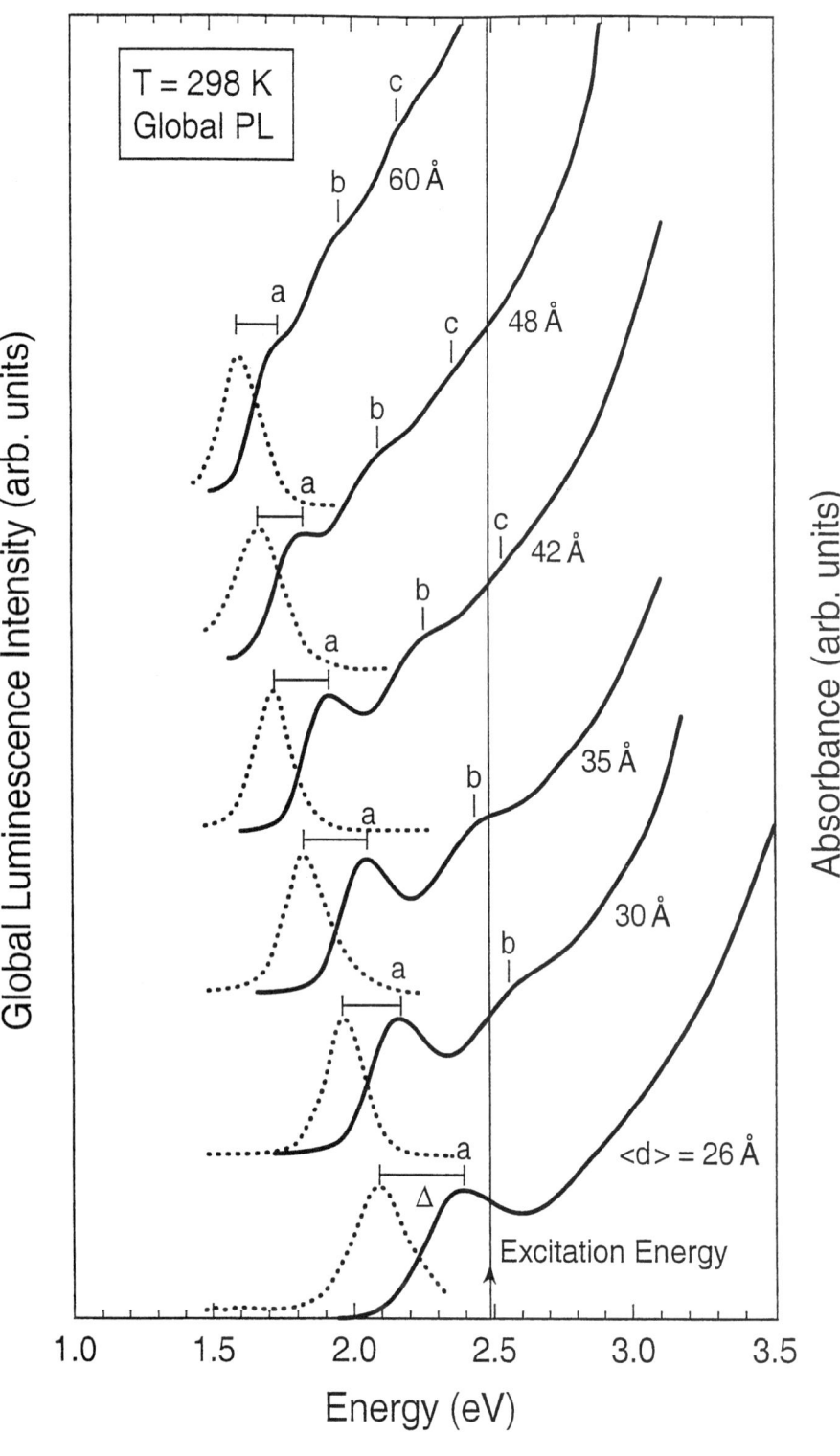

is from an intrinsic, spin-forbidden state, split from its singlet counterpart owing to screened electron-hole exchange [dark exciton model (101, 102)].

The line widths of FLN spectra are typically 15–30 meV; although these line widths are significantly narrower than the 175–225 meV line widths typically obtained from global (nonresonant) PL excitation, they are still much broader than line widths reported from PL measurements on single dots. For a variety of II-VI and III-V QDs, single-dot PL line widths have been reported to range from 40–1000 μeV (103–108); the experimental PL line widths observed for single QDs are smaller than the line widths obtained in Reference 88 by a factor of about 2 to 50. The broader line widths in the FLN spectra are attributed to the significant QD size and shape distribution that still remains in the FLN experiment. Also, the absorption peak moves about 35 to 45 meV for every 1-Å change in QD diameter. Thus, the PL line width of 15–30 meV for FLN spectra reflects a QD-diameter variation of less than 1 Å! It is apparent that the PL line width is extremely sensitive to the spread in QD diameters, and that the determination of true line widths requires PL data obtained from single dots.

Phonon Bottleneck and Slowed Hot-Electron Cooling in Quantum Dots

The first prediction of slowed cooling at low light intensities in quantized structures was made by Boudreaux et al (10). They anticipated that cooling of carriers would require multiphonon processes when the quantized levels are separated in energy by more than phonon energies. They analyzed the expected slowed cooling time for hot holes at the surface of highly doped n-type TiO_2 semiconductors, in which quantized energy levels arise because of the narrow space-charge layer (i.e. depletion layer) produced by the high doping level. The carrier confinement in this case is produced by the band bending at the surface; for a doping level of 1×10^{19} cm^{-3} the potential well can be approximated as a triangular well extending 200 Å from the semiconductor bulk to the surface and with a depth of 1 eV at the surface barrier. The multiphonon relaxation time was estimated from

$$\tau_c \sim \omega^{-1} \exp(\Delta E/kT), \qquad 24.$$

where τ_c is the hot-carrier cooling time, ω is the phonon frequency, and ΔE is the energy separation between quantized levels. For strongly quantized electron levels, with $\Delta E > 0.2$ eV, τ_c could be > 100 ps according to Equation 24.

However, carriers in the space charge layer at the surface of a heavily doped semiconductor are only confined in one dimension, as in a quantum film. This quantization regime leads to discrete energy states that have dispersion in k-space

Figure 9 Absorption (solid line) and global photoluminescence (dotted line) spectra at 298 K for colloidal ensembles of InP quantum dots with different mean diameters. All quantum dots colloidal samples were photoexcited at 2.48 eV (88).

(109). This means the hot carriers can cool by undergoing interstate transitions that require only one emitted phonon followed by a cascade of single phonon intrastate transitions; the bottom of each quantum state is reached by intrastate relaxation before an interstate transition occurs. Thus, the simultaneous and slow multiphonon relaxation pathway can be bypassed by single phonon events, and the cooling rate increases correspondingly.

More complete theoretical models for slowed cooling in QDs have been proposed recently by Bockelmann and co-workers (19, 110) and Benisty and co-workers (18, 20). The proposed Benisty mechanism (18, 20) for slowed hot-carrier cooling and phonon bottleneck in QDs requires that cooling only occurs via longitudinal optical (LO) phonon emission. However, there are several other mechanisms by which hot electrons can cool in QDs. Most prominent among these is the Auger mechanism (111). Here, the excess energy of the electron is transferred via an Auger process to the hole, which then cools rapidly because of its larger effective mass and smaller energy-level spacing. Thus, an Auger mechanism for hot-electron cooling can break the phonon bottleneck (111). Other possible mechanisms for breaking the phonon bottleneck include electron-hole scattering (112), deep-level trapping (113), and acoustical-optical phonon interactions (114, 115).

Experimental Determination of Relaxation/Cooling Dynamics and a Phonon Bottleneck in Quantum Dots

Over the past several years many investigations have been published that explore hot-electron cooling/relaxation dynamics in QDs and the issue of a phonon bottleneck in QDs. The results are controversial, and it is quite remarkable that there are so many reports that both support (116–130) and contradict (113, 131–144) the prediction of slowed hot-electron cooling in QDs and the existence of a phonon bottleneck. One element of confusion that is specific to the focus of this review is that although some of these publications report relatively long hot-electron relaxation times (tens of ps) compared with what is observed in bulk semiconductors, the results are reported as being not indicative of a phonon bottleneck because the relaxation times are not excessively long and PL is observed (145–147). (Theory predicts an infinite relaxation lifetime of excited carriers for the extreme, limiting condition of a phonon bottleneck; thus, the carrier lifetime would be determined by nonradiative processes and PL would be absent.) However, because the interest here is on the rate of relaxation/cooling compared to the rate of electron transfer, we consider that slowed relaxation/cooling of carriers has occurred in QDs if the relaxation/cooling times are greater than 10 ps (about an order of magnitude greater than for bulk semiconductors). This is because previous work that measured the time of electron transfer from bulk III-V semiconductors to redox molecules (metallocenium cations) adsorbed on the surface found that electron transfer times can be sub-ps to several ps (28, 148–150); hence photoinduced hot electron transfer can be competitive with electron cooling and relaxation if the latter is greater than tens of ps.

In a series of papers Sugawara and colleagues (118, 119, 121) reported slow hot-electron cooling in self-assembled InGaAs QDs produced by Stranski-Krastinow (SK) growth on lattice-mismatched GaAs substrates. Using time-resolved PL measurements, the excitation-power dependence of PL, and the current dependence of electroluminescence spectra, these researchers reported cooling times ranging from 10 ps to 1 ns. The relaxation time increased with electron energy up to the fifth electronic state. Also, Mukai & Sugawara (151) recently published an extensive review of phonon bottleneck effects in QDs, which concludes that the phonon bottleneck effect is indeed present in QDs.

Gfroerer et al reported slowed cooling of up to 1 ns in strain-induced GaAs QDs formed by depositing tungsten stressor islands on a GaAs QW with AlGaAs barriers (130). A magnetic field was applied in these experiments to sharpen and further separate the PL peaks from the excited state transitions and thereby determine the dependence of the relaxation time on level separation. The authors observed hot PL from excited states in the QD, which could only be attributed to slow relaxation of excited (i.e. hot) electrons. Because the radiative recombination time is about 2 ns, the hot-electron relaxation time was found to be of the same order of magnitude (about 1 ns). With higher excitation intensity sufficient to produce more than one electron-hole pair per dot the relaxation rate increased.

A lifetime of 500 ps for excited electronic states in self-assembled InAs/GaAs QDs under conditions of high injection was reported by Yu et al (125). PL from a single GaAs/AlGaAs QD (128) showed intense high-energy PL transitions that were attributed to slowed electron relaxation in this QD system. Kamath et al (129) also reported slow electron cooling in InAs/GaAs QDs.

QDs produced by applying a magnetic field along the growth direction of a doped InAs/AlSb QW showed a reduction in the electron relaxation rate constant from 10^{12} s^{-1} to 10^{10} s^{-1} (120, 152).

In addition to slow electron cooling, slow hole cooling was reported by Adler et al (126, 127) in SK InAs/GaAs QDs. The hole relaxation time was determined to be 400 ps, based on PL rise times, whereas the electron relaxation time was estimated to be less than 50 ps. These QDs only contained one electron state but several hole states; this explained the faster electron cooling time because a quantized transition from a higher quantized electron state to the ground electron state was not present. Heitz et al (122) also report relaxation times for holes of about 40 ps for stacked layers of SK InAs QDs deposited on GaAs; the InAs QDs are overgrown with GaAs, and the QDs in each layer self-assemble into an ordered column. Carrier cooling in this system is about two orders of magnitude slower than in higher dimensional structures.

All of the above studies on slowed carrier cooling were conducted on self-assembled SK-type QDs. Studies of carrier cooling and relaxation have also been performed on II-VI CdSe colloidal QDs by Klimov et al (137, 153) and Guyot-Sionnest et al (116). The former group first studied electron relaxation dynamics from the first-excited 1P to the ground 1S state using interband pump-probe spectroscopy (137). The CdSe QDs were pumped with 100 fs pulses at 3.1 eV to create high-energy electrons and holes in their respective band states, and then probed with fs white light continuum pulses. The dynamics of the interband bleaching and induced absorption caused by state filling was monitored to determine the electron relaxation time from the 1P to the 1S state. The results showed very fast 1P to 1S relaxation, on the order of 300 fs, and was attributed to an Auger process for electron relaxation that bypassed the phonon bottleneck. However, this experiment cannot separate the electron and hole dynamics from each other. Guyot-Sionnest et al (116) followed up these experiments using fs infrared pump-probe spectroscopy. A visible pump beam creates electrons and holes in the respective band states and a subsequent infrared beam is split into an infrared pump and an infrared probe beam; the infrared beams can be tuned to monitor only the intraband transitions of the electrons in the electron states and thus can separate electron dynamics from hole dynamics. The experiments were conducted with CdSe QDs that were coated with different capping molecules (TOPO, thiocresol, and pyridine), which exhibit different hole-trapping kinetics. The rate of hole trapping increased in the order: TOPO, thiocresol, and pyridine. The results generally show a fast relaxation component (1–2 ps) and a slow relaxation component ($\approx$200 ps). The relaxation times follow the hole-trapping ability of the different capping molecules and are longest for the QD systems having the fastest hole-trapping caps; the slow component dominates the data for the pyridine cap, which is attributed to its faster hole-trapping kinetics.

These results (116) support the Auger mechanism for electron relaxation, whereby the excess electron energy is rapidly transferred to the hole, which then relaxes rapidly through its dense spectrum of states. When the hole is rapidly removed and trapped at the QD surface the Auger mechanism for hot electron relaxation is inhibited and the relaxation time increases. Thus, in the above experiments, the slow 200 ps component is attributed to the phonon bottleneck, most prominent in pyridine-capped CdSe QDs, whereas the fast 1–2 ps component reflects the Auger relaxation process. The relative weight of these two processes in a given QD system depends upon the hole-trapping dynamics of the molecules surrounding the QD.

Klimov et al further studied carrier relaxation dynamics in CdSe QDs and published a series of papers on the results (153, 154); a review of this work was also recently published (155). These studies also strongly support the presence of the Auger mechanism for carrier relaxation in QDs. The experiments used ultrafast pump-probe spectroscopy with either two beams or three beams. In the former, the QDs were pumped with visible light across their band gaps (hole states to electron states) to produce excited state (i.e. hot) electrons; the electron relaxation was monitored by probing the bleaching dynamics of the resonant HOMO to LUMO transition with visible light or by probing the transient infrared absorption of the 1S to 1P intraband transition, which reflects the dynamics of electron occupancy in the LUMO state of the QD. The three-beam experiment was similar to that of Guyot-Sionnest et al (116) except that the probe in the experiments of Klimov et al is a white-light continuum. The first pump beam is at 3 eV and creates electrons and holes across the QD band gap. The second beam is in the infrared and is delayed with respect to the optical pump; this beam repumps electrons that have relaxed to the LUMO back up in energy. Finally, the third beam is a broad-band white-light continuum probe that monitors photoinduced interband absorption changes over the range of 1.2 to 3 eV. The experiments were done with two different caps on the QDs: a ZnS cap and a pyridine cap. The results showed that with the ZnS-capped CdSe the relaxation time from the 1P to 1S state was about 250 fs, whereas for the pyridine-capped CdSe, the relaxation time increased to 3 ps. The increase in the latter experiment was attributed to a phonon bottleneck produced by rapid hole trapping by the pyridine, as also proposed by Guyot-Sionnest et al. However, the time scale of the phonon bottleneck induced by hole trapping by pyridine caps on CdSe that was reported by Klimov et al was not as great as that reported by Guyot-Sionnest et al.

In contradiction to the results discussed above, many other investigations exist in the literature in which a phonon bottleneck was apparently not observed. These results were reported for both self-organized SK QDs (113, 131–144) and II-VI colloidal QDs (137, 139, 141). However, in several cases (122, 145, 147), hot-electron relaxation was found to be slowed, but not sufficiently to enable the authors to conclude that this was evidence of a phonon bottleneck. For the issue of hot-electron transfer this conclusion may not be relevant because, in this case, one is not interested in the question of whether the electron relaxation is slowed so drastically that nonradiative recombination occurs and quenches photoluminescence, but rather whether the cooling is slowed sufficiently so that excited-state electron transport and transfer can occur across the semiconductor-molecule interface before cooling. For this purpose, the cooling time need only be increased above about 10 ps because electron transfer can occur within this time scale (28, 148–150).

The experimental techniques used to determine the relaxation dynamics in the above experiments showing no bottleneck were all based on time-resolved PL or transient absorption spectroscopy. The SK QD systems that were studied and that exhibited no apparent phonon bottleneck include $In_xGa_{1-x}As/GaAs$ and GaAs/AlGaAs. The colloidal QD systems were CdSSe QDs in glass (750 fs relaxation time) (141) and CdSe (133). Thus, the same QD systems studied by different researchers showed both slowed cooling and nonslowed cooling in different experiments. This suggests a strong sample-history dependence for the results; perhaps the samples differed in their defect concentration and type, surface chemistry, and other physical parameters that affect carrier cooling dynamics. Much additional research is required to sort out these contradictory results.

Quantum Dot Solar Cells

As mentioned above in the INTRODUCTION and "RELAXATION DYNAMICS OF HOT ELECTRONS IN QUANTUM DOTS," slowed hot-carrier cooling in QDs can possibly be utilized to produce enhanced efficiencies for the conversion of radiant energy into electrical or chemical potential energy. The two fundamental pathways for enhancing the conversion efficiency [increased photovoltage (9, 10) or increased photocurrent (11, 12)] can be accessed, in principle, in three different QD solar cell configurations. These are described below; however, it is emphasized that these potential high-efficiency configurations are speculative and that there is no experimental evidence yet that demonstrates actual enhanced conversion efficiencies in any of these systems.

Photoelectrodes Composed of Quantum Dot Arrays In this configuration the QDs are formed into an ordered 3-D array with inter-QD spacing sufficiently small that strong electronic coupling occurs and minibands are formed to allow long-range electron transport. The system is a 3-D analog to the 1-D superlattice, and miniband structures discussed in "Energy Levels and Density of States in Quantum Wells and Superlattices" and shown in Figure 3. The delocalized quantized 3-D miniband states could be expected to slow the carrier cooling and permit the transport and collection of hot carriers to produce a higher photopotential in a photovoltaic cell or in a photoelectrochemical cell in which the 3-D QD array is the photoelectrode (156). Also, impact ionization might be expected to occur in the QD arrays, enhancing the photocurrent. However, hot-electron transport/collection and impact ionization cannot occur simultaneously; they are mutually exclusive, and only one of these processes can be present in a given system.

Significant progress has been made in forming 3-D arrays of both colloidal (157) and SK (94) II-VI and III-V QDs. The former have been formed via evaporation and crystallization of colloidal QD solutions containing a uniform QD size distribution; crystallization of QD solids from broader size distributions leads to close-packed QD solids, but with a high degree of disorder. Concerning the latter, arrays of SK QDs have been formed by successive epitaxial deposition of SK QD layers; after the first layer of SK QDs is formed, successive layers tend to form with the QDs in each layer aligned on top of each other (94, 158). Theoretical and experimental studies of the properties of QD arrays are currently under way. Major issues are the nature of the electronic states as a function of interdot distance, array order versus disorder, QD orientation and shape, surface states, surface structure/passivation, and surface chemistry. Transport properties of QD arrays are also of critical importance and are under investigation.

Quantum Dot–Sensitized Nanocrystalline TiO_2 Solar Cells This configuration is a variation of a recent promising new type of photovoltaic cell that is based on dye sensitization of nanocrystalline TiO_2 layers (159–162). In this latter photovoltaic cell dye molecules are chemisorbed onto the surface of 10–30-nm TiO_2 particles that have been sintered into a highly porous nanocrystalline 10–20 μm TiO_2 film. Upon photoexcitation of the dye molecules, electrons are very efficiently injected from the excited state of the dye into the conduction band of the

TiO$_2$, affecting charge separation and producing a photovoltaic effect. The cell circuit is completed using a nonaqueous redox electrolyte that contains I$^-$/I$_3^-$ and a Pt counter electrode to allow reduction of the adsorbed photooxidized dye back to its initial nonoxidized state (via I$_3^-$ produced at the Pt cathode by reduction of I$^-$).

For the QD-sensitized cell QDs are substituted for the dye molecules; they can be adsorbed from a colloidal QD solution (163) or produced in situ (164–167). Successful photovoltaic effects in such cells have been reported for several semiconductor QDs including InP, CdSe, CdS, and PbS (163–167). Possible advantages of QDs over dye molecules are the tunability of optical properties with size and better heterojunction formation with solid hole conductors. Also a unique potential capability of the QD-sensitized solar cell is the production of quantum yields greater than one by the inverse Auger effect (impact ionization) (21). Dye molecules cannot undergo this process. Efficient inverse Auger effects in QD-sensitized solar cells could produce much higher conversion efficiencies than are possible with dye-sensitized solar cells

Quantum Dots Dispersed in Organic Semiconductor Polymer Matrices Recently, photovoltaic effects have been reported in structures consisting of QDs forming junctions with organic semiconductor polymers. In one configuration, a disordered array of CdSe QDs is formed in a hole-conducting polymer—MEH-PPV [poly(2-methoxy, 5-(2′-ethyl)-hexyloxy-p-phenylenevinylene] (168). Upon photoexcitation of the QDs, the photogenerated holes are injected into the MEH-PPV polymer phase and are collected via an electrical contact to the polymer phase. The electrons remain in the CdSe QDs and are collected through diffusion and percolation in the nanocrystalline phase to an electrical contact to the QD network. Initial results show relatively low conversion efficiencies (168, 169), but improvements have been reported with rod-like CdSe QD shapes (170) embedded in poly(3-hexylthiophene) (the rod-like shape enhances electron transport through the nanocrystalline QD phase). In another configuration (171), a polycrystalline TiO$_2$ layer is used as the electron-conducting phase, and MEH-PPV is used to conduct the holes; the electron and holes are injected into their respective transport mediums upon photoexcitation of the QDs.

A variation of these configurations is to disperse the QDs into a blend of electron- and hole-conducting polymers (172). This scheme is the inverse of light-emitting diode structures based on QDs (173–177). In the photovoltaic cell, each type of carrier-transporting polymer would have a selective electrical contact to remove the respective charge carriers. A critical factor for success is to prevent electron-hole recombination at the interfaces of the two polymer blends; prevention of electron-hole recombination is also critical for the other QD configurations mentioned above.

All of the possible QD–organic polymer photovoltaic cell configurations would benefit greatly if the QDs could be coaxed into producing multiple electron-hole pairs by the inverse Auger/impact ionization process (21). This is also true for all the QD solar cell systems described in "Quantum Dot Solar Cells." The most important process in all the QD solar cells for reaching very high conversion efficiency is the multiple electron-hole pair production in the photoexcited QDs; the various cell configurations simply represent different modes of collecting and transporting the photogenerated carriers produced in the QDs.

ACKNOWLEDGMENTS

I acknowledge and thank my present and former National Renewable Energy Laboratory (NREL) collaborators who have performed the NREL work reviewed here: OI Mićić, R Ellingson, M Hanna, BB Smith, DC Selmarten, Y Rosenwaks, K Jones, D Levi, MW Peterson, D Szmyd, and HM Cheong; I also thank A Zunger and H Fu for discussions on these topics. Finally, I am very grateful for the long-term and generous support of the US Department of Energy, Office of Science, Office of Basic Energy Sciences, Division of Chemical Sciences.

Visit the Annual Reviews home page at www.AnnualReviews.org

LITERATURE CITED

1. Pankove JI. 1975. *Optical Processes in Semiconductors.* New York: Dover
2. Sze S. 1981. *Physics of Semiconductor Devices.* New York: Wiley
3. Stanton CJ, Bailey DW, Hess K, Chang YC. 1988. *Phys. Rev. B* 37:6575
4. Pelouch WS, Ellingson RJ, Powers PE, Tang CL, Szmyd DM, Nozik AJ. 1992. *Phys. Rev. B* 45:1450
5. Pelouch WS, Ellingson RJ, Powers PE, Tang CL, Szmyd DM, Nozik AJ. 1992. *Semicond. Sci. Technol. B* 7:337
6. Green MA. 1992. *Solar Cells.* Kensington, Aust.: Univ. New South Wales
7. Nozik AJ. 1978. *Annu. Rev. Phys. Chem.* 29:189
8. Sugawara M. ed. 1999. In *Semiconductors and Semimetals.* See Ref. 179, p. 86
9. Ross RT, Nozik AJ. 1982. *J. Appl. Phys.* 53:3813
10. Boudreaux DS, Williams F, Nozik AJ. 1980. *J. Appl. Phys.* 51:2158
11. Landsberg PT, Nussbaumer H, Willeke G. 1993. *J. Appl. Phys.* 74:1451
12. Kolodinski S, Werner JH, Wittchen T, Queisser HJ. 1993. *Appl. Phys. Lett.* 63:2405
13. Nozik AJ, Boudreaux DS, Chance RR, Williams F. 1980. In *Advances in Chemistry*, ed. M Wrighton, 184:162. New York: ACS
14. Williams FE, Nozik AJ. 1984. *Nature* 311:21
15. Nozik AJ. 1980. *Philos. Trans. R. Soc. London. Ser. A* 295:453
16. Rosenwaks Y, Hanna MC, Levi DH, Szmyd DM, Ahrenkiel RK, Nozik AJ. 1993. *Phys. Rev. B* 48:14675
17. Williams F, Nozik AJ. 1978. *Nature* 271:137
18. Benisty H, Sotomayor-Torres CM, Weisbuch C. 1991. *Phys. Rev. B* 44:10945
19. Bockelmann U, Bastard G. 1990. *Phys. Rev. B.* 42:8947
20. Benisty H. 1995. *Phys. Rev. B* 51:13281
21. Nozik AJ. 1997. Unpublished manuscript
22. Dingle R, Wiegmann W, Henry CH. 1974. *Phys. Rev. Lett.* 33:827
23. Weisbuch C, Vinter B. 1991. *Quantum*

Semiconductor Structures. New York: Academic
24. Bastard G. 1988. *Wave Mechanics Applied to Semiconductor Heterostructures.* New York: Halsted
25. Jaros M. 1989. *Physics and Applications of Semiconductor Microstructures.* Oxford: Oxford Univ. Press
26. Yoffe AD. 1993. *Adv. Phys.* 42:173
27. Dingle R, ed. 1987. *Semiconductors and Semimetals*, Vol. 24. New York: Academic
28. Miller RDJ, McLendon G, Nozik AJ, Schmickler W, Willig F. 1995. *Surface Electron Transfer Processes.* New York/Weinheim/Cambridge: VCH
29. Peterson MW, Turner JA, Parsons CA, Nozik AJ, Arent DJ, et al. 1988. *Appl. Phys. Lett.* 53:2666
30. Parsons CA, Thacker BR, Szmyd DM, Peterson MW, McMahon WE, Nozik AJ. 1990. *J. Chem. Phys.* 93:7706
31. Bastard G. 1981. *Phys. Rev. B* 24:5693
32. Bastard G. 1982. *Phys. Rev. B* 25:7594
33. Altarelli M. 1985. *J. Lumin.* 30:472
34. Bastard G, Brum JA. 1986. *J. Quantum Electron.* 22:1625
35. Dingle R, Gossard AC, Wiegmann W. 1975. *Phys. Rev. Lett.* 34:1327
36. Jung H, Fisher A, Ploog K. 1984. *Appl. Phys.* A33:97
37. Reynolds DC, Bajaj KK, Litton CW. 1985. *Appl. Phys. Lett.* 46:51
38. Aspnes DE. 1980. *Handbook on Semiconductors.* New York: North Holland
39. Edelstein DC, Tang CL, Nozik AJ. 1987. *Appl. Phys. Lett.* 51:48
40. Xu ZY, Tang CL. 1984. *Appl. Phys. Lett.* 44:692
41. Rosker MJ, Wise FW, Tang CL. 1986. *Appl. Phys. Lett.* 49:1726
42. Ryan JF, Taylor RA, Tuberfield AJ, Maciel A, Worlock JM, et al. 1984. *Phys. Rev. Lett.* 53:1841
43. Christen J, Bimberg D. 1990. *Phys. Rev. B* 42:7213
44. Cai W, Marchetti MC, Lax M. 1986. *Phys. Rev. B* 34:8573
45. Nozik AJ, Parsons CA, Dunlavy DJ, Keyes BM, Ahrenkiel RK. 1990. *Solid State Commun.* 75:297
46. Lugli P, Goodnick SM. 1987. *Phys. Rev. Lett.* 59:716
47. Campos VB, Das Sarma S, Stroscio MA. 1992. *Phys. Rev. B* 46:3849
48. Joshi RP, Ferry DK. 1989. *Phys. Rev. B* 39:1180
49. Nirmal M, Dabbousi BO, Bawendi MG, Macklin JJ, Trautman JK, et al. 1996. *Nature* 383:802
50. Efros AL, Rosen M. 1997. *Phys. Rev. Lett.* 78:1110
51. Brus LE. 1984. *J. Chem. Phys.* 80:4403
52. Brus LE. 1986. *J. Chem. Phys.* 90:2555
53. Schmidt HM, Weller H. 1986. *Chem. Phys. Lett.* 129:615
54. Ekimov AI, Efros AL, Ivanov MG, Onushenko AA, Shumilov SK. 1989. *Solid State Commun.* 69:565
55. Kagan CR, Murray CB, Nirmal M, Bawendi MG. 1996. *Phys. Rev. Lett.* 76:1517
56. Kayanuma Y, Momiji H. 1990. *Phys. Rev. B* 41:10261
57. Kayanuma Y. 1988. *Phys. Rev. B* 38:9797
58. Takagahara T. 1993. *Phys. Rev. B* 47:4569
59. Takagahara T. 1993. *Phys. Rev. Lett.* 71:3577
60. Hu YZ, Lindberg M, Koch SW. 1990. *Phys. Rev. B* 42:1713
61. Park SH, Morgan RA, Hu YZ, Lindberg M, Koch SW, Peyghambarian N. 1990. *J. Opt. Soc. Am.* 7:2097
62. Pollock EL, Koch SW. 1991. *J. Chem. Phys.* 94:6766
63. Nomura S, Kobayashi T. 1991. *Solid State Commun.* 78:677
64. Sercel PC, Vahala KJ. 1990. *Phys. Rev. B* 42:3690
65. Efros AL. 1992. *Phys. Rev. B.* 46:7448
66. Ekimov AI, Hache F, Schanne-Klein MC, Ricard D, Flytzanis C, et al. 1993. *J. Opt. Soc. Am. B* 10:100
67. Grigoryan B, Kazaryan EM, Efros AL,

Yazeva TV. 1990. *Sov. Phys. Solid State* 32:1031
68. Koch SW, Hu YZ, Fluegel B, Peyghambarian N. 1992. *J. Cryst. Growth* 117:592
69. Banyai L, Gilliot P, Hu YZ, Koch SW. 1992. *Phys. Rev. B* 45:14136
70. Banin U, Lee JC, Guzelian AA, Kadavanich AV, Alivisatos AP, et al. 1998. *J. Chem. Phys.* 109:2306
71. Fu H, Wang L-W, Zunger A. 1997. *Appl. Phys. Lett.* 71:3433
72. Fu H, Wang L-W, Zunger A. 1998. *Phys. Rev. B* 57:9971
73. Fu H, Wang L-W, Zunger A. 1998. *Appl. Phys. Lett.* 73:1157
74. Efros AL, Rosen M. 1998. *Appl. Phys. Lett.* 73:1155
75. Efros AL, Rosen M. 2000. *Annu. Rev. Mat. Sci.* 30:475
76. Fu H, Zunger A. 1998. *Phys. Rev. B* 57:R15064
77. Woggon U. 1997. *Optical Properties of Semiconductor Quantum Dots.* Berlin/Heidelberg: Springer-Verlag
78. Xia J-B. 1989. *Phys. Rev. B* 40:8500
79. Weller H, Eychmüller A. 1997. In *Semiconductor Nanoclusters—Physical, Chemical and Catalytic Aspects.* ed. PV Kamat, D Meisel, 103:5. Amsterdam: Elsevier
80. Overbeek JTG. 1982. *Adv. Colloid Interface Sci.* 15:251
81. Murray CB. 1995. *Synthesis and characterization of ii-vi quantum dots and their assembly into 3D quantum dot superlattices.* PhD thesis. MIT. Cambridge, MA
82. Murray CB, Norris DJ, Bawendi MG. 1993. *J. Am. Chem. Soc.* 115:8706
83. Mićić OI, Sprague JR, Curtis CJ, Jones KM, Machol JL, Nozik AJ. 1995. *J. Phys. Chem.* 99:7754
84. Banin U, Cerullo G, Guzelian AA, Bardeen CJ, Alivisatos AP, Shank CV. 1997. *Phys. Rev. B* 55:7059
85. Guzelian AA, Katari JEB, Kadavanich AV, Banin U, Hamad K, et al. 1996. *J. Phys. Chem.* 100:7212
86. Mićić OI, Curtis CJ, Jones KM, Sprague JR, Nozik AJ. 1994. *J. Phys. Chem.* 98:4966
87. Mićić OI, Sprague JR, Lu Z, Nozik AJ. 1996. *Appl. Phys. Lett.* 68:3150
88. Mićić OI, Cheong HM, Fu H, Zunger A, Sprague JR, et al. 1997. *J. Phys. Chem. B* 101:4904
89. Mićić OI, Nozik AJ. 1996. *J. Lumin.* 70:95
90. Olshavsky MA, Goldstein AN, Alivisatos AP. 1990. *J. Am. Chem. Soc.* 112:9438
91. Uchida H, Curtis CJ, Kamat PV, Jones KM, Nozik AJ. 1992. *J. Phys. Chem.* 96:1156
92. Nozik AJ, Uchida H, Kamat PV, Curtis C. 1993. *Isr. J. Chem.* 33:15
93. Guzelian AA, Banin U, Kadavanich AV, Peng X, Alivisatos AP. 1996. *Appl. Phys. Lett.* 69:1432
94. Sugawara M, ed. 1999. See Ref. 179, pp. 1–350
95. Hanna MC, Lu ZH, Cahill AF, Heben MJ, Nozik AJ. 1997. *J. Cryst. Growth.* 174:605
95a. Bimberg D, Grundmann M, Ledentsov NN. 1999. *Quantum Dot Heterostructures*, pp. 82–83. Chichester: Wiley
96. Ruminov S, Scheerschmidt K. 1995. *Phys. Status Solidi A* 150:471
97. Yang W, Lee H, Johnson TJ, Sercel PC, Norman AG. 2000. *Phys. Rev. B* 61:2784
98. Lee H, Yang W, Lowe-Webb R, Sercel PC. 1998. See Ref. 178, p. 205
99. Fu H, Zunger A. 1997. *Phys. Rev. B* 55:1642
100. Fu H, Zunger A. 1997. *Phys. Rev. B* 56:1496
101. Nirmal M, Norris DJ, Kuno M, Bawendi MG, Efros AL, Rosen M. 1995. *Phys. Rev. Lett.* 75:3728
102. Efros AL, Rosen M, Kuno M, Nirmal M, Norris DJ, Bawendi M. 1996. *Phys. Rev. B* 54:4843
103. Empedocles SA, Norris DJ, Bawendi MG. 1996. *Phys. Rev. Lett.* 77:3873

104. Forchel A, Steffen R, Koch T, Michel M, Albrecht M, Reinecke TL. 1996. *Semicond. Sci. Technol.* 11:1529
105. Gammon D, Snow ES, Katzer DS. 1995. *Appl. Phys. Lett.* 67:2391
106. Grundmann M, Christen J, Ledentsov NN, Bohrer J, Bimberg D, et al. 1995. *Phys. Rev. Lett.* 74:4043
107. Samuelson L, Carlsson N, Castrillo P, Gustafsson A, Hessman D, et al. 1995. *Jpn. J. Appl. Phys.* 34:4392
108. Nagamune Y, Watabe H, Nishioka M, Arakawa Y. 1995. *Appl. Phys. Lett.* 67:3257
109. Jaros M. 1989. In *Physics and Applications of Semiconductor Microstructures*, p. 83. New York: Oxford Univ. Press
110. Bockelmann U, Egeler T. 1992. *Phys. Rev. B* 46:15574
111. Efros AL, Kharchenko VA, Rosen M. 1995. *Solid State Commun.* 93:281
112. Vurgaftman I, Singh J. 1994. *Appl. Phys. Lett.* 64:232
113. Sercel PC. 1995. *Phys. Rev. B* 51:14532
114. Inoshita T, Sakaki H. 1992. *Phys. Rev. B* 46:7260
115. Inoshita T, Sakaki H. 1997. *Phys. Rev. B* 56:R4355
116. Guyot-Sionnest P, Shim M, Matranga C, Hines M. 1999. *Phys. Rev. B* 60:R2181
117. Wang PD, Sotomayor-Torres CM, McLelland H, Thoms S, Holland M, Stanley CR. 1994. *Surf. Sci.* 305:585
118. Mukai K, Sugawara M. 1998. *Jpn. J. Appl. Phys.* 37:5451
119. Mukai K, Ohtsuka N, Shoji H, Sugawara M. 1996. *Appl. Phys. Lett.* 68:3013
120. Murdin BN, Hollingworth AR, Kamal-Saadi M, Kotitschke RT, Ciesla CM, et al. 1999. *Phys. Rev. B* 59:R7817
121. Sugawara M, Mukai K, Shoji H. 1997. *Appl. Phys. Lett.* 71:2791
122. Heitz R, Veit M, Ledentsov NN, Hoffmann A, Bimberg D, et al. 1997. *Phys. Rev. B* 56:10435
123. Heitz R, Kalburge A, Xie Q, Grundmann M, Chen P, et al. 1998. *Phys. Rev. B* 57:9050
124. Mukai K, Ohtsuka N, Shoji H, Sugawara M. 1996. *Phys. Rev. B* 54:R5243
125. Yu H, Lycett S, Roberts C, Murray R. 1996. *Appl. Phys. Lett.* 69:4087
126. Adler F, Geiger M, Bauknecht A, Scholz F, Schweizer H, et al. 1996. *J. Appl. Phys.* 80:4019
127. Adler F, Geiger M, Bauknecht A, Haase D, Ernst P, et al. 1998. *J. Appl. Phys.* 83:1631
128. Brunner K, Bockelmann U, Abstreiter G, Walther M, Böhm G, et al. 1992. *Phys. Rev. Lett.* 69:3216
129. Kamath K, Jiang H, Klotzkin D, Phillips J, Sosnowski T, et al. 1998. *Inst. Phys. Conf. Ser.* 156 (*Compound Semicond. Symp.* 1997):525
130. Gfroerer TH, Sturge MD, Kash K, Yater JA, Plaut AS, et al. 1996. *Phys. Rev. B* 53:16474
131. Li X-Q, Nakayama H, Arakawa Y. 1998. See Ref. 178, p. 845
132. Bellessa J, Voliotis V, Grousson R, Roditchev D, Gourdon C, et al. 1998. See Ref. 178, p. 763
133. Lowisch M, Rabe M, Kreller F, Henneberger F. 1999. *Appl. Phys. Lett.* 74:2489
134. Gontijo I, Buller GS, Massa JS, Walker AC, Zaitsev SV, et al. 1999. *Jpn. J. Appl. Phys.* 38:674
135. Li X-Q, Nakayama H, Arakawa Y. 1999. *Jpn. J. Appl. Phys.* 38:473
136. Kral K, Khas Z. 1998. *Phys. Status Solidi B* 208:R5
137. Klimov VI, McBranch DW. 1998. *Phys. Rev. Lett.* 80:4028
138. Bimberg D, Ledentsov NN, Grundmann M, Heitz R, Boehrer J, et al. 1997. *J. Lumin.* 72–74:34
139. Woggon U, Giessen H, Gindele F, Wind O, Fluegel B, Peyghambarian N. 1996. *Phys. Rev. B* 54:17681
140. Grundmann M, Heitz R, Ledentsov N,

Stier O, Bimberg D, et al. 1996. *Superlattices Microstruct.* 19:81
141. Williams VS, Olbright GR, Fluegel BD, Koch SW, Peyghambarian N. 1988. *J. Mod. Opt.* 35:1979
142. Ohnesorge B, Albrecht M, Oshinowo J, Forchel A, Arakawa Y. 1996. *Phys. Rev. B* 54:11532
143. Wang G, Fafard S, Leonard D, Bowers JE, Merz JL, Petroff PM. 1994. *Appl. Phys. Lett.* 64:2815
144. Sandmann JHH, Grosse S, von Plessen G, Feldmann J, Hayes G, et al. 1997. *Phys. Status Solidi B* 204:251
145. Heitz R. Veit M, Kalburge A, Zie Q. Grundmann M, et al. 1998. *Physica E* 2:578
146. Li X-Q, Arakawa Y. 1998. *Phys. Rev. B* 57:12285
147. Sosnowski TS, Norris TB, Jiang H, Singh J, Kamath K, Bhattacharya P. 1998. *Phys. Rev. B* 57:R9423
148. Meier A. Selmarten DC, Siemoneit K, Smith BB, Nozik AJ. 1999. *J. Phys. Chem. B* 103:2122
149. Meier A, Kocha SS, Hanna MC, Nozik AJ, Siemoneit K, et al. 1997. *J. Phys. Chem. B* 101:7038
150. Diol SJ, Poles E, Rosenwaks Y, Miller RJD. 1998. *J. Phys. Chem. B* 102:6193
151. Mukai K, Sugawara M. 1999. See Ref. 179, p. 209
152. Murdin BN, Hollingworth AR, Kamal-Saadi M, Kotitschke RT, Ciesla CM. et al. 1998. See Ref. 178, p. 1867
153. Klimov VI, Mikhailovsky AA, McBranch DW, Leatherdale CA, Bawendi MG. 2000. *Phys. Rev. B* 61:R13349
154. Klimov VI, McBranch DW, Leatherdale CA, Bawendi MG. 1999. *Phys. Rev. B* 60:13740
155. Klimov VI. 2000. *J. Phys. Chem. B* 104:6112
156. Nozik AJ. 1996. Unpublished manuscript
157. Murray CB, Kagan CR, Bawendi MG. 2000. *Annu. Rev. Mater. Sci.* 30:611
158. Nakata Y, Sugiyama Y, Sugawara M. 1999. In *Semiconductors and Semimetals*, Vol. 60, ed. M Sugawara, 60:117. San Diego: Academic
159. Hagfeldt A, Grätzel M. 2000. *Acc. Chem. Res.* 33:269
160. Moser J, Bonnote P, Grätzel M. 1998. *Coord. Chem. Rev.* 171:245
161. Grätzel M. 2000. *Prog. Photovoltaics* 8:171
162. Kalyanasundaram K, Grätzel M. 1999. In *Optoelectronic Properties of Inorganic Compounds*, ed. DM Roundhill, JP Fackler Jr, p. 169. New York: Plenum
163. Zaban A, Mićić OI, Gregg BA, Nozik AJ. 1998. *Langmuir* 14:3153
164. Vogel R, Weller H. 1994. *J. Phys. Chem.* 98:3183
165. Weller H. 1991. *Ber. Bunsen-Ges. Phys. Chem.* 95:1361
166. Liu D, Kamat PV. 1993. *J. Phys. Chem.* 97:10769
167. Hoyer P, Könenkamp R. 1995. *Appl. Phys. Lett.* 66:349
168. Greenham NC, Peng X, Alivisatos AP. 1996. *Phys. Rev. B* 54:17628
169. Greenham NC, Peng X, Alivisatos AP. 1997. In *Future Generation Photovoltaic Technologies: First NREL Conference*, ed. R McConnell, p. 295. Am. Inst. Phys.
170. Huynh WU, Peng XG, Alivisatos AP. 1999. *Adv. Mater.* 11:923
171. Arango AC, Carter SA, Brock PJ. 1999. *Appl. Phys. Lett.* 74:1698
172. Nozik AJ, Rumbles G, Selmarten DC. 2000. Unpublished manuscript
173. Dabbousi BO, Bawendi MG, Onitsuka O, Rubner MF. 1995. *Appl. Phys. Lett.* 66:1316
174. Colvin V, Schlamp M, Alivisatos AP. 1994. *Nature* 370:354
175. Schlamp MC, Peng X, Alivisatos AP. 1997. *J. Appl. Phys.* 82:5837
176. Mattoussi H, Radzilowski LH, Dabbousi

BO, Fogg DE, Schrock RR, et al. 1999. *J. Appl. Phys.* 86:4390
177. Mattoussi H, Radzilowski LH, Dabbousi BO, Thomas EL, Bawendi MG, Rubner MF. 1998. *J. Appl. Phys.* 83:7965
178. Gershoni D, ed. 1998. *Proc. Int. Conf. Phys. Semicond. 24th* Singapore: World Sci.
179. Willardson RK, Weber ER, eds. 1999. *Semiconductors and Semimetals*, San Diego: Academic. Vol. 60

RATIOMETRIC SINGLE-MOLECULE STUDIES OF FREELY DIFFUSING BIOMOLECULES

Ashok A Deniz[*,1], Ted A Laurence[2,3], Maxime Dahan[3,†], Daniel S Chemla[2,3], Peter G Schultz[*,1], and Shimon Weiss[*,3,4]

[1]Department of Chemistry, The Skaggs Institute for Chemical Biology, The Scripps Research Institute, La Jolla, California 92037; e-mail: deniz@scripps.edu; schultz@scripps.edu
[2]Department of Physics, University of California, Berkeley, California 94720; e-mail: tedal@uclink4.berkeley.edu
[3]Materials Sciences; and [4]Physical Biosciences Divisions, Lawrence Berkeley National Laboratory, Berkeley, California 94720; e-mail: sweiss@lbl.gov, maxime.dahan@lkb.ens.fr, dschemla@lbl.gov

Key Words   protein folding, polymer physics, fluorescence resonance energy transfer, subpopulations, fluorescence lifetime

■ **Abstract**   We outline recent developments in biological single-molecule fluorescence detection with particular emphasis on observations by ratiometric fluorescence resonance energy transfer (FRET) of biomolecules freely diffusing in solution. Single-molecule-diffusion methodologies were developed to minimize perturbations introduced by interactions between molecules and surfaces. Confocal microscopy is used in combination with sensitive detectors to observe bursts of photons from fluorescently labeled biomolecules as they diffuse through the focal volume. These bursts are analyzed to extract ratiometric observables such as FRET efficiency and polarization anisotropy. We describe the development of single-molecule FRET methodology and its application to the observation of the Förster distance dependence and the study of protein folding and polymer physics problems. Finally, we discuss future advances in data acquisition and analysis techniques that can provide a more complete picture of the accessible molecular information.

[†]Present address: Laboratoire Kastler Brossel, Ecole Normale Superieure, 75005 Paris, France.
[*]Corresponding authors.

INTRODUCTION

Single-Molecule Fluorescence Detection and Biological Applications

Although we most often think about and model molecular systems individually, experimental science has been dominated by measurements that result in ensemble averages. This approach has traditionally hidden much of the rich variety present at microscopic and mesoscopic scales. However, over the last decade, single-molecule methods have rapidly become an important tool in the repertoire of the experimentalist, with implications in a host of scientific disciplines (1–8). Single-molecule measurements are especially useful for the study of complex systems, which are ubiquitous in biology. For systems involving static or dynamic heterogeneity, single-molecule methods provide the unique ability to probe distributions of static properties and dynamics of interconversion between different states of the system directly, without the need to synchronize or trigger the interconversion. This ability is crucial in many biological contexts, in which triggering is not possible or ensembles of molecules moving stochastically on complex reaction landscapes quickly lose their coherence, resulting in averaged behavior. Examples range from enzyme reactions with proposed multiple conformational substates (9, 10) and protein folding with multiple unfolded states, pathways, intermediates, and transition regions (11–15) to interactions between cell surface receptors or more downstream components of signal transduction pathways (16, 17). Other examples include interactions among proteins, DNA, and RNA during cellular functions such as recombination, transcription, and translation and very complex aggregation processes that lead to diseases such as Alzheimer's and Creutzfeldt-Jakob diseases (18–21). Single-molecule methods also permit the observation of processes at extremely low molecular concentrations; for example, conformational properties of individual prion proteins under aggregation conditions may be studied. A very intriguing application of single-molecule detection is the simultaneous observation of transitions occurring in different parts of a system, allowing the direct evaluation of synergistic effects during biopolymer structural transitions, assembly, or enzyme catalysis. Finally, these methods may lead to significant technical advances in areas such as high-throughput screening and DNA sequencing (5, 22–24). Although very young, the field of biological single-molecule spectroscopy has already been elaborated in excellent reviews (2, 8, 25–27). This review focuses on the development of ratiometric single-molecule methods during the last couple of years, aimed at studying freely diffusing fluorescent biological systems that are minimally perturbed by surfaces.

The detection of fluorescence from single molecules involves repeated cycling of the molecule between ground and excited states and detection of the series of emitted photons; nonradiative relaxation of the excited state results in a reduction of the maximum photon flux. Site-specific labeling of biomolecules with appropriate dyes is a prerequisite for these experiments (2). Fluorescently labeled single biomolecules can be detected in two configurations, either immobilized (in a matrix or by surface attachment) or freely diffusing in solution. These two detection formats provide different and complementary kinds of information. With immobilized molecules, a particular molecule can be observed over an extended period, and a time trace of its properties can be continuously recorded. Such an experiment can result in the observation of its stochastic fluctuations under equilibrium or nonequilibrium conditions, and it represents the most complete characterization of the molecule. However, care must be taken to ensure minimal perturbation due to the immobilization process. In the other format, when diffusing or flowing single molecules in a liquid traverse the laser excitation volume, fluorescence photon bursts are generated. Such bursts can be analyzed for their duration, brightness, spectrum, and fluorescence lifetime, thereby providing molecular information on identity, size, diffusion coefficient, and concentration (5, 22). These bursts are short (i.e. typically on a millisecond time scale) and provide little information on slower fluctuations. However, they can provide invaluable information about the distributions of molecular properties of interest, undisturbed by surface effects, and about changes in these distributions under nonequilibrium conditions. Because a large number of events (photon bursts) can be collected in a relatively short time, statistical analyses of these data are possible, and histograms can be constructed. Most notably, subpopulations of molecules in heterogeneous ensembles can be identified (28–30), and the properties of these subpopulations can be individually investigated. Until recently, most studies of single-molecule fluorescence bursts were limited to measuring distributions in burst size or in fluorescence lifetime. For example, two-color burst analysis (31) and multiple-lifetime (28) approaches for identification and separation were suggested and implemented. The related techniques, fluorescence correlation spectroscopy (22, 32–34) and two-color cross-correlation spectroscopy (31, 34, 35), have been used to analyze, sort, and detect conformational states of a few or single molecules in the excitation volume.

Ratiometric Diffusion Methods

In 1999, two groups demonstrated the ability to make ratiometric measurements on freely diffusing molecules; our group used single-pair fluorescence (or Förster) resonance energy transfer (spFRET) to identify and record changes in subpopulations in a mixture (29, 36), while the Seidel group used polarization anisotropy to increase the classification confidence for components in a mixture (30). The ability to distinguish subpopulations and to classify collected photons or photon bursts according to these subpopulations is a key feature of the methodology. Owing to their low intensities, single-molecule signals suffer from intrinsic fluctuations. Such fluctuations are enormously amplified in diffusion measurements due to the variation in the excitation intensity the molecule is subjected to as it diffuses through various regions of the focal volume. The ratiometric method, in effect, provides a way to normalize the data with respect to many of these fluctuations. This vastly simplifies the data interpretation, and it is possible to extract several parameters of interest with a minimum of complex modeling and data analysis. Hence, ratiometric diffusion methodologies have been developed to study unperturbed biological molecules in their normal solution environment. Using the ability provided by these methods to quickly resolve static structural distributions and subpopulations in mixtures, it is possible to measure submillisecond structural dynamics of individual subpopulations, measure slower changes in properties of subpopulations, and perform appropriate controls during the development of immobilization methods. In the following sections, we describe the development of ratiometric-diffusion single-molecule methods with a focus on spFRET and its application to protein folding and polymer physics problems. We conclude with a discussion of advances in data collection and analysis, limitations of the methodologies, and directions for the future.

The Observables

Fluorescence experiments provide many observables that can be recorded in a two-channel, ratiometric fashion. The most common among these are fluorescence (or Förster) resonance energy transfer (FRET), polarization anisotropy, and spectral peak position.

FRET is the nonradiative transfer of electronic excitation energy from donor to acceptor dye molecules via a weak dipole-dipole coupling mechanism. The transfer efficiency was predicted by Förster to decrease with R, the distance between the two dyes, as $1/[1 + (R/R_0)^6]$ (37). R_0, the Förster radius, is the distance corresponding to 50% energy transfer and depends on the photophysical properties of the dyes and their relative orientations. Stryer & Haugland demonstrated this distance dependence in 1967 (38), and, since then, FRET has been used as a spectroscopic ruler in ensemble experiments (37, 39). It allows distance measurements on the 20- to 80-Å scale and is a tool well suited to studying conformations of biological macromolecules. FRET was first observed at the single-molecule level in 1996 by Ha et al, who used near-field optical microscopy to observe energy transfer between two dyes on a dry surface (40). More recently, Deniz et al (29) used the diffusion FRET methodology to inspect the distance dependence of FRET at single-molecule resolution. During the past one and a half years, spFRET has been used in a variety of biological applications, including DNA (29; JL Glass, JR Grumwell, TD Lacoste, AA Deniz, DS Chemla, PG Schultz, unpublished data), RNA (42, 43), protein-folding (44–46), enzyme reactions (47), protein-calcium binding (41), cell surface receptor-ligand binding (48), and signal transduction (49). spFRET measurements can be used not only to probe average distances as in an ensemble experiment but also to observe distributions and the time evolution of conformational (distance) properties directly. The FRET efficiency E (quantum yield for energy transfer) can be expressed in terms of k_D, the "donor-only fluorescence," and k_T, the "energy transfer" rate constants. Because the transfer process results in an excited state for the acceptor, the efficiency can be recast in terms of the detected donor (I_d) and acceptor (I_a) fluorescence intensities as

$$E = \frac{k_T}{k_D + k_T} = \frac{1}{1 + (k_D/k_T)} = \frac{1}{1 + \gamma(I_d/I_a)}, \qquad 1.$$

where γ is a correction factor that contains components from detection efficiencies, and fluorescence quantum yields. Hence, measurement of the ratio (I_d/I_a) allows ratiometric determination of E in single-molecule experiments.

Polarization properties of the emitted fluorescence provide information about the reorientational properties of the fluorophore on the time scale of the fluorescence lifetime. This observable has been used extensively in both steady-state and time-resolved formats in biological applications such as protein folding and drug-protein interactions (37). At the single-molecule level (50), polarization properties have been used to distinguish components of mixtures (30), to measure changes in myosin light-chain orientation (51), and to observe the rotation of single F1-ATPase molecules (52). A striking example is the F1-ATPase work by Adachi et al (53), in which the 120° rotation steps of the central γ-subunit during the hydrolysis of ATP were directly monitored in real time. They achieved this by monitoring the fluorescence polarization properties of single dye molecules attached to the γ-subunit, and they were able to demonstrate that the rotation rate was load independent. Polarization methods might find their most powerful applications in monitoring the activity of single motor proteins as they move along DNA. From a synthesis and data collection perspective, polarization measurements are advantageous over FRET, because they often require labeling with only a single dye and suffer less from interpretation ambiguities. However these measurements do not provide distance information, and they are best used when angular conformational changes are expected, such as in F1-ATPase (vide infra). For the remainder of this review, we confine much of our discussion to the FRET ratiometric observable. Many of the fundamental considerations also apply to other ratiometric observables.

DATA ACQUISITION AND ANALYSIS

Ratiometric Data Acquisition and Histogram Analysis

Briefly, our experimental setup for the ratiometric experiments (29, 36) consists of laser excitation and detection via an inverted confocal microscope (Zeiss Axiovert S100 TV) coupled to a high-sensitivity detection setup. An air-cooled argon ion laser is used for most experiments, whereas a mode-locked titanium-sapphire laser tuned to 1 μm and doubled to 500 nm is used for time-correlated single-photon counting. A tight focal spot is created by focusing the laser excitation through a high numerical aperture (NA 1.3) oil immersion objective. To detect freely diffusing single molecules, the focal point is placed within the low concentration (<100 pM) sample solution, and fluorescence emission is collected through the same objective and imaged onto a pinhole to reject out-of-focus light. The effective confocal volume is about 1 fl. The fluorescence is then split into two parts via a dichroic mirror (FRET, spectral fluctuations) or a polarizing-beam-splitter cube (polarization anisotropy), and these two parts are separately focused onto two avalanche photodiode detectors (APDs), which are used for single-photon counting. The output pulses of the APDs are timed by using a counter-timer board (model PCI-6602; National Instruments). The overall detection efficiency is estimated to be a few percent (36), and with typical dyes, peak count rates of 10 kHz–1 MHz are observed. For FRET measurements, the dye pair is typically composed of tetramethylrhodamine (TMR) (donor, peak emission at 575 nm) and Cy5 (acceptor, peak emission at 670 nm); the large extinction coefficient of Cy5 (250,000) allows a good spectral separation while maintaining a large R_0 of ∼65 Å (PR Selvin, personal communication). For time-correlated single-photon counting, the APD signals are routed through a time-to-amplitude converter (EG&G Ortec 567) to an analog-to-digital converter board (PCI-6111E; National Instruments) to measure the time between the laser pulse and the photon arrival. A single commercial board can be used for the entire process (PicoQuant TimeHarp 100), which simultaneously measures the macroscopic timing of the photon (100-ns time scale) and the microscopic arrival time with respect to the laser pulse (100-ps time scale). In both cases, the synchronization pulses from the laser are electronically gated by the pulses from the APD. A different gating delay for each APD channel allows multiple channel detection with a single time-to-amplitude converter or TimeHarp board.

For FRET experiments, the donor (TMR) is excited, and the donor and acceptor (Cy5) emission are separately and simultaneously detected. Pairs of data points corresponding to the number of detected photons in the two channels, I_d and I_a, are recorded with integration times ranging from 0.2 to 1 ms; these data can be used to plot FRET histograms (29). A better method is to record the time lag between successive photons with submicrosecond time resolution for each channel. This permits the construction of FRET histograms and also analysis by burst and correlation methods (vide infra). Figure 1a (see color insert) shows a time trace of dual-detector (0.2 ms integration time) data for donor-acceptor-labeled DNA molecules. Due to the low concentration, most of the points consist of background signals, resulting predominantly from Raman and Rayleigh scattering, whereas the occasional peaks correspond to single donor-acceptor-labeled DNA molecules diffusing through the focal volume and emitting bursts of photons. The durations of these bursts (mean of ∼200 μs) are determined by the molecular diffusion rate, whereas their amplitudes depend on several factors, such as the diffusion path of the molecule through the focal volume and the photophysical properties of the dyes.

The time trace shows fluorescence bursts on both the donor and acceptor channels. Since direct excitation of the acceptor is negligible and leakage of the donor emission into the acceptor channel is small, large bursts on the acceptor channel result from Cy5 emission caused by energy transfer. A simple approach is used to extract FRET information from these photon bursts. First, a threshold is used to discriminate dye signals from the background noise. That is, pairs of data points from a time bin are accepted only when the sum of the signals ($I_a + I_d$) is above a given threshold T. The threshold value is chosen to reject background effectively while retaining as many points as possible. We term this the "SUM" rejection criterion. For the accepted events, the FRET efficiencies are calculated according to Equation 1 and collected in a histogram (Figure 1b). The correction factor γ is estimated to be $\sim$1 from surface-immobilized molecules (47), and a value of 1 was used for the data presented here (also see following section).

The above background rejection procedure is chosen to allow quick and simple processing of data while minimizing the bias of the thresholding criterion. Equation 1 can be rewritten as follows:

$$I_a + \gamma I_d = I_a/E. \qquad 2.$$

Substituting $I_a = E N_d \eta_a \phi_a$, where N_d is the number of donor excited states created during the integration period and η_a and ϕ_a are the acceptor detection efficiency and fluorescence quantum yield respectively, we can write

$$I_a + \gamma I_d = I_a/E = N_d \eta_a \phi_a, \qquad 3.$$

showing that the sum ($I_a + \gamma I_d$) is independent of the FRET efficiency. Our intention is to identify and study the properties of subpopulations or distributions of molecules based on their FRET efficiencies. Hence, using the SUM criterion is the simplest procedure for background rejection that provides representative data sampling without a bias with respect to E. Alternate "AND" ($I_a > T_a$, AND $I_d > T_d$) and "OR" ($I_a > T_a$, OR $I_d > T_d$) rejection criterion have recently been suggested and reported to enhance the ability to detect subpopulations in some cases (55). However, since "AND" accepts selectively close to $E = 0.5$, whereas "OR" accepts selectively at low or high E, these criteria are biased and less appropriate for extraction of FRET efficiencies. Finally, a more global analysis for FRET can be considered in which the SUM criterion is used in conjunction with a range of threshold values weighted using the signal value or the photon count histogram.

Figure 1b (see color insert) shows a representative FRET efficiency histogram from data for a double-stranded DNA 7 molecule, in which the donor and acceptor dyes were separated by 7 bp. Two peaks are observed, one at an efficiency greater than 0.95 and the other E at close to zero. The high efficiency peak is consistent with expectations based on the calculated distance between the dyes. The peak near zero E implies fluorescence bursts with emission from the donor only. It has contributions from several sources, including solvent impurities, molecules with donor dyes only, accelerated photobleaching of the acceptor during the single-molecule experiment, and cis-isomers of the acceptor dye (56). Although this peak is problematic for analysis of histograms with low FRET peaks ($E < 0.3$), it is less relevant for the $E > 0.4$ portion of the histogram, where FRET peaks (distributions) may be analyzed to reveal their various moments, mean E, area, and widths. Although Gaussian functions are often used for fitting the histograms, these are appropriate only between ~0.2 and 0.8 E, especially for the extraction of mean E and width (36). Outside this range, the nonsymmetric shape of the distributions is fit better by using beta functions. Finally, these histograms may be used to identify and follow subpopulations of molecules individually based on differences in E.

Moments of the Histograms

Although the moments extracted from the E histograms give an approximate measure of populations, distance means, and distributions, they are skewed owing to contributions from other factors. Next, we discuss the γ factor, shot noise, orientational distributions, and fast distance fluctuations and some methods that may be used to quantify and separate out their contributions. Given the uncertainties in the estimated E values, some authors refer to them as proximity ratios (42, 43, 55, 57).

The γ-scaling factor can consist of a nonsymmetric distribution centered off unity, and thus it can influence both the mean and width of E distributions. Although this factor is hard to pin down accurately, lifetime-based methods to measure E independently will prove very useful for this purpose. Another major contributor to the widths is shot noise. The emissions of both fluorophores (I_a and I_d) exhibit approximately Poisson distributions, with mean values that depend on the excitation intensity and the photophysical characteristics of the dyes (36). For the observed low signals, relative variations play a significant role. This results in distributions in the computed E, which in turn result in lower limits for the widths of the E histogram peaks. To estimate this limit, a simple model was first used in which both emission channels I_a and I_d were described by Poisson variables (36). Their mean values were $\mu_a = E_m \bullet T$ and $\mu_d = (1 - E_m) \bullet T$, respectively, where E_m is the mean transfer efficiency and T is the SUM threshold used in the analysis. Because only signals above this threshold are processed, they have a smaller relative shot noise, and, hence, this calculation places an upper bound on the calculated values. The derivation for the standard deviation in (36) should be revised (CAM Seidel, personal communication), because the variances of the statistical quantities should be added, not the standard deviations. The formula should read:

$$(\Delta d)^2 = \left(\frac{\partial d}{\partial I_a}(I_a = \mu_a, I_d = \mu_d)\right)^2 \Delta I_a^2 + \left(\frac{\partial d}{\partial I_d}(I_a = \mu_a, I_d = \mu_d)\right)^2 \Delta I_d^2,$$
4.

where d is the ratiometric observable, and for FRET efficiency, $d = E = I_a/(I_a + I_d)$.

Using this formula, the FRET width (standard deviation) is given by $\Delta E = \sqrt{E_m(1 - E_m)/T}$. The values calculated by using this formula are significantly lower than those calculated in reference 36. The solid line in Figure 2 (see color insert) displays this upper bound estimate for $T = 20, 50,$ and 100. The shot noise induced fluctuations strongly depend on the FRET mean value with a maximum at $E = 0.5$, and they clearly show the expected decrease with increase in signal intensity. More accurate models will include the effects of distributions of photon counts and diffusion of the molecules in the focal volume (36). Besides shot noise, distributions in E depend on distributions in the Förster radius R_0 and, more interestingly, in the distance between the two dyes. The dominant factor in R_0 fluctuations is κ^2, the orientational factor, which depends on the relative orientation of the two dipoles. A common assumption is that the two dyes are freely rotating on a time scale comparable to or faster than the fluorescence lifetime and that κ^2 can be dynamically averaged to a fixed value of two-thirds (37). However, a combination of results from steady-state polarization measurements and simulations shows that the presence of hindered rotation can cause both peak broadening and small shifts in mean E (44).

Fluctuations or distributions of distances between the two dyes can give rise to shifts and distributions in E. From a structural-biology point of view, this is the most interesting case, as distance measurements are of interest in conformational fluctuations in proteins, in end-to-end fluctuations of polymers, and in protein folding. Haas and coworkers first developed ensemble time-resolved FRET techniques, data analysis, and modeling to extract such distributions from ensemble data (59, 60). Single-molecule methods might provide new insights in this context. It is noted that fast fluctuations are averaged out within the measurement period and will not contribute to the observed widths, although they will shift the mean E to a higher value (60). However, slower distance fluctuations from conformational or dye-biomolecule interactions can contribute significantly.

Relative areas of subpopulations are particularly useful, because they provide a measure for the occupation probabilities of a particular state by the molecular system; this ability has been used to calculate protein denaturation curves (44). Relative areas can potentially be influenced by biased background rejection algorithms and also by different photophysics, photochemistry, or diffusion time between the different subpopulations.

In summary, many contributions affect the parameters that are extracted from the E histograms. Although this circumstance can be problematic for accurate measurements, the studies we describe in the next sections demonstrate that it is nevertheless possible to extract new and meaningful information from these histograms. These issues need to be investigated in greater detail by using improved spacer constructs and more advanced data collection and analysis methods, which can rule out some of the above contributions and correct for others.

THE FÖRSTER DISTANCE DEPENDENCE AND SUBPOPULATIONS

Before the spFRET methodology was used for biological applications, the ability to measure the FRET distance dependence at single-molecule resolution had to be established. Hence, Deniz et al used double-stranded DNA as a conveniently synthesized rigid donor-acceptor spacer molecule (29) to validate the spFRET methodology, in experiments similar to the pioneering ensemble ones of Stryer & Haugland (38). A series of seven DNAs were synthesized, with donor-acceptor separations that spanned a range of distances required for E from <0.3 to >0.95. Diffusion experiments were carried out by using these constructs to generate FRET histograms as described above. The mean FRET efficiencies are plotted vs distance in Figure 3a (see color insert). The *solid line* is the theoretical curve with an R_0 of 65 Å calculated from available photophysical data (PR Selvin, personal communication). Clearly, there is a monotonic decrease in mean E as a function of distance, showing that the technique is useful for monitoring distance changes with a resolution of ± 5 Å under favorable conditions.

Deniz et al (29) then carried out spFRET measurements on a mixture of DNA molecules to demonstrate the ability to resolve subpopulations. Figure 3b (see color insert) shows that two FRET peaks corresponding to DNA 7 (7-bp dye separation; $E = 0.95$) and DNA 14 (14-bp dye separation; $E = 0.75$) are clearly resolved from a mixture of these molecules. Furthermore, the area ratio of the fitted Gaussians is 1.3:1, in good agreement with the composition of the mixture. The resolution that can be achieved is directly related to the mean FRET efficiency E and the widths of the distributions. Using a Rayleigh type criterion, it is estimated that two subpopulations separated by 0.2 to 0.3 E units can be distinguished with this approach. The utility of this resolving power is evident in the following study of chymotrypsin inhibitor 2 (CI2) folding.

PROTEIN FOLDING

Although much has been learned about protein folding during the past few decades, several mechanistic and dynamical aspects of this fascinating process remain uncharted (61–66). A major hurdle impeding progress is the complexity of the protein folding reaction. Although fairly general submillisecond triggering methods have been developed during the past decade (67), complex folding landscapes lead to loss of coherence and consequent averaging and loss of detail for ensembles of folding/unfolding proteins. Single-molecule methods provide a set of tools ideally suited to address this experimental challenge, and they promise to be invaluable in detailed studies of proteins and other biopolymers that show complex folding behavior, such as shifts in properties of states and multiple pathways or intermediates (66, 68–70). Questions regarding cooperativity of folding transitions and on- versus off-pathway characteristics of intermediates can be resolved most directly at single-molecule resolution. The validity of the "funnel model," (61, 63, 65, 71) and the origins of "strange kinetics" (72, 73) of protein folding may also be tested. Furthermore, single-molecule methods offer the intriguing possibility of observing correlations between fluctuations or movements occurring in two or more parts of a protein during the folding process. Such correlations could provide the most definitive experimental pictures of the folding energy landscape (74). Finally, we have seen a dramatic increase in theoretical work in this field during recent years

(61, 63, 71, 75–80). Single-molecule folding experiments can be used to benchmark and test theoretical methods, which in turn may allow a better interpretation of the single-molecule results.

While single-molecule approaches to the folding problem appear very promising, the field is in its infancy, and few reports of such work have appeared so far (44–46). The spFRET distance measure is especially useful for monitoring folding reactions. It provides an effective reaction coordinate that affords a global view of the conformational distributions and dynamics during the folding reaction. In 1999, Jia et al used spFRET to monitor fluctuations in single surface-immobilized GCN4 peptide molecules as a function of denaturant (45). Immobilization was achieved by electrostatic adsorption via a negatively charged tail to a positively charged silanized glass surface. This work reported exceptionally well-resolved fluctuations in the FRET efficiency for these single coiled-coil peptides, presumably as they folded and unfolded. The time trace data were used to plot E histograms, distance distributions, and potentials, as a function of denaturant. Unfortunately, as the authors point out, surface interactions could have contributed significantly to changing the energy landscape, reflected in broad histograms.

The recent study of CI2 by Deniz et al (44) demonstrated that the spFRET diffusion methodology provides results that are consistent with ensemble measurements and furthermore that novel insights may be extracted from the data. The protein system was chosen for two major reasons, the wealth of existing ensemble folding data for this protein (81–83) and the relative ease of synthesis of site-specific donor acceptor-labeled protein (44). The inset in Figure 4 (see color insert) shows the structure of the protein (84) and the positions of the labels; the presence of many common elements of secondary structure and the simple two-state folding behavior exhibited make the protein an ideal model system for these first single-molecule folding investigations. Guanidinium chloride (GdmCl) denaturation curves for the protein at single-molecule resolution [Figure 4 (see color insert)] show the cooperative transition at around 4 M GdmCl for the pseudo-wild-type protein (81). To understand the single-molecule denaturation curves, consider the spFRET histograms in Figure 5a (see color insert), at three different concentrations along the denaturation curve. Below 3 M GdmCl, a peak at $\sim$0.95 E is observed, corresponding to the folded state. At a high denaturant concentration (6 M), only a peak at around 0.65 E is detected, corresponding to the unfolded state. As intermediate denaturant concentrations are scanned, various ratios of the two peaks are observed; the sample histogram at 4 M GdmCl shows approximately equal 0.65 and 0.95 E peaks. The assignments of the peaks were confirmed by comparison to data from a destabilized mutant of CI2 (K17G), which shows the same peaks in the histogram and whose relative ratio is a different function of the denaturant concentration, consistent with its lowered stability. Hence, using these spFRET histograms, the folded and unfolded states of CI2 can be directly observed and quantified, and this work provides a direct confirmation of the two-state folding model previously inferred from ensemble data (81).

Because the areas of the peaks are a measure of the occupation probabilities for the two states, a fractional occupation for the folded state can be extracted. A plot of this fraction versus denaturant concentration results in a single-molecule denaturation curve. Such curves derived from ensemble data are routinely used to quantify protein stability (85). GdmCl denaturation curves for pseudo wild-type CI2 and the destabilizing mutant are shown in Figure 4 (see color insert). It is noteworthy that the single-molecule curves are indistinguishable, within error (1 std. dev.), from the ensemble curves, showing that the equilibrium properties of the system are faithfully reproduced by the single-molecule diffusion experiment, providing a validation of the methodology for such studies. An interesting distinction may be drawn between the ensemble and single-molecule curves. In equilibrium ensemble experiments, direct information concerning the fraction of the protein in a particular state is not available because only average properties are measured. As an example, consider protein tryptophan fluorescence, which is commonly used to monitor folding/unfolding transitions (85). In such experiments, tryptophan fluorescence intensities for the folded and unfolded states are usually assumed to be linear functions of the denaturant concentration. Then, assuming a two-state model, the total fluorescence intensity may be expressed as a linear combination of folded- and denatured-state intensities. Such a model-dependent extrapolation procedure must be used to calculate the fraction of native states in the transition region for most ensemble denaturation experiments. This procedure becomes problematic when the fluorescence intensities of the two states are not linear functions of denaturant concentration, especially in the transition region, as pointed out by Dill & Shortle (86). For example, a parabolic baseline model was used recently to explain discrepancies between the van't Hoff denaturation enthalpy measured by circular dichroism, and the calorimetric denaturation enthalpy measured by differential scanning calorimetry (87). In contrast, the single-molecule approach analyzes the occupation probabilities of the individual states directly, providing a solution to this problem. This ability should prove useful in more accurate stability studies of proteins or other biomolecules, especially with more complex multistep folding reactions.

Potentials calculated from E histograms [44, 45; Figure 5b (see color insert)] show the presence of a double-well shape in the transition region. The figure depicts upper bounds on widths of these wells and lower bounds on the barriers between them, as a consequence of the width contributions described earlier. Such potentials can be used for comparisons to theory. While the reported potentials are crude approximations, better methods of analysis and future experiments on immobilized molecules will allow the calculation of more accurate potentials. Intriguingly, the mean E measured for the denatured peak shows a poorly resolved shift close to the transition region, which might correspond to changes in the overall dimensions of the protein as a function of denaturant. The fast collapse from an expanded to a compact denatured state has been observed during protein folding, and a recent study by Hagen & Eaton (88) provided evidence for a cooperative collapse process. The implications, generality, and detailed features of such transitions are currently under debate (89–91). In this context, the measured E shift for CI2 is being re-examined with lifetime methods described below, which could result in the resolution of two denatured states, expanded and compact, and their interconversion dynamics.

Related spFRET studies have targeted conformational changes in RNA and DNA (29, 42, 43; JL Glass, JR Grumwell, TD Lacoste, AA Deniz, DS Chemla, PG Schultz, unpublished data). The most outstanding example is the recent study of surface immobilized *Tetrahymena* ribozyme molecules by Zhuang et al (42). spFRET was used to follow the docking and undocking of the P1 subunit on the ribozyme. The buffer exchange-triggered folding of single-ribozyme molecules was also monitored. An extremely exciting result from this work was the observation of a sparsely populated intermediate during triggered folding, clearly demonstrating a unique capability of single-molecule experiments. Furthermore, the authors confirmed that the kinetic properties of these molecules were consistent with ensemble work, showing that surface immobilization can indeed be minimally perturbing under appropriate conditions. No doubt, we will soon see work reporting similar observations of the folding and unfolding of single proteins. Diffusion methods will play a vital role as controls in these immobilization developments, providing benchmark histograms and submillisecond fluctuation kinetics for comparison.

POLYMER PHYSICS

Using single-stranded DNA as a model polymer system, spFRET was used to study polyelectrolyte chain distributions, dynamics, and scaling properties in the short chain regime (92). There are several outstanding fundamental questions regarding polyelectrolytes in solution, especially due to the complexity resulting from Coulomb interactions (93, 94). Single-molecule studies on relatively isolated polyelectrolytes could help answer these questions. To study scaling issues in this length regime, a series of short poly-dT oligonucleotides [$(dT)_{15}$, $(dT)_{20}$, $(dT)_{30}$, $(dT)_{40}$, and $(dT)_{50}$], end labeled with donor and acceptor dyes, were synthesized. Poly-dT sequences were chosen to minimize specific interactions between residues in the chain for this first set of experiments, and lengths were chosen by considering isomeric-state simulations (RIS) based on the model of Yevich & Olson (95). Figure 6a (see color insert) shows the characteristic ratio (C_N) for single-stranded DNA as a function of length in bases (N). The characteristic ratio provides a measure of the polymer stiffness. The points are averages from simulations based on the rotational isomeric model, while the line is a fit of this simulation data to the wormlike chain model. This model assumes a continuum form for the polymer chain, with a certain amount of rigidity. The fit gives a persistence length of 15 Å for these idealized conditions, which is within the range of recent experimental estimates (96, 97). An important point here is that, although C_N eventually levels out for single-stranded DNA when N is >100, it has a strong, nonlinear dependence for shorter lengths.

Figure 6b (see color insert) shows E histograms for $(dT)_{20}$ as a function of salt concentration, showing the collapse of the DNA molecule at higher concentrations where electrostatic screening between the DNA backbone negative charges becomes more effective. Fits to the histograms allow the calculation of modified characteristic ratios ($<R^2>/N$), which are shown in Figure 6c (see color insert) as a function of length, for several salt concentrations. The downward slope in 100 mM and 1 M salt means that the size of the single-stranded DNAs increases more slowly than one would expect for a Gaussian chain. This could indicate poor solvent conditions and a collapsed structure for high salt. Several corrections must be included to improve the accuracy of the above results—these issues are under current investigation. Thus, the spFRET method allows measurement of polymer end-to-end distances at low concentrations, which minimizes any intermolecular interaction between strands. Furthermore, it provides a simple test for heterogeneity in the sample caused by such aggregation effects. It is most interesting that the lifetime methods described later are currently being used to investigate the distance distributions and fluctuation time scales as a function of length scaling and solution conditions in this short-length chain regime. These experiments will be extended to study sequence dependence of these distance and interaction properties, which are of interest in several biological systems (96, 98).

ADVANCED DATA ACQUISITION AND ANALYSIS

Every photon emitted by a fluorophore contains a variety of information; hence the single-photon counting data generated in single-molecule experiments can be analyzed in powerful ways to extract several physical parameters of the emitting system. This idea is summarized in Figure 7 (see color insert). Each emitted photon can be characterized with respect to excited-state duration, polarization, spectral position, and the time delay from the previous emitted photon. Integration over several such single-photon values can result in fluorescence intensity, lifetime, and related information. The photon information can be used to generate ratiometric histograms, lifetime curves, correlation information, and distributions from burst analysis, such as diffusion and intensity distributions. A key point here is the ability to draw correlations between the results from these different kinds of analyses. VanOrden et al (99) and Fries et al (30) have pioneered such multidimensional analyses methods, mainly to minimize the error rates in identification of subpopulations during single-molecule DNA sequencing and other proposed applications (30, 99). In the current context, the ratiometric methods discussed in this review offer the ability to monitor separately and carry out the above analyses on photons emitted by individual subpopulations in a mixture. The following sections discuss a few illustrative examples of such analytical methods.

Methods of Burst and Fluctuation Analysis

Bursts of fluorescence photons in diffusion experiments may be identified and further analyzed either at the pointwise or the full-burst levels. For the pointwise analysis, the signal is merely integrated over a specified sampling interval, as for much of the spFRET analysis presented. Alternatively, for the burst analysis, the data are first processed to identify groups of photons emitted by a specific molecule as it traverses the focal volume. Although the pointwise analysis is simpler and provides the possibility to control the averaging time scale, burst analysis can offer a larger signal and consequently lower relative shot noise. However, burst analysis can have the disadvantage of greater averaging over interconverting subpopulations during a burst, and these two analytical methods should be considered complementary. Identification of bursts followed by analysis of subdivisions of these bursts is even more powerful, because it incorporates the advantages of both methods.

Fries et al (30) have described the burst integrated fluorescence lifetime method, in which they analyze distributions of bursts with respect to size, duration, and lifetime, allowing them to resolve mixtures of dyes (30). Chen et al (100) have described photon count histogram analysis, which is a related fluctuation analysis method. By fitting the photon count histogram to a theoretical model, they are able to identify subpopulations in a mixture by differences in their brightness. A similar approach has been developed by Kask et al (101), which they have termed fluorescence intensity distribution analysis; this approach takes into account the shape of the confocal spot, as well as providing a more efficient calculation algorithm (101). These fluctuation methods have been used to resolve subpopulations in binary dye mixtures (102) and solutions of proteins that have been statistically labeled with one or more fluorophores (102) and to study DNA hybridization and cleavage (103).

Fluorescence autocorrelation analysis was introduced by Magde et al nearly three decades ago (104) and more recently has received renewed attention by Eigen & Rigler (22) as a method with broad implications for the modern biotechnology industry. Although the autocorrelation function itself disregards a considerable amount of temporal information, fluorescence correlation spectroscopy is a very powerful method that can provide information on fluorescence fluctuation time scales and amplitudes at equilibrium, over a wide range of time scales (33). Recently, two-color cross-correlation analysis has been used to study DNA cleavage (34), binding of DNA duplexes to transcription activator protein complexes (35), and detection of pathological prion protein aggregates (20, 21). A drawback of this technique, as conventionally practiced, is that it results in averaged information, from which it is hard to deconvolute contributions from individual subpopulations of molecules. The ratiometric single-molecule approach allows us to add this capability to such correlation analyses.

For FRET experiments, donor and acceptor photon data can be acquired in a time lag mode; that is, the time difference between each successive photon is measured using a counter-timer board. Then a cross correlation between donor and acceptor channels affords the ability to measure kinetics of processes that result in changes in E. In contrast to the earlier two-color cross correlation, such fluctuations result in negative correlations, which allow them to be distinguished simply from other fluctuation processes that result in a positive correlation (41). Furthermore, the same time lag data may be used to plot a FRET histogram, followed by a correlation analysis of signals that correspond to subsets of this histogram. In addition to first-order correlations such as the above, higher-order correlations may be used. These afford the ability to investigate the contributions of non-Markovian dynamics, as recently discussed in the analysis of data for flavoenzyme (105, 106) and horseradish peroxidase catalysis (107, 108). These authors interpret their observations as evidence for the presence of multiple conformational substates indistinguishable by their fluorescence. Finally, methods of burst and correlation analysis, in conjunction with ratiometric separation of subpopulations, might prove useful for detection of aggregation processes, such as between cell surface receptors during signaling, between misfolded proteins resulting in disease, or between proteins and DNA during cellular processes, in which differences in molecular brightness and/or diffusion rates are expected.

Time-Correlated Single-Photon Counting

Even more information can be extracted by collecting data in a time-correlated manner, a method commonly used in ensemble fluorescence lifetime measurements. In this method, excitation is via a pulsed laser (femtosecond or picosecond pulse widths; 50- to 100-MHz repetition rates), and the time delay between the excitation pulse and the emission of the corresponding fluorescence photon is recorded. Single-molecule detection and analysis by this method has been used by several groups in a single-channel detection mode (28, 99). This method can be used to classify photon bursts according to lifetime, allowing the identification and quantification of components in a mixture with different lifetimes. This approach has specifically been developed as a tool for single-molecule DNA sequencing applications using a single excitation source and single detector. VanOrden et al (99) have shown that simultaneously measuring burst size and intraburst fluorescence lifetime increases the accuracy of identification of subpopulations in a mixture. Lamb et al (109) have used the arrival time information in conjunction with fluorescence correlation spectroscopy experiments to study heterogeneity of solutions. The group of Seidel first used two-channel lifetime detection in combination with ratiometric polarization anisotropy measurements, in an approach termed the multidimensional burst integrated fluorescence lifetime technique (30). Here, the authors recorded each photon by using both the time lag and time-correlated data collection approaches and then analyzed the photons emitted during single bursts to classify molecules based on burst size, and steady-state and time-resolved anisotropies.

Using this lifetime capability, an interesting analysis may be carried out with ratiometric data collection. Ratiometric diffusion histograms are first plotted to identify subpopulations, and then the arrival time histogram is plotted for all of

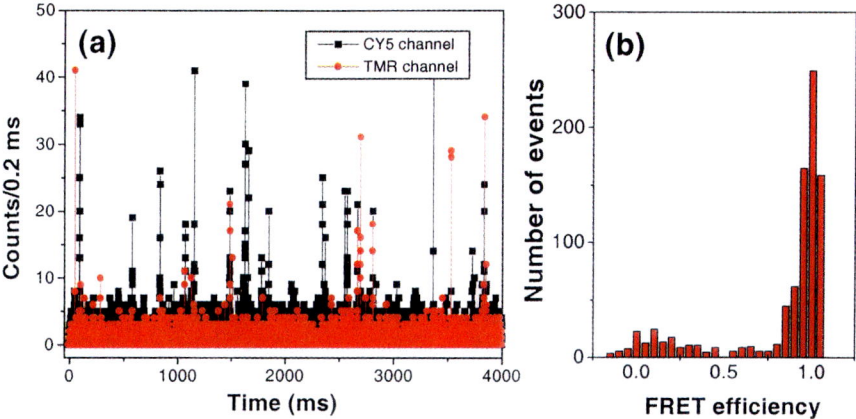

Figure 1 (*a*) Dual-channel time trace showing acceptor (*black, squares*) and donor (*red, circles*) data for DNA 7 (see text). (*b*) FRET efficiency histogram generated from dual-channel data, using a threshold of 20 counts.

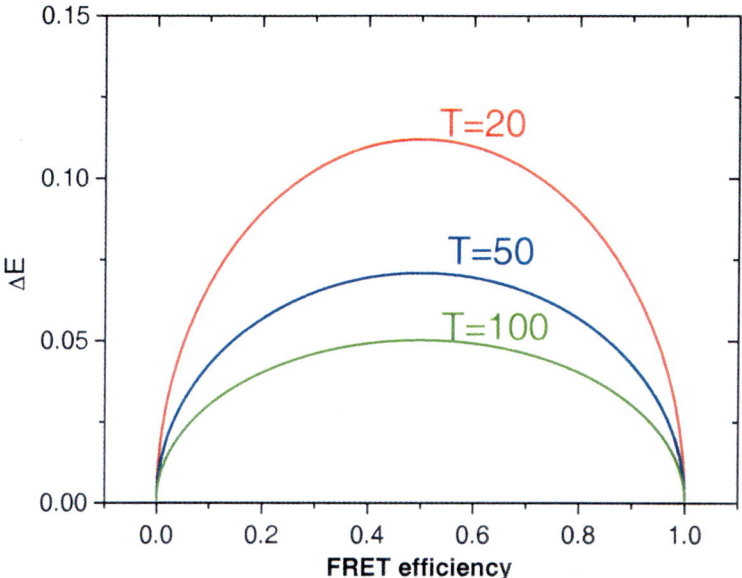

Figure 2 Shot noise width contributions to *E* histogram peaks, for $T = 20$, 50, and 100 counts, calculated using the simple model described in the text.

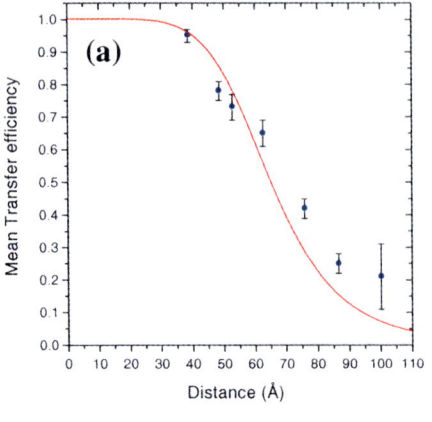

Figure 3 (*a*) Plot of mean FRET efficiency vs distance. Error bars are standard deviations from multiple experiments. The solid line is the theoretical curve for $R_0 = 65$ Å, and it is presented as a reference. (*b*) FRET histogram of a 1:1 mixture of DNA 7 and DNA 14 (see text). (Modified with permission from reference 29.)

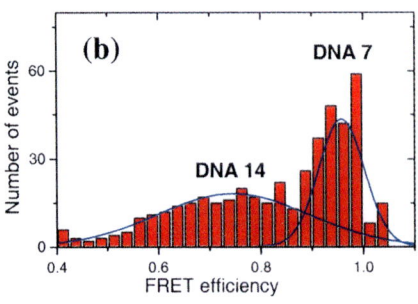

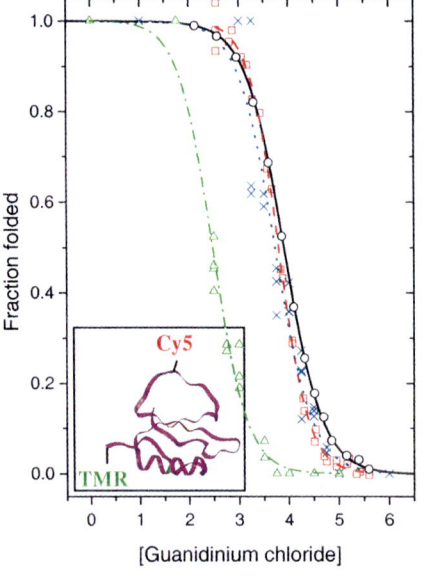

Figure 4 GdmCl denaturations curves for CI2 (symbols and lines represent data and sigmoidal fits respectively). Pwt CI2: ensemble tryptophan (*black*), ensemble FRET (*red*), and spFRET (*blue*). Destabilized mutant CI2: spFRET (*green*). Inset shows structure of CI2 and attachment positions of dyes. (Modified with permission from (44).)

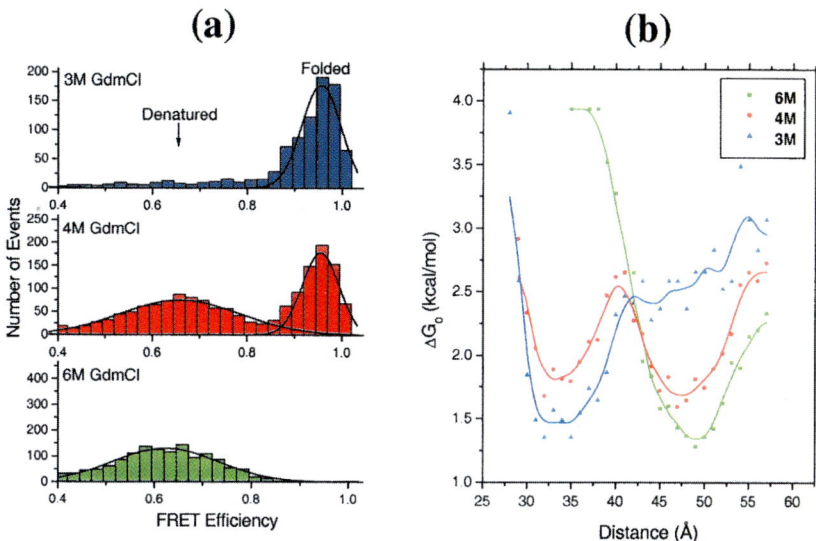

Figure 5 (*a*) FRET histograms for pwt CI2 at 3M (*blue*), 4M (*red*), and 6M (*green*) GdmCl. (*b*) Distance potentials for pwt CI2 at 3M (*blue*), 4M (*red*), and 6M (*green*) GdmCl. (Modified with permission from (44).)

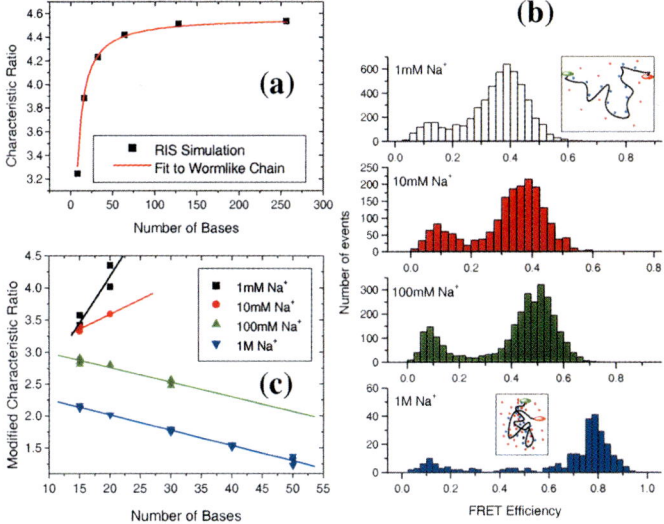

Figure 6 (*a*) Plot of characteristic ratio versus number of bases. Symbols show results of RIS simulations, while the solid line shows a fit to a wormlike chain model. (*b*) FRET histograms for $(dT)_{20}$ as a function of salt concentration. The insets show highly simplified cartoons of ss-DNA at low (expanded) and high (collapsed) salt. Blue dots depict DNA negative charges and red the sodium counterions. (*c*) Modified characteristic ratio (see text) as a function of number of bases for different salt concentrations.

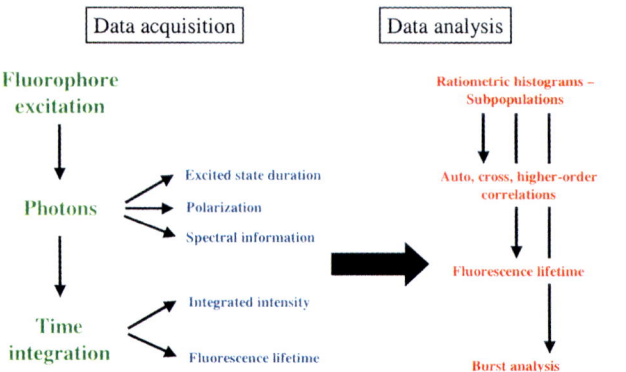

Figure 7 Scheme showing available photon information, and various possibilities for data analyses.

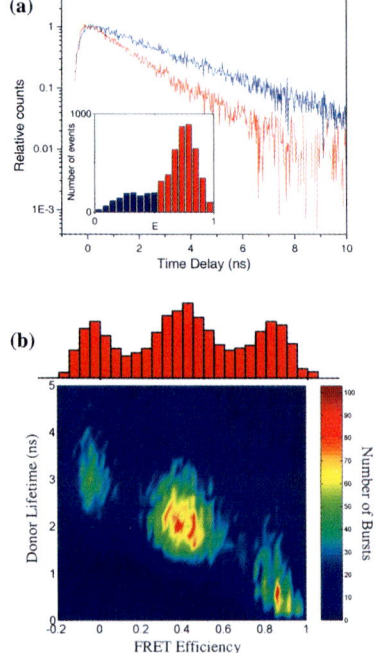

Figure 8 (*a*) Donor channel lifetime decays for "low FRET" (*blue*) and "high FRET" (*red*) subpopulations in a mixture. The inset shows a FRET histogram with the above subpopulations in blue and red respectively. (*b*) The plot is a two-dimensional histogram, showing the number of bursts with a given donor lifetime and FRET efficiency. Along the top axis is a projection of the histogram onto the E co-ordinate. The sample is a mixture of two doubly labeled FRET subpopulations. The high FRET subpopulation is a ds-DNA with 7 base pair separation between donor and acceptor, and the low FRET subpopulation is a donor-acceptor end labeled $(dT)_{40}$.

the photons corresponding to a particular subpopulation of molecules. A simple example of such a separation is illustrated in Figure 8a (see color insert), in which a donor-acceptor DNA molecule is studied. Here, a FRET histogram is plotted in the inset, showing the separation of subpopulations corresponding to the "zero peak" (*blue*), and doubly labeled DNA molecules (*red*); the figure shows donor lifetime for these two subpopulations. The high FRET component (*red*) clearly shows the accelerated decay due to energy transfer. The acceptor lifetime decay also showed a similar change. Although lifetime analysis may be carried out on individual bursts, the advantage of the current method is that it provides a larger number of photons and can be used for identifying multiple or distributed exponential decays. This is particularly interesting for the use of FRET measurements to look at subpopulations that interconvert on the submicrosecond time scale and fluctuations of proteins, DNA, and other polymers, as first demonstrated by the pioneering ensemble work of Haas et al (59) and Haas & Steinberg (60). End-to-end distance distributions or fluctuations in donor-acceptor-labeled polymers results in nonexponential decays of donor and acceptor fluorescence lifetime. Several groups have used analyses of such nonexponential donor decays to identify peptide, protein, DNA, and RNA conformational distributions (37). Because the analysis of such decays relies on complex models, it is of great utility to be able to separate out lifetime components of subpopulations, using the ratiometric single-molecule approach first, followed by further analyses of these separate decays. In particular, this approach is being pursued to investigate the unfolded state of CI2 in the region showing possible changes in the protein dimensions and for the study of fluctuations in single-stranded DNA.

Figure 8b (see color insert) demonstrates how lifetime information can be used in conjunction with ratiometric observables to facilitate separation of fluorescence bursts into subpopulations. The data are taken from a mixture of DNA molecules consisting of high-FRET and lower-FRET subpopulations. The plot is a two-dimensional histogram, showing the number of fluorescence bursts with a given donor lifetime and FRET efficiency E. The projection along the top axis is the one-dimensional histogram of E. For both graphs, the peak corresponding to the high-FRET subpopulation is on the right. The peak of the lower-FRET subpopulation is in the middle, and the "zero FRET" peak is on the left. Because the donor lifetime decreases as E increases, the separation between subpopulations is improved when using the additional information from lifetime data. An extension of these lifetime methods is to have four detection channels, each recording both time lag and lifetime information. Such a setup could monitor polarization properties of each FRET channel, resulting in the most efficient use of the available photon information (CAM Seidel, personal communication).

CONCLUDING REMARKS

The ratiometric-diffusion, single-molecule methods discussed in this review provide powerful means for the study of complex structural distributions and fluctuation processes in biology, over time scales ranging from nanoseconds to milliseconds. Changes in distributions caused by nonequilibrium conditions may also be monitored on slower time scales. We have discussed selected applications of these methods to folding and polymer physics, providing the first glimpses of the future. Progress will require developments in both spectroscopic and biochemical methodologies. Current efforts include incorporation of the diffusion methods into studies of immobilized molecules, both to acquire an overview of the single-molecule properties of the system rapidly, and to develop nonperturbative methods for gel or surface immobilization. Furthermore, fluorophores with better photophysical and chemical properties need to be developed, potentially by the use of combinatorial-synthesis methods. Optimized immobilization methods, in conjunction with improved fluorophores, labeling, and detection methods, will allow the detailed study of full conformational time trajectories during protein folding or enzymatic catalysis. Ultimately, such fluctuations might be directly viewed as they take place during cellular activity. The methods described will be useful in the exploration of a range of biological phenomena at high resolution, especially in the new frontiers of cell biology, medicine, and nanotechnology.

ACKNOWLEDGMENTS

We thank Ganga Beligere, Philip Dawson, AnnElise Faulhaber, Jen Glass, Jocelyn Grunwell, Taekjip Ha, David King, Philo Lacoste, Andy Martin, Alice Ting, and David Wemmer for their contributions to the ratiometric-diffusion, single-molecule work. Financial support is acknowledged, from the Laboratory Directed Research and Development Program of Lawrence Berkeley National Laboratory under U.S. Department of Energy contract No. DE-AC03-76SF00098, from the Office of Naval Research (contract N0001498F0402) and the National Science Foundation under grant No.CHE-971-4390, by the National Institutes of Health (grants No. RO1 GM49220-09 and GM59380), and by the Skaggs Institute for Chemical Biology.

Visit the Annual Reviews home page at www.AnnualReviews.org

LITERATURE CITED

1. Tamarat P, Maali A, Lounis B, Orrit M. 2000. *J. Phys. Chem. A* 104:1–16
2. Weiss S. 1999. *Science* 283:1676–83
3. Rigler R, Wolynes PG. 1999. *Chem. Phys.* 247:vii–viii
4. Moerner WE, Orrit M. 1999. *Science* 283:1670–76
5. Ambrose WP, Goodwin PM, Jett JH, Van Orden A, Werner JH, Keller RA. 1999. *Chem. Rev.* 99:2929–56
6. Xie XS, Trautman JK. 1998. *Annu. Rev. Phys. Chem.* 49:441–80
7. Basche T, Moerner WE, Orrit M, Wild UP. 1997. *Single Molecule Optical Detection, Imaging and Spectroscopy.* Cambridge, United Kingdom: Wiley-VCH
8. Nie S, Zare RN. 1997. *Annu. Rev. Biophys. Biomol. Struct.* 26:567–96
9. Frauenfelder H, Sligar SG, Wolynes PG. 1991. *Science* 254:1598–603

10. Rader SD, Agard DA. 1997. *Protein Sci.* 6:1375–86
11. Lu H, Booth PJ. 2000. *J. Mol. Biol.* 299:233–43
12. Leeson DT, Gai F, Rodriguez HM, Gregoret LM, Dyer RB. 2000. *Proc. Natl. Acad. Sci. USA* 97:2527–32
13. Goldbeck RA, Thomas YG, Chen E, Esquerra RM, Kliger DS. 1999. *Proc. Natl. Acad. Sci. USA* 96:2782–87
14. Moran LB, Schneider JP, Kentsis A, Reddy GA, Sosnick TR. 1999. *Proc. Natl. Acad. Sci. USA* 96:10699–704
15. Shastry MC, Udgaonkar JB. 1995. *J. Mol. Biol.* 247:1013–27
16. Deller MC, Jones EY. 2000. *Curr. Opin. Struct. Biol.* 10:213–19
17. Pawson T, Nash P. 2000. *Genes Dev.* 14:1027–47
18. Cohen FE, Prusiner SB. 1998. *Annu. Rev. Biochem.* 67:793–819
19. Cohen FE. 1999. *J. Mol. Biol.* 293:313–20
20. Bieschke J, Giese A, Schulz-Schaeffer W, Zerr I, Poser S, et al. 2000. *Proc. Natl. Acad. Sci. USA* 97:5468–73
21. Tjernberg LO, Pramanik A, Björling S, Thyberg P, Thyberg J, et al. 1999. *Chem. Biol.* 6:53–62
22. Eigen M, Rigler R. 1994. *Proc. Natl. Acad. Sci. USA* 91:5740–47
23. Rigler R. 1995. *J. Biotechnol.* 41:177–86
24. Fernandes PB. 1998. *Curr. Opin. Chem. Biol.* 2:597–603
25. Bai C, Wang C, Xie XS, Wolynes PG. 1999. *Curr. Opin. Struct. Biol.* 96:11075–76
26. Fisher TE, Marszalek PE, Fernandez JM. 2000. *Nat. Struct. Biol.* 7:719–24
27. Weiss S. 2000. *Nat. Struct. Biol.* 7:724–29
28. Zander C, Drexhage KH, Han KT, Wolfrum J, Sauer M. 1998. *Chem. Phys. Lett.* 286:457–65
29. Deniz AA, Dahan M, Grunwell JR, Ha T, Faulhaber AE, et al. 1999. *Proc. Natl. Acad. Sci. USA* 96:3670–75
30. Fries JR, Brand L, Eggeling C, Kollner M, Seidel CAM. 1998. *J. Phys. Chem. A* 102:6601–13
31. Schwille P, Meyer-Almes FJ, Rigler R. 1997. *Biophys. J.* 72:1878–86
32. Maiti S, Haupts U, Webb WW. 1997. *Proc. Natl. Acad. Sci. USA* 94:11753–57
33. Widengren J, Rigler R. 1998. *Cell. Mol. Biol.* 44:857–79
34. Kettling U, Koltermann A, Schwille P, Eigen M. 1998. *Proc. Natl. Acad. Sci. USA* 95:1416–20
35. Rippe K. 2000. *Biochemistry* 39:2131–39
36. Dahan M, Deniz AA, Ha T, Chemla DS, Schultz PG, Weiss S. 1999. *Chem. Phys.* 247:85–106
37. Lakowicz JR. 1999. *Principles of Fluorescence Spectroscopy.* New York: Plenum
38. Stryer L, Haugland RP. 1967. *Proc. Natl. Acad. Sci. USA* 58:719–26
39. Selvin PR. 2000. *Nat. Struct. Biol.* 7:730–34
40. Ha T, Enderle T, Ogletree DF, Chemla DS, Selvin PR, Weiss S. 1996. *Proc. Natl. Acad. Sci. USA* 93:6264–68
41. Brasselet S, Peterman EJG, Miyawaki A, Moerner WE. 2000. *J. Phys. Chem. B* 104:3676–82
42. Zhuang X, Bartley LE, Babcock HP, Russell R, Ha T, et al. 2000. *Science* 288:2048–51
43. Ha T, Zhuang X, Kim HD, Orr JW, Williamson JR, Chu S. 1999. *Proc. Natl. Acad. Sci. USA* 96:9077–82
44. Deniz AA, Laurence TA, Beligere GS, Dahan M, Martin AB, et al. 2000. *Proc. Natl. Acad. Sci. USA* 97:5179–84
45. Jia YW, Talaga DS, Lau WL, Lu HSM, DeGrado WF, Hochstrasser RM. 1999. *Chem. Phys.* 247:69–83
46. Ishii Y, Yoshida T, Funatsu T, Wazawa T, Yanagida T. 1999. *Chem. Phys.* 247:163–73
47. Ha TJ, Ting AY, Liang J, Caldwell WB, Deniz AA, et al. 1999. *Proc. Natl. Acad. Sci. USA* 96:893–98
48. Schütz GJ, Trabesinger W, Schmidt T. 1998. *Biophys. J.* 74:2223–26
49. Sako Y, Minoghchi S, Yanagida T. 2000. *Nat. Cell. Biol.* 2:168–72

50. Ha T, Laurence TA, Chemla DS, Weiss S. 1999. *J. Phys. Chem. B* 103:6839–50
51. Warshaw DM, Hayes E, Gaffney D, Lauzon AM, Wu JR, et al. 1998. *Proc. Natl. Acad. Sci. USA* 95:8034–39
52. Kinosita K, Jr. 1999. *FASEB J.* 13(Suppl. 2):S201–8
53. Adachi K, Yasuda R, Noji H, Itoh H, Harada Y, et al. 2000. *Proc. Natl. Acad. Sci. USA* 97:7243–47
54. Deleted in proof
55. Ying LM, Wallace MI, Balasubramanian S, Klenerman D. 2000. *J. Phys. Chem. B* 104:5171–78
56. Widengren J, Schwille P. 2000. *J. Phys. Chem. A* 104:6416–28
57. Brasselet S, Peterman EJG, Miyawaki A, Moerner WE. 2000. *J. Phys. Chem. B* 104:3676–82
58. Deleted in proof
59. Haas E, Wilchek M, Katchalski-Katzir E, Steinberg IZ. 1975. *Proc. Natl. Acad. Sci. USA* 72:1807–11
60. Haas E, Steinberg IZ. 1984. *Biophys. J.* 46:429–37
61. Onuchic JN, Luthey-Schulten Z, Wolynes PG. 1997. *Annu. Rev. Phys. Chem.* 48:545–600
62. Dobson CM, Karplus M. 1999. *Curr. Opin. Struct. Biol.* 9:92–101
63. Dill KA. 1999. *Protein Sci.* 8:1166–80
64. Brooks CL III, Gruebele M, Onuchic JN, Wolynes PG. 1998. *Proc. Natl. Acad. Sci. USA* 95:11037–38
65. Pande VS, Grosberg AY, Tanaka T, Rokhsar DS. 1998. *Curr. Opin. Struct. Biol.* 8:68–79
66. Onuchic JN, Nymeyer H, García AE, Chahine J, Socci ND. 2000. *Adv. Protein Chem.* 53:87–152
67. Eaton WA, Munoz V, Hagen SJ, Jas GS, Lapidus LJ, et al. 2000. *Annu. Rev. Biophys. Biomol. Struct.* 29:327–59
68. Bilsel O, Matthews CR. 2000. *Adv. Protein Chem.* 53:153–207
69. Brockwell DJ, Smith DA, Radford SE. 2000. *Curr. Opin. Struct. Biol.* 10:16–25
70. Wolynes PG. 1997. *Proc. Natl. Acad. Sci. USA* 94:6170–75
71. Lazaridis T, Karplus M. 1997. *Science* 278:1928–31
72. Sabelko J, Ervin J, Gruebele M. 1999. *Proc. Natl. Acad. Sci. USA* 96:6031–36
73. Eaton WA. 1999. *Proc. Natl. Acad. Sci. USA* 96:5897–99
74. Dill KA, Chan HS. 1997. *Nat. Struct. Biol.* 4:10–19
75. Alm E, Baker D. 1999. *Curr. Opinion Struct. Biol.* 9:189–96
76. Brooks CL III. 1998. *Curr. Opin. Struct. Biol.* 8:222–26
77. Galzitskaya OV, Finkelstein AV. 1999. *Proc. Natl. Acad. Sci. USA* 96:11299–304
78. Ladurner AG, Itzhaki LS, Daggett V, Fersht AR. 1998. *Proc. Natl. Acad. Sci. USA* 95:8473–78
79. Pande VS, Rokhsar DS. 1999. *Proc. Natl. Acad. Sci. USA* 96:1273–78
80. Thirumalai D, Klimov DK. 1999. *Curr. Opin. Struct. Biol.* 9:197–207
81. Itzhaki LS, Otzen DE, Fersht AR. 1995. *J. Mol. Biol.* 254:260–88
82. Jackson SE, Fersht AR. 1991. *Biochemistry* 30:10428–35
83. Jackson SE, Fersht AR. 1991. *Biochemistry* 30:10436–43
84. Mcphalen CA, James MNG. 1987. *Biochemistry* 26:261–69
85. Fersht AR. 1999. *Structure and Mechanism in Protein Science*. New York: Freeman
86. Dill KA, Shortle D. 1991. *Annu. Rev. Biochem.* 60:795–825
87. Yadav S, Ahmad F. 2000. *Anal. Biochem.* 283:207–13
88. Hagen SJ, Eaton WA. 2000. *J. Mol. Biol.* 297:781–89
89. Shastry MC, Roder H. 1998. *Nat. Struct. Biol.* 5:385–92
90. Pollack L, Tate MW, Darnton NC, Knight JB, Gruner SM, et al. 1999. *Proc. Natl. Acad. Sci. USA* 96:10115–17
91. Plaxco KW, Millett IS, Segel DJ, Doniach

S, Baker D. 1999. *Nat. Struct. Biol.* 6:554–56
92. Laurence TA, Deniz AA, Faulhaber AE, Dahan M, Chemla DS, et al. 2000. Presented at 44th Ann. Biophys. Soc. Meet. New Orleans
93. Foster S, Schmidt M. 1995. *Adv. Polym. Sci.* 120:51–133
94. Barrat J-L, Joanny J-F. 1996. *Adv. Chem. Phys.* 94:1–66
95. Yevich R, Olson WK. 1979. *Biopolymers* 18:113–45
96. Mills JB, Vacano E, Hagerman PJ. 1999. *J. Mol. Biol.* 285:245–57
97. Smith SB, Cui Y, Bustamante C. 1996. *Science* 271:795–99
98. Wuite GJ, Smith SB, Young M, Keller D, Bustamante C. 2000. *Nature* 404:103–6
99. VanOrden A, Machara NP, Goodwin PM, Keller RA. 1998. *Anal. Chem.* 70:1444–51
100. Chen Y, Müller JD, So PT, Gratton E. 1999. *Biophys. J.* 77:553–67
101. Kask P, Palo K, Ullmann D, Gall K. 1999. *Proc. Natl. Acad. Sci. USA* 96:13756–61
102. Müller JD, Chen Y, Gratton E. 2000. *Biophys. J.* 78:474–86
103. Kask P, Palo K, Fay N, Brand L, Mets U, et al. 2000. *Biophys. J.* 78:1703–13
104. Magde D, Elson E, Webb WW. 1972. *Phys. Rev. Lett.* 29:705–8
105. Lu HP, Xun L, Xie XS. 1998. *Science* 282:1877–82
106. Xie XS, Lu HP. 1999. *J. Biol. Chem.* 274:15967–70
107. Edman L, Rigler R. 2000. *Proc. Natl. Acad. Sci. USA* 97:8266–71
108. Edman L, Foldes-Papp Z, Wennmalm S, Rigler R. 1999. *Chem. Phys.* 247:11–22
109. Lamb DC, Schenk A, Rocker C, Scalfi-Happ C, Nienhaus GU. 2000. *Biophys. J.* 79:1129–38

TIME-RESOLVED PHOTOELECTRON SPECTROSCOPY OF MOLECULES AND CLUSTERS

Daniel M Neumark

Department of Chemistry, University of California, Berkeley, California 94720, and Chemical Sciences Division, Lawrence Berkeley National Laboratory, Berkeley, California 94720; e-mail: neumark@cchem.berkeley.edu

Key Words photoelectron spectroscopy, picosecond and femtosecond lasers, clusters, negative ions

■ **Abstract** Time-resolved photoelectron spectroscopy (TRPES) has become a powerful new tool in studying the dynamics of molecules and clusters. It has been applied to processes ranging from energy flow in electronically excited states of molecules to electron solvation dynamics in clusters. This review covers experimental and theoretical aspects of TRPES, focusing on studies of neutral and negatively charged species.

INTRODUCTION AND OVERVIEW

The past fifteen years have witnessed an explosion in the use of ultrafast lasers to follow chemical dynamics in real time with femtosecond resolution (1–3). This methodology has been applied to chemical reactions ranging in complexity from bond-breaking in diatomic molecules to dynamics in large inorganic and biological molecules, and has led to breakthroughs in our understanding of fundamental chemical processes.

Most of these experiments involve a pump-probe configuration in which an ultrafast pump pulse initiates a reaction or, more generally, creates a nonstationary state, and the evolution of this state is monitored by means of a probe pulse. The amount of information obtained from these experiments is very much dependent on the probe scheme. In condensed phase work, transient absorption is often the method of choice because of its generality. In studies of molecules and cluster in the gas phase, number densities are generally too low for transient absorption measurements, so other techniques must be used. The most popular methods, laser-induced fluorescence and resonant multiphoton ionization, require the probe laser to be resonant with an electronic transition in the species being monitored. However, as a chemical reaction initiated by the pump pulse evolves toward products, one expects the nature of the species under observation to change. Hence these probe methods are restricted to observation of the dynamics within a small region of the reaction coordinate. For example, in a seminal experiment by Zewail

& Bernstein (4) the reaction

$$H + CO_2 \rightarrow OH + CO \qquad 1.$$

was initiated by pump laser photolysis of the HI moiety in the van der Waals complex $CO_2 \cdot HI$, and the probe laser monitored OH fluorescence as a function of time. Thus, the reaction could be "clocked", in the sense that one could determine the time interval between initiation of the reaction and product formation, but not the detailed dynamics in the transition state region between reactants and products.

The development of more powerful probe methods in ultrafast experiments that could be used to monitor a reaction along the entire reaction coordinate has become a very active area in recent years. One approach is the generation of ultrafast X-ray or electron pulses to perform time-resolved X-ray or electron diffraction experiments, the goal being to be able to directly determine nuclear positions as a function of time once a reaction has been initiated (5–8). This method holds considerable promise but requires further development before it can be readily applied to chemical dynamics.

This article focuses on a different approach, time-resolved photoelectron spectroscopy, a probe methodology that has been demonstrated to be able to follow dynamics along the entire reaction coordinate and is considerably easier to implement (at present) than time-resolved diffraction experiments. In these experiments, the probe laser generates electrons through photoionization or photodetachment, and the electron kinetic energy distribution is followed as a function of time. Time-resolved photoelectron spectroscopy (TRPES) was first developed and applied to electronic dynamics on semiconductor surfaces in the mid 1980's (9) and indeed, is still a powerful tool in the investigation of metal and semiconductor surfaces (10–13). However, in this article the focus is on applications to isolated molecules and clusters, both neutral and negatively charged, in the gas phase.

TRPES experiments have been performed on neutral molecules for several years using nanosecond (ns), picosecond (ps), and femtosecond (fs) lasers to generate the pump and probe pulses. The type of dynamics one can follow depends on the temporal resolution of the laser system. TRPES experiments with ns or ps resolution have been used to probe lifetimes and radiationless decay pathways of excited electronic states. With fs resolution, experiments of this type can be done on very short-lived electronic states. In addition, one can follow a host of vibrational dynamics including dissociation, vibrational relaxation, and coherent wavepacket motion. The results of ns or ps TRPES experiments often complement information that can be obtained with other techniques such as laser-induced fluorescence and dispersed emission. Femtosecond TRPES experiments are complementary to experiments in which mass-selected ion yields are measured as a function of pump-probe delay. For example, in photodissociation, ion-yield experiments are very useful for identifying any transient species and monitoring the production of products (14), whereas TRPES provides an additional level of detail concerning the energy content of these species as a function of time.

TRPES experiments have also been performed on mass-selected negative ions. While the more complex sources and lower number densities in negative ion experiments present challenges absent in the neutral experiments, detachment energies are generally significantly lower than ionization energies in neutral species, so it is easier to generate probe laser pulses with sufficient energy to eject an electron. In addition, studies of clusters are straightforward in negative ion experiments because the ions can be mass-selected prior to their interaction with the laser pulses; analogous neutral studies require collecting photoions in coincidence with photoelectrons so that the identity of the ionized species can be ascertained.

In a typical instrument for a gas phase experiment, a skimmed neutral molecular beam or mass-selected negative ion beam interacts with pump and probe laser pulses in the interaction region of a photoelectron (PE) energy analyzer. Because these experiments involve pulsed lasers, time-of-flight electron energy analyzers have been used in virtually every time-resolved PE experiment to date. High collection efficiency analyzers are particularly useful to increase the data acquisition rate; parabolic reflectors (15) or more commonly, "magnetic bottle" analyzers (16, 17) with a collection efficiency of 50% or better have been incorporated into many of the instruments currently in use. The recent development of high power femtosecond lasers with repetition rates of 1 kHz or higher has also greatly facilitated time-resolved PE experiments with femtosecond temporal resolution. The combination of photoelectron spectroscopy and femtosecond lasers is particularly appealing because the energy resolution of electron energy analyzers (10–100 meV) is comparable to the energy spread of femtosecond laser pulses (20–30 meV for a 100 fs pulse).

An example of the principle and power of time-resolved PES is shown in Figure 1, in which femtosecond lasers (80–120 fs wide) are used to probe the dissociation of the negative ion I_2^- on its $A'^2\Pi_{g,1/2}$ excited state (18, 19). In this experiment, a pump pulse at 780 nm excites I_2^- from its ground $X^2\Sigma_u^+$ state to the repulsive $A'^2\Pi_{g,1/2}$ state. The pump pulse creates a localized wavepacket that evolves toward $I + I^-$ products. The evolution of this wavepacket is monitored by electron photodetachment with a pulse at 260 nm, and the resulting photoelectron spectrum is measured as a function of pump-probe delay. At each delay, the PE spectrum represents a Franck–Condon (FC) mapping of the dissociating–wavepacket onto the various I_2 potential energy curves, thereby providing a series of "snapshots" of the wavepacket as it moves from the initial region of excitation to asymptotic products.

Two sample PE spectra at delay times of 70 and 320 fs are shown. In the 70 fs spectrum, dissociation on the $A'^2\Pi_{g,1/2}$ state is underway but not complete. As a consequence, the PE spectrum reflects the FC overlap of the excited state wavepacket with the vibrational eigenstates supported by the various I_2 potential energy curves, and it therefore has the appearance of a molecular PE spectrum. By 300 fs, dissociation is essentially complete, and the PE spectrum represents photodetachment of the I^- photoproduct to the well-separated $^2P_{3/2}$ and $^2P_{1/2}$ spin-orbit states of the I atom. The FPE spectra of I_2^-, discussed in more detail later in

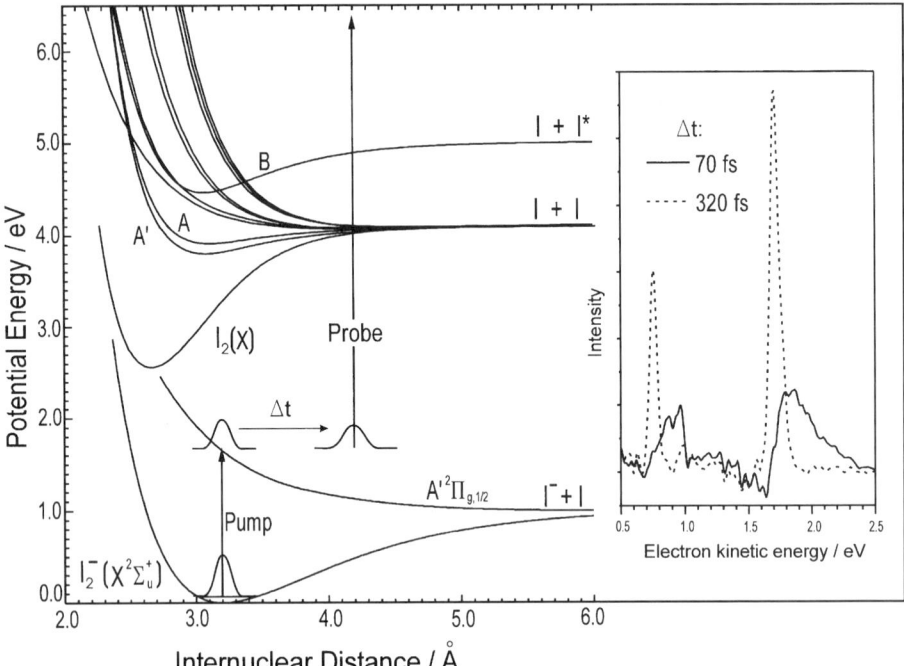

Figure 1 Time-resolved photoelectron spectroscopy of I_2^-. Relevant potential energy curves for I_2^- and I_2 are shown, along with PE spectra taken at pump-probe delay times of 70 and 320 fs.

this article, thus illustrate a simple yet fundamental process, namely the breaking of a chemical bond. They also emphasize several general advantages of TRPES: (*a*) one can indeed monitor evolution of the excited state dynamics along the entire reaction coordinate, and (*b*) the probe laser does not have to be tuned, since the PE spectrum at each delay provides the full FC mapping of the evolving wavepacket onto those electronic states accessible by photodetachment (or photoionization, in the case of neutrals).

In several time-resolved ionization experiments, zero-electron-kinetic-energy (ZEKE) spectroscopy rather than fixed-frequency PE spectroscopy was used to analyze the electrons produced by the probe pulse. ZEKE spectroscopy (20) offers considerably better resolution (0.1 meV) than PE spectroscopy but requires a tunable probe laser. This causes some experimental difficulties when dealing with ultrafast (particularly fs) lasers. In addition, while PE spectroscopy yields a complete FC mapping of the evolving state at a single probe wavelength, the equivalent information from a ZEKE experiment requires taking data at many probe wavelengths, negating the multiplexed energy analysis offered by PE spectroscopy. Moreover, while the additional energy resolution of ZEKE spectroscopy is useful in ps experiments, since the linewidths of ps lasers are typically a few wavenumbers, this advantage is lost in fs experiments. For these reasons, relatively few time-resolved ZEKE studies have been carried out; these are discussed below along with TRPES experiments.

With this overview in mind, the rest of this article reviews past progress in the field of TRPES. Studies of neutral molecules and clusters are considered first, followed by a discussion of negative ion experiments. Theoretical treatments of both classes of experiments are discussed. Finally, experimental and theoretical developments in the area of time-resolved photoelectron angular distributions are considered.

STUDIES OF NEUTRAL MOLECULES AND CLUSTERS

In this section, TRPES experiments with nanosecond and picosecond lasers are discussed first, followed by femtosecond PES studies. This division, while somewhat arbitrary, is approximately chronological and also serves to emphasize how the nature of the dynamics that can be probed depends on the laser pulse length.

The first gas phase TRPES experiments were demonstrated in the 1980's. Using either 2 ns or 5 ps laser pulses, Pallix & Colson (21) measured the photoelectron spectra from one-color 1+2 resonant multiphoton ionization of sym-triazine via the excited $^1E''$ state. Although these were not pump-probe experiments, the photoelectron spectra showed a clear dependence on the width of the laser pulse; spectra using the ps laser showed two broad peaks, while in those obtained with the ns laser, the peak at higher electron kinetic energy (eKE) was much diminished in intensity. These observations were explained in terms of intersystem crossing (ISC) from the initially excited $^1E''$ state, responsible for the higher energy peak, to the lowest triplet (T_1) state, on a timescale between 2 ns and 5 ps. The first pump-probe time-resolved PES experiment was performed by Sekreta & Reilly (22) on benzene using ns laser pulses. In these experiments, the pump laser excited the 6_0^1 level of the $^1B_{2u}$ state. The pump-probe spectra showed the contribution of this state to the PE spectrum diminishing with time while a peak at lower eKE grew in on a time scale of ∼50 ns. These results were attributed to ISC from the $^1B_{2u}$ state to the lower-lying $^3B_{1u}$ (T_1) state.

These two experiments illustrate several principles of time-resolved PES as applied to radiationless transitions. First, optically dark states (the T_1 states in these examples) are detected as easily as optically allowed excited states. Hence, one can track ISC and other radiationless processes that might occur subsequent to electronic excitation. In addition, although total energy is conserved in radiationless transitions, the photoelectron spectra of the excited singlet and triplet states are easily distinguished from one another. This occurs because the difference in electronic term values between the initial and final states generally appears as vibrational excitation in the lower energy state. As long as changes in geometry upon ionization from either state are relatively small, which is often the case for aromatic molecules, FC factors dictate that $\Delta v = 0$

transitions dominate the PE spectra from both states. Hence, vibrational energy is approximately conserved upon ionization, and the bands in the PE spectra are separated by the difference in the electronic term energies.

Starting in 1990, other applications of time-resolved PES were developed. In experiments performed with picosecond lasers, Reilly (23), Knee (24), and their co-workers showed that time-resolved PES could be used to follow intramolecular vibrational relaxation (IVR) in excited electronic states. In IVR, the vibrational character of the initially excited vibronic level evolves with time, but the molecule remains in the same electronic state. In Reilly's experiments, the $^1B_2 \leftarrow {}^1A_1 1_0^1$ transition was pumped, and vibrational structure in the photoelectron spectrum in the band associated with ionization of the $^1B_2$ state exhibited a distinct time-dependence due to mixing of the $v_1 = 1$ level with the low-frequency vibrational modes associated with the alkyl chain. This result, while consistent with previous fluorescence studies by Smalley (25) and Zewail (26) nicely demonstrated how IVR could be readily distinguished from electronically nonadiabatic transitions in time-resolved PES; IVR is manifested as changes in vibrational structure within a single photoelectron band, as distinct from electronically nonadiabatic transitions in which new bands appear.

In picosecond ZEKE experiments by Knee and coworkers on fluorene (24), quantum beats were observed at a low level of vibrational excitation (800 cm^{-1}) in the S_1 state, and irreversible IVR was seen when higher vibrational levels ($\sim$ 1700 cm^{-1}) were excited. The signature of IVR was quite clear as the ZEKE spectrum evolved from a relatively sparse band at early time to a more extended and congested band by 500 ps. In related experiments on the aniline $\cdot$ CH$_4$ van der Waals complex (27), both IVR and predissociation were followed as a function of time and of vibrational excitation in the S_1 state of the aniline chromophore. This group also applied picosecond ZEKE spectroscopy to the S_1 state of benzene to distinguish between IVR within the S_1 state and electronic relaxation to the lower-lying T_1 and S_0 states (28).

While the above studies focused on dynamics of the S_1 state in aromatics, Weber and co-workers (29, 30) used picosecond PES to investigate electronic state mixing and internal conversion (IC) dynamics in higher-lying singlet states, in particular the S_3 and S_4 states of azulene and the S_2 state of phenanthrene, all of which exhibited IC to lower lying singlet states. Weber has also used picosecond PES to follow ISC from the S_1 state in aniline and 2- and 3-aminopyridine, finding in all cases that only the T_1 state was populated even though higher-lying triplet states were accessible (31).

Syage (32, 33) demonstrated that picosecond time-resolved PES could be used to follow reactions and solvation dynamics within clusters, the first example of a particularly powerful application of the technique. This experiment was applied to excited state proton transfer in phenol$\cdot$(NH$_3$)$_n$ clusters. The phenol chromophore is electronically excited with a 266 nm pump pulse, and proton transfer to the solvent molecules occurs, resulting in a PhO$^-\cdots$H$^+$(NH$_3$)$_n$ transient species in which the NH$_3$ molecules rearrange to better solvate the proton. Although some of these dynamics were inferred previously from time-resolved ionization mass spectrometry on these and naphthol·$(NH_3)_n$ clusters (34–36), time-resolved PES yields a more detailed picture. In particular, the PE spectra of the n = 5 cluster showed a transient feature appearing on a time scale of 50 ps followed by a shifting of this feature toward higher eKE on a time scale of 300 ps; these observations were attributed to proton transfer and solvation, respectively.

In other applications of picosecond TRPES, de Lange and co-workers used picosecond TRPES to follow dissociation in the repulsive A band of CH_3I (37) and to measure lifetimes of the predissociative $\tilde{B}^1 E''$ and $\tilde{C}'^1 A_1'$ Rydberg states of NH_3 (38). Fischer and co-workers (39–41) used this method to measure lifetimes of several vibrational levels of the B, C, and D excited electronic states of the allyl radical, all of which lie about 5 eV above the ground state. The PE bands associated with the excited states disappeared on a time scale of 7–20 ps, with no other bands growing in. The observed decays were therefore attributed to IC to the allyl ground state, which could not be ionized with the probe pulse. The variation of decay rates with excitation energy approximately followed the density of ground state vibrational levels, further supporting the proposed mechanism.

The first experiments at femtosecond time resolution incorporating electron energy analysis were reported in the mid-1990s. Gerber (42), Stolow (43, 44), and their co-workers applied femtosecond ZEKE spectroscopy to vibrational wavepacket dynamics in Na_3 and I_2, respectively while Cyr & Hayden (45) used femtosecond TRPES to investigate the dynamics of internal conversion in 1,3,5-hexatriene. Gerber used a femtosecond pump pulse to create a superposition of vibrational levels in the B state of Na_3, and measured both total ion and ZEKE electron yield as a function of pump-probe delay. Both spectra showed clear recurrences due to coherent wavepacket dynamics on the B state. Stolow's experiments applied to the I_2 B state were similar in principle but more extensive. They yielded beautiful oscillatory structure, the phase of which varied with probe wavelength, and which showed dephasing and rephasing characteristic of wavepacket motion in an anharmonic potential.

In Hayden's work, the S_2 state of 1,3,5-hexatriene was excited, and a combination of time-resolved ion yield and PE spectroscopy measurements was used to unravel the ensuing dynamics. The PE spectra for the *cis* isomer showed very rapid (∼20 fs) internal conversion from the S_2 to S_1 state, followed by IVR within the S_1 state on a time scale of 300 fs. The ion yield measurements indicated lifetimes of 750 and 270 fs for the S_1 states in the *cis* and *trans* isomers, respectively, with respect to IC to the ground electronic state. This experiment represents a clear example of how TRPES can follow very fast dynamics involving multiple electronic states.

These first fs-resolved experiments were soon followed by a flurry of related activity in other laboratories, including the first TRPE experiments by Neumark (18) on negative ions (see following section) and experiments by Chen (46) and Baumert (47) on NO and Na_2, respectively. These and all subsequent experiments used PE rather than ZEKE detection of the electrons.

The work by Baumert and co-workers (47) on Na_2 exemplifies the multiplexed nature of femtosecond PE experiments. The principle of the experiment is shown in Figure 2a. Two-photon excitation was used to create a wavepacket on the excited $2^1\Pi_g$ state of Na_2, and the PE spectrum after ionization with a probe pulse was measured. The eKE changes with the phase of the wavepacket because of the difference between the potentials for the $2^1\Pi_g$ state and the $X^2\Sigma_g^+$ state of Na_2^+, with higher (lower) eKE at the inner (outer) turning point. As a result, a plot of

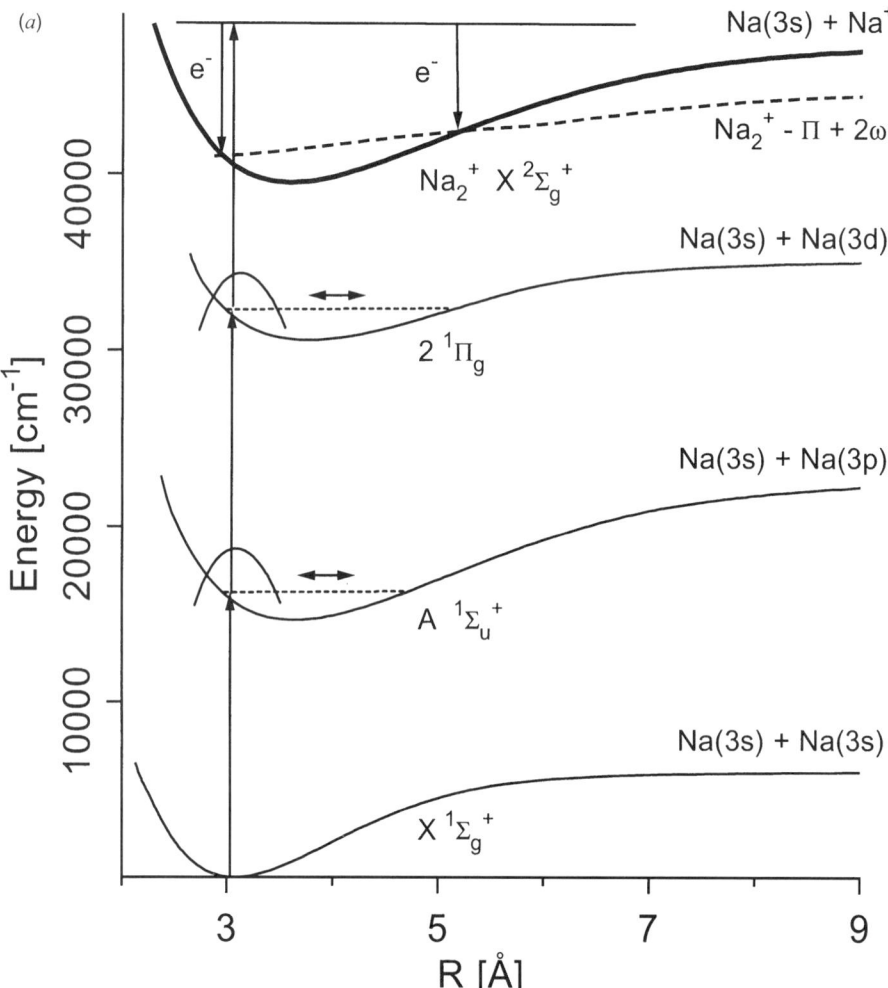

Figure 2 Femtosecond PE spectroscopy of Na_2. Potential energy curves of Na_2 and Na_2^+ are shown in (a), while PE spectra showing coherent wavepacket motion on the excited Na_2 states are shown in (b). Adapted from Reference (47).

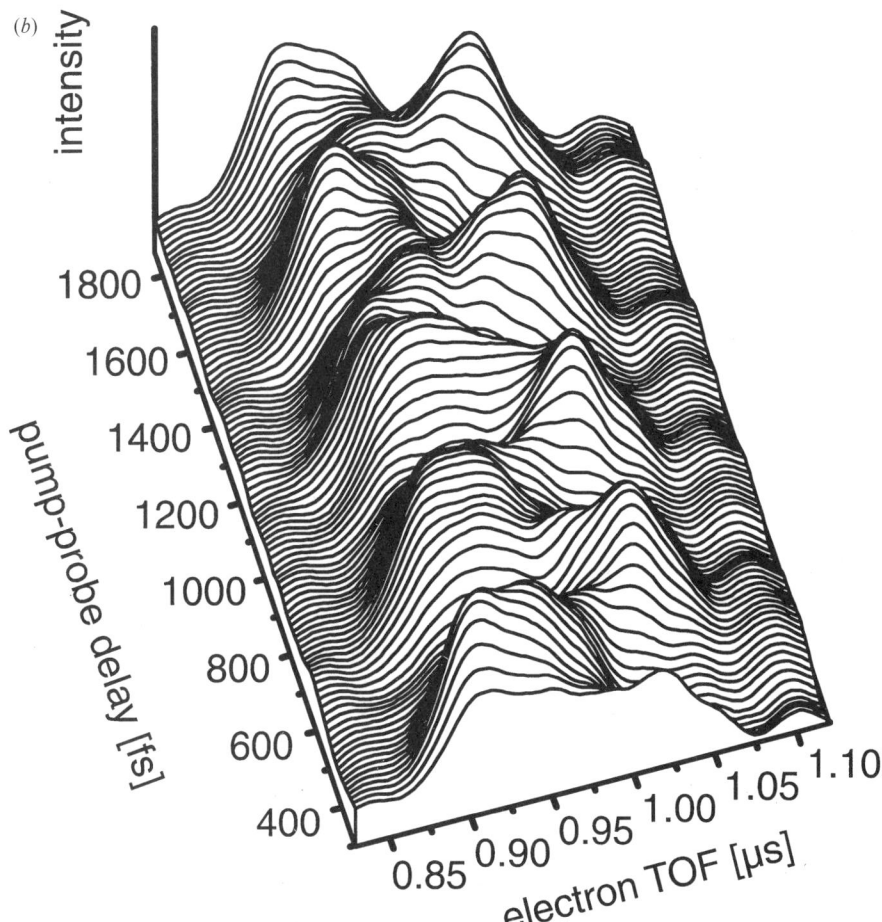

Figure 2 (*Continued*)

the photoelectron spectra versus time Figure 2*b* shows the complete excited state wavepacket dynamics. Note that a wavepacket is also created on the lower-lying $A^1\Sigma_u^+$ state through absorption of one pump photon, but because the $2^1\Pi_g \rightarrow A^1\Sigma_u^+$ transition was resonant with the pump laser, two-photon ionization out of the $A^1\Sigma_u^+$ occurs only at the inner turning point. The experimental results are in good agreement with an earlier theoretical treatment by Meier & Engel (48) and nicely complement the time-resolved total ion yield experiments by Schreiber & Wöste (49, 50).

Chen's work on NO (46) was the first in which intensity effects on time-resolved PE spectra were investigated. The experiment was in principle quite straightforward: measurement of the pump-probe photoelectron spectrum from

2 + 1 ionization of NO through its $C^2\Pi$ Rydberg state. At low intensity, the PE spectra with a 400 fs pump-probe delay was dominated by a single peak corresponding to the $\Delta v = 0$ transition from the $v = 0$ level of the C state. At higher pump intensities (10^{13} W/cm^2), a second feature appeared that was not resonant with any ionization transition. This feature was attributed to the $v = 3$ level of the $A^2\Sigma^+$ state being ac-Stark-shifted about 200 meV by the pump pulse into near-resonance with the C ($v = 0$) level and then returning to its zero-field value after the pump pulse terminated. More complex Stark-shifting and state mixing involving the nearby $B^2\Pi$ state were observed in the PE spectra at zero pump-probe delay.

Related experiments in which the probe laser intensity was varied were carried out by Baumert (51) on Na$_2$, showing enhanced contributions from ionization out of the $A^1\Sigma_u^+$ state to the femtosecond PE spectra at high probe powers. This was attributed to perturbations of the A and $2^1\Pi_g$ states of Na$_2$ by the probe pulse, so that the transition between the two states was near-resonant with the probe photon energy over the entire range of bond distances spanned by the wavepacket on the $A^1\Sigma_u^+$ state, as opposed to the inner turning point only (see above). Note that atomic and molecular PE spectra taken at even higher photon intensities show a whole new class of processes, such as above threshold ionization and tunneling/barrier suppression ionization. These "strong field" effects are outside the scope of this article and have been reviewed elsewhere (52).

Following the work of Hayden (45), several other groups have used femtosecond TRPES to follow very fast nonradiative decay dynamics in excited electronic states. For example, Stolow's group investigated the dynamics of (NO)$_2$ following electronic excitation at 210 nm (53). The parent (NO)$_2^+$ ion signal decayed with a time constant of 322 fs. In the absence of other information, this decay might be attributed to dissociation. However, time-resolved PE spectra showed a sharp peak associated with ionization of the NO ($A^2\Sigma^+$) photoproduct growing with a considerably slower time-constant of 730 fs. Hence, there appears to be an intermediate electronic state of the dimer that is dark with respect to photoionization at the probe wavelength (287 nm). This group also studied $S_2 \rightarrow S_1$ IC dynamics in decatetraene (54). Although the two excited states are nearly degenerate, photoionization of the S_2 state can access the cation ground state by a one-electron transition, whereas photoionization of the S_1 state leads primarily to an excited cation state lying 1.2 eV higher. Hence, the S_2 and S_1 states are readily distinguished in TRPE spectroscopy which shows that IC between the two excited states occurs on a time scale of 386 fs. Related work on the decay rates of the S_2 states in benzene, phenol and pyrazine has been reported by Radloff et al (55), Weber (56), and Stert et al (57), respectively.

Femtosecond TRPES has also been used to study dynamics in clusters. Soep and co-workers (58) examined the excited state double proton transfer reaction in the 7-azaindole dimer, comparing PE spectra obtained by 1 + 1 ionization with 0.8 ps and 5.0 ps laser pulses. They found significantly higher PE yield with shorter pulse ionization, indicating a subpicosecond excited state lifetime consistent with previous time-resolved ion yield measurements by Zewail and co-workers (59).

The 7-azaindole dimer is a very stable species, due to its double hydrogen bond, and can be generated as the dominant species in a free jet expansion. This is not the case for most neutral clusters, however, and to ensure that photoelectrons come from a cluster of particular size, it is desirable to collect photoelectrons in coincidence with ions. The first time-resolved coincidence experiment of this type, performed by Radloff et al (55) compared the excited state dynamics of benzene dimer with those of the monomer. They found that the decay of the S_2 to S_1 states was very rapid for both species (about 50 fs), whereas the $S_1 \rightarrow S_0$ IC decay was significantly faster in the monomer than in the dimer, 7.6 versus 100 ps. Measurement of the PE spectra made it straightforward to distinguish the two processes; $S_2 \rightarrow S_1$ IC was indicated by changes in the shape of the TRPE spectra, while $S_1 \rightarrow S_0$ IC rate was followed by the decay of the integrated PE signal. This group also investigated the dynamics of hydrogen transfer in electronically excited $(NH_3)_2$ (60). By monitoring the PE signal associated with the $(NH_3)_2^+$ and NH_4^+ ions, they were able to sort out the rather complicated dynamics associated with this system, finding time constants of 170 fs for the decay of $(NH_3)_2^*$ to $NH_4 \cdot\cdot NH_2$, and 4 ps for dissociation of the latter complex to $NH_4 + NH_2$.

Two recent important experimental developments in TRPES are likely to play a major role in the near future. Suzuki et al (61) have applied femtosecond time-resolved photoelectron imaging to ISC from the S_1 state of pyrazine. In this experiment, the electron distribution is imaged onto a position-sensitive detector, enabling one to follow the time-dependence of the photoelectron angular distribution as well as the energy distribution, and indeed Suzuki observed changes in the photoelectron angular distribution accompanying ISC. Davies et al (62) have developed an even more sophisticated experiment in which the electrons and ions produced from the pump and probe lasers are collected in coincidence and imaged. This instrument was used to follow dissociative multiphoton ionization of NO_2. At each delay time, one can measure how the electron kinetic energy varies with fragment recoil energy, thereby providing a very detailed picture of the excited state dissociation dynamics.

The other key development is the incorporation of femtosecond laser pulses with photon energies in the vacuum ultraviolet and beyond into TRPES experiments. The choice of neutral systems studied with TRPES has been limited by the requirement of a low-lying electronic state accessible via the pump pulse, and more importantly, the preference to be able to ionize the excited electronic state and its decay or dissociation products with a single probe photon. Femtosecond laser wavelengths around 200 nm have been achieved through fourth harmonic generation in nonlinear crystals, but in order to go beyond this one has to generate higher harmonics in pulsed free jets of rare gas atoms or in optical fibers (63). Recently, Sorensen et al (64) have performed TRPES on a high Rydberg state of C_2H_2 using the 6th harmonic (9.42 eV) of a Ti:sapphire laser as the pump pulse. Even higher harmonics have been incorporated into TRPES experiments by the Leone group. These were used as the probe pulse to follow the dissociation of Br_2 in real time. As these techniques become more commonplace, the range of applicability of TRPES will be increased significantly.

TRPES OF NEGATIVE IONS

Most of the neutral TRPES experiments described above have focused on radiationless, nondissociative decay in electronically excited states. Excited state dynamics are also probed in negative ion experiments, although the specific problems have been quite different, including photodissociation of bare and solvated ions, autodetachment dynamics, and electron solvation dynamics in clusters. In contrast to neutrals, experiments on negative ion clusters are straightforward in principle because the ions can be size-selected prior to interacting with the laser pulses. In addition, because detachment thresholds for negative ions are significantly lower than neutral ionization potentials, one can probe ground state dynamics induced by the pump pulse (or pulses) in negative ion experiments, whereas ground state ionization has generally been out of range in neutral TRPES experiments.

The first TRPES on a negative ion was performed with femtosecond laser pulses on I_2^-, in which the $A'^2\Pi_{g,1/2} \leftarrow X^2\Sigma_u^+$ transition (Figure 1) was pumped and the resulting dynamics on the repulsive upper state was probed (18). While the original study showed dissociation to be complete by 300 fs, as signified by the appearance of peaks associated with I^- photodetachment, subsequent work at higher electron energy resolution showed that the I^- peaks shifted 10 meV toward higher eKE from 320–720 fs (19). This was attributed to an attractive well at long range on the $A'^2\Pi_{g,1/2}$ state potential due to the charge-polarizability interaction between the separating I and I^- fragments. From these data an accurate potential for the $A'^2\Pi_{g,1/2}$ was obtained.

One also obtains information on ground state spectroscopy and dynamics from these experiments. In the process of exciting the $A'^2\Pi_{g,1/2} \leftarrow X^2\Sigma_u^+$ transition, the pump pulse generates a wavepacket on the $\tilde{X}^2\Sigma_u^+$ state of I_2^- composed primarily of the $v = 0$ and 1 states by resonant impulsive stimulated Raman scattering (RISRS) (66). Wavepacket motion appears as oscillatory structure at particular electron energies in the time-dependent PE spectra; from these oscillations the vibrational frequency of the anion, 110 ± 2 cm^{-1}, is obtained (67). This oscillatory structure persists long after the upper state dissociation dynamics are over, making it straightforward to distinguish ground and excited state dynamics. As in Baumert's experiments on Na_2 (47), different electron kinetic energies correspond to different phases of the wavepacket motion.

While the RISRS experiment probes vibrational motion only near the minimum of the $I_2^-(X^2\Sigma_u^+)$ ground state, one can access much higher vibrational levels, up to within 2% of the dissociation limit, in a three-pulse, pump-dump-probe experiment, in which the pump and dump pulses create a superposition of ground state vibrational levels via femtosecond stimulated emission pumping (SEP) (68).

The excitation energy of the wavepacket is varied by tuning the dump pulse. The wavepacket oscillates between the inner and outer turning points of the ground state potential, and these oscillations are readily detected through measurement of the photoelectron spectrum with a femtosecond probe pulse (at 263 nm) at a series of delay times. As the excitation energy is increased, the oscillation frequency drops, and the anharmonicity (measured by the rephasing time of the oscillations) increases. By measuring the frequency and anharmonicity as a function of excitation energy, a high quality potential for the I_2^- ground state was developed that should be accurate all the way up to the dissociation limit.

The above work on the ground and excited states of bare I_2^- was complemented by studies of how the excited state dynamics and ground state spectroscopy of I_2^- are affected by clusters with a weakly interacting solvating species (Ar) and a more strongly interacting solvent (CO_2). Based on previous experimental work by Lineberger (69–71), and molecular dynamics simulations by Parson (72–75), and Coker (76, 77), the excited state dynamics are strongly affected by solvation. Instead of undergoing direct dissociation on the A' state, the recoiling fragments interact with the surrounding solvent species, resulting in recombination, vibrational relaxation, and solvent evaporation. The femtosecond PE spectra enable one to follow these dynamics in considerable detail.

Femtosecond PES studies of $I_2^-(Ar)_n$ clusters ($n = 6$–20) (78, 79) showed dissociation of the I_2^- within the cluster is complete by 300 fs. From 300 fs–1 ps, the spectra yield the number of Ar atoms interacting with the I^- fragment. At later times, recombination of I_2^- occurs in $I_2^-(Ar)_{n \geq 12}$ on both the X and A states; the latter is a weakly bound excited state lying between the X and A' states. Analysis of the spectra yields the time scale for X state vibrational relaxation and solvent evaporation. In $I_2^-(Ar)_{20}$, energy transfer from I_2^- to Ar atoms through vibrational relaxation is slightly faster than energy loss from the cluster through Ar evaporation, indicating the temporary storage of energy within Ar cluster modes.

Similar experiments were performed on $I_2^-(CO_2)_n$ clusters with n = 4–16 (80, 81). At short times (<1 ps), the PE spectra show evidence for rearrangement of the solvent molecules around the separated I and I^- fragments. At longer times, features associated with recombination, vibrational relaxation, and solvent evaporation are observed for clusters with n $\geq$ 6. These dynamics occur more rapidly as the cluster size increases, and in all cases are much faster than in the Ar clusters. In addition, substantial trapping in a solvent-separated state is seen for clusters with n > 9; this state, in which I and I^- are trapped within the cluster but have not recombined, persists for at least 200 ps.

In order to probe the effect of solvation on the I_2^- ground state potential, excitation of coherent ground state vibrational motion via RISRS was used to measure vibrational frequencies of the I_2^- chromophore in size selected $I_2^-(Ar)_n$ and $I_2^-(CO_2)_n$ clusters to wavenumber accuracy (82). Blue shifting of the frequency occurs upon solvation, with larger shifts observed for solvation with CO_2 (i.e. 4.5 cm^{-1} for $I_2^-(CO_2)_9$). The blue-shifting in the Ar clusters appears to result from repulsive solute-solvent interactions. The situation in CO_2 clusters is more complicated and may result from a small amount of electron delocalization into the solvent, increasing the I_2^- vibrational frequency since the highest occupied molecular orbital in I_2^- is anti-bonding. The effect of solvation on much higher vibrational levels of I_2^- was probed by using femtosecond SEP on $I_2^-(CO_2)_4$ to generate a highly excited, coherent wavepacket within a cluster (83). The FPE spectra showed significant coherent relaxation of this wavepacket over 3 ps, in which one observed oscillatory motion, the period of which decreased with time as the I_2^- chromophore lost vibrational energy to the surrounding solvent molecules.

Femtosecond TRPES was also used to study the photodissociation dynamics of polyatomic anions. The first study of this type was performed by Gantefor et al (84) on Au_3^-. When this species was excited at a photon energy of 3.0 eV, a long-lived excited state was formed that decayed to $Au^- + Au_2$ on a time scale of 1500 ps; the PE spectra showed a broad feature at early times coalescing to a narrow feature corresponding to the Au^- photoproduct. At a slightly higher photon energy (3.14 eV), dissociation was far more rapid, with both Au^- and Au_2^- formed in less than 50 ps.

A similar experiment in principle was performed on I_3^- at 390 nm (85). The FPE spectra in Figure 3 show depletion of the I_3^- reactant and growth of the I_2^- and I^- products; these are clearly distinguishable since the time-evolution of the three species occurs at different electron energy ranges. Dissociation is complete after several hundred femtoseconds. In addition, the I_2^- exhibits coherent oscillations with a period of 550 fs corresponding to $\sim$0.70 eV of vibrational excitation, or $\langle v \rangle = 67$. The oscillations dephase by 4 ps and rephase at 45 and 90.5 ps on the anharmonic I_2^- potential. The gas phase frequency of ground state I_3^- was determined from oscillations in the photoelectron spectrum induced by RISRS. The dissociation dynamics were modeled using one- and two-dimensional wavepacket simulations, from which we attributed the formation of I^- to three body dissociation to $2I + I^-$ along the symmetric stretching coordinate of the excited anion potential. The photodissociation dynamics of gas phase I_3^- differ considerably from those observed previously in EtOH solution by Ruhman (86) and Vohringer (87) using transient absorption spectroscopy. In solution, substantially less I_2^- vibrational excitation is observed ($\langle v \rangle = 12$) at time delays as short as 500 fs, and no I^- is produced except as a minor channel at 266 nm (88).

Anion TRPES experiments have also focused on electron solvation dynamics in clusters. When a halide anion is dissolved in water and many other solvents, one observes absorption bands in the ultraviolet corresponding to ejection of the excess electron into the surrounding solvent molecules (89, 90). Recently, the cluster analog of these charge-transfer-to-solvent (CTTS) bands was spectroscopically observed by Johnson (91) and Cheshnovsky (92, 93) in $I^-(H_2O)_n$ and $I^-(Xe)_n$ clusters, respectively. Excitation of the CTTS bands in solution produces solvated electrons (94), so TRPES measurements on $I^-(S)_n$ clusters (S = Xe, H_2O, NH_3, CH_3OH) were undertaken to follow the electron solvation dynamics in the finite solvent networks on these clusters subsequent to excitation of the cluster CTTS band.

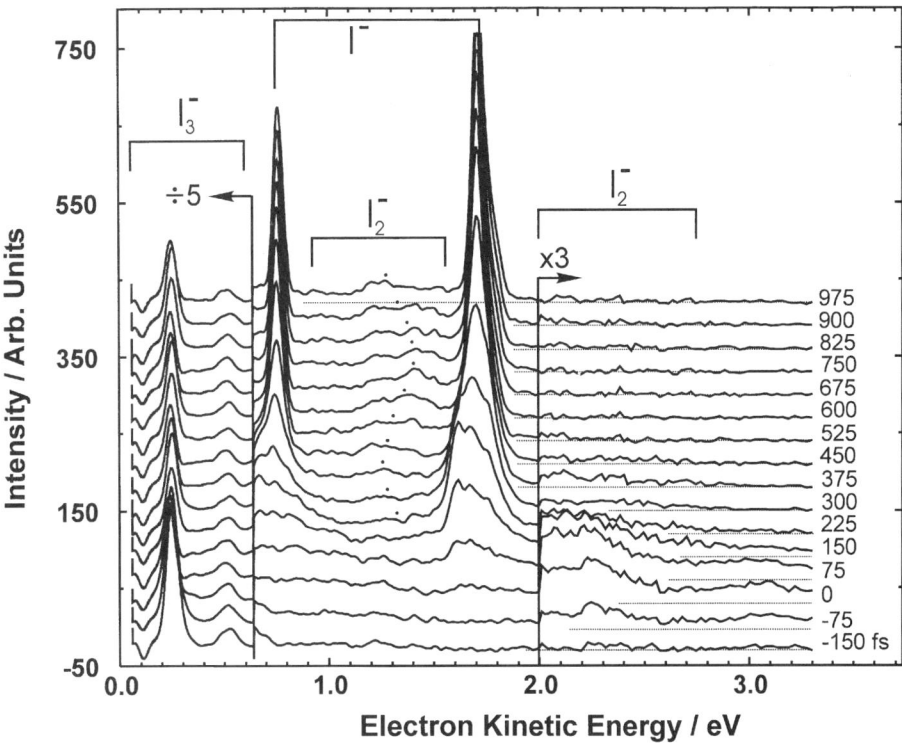

Figure 3 Femtosecond PE spectra of I_3^- using a pump and probe wavelengths of 390 nm and 260 nm, respectively. Contributions to the spectra from I_3^- reactant and the I_2^- and I^- photoproducts are indicated. Adapted from Reference (85).

In $I^-(Xe)_n$ clusters (95) the photoelectron spectra rapidly build up in intensity and then decay on a time scale that increases from 550 fs for the n = 11 cluster to 1,550 fs for n = 38 cluster. The electron kinetic energy distribution does not shift with time; only the integrated intensity varies. In contrast, for the molecular solvent species H_2O, NH_3, and CH_3OH (96–98), a shift toward lower eKE by as much as 0.3 eV was observed after several hundred femtoseconds. This shift was ascribed to reorganization of the solvent network as the solvent molecules rearrange to stabilize and partially solvate the excess electron. The absence of this shift for Xe indicates that no solvent rearrangement occurs in the $I^-(Xe)_n$ clusters.

There are distinct differences among the solvent species that do show partial solvation. Clusters with NH_3 and CH_3OH show a gradual buildup of the solvation shift as the number of solvent molecules is increased from n = 4 to n = 8, whereas water clusters show no shift for n = 4, and the full shift of ∼0.3 eV occurs for clusters with n ≥ 5. Hence, the onset of solvation appears far more abrupt in clusters with water. The lifetime of the partially solvated state is also strongly solvent dependent. For example, in I$^-$(NH$_3$)$_{15}$, the largest NH$_3$ cluster studied, this state decays 53 ps, whereas a much longer decay time of 440 ps was seen for I$^-$(H$_2$O)$_8$.

Finally, Minemoto et al (99) have used TRPES to measure the lifetime of an excited state of C$_3^-$ that lies above the threshold for autodetachment to C$_3$ + e^-. This state, first observed in a frequency-resolved experiment by Tulej et al (100), has been tentatively assigned as the $^2\Delta_u$ state and was found to have a lifetime of 2.6 ps with respect to autodetachment. Measurements of this type on larger carbon cluster anions are likely to be of significant interest given the large number of low-lying electronic states in these species.

THEORETICAL TREATMENT OF TRPES

The theoretical treatment of TRPES involves a straightforward but somewhat computationally demanding application of time-dependent quantum mechanics. In 1991, Seel & Domcke (101) wrote the first systematic theoretical paper specifically addressing TRPES and how it could be applied to follow $S_2 \to S_1$ IC dynamics in pyrazine, a problem that was not addressed experimentally until 2000 (57). In 1993, Meier & Engel (48) applied a similar treatment to TRPES of Na$_2$, predicting many of the features associated with coherent wavepacket motion that were later observed experimentally by Baumert (47).

Both papers described a nonperturbative treatment of the problem, the simplest version of which involves the ground state, a single excited state, and a single ionized (or detached) state. One then solves the time-dependent Schrodinger equation in the presence of the pump and probe laser fields, $E_{pu}(t)$ and $E_{pr}(t-\Delta t)$, to obtain the time-dependent wavefunction:

$$|\psi(t)\rangle = \sum_{n=0,1} \chi_n(R,t) |n\rangle + \int_0^{E_{max}} dE_k \chi_2(R, E_k, t) |k\rangle. \qquad 2.$$

Here the $\chi_n(t)$ (n = 0, 1, 2) are nuclear wavefunctions associated with the ground, excited, and ionized electronic states $|0\rangle$, $|1\rangle$, and $|k\rangle$, respectively, with $\chi_2(R, E_k, t)$ the nuclear wavefunction corresponding to electron kinetic energy E_k. The photoelectron spectrum $P(E_k)$ is then given by:

$$P(E_k, \Delta t) = \lim_{t \to \infty} \int dR |\chi_2(R, E_k, t)|^2. \qquad 3.$$

In this exact treatment, the photoelectron spectrum has to be calculated for each delay time Δt and each electron kinetic energy E_k, a somewhat tedious procedure.

A considerable simplification is obtained by solving for $\chi_2(R, t)$ exactly considering the pump pulse alone, and treating the probe pulse perturbatively, as first proposed by Meier & Engel (102, 103). If $\chi_2(R, E_k, t)$ is calculated to first order and substituted into (3), one finds (18, 102)

$$P(E_k, \Delta t) = \frac{|\mu_{12}|^2}{\hbar^2} \left| \int_{-\infty}^{\infty} dt' e^{iE_k t'} \left[e_{pr}(t' - \Delta t) e^{iH_2 t'} \chi_1(R, t') \right] \right|^2, \qquad 4.$$

where μ_{12} is the transition dipole for photoionization or photodetachment and H_2 is the nuclear hamiltonian for the ionized or detached potential energy surface. The advantage of Equation 4 is that it generates the entire photoelectron spectrum at time delay Δt by a Fourier transform of the term in brackets. The perturbative treatment of TRPES is discussed in detail by Batista et al (104) who also propose a semiclassical method that gave excellent agreement with the full quantum mechanical treatment in simulations of the I_2^- TRPE spectra.

The assumption of a constant transition dipole for photoionization or photodetachment in all the above treatments is problematic when the state created by the pump pulse undergoes large amplitude nuclear motion or, for that matter, dissociation. Arasaki et al (105, 106) have carried out a more sophisticated calculation on the TRPES of Na_2 in which the photoionization transition amplitudes are explicitly calculated as a function of internuclear distance. In their calculation, the intermediate excited state was the $2^1\Sigma_u^+$ state of Na_2, a double-minimum state in which the character of the electronic wavefunction changes significantly on either side of the barrier; Arasaki et al. show that this affects both the photoelectron energy and angular distributions.

TIME-RESOLVED PHOTOELECTRON ANGULAR DISTRIBUTIONS

It has long been recognized in conventional (i.e. nontime-resolved) photoelectron spectroscopy that the measurement of photoelectron angular distributions provides information complementary to that obtained by photoelectron energy distributions. The angular distributions are often more difficult to interpret than the energy distributions, but they often shed light on aspects of the ionization/detachment process such as the symmetries of the electronic states involved and the presence of vibronic coupling in the ionized species. In addition, the measurement of photoelectron angular distributions can help in making assignments if two or more ionization/detachment bands are overlapped. In time-resolved experiments, there has been relatively little work on the measurement of time-resolved angular distributions compared to energy distributions, and much of this has been theoretical rather than experimental. Nonetheless it is clear from the work done so far that this type of measurement can provide significant insight into the dynamics induced by the pump pulse.

The photoelectron angular distribution from one-photon ionization of a randomly oriented ensemble of molecules is given by

$$\frac{d\sigma}{d\Omega} = \frac{\sigma}{4\pi}(1 + \beta P_2(\cos\theta)), \qquad 5.$$

where θ is the angle between the electron velocity and the polarization direction of the ionizing light. The anisotropy parameter β lies between -1 and 2 and depends on the symmetry of the initial and final states as well as the electron kinetic energy. Zare and co-workers (107) observed deviations from Equation 5 in resonant multiphoton ionization of NO because of a net alignment (or orientation) created in the intermediate electronic state through absorption of the first photon. Reid (108) was the first to show that this sensitivity of the angular distribution on the rotational wavefunction in the intermediate state could lead to time-evolution of the photoelectron angular distribution if the pump pulse in a TRPES experiment created a coherent superposition of angular momentum levels, and demonstrated this in a nanosecond-resolution experiment using this time-dependence to follow hyperfine depolarization in the NO ($A^2\Sigma^+$) state (109).

On a faster time scale, theoretical work by Reid (108, 110) McKoy (105), and Seideman (111, 112) has shown that if an ultrafast pump pulse is used to generate an aligned, coherent superposition of rotational levels, then one should observe periodicity in the photoelectron angular distribution on a picosecond time scale analogous to the recurrences seen in rotational coherence spectroscopy. Reid (113) has carried out experiments of this type through excitation and ionization of the S_1 state of para-difluorobenzene using picosecond laser pulses, although the level of vibrational excitation in the S_1 state was sufficiently high so that the time-dependent angular distributions were governed more by intramolecular vibrational energy redistribution than by pure rotational coherences.

Finally, time-dependent angular distributions measured on a femtosecond time scale can also be used to probe nonadiabatic transitions and photodissociation dynamics in polyatomic molecules following electronic excitation with ultrafast pulses. Effects of this type were seen in the TRPE imaging experiments of Suzuki (61), discussed previously, in which ISC in pyrazine was studied. Theoretical work by Seideman and co-workers (114, 115) has shown that the photoelectron angular distributions are very sensitive to changes in the shape and symmetry of the excited state electronic wavefunction associated with nonadiabatic dynamics, and these distributions should provide complementary information to energy-resolved measurements. The significance of measuring time-dependent PE angular distributions was demonstrated in an experiment performed very recently by Hayden and co-workers (116) on the photodissociation of NO_2. By measuring time-dependent photoelectron angular distributions in coincidence with NO^+ product scattered in a particular direction (parallel to the laser polarization), they could follow the evolution of the photoelectron angular distribution in the molecular frame of reference. As shown in Figure 4, the distribution evolves from a highly asymmetric distribution at short times to a forward-backward symmetric distribution by 1 ps. The short-time dynamics were attributed to the effect of the departing O atom on the electronic wavefunction of the NO photoproduct.

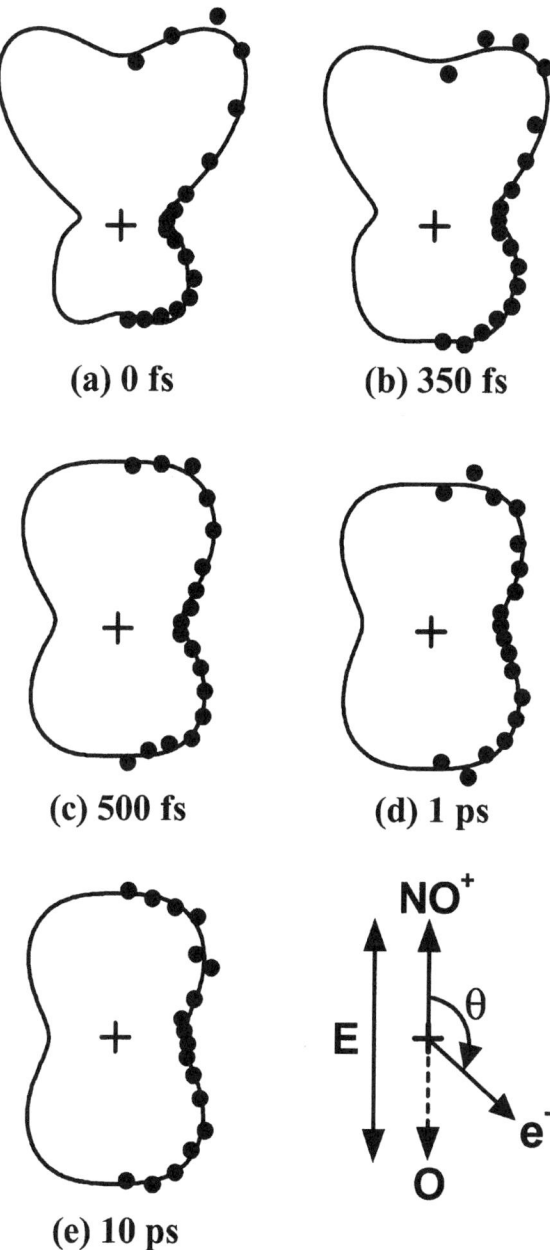

Figure 4 Time-resolved photoelectron angular distributions from NO_2 photodissociation. These are molecular frame distributions in electrons are collected in coincidence with NO^+ fragments scattered in a particular direction. Reproduced from Reference (116) with permission.

SUMMARY AND OUTLOOK

This review has presented a comprehensive discussion of time-resolved photoelectron spectroscopy as applied to gas phase molecular and cluster dynamics. A wide range of dynamical processes in neutral and negatively charged species had already been investigated using this technique, and TRPES is now recognized as a powerful new tool for the study of chemical dynamics. Two new developments described in this article hold considerable promise in the near future. First, as high harmonic generation of ultrafast laser pulses becomes more routine, it should be possible to perform TRPES of the valence electrons of many more neutral species, and also to investigate time-dependent PE spectra of the core electrons, thereby providing a complementary perspective of dissociatve and reactive processes initiated by the pump pulse. The recent incorporation of electron imaging techniques in TRPES experiments is also very appealing as it enables the measurement of time-dependent photoelectron angular distributions.

While TRPES experiments have thus far been limited to the gas phase and semiconductor or metal surfaces, there is no reason why they cannot be extended to other media. In principle, TRPES can be applied to any type of system accessible to conventional PES. For example, conventional photoelectron spectroscopy has been applied to liquid jets in order to probe structure and dynamics at the gas-liquid interface (117), and this would clearly be an interesting application of TRPES. Experiments of this type would provide a vital link with some of the cluster dynamics experiments described in this article.

ACKNOWLEDGMENTS

Support from the National Science Foundation under Grant No. CHE-9710243 is gratefully acknowledged. The author is indebted to his many co-workers on the anion femtosecond photoelectron spectroscopy project in his laboratory: B Jefferys Greenblatt, Martin Zanni, Alison Davis, Christian Frischkorn, Rainer Weinkauf, Leo Lehr, Benoit Soep, Mohammed Elhanine, Arthur Bragg, and Roland Wester.

Visit the Annual Reviews home page at www.AnnualReviews.org

LITERATURE CITED

1. Khundkar LR, Zewail AH. 1990. *Annu. Rev. Phys. Chem.* 41:15–60
2. Polanyi JC, Zewail AH. 1995. *Accts. Chem. Res.* 28:119–32
3. Zewail AH. 1996. *J. Phys. Chem.* 100:12,701–24
4. Scherer NF, Khundkar LR, Bernstein RB, Zewail AH. 1987. *J. Chem. Phys.* 87:1451–53
5. Williamson JC, Cao JM, Ihee H, Frey H, Zewail AH. 1997. *Nature* 386:159–62
6. Rischel C, Rousse A, Uschmann I, Albouy PA, Geindre JP, et al. 1997. *Nature* 390:490–92
7. Chin AH, Schoenlein RW, Glover TE, Balling P, Leemans WP, Shank CV. 1999. *Phys. Rev. Lett.* 83:336–39
8. Siders CW, Cavalleri A, Sokolowski-Tinten

K, Toth C, Guo T, et al. 1999. *Science* 286:1340–42
9. Haight R, Bokor J, Stark J, Storz RH, Freeman RR, Bucksbaum PH. 1985. *Phys. Rev. Lett.* 54:1302–5
10. Hertel T, Knoesel E, Wolf M, Ertl G. 1996. *Phys. Rev. Lett.* 76:535–38
11. Hofer U, Shumay IL, Reuss C, Thomann U, Wallauer W, Fauster T. 1997. *Science* 277:1480–82
12. Ge NH, Wong CM, Lingle RL, McNeill JD, Gaffney KJ, Harris CB. 1998. *Science* 279:202–5
13. Petek H, Weida MJ, Nagano H, Ogawa S. 2000. *Science* 288:1402–4
14. Pedersen S, Herek JL, Zewail AH. 1994. *Science* 266:1359–64
15. Steadman J, Syage JA. 1993. *Rev. Sci. Instrum.* 64:3094–3103
16. Kruit P, Read FH. 1983. *J. Phys. E* 16:313–24
17. Cheshnovsky O, Yang SH, Pettiette CL, Craycraft MJ, Smalley RE. 1987. *Rev. Sci. Instrum.* 58:2131–37
18. Greenblatt BJ, Zanni MT, Neumark DM. 1996. *Chem. Phys. Lett.* 258:523–29
19. Zanni MT, Batista VS, Greenblatt BJ, Miller WH, Neumark DM. 1999. *J. Chem. Phys.* 110:3748–55
20. Muller-Dethlefs K, Schlag EW. 1991. *Annu. Rev. Phys. Chem.* 42:109–36
21. Pallix JB, Colson SD. 1985. *Chem. Phys. Lett.* 119:38–41
22. Sekreta E, Reilly JP. 1988. *Chem. Phys. Lett.* 149:482–86
23. Song XB, Wilkerson CW, Lucia J, Pauls S, Reilly JP. 1990. *Chem. Phys. Lett.* 174:377–84
24. Smith JM, Lakshminarayan C, Knee JL. 1990. *J. Chem. Phys.* 93:4475–76
25. Powers DE, Hopkins JB, Smalley RE. 1980. *J. Chem. Phys.* 72:5721–30
26. Baskin JS, Dantus M, Zewail AH. 1986. *Chem. Phys. Lett.* 130:473–81
27. Smith JM, Zhang X, Knee JL. 1993. *J. Chem. Phys.* 99:2550–59
28. Smith JM, Zhang X, Knee JL. 1995. *J. Phys. Chem.* 99:1768–75
29. Weber PM, Thantu N. 1992. *Chem. Phys. Lett.* 197:556–61
30. Thantu N, Weber PM. 1993. *Chem. Phys. Lett.* 214:276–80
31. Kim B, Schick CP, Weber PM. 1995. *J. Chem. Phys.* 103:6903–13
32. Syage JA. 1993. *Chem. Phys. Lett.* 202:227–32
33. Syage JA. 1995. *J. Phys. Chem.* 99:5772–86
34. Breen JJ, Willberg DM, Gutmann M, Zewail AH. 1990. *J. Chem. Phys.* 93:9180–84
35. Kim SK, Breen JJ, Willberg DM, Peng LW, Heikal A, et al. 1995. *J. Phys. Chem.* 99:7421–35
36. Hineman MF, Brucker GA, Kelley DF, Bernstein ER. 1992. *J. Chem. Phys.* 97:3341–47
37. Dobber MR, Buma WJ, Delange CA. 1993. *J. Chem. Phys.* 99:836–53
38. Dobber MR, Buma WJ, Delange CA. 1995. *J. Phys. Chem.* 99:1671–85
39. Schultz T, Fischer I. 1997. *J. Chem. Phys.* 107:8197–8200
40. Schultz T, Fischer I. 1998. *J. Chem. Phys.* 109:5812–22
41. Schultz T, Clarke JS, Gilbert T, Deyerl HJ, Fischer I. 2000. *Faraday Discuss.* 115:17–31
42. Baumert T, Thalweiser R, Gerber G. 1993. *Chem. Phys. Lett.* 209:29–34
43. Fischer I, Villeneuve DM, Vrakking MJJ, Stolow A. 1995. *J. Chem. Phys.* 102:5566–69
44. Fischer I, Vrakking MJJ, Villeneuve DM, Stolow A. 1996. *Chem. Phys.* 207:331–54
45. Cyr DR, Hayden CC. 1996. *J. Chem. Phys.* 104:771–74
46. Ludowise P, Blackwell M, Chen Y. 1996. *Chem. Phys. Lett.* 258:530–39
47. Assion A, Geisler M, Helbing J, Seyfried V, Baumert T. 1996. *Phys. Rev. A* 54:R4605–7
48. Meier C, Engel V. 1993. *Chem. Phys. Lett.* 212:691–96
49. Rutz S, Greschik S, Schreiber E, Wöste

L. 1996. *Chem. Phys. Lett.* 257:365–73
50. Schreiber E. 1998. *Femtosecond Real-Time Spectroscopy of Small Molecules and Clusters*. New York: Springer
51. Frohnmeyer T, Hofmann H, Strehle M, Baumert T. 1999. *Chem. Phys. Lett.* 312:447–54
52. Levis RJ, DeWitt MJ. 1999. *J. Phys. Chem. A* 103:6493–6507
53. Blanchet V, Stolow A. 1998. *J. Chem. Phys.* 108:4371–74
54. Blanchet V, Zgierski MZ, Seideman T, Stolow A. 1999. *Nature* 401:52–54
55. Radloff W, Stert V, Freudenberg T, Hertel IV, Jouvet C, DedonderLardeux C, Solgadi D. 1997. *Chem. Phys. Lett.* 281:20–26
56. Schick CP, Carpenter SD, Weber PM. 1999. *J. Phys. Chem. A* 103:10,470–76
57. Stert V, Farmanara P, Radloff W. 2000. *J. Chem. Phys.* 112:4460–64
58. LopezMartens R, Long P, Solgadi D, Soep B, Syage J, Millie P. 1997. *Chem. Phys. Lett.* 273:219–26
59. Douhal A, Kim SK, Zewail AH. 1995. *Nature* 378:260–63
60. Stert V, Radloff W, Schulz CP, Hertel IV. 1999. *Eur. Phys. J. D* 5:97–106
61. Suzuki T, Wang L, Kohguchi H. 1999. *J. Chem. Phys.* 111:4859–61
62. Davies JA, LeClaire JE, Continetti RE, Hayden CC. 1999. *J. Chem. Phys.* 111:1–4
63. Durfee CG, Rundquist AR, Backus S, Herne C, Murnane MM, Kapteyn HC. 1999. *Phys. Rev. Lett.* 83:2187–90
64. Sorensen SL, Bjorneholm O, Hjelte I, Kihlgren T, Ohrwall G, et al. 2000. *J. Chem. Phys.* 112:8038–42
65. Deleted in proof
66. Banin U, Bartana A, Ruhman S, Kosloff R. 1994. *J. Chem. Phys.* 101:8461–81
67. Zanni MT, Taylor TR, Greenblatt BJ, Soep B, Neumark DM. 1997. *J. Chem. Phys.* 107:7613
68. Zanni MT, Davis AV, Frischkorn C, Elhanine M, Neumark DM. 2000. *J. Chem. Phys.* 112:8847–54
69. Papanikolas JM, Vorsa V, Nadal ME, Campagnola PJ, Buchenau HK, Lineberger WC. 1993. *J. Chem. Phys.* 99:8733–50
70. Vorsa V, Campagnola PJ, Nandi S, Larsson M, Lineberger WC. 1996. *J. Chem. Phys.* 105:2298–308
71. Vorsa V, Nandi S, Campagnola PJ, Larsson M, Lineberger WC. 1997. *J. Chem. Phys.* 106:1402–10
72. Faeder J, Delaney N, Maslen PE, Parson R. 1997. *Chem. Phys. Lett.* 270:196–205
73. Delaney N, Faeder J, Maslen PE, Parson R. 1997. *J. Phys. Chem. A* 101:8148–51
74. Faeder J, Delaney N, Maslen PE, Parson R. 1998. *Chem. Phys.* 239:525
75. Faeder J, Parson R. 1998. *J. Chem. Phys.* 108:3909–914
76. Batista VS, Coker DF. 1997. *J. Chem. Phys.* 106:7102–16
77. Margulis CJ, Coker DF. 1999. *J. Chem. Phys.* 110:5677–90
78. Greenblatt BJ, Zanni MT, Neumark DM. 1997. *Science* 276:1675
79. Greenblatt BJ, Zanni MT, Neumark DM. 1999. *J. Chem. Phys.* 111:10,566–77
80. Greenblatt BJ, Zanni MT, Neumark DM. 1997. *Faraday Discuss.* 108:101–13
81. Greenblatt BJ, Zanni MT, Neumark DM. 2000. *J. Chem. Phys.* 112:601–12
82. Zanni MT, Greenblatt BJ, Neumark DM. 1998. *J. Chem. Phys.* 109:9648–51
83. Davis AV, Zanni MT, Frischkorn C, Elhanine M, Neumark DM. 2000. *J. Electron Spectrosc. Rel. Phenom.* 112:221–30
84. Gantefor G, Kraus S, Eberhardt W. 1998. *J. Electron Spectrosc. Rel. Phenom.* 88:35–40
85. Zanni MT, Greenblatt BJ, Davis AV, Neumark DM. 1999. *J. Chem. Phys.* 111:2991–3003
86. Banin U, Ruhman S. 1993. *J. Chem. Phys.* 98:4391–4403
87. Kuhne T, Vohringer P. 1996. *J. Chem. Phys.* 105:10,788–802
88. Kuhne T, Kuster R, Vohringer P. 1998. *Chem. Phys.* 233:161–78

89. Franck J, Scheibe G. 1928. *Z. Phys. Chem. A* 139:22–31
90. Blandamer MJ, Fox MF. 1970. *Chem. Rev.* 70:59–93
91. Serxner D, Dessent CEH, Johnson MA. 1996. *J. Chem. Phys.* 105:7231–34
92. Becker I, Markovich G, Cheshnovsky O. 1997. *Phys. Rev. Lett.* 79:3391–94
93. Becker I, Cheshnovsky O. 1999. *J. Chem. Phys.* 110:6288–97
94. Jortner J, Ottolenghi M, Stein G. 1964. *J. Phys. Chem.* 68:247–55
95. Zanni MT, Frischkorn C, Davis AV, Neumark DM. 2000. *J. Phys. Chem. A* 104:2527–30
96. Lehr L, Zanni MT, Frischkorn C, Weinkauf R, Neumark DM. 1999. *Science* 284:635–38
97. Davis AV, Zanni MT, Frischkorn C, Neumark DM. 2000. *J. Electron Spectrosc. Relat. Phenom.* 108:203–11
98. Frischkorn C, Zanni MT, Davis AV, Neumark DM. 2000. *Faraday Discuss.* 115:49–62
99. Minemoto S, Muller J, Gantefor G, Munzer HJ, Boneberg J, Leiderer P. 2000. *Phys. Rev. Lett.* 84:3554–57
100. Tulej M, Fulara J, Sobolewski A, Jungen M, Maier JP. 2000. *J. Chem. Phys.* 112:3747–53
101. Seel M, Domcke W. 1991. *J. Chem. Phys.* 95:7806–22
102. Meier C, Engel V. 1994. *J. Chem. Phys.* 101:2673–77
103. Meier C, Engel V. 1994. *Phys. Rev. Lett.* 73:3207–10
104. Batista VS, Zanni MT, Greenblatt BJ, Neumark DM, Miller WH. 1999. *J. Chem. Phys.* 110:3736–47
105. Arasaki Y, Takatsuka K, Wang K, McKoy V. 1999. *Chem. Phys. Lett.* 302:363–74
106. Arasaki Y, Takatsuka K, Wang K, McKoy V. 2000. *J. Chem. Phys.* 112:8871–84
107. Leahy DJ, Reid KL, Zare RN. 1991. *J. Chem. Phys.* 95:1757–67
108. Reid KL. 1993. *Chem. Phys. Lett.* 215:25–30
109. Reid KL, Duxon SP, Towrie M. 1994. *Chem. Phys. Lett.* 228:351–56
110. Reid KL, Underwood JG. 2000. *J. Chem. Phys.* 112:3643–49
111. Althorpe SC, Seideman T. 1999. *J. Chem. Phys.* 110:147–55
112. Seideman T. 1999. *Phys. Rev. Lett.* 83:4971–74
113. Reid KL, Field TA, Towrie M, Matousek P. 1999. *J. Chem. Phys.* 111:1438–45
114. Seideman T. 2000. *J. Chem. Phys.* 113:1677–80
115. Blanchet V, Lochbrunner S, Schmitt M, Shaffer JP, Larsen JJ, et al. 2000. *Faraday Discuss.* 115:33–48
116. Davies JA, Continetti RE, Chandler DW, Hayden CC. 2000. *Phys. Rev. Lett.* 84:5983–86
117. Faubel M, Steiner B, Toennies JP. 1997. *J. Chem. Phys.* 106:9013–31

PULSED EPR SPECTROSCOPY: Biological Applications

Thomas Prisner, Martin Rohrer, and Fraser MacMillan

Institute for Physical and Theoretical Chemistry, J. W. Goethe-University Frankfurt, Marie-Curie-Strasse 11, D-60439 Frankfurt am Main, Germany; e-mail: prisner@chemie.uni-frankfurt.de

Dedicated to the memory of Jerry Babcock

Key Words pulsed ENDOR, ESEEM, high-field EPR, enzymes, paramagnetic centers

■ **Abstract** Pulsed electron paramagnetic resonance (EPR) methods such as ESEEM, PELDOR, relaxation time measurements, transient EPR, high-field/high-frequency EPR, and pulsed ENDOR, have been used successfully to investigate the local structure and dynamics of paramagnetic centers in biological samples. These methods allow different contributions to the EPR spectra to be distinguished and can help unravel complicated EPR spectra consisting of overlapping resonance lines, as are often found in disordered protein samples. The basic principles, specific potentials, technical requirements, and limitations of these advanced EPR techniques will be reviewed together with recent applications to metal centers, organic radicals, and spin labels in proteins.

INTRODUCTION

Electron paramagnetic resonance (EPR) spectroscopy is a well-known method for the examination of the local structure of paramagnetic molecules. In proteins such paramagnetic molecules can be naturally stable cofactors (e.g. heme molecules, iron-sulfur clusters, or metal ions); transiently generated radicals within a reaction cycle (chromophores, cofactors, or amino acid radicals); or artificial nitroxide spin labels attached to the protein. In contrast to nuclear magnetic resonance (NMR), EPR is not restricted by the size of the protein because only the paramagnetic centers and their interaction with the protein are spectroscopically visible. In cases in which only one paramagnetic center is located within the protein a simple one-dimensional (1D) continuous wave (cw) EPR experiment can give detailed information about the properties of the paramagnetic species, such as its oxidation state, ligand symmetry, and concentration. In more realistic cases, as in enzymes, the situation is far more complex: Often more than one paramagnetic species is involved (some of them may be transiently formed within the catalytic cycle of the enzyme), which are also subjected to internal dynamics. In these cases the simple cw-EPR spectra will suffer from the same difficulties as NMR for large molecules: Spectral lines overlap and broaden. Therefore, it becomes very difficult to analyze the spectra quantitatively and to obtain a unique solution; the problem is ill defined. As in NMR, pulse methods help unravel the different contributions to 1D-EPR spectra. Overlapping spectral components of different paramagnetic centers may be separated by pulsed EPR methods, for example by their relaxation times or by their spin-magnetic moment.

Whereas the methodological principles for the manipulation of the spin system (electron spin S in EPR, nuclear spin I in NMR) are very similar in NMR and EPR spectroscopy, the technical requirements and therefore the practical realization is quite different in both fields. Owing to the much larger magnetic moment of the electron spin S (nearly a factor of 1000 larger compared with a proton nuclear spin) the technical requirements (resonance frequency, relaxation times, pulse lengths) for EPR are much more demanding as compared with NMR spectroscopy. Nevertheless, most of the technical restrictions have been overcome by the development of specific pulse methods and techniques. These methods can, similar to heteronuclear NMR spectroscopy, not only affect the unpaired electron of a single paramagnetic species, but at the same time affect magnetically coupled nuclear spins as in pulsed electron nuclear double resonance (ENDOR) or the electron spin of another paramagnetic center in pulsed electron double resonance (PELDOR) experiments. Whereas in the experiments mentioned above the nuclear spin or the additional electron spin are excited by a second radiofrequency (RF) or microwave (MW) pulse, in many cases both types of couplings can also be examined with a single MW pulse in resonance with the unpaired electron spin. These methods are called electron spin echo envelope modulation (ESEEM) for the investigation of nuclear couplings and "2 + 1" for the examination of coupling to other paramagnetic centers. A further dimension can be added if they are performed at different external magnetic fields. Experiments at high magnetic fields (>2 T), especially, have proven to significantly enhance the amount of information that can be gathered from EPR and ENDOR spectra of aromatic organic radicals in proteins. All these advanced EPR techniques have dramatically increased the potential of EPR spectroscopy, especially in the field of biochemical applications. Nevertheless, it should be mentioned that for all biological systems discussed in this review, cw-EPR and cw-ENDOR methods have preceded the pulsed EPR investigations and have built a most valuable starting base for the advanced investigations reviewed herein.

PULSED METHODS AND RECENT APPLICATIONS TO BIOLOGICAL SYSTEMS

Technical Aspects

Unfortunately, one of the main advantages of pulsed NMR spectroscopy—the enhanced sensitivity of pulsed Fourier transform (FT) spectroscopy—does not apply to EPR spectroscopy on biological samples. In contrast to NMR spectroscopy, the rotational correlation time of biological macromolecules at room temperature is in most cases much too slow to effectively average the large anisotropic interactions of the unpaired electron spin. This leads to broad unresolved lineshapes—in many cases much broader than the available pulse excitation width, so that only a part of a spectrum can be recorded at a given magnetic field value. Therefore, the multiplex advantage of the pulsed experiments—recording of several lines at the same time—does not apply to EPR on macromolecules. Additionally, because of these broad linewidths and the very short relaxation times of paramagnetic centers, e.g. for transition metal ions, the pulse lengths have to be very short. To allow the short MW pulses to pass through the cavity circuit the bandwidth of the resonant MW cavity usually has to be lowered, as compared with cw-applications. This reduction of the Q value of the pulsed resonant circuits leads to a lower sensitivity.

Consequently, the only reason to perform pulsed experiments on proteins is the possibility of enhancing the spectral information on spin systems, as mentioned above. In many cases the lineshape of the paramagnetic center is inhomogeneously broadened in protein samples, owing to orientationally disordered samples or other inhomogeneities in the protein. Thus, most of the interesting interactions of the paramagnetic center with its surrounding, as for example the hyperfine coupling (hfc) to magnetic nuclei in the close surrounding, are hidden under these broadening mechanisms and cannot be observed directly in cw-EPR experiments. Pulsed experiments can refocus such static inhomogeneous broadening contributions and dramatically increase the spectral resolution with respect to other interactions, such as hfcs to nuclei or dipolar couplings to other paramagnetic species. Despite the fact that the MW pulses in pulsed EPR are by far less ideal than in NMR spectroscopy, the possibility of manipulating the spin system is still of great advantage in unraveling the information content of spectra in complex spin systems, by suppressing specific interactions and by diluting the spectra in a more-dimensional spectral space.

As already mentioned, the requirements on the equipment for pulsed EPR measurements are demanding. Pulse lengths have to be well below 100 ns, typically 5–10 ns for a $\pi/2$-pulse at X-band MW frequencies with a 1 kW MW amplifier. Care has to be taken that the bandwidth of all components in the excitation and detection channel support these short pulses without distortion. The important limitation for the observation of species with short transverse relaxation times is the dead time of the receiver after intense MW pulses. Typically, this time is about 50 ns and limits the detection to systems with longer transverse relaxation times T_2. It also makes it difficult to observe the free induction decay signal of most samples. For many paramagnetic ions the longitudinal relaxation time T_1 is also short even at low temperature. Only very recently have fast signal digitizers ($>10^8$ points/s, 8 bit resolution) and averagers ($>10,000$ acquisition/s, 1024 data points) been available for effective data collection with an optimum duty cycle (1). Fortunately, pulsed experiments put less stringent requirements on MW source noise, cavity stability, microphonics, and other sources of noise and baseline drift. This allows the less sensitive and therefore more time consuming pulsed experiments to be performed even on biological systems with low spin concentrations.

Spin Hamilton Operator

The magnetic properties of the unpaired electron spin of a paramagnetic center can be described by a spin Hamilton operator of the form (2):

$$H = \beta_e \vec{S}_p \cdot \hat{g} \cdot \vec{B}_0 + \sum_j \gamma_j \cdot \vec{I}_j \cdot \vec{B}_0 + \sum_j \vec{S}_p \cdot \hat{A}_j \cdot \vec{I}_j$$
$$+ \sum_j \vec{I}_j \cdot \hat{Q}_j \cdot \vec{I}_j + \sum_k \vec{S}_p \cdot \hat{D}_{pk} \cdot \vec{S}_k + \vec{S}_p \cdot \hat{D}_{pp} \cdot \vec{S}_p \qquad 1.$$

The first term represents the interaction of the unpaired electron spin $\vec{S}_p$ with the external magnetic field $\vec{B}_0$ via the anisotropic g-tensor of the paramagnetic molecule. The next two terms represent the magnetic interactions of close-by nuclear spins $\vec{I}_j$ with the external magnetic field and with the electron spin via the hfc tensor $\hat{A}$. The fourth term is the quadrupolar interaction $\hat{Q}$ for nuclear spins with $I > 1/2$. The remaining terms are the interaction of the paramagnetic center with other unpaired electron spins $\vec{S}_k$, including dipolar and exchange contributions and the so-called zero field splitting for electron spins $S > 1/2$. This Hamilton operator can be simplified for specific types of centers under many conditions, where one or two terms in the Hamilton operator dominate the spectra. Although ENDOR or ESEEM experiments are only sensitive to the interaction to other close-by nuclei within a distance range of up to 0.8 nm, PELDOR and pulsed EPR relaxation experiments are sensitive to dipolar interaction to other paramagnetic centers with distances up to approximately 5 nm. High magnetic field experiments can resolve the anisotropy of the electronic g-tensor, whereas this anisotropy is suppressed for most aromatic organic radicals at lower magnetic fields. Therefore, these different experiments can help distinguish between these contributions to the spin Hamilton operator and allow, even for complex systems, a detailed understanding of the paramagnetic center.

For a theoretical description of pulsed experiments, the time evolution of the quantum mechanical system under nonstationary perturbations (by MW and/or RF excitation) has to be considered. These can be described with the density matrix of the spin system including the excitation fields and spin interactions. The time evolution of the spin system is given by the stochastic Liouville equation, which has to be solved. Nevertheless, the basics of most of the pulsed experiments can be understood on the basis of simple energy level diagrams, in which only the eigenfunctions and eigenvalues of the spin Hamilton operator without excitation perturbation are considered.

In Figure 1 the resulting energy level scheme and spin eigenfunctions are shown for two simple cases: (a) the coupling of an unpaired electron spin $S = 1/2$ to a single nuclear spin $I = 1/2$ and (b) the coupling of two unpaired electron spins $S_1 = 1/2$ and $S_2 = 1/2$. The spin Hamiltonian eigenfunctions, when expressed in the

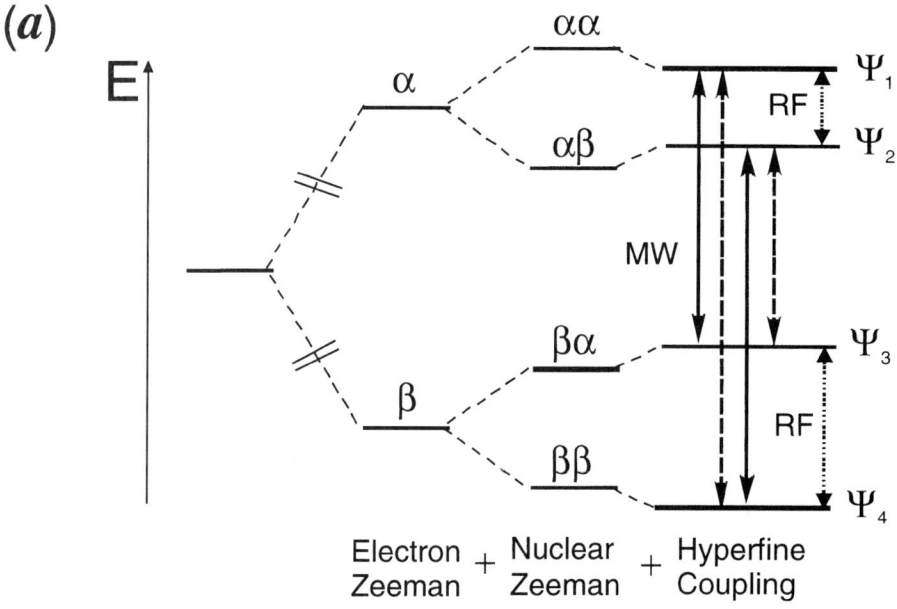
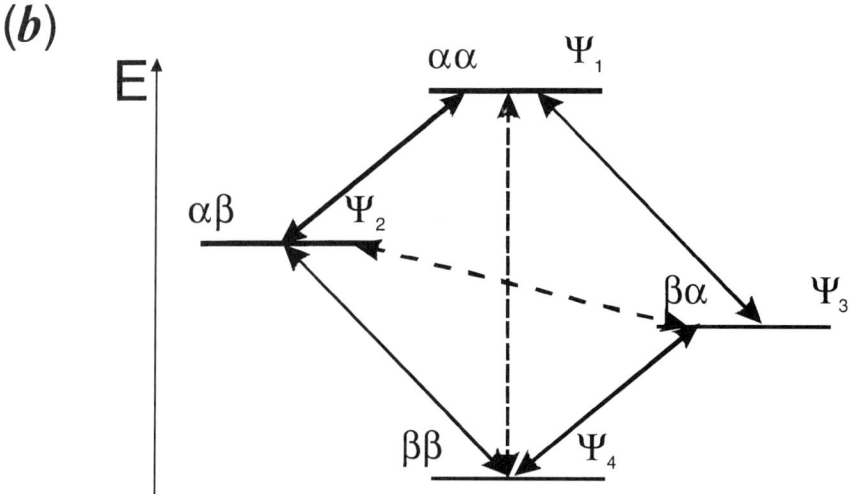

Figure 1 Energy level diagram for (a) an electron spin S = 1/2 coupled to a nuclear spin I = 1/2 and (b) two electron spins $S_1 = 1/2$ and $S_2 = 1/2$. The high-field basis functions α (m_S or m_I = +1/2) and β (m_S or m_I = −1/2) for the eigenfunctions $\Psi_1 - \Psi_4$ are valid only without coupling of the two spins. Allowed (solid arrows) and 'forbidden' EPR (dashed arrows) and NMR (dotted arrows) transitions between the spin states are included.

high-field eigenfunctions, α and β, are mixed states owing to anisotropic tensor interactions [hfc tensor $\hat{A}$ in (a) and dipolar and exchange coupling tensor $\hat{D}$ in (b)]. As a consequence of this mixing of the eigenfunctions, all transitions between the four eigenstates are allowed to a different extent under pulsed irradiation of the spin system. However, in case (a) the energy spacing by the nuclear and the electronic magnetic moment differs by approximately three orders of magnitude (much larger than shown in the diagram); hence, MW pulses only affect EPR transitions and RF pulses only affect NMR transitions of the coupled spin system. Because of the large hfc, RF pulses select a single NMR transition in only one electron spin manifold (α or β) in most cases, e.g. will induce transitions either between $\Psi_1 \leftrightarrow \Psi_2$ or $\Psi_3 \leftrightarrow \Psi_4$. In contrast, MW pulses select a single EPR transition (e.g. only $\Psi_1 \leftrightarrow \Psi_3$) or semiselective ($\Psi_1 \leftrightarrow \Psi_3$ and $\Psi_2 \leftrightarrow \Psi_4$). Two further EPR transitions will be partially allowed because of the mixing of the nuclear spin states by the anisotropic hfc (between state $\Psi_1 \leftrightarrow \Psi_4$ and $\Psi_2 \leftrightarrow \Psi_3$), as shown by the dashed lines in Figure 1.

Pulsed Electron Nuclear Double Resonance (Pulsed ENDOR)

Cw-ENDOR was first introduced by Feher (3). It was rapidly demonstrated to be a powerful tool to significantly increase spectral resolution as compared with cw-EPR. Analogous pulsed methods have been introduced by Mims (4) and Davies (5). Today cw and pulsed ENDOR techniques are well-established advanced EPR methods, being commonly applied in biology and organic chemistry.

Regardless of whether they are performed in a cw or pulsed mode, all ENDOR experiments have an EPR signal that is monitored, during which NMR transitions are induced that lead to changes of the EPR signal. These changes of an EPR signal are recorded as the ENDOR spectrum.

A thorough theoretical description of ENDOR methods is complex and requires significant mathematical and physical background knowledge (for review see 6, 7). Nevertheless, a simple energy level consideration (see Figure 1) allows a phenomenological understanding that is sufficient to explain the advantages and limitations of the method. In particular, the pulsed Davies ENDOR experiment can be explained in this picture quite easily. The description is based on a system characterized solely by an electron spin S = 1/2 and a nuclear spin I = 1/2. Owing to the applied static magnetic field B_0, the energy levels of the two quantum spin states α and β, both for S and I, are no longer degenerate. The resulting electron and nuclear Zeeman energy splittings differ by roughly three orders of magnitude as the result of the different electronic and nuclear gyromagnetic ratios. Furthermore, the hfc between the electronic and nuclear spins leads to an increase/decrease of the nuclear Zeeman splitting between the α and β electronic spin states. According to quantum mechanical selection rules, EPR transitions with $\Delta m_S = 1$ and $\Delta m_I = 0$ and NMR transitions with $\Delta m_I = 1$ and $\Delta m S = 0$ are allowed. During the basic pulsed Davies ENDOR experiment one of the EPR transitions is selectively induced (by a MW π-pulse) and monitored (by a MW Hahn echo pulse sequence) as shown in Figure 2. Under ideal conditions, the echo signal is thereby inverted as a result of the inversion pulse at the beginning of the sequence. This holds if the longitudinal relaxation time T_1^S is much longer than the pulse separation time T (see Figure 2). Selective excitation of one of the NMR transitions by the additional RF pulse will now change the detected EPR echo signal, which is the ENDOR effect. The echo can disappear completely if the RF matches exactly one of the two allowed NMR transition frequencies and if

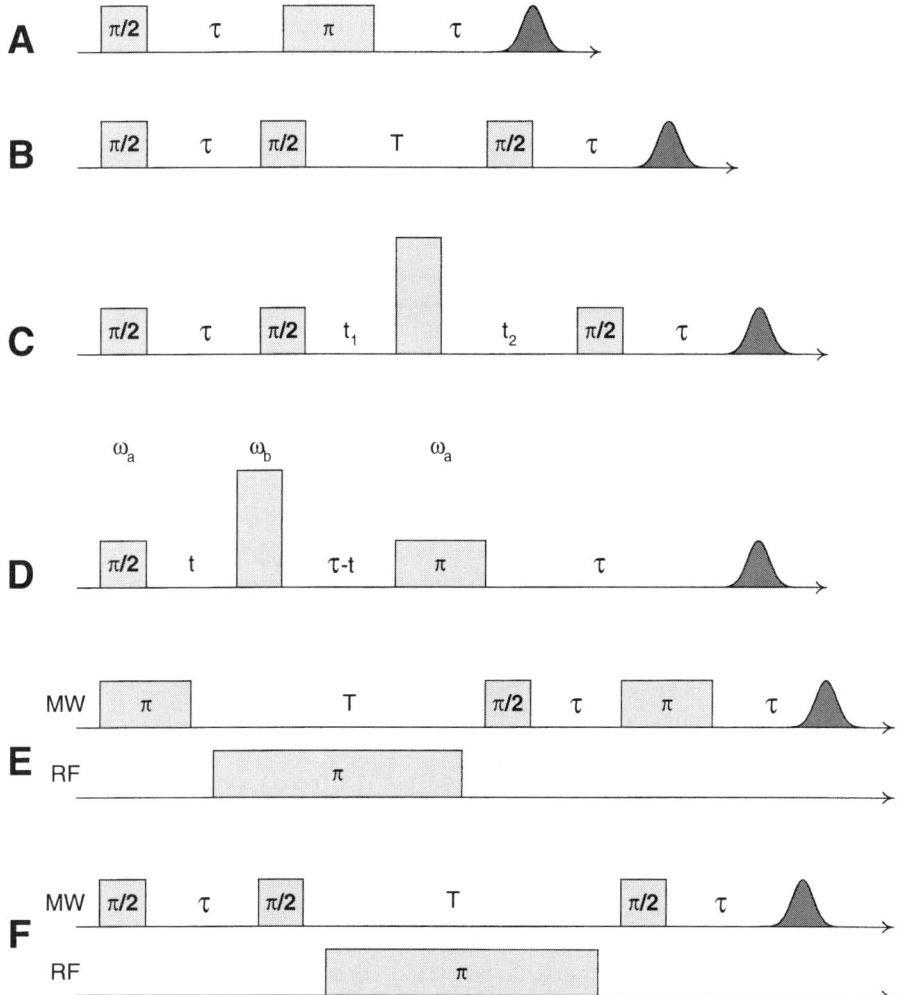

Figure 2 Pulsed-EPR and ENDOR experiments described in text. (*A*) Two-pulse (Hahn) echo, (*B*) three-pulse (stimulated) echo, (*C*) four-pulse echo (HYSCORE), (*D*) PELDOR pulse (for $\omega_a = \omega_b$, which corresponds to the "2 + 1" pulse sequence), (*E*) Davies ENDOR, (*F*) Mims ENDOR.

the pulse flip angle corresponds to a RF π-pulse for the nuclear spin. During the experiment the RF is swept linearly in time and the ENDOR spectra are recorded as a function of the RF. As can be rationalized from Figure 1, the magnitude of the observed hyperfine splitting is directly reflected in the frequency spacing between the two ENDOR signals. This applies for cases in which the hfc is small compared with the nuclear Zeeman energy at the chosen magnetic field if no further interactions such as zero field splitting or quadrupolar interactions have to be considered.

The technical effort required to generate strong RF fields over a broad frequency range to the cavity is rather high and quite expensive. Fortunately, over the past two decades, the instrumentation for extending EPR experiments to ENDOR capability has become readily available.

The basic advantages of ENDOR are the drastic increase of hyperfine resolution. Additionally, when a large number of equivalent nuclei interact with the electronic spin, the ENDOR spectra are significantly simplified compared with the corresponding EPR spectra. Drawbacks of ENDOR, especially when powder-type ENDOR spectra are to be interpreted, arise from overlapping lines originating from different interacting nuclei. These can be overcome by high-field ENDOR (see "Pulsed High-Field/High-Frequency EPR") or by application of advanced ENDOR methods (8, 9).

Another difficulty in the interpretation is often encountered if the observed hfc is larger than twice the nuclear Zeeman interaction. The ENDOR frequencies ν_{ENDOR} are given (for the simple cases under consideration) by:

$$\nu_{ENDOR} = \left| \nu_n \pm \frac{A}{2} \right|, \quad \nu_n = \frac{g_n \mu_n B_0}{h} \qquad 2.$$

where g_n and μ_n are the nuclear g-factor and the nuclear magneton, respectively, h is the Planck's constant, and A is the hfc. In cases in which the free nuclear ENDOR frequency ν_n is greater than $|A/2|$, the ENDOR lines are centered symmetrically around ν_n. In contrast, if ν_n is smaller than $|A/2|$ the ENDOR lines are centered around $|A/2|$ with a spacing corresponding to $2\nu_n$. This complicates the distinct assignment of ENDOR lines and makes the interpretation more difficult. One obvious way around this problem is the performance of ENDOR at different magnetic fields and frequencies because ν_n depends on the magnetic field B_0, whereas hfc does not.

Further important parameters to be considered concern the spin relaxation times of the system under investigation: For cw experiments, the most pronounced ENDOR effect is normally obtained when the longitudinal electronic (T_1^S) and nuclear spin relaxation (T_1^I) constants are equal, leading to the largest changes of the saturated EPR transition under observation. This condition can often be adjusted by means of sample temperature and solvent viscosity. It can also be overcome by advanced methods such as electron-nuclear-nuclear resonance (TRIPLE resonance), which has been applied successfully to measurements of organic radicals in liquid solution under physiological conditions (10). In contrast to cw-ENDOR, in pulsed-ENDOR experiments the ratio T_1^S/T_1^I is not of importance for the ENDOR effect. Instead, the dominant requirement is that T_1^S must be long enough to allow for the application of an RF π-pulse within the pulse separation time T. Another limitation results from the transversal electronic spin relaxation time T_2^S, which has to be longer than the pulse separation time τ if an echo detection of the ENDOR effect is used (see Figure 2). Direct detection of the free induction decay may overcome this problem. Another limitation of ENDOR applications is found for very small hfcs: It is caused by the strength of the RF fields incident to the sample, which must not exceed the magnitude of the hfc under observation. If this happens NMR pulses are no longer selective, which reduces the ENDOR effect. Reducing the RF power, which however, also reduces the ENDOR effect if the pulse length cannot be prolonged at the same time, may eliminate this handicap. Further limitations of the ENDOR method arise for very small nuclear-level splittings (as is often the case for ^{2}H or ^{14}N at X-band frequencies) because of RF excitation problems at very low frequencies. A way around this problem is to apply ESEEM techniques, as described below.

Applications of ENDOR spectroscopy to biological systems have concentrated on organic radicals and metal ion centers in proteins. Both types of radicals are usually investigated at low temperature. Especially for metal centers the temperature has to be below 20 K in most cases to obtain long enough relaxation times allowing a successful ENDOR experiment.

Within the group of organic radicals the chromophores involved in the electron transfer reaction of photosynthetic proteins in particular have been extensively studied, not only by cw- but also by pulsed-ENDOR methods (for review see 11). The proton couplings of the two semiquinone electron acceptor radicals $Q_A^{-\bullet}$ and $Q_B^{-\bullet}$ in bacterial reaction centers (bRC) (12), of the first quinone acceptor $A_1^{-\bullet}$ in photosystem I (PSI) (13) and of the quinone $Q_A^{-\bullet}$ in photosystem II (PSII) have been investigated by pulsed ENDOR spectroscopy. Pulsed W-band (95 GHz, 3.4 T) ENDOR experiments on $Q_A^{-\bullet}$ of bRC (14) allowed full proton hfc tensorial information to be obtained, owing to the possibility of performing orientation-selective experiments at high fields, as explained below. In PSII proteins the two tyrosine radicals, redox active tyrosine 160 of the D2 protein of PSII ($Y_D^{-\bullet}$) and redox active tyrosine 161 of the D1 protein of PSII ($Y_Z^{-\bullet}$) have been actively investigated by a number of groups (15–18). Transient ENDOR (19) and pulsed ENDOR (20) have been used to investigate the triplet state of the primary chlorophyll donor $P_{865}^{+\bullet}$ of bRC. Pulsed-ENDOR spectroscopy was also applied to the transient correlated-radical pair $P_{700}^{+\bullet}A_1^{-\bullet}$ in PSI (21). Very recently both chlorophyll and carotene cation radicals in PSII were characterized by pulsed ENDOR (22).

Similar characterizations have been undertaken concerning protein-bound organic molecules. Pulsed ENDOR was used to examine the semiquinone of quinol oxidase (23), a methylamine dehydrogenase complex (24), a flavin radical in monoamine oxidase (25, 26), a spin-coupled tryptophan radical in cytochrome c peroxidase (27, 28), and the tyrosine radical of ribonucleotide reductase (29). On the latter radical orientation-selective D-band (140 GHz, 5 T) ENDOR was also successfully applied to obtain tensorial information of the hfc tensors (30). The capability of the ENDOR method to distinguish different protein environments can be seen in Figure 3, in which pulsed-ENDOR spectra of such tyrosine radicals from different organisms are compared.

Applications of pulsed ENDOR to metalloenzymes, especially for Cu-proteins, FeS-centers, and binuclear Cu complexes, have been reviewed (31–35). ENDOR has been successfully used to determine ^{14}N-hfcs of the Cu ligands and ^{1}H/^{2}H couplings to ambient water molecules in the vicinity of the metal site (36, 37). Higher fields were used to gain more spectral information with Q-band ENDOR on azurin (38) and with W-band ENDOR on the Cu$_A$ site of cytochrome c oxidase (39). The interaction of a Mn^{2+}-ion center with ATP has been investigated by pulsed ENDOR in pyruvate kinase (40) and compared with the ^{1}H-, ^{31}P-, and ^{17}O-hfc determined in model systems (41). The manganese cluster in PSII was also investigated by pulsed-ENDOR methods in its different catalytic states (42). High-field W-band ENDOR was recently used to study the manganese center in a single crystal of concavalin A (43). Also, the local surroundings of binuclear iron centers in uteroferrin and methane monooxygenases (44–47), the iron-molybdenum center of molybdoferredoxin (35, 48), the iron-sulfur centers of a number of iron-sulfur proteins (34, 49–52), and the heme centers of cytochrome P450 cam and chloroperoxidase (53–55) have been investigated by pulsed ENDOR.

Electron Spin Echo Envelope Modulation

The electron spin echo envelope modulation (ESEEM) effect is induced by nuclei hyperfine-coupled to the unpaired electron spin. These couplings manifest themselves in a periodic modulation of the electron spin echo intensity as a function of the pulse separation time. Fourier transformation of the time-domain echo envelope leads to a hyperfine spectrum in the frequency domain similar to the corresponding ENDOR spectrum. These modulations of the electron spin echo intensity due to the coupled nuclei were already observed in early electron spin echo experiments (56, 57) and later theoretically understood and analytically described (58–60).

For nuclei coupled to the electron spin via an anisotropic hyperfine interaction, 1D- and 2D-pulsed ESEEM experiments have proven to be powerful tools to examine these interactions, especially for weak couplings and for nuclei with a small Zeeman splitting (61). In contrast to the ENDOR experiment, no RF irradiation is needed; the nuclear spins are only indirectly affected via their hfc to the electron spin.

Figure 3 Pulsed Davies X-band ENDOR on tyrosyl radicals from three different organisms (F MacMillan, F Lendzian, A Boussac, G Lassmann, W Lubitz 2000, unpublished). (A) Y$_D^{ox}$ of PSII, (B) Y$_{177}$ of ribonucleotide reductase in mouse, (C) Y$_{122}$ of ribonucleotide reductase in *Escherichia coli*.

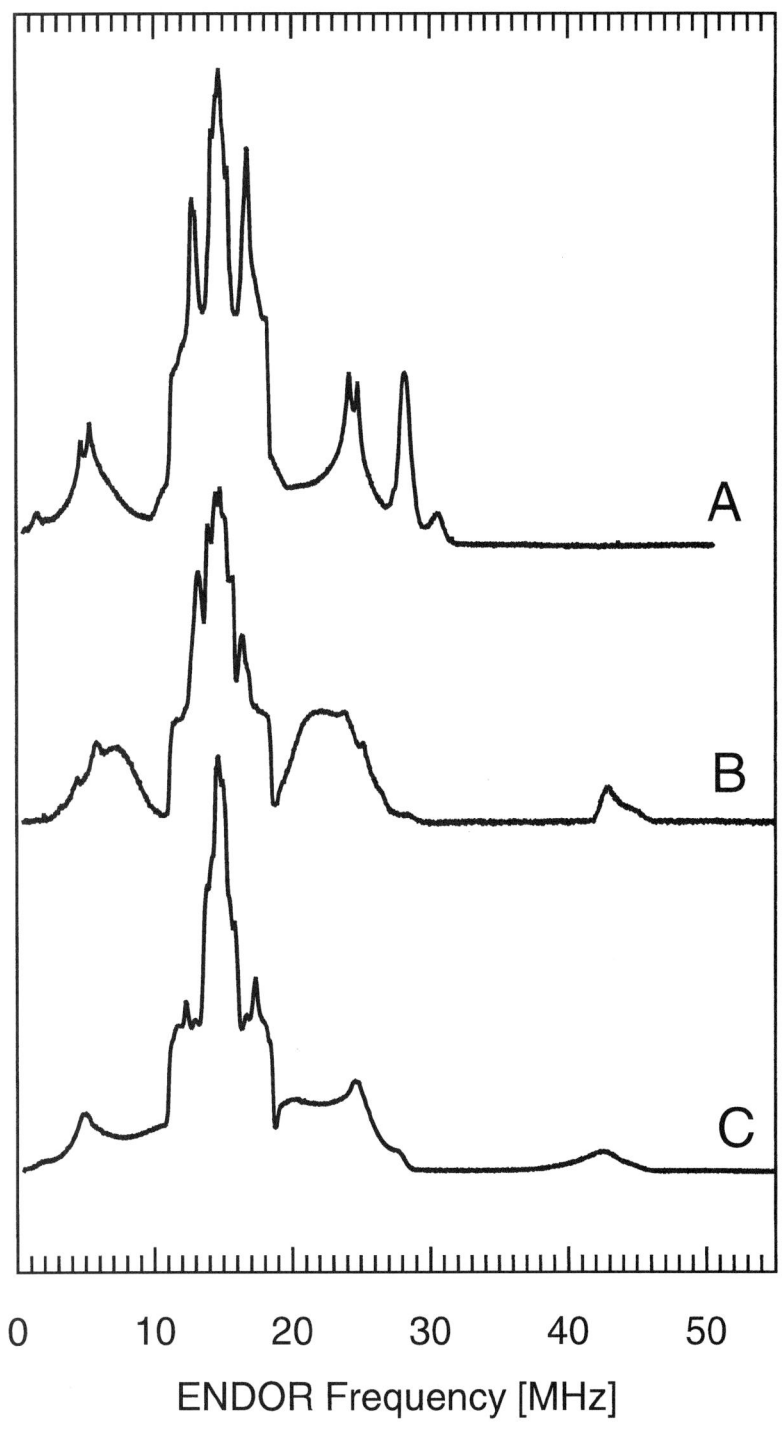

Necessary conditions for the observation of strong modulation effects are (*a*) anisotropic hyperfine or quadrupolar interaction, (*b*) strong MW pulses, and (*c*) cancellation of nuclear splitting for one electronic manifold. Because of condition (*a*) ESEEM effects cannot be directly observed in liquid samples of small molecules. Condition (*b*) means that at least one MW pulse must be strong enough to drive simultaneously an allowed and a forbidden EPR transition (for example $\Psi_1 \leftrightarrow \Psi_3$ and $\Psi_1 \leftrightarrow \Psi_4$). If the transition probabilities of the "allowed" and "forbidden" transition are similar, the ESEEM effect will be most pronounced. This can be achieved by tuning the nuclear level splitting of one electronic manifold by proper choice of the external magnetic field B_0. This is expressed in condition (*c*) and will depend on the nuclear spin and the specific coupling strength (62).

In the 1D ESEEM experiment the two-pulse Hahn echo intensity is monitored as a function of the pulse separation time τ (Figure 2). The oscillation of the echo intensity arises from spins that evolve on different electronic coherences before and after the refocusing pulse (for example coherence $\Psi_1 \leftrightarrow \Psi_3$ in the first time interval τ between the MW pulses and coherence $\Psi_1 \leftrightarrow \Psi_4$ within the second time interval τ after the second MW pulse). Because the respective energy spacings are different for both transitions the refocusing is not 100% but is modulated with the energy difference between the two transitions, which is in fact the nuclear spin–level splitting. In this experiment the modulation can only be observed on a time scale of the transverse electron relaxation time T_2^S, which is in many cases too short to obtain high enough resolution in complex spectra with several coupled nuclei.

A 2D version of the experiment is the stimulated echo pulse sequence, in which both pulse spacing times τ and T are varied (Figure 2). In this experiment nuclear coherence (for example between $\Psi_1 \leftrightarrow \Psi_2$) created by the first two pulses evolves within the time T and is then transferred back into observable electron coherence by the last pulse. In this case the modulation is typically observable on the time scale of the transverse nuclear relaxation time T_2^I, which is often longer than T_2^S but shorter than T_1^S.

In a hyperfine sublevel correlation experiment (HYSCORE) the nuclear coherence within one electronic manifold (for example $\Psi_1 \leftrightarrow \Psi_2$) is transferred by an additional MW pulse into the corresponding nuclear coherence within the other electronic manifold ($\Psi_3 \leftrightarrow \Psi_4$) (63). A 2D-FT leads to a 2D spectrum in frequency space with off-diagonal correlation peaks between hyperfine lines belonging to the same nucleus in both electronic manifolds. This is very helpful in unraveling complex hyperfine spectra with overlapping line contributions from different nuclei.

ESEEM spectra can be analyzed in the time domain as well as in the frequency domain (64). The frequencies in the Fourier transformed ESEEM time traces are the same as in the ENDOR experiment. In contrast, the amplitudes are more complicated functions of spin Hamilton operator parameters (such as hyperfine and quadrupole tensor elements and number of equivalent coupled spins) and experimental parameters (such as MW excitation power, frequency offset, external magnetic field value, pulse lengths, and dead times). This can be a problem, especially for the assignment of broad anisotropic hfcs in disordered powder samples (61). In disordered samples specific orientations with effective hfcs close to the cancellation condition will be strongly enhanced in the modulation intensity, leading to strong distortions for anisotropic hfc powder patterns. Additionally, blind-spot artifacts for pulse sequences with more than two pulses will modulate the powder pattern lineshape. Alternatively, the modulation depth as well as the damping of the modulation contains additional and independent information on the coupled spin system.

If several nuclei couple to the paramagnetic center, the experimental time trace will be the product of the modulations of the individual nuclei and may become very complex. In order to highlight specific modulations, an isotope-labeling and waveform-division method can be used, as proposed by Mims and coworkers (65, 66). For example, an isotope labeling of ^{12}C (I = 0) → ^{13}C (I = 1/2) results in time traces that are identical, but the isotope labeled time trace is multiplied by the ^{13}C echo modulation. Therefore, a division of the time traces of the two experiments, $^{13}C/^{12}C$, only leaves the specific ^{13}C modulation, if all the other parameters for the sample and experiments are kept constant (67). Although this method is mathematically not exact for disordered powder samples and stimulated echo experiment, as was clearly stated by the authors, it is very successfully applied to disordered protein samples to obtain quantitative information, especially on the number of equivalent nuclei (e.g. water molecules) coupled to the paramagnetic center under investigation.

Nuclei with couplings close to the cancellation condition are strongly enhanced in the ESEEM spectra. Different nuclei can be tuned to this condition by varying the external magnetic field (and thereby the nuclear Zeeman splitting). Therefore, a careful tuning of the MW highlights specific regions of the paramagnetic surrounding (Figure 4). For such nuclei close to the cancellation condition the adiabatic precession of the electron spin in the resulting effective field (external magnetic field plus hyperfine field) under RF irradiation negatively interferes with the excitation field and thereby diminishes the ENDOR effect (65). This shows nicely the complementarities of these two methods. If the modulation depth is small for nuclei not close to the cancellation condition, applying matched pulses can enhance the modulation intensity (68), thereby increasing the sensitivity for such nuclei. This method has not been applied to biological questions yet.

The same groups have often applied ESEEM and pulsed-ENDOR spectroscopy to similar biological systems and paramagnetic centers. Whereas ENDOR investigations concentrated more on the nuclei with higher nuclear Zeeman frequencies, such as ^1H, ^{13}C, or ^{31}P, the ESEEM method was mostly used to obtain the complementary information on low frequency nuclei, such as ^2H, ^{14}N, and ^{15}N.

A very recent review gives a comprehensive listing of ESEEM spectroscopy applied to metal centers in enzymes and model systems (69). Applications on chromophores in photosynthetic reaction centers have also been reviewed recently (11, 61). Therefore, only a few applications highlighting the potential of ESEEM

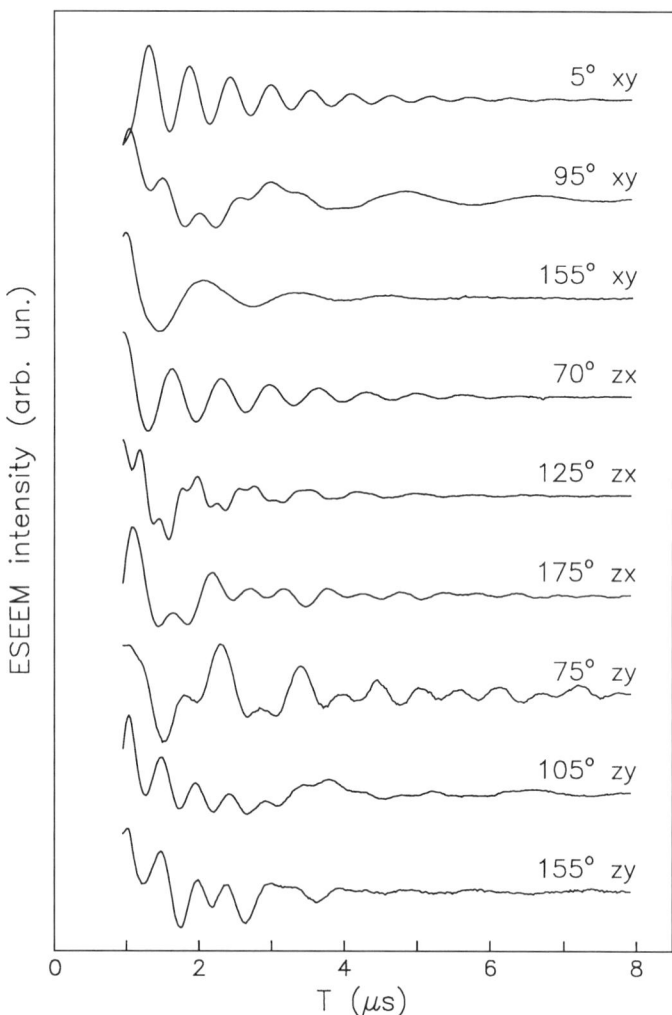

Figure 4 Stimulated echo envelope for a single crystal of azurin. Different orientations of the magnetic field with respect to the g-tensor principal axis system are shown. The decay of the echo intensity has been removed to emphasize the modulation-frequency dependence on the crystal orientation [Reprinted with permission from (80).]

spectroscopy for the investigation of the local structure of paramagnetic centers will be described here in more detail.

Cu^{2+}-metal centers ($S = 1/2$, $I = 3/2$) in enzymes have been extensively studied by ESEEM methods. One of the first ESEEM applications on biological systems was done on the Cu^{2+}-center in stellacyanin (70). For the ^{14}N-nucleus close to cancellation the quadrupole tensor $\hat{Q}$ can be determined by analyzing the echo modulation. Comparison of the data obtained in these experiments with those from model systems and nuclear quadrupole resonance data allows the identification of the ligating molecule (mostly histidine), the ligand positions and orientations, and also the protonation of the nitrogens (72, 73, 246). Two to four histidine ligands were detected by ESEEM spectroscopy for different Cu^{2+}-proteins by comparison of the experimental ESEEM time traces with simulations (74–79). Multifrequency ESEEM is especially helpful for tuning different nitrogens with distinguishable hfcs to the cancellation condition. Whereas at X-band frequency the ESEEM experiment is especially sensitive to the remote imidazole nitrogen, the situation is different at higher frequencies. With W-band ESEEM experiments the directly ligated nitrogen of the imidazole ring in the blue copper protein azurin has been observed in single crystal studies (80) (Figure 4), whereas X-band ESEEM detected the remote nitrogens (81, 82). Isotope labeling ($^1H/^2H$ or $^{12}C/^{13}C$) and the spectra division method were used to obtain quantitative information about other ligands of the Cu^{2+}-ion, such as water molecules (67, 83–86). The replacement of a histidine nitrogen ligand by the enzyme inhibitor cyanide was proven in superoxide dismutase by isotope-labeled ($^{14}N/^{15}N$) CN^- (87) and the replacement of a water ligand by N_3^- in laccase (88). Recently, Cu^{2+} was used to replace nonheme Fe^{2+} in bRC and in a dioxygenase. The histidine couplings observed on the Cu-ESEEM spectra were used to identify possible ligands to the native Fe^{2+}-ion (89, 90).

Mn^{2+} is another metal ion that was investigated by ESEEM spectroscopy in several protein complexes. In many cases Mn^{2+} can substitute for other naturally abundant metal ions, as for example the diamagnetic Mg^{2+}-ion, and can therefore serve as a paramagnetic probe in the investigation of these metal sites. The analysis of field swept spectra and the quantitative analysis of ESEEM experiments is rather complicated (91), owing to the high spin state ($S = 5/2, I = 5/2$), the large manganese hfc ($\sim$9 mT), and the zero field splitting tensor (typically coupling strengths D of 10–100 mT). Because of the high spin state, the different allowed and forbidden electronic transitions overlap, leading to complicated EPR and ENDOR spectra. In principle, pulsed experiments will allow these different transitions to be distinguished by their transition moments (92), but this has been used only recently to simplify such spectra (93). Nevertheless, quantitative information on the ligand sphere could be obtained by either isotope exchange of the water solvent (e.g. ^2H_2O or $H_2^{17}O$) or isotope labeling of specific amino acids ($^{14}N/^{15}N$, ^{13}C, ^{17}O, ^2H). This allowed the determination of the number of water ligands (94) and the identification of the peptide ligands (95). What can be achieved by ESEEM spectroscopy on such a metal center in a protein complex was demonstrated very impressively on the p21$^{ras.}$ Mn·GppNHp complex (96, 97). Mn^{2+} replaced the natural diamagnetic Mg^{2+}-ion in this complex for the EPR investigations. By specific mutations of amino acids in the close neighborhood of the metal center (^{15}N-Ser17, $^2H/^{13}C/^{15}N$-Thr35, ^{13}C-Asp57) and by detailed analysis and simulation of the high spin ESEEM effects (91) an EPR structure of the metal site (up to 0.5 nm) could be obtained in good agreement with X-ray results (98). In the same way the metal binding sites of isolated F1-ATPase from spinach chloroplasts and from *Bacillus PS3* have been studied by ESEEM spectroscopy using Mn^{2+} (99, 100) or vanadium (101, 102) as a paramagnetic substitute for Mg^{2+} and Ca^{2+}. The protein and ATP nucleotide binding ligands to the metal ion and the change in the metal binding site upon addition of ATP have been identified by EPR spectroscopy. In cytochrome *c* oxidase ESEEM spectroscopy was used to identify the amino acid ligands of the Mn^{2+}-site, which does not participate in the electron transfer reaction but occurs naturally in several organisms (103). In a recent work $^{14}N/^{15}N$ and $^1H_2O/^2H_2O$ ESEEM experiments have been used to identify the interaction of Mn^{2+}-ions with guanosine in the hammerhead ribozyme complex (104).

Multifrequency ESEEM, ranging from S-band (3 GHz) to P-band (15 GHz), was used to distinguish different states of Mo^{5+} (S = 1/2) in molybdoenzymes such as xanthine or sulfide oxidase. Additionally, their ligation sphere and geometry was postulated on the basis of these EPR and EXAFS measurements (105–109).

Photosynthetic proteins have been intensively and competitively examined by ESEEM spectroscopy. The electron spin density distribution on the primary electron donor $P^{+\bullet}$, a chlorophyll dimer, was intensively studied by nitrogen 1D stimulated echo and 2D HYSCORE experiments (61, 110–112). Complementary to ENDOR investigations, ESEEM experiments were used for the different quinone acceptors to detect hfcs of nitrogens in amino acid residues. Recent publications have appeared on the electron acceptor $A_1^{-\bullet}$ of PSI (113), on $Q_A^{-\bullet}$ in bRCs (114–116), on $Q_A^{-\bullet}$ in PSII (117) (attempting to resolve some earlier controversial results), and on $Q_H^{-\bullet}$ in quinol oxidase (118). In PSII the tyrosyl radical $Y_D^{-\bullet}$ was investigated by $^1H/^2H$ ESEEM (119, 120), the manganese cluster (121–126), and very recently, the carotene cation radical (127), in which a close-by tryptophan residue was assigned.

Further HYSCORE spectroscopy has been done to analyze nitrogen and proton/deuteron hfcs of flavin radicals in ferredoxin-$NADP^+$ reductase and in flavodoxin (26, 128). In the proteins amine oxidase and methylamine dehydrogenase intermediate substrate radicals within the catalytic cycle could be analyzed by ESEEM spectroscopy (129–131). A recent HYSCORE experiment on the bound quinone $Q_H^{-\bullet}$ in quinol oxidase is shown in Figure 5. The obtained nitrogen quadrupole tensor Q shows that the semiquinone is most strongly coupled to a nitrogen nucleus of the protein backbone (118).

Pulsed Electron Double Resonance (PELDOR)

Because the interaction of two electron spins with each other is formally identical to the interaction of an electron spin with a nuclear spin (see Equation 1 and Figure 1), the conceptual principles of pulse methods to observe small magnetic dipolar couplings between distant paramagnetic centers are very similar to the ESEEM experiment. PELDOR experiments (132), also called double electron electron resonance (DEER) (133), have been used to measure interactions between radicals for distances of up to 5 nm (134, 135). For this method two different MW

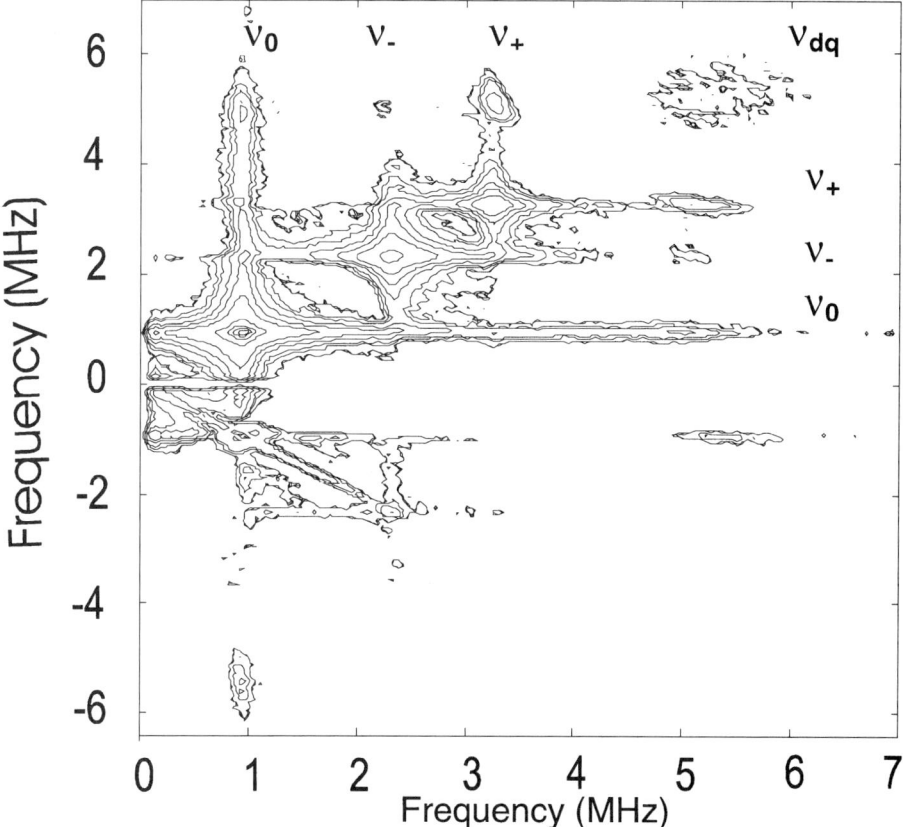

Figure 5 X-band HYSCORE spectra of the semiquinone $Q_H^{-\bullet}$ of bo_3 ubiquinol oxidase from *E. coli* after echo decay subtraction and magnitude Fourier transformation of the 2D time traces. The correlation peaks of the three quadrupolar split levels (ν_0, ν_-, ν_+) of the α electronic manifold with the double quantum peak (ν_d) of the β electronic manifold are well resolved (118).

frequencies (ω_a in resonance with electron spin A and ω_b in resonance with electron spin B; see Figure 2) have to be applied to the sample via the MW resonator. An example of a new four-pulse version of this experiment applied to two biradical model systems (with interspin distances of 1.9 and 2.8 nm) is shown in Figure 6, in which the dipolar oscillations, containing the distance information, are easily observable (135). Instead of a pulsed excitation with two distinct MW frequencies, the magnetic field can be jumped between pulses to get into resonance condition with the second paramagnetic species (136, 137). If the spectra of the two paramagnetic centers overlap, the experiment can also be performed with a single MW frequency; this is the so-called 2 + 1 pulse sequence (138, 139). In this experiment care has to be taken to distinguish the dipolar coupling of the two paramagnetic centers from

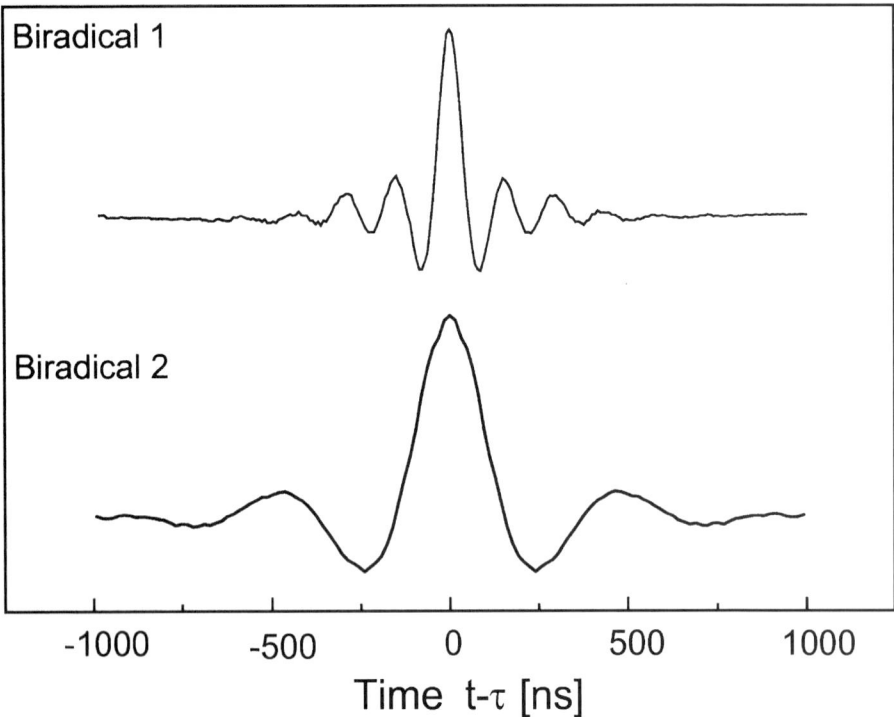

Figure 6 Four-pulse double electron electron resonance time domain signal on two biradical model systems after subtraction of the background signal owing to intermolecular interactions. From the oscillations an electron-electron distance of 1.94 ± 0.05 nm for biradical 1 (upper trace) and of 2.83 ± 0.05 nm for biradical 2 (lower trace) could be determined. [Adapted with permission from (135).]

hfc to nuclei, as observed by ESEEM (140). Another pulse scheme, the DQ 2D-EPR experiment (141), is a direct EPR analogue of a classical nuclear magnetic resonance (NMR) double quantum coherence experiment, and was demonstrated on a nitroxide biradical sample with an interspin distance of 3 nm (142).

The possibility of measuring long distances with pulsed-EPR methods is extremely attractive, especially for biological samples. It can be used to identify electron transfer pathways in proteins and measure the distances and relative orientations of the involved centers. In combination with site-specific double-spin labeling techniques it will allow the determination of long range structural constraints of the protein backbone and information on conformational mobility in specific loop regions. A more comprehensive overview of cw-EPR, PELDOR, and pulsed-EPR relaxation methods to measure distances between paramagnetic centers and some first applications will appear soon (143).

An early application of the "2 + 1" pulse method to biological systems was the measurements of the distance between two spin-labeled cysteine residues of a hemoglobin tetramer (139). The measured distance of 3.5 ± 0.15 nm was in good agreement with the X-ray structure. PELDOR and the "2 + 1" pulse sequence were used to measure the distances between the tyrosine radicals $Y_D^\bullet$ and $Y_Z^\bullet$, the semiquinone radical $Q_A^{-\bullet}$, and the Mn_4-cluster in PSII samples (144–146). Distances of up to 4 nm and relative orientations have been determined between the different chromophores. Together with pulsed-EPR relaxation measurements, out-of-phase-echo measurements (both described in the next two sections), and cw-EPR power saturation measurements, these data allowed prediction of a skeleton of the chromophore locations within the protein complex. Very recently, an application of the four-pulse double electron electron resonance experiment (135) was performed in double spin-labeled human carbonic anhydrase II to measure distances and mobility in the range of 2 nm (147).

Pulsed EPR/Relaxation Measurements

If one of the two paramagnetic species is a fast relaxing species, as for example a metal center, the PELDOR method will not be applicable anymore. However, information on the interspin distance can be gained by measuring the relaxation enhancement on the slowly relaxing paramagnetic species, caused by the dipolar coupling to the fast relaxing species (148). Usually the relaxation behavior of a paramagnetic center will be a complex temperature-dependent process with contributions from different physical mechanisms. Therefore, to obtain a quantitative value for the interspin distance, the relaxation rates of both paramagnetic species without the dipolar coupling but otherwise identical conditions have to be known. This is difficult to achieve in biological samples, but often model systems for the slowly relaxing species are used to obtain their unperturbed rates. In some proteins one of the paramagnetic species can be switched into a diamagnetic state. The rates of the fast relaxing species will not be perturbed by the dipolar coupling but are often difficult to measure.

Applications of these pulsed-EPR techniques to determine interspin distances in biological samples have been demonstrated most impressively in PSII samples. Already, changes of the relaxation time T_1 of the tyrosine radical $Y_D^\bullet$ depending on the redox state of the Mn_4-cluster have been interpreted as arising from the dipolar coupling of the two paramagnetic species (149). Thereafter, distances between the Mn_4-cluster (or for Mn_4-cluster-depleted samples to the Fe^{+2} ion) and the paramagnetic species $Y_D^\bullet$, $Y_Z^\bullet$ $P_{680}^{+\bullet}$, $Chl_Z^{+\bullet}$, and $Pheo^{-\bullet}$ were estimated by saturation recovery relaxation measurements (150–154). Similar to the PSII experiments, the coupling of the tyrosyl radical $Y_{122}^\bullet$ with the diferric center in ribonucleotide reductase was also detected by saturation recovery experiments (155).

A variant of this method is the hole-burning method (156), in which a selective excitation hole is burned into the inhomogeneous broadened EPR line, and the broadening of the hole by relaxation effects is studied. This method was applied to PSII and PSI chromophores to measure distances (157, 158).

Other examples for the use of this relaxation experiment are measurements on spin-labeled methemoglobin and methmyoglobin, in which different nitroxide spin-label to heme-iron distances (in the range of 1.5–3 nm) were determined (159, 160).

A different relaxation source for paramagnetic centers beside dipolar coupling to fast relaxing electron or nuclear spins is the motional modulation of the Larmor resonance frequency by orientation-dependent anisotropic interactions, as determined by the hfc or the g-tensor. In this case the measurement of the relaxation times over the whole spectral width allow determination of not only information on the rotational correlation time τ_c of the molecule but also detailed information on the type of motion the molecule undertakes. In proteins this motion of a bound nitroxide spin label is non-Brownian and hindered, depending on the specific position of the spin label. Therefore, these measurements can help obtain detailed information not only on the molecule itself, but also on the protein backbone. The time windows of motional processes observable on spin labels by cw-EPR have been drastically extended by cw-saturation-transfer-EPR (161) and pulsed EPR methods, such as saturation recovery EPR and 2D-electron-spin-echo (2D-ESE) spectroscopy (162, 163). Whereas the very slow motional region (with correlation times of up to 1 ms) can be detected by saturation-transfer–EPR (164, 165), pulsed EPR allows investigation of motional processes lying in the correlation time range between ns and several hundred μs. ESE spectroscopy is one of the basic tools that allows probing of slow molecular motions of spin labels, for example in membrane model systems (166, 167). Furthermore, the examination of molecular motions can be done by 2D-FT EPR; a comprehensive treatment including descriptions of free induction decay–based correlation spectroscopy (COSY) and spin echo correlation spectroscopy (SECSY) can be found in (168).

Characterizing protein-protein and protein-lipid interactions by observation of changes in rotational diffusion of the spin labels involved requires the use of comprehensive mathematical and rotational diffusion models (169–172). As mentioned above, application of pulsed-EPR techniques simplifies the interpretation by distinct separation of time-dependent and static-spin Hamiltonian parameters.

An early investigation of echo-detected motional dynamics was performed on a maleimide spin-labeled deoxygenated hemoglobin, in which the motion of the polymerized molecule could be differentiated from monomeric molecules in solution under physiological conditions (173). Most recently, this method has been used for probing temperature-dependent librational motion in glass formed from the intracellular medium at low temperature in seed and pollen. Correlation between the water content in *Typha latifolia* pollen and the librational motion of the spin label in the cytoplasma could be shown (174). Echo-detected EPR has further been used to characterize the molecular motion of the cholestane spin label in a multibilayer in the gel phase (167). With a model of uniaxial librations, the projection angle of the libration axis in the molecular plane was determined relative to the N–O bond direction of the spin label. Motional states of spin labels were also characterized with pulsed-EPR in cardiolipin-cytochrome c bilayers (175) and in spin-labeled Ca^{2+}-ATPase in the sarcoplasmatic reticulum membrane (176). Interactions of stearic acid spin-label pairs in multilamellar liposomal dispersions were characterized with saturation recovery-EPR methods and by measurements of the spin lattice relaxation time T_1 (177). Site-specific spin labeling and pulsed-EPR techniques have been used to characterize melittin at membrane surfaces (178), most recently using pulsed high-field (HF)-EPR on bacteriorhodopsin (179). Thereby, different modes of molecular motions were characterized in dependence on the position of the spin label. Pulsed-HF-EPR allows these investigations to be extended to other classes of molecules, such as organic radicals with only small anisotropic hfc and g-tensors, which are not resolved at X-band frequencies. At high fields these radicals' g-tensor anisotropy can be resolved (as described below), and the contribution of this tensor to the motional relaxation will be strongly enhanced (180). W-band pulsed-EPR experiments were used to investigate the librational motion of the semiquinone $Q_A^{-\bullet}$ in bRC (181). The uniaxial librational motion of the protein-bound semiquinone along the C-O axis (in contrast to the more isotropic Brownian tumbling of semiquinone in frozen solution) could clearly be resolved by the 2D-ESE experiments (Figure 7).

Transient EPR/Correlated Radical Pairs

Photosynthetic proteins (especially bRC in which high resolved X-ray structures are available) have not only been extensively studied with all sorts of advanced EPR methods, but have also helped in the development of new methods and concepts of EPR spectroscopy (182, 183).

Transient EPR methods, first developed for the detection of photo-excited transient triplet states of aromatic molecules (184), have been used extremely successfully to obtain unique information on the distances and relative orientations of the chromophores involved in the fast electron-transfer reaction in photosynthetic proteins. In these proteins a fast electron transfer reaction is initiated by photoexcitation, leading to a transient radical pair of an electron donor molecule $P^{+\bullet}$ and an electron acceptor molecule $Q^{-\bullet}$ on a sub-ns time scale. Because the electron transfer reaction is started from the excited singlet state of the primary donor $^1P^*$, the population of the dipolar coupled radical pair energy levels (see Figure 1b) does not obey Boltzmann distribution law. Instead, only the Ψ_2 and Ψ_3 levels are selectively populated via intersystem crossing, which leads to an

---

Figure 7 Two-dimensional-electron-spa-echo (2D-ESE) (W-band) experiments on ubisemiquinone radical in frozen isopropanol solution (upper trace) and in bRC (lower trace). Both measurements were performed at 120 K. The anisotropy of the relaxation time T_2 within the powder spectra is clearly visible and very different for the two environments.

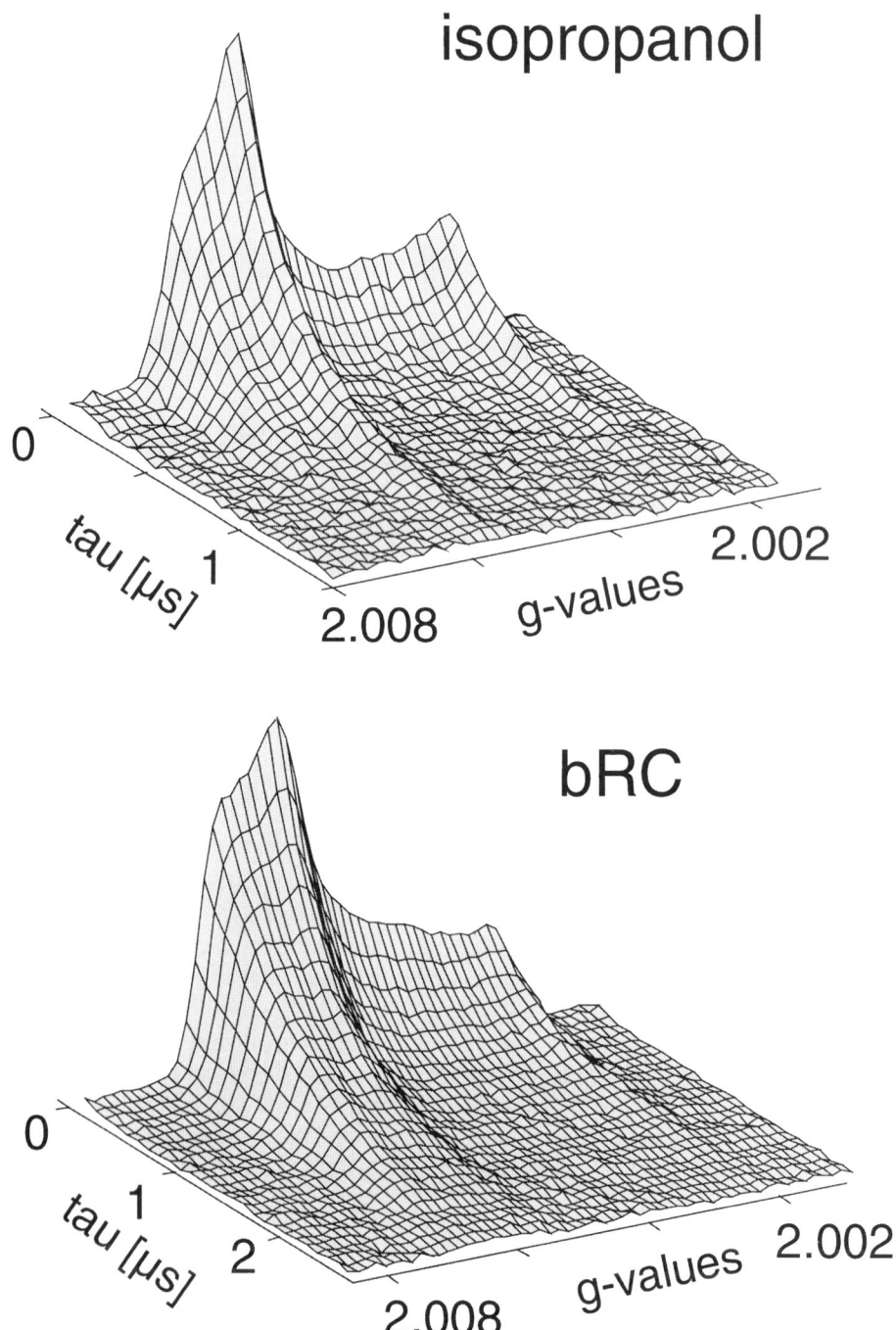

unusual spin polarization, described as longitudinal two-spin order, resulting in the specific features of the EPR signals.

These EPR spectra, detected as time-resolved MW responses directly after the laser flash, are therefore highly spin polarized with emissive and absorptive components. The first applications of the transient EPR method to biological systems were performed on bRC (185). The spin polarized correlated radical pair transient was recorded for different resonant field values B_0 and could be used to deduce the relative orientation and, to some extent, the distance of the dipolar-coupled radical pair $P^{+\bullet}$-$Q^{-\bullet}$ (183, 186, 187). High-field two-pulse echo-detected polarized spectra were used to get enhanced information on the relative orientation of the $P_{865}^{+\bullet}$-$Q_A^{-\bullet}$ pair in bRC (188) and on the pair $P_{700}^{+\bullet}$-$A_1^{-\bullet}$ in PSI (189) because the anisotropy of the semiquinone g-tensor could be resolved at such field values (3.4 T).

In addition to the nonthermal polarization created by the photo-induced electron transfer, zero-quantum coherence between the charge-separated spin states is created, which are (because of the dipolar and exchange coupling) not eigenstates of the coupled radical pair. Because of the large distance of the two radicals (~ 2 nm) and the fast electron transfer (<1 ns), coherence between the populated states Ψ_2 and Ψ_3 exists after a short laser pulse excitation. This was theoretically predicted (190) and experimentally observed for short delay times after the laser flash PSI (191). The coherent quantum beat signals, observed as a function of the magnetic field position, could be used to obtain detailed structural information on the radical pair (192). Finally, the unusual out-of-phase-echo signal (193) (observed in phase with respect to the excitation pulses) of this coupled radical pair can be used to measure with high accuracy the distance between the two paramagnetic species because the echo amplitude is periodically modulated as a function of the pulse separation time τ (see Figure 2) by the dipolar (and exchange) interaction (194–196). This method is just as sensitive to distances as the more complicated PELDOR or double quantum experiments and has been used to measure the distances between the different cofactors in bRC, PSI, and PSII with high accuracy (182, 197–205). Changes of these distances in light-adapted frozen samples were also investigated by these methods (206, 207). Figure 8 shows the Fourier transformations of time traces obtained from the radical pair $P_{865}^{+\bullet}$-$Q_A^{-\bullet}$ of bRC and from $P_{700}^{+\bullet}$-$A_1^{-\bullet}$ in PSI. The distances of 2.82 nm and 2.54 nm could be deduced by simulations (dotted lines) (202). Finally, it has also been shown theoretically (208) as well as experimentally in bRC (209) that double quantum coherences can be created under these specific starting conditions by just a single MW pulse.

High-Field/High-Frequency EPR

As can be seen from the first term of the spin Hamiltonian (Equation 1), the electron Zeeman splitting scales linearly with the magnetic field B_0. In first order, an increase of B_0 requires an equivalent increase of the Larmor frequency ν_{MW}. This applies for cases in which a particular spin system is studied at different magnetic

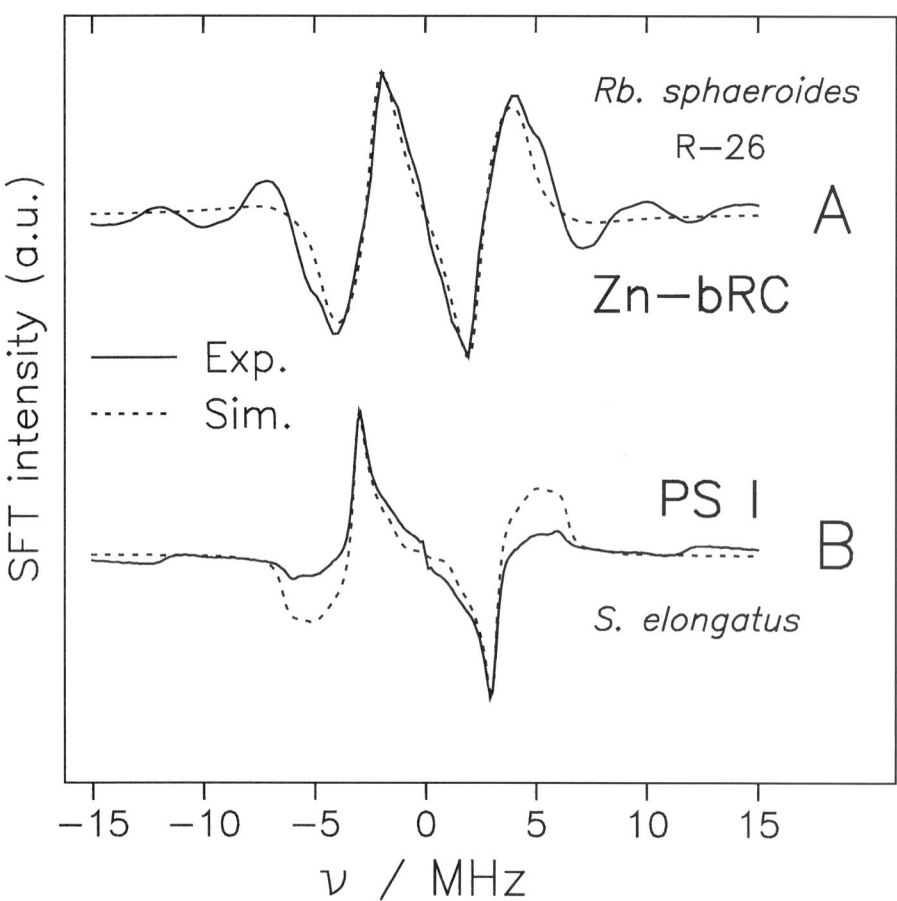

Figure 8 Sine Fourier transforms of the out-of-phase echo modulation for the radical pair $P_{865}^{+\bullet}$-$Q_A^{-\bullet}$ of bRC and the pair $P_{700}^{+\bullet}$-$A_1^{-\bullet}$ of PSI. The dead time of the traces was reconstructed with maximum entropy methods. The simulations (dotted lines) of the experimental patterns yield 2.84 ± 0.03 nm for the distance in bRC and 2.54 ± 0.03 nm in PSI. A small exchange coupling of $J = 1$ μT was obtained for both systems. [Adapted with permission from (182).]

fields if there are no additional field-dependent parameters. Consequently, "high-field EPR" and "high-frequency EPR" are effectively synonyms in most cases.

In principle, the motivation for HF-EPR is the same as for the comparable development in NMR. As in NMR at increasing Zeeman energy, the spectral resolution increases in most cases and the population difference or polarization of the spin states (α and β) increases according to the Boltzmann distribution as well. This leads, together with the higher quantum energy of the observed transitions, to higher sensitivity of HF-EPR experiments. Besides these very basic motifs,

HF-EPR opens new frontiers for systems with large zero field splitting, which may be totally EPR-silent at normal EPR frequencies (210), and for systems with very small g anisotropies and unresolved hfcs, as in many aromatic organic radicals (211–217).

In HF-EPR the magnetic field is usually higher than 2 Tesla and thus requires, typically, the use of a superconducting magnet. The first corresponding frequency band is called W-band and ranges from 70–110 GHz. Many technical requirements substantially change for such high-MW frequencies close or above the limit for conventional semiconductor MW technology. "Classical" spectrometer designs with a MW resonator in reflection mode and using commercial MW components such as a waveguide MW circulator can still be realized at frequencies up to approximately 140 GHz (D-band) (1, 218–221). At higher frequencies the small wavelengths become prohibitive for the effective operation of rectangular waveguides, and quasi-optical methods have to be used to replace classical MW components. By using such techniques, or by combining them with classical MW components, sensitive cw spectrometers can be built at frequencies even in the low-mm and in the sub-mm range (222–224).

Cw HF-EPR was first developed by the group of Y Lebedev in Moscow (225), who used a MW frequency of 140 GHz. The first pulsed HF-EPR setups were built in Leiden (226), followed by spectrometers in Moscow (218), at MIT (227), and in Berlin (228) at W-band or D-band frequencies. By using pulsed far-infrared lasers, pulsed EPR was demonstrated at 604 GHz in Grenoble (229, 230). Since 1996, HF-EPR instrumentation at W-band has been offered commercially by Bruker Analytik (1) for cw experiments as well as pulsed-EPR experiments.

When half-integer high-spin systems ($S > 1/2$) are studied, the linewidth of the central EPR transition ($m_S = +1/2$ to $m_S = -1/2$) is broadened by higher order zero field splitting contributions. Because of the inverse field dependency of these contributions, HF-EPR often leads to narrower lines and thus to increased information content for small interactions such as hfc of ligands (231–233) or dipolar coupling to other paramagnetic centers (234). The spectral resolution for systems with large hfc is often further increased as a result of the suppression of forbidden m_I transitions at higher field values. Finally, as already mentioned above, systems with very large zero field splitting (in the order of the electron Zeeman splitting) even become EPR-silent at low field values (210).

As a result of the increased Boltzmann spin polarization, magnetic resonance experiments become more sensitive at higher fields. However, this effect is partially compensated by technical restrictions, limiting the performance of many MW components at the smaller wavelengths (180). Another reason for the drastic increase in absolute sensitivity (in spins/mT) is found for HF-EPR spectrometers utilizing a MW-resonator. Owing to the smaller wavelength, the resonator dimensions decrease, leading to smaller sample volumes (<1 μl) with still high filling factors (219, 220, 228). For simple transmission mode setups without resonator (217, 235, 236), the latter argument does not apply. Instead, the concentration sensitivity (in spin/mT/cm^3) has to be considered in this case because large sample volumes (several hundred μl) can and must be used. Despite this, cw-EPR experiments can be performed quite successfully in broadband setups without resonators. Alternatively, for pulsed HF-EPR the incorporation of a MW resonator, preferably in the reflective operation mode, is indispensable because of the limited MW power available. The bandwidth of the MW resonator is broad enough to support the MW pulses even for optimal large Q values at high MW frequencies (whereas the Q value and therefore the sensitivity has to be lowered drastically in the case of pulsed X-band EPR applications). This leads to good conversion factors for the MW B_1-field, a short ringing and dead time after the pulses, and a high sensitivity (228).

The performance of ENDOR and related double resonance experiments at high magnetic fields offers additional specific advantages, for example for studying organic radicals with small g-anisotropies: The improved spectral resolution obtained in the EPR dimension allows one to record orientation-selective ENDOR spectra and thereby to improve the information content of ENDOR spectroscopy on powder-type samples with randomly oriented radicals (14, 30, 237). Another significant advantage of HF-ENDOR is the increased nuclear Zeeman interaction, which separates the free nuclear Larmor frequencies of different nuclei and thereby enhances the spectral resolution in the NMR dimension. Accordingly, spectral overlap can be reduced and spectra can be simplified (39, 238). Considerable simplifications of ENDOR spectra are also observed owing to the increased ratio of nuclear Zeeman frequency/hfc, as pointed out above. In analogy to the advantages of HF-EPR on metalloproteins, contributions of higher m_S-transitions are suppressed in the HF-ENDOR experiments as well (239). Furthermore, the advantages of HF-ENDOR can be combined with those of pulsed TRIPLE resonance methods (240). Thereby, spectral assignments of different nuclei and metal centers can be mapped in 2D spectra (241).

Applications of pulsed HF-EPR in biological systems are mentioned and cited above. Whereas ESEEM spectroscopy at high magnetic fields is limited to few nuclei with specific hfc (228, 242), the application of pulsed HF-ENDOR (220, 228, 237) and HF-2D-ESE (228) offers several advantages, as mentioned above. First, applications to biological systems include the W-band ESEEM spectroscopy on the direct coordinated histidine ligands (80, 243) (Figure 4) and W-band ENDOR spectroscopy on two remote histidine nitrogens and three distant protein backbone nitrogens (238) in small single-crystal azurin samples. The ubisemiquinone $Q_A^{-\bullet}$ in bRC was investigated by pulsed W-band ENDOR (14), 2D-ESE (181) and transient EPR (188) experiments. The relative orientation of the quinone molecule with respect to the bacteriochlorophyll donor molecule $P_{865}^{+\bullet}$, its hydrogen bonding to the protein and its uniaxial librational mobility in the protein pocket could be deduced from these experiments. Similar transient W-band EPR experiments were performed on the correlated radical pair $P_{700}^{+\bullet}$-$A_1^{-\bullet}$ in PSI samples (189). Other applications of W-band pulsed ENDOR are the determination of nitrogen and proton hfcs to the Cu_A center in cytochrome c oxidase (39) and to the Mn^{2+}-ion in

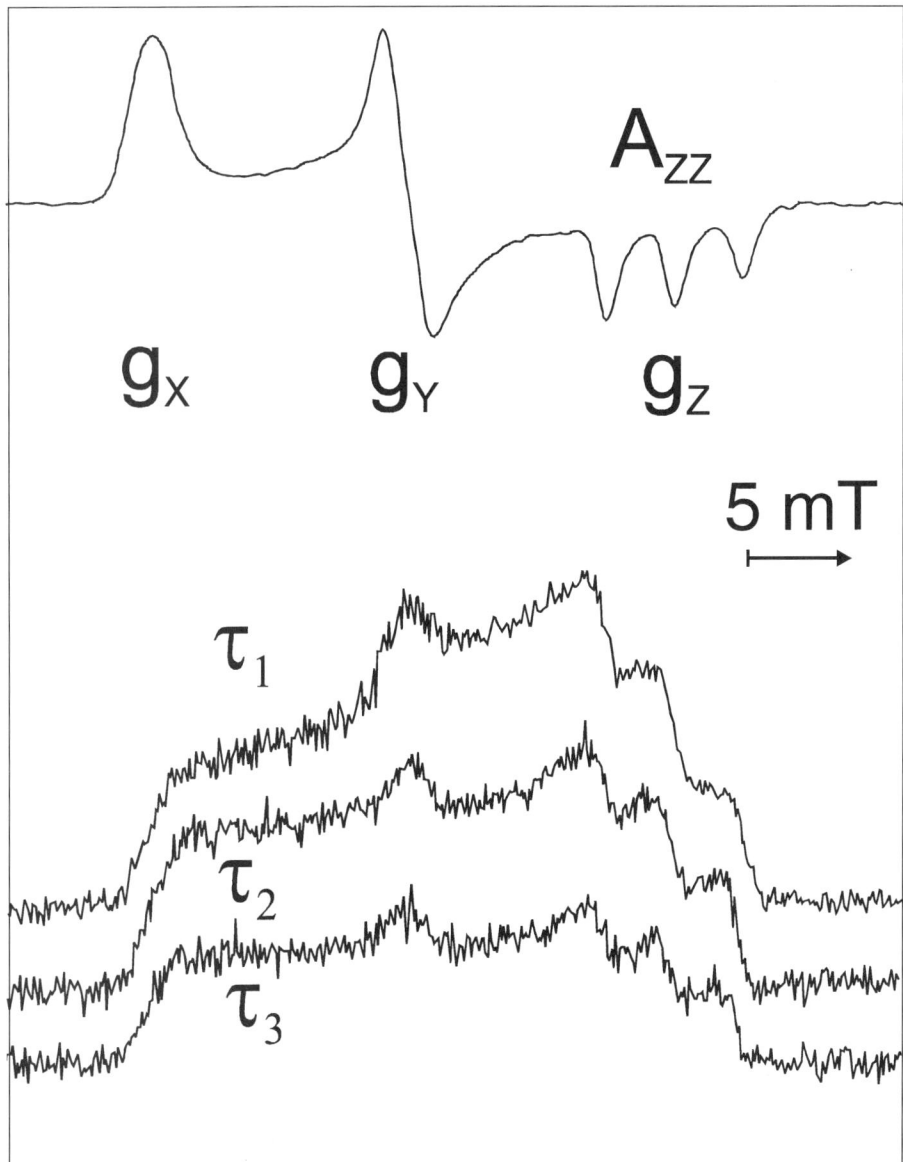

Figure 9 Field swept two-pulse echo spectrum of TEMPO spinlabel in a polystyrene matrix at 180 GHz MW frequency. In the cw spectra (upper trace) the well-resolved canonical peaks of the anisotropic g-tensor and the resolved nitrogen hfc A_{zz} component can be easily seen. The echo-detected spectra recorded for different τ values ($\tau_1 = 100$ ns, $\tau_2 = 200$ ns, $\tau_3 = 400$ ns) show the anisotropic T_2 relaxation time.

concavalin A (43) and the determination of the tyrosine radical in ribonucleotide reductase (30). Field-swept echo-detected D-band (140 GHz) EPR spectra of the apogalactose oxidase tyrosyl radical at low temperatures showed an increased hfc resolution compared to the saturation-broadened cw-EPR spectrum (244). In a recent publication first W-band field-swept echo experiments for the investigation of motional dynamics of nitroxide spin-labeled bacteriorhodopsin at the dynamic glass transition temperature were presented (179). The anisotropic transverse relaxation time T_2, measured over the whole powder pattern, can be used to analyze this motion in detail. An example of such an experiment, performed on our home-built pulsed 180 GHz EPR spectrometer (245), on a nitroxide spinlabel in polystyrene sample is shown in Figure 9. At these high magnetic fields (6.4 T) the anisotropic g-tensor is well resolved and dominates the spectra, in contrast to X-band frequencies in which the anisotropic nitrogen hfc tensor is dominant. Therefore, by HF-EPR orientational selection can be observed along g_x, g_y, and g_z, whereas at X-band only the z orientation is well resolved. The differences of the relaxation times T_2 for the different g-values can be seen by the change of the relative spectral intensities for different pulse separation times τ the well.

OUTLOOK

Pulsed-EPR and ENDOR methods extend the possibilities of cw-EPR to extract structural information from paramagnetic centers. This is especially true for disordered biological samples with several spectral overlapping paramagnetic species. The basic pulse sequences and experiments designed to unravel complex systems as described in this review have been developed and calibrated on model systems. As can be seen, they have been already successfully applied to study the local structure of some paramagnetic centers in biological systems. Their use in biological systems will, to our belief, further increase in the near future. One of the main reasons for this is that molecular quantum theoretical calculations have rapidly improved and the link from EPR spectroscopic data to a molecular structure will be much more direct in the future.

ACKNOWLEDGMENTS

We are indebted to all members of our group for their helpful suggestions and assistance in preparing the manuscript, especially Hildegard Mathis for getting the references in the correct order. Furthermore, we would like to thank all the authors who agreed to contribute a figure from their own research work to this review, especially Wolfgang Lubitz, who contributed unpublished work performed in his laboratory (Figure 3). Financial support of our work by the Deutsche Forschungsgemeinschaft (SFB472 and SP HF-EPR), the county of Hesse and the Fond der Chemischen Industrie is gratefully acknowledged.

Visit the Annual Reviews home page at www.AnnualReviews.org

LITERATURE CITED

1. Höfer P, Maresch GG, Schmalbein D, Holczer K. 1996. *Bruker Rep.* 142:15–21
2. Poole CP, Farach HA. 1972. *The Theory of Magnetic Resonance*. New York: Wiley
3. Feher G. 1956. *Phys. Rev.* 103:834–35
4. Mims WB. 1965. *Proc. R. Soc. A* 283:452
5. Davies ER. 1972. *Phys. Lett.* 47a:1–2
6. Grupp A, Mehring M. 1990. In *Modern Pulsed and Continuous Wave Electron Spin Spectroscopy*, ed. L Kevan, MK Bowman, pp. 195–229. New York: Wiley
7. Schweiger A. 1991. *Angew. Chem. Int. Ed. Engl.* 30:265–92
8. Jeschke G, Schweiger A. 1995. *J. Magn. Reson. A* 119:45–52
9. Jeschke G, Schweiger A. 1997. *J. Chem. Phys.* 106:9979–91
10. Dinse KP, Biehl R, Möbius K. 1974. *J. Chem. Phys.* 61:4335–41
11. Angerhofer A, Bittl R. 1996. *Photochem. Photobiol.* 63:11–38
12. Lubitz W, Feher G. 1999. *Appl. Magn. Reson.* 17:1–48
13. Teutloff C, MacMillan F, Bittl R, Lendzian F, Lubitz W. 1998. See Ref. 246, pp. 607–10
14. Rohrer M, MacMillan F, Prisner TF, Gardiner AT, Möbius K, Lubitz W. 1998. *J. Phys. Chem. B* 102:4648–57
15. Gilchrist ML, Ball JA, Randall DW, Britt RD. 1995. *Proc. Natl. Acad. Sci. USA* 92:9545–49
16. Force DA, Randall DW, Britt RD, Tang XS, Diner BA. 1995. *J. Am. Chem. Soc.* 117:12643–44
17. Campbell KA, Peloquin JM, Diner BA, Tang XS, Chisholm DA, Britt RD. 1997. *J. Am. Chem. Soc.* 119:4787–88
18. Mino H, Kawamori A, Ono TA. 2000. *Biochim. Biophys. Acta* 1457:157–65
19. DiValentin M, Kay CWM, Giacometti G, Möbius K. 1996. *Chem. Phys. Lett.* 824:434–41
20. Lendzian F, Bittl R, Lubitz W. 1998. *Photosynth. Res.* 55:189–97
21. Bittl R, Zech SG, Teutloff C, Krabben L, Lubitz W. 1998. See Ref. 246, pp. 509–14
22. Faller P, Maly T, Rutherford AW, MacMillan F. 2001. *Biochemistry.* 39 40:320–26
23. Veselov AV, Osborne JP, Gennis RB, Scholes CP. 2000. *Biochemistry* 39:3169–75
24. Warncke K, McCracken J. 1995. *J. Chem. Phys.* 103:6839–40
25. Medina M, Cammack R. 1996. *J. Chem. Soc. Perkin Trans.* 2:633–38
26. Medina M, Lostao A, Sancho J, Gomez-Moreno C, Cammack R, et al. 1999. *Biophys. J.* 77:1712–20
27. Hoffman BM, Roberts JE, Brown TG, Kang CH, Margoliash E. 1979. *Proc. Natl. Acad. Sci. USA* 76:6132–36
28. Sivaraja M, Goodin DB, Smith M, Hoffman BM. 1989. *Science* 245:738–40
29. van Dam PJ, Willems JP, Schmidt PP, Pötsch S, Barra AL, et al. 1998. *J. Am. Chem. Soc.* 120:5080–85
30. Bennati M, Farrar CT, Bryant JA, Inati SJ, Weis V, et al. 1999. *J. Magn. Reson.* 138:232–43
31. Thomann H, Mims WB. 1992. *Pulsed Magnetic Resonance: NMR, ESR, Optical*, ed. DMS Bagguley, DM Slingsby, pp. 362–89. London/New York: Oxford Univ
32. Kroneck PMH, Kastrau DHW macher W, Hole UH, Neese l WG. 1997. In *Bioinorganic Chem.* AX Trautwein, pp. 710–24. New York: Wiley—VCH Verlag GmbH
33. Huttermann J. 1997. In *Bioinorganic Chemistry*, ed. AX Trautwein, pp. 696–709. New York: Wiley—VCH Verlag GmbH
34. Cammack R, Gay E, Shergill JK. 1999. *Coord. Chem. Rev.* 190–192:1003–22
35. Lee HI, Doan PE, Hoffman BM. 1999. *J. Magn. Reson.* 140:91–107

36. Gurbiel RJ, Peoples R, Doan PE, Cline JF, McCracken J, et al. 1993. *Inorg. Chem.* 32:1813–19
37. Hansen AP, Britt RD, Klein MP, Bender CJ, Babcock GT. 1993. *Biochemistry* 32:13718–24
38. Fan C, Doan PE, Davoust CE, Hoffman BM. 1992. *J. Magn. Reson.* 98:62–72
39. Gromov I, Krymov V, Manikandan P, Arieli D, Goldfarb D. 1999. *J. Magn. Reson.* 139:8–17
40. Tan X, Poyner R, Reed GH, Scholes CP. 1993. *Biochemistry* 32:7799–810
41. Tan X, Bernardo M, Thomann H, Scholes CP. 1993. *J. Chem. Phys.* 98:5147–57
42. Peloquin JM, Campbell KA, Britt RD. 1998. *J. Am. Chem. Soc.* 120:6840–41
43. Manikandan P, Carmieli R, Shane T, Kalb AJ, Goldfarb D. 2000. *J. Am. Chem. Soc.* 122:3488–94
44. Doi K, McCracken J, Peisach J, Aisen P. 1988. *J. Biol. Chem.* 263:5757–63
45. Thomann H, Bernardo M, McCormick JM, Pulver S, Andersson KK, et al. 1993. *J. Am. Chem. Soc.* 115:8881–82
46. Hoffman BM. 1994. *J. Phys. Chem.* 98:11657–65
47. DeRose VJ, Woo JCG, Hawe WP, Hoffman BM, Silverman RB. 1996. *Biochemistry* 35:11085–91
48. Pollock RC, Lee HI, Cameron LM, DeRose VJ, Hales B, et al. 1995. *J. Am. Chem. Soc.* 117:8686–87
49. Thomann H, Bernardo M, Adams MWW. 1991. *J. Am. Chem. Soc* 13:7044–46
50. Fan C, Kennedy MC, Beinert H, Hoffman BM. 1992. *J. Am. Chem. Soc.* 114:374–75
51. Doan PE, Fan C, Hoffman BM. 1994. *J. Am. Chem. Soc.* 116:1033–41
52. Shergill JK, Butler CS, White AC, Cammack R, Mason JR. 1994. *Biochem. Soc. Trans.* 22:288S
53. Thomann H, Bernardo M, Goldfarb D, Kroneck PMH, Ullrich V. 1995. *J. Am. Chem. Soc.* 117:8243–51
54. Goldfarb D, Bernardo M, Thomann H, Kroneck PMH, Ullrich V. 1996. *J. Am. Chem. Soc.* 118:2686–93
55. Lee HI, Dexter AF, Fann YCh, Lakner FJ, Hager LP, Hoffman BM. 1997. *J. Am. Chem. Soc.* 119:4059–69
56. Mims WB, Nassau K, McGee JD. 1961. *Phys. Rev.* 123:2059
57. Cowen JA, Kaplan DE. 1961. *Phys. Rev.* 124:1098–101
58. Mims WB. 1972. *Phys. Rev. B* 5:2409–19
59. Shubin AA, Dikanov SA. 1983. *J. Magn. Reson.* 52:1
60. Ponti A. 1997. *J. Magn. Reson.* 127:87–104
61. Dikanov SA, Tsvetkov YD. 1992. In *Electron Spin Echo Envelope Modulation (ESEEM) Spectroscopy*, ed. SA Dikanov, YD Tsvetkov, pp. 334–58. Boca Raton, FL: CRC
62. Flanagan HL, Singel D. 1988. *J. Chem. Phys.* 87:25606–16
63. Höfer P, Grupp A, Mehring M. 1986. *Phys. Rev.* 33:3519–22
64. Mims WB. 1972. *Phys. Rev. B* 6:3543–45
65. Mims WB, Peisach J. 1989. In *Advanced EPR: Applications in Biology and Biochemistry*, ed. AJ Hoff, pp. 1–58. Amsterdam: Elsevier
66. Mims WB, Davis JL, Peisach J. 1990. *J. Magn. Reson.* 86:273–92
67. Zweier JJ, Peisach J, Mims WB. 1982. *J. Biol. Chem.* 257:10314–16
68. Jeschke G, Rakhmatullin R, Schweiger A. 1998. *J. Magn. Reson.* 131:261–71
69. Deligiannakis Y, Loudoudi M, Hadjiliadis N. 2000. *Coord. Chem. Rev.* 204:1–112
70. Mims WB, Peisach J. 1978. *J. Chem. Phys.* 69:4921–30
71. Deleted in proof
72. Colaneri MJ, Peisach J. 1995. *J. Am. Chem. Soc.* 117:6308–15
73. Place C, Zimmermann JL, Mulliez E, Guillot G, Bois C, Chottard JC. 1998. *Inorg. Chem.* 37:4030–39
74. McCracken J, Desai PR, Papadopoulos NJ, Villafranca JJ, Peisach J. 1988. *Biochemistry* 27:4133–37

75. Goldfarb D, Fauth JM, Farver O, Pecht I. 1992. *Appl. Magn. Reson.* 3:333–51
76. Balasubramanian S, Carr RT, Bender CJ, Peisach J, Benkovic SJ. 1993. *Adv. Exp. Med. Biol.* 338:67–70
77. Balasubramanian S, Carr RT, Bender CJ, Peisach J, Benkovic SJ. 1994. *Biochemistry* 33:8532–37
78. Crowder MW, Stewart JD, Roberts VA, Bender CJ, Tevelrakh E, et al. 1995. *J. Am. Chem. Soc.* 117:5627–34
79. Bender CJ, Casimiro DR, Peisach J, Jane-Dyson H. 1997. *J. Chem. Soc.* 93:3967–80
80. Coremans JWA, Poluektov OG, Groenen EJJ, Canters GW, Nar H, Messerschmidt A. 1997. *J. Am. Chem. Soc.* 119:4726–31
81. Kofman V, Farver O, Pecht I, Goldfarb D. 1996. *J. Am. Chem. Soc.* 118:1201–6
82. van Gastel M, Coremans JWA, Jeuken LJC, Canters GW, Groenen EJJ. 1998. *J. Phys. Chem.* 102:4462–70
83. McCracken J, Peisach J, Dooley DM. 1987. *J. Am. Chem. Soc.* 109:4064–72
84. Eaton SS, Dubach J, More KM, Eaton GR, Thurman G, Ambruso DR. 1989. *J. Biol. Chem.* 264:4776–81
85. Lu J, Bender CJ, McCracken J, Peisach J, Severns JC, McMillin DR. 1992. *Biochemistry* 31:6265–72
86. Bubacco L, Magliozzo RS, Wirt MD, Beltramini M, Salvato B, Peisach J. 1995. *Biochemistry* 34:1524–33
87. Dikanov SA, Burgard C, Hüttermann J. 1993. *Chem. Phys. Lett.* 212:493
88. Lu J, Bender CJ, McCracken J, Peisach J, Severns JC, McMillin DR. 1992. *Biochemistry* 31:6265–72
89. Utschig LM, Poluektov O, Tiede DM, Thurnauer MC. 2000. *Biochemistry* 39:2961–69
90. Hogan DA, Smith SR, Saari EA, McCracken J, Hausinger RP. 2000. *J. Biol. Chem.* 275:12400–9
91. Larsen RG, Halkides CJ, Singel DJ. 1993. *J. Chem. Phys.* 98:6704–21
92. Stoll S, Jeschke G, Willer M, Schweiger A. 1998. *J. Magn. Res.* 130:86–96
93. Hofbauer W, Bittl R. 2000. *J. Magn. Reson.* 147:226–31
94. Serpersu EH, McCracken J, Peisach J, Mildvan AS. 1988. *Biochemistry* 27:8034–44
95. LoBrutto R, Smithers GW, Reed GH, Orme-Johnson WH, Tan SL, Leigh JS Jr. 1986. *Biochemistry* 25:5654–60
96. Halkides CJ, Farrar CT, Larsen RG, Redfield AG, Singel DJ. 1994. *Biochemistry* 33:4019–35
97. Farrar CT, Halkides CJ, Singel DJ. 1997. *Structure* 5:1055–66
98. Tong L, de Vos AM, Milburn MV, Kim SH. 1991. *J. Mol. Biol.* 217:503–16
99. Buy C, Girault G, Zimmermann JL. 1996. *Biochemistry* 35:9880–91
100. Girault G, Berger G, Zimmermann JL. 1998. *Photosynth. Res.* 57:253–66
101. Houseman ALP, Morgan L, LoBrutto R, Frasch WD. 1994. *Biochemistry* 33:4910–17
102. Houseman ALP, LoBrutto R, Frasch WD. 1995. *Biochemistry* 34:3277–85
103. Espe MP, Hosler JP, Ferguson-Miller S, Babcock GT, McCracken J. 1995. *Biochemistry* 34:7593–602
104. Morrissey SR, Horton TE, Grant CV, Hoogstraten CG, Britt RD, DeRose VJ. 1999. *J. Am. Chem. Soc.* 121:9215–18
105. Lorigan GA, Britt RD, Kim JH, Hille R. 1994. *Biochim. Biophys. Acta* 1185:284–94
106. Kurshev VV, Kevan L, Basu P, Enemark J. 1995. *J. Phys. Chem.* 99:11288–91
107. Pacheco A, Basu P, Borbat P, Raitsimring AM, Enemark JH. 1996. *Inorg. Chem.* 35:7001–8
108. Raitsimring AM, Pacheco A, Enemark JH. 1998. *J. Am. Chem. Soc.* 120:11263–78
109. Astashkin AV, Mader ML, Pacheco A, Enemark JH, Raitsimring AM. 2000. *J. Am. Chem. Soc.* 122:5294–302
110. De Groot A, Hoff AJ, De Beer R, Scheer H. 1985. *Chem. Phys. Lett.* 113:286–90

111. Käß H, Lubitz W. 1996. *Chem. Phys. Lett.* 251:193–203
112. Mac M, Bowlby NR, Babcock GT, McCracken J. 1998. *J. Am. Chem. Soc.* 120:13215–23
113. Hanley J, Deligiannakis Y, MacMillan F, Bottin H, Rutherford AW. 1997. *Biochemistry* 36:11543–49
114. Bosch MK, Gast P, Hoff AJ, Spoyalov AP, Tsvetkov YuD. 1995. *Chem. Phys. Lett.* 239:306–12
115. Lendzian F, Rautter J, Kaess H, Gardiner A, Lubitz W. 1996. *Ber. Bunsen-Ges.* 100:2036–40
116. Gardiner AT, Zech SG, MacMillan F, Kaess H, Bittl R, et al. 1999. *Biochemistry* 38:11773–87
117. Deligiannakis Y, Hanley J, Rutherford AW. 2000. *J. Am. Chem. Soc.* 121:7653–64
118. Grimaldi S, MacMillan F, Ostermann T, Michel H, Ludwig B, Prisner T. 2001. *Biochemistry* 40:1037–43
119. Warncke K, McCracken J, Babcock GT. 1994. *J. Am. Chem. Soc.* 116:7332–40
120. Tang XS, Randall DW, Force DA, Diner BA, Britt RD. 1996. *J. Am. Chem. Soc.* 118:7638–39
121. Britt RD, Zimmermann JL, Sauer K, Klein MP. 1989. *J. Am. Chem. Soc.* 111:3522–32
122. DeRose VJ, Yachandra VK, McDermott AE, Britt RD, Sauer K, Klein MP. 1991. *Biochemistry.* 30:1335–41
123. Tang XS, Diner BA, Larsen BS, Gilchrist ML, Lorigan GA, Britt RD. 1994. *Proc. Natl. Acad. Sci. USA* 91:704–8
124. Ivancich A, Barynin VV, Zimmermann JL. 1995. *Biochemistry* 34:6628–39
125. Dorlet P, Valentin MD, Babcock GT, McCracken JL. 1998. *J. Phys. Chem. B* 102:8239–47
126. Britt RD, Peloquin JM, Campbell KA. 2000. *Annu. Rev. Biophys. Biomol. Struct.* 29:463–95
127. Deligiannakis Y, Hanley J, Rutherford AW. 2000. *J. Am. Chem. Soc.* 122:400–1
128. Martinez JI, Alonso PJ, Gomez-Moreno C, Medina M. 1997. *Biochemistry* 36:15526–37
129. Warncke K, Brooks HB, Babcock GT, Davidson VL, McCracken J. 1993. *J. Am. Chem. Soc.* 115:6464–65
130. Warncke K, Brooks HB, Lee HI, McCracken J, Davidson VL, Babcock GT. 1995. *J. Am. Chem. Soc.* 117:10063–75
131. Ke SC, Warncke K. 1999. *J. Am. Chem. Soc.* 121:9922–27
132. Milov AD, Ponomarev AB, Tsvetkov YuD. 1984. *Chem. Phys. Lett.* 110:67–72
133. Larsen RG, Singel DJ. 1993. *J. Chem. Phys.* 98:5134–46
134. Martin RE, Pannier M, Diederich F, Gramlich V, Hubrich M, Spiess HW. 1998. *Angew. Chem. Int. Ed.* 37:2834–37
135. Pannier M, Veit S, Godt A, Jeschke G, Spiess HW. 2000. *J. Magn. Res.* 142:331–40
136. Rangan SK, Bhagat VR, Sastry VSS, Venkataraman B. 1979. *J. Magn. Res.* 33:227–40
137. Maresch GG, Weber M, Dubinskii AA, Spiess HW. 1992. *Chem. Phys. Lett.* 193:134–40
138. Kurshev VV, Raitsimring AM, Tsvetkov YuD. 1989. *J. Magn. Reson.* 81:441–54
139. Raitsimring A, Peisach J, Lee HC, Chen X. 1992. *J. Phys. Chem.* 96:3526–31
140. Raitsimring A, Crepeau RH, Freed JH. 1995. *J. Chem. Phys.* 102:8746–62
141. Saxena S, Freed JH. 1996. *Chem. Phys. Lett.* 251:102–10
142. Borbat PP, Freed JH. 1999. *Chem. Phys. Lett.* 313:145–54
143. Berliner LJ. 2000. *Biological Magnetic Resonance: Distance Measurements in Biological Systems by EPR*, ed. GR Eaton, SS Eaton, LJ Berliner. New York: Kluwer Academic
144. Astashkin AV, Kodera Y, Kawamori A. 1994. *Biochim. Biophys. Acta* 1187:89–93

145. Shigemori K, Hara H, Kawamori A, Akabori K. 1998. *Biochim. Biophys. Acta* 1363:187–98
146. Astashkin AV, Hara H, Kawamori A. 1998. *J. Chem. Phys.* 108:3805–12
147. Persson M, Harbridge JR, Hammarström P, Mitri R, Mårtensson LG, et al. *Biophys. J.* Submitted
148. Kulikov AV, Likhtenstein GI. 1977. *Adv. Mol. Relax. Interact. Process* 10:47
149. Evelo RG, Styring S, Rutherford AW, Hoff AJ. 1989. *Biochim. Biophys. Acta* 973:428–42
150. Hirsh DJ, Beck WF, Innes JB, Brudvig GW. 1992. *Biochemistry* 31:532–41
151. Kodera Y, Takura K, Kawamori A. 1992. *Biochim. Biophys. Acta* 1101:23–32
152. Koulougliotis D, Innes JB, Brudvig GW. 1994. *Biochemistry* 33:11814–22
153. Koulougliotis D, Tang XS, Diner BA, Brudvig GW. 1995. *Biochemistry* 34:2850–56
154. Deligiannakis Y, Rutherford AW. 1996. *Biochemistry* 35:11239–46
155. Hirsh DJ, Beck WF, Lynch JB, Que L, Brudvig GW. 1992. *J. Am. Chem. Soc.* 114:7475–81
156. Dzuba SA, Kodera Y, Hara H, Kawamori A. 1993. *J. Magn. Reson. A* 102:257–60
157. Kodera Y, Dzuba SA, Hara H, Kawamori A. 1994. *Biochim. Biophys. Acta* 1186:91–99
158. Hara H, Kawamori A. 1997. *Appl. Magn. Reson.* 13:241–57
159. Budker V, Du JL, Seiter M, Eaton GR, Eaton SS. 1995. *Biophys. J.* 68:2531–42
160. Zhou Y, Bowler BE, Lynch K, Eaton SS, Eaton GR. 2000. *Biophys. J.* 79:1039–52
161. Hyde JS, Dalton LR. 1972. *Chem. Phys. Lett.* 16:568–72
162. Freed JH. 1979. In *Time Domain Electron Spin Resonance*, ed. L Kevan, RN Schwartz, pp. 31–66. New York: Wiley
163. Millhauser GL, Freed JH. 1984. *J. Chem. Phys.* 81:37–48
164. Marsh D. 1980. *Biochemistry* 19:1632–37
165. Beth AH, Robinson BH. 1989. In *Biological Magnetic Resonance: Spin Labeling: Theory and Applications*, ed. LJ Berliner, J Reubens, 8:179–253. New York: Plenum
166. Kar L, Millhauser GL, Freed JH. 1984. *J. Phys. Chem.* 88:3951–56
167. Dzuba SA, Watari H, Shimoyama Y, Maryasov AG, Kodera Y, Kawamori A. 1995. *J. Magn. Reson. A* 115:80
168. Lee S, Budil DE, Freed JH. 1994. *J. Chem. Phys.* 101:5529–58
169. Saffman PG, Delbrück M. 1975. *Proc. Natl. Acad. Sci. USA* 72:3111–13
170. Freed JH. 1964. *J. Chem. Phys.* 41:2077–83
171. Ge M, Field KA, Aneja R, Holowka D, Baird B, Freed JH. 1999. *Biophys. J.* 77:925–33
172. Liang Z, Freed JH, Keyes RS, Bobst AM. 2000. *J. Phys. Chem. B* 104:5372–81
173. Kar L, Johnson ME, Bowman MK. 1987. *J. Magn. Reson.* 75:397–413
174. Buitink J, Dzuba SA, Hoekstra FA, Tsvetkov YD. 2000. *J. Magn. Reson.* 142:364–68
175. Pinheiro TJT, Bratt PJ, Davis IH, Doetschman DC, Watts A. 1993. *J. Chem. Soc. Perkin Trans.* 211:2113–17
176. van der Struijf C, Pelupessy TPH, van Faassen EE, Levine YK. 1996. *J. Magn. Reson. B* 111:158–67
177. Yin JJ, Feix JB, Hyde JS. 1990. *Biophys. J.* 58:713–20
178. Altenbach C, Froncisz W, Hyde JS, Hubbell WL. 1989. *Biophys. J.* 56:1183–91
179. Steinhoff HJ, Savitsky A, Wegener C, Pfeiffer M, Plato M, Möbius K. 2000. *Biochim. Biophys. Acta* 1457:253–62
180. Prisner TF. 1997. In *Advances in Magnetic and Optical Resonance*, ed. WS Warren, 20:245–300. San Diego: Academic
181. Rohrer M, Gast P, Möbius K, Prisner TF. 1996. *Chem. Phys. Lett.* 259:523–30
182. Bittl R, Zech SG. 1997. *J. Phys. Chem. B* 101:1429–36

183. Stehlik D, Möbius K. 1997. *Annu. Rev. Phys. Chem.* 48:739–78
184. Kim SS, Weisman SI. 1976. *J. Magn. Reson.* 24:167–69
185. Hoff AJ, Gast P, Romijn JC. 1977. *FEBS Lett.* 73:185–90
186. Stehlik D, Bock CH, Thurnauer MC. 1989. In *Advanced EPR: Applications in Biology and Biochemistry*, ed. AJ Hoff, pp. 371–404. Amsterdam: Elsevier
187. Snyder SW. 1993. In *The Photosynthetic Reaction Center*, ed. J Deisenhofer, J Norris, 2:285–330. New York: Academic
188. Prisner TF, van der Est A, Bittl R, Lubitz W, Stehlik D, Möbius K. 1995. *Chem. Phys.* 194:361–79
189. Van der Est A, Prisner TF, Bittl R, Fromme P, Lubitz W, Möbius K, Stehlik D. 1997. *J. Phys. Chem. B* 101:1437–43
190. Salikhov KM, Bock CH, Stehlik D. 1990. *Appl. Magn. Res.* 1:195–211
191. Kothe G, Weber S, Bittl R, Ohmes E, Thuranuer MC, Norris JR. 1991. *Chem. Phys. Lett.* 186:474–80
192. Weber S, Ohmes E, Thurnauer MC, Norris JR, Kothe G. 1996. In *Reaction Centres of Photosynthetic Bacteria: Structure and Dynamics*, ed. ME Michel-Beyerle, pp. 341–51. Berlin: Springer
193. Thurnauer MC, Clark C. 1984. *Photochem. Photobiol.* 40:381–86
194. Salikhov KM, Kandrashkin YE, Salikhov AK. 1992. *Appl. Magn. Reson.* 3:199–216
195. Zwanenburg G, Hore PJ. 1995. *J. Magn. Reson.* 114:139–46
196. Jeschke G, Bittl R. 1998. *Chem. Phys. Lett.* 294:323–31
197. Dzuba SA, Gast P, Hoff AJ. 1995. *Chem. Phys. Lett.* 236:595–602
198. Zech SG, Lubitz W, Bittl R. 1996. *Ber. Bunsen-Ges.* 100:2041–44
199. Dzuba SA, Hoff AJ. 1997. *Chem. Phys. Lett.* 268:273–79
200. Hara H, Dzuba SA, Kawamori A, Akabori K, Tomo T, et al. 1997. *Biochim. Biophys. Acta* 1322:77–85
201. Zech SG, Kurreck J, Eckert HJ, Renger G, Lubitz W, Bittl R. 1997. *FEBS Lett.* 414:454–56
202. Bittl R, Zech SG, Fromme P, Witt HT, Lubitz W. 1997. *Biochemistry* 36:12001–4
203. Zech SG, Kurreck J, Renger G, Lubitz W, Bittl R. 1999. *FEBS Lett.* 442:79–82
204. Yoshii T, Hara H, Kawamori A, Akabori K, Iwaki M, Itoh S. 1999. *Appl. Magn. Reson.* 16:565–80
205. Yoshii T, Kawamori A, Tonaka M, Akabori K. 1999. *Biochim. Biophys. Acta* 1413:43–49
206. Borovykh IV, Dzuba SA, Proskuryakov II, Gast P, Hoff AJ. 1998. *Biochim. Biophys. Acta* 1363:182–86
207. Zech SG, Bittl R, Gardiner AT, Lubitz W. 1997. *Appl. Magn. Reson.* 13:517–29
208. Tang J, Norris JR. 1995. *Chem. Phys. Lett.* 233:192–200
209. Dzuba SA, Bosch MK, Hoff AJ. 1996. *Chem. Phys. Lett.* 248:427–33
210. Dei A, Gatteschi D, Pardi LA, Barra AL, Brunel LC. 1990. *Chem. Phys. Lett.* 175:589–91
211. Burghaus O, Plato M, Rohrer M, Möbius K, MacMillan F, Lubitz W. 1993. *J. Phys. Chem.* 97:7639–47
212. Prisner TF, McDermott AE, Un S, Norris JR, Thurnauer MC, Griffin RG. 1993. *Proc. Natl. Acad. Sci. USA* 90:9485–88
213. Gerfen GJ, Bellew BF, Un S, Bollinger JM, Stubbe JA, et al. 1993. *J. Am. Chem. Soc.* 115:6420–21
214. Möbius K. 1993. In *Biological Magnetic Resonance: EMR of Paramagnetic Molecules*, ed. LJ Berliner, J Reubens, 13:253–74. New York: Plenum
215. Bratt PJ, Rohrer M, Krzystek J, Evans MCW, Brunel LC, Angerhofer A. 1997. *J. Phys. Chem. B* 101:9686–89
216. Möbius K. 2000. *Chem. Soc. Rev.* 29:129–39
217. Dorlet P, Rutherford AW, Un S. 2000. *Biochemistry* 39:7826–34
218. Bresgunov AY, Dubinskii AA, Krimov VN, Petrov YG, Poluektov OG, Lebedev YS. 1991. *Appl. Magn. Reson.* 2:715–28

219. Burghaus O, Rohrer M, Götzinger T, Plato M, Möbius K. 1992. *Meas. Sci. Technol.* 3:765–74
220. Disselhorst JAJM, van der Meer H, Poluektov OG, Schmidt J. 1995. *J. Magn. Reson. A* 115:183–88
221. Becerra LR, Gerfen GJ, Bellew BF, Bryant JA, Hall DA, et al. 1995. *J. Magn. Reson. A* 117:28–40
222. Earle KA, Tipikin DS, Freed JH. 1996. *Rev. Sci. Instrum.* 67:2502–13
223. Smith GM, Lesurf JCG, Mitchell RH, Riedi PC. 1998. *Rev. Sci. Instrum.* 69:3924–37
224. Fuchs M, Prisner T, Möbius K. 1999. *Rev. Sci. Instrum.* 70:3681–83
225. Grinberg OY, Dubinskii AA, Shuvalov VF, Oranskii LG, Kurochkin VI, Lebedev YS. 1976. *Dokl. Phys. Chem.* 230:923–30 (In English)
226. Weber RT, Disselhorst JAJM, Prevo LJ, Schmidt J, Wenckebach WT. 1988. *J. Magn. Reson.* 81:129–44
227. Prisner TF, Un S, Griffin RG. 1992. *Isr. J. Chem.* 32:357–63
228. Prisner TF, Rohrer M, Möbius K. 1994. *Appl. Magn. Reson.* 7:167–83
229. Kutter C, Moll HP, van Tol J, Zuckermann H, Maan JC, Wyder P. 1995. *Phys. Rev. Lett.* 74:2925–28
230. Moll HP, Kutter C, van Tol J, Zuckermann H, Wyder P. 1999. *J. Magn. Reson.* 137:46–58
231. Bellew BF, Halkides CJ, Gerfen GJ, Griffin RG, Singel D. 1996. *Biochemistry* 35:12186–93
232. Geyer M, Schweins T, Herrmann C, Prisner T, Wittinghofer A, Kalbitzer HR. 1996. *Biochemistry* 35:10308–20
233. Rohrer M, Prisner TF, Brügmann O, Käß H, Spörner M, et al. 2001. *Biochemistry* In press
234. Käß H, MacMillan F, Ludwig B, Prisner TF. 2000. *J. Phys. Chem. B* 104:5362–71
235. Müller F, Hopkins MA, Coron N, Grynberg M, Brunel LC, Martinez G. 1989. *Rev. Sci. Instrum.* 60:3681–84
236. Hassan AK, Pardi LA, Krzystek J, Sienkiewicz A, Goy P, et al. 2000. *J. Magn. Reson.* 142:300–12
237. Rohrer M, Plato M, MacMillan F, Grishin Y, Lubitz W, Möbius K. 1995. *J. Magn. Reson. A* 116:59–66
238. Coremans JWA, Poluektov OG, Groenen EJJ, Canters GW, Nar H, Messerschmidt A. 1996. *J. Am. Chem. Soc.* 118:12141–53
239. Goldfarb D, Strohmaier KG, Vaughan DEW, Thomann H, Poluektov OG, Schmidt J. 1996. *J. Am. Chem. Soc.* 118:4665–71
240. Mehring M, Höfer P, Grupp A. 1987. *Ber. Bunsenges. Phys. Chem.* 91:1132–37
241. Epel B, Goldfarb D. 2000. *J. Magn. Reson.* 146:196–203
242. Bloeß A, Möbius K, Prisner TF. 1998. *J. Magn. Reson.* 134:30–35
243. Coremans JWA, van Gastel M, Poluektov OG, Groenen EJJ, den Blaauwen T, et al. 1995. *Chem. Phys. Lett.* 235:202–10
244. Gerfen GJ, Bellew BF, Griffin RG, Singel DJ, Ekberg CA, Whittaker JW. 1996. *J. Phys. Chem.* 100:16739–48
245. Rohrer M, Brügmann O, Kinzer B, Prisner TF. In preparation
246. Garab G. 1998. *Photosynthesis: Mechanisms and Effect*, Vol. 1. Dordrecht, The Netherlands: Kluwer Academic
247. Peisach J. 1995. *Proc. Int. Conf. Bioradicals Detected ESR Spectroscopy*, ed. H Ohya-Nishiguchi, L Packer, pp. 203–15 Basel: Birkhauser

FAST PROTEIN DYNAMICS PROBED WITH INFRARED VIBRATIONAL ECHO EXPERIMENTS

Michael D Fayer
Department of Chemistry, Stanford University, Stanford, California 94305; e-mail: fayer@fayerlab.stanford.edu

Key Words pure dephasing, protein glass transition, protein viscosity dependence, protein surface fluctuations

■ **Abstract** IR vibrational echo experiments are used to study dynamics in myoglobin (Mb) by investigating the dephasing of the CO-stretching mode of CO bound at the active site of the protein (Mb-CO). The temperature dependence and the viscosity dependence of Mb-CO pure dephasing have been measured in several solvents. In low-temperature, glassy solvents, the pure dephasing has a power law temperature dependence, $T^{1.3}$, that reflects glasslike protein dynamics. In liquids, the temperature dependence is much steeper and arises from a combination of pure temperature dependence and the influence of decreasing solvent viscosity with increasing temperature. As the solvent viscosity decreases, the ability of the protein's surface to undergo topological fluctuations increases, which in turn increases the internal protein-structural fluctuations. The protein-structural motions are coupled to the CO bound at the active site by electric field fluctuations that accompany movements of polar residues. The dynamic electric field-coupling mechanism is tested by observing differences in the temperature dependence of the pure dephasing of Mb-CO mutations.

INTRODUCTION

In 1993, the first ultrafast vibrational echo experiments were performed on condensed-matter systems, using tunable IR pulses (1). The development and application of ultrafast IR vibrational echoes and other IR coherent-pulse sequences are providing a new approach to the study of the structural states of molecules in complex molecular systems such as liquids, glasses, and proteins (2–11). The vibrational echo experiments and related ultrafast vibrational-coherence experiments are outgrowths of advances made in magnetic resonance methods and coherent optical spectroscopy over many decades.

NMR spin echo experiments, first performed in 1950, began a new era in spectroscopy (12). The spin echo was the first spectroscopic experiment to take advantage of coherent interactions of a radiation field with the system to obtain information not available in an absorption measurement. The spin echo is the simplest of all pulsed magnetic resonance experiments. It involves the application of two radio frequency pulses and observation of the time-dependent response of the sample. Since 1950, a large number of complex pulse sequences have been developed and applied to the study of magnetic spin systems (13). All of these have direct lineage to the earlier spin echo experiments.

In 1964, photon echo experiments extended spin echo experiments to electronically excited states (14, 15). The photon echo began the application of coherent-pulse techniques in the visible and UV portions of the electromagnetic spectrum. Since its development, the photon echo and related pulse sequences have been applied to a wide variety of problems including dynamics and intermolecular interactions in crystals, glasses, proteins, and liquids (16–19). Like the spin echo, the photon echo and other optical, coherent-pulse sequences provide information that is not available from absorption or fluorescence spectroscopy.

The spin echo, the photon echo, and the vibrational echo are, in many respects, similar experiments. The term vibrational echo is used to distinguish IR experiments on vibrations from radio frequency experiments on spins or visible and UV experiments on electronic states. In this chapter, recent vibrational echo experiments on the proteins myoglobin (Mb) and hemoglobin (Hb) are described.

The dynamics of proteins on a wide variety of time scales are intimately related to protein function. Fast and moderately fast fluctuations of protein structure enable a protein to sample a complex conformational-energy landscape. These rapid motions give rise to the slower processes associated with protein function. Molecular-dynamics simulations have shown that a protein can sample thousands of conformations within a very short time (20, 21). Understanding these dynamics provides an important connection between protein function and protein structure, as determined by X ray (22, 23), NMR (24, 25), and other experimental techniques (26–30) and theory (20, 21).

The importance of dynamic fluctuations in proteins is illustrated by Mb, which is a 153-amino-acid protein with the primary biological function of reversibly binding and transporting O_2 in muscle tissue. The ability of Mb to bind O_2 and other biologically relevant ligands, such as CO or NO, results from a nonpeptide prosthetic group, heme, which is located in the protein's "pocket" and is covalently bound at the proximal histidine (H93) of the globin protein. The X-ray crystal structure of Mb indicates that there are no static gaps for ligands to pass through (31). For ligands to move in and out of the pocket, they must traverse the intervening protein. Traversing the protein is made possible at room temperature by dynamic fluctuations in the protein structure that open paths for ligand diffusion through the protein (32). Because of the ability of ligands to move through the protein, in some sense, the protein has liquidlike character at room temperature (32).

Vibrational echo experiments can be applied to Mb by examining the dynamics of the CO ligand bound at the active site of the protein (Mb-CO). The vibrational levels of a molecular oscillator in a condensed-matter system are influenced by the surrounding medium through intermolecular interactions. The time-averaged forces exerted by the solvent on an oscillator cause a static shift in the vibrational-absorption frequency relative to its frequency in the gas phase. The frequency shifts of the vibrational transitions of a molecule between the gas phase and a condensed-matter environment are indicators of the effect of the solvent on the internal mechanical degrees of freedom of a solute.

The fluctuating forces exerted by a medium also produce fluctuations in the molecular structure. Such structural fluctuations cause the vibrational eigenstates to be time dependent, and, thus, the vibrational-energy eigenvalues are time dependent. Time evolution of the vibrational-energy eigenvalues produces fluctuations in the vibrational-transition energies. Structural fluctuations and the associated energy fluctuations are not generally observable in either an IR or Raman vibrational spectrum. The extent and time dependence of the fluctuations of a molecular oscillator's vibrational-energy levels are sensitive to the nature of the dynamics of the condensed-matter environment and the strength of intermolecular interactions. In Mb-CO, protein-structural fluctuations are responsible for the time dependence of the CO energy levels.

In principle, information on dynamical intermolecular interactions of an oscillator with its environment can be obtained from vibrational-absorption spectra. The forces experienced by the oscillator determine the vibrational line shape and width. The IR absorption line shape is related to these microscopic dynamics through the Fourier transform of the two-time transition dipole correlation function (33–36), which includes any inhomogeneous broadening (33, 34, 36). The line shape and width depend on temperature and other properties of the environment that influence the oscillator. However, a vibrational-absorption spectrum reflects the full range of mechanisms that broaden the vibrational line shape, including dynamical contributions and static or essentially static inhomogeneous contributions to the line shape (33–36). In proteins such as Mb-CO and Hb-CO, inhomogeneous broadening of the CO vibrational transition exceeds the dynamical line width (10, 37). Under these circumstances, measurements of absorption spectra do not provide information on protein dynamics.

The vibrational echo experiment is a coherent IR pulse sequence that removes inhomogeneous broadening from the vibrational transition and reveals the underlying dynamical line shape. In contrast to an IR absorption spectrum, the vibrational echo is described by a four-time correlation function (18, 38, 39). The vibrational echo is one of a class of spectral line-narrowing experiments described by a four-time correlation function (18).

In a vibrational echo experiment, a source of short IR pulses is tuned to the vibrational transition of interest. The vibrational echo employs a two-pulse excitation sequence. The first pulse places each solute molecule's vibration into a superposition state, which is a mixture of the $v=0$ and $v=1$ vibrational levels. Each vibrational superposition has a microscopic electric dipole associated with it. This dipole oscillates at the vibrational-transition frequency. Immediately after the first pulse, all of the microscopic dipoles in the sample oscillate in phase. Because there is an inhomogeneous distribution of vibrational-transition frequencies, the individual dipoles oscillate with some distribution of frequencies. Thus, the initial-phase relationship is very rapidly lost. This is referred to as the free-induction decay. After a time τ, a second pulse, traveling along a path that makes an angle θ with that of the first pulse, passes through the sample. This second pulse changes the phase factors of each vibrational-superposition state in a manner that initiates a rephasing process. At time τ after the second pulse, the sample emits a third coherent pulse of light. The emitted pulse propagates along a path that makes an angle of 2θ with the path of the first pulse. The third pulse is the vibrational echo. It is generated when the ensemble of microscopic dipoles (one for each vibrational oscillator) is rephased at time 2τ. The phased array of microscopic dipoles behaves as a macroscopic oscillating dipole, which generates an IR pulse of light. A free-induction decay (inhomogeneous frequency distribution) again destroys the phase relationships, so only a short pulse of light is generated.

The rephasing at 2τ has removed the effects of the inhomogeneous broadening. However, fluctuations due to coupling of the vibrational mode (CO vibration) to the heat bath (protein-structural dynamics) cause the oscillation frequencies also to fluctuate. Thus, at 2τ there is not perfect rephasing. As τ is increased, the fluctuations produce increasingly large accumulated phase errors among the microscopic dipoles, and the size of the vibrational echo is reduced. A measurement of the vibrational echo intensity vs τ, the delay time between the pulses, is called a vibrational echo decay curve. Thus, the vibrational echo decay is related to the fluctuations in the vibrational frequencies, not the inhomogeneous spread in frequencies. The Fourier transform of the vibrational echo decay yields the underlying dynamical line shape (40, 41). For example, if the vibrational echo decay is exponential, the line shape is Lorentzian, and the line width is determined by the exponential decay constant. The vibrational echo makes the vibrational-dynamic line shape experimentally observable. In fact, the vibrational echo decay measures directly the decay of the system's off-diagonal density matrix elements and is the fundamental observable.

To obtain a physical feel for the manner in which the vibrational echo experiment can reveal homogeneous fluctuations despite a broad inhomogeneous spread of transition frequencies, consider the following foot race scenario (42, 43). Initially, all of the runners in this scenario are lined up at the starting line (see Figure 1). At $t = 0$, the starting gun (analogous to the first IR pulse) is fired, and the runners take off down the track. After running for some time, the faster runners are out in front, and the slower runners are somewhat behind. The runners are no longer in a line because of the inhomogeneity of their speeds. At time τ, the gun is again fired (analogous to the second IR pulse), and all of the runners turn around and run back toward the starting line. If each runner maintains a constant speed out and back, then all of the runners will cross the starting line exactly in line again (see Figure 1, second panel from bottom). When the second gun is fired, the faster runners are farther away from the starting line than the slower runners, but because fast runners run more quickly, the differences in distances are exactly made up for by the differences in speeds. At the starting line, the group is "rephased"; the inhomogeneity in speeds is nullified. If the runners do not run at exactly constant

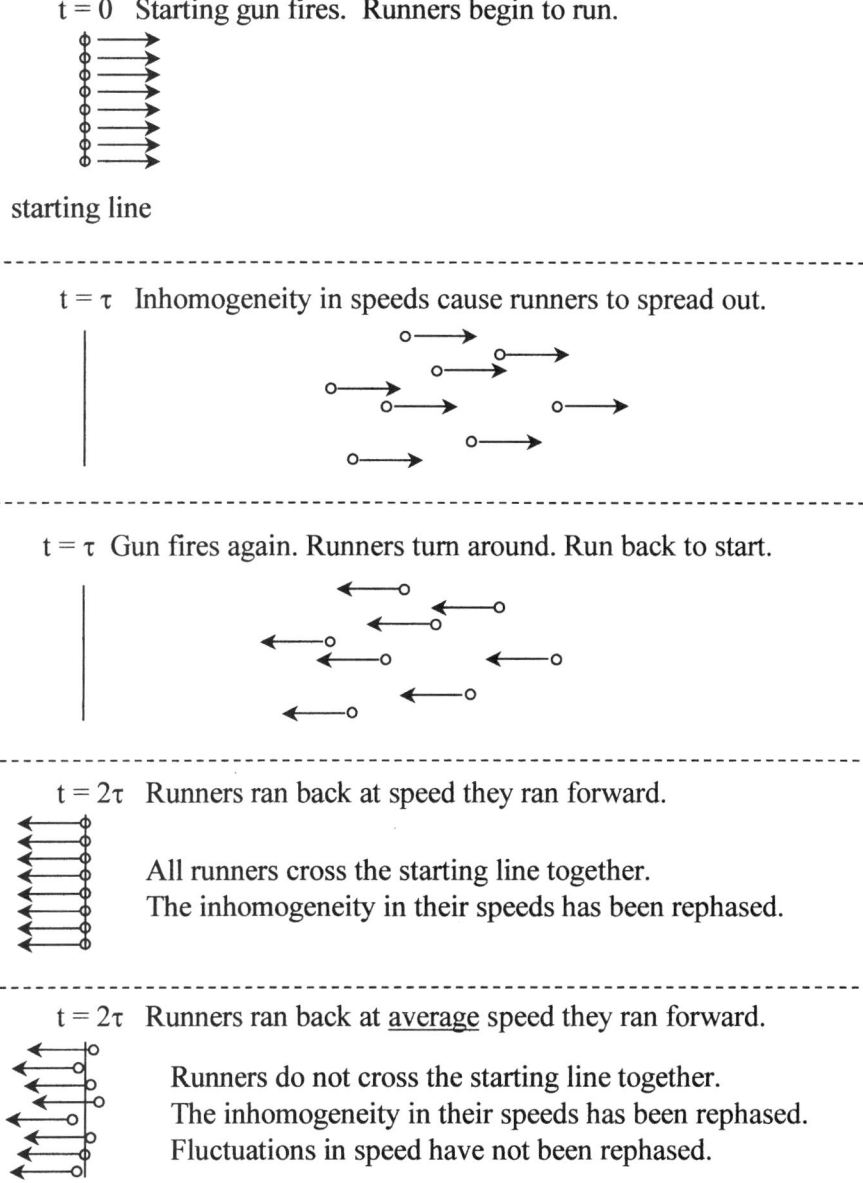

Figure 1 Analogous use of runners on a racetrack to illustrate the mechanism by which a vibrational echo experiment eliminates inhomogeneous broadening and permits measurement of pure dephasing (energy level fluctuations). The first and second firings of the starting gun correspond to the first and second pulses in the vibrational echo pulse sequence. See text for other details.

speeds, but each runner has some fluctuation in speed about his average (dynamical fluctuations), then the runners will not cross the starting line exactly in a line (see Figure 1, bottom panel). The rephasing will be imperfect. A snapshot of the group as it crosses the starting line on the return leg of the race will show a spread in positions about the starting line, revealing the small fluctuations in the runners' speeds. If the runners run out for a longer time (τ is increased), then the fluctuations will produce a greater spread in positions about the starting line. The increase in the spread as τ is increased is a measure of the fluctuations in the speeds, and the measurement is not influenced by the inhomogeneity in the speeds. In the same manner, the vibrational echo experiment reveals the fluctuations in the vibrational-transition frequency despite a large inhomogeneous distribution of vibrational energies.

Recently, the ultrafast IR vibrational echo technique has been applied to the study of Mb-CO dynamics (4, 5, 7, 10, 44) and those of the closely related protein Hb-CO (37). The vibrational echo measurements of the pure dephasing of the CO-stretching mode are sensitive to the complex protein dynamics communicated to the CO ligand bound at the active site of the protein. Unlike other ultrafast techniques (45, 46), which involve electronic excitation of chromophores, the vibrational echo experiments directly examine effects of fluctuations of protein structure on the ground state potential-energy surface.

In this review, first some details of the vibrational echo experiments and experimental procedures are presented. Then some of the basic data taken on Mb-CO are discussed, and a model is presented for the mechanism that couples protein fluctuations to CO bound at the active site and causes CO vibrational dephasing. Experiments on mutant Mbs are used to support the model (5, 7). Then data on the temperature dependence of the dephasing of Mb-CO are presented. At low temperatures in glassy solvents, the temperature dependence is a power law, $T^{1.3}$. This temperature dependence is suggestive of glasslike dynamics, and there is evidence of a protein "glass transition" at $\sim$200 K (10). The solvent viscosity dependence of the Mb-CO dephasing is analyzed. A viscoelastic model is used to describe the data (47). The results demonstrate the importance of fluctuations of the protein's surface in determining the extent of internal-protein dynamics. Finally, results of experiments on Hb-CO are compared with the results on Mb-CO.

THE VIBRATIONAL ECHO METHOD AND EXPERIMENTAL PROCEDURES

The Vibrational Echo Method

The Mb-CO absorption spectrum (1945 cm^{-1}) is inhomogeneously broadened, even at room temperature. The line shape reflects the distribution of quasistatic protein configurations that give rise to a range of CO frequencies, but the line shape provides no information about the protein dynamics. The vibrational echo is a two-pulse time domain technique that is sensitive to the dynamics of the CO frequency and thus can provide information on the protein dynamics.

Standard treatments of echo spectroscopy (or line shapes) describe two extreme limits (40, 48, 49)—very fast processes, which produce homogeneous dephasing and a motionally narrowed spectroscopic line, and very slow or static processes, which produce an inhomogeneously broadened spectroscopic line. In standard applications, the echo experiment extracts the homogeneous line width (dephasing rate) from a transition dominated by inhomogeneous broadening. In this case, the echo decay is given by

$$S_E(\tau) = S_E(0)e^{-4\tau/T_E}, \qquad 1.$$

and the experimental echo decay time T_E is equal to the ensemble averaged homogeneous dephasing time T_2 in the sense of the Bloch equations. The homogeneous line width is $(\pi T_2)^{-1}$ (40).

A common source of homogeneous dephasing is continuous frequency modulations that produce motional narrowing. For processes that cause motional narrowing, the modulation time τ_m of the fast process is fast compared with the typical size of the frequency perturbation Δ_m (the rms range of frequencies sampled); that is, $\Delta_m \tau_m \ll 1$ (50, 51). Under these conditions, the echo measures the dephasing time $T_E^{-1} = T_2^{-1} = \Delta_m^2 \tau_m$ (49). It does not measure the modulation time directly, although T_E and τ_m can be related by an appropriate model.

At the other limit, the slow process is assumed to be essentially static. It has no effect on the echo decay, but it does contribute to the absorption line width. If the typical size of the quasistatic frequency variation is Δ_I, the Fourier transform of the absorption line shape, often called a free-induction decay, is

$$S_{FID}(\tau) = S_{FID}(0)e^{-\Delta_I^2 \tau^2/2}e^{-4\tau/T_2}. \qquad 2.$$

If the inhomogeneous line width Δ_I is large, no information on the dephasing time or the modulation times of the system can be obtained from the line shape. However, the echo can extract the value of T_2, which is related to the system's dynamics.

The effect of the vibrational population lifetime T_1 is removed by combining the vibrational echo time T_E with a pump probe measurement (transient absorption) of T_1 to yield the pure dephasing time T_E^* through

$$\frac{1}{T_E} = \frac{1}{T_E^*} + \frac{1}{2T_1}. \qquad 3.$$

The pure dephasing time T_E^* is caused only by CO transition frequency fluctuations, which in turn are caused by the protein dynamics.

A variety of physical processes conform to this standard model, that is, there is a vast separation of time scales that produces a homogeneous line (not necessarily motionally narrowed) that underlies an inhomogeneously broadened absorption spectrum. Examples include absorption spectra and photon echo experiments on chromophores in low-temperature glasses (19), in which the pure dephasing is caused by two-level system (TLS) dynamics (52, 53) and possibly IR absorption spectra and vibrational echo experiments on proteins at moderate to low temperatures (10). The low-temperature and viscosity-independent processes seen in the experiments below fit this model. However as discussed below, the viscosity-dependent portion of the dephasing does not fit the model (54).

The standard model is adequate if the dynamics are either very fast or very slow. However, there is also a broad range of modulation times between these limits, in which the vibrational echo behaves quite differently from these standard treatments. In this intermediate range, the modulation time is long enough to prevent motional narrowing ($\Delta_m \tau_m \gg 1$), but it is not slow enough for its effects to be eliminated from the two-pulse echo experiment; that is, Δ_m is not part of the inhomogeneous line. A process that occurs in this intermediate time range produces what is called spectral diffusion, that is, relatively slow evolution of the vibrational frequency. Spectral diffusion can be studied using three-pulse stimulated echoes (40, 48, 49, 55) or the time-dependent versions of hole burning or fluorescence line narrowing (18, 19, 56). However, with proper interpretation, the two-pulse echo also becomes a powerful tool for measuring spectral diffusion (47, 54).

Relatively little attention has been paid to two-pulse echoes in the spectral-diffusion range. Older work focused on models appropriate for spin resonance (57–59). After the basic results of Yan & Mukamel (60), a more detailed examination of the behavior of the two-pulse echo as the modulation time is varied has been presented (54). When τ_m is in the spectral-diffusion range, the echo decay time T_E is no longer equal to a dephasing time T_2 in the sense of the Bloch equations, and the echo decay time is related to an inverse dynamic line width that is not the equivalent of the standard homogeneous line width. In the spectral-diffusion regime, the echo decay is more directly related to the underlying modulation time τ_m. In general, if the initial decay of the frequency-frequency correlation function has the form $1 - \tau_m^\beta$ (the short time expansion of a variety of time-dependent functions, for example, an exponential, a Gaussian, and a stretched exponential), then (54)

$$\Delta_m T_E = B_\beta \Gamma\left(\frac{1}{\beta+2}\right)(\Delta_m \tau_m)^{\beta/(\beta+2)}, \qquad 4.$$

and

$$B_\beta = \frac{1}{2}\left(\frac{8}{\beta+2}\right)^{\frac{\beta+1}{\beta+2}}\left(\frac{\beta+1}{2^\beta-1}\right)^{\frac{1}{\beta+2}}. \qquad 5.$$

In the specific case of an exponentially decaying frequency-frequency correlation function, $\beta = 1$, and Yan & Mukamel's result is recovered (60):

$$T_E = \left(\frac{4}{3}\right)^{2/3} \Gamma(1/3) \left(\frac{\tau_m}{\Delta_m^2}\right)^{1/3}. \qquad 6.$$

In these equations, the standard gamma function Γ is used, and T_E is defined as the integral decay time (54). The weak cube-root dependence, $T_E \propto \tau_m^{1/3}$, makes the echo useful over a very wide range of modulation times. It is also very different from the $T_E \propto \tau_m^{-1}$ dependence expected in the motionally narrowed limit. The data presented below on the viscosity-dependent pure dephasing of Mb-CO follow $T_E \propto \tau_m^{1/3}$, implying that the protein-structural fluctuations occur on intermediate time scales and cause viscosity-dependent spectral diffusion of the CO vibrational frequency.

Experimental Procedures

The IR vibrational echo experiments on Mb-CO (most of the experiments) were performed by using the Stanford free-electron laser (FEL). The FEL produces tunable, ps-duration, mid-IR pulses. The experiments on Hb-CO were performed using a Ti:Sapphire-based optical-parametric-amplifier (OPA) system to generate tunable sub-ps-duration mid-IR pulses (37). The IR pulses from the FEL had energies of ~0.5 µJ and were nearly transform-limited Gaussian distributions of 1.2 ps in duration. Both the autocorrelation and the spectrum of the IR pulse were monitored continuously during the experiments. The spot size in the sample was ~100 µm. The energies at the sample in the two pulses of the echo sequence were ~50 and ~150 nJ, respectively. The vibrational lifetime T_1 was measured by pump probe experiments (transient absorption).

The FEL experimental apparatus is shown in Figure 2 (61, 62). The IR beam enters the experimental area roughly collimated. L1 and L2 reduce the beam size. At the focus of the telescope is a Ge acousto-optic modulator (AOM) for pulse selection, within a 1:1 cylindrical telescope using CaF$_2$ lenses. Micropulses are selected out of each FEL macropulse (~2 ms in duration at a 10 Hz repetition rate

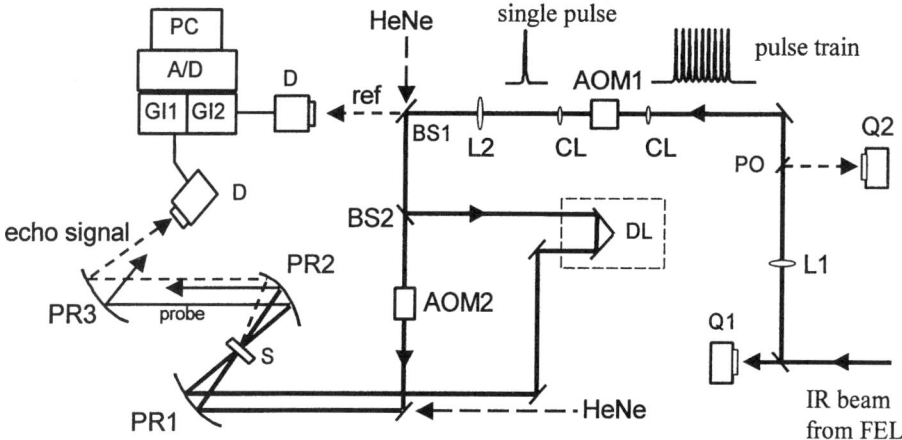

Figure 2 Vibrational echo experimental setup when using the Stanford free-electron laser (*FEL*) as the source of IR pulses. Pulse selection is accomplished with an acousto-optic modulator 1 (*AOM1*), while *AOM2* is used for chopping of pulse 2. *A/D*, analog-to-digital converter; *BS*, ZnSe beam splitter; *CL*, cylindrical lens; *D*, detector; *DL*, optical delay line; *GI*, gated integrator; *L*, lens; *PC*, computer; *PO*, pick-off; *PR*, off-axis parabolic reflector; *Q*, position-sensitive detector; *S*, sample.

with micropulses occurring at 12 MHz) at a repetition rate of 50 kHz by the AOM single-pulse selector. The cylindrical telescope makes the AOM rise time less than the interpulse separation. The pulse selection yields an effective experimental repetition rate of 1 kHz, and an average power of <0.5 mW. A ZnSe beam splitter allows 1% of the IR beam to be directed into an HgCdTe reference detector. All data from pulses with intensities outside a 10% window were discarded.

The two pulses for vibrational echo or pump probe experiments were obtained with a 10% ZnSe beam splitter. The 10% beam (first pulse in the vibrational echo sequence and probe pulse) is sent through a computer-controlled stepper motor delay line. The remaining portion (second vibrational echo pulse or pump pulse) is chopped at 25 kHz with a second Ge AOM. A HeNe beam is made collinear with each IR beam for alignment purposes. The two pulses were focused using an off-axis parabolic reflector, for achromatic focusing of the IR and HeNe. The beams and vibrational echo signal were recollimated with a second parabolic reflector and focused into a HgCdTe signal detector with a third parabolic reflector. By selecting the desired beam with an iris between the second and third parabolic reflectors, either the vibrational echo or pump probe signal could be observed. The vibrational echo signal and the intensity reference signal were sampled by two gated integrators and digitized for collection by computer. Careful studies of power dependence and repetition rate dependence of the data were performed. It was determined that there were no heating or other unwanted effects.

The measurements were made on native horse heart Mb (Sigma; used without further purification). Samples were prepared by adding 15 mM lyophilized Mb-Fe^{+3} to either 50:50, 80:20, or 95:5 wt% glycerol–0.1 M phosphate buffer (pH 7) or 50:50 (vol/vol) ethylene glycol–0.1 M phosphate buffer (pH 7). The resulting solutions were then stirred under CO atmosphere for 8 h before being reduced by a 10-fold molar excess of dithionite.

The water sample was prepared in the same manner except that the concentration was 30 mM in 0.1 M phosphate buffer (pH 7). The trehalose sample was prepared by making an $\sim$10 wt% solution of trehalose in 0.1 M phosphate buffer (pH 7) and dissolving Mb-Fe^{+3} in it to a final concentration of 1–2 mM. A few drops of the resulting solution were placed on a sapphire window and allowed to dry under CO atmosphere for a few days. Another sapphire window was pushed against the first one to give a thickness of $\sim$100 μm.

The viscosity measurements were performed with Cannon-Ubbelohde viscometers. For the Mb-CO in 50:50 ethylene glycol-water, temperature-dependent-viscosity measurements were made in a variety of low temperature baths (47). To obtain the viscosity, η, at lower temperatures, a Vogel-Tammann-Fulcher (VTF) equation (63–65) was used to model the viscosity at all temperatures. A fit was made to the measured points to yield (47)

$$\eta = (2.0 \times 10^{-4} \text{ cP}) \exp\left(-\frac{2500 \text{ K}}{T - 70 \text{ } K}\right). \qquad 7.$$

However, a single VTF curve cannot accurately emulate the viscosity over such a large temperature range. Therefore, the viscosities at the lowest temperatures are approximate.

RESULTS AND DISCUSSION

The Data on Mb-CO

Figure 3a shows a vibrational echo decay of Mb-CO in the solvent trehalose glass at 11 K (10). The signal-to-noise ratio is excellent despite a large background absorption by the protein and the trehalose. The Mb-CO peak has an optical density ~0.2 on a background of optical density 1. The data are fit well by a 7.0-ps decay constant exponential curve, which corresponds to a homogeneous dephasing time of 28 ps and a homogeneous width of 0.38 cm^{-1}. At room temperature, the dynamical width increases to ~2 cm^{-1}. The full absorption line width at half maximum is ~15 cm^{-1} at all temperatures. Therefore, the Mb-CO absorption line is inhomogeneously broadened at all temperatures studied, and dynamical information could not be obtained from an absorption spectrum.

Figure 3b is a semilog plot of the temperature dependence in trehalose of T_E (circles), $2T_1$ (squares), and T_E^* (triangles). T_E^* was computed from the other two quantities, using Equation 3. T_1 has a slight, essentially linear temperature dependence. It can be seen that, at low temperatures, the vibrational echo decay is nearly lifetime limited. By room temperature, the vibrational echo decay is dominated by pure dephasing.

Figure 4 displays temperature-dependent pure dephasing line widths, $1/\pi T_E^*$, for Mb-CO in glycerol/water. At low temperature, the data were a power law with temperature dependence $T^{1.3}$. The power law is shown more clearly below. At higher temperature, the functional form of the data changes; the temperature dependence becomes steeper. A detailed discussion of the temperature and solvent viscosity dependence is given below. First, the mechanism that couples time dependent structural fluctuations of the protein to the CO vibrational frequency to produce pure dephasing is described.

Protein Structural Dynamics and CO Dephasing: Mutant Studies

In the various experiments, Mb-CO is dissolved in a variety of liquid and glassy solvents. The solvent viscosity plays an important role in determining the nature and extent of the Mb structural fluctuations that contribute to pure dephasing. However, the solvent does not directly couple to the CO, which is bound at the active site. The Mb pocket protects the CO from direct interaction with the solvent.

For the solvent to cause dephasing, its motions must couple to the transition frequency of the CO. When molecules go from the gas phase to a condensed phase, there is a shift of electronic and vibrational-transition frequencies. This effect is referred to as the solvent shift. Intermolecular interactions with the

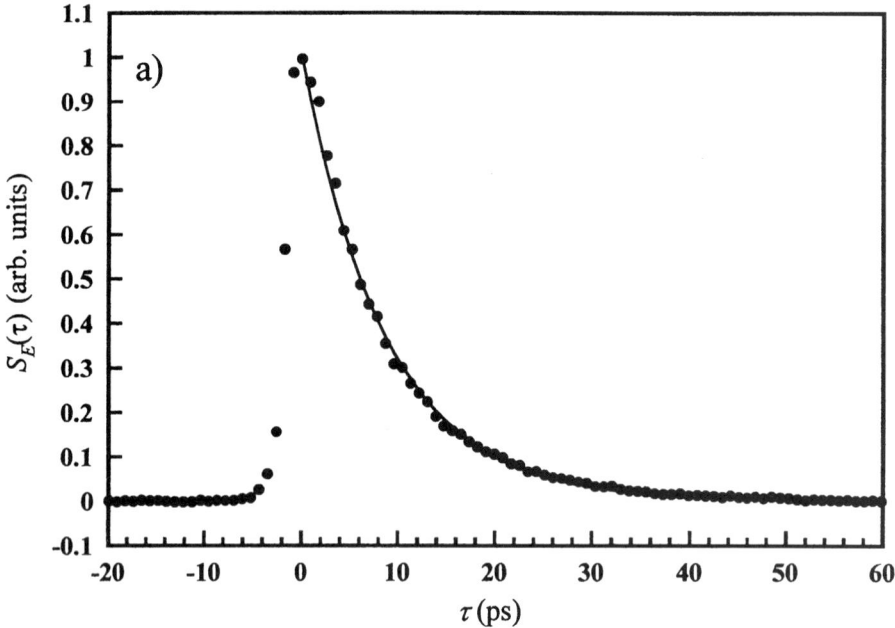

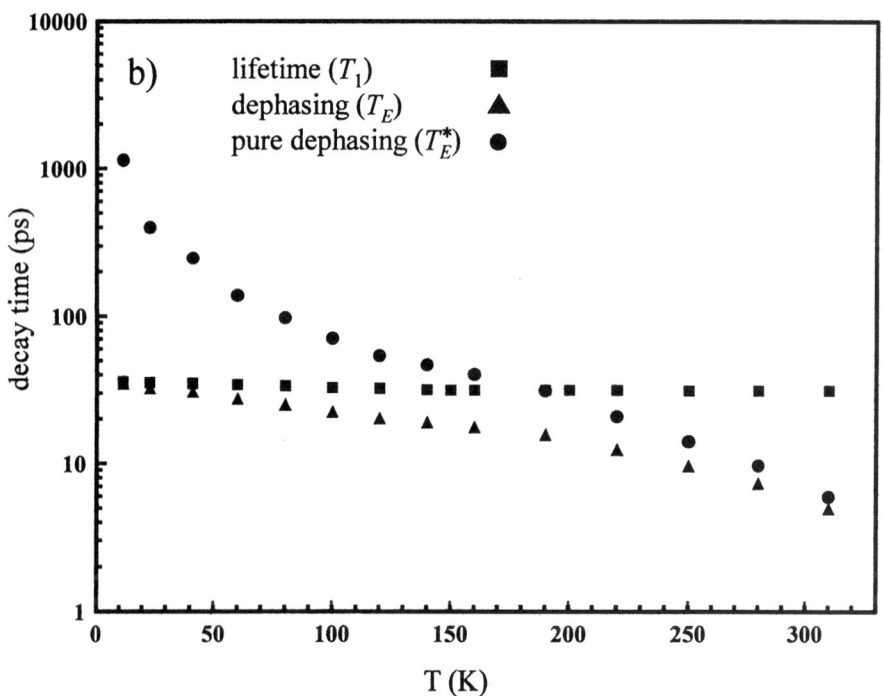

condensed-matter environment are responsible for line broadening as well as the solvent shift. These phenomena are closely related. The line broadening can be static, giving rise to an inhomogeneous line, or dynamic. In either case, it is the variations in the solvent shift that cause line broadening.

In Mb-CO, the nature of the solvent itself has little effect on the CO vibrational-transition frequency. The Mb-CO transition frequency is virtually identical in a wide variety of liquids and glasses (66, 67). The solvent shift is unaffected by the medium surrounding the protein, even when the change is from a liquid solvent such as water or a mixture of glycerol and water to a protein crystal (67). In contrast, the frequency difference between the A_0 and A_1 CO absorption lines in Mb-CO is 24 cm^{-1} (66). This difference in the CO absorption frequency is caused by a change in conformation of the protein, particularly the position of the distal histidine, an amino acid close to the CO. Changes in the protein structure can have a major influence on the CO vibrational frequency, whereas changes in the solvent have a negligible influence. This spectroscopic information leads to the reasonable conclusion that fluctuations of the protein structure will cause pure dephasing, while fluctuations of the solvent structure will not. The solvent does provide a heat bath and a boundary condition that are intimately involved in the protein fluctuations and the dephasing, but the argument made above strongly supports the idea that the dephasing does not arise from direct coupling of the solvent dynamics to the CO transition frequency.

The $T^{1.3}$ power law pure-dephasing temperature dependence displayed in Figure 4 is also observed in several other solvents (10; see below). The power laws in the various solvents have the same slopes and the same values of the pure dephasing rates, showing that the dephasing is not affected by the specific chemical composition of the solvent. This fact is another demonstration that the solvent does not directly interact with the CO to produce pure dephasing. Rather, the vibrational dephasing is caused by protein-structural fluctuations (5, 7, 62).

The manner in which the Mb protein fluctuations are communicated to the CO bound at the active site and in which they produce pure dephasing has been described previously (5, 7, 47, 62). A model was developed that ascribes the CO vibrational pure dephasing to global structural fluctuations of the protein. The protein-structural dynamics produce fluctuating electric fields because polar groups throughout the protein are moving. The fluctuating electric fields cause the CO vibrational frequency to fluctuate via the Stark effect (68). A number of detailed quantum chemical studies (69, 70) and experimental studies (68) demonstrate that changes in the electric field act directly on the CO, changing its frequency through the Stark effect. The theoretical and experimental studies are for static changes in

←

Figure 3 (a) Vibrational echo decay data for Mb-CO in trehalose. The line through the data is a fit to a single exponential. (b) Mb-CO lifetime data ($2T_1$), vibrational echo data (T_E), and pure dephasing data (T_E^*) as a function of temperature. The pure-dephasing data are obtained by removing the T_1 contribution from T_E, using Equation 3.

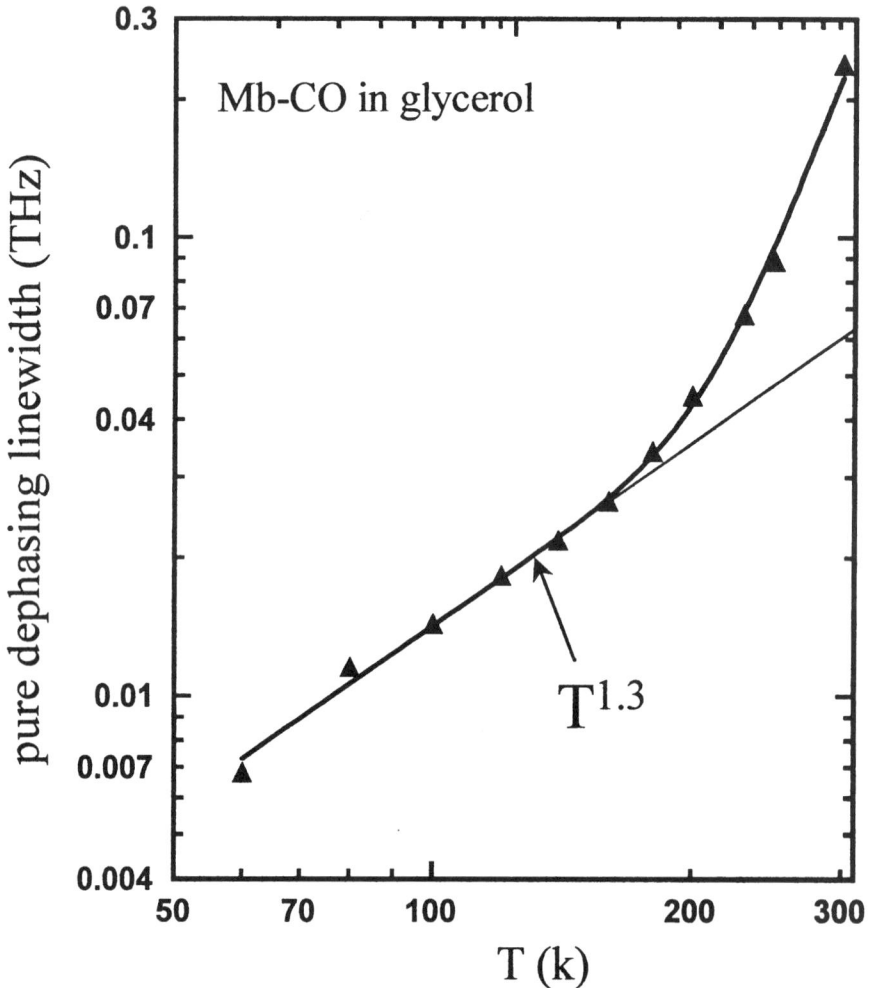

Figure 4 Vibrational echo pure-dephasing line width ($1/\pi T_E^*$) data vs temperature for Mb-CO in glycerol-water. At low temperatures, the temperature dependence is $T^{1.3}$. At higher temperatures, the change in the data with temperature becomes steeper.

the electric field (68–70), whereas the pure dephasing is caused by time-dependent electric fields that arise from the protein's structural dynamics. In either case, the Stark effect is the primary mechanism causing frequency shifts in the CO vibration (47).

To test the fluctuating electric field induced pure dephasing model, experiments were conducted on mutants of Mb-CO (5, 7). Figure 5 displays the temperature-dependent pure dephasing line widths of a mutant of Mb-CO in which the polar

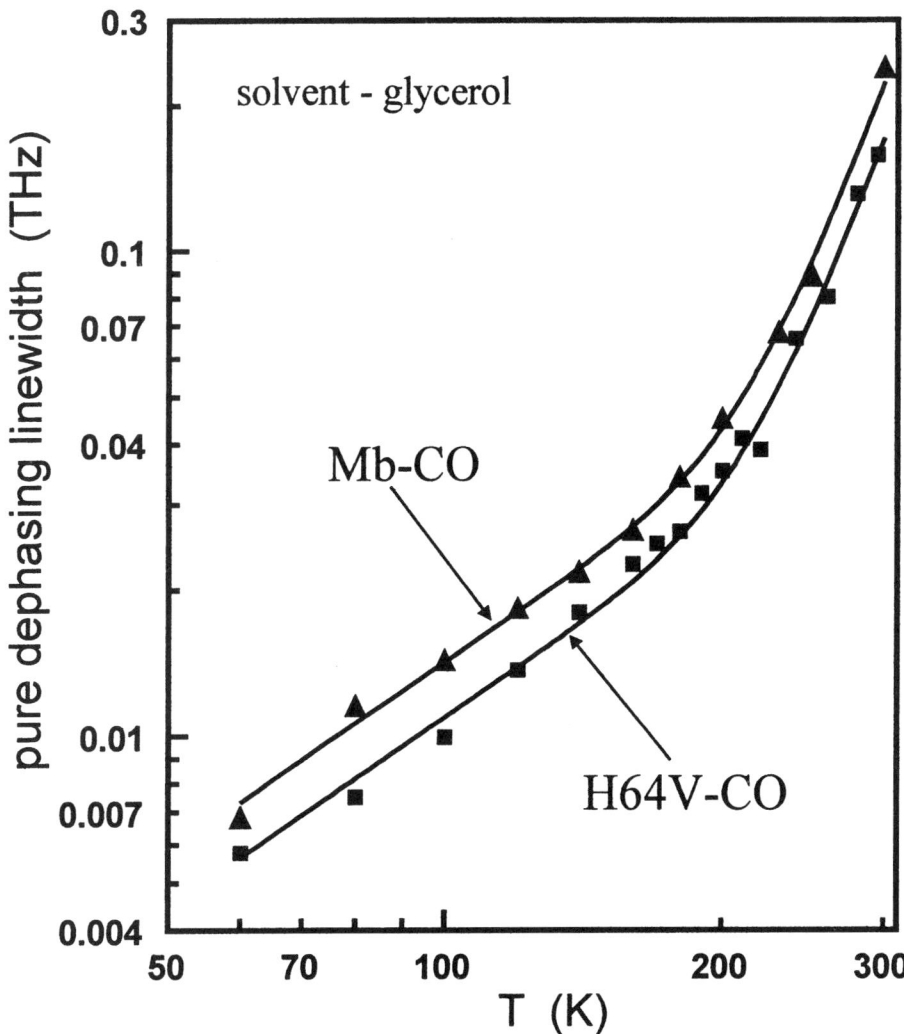

Figure 5 A comparison of the vibrational echo pure-dephasing line width ($1/\pi T_E^*$) data vs temperature for myoglobin Mb-CO and H64V-CO (the mutant with the polar distal histidine replaced with a nonpolar valine) in glycerol-water. The functional forms of the two data sets are identical, but the pure-dephasing line width is ~20% narrower in H64V-CO at all temperatures. These data support the model stating that pure dephasing is caused by global protein-structural fluctuations that produce a fluctuating electric field at the CO bound at the active site of Mb.

distal histidine is replaced with nonpolar valine (H64V-CO). Also shown in the Figure are the Mb-CO data displayed in Figure 4. The dephasing rate for H64V-CO is slower than for Mb-CO over the entire temperature range. Comparing the data points at each temperature reveals that the H64V-CO rates are ~20% slower than the Mb-CO rates with no systematic variation in the form of the temperature dependence. A solid line was drawn through the H64V-CO data. The line is the same line that runs through the Mb-CO data. The line was translated down with no change in shape. Translation downward on the log plot is the same as multiplying the function that passes through the Mb-CO data by a constant. It can be seen that, within experimental error, the functional forms of the two data sets are identical.

In the H64V-CO experiments, everything about the protein pocket and the experimental procedures is the same as in the previous Mb experiments except for the substitution of the distal histidine with a valine. This amino acid change causes an ~20% reduction in the pure-dephasing rate. Thus, a change in the protein amino acid sequence produces a change in the coupling of the protein fluctuations to the CO ligand bound at the active site. These results strongly support the model that CO pure dephasing is caused by coupling to the protein fluctuations, because a change in the protein with no change in the solvent produces a substantial change in the rate of pure dephasing.

In the fluctuating-electric-field dephasing mechanism, protein motions result in motions of the amino acids. The dynamics of these amino acids, particularly the polar ones, produce fluctuating electric fields at the CO bound to the active site. The fluctuating fields generate time-dependent variations in the CO vibrational frequency via the Stark effect. The fluctuating vibrational frequencies cause the vibrational pure dephasing measured by the vibrational echo.

X-ray crystallographic data of the H64V mutant show that the equilibrium structure of the protein is not significantly different from that of Mb (22). Therefore, it is unlikely that there is a significant change in the global dynamics of the protein. Thus, the functional form of the temperature dependence of the H64V-CO vibrational pure dephasing is unchanged from that of Mb-CO, because it reflects the spectrum of protein fluctuations that are coupled to the CO. However, the strength of the coupling of the protein fluctuations to the CO is reduced because one of the closest sources of the fluctuating electric field has been removed.

As a further test of the electric field model, another mutant, H93G (N-methyl-imidazole) [H93G(N-MeIm)]-CO (71, 72) was studied (7). The heme at the active site of Mb has only one covalent linkage to the protein, the iron-H93 bond. H93 is contained in the F α-helix of the globin. In the mutant, the proximal H93 is replaced by glycine, leaving a cavity on the proximal side of the heme (23). Many exogenous ligands (L) to the heme iron, such as imidazoles and pyridines, can be substituted into this cavity, producing a series of proteins—H93G(L)s (71). Although these proteins retain a covalent linkage between the heme iron and the proximal ligand, the covalent connection to the protein backbone is severed. When N-methyl-imidazole is used as the exogenous proximal ligand, the hydrogen bond between the proton on the imidazole imino-nitrogen and the hydroxyl group on serine 92 is also absent (72). As with native Mb, the open heme coordination site on the distal side binds biologically important diatomic molecules such as O_2, CO, and NO. The H93G(N-MeIm) protein has been shown to have a structure very similar to that of native Mb (7, 72). The CO transition frequency and room temperature vibrational CO lifetime of the mutant are almost identical to the native value (72–74).

H93G(N-MeIm)-CO has the only covalent bond between the active site and the protein removed. However, N-methylimodazole is essentially the side group as histidine. Therefore, the electrostatic environment is unchanged although the mechanical linkage to the protein is broken, removing the ability of local protein-conformational fluctuations to move the proximal histidine. Such motions could push and pull the Fe in and out of the plane of the heme ring system, changing the CO vibrational frequency and, therefore, causing pure dephasing.

Figure 6 displays temperature dependent, pure dephasing line widths for H93G(N-MeIm)-CO and Mb-CO. Within experimental error, eliminating the covalent bond to the Fe but leaving the electrostatic nature of the protein unchanged does not influence the magnitude or the temperature dependence of the pure dephasing. These vibrational echo measurements give further support to the fluctuating global electric field mechanism. When the polar histidine is replaced by nonpolar valine, H64V-CO, the pure dephasing is reduced (see Figure 5) because one contributor to the fluctuating electric field is gone. The H93G(N-MeIm)-CO experiments show that pure dephasing does not arise from local motions of the Fe-CO caused by movement of H93, but rather that fluctuating electric fields are the dominate cause of pure dephasing in Mb-CO.

Temperature Dependence and the Protein-Glass Transition

Figure 7 shows the pure-dephasing contribution to the line width, $1/\pi T_E^*$, vs temperature on a log plot for Mb-CO in trehalose (10). Trehalose is a glassy solid at all of the experimental temperatures. On a log plot, a power law is a straight line. As can be seen in Figure 7, between 11 and ~200 K, the functional form of the data is a power law,

$$\frac{1}{\pi T_E^*} = aT^{1.3}, \qquad 8.$$

where the prefactor $a = 3.5 \times 10^7 \pm 0.1 \times 10^7$ Hz/(degree K)$^{1.3}$. The error bar on the power law exponent is ± 0.1. This power law is identical to the one in Figure 4. However, the trehalose data are for a much broader range of temperatures and leave little doubt as to the functional form of the data. The same power law was observed in ethylene glycol-water (see below).

In Figure 7, there is a change in the functional form of the data at ~200 K. The points above ~200 K can be fit with

$$\frac{1}{\pi T_E^*} = 3.3 \times 10^{12} e^{\frac{-650}{k_B T}} \text{ Hz}, \qquad 9.$$

where k_B is Boltzmann's constant, $k_B T$ has units of cm^{-1}, and the error bars on the prefactor and activation energy are $\pm 0.2 \times 10^{12}$ Hz and ± 25 cm^{-1}, respectively.

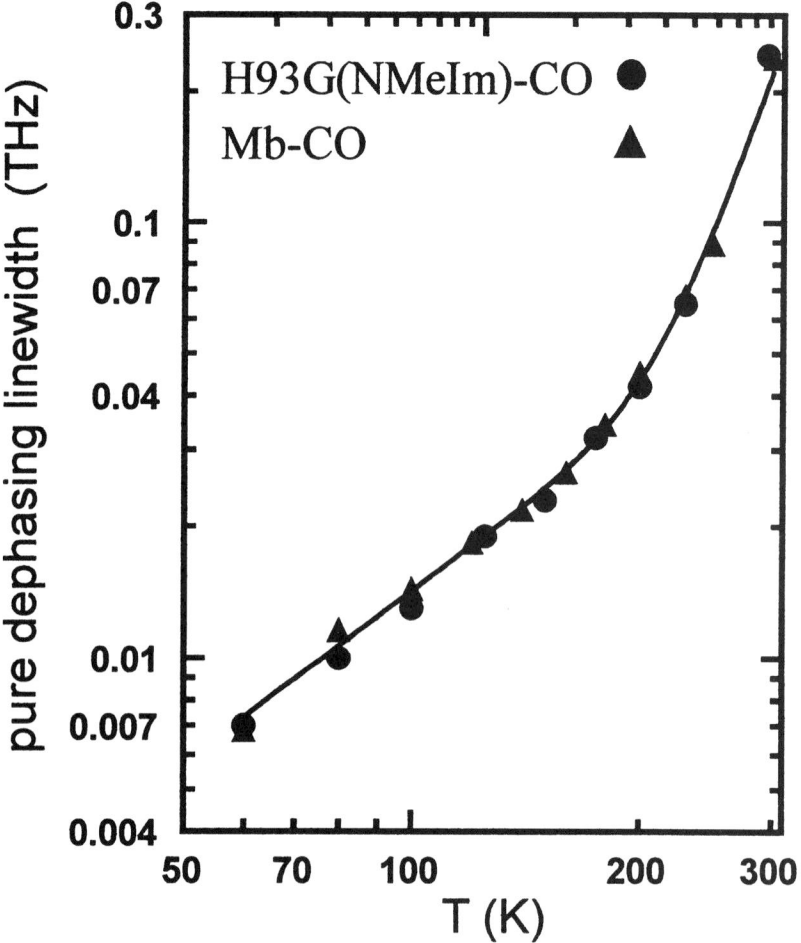

Figure 6 A comparison of the vibrational echo pure-dephasing line width ($1/\pi T_E^*$) data vs temperature for myoglobin Mb-CO and H93G (*N*-methylimidazole)-CO (the proximal histidine is replaced by glycine, and *N*-methylimidazole is the exogenous ligand) in glycerol-water. Within experimental error, the two data sets are identical. Breaking the only covalent bond between the heme and the protein does not change the pure dephasing when the electrostatic environment is unchanged because of the substitution of the histidine side chain with the equivalent *N*-methylimidazole.

It is clear that there is a change in the functional form of the temperature dependence at ∼200 K. However, it is important to emphasize that the form of Equation 9 is not unique, given the small number of points. A very good fit is obtained if the data are fit to a power law plus a VTF-type equation (63–65). A VTF equation describes many processes, such as viscosity in super-cooled liquids as they approach the

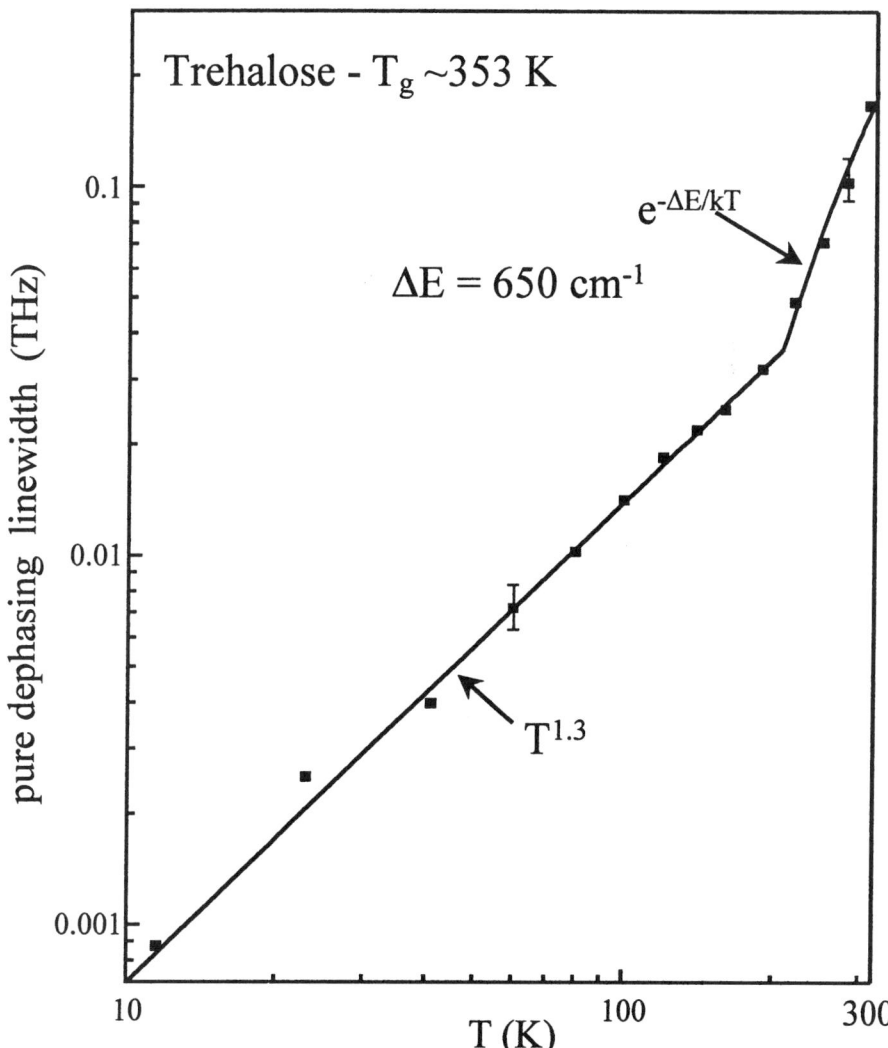

Figure 7 Temperature-dependent pure-dephasing line width ($1/\pi T_E^*$) of Mb-CO in trehalose. Trehalose is a glass at all of the experimental temperatures shown. The data are fit with a power law $T^{1.3}$ below ~200 K and with an exponentially activated process ($\Delta E = 650\,\text{cm}^{-1}$) at >200 K. The $T^{1.3}$ temperature dependence is indicative of glasslike behavior, and the break in the temperature dependence may indicate the "protein-glass transition" at ~200 K.

glass transition temperature. The VTF equation for the dynamic line width is

$$\frac{1}{\pi T_E^*} = b\mathrm{e}^{\frac{-E}{k_\mathrm{B}(T-T_0)}}.\qquad 10.$$

For a true glass-forming liquid, T_0 is the "ideal" glass transition temperature. It typically has a value a few tens of degrees below the laboratory T_g (63–65). A fit to the data with the combination of Equation 8 and Equation 10 yields a T_0 of ~180 K and an E corresponding to a temperature of ~230 K. These parameters can vary somewhat about the given values because of the wide range of fits that can be achieved when fitting four points with three parameters. However, the power law is always identical, independent of the form used to fit the points above ~200 K. If the exponential fit and the VTF fit are extended to higher temperatures, they do not become distinguishable below 500 K. Therefore, experiments at temperatures below the Mb denaturation temperature cannot distinguish these two forms. Regardless of the form that is used to fit the data, it is clear that there is a sudden change in the nature of the temperature dependence of the pure dephasing at ~200 K, and below ~200 K, the temperature dependence is $T^{1.3}$.

It has been observed in a wide variety of low-temperature glasses, far below T_g, that the temperature dependence of optical pure dephasing of electronic transitions, measured by photon echo and hole-burning experiments, is a power law and that the typical power law is $T^{1.3}$ (18, 75–78). Heat capacities of low-temperature glass also display power law temperature dependence, again with the value of the exponent being somewhat greater than 1 (53, 79).

Power law temperature dependences of heat capacities and optical dephasing, as well as other observable properties have been explained in terms of the tunneling TLS model of glasses (52, 80). Generally, the TLS model is invoked only below a few K. A power law temperature dependence of the vibrational pure-dephasing line width was observed for a solute in an organic glass up to ~20 K (9). TLSs in glasses arise from slight differences in local structures. The complex structural energy landscape is modeled as double wells having a broad distribution of energy differences between the two sides of the double well. The dynamics in glasses at low temperatures are caused by phonon-assisted tunneling among local structures modeled as transitions between the two sides of the double wells.

In the Mb-CO vibrational dephasing, the $T^{1.3}$ temperature dependence is observed at much higher temperatures than in true glasses. One possible explanation of the power law temperature dependence is thermally assisted tunneling among slightly different protein configurations. Small internal protein structural changes might be described in terms of protein TLSs (5, 10, 62). The protein TLSs are akin to the TLSs of very low-temperature glasses except that the protein energy landscape would have to be such that tunneling is the dominant process, even at temperatures of ≤200 K. If this is the case, the same statistical mechanics used to describe the low-temperature (~1 K) optical dephasing of electronic transitions of chromophores in low-temperature glasses (19, 81) can be used to describe the protein TLS-induced vibrational dephasing of Mb-CO at much higher temperatures

(~100 K). Alternatively, the power law temperature dependence could arise from activation over barriers rather than tunneling, if there is the appropriate energy landscape to provide the necessary broad distribution of activation energies (82, 83). In either case, the $T^{1.3}$ temperature-dependent pure dephasing can occur because of motion on a broad protein energy landscape in a manner analogous to dynamics in true glasses. Although an entirely different mechanism cannot be ruled out, the similarity of the vibrational echo data for Mb-CO to dephasing and other measurements on low-temperature glasses is suggestive of a type of glasslike state of the protein.

Switching from a power law to an activated process or the power law to the VTF function is appropriate if there is a transition in the fundamental nature of the dynamics. There has been considerable discussion in the literature concerning a "protein-glass transition." A glass transition for a bulk material is recognized to be dynamical in nature (64). Below T_g, the system can no longer interconvert on a reasonable time scale among the full range of structural configurations accessed in the liquid. Previous temperature-dependent experiments on Mb-CO have revealed a type of dynamical transition in the protein at ~200 K that has been referred to as a protein-glass transition (84–86). This is not a glass transition in the normal sense because a transition from a liquid to a glass as temperature is decreased is a phenomenon associated with a bulk material, whereas a protein is a single molecule. However, its complexity is so great that it undergoes continual structural evolution among a vast number of configurations. Below the protein T_g (T_g^P), sampling of protein configurations is greatly slowed. Thus, like a true liquid-glass, a protein may undergo a type of dynamical transition.

The Mb dynamics near ~200 K, which suggest a protein-glass transition, have been the subject of a considerable number of investigations. For example, inelastic neutron scattering of hydrated Mb at <180 K measures only vibrational motion, and at >180 K there is a dynamical transition, which is interpreted as the onset of torsional jumps between states (87, 88). Molecular dynamics simulations of the torsional transitions of the dihedral angles of Mb in water show that the anharmonic mean square displacements change at 200 K, which is indicative of a glasslike transition (89). In addition, IR (90–92) and visible (93) transition frequencies, dielectric relaxation (90), specific heat of water in Mb crystals (90), and Mössbauer spectra of ^{57}Fe-Mb (94) all show breaks near 200 K. Some experiments have suggested that this is a "slaved" glass transition; that is, the protein undergoes a transition induced by the true glass transition of the solvent (84–86).

The vibrational echo data presented in Figure 7 may be the strongest evidence for a protein-glass transition at ~200 K. The solvent is a glass. Therefore, the break in the functional form of the dynamics cannot be ascribed to a change in the nature of the solvent dynamics (see below). The $T^{1.3}$ temperature dependence observed at <200 K is the typical temperature dependence observed for photon echo-dephasing measurements, heat capacities, and other experiments in true glasses at low temperatures. If the power law pure vibrational-dephasing temperature dependence below ~200 K is caused by motion on a glasslike energy landscape, then data above ~200 K could result from activation above the top of the landscape (64).

Solvent Viscosity and Protein Dynamics

In the experiments with Mb-CO in trehalose, the solvent is a glass at all temperatures studied (10–310 K). Therefore, the dephasing arises from the protein dynamics with a rigid, essentially infinite viscosity solvent. Figure 8 displays data taken in a 50:50 (vol/vol) mixture of ethylene glycol-water (EgOH:H$_2$O) along with the trehalose data displayed in Figure 7 (10, 47). The EgOH:H$_2$O sample has a glass transition at ~140 K and a rapidly decreasing viscosity as temperature increases. The dephasing rates in the two solvents are identical ($T^{1.3}$) below ~150 K, at which temperature both solvents have extremely high or infinite viscosity. In contrast, the dynamics are dramatically different above ~150 K, where

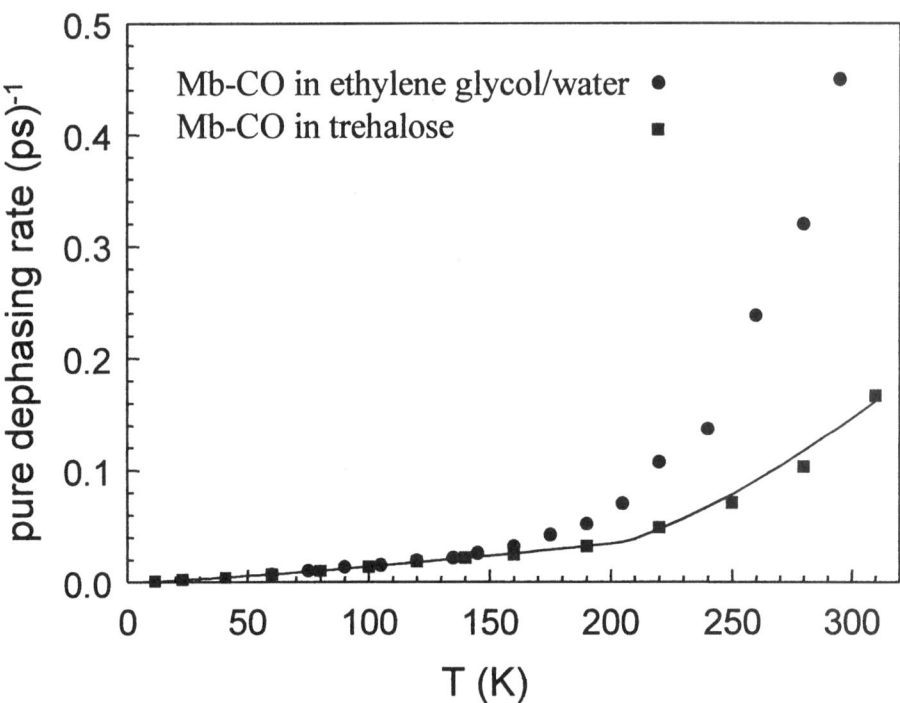

Figure 8 Temperature-dependent pure-dephasing rates ($1/T_E^*$) of Mb-CO in trehalose (*boxes*) and 50:50 ethylene glycol-water [EgOH:H$_2$O (*circles*)]. The line through the trehalose data is the same as that shown in Figure 7. Trehalose is a glass at all of the experimental temperatures. The EgOH:H$_2$O with protein solution goes through its glass transition at ~140 K. The data in the two solvents have the identical $T^{1.3}$ temperature dependence at low temperatures, at which both solvents are glasses. Once EgOH:H$_2$O becomes a liquid, the Mb-CO pure-dephasing rate in this solvent increased rapidly with temperature, compared with the rate measured in trehalose.

the viscosities of the solvents differ substantially. With increasing temperature, the pure-dephasing rate of Mb-CO in the fluid EgOH:H_2O increases much more rapidly than it does in the solid trehalose. The additional dephasing in EgOH:H_2O is consistent with a solvent viscosity effect. However, the viscosity of EgOH:H_2O changes many orders of magnitude between 150 and 295 K, whereas pure dephasing changes only somewhat more than 1 order of magnitude. Given the large change in viscosity and the comparatively small change in the rate of pure dephasing, the relationship between the pure dephasing and solvent viscosity is not immediately apparent.

As discussed below, a comparison of the solvent-dependent and temperature-dependent data shows that the change in the pure dephasing rate with temperature is caused by a combination of viscosity dependence and pure temperature dependence. The dephasing in glassy trehalose (infinite viscosity) is due only to pure temperature-dependent processes. When this contribution is removed from data in fluid solvents, the remaining "reduced" dephasing rate is dependent on only the viscosity of the solvent, whether the solvent viscosity is varied by changing the temperature of a single solvent or by changing the solvent composition at a fixed temperature.

A viscoelastic continuum model is presented to analyze the data (47). The key idea in this model is that the internal dynamics of the protein are strongly constrained by the ability of the protein's surface to move (10, 47). When the protein is embedded in a disordered glass, a variety of surface topologies occur, but each protein molecule is fixed (or nearly fixed) with a single surface topology. Internal fluctuations are restricted to those protein motions that involve little motion at the surface. The vibrational echo decay time of the CO, which is sensitive to the magnitude and rate of the internal structural fluctuations of an individual protein molecule, is slow. When the protein is in a liquid, its surface can move more freely, allowing protein molecules to change their surface topologies. However, just above the glass transition temperature, the solvent is very viscous, and the surface motion is very slow. Internal dynamics connected with surface motion are also slow, and their effect on the echo decay rate remains small. Other dephasing processes that occur in the glass, namely, the vibrational lifetime and the temperature-induced dephasing, overwhelm the contribution of the structural fluctuations enhanced by surface motions, even though the solvent is a liquid. As the solvent viscosity is decreased further (and the temperature is raised), more rapid surface fluctuations permit faster internal protein structural fluctuations. This viscosity-dependent contribution to the echo decay becomes observable and is increasingly important as the viscosity is decreased.

In the model, the protein is taken to be a compressible breathing sphere with surface motions that are constrained by the viscoelastic properties of the solvent. Both the magnitude and correlation time of the thermal fluctuations of the protein's size are calculated. Taking the motions of the surface to be linearly coupled to the CO frequency, the echo decay time is calculated as a function of the solvent viscosity and temperature. A detailed analysis of the properties of the echo experiment in the presence of slow modulation of the transition frequency (54), that is, spectral diffusion, which is briefly discussed above, is essential to applying the viscoelastic model.

Figure 9 displays isothermal pure-dephasing data as a function of the solvent viscosity (47). The viscosities are varied by changing the solvent. The compositions of the mixtures are given earlier in this review. From Figure 9, it is clear that the Mb-CO pure dephasing is sensitive to the solvent viscosity. These data and the data taken in trehalose support the idea that there is both a temperature-dependent contribution to the dephasing, which does not depend on changing viscosity, and a viscosity dependence that does not depend on changing temperature.

It is possible to quantify the influence of changing viscosity on pure dephasing. At room temperature, the trehalose sample displays significant pure dephasing although the viscosity is essentially infinite. Thus, the pure-dephasing rate in trehalose represents the infinite-viscosity point. The pure dephasing in this sample is due only to temperature-induced structural fluctuations of the protein that do not require participation of solvent motion to any significant extent. To obtain the viscosity dependence, the room temperature infinite-viscosity, pure-dephasing rate, that is, the rate in trehalose, is subtracted from the pure-dephasing rates measured

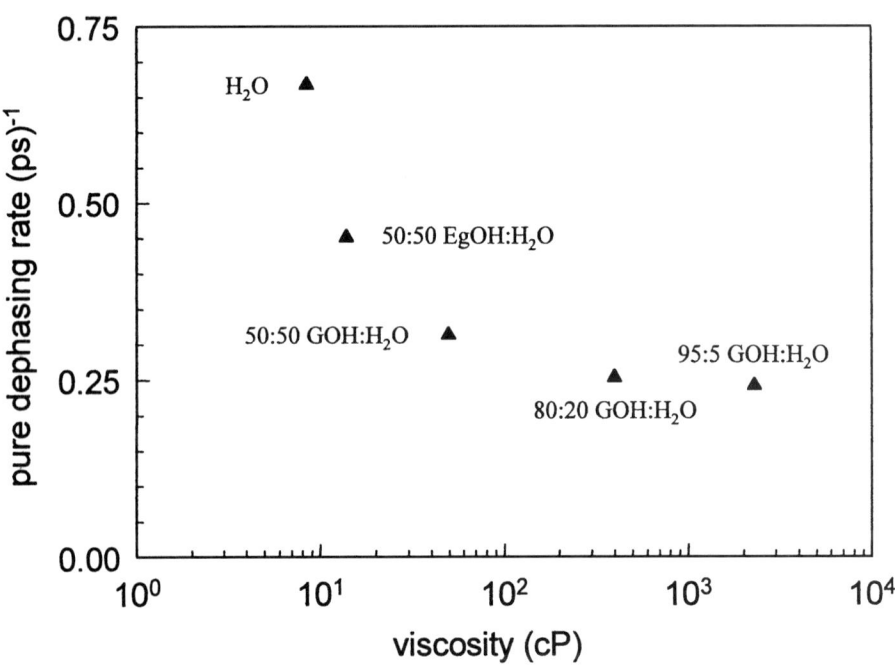

Figure 9 Isothermal (295 K) viscosity-dependent Mb-CO pure-dephasing rates ($1/T_E^*$) in solvents of various compositions (see text). The data demonstrate that the Mb-CO pure dephasing is significantly dependent on viscosity at constant temperature. EgOH, ethylene glycol; GOH, glycerol.

at finite viscosities,

$$\frac{1}{T_E^r(\eta)} = \frac{1}{T_E(\eta, T)} - \frac{1}{T_E(\eta = \infty, T)}, \qquad 11.$$

where T_E^r is the reduced pure dephasing time.

The model in Equation 11 assumes that the temperature-dependent contribution to pure dephasing at infinite viscosity and the viscosity-dependent contribution are additive. The additivity feature of the model can be tested experimentally. Figure 10 is a plot of the reduced isothermal-viscosity-dependent data from Figure 9 (47). The room temperature, infinite-viscosity, pure-dephasing rate was subtracted using Equation 11. Also shown in Figure 10 are data obtained as a function of temperature in 50:50 (vol/vol) EgOH:H_2O (Figure 8). For each EgOH:H_2O

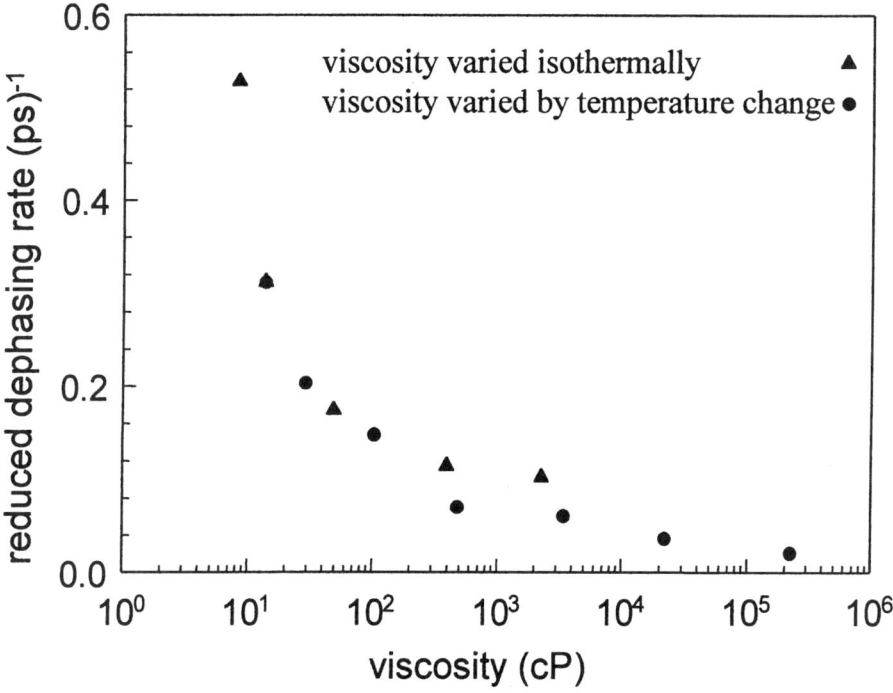

Figure 10 Pure-dephasing rates at a given viscosity minus the 295 K rate at infinite viscosity (see Equation 11). *Triangles*, isothermal (295 K) viscosity-dependent Mb-CO reduced pure-dephasing rates; *circles*, temperature-dependent Mb-CO in 50:50 ethylene glycol-water (EgOH:H_2O) pure-dephasing rates minus the corresponding infinite-viscosity (trehalose data) pure-dephasing rates at each temperature plotted against the viscosity of the EgOH:H_2O protein solution. The *circles* are the viscosity-induced part of the pure dephasing rate at various temperatures. (Note that the point at ~10 cP is an *overlapping circle and triangle*.). The fact that the circles and the triangles are intermixed demonstrates that the viscosity component and the pure-temperature component of the pure-dephasing rates are additive within experimental error.

point, the pure-dephasing rate in trehalose was subtracted. For the points between room temperature and 210 K, the viscosity is known (47), and the differences between the EgOH:H$_2$O data and the trehalose data are plotted at the appropriate viscosity. Notice that, within experimental error, the isothermal-viscosity points and the viscosity points obtained at various temperatures are intermingled. The trend is the same. Therefore, the infinite-viscosity, pure-dephasing rate and the viscosity-dependent, pure-dephasing rate are additive within experimental error.

The viscosity dependence of the experimental data displays two qualitative features. First, the reduced dephasing rate is almost entirely determined by the solvent viscosity. The detailed chemical properties of the solvent do not make much difference as shown by the isothermal data taken in a variety of solvents (Figure 9). The apparent difference in the temperature dependence between EgOH:H$_2$O and trehalose (Figure 8) is almost entirely caused by the temperature dependence of the viscosity (Figure 10). It is not obvious that the CO pure dephasing should have a strong sensitivity to the solvent viscosity. The CO is located internally in Mb, precluding a direct interaction between the CO and the bulk solvent. The dephasing is not sensitive to the particular solvent when the solvents are all glasses. The frequency of the Mb-CO stretch is insensitive to the solvent, demonstrating that there is no direct interaction of the solvent with the CO.

The second qualitative result is that the viscosity dependence of the dephasing rate is relatively weak. The data summarized in Figure 10 cover changes of viscosity of >5 orders of magnitude, but the reduced dephasing rate changed by $\sim$1.5 orders of magnitude.

These facts can be explained using the following model (47). As discussed above, the change in the frequency, ω, of the CO is directly proportional to the electric field, E, at its site owing to a vibrational Stark shift (47, 68). Fluctuations in the instantaneous configuration of the protein cause fluctuations in this electric field, δE. Taking $\omega°$ as the frequency at the time-averaged protein configuration, the time-dependent CO frequency is calculated as

$$\omega(t) = \omega° + \frac{\delta \mu_{01} \delta E(t)}{\hbar}. \qquad 11.$$

The value of the change in dipole moment from the ground to first excited vibrational level, $\delta_{01} = 0.14 \, \text{Debye} = 2.4 \, \text{cm}^{-1}/(\text{MV/cm})$, has been measured recently (68). When the protein is in a solid, glassy solvent, the surface topology of the protein is essentially fixed. The solvent resists any shearing motions, so the surface of the protein can change only by elastic distortions of the glassy solvent. Because the compressibility of a glass is very low, surface motions of the protein are severely limited. If the internal motions of the protein are strongly coupled to the motion of the surface, the internal protein dynamics are also tightly constrained. The protein can still undergo internal structural fluctuations, but only those fluctuations are permitted that do not move the protein's surface very far. These motions, as sensed by CO bound at the active site of Mb, are reflected in the temperature-dependent, pure-dephasing measured in trehalose. The only feature of the solvent that is important is that it keeps the protein surface fixed. Thus, in EgOH:H$_2$O below its T_g, the behavior of Mb-CO is identical to its behavior in trehalose. The pure dephasing of Mb-CO in glycerol-water is also identical at temperatures below that solvent's glass transition (10; see Figure 4).

As the temperature is increased above T_g, a new contribution to the Mb-CO echo decay measurements comes into play. When the solvent is a fluid, the range of motion of the protein's surface and, as a result, the range of its internal motions are much greater. The additional amplitude of the protein motion increases the magnitude of dynamic fluctuations in the electric field at the CO bound in the interior of the protein and thus in the magnitude of dynamic fluctuations in the CO vibrational frequency. However, the viscosity of the solvent determines the rate of this increased motion. Just above T_g, the liquid is extremely viscous, and the increased protein motions are very slow. The effect on the CO echo decay is correspondingly weak. The increased dephasing caused by motions allowed by the solvent is undetectable over the other contributions to the dephasing rate produced by the solvent-independent processeses. That is, just above T_g, the solvent-induced dephasing is still in the quasistatic limit and is eliminated by the echo experiment. In EgOH:H$_2$O, this is the situation from ~136 to 150 K. The dephasing rate remains almost identical to that in the true glass, trehalose (Figure 8).

As the temperature is increased further, the viscosity of the solvent drops rapidly. At some point, the protein-structural fluctuations dependent on moving its surface become fast enough to have a measurable contribution to the echo decay. The system enters the spectral-diffusion regimen. Further increases in temperature reduce the solvent viscosity and increase the internal protein fluctuation rate by many orders of magnitude. The echo decay rate also increases. However, because of the weak power law connecting the fluctuation time, τ_m, and the echo decay time in the spectral-diffusion region (Equation 6), the echo decay rate increases only weakly with decreasing viscosity, a little over 1 order of magnitude. This region covers ~150–295 K in EgOH:H$_2$O (Figure 8).

The model gives a good qualitative account of the dephasing rate vs temperature in EgOH:H$_2$O and in trehalose (47). Moreover, it explains how the CO vibration can be sensitive to the solvent viscosity without any direct coupling between the solvent and the CO. The disparity in the magnitude of change in the viscosity and the echo decay rate is also accounted for.

A recently developed viscoelastic-continuum theory of solvent dynamics (95–97) can be used to make the model quantitative (47). The protein is modeled as a sphere of radius r_p embedded in a viscoelastic and continuous solvent (47). The solvent's viscoelastic behavior is characterized by a decaying shear modulus $G(t)$. For sufficiently short periods, the solvent behaves as a solid with a short-time (infinite-frequency) shear modulus G_∞. At times comparable to or longer than the decay time of $G(t)$, the solvent can flow, allowing additional fluctuation of the protein's surface. This decay time is directly related to the solvent viscosity. The protein is also treated as a continuous material with a bulk modulus K_p. Viscous relaxation or "flow" of the protein in response to its distortion is irrelevant and not included. With these assumptions, the protein will almost instantaneously transmit changes in its surface to its interior.

Figure 10 shows that the temperature-induced dephasing and the viscosity-induced dephasing are separable within experimental error. In the model, it is assumed that the corresponding protein motions with a frozen surface and the additional motion allowed by moving the surface can be treated as independent processes. The two types of motions need not involve different protein coordinates. Both processes may involve the same internal motions of the protein. The freeing of the surface only increases the amplitude of the motion along the same coordinates that cause the viscosity-independent dephasing. It is also possible that the protein has conformational changes that relax much more slowly than the solvent. Any change in the CO frequency due to these conformations will be in the quasistatic limit and will not contribute to the vibrational echo decay. The viscoelastic model treats only the viscosity-dependent dephasing reflected in the reduced dephasing time $T_E^r(\eta)$.

The protein's fluctuations are fully characterized by its change in radius δr_p (47). Changes in radius are linked to changes in the electric field at the CO, because of displacements of charged groups within the protein. The details of the coupling are not treated. Rather, a phenomenological proportionality constant b is used, and

$$\delta E(t) = b \delta r_p(t). \qquad 12.$$

The viscosity-dependent protein dynamics in the model (47) are now equivalent to the structural dynamics in other viscoelastic models (95–97). The magnitude of the thermal fluctuations in the protein radius is given by

$$\langle \delta r_p^2 \rangle = \frac{kT}{12\pi K_p r_p}, \qquad 13.$$

where k is the Boltzmann constant. Combining Equations 11, 12, and 13, the magnitude of the CO frequency modulation is

$$\Delta_m = \langle (\omega(0) - \omega^\circ)^2 \rangle^{1/2} = \frac{b\delta\mu}{\hbar\sqrt{4\pi}} \left(\frac{kT}{3K_p r_p} \right)^{1/2} = \Delta_0 \left(\frac{T}{T_0} \right)^{1/2}, \qquad 14.$$

where $\omega(0)$ is the frequency at $t = 0$. The most important feature of Equation 14 is the temperature dependence of Δ_m. The temperature-independent constants are collected into Δ_0, the modulation amplitude at the reference temperature T_0. Here, $T_0 = 295$ K.

The relaxation time of the thermal fluctuations is proportional to the solvent viscosity, with the proportionality constant determined by the solvent and protein moduli, as follows:

$$\tau_m = \alpha \frac{\eta}{G_\infty}, \qquad 15.$$

and

$$\alpha = 1 + \frac{4G_\infty}{3K_p}. \qquad 16.$$

As an estimate, $K_p \approx K_\infty$, and the Cauchy relation for simple solids can be used (98), such that $G_\infty = (3/5)K_\infty$, giving $\alpha \approx 9/5$. Using the value for water (99) gives an estimate $G_\infty \approx 10 \times 10^{10}$ dyne/cm^2. Over the experimental range of viscosities of $\sim 10^1$–10^6 cP, modulation times in the approximate range 2 ps–200 ns are expected.

The echo response can be calculated using Equations 14–16 under a wide variety of conditions. Given the long modulation times predicted, it is reasonable to take the system to be in the spectral-diffusion regime $\tau_m \Delta_m \gg 1$. Using Equations 4, 5, 14, and 15, gives

$$T_E \sqrt{\frac{T}{T_0}} = C \left(\eta(T) \sqrt{\frac{T}{T_0}} \right)^{\frac{\beta}{\beta+2}}, \qquad 17.$$

and

$$C = B_\beta \Gamma \left(\frac{1}{\beta+2} \right) \left(\frac{\alpha^\beta}{G_\infty^\beta \Delta_0^2} \right)^{\frac{1}{\beta+2}}. \qquad 18.$$

The dominant source of temperature dependence in Equation 17 is the viscosity, which has a steep VFT (63–65) dependence on temperature. The constants in Equation 18 can be approximated as temperature independent. The moduli are actually temperature dependent, but this dependence is relatively weak (100). In Equation 17, other than the viscosity, the only temperature dependence is the explicit square root of the temperature, which is weak.

Thus, the viscoelastic theory explains why the solvent viscosity is the dominant factor in determining the dephasing rate (47). Furthermore, for reasonable values of β, that is, near unity, the dephasing rate varies with the viscosity raised to a small fractional power. Again, the theory is in accord with the relatively weak dependence on viscosity observed in the experiments.

The viscoelastic theory was tested quantitatively as shown in Figures 11 and 12 (47). In Figure 11, both the isothermal and temperature-dependent data are included. Equation 17 predicts a linear relationship on this log plot, with a positive slope. If the system were in the fast modulation limit, then the slope would be negative, and if there were a shift between the fast modulation limit (motional narrowing) and the slow modulation limit (spectral diffusion) as the viscosity changes, then there would be a change in the sign of the slope and a minimum in the curve (54). The slope remains positive over the entire viscosity range, so the assumption that the system is in the spectral diffusion regime is valid.

The line in Figure 11 has the slope predicted by Equation 17 for $\beta = 1$, which in turn corresponds to a cube-root dependence of the dephasing rate on the viscosity. The line shows quite good agreement with the data. The points at the highest viscosities fall off the line, but these are the points with the greatest uncertainty in the viscosities. $\beta = 1$ corresponds to an exponential decay of the CO frequency-frequency correlation function and, consequently, of the solvent shear modulus

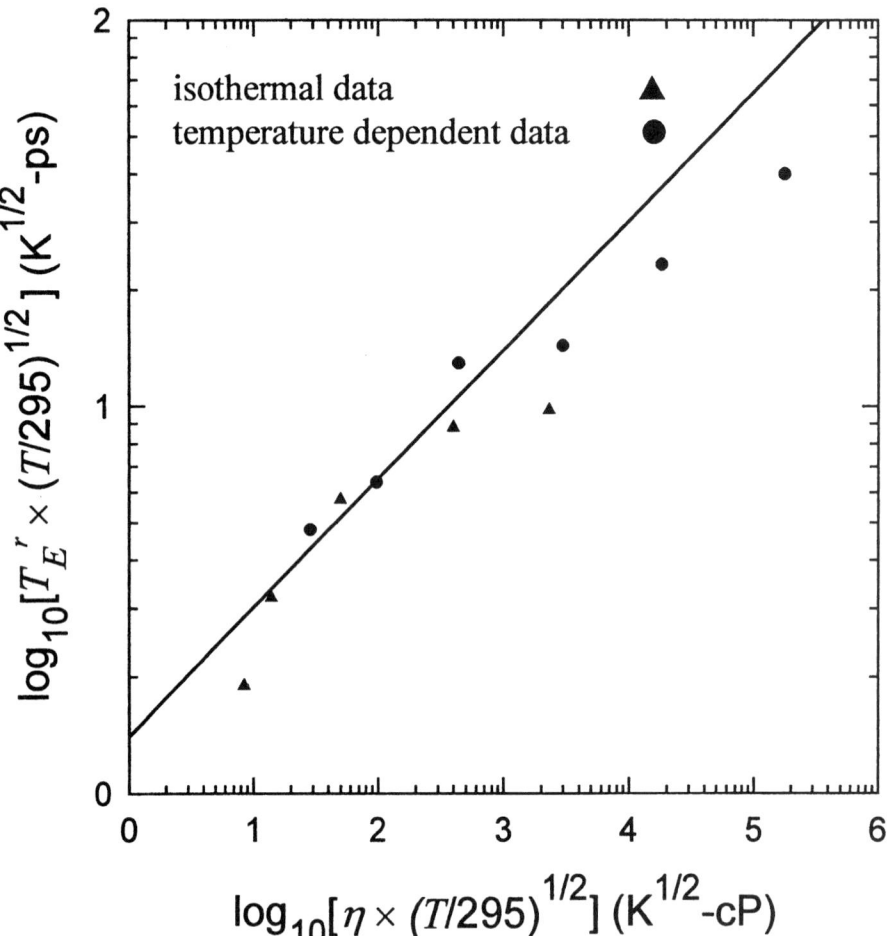

Figure 11 The reduced-echo pure-dephasing time from both the isothermal (*triangles*) and temperature-dependent (*circles*) measurements are plotted against viscosity in accord with Equation 17. The line corresponds to the cube root dependence on viscosity predicted for an exponentially relaxing shear modulus ($\beta = 1$). The line also corresponds to the fit shown in Figure 12.

(54). It is common for shear modulus relaxation functions to become nonexponential at high viscosity, which would lead to a smaller value of β. This effect could also contribute to the deviations seen at the highest viscosities.

The solid line in Figure 12 displays the fit of the reduced dephasing rate from Figure 11, combined with the viscosity-independent dephasing data taken in trehalose (Equations 8 and 9) to recreate the full temperature-/viscosity-dependent dephasing rate in EgOH:H$_2$O (47). The theory does a remarkable job of reproducing the data qualitatively and essentially quantitatively. It misses the lowest

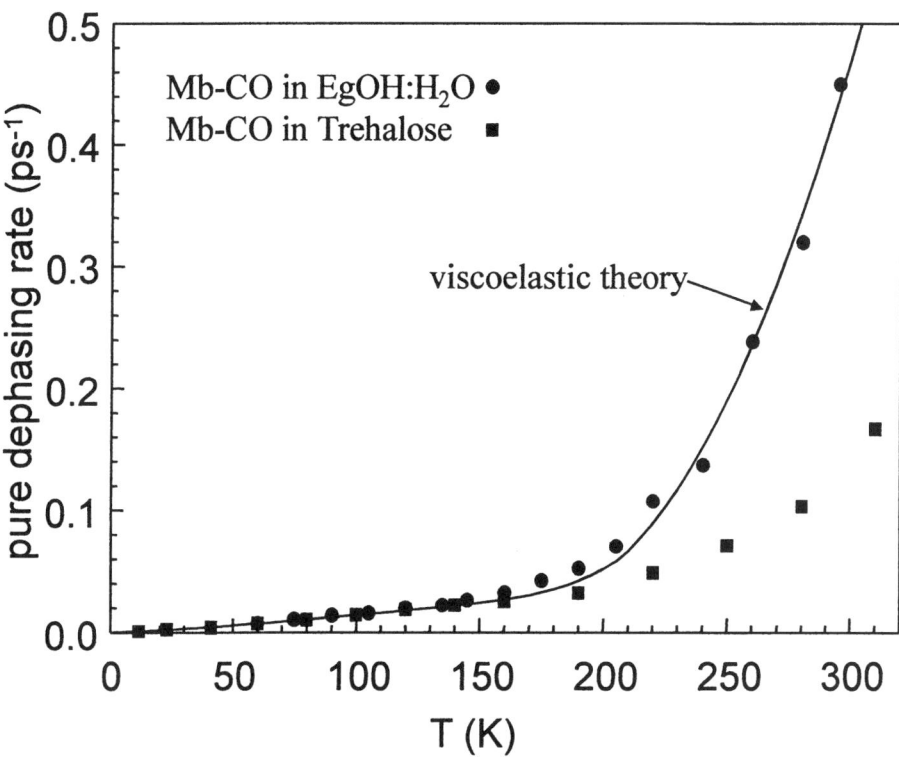

Figure 12 Temperature-dependent Mb-CO pure-dephasing rates ($1/T_E^*$) in trehalose (*squares*) and in 50:50 ethylene glycol-water (EgOH:H_2O; *circles*). The line through the EgOH:H_2O data is the fit to the viscoelastic theory (Equation 17) added to the temperature dependence in trehalose. The viscoelastic theory attributes the difference between the trehalose and EgOH:H_2O rates to fluctuations of the protein surface that are governed by the solvent viscosity. The theory does a remarkable job of reproducing the data. These results confirm the importance of the role of protein surface fluctuations in the internal protein structural fluctuations responsible for CO pure dephasing.

temperature points somewhat, possibly either for the reasons discussed above or because of an inherent limitation of the theory. The calculated curves in Figure 11 and Figure 12 are different methods of comparing theory and experiment.

The viscosity-dependent dephasing in EgOH:H_2O is active at temperatures below the putative protein-glass transition (see Figures 7 and 8). As discussed above, the Mb-CO pure-dephasing data in trehalose have a break at ~200 K that may be caused by the protein-glass transition (10). Although the viscosity-independent dephasing in trehalose shows a break, the additional viscosity-dependent dephasing that occurs in EgOH:H_2O does not show an observable break at the protein-glass transition. This observation is compatible with the viscoelastic theory. Within the theory, the response of the solvent is governed by its shear relaxation, but the protein responds only through its compressibility (J Jiang and MA Berg, personal communication). For typical liquids, the compressibility changes only ~20% at the glass transition (102, 103). If the change in protein compressibility at the protein-glass transition were similarly small, no observable break in the internal dynamics connected with surface motion would be expected. This conclusion is supported by computer simulations, which have shown that there is not a dramatic change in the dynamical behavior of Mb across the protein-glass transition (20, 21). Thus, the viscoelastic theory predicts that the viscosity-dependent dephasing mechanism should remain active below the protein-glass transition, as is observed experimentally. In general, the solvent viscosity can influence a protein's dynamics even below its glass transition. Another possible reason that viscosity-dependent dephasing in EgOH:H_2O is observed at <200 K is a shift in the protein-glass transition to a lower temperature in the liquid solvent (10). This idea is supported by the fact that, in a system such as a thin polymer film, which has a large surface-to-volume ratio in which the surface is free to move, the T_g shifts below that of the bulk material (104).

The fits in Figures 11 and 12 are very good considering that there is only one adjustable parameter, C (Equations 17 and 18). The fit gives $C = 1.4$ ps/cP$^{1/3}$ (47). Using Equations 14 and 18, the value of b corresponding to the fit can be obtained. With $\delta\mu_{01} = 0.14$ Debye, an average protein radius of 3.5 nm, and the previous estimate $\alpha \approx 9/5$, it is found that $b \approx 8.3 \times 10^3$ dyne/Debye.

The fit value of C can be used to place a bound on Δ_0 (47). Detailed analysis (47) gives 1.1 ps$^{-1} \leq \Delta_0 \leq 1.2$ ps$^{-1} = \Delta_I$ where Δ_I is the inhomogeneous line width. This is an important result. The process responsible for the viscosity-dependent spectral diffusion accounts for almost all of the IR line width. Therefore, the viscosity-dependent protein fluctuations responsible for the vibrational echo decay on a relatively fast time scale span a very broad range of time scales. The results suggest that there are no other distinctly different slow processes responsible for the inhomogeneous line width. Increasingly slower spectral diffusion of the same nature as that measured on fast time scales by the vibrational echo experiments accounts for essentially the entire absorption line.

Taking $\Delta_0 \approx \Delta_I$ along with Equation 18, the previous estimate of $\alpha \approx 9/5$ and the measured value of C gives a value for the solvent shear modulus: $G_\infty = 15 \times 10^{10}$ dyne/cm^2. For comparison, the value for pure water is shown as $G_\infty = 11 \times 10^{10}$ dyne/cm^2 (99). Given the errors in the experimental fit and in the estimated parameters, this inferred value of G_∞ is very reasonable. (Implicit in the analysis is an assumption that the value of this modulus is independent of solvent composition and temperature. Errors in this approximation can lead to some of the scatter around the fit.). The fit value of G_∞ then leads to an estimate of the protein modulation time in the various solvents through Equation 15, $\tau_m \approx (0.12$ ps/cP$)\eta$. Thus, a quantitative estimate of the viscosity-dependent modulation time is obtained from the vibrational echo experiments.

A number of imprecisely known factors contribute to a moderate level of uncertainty in all of the above estimates. However, they should be accurate enough to establish with some certainty the parameters involved in solvent-induced protein dynamics. The fact that all of these numbers are internally self-consistent as well as compatible with physical expectations strongly supports the model and the values of parameters obtained from the data analysis.

The viscosity-dependent vibrational echo results reveal a fundamentally important property of proteins. The internal structural fluctuations of proteins are intimately related to the surface fluctuations and the viscoelastic properties of the medium in which the protein is embedded. The data in the glassy solvent trehalose shows that a protein's structure fluctuates even when the surface topology of the protein is fixed. However, in a liquid solvent like ethylene glycol-water, structural fluctuations of the protein are greatly enhanced because internal evolution of the protein's structure requiring changes in the protein's surface topology is permitted.

The viscoelastic model fits the data using physically reasonable parameters. The viscosity-dependent echo decay time is proportional to $\eta^{1/3}$, where η is the solvent viscosity. This behavior is characteristic of dephasing caused by spectral diffusion (relatively slowly evolving CO frequency) with an exponential frequency-frequency correlation function (54). The entire experimental region lies within the spectral diffusion and quasistatic regions; that is, motional narrowing of the CO vibrational line is not observed. The success of the model in describing both the isothermal (solvent-dependent) viscosity data and the temperature-dependent viscosity data supports the underlying concept that the solvent plays an important role in protein structural dynamics through its influence on protein surface motions, which in turn have a substantial affect on the internal structural dynamics of the protein (47).

Vibrational Echo Experiments on Hemoglobin-CO

Human Hb A is a tetrameric protein composed of two α and two β subunits. The intrinsic differences in the subunits' ligand affinities and the interactions between them are necessary for the proper functioning of Hb (105). When isolated, the subunits exhibit different behavior from that of the $\alpha_2\beta_2$ tetramer (106). The CO stretch of carbonmonoxy Hb (Hb-CO) at room temperature displays a dominant peak, C_{III} at $\sim$1951 with a full width at half maximum of $\sim$8 cm^{-1}. From difference absorption spectroscopy, it is known that the subunits have slightly different absorbencies. The α and β subunits' maximum absorbencies are at 1950.5 cm^{-1} and 1951.6 cm^{-1}, respectively (106). This difference indicates that the CO ligands' environments are not identical in the two subunits. The Mb protein has three distinct CO-stretching modes, which have relative amplitudes that are sensitive to changes in pH, temperature, and amino acid sequence (107). In Mb-CO, the dominant line, A_1, is $\sim$75% of the integrated band area, whereas in Hb, the C_{III} line is >96% (106, 108).

The secondary and tertiary structures of Mb are similar to those of the α and β subunits of intact Hb (109). The similarities in structures are noteworthy because the primary sequences of the three chains are the same at only 24 of 141 positions

(31). Most of these differences are subtle changes of size or polarity, such as a leucine in Hb$_\alpha$ and Mb at F1, where a phenylalanine is found in Hb$_\beta$. Some differences are more dramatic, such as F7 in which a serine occurs in Mb but an alanine occurs in Hb$_\alpha$ and Hb$_\beta$ (31). These changes in amino acid polarity do not significantly change the secondary or tertiary structure of Hb. However, changes in the polarity will change the electrostatic nature of the protein.

Vibrational echo studies were performed on Hb-CO using an IR OPA pumped by a regenerative amplified Ti:Sapphire system rather than the FEL. The advent of Ti:Sapphire-based commercial equipment capable of generating tunable mid-IR pulses that are useful for performing detailed vibrational echo studies of proteins and other systems is expanding the application of vibrational echo methods to a wide variety of problems.

For the experiments on Hb-CO, the bandwidth of the IR pulses was limited to avoid pumping to higher vibrational levels of the CO mode (2, 110) and to have the bandwidth of the IR pulses approximately the same as the line width. The bandwidth is limited by placing a slit in the stretcher. The output of the regenerative amplifier has wings generated by the sharp cutoff of the bandwidth caused by the slit. However, the bandwidth is narrowed slightly, and the wings are eliminated by the use of a grating and four consecutive nonlinear processes in the OPA used to generate the IR. The IR pulses are 900 fs in duration and have nearly Gaussian shape as determined by IR autocorrelation. The IR bandwidth was $\sim$18 cm^{-1}. At 5 μm, the OPA typically produces 6–7 μJ/pulse at 1 kHz.

The experimental vibrational echo setup used with the OPA system is very similar to that used with the FEL (see Figure 2). However, single-pulse selection is not necessary since the Ti:Sapphire-OPA system has a repetition rate of 1 KHz. One of the beams was chopped at 500 Hz, and a lock-in amplifier was used in the detection. Pulse energies up to $\sim$4 μJ/pulse were available at the sample. Power studies were performed at high and low temperatures to ensure that there were no intensity or heating artifacts.

The sample was prepared from human Hb in a manner similar to that used for the Mb-CO samples (37). The concentration was $\sim$18 mmol in a 50:50 (vol/vol) ethylene glycol (EgOH)-phosphate buffer (pH 7) mixture. Thermal stability of proteins is not greatly affected by relatively high concentrations of EgOH (111–113), and concentrations $\leq$22 mol% are not thought to cause structural perturbations.

Fourier transform IR spectra were recorded as a function of temperature across the temperature range of the echo experiments. The room temperature peak was centered at 1951.8 cm^{-1} and had a width of 8.0 cm^{-1} and an absorbance of 0.5 on a 1.2 background. The peak center and width changed slightly and monotonically as the temperature was lowered. By 50 K, the peak center had shifted to 1949.3 cm^{-1}, and the width had grown to 8.1 cm^{-1}. The OPA wavelength was tuned to the absorption center at each temperature, so the α and β CO modes were both driven (37).

Figure 13 displays a vibrational echo decay measurement taken with the Ti:Sapphire/OPA system. These data are for Hb-CO in EgOH-H$_2$O, at 40 K. Also shown is an exponential fit to the decay data. The decay time is 11.0 ps, yielding

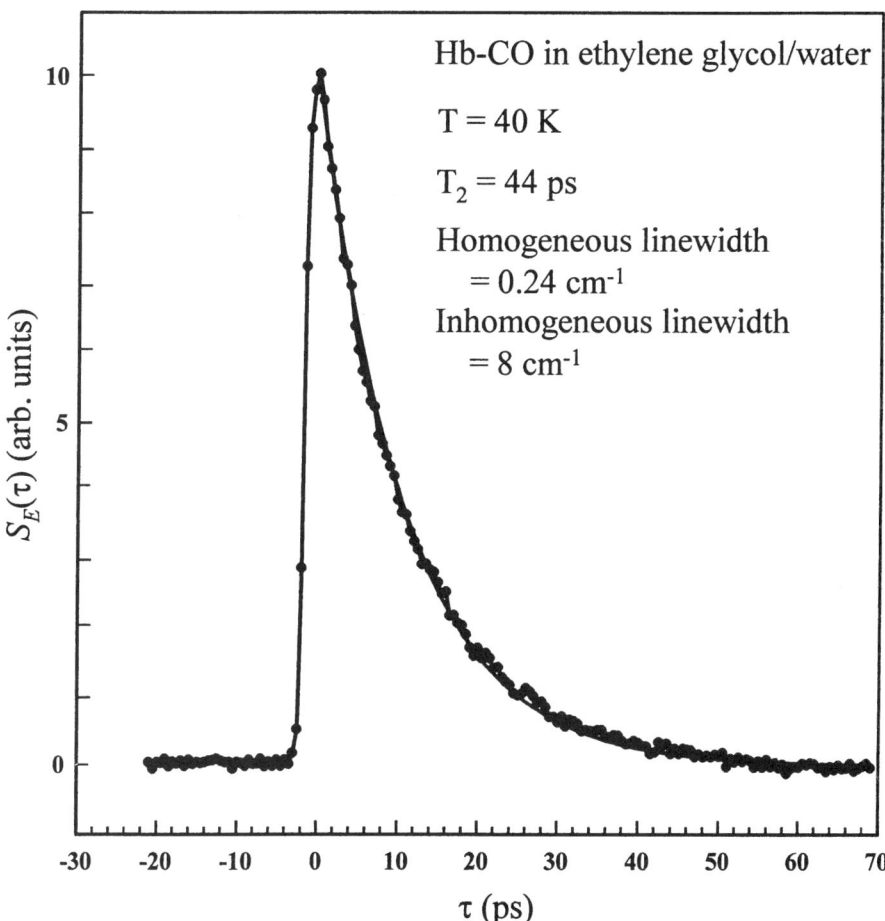

Figure 13 An example of vibrational echo data taken on Hb-CO in ethylene glycol-water and a single exponential fit to the data. The data were taken at 40 K. The decay time was 11.0 ps, giving a T_2 of 44.0 ps; the corresponding homogeneous line width was 0.24 cm^{-1}. The inhomogeneous line width was 8 cm^{-1}. The absorption line was inhomogeneously broadened at all temperatures studied.

a T_2 of 44 ps. These echo data were taken on a sample in which the protein had a very strong background absorption compared with that of the CO peak under study. Nonetheless, it is possible to obtain high-quality vibrational echo data. The data took approximately 10 min to acquire.

Figure 14 displays the pure-dephasing rates for Hb-CO as a function of temperature and a comparison to the pure dephasing rates of Mb-CO, both in EgOH-H$_2$O. As can be seen from the figure, Hb-CO- and Mb-CO-dephasing data have the same functional form and differed only in the magnitudes of their pure-dephasing rates. The Mb-CO data are the same as shown in Figure 8 (10). The line through the

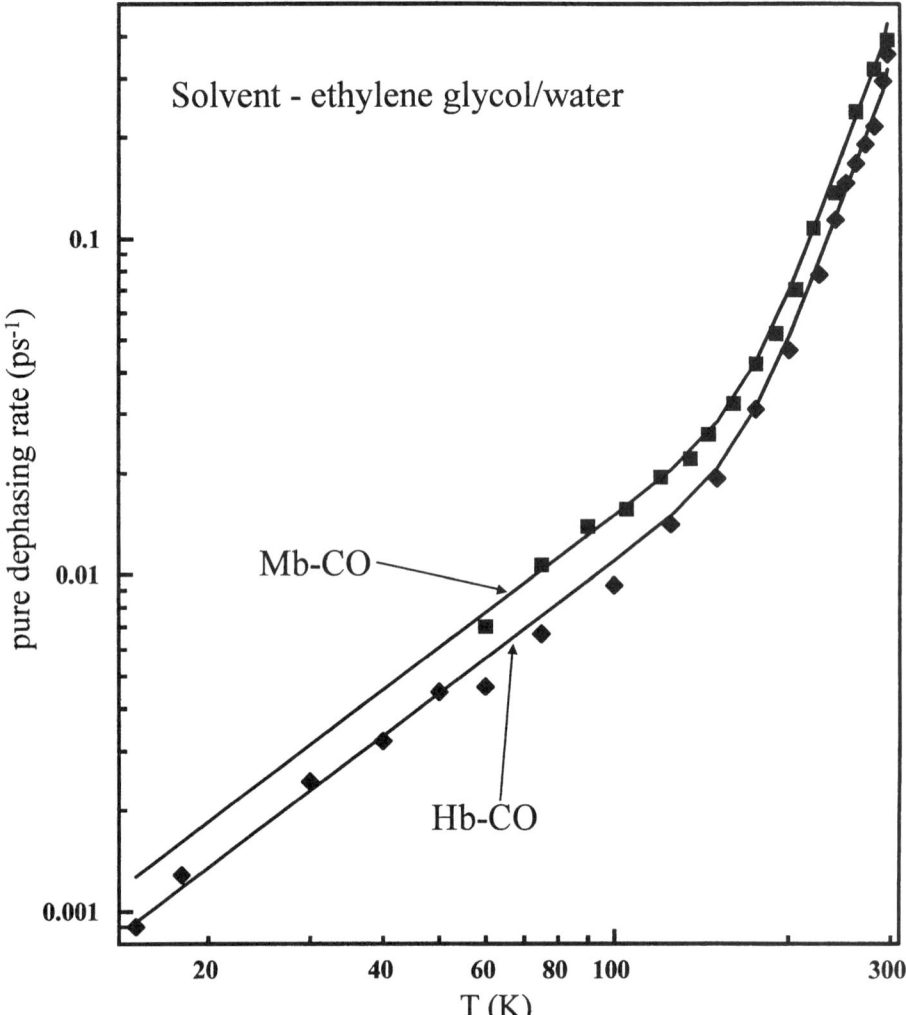

Figure 14 Pure-dephasing rates of Hb-CO and Mb-CO, both in glycerol-water, on a log plot. The data from Mb-CO were the same as in Figure 8. The line through the Hb-CO data is the line though the Mb-CO data multiplied by the constant 0.73. The functional forms of the temperature dependences were identical.

Mb-CO data is not from the viscoelastic theory (Figure 12). It is an empirical fit of the form given in Equations 8 and 9, that is, a $T^{1.3}$ power law at low temperature and an exponentially activated function at high temperature (10). This form is used to put a line through the Mb-CO data to make possible a direct comparison to the Hb-CO data. The line through the Hb-CO data is not a fit; rather, it is the Mb-CO fit multiplied by a constant factor 0.73. These results show that the functional form of the temperature-dependent pure-dephasing dynamics of Hb-CO is identical to Mb-CO, but the pure dephasing in Hb-CO is 27% slower across the entire temperature range.

As proposed above, the source of the pure dephasing in Mb-CO involves motions of the polar groups in the protein that generate fluctuating electric fields at the CO (47). This model is supported by mutant studies and analysis of data using the viscoelastic theory. A possible explanation for the reduction in the rate of pure dephasing in Hb-CO compared with Mb-CO (Figure 14) is that the magnitudes of the fluctuating electric field at the CO in Hb-CO are less than in Mb-CO. The fluctuating electric field in Hb-CO may be reduced either because of the differences in the locations of polar groups or because the protein dynamics on the time scales observed by the vibrational echo experiment are reduced in Hb-CO compared with Mb-CO. Hb is approximately fourfold greater in volume than Mb. The change in surface-to-volume ratio could influence the viscoelastic response of the protein. The fact that, within experimental error, the functional form of the pure-dephasing temperature dependence in Hb-CO is identical to that in Mb-CO demonstrates that the fast global dynamics of the two proteins, as sensed by the CO ligand bound at the active sites of the proteins, are very similar. This suggests that the distributions of barriers controlling structural fluctuations have the same nature in the two proteins.

CONCLUDING REMARKS

The application of ultrafast IR vibrational echo experiments to the study of proteins is a new approach for the investigation of protein dynamics. To date, Mb-CO has been studied extensively, and initial experiments have been conducted on Hb-CO. For both proteins, pure vibrational dephasing of the CO ligand bound at the active sites of the protein provides information on the global structural fluctuations of the protein. Temperature-dependent, vibrational echo, pure-dephasing measurements have been made on Mb-CO in a variety of solvents and on two mutants of Mb-CO. In addition, the isothermal (300 K) viscosity dependence of Mb-CO pure dephasing has been measured.

The temperature-dependent, vibrational echo results show that the pure dephasing of the Mb mutant, H64V-CO, is ~21% slower than that of native Mb-CO with no change in the functional form of the temperature dependence. The temperature dependence of the pure dephasing of the mutant H93G(N-MeIm)-CO is identical to that of the native Mb-CO. The general mechanism proposed to explain the coupling of conformational fluctuations of the protein to the vibrational-transition energy of CO bound at the active site is supported by these results. The model states that global protein motions produce a fluctuating electric field that is responsible for the CO pure dephasing via a direct Stark effect (68) on the CO vibrational frequency (47). Replacing the polar distal histidine with the nonpolar valine (H64V-CO) removes one source of the fluctuating electric field, thus reducing the coupling between the protein fluctuations and the measured pure dephasing. The picture that emerges is that the CO bound at the active site acts as an antenna for the fluctuating electric fields produced by the protein dynamics, providing an avenue for study of protein-structural fluctuations through vibrational echo measurements of pure dephasing.

At low temperature, the pure vibrational dephasing of Mb-CO in several solvents and Hb-CO have the same $T^{1.3}$ power law temperature dependence. This power law temperature dependence is the signature of glass dynamics in low-temperature glasses and suggests that the protein has passed through a "protein-glass transition." In the solvent trehalose, which is a glass at all of the experimental temperatures, the pure dephasing of Mb-CO displays a transition from $T^{1.3}$ to a steeper temperature dependence above $\sim$200 K. The break in the temperature dependence is consistent with a protein-glass transition at $\sim$200 K.

Mb-CO in liquid solvents displays much steeper pure-dephasing temperature dependence than it does in the glassy-solvent trehalose. The increased temperature dependence is a combination of pure temperature dependence and viscosity dependence. As the temperature is increased, the solvent viscosity decreases, leading to an additional contribution to pure dephasing. The existence of viscosity dependence is confirmed by an isothermal viscosity-dependent pure-dephasing study. The influence of viscosity on protein dynamics arises from the coupling of the protein internal structure to its surface topology. In a glassy solvent, the surface topology is essentially fixed. Only those protein motions that can occur without changing the surface topology are permitted. As the solvent viscosity decreases, the protein's surface becomes increasingly free to move, enabling protein-structural fluctuations that require changes in the surface topology. The viscosity-dependent behavior of Mb-CO pure dephasing was described quantitatively using a viscoelastic theory of the pure dephasing (47). Nearly quantitative agreement was found between theory and experiment. The theory permits the time scales and nature of viscosity-dependent protein structural fluctuations to be delineated.

The application of ultrafast IR vibrational echoes, related pulse sequences, and multidimensional IR vibrational echo methods to problems of biological interest is just beginning. Most biological and chemical processes occur on the electronic ground state potential surface. They involve structural transformations induced by structural fluctuations. The structural, that is, mechanical, degrees of freedom of molecules are described in terms of the molecular vibrations. Structural dynamics are vibrational dynamics. Vibrational echo techniques, which directly probe vibrational dynamics, will be an increasingly important probe for fundamental biological and chemical processes.

ACKNOWLEDGMENTS

Many people contributed to this work. In particular, I thank Dr. Kirk Rector, currently at Los Alamos National Laboratory, Los Alamos, NM, who was the driving force behind the experiments. I give special thanks to Professor Mark A. Berg, Department of Chemistry and Biochemistry, University of South Carolina,

Columbia, SC, who was primarily responsible for the viscoelastic theory of the protein pure dephasing. All of the Mb-CO experiments were conducted at the Stanford Free Electron Laser Center. I thank Professor H. Alan Schwettman and Professor Todd I. Smith and their students and staff at the Stanford FEL Center for their help and collaboration on many of the experiments. I also thank Professor Dana D. Dlott, Department of Chemistry, University of Illinois, Urbana, IL, who was a major participant in a number of the experimental studies described here. The research described in this review was supported by the National Institutes of Health (grant 1RO1-GM61137) and by the National Science Foundation (grant DMR-0088942).

Visit the Annual Reviews home page at www.AnnualReviews.org

LITERATURE CITED

1. Zimdars D, Tokmakoff A, Chen S, Greenfield SR, Fayer MD. 1993. *Phys. Rev. Lett.* 70:2718–21
2. Tokmakoff A, Kwok AS, Urdahl RS, Francis RS, Fayer MD. 1995. *Chem. Phys. Lett.* 234:289–95
3. Tokmakoff A, Fayer MD. 1995. *J. Chem. Phys.* 102:2810–36
4. Rella CW, Kwok A, Rector KD, Hill JR, Schwettman HA, Dlott DD, Fayer MD. 1996. *Phys. Rev. Lett.* 77:1648–51
5. Rector KD, Rella CW, Kwok AS, Hill JR, Sligar SG, Chien EYP, Dlott DD, Fayer MD. 1997. *J. Phys. Chem. B* 101:1468–75
6. Rector KD, Kwok AS, Ferrante C, Francis RS, Fayer MD. 1997. *Chem. Phys. Lett.* 276:217–23
7. Rector KD, Engholm JR, Hill JR, Myers DJ, Hu R, Boxer SG, Dlott DD, Fayer MD. 1998. *J. Phys. Chem. B.* 102:331–33
8. Rector KD, Zimdars DA, Fayer MD. 1998. *J. Chem. Phys.* 109:5455–65
9. Rector KD, Fayer MD. 1998. *J. Chem. Phys.* 108:1794–1803
10. Rector KD, Engholm JR, Rella CW, Hill JR, Dlott DD, Fayer MD. 1999. *J. Phys. Chem. A* 103:2381–87
11. Rector KD, Fayer MD, Engholm JR, Crosson E, Smith TI, Schwettmann HA. 1999. *Chem. Phys. Lett.* 305:51–56
12. Hahn EL. 1950. *Phys. Rev.* 80:580–87
13. Schmidt-Rohr K, Spiess HW. 1994. *Multidimensional Solid-State NMR*. London: Academic
14. Kurnit NA, Abella ID, Hartmann SR. 1964. *Phys. Rev. Lett.* 13:567–70
15. Abella ID, Kurnit NA, Hartmann SR. 1966. *Phys. Rev. Lett.* 14:391–94
16. Molenkamp LW, Wiersma DA. 1985. *J. Chem. Phys.* 83:1–9
17. Vainer YG, Personov RI, Zilker S, Haarer D. 1996. In *Int. Meet. Hole Burning Relat. Spectrosc. Sci. Appl., 5th*, Brainerd, Minn, ed. GJ Small, 291:51–55. Brainerd, Minn: Gordon & Breach
18. Berg M, Walsh CA, Narasimhan LR, Littau KA, Fayer MD. 1988. *J. Chem. Phys.* 88:1564–87
19. Narasimhan LR, Littau KA, Pack DW, Bai YS, Elschner A, Fayer MD. 1990. *Chem. Rev.* 90:439–57
20. Elber R, Karplus M. 1987. *Science* 235:318–21
21. Vitkup D, Ringe D, Petsko GA, Karplus M. 2000. *Nat. Struct. Biol.* 7:34–38
22. Quillin ML, Arduini RM, Olson JS, Phillips GN Jr. 1993. *J. Mol. Biol.* 234:140–55
23. Barrick D. 1994. *Biochemistry* 33:6546–54
24. Tsuda S. 1996. *Crystallogr. Soc. Jpn.* 38:84–92

25. Clore GM, Gronenborn AM. 1991. *Prog. Nucl. Magn. Reson. Spectrosc.* 23:43–92
26. Braunstein DP, Chu K, Egeberg KD, Frauenfelder H, Mourant JR, Nienhaus GU, Ormos P, Sligar SG, Springer BA, Young RD. 1993. *Biophys. J.* 65:2447–54
27. Jackson TA, Lim M, Anfinrud PA. 1994. *Chem. Phys.* 180:131–40
28. Janes SM, Dalickas GA, Eaton WA, Hochstrasser RM. 1988. *Biophys. J.* 54:545–49
29. Oldfield E, Guo K, Augspurger JD, Dykstra CE. 1991. *J. Am. Chem. Soc.* 113:7537–41
30. Surewicz WK, Mantsch HH. 1996. In *Spectroscopic Methods for Determining Protein Structure in Solution*, ed. HA Havel, pp. 135–62. New York: VCH Publ.
31. Stryer L. 1988. *Biochemistry.* New York: Freeman
32. Beece D, Eisenstein L, Frauenfelder H, Good D, Marden MC, Reinisch L, Reynolds AH, et al. 1980. *Biochemistry* 19:5147–57
33. Gordon RG. 1968. *Adv. Magn. Reson.* 3:1–17
34. Gordon RG. 1965. *J. Chem. Phys.* 43:1307–12
35. Tokmakoff A, Zimdars D, Urdahl RS, Francis RS, Kwok AS, Fayer MD. 1995. *J. Phys. Chem.* 99:13310–320
36. Oxtoby DW. 1981. *Annu. Rev. Phys. Chem.* 32:77–101
37. Rector KD, Thompson DE, Merchant K, Fayer MD. 2000. *Chem. Phys. Lett.* 316:122–28
38. Loring RF, Mukamel S. 1985. *J. Chem. Phys.* 83:2116–28
39. Bai YS, Fayer MD. 1989. *Phys. Rev. B.* 39:11066–84
40. Farrar TC, Becker DE. 1971. *Pulse and Fourier Transform NMR.* New York: Academic
41. Skinner JL, Andersen HC, Fayer MD. 1981. *J. Chem. Phys.* 75:3195–202
42. Hahn EL. 1953. *Phys. Today* 6:4–7
43. Brewer RG, Hahn EL. 1984. *Sci. Am.* 251:50–57
44. Rector KD, Fayer MD. 1998. *Int. Rev. Phys. Chem.* 17:261–306
45. Leeson DT, Wiersma DA. 1995. *Phys. Rev. Lett.* 74:2138–41
46. Thijssen HPH, Dicker AIM, Völker S. 1982. *Chem. Phys. Lett.* 92:7–12
47. Rector KD, Jiang J, Berg M, Fayer MD. 2001. *J. Phys. Chem.* In press
48. Levenson MD. 1982. *Introduction to Nonlinear Laser Spectroscopy.* San Jose, Calif.: Academic
49. Slichter CP. 1989. *Principles of Magnetic Resonance.* Berlin: Springer-Verlag
50. Anderson PW. 1954. *J. Phys. Soc. Jpn.* 9:316–39
51. Kubo R. 1961. In *Fluctuation, Relaxation and Resonance in Magnetic Systems*, ed. D Ter Haar. London: Oliver & Boyd
52. Anderson PW, Halperin BI, Varma CM. 1972. *Philos. Mag.* 25:1–7
53. Phillips WA. 1981. *Amorphous Solids: Low Temperature Properties, Topics in Current Physics.* Berlin: Springer-Verlag
54. Berg MA, Rector KD, Fayer MD. 2000. *J. Chem. Phys.* 113:3233–42
55. Mukamel S. 1995. *Principles of Nonlinear Optical Spectroscopy.* New York: Oxford Univ. Press
56. Ma J, Fourkas JT, Vanden Bout DA, Berg M. 1997. In *Supercooled Liquids: Advances and Novel Applications*, Vol. 676, ed. JT Fourkas, D Kivelson, U Mohanty, KA Nelson, pp. 199–211. Washington, DC: Am. Chem. Soc.
57. Klauder JR, Anderson PW. 1962. *Phys. Rev.* 125:912–22
58. Hu P, Hartmann SR. 1974. *Phys. Rev. B* 9:1–13
59. Hu P, Walker LR. 1978. *Phys. Rev. B* 18:1300–1305
60. Yan YJ, Mukamel S. 1991. *J. Chem. Phys.* 94:179–90
61. Tokmakoff A, Fayer MD. 1995. *J. Chem. Phys.* 103:2810–36
62. Rella CW, Rector KD, Kwok AS, Hill JR,

Schwettman HA, Dlott DD, Fayer MD. 1996. *J. Phys. Chem.* 100:15620–29
63. Angell CA, Smith DL. 1982. *J. Phys. Chem.* 86:3845–52
64. Angell CA. 1988. *J. Phys. Chem. Solids* 49:863–71
65. Fredrickson GH. 1988. *Annu. Rev. Phys. Chem.* 39:149–80
66. Ansari A, Beredzen J, Braunstein D, Cowen BR, Frauenfelder H, et al. 1987. *Biophys. Chem.* 26:337–55
67. Ivanov D, Sage JT, Keim M, Powell JR, Asher SA, Champion PM. 1994. *J. Am. Chem. Soc.* 116:4139–40
68. Park E, Andrews S, Boxer SG. 1999. *J. Phys. Chem.* 103:9813–17
69. Oldfield E, Guo K, Augspurger JD, Dykstra CE. 1991. *J. Am. Chem. Soc.* 113:7537–41
70. Augspurger JD, Dykstra CE, Oldfield E. 1991. *J. Am. Chem. Soc.* 113:2447–51
71. DePillis GD, Decatur SM, Barrick D, Boxer SG. 1994. *J. Am. Chem. Soc.* 116:6981–51
72. Decatur SM, DePillis GD, Boxer SG. 1996. *Biochemistry* 35:3925–32
73. Hill JR, Tokmakoff A, Peterson KA, Sauter B, Zimdars DA, Dlott DD, Fayer MD. 1994. *J. Phys. Chem.* 98:11213–19
74. Hill JR, Dlott DD, Rella CW, Peterson KA, Decatur SM, Boxer SG, Fayer MD. 1996. *J. Phys. Chem.* 100:12100–107
75. Walsh CA, Berg M, Narasimhan LR, Fayer MD. 1986. *Chem. Phys. Lett.* 130:6–11
76. Walsh CA, Berg M, Narasimhan LR, Fayer MD. 1987. *J. Chem. Phys.* 86:77–87
77. Thijssen HPH, Volker S. 1985. *Chem. Phys. Lett.* 120:496–502
78. Thijssen HPH, van den Berg R, Volker S. 1985. *Chem. Phys. Lett.* 120:503–8
79. Vohringer P, Arnett DC, Yang TS, Scherer NF. 1995. *Chem. Phys. Lett.* 237:387–98
80. Phillips WA. 1972. *J. Low Temp. Phys.* 7:351–58
81. Leeson DT, Wiersma DA, Fritsch K, Friedrich J. 1997. *J. Phys. Chem. B* 101:6331–40
82. Gilroy KS, Phillips WA. 1981. *Philos. Mag. B.* 43:735–46
83. Kohler W, Zollfrank J, Friedrich J. 1989. *Phys. Rev. B* 39:5414–24
84. Iben IET, Basunstein D, Doster W, Frauenfelder H, Hong MK, et al. 1989. *Phys. Rev. Lett.* 62:1916–19
85. Parak F, Frauenfelder H. 1993. *Physica A* 201:332–45
86. Frauenfelder H, Sligar SG, Wolynes PG. 1991. *Science* 254:1598–603
87. Doster W, Cusack S, Petry W. 1989. *Nature* 337:754–56
88. Loncharich RJ, Brooks BR. 1990. *J. Mol. Biol.* 215:439–55
89. Steinbach PJ, Brooks BR. 1993. *Proc. Natl. Acad. Sci. USA* 90:9135–39
90. Doster W, Bachleitner A, Dunau R, Hiebl M, Luscher E. 1986. *Biophys. J.* 50:213–19
91. Hong MK, Draunstein D, Cowen BR, Frauenfelder H, Iben IET, et al. 1990. *Biophys. J.* 58:429–36
92. Mayer E. 1994. *Biophys. J.* 67:862–73
93. Cordone L, Cupane A, Leone M, Vitrano E. 1988. *J. Mol. Biol.* 199:213–18
94. Parak F, Knapp EW, Kucheida D. 1982. *J. Mol. Biol.* 161:177–94
95. Berg MA, Hubble HW. 1998. *Chem. Phys.* 233:257–66
96. Berg M. 1998. *J. Phys. Chem.* 102:17–30
97. Berg MA. 1999. *J. Chem. Phys.* 110:8577–88
98. Zwanzig R, Mountain RD. 1965. *J. Chem. Phys.* 43:4464–69
99. Bertolini D, Tani A. 1995. *Phys. Rev. E* 52:1699–710
100. Piccirelli R, Litovitz TA. 1957. *J. Acoust. Soc. Am.* 29:1009–115
101. Deleted in proof
102. Litovitz TA, Davis DM. 1965. In *Physical Acoustics*, Vol. IIA, ed. WP Mason. New York: Academic

103. Harrison G. 1976. *The Dynamic Properties of Supercooled Liquids.* New York: Academic
104. Forrest JA, Dalnoki-Veress K, Dutcher JR. 1997. *Phys. Rev. E.* 56:5705–16
105. Weber G. 1982. *Nature* 300:603–7
106. Potter WT, Hazzard JH, Kawanishi S, Caughey WS. 1983. *Biochem. Biophys. Res. Commun.* 116:719–25
107. Caughey WS, Shimada H, Choc MC, Tucker MP. 1981. *Proc. Natl. Acad. Sci. USA* 78:2903–7
108. Karavitis M, Fronticelli C, Brinigar WS, Vasquez GB, Militello V, Leone M, Cupane A. 1998. *J. Biol. Chem.* 273:23740–49
109. Antonini E, Brunori M. 1971. *Hemoglobin and Myoglobin in Their Reactions with Ligands.* Amsterdam: Elsevier/North-Holland Biomed.
110. Rector KD, Kwok AS, Ferrante C, Tokmakoff A, Rella CW, Fayer MD. 1997. *J. Chem. Phys.* 106:10027–36
111. Haire RN, Hedlund BE. 1983. *Biochemistry* 22:327–34
112. Back JF, Oakenfull D, Smith MD. 1979. *Biochemistry* 18:5191–96
113. Gerlsma SY, Stuur ER. 1972. *Proc. Natl. Acad. Sci. USA* 74:4135–38

STRUCTURE AND BONDING OF MOLECULES AT AQUEOUS SURFACES

GL Richmond
Department of Chemistry, University of Oregon, Eugene, Oregon 97403;
e-mail: Richmond@oregon.uoregon.edu

Key Words vibrational spectroscopy, nonlinear optics, hydrogen bonding, water

■ **Abstract** Significant advances toward understanding the structure of aqueous surfaces on a molecular level have been made in recent years. This review focuses on the recent contributions of surface vibrational sum frequency spectroscopy (VSFS) to this field of study. An overview of recent VSFS studies of the molecular structure and orientation of molecules at the vapor-water interface and the interface between water and an immiscible organic liquid is presented, with particular emphasis on studies that compare the molecular properties and adsorbate behavior at these two different but related interfaces. This discussion is preceded by a general introduction to VSFS studies at aqueous surfaces and a description of the fundamental principles underlying the technique.

INTRODUCTION

Aqueous surfaces are central to our existence on this planet. The water-vapor interface provides a template for many reactive processes in the atmosphere and aqueous surfaces on the earth. Protein folding, membrane formation, and micellar assembly are a few of the many processes for which the interaction of water with a hydrophobic macromolecular surface is key. Ion or solute transport across a water-hydrophobic liquid interface relies upon the differing interactions between the transferring species and the water and organic solvent molecules in the interfacial region. Our understanding of interactions at water surfaces has largely been derived from theoretical efforts in the past several decades (1–12). This is particularly true for buried liquid interfaces such as the junction between two immiscible liquids or water-hydrating macromolecule and organic molecular assemblies (13–21). Recent advances in experimental methods that can probe aqueous surfaces with molecular specificity have made possible the first direct measurements of the molecular properties of water surfaces (22–38). These studies demonstrate the unique properties of liquid surfaces relative to solid surfaces. They also illustrate the complex interactions that are present at these interfaces, interactions that will require a combination of advanced experimental and theoretical approaches to understand. However, the challenge is worth pursuing with vigor, given the key role that these surfaces play in chemical, physical, and biological processes.

One particularly promising method for studying liquid surfaces and interfaces is vibrational sum frequency spectroscopy (VSFS). This review provides an overview of some of the recent studies in which VSFS has been used to measure the molecular properties of vapor-water and organic-liquid-water interfaces. The review begins with a description of VSFS, including examples of optical systems used for VSFS measurements. Information gained from selected studies in this field is then described, beginning with studies on the structure and hydrogen bonding of interfacial water molecules. This is followed by a summary of studies that examine how interfacial water is perturbed by the presence of surfactants and adsorbates. The remaining portion of the review focuses on studies of adsorbate structure and conformation measured at both types of interfaces. This review places particular emphasis on recent findings concerning the similarities and differences in the behavior of molecules at aqueous vapor-liquid and aqueous liquid/hydrophobic liquid interfaces. The comparison provides some fascinating new insights about differences in hydrogen bonding of water and the assembly of monolayers at these two interfaces.

VIBRATIONAL SUM FREQUENCY SPECTROSCOPY

Theoretical Considerations

VSFS is a second-order nonlinear optical process that directly measures the vibrational spectrum of molecules at an interface. Pioneered by Shen in the mid 1980s (39), this technique is uniquely suited for surface studies because of its inherent surface sensitivity. Under the dipole approximation, this second-order process is forbidden in media that possess inversion symmetry (40, 41). At the interface between two centrosymmetric media there is no inversion center and thus vibrational sum frequency (VSF) is allowed in the interfacial region. Consequently, the asymmetric nature of interfaces allows a regional selectivity to interfacial properties on a molecular level that is not inherent to other linear surface vibrational spectroscopies.

In a VSFS experiment, a visible laser beam and a tunable IR laser beam are coincident at the interface. The energy range of the tunable IR laser is chosen to overlap with the energies of vibrational modes of molecules present at the interface. By scanning the energy of the IR laser and monitoring the generated sum frequency (SF) signal, a vibrational spectrum of the interfacial molecules can be measured. The VSFS intensity is proportional to the square of the surface nonlinear susceptibility $\chi_s^{(2)}(\omega_{sfg} = \omega_{vis} + \omega_{ir})$ as

$$I_{sfg} \propto |P_{sfg}|^2 \propto \left| \chi_{NR}^{(2)} + \sum_\nu |\chi_{R_\nu}^{(2)}| e^{i\gamma_\nu} \right|^2 I_{vis} I_{ir}, \qquad 1.$$

where P_{sfg} is the nonlinear polarization at ω_{sfg}; χ_{NR} and χ_{R_ν} are the nonresonant and resonant parts of $\chi_s^{(2)}$, respectively; γ_ν is the relative phase of the νth vibrational mode; and I_{vis} and I_{ir} are the visible and IR intensities, respectively. Because the susceptibility is generally complex, the resonant terms in the summation are associated with a relative phase γ_ν, which is used to account for any interference between two modes that overlap in energy. $\chi_{R_\nu}^{(2)}$ is also proportional to the number density of molecules, N, and the orientationally averaged molecular hyperpolarizability, β_ν, as follows:

$$\chi_{R_\nu}^{(2)} = \frac{N}{\varepsilon_0}\langle\beta_\nu\rangle. \qquad 2.$$

Thus the square root of the measured SF intensity is proportional to the number density of molecules at the surface or interface. The molecular hyperpolarizability, β_ν, is enhanced when the frequency of the IR field is resonant with a SF-active vibrational mode of a molecule at the surface or interface. This enhancement in β_ν leads to an enhancement in the nonlinear susceptibility $\chi_{R_\nu}^{(2)}$, which can be expressed as

$$\chi_{R_\nu}^{(2)} \propto \frac{A_\nu}{\omega_\nu - \omega_{ir} - i\Gamma_\nu}, \qquad 3.$$

where A_ν is the intensity of the νth mode and is proportional to the product of the Raman and IR transition moments, ω_ν is the resonant frequency, and Γ_ν is the line width of the transition. Because the intensity term, A_ν, is proportional to both the IR and Raman transition moments, only vibrational modes that are both IR and Raman active will be SF active. Molecules or vibrational modes that possess an inversion center will not be SF active.

The surface susceptibility $\chi_s^{(2)}$ in general is a 27-element tensor. It can often be reduced to several nonvanishing elements by invoking symmetry constraints. Liquid surfaces and interfaces of liquid surfaces are isotropic in the plane of the surface. The symmetry constraints for an in-plane isotropic surface ($C_{\infty v}$) reduce $\chi_s^{(2)}$ down to the following four independent nonzero elements:

$$\chi_{zzz}^{(2)}; \chi_{xxz}^{(2)} = \chi_{yyz}^{(2)}; \chi_{xzx}^{(2)} = \chi_{yzy}^{(2)}; \chi_{zxx}^{(2)} = \chi_{zyy}^{(2)}, \qquad 4.$$

where z is the direction normal to the surface. These four independent elements contribute to the VSFS signal under the four different polarization conditions, *SSP*, *SPS*, *PSS*, and *PPP* (explained below), which are listed in the order of decreasing frequency (SF, Vis, IR). Light polarized parallel to the incident plane is referred to as *P* polarization, whereas light polarized perpendicular to the incident plane is *S* polarized. Which vibrational modes contribute to a particular polarization combination depends on the polarization of the IR field and the direction of the IR and Raman transition moments. The *SSP* polarization combination accesses vibrational modes with dipole transition moments that have components perpendicular to the surface plane. *SPS* and *PSS* polarization combinations access modes that have transition moments with components parallel to the surface plane. The intensity under *PPP* polarization conditions contains contributions from all of the tensor elements. Thus vibrational modes with components that are both perpendicular and parallel to the surface plane will be present in *PPP*-polarized VSFS spectra. *SSP* polarization has been used in most studies to date. *PSS* polarization has been used less frequently because of low signal levels for most liquid surfaces. Studies conducted with *SSP*, *SPS*, and *PPP* have been very useful in verifying peak assignments and orientation in VSFS (38).

Experimental Considerations

As mentioned above, VSFS involves overlapping two pulsed laser beams at the surface, one of fixed frequency in the visible and another of variable frequency in the IR spectrum. The SF process at liquid surfaces is very weak, so pulsed lasers are necessary to attain detectable SF signals. Because the SF signal increases with peak intensity, shorter pulses (i.e. picosecond or femtosecond) are optimal, although shorter pulses translate into broader IR bandwidths. Nanosecond lasers are often easier to operate and have narrower bandwidths but can lead to unwanted heating unless an optical coupling scheme such as total internal reflection (TIR) (38) or other mechanisms such as rotation of the sample (29) are used.

Most VSF studies to date have focused in IR regions where nanosecond and picosecond lasers operate with the highest power densities, for example, in the 3-μm region. IR light has been produced by a number of optical parametric processes, such as generation (OPG), oscillation (OPO), and amplification (OPA) (38, 42) as well as difference frequency mixing (43) and stimulated Raman scattering (44). In our laboratory we use both nanosecond and picosecond systems. The nanosecond systems are used primarily for liquid-liquid studies with the light coupled to the interface in a TIR geometry to attain sensitivity levels comparable to picosecond laser studies (38). For such TIR studies, the incident beams are sent through the higher-index medium at the critical angle for each particular beam. The VSF response is collected in reflection at the corresponding critical angle for the SF signal. The enhancement in the SF response using this TIR geometry is several orders of magnitude higher than that obtained using an external reflection geometry. The nanosecond system relies upon the 1064-nm output of an injection-seeded neodymium: yttrium/aluminum/garnet laser to pump a potassium titanyl phosphate (KTP) OPO/OPA assembly. This system produces tunable IR from 2.5 μm (4000 cm^{-1}) to 5 μm (1975 cm^{-1}), with usable energies ranging from 4 mJ to 1 mJ at the two respective limits. It also operates at 1 cm^{-1} resolution and has a variable repetition rate (1–100 Hz). For air-liquid studies, a picosecond laser is used (42). This picosecond system produces 800-nm light (kilohertz repetition rate, 2 ps, 1.6 W) with a titanium:sapphire (Coherent Mira) passively mode-locked laser pumped with 5.5 W of 532-nm laser light from a Coherent Verdi laser. The 800-nm light used is temporally stretched, amplified with a Quantronix regenerative amplifier and double-pass titanium:sapphire amplifier, and recompressed. The IR light (2700–4000 cm^{-1}) is produced via a home-built optical parametric amplifier pumped by the 800-nm light. This amplifier consists of two angle-tuned KTP crystals. Recent changes to this system over previous work (45) include the incorporation of a 2-cm-long MgO:LiNbO$_3$ crystal that is used as the optical parametric generator (OPG) from which the resultant 1- to 1.2-μm wavelengths are used to seed the KTP crystal. A grating for the selection of the seed wavelength is used. The IR light is generated by seeding the first KTP crystal and then amplifying it through the second KTP crystal. For the air-water studies, both beams are passed though a closed cell to minimize contamination. This system is currently being upgraded to produce IR as low as 8 μm.

Other variations on the traditional means of acquiring SF spectra in a scanning IR mode have recently appeared, both using femtosecond pulses. McGuire et al (46) have demonstrated a Fourier-transform spectroscopic technique based on VSFS. This method results in nearly unlimited spectral resolution and is based on Fourier transform of an SF-upconverted interferogram of an IR-induced polarization on the surface of the sample. Another method capitalizes on the variation of the SF exit angle with frequency, with the SF signal collected by different elements of a multielement detector (47). Here, the spectral resolution depends on the beam divergence. The third method simply disperses the broadband SF signal that is generated by femtosecond IR and narrow-band visible pulses by a monochromator and records the signal on a CCD camera (48).

HYDROGEN BONDING OF INTERFACIAL WATER

The unique properties of water surfaces have fascinated scientists for centuries. Strong hydrogen bonding between water molecules at a water surface is generally recognized to be responsible for the anomalously high surface tension present at the vapor/water interface (3). This section summarizes VSF measurements of the structure and hydrogen bonding of water molecules at both the vapor-water and organic-matter–water interfaces and discusses differences found in the behavior of water at these two interfaces. The summary includes recent developments in analysis of VSF spectral data of water. Results of a series of papers that have investigated the effect of different charged surfactants on the hydrogen bonding of water at both interfaces are also summarized. The section concludes with a description of vapor-water studies that are pertinent to important environmental issues.

Vibrational Sum Frequency Spectrum of the Vapor-Water Interface

The vibrational spectrum of water is of particular interest because hydrogen bond interactions are highly sensitive to the local molecular environment (3, 49–53). The vibrational spectrum of water therefore provides a sensitive probe of the structure and energetics of the hydrogen bond network at the water interface. The sensitivity of VSFS to surface water vibrational modes is accompanied by a complexity in

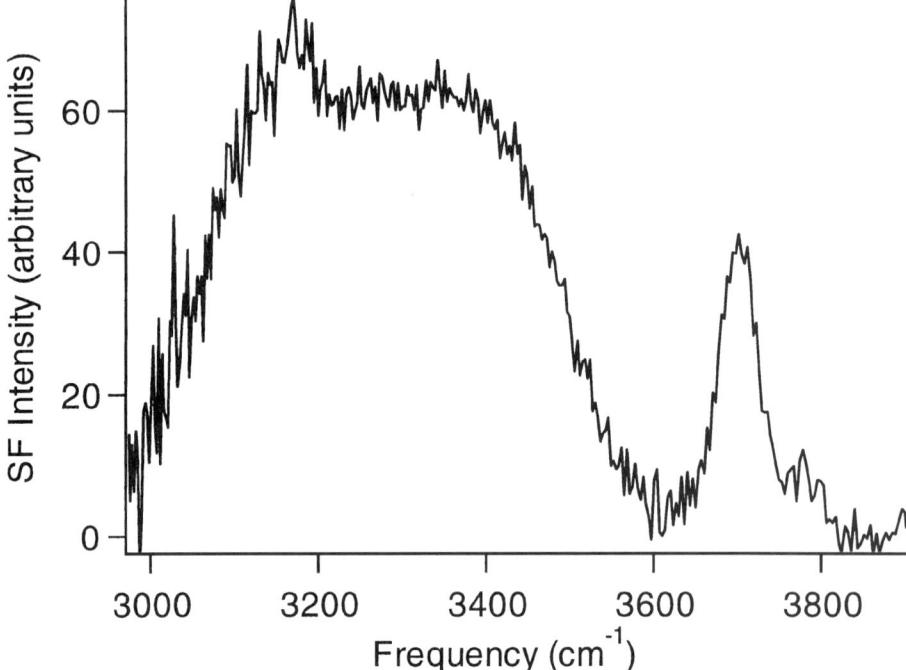

Figure 1 Vibrational sum frequency spectrum of the vapor/water interface using *SSP* polarization. A picosecond laser system described in the text was used. From LF Scatena & GL Richmond, submitted manuscript.

spectral interpretation that is only beginning to be explored by researchers in depth, owing to the variety of environments of the interfacial water molecules and various degrees of hydrogren bonding. This difficulty in interpretation is not unique to surface vibrational spectroscopy but has been a major point of controversy over past decades in the interpretation of bulk water spectra (52–54).

Du et al (55) reported the first VSF spectrum of the vapor-water interface. Figure 1 shows the VSF spectrum from this interface that was obtained in this laboratory (56, 57). The data have been taken with *SSP* polarization, which accesses vibrational modes that have components of their transition dipole in the plane perpendicular to the interface. The experiments were conducted in a purged cell as well as under ambient conditions. The spectra give identical results as long as the ambient experiments are conducted in such a way as to minimize any contamination problems. The 2-ps titanium:sapphire-based system described above was used. The spectrum shares the general features of earlier work of Du et al (55). Assignment of spectral features in this spectrum relies heavily on IR and Raman assignments of OH stretching modes taken from bulk water measurements. The surface water spectrum has the general shape of an isotropic Raman bulk water OH spectrum, yet there are important differences. Consistent with Raman and IR pH data and cluster studies, the VSF intensity in the 2900 cm^{-1} region is attributed to the very strong symmetric hydrogen bonding found in various water cluster distributions (58). Cationic and anionic water species that tend to partition to an aqueous surface could also be contributing to this intensity (59, 60). The broad band from ~3000 to 3600 cm^{-1} is assigned to the large distribution of OH hydrogen-bonding stretching modes in which the oxygen is tetrahedrally coordinated (49, 61–64). The energy region from ~3000 to 3250 cm^{-1} is attributed to strong intermolecular in-phase hydrogen bonds of water molecules that give rise to a highly correlated hydrogen-bonding network. This region is dominated by a continuum of OH symmetric stretches, v_1. The higher-energy broadband region (~3250 to 3500 cm^{-1}) is assigned to more weakly correlated hydrogen-bonded stretching modes of molecular water that encompass both v_1 (OH symmetric stretch) and, to a lesser extent, v_3 (OH asymmetric stretch) vibrational modes. The distinct peak at 3702 cm^{-1} is assigned to the dangling OH bond of interfacial water molecules that straddle the interface. This bond projects into the vapor phase. The relatively high energy of this mode reflects its intramolecular uncoupling with the other OH bond in the molecule, the donor bond. The donor bond is expected to appear at lower energies owing to its interactions in the surface plane with other interfacial water molecules. Interactions with the free OH bond mode are largely through the oxygen atom. Spectral corrections for the dispersion in the Fresnel coefficients and normalization in the SF response for a nonresonant substrate are not shown. When these corrections are made, there is minimal change in the spectrum except to enhance the intensity of the free OH mode relative to the broad hydrogen-bonded peaks. Schnitzer et al (65), Simonelli et al (66), and Schnitzer et al (67) have made similar measurements of the air/water spectrum using a nanosecond system rather than the picosecond laser systems used in the work by Allen et al (68) and Du et al (22). Schnitzer et al and Simonelli et al observed the free OH peak and a broad distribution of OH modes in the 3000–3500 cm^{-1} region, but their spectra also show an intensity drop to nearly zero around 3200 cm^{-1} (65–67).

Overall, the vapor-water interfaces suggest very strong hydrogen bonding at the surface. The strong intensity in the lower-energy region is consistent with a highly coordinated water structure, often referred to as "icelike" hydrogen bonding between water molecules because intensity in the 3200 cm^{-1} region is characteristic of IR peaks found in bulk ice. IR absorption in the 3400 cm^{-1} region of the spectrum is more characteristic of liquid water. Du et al (69) estimate that ~20% of surface water molecules have one free OH projecting into the vapor phase.

Water Hydrogen Bonding at the Organic-Liquid/Water Interface

The interfacial tension of an organic-liquid/water interface is known to decrease with the increased polarity of the organic phase until both phases are miscible. Most of what is known on a molecular level about water structure next to a hydrophobic fluid or solid surface has come from theoretical efforts (13–21). The difficulty in accessing this interface with a technique that can selectively probe interfacial water is the reason that so few experimental studies have been conducted. Nearly all studies to date have used VSFS and, of these, most have been conducted at the CCl_4/H_2O interface (70–74).

The CCl_4/H_2O system has some attractive features that make it a model system for understanding the hydrogen bonding of water near a hydrophobic liquid. In addition to being largely immiscible in water, it has an interfacial tension (48 mN m^{-2}) similar to alkane/water (∼51 mN m^{-2} for C_6–C_8). CCl_4 has a electric dipole polarizability that is nearly identical to alkanes such as n-hexane but does not have CH stretch vibrational modes that energetically overlap and complicate the spectral interpretation of the OH water modes used to study hydrogen bonding at interfaces.

Figure 2a shows the VSF spectra of water measured at the CCl_4/H_2O interface. The data are taken in a TIR optical geometry using the 3.5-ns neodymium: yttrium/aluminum/garnet-pumped laser system described above (LF Scatena & GL Richmond, submitted manuscript). Because the two experiments used different laser systems, peak positions in this spectrum and that for the vapor-water interface shown in Figure 1 can be compared, but only qualitative comparisons with peak intensities are appropriate. The most obvious difference between the two spectra is the remarkable shift in overall intensity to higher energies for CCl_4/H_2O, which is indicative of significantly weaker hydrogen bonding at this liquid-liquid interface. This observation is consistent with a much lower interfacial tension relative to air-water (71 mN m^{-2}). These VSF results for CCl_4/H_2O are a refinement on earlier work for water at this interface (75) in which the dangling bond OH mode (free OH) could not be probed because of laser limitations. These later results were obtained in a special cell designed to minimize solvent volume.

The weaker nature of the hydrogen bonding at this interface has facilitated assignment of spectral features in the CCl_4/H_2O spectrum. The free-OH mode for water is clearly apparent at the CCl_4/H_2O interface, but its energy is red shifted (3662 ± 1 cm^{-1}) relative to the vapor-water interface (3702 cm^{-1}) (55–57). From studies of HOD monomers in CCl_4, LF Scatena & GL Richmond (submitted manuscript) have determined that this red shift is the result of an attractive interaction between the dangling OH bond and the surrounding CCl_4 molecules. The binding energy for the H_2O-CCl_4 complex is reported to be −1.4 kcal mol^{-1} (76, 77). (This mode is therefore referred to as the dangling bond mode because it is not as free as in the vapor phase.) The results are consistent with previous simulations of this interface that suggest a locally sharp transition between phases (13, 21). VSF studies of HOD at this interface provide further assignments. In addition to the OH dangling bond and OD dangling bond of HOD at the interface (measured at 3664 cm^{-1} and 2712 cm^{-1}, respectively), a broader peak is observed near 3450 cm^{-1} that is attributed to the donor OH mode of the water molecules that straddle the interface. Because intensity is detected with I_{SSP} polarization, this H-bonded mode, which is largely uncoupled from the OH dangling-bond mode

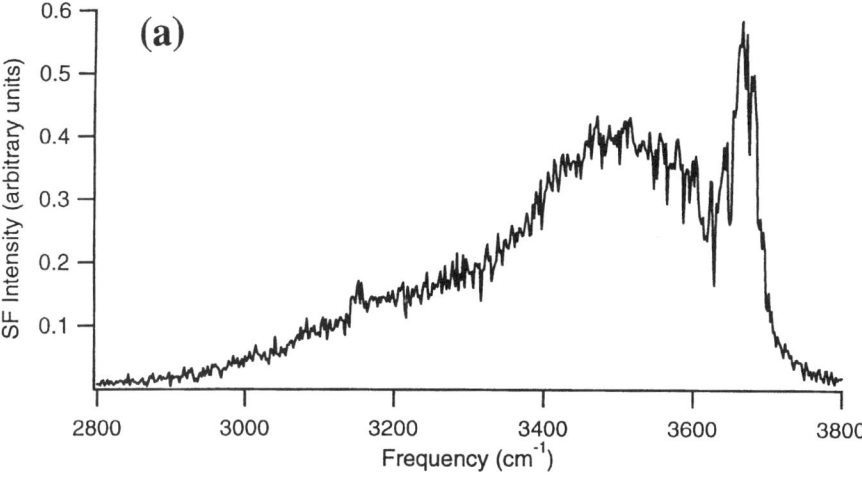
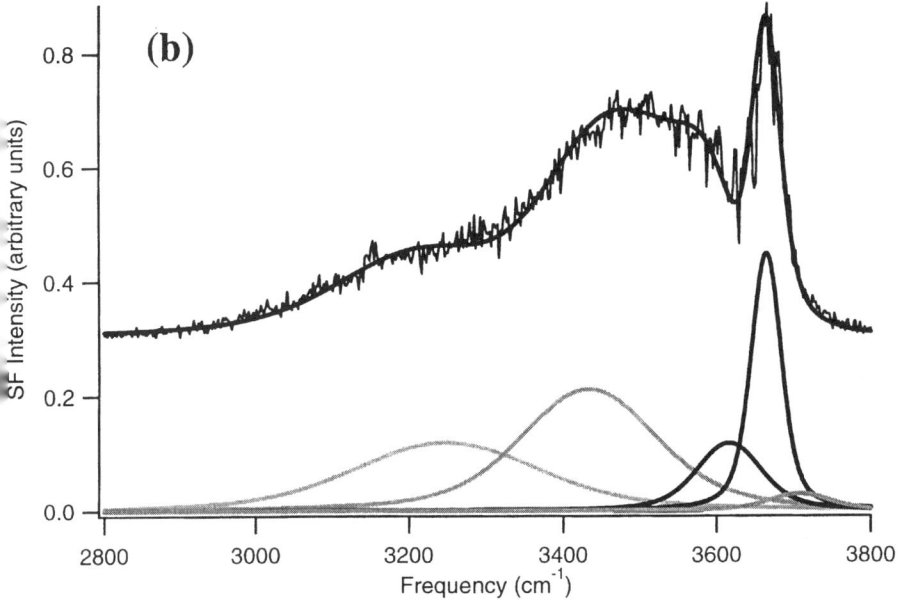

Figure 2 (*a*) Vibrational sum frequency spectrum of the CCl$_4$/H$_2$O interface using the 1-ns laser system described in the text (LF Scatena & GL Richmond, submitted manuscript). The spectra were collected in a total internal reflection geometry with *SSP* polarization used. This spectrum was achieved without the trace impurities that were apparent in the previously published spectra (75). (*b*) Spectral analysis of the CCl$_4$/H$_2$O interface using the procedures outlined in the text. The upper curve shows the spectral fit; the lower curves show contributions from each of the vibrational modes. The gray peaks have a phase opposite to that of the black peaks, as determined from fits to the data using Voigt profiles. Published with permission from Reference 57.

in the molecule, does not lie exactly in-the-plane of the interface but has some perpendicular component.

The remaining portion of the H$_2$O spectrum of Figure 2a between 3450 and 3700 cm^{-1} is attributed to weakly interacting water molecules in the interfacial region, such as water monomers (LF Scatena & GL Richmond, submitted manuscript). The OH stretch of water monomers in bulk CCl$_4$ shows two characteristic IR peaks, one associated with the symmetric OH stretch (SS or v_1) of water monomers (3616 cm^{-1}) and the other (3708 cm^{-1}) associated with the asymmetric OH stretch (AS or v_3). These two peaks energetically bracket the dangling bond stretch mode in the VSF spectrum. Careful analysis (described below) of the data shows that both modes are present in the VSF spectrum, the SS and the AS of water monomers respectively, which constructively and destructively interfere with the neighboring dangling OH bond mode (57).

Overall, the results suggest that, unlike the air-water interface, the hydrogen bonding at the CCl$_4$/H$_2$O interface is very weak. Intensity is observed in the strongly hydrogen-bonded region (3200–3400 cm^{-1}), but overall the spectra are dominated by water species that have only weak interactions with other water molecules and CCl$_4$ molecules. This is the first spectroscopic evidence for the existence of water molecules at an oil-water interface that have the spectroscopic characteristics of water monomers in the organic phase.

Vibrational Sum Frequency Spectral Analysis of Interfacial H$_2$O

Obtaining spectral fits to VSF water spectra is difficult because of the wide range of contributing vibrational modes, the breadth of the spectral peaks for hydrogen-bonded water molecules, and the complex interference effects that can arise between adjacent vibrational modes. Because VSFS is a coherent nonlinear spectroscopic technique, each resonant vibrational mode has an inherent phase for a fixed orientation. Resonant modes that change phase with the orientation of the molecule may interfere constructively or destructively when overlapped in frequency. The most detailed analysis to date for water spectra has recently appeared (57). This analysis has been applied to the water-vapor and the CCl$_4$/H$_2$O data described above. The work considers a range of water species present at a water surface and the possible interference between these contributing modes, and it takes into account the phase of the SF response from contributing vibrational modes. This phase is useful in obtaining an average orientation of molecules at the surface by relating the macroscopic second-order susceptibility, $\chi^{(2)}$, of the system to the molecular hyperpolarizabilities, β_v, of the individual molecules at the interface (78, 79). The molecular hyperpolarizability is often described as

$$\beta_{lmn,v} = \frac{\langle g|\alpha_{lm}|v\rangle \langle v|\mu_n|g\rangle}{\omega_{IR} - \omega_v + i\Gamma_v}, \qquad 4.$$

where l, m, and n represent the molecular inertial axes (a, b, and c); μ_n and α_{lm} represent the dipole and Raman vibrational transition moments, respectively; and a Lorentzian distribution of resonant transition energies is assumed (80). The measured macroscopic property $\chi^{(2)}$, is a sum of the molecular hyperpolarizability over all of the molecules at the interface, taking into account the orientation of the different molecular species. To obtain information on the orientation of the molecules, the observed Cartesian components of the macroscopic second-order susceptibility $\chi^{(2)}_{IJK,\nu}$ must be derived from the corresponding spectroscopically active components of the molecular hyperpolarizability β_{lmn} through an Euler angle rotation of the molecular axis system into the laboratory axis system. Hirose et al (78, 79) derived a general expression for the transformation from a molecular fixed-axis system to a laboratory fixed-axis system as

$$\chi^{(2)}_{IJK,\nu} = \sum_{lmn} \mu_{IJK:lmn} \cdot \beta_{lmn,\nu}. \qquad 5.$$

The indices I, J, and K are replaced by the laboratory frame coordinates X, Y, and Z, respectively, observed in a specific experiment. The indices l, m, and n run through the molecular coordinates a, b, and c. The orientation of the molecular axis system in the laboratory frame is defined by the transformation tensor, $\mu_{IJK:lmn}$ through the Euler angles (θ, ϕ, and χ). If the signs of $\beta_{lmn,\nu}$ and $\chi^{(2)}_{IJK,\nu}$ are known, the average orientation of the molecules can be constrained by analyzing how the sign of the transformation tensor changes with respect to the angles θ, ϕ, and χ. The signs of the $\chi^{(2)}_{IJK,\nu}$ terms in Equation 5 are known through a comprehensive fit of the observed SF spectra to Equations 2 and 3, and the signs of the $\beta_{lmn,\nu}$ components are known through ab initio calculations (81).

In many respects, the CCl_4/H_2O spectrum is a good test of the analysis, given the weakly interacting nature of the observed water molecules and hence the narrow bandwidths. For example, under I_{SSP} polarization, the SS and AS of monomeric H_2O would be expected to have opposite sign conventions (plus and minus, respectively), meaning that they are nearly 180° out of phase (57). Given that the OH dangling bond (plus sign convention) has a significant contribution perpendicular to the interface, if the water monomers are oriented with their dipoles in the same direction as the dangling bond, the SS(+) and AS(−) modes should interfere constructively and destructively, respectively, with the dangling bond mode. This is verified by fitting the data with the appropriate phase relationships using Voigt profiles (Figure 2b). Brown et al conclude from the fits and the derived sign of the phases that the SS and AS observed intensities represent water monomers at the interface that have a net orientation with their hydrogen atoms pointed into the CCl_4. The peak energies and widths determined from the fits agree well with Fourier transform IR data of water monomers in bulk water. Spectral fits place the AS and SS of these monomer and monomer-like waters with their hydrogen atoms oriented into the CCl_4 phase at 3616 ± 2 cm^{-1} and 3706 ± 2 cm^{-1}, respectively, compared with the Fourier transform IR-measured peaks at 3616 cm^{-1} and 3708 cm^{-1}. The experimental results are in agreement with the molecular dynamics calculation of Chang & Dang for the CCl_4/H_2O interface (77).

EFFECT OF SURFACTANT ADSORPTION

Characterizing the interaction between water and a charged surfactant is important for understanding surfactant behavior in commercial soaps and detergents. However, it also is relevant to our understanding of many biological processes, including the folding of proteins where water interacts with charged groups on a hydrocarbon backbone. The series of studies described below focused on understanding how the hydrogen bonding of water is altered in the presence of increasing concentrations of simple charged alkyl surfactants. These studies were conducted at both the CCl_4/H_2O and air-water interfaces.

The hydrogen bonding of water is highly sensitive to the presence of charged surfactants, with changes occurring in the water spectrum at trace concentrations in the aqueous phase (LF Scatena & GL Richmond, manuscript in preparation). Figure 3 demonstrates this effect for the CCl_4/H_2O interface. When sodium dodecyl sulfate (SDS) is added in nanomolar concentrations to the aqueous phase, the trace SDS that adsorbs at the interface causes a dramatic change in the water spectrum that continues as the interfacial concentration increases to a monolayer, as seen by a comparison of Figure 2a with Figure 3a. As small amounts of SDS adsorb at the interface, the spectrum shows a strengthening in the hydrogen bonds between water molecules. The dangling OH bond essentially disappears as interfacial concentrations increase toward fractions of a monolayer. Further detailed studies of this effect at low concentrations in the nanomolar range will appear in a later publication. At higher concentrations, there is a multifold enhancement in the OH intensity, as shown by the difference in intensities displayed in Figure 3b.

The effect on the water spectrum of adding surfactants at aqueous-phase concentrations in the $>10^{-4}$ M range has been the focus of studies from this laboratory (82–84). In the absence of a charged surfactant or a mixture of anionic and cationic surfactants at the interface, this enhancement is not present. Figure 4 demonstrates this effect for SDS, dimethylammonium chloride (DAC), and a mixture of the two compounds at the air-water interface. The data were taken with *SSP* polarization that samples water dipoles oriented perpendicular to the interface. As shown, increased intensity is observed in the OH stretch region, which corresponds to strong hydrogen-bonding modes in the 3100–3500 cm^{-1} region of the water spectrum. For both interfaces, this intensity increases with added interfacial surfactant concentration, with a leveling off in the intensity that occurs well before monolayer formation (84). Figure 5 demonstrates this effect for a series of increasing concentrations of SDS at the CCl_4/H_2O interface. In these studies, the free OH mode was not studied owing to limited wavelength capabilities in the 3600 cm^{-1} region.

A series of detailed studies has been conducted to identify the factors that contribute to the enhanced VSF signal from water in the presence of charged surfactants

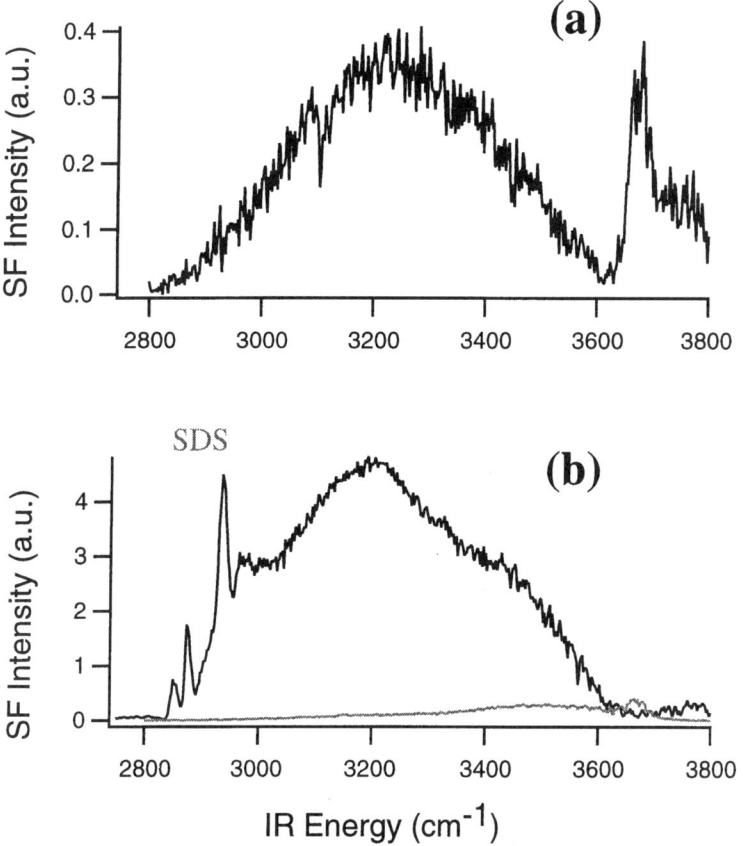

Figure 3 Effect of addition of sodium dodecyl sulfate (SDS) on the spectrum of water at the CCl_4/H_2O interface using *SSP* polarization. (*a*) Vibrational sum frequency (VSF) spectrum of the interface with 292-nM concentration of SDS in the aqueous phase; head group area $\gg 10^4$ A^2 molecule^{-1}. (*b*) VSF spectrum with 5 mM SDS in the aqueous phase. This corresponds to approximately a monolayer of SDS at the interface. The CH stretch modes are apparent near 2900 cm^{-1}. The gray line corresponds to the signal from the neat CCl_4/H_2O interface. From L. Scatena and G. L. Richmond, unpublished data.

(71, 84–86). Two factors have been considered. The simplest is the increased orientation of water molecules perpendicular to the interface that is caused by the large electrostatic field, E_0, created by the surfactant and its counterion. A significant surface charge exists at an interface where charged surfactant is adsorbed; this charge produces the large electrostatic field. An additional contribution to the *SF* polarization arises from a third-order polarization in Equation 6,

$$P_{\text{sfg}} = \chi^{(2)} : E_{\text{vis}}E_{\text{ir}} + \chi^{(3)} : E_{\text{vis}}E_{\text{ir}}E_0, \qquad 6.$$

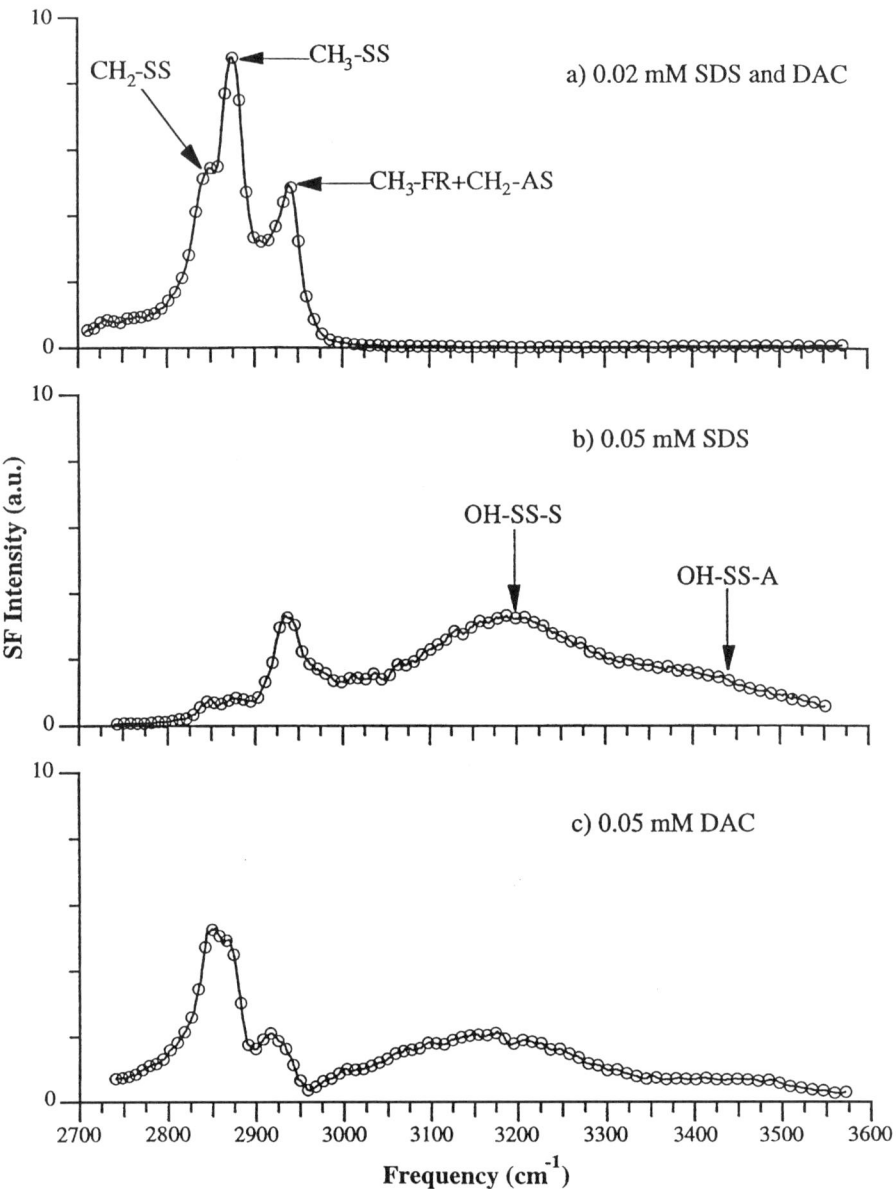

Figure 4 Vibrational sum frequency spectra under *SSP* polarization conditions from the air-water interface of an aqueous solution of (*a*) a mixture of 0.02 mM dimethylammonium chloride (DAC) and 0.02 mM sodium dodecyl sulfate (SDS), (*b*) 0.05 mM SDS, and (*c*) 0.05 mM DAC. Solid curves are a guide to the eye. Published with permission from Reference 89.

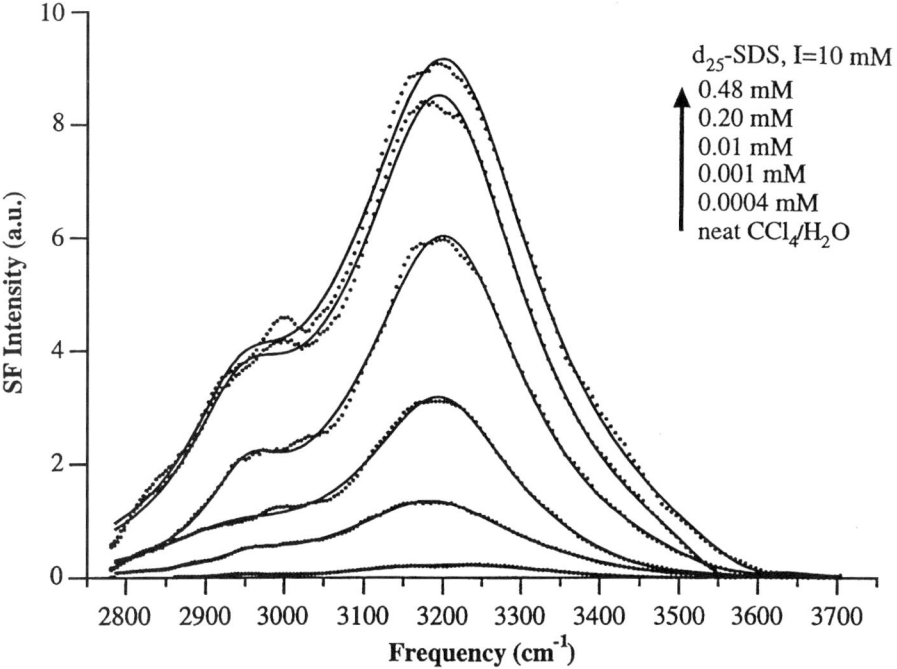

Figure 5 Vibrational sum frequency spectra from the CCl_4/SDS/H_2O interface under *SSP* polarization conditions for various bulk concentrations of sodium dodecyl sulfate (SDS) and an ionic strength of 10 mM. Solid curves are least squares fit to the data. Deuterated SDS (d_{25}) was used. Vibrational modes assigned include the methyl symmethic stretch (CH_3SS), methylene symmetric stretch (CH_2SS), methylene asymmetric stretch (CH_2AS) and Fermi resonance (FR). Published with permission from Reference 84.

which contains the electrostatic field dependence of the nonlinear polarization induced at the interface. In the absence of a large electrostatic field, the interfacial water molecules are randomly oriented after a few water layers and thus do not contribute to the nonlinear polarization. The presence of a large electrostatic field aligns the interfacial water molecules beyond the first few water layers and thus removes the centrosymmetry in this region, allowing more water molecules to contribute to the nonlinear polarization. The depth of the asymmetric region is on the order of the Debye length or 3 nm at an ionic strength of 10 mM, and 10 nm at an ionic strength of 1.0 mM. This depth corresponds to $\sim$10–30 water layers, respectively (84). As a consequence, the second factor in the observed enhancement in the OH-stretching region is the increased number of water molecules that are sampled, owing to the contribution from the $\chi^{(3)}$ term in Equation 6.

To confirm this enhancement, studies comparing the VSF response with the interfacial potential have been conducted at a series of surfactant concentrations and ionic strengths (84). Assuming that $\chi^{(3)}$ is constant over the interfacial region, the interfacial potential can be shown to have a linear relationship with P_{sfg}. Using the Gouy-Chapman model (87, 88), interfacial potential can be expressed as a function of surface charge density and ionic strength as

$$\Phi(0) = \frac{2kT}{ze}\sinh^{-1}\left(\sigma\sqrt{\frac{\pi}{2\varepsilon kTI}}\right), \qquad 7.$$

where σ is the surface charge density, z is the sign of the charged surfactant molecule, and I is the ionic strength of the bulk solution. The concentration and ionic strength studies confirm that a significant portion of the observed enhancement in the OH-stretching region is caused by the increased volume of water molecules, which gives rise to the VSF response (84). However, the lack of linearity between P_{sfg} and the surface potential for both interfaces suggests that orientation also contributes to the enhancement, that is, that there is alignment in the interfacial water molecules that accompanies the interfacial electrostatic field produced by the charged surfactant.

As in the neat water interfacial studies described above, interference between adjacent modes assists in determining the orientation of the water molecules at these charged interfaces. Analysis of the interference between the CH stretch modes and the water bands indicates that for cationic surfactants the water dipoles are oriented with their hydrogen atoms pointed toward the bulk aqueous solution. For anionic surfactants, water molecules are oriented with their oxygen atoms pointed toward the bulk aqueous solution (89).

STUDIES OF ENVIRONMENTALLY RELEVANT AQUEOUS-PHASE–AIR INTERFACES

The chemistry that occurs in the atmosphere involves both homogeneous and heterogeneous processes. Atmospherically important reactions have been studied extensively in the gas phase, but few published studies have examined in molecular detail the adsorption, molecular properties, and reactivity of atmospherically relevant molecules at water surfaces. Several studies dealing with organosulfur adsorbates on water surfaces have recently appeared, specifically studies of dimethyl sulfoxide (DMSO) and methane sulfonic acid (MSA) (56, 68, 90). Both molecules are present as trace constituents in the atmosphere and, although water soluble, have significant surface activities. DMSO has been proposed as a heterogeneous precursor to atmospheric condensed-phase MSA through an atmospheric cycle originating with dimethyl sulfide, a phytoplankton degradation product (91–93). Aerosol particles containing MSA are thought to contribute to the class of aerosols that effectively scatters radiation out of the atmosphere (92, 94). The molecular structure and orientation of DMSO have been examined in two studies, one of which looked at how the molecule changes as a function of surface concentration (90) and a second that examined the DMSO-water interaction at vapor-water interfaces. For neat DMSO studies, the methyl transition dipole moment is oriented a maximum of 55° from the surface normal on average (90). Aqueous solutions of DMSO show that this molecule partitions to the surface. With decreasing DMSO concentration in an aqueous phase, observed VSF spectral shifts in the methyl-stretch mode have been attributed to increased electronic interactions between sulfur and the methyl groups of DMSO. Evidence is provided for clustering of DMSO molecules at the surface at higher interfacial concentrations. The second study of DMSO was conducted with improved spectral resolution and VSF detection to allow a more detailed analysis of the VSF spectrum of DMSO at the water surface, particularly interference between the methyl SS and AS modes, and the OH stretch modes of water in this region (68). With increased DMSO adsorption, a decrease in the intensity of the OH stretch mode of highly coordinated water molecules is observed. Allen et al attributed this effect to the strength of the hydrogen bonds that DMSO forms with water, which are stronger than those between water molecules and thus effectively disrupt the long-range order of the hydrogen-bonding network. MSA also partitions at the surface and has a preferred orientation in which the MSA methyl group points away from the liquid surface (56). The surrounding surface water structure is significantly affected by the adsorption of MSA, with small amounts of MSA found to enhance the intermolecular hydrogen bonding between interfacial water molecules. Studies with mixtures of water and sulfuric acid show that MSA is effectively displaced by sulfuric acid at an aqueous surface (56). Other VSF studies have examined the water structure in the presence of different acids, including sulfuric and nitric acid (65, 95–97). The VSF of surface water containing a series of salts has also been examined and shows how the surface is perturbed by the presence of these species in the bulk and surface phase (67).

SURFACTANT STUDIES AT LIQUID SURFACES

Interest in the adsorption of surfactants at liquid surfaces stems from the wide application of surfactants in commercial products. Although most surfactants are still used for conventional cleaning and hygiene purposes, they are also widely used as stabilizing foams and emulsions in food processing and in beverages, in the stabilization of particulate dispersions, and in the secondary recovery of oil from porous rock beds (98, 99). All of these uses depend on the amphiphilic character of the molecule—part nonpolar hydrophobic hydrocarbon and part polar hydrophilic moiety.

Monolayers of surfactants assembled at solid surfaces have been extensively studied during the last decade. What has been learned from these studies is that the assembly and ordering of these surfactant monolayers are affected by van der Waals interactions between the alkyl chains of adjacent surfactants. For liquid surfaces where the surfactants are not covalently bound to the surface and the interface is more dynamic, head group–head group and head group–solvent interactions have the potential to play a significant if not more important role than that played by chain-chain interactions in the assembly process. In the presence of a solvent that can penetrate between the alkyl chains of the surfactants, such as for liquid-liquid interfaces, the picture is often even more complicated. This section describes the results of a number of studies of surfactant structure and conformation that were conducted at both the air-water and liquid-liquid interfaces. This summary is not intended to be comprehensive but merely provides an overview of the types of studies conducted thus far with VSFS. An example that demonstrates the difference in the assembly of surfactants at these two interfaces is highlighted.

VSF studies of surfactants have largely examined the CH stretch modes of the alkyl chains of various surfactants as a means of determining conformational ordering. Conformational ordering is qualitatively monitored by measuring the ratio of peak areas for the methyl SS and methylene SS modes in the VSF spectra (38). Under *SSP* polarization, chains that are relatively ordered and have few gauche defects should show little if any signal from the methylene CH stretch modes, owing to the symmetry of the molecule. In contrast, gauche defects owing to conformational disordering cause the VSF intensity of the methylene modes to increase as local symmetry constraints are relaxed. The terminal methyl group, which possesses both IR- and Raman-active vibrational modes, is by nature in a noncentrosymmetric environment and is further used in polarization experiments to determine the tilt angle of the hydrocarbon chain in these systems (100).

Surfactants at the Air-Water Interface

Some of the first VSF studies of surfactants measured at the air-water interface involved pentadecanoic acid (100). By monitoring the CH stretch region of the alkyl chains, it was found that, in the condensed phase, the alkyl chains extend and orient nearly normal to the surface. In the liquid expanded phase, the chains are highly disordered (100, 101). In a later related study, a carboxylic acid film of hexacosanaoic acid was studied at the air-water interface (102). The CH modes of the alkyl chains, the CO stretch in the head group, and the OH stretches of surface water have been monitored as a function of pH. The alkyl chains remained conformationally ordered at all pH values. At low pH values, at which the monolayer is neutral, the surface water is disordered by hydrogen bonding of water molecules with acid head groups. At high pH values, at which the head groups are ionized, the resulting surface fields lead to a more ordered hydrogen-bonding network.

Bell et al (103) have investigated the structure of monolayers of cationic surfactant, hexadecyltrimethylammonium *p*-tosylate, at the surface of water. They find that the *p*-tosylate ions are oriented with their methyl groups pointing away from the aqueous subphase and with the C_2 axis tilted an average of 30 to 40° from the surface normal. The vibrational spectrum of the cationic surfactant indicates that the number of gauche defects in the monolayer does not change dramatically when the counterion is changed from *p*-tosylate to bromide. However, the ends of the surfactant's hydrocarbon chains are tilted much farther from the surface normal in the presence of *p*-tosylate than in the presence of bromide. Alkyl cationic surfactant (tetradecyl-trimethylammonium) with an aromatic anionic counterion (benzoate) has also been examined by Ward et al (104). Other studies of the effect of different halide counterions on monolayers of cationic surfactants at the air-water interface found no evidence for specific effects on the structure of the surfactant monolayers (105).

A series of soluble surfactants has also been examined and compared at the air-water interface by Bell et al (106). Bell et al find that, in general, the number of gauche conformations increases as the area per chain increases. Comparison of surfactants with the same chain length and area per molecule shows that the structure of the chain region of the monolayer is sensitive to the nature of the head group and not just to the packing density.

Monolayers of nonionic surfactants poly(ethylene glycol) monodecyl ethers ($C_{12}E_m$; $m = 2$–8) have been examined at the air-water interface (107). The results show an increase in conformational disorder with increasing area per molecule and an apparent decrease in the angle of tilt of the methyl group. The study also suggests that the value of m does not affect the structure of the monolayer for $m = 4$–8. In another study of uncharged surfactants, Zhang et al (108) have examined the adsorption of long-chain amphiphiles containing the nitrile (CN) head group. These air-water studies monitored the spectroscopy of the CN stretch modes and the CD stretch modes of the terminal CD_3 moiety on the alkyl chains of these Langmuir monolayers. The results indicate that the orientations of the head group and terminal methyl group vary in a markedly different manner with amphiphile surface density. For the CN head group, Zhang et al observe a sharp change in the orientation angle of that group, with surface density at the density corresponding to a phase transition from the gas-liquid coexistence region to the liquid region. This change is discussed in terms of the difference in solvating water environment during the phase transition. The orientation of the tail is quite sensitive to the monolayer density, with the tail continuously becoming more upright upon compression.

Mixed monolayers of SDS and dodecanol at the air-water interface have been studied by Casson & Bain (109). They find that when a trace of dodecanol is introduced into a millimolar solution of SDS, the monolayer is transformed from a loosely packed structure to a densely packed structure like that of a pure dodecanol monolayer. Temperature-dependent studies suggest that this mixed monolayer undergoes a phase transition at 16°C, with a decrease in packing and an increase in conformational disorder. This behavior is analogous to that observed in a pure dodecanol monolayer. From these results Casson & Bain conclude that previously reported phase transitions for SDS at the air-water interface are probably a result of the presence of trace amounts of dodecanol in the SDS. Later studies provide evidence for a liquid-gas phase transition in monolayers of dodecanol adsorbed at the air-water interface (110).

Stanners et al (111) have examined the adsorption at the air-water interface of a series of alcohols from C_1 to C_8 in the CH and OH stretch regions. They find that the alkyl chains point away from the liquid for all alcohols studied. For C_6–C_8, the spectra show the presence of *trans*-gauche defects in the alkyl chain. The shift of spectra to lower frequency in the OH stretch region that was found for all alcohols indicates a well-ordered hydrogen-bonding network at the interface. Longer-chain alcohols have been pursued in other studies (112, 113).

Surfactants at Liquid-Liquid Interfaces

Surfactants studied at a liquid-liquid interface pose a challenge owing to the buried nature of the interface. Both the IR and visible beams must transmit through one of the solvents, but the number of solvents that are transparent in the spectral region of the surfactant vibrational modes is few. Even the weak IR absorbance caused by a thin layer of solvent can greatly diminish the incident IR beam. The most common organic solvent that has minimal absorption in the IR and is also insoluble in water is CCl_4. Therefore, all of the studies thus far of surfactants at liquid-liquid interfaces have been conducted at the CCl_4-water interface.

The first experiments in this area demonstrated the feasibility of using VSFS to measure the spectroscopy of charged alkyl surfactants at a liquid-liquid interface with a TIR geometry. D_2O has been used as the aqueous phase to minimize OH interference in these experiments. The studies demonstrate the ability to use TIR VSFS to measure interfacial concentrations of SDS below 0.01 monolayers. The studies have monitored the CH stretch modes of this surfactant and have shown that, whereas there is increased conformational ordering with interfacial concentration, the surfactants show significant gauche defects in the alkyl chains even at the highest concentrations. This result is attributed to interactions with the solvent that disrupt the van der Waals interactions between adjacent alkyl chains (38, 114). Polarization studies indicate that the terminal methyl group on the alkyl chain is oriented, on average, along the surface normal. Deuteration studies have been conducted to confirm the assignment of several CH vibrational modes. Later related studies examined a series of cationic and anionic surfactants of similar chain length, SDS, sodium dodecyl sulfonate (DDS), dodecyltrimethylammonium chloride (DTAC), and DAC, to understand how the different head groups alter the molecular conformation of the alkyl chains (70, 115). Figure 6 shows examples of the spectra of these molecules measured at the CCl_4/H_2O interface using *SSP* (see Figures 6*a*, *c*, *e*, *g*) and *SPS* (see Figures 6*b*, *d*, *f*, *h*) polarizations. The alkyl chains of the cationic surfactants possess the fewest gauche defects. Mixed cationic and anionic surfactants present at the interface lead to an increase in

Figure 6 Vibrational sum frequency spectra acquired with *P*-polarized IR and *S*-polarized visible light: (*a*) sodium dodecyl sulfate (SDS), (*c*) sodium dodecyl sulfonate (DDS), (*e*) dodecyltrimethylammonium chloride (DTAC), and (*g*) dodecylammonium chloride (DAC). Sum frequency (SF) spectra were acquired with *S*-polarized IR and *P*-polarized visible light: (*b*) SDS, (*d*) DDS, (*f*) DTAC, and (*g*) DAC. Spectra were obtained at the CCl_4/D_2O interface with a bulk aqueous-phase concentration of 5.0 mM for all of the surfactants studied. The generated SF was *S*-polarized in all cases. The solid curves represent a fit to the spectra using a combination of Gaussian and Lorentzian functions for each peak. Reproduced with permission from Reference 115.

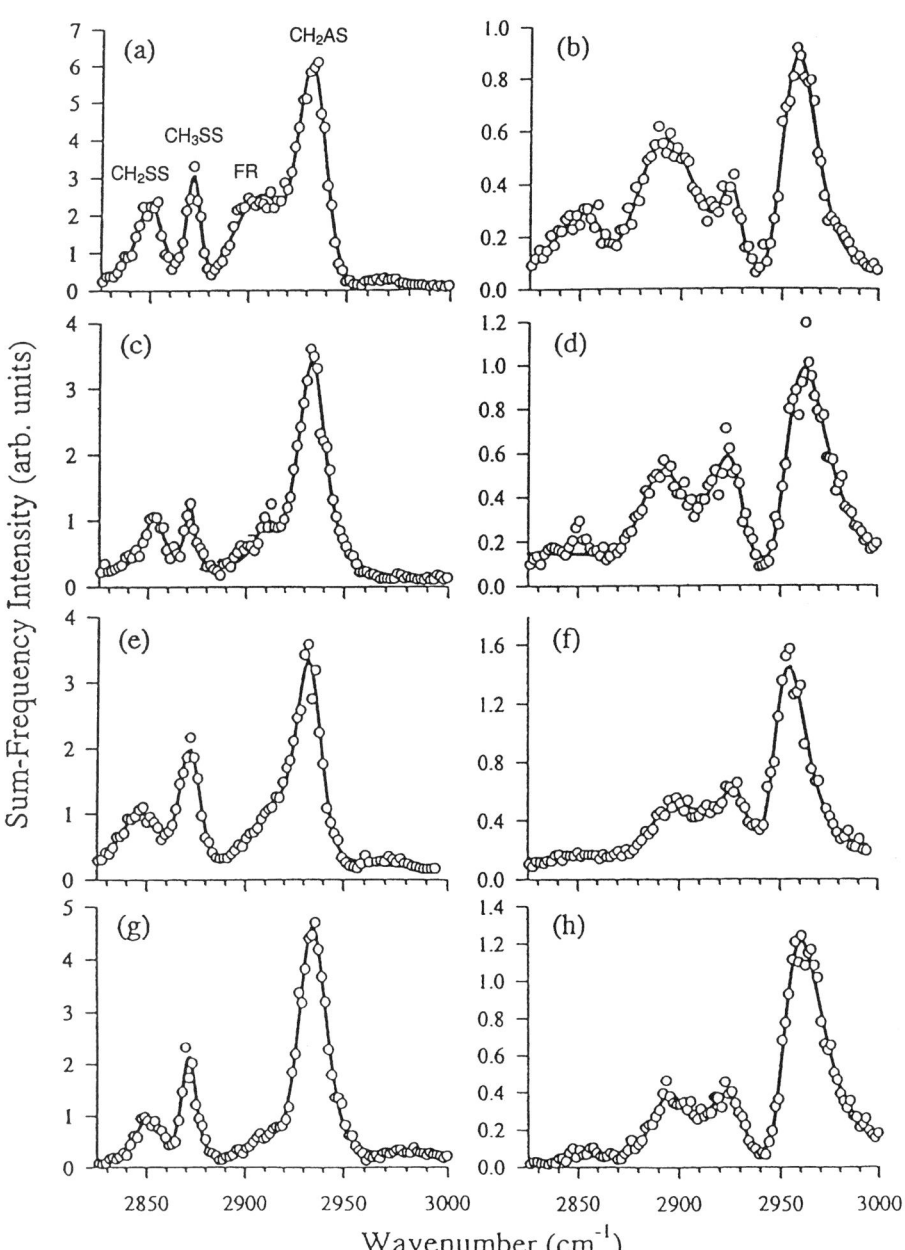

ordering of the chains, which is attributed to a reduction in head group repulsion at the interface. Studies of sulfonate-containing surfactants of different chain lengths find that the degree of disorder varies greatly with chain length for the surfactants at the CCl_4/D_2O interface (116). The shortest alkyl chain, C_6, displays the fewest gauche defects, and the longest chains, C_{11} and C_{12}, show the most disorder. No preference in ordering like that often found for monolayers on solid surfaces is observed between even and odd chain surfactants.

Comparative Studies at Both Liquid-Liquid and Liquid-Air Interfaces

One of the more interesting studies that demonstrates the differences and similarities in how surfactants adsorb, orient, and order themselves at a liquid-liquid versus an air-water interface has been conducted with alkyl and aryl sulfonate surfactants by Watry & Richmond (117). Linear alkane sulfonate surfactants and linear alkylbenzenesulfonates (LABS) make up a large fraction of the surfactants used in commercial detergents and cleansers, with LABS being one of the most widely used. The two surfactants studied by Watry & Richmond were DDS and sodium dodecylbenzenesulfonate (DBS). Figure 7 shows a spectrum of DBS measured at the CCl_4/H_2O interface (Figure 7a) and the air-water interface (Figure 7b), both with *SSP* polarizations. Because the two experiments were conducted with different laser systems, peak positions can be compared but not absolute intensities. The CH stretch modes of the alkyl chains are clearly observed, as are the expected modes ν_{7b} and ν_2 of the aromatic ring. Polarization studies have been used to monitor how the orientation of the benzene ring in DBS changes with interfacial concentration. Figure 8 shows a plot of the ν_2 intensity as a function of surface concentration for DBS at the air-water interface (Figure 8a) and the CCl_4/D_2O interface (Figure 8b). The liquid-liquid interface shows a monotonic increase in intensity with surface concentration. The benzene rings orient perpendicularly to the interface, and this orientation is not affected by changes in interfacial concentration even at surface concentrations up to a monolayer. This is in stark contrast to the effects observed at the air-water interface (Figure 8a). A sharp increase in intensity is observed as monolayer coverage is approached, which suggests that the increased interaction between the benzene rings as the surface concentration increases causes a change of orientation from a more planar to a perpendicular orientation relative to the surface plane. The different behaviors of the aryl group at the two interfaces are attributed to different interfacial potentials and the presence of polarizable CCl_4, which causes the benzene rings to continue their initial orientation at the liquid-liquid interface but to change in orientation at the air-water interface once crowding and increased benzene-benzene interactions occur.

The alkyl chain conformation for these two molecules at the different interfaces has also been compared. For DDS, the alkyl chains increase in conformational ordering as the interfacial concentration is increased at both interfaces. For DBS, a high degree of gauche defects is apparent in the chains at all concentrations at both

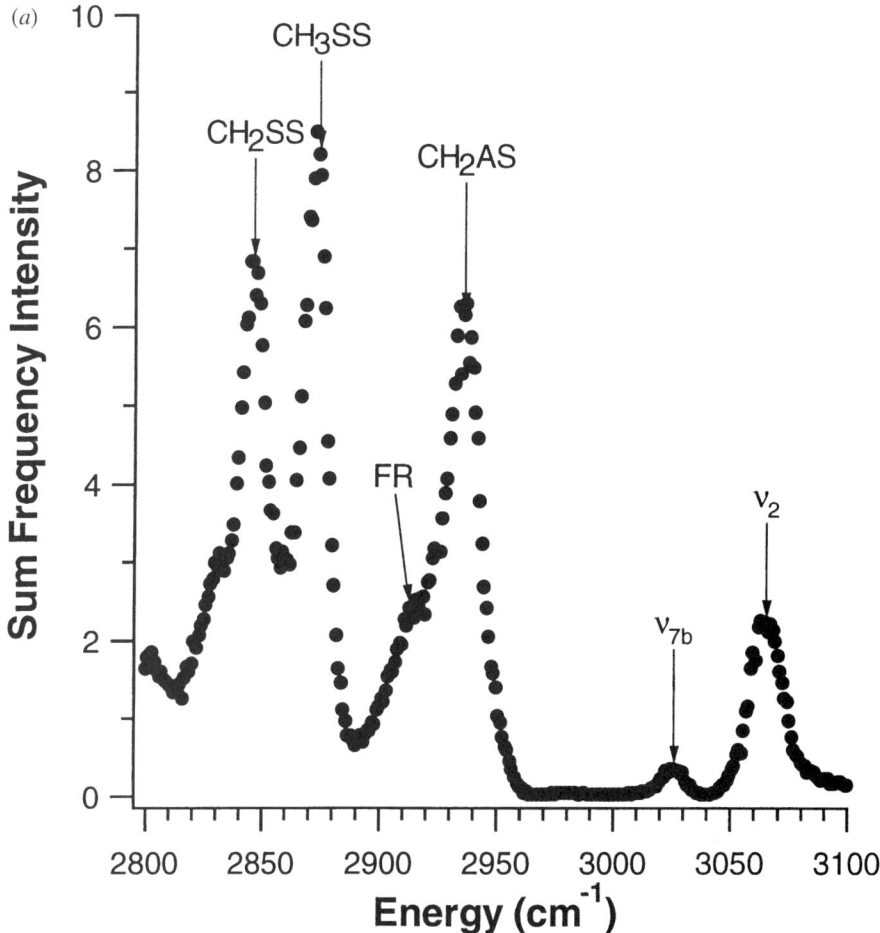

Figure 7 Vibrational sum frequency spectrum of (*a*) sodium dodecylbenzenesulfonate at the CCl_4/D_2O interface under conditions of 0.1 M NaCl, *SSP* polarization, and 1-ns laser pulses, and (*b*) sodium dodecylbenzenesulfonate at the air-D_2O interface, under conditions of 0.1 M NaCl, *SSP* polarization, and 2-ps laser pulses. Solid curves are fits to the data assuming a Voigt functional form for the peaks. Reproduced with permission from Reference 117.

interfaces. At the air-water interface, the benzene reorientation of DBS appears to have minimal effect on alkyl chain ordering. A key to understanding these differences comes from measurements of the limiting surface area per molecule. These studies measure nearly identical surface head group areas for DDS and DBS at monolayer coverages (~ 60 A^2 molecule^{-1}) even though the geometric head group of DBS is much larger than DDS. This suggests that DBS surfactants exist in a staggered head group geometry, a picture that is consistent with the disruption

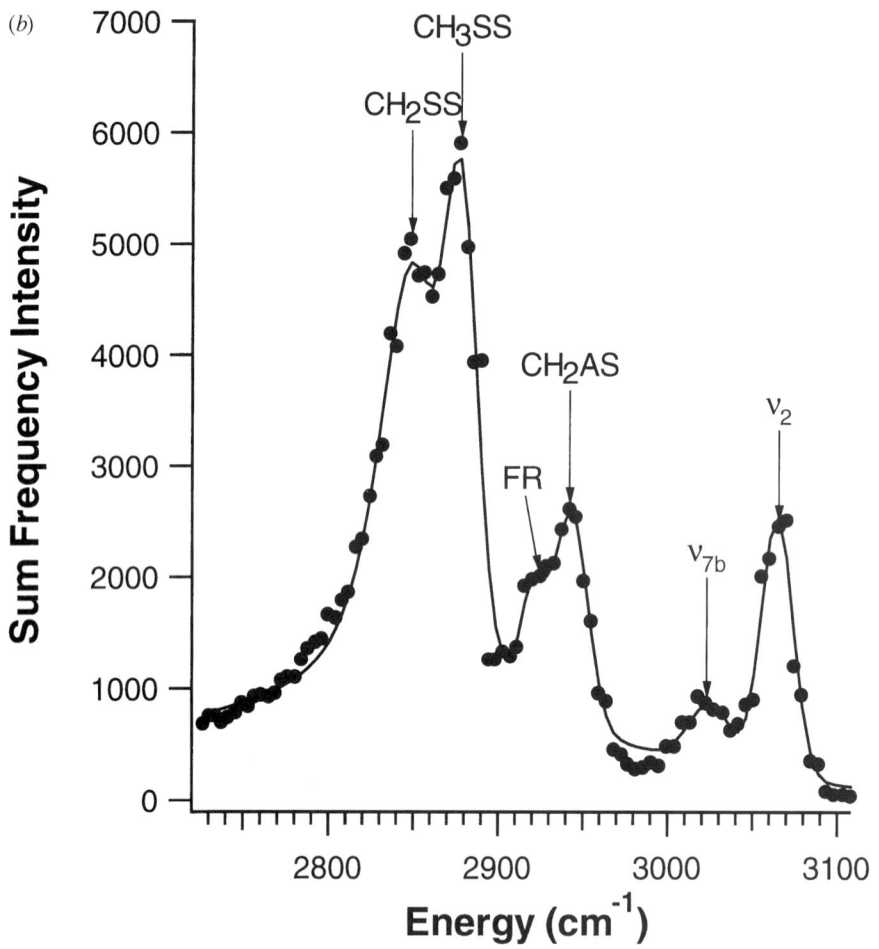

Figure 7 (*Continued*)

by the benzene rings of chain-chain interactions for the first few methylene groups adjacent to the benzene ring for DBS. There is no evidence that increased surface concentration and subsequent packing of alkyl chains can overcome this disruption.

BIOLOGICALLY RELEVANT SYSTEMS

Phospholipid Assembly

Phosphatidylcholines (PCs) are amphiphilic molecules possessing two long acyl chains connected to a zwitterionic head group by means of a three-carbon glycero backbone. These phospholipids are a major component of most cell membranes

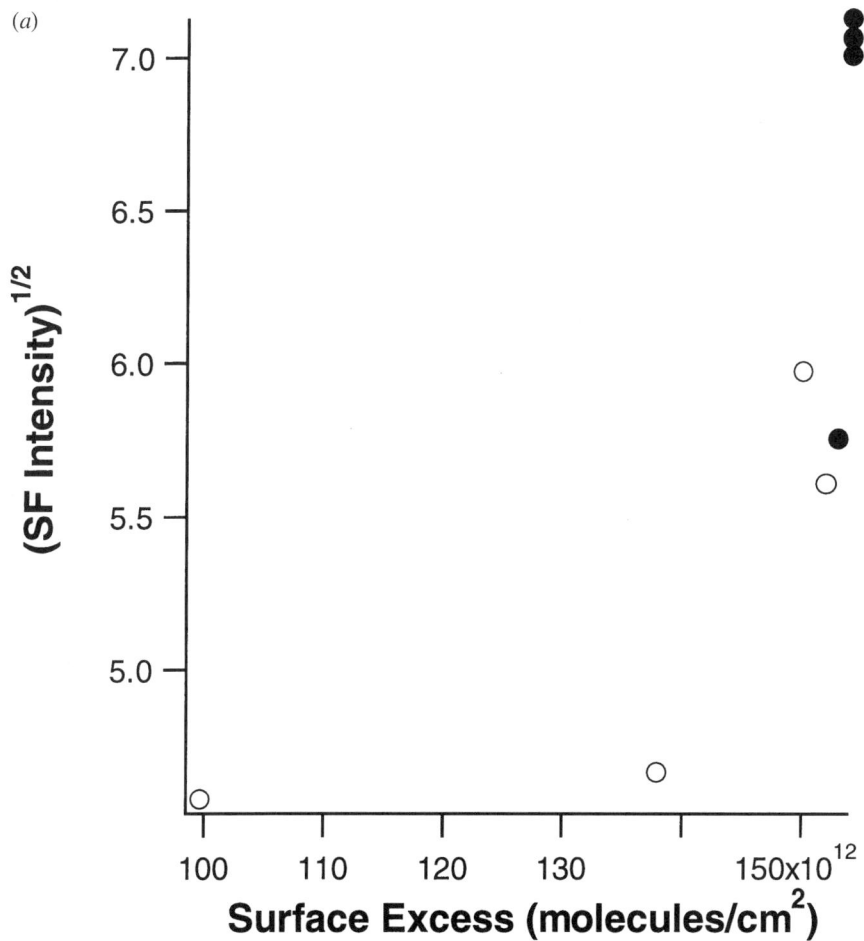

Figure 8 Square root of sum frequency intensity of the mode ν_2 as a function of surface concentration for (a) sodium dodecylbenzenesulfonate (DBS) at the air-water interface and (b) DBS at the CCl_4-water interface with 0.1 M NaCl. The line is a fit to the data. Solid circles indicate data points that correspond to concentrations at which the surfactant interfacial concentration is near a monolayer of coverage. Reproduced with permission from Reference 117.

and, consequently, have been the subject of intense scientific scrutiny over the past three decades. Cell membranes generally consist of a lipid bilayer structure, and PC monolayers have served as simple model membranes (118). VSFS studies of PC monolayers adsorbed at the CCl_4/D_2O interface and the air-water interface have been conducted to ascertain how the molecular structure of the monolayer depends on such variables as acyl chain length, surface concentration, temperature, and, by inference, the degree of chain solvation by the organic solvent (119). Some examples of these studies are given below.

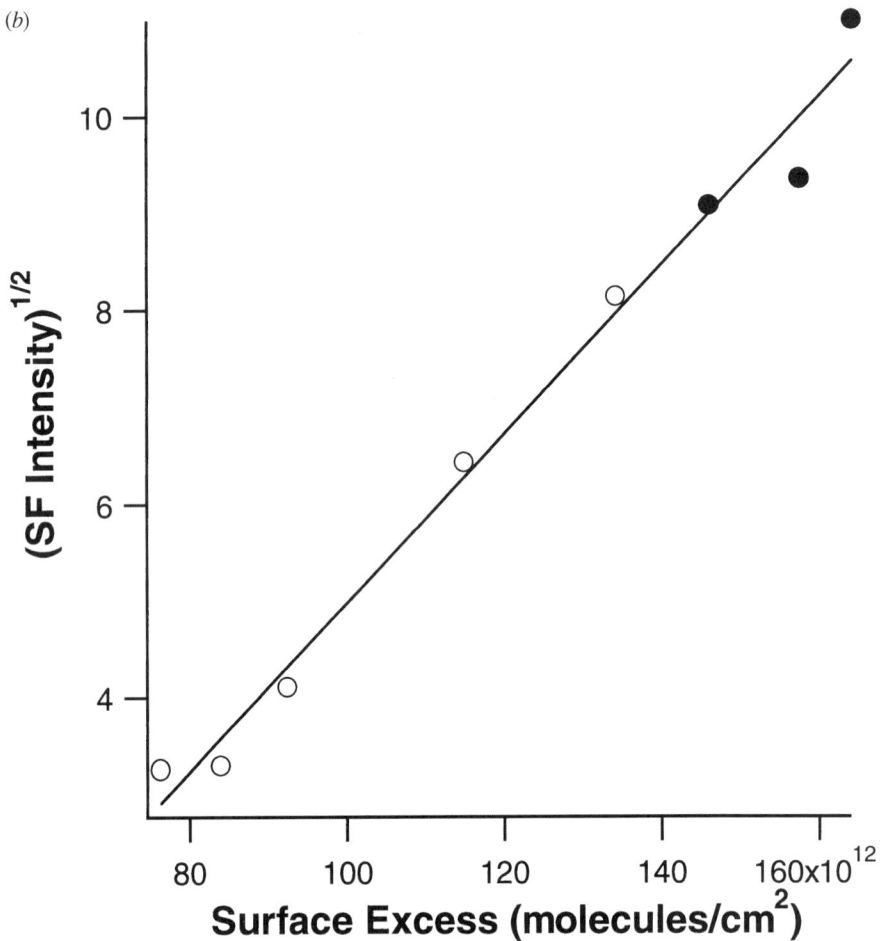

Figure 8 (*Continued*)

The first measurements of the vibrational spectroscopy of PCs at a liquid surface have been conducted with dialkylphosphocholines (Figure 9) (120). The four PCs examined included dilauroyl-PC (DLPC), dimyristoyl-PC (DMPC), dipalmitoyl-PC (DPPC) and distearoyl-PC (DSPC), with acyl chain units of C_{12}, C_{14}, C_{16}, and C_{18}, respectively. The CH stretch modes were used to determine conformational ordering of the chains in a manner similar to the surfactant studies described above. Figure 10 shows the spectra of the four PCs examined at the CCl_4/D_2O interface under *SSP* polarization. Conformational ordering of the chains has been determined by the ratio of areas under the d^+ (methylene symmetric stretch mode) and r^+ (methyl symmetric stretch mode). In the first set of studies of these PCs at the liquid-liquid interface it was found through temperature-controlled experiments that the lipid bilayer gel to liquid crystalline phase transition temperature plays

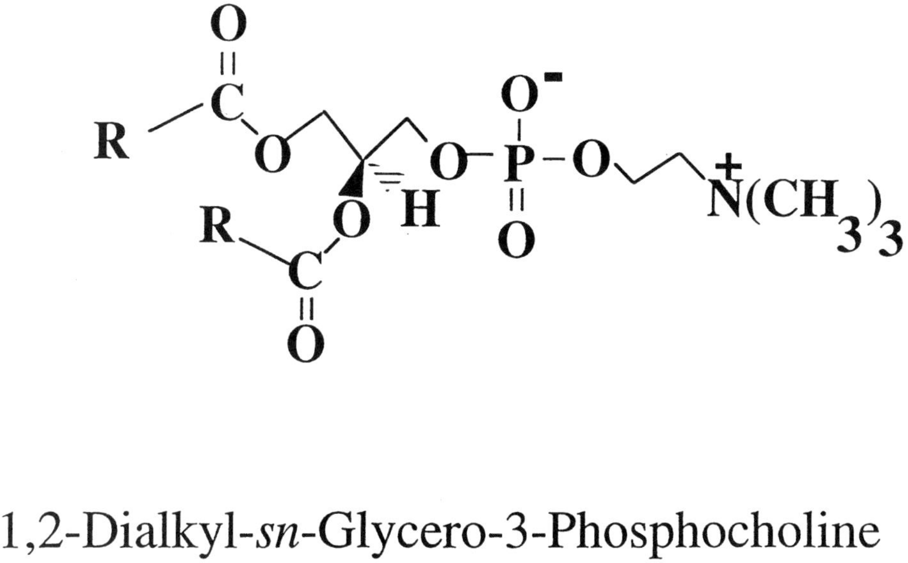

1,2-Dialkyl-*sn*-Glycero-3-Phosphocholine

Figure 9 Molecular structure of dialkyl phosphocholines used in this study. $R = C_{n-1}H_2CH_3$ where $N = 12$ is dilauroyl-phosphocholine, $N = 14$ is dimyristoyl-phosphocholine, $N = 16$ is dipalmitoyl-phosphocholine, and $N = 18$ is distearoyl-phosphocholine.

a pivotal role in determining the interfacial coverage and subsequent alkyl chain structure. This and other aspects of PC adsorption have been studied in further detail by Walker et al (73). In these liquid-liquid studies, polarization experiments showed that the acyl chains do not exhibit long-range order. Acyl chains within a tightly packed monolayer are found to stand up with their methyl C_3 axes aligned perpendicular to the interface. Temperature studies of the monolayer order suggest that a barrier exists to organic solvent penetration of the acyl chain network of a tightly packed, adsorbed monolayer. At the liquid-liquid interface, it is found that shorter-chain PC species form monolayers that are more ordered than those of longer-chain species, although the dependence of monolayer order on acyl chain length is small. When identical studies are conducted at the air-water interface, the opposite trend is observed. That is, the longer-chain PCs form monolayers that are dramatically more ordered than those of their shorter-chain counterparts. This difference can be seen in Figure 11, where ratios of r^+/d^+ for the PCs at the different interfaces are compared. This disparity provides evidence for acyl chain solvation by the organic CCl_4 solvent.

A study by Smiley & Richmond (121) has examined alkyl chain ordering of asymmetric PCs adsorbed at the CCl_4/D_2O interface. The large majority of biological phospholipids contain two dissimilar hydrocarbon chains per molecule. The two chains may differ in length and degree of unsaturation in such a way that a highly complex mixture of phospholipid structures is present in a particular bilayer membrane, allowing other structural elements to be accommodated without disruption

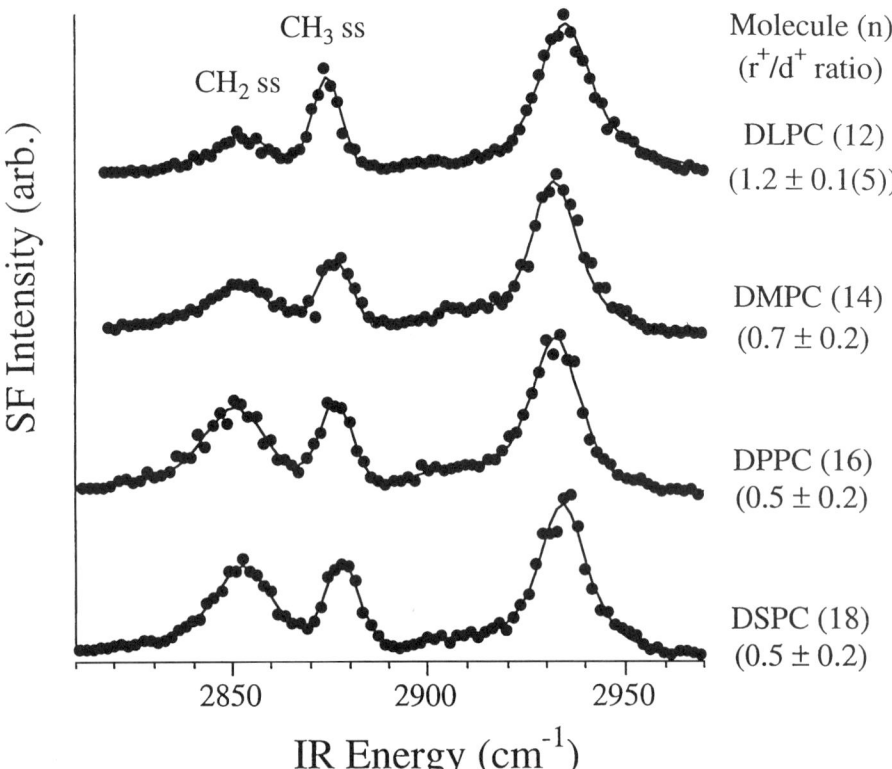

Figure 10 Vibrational sum frequency spectra of tightly packed monolayers composed of dilauroyl-phosphocholine (DLPC), dimyristoyl-phosphocholine (DMPC), dipalmitoyl-phosphatidylcholine (DPPC), and distearoyl-phosphatidylcholine (DSPC) at the water-CCl_4 interface. Ratios of the methylene SS intensity/methy SS intensity (r^+/d^+) appear underneath the PC acronyms. Spectral features were fit to Voigt profiles. Reproduced with permission from Reference 73.

of bilayer integrity (122). Smiley & Richmond have used a somewhat different preparation procedure than that used by Walker et al (73), owing to the insolubility of several of the PCs in water. These studies of a series of saturated symmetric and asymmetric chain PCs find that both symmetric PCs with 16 or fewer carbons per acyl chain and highly asymmetric PCs produce relatively disordered films at the liquid-liquid interfaces. The longest PCs studied, C_{18}/C_{18}, C_{18}/C_{16}, and C_{16}/C_{18}, form well-ordered monolayers at room temperatures. The highly disordered nature of the chains for the highly asymmetric PCs is consistent with a picture of reduced chain-chain interactions among mismatched portions of the longer chains. The greater disorder seen in the shorter-chain PCs, irrespective of chain mismatch, is attributed to reduced chain-chain interactions over a smaller chain length. This study has been further expanded to longer-chain PCs, up to C_{22} (123). The results show a strong increase in relative ordering for the longer-chain PCs, including

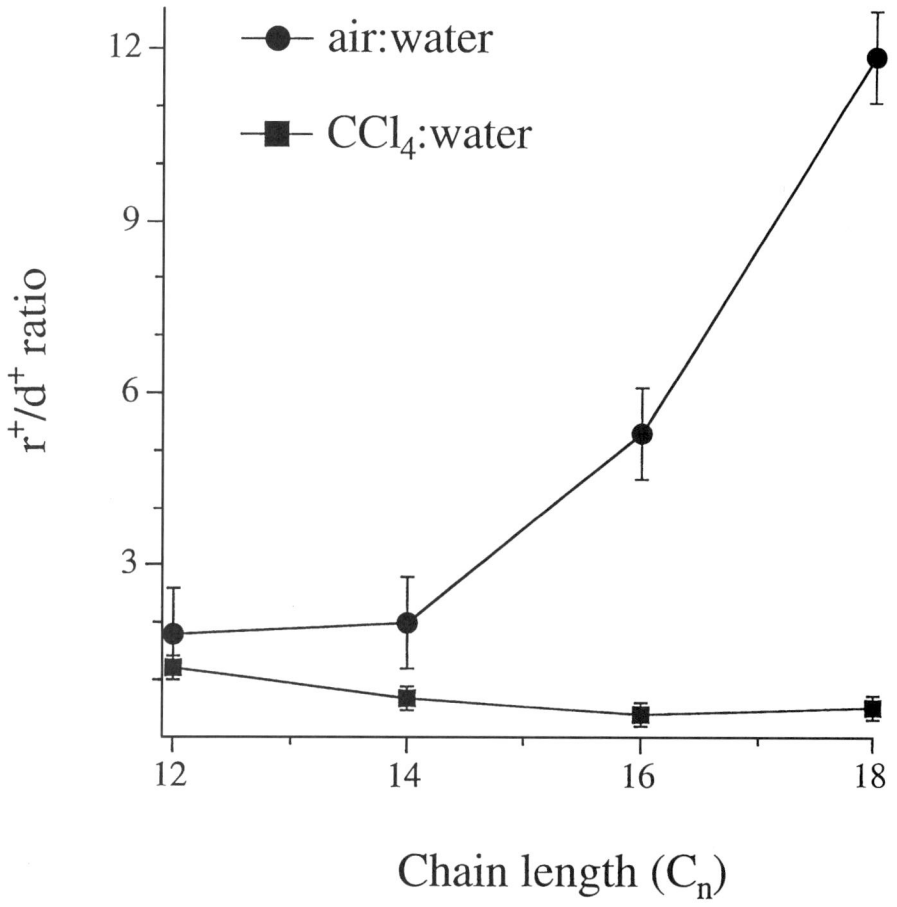

Figure 11 A comparison of the r^+/d^+ ratios for the different phosphatidylcholine monolayers at the air-water and CCl_4-water interfaces. Error bars reflect the uncertainty limits. Reproduced with permission from Reference 73.

those with an odd number of carbons per chain. The chain ordering of selected mixtures of adsorbed PCs is seen to be composition dependent, indicative of natural structural variances present in the large variety of functional biological membrane assemblies suited to a multitude of modes.

CONCLUSIONS

Obtaining a molecular level picture of liquid surfaces is a challenging area of surface science. From an experimental perspective, it is an exciting time to be delving into issues of this nature because of the emergence of many new experimental techniques that measure properties in the molecular domain. VSFS will clearly play an important role in this field as has already been demonstrated over the past decade. This summary has provided an overview of many of the studies of liquid surfaces by VSFS that have been conducted since the first results were reported in the mid-1980s. The examples described include measurements of the molecular interactions between water molecules at the vapor-water and organic-matter–water interfaces, the effect of adsorbed species on water structure and hydrogen bonding, and studies of the conformation and assembly of surfactants, macromolecules, biomolecules, and atmospherically relevant molecules at these interfaces.

There are numerous opportunities and challenges for this technique in the future. Studies are beginning to emerge that go beyond the traditionally probed 3-μm region to longer wavelengths at which a variety of additional modes can be measured. Extension to these longer-wavelength regions will require continued development of IR generation methods and pulsed lasers. As other modes become spectroscopically accessible, challenges will arise in spectral interpretation owing to the coherent nature of the VSF response, similar to the challenges demonstrated in this review for water spectra. Close coupling of the knowledge base derived from IR and Raman molecular spectroscopy with nonlinear optical principles will be important in interpretation of the spectral data to derive meaningful and rigorous interpretation of the observed spectral response. VSF studies of simple systems will be very valuable as the field moves toward understanding of the structure of more complex biological and polymeric systems. Time-resolved measurements that take advantage of the short pulsed nature of the lasers used in VSFS offer additional exciting opportunities.

ACKNOWLEDGMENTS

The author gratefully acknowledges the support of the National Science Foundation (CHE-9725751), the Office of Naval Research, and the Department of Energy's Office of Basic Energy Sciences for various studies described in this review that originated from this laboratory.

Visit the Annual Reviews home page at www.AnnualReviews.org

LITERATURE CITED

1. Stillinger FH. 1980. *Science* 209:451
2. Stillinger FH. 1973. *J. Solut. Chem.* 2:141–58
3. Tanford C. 1973. *The Hydrophobic Effect: Formation of Micelles and Biological Membranes.* New York: Wiley-Intersci.
4. Wilson MA, Pohorille A, Pratt LR. 1987. *J. Phys. Chem.* 91:4873–78
5. Hummer G, Garde S, García AE, Paulaitis ME, Pratt LR. 1988. *J. Phys. Chem. B* 102:10469–82
6. Lum K, Chandler D, Weeks JD. 1999. *J. Phys. Chem. B* 103:4570–77
7. Sokhan VP, Tildesley DJ. 1997. *Mol. Phys.* 92:625–40
8. Taylor RS, Dang LX, Garrett BC. 1996. *J. Phys. Chem.* 100:11720–25

9. Stillinger FH, Ben-Naim A. 1967. *J. Chem. Phys.* 47:4431–37
10. Townsend RM, Rice SA. 1991. *J. Chem. Phys.* 94:2207–18
11. Weeks JD. 1997. *J. Chem. Phys.* 67:3106–21
12. Croxton CA. 1986. *Fluid Interfacial Phenomena*. New York: Wiley & Sons
13. Benjamin I. 1992. *J. Chem. Phys.* 97:1432–45
14. Schweighofer K, Benjamin I. 2000. *J. Chem. Phys.* 112:1474–82
15. Schweighofer KJ, Essmann U, Berkowitz M. 1997. *J. Phys. Chem. B* 101:3793–99
16. Gao J, Jorgensen WL. 1988. *J. Phys. Chem.* 92:5813–22
17. Carpenter IL, Hehre WJ. 1990. *J. Phys. Chem.* 94:531–36
18. Feller SE, Zhang Y, Pastor RW. 1995. *J. Chem. Phys.* 103:10267–76
19. vanBuuren AR, Marrink S-J, Berendsen HJC. 1993. *J. Phys. Chem.* 97:9206–12
20. Napari I, Laaksonen A, Talanquer V, Oxtoby DW. 1999. *J. Chem. Phys.* 110:5906–12
21. Linse P. 1987. *J. Chem. Phys.* 86:4177–87
22. Du Q, Freysz E, Shen YR. 1994. *Phys. Rev. Lett.* 72:238–41
23. Ren Y, Meuse CW, Hsu SL, Stidham HD. 1994. *J. Phys. Chem.* 98:8424–30
24. Willard DM, Riter RE, Levinger NE. 1998. *J. Am. Chem. Soc.* 120:4151–68
25. Chamberlain J, Pemberton JE. 1997. *Langmuir* 13:3074–79
26. Dluhy RA. 1986. *J. Phys. Chem.* 90:1373–79
27. Buontempo JT, Rice SA. 1993. *J. Chem. Phys.* 98:5835–46
28. Wolfrum K, Graener H, Laubereau A. 1993. *Chem. Phys. Lett.* 214:41–46
29. Bell GR, Bain CD, Ward RN. 1996. *J. Chem. Soc. Faraday Trans.* 92:515–23
30. Knobler CM. 1990. *Advances in Chemical Physics*. New York: Wiley & Sons
31. Fiehrer KM, Nathanson GM. 1997. *J. Am. Chem. Soc.* 119:251–52
32. Brezesinski G, Thoma M, Struth B, Mohwald H. 1996. *J. Phys. Chem.* 100:3126–30
33. Lu JR, Li ZX, Thomas RK, Penfold J. 1996. *J. Chem. Soc. Faraday Trans.* 92:403–8
34. Weinbach SP, Kjaer K, Bouwman W, Als-Nielsen J, Leiserowitz L. 1996. *J. Phys. Chem.* 100:8356–62
35. Israelachvili JN. 1996. *Intermolecular and Surface Forces*. New York: Academic
36. Eisenthal KB. 1996. *Chem. Rev.* 96:1343–60
37. Wirth MJ, Burbage JD. 1992. *J. Phys. Chem.* 96:9022–25
38. Conboy JC, Messmer MC, Richmond GL. 1996. *J. Phys. Chem.* 100:7617–22
39. Shen YR. 1989. *Nature* 337:519–25
40. Bloembergen N. 1966. *Opt. Acta* 13:311–22
41. Bloembergen N, Simmon HJ, Lee CH. 1969. *Phys. Rev.* 181:1261–71
42. Gragson DE, McCarty BM, Richmond GL, Alavi DS. 1996. *J. Opt. Soc. B* 13:2075–83
43. Tadjeddine A, Peremans A, Guyot-Sionnest P. 1995. *Surf. Sci.* 335:210–20
44. Bain CD. 1995. *J. Chem. Soc. Faraday Trans.* 91:1281–96
45. Gragson DE, Alavi DS, Richmond GL. 1995. *Opt. Lett.* 20:1991–93
46. McGuire JA, Beck W, Wei X, Shen YR. 1999. *Opt. Lett.* 24:1877–80
47. van der Ham EWM, Vrehen QHR, Eliel ER. 1996. *Opt. Lett.* 21:1448–50
48. Richter LJ, Petralli-Mallo TP, Stephenson JC. 1998. *Opt. Lett.* 23:1594–97
49. Scherer JR. 1978. In *Advances in Infrared and Raman Spectroscopy*, Vol. 5, ed. RJH Clark, RE Hester, pp. 149–216. Philadelphia: Heyden
50. Lippincott ER, Finch JN, Schroeder R. 1959. In *Hydrogen Bonding*. New York: Pergamon
51. Schuster P, Zundel G, Sandorfy C. 1976. *The Hydrogen Bond. Recent Developments in Theory and Experiment*. Amsterdam: North Holland
52. Eisenberg D, Kauzmann W. 1969. *The*

Structure and Properties of Water. New York: Oxford Univ. Press
53. Franks F. 1972. *Water: A Comprehensive Treatise.* New York: Plenum
54. Pimental GC, McClellan AL. 1960. *The Hydrogen Bond.* San Francisco: Freeman
55. Du Q, Superfine R, Freysz E, Shen YR. 1993. *Phys. Rev. Lett.* 70:2313–16
56. Allen HC, Raymond EA, Richmond GL. 2001. *J. Phys. Chem.* In press
57. Brown MG, Raymond EA, Allen HC, Scatena LF, Richmond GL. 2000. *J. Phys. Chem.* 104:10220–26
58. Goss LM, Sharpe SW, Blake TA, Vaida V, Brault JW. 1999. *J. Phys. Chem.* 103:8620–24
59. Falk M, Giguere PA. 1957. *Can. J. Chem.* 35:1195–204
60. Zundel G. 1976. In *The Hydrogen Bond*, Vol. II, *Structure and Spectroscopy*, Chap. 15, ed. P Schuster, G Zundel, C Sandorfy, pp. 687–766. New York: North-Holland
61. Walrafen GE, Yang WH, Chu YC. 1997. *ACS Symp. Ser.* 676:287–308
62. Hare DE, Sorenson CM. 1990. *J. Chem. Phys.* 84:25–33
63. Buch V, Devlin JP. 1999. *J. Phys. Chem.* 110:3437–43
64. Whalley E, Klug DD. 1986. *J. Chem. Phys.* 84:78–80
65. Schnitzer C, Baldelli S, Shultz MJ. 1999. *Chem. Phys. Lett.* 313:416–20
66. Simonelli DM, Baldelli S, Shultz MJ. 1998. *Chem. Phys. Lett.* 298:400–404
67. Schnitzer C, Baldelli S, Shultz MJ. 2000. *J. Phys. Chem. B* 104:585–89
68. Allen HC, Raymond EA, Richmond GL. 2000. *Curr. Opin. Colloids Surf.* 5:74–80
69. Du Q, Freysz E, Shen YR. 1994. *Science* 264:826–28
70. Conboy JC, Messmer MC, Walker R, Richmond GL. 1997. *Prog. Colloids Polym. Sci.* 103:10–20
71. Gragson DE, Richmond GL. 1997. *J. Chem. Phys.* 107:9687–90
72. Richmond GL. 1997. *Anal. Chem. News Views* 69:A536–43
73. Walker RA, Gruetzmacher JA, Richmond GL. 1998. *J. Am. Chem. Soc.* 120:6991–7003
74. Walker RA, Smiley BE, Richmond GL. 1999. *Spectrscopy* 14:18–22
75. Gragson DE, Richmond GL. 1997. *Langmuir* 13:4804–6
76. Ahlström P, Wallqvist A, Engström S, Jonsson B. 1989. *Mol. Phys.* 68:563–68
77. Chang T-M, Dang LX. 1996. *J. Chem. Phys.* 104:6772–83
78. Hirose C, Akamatsu N, Domen K. 1992. *Appl. Spectrosc.* 46:1051–72
79. Hirose C, Akamatus N, Domen K. 1992. *J. Chem. Phys.* 96:997–1004
80. Shen YR. 1984. *The Principles of Nonlinear Optics.* New York: Wiley & Sons
81. Fredkin DR, Komornicki A, White SR, Wilson KR. 1983. *J. Chem. Phys.* 78:7077–92
82. Walker RA, Richmond GL. 1999. *Colloids Surf. A* 154:175–80
83. Gragson DE, McCarty BM, Richmond GL. 1997. *J. Am. Chem. Soc.* 119:6144–52
84. Gragson DE, Richmond GL. 1998. *J. Am. Chem. Soc.* 120:366–75
85. Gragson DE, Richmond GL. 1998. *J. Phys. Chem. B* 102:569–76
86. Gragson DE, McCarty BM, Richmond GL. 1996. *J. Phys. Chem.* 100:14272–75
87. Zhao X, Ong S, Eisenthal KB. 1993. *Chem. Phys. Lett.* 202:513–20
88. Chattoraj DK, Birdi KS. 1984. *Adsorption and the Gibbs Surface Excess.* New York: Plenum
89. Gragson DE, Richmond GL. 1998. *J. Phys. Chem.* 102:3847–61
90. Allen HC, Gragson DE, Richmond GL. 1999. *J. Phys. Chem.* 103:660–66
91. Barone SB, Turnipseed AA, Ravishankara AR. 1995. *Faraday Discuss.* 100:39–54
92. Davis D, Chen G, Kasibhatla P, Jefferson A, Tanner D, et al. 1998. *J. Geophys. Res.* 103:1657–78
93. Jefferson A, Tanner DJ, Eisels FL, Davis DD, Chen G, et al. 1998. *J. Geophys. Res.* 103:1647–56

94. Charlson RJ, Lovelock JE, Andrae MO, Warren SG. 1987. *Nature* 326:655–61
95. Raduge C, Pflumio V, Shen YR. 1997. *Chem. Phys. Lett.* 274:140–44
96. Baldelli S, Schnitzer C, Shultz MJ, Campbell DJ. 1997. *J. Phys. Chem. B* 49:10435–41
97. Baldelli S, Schnitzer C, Shultz MJ, Campbell DJ. 1998. *Chem. Phys. Lett.* 287:143–47
98. Morse PM. 1999. *Chem. Eng. News* 35–48
99. Porter MR. 1994. *Handbook of Surfactants.* 2nd ed. Glasgow, UK: Chapman & Hall.
100. Guyot-Sionnest P, Hunt JH, Shen YR. 1987. *Phys. Rev. Lett.* 59:1597–600
101. Hunt JH, Guyot-Sionnest P, Shen YR. 1987. *Chem. Phys. Lett.* 133:189–92
102. Miranda PB, Du Q, Shen YR. 1998. *Chem. Phys. Lett.* 286:1–8
103. Bell GR, Li ZX, Bain CD, Fischer P, Duffy DC. 1998. *J. Phys. Chem. B* 102:9461–72
104. Ward RN, Duffy DC, Bell GR, Bain CD. 1996. *Mol. Phys.* 88:269–80
105. Knock MM, Bain CD. 1999. *Langmuir* 16:2857–65
106. Bell GR, Bain CD, Ward RN. 1996. *J. Chem. Soc. Faraday Trans.* 92:515–23
107. Goates SR, Schofield DA, Bain CD. 1999. *Langmuir* 15:1400–9
108. Zhang D, Gutow J, Eisenthal KB. 1994. *J. Phys. Chem.* 98:13729–34
109. Casson BD, Bain CD. 1998. *J. Phys. Chem. B* 102:7434–41
110. Casson BD, Bain CD. 1999. *J. Am. Chem. Soc.* 121:2615–16
111. Stanners CD, Du Q, Chin RP, Cremer P, Somorjai GA, Shen YR. 1995. *Chem. Phys. Lett.* 232:407–13
112. Wolfrum K, Laubereau A. 1994. *Chem. Phys. Lett.* 228:83–88
113. Braun R, Casson BD, Bain CD. 1995. *Chem. Phys. Lett.* 245:326–34
114. Messmer M, Conboy JC, Richmond GL. 1995. *J. Am. Chem. Soc.* 117:8039–40
115. Conboy JC, Messmer MC, Richmond GL. 1997. *J. Phys. Chem.* 101:6724–33
116. Conboy JC, Messmer MC, Richmond GL. 1998. *Langmuir* 14:6722–27
117. Watry M, Richmond GL. 2000. *J. Am. Chem. Soc.* 122:875–83
118. Gershfeld NL. 1976. *Annu. Rev. Phys. Chem.* 27:349–60
119. Smiley BE, Walker RA, Gragson DE, Hannon TE, Richmond GL. 1998. *Proc. SPIE* 3273:134–44
120. Walker RA, Conboy JC, Richmond GL. 1997. *Langmuir* 13:3070–73
121. Smiley BE, Richmond GL. 1999. *J. Phys. Chem.* 103:653–59
122. Huang C, Mason JT. 1986. *Biochim. Biophys. Acta* 864:423–70
123. Smiley BE, Richmond GL. 2000. *Biopolym. Biospectrosc.* 57:111–16

LIGHT-EMITTING ELECTROCHEMICAL PROCESSES

Neal R Armstrong,[1] R Mark Wightman,[2] and Erin M Gross[2]

[1]Department of Chemistry, University of Arizona, Tucson, Arizona 85721;
e-mail: nra@u.arizona.edu
[2]Department of Chemistry, University of North Carolina–Chapel Hill, Chapel Hill,
North Carolina 27599; e-mail: rmw@unc.edu; erinm@email.unc.edu

Key Words electrogenerated chemiluminescence, microelectrode, organic light-emitting diode, light-emitting polymer, electroluminescence

■ **Abstract** Electrochemical processes leading to light emission are reviewed, with emphasis on aspects of this subject relevant to the understanding and optimization of electrogenerated luminescence (EL) in organic thin-film materials. The basic energetic requirements of light emission from electrochemically initiated solution redox reactions [electrogenerated chemiluminescence (ECL)] are reviewed first. This review is followed by a discussion of light-emitting electrochemical processes that have been observed in hybrids of ionically conducting polymers and electronically conducting polymers. Finally, the features of EL in insulating polymers and molecular thin films are reviewed, along with recent electrochemical and ECL studies of the small-molecule components of certain organic light-emitting diodes. These studies provide a conceptual framework for understanding and optimizing these materials and the EL process.

INTRODUCTION

Light-emitting electrochemical processes have interested a broad spectrum of the scientific community for at least the last four decades (1–8). The initial interest in these processes arose from the fact that they provided a spectroscopic means of probing the molecular aspects of homogeneous and heterogeneous electron transfer, but also because it was believed that they might eventually be exploited for displays and even lasers (9).

Beginning 14 years ago, organic thin-film technologies were introduced that finally showed the possibility of display and lighting applications arising from electrogenerated luminescence (EL). As shown in Figure 1a, these thin-film devices are typically arranged in single or multiple layers, sandwiched between an anode [typically a transparent oxide film such as indium–tin oxide (ITO)] and a cathode such as Al, Mg, Ag/Mg, or Li/Al alloys. Tang & Van Slyke (10) introduced a near-display-brightness technology in 1987, the organic light-emitting diode (OLED, shown schematically in Figure 1c), which consisted of vacuum-deposited layers of triarylamines and a luminescent metal complex, Alq$_3$ [aluminuminato-tris (8-hydroxyquinolate)] (10, 11). Burroughes et al next introduced display-quality polymeric thin-film technologies, light-emitting polymers (LEPs), as single-layer [poly(p-phenylenevinylene), PPV] devices (Figure 1b) (12). Heeger and coworkers followed with LEPs of comparable composition (13). LEP materials were later modified to add a second layer (a cyano-derivative of PPV) to create a system conceptually close to the vacuum-deposited two-layer OLEDs (14). Recent reviews summarize much of this early developmental work (15–17).

These initial reports dealt with organic thin films that were insulators, requiring high internal fields to sustain charge injection and recombination events leading to luminescence. Luminescent polymer thin films incorporating ion-conducting polymers and high concentrations of mobile ions were introduced soon after [designated as light-emitting electrochemical cells (LECs)], which also showed impressive electroluminescent responses (18–21). These have been reported as single-layer devices (Figure 1d) and as single-layer devices divided into two domains by virtue of differences in mobilities of charge-compensating ions and the "freezing out" of their motions after device activation (Figure 1e) (20–22). More recently, these materials have appeared as multilayer films, grown by self-assembly in a stepwise fashion (23, 24). Electroluminescent materials have been optimized to the point that entry-level displays are now being sold (25) although significant work remains to optimize their brightness, luminous efficiency, color range, and long-term stabilities (17, 26).

For small-molecule-based OLED materials, consisting of weakly interacting molecular components, solution electrochemical studies can be directly related to both the frontier orbital energies of the molecular components of these devices and the energetic requirements for charge recombination events that lead to luminescent states (27, 28). For polymeric EL materials, this "molecular view" of their electronic and luminescent properties has been useful as well, especially because for many LEP materials, the polymer layers generally contain high concentrations of smaller oligomeric components, which may often be the critical components in light emission (26).

Figure 1 Schematic views of organic light-emitting diodes and light-emitting polymers. (*a*) A two-layer device consisting of an indium–tin oxide (ITO) anode: layer 1, a hole-transporting organic layer (HTL) (∼40–60 nm); layer 2, an electron-transporting (and generally emissive) organic layer (ETL) (∼40–60 nm) topped by a cathode (typically Mg, Ca, Al, Li/Al alloys, etc). (*b, c*) Energy versus distance profiles in a single-layer (panel *b*) and a two-layer organic light-emitting diode (OLED) or light-emitting polymer (panel *c*). (*d*) Energy versus distance profiles in a single-layer light-emitting electrochemical cell (LEC). (*e*) Relative concentration versus distance profiles for a "two-layer" LEC, e.g. a single-layer LEC based on a polymer of ruthenium trisbipyridyl [Ru(bipy)$_3^{2+}$] after electrolysis has established concentration gradients for the Ru^{2+}, Ru^{3+}, and Ru$^+$ states, providing for rectification in the current/voltage response (adapted from Figure 1 of Reference 20, by permission).

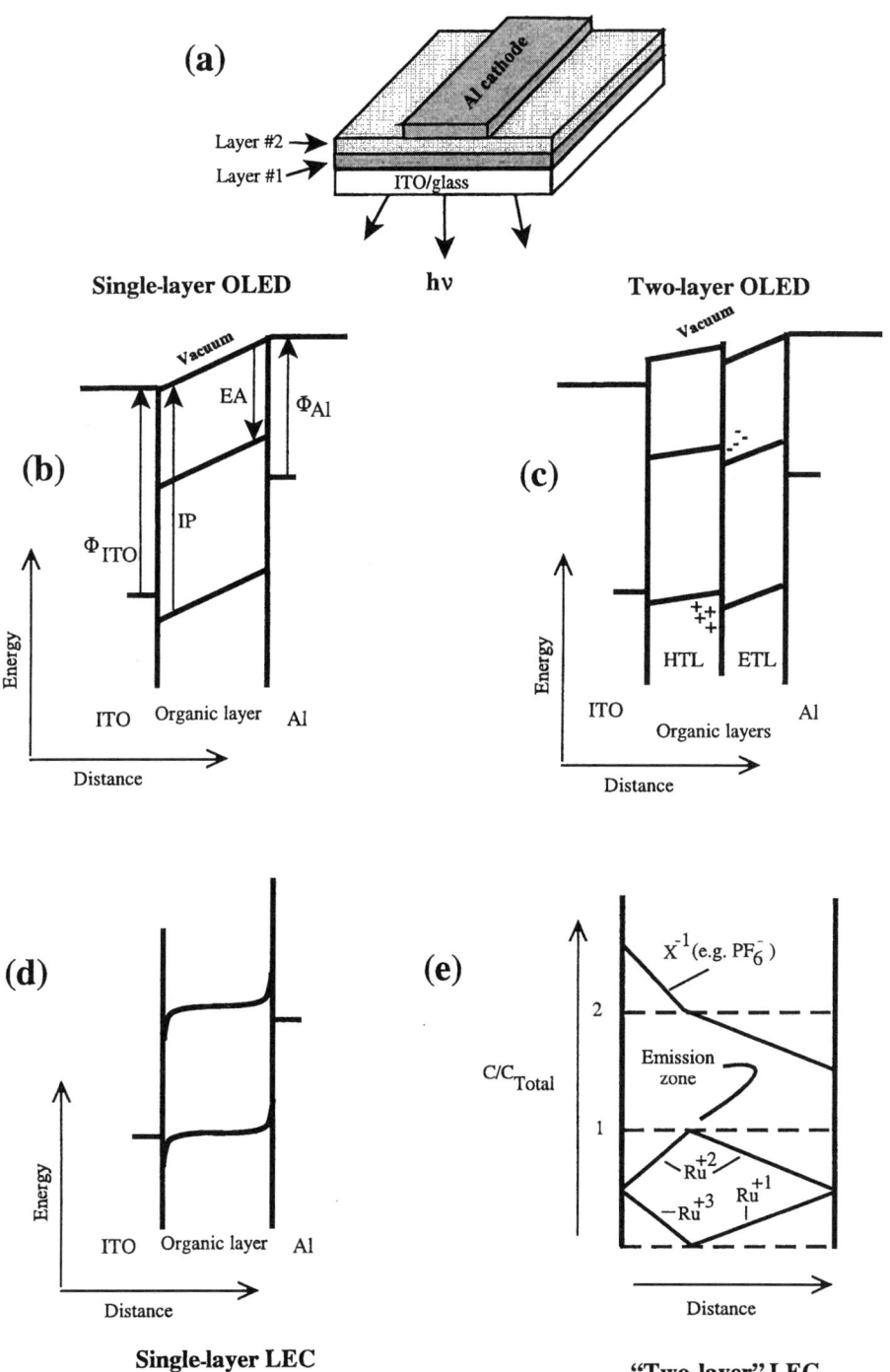

The first section of this review therefore deals with recent studies of solution electrogenerated chemiluminescence (ECL) of probe molecules such as diphenylanthracene (DPA), which illustrate how the basic steps in electron transfer determine the efficiency of light emission. The second section reviews the recent development of LECs, which show emission most closely related to solution ECL events. The third section briefly reviews the operation of OLED and LEP displays, based on insulating thin-film materials, the role played by electrochemical characterization of the molecular components of these technologies, and the way ECL studies of those molecular components can provide direction for optimizing these technologies.

LIGHT EMISSION FROM RADICAL RECOMBINATION REACTIONS OF MODEL COMPOUNDS IN SOLUTION: ELECTROGENERATED CHEMILUMINESCENCE

Principles of Solution Electrogenerated Chemiluminescence

Heterogeneous electrochemical reduction of an acceptor species (A) and oxidation of a donor (D) are the first steps in ECL:

$$A + e^- \rightarrow A^{\cdot -}, \qquad 1.$$

$$D \rightarrow D^{\cdot +} + e^-. \qquad 2.$$

In the simplest ECL processes, $A^{\cdot -}$ and $D^{\cdot +}$ are the anion radical and the cation radical, respectively, of the same aromatic molecule (8, 29, 30). Subsequent electron transfer reactions between $A^{\cdot -}$ and $D^{\cdot +}$ include the formation of the singlet state ($^1A^*$) with rate coefficient k_S' (Equation 3), the formation of the triplet state ($^3A^*$) with rate coefficient k_T' (Equation 4), or the formation of the neutral ground states of both A and D with rate coefficient k_g' (Equation 5):

$$A^{\cdot -} + D^{\cdot +} \xrightarrow{k_S'} {}^S A^* + D \qquad 3.$$

$$\xrightarrow{k_T'} {}^T A^* + D \qquad 4.$$

$$\xrightarrow{k_g'} A + D. \qquad 5.$$

The prime notation in each rate coefficient arises from the fact that formation of $D^{\cdot +}/A^{\cdot -}$ "encounter complexes" (discussed later) may precede these bimolecular electron transfer events, with rate coefficients for formation (k_d) (by diffusion) and dissociation (k_{-d}) of these complexes, such that $k' = [k \cdot k_d]/[k + k_{-d}]$, where the electron transfer rate coefficient $k = k_S$, k_T, or k_g.

Marcus' theory for electron transfer predicts that k_S, k_T, and k_g are controlled by the excess free energy in the $A^{\cdot -}/D^{\cdot +}$ redox reaction according to the following (31, 32):

$$k_{(S,T,g)} \propto \exp\left[-\left(\Delta G_{(S,T,g)} - \lambda\right)^2/4\lambda k_B T\right], \qquad 6.$$

where $\Delta G_{(S,T,g)}$ is the excess free energy in the redox process related to the difference in redox potentials for the reduction of A and oxidation of D ($\Delta E^\circ = E^\circ_{\text{oxidation,D}} - E^\circ_{\text{reduction,A}}$), and the energy of the final state achieved (S, T, or g); λ is the overall reorganization energy involved in reaching the activated state; k_B is the Boltzmann constant; and T is the absolute temperature. For many ECL reactions, $\Delta G_{(g)}$ is large enough to place the reaction in Equation 5 in the "Marcus inverted region," making k_g much smaller than k_S or k_T. Luminescent states are therefore the preferred final product of the reaction between $D^{\cdot +}$ and $A^{\cdot -}$. Predictions that ECL processes could be described in this fashion were among the first made by Marcus and others in the development of modern electron-transfer theory.

The ionization potentials (IP) of D and the electron affinity (EA) of A are related, respectively, to the first solution oxidation potential of D and the first reduction potential of A (28, 33, 34). They can be used to estimate the excess free energy of these redox systems in any dielectric material, relative to the ground state, according to the following:

$$\Delta G = -\text{IP}_D + \text{EA}_A - \Delta G^\circ_{\text{solv,D}} - \Delta G^\circ_{\text{solv,A}} - w_{a,\mu}, \qquad 7.$$

where $\Delta G^\circ_{\text{solv}}$ is the Born solvation energy of the charged species ($D^{\cdot +}$ or $A^{\cdot -}$), and $w_{a,\mu}$ is the work associated with bringing the two ions into close proximity. In high-dielectric-constant solvents with high concentrations of inert electrolytes $w_{a,\mu}$ is generally in the range from 0.001 to 0.01 eV and can be neglected. Confirmation that the rates of singlet or triplet formation could be controlled by controlling $\Delta E^\circ = E^\circ_{\text{oxidation,D}} - E^\circ_{\text{reduction,A}}$ have been carried out for several systems, a few of which are discussed below.

Reactions where the excess free energy is large enough to allow for direct creation of $^1A^*$ (Equation 3) are termed energy-sufficient or S-route processes. When the excess free energy in the electron exchange process is not enough to create $^1A^*$ directly $^3A^*$ may form (Equation 4) and triplet–triplet annihilation reactions can then lead to the formation of $^1A^*$:

$$^3A^* + {}^3A^* \rightarrow {}^1A^* + A. \qquad 8.$$

Equations 4 and 8 represent energy-deficient or T-route processes. For a few systems, there is also evidence of an "exciplex" species as the emissive product of the electron transfer reaction; that is, emission occurs from a complex involving $A^{\cdot -}$ ••• $D^{\cdot +}$, typically at much lower energies than the emission spectra of either

A* or D* (35–38). ECL processes involving such exciplex states, however, are typically less efficient than those leading to direct formation of either singlet or triplet states of A*.

For energy-sufficient ECL processes it has been assumed that spin statistics dictate the formation of 25% singlet states of A* and 75% triplet states (2, 7, 8). Such partitioning of energy between spin states can be rationalized by a simple two-level model (Figure 2) with the assumption that the excited states of A are lower in energy than those for D thus making it the energy acceptor in this system and thus the emissive species (38). A similar scheme can be written when D* is the emissive species. Whether these spin statistics hold over a wide range of small-molecule and polymeric materials is still an area of active exploration and recent publications suggest that singlet yields much greater than 25% are possible (39–42).

Provided this singlet/triplet ratio holds, consideration of the relative rates of bimolecular electron transfer for singlet state production $^S A^*$, the triplet state $^T A^*$, and the ground state of A yields the efficiency of the ECL process, ϕ_{ECL}, which depends on both the relative rates for each of the processes described in Equations 3 through 5 and the luminescence efficiency, ϕ_f:

$$\phi_{ECL} = \phi_f k_s'/(k_s' + 3k_T' + k_g'). \qquad 9.$$

"Encounter complexes" (Figure 3) are likely to play a key role in modulating the rates of formation of emissive states. This concept has been extensively developed for photo-induced electron-transfer reactions in solution (43) which are viewed as the inverse of the ECL process, as shown in Figure 3, wherein there is separation of $D^{·+}$ and $A^{·-}$ by a single layer of solvent (solvent-separated radical-ion pairs, SSRIP), or simple contact radical-ion pairing (CRIP). The rate coefficients k_d and k_{-d}, which affect k_s', k_T', and k_g', are therefore expected to be strongly dependent on the polarity and polarizability of the solvent environment. This concept has not been thoroughly explored, however, for light-emitting electrochemical processes, and the impact of these encounter complexes on EL in condensed-phase environments is not yet clear.

Ionic species also exert control over the energetics of light-emitting electrochemical processes (33, 34, 44). In concentrated electrolyte solutions, $D^{·+}$ and $A^{·-}$ may exist in ion-paired forms with counter ions from the supporting electrolyte especially in low-dielectric-constant media with high concentrations of added electrolyte:

$$D^{·+} + X^- \rightleftarrows D^{·+} \cdots X^-, \qquad 10.$$

$$K_{DX} = [D^{·+} \cdots X^-]/[D^{·+}][X^-], \qquad 11.$$

$$A^{·-} + M^+ \rightleftarrows A^{·-} \cdots M^+, \qquad 12.$$

$$K_{AM} = [A^{·-} \cdots M^+]/[A^{·-}][M^+]. \qquad 13.$$

K_{AM} and K_{DX} are defined as the ion-pairing equilibrium constants for $A^{·-}$ and $D^{·+}$ with solution cations and anions, respectively.

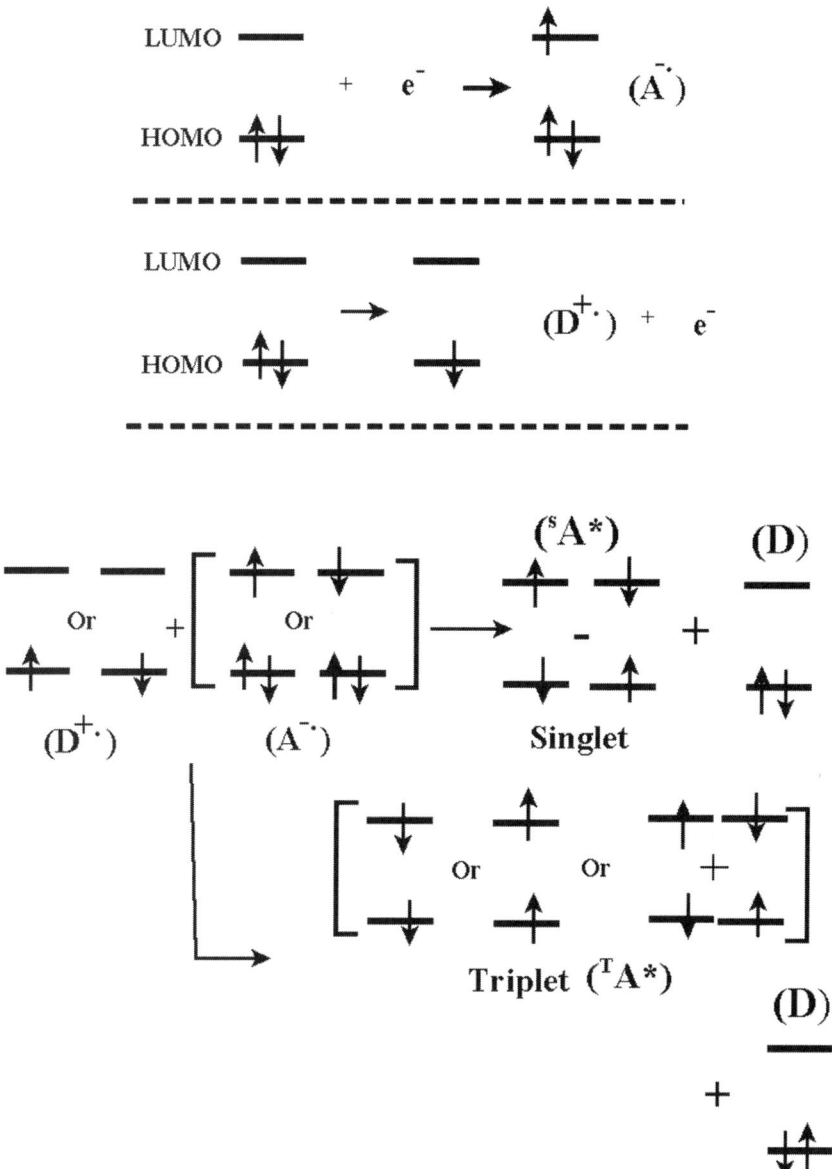

Figure 2 Schematic view of the spin states of radical species ($A^{\cdot -}$ and $D^{\cdot +}$) before and after bimolecular electron transfer, leading to 25% singlet states (spins antiparallel with zero resultant spin angular momentum) and 75% triplet states (spins parallel, with three resulting nonzero spin angular momentum vectors).

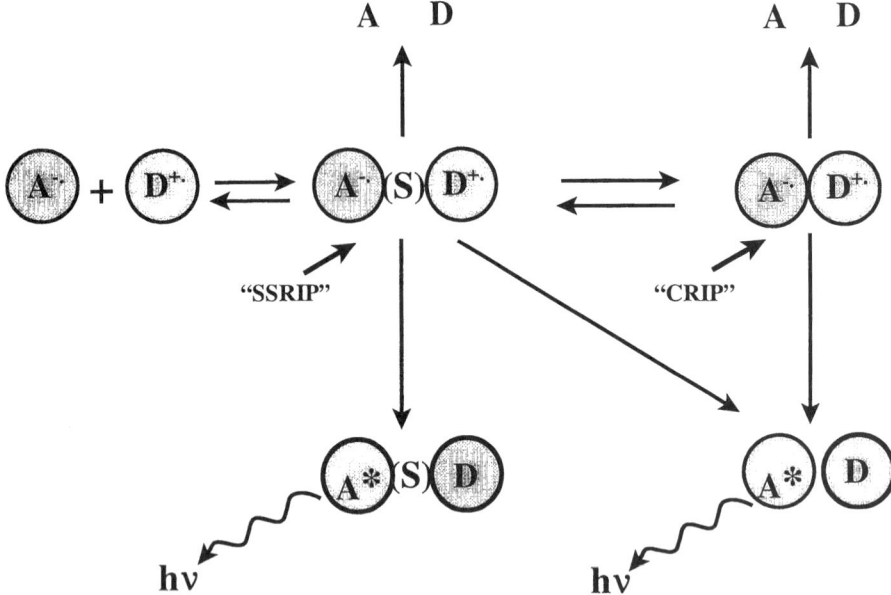

Figure 3 Schematic view of solvent-separated radical-ion pairs (SSRIP) and contact radical-ion pairs (CRIP) as they may exist during bimolecular electron transfer between $A^{·-}$ and $D^{·+}$ (adapted from Figure 5, Reference 43).

Such ion-pairing processes lower $\Delta E^o = E^o_{oxidation,D} - E^o_{reduction,A}$ through stabilization of the ionic states, lowering the magnitude of excess free energy available for the recombination event and the subsequent rates of bimolecular electron transfer and ϕ_{ECL}. Low-dielectric-constant environments, without added ions, are therefore expected to provide for higher excess free energies controlling the $A^{·-}/D^{·+}$ electron-transfer process, and therefore higher ϕ_{ECL}.

Electrogenerated Chemiluminescence in Model Systems

The recombination reactions of the electrogenerated anion and cation radicals of molecules such as 9,10-diphenylanthracene ($DPA^{·-}/DPA^{·+}$) and rubrene ($rubrene^{·-}/rubrene^{·+}$) (shown in Figure 4) are among the most studied light-emitting electrochemical processes (8, 33, 34, 45–47). DPA and rubrene both demonstrate rapid heterogeneous electrochemical reduction and oxidation (at Pt and Au electrodes) to form stable anion radicals and cation radicals in a wide variety of solvents. Perhaps as importantly, these molecules have high values of ϕ_F, the quantum yield for fluoresence. Organometallic compounds such as ruthenium(II)trisbipyridyl [$Ru(bipy)_3^{2+}$] (Figure 4) also show good ECL efficiencies following electrogeneration of $Ru(bipy)_3^{3+}$ and $Ru(bipy)_3^{+}$ and recombination to form $Ru(bipy)_3^{2+*}$ (8, 47, 48).

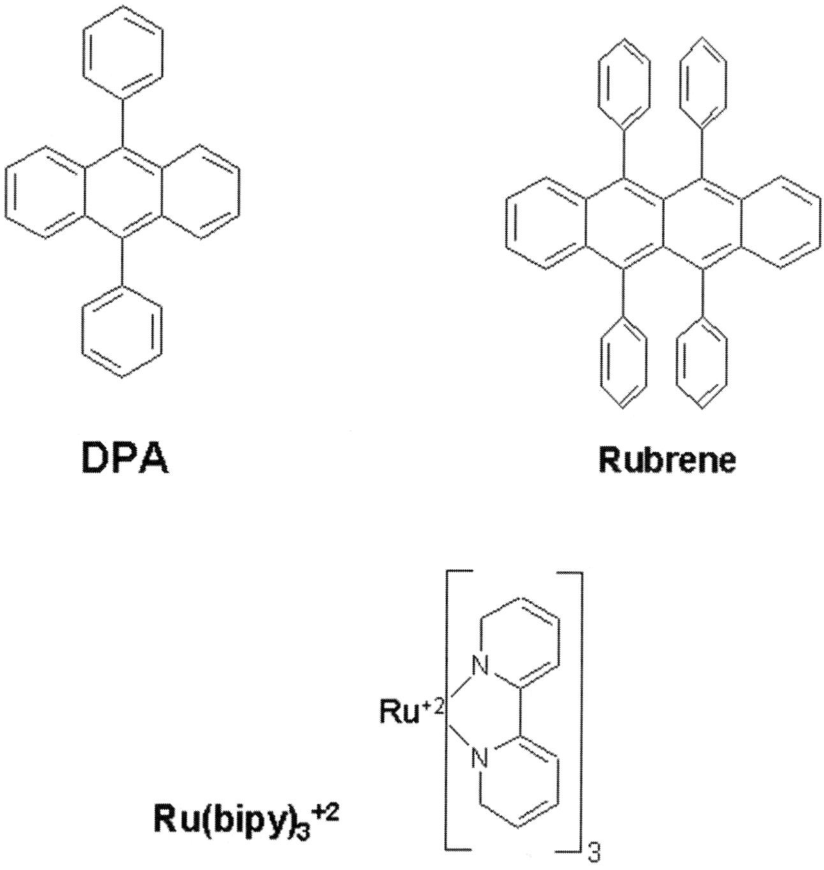

DPA   **Rubrene**

Ru(bipy)$_3$$^{+2}$

Figure 4 Schematic drawings of diphenylanthracene (DPA), rubrene, and the emissive metal complex Ru(bipy)$_3^{2+}$.

ECL processes can now be studied on short time scales, at high sensitivity, and in a variety of solvent media. Using ultramicroelectrodes, both the cation and anion radical states of DPA, as well as related donors or acceptors, have been created on time scales of 10^{-6} to 10^{-3} s (33, 34, 49–51). Electrodes with areas as small as 10^{-7} cm^2 make it possible to conduct two-electrode or three-electrode voltammetry or potential-step experiments in small solution volumes, with picoampere to nanoampere currents generated on each voltammetric sweep or voltage pulse. Voltammetric sweep rates of $\geq 100,000$ V s^{-1} can be employed, or voltage pulses can be used at frequencies from 1 to 100 kHz (to form the cation radical on the forward pulse and the anion radical on the reverse pulse) (Figure 5A). Radical species are therefore created under diffusion control in a solution region that extends out from the indicator electrode from 100 to 1000 nm

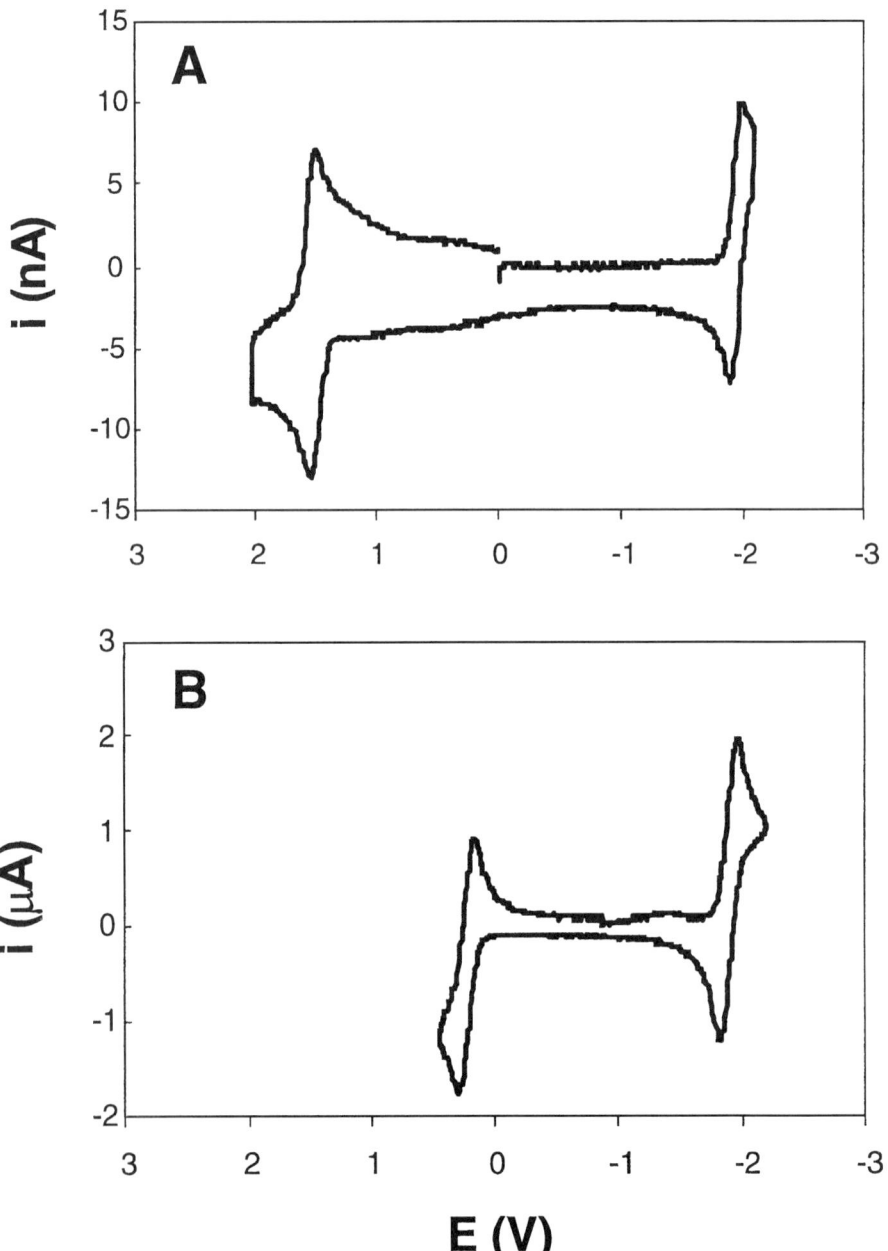

Figure 5 (A) Fast-scan cyclic voltammogram of 5-mM diphenylanthracene in benzonitrile with 0.1 M tetrabutylammonium hydroxide at a 2.5-μm (radius) Au disk; sweep rate = 410 V s^{-1}. (B) Fast-scan cyclic voltammogram of 10-mM diphenylanthracene and 10-mM tetramethylphenylenediamine in 50/50 acetonitrile/benzonitrile at a 52-μm (radius) Au disk. Both voltammograms used a silver pseudo reference electrode.

assuming diffusion coefficients of ca. $10^{-6}-10^{-5}$ cm^2/sec. These low currents allow for experiments to be conducted in low-dielectric-constant solvent media, with low supporting-electrolyte concentrations, approaching the conditions of the condensed-phase environment (34). Sensitivities in these experiments have been extended to levels that even allow for the measurement of light emission from single-molecule ECL events (50).

The ECL processes from DPA and related redox couples were recently explored in a low-dielectric-constant solvent (dimethoxyethane) at supporting electrolyte concentrations ranging from 10^{-1} to 10^{-4} M (33, 34). As supporting electrolyte concentrations were lowered by four orders of magnitude, the ΔE° values for the DPA/DPA$^{\cdot -}$ and DPA/DPA$^{\cdot +}$ processes systematically increased from -3.33 to -3.53 V, as predicted by Equations 10 through 13. The values for ϕ_{ECL} more than doubled, from 0.096 to 0.234 (34). Similar increases were seen for ECL processes for rubrene and for cross-reactions (see below) between DPA$^{\cdot +}$ and two different anion radical species. Extrapolating from these studies, similar processes occurring in low-dielectric-constant condensed-phase materials will likely possess excess free energies in the recombination process at least 0.2 eV larger than those observed for the isolated molecular components in concentrated electrolyte solutions (28) and higher electroluminescence efficiencies than seen in solution studies.

The excess free energy in these radical annihilation reactions has also been systematically varied using other redox couples to create either radical anions (A$^{\cdot -}$) or radical cations (D$^{\cdot +}$) to participate in "cross-reactions" with DPA$^{\cdot +}$ (D$^{\cdot +}$) or DPA$^{\cdot -}$ (A$^{\cdot -}$), respectively (8, 51). DPA$^{\cdot -}$ formed on the cathodic potential step can participate in cross-reactions with the radical cations of donor molecules such as tetramethylphenylenediamine, (TMPD$^{\cdot +}$), which are formed on the anodic potential step at potentials less positive of those needed to form DPA$^{\cdot +}$ (Figure 5B). Alternatively, radical anions of molecules such as benzophenones (BP$^{\cdot -}$) can be generated at potentials positive of those needed to form DPA$^{\cdot -}$, and BP$^{\cdot -}$ can then react with electrogenerated DPA$^{\cdot +}$. Recombination reactions with ΔE° values less than those seen for the DPA$^{\cdot +}$/DPA$^{\cdot -}$ reaction, but which have excess free energies that exceed those needed to create ^{1}DPA*, are still S-route and proceed according to Equation 3. Cross-reactions not possessing adequate energy to create ^{1}DPA* directly can nevertheless give rise to singlet state production through T-route processes (Equations 4 and 8). The DPA$^{\cdot -}$/TMPD$^{\cdot +}$ cross-reactions, which follow the steady-state production of these two species, as shown in Figure 5B, are typical of such T-route systems.

Creating D$^{\cdot +}$ and A$^{\cdot -}$ at the same ultramicroelectrode using alternating voltage pulses at high frequencies allows for a clear differentiation of the temporal nature of the ECL response for energy-sufficient and energy-deficient reactions (51). For S-route cross-reactions, the time dependence for the production of the singlet state emission response is qualitatively similar to that seen at high concentrations of DPA when only the DPA$^{\cdot -}$/DPA$^{\cdot +}$ process is occurring (Figure 6). ECL emission is seen to rise sharply immediately following application of the voltage pulse that creates the critical species for the cross-reaction. For T-route reactions, the time dependence for the production of the DPA singlet

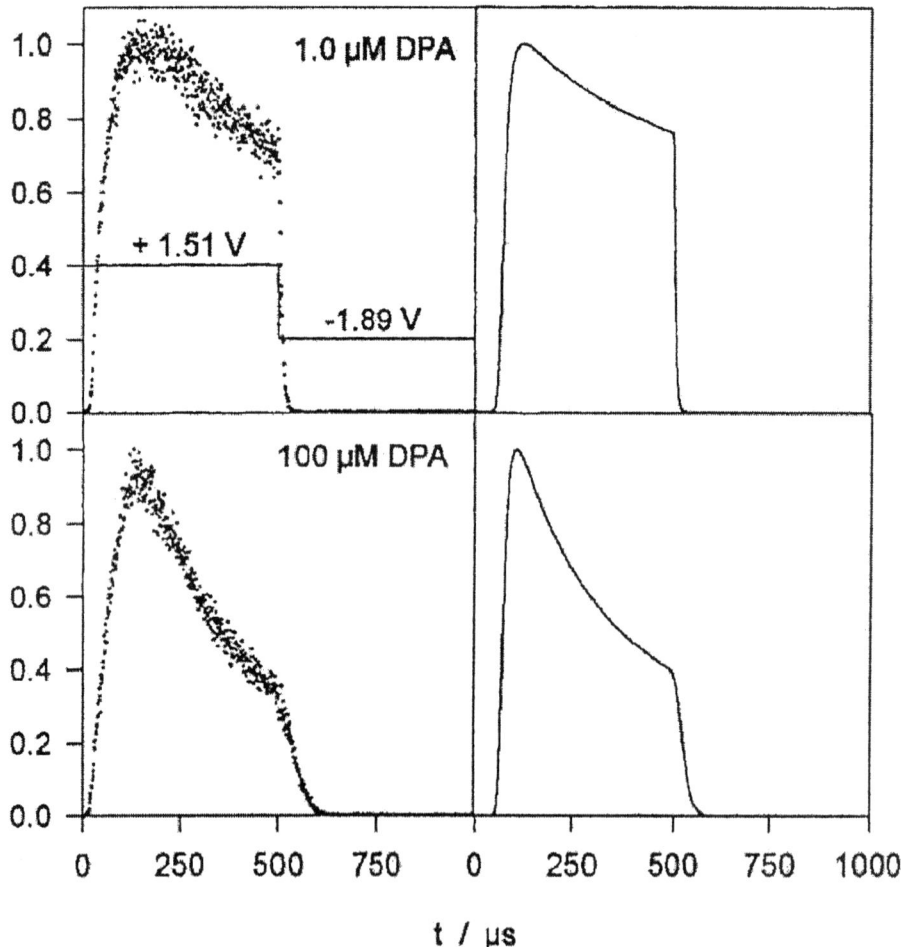

Figure 6 Normalized electrogenerated chemiluminescence responses seen for energy-sufficient systems: 10^{-6} M (upper panels) and 10^{-4} M (lower panels) diphenylanthracene (DPA) in acetonitrile with 10^{-3} M benzophenones. Data are plotted on the left, and simulations are shown on the right, assuming a diffusion-controlled S-route process (from Reference 51, by permission).

state changes dramatically at low vs high DPA concentrations (Figure 7). At low DPA concentrations in the T-route reactions, the creation of DPA singlet states lags substantially behind the application of the voltage pulse, owing to the need to build up the solution concentration of ^{3}DPA* near the electrode to maximize production of ^{1}DPA* through triplet-triplet annihilation. As the concentration of DPA is increased, the population of ^{3}DPA* created on each voltage pulse increases, and the time dependence of emission takes on the appearance of plots seen for S-route reactions.

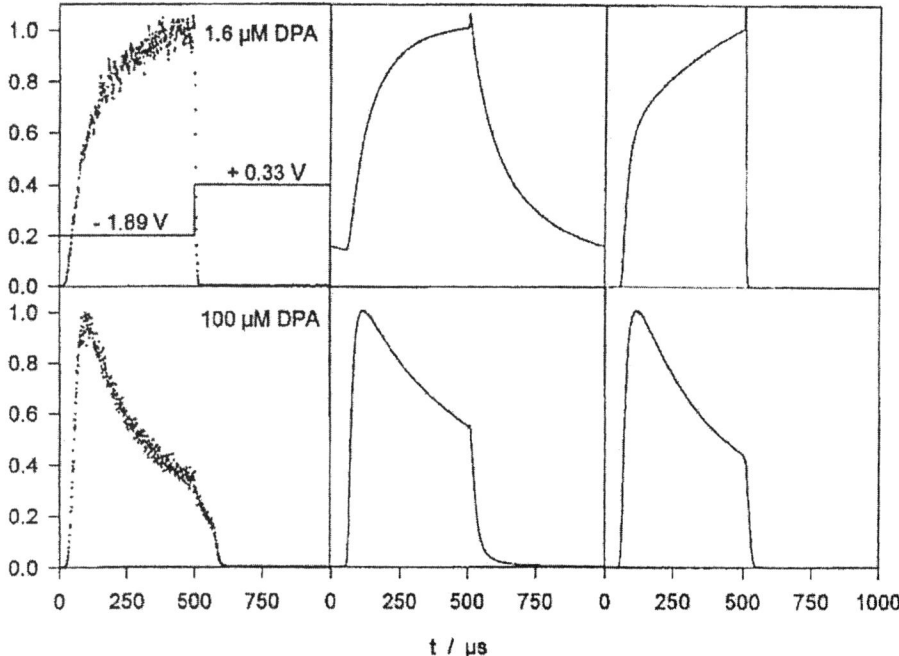

Figure 7 Normalized electrogenerated chemiluminescence responses seen for energy-deficient systems: $\sim 2 \times 10^{-6}$ M (*upper panels*) and 10^{-4} M (*lower panels*) diphenylanthracene (DPA) in acetonitrile, with 10^{-3} M tetramethylphenylenediamine. Data are shown on the left, and simulations are shown in the center and on the right, assuming a diffusion-controlled T-route process (from Reference 51, by permission).

Exploring the concentration dependence of ECL response for such systems also allows differentiation of S-route vs T-route pathways. As shown in Figure 8, a log ECL emission vs log DPA concentration plot for several different redox couples (acting either as $A^{\cdot -}$ or $D^{\cdot +}$, with $DPA^{\cdot +}$ or $DPA^{\cdot -}$, respectively) produces two distinct slopes. For S-route cases, the slope of this plot is 1.0, whereas for T-route cases, where second-order triplet-triplet annihilation reactions create the $^1DPA^*$ state, the slope of this plot is 2.0. For redox reactions that have excess free energies that lie close to the minimum needed for direct creation of the singlet state (e.g. npk in Figure 8), a log ECL emission/log DPA concentration plot is observed with a slope intermediate between 1.0 and 2.0.

Figure 8 shows that production of singlet states through T-route processes approaches the efficiencies for S-route formation of singlet states at the highest concentrations of DPA. Although it is not likely that T-route efficiencies will surpass those of the S-route, the T-route will produce the same emissive state with less energy required from the electrochemical processes leading to $A^{\cdot -}$ and $D^{\cdot +}$. In condensed phase devices these lower energies may translate into lower drive voltages and lower input powers (see below).

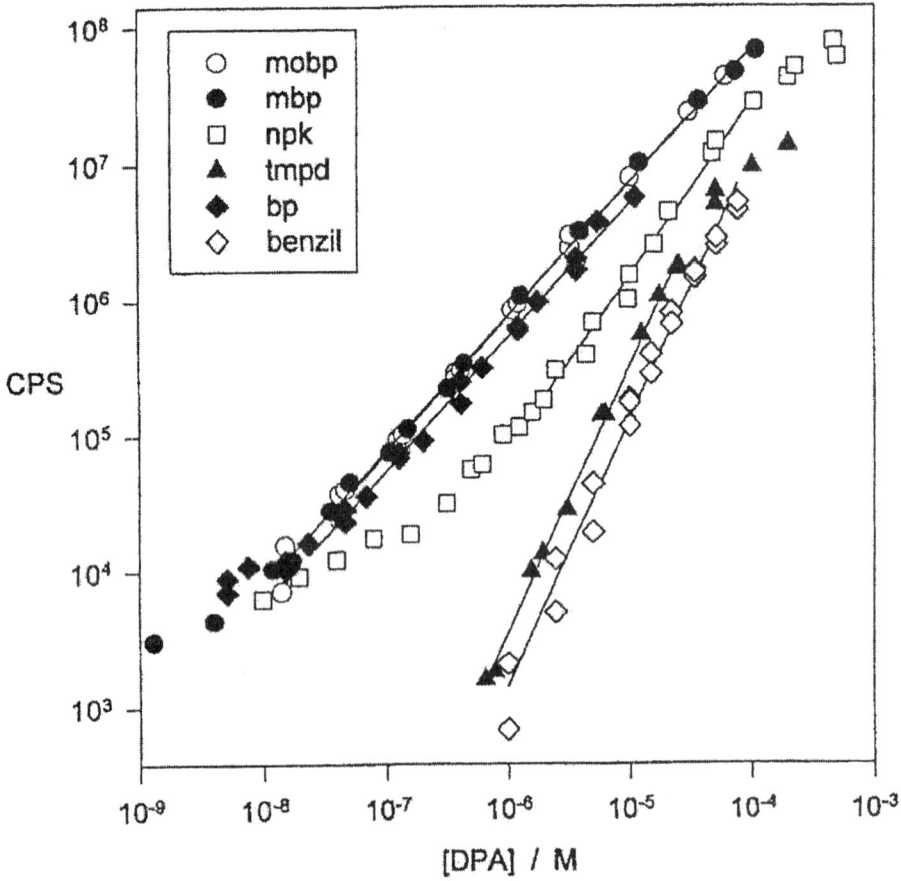

Figure 8 Logarithmic plot of electrogenerated chemiluminescence intensity vs diphenylanthracene (DPA) concentration for a series of S-route and T-route processes. mobp = 4-methoxybenzophenone; mbp = 4-methylbenzophenone; npk = naphthyl phenyl ketone; tmpd = N,N,N',N'-tetramethyl-1,4-phenylendiamine; bp = benzophenone (from Reference 51, by permission).

LIGHT-EMITTING ELECTROCHEMICAL (THIN-LAYER) CELLS

Light-Emitting Electrochemical Cells Without Rectifying Current/Voltage Behavior

ECL processes in thin-layer solution cells (LECs) were examined over two decades ago as potential light-emitting devices. Keszthelyi et al (52), Laser & Bard (53), and Brilmyer & Bard (54) examined the limitations to light emission in the context of the DPA, rubrene, and Ru(bipy)$_3^{2+}$ systems, making predictions relevant to light-emitting devices created more recently. DPA$^{·+}$/DPA$^{·-}$, rubrene$^{·-}$/rubrene$^{·+}$, or Ru(bipy)$_3^{3+}$/Ru(bipy)$_3^{+}$ redox couples were created in thin films of low-viscosity, high-dielectric solvents such as acetonitrile, including supporting electrolyte concentrations up to 0.1 M sandwiched between semitransparent cathodes and anodes of identical composition (ITO) (Figure 9).

In such cells, a solution thickness of 1 to 5 μm allows the radical species to diffuse together in times less than the half-lives of A$^{·-}$ and D$^{·+}$, thus allowing for light emission from near the middle of the thin solution layer. Those initial studies revealed several processes that limit the efficiency of light emission and lifetime of such a device and of more modern LECs based on polymeric ionic conductors:

1. The donor (anion radical) and acceptor (cation radical) may be chemically unstable due to coupling reactions, reactions with impurities in the solvent

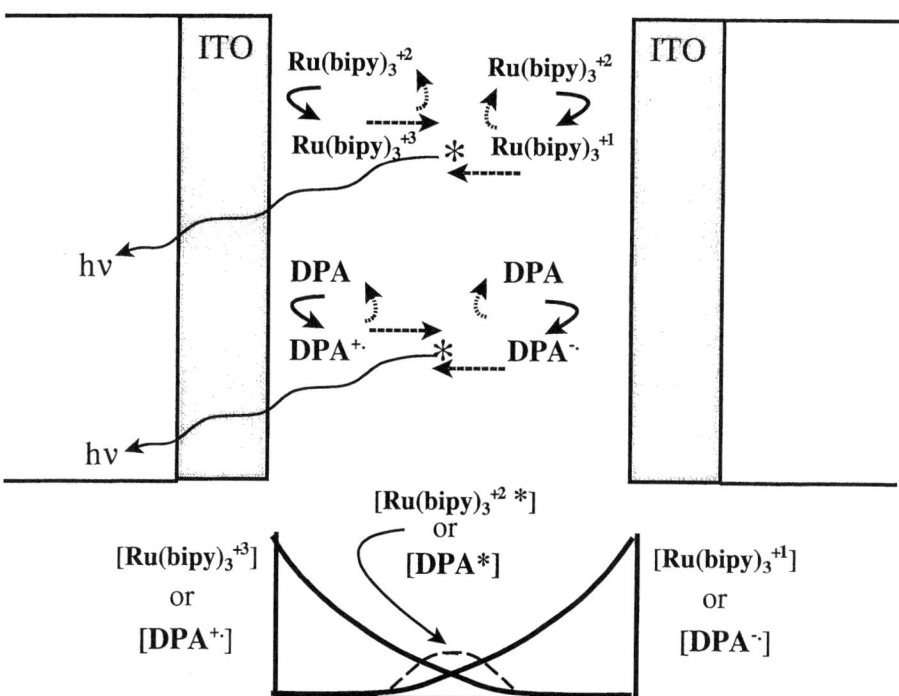

Figure 9 Schematic view of a solution thin-layer, light-emitting electrochemical cell, using either diphenylanthracene (DPA) or Ru(bipy)$_3^{2+}$ as the emitting species. The upper region shows the electrochemical and diffusion/migration processes that take place to provide the reactive intermediates in the center of the thin film. The lower region shows expected concentration profiles of each species provided that the mobilities of these species and the corresponding counter ions are all equal.

(especially proton donors and acceptors), and creation of insoluble (polymeric) films on the anode and cathode surfaces.

2. Direct electrolysis of impurities such as O_2, which lead to reactive radical species such as $HO_2^{\cdot}$, leads to the destruction of either D or A.
3. Diffusion/migration of $D^{\cdot-}$ and $A^{\cdot+}$ to the recombination region in the device is much slower than charge hopping between closely spaced molecules (as in condensed-phase LEPs or OLEDs).
4. The rates of recombination of these radical species may be lowered by the presence of high concentrations of supporting electrolyte.

The issue of formation of insoluble, electroinactive films on either the anode or cathode surface can be troublesome, especially for systems such as $Ru(bipy)_3^{3+}/Ru(bipy)_3^{+}$, since reduction of $Ru(bipy)_3^{+}$ by an additional one-electron process occurs at potentials close to that used for formation of $Ru(bipy)_3^{+}$ from $Ru(bipy)_3^{2+}$. This further reduction product is unstable, and its formation generally leads to deterioration of the ECL performance of the thin-layer cell described above.

Concentration profiles for an idealized solution LEC of any thickness are also shown in Figure 9. Each charge transfer event is accompanied by a local increase or decrease of the concentration of counter ions, which maintains the necessary local electroneutrality in the low-field regions in the center of the device (see below). The mobilities of these ions therefore play a critical role in determining the exact shape of these concentration profiles. Provided these diffusion and migration rates are equal for all species in solution, which may be approximately true for devices based on low-viscosity, high-dielectric-constant solvents, the concentration vs distance profiles for donor and acceptor species are expected to be mirror images of each other. The maximum probability for recombination and emission of light should therefore occur in the center of the thin film of solution.

The absolute magnitude of the concentration profile for the emissive species is subsequently controlled by rates of recombination. Recombination events can still take place at significant distances away from the center of the device, as the probability of formation of the requisite encounter complexes may be low in solutions where the maximum concentrations of reactive species is limited (10^{-3} to $\sim 10^{-1}$ M). Creation of emissive states within ~ 10 nm of either cathode or anode may lead to strong quenching by creating image dipoles in these conductive materials. Also, if $A^{\cdot-}$ or $D^{\cdot+}$ migrate/diffuse to the opposite electrode without recombination, their corresponding oxidation/reduction represents another major energy-loss pathway. It is clearly desirable to ensure that high concentrations of these reactive species are produced in confined volumes of material and that recombination events leading to emission take place near the center of the thin-layer device. This is something that is achievable in viscous, polymeric environments and difficult to achieve using concentrated low-viscosity solutions.

For these simple solution ECL cells with identical cathodes and anodes, the current–voltage (i/V) response will be the same regardless of which electrode is made the cathode or anode; that is, there is no rectification in the i/V response, and no net current is seen until the potential difference between the anode and cathode exceeds the difference in formal potentials for $A/A^{·-}$ and $D^{·+}/D$.

Light-Emitting Electrochemical Cells Displaying Rectifying Current/Voltage Behavior

Cells produced from polymeric versions of D and A (typically D and A are the same molecule type) can show the low turn-on voltages of the more conventional thin-layer electrochemical cells and may show rectification in the i/V behavior, depending on the mismatch of work functions of anode and cathode and the coupled rates of migration of electrical and ionic charge (18–20, 22, 55–61). Two basic types of LECs that were recently introduced demonstrate these rectification properties: (*a*) those comprised of extended conjugation polymers based on PPV, or ladder polymers based on poly(*p*-phenylenes) (PPP), examples of which are shown in Figure 10 (18, 19, 54–59); and (*b*) those based on redox active monomers incorporated as side chains in a polymeric network (e.g. polymers based on the $a\text{Ru(bipy)}_3^{2+}$ system and related ruthenium-ligand complexes) (20, 21, 23, 24, 60, 61). These polymers are typically mixed with molar equivalents of polyethylene oxide (PEO) or comparable ion-conducting polymer and high concentrations (0.1 M) of an inert electrolyte (e.g. LiCF_3SO_3). In the case of PPV-based LECs, the monomeric precursor to PPV was initially mixed with high-molecular-weight PEO and then thermally converted to the PPV polymer, interspersed with the ion conducting polymer (18, 19). Cathodes and anodes can be made from the same electrode material (e.g. gold) or may be materials of differing work functions (e.g. gold versus aluminum), which helps to establish the rectifying behavior of these cells.

The low turn-on voltages of these devices, relative to their condensed-phase (insulator) counterparts, have been one of their most attractive features. Reasonable light-emission intensities are achieved with drive voltages less than 5 V, as was the case for the solution-based thin-layer electrochemical cells. These low onset voltages arise because of the intense electric field gradients that develop in the polymeric thin film within 10 nm of each electrode/organic material interface (56, 62). At high concentrations of incorporated supporting electrolyte, with constant applied potentials in excess of $\Delta E^\circ = E^\circ_{\text{oxidation,D}} - E^\circ_{\text{reduction,A}}$, an approximation to the dependence of electric field with distance from the electrode interface ($d\phi/dx$) can be extrapolated from models originally developed for electrical double-layer formation in solution:

$$d\phi/dx = -(8k_B T n^\circ/\varepsilon\varepsilon_0)^{1/2}\sinh(ze\phi/2k_B T), \qquad 14.$$

where k_B is the Boltzmann constant, T is the absolute temperature, z is the absolute charge on each of the ions of the electrolyte (assuming a symmetric 1:1 electrolyte), e is the charge on the electron, ε is the dielectric constant of the medium, ε_0 is the permittivity of free space, and n° is the bulk concentration of ionic species in the organic thin film (56, 62, 63). The value of $d\phi/dx$ is therefore directly related to

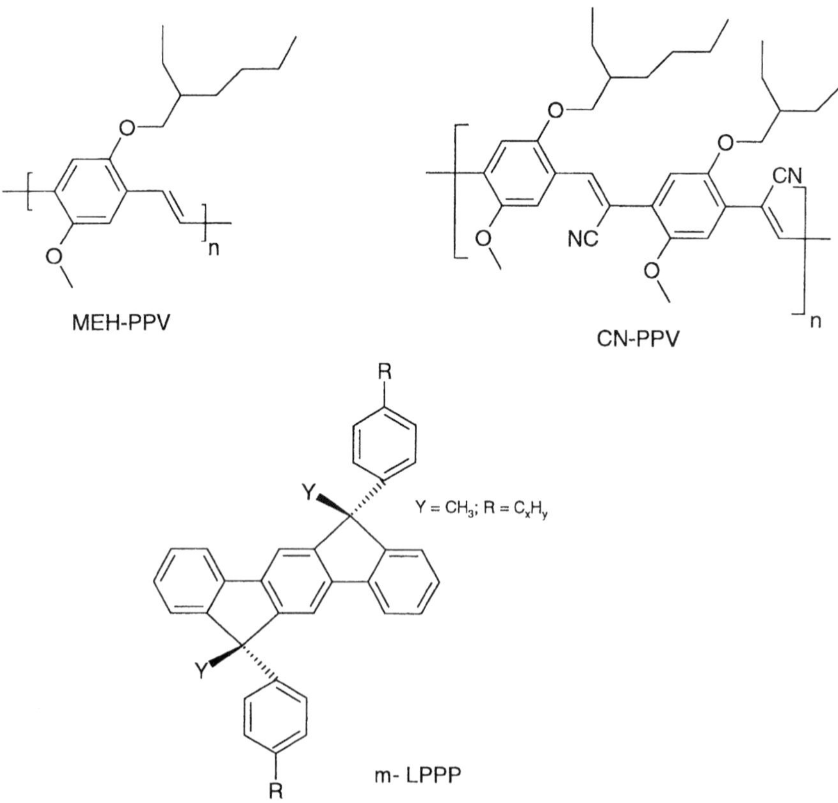

Figure 10 Molecular structures for three common polymers that have shown efficient emission as condensed-phase light-emitting polymers or in light-emitting electrochemical cells. The upper two structures are derivatives of poly(p-phenylenevinylenes), whereas the lower structure is a ladder polymer variant of poly(phenylene).

the dielectric properties of the polymeric medium and the concentrations of ionic species in the organic material. $d\phi/dx$ can reach 10^6 V cm^{-1} for concentrated ionic solutions.

When mixed PPV/PEO thin films loaded with high concentrations of LiCF$_3$SO$_3$ and identical anode and cathode materials (gold) were used, it was noticed that the zone of maximum emission moved from the middle of the polymeric thin film toward the cathode (18, 19). It was later argued that the rate of hole migration along the PPV chain is higher than the rate of electron migration (which is generally true for most conjugated polymeric systems). The shift in the zone of maximum emission toward the cathode is simply a reflection of the enhanced ability of holes to migrate toward the cathode vs electrons moving toward the anode. This process has been extensively modeled with respect to the dependence on the applied potential, concentration of inert electrolyte, rate coefficient for charge recombination leading to the emissive state, and time after application of the initial electric field (56).

Asymmetry in the location of the emissive zone is also seen in LECs based on side-chain polymers, such as those derived from the ruthenium trisbipyridyl complex, linked through vinyl groups, poly[Ru(vbpy)$_3$](PF$_6$)$_2$ (20, 21). In this type of polymer system, the ruthenium trisbipyridyl groups have a nominal 2+ charge and two counter anions per site prior to the onset of redox activity. The reduction of the ruthenium trisbipyridyl sites at the cathode (Ru^{2+} → Ru$^+$) and the oxidation of the ruthenium trisbipyridyl sites at the anode (Ru^{2+} → Ru^{3+}) is coupled to the hopping of electrons from Ru(vbpy)$_3^{(n-1)+}$ to Ru(vbpy)$_3^{n+}$ sites and the corresponding movement of the counter anions in the opposite direction. These charge-transport rates are often expressed as electron diffusion coefficients, $D_{E3+/2+}$ and $D_{E2+/1+}$, which for the poly[Ru(vbpy)$_3$](PF$_6$)$_2$ system show a $D_{E3+/2+}/D_{E2+/1+}$ ratio of ~4.0. Following a few seconds of electrolysis in solvent-swollen films of this polymer, the emissive zone is established closer to the anode than the cathode by the same ratio of distances. The concentration profiles for each of the ruthenium oxidation states and the corresponding concentration profile for the counter ion are shown in Figure 1e for this type of cell.

Once this type of asymmetric concentration profile is established, it is possible to "freeze" these profiles in place within the film, either by withdrawing the solvent that initially swelled the polymer matrix and/or by cooling the entire device to well below the glass transition temperature for the polymer matrix (21, 22, 57, 60, 61). The i/V behavior of the device then takes on the characteristics of a truly rectifying cell, and it is possible to apply significantly higher fields, thus increasing the output brightness. When "frozen-junction" conditions are met, counter-ion movement is effectively stopped, and emission processes are limited by the hopping of electrons from site-to-site (i.e. electroneutrality at the molecular level is no longer maintained). These devices then take on the operating characteristics of the more conventional small-molecule or polymeric devices.

These devices have many advantages over other polymeric or small-molecule devices. The low turn-on voltages of these devices are attractive, processing to achieve pin-hole-free films may be easier for these nearly liquid thin films, and high brightness has been achieved with the frozen-junction version of these LECs. Nevertheless, added solvents and supporting electrolytes in the thin film represent sources of impurities that are difficult to remove before these thin films can withstand high fields for long times, and the stability problems noted two decades ago in solution-based cells appear to be very significant.

Recent studies reported by Rubner and coworkers, however, show that good light intensity can be achieved by a simple device based on "neat" thin films of Ru(bipy)$_3^{2+}$/PF$_6^-$ using various hydroxyl- and ester-substituted ruthenium bipyridyl complexes with no added polymer or solvent (24). These devices produce the emission from the Ru(bipy)$_3^{2+}$ excited state at low voltages with good stability and reasonable external efficiencies (approaching 1%). Steady-state concentrations of all of the essential electroactive species are quickly built up within the thin film, despite the absence of ion-conducting polymers and supporting electrolyte, although an activation process is still required. The time needed for this activation, however, is shortened by using a ~5-V pulse (5–10 s), followed by sustained voltages of 2.5 to 3 V for the duration of the experiment. Although it is not clear that high brightness levels will be achievable by this approach, the apparent stability and ease of creation of such thin films may make them attractive for low-light-level applications.

Friend and coworkers also recently showed that the advantage of low turn-on voltages can be realized in single-layer LEPs (e.g. PPV) in which low concentrations of electrolyte (e.g. $LiCF_3SO_3$) have been added, even without adding an ion-conducting polymer (62). Apparently, even in such PPV films there is sufficient ion mobility to establish the high interfacial fields that accelerate the necessary redox processes leading to emissive state production, and one can realize the advantages of the LEC without added solvent or PEO-like polymers.

LIGHT-EMITTING POLYMERS AND ORGANIC LIGHT-EMITTING DIODES

Electrochemical Characterization of Single-Layer Light-Emitting Polymer Films

The first light-emitting polymer films were based on spin-coated thin films of either PPV or MEH–PPV (Figure 10) (e.g. ITO/PPV/Ca). These first devices depended on work function differences between the anode and cathode and differences in charge mobilities to provide the rectification response (Figure 1b) (12–14). The relationship between charge injection and recombination events and the electrochemical properties of these materials were at first not clear. By using solution contacts for one of the electrodes (e.g. ITO/MEH–PPV/acetonitrile, 0.1 M tetrabutylammonium tetrafluorborate, Bu_4NBF_4), studies could be conducted that showed the connection between ECL and the condensed-phase properties of these materials (64). These studies were made possible by the fact that (a) ultrathin films of MEH–PPV rapidly equilibrated with changing electrode potentials and (b) the polymer thin film and the cation and anion radical forms of its redox products were insoluble in acetonitrile.

The reduction of MEH–PPV to its radical anion (negative polaron) and radical cation (positive polaron) states was easily seen in voltammetric scans, which showed both states to be chemically stable on the time scale of seconds. The separation in apparent formal potentials for both electrochemical processes (ΔE) was ~2.4 V, which is close to the absorption energy for the $\pi \rightarrow \pi^*$ transition in this polymer. By pulsing the MEH–PPV/Pt electrode back and forth between the potentials needed to create both the reduced and oxidized forms of MEH–PPV, both species were generated within the thin film at sufficient concentrations that their recombination led to the emissive states normally seen in displays based on this same molecule, that is, red–orange emission at λ_{max} of ~625 nm. Of course, these electrochemical events were accompanied by counter-ion incorporation during both reduction and oxidation scans, so that after a single voltammetric cycle these thin films must have taken on compositions reminiscent of the LECs discussed above. Nevertheless, the fact that the emission event in these hybrid condensed-phase/solution experiments was formed by processes recognizable as electrochemical in nature suggested that electrochemical studies of other components of LEPs and OLEDs would be relevant to understanding and optimizing their operation.

Electrochemical Characterization of the Hole- and Electron-Transport Materials of Multilayer Organic Light-Emitting Diodes

In a single-layer light-emitting thin film with equal rates of hole and electron injection and equal rates of hole and electron transport, the highest concentration of $A^{·-}/D^{·+}$ recombination events will occur in the center of the thin film, as with the thin-layer electrochemical cells described above. In general, however, hole mobilities in organic materials are ~10^3 higher than electron mobilities, and hole injection can also be more facile than electron injection (15). In a single-layer device, this leads to a significant asymmetry in the $A^{·-}$ and $D^{·+}$ populations within the film. Consequently, recombination events occur predominantly near the cathode, where luminescence quenching represents a significant efficiency loss.

Two-layer devices, consisting of vacuum-deposited hole-transport layers (HTL) and electron-transport (luminescent) layers (ETL), solve some of these problems and are represented by the ITO/TPD/Alq$_3$/Mg(Ag) devices first introduced by Tang and Van Slyke [TPD = 4,4'-bis(m-tolylphenylamino)biphenyl, Alq$_3$ = tris (8-hydroxyquinolato)aluminum] (Figure 11) (10, 11). Similar solutions to this problem in LEPs were achieved by PPV/CN–PPV devices or MEH–PPV/CN–MEH–PPV (structures shown in Figure 10), where –CN substituents on the polymer chain increase the EA of this material, making it usable in device geometries such as ITO/PPV/CN–PPV/Al (14, 15).

In such thin films, hole injection (oxidation) into the HTL is followed by charge transport to the HTL/ETL interface (Figure 1c), where the differences in IP of the two materials lead to an energetic barrier to hole injection into the ETL. Electron injection (reduction) into the ETL is followed by slower transport of charge to the ETL/HTL interface. Differences in EA of the ETL and HTL lead to an energetic barrier to electron injection into the HTL. Two possible routes are then available for recombination leading to light emission, described as follows:

1. Holes collected at the HTL/ETL interface can "diffuse" across this barrier under the influence of the high fields used in these devices, thereby creating both the oxidized and reduced versions of the ETL within a few nanometers of this interface, which can then recombine to form the

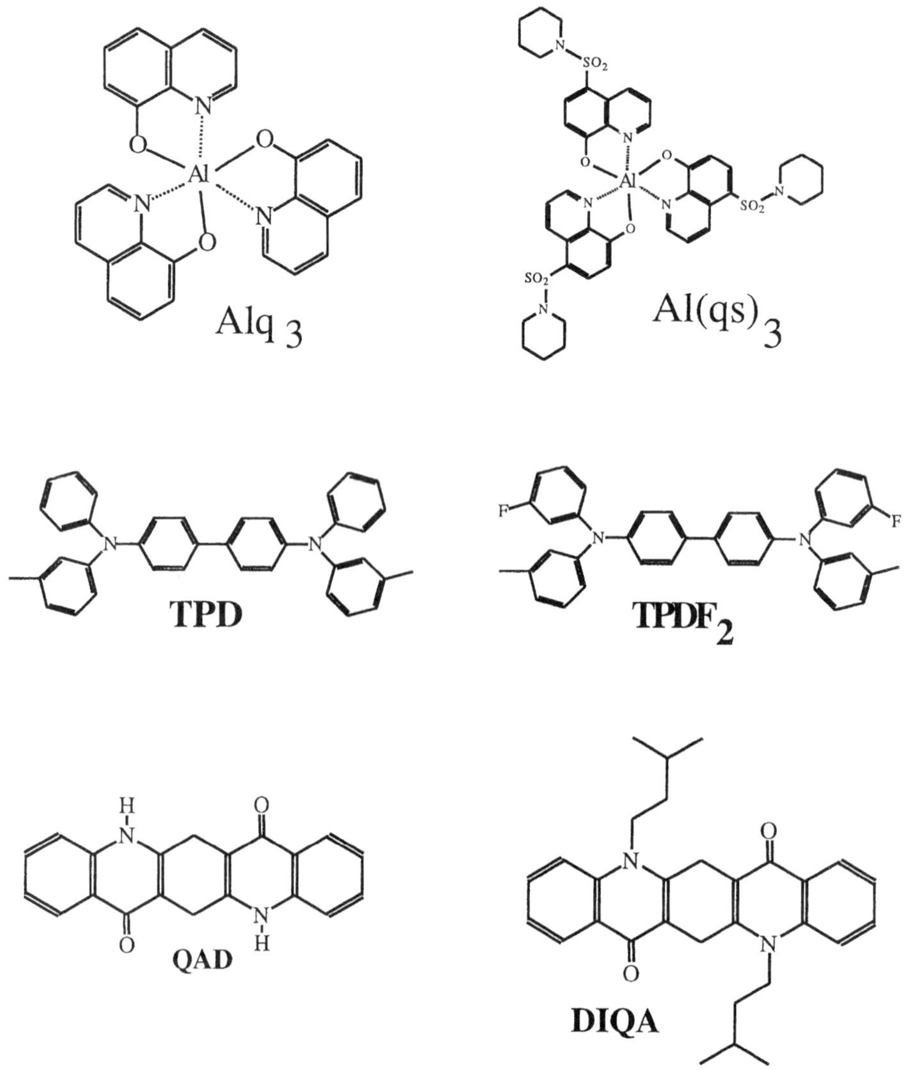

Figure 11 Molecular structures for small-molecule components of green-emitting organic light-emitting diodes. Alq$_3$ and Al(qs)$_3$ are emitting and electron-transporting species, while 4,4'-bis(*m*-tolylphenylamino)biphenyl (TPD) and TPDF$_2$ are hole-transporting species. QAD and DIQA are quinacridone and its *N,N'*-diisoamyl derivative, respectively, used as dopants in Alq$_3$-based organic light-emitting diodes (References 27 and 28).

emissive state (10, 11, 65). The products of hole injection into the ETL and electron injection into this same layer lead to recombination events such as the following for the Alq$_3$/TPD OLED:

$$\mathrm{Alq_3^{\cdot-} + Alq_3^{\cdot+} \rightarrow {}^{S\ or\ T}Alq_3^* + Alq_3}. \qquad 15.$$

Hole injection into the ETL and recombination reactions as shown in Equation 15 are predicted to be more efficient as the IP of the hole-transporting molecules increases and becomes comparable to the IP of the electron-transporting molecules.

2. Holes and electrons collected on either side of the HTL/ETL interface may recombine across that interface through cross-reaction schemes such as the following:

$$\mathrm{Al_{q3}^{\cdot-} + TPDX^{\cdot+} \rightarrow {}^{S\ or\ T}Al_{q3}^* + (TPDX)} \qquad 16.$$

$$\mathrm{Al(qs)_3^{\cdot-} + (TPDX^{\cdot+}) \rightarrow {}^{S\ or\ T}Al(qs)_3^* + (TPDX)}. \qquad 17.$$

where Al(qs)$_3$ is tris(5-piperidinylsulfonamide-8-quinolato-N$_1$O$_8$) aluminum (66) (Figure 11).

This reaction is likely to be limited to charged species located within one molecular diameter of the HTL/ETL interface. It could become important for thin films with intentionally roughened HTL/ETL interfaces or in systems where there is an intentional mixing of these two components in the center of the thin-film device (68).

The experimental protocols developed to study solution ECL processes for model systems such as DPA were recently extrapolated to TPD, Alq$_3$, and their derivatives. These studies assessed the energetic requirements for cross-reactions involving the anion radical form of the electron-transporting molecule and the cation radical form of the hole-transporting molecule in OLEDs (option No. 2 above) (27, 28). In mixed acetonitrile/toluene solvents, radical anion/radical cation annihilation reactions (Equation 15) could be initiated in solution, which showed only the formation of ^{S}Alq$_3^*$ with an emission wavelength within 1 to 2 nm of that seen for this species in condensed-phase device environments ($\lambda_{max} = \sim$520 nm).

The radical anion and radical cation forms of Alq$_3$, however, are chemically unstable even in dry, deoxygenated solvents (shown by the voltammetry in Figure 12), which may be correlated with chemical instabilities of devices based on this molecule (67). Also, the low solubility of this molecule in most solvents precluded a thorough investigation of its electrochemical properties over a wide range of concentrations. The more soluble trisulfonamide derivative of this molecule, Al(qs)$_3$ (Figure 11), shows three successive chemically reversible reduction processes in solution (Figure 12), with a first reduction potential consistent with an increase in EA of this molecule of $\sim$0.3 eV relative to the parent Alq$_3$ (27, 28).

TPD and TPDX derivatives with electron-donating or electron-withdrawing substituents (alkoxy groups and fluoro, difluoro, and trifluoromethyl groups

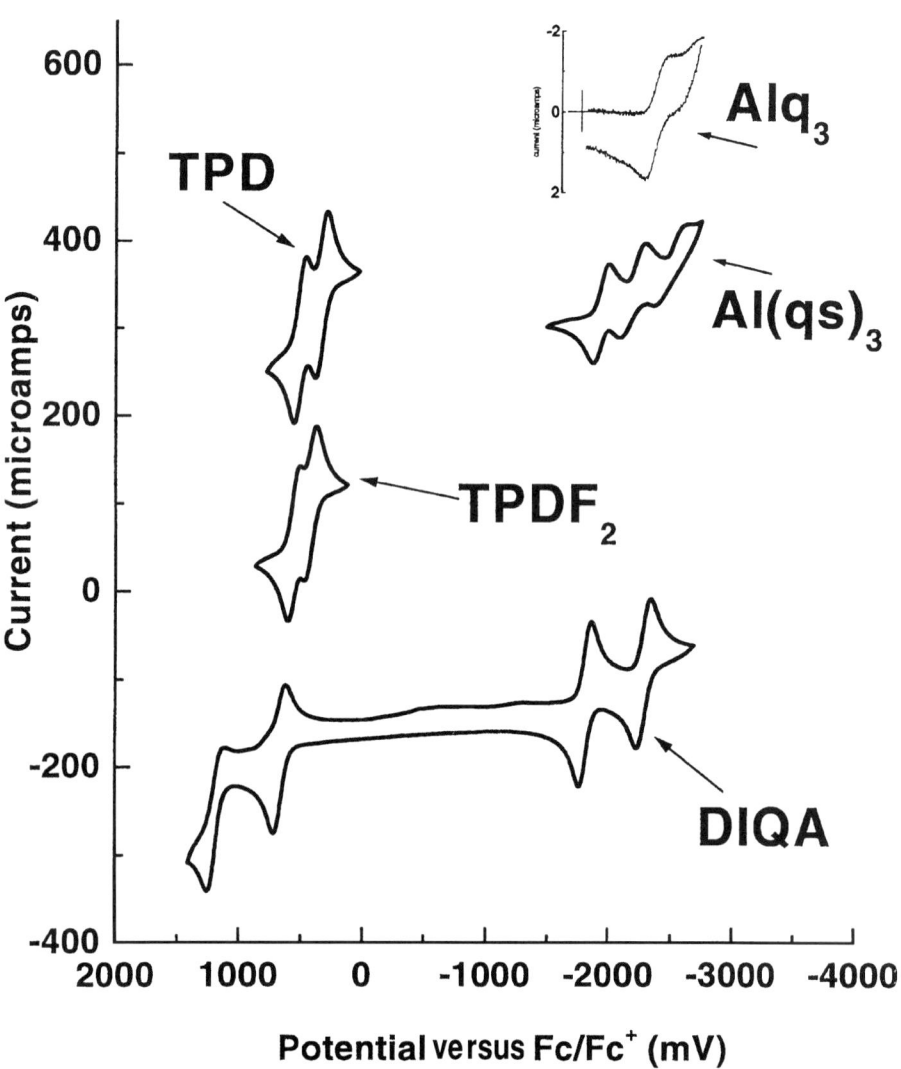

Figure 12 Electrochemical responses of $\sim 10^{-3}$ M acetonitrile/toluene solutions of the organic light-emitting diode small molecules shown in Figure 11. The cyclic voltammogram shown for Alq$_3$ was obtained at a microelectrode at a sweep rate of 2000 V s^{-1} because of the instability of the Alq$_3$ radical anion. All other voltammograms were obtained at 100 mV s^{-1} (adapted from Reference 27 by permission).

appended to these molecules at ortho, meta, and para positions) (68) show two closely spaced one-electron oxidation events (Figure 12) (27, 28). These oxidation processes lead to chemically stable cation radicals (on the voltammetric time scale). The first oxidation potential varies linearly by more than $\sim$0.5 V with the Hammett acidity of the substituents (JA Anderson, S Thayumanavan, S Barlow, S Marder, PA Lee, & NR Armstrong, submitted, *Chemistry of Materials*).

The excess free energy in $AL_{q3}^{\cdot-}/TPDX^{\cdot+}$ and $Al(qs)_3^{\cdot-}/TPDX^{\cdot+}$ cross-reactions (Equation 16 and 17) can therefore also be varied by $\sim$0.5 eV. A series of microelectrode ECL studies have been conducted to confirm the dependence of ϕ_{ECL} on this excess free energy. Regardless of the magnitude of the excess free energy for these solution redox processes, the observed luminescence was from the singlet, $^SAl(qs)_3^*$, whose emission spectrum closely resembled that seen for OLEDs based on this molecule (66). For most of the triarylamines explored, however, the excess free energy in the $Al(qs)_3^{\cdot-}/TPDX^{\cdot+}$ cross-reaction was not sufficient to populate the $^SAl(qs)_3^*$ state directly. Indeed, the temporal nature of the ECL response suggested that the triplet $^TAl(qs)_3^*$ formed first, followed by triplet-triplet annihilation, as was discussed above for certain cross-reactions with the radical forms of DPA. The efficiency of the ECL process, ϕ_{ECL}, increased exponentially with increasing first oxidation potential in the TPDX system. For the highest oxidation potential TPDX molecule explored (the *m*-difluoro derivative), ϕ_{ECL} is about fivefold higher than that seen for the $DPA^{\cdot-}/TMPD^{\cdot+}$ cross-reaction.

Electrochemical Characterization of Molecular Dopants for Multilayer Organic Light-Emitting Diodes

Later versions of Alq$_3$/TPD OLEDs have incorporated an *N*-methylquinacridone dye in the Alq$_3$ layer, at $\sim$0.5 to 2.0 wt%, which greatly enhances both the green-emission efficiency and stability of these devices (70). Recent studies have suggested that in addition to being an efficient energy acceptor in the Alq$_3$ layer, these quinacridones may also act as charge traps, further enhancing the EL response of the OLED (71). Experiments were conducted on varying concentrations of a quinacridone derivative doped into a polymer thin film with a high concentration of Alq$_3$. These studies showed that the Alq$_3^*$ luminescence response was rapidly quenched by increasing quinacridone concentrations, becoming undetectable at quinacridone concentrations greater than $\sim$2% (wt/wt). The Alq$_3^*$ electroluminescent response of these same thin films, however, decayed at a much higher rate with increasing quinacridone concentration, suggesting an additional pathway for loss of Alq$_3^*$ luminescence, that is, capture of charges from either $Alq_3^{\cdot-}$ and/or $Alq_3^{\cdot+}$. Our electrochemical and ECL characterization of these molecules confirms that such charge-capture reactions are possible (27, 28).

Figure 11 shows the structure and Figure 12 shows the voltammetric reduction and oxidation of *N,N'*-diisoamylquinacridone (DIQA). The one-electron reduction and oxidation processes lead to stable radical anions and radical cations, respectively. Formation of these species occurs at potentials that suggest that DIQA and comparable N,N'-dialkyl derivatives of quinacridone will act as electron acceptors toward $\text{Alq}_3^{\cdot-}$ and as weak electron donors toward $\text{TPD}^{\cdot+}$, the cation radical states of certain TPD derivatives, and the cation radical states of other hole-transporting polymers such as poly(vinylcarbazole) (PVK):

$$\text{Alq}_3^{\cdot-} + \text{DIQA} \rightarrow \text{Alq}_3 + \text{DIQA}^{\cdot-}, \qquad 18.$$

$$\text{DIQA} + \text{PVK}^{\cdot+} \rightarrow \text{DIQA}^{\cdot+} + \text{PVK}. \qquad 19.$$

Assuming that the interactions between molecular species in the condensed phase are weak, it is possible to extrapolate from these solution experiments to the molecular solid environment and to conclude that these alkyl-substituted quinacridones can indeed act as sites for charge trapping. The relative stability of the cation and anion radicals of these quinacridones, may contribute to the increased stability seen in OLEDs doped with these materials.

Our solution ECL studies showed that the tendency to form emissive states from these DIQA radicals was also consistent with their ability to enhance OLED efficiencies (28). The $\text{DIQA}^{\cdot-}/\text{DIQA}^{\cdot+}$ cross-reaction was first shown to produce the $^S\text{DIQA}^*$ emissive state with an efficiency ($\phi_{\text{ECL}} = \sim 7\%$), exceeding that seen for creation of $^S\text{DPA}^*$ from the $\text{DPA}^{\cdot-}/\text{DPA}^{\cdot+}$ which is an indication of both its high fluorescence efficiency and the significant rate of emissive state formation from the $\text{DIQA}^{\cdot-}/\text{DIQA}^{\cdot+}$ cross-reaction.

$\text{DIQA}^{\cdot-}/\text{TPDX}^{\cdot+}$ cross-reactions all proved to follow T-route pathways, leading to $^T\text{DIQA}^*$ and $^S\text{DIQA}^*$ through triplet-triplet annihilation, as indicated by both the temporal dependence of the ECL response and the second-order DIQA concentration dependence on ϕ_{ECL}. An S-route pathway to $^S\text{DIQA}^*$ was found when TPDX was replaced by a di-t-butyl derivative of N-ethylcarbazole (ETBC) ($\text{DIQA}^{\cdot-}/\text{ETBC}^{\cdot+}$ cross-reaction), where the first oxidation potential of ETBC was sufficiently positive as to increase the excess free energy in the cross-reaction above the threshold needed for S-route formation of $^S\text{DIQA}^*$.

When the efficiency of T-route ECL is plotted as a function of the excess free energy ($\Delta^T G$) in the $\text{DIQA}^{\cdot-}/\text{TPDX}^{\cdot+}$ reaction (Figure 13), it is observed that ϕ_{ECL} increases exponentially over several orders of magnitude, as would be predicted by the Marcus relationship (Equation 6) for such redox processes. This increase in ϕ_{ECL} continues until $\Delta^T G$ reaches ~ 0.5 eV, whereupon it plateaus, consistent with the rate-limiting reaction reaching diffusion control. As with the DPA cross-reactions discussed in the first section, plots of ϕ_{ECL} vs DIQA concentration, extrapolated to molar concentration levels of DIQA for both the S-routes and T-routes, appear to approach one another. Although indirect formation of singlet states through the T-route is a less efficient process in solution, it requires less input power and may prove to be a viable pathway toward creation of light in condensed-phase environments. This finding may prove to be important in systems where the target molecule has high photoluminescence efficiency and shows stable cation

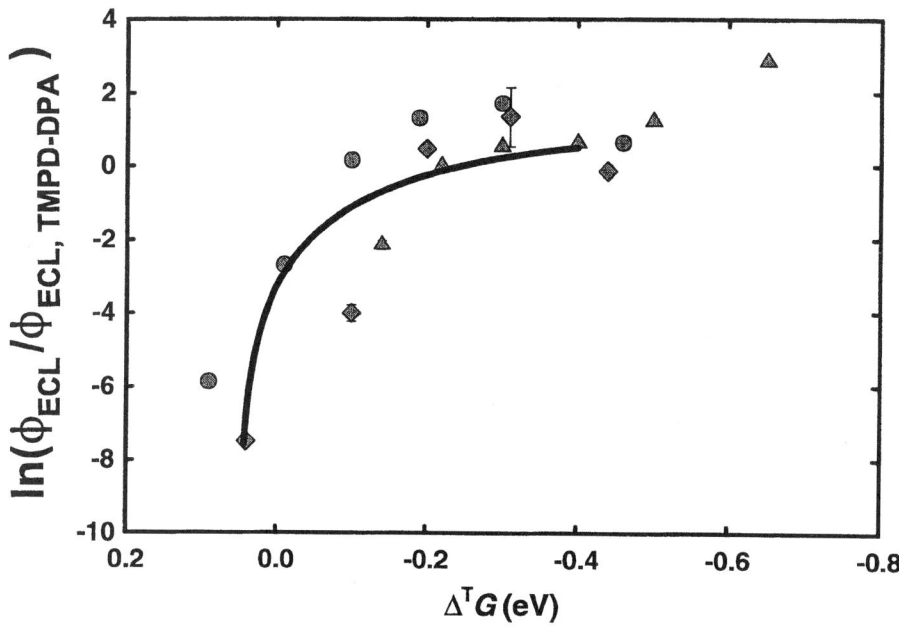

Figure 13 Natural log of normalized ratios of electrogenerated chemiluminescence (ECL) response vs the excess free energy ($\Delta^T G$) in T-route DIQA$^{\cdot -}$/TPDX$^{\cdot +}$ cross-reactions. The ECL response is normalized to the response seen for the T-route DPA$^{\cdot -}$/TMPD$^{\cdot +}$ cross-reaction (from Reference 28, by permission).

and anion radical states, but must use the T-route D$^{\cdot +}$/A$^{\cdot -}$ cross-reaction to form the singlet state (30).

Optimization of Two-Layer Small-Molecule Organic Light-Emitting Diodes

The initial electrochemical studies of Alq$_3$ and TPDX materials suggested that a systematic increase in the IP of the HTL would increase the efficiency of the Alq$_3$/TPD OLED, either by lowering the barrier to hole injection into the Alq$_3$ layer or by increasing the reaction rate for Alq$_3$$^{\cdot -}$/TPDX$^{\cdot +}$ cross-reactions at the organic/organic' interface (72). As shown in Figure 14, side-chain polymers of TPD and its *p*-methoxy and *m*-fluoro derivatives were created (68) and were spun cast onto the ITO anode (~40–50 nm), followed by vapor deposition of the Alq$_3$$^{\cdot -}$ layer (~40 nm), followed by a Mg cathode [ITO/poly(TPDX)/Alq$_3$/Mg]. The external quantum efficiency showed a systematic increase for the three polymers chosen for study: $\phi_{EL,TPDF2} > \phi_{EL,TPD} > \phi_{EL,p\text{-}CH3O\text{-}TPD}$, as would have been predicted from the solution electrochemical characterization of these molecules and from the ECL studies of related derivatives described above. Additional

P1
H_3CO
OCH_3
IP = 5.06 eV
P2
IP = 5.38 eV
P3
F
F
IP = 5.56 eV

Current Density (mA/cm²)
ITO / polymer P1 / Alq₃ / Mg
ITO / polymer P2 / Alq₃ / Mg
ITO / polymer P3 / Alq₃ / Mg
P1
P2
P3
Luminance (cd/m²)
Ext. Quantum Efficiency (%)
P3
P2
P1
Bias Voltage (V)

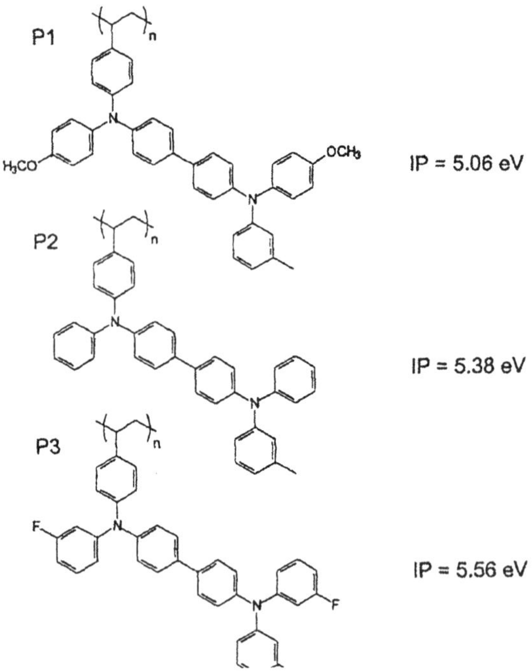

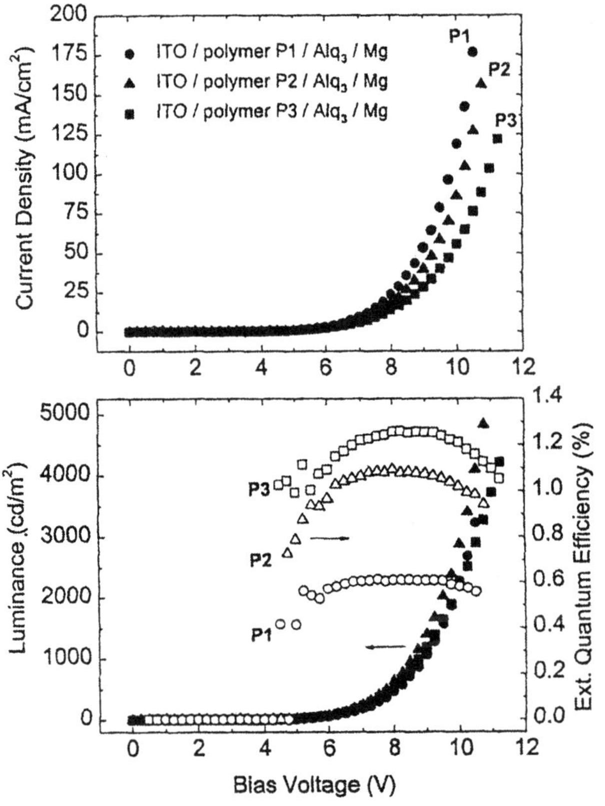

modifications to the most efficient of these OLEDs were made by incorporation of a quinacridone derivative into the Alq_3 layer and by using a modified Al cathode [ITO/poly(TPDX)/Alq_3:quinacridone/LiF (0.8 nm)/Al], producing a device that at an applied voltage of 3.0 V gave a luminance of 15 cd m^{-2}, an external quantum efficiency of 4.5%, and a luminous efficiency of $\sim$20 lm W^{-1} (lumens per watt). This last efficiency is considered a benchmark of EL devices, corresponding to the approximate efficiency of an incandescent light bulb, and is considered an acceptable level for device applications (17).

Although it is tempting to correlate this optimization of OLED efficiency directly with the increases in ECL efficiency arising from increases in excess free energy in a $D^{·+}/A^{·-}$ cross-reaction, we must consider as well the processes noted above that can also contribute to enhanced EL efficiency. First, increases in IP of the TPD layer, while clearly accelerating the rate of the $TPD^{·+}/Alq_3^{·-}$ cross-reaction, also decrease the energy barrier to direct-hole injection into the Alq_3 layer, which may also lead to enhanced emission (73). In addition, increases in IP for the TPD layer will increase the energy barrier to hole injection into this layer, which might not change the rate of that process in solution but could decelerate it in the condensed phase (74). Such a deceleration of hole injection would serve to help balance the rates of arrival of holes and electrons at the TPD/Alq_3 interface, thereby increasing slightly the probability for recombination leading to emission. Lastly, hole-transport mobilities in the TPD layers with the highest IP may be lower than for the parent molecule, again aiding in balancing the arrival of electrons and holes at the organic/organic' interface. All of these possibilities are under active exploration. Side-chain and conjugated polymers are being designed, however, to maximize direct contact of a hole-transporting molecule and an emissive electron-transporting molecule in the center of a three-layer OLED. The ECL reactions described above are therefore likely to provide a direct means for characterizing and optimizing the condensed-phase electrochemical reactions that lead to emissive state production.

Optimization of these condensed-phase redox processes has now been coupled to better harvesting of the emissive energy from both singlet and triplet states using tailored composition dopants, which have more than doubled the efficiencies of both OLEDs and LEPs in just the past 2 years (39–41, 75). It is not unreasonable to predict that OLEDs and LEPs will soon routinely surpass the efficiencies of incandescent lighting and may even approach the external efficiencies of fluorescent lighting. Studies of light-emitting electrochemical processes will certainly play a key role in these developments.

←

Figure 14 Molecular structures (*top*) and current/voltage, luminance/voltage, and external quantum efficiency/voltage curves (*bottom*) for three different Alq_3/TPDX organic light-emitting diodes (adapted by permission from Reference 72).

ACKNOWLEDGMENTS

NRA and RMW research activities described in this review were supported in part by the National Science Foundation (Chemistry) and by the Office of Naval Research (MURI Program). We also gratefully acknowledge a careful review of this manuscript by Sumit Mazumdar.

Visit the Annual Reviews home page at www.AnnualReviews.org

LITERATURE CITED

1. Dresner J. 1969. *RCA Rev.* 30:322–34
2. Pope M, Swenberg CE. 1999. In *Electronic Processes in Organic Crystals and Polymers*, pp. 1193–215. New York: Oxford Univ. Press
3. Mehl W, Hale JM. 1967. In *Advances in Electrochemistry and Electrochemical Engineering*, ed. P Delahay, New York: Interscience. 339 pp.
4. Visco RE, Chandross E. 1964. *J. Am. Chem. Soc.* 86:5350
5. Marcus RA. 1965. *J. Chem. Phys.* 43:2654–57
6. Marcus RA. 1970. *J. Chem. Phys.* 52:2803–4
7. Feldberg SW. 1966. *J. Phys. Chem.* 70:3928–30
8. Faulkner LR, Bard AJ. 1977. In *Electroanalytical Chemistry, A Series of Advances*, ed. AJ Bard, pp. 2–95. New York: Marcel Dekker
9. Measures RM. 1974. *Appl. Opt.* 13:1121–33
10. Tang CW, Van Slyke SA. 1987. *Appl. Phys. Lett.* 51:913–15
11. Tang CW, Van Slyke SA, Chen CH. 1989. *J. Appl. Phys.* 65:3610–16
12. Burroughes JH, Bradley DDC, Brown AR, Marks RN, Mackay K, et al. 1990. *Nature* 347:539–41
13. Gustafsson G, Cao Y, Treacy GM, Klavetter F, Colaneri N, Heeger AJ. 1992. *Nature* 357:477–79
14. Greenham NC, Moratti SC, Bradley DDC, Friend RH, Holmes AB. 1993. *Nature* 365:628–30
15. Friend RH, Greenham NC. 1998. In *Handbook of Conducting Polymers*, ed. TA Skotheim, RL Elsenbaumer, JR Reynolds, pp. 823–47. New York: Marcel Dekker
16. Friend RH, Gymer RW, Holmes AB, Burroughes JH, Marks RN, et al. 1999. *Nature* 397:121–28
17. Sheats JR, Antoniadis H, Hueschen M, Leonard W, Miller J, et al. 1996. *Science* 273:884–88
18. Pei Q, Yu G, Zhang C, Yang Y, Heeger AJ. 1995. *Science* 269:1086–88
19. Pei Q, Yang Y, Yu G, Zhang C, Heeger AJ. 1996. *J. Am. Chem. Soc.* 118:3922–29
20. Maness KM, Terrill RH, Meyer TJ, Murray RW, Wightman RM. 1996. *J. Am. Chem. Soc.* 118:10609–16
21. Maness KM, Masui H, Wightman RM, Murray RW. 1997. *J. Am. Chem. Soc.* 119:3987–93
22. Gao J, Yu G, Heeger AJ. 1997. *Appl. Phys. Lett.* 71:1293–95
23. Wu AP, Lee J, Rubner MF. 1998. *Thin Solid Films.* 329:663–67
24. Wu AP, Yoo D, Lee J-K, Rubner MF. 1999. *J. Am. Chem. Soc.* 121:4883–91
25. Several examples of this display technology can be found on web sites maintained by Pioneer, Phillips, IBM, Cambridge Display Technolgies, Uniax, etc., and are current as of November, 2000. The following appear to be good links to general information about these displays: http://164.109.152.35/features/0003_OEL1.asp (Pioneer Electronics); http://www.almaden.ibm.com/st/projects/oleds/ (IBM); http://

www.cdtltd.co.uk/ (Cambridge Display Technologies); http://www.uniax.com/ (Uniax)
26. Sheats JR, Barbara PF. 1999. *Acc. Chem. Res.* 32:191–92
27. Anderson JD, McDonald EM, Lee PA, Anderson ML, Ritchie EL, et al. 1998. *J. Am. Chem. Soc.* 120:9646–55
28. Gross EM, Anderson JD, Slaterbeck AF, Thayumanavan S, Barlow S, et al. 2000. *J. Am. Chem. Soc.* 122:4972–79
29. Debad JD, Morris JC, Lynch V, Magnus P, Bard AJ. 1996. *J. Am. Chem. Soc.* 118:2374–79
30. Lee SK, Zu YB, Herrmann A, Geerts Y, Mullen K, Bard AJ. 1999. *J. Am. Chem. Soc.* 121:3513–20
31. Marcus RA, Sutin N. 1985. *Biochim. Biophys. Acta* 811:265–322
32. Marcus RA. 1997. *J. Electroanal. Chem.* 438:251–59
33. Maness KM, Bartlet JE, Wightman RM. 1994. *J. Phys. Chem.* 98:3993–98
34. Maness KM, Wightman RM. 1995. *J. Electroanal. Chem.* 396:85–95
35. Park S-M. 1978. *Photochem. Photobiol.* 28:83–90
36. Oyama M, Mitani M, Okazaki S. 2000. *Electrochem. Commun.* 2:363–66
37. Wang JF, Kawabe Y, Shaheen SE, Morrell MM, Jabbour GE, et al. 1998. *Adv. Mater.* 10:230–33
38. Weller A. 1968. *Pure Appl. Chem.* 16:115–23
39. Baldo MA, O'Brien DF, You Y, Shoustikov A, Sibley S, et al. 1998. *Nature* 395:151–54
40. Baldo MA, O'Brien DF, Thompson ME, Forrest SR. 1999. *Phys. Rev. B Conds. Matt.* 60:14422–28
40a. Chawdhury N, Kohler A, Friend RH, Wong WY, Lewis J, et al. 1999. *J. Chem. Phys.* 110:4963–70
40b. Kim J, Ho PKH, Greenham NC, Friend RH. 2000. *J. Appl. Phys.* 88:1073–81
40c. Cao Y, Parker ID, Yu G, Heeger AJ. 1999. *Nature* 397:414–17
41. Shuai Z, Beljonne D, Silbey RJ, Bredas J-L. 2000. *Phys. Rev. Lett.* 84:131–34
42. Cleave V, Yahioglu G, Le Barny P, Friend RH, Tessler N. 1999. *Adv. Mater.* 11:285–88
43. Gould IR, Farid S. 1996. *Acc. Chem. Res.* 29:522–28
44. Marcus RA. 1998. *J. Phys. Chem. B* 102:10071–77
45. Faulkner LR, Tachikawa H, Bard AJ. 1972. *J. Am. Chem. Soc.* 94:691–99.
46. Beideman FE, Hercules DM. 1979. *J. Phys. Chem.* 83:2203–9
47. Collinson MM, Wightman RM, Pastore P. 1994. *J. Phys. Chem.* 98:11942–47
48. Tokei-Takvoryan NE, Hemingway RE, Bard AJ. 1973. *J. Am. Chem. Soc.* 95:6582–89
49. Collinson MM, Wightman RM. 1993. *Anal. Chem.* 65:2576–82
50. Collinson MM, Wightman RM. 1995. *Science* 268:1883–85
51. Ritchie EL, Pastore P, Wightman RM. 1997. *J. Am. Chem. Soc.* 119:11920–25
52. Keszthelyi CP, Laser D, Bard AJ. 1975. *J. Electrochem. Soc.* 122:1642
53. Laser D, Bard AJ. 1975. *J. Electrochem. Soc.* 122:632–40
54. Brilmyer GH, Bard AJ. 1980. *J. Electrochem. Soc.* 127:104–10
55. Li YF, Gao J, Wang DL, Yu G, Cao Y, Heeger AJ. 1998. *Synth. Met.* 97:191–94
56. Manzanares JA, Reiss H, Heeger AJ. 1998. *J. Phys. Chem.* 102:4327–36
57. Yu G, Cao Y, Andersson M, Gao J, Heeger AJ. 1998. *Adv. Mater.* 10:385–88
58. Tasch S, Gao J, Wenzl FP, Holzer L, Leising G, et al. 1999. *Electrochem. Solid State Lett.* 2:303–5
59. Tasch S, Holzer L, Wenzl FP, Gao J, Winkler B, et al. 1999. *Synth. Metals* 102:1046–49
60. Elliott CM, Pichot F, Bloom CJ, Rider LS. 1998. *J. Am. Chem. Soc.* 120:6781–84
61. Pichot F, Bloom CJ, Rider LS, Elliott CM.

1998. *J. Phys. Chem. B* 102:3523–30
62. deMello JC, Tessler N, Graham SC, Friend RH. 1998. *Phys. Rev. B* 57:12951–63
63. Bard AJ, Faulkner LR. 1980. In *Electrochemical Methods: Fundamentals and Applications*, pp. 500–14. New York: Wiley
64. Richter MM, Fan F-RF, Klavetter F, Heeger AJ, Bard AJ. 1994. *Chem. Phys. Lett.* 226:115–20
65. Burrows PE, Shen Z, Bulovic V, McCarty DM, Forrest SR, et al. 1996. *J. Appl. Phys.* 79:7991–8006
66. Hopkins TA, Meerholz K, Shaheen S, Anderson ML, Schmidt A, et al. 1996. *Chem. Mater.* 8:344–51
67. Aziz H, Popovic ZD, Hu N-X, Hor A-M, Xu G. 1999. *Science* 283:1900–2
68. Bellmann E, Shaheen SE, Thayumanavan S, Barlow S, Grubbs RH, et al. 1998. *Chem. Mater.* 10:1668–76
69. Deleted in proof
70. Shi JM, Tang CW. 1997. *Appl. Phys. Lett.* 70:1665–67
71. Shaheen SE, Kippelen B, Peyghambarian N, Wang JF, Anderson JD, et al. 1999. *J. Appl. Phys.* 85:7939–45
72. Shaheen SE, Jabbour GE, Kippelen B, Peyghambarian N, Anderson JD, et al. 1999. *Appl. Phys. Lett.* 74:3212–14
73. Giebeler C, Antoniadis H, Bradley DDC, Shirota Y. 1999. *J. Appl. Phys.* 85:608–15
74. Jabbour GE, Shaheen SE, Morrell MM, Anderson JD, Lee P, et al. 2000. *IEEE J. Quantum Electron.* 36:12–17
75. Kwong RC, Sibley S, Dubovoy T, Baldo M, Forrest SR, Thompson ME. 1999. *Chem. Mater.* 11:3709–13

REACTIONS AND THERMOCHEMISTRY OF SMALL TRANSITION METAL CLUSTER IONS

PB Armentrout

Chemistry Department, University of Utah, Salt Lake City, Utah 84112;
e-mail: armentrout@chemistry.utah.edu

Key Words bond dissociation energies, collision-induced dissociation, kinetic energy, metal hydrides, metal oxides

■ **Abstract** This review discusses the reactivities and thermodynamics of small-size-specific transition metal clusters and focuses on thermodynamic information, which has not been comprehensively discussed before. Because of this focus, guided-ion-beam mass spectrometry was used to acquire much of the data. The details of this technique and the associated data analysis methods are provided. Results on the stabilities of bare transition metal clusters are provided for neutral, cationic, and anionic species. Implications for the electronic and geometrical structures are discussed, as well as the extrapolation of these values to bulk phase behavior. Detailed results for reactions of transition metal clusters with D_2 and the oxygen donors O_2 and CO_2 are reviewed. Available bond energies between size-specific clusters and one D atom and one and two O atoms are compiled, and their implications are evaluated and favorably compared with bulk phase analogs. Several additional thermodynamic studies of various cluster systems are also discussed.

INTRODUCTION

Motivations

Four principal motivations lie behind research on the gas phase chemistry of metal clusters: (*a*) Clusters may serve as effective and experimentally tractable models for surfaces and heterogeneous catalysts; (*b*) some clusters may exhibit unique reactivity and properties, making them intriguing chemical reagents; (*c*) cluster research serves as an interface between theory and experiment; and (*d*) the investigation of any new form of matter presents a challenge and opportunity for novel discoveries. The first of these factors is arguably the most important from a technological standpoint. The cluster model for catalysts is plausible, given that many industrial catalysts involve highly dispersed metals and that surface defect sites are often the active sites for chemistry (1). A nice verification of this hypothesis is the recent demonstration that coordinatively unsaturated ruthenium atoms are the catalytic sites for carbon monoxide oxidation on RuO_2 [1,1,0] (2). Further, studies of transition metal clusters in the gas phase have found that as the cluster size increases, the chemistry rapidly reaches a limiting behavior, suggestive of an approach to bulk phase behavior. Several examples of such behavior are provided in the remainder of this review. Such observations are consistent with the idea that chemical bonds are phenomena more local than quantities such as ionization energies and electron affinities.

Clearly, more extensive research is required in both the cluster and surface science fields to determine how literally the cluster-surface analogy may be taken. However, the second rationale for cluster studies is that the analogy fails for at least some clusters, such that specific-sized clusters may be found with very high chemical specificity. For example, studies by Irion and coworkers found that Fe_4^+ is unique in its ability to form benzene from smaller hydrocarbons (3) and in its thermal reactivity with ammonia (4) and ethane (5). Similarly, the tetramer and pentamer cations of cobalt have been observed to be particularly reactive with ethene (6), methanol (Oiestad A, Uggerud E. unpublished results), and several other neutral species (8). In addition, supported iridium clusters show a dependence on cluster size in the hydrogenation of toluene (9), and nickel clusters supported on MgO exhibit size-dependent chemical reactivity toward CO dissociation (10).

Gas phase cluster research also plays an increasingly important role as an interface with theory, including both electronic structure theory (at many levels from empirical to ab initio) and dynamics calculations. Advances have allowed theory to become a primary tool in helping to explain and understand the complex chemistry occurring at surfaces, but of necessity, the surfaces are usually modeled by a limited number of surface atoms. In contrast, clusters are sufficiently small that they can be modeled directly by theory. Quantitative experimental results on the reactivity and thermodynamics of clusters thereby provide a benchmark for theory, which can then examine surface chemistry with increased reliability.

Focus of the Review

Various detailed measurements of the physical and chemical properties of clusters can be made, most of which have a ready analogy in surface science and catalysis. Such properties include ionization energies (11–22), electron affinities (23–27), and magnetic properties (28–32). The chemistry of transition metal clusters at thermal energies has been examined in fast-flow-tube reactors (33–48), ion cyclotron resonance mass spectrometers (3–5, 49–54), and drift cells (55). In most of these methods, mass-selection techniques are used, thereby permitting size-specific cluster studies and directly addressing questions relating to possible fragmentation during reaction and ionization.

Despite the considerable progress made in studies of metal clusters since the early days of magic numbers (clusters of a particular size showing enhanced stability or reactivity), the complexity of metal clusters means that a quantitative understanding of these species is still developing, especially for the open-shell transition metal systems. Detailed electronic, structural, and thermodynamic properties of clusters are largely unavailable. These three aspects are intimately intertwined and related. In this review, I examine progress made in the examination of transition metal cluster reactivity and thermochemistry. In contrast to previous reviews that concentrated on the electronic structure of small metal clusters (56, 57), the reactivities of clusters of neutral transition metals (58), and ionic metal and semimetal clusters (57, 59), the focus of this review is on the thermodynamics of the clusters and their reactions.

The emphasis on thermodynamics directs much of this review to studies using guided-ion-beam mass spectrometry. This technique shares the capability of mass selectivity with many of the methods noted above. However, it has the advantage of being able to study cluster chemistry at both thermal and hyperthermal energies (the latter is equivalent to activated conditions on surfaces). Flow-tube reactors, ion cyclotron resonance techniques, and drift cells are typically limited to thermal and near-thermal conditions. Although ion cyclotron resonance techniques can be used to examine hyperthermal reactions, the derivation of quantitative information from such experiments is more difficult than from beam techniques (see references 60 and 60a). The ability to examine the chemistry as a function of kinetic energy allows the acquisition of thermochemical information; and such information is a powerful probe of the mechanisms and dynamics of the reactions. This power has been illustrated in work on transition metal clusters, as discussed in this review, and also by Jarrold (61–63) and Anderson (64–64e) for silicon, aluminum, and boron clusters.

Ion versus Neutral Studies

Much of the work discussed in this review concerns reactions and properties of transition metal clusters in the gas phase. A pertinent question with regard to studies of such species is their relevance to real-world catalysts. Although it might seem that neutral clusters are needed to accurately model heterogeneous catalysis on a metal surface, several reasons indicate that investigations of transition metal cluster ions are entirely appropriate for research aimed at understanding and ultimately designing transition-metal-based catalysts.

First, for clusters larger than five or six atoms, metal cluster cations of Nb_n, Co_n, and Fe_n were found to react with patterns of size-dependent reaction rates similar to those of the neutral clusters (42, 43, 65, 66). Bondybey and coworkers (53, 53a, 54) found similar reactivities for the larger anionic and cationic niobium and rhodium clusters interacting with benzene. The close correspondence of the reaction rate patterns between negatively charged, neutral, and positively charged clusters demonstrates that the fundamental chemistry is relatively insensitive to the charge state of the metal. This result occurs possibly because the charge is delocalized over the metal cluster, resulting in no strong charges at any given point on the surface of the cluster. Also, as reviewed below, we have found that the thermodynamic stabilities of neutral and cationic transition metal clusters are virtually the same for clusters of six or more atoms.

This is not to say that no differences occur in the chemistry of neutral and ionic clusters. For example, although the broad reactivity patterns of neutral and cationic iron clusters resemble one another, differences in reactivity as large as two orders of magnitude are found for specific-sized clusters (43). Similarly, the relative reactivity of some cationic cobalt clusters is about an order of magnitude higher than that of the neutral clusters (66). Rather than being a detriment, it seems likely that such differences (imbedded in an otherwise similar reactivity pattern) can probably be used advantageously to better understand details of the chemistry of both charge states.

Second, the actual use of small metal clusters as catalysts requires them to be supported on some type of inert material such as silica, alumina, MgO, or possibly graphite. Regardless of the nature of the support, some electronic rearrangement and charge transfer to the support occur (67), potentially leaving the metal cluster with at least a partial charge. Further, defect or unsaturated sites of surfaces may be more reactive because they are electron-deficient (or rich) sites. These ideas suggest that cationic (or anionic) clusters may, in fact, be better models for catalysts than neutral clusters are.

Finally, the thermodynamic studies discussed in this review demonstrate a good correspondence between ligand bonds to metal cluster cations and those to single-crystal surfaces. This good agreement suggests that chemical binding is largely a local phenomenon and indicates that clusters, even charged ones, can act as a quantitative model for surface chemistry and thermodynamics.

EXPERIMENTAL APPROACHES

Transition Metal Cluster Ion Sources

The study of metal cluster cations in the gas phase has been enabled by the development of the laser vaporization/supersonic expansion source (68, 69). In our version (70), a copper vapor laser with a 7-kHz repetition rate is used for sample vaporization. It is operated at sufficiently high power (3–4 mJ pulse^{-1}, 22–28 W) that atomic metal cations are also formed. Both ionic and neutral clusters of metal atoms are formed by three-body collisions in a high pressure of continuously flowing He. We estimate that an average ion undergoes $\sim 10^5$ collisions with He in our 2-mm-diameter, 6.4-cm-long source channel. The clusters are then cooled further in a mild supersonic expansion that occurs in a field-free region. No post-ionization of neutral clusters is used. Ion-focusing lenses in two differentially pumped regions that follow the source chamber are kept at low potentials. Under these conditions, we find no evidence of collisional reheating of the cluster ions (71). Studies show that cluster cations generated in this source are much colder than those generated in electron impact sources (72); however, there is no direct quantitative measure of the absolute internal temperature of our cluster ions. The high-pressure conditions in this source are sufficient to suggest that the clusters are fully thermalized and perhaps colder because of the supersonic expansion. All of our work proceeds with this assumption, and the results appear to be consistent with it as well.

Anionic metal clusters can also be formed by using a laser vaporization/supersonic expansion source (45), but this source is no longer a requirement. A simpler dc discharge source in which a cathode made of the desired metal is sputtered by Ar^+ ions generates metal cluster anions in abundance (46, 46a). Thermalization of the clusters is again achieved by collisions with the bath gas. Such a source produces few cationic clusters, presumably because these undergo energetic collisions with the cathode, leading to dissociation.

Instrumentation and Experimental Methods

In our work, the metal cluster cations are studied with a guided-ion-beam tandem mass spectrometer built especially for this purpose (71). Briefly, our instrument comprises the cluster ion source described above, a magnetic sector mass spectrometer for selection of a particular cluster size, a well-defined zone where reactions occur at variable kinetic energies under single-collision conditions, a quadrupole mass filter for product analysis, and a sensitive ion detector. This Daly detector (73) incorporates a 28-kV conversion dynode that provides very good detection sensitivity for all masses studied. The reaction region is surrounded by an octopole ion beam guide (74) that permits very low ion energies (<0.1 eV) to be reached and routine determination of the absolute energy scale using retarding techniques (75). By varying the voltage difference between the source and the reaction zone, the kinetic energies of the ions are easily varied from thermal ($\sim$0.03 eV) to more than four orders of magnitude higher. Reactant and product ion intensities are measured as a function of the ion kinetic energy, and these are converted to absolute reaction cross sections (75). Uncertainties in these cross sections are estimated as $\pm 30\%$.

Data Analysis

To obtain thermodynamic information from our studies, we rely on varying the kinetic energy available to the reaction system and then modeling the cross sections for endothermic reactions. The threshold analysis procedure for reactivity studies of transition metal clusters was described previously (76, 77). Briefly, the energy dependence of cross sections for endothermic processes in the threshold region is modeled as follows:

$$\sigma(E) = \sigma_0 \sum g_i (E + E_i - E_0)^N / E, \qquad 1.$$

where σ_0 is an energy-independent scaling parameter, N is an adjustable parameter, E is the relative kinetic energy, and E_0 is the threshold for the reaction at 0 K. The summation is over the ro-vibrational states i having energies E_i and populations g_i, where $\Sigma g_i = 1$. We assume that the relative reactivity, as reflected by σ_0 and N, is the same for all vibrational states. The Beyer-Swinehart algorithm (78–81) is used to evaluate the density of the ion vibrational states, and the relative populations

g_i are calculated by the appropriate Maxwell-Boltzmann distribution at 300 K. Vibrational frequencies for the bare-cluster ions are obtained by using a Debye model suggested by Jarrold and Bower (62). Equation 1 has been used successfully in reproducing the cross sections of various ion-molecule reactions (82) as well as collision-induced dissociation (CID) processes (83–85).

To reproduce the experimental data using Equation 1, several additional experimental effects must be considered. First, the thermal motion of the target gas and the kinetic energy distribution of the parent ion beam are both convoluted into Equation 1 as described previously (75). Second, CID processes and some reactions are sensitive to multiple collisions between the ions and the collision gas. This sensitivity occurs because even at low neutral reagent pressures, a small but finite probability exists for an ion to undergo two collisions with the neutral species in the gas cell. In CID systems, the extra collision deposits additional energy into the cluster ion, resulting in two effects. First, this extra energy lowers the observed experimental threshold for CID, an effect that becomes more obvious as the threshold increases. Second, the extra energy shortens the ion lifetime and thus yields much more efficient dissociation, an effect that is most prominent for large parent ions that have longer dissociation lifetimes. In reaction systems, the secondary collision can lead to further reaction, altering the shape of the cross section under study. To remove these pressure-dependent effects, cross sections are acquired at multiple pressures and extrapolated to zero neutral reactant pressure (84, 86, 87). This method yields a reaction cross section that is rigorously a result of single-collision events. The cross sections extrapolated to zero pressure are analyzed to determine thermochemistry.

We also account for the possibility that the processes being modeled occur more slowly than the experimental time window available, which is $\sim 10^{-4}$ in our apparatus. This possibility is considered by incorporating Rice-Ramsperger-Kassel-Marcus (RRKM) theory (81, 88) into Equation 1, as outlined elsewhere (71, 89, 90). All threshold analyses using Equation 1 and discussed below include this lifetime analysis. For these calculations, the transition state and its molecular constants are chosen as described in our work on the reactions of Fe_n^+ (76, 77). We believe that the most reasonable choice for the transition state places it at the point where the last species is lost from the transiently formed intermediate. This choice presumes that dissociations of atoms lost prior to this event are facile and occur much more rapidly than this final atom loss step, which is a reasonable approximation as discussed previously (76, 77). In most cases, we assume that the transition state is a loose one, having molecular constants similar to the dissociated products. Most of the vibrational frequencies are chosen to equal those of the products, whereas the transitional modes (those turning into translations and rotations of the products) are chosen as outlined in recent work (91, 92).

Two types of endothermic reactions are used to acquire thermodynamic information regarding transition metal clusters. In a CID process,

$$M_n^+ + Xe \rightarrow M_{n-1}^+ + M + Xe, \qquad 2.$$

the threshold for reaction equals the desired bond energy, $E_0(2) = D_0(M_{n-1}^+ - M)$. In the following reaction,

$$M_n^+ + AB \rightarrow M_n A^+ + B, \qquad 3.$$

the desired bond energies are given by $D(M_n^+ - A) = D(A-B) - E_0(3)$. In both cases, it is assumed that no activation barriers exceed the reaction endothermicity. This assumption is generally true for ion-molecule reactions and has been tested explicitly many times (82, 84, 85, 93). Reverse activation barriers are often absent in ion-molecule processes because of the long-range ion-induced dipole or higher-order interactions. Exceptions have generally involved spin or orbital angular momentum restrictions (82, 93), but for transition metal systems, even these restrictions may not apply. Because transition metal systems often have a high density of electronic states and appreciable spin-orbit coupling, it is feasible for the many potential energy surfaces to mix, allowing ground state parent ions to dissociate to ground state products via adiabatic processes.

STABILITIES AND STRUCTURES OF TRANSITION METAL CLUSTERS

The stabilities of bare transition metal clusters have been studied using both CID and photodissociation (PD) methods. In our laboratory, we have used CID to study the cationic clusters of 10 different transition metal elements. This work includes the first-row transition metal clusters of Ti_n^+ ($n = 2$–22) (94), V_n^+ ($n = 2$–20) (95), Cr_n^+ ($n = 2$–21) (96, 97), Mn_2^+ (98, 99), Fe_n^+ ($n = 2$–19) (71, 100, 101), Co_n^+ ($n = 2$–18) (72, 87), Ni_n^+ ($n = 2$–18) (102, 103), and Cu_2^+ (Lian L, Schultz RH, Armentrout PB. unpublished results); the second-row transition metal clusters of Nb_n^+ ($n = 2$–11) (86, 105); and the third-row transition metal clusters of Ta_n^+ ($n = 2$–4) (106). Using similar techniques, Ervin's group performed comparable experiments on anionic clusters of the coinage metals [Cu_n^- ($n = 2$–8) (107), Ag_n^- ($n = 2$–11) (108), and Au_n^- ($n = 2$–7) (109)] and group 10 metals [Pd_3^- (110) and Pt_n^- ($n = 3$–6) (111)]. A multiple CID experiment in an ion cyclotron resonance trap was recently used to measure dissociation energies for Ag_n^+ ($n = 2$–25) (112) and Ag_n^{2+} ($n = 9$–25) (113). PD methods have been used for very accurate measurements of the bond energies for a limited number of small cluster cations (114–122), a more comprehensive set of Ag_n^+ ($n = 8$–21) clusters (123), and several cluster anions [Ag_n^- ($n = 7$–11) (124) and Au_n^- ($n = 6, 7$) (109)]. Results and conclusions from this work are overviewed in this section.

Dissociation Pathways

CID and PD studies of transition metal cluster systems find that cluster ions dissociate by two general pathways: fission, which is the dissociation into bound molecular fragments, and evaporation, which is the sequential loss of atoms (analogous to vaporization of bulk phase metals). We find that most metal cluster cations preferentially dissociate by evaporation, and the behavior of Cr_{13}^+ shown in Figure 1 is typical. Indeed, titanium (94), iron (101), cobalt (87), and tantalum (106) cluster ions do so exclusively. Most chromium and nickel clusters also dissociate by evaporation, although the trimer and pentamer cations emit neutral dimers as a primary dissociation pathway (97, 103). This finding is apparent in the results for Cr_5^+ shown in Figure 1. Small vanadium cluster cations also dissociate by dimer elimination, but atom loss is the predominant dissociation pathway for all but the tetramer (95). Niobium is the only metal that we have studied that undergoes fission for all cluster sizes (86). For silver cluster cations, monomer evaporation is again predominant, although Ag_{11}^+ and Ag_{13}^+ are also observed to expel a dimer (123). In contrast, the anions of the coinage metals experience substantial dimer loss, and this channel dominates for the odd-sized clusters (107–109).

For most metal clusters, the lowest energy dissociation channel is atom loss, and therefore evaporation is anticipated to be the dominant dissociation pathway. However, it is somewhat surprising that many clusters show no evidence of fission, even though the loss of a dimer (or higher-order cluster) is always thermodynamically preferred in comparison with the sequential loss of two (or more) atoms. Therefore, sequential atom loss (evaporation) must dominate the dissociation pathways of most metal clusters because of favorable kinetics. Clearly, cluster fission generally requires rearrangement of the cluster to form the ground state structures of the fragment clusters, whereas cluster evaporation requires less rearrangement. The clusters that do dissociate by fission processes can be correlated with events that are thermodynamically very favorable and sufficient to overcome the kinetic constraint. Thus, the odd-sized anionic coinage metal clusters, which are even-electron, often closed-shell species, are much more stable than the even-sized clusters, which have open-shell electronic configurations. Thus, whereas even-sized clusters lose atoms to generate closed-shell odd-sized cluster anions, the odd-sized clusters lose neutral dimers to form two stable closed-shell products. For niobium, the binding energies of the cationic clusters are sufficiently large (see the next section) that the driving force for fission is much higher than that for clusters of the first-row transition metals.

Bond Energies of Cationic Clusters

Among the most accurately known bond dissociation energies (BDEs) of transition metal clusters are those obtained in PD studies. In several of these systems, namely, Ti_2^+, V_2^+, Co_2^+, and Co_3^+ (114, 115), a step function in the dissociation probability is observed at a specific photon energy. This finding demonstrates that an extremely high density of states occurs in these molecules at the dissociation limit. This

Figure 1 Collision-induced dissociation cross sections of Cr_5^+ (*panel a*) and Cr_{13}^+ (*panel b*) with Xe as a function of collision energy in the center-of-mass frame (adapted from Reference 97).

THERMOCHEMISTRY OF TRANSITION METAL CLUSTERS 431

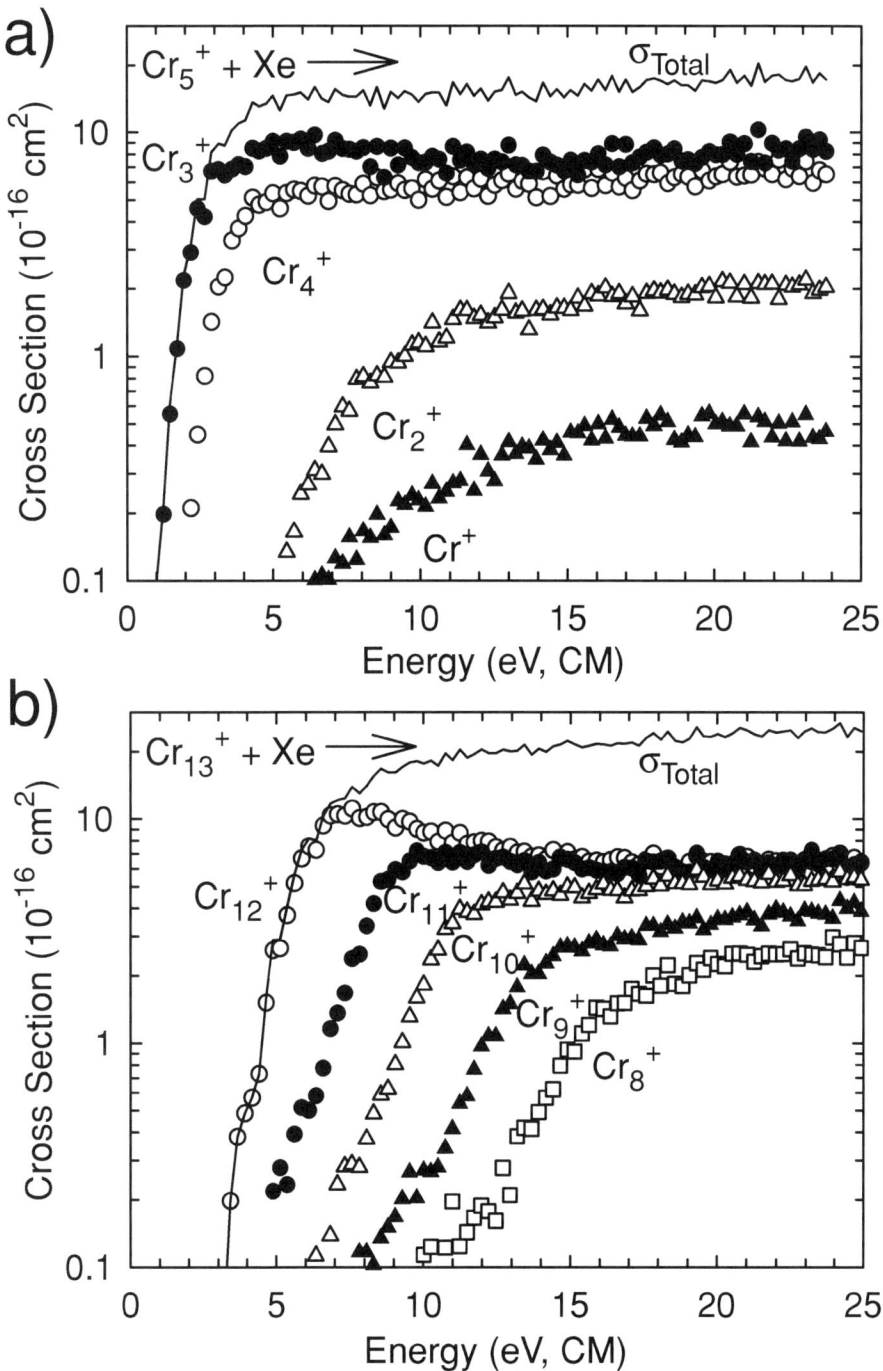

conclusion is confirmed by the large number (>3000) of adiabatic potential energy curves calculated to evolve from separated atom limits within 1 eV above the ground-separated atom limits. Confirmation that the observed predissociation limit is the adiabatic bond energy comes from good agreement between the PD values and those from CID experiments [e.g. for Ti_2^+, $D(PD) = 234.9 \pm 0.2$ kJ mol^{-1}, whereas $D(CID) = 229 \pm 7$ kJ mol^{-1}]. In contrast, PD studies on Cr_2^+ and Ni_2^+ by Brucat and coworkers (116, 118) found only discrete lines, such that only an upper limit to the BDE can be derived. This result appears to be related to the lower density of states for these two molecules (<800 curves calculated to lie within 1 eV of the ground-separated atom limits). In these cases, the PD limits observed are considerably above the CID bond energies (96, 101).

An interesting example is V_3^+, for which we observe a sharp predissociation threshold (116), but it occurs at a photon energy significantly higher than the BDE of V_3^+ measured by CID methods (219 ± 9 kJ mol^{-1}) (95). Further, the predissociation onset lies within a weaker dissociation continuum that is induced by absorption of radiation at lower photon energies. The explanation for this behavior starts with the assignment of the ground electronic state of V_3^+ as $^9A_2''$, as suggested in a theoretical study of the $3d$ transition metal trimers (125). If the total electronic spin remains a nearly good quantum number upon dissociation, then the first spin-allowed dissociation asymptote is, $V_2^+(X^4\Sigma_g^-) + V(3d^4 4s^1, {}^6D)$, where the ^6D is an excited state of V. This possibility explains how a nonzero PD signal may be observed at energies below the predissociation threshold. Correcting the predissociation threshold by the energy of the $V(3d^4 4s^1, {}^6D_{1/2})$ excited state then provides $D_0(V_2^+-V) = 224.1 \pm 0.1$ kJ mol^{-1}, in good agreement with the value from CID experiments.

As these results demonstrate, PD methods can yield extremely accurate and precise BDE measurements, but unfortunately they are very difficult to extend to clusters larger than trimers. In contrast, CID methods are straightforwardly applied to clusters of ~20 atoms. Values obtained from our CID studies for larger clusters, in some cases revised to 0 K values (126), are summarized in Table 1. Figure 2 shows the results as a function of cluster size for iron and chromium clusters. Comparisons to PD results for small clusters provide a critical test of the CID method for determining accurate thermochemistry for transition metal cluster cation systems. Although comparisons to theory are not available for most of these values, good agreement between the CID values and the calculated BDEs has been obtained for the smaller vanadium cluster cations (127, 128). Table 1 also includes values from multiple CID experiments on Ag_n^+ clusters (112), which have been verified for the larger clusters ($n = 8$–21) by time-resolved PD experiments (TRPD) (123). Theoretical calculations (129) on the silver clusters agree reasonably well with the experimental values except for $n = 6$ and 7, where large deviations occur. Other TRPD measurements have included measurements of $D(Cu_5^+-Cu) = 217 \pm 10$ kJ mol^{-1} and $D(Cu_7^+-Cu) = 230 \pm 5$ kJ mol^{-1} (122).

In addition to the values in Table 1, CID methods have been used to measure the BDEs of dimer cations of copper and manganese: $D(Cu^+-Cu) = 177 \pm$

TABLE 1 Bond energies of M_{n-1}^+–M (kJ mol^{-1})[a]

$n\backslash M$	Ti[b,c]	V[d]	Cr[e]	Fe[c,f]	Co[g]	Ni[c,h]	Nb[c,i]	Ag[j]
2	229	302	125	268	265	204	569	168
3	230	219	194	169	185	235	490	261
4	338	341	100	216	205	199	590	101
5	340	313	215	260	274	223	557	190
6	354	398	171	315	319	275	568	124
7	400	372	246	320	283	296	650	218
8	278	385	217	252	303	265	592	211
9	346	354	248	281	283	283	578	285
10	335	382	232	284	285	283	604	152
11	344	382	248	308	314	293	600	206
12	406	398	258	334	329	335		180
13	470	449	289	408	351	352		233
14	321	388	294	281	303	299		193
15	400	452	267	377	380	334		242
16	360	376	282	319	355	338		198
17	344	404	273	323	346	354		246
18	318	381	208	311	371	341		227
19	450	408	233	373				252
20	409	443	289					225
21	395		253					249
22	410							181
23								211
24								186
25								210

[a] Values are at 0 K. Absolute uncertainties gradually increase from $\sim$10 kJ mol^{-1} for small-cluster ions to 50 kJ mol^{-1} for larger-cluster ions. Relative uncertainties are <15 kJ mol^{-1} for all clusters.
[b] Data from Reference 94.
[c] Values have been adjusted to 0 K as discussed in Reference 126.
[d] Data from Reference 95.
[e] Data from Reference 97.
[f] Data from Reference 101.
[g] Data from Reference 87.
[h] Data from Reference 103.
[i] Data from Reference 105.
[j] Data from Reference 112. Absolute uncertainties are <20 kJ mol^{-1}.

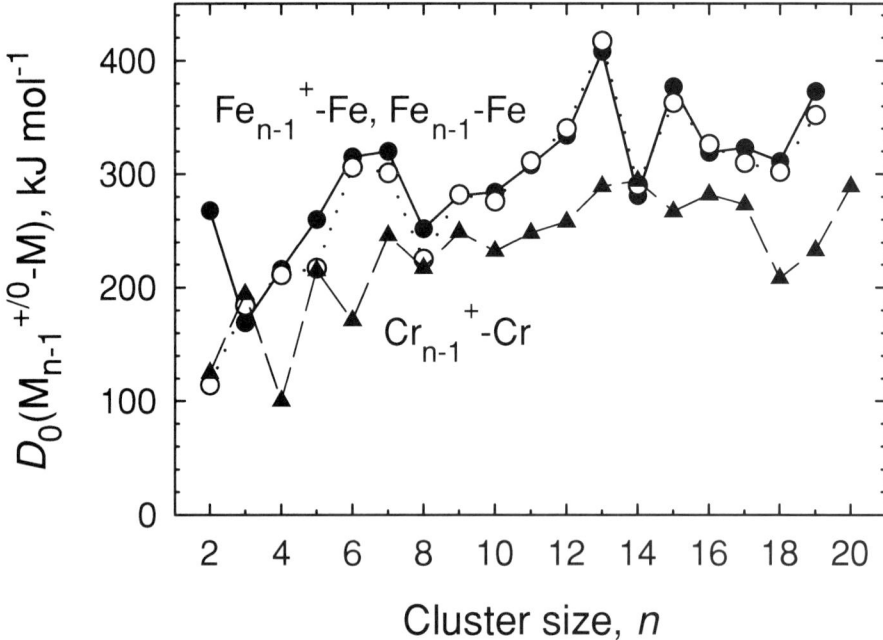

Figure 2 Metal cluster bond dissociation energies for iron cations (●), iron neutral species (○), and chromium cations (▲) as a function of cluster size (data are taken from Tables 1 and 2).

10 kJ mol^{-1} (L Lian, RH Schultz & PB Armentrout, unpublished results) and $D(Mn^+-Mn) = 109 \pm 20$ kJ mol^{-1} (99). The former value agrees well with the best literature value, 178 ± 8 kJ mol^{-1} (130). The CID value for $D(Mn^+-Mn)$ is just below a lower limit of 134 kJ mol^{-1} determined by PD (131). This discrepancy is attributed to internal excitation left by the electron impact source used to generate Mn_2^+ from $Mn_2(CO)_{10}$, a problem not encountered with the laser vaporization/supersonic expansion source used for all other clusters in our studies.

Bond Energies of Neutral Clusters

The BDEs of cationic transition metal clusters measured in our CID studies can be converted to BDEs for the neutral clusters when the adiabatic ionization energies are available. This conversion is done with the following equation:

$$D(M_{n-1}-M) = D(M_{n-1}^+-M) + IE(M_n) - IE(M_{n-1}), \qquad 4.$$

where IE is the ionization energy. Ionization energies, which are probably adiabatic rather than vertical values, are available for most clusters of vanadium (11, 132), iron (12–14), cobalt (12, 14), nickel (14, 15, 133), and niobium (11, 16). Neutral transition metal BDEs calculated from these ionization energies and the cationic BDEs in Table 1 are listed in Table 2. Additional BDEs for many smaller neutral

TABLE 2 Bond energies for M_{n-1}–M (kJ mol^{-1})[a]

n\M	V[b]	Fe[c]	Co[c]	Ni[d]	Cu[e]	Nb[f]
2	266[g]	114[h]	≤127[i]	200[j]	197	509
3	137	184[h]	≥140[k]	87	111	448
4	354	211[h]	233[l]	158	262	571
5	297	217[h]	274[l]	272	178	539
6	389	306	319[l]	334[l]	247	561
7	360	301	256	227	257	647
8	397	225	283	271	288	610
9	339	282	279	274		546, 519
10	379	276	294	275		631
11	365	311	301	283		528
12	395	340	329	334		
13	447	417	355	353		
14	391	290	301	298		
15	<426	363	371	333		
16	>404	326	369	340		
17	405	310	332	355		
18	372	302	371	339		
19	<388	352				

[a]Values are at 0 K and, except where noted, are calculated using Equation 4 with cluster ion BDEs from Table 1 and ionization energies (IEs) from the references given. Absolute uncertainties gradually increase from ~10 kJ mol^{-1} for small cluster ions to 50 kJ mol^{-1} for larger clusters. Relative uncertainties are <15 kJ mol^{-1} for all clusters.
[b]Data are from Reference 11.
[c]Data from Reference 12, except as noted.
[d]Data from Reference 15, except as noted.
[e]Calculated from the $D(Cu_{n-1}$–Cu) bond energies in Table 3 combined with electron affinities from Reference 25.
[f]Data from Reference 16. For Nb$_9$, two isomers have been identified.
[g]Data from Reference 132.
[h]Data from Reference 13.
[i]IE(Co$_2$) ≤ 6.42 eV, MD Morse & ZW Fu, personal communication.
[j]Data from Reference 133.
[k]$D(Co$–Co$) + D(Co_2$–Co$) = D(Co^+$–Co$) + D(Co_2^+$–Co$) + IE(Co_3)$–IE(Co$) = 281 \pm 14$ kJ mol^{-1}.
[l]Data from Reference 14.

transition metal dimers obtained using a variety of techniques have been reviewed comprehensively by Morse (56) and are not included here. One test of the accuracy of these results comes from a comparison of the BDEs for neutral nickel clusters, Ni_n ($n = 5$–23), with those calculated by Stave & DePristo (134). Here, experiment and theory agree nicely with respect to the patterns in the BDEs as the cluster size varies, and the deviation in the absolute values (theoretical minus experimental values) is 20 ± 29 kJ mol^{-1}.

Comparison of the cationic and neutral cluster BDEs reveals that large differences do not exist for most clusters, as shown in the example for iron in Figure 2. For this metal, only the dimer and pentamer show appreciable differences between the cationic and neutral BDEs. For vanadium, neutral and cationic cluster BDEs are similar throughout the range studied. For cobalt, the dimer and trimer show large differences. Nickel clusters show large changes between neutral and cationic BDEs for clusters smaller than the octamer, but they show very similar values for larger clusters. Niobium clusters have neutral and cationic BDEs that parallel one another fairly closely from the dimer to the octamer, but large oscillations are found for Nb_9, Nb_{10}, and Nb_{11}. This unusual behavior is probably attributable to the various isomers that occur for niobium clusters (neutral and cationic) in this size range (39, 40, 42, 65, 135).

These results indicate that the removal of a single electron has little effect on the thermodynamic stability of most clusters, except for the smallest clusters. This conclusion could mean that the charge in a cationic cluster is delocalized over the cluster such that as the cluster increases in size, the effect of the "missing" electron is mediated. Because all atoms in the cluster have only a fractional charge, the energetics for dissociation of the cluster cation are approximately the same as for the neutral cluster. A second possible explanation is that the charge is highly localized and provides a site of strong bonding. In this scenario, dissociation in charged clusters occurs at a site remote from this charge localization, that is, at a site neutral in character.

Bond Energies of Anionic Clusters

BDEs have been measured for fewer anionic transition metal clusters in comparison with cationic and neutral clusters and come exclusively from the work of Ervin (107–111, 124, 136, 137). These values are summarized in Table 3, except for $D(Pd_2^- - Pd) = 218 \pm 35$ kJ mol^{-1} (110), and they generally come from CID measurements. For Ag_n^- ($n = 7$–11), Au_6^-, and Au_7^-, results from TRPD studies are also available and are generally preferred. Ervin and coworkers found that the CID and TRPD values agree within experimental error but that the CID values are systematically slightly high (less than the experimental error, however). The origins of these small differences have been discussed in some detail (109), but no firm conclusions can be reached at this time. For the silver clusters, experimental BDEs are found to agree reasonably well with theoretical calculations (138, 139).

TABLE 3 Bond energies for M_{n-1}^{-}–M (kJ mol^{-1})[a]

$n\backslash M$	Cu[b]	Ag[c]	Au[d]	Pt[e]
2	158	177	≤323	
3	233	233	≤382	425
4	186	131	≤212	415
5	227	187	≤253	473
6	246	182	214[f]	
7	277	239[f]	269[f]	
8	233	159[f]		
9		219[f]		
10		176[f]		
11		230[f]		

[a]Values are at 0 K taken from CID measurements except where noted. Absolute uncertainties range from ~10–50 kJ mol^{-1}. Relative uncertainties are probably <15 kJ mol^{-1} for all clusters.
[b]Data from Reference 107.
[c]Data from References 108 and 124.
[d]Data from Reference 109.
[e]Data from Reference 111.
[f]TRPD results.

Electronic and Geometric Structures of Metal Clusters

Our understanding of the variations in transition metal cluster BDEs, such as those shown in Figure 2, is still in its infancy. In some cases, it is clear that electronic effects dominate the stability patterns. The best examples are clusters of the coinage metals (Cu, Ag, and Au), which all have atoms with s^1d^{10} ground state electron configurations. For cationic, neutral, and anionic clusters of these elements, the BDEs show strong even-odd oscillations (see Tables 1–3). Odd-sized ionic clusters are most stable, whereas even-sized clusters have the strongest BDEs for the neutral clusters. These variations are clearly related to the number of valence s electrons; that is, all clusters having an even number of s electrons are relatively stable. Of the small clusters, scrutiny of the results shows that the most stable are M_3^+, M_9^+, M_8, and M_7^-. This finding is consistent with the electron shell closings predicted by the jellium model (140, 141) for two and eight electrons. The next shell closings are predicted to occur at 18 and 20 electrons, consistent with the large BDEs observed for Ag_{19}^+ and Ag_{21}^+ (see Table 1) (112).

Small vanadium and chromium clusters also show even-odd oscillations in their stabilities. As for the coinage metals, the stable cationic clusters of chromium have an odd number of atoms (see Figure 2 and Table 1). The ground states of Cr and Cr$^+$ are $^7S(4s^1 3d^5)$ and $^6S(3d^5)$, respectively, such that odd-sized Cr_n^+ clusters have an even number of 4s electrons, and even-sized clusters have an odd number of 4s electrons. This result suggests that the binding of these clusters (up to about $Cr_{10}{}^+$) is dominated by interactions among the 4s electrons. In contrast to chromium and the coinage metals, the even-sized clusters of vanadium are the most stable for both cationic and neutral clusters (see Tables 1 and 2). The ground states of V and V^+ are $^4F(4s^2\ 3d^3)$ and $^5D(3d^4)$, respectively. The ground state of $V_2{}^+$ is probably formed by combining these two species, such that there are two electrons in a σ bond formed from the 4s orbitals. To generate the ground state of V_2, however, two atoms of V in its $^6D(4s^1\ 3d^4)$ low-lying excited state are probably combined, again forming a species with two electrons in a 4s-orbital-based σ bond (115, 142). The even-odd oscillations can then be understood if larger cationic and neutral vanadium clusters are both formed by the addition of vanadium atoms in the ^6D state, leading to an odd number of 4s electrons for odd numbers of nuclei and to an even number of 4s electrons for even numbers of nuclei.

For other transition metals, the exact number of electrons does not grossly influence the stability of metal clusters (except for the smallest ones). This finding suggests that geometric structures control the relative stability of different sized clusters of these transition metals. Compact and symmetric geometries are anticipated for metals, as demonstrated by the chemical probe technique of Riley & Parks (58, 143, 144) and by calculations (127, 128, 134). To obtain structural information from the stabilities of clusters, we examine the most stable clusters (excluding dimers and trimers). From our work, these are as follows: $Ti_n{}^+$ at $n = 7, 13, 15$, and 19; $V_n{}^+$ at $n = 6, 13$, and 15; $Cr_n{}^+$ at $n = 5, 7, 9, 13, 14$, and 20; $Fe_n{}^+$ at $n = 6, 7, 13, 15$, and 19; $Co_n{}^+$ at $n = 6$; $Ni_n{}^+$ at $n = 7$ and 13; and $Nb_n{}^+$ at $n = 4$ and 7. For neutral clusters, the most stable clusters are V_n at $n = 4, 6$, and 13; Fe_n at $n = 6, 13, 15$, and 19; Co_n at $n = 6$; Ni_n at $n = 6$ and 13; and Nb_n at $n = 4$ and 7. Note that the most stable neutral and cationic clusters are the same sizes in most cases.

One prominent sequence is 7, 13, and 19 (e.g. for $Ti_n{}^+$ and $Ni_n{}^+$), which suggests the icosahedral structures of a pentagonal bipyramid, icosahedron (one central atom with 12 equivalent surrounding atoms), and double icosahedron (with two central atoms), respectively. This type of numerology is equivocal, however, because compact and octahedrally symmetric structures based on a face-centered-cubic closest-packed lattice also exist for $n = 13$ and 19. However, this type of structure should be correlated with a stable hexamer ($n = 6$). Thus, when the sequence of 7, 13, and 19 is dominant, as for $Ti_n{}^+$ and $Ni_n{}^+$, icosahedral packing may be implied. This finding is consistent with the conclusions drawn from chemical probe methods for neutral nickel clusters (58, 143), except that Ni_6 is believed to be octahedral. When the sequence is 6, 13, 15, and 19, as for $V_n{}^+$ and Fe_n, octahedral face-centered-cubic-like packing may be implied. This result parallels the conclusions of the chemical probe method for neutral clusters of iron (34). Of course, clusters in the size range studied here need not follow a single pattern of packing, as noted for Ni_n clusters. In addition, there may be multiple isomers having nearly degenerate energies (134, 145).

Approach to Bulk Phase Thermodynamics

As noted in the introduction, a key interest in cluster studies is the approach to bulk phase behavior. For cluster BDEs, this behavior is studied by comparing the cluster cohesive energy, $E_c(n)$, with the heat of vaporization at 0 K, $\Delta_{vap}H_0°$ (146). The variable $E_c(n)$ is defined as the atomization energy per atom, $E_A(n)/n$, where $E_A(n)$ is given by $\Sigma D(M_i^+–M)$ over the range $i = (n-1)$ to 1. For cluster ions of about 20 atoms, the cohesive energies are about 70% of $\Delta_{vap}H_0°$. Thus, these clusters are approaching but have not arrived at bulk phase behavior, which seems reasonable when it is remembered that nearly all of the atoms in these clusters are surface atoms.

The deviation between cluster and bulk phase thermodynamics can be quantified with the spherical drop model of Miedema (147, 148). This model assumes that the cohesive energy of a spherical metal cluster and that of the same group of atoms in the bulk differ by the surface energy of the cluster, as expressed in the following equation:

$$E_c(n) = \Delta_{vap}H_0° - (36\pi/n)^{1/3}\gamma° V_a^{2/3}, \qquad 5.$$

where $\gamma°$ is the surface energy of the bulk metal and $V_a^{2/3}$ is the atomic size in the bulk. The value $(36\pi)^{1/3}$ in this equation is the dimensionless ratio between the surface area of a sphere (πd^2, where d is the diameter) and its volume ($\pi d^3/6$) to the two-thirds power. Plots of our values for $E_c(n)$ vs $n^{-1/3}$ extrapolate to the bulk phase value of $\Delta_{vap}H_0°$ within 10% at $n = \infty$ for all eight cationic and five neutral metal systems (142) obtained from our work. This result is shown for a few examples in Figure 3. This figure includes data for silver cluster cations (112), which also extrapolate very well to the bulk phase heat of vaporization, as do comparable data for silver cluster anions (108). For our data, the slopes of such plots for cationic clusters yield values for $\gamma° V_a^{2/3}$ that are $19 \pm 7\%$ larger than the bulk phase values recommended by Miedema (147, 148), whereas the data for neutral clusters yield values that are $41 \pm 11\%$ larger. As discussed elsewhere (126), the deviations can be explained if these very small clusters have a surface area larger than that of a smooth sphere, which seems appropriate for the cluster surface at this atomic scale. Further, this deviation was also observed by Miedema, who found that the experimental data for neutral dimers of various metals were best described by a coefficient in Equation 5 of 5.82, 20% higher than $(36\pi)^{1/3} = 4.84$. The observation that $\gamma° V_a^{2/3}$ values for the cationic clusters are smaller than those for the neutral clusters may be simply the effect of a smaller volume that results from contraction of the electron orbitals with increased nuclear charge. Overall, this comparison shows that the properties of the transition metals that control the binding energies of small clusters are the same properties that control the bulk phase thermodynamic stabilities.

The spherical drop model with no adjustable parameters reproduces the trends in the cluster binding energies remarkably well and successfully explains the large increase in BDEs observed for the second-row transition metal Nb in comparison

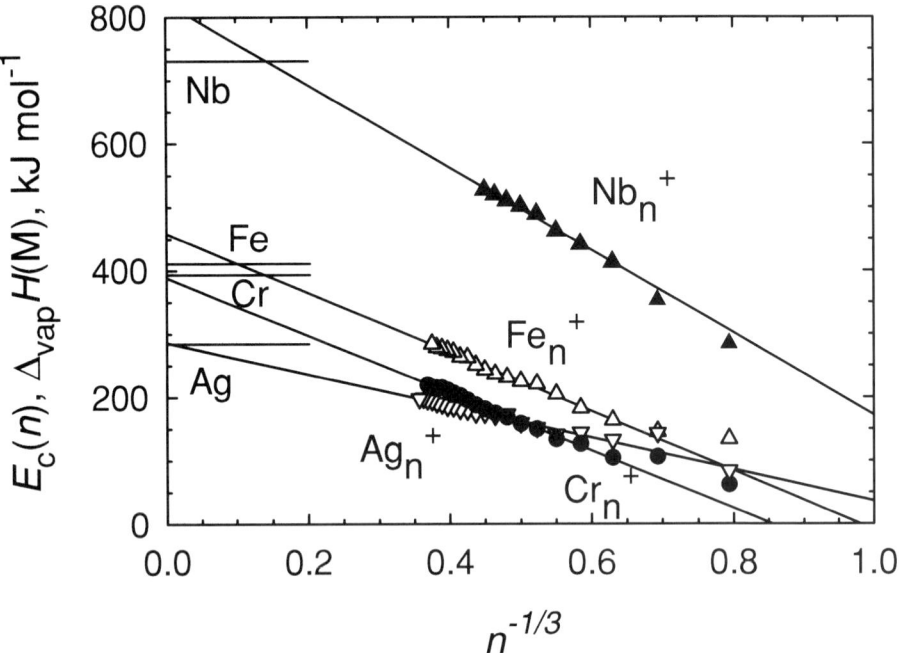

Figure 3 Cluster cohesive energies as a function of $n^{-1/3}$ for cationic clusters of chromium (●), iron (△), niobium (▲), and silver (▽) calculated from the data in Table 1. Heats of vaporization for each element are shown as horizontal lines on the left axis.

with the first-row metals (see Table 1 and Figure 3) (142). This agreement helps verify that the absolute BDEs obtained in CID, PD, and TRPD studies are reasonable. Despite this utility, it should be realized that the spherical drop model yields only the average approach of cluster cohesive energies to bulk phase vaporization. This simple model cannot predict the absolute binding energies of specific clusters or which clusters are more stable than their neighbors, that is, the nonmonotonic behavior observed in Figure 2.

REACTIONS OF CLUSTERS WITH D_2

Reactions of transition metal clusters with dihydrogen are among the best-studied cluster chemistry (58), dating from early work of Smalley (36) and the intriguing observation of an anticorrelation between the reactivity of iron clusters and their ionization energy by the Exxon group (149). From these earliest studies, it was obvious that reactions with H_2 are quite sensitive to the size of the metal clusters and hence to their structure, both electronic and geometric. Despite this activity, studies in which the thermochemistry of dihydrogen reactions is quantitatively assessed are scarce. Temperature-dependent studies of the reactions of deuterium with iron (33), cobalt (150), and nickel (151) clusters have allowed the measurement of activation barriers for the initial step in reactions with specific-sized clusters. One intriguing study is that of Liu et al (162) in which they determined the average cluster H_2 desorption energy in hydrogen-saturated clusters by examining the laser-induced fragmentation patterns. Significantly, this method yields the average BDE but does not provide information about the energetics of ligand binding to bare clusters, that is, the first and generally rate-limiting reaction step.

Reaction Pathways

Our experimental studies of the thermochemistry of cluster reactions with D_2 include studies of Cr_n^+ ($n = 2$–14) (152) and Fe_n^+ ($n = 2$–15) (153) cluster cations. Preliminary work for V_n^+ ($n = 2$–13) is also available (R Liyanage, J Conceição & PB Armentrout, unpublished results). Figure 4 shows examples of the energy dependence of the cross sections for reactions of Fe_n^+ ($n = 4$ and 13) with D_2. Directly analogous results are observed for other iron cluster reactants as well as vanadium and chromium cluster reactants. For all three metal systems, the only products observed were those formed in Reactions 6 and 7:

$$M_n^+ + D_2 \rightarrow M_nD^+ + D, \qquad 6.$$

$$\rightarrow M_nD_2^+. \qquad 7.$$

Only Reaction 6 is observed for the smallest clusters of each metal, whereas both reactions are observed for V_n^+ with $n \geq 5$, Cr_n^+ with $n \geq 6$, and Fe_n^+ with $n \geq 9$. Despite a careful search for products with fewer metal atoms, none were observed. The failure to see CID of the cluster ions with D_2 can be rationalized on the basis of the small mass and low polarizability of the D_2 target gas (70, 72). In addition, the failure to observe M_mD^+ and $M_mD_2^+$ products where $m < n$ indicates that the M_nD^+ and $M_nD_2^+$ products dissociate exclusively by D and D_2 loss, respectively. This conclusion is consistent with the thermochemistry derived in this work.

The formation of $M_nD^+ + D$ is observed to be endothermic for all of the clusters studied. All cross sections exhibit similar energy dependences in which they rise from apparent thresholds at about 2 eV and peak at 4.5–6.5 eV (see Figure 4). The decline in the cross sections at higher energies is a result of the dissociation process,

$$M_n^+ + D_2 \rightarrow M_nD^+ + D \rightarrow M_n^+ + D + D, \qquad 8.$$

which has a thermodynamic threshold of 4.56 eV $= D_0(D_2)$ (155). For smaller cluster ions, the cross sections peak at energies close to this value (Figure 4a). As the clusters get larger, the peaks gradually shift to higher energies because the larger clusters are able to accommodate the excess energy (Figure 4b). At any particular kinetic energy above the onset of Reaction 8, the lifetime of the cluster increases and the dissociation probability decreases as the clusters get larger.

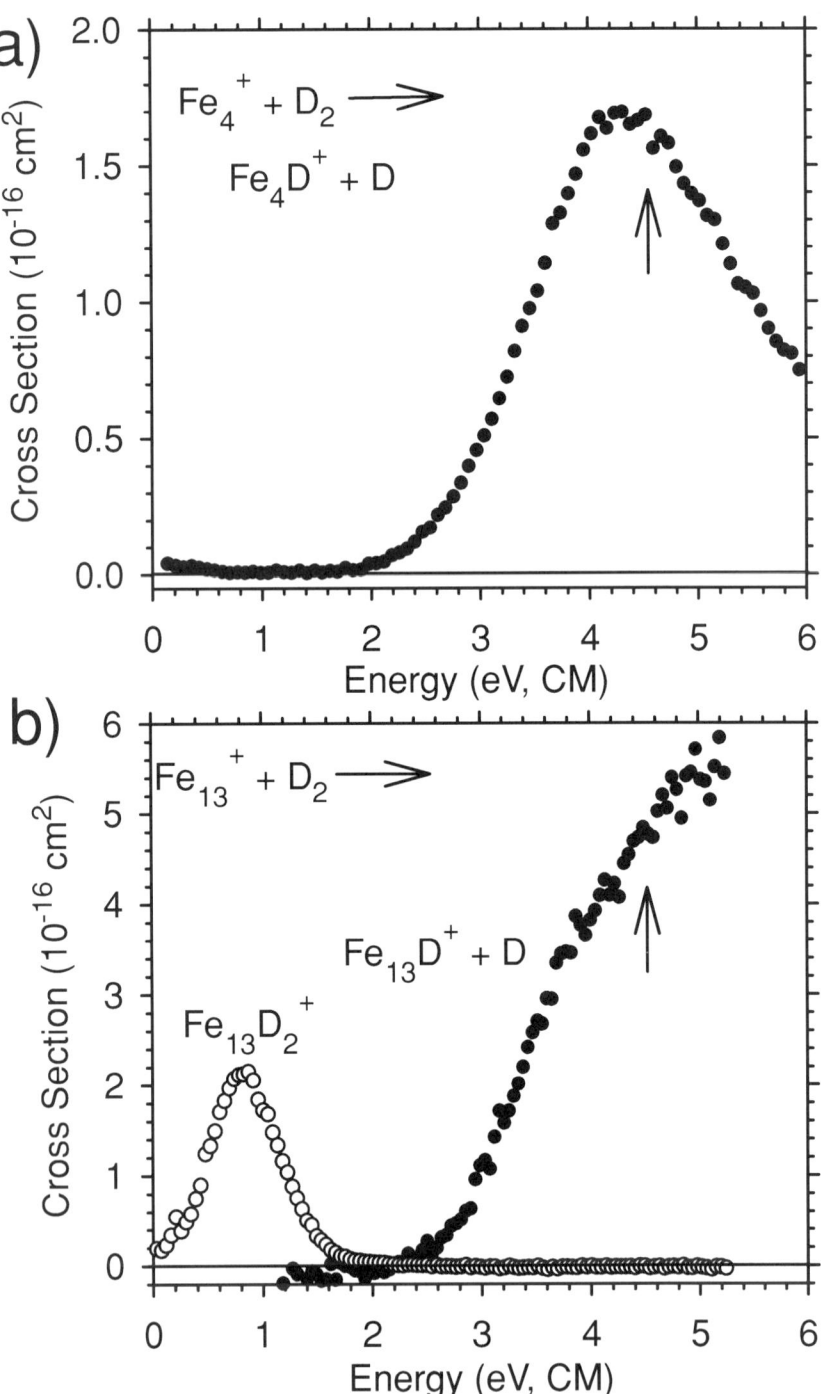

For most clusters, the cross sections for the formation of $M_nD_2^+$ increase with increasing energy (Figure 4b). This behavior is characteristic of endothermic reactions or those with barriers. Reaction 7 exhibits no barrier only for Fe_9^+ and Cr_n^+ ($n = 6$–8). The cross sections for Reaction 7 reach maxima near 0.8–1.4 eV and then decline. The only process that can account for this decline is dissociation by loss of molecular D_2 in the overall reaction,

$$M_n^+ + D_2 \to M_nD_2^+ \to M_n^+ + D_2. \qquad 9.$$

The $M_nD_2^+$ product ions can conceivably have one of two forms: (a) a weakly bound adduct held together by ion-induced dipole and ion-quadrupole attraction, that is, a physisorbed state, or (b) a strongly bound species in which the deuterium atoms are chemically bonded to the cluster, that is, a dissociatively chemisorbed state. Experimental evidence appears most consistent with the latter form. The thresholds observed for Reaction 7 with iron cluster cations are comparable to the estimates made by Richtsmeier et al (33) for the activation energies of H_2 chemisorption on neutral iron clusters. Although not definitive, it appears that these thresholds must correspond to a barrier in the entrance channel for the reaction, in between the physisorbed and chemisorbed states. As discussed elsewhere (156), the presence of such a barrier depends critically on the amount of s character at the metal center where chemisorption is occurring. The s character can vary from one cluster size to another and from cationic to neutral to anionic states of the clusters. In this regard, it is interesting to note that the probability of Reaction 7 is somewhat larger for even-sized chromium clusters when $n \leq 10$. (This pattern breaks down for larger clusters because Cr_{11}^+ has the highest probability for undergoing Reaction 7.) As noted above, these clusters are believed to have open-shell s character rather than the closed-shell s character for the odd-sized clusters.

Bond Energies of Cluster-Deuterides

Analyses of the threshold for Reaction 6 using Equation 1 yield M_n^+–D BDEs. These are listed in Table 4 and shown in Figure 5 and can be compared to values for H atom binding to bulk surfaces. For chromium, the binding energy to polycrystalline surfaces is about 310 kJ mol^{-1} (157, 158). The largest Cr_n^+–D BDEs approach this limit, whereas smaller clusters (except Cr_2D^+) have considerably weaker BDEs. The bulk phase BDE is 270 ± 10 kJ mol^{-1} for Fe (1,0,0), Fe (1,1,0), and Fe (1,1,1) (159–161), which is comparable to the highest of the Fe_{n-1}^+–D BDEs (those for $n = 7$ and $n \geq 11$). However, the lowest BDEs (for

Figure 4 Reactions of Fe_4^+ (*panel a*) and Fe_{13}^+ (*panel b*) with D_2 as a function of collision energy in the center-of-mass frame. Arrows indicate the D_2 bond energy of 4.55 eV (adapted from Reference 153).

TABLE 4 Bonds Energies of M_n^+–L_x (kJ mol^{-1})[a]

n	Cr_n^+–D[b]	Fe_n^+–D[c]	V_n^+–O[d]	V_n^+–2O[d]	Cr_n^+–O[e]	Cr^+–2O[f]	Fe_n^+–O[g]	Fe_n^+–2O[h]
1	135[i]	208[j]	578[k]	882	359[k]		341[k]	
2	276	140	492	1254	542	1119	492	907
3	175	171	704	1331	584	1167	540	1081
4	206	208	724	1416	669	1312	569	1148
5	194	183	704	1472	615	1293	550	1148
6	223	264	714	1447	619	1360	531	1158
7	206	204	675	1374	622	1341	502	1081
8	235	190	540	1289	652	1322	540	1100
9	228	230	579	1324	628	1312	550	1119
10	252	259	646	1220	648	1245	560	1110
11	238	243	511	1210	624	1264	579	1071
12	243	245	434	1196	632	1274	598	1052
13	274	243	492	1223	656	1283	569	1061
14	263	243	521	1131	611	1187	627	1090
15		277	569	1158	610	1129	589	1032
16					577		589	965
17					512		560	

[a]Value are at 0 K. Absolute uncertainties range from ~10 to 80 kJ mol^{-1}.
[b]Data from Reference 152.
[c]Data from Reference 153.
[d]Data from Reference 165.
[e]Data from Reference 169.
[f]Data from Reference 166.
[g]Data from Reference 77.
[h]Data from Reference 76.
[i]Data from Elkind JL & Armentrout PB. 1986. *J. Phys. Chem.* 90:5736–45.
[j]Data from Elkind JL & Armentrout PB. 1987. *J. Chem. Phys.* 86:1868–77.
[k]Data from Fisher ER, Elkind JL, Clemmer DE, Georgiadis R, Loh SK, et al 1990. *J. Chem. Phys.* 93:2676–91.

$n = 3$ and 4) are almost half the bulk value. These variations must be related to the geometric and electronic structures of the metal cluster ions, as discussed further below.

It is also useful to compare these values to related thermochemistry for neutral cluster systems. In this case, such information is available only for iron clusters. Liu et al (162) studied the energetics of D_2 desorption from neutral iron clusters saturated with D_2 (i.e. Fe$_n$D$_m$), as induced by multiphoton laser irradiation. They found that the loss of one D_2 molecule from such clusters requires a near constant energy of about 125 ± 20 kJ mol^{-1} over the size range of $n = 10$–32. This value corresponds to an upper limit to the average BDE of the D atom to the cluster of about 280 kJ mol^{-1}, in reasonable agreement with our BDEs of Fe$_{n-1}^+$–D ($n = 11$–16). This concordance suggests that the BDEs of D atoms to iron clusters are not sensitive to the charge state or to the coverage. The latter finding implies that similar sites are available for binding D atoms to the clusters up to the saturation limit reached in the studies of Liu et al, that is, essentially one D atom per Fe atom on the surface.

Probes of Electronic Structure

Several important insights into the bonding of bare clusters and of D atoms to these clusters are revealed by comparing the M$_{n-1}^+$–D BDEs with those previously determined for M$_{n-1}^+$–M (97). These comparisons take the point of view that the D atom binding energies act as a probe of the electronic structure of the metal clusters, specifically in regard to the accessibility of an open-shell binding site. In the iron system, shown in Figure 5b, this comparison reveals that the trends for many cluster sizes are similar; that is, the Fe$_{n-1}^+$–Fe BDEs are slightly larger than the Fe$_{n-1}^+$–D BDEs. This concordance suggests that iron clusters bind both D and Fe atoms largely using the 4s-like orbitals. However, Fe$_{n-1}^+$–Fe BDEs are enhanced in comparison with the Fe$_{n-1}^+$–D BDEs by an average of 29 ± 7 kJ mol^{-1}, excluding the exceptional sizes noted below, which indicates some 3d orbital participation in the Fe$_n^+$–Fe bonds. Although most BDEs are comparable, there are clearly several clusters for which the metal-metal bond is much stronger, specifically for Fe$_6^+$, Fe$_9^+$, Fe$_{13}^+$, and Fe$_{15}^+$. As noted above, the $n = 6$, 13, and 15 pattern of stability suggests octahedral face-centered-cubic atom packing, which implies that these clusters have particularly symmetric geometries (101). This conclusion is consistent with work by Riley and coworkers on molecular adsorption on neutral Fe clusters (34). However, when a D atom is substituted for one of the Fe atoms in an octahedrally symmetric and stable cluster (e.g. Fe$_6^+$ to form Fe$_5$D$^+$), the symmetry is removed. This substitution apparently changes the energy of the electronic states considerably, leading to much weaker Fe$_{n-1}^+$–D BDEs for these symmetric clusters.

As shown in Figure 5a, the Cr$_{n-1}^+$–D and Cr$_{n-1}^+$–Cr BDEs exhibit strong even-odd oscillations up to about $n = 10$. The correspondence of these oscillations can be understood by noting that D(1s^1) and Cr(4s^1 3d^5) atoms both have singly occupied s orbitals as their frontier orbital. The oscillations are a strong confirmation of the hypothesis given above that the 4s electron dominates the bonding interaction in the bare chromium clusters, and the half-filled 3d^5 shell remains largely unperturbed. For larger chromium clusters, the even-odd oscillations wash out, but the difference in metal-metal and metal-deuteride BDEs shows a notable maximum at Cr$_{13}^+$. As for the iron systems, we speculate that this result indicates a particularly stable cluster. This cluster is plausibly assigned as an icosahedron, but an octahedrally symmetric Cr$_{13}^+$ is also feasible.

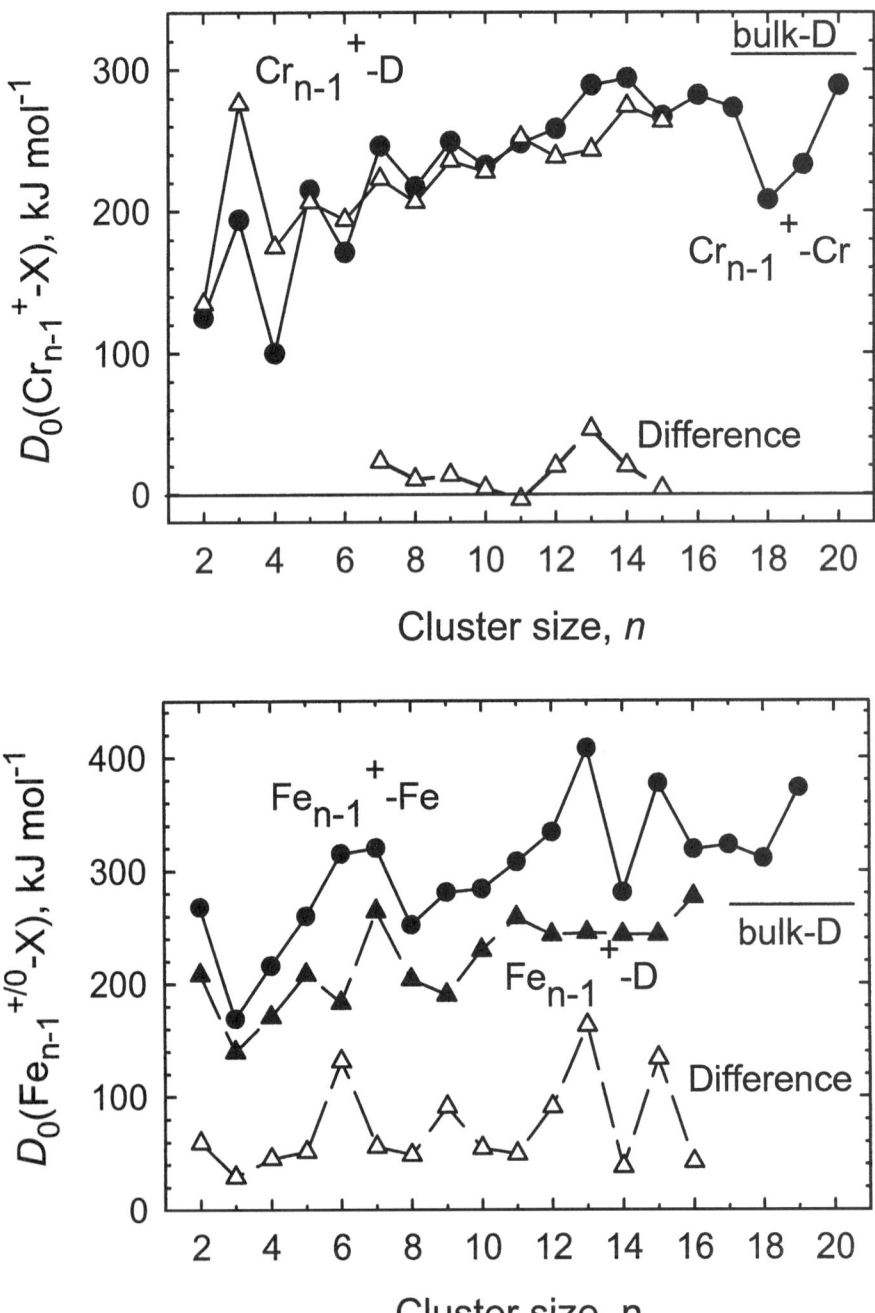

REACTIONS OF CLUSTERS WITH O_2

Early studies (41, 163, 164) of the oxidation of transition metal clusters by dioxygen at thermal energies showed efficient reactivity that is relatively indiscriminate; that is, little change in reactivity occurs as a function of cluster size, in contrast to the results with D_2 (58). The qualitative observations made in these and subsequent works are quite consistent with the picture that emerges from quantitative studies of these systems using guided-ion-beam mass spectrometry. Our experimental studies of the kinetic energy dependence of reactions of cluster ions with O_2 include V_n^+ ($n = 2$–17) (165), Cr_n^+ ($n = 2$–18) (166), and Fe_n^+ ($n = 2$–18) (76).

Reaction Pathways

As illustrated in Figure 6, the reactions of metal cluster cations with O_2 are complicated because many products are observed, in contrast to the simplicity of the D_2 reactions, as shown in Figure 4. Although smaller clusters show some variability in the products observed, larger clusters ($n \geq 6$) exhibit strong similarities in their product distributions. Figure 6 shows results for the reactions of Cr_{13}^+ and Fe_{13}^+ with dioxygen, representative of the behavior of clusters larger than the pentamers of both metals. The magnitudes of the total reaction cross sections at all energies are comparable to the collision cross section (dominated by the long-range ion-induced dipole interaction at the lowest energies and by the hard-sphere cross section at elevated energies). This finding indicates that the reactions are efficient and exothermic. Cluster dioxide ions are the dominant products at all energies. As revealed in this work, metal-oxygen bonds are stronger than metal-metal bonds for all three metal systems (V_n^+, Cr_n^+, and Fe_n^+); hence, primary products dissociate by metal atom loss to form smaller cluster dioxide ions. Oxygen atom loss to form cluster monoxide ions (M_mO^+) is much less efficient. The sequential nature of the metal atom loss processes is evident from the observation that the cross sections for $M_mO_2^+$ products decline as the $M_{m-1}O_2^+$ cross sections rise, as shown in Figure 6. It is also useful to note that the dissociation process corresponds almost exclusively to $M_mO_2^+ \rightarrow M_{m-1}O_2^+ + M$, which indicates that the ionization energies of the cluster dioxides must be less than the ionization energy of the atomic metal.

The largest $M_mO_2^+$ product observed varies as a function of cluster size and metal identity. Thus, at thermal energies, the dominant product for the reaction of small Cr_n^+ ($n < 12$) clusters is $Cr_{n-3}O_2^+$, whereas medium-sized clusters ($n =$

Figure 5 Size-dependent bond dissociation energies for metal cluster cations (taken from Table 1) compared with metal cluster cation-deuterium bond energies (taken from Table 4) for chromium (*panel a*) and iron (*panel b*). The differences between these bond energies are also shown. Values for the bulk metal-D bonds are indicated by horizontal lines and are taken from References 157 to 161.

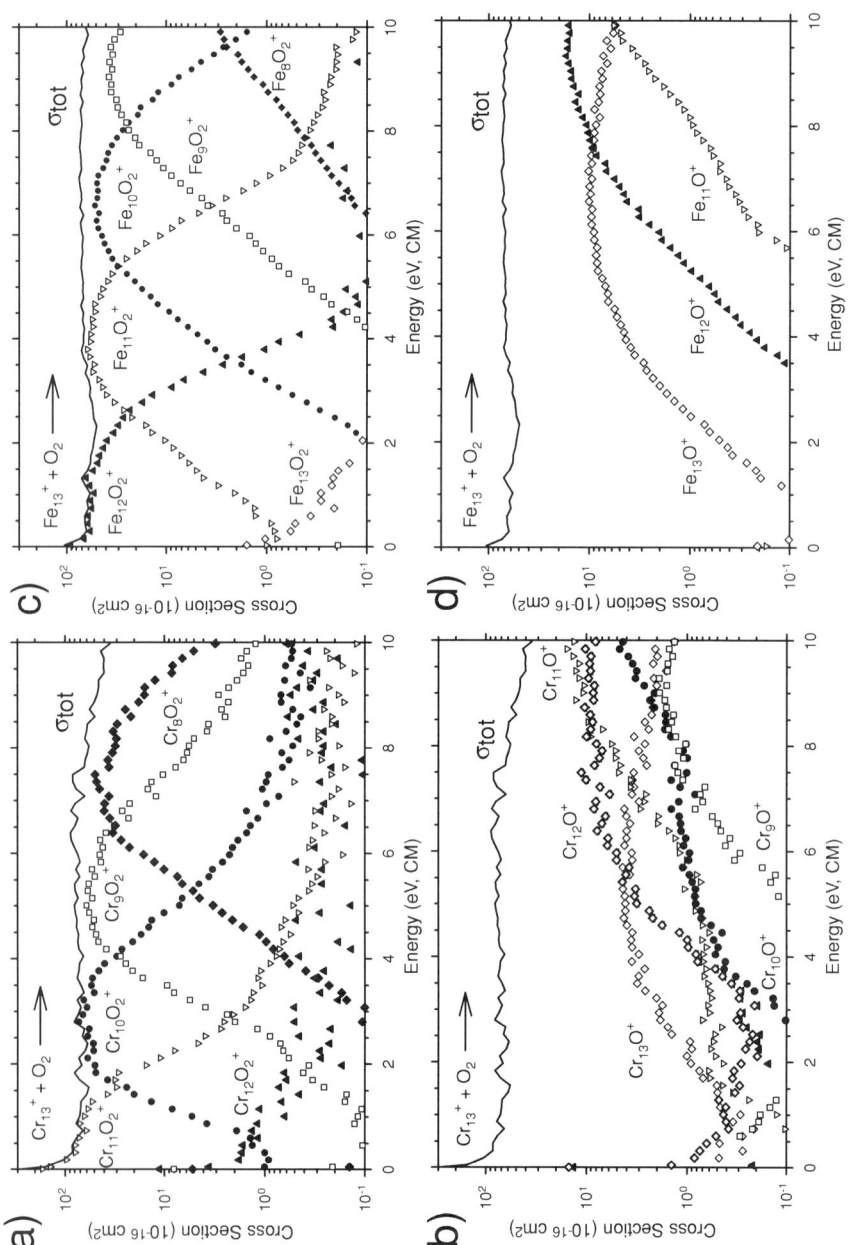

12–17) produce $Cr_{n-2}O_2^+$; the largest cluster studied (Cr_{18}^+) yields $Cr_{n-1}O_2^+$. For vanadium, small ($n < 10$) clusters react to yield $V_{n-2}O_2^+$ preferentially at low energies, whereas larger clusters produce $V_{n-1}O_2^+$. For iron, small clusters react to produce $Fe_{n-2}O_2^+$ primarily; whereas medium-sized clusters (Fe_9^+ to Fe_{15}^+) form $Fe_{n-1}O_2^+$; the $Fe_nO_2^+$ adduct, which is not observed until Fe_{11}^+, dominates the product spectrum for Fe_{16}^+ and larger clusters. Unlike for the iron system, $Cr_nO_2^+$ and $V_nO_2^+$ adducts are never major products. These changes in the dominant $M_mO_2^+$ product observed can be attributed to changes in the lifetime of the transient $M_nO_2^+$ intermediate as the cluster size increases and to the difference in the exothermicity of reaction, which is much higher for the early transition metals than for iron (see below).

Figure 6 shows that cluster monoxide ion products are formed in endothermic processes and with much lower efficiencies than the cluster dioxide products. The M_nO^+ product cross sections generally exhibit thresholds in the vicinity of 1 eV, although in the vanadium system, we also observe exothermic formation of $V_{n-1}O^+ + VO$. This exothermic process is driven by the very large VO BDE of 6.5 eV (167), in comparison with 4.5 eV for CrO and 4.2 eV for FeO (168). A similar process is also apparent in the chromium reactivity shown in Figure 6, where obvious features at low energies in the $Cr_{11}O^+$ and $Cr_{10}O^+$ cross sections are consistent with neutral CrO formation. As the energy is increased, the monoxide cation products also dissociate by sequentially losing metal atoms. For most clusters, simple fragmentation to form $M_{n-1}^+ + M$ products is an inefficient process. Such products are formed with thresholds largely characteristic of simple CID processes, although they have much smaller magnitudes than those observed with Xe as the collision gas.

Bond Energies of Cluster-Oxides

Because the primary reaction products in oxidation reactions by O_2 are exothermic, it is difficult to glean thermodynamic information from these systems. Limits to the thermochemistry are obtained straightforwardly from observations of exothermic reactivity, but analysis of product cross sections exhibiting thresholds could be severely affected by competition with the much more intense products formed in exothermic processes. This difficulty is especially prevalent in the present cases where the channels are closely coupled because all products are formed by sequential loss of metal atoms, O atoms, or MO molecules from the transiently formed $M_nO_2^+$ complex. Despite this concern, we believe that accurate thermochemistry

←

Figure 6 Reactions of Cr_{13}^+ (*panels a and b*) and Fe_{13}^+ (*panels c and d*) with O_2 as a function of collision energy in the center-of-mass frame. *Upper* and *lower frames* show cluster dioxide and cluster monoxide cation products, respectively (adapted from Reference 166 (Cr data) and Reference 76 (Fe data)).

can be obtained by direct measurements of the thresholds for some (but not all) endothermic processes and by relative measurements of the thresholds for $M_mO_p^+$ and $M_{m-1}O_p^+$ ($p = 1$ and 2) products. The thermochemistry obtained agrees well with limits imposed by the observation of exothermic pathways. For the monoxide product ions, thresholds for the formation of the $M_nO^+ + O$ product channels do not correspond to the thermodynamic limit. Although this observation could be a consequence of barriers to this process, it seems more likely that the thresholds are delayed because of the severe competition with the dioxide product ions. In the O_2 reaction systems, M_n^+–O BDEs were determined primarily from the thresholds for the secondary process forming $M_{n-1}O^+ + M + O$. An important check of these results is our study of the reactivity of chromium and iron cluster cations with CO_2. These are discussed in the next section along with the final BDEs obtained for cluster oxides.

REACTIONS OF CLUSTERS WITH CO_2

Reactions of metals with CO_2 are of interest in understanding the activation of CO_2 for use as a feedstock for catalysis. In addition, it is an ideal donor of a single oxygen atom because the carbon monoxide bond is so strong ($D = 11.1$ eV). Indeed, the O–CO BDE is similar in strength (5.4 eV) to that for O_2 (5.1 eV), such that nearly direct comparisons can be made between the cross sections for formation of M_mO^+ products in the two systems. Therefore, we have used CO_2 to augment our studies of the oxidation thermochemistry of both Cr_n^+ ($n = 1$–18) (169) and Fe_n^+ ($n = 1$–18) (77) cluster ions.

Reaction Pathways

Figure 7 illustrates the reactivity observed for reactions of Fe_n^+ ($n = 1$–18) clusters with CO_2. Similar results are obtained for chromium clusters. The primary product ions observed are M_mO^+ and M_m^+, where $m \leq n$, although $M_nCO_2^+$ adducts are formed for the largest clusters with $n \geq 9$ ($n \geq 10$ for Cr). The M_nO^+ product cross sections show an unusual energy dependence having both a barrierless exothermic component that declines with increasing kinetic energy and a feature at higher kinetic energies that exhibits a threshold, as shown in Figure 7. We considered two possible interpretations for this observation. In the first interpretation, the reaction could occur by a trapping-mediated (resonant) pathway at low energies and a direct reaction pathway at higher energies. Theoretical (170, 171) and surface (172–174) studies of the trapping-mediated pathway imply that it should show a distinct dependence on the temperature of the cluster ions. In the iron system (77), several means were used to heat the clusters, but no such dependence could be observed. In the second interpretation, the two features of the cross section could be associated with adiabatic and diabatic pathways that result from the spin-forbidden character of the CO_2 dissociation. Specifically, ground state

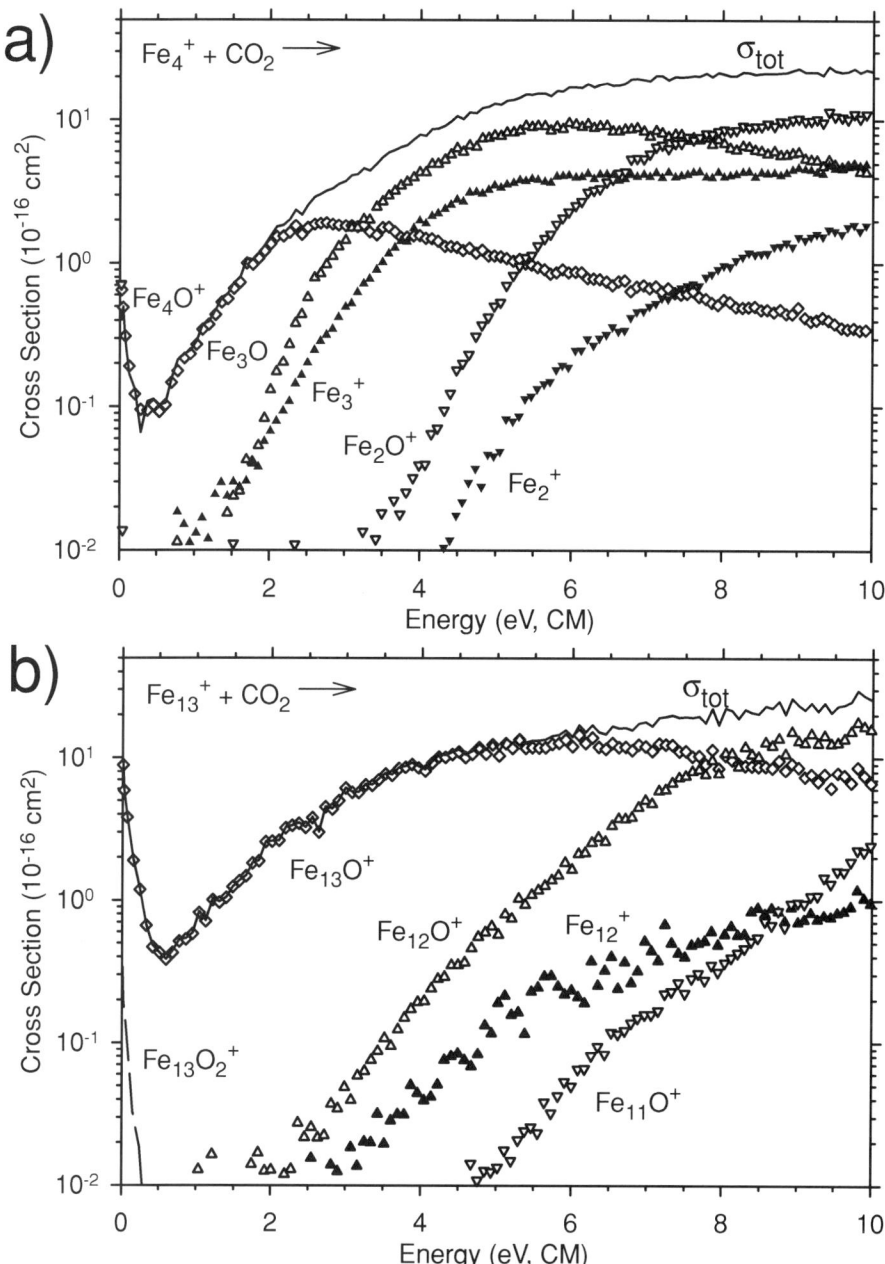

Figure 7 Reactions of Fe_4^+ (*panel a*) and Fe_{13}^+ (*panel b*) with CO_2 as a function of collision energy in the center-of-mass frame (adapted from Reference 77).

$CO_2(^1\Sigma_g^+)$ does not diabatically dissociate to ground state products, $CO(^1\Sigma^+) + O(^3P)$, in a spin-allowed process. The spin-allowed asymptote, $CO(^1\Sigma^+) + O(^1D)$, requires an additional 1.97 eV of energy. In this interpretation, interactions between the singlet and triplet CO_2 surfaces are allowed because of the intervention of the metal cluster cations. Despite being exothermic and barrierless, the residual spin-forbidden character of the ground state surface explains the relative inefficiency of these processes. As the excited-state asymptote is reached, a fully spin-allowed process can occur, but it is thermodynamically more demanding.

Bond Energies of Cluster-Oxides

Remembering that $D(OC–O)$ is actually greater than $D(O_2)$, comparison of the $Fe_{13}O^+$ products in Figures 6 and 7 shows unequivocally that the threshold observed in the O_2 system for this product ion cannot correspond to the thermodynamic limit. As noted above, we attribute this finding to the severe competition with the formation of the dominant $Fe_mO_2^+$ products. In contrast, the apparent thresholds for the secondary $Fe_{12}O^+$ products are comparable in both systems. (In making this comparison, note the difference in the logarithmic scale of the two figures.) Hence, analysis of the thresholds for formation of the $M_{n-1}O^+ + M + CO$ product channel should lead to reliable thermochemistry. Indeed, because of the complicated low-energy behavior of the $M_nO^+ + CO$ cross sections, these primary reaction pathways cannot be used for reliable thermochemical measurements.

The $M_n^+–O$ BDEs derived from the CO_2 systems are shown in Figure 8 and listed in Table 4. For iron clusters, these values agree reasonably well with the thermochemistry determined from the thresholds for formation of the $M_{n-1}O^+ + M + O$ product channel in the O_2 system, also shown in Figure 8. For chromium, BDEs obtained from the O_2 system are within experimental error but are systematically lower (by about 50 kJ mol^{-1}) than the values obtained from the CO_2 systems. Again, we attribute this result to effects associated with competition with the dominant $Cr_mO_2^+$ products. Thus, the $M_n^+–O$ BDEs determined in the CO_2 systems are believed to be more reliable than those from our O_2 studies. We can also compare these results with those for binding two oxygen atoms to the clusters ($M_n^+–2O$ BDEs, Table 4), as determined from an analysis of the cross sections for formation of $M_mO_2^+$ products in endothermic reactions in the O_2 systems. Figure 8 shows that the $M_n^+–O$ BDEs are comparable to half the $M_n^+–2O$ BDEs for both metals. As these determinations are completely independent, this agreement substantiates both measurements. The pattern in these BDEs mimics the relative reactivity at thermal energies observed for neutral iron clusters (41); that is, the monomer and dimer react slowly, whereas larger clusters have similarly large rate constants.

We can compare the metal cluster oxygen BDEs measured in our work with those obtained on bulk phase surfaces. The adsorption of molecular oxygen on

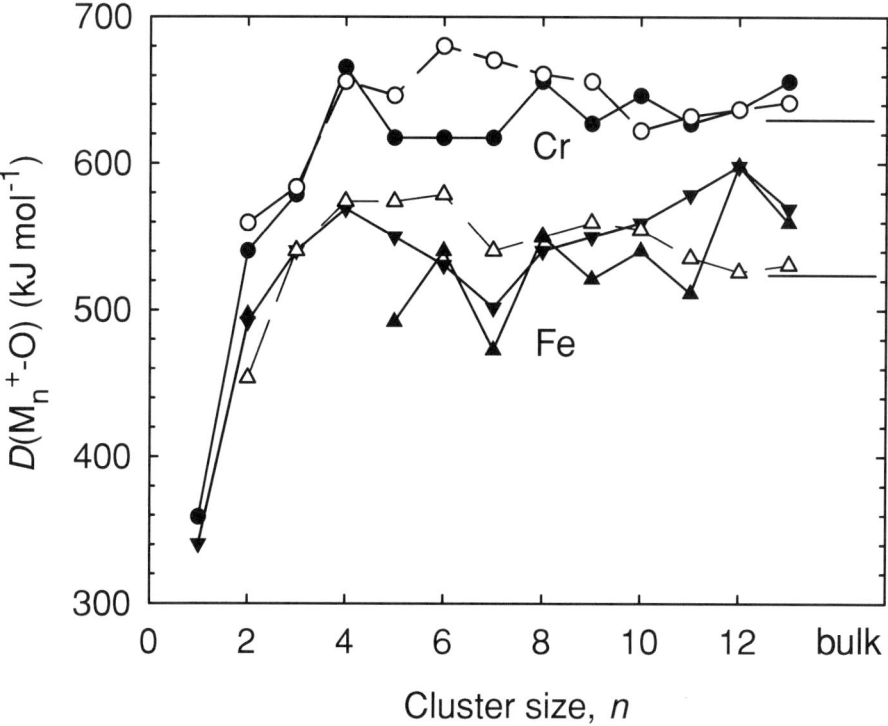

Figure 8 Size-dependent metal cluster cation-oxide bond energies for chromium (○, ●) and iron (△, ▼) taken from Table 4, and for iron (▲) taken from the O_2 reaction system (reference 76). The open symbols represent $D(M_n^+-2O)/2$. Values for the bulk metal-O bonds are indicated by horizontal lines and are taken from Reference 161.

chromium and iron surfaces occurs by a dissociative chemisorption mechanism (see reference 175 and references therein). This observation is consistent with the most current thermodynamic measurements, which come from calorimetry experiments (176–178). For chromium and iron films, desorption of O_2 requires 730 ± 30 kJ mol^{-1} and 300–600 kJ mol^{-1}, respectively; this thermochemistry is equivalent to M_n^+–O bond energies of 610 kJ mol^{-1} and 400–550 kJ mol^{-1}, respectively. Benziger (161) lists atomic adsorption BDEs for oxygen of 630 kJ mol^{-1} and 523 kJ mol^{-1} for chromium and iron surfaces, respectively. These various values are directly comparable to the values found in our work for larger clusters, as shown in Figure 8. In addition, $\Delta_{vap}H^0(Cr)$ and $\Delta_{vap}H^0(Fe)$ are 400 kJ mol^{-1} and 415 kJ mol^{-1}, respectively, suggesting that the metal should vaporize before oxygen desorbs for oxidized metal surfaces. Thus, the thermochemistry obtained here for small clusters is comparable to that for the bulk phase. In addition, the fact that the average chromium cluster oxide BDE exceeds that for iron cluster oxides clearly reflects the driving force for the selective oxidation of chromium in Fe-Cr alloys. This result is another indication that the use of clusters to model bulk phase reactivity may be reasonable (1).

OTHER THERMODYNAMIC STUDIES

Using methods similar to those employed in our group, Ervin and coworkers measured CO binding energies to clusters of group 10 cluster anions (107, 110, 111, 179). Significantly, this work was able to elucidate the differences in the binding of CO at two different binding sites (111, 179). High BDEs (220–250 kJ mol^{-1}) are assigned to bridging positions, whereas low values (80–110 kJ mol^{-1}) are believed to correspond to binding in terminal positions. The latter values are comparable to binding energies on platinum surfaces (180, 181). In recent work on the binding of CO to copper cluster anions, it was found that Cu_5CO^- has the strongest cluster-carbonyl bond energy among the Cu_nCO^- ($n = 3$–7) clusters. This cluster has eight valence electrons, such that it is a closed-shell species in the jellium model (140, 141).

Recent work by Castleman and coworkers used kinetic-energy-dependent methods similar to the those used in our work to examine the thermochemistry of niobium oxide cluster cations (182) and vanadium carbon cluster cations (183). In the former study, $Nb_3O_7^+$ and $Nb_4O_{10}^+$ were the most strongly bound clusters, and $Nb_4O_9^+$ and $Nb_5O_{12}^+$ were other stable species observed. Loss of NbO_2 units is a common dissociation pathway consistent with ab initio calculations, indicating that these clusters have oxygen atoms bridging the niobium atoms, which are terminated by two oxygen atoms apiece. In the latter work, the $V_8C_{12}^+$ cluster was shown to be substantially more stable than $V_8C_{11}^+$. This finding is consistent with the proposal that the former species has a cagelike structure; that is, it is a metallocarbohedrene or met-car (184).

The chemical probe studies of Riley & Parks have provided some of the best structural information about transition metal clusters (58, 143, 144, 150, 151). As part of these studies, however, detailed thermodynamic information was obtained by equilibrium measurements of the binding of stable molecules to clusters. These include D_2O binding energies to iron clusters ($n = 7$–27) (185), NH_3 binding energies to selected iron clusters ($n = 19, 23, 26, 29, 32, 34$) (186), and the binding of water and ammonia to bare and hydrogenated cobalt and nickel clusters (187). The former reaction system is interesting because a correlation between the Fe_n–OD_2 BDEs and the reactivity of Fe_n with H_2 (185) is found; for example, weak BDEs and very slow reaction rates are found for Fe_n ($n = 15$–17). This result suggests that the electron acceptor properties of the metal clusters are an important aspect of the ability to activate the H_2 bond. The study of the thermochemistry of selected $Fe_n(NH_3)_m$ ($m = 1$–12) clusters shows a gradual decline in the BDEs with increasing m. When the bond energies are corrected for the surface coverage (m divided by the number of surface atoms in the cluster), the trends for all clusters are basically identical. The BDEs are found to be about twice as high as those measured for Fe (1,1,1) and Fe (1,0,0) surfaces (189). This result is attributed to the small size of the cluster and its noncrystalline packing.

OUTLOOK

The early days of gas phase cluster research were rife with reports of magic numbers. Such observations were stimulating and led to several useful insights such as the jellium model and fullerene research. However, continued progress in the field now requires a more systematic and rigorous experimental approach. Quantitative information that can be compared with results in the areas of surface science and catalysis is needed to identify clusters serving as good models and those that are exceptional. This review has attempted to demonstrate that quantitative thermodynamic information on the stability and reactivity of small transition metal clusters in the gas phase is available and applicable to the bulk phase limit. Guided-ion-beam mass spectrometry appears to be one of a handful of experimental methods capable of providing such information. Although the acquisition of such data is difficult, the results seem to be reasonable and capable of providing useful insight into surface chemistries. Many experiments are possible by varying the metal and chemical reactant. Periodic trends in such chemistry would be particularly interesting to investigate systematically.

As a guide to future prospects, two experiments are in progress in our laboratory. We recently examined the reactions of size-specific iron cluster cations with methane and ammonia. Many products were observed, including species such as C, CH, CH_2, N, NH, and NH_2 bound to the clusters. As with the systems reviewed above, bond energies can be extracted from an analysis of the kinetic energy dependence of the cross sections for these products. As for the D and O atoms bound to clusters, the bond energies for C and N bound to the larger clusters agree nicely with bulk phase values. Such comparisons for the molecular fragments (CH, CH_2, NH, and NH_2) cannot be made because bond energies of such species bound to surfaces are unavailable experimentally. Surface studies are able to measure the heat of chemisorption of atomic species and stable molecules such as CO, ethene, water, or ammonia to surfaces, but there is essentially no experimental information on the thermochemistry of surface species inbetween these two limits (see Reference 190). Thus, the cluster binding energies emerging from our studies will provide some of the first quantitative experimental information on these molecular fragments. As such fragments are critical intermediates in a host of industrially important catalytic chemistries, we anticipate that cluster thermodynamics may be particularly useful in this area.

ACKNOWLEDGMENT

Financial support from the Chemical Sciences, Geosciences, and Biosciences Division, Office of Basic Energy Sciences, Office of Science, U.S. Department of Energy, is gratefully acknowledged.

Visit the Annual Reviews home page at www.AnnualReviews.org

LITERATURE CITED

1. Somorjai G. 1981. *Chemistry in Two Dimensions: Surfaces*. Ithaca: Cornell Univ. Press
2. Over H, Kim YD, Seitsonen AP, Wendt S, Lundgren E, et al. 2000. *Science* 287:1474–76
3. Schnabel P, Irion MP, Weil KG. 1991. *J. Phys. Chem.* 95:9688–94
4. Irion MP, Schnabel P. 1991. *J. Phys. Chem.* 95:10596–99
5. Schnabel P, Irion MP. 1992. *Ber. Bunsenges Phys. Chem.* 96:1101–3
6. Irion MP, Schnabel P, Selinger A. 1990. *Ber. Bunsenges Phys. Chem.* 94:1291–95
7. Deleted in proof
8. Nakajima A, Kishi T, Sone Y, Nonose S, Kaya K. 1991. *Z. Phys. D* 19:385–87
9. Xu Z, Xiao FS, Purnell SK, Alexeev O, Kawi S, et al. 1994. *Nature* 372:346–48
10. Heiz U, Vanolli F, Sanchez A, Schneider WD. 1998. *J. Am. Chem. Soc.* 120:9668–71
11. Cox DM, Whetten RL, Zakin MR, Trevor DJ, Reichmann KC, Kaldor A. 1986. In *Advances in Laser Science, Optical Science and Engineering, Ser. 6*, Vol. I, ed. WC Stwalley, M Lapp, p.527–30. *AIP Conf. Proc.* 146. New York: Am. Inst. Phys.
12. Yang S, Knickelbein MB. 1990. *J. Chem. Phys.* 93:1533–39
13. Rohlfing EA, Cox DM, Kaldor A, Johnson KH. 1984. *J. Chem. Phys.* 81:3846–51
14. Parks EK, Klots TD, Riley SJ. 1990. *J. Chem. Phys.* 92:3813–26
15. Knickelbein MB, Yang S, Riley SJ. 1990. *J. Chem. Phys.* 93:94–104
16. Knickelbein MB, Yang S. 1990. *J. Chem. Phys.* 93:5760–67
17. Collings BA, Rayner DM, Hackett PA. 1993. *Int. J. Mass Spectrom. Ion Proc.* 125:207–214
17a. Athanassenas K, Kreisle D, Collings BA, Rayner DM, Hackett PA. 1993. *Chem. Phys. Lett.* 213:105–110
18. Alameddin G, Hunter J, Cameron D, Kappes MM. 1992. *Chem. Phys. Lett.* 192:122–28
19. Homer ML, Persson JL, Honea EC, Whetten RL. 1991. *Z. Phys. D* 22:441–47
20. Limberger HG, Martin TP. 1989. *J. Chem. Phys.* 90:2979–91
21. Dugourd P, Rayane D, Labastie P, Vezin B, Chevaleyre J, Broyer M. 1992. *Chem. Phys. Lett.* 197:433–37
22. Simard B, Lebeault-Dorget MA, Marijnissen A, ter Meulen JJ. 1998. *J. Chem. Phys.* 108:9668–74
23. Taylor KJ, Pettiette-Hall CL, Cheshnovsky O, Smalley RE. 1992. *J. Chem. Phys.* 96:3319–29
24. Leopold DG, Lineberger WC. 1986. *J. Chem. Phys.* 85:51–55
25. Ho J, Ervin KM, Lineberger WC. 1990. *J. Chem. Phys.* 93:6987–7002
26. Gausa M, Gantefor G, Lutz HO, Meiwes-Broer KH. 1990. *Int. J. Mass Spectrom. Ion Proc.* 102:227–37
27. Kietzmann H, Morenzin J, Bechthold PS, Gantefor G, Eberhardt W. 1998. *J. Chem. Phys.* 109:2275–78
28. Whetten RL, Cox DM, Trevor DJ, Kaldor A. 1985. *Surf. Sci.* 156:8–35
29. Billas IML, Becker JA, Chatelain A, deHeer WA. 1993. *Phys. Rev. Lett.* 71:4067–70
30. Holczer K, Klein O, Gruner G, Thompson JD, Diederich F, Whetten RL. 1993. *Phys. Rev. Lett.* 67:271–74
31. Louderback JG, Cox AJ, Lising LJ, Douglass DC, Bloomfield L. 1993. *Z. Phys. D* 26:301–3
32. Feng L, Khanna SN, Jena P. 1991. *Phys. Rev. B* 43:8179–82
33. Richtsmeier SC, Parks EK, Liu K, Pobo

33. LG, Riley SJ. 1985. *J. Chem. Phys.* 82:3659–65
34. Parks EK, Weiller BH, Bechthold PS, Hoffman WF, Nieman GC, et al. 1988. *J. Chem. Phys.* 88:1622–32
35. Geusic ME, Morse MD, O'Brien SC, Smalley RE. 1985. *Rev. Sci. Instrum.* 56:2123–30
36. Geusic ME, Morse MD, Smalley RE. 1985. *J. Chem. Phys.* 82:590–91
37. Geusic ME, Morse MD, Heath JR, Smalley RE. 1985. *J. Chem. Phys.* 83:2293–2304
38. Song L, Eychmuller A, St Pierre RJ, El-Sayed MA. 1989. *J. Phys. Chem.* 93:2485–90
39. Hamrick Y, Taylor S, Lemire GW, Fu ZW, Shui JC, Morse MD. 1988. *J. Chem. Phys.* 88:4095–98
40. Hamrick YM, Morse MD. 1989. *J. Phys. Chem.* 93:6494–6501
41. Whetten RL, Cox DM, Trevor DJ, Kaldor A. 1985. *J. Chem. Phys.* 89:566–69
42. Zakin MR, Brickman RO, Cox DM, Kaldor A. 1988. *J. Chem. Phys.* 88:3555–60
43. Zakin MR, Brickman RO, Cox DM, Kaldor A. 1988. *J. Chem. Phys.* 88:6605–10
44. Leuchtner RE, Harms AC, Castleman AW. 1990. *J. Chem. Phys.* 92:6527–37
44a. Bell RC, Zemski KA, Castleman AW. 1998. *J. Phys. Chem. A* 102:8293–99
44b. Bell RC, Zemski KA, Castleman AW. 1999. *J. Phys. Chem. A* 103:1585–91
44c. Bell RC, Zemski KA, Castleman AW. 1999. *J. Phys. Chem. A* 103:2992–98
44d. Vann WD, Bell RC, Castleman AW. 1999. *J. Phys. Chem.* 103:10846–50
44e. Zemski KA, Bell RC, Castleman AW. 2000. *J. Phys. Chem. A* 104:5732–41
45. Vann WD, Wagner RL, Castleman AW. 1998. *J. Phys. Chem.* 102:8804–11
46. Hintz PA, Ervin KM. 1995. *J. Chem. Phys.* 103:7897–906
46a. Kapiloff E, Ervin KM. 1997. *J. Phys. Chem. A* 101:8460–69
47. Holmgren L, Andersson M, Rosen A. 1998. *J. Chem. Phys.* 109:3232–39
48. Wu Q, Lu W, Yang S. 1998. *J. Chem. Phys.* 109:8935–39
49. Alford JM, Williams PE, Trevor DJ, Smalley RE. 1986. *Int. J. Mass Spectrom. Ion Proc.* 72:33–51
50. Buckner SW, Gord JR, Freiser BS. 1988. *J. Chem. Phys.* 88:3678–81
51. Irion MP, Selinger A, Wendel R. 1990. *Int. J. Mass Spectrom. Ion Proc.* 96:27–47
52. Dietrich G, Dasgupta K, Kuznetsov S, Lützenkirchen K, Schweikhard L, Ziegler J. 1996. *Int. J. Mass Spectrom. Ion Proc.* 157/158:319–28
53. Berg C, Schindler T, Lammers A, Niedner-Schatteburg G, Bondybey VE. 1995. *J. Phys. Chem.* 99:15497–501
53a. Albert G, Berg C, Beyer M, Achatz U, Joos S, et al. 1997. *Chem. Phys. Lett.* 268:235–41
54. Berg C, Beyer M, Achatz U, Joos S, Niedner-Schatteburg G, Bondybey VE. 1998. *J. Chem. Phys.* 108:5398–403
55. Radi PP, von Helden G, Hsu MT, Kemper PR, Bowers MT. 1991. *Int. J. Mass Spectrom. Ion Proc.* 109:49–73
56. Morse MD. 1986. *Chem. Rev.* 86:1049–1109
57. Castleman AW, Bowen KH. 1996. *J. Phys. Chem.* 100:12911–44
58. Knickelbein MB. 1999. *Annu. Rev. Phys. Chem.* 50:79–115
59. Parent DC, Anderson SL. 1992. *Chem. Rev.* 92:1541–65
60. Forbes RA, Lech LM, Freiser BS. 1987. *Int. J. Mass Spectrom. Ion Proc.* 77:107–21
60a. Beyer M, Bondybey VE. 1997. *Rapid Commun. Mass Spectrom.* 11:1588
61. Jarrold MF, Bower JE, Kraus JS. 1987. *J. Chem. Phys.* 8:3876–85
61a. Jarrold MF, Bower JE, Creegan K. 1989. *J. Chem. Phys.* 90:3615–28
62. Jarrold MF, Bower JE. 1986. *J. Chem. Phys.* 85:5373–75

62a. Jarrold MF, Bower JE. 1987. *J. Chem. Phys.* 87:1610–19
62b. Jarrold MF, Bower JE. 1988. *J. Am. Chem. Soc.* 110:70, 6706–16
62c. Jarrold MF, Bower JE. 1988. *J. Phys. Chem.* 92:5702–5
63. Jarrold MF, Bower JE. 1987. *J. Chem. Phys.* 87:5728–38
64. Hanley L, Ruatta SA, Anderson SL. 1987. *J. Chem. Phys.* 87:260–68
64a. Ruatta SA, Anderson SL. 1988. *J. Chem. Phys.* 89:273–86
64b. Ruatta SA, Hanley L, Anderson SL. 1987. *Chem. Phys. Lett.* 137:5–9
64c. Hanley L, Whitten JL, Anderson SL. 1988. *J. Phys. Chem.* 92:5803–12
64d. Hanley L, Anderson SL. 1987. *J. Phys. Chem.* 91:5161–63
64e. Hanley L, Anderson SL. 1988. *J. Chem. Phys.* 89:2848–60
65. Elkind JL, Weiss FD, Alford JM, Laaksonen RT, Smalley RE. 1988. *J. Chem. Phys.* 88:5215–24
66. Brucat PJ, Pettiette CL, Yang S, Zhang LS, Craycraft MJ, Smalley RE. 1986. *J. Chem. Phys.* 85:4747–48
67. Pacchioni G, Rosch N. 1996. *J. Chem. Phys.* 104:7329–37
68. Dietz TG, Duncan MA, Powers DE, Smalley RE. 1981. *J. Chem. Phys.* 74:6511–12
69. Hopkins JB, Langridge-Smith PRR, Morse MD, Smalley RE. 1983. *J. Chem. Phys.* 78:1627–37
70. Loh SK, Hales DA, Armentrout PB. 1986. *Chem. Phys. Lett.* 129:527–32
71. Loh SK, Hales DA, Lian L, Armentrout PB. 1989. *J. Chem. Phys.* 90:5466–85
72. Hales DA, Armentrout PB. 1990. *J. Cluster Sci.* 1:127–42
73. Daly NR. 1959. *Rev. Sci. Instrum.* 31:264–67
74. Teloy E, Gerlich D. 1974. *Chem. Phys.* 4:417–27
75. Ervin KM, Armentrout PB. 1985. *J. Chem. Phys.* 83:166–89
76. Griffin JB, Armentrout PB. 1997. *J. Chem. Phys.* 106:4448–62
77. Griffin JB, Armentrout PB. 1997. *J. Chem. Phys.* 107:5345–55
78. Beyer T, Swinehart DF. 1973. *Comm. Assoc. Comput. Mach.* 16:379
79. Stein SE, Rabinovitch BS. 1973. *J. Chem. Phys.* 58:2438–45
80. Stein SE, Rabinovitch BS. 1977. *Chem. Phys. Lett.* 49:183–88
81. Gilbert RG, Smith SC. 1990. *Theory of Unimolecular and Recombination Reactions*. Oxford, UK: Blackwell Sci.
82. Armentrout PB. 1992. In *Advances in Gas Phase Ion Chemistry*, Vol. 1, ed. NG Adams, LM Babcock, pp. 83–119. Greenwich, Conn.: JAI
83. More MB, Ray D, Armentrout PB. 1999. *J. Am. Chem. Soc.* 121:417–23
84. Schultz RH, Crellin K, Armentrout PB. 1991. *J. Am. Chem. Soc.* 113:8590–601
85. Dalleska NF, Honma K, Armentrout PB. 1993. *J. Am. Chem. Soc.* 115:12125–31
86. Hales DA, Lian L, Armentrout PB. 1990. *Int. J. Mass Spectrom. Ion Proc.* 102:269–301
87. Hales DA, Su CX, Lian L, Armentrout PB. 1994. *J. Chem. Phys.* 100:1049–57
88. Robinson PJ, Holbrook KA. 1972. *Unimolecular Reactions*. London: Wiley-Intersci.
89. Khan FA, Clemmer DE, Schultz RH, Armentrout PB. 1993. *J. Phys. Chem.* 97:7978–87
90. Rodgers MT, Ervin KM, Armentrout PB. 1997. *J. Chem. Phys.* 106:4499–508
91. Rodgers MT, Armentrout PB. 1997. *J. Phys. Chem. A* 101:2614–25
92. Rodgers MT, Armentrout PB. 1997. *J. Phys. Chem. A* 101:1238–49
93. Armentrout PB. 1987. In *Structure/Reactivity and Thermochemistry of Ions*, ed. P Ausloos, SG Lias, pp. 97–164. Dordrecht, Netherlands: Reidel
94. Lian L, Su CX, Armentrout PB. 1992. *J. Chem. Phys.* 97:4084–93

95. Su CX, Hales DA, Armentrout PB. 1993. *J. Chem. Phys.* 99:6613–23
96. Su CX, Hales DA, Armentrout PB. 1993. *Chem. Phys. Lett.* 201:199–204
97. Su CX, Armentrout PB. 1993. *J. Chem. Phys.* 99:6506–16
98. Ervin K, Loh SK, Aristov N, Armentrout PB. 1983. *J. Phys. Chem.* 87:3593–96
99. Armentrout PB. 1986. In *Laser Applications in Chemistry and Biophysics*, ed. M El-Sayed. *Proc. Int. Soc. Opt. Eng.* 620:38–45
100. Loh SK, Lian L, Hales DA, Armentrout PB. 1988. *J. Phys. Chem.* 92:4009–12
101. Lian L, Su CX, Armentrout PB. 1992. *J. Chem. Phys.* 97:4072–83
102. Lian L, Su CX, Armentrout PB. 1991. *Chem. Phys. Lett.* 180:168–72
103. Lian L, Su CX, Armentrout PB. 1992. *J. Chem. Phys.* 96:7542–54
104. Deleted in proof
105. Loh SK, Lian L, Armentrout PB. 1989. *J. Am. Chem. Soc.* 111:3167–76
106. Hales DA. 1990. *CID of Transition Metal Cluster Ions*. PhD thesis. Univ. Utah, Salt Lake City. 204 pp.
107. Spasov VA, Lee TH, Ervin KM. 2000. *J. Chem. Phys.* 112:1713–20
108. Spasov VA, Lee TH, Maberry JP, Ervin KM. 1999. *J. Chem. Phys.* 110:5208–17
109. Spasov VA, Shi Y, Ervin KM. 2000. *Chem. Phys.* 262:75–91
110. Spasov VA, Ervin KM. 1998. *J. Chem. Phys.* 109:5344–50
111. Grushow A, Ervin KM. 1997. *J. Chem. Phys.* 106:9580–93. Erratum. 1997. *J. Chem. Phys.* 107:8310
112. Krückeberg S, Dietrich G, Lützenkirchen K, Schweikhard L, Walther C, Ziegler J. 1999. *J. Chem. Phys.* 110:7216–27
113. Krückeberg S, Dietrich G, Lützenkirchen K, Schweikhard L, Ziegler J. 1999. *Phys. Rev. A* 60:1251–57
114. Russon LM, Heidecke SA, Birke MK, Conceição J, Armentrout PB, Morse MD. 1993. *Chem. Phys. Lett.* 204:235–40
115. Russon LM, Heidecke SA, Birke MK, Conceição J, Morse MD, Armentrout PB. 1994. *J. Chem. Phys.* 100:4747–55
116. Fu Z, Russon LM, Morse MD, Armentrout PB. 2000. *Int. J. Mass Spectrom. Ion Proc.* In press
117. Lessen DE, Asher RL, Brucat PJ. 1991. *Chem. Phys. Lett.* 182:412–14
118. Asher RL, Bellert D, Buthelezi T, Brucat PJ. 1994. *Chem. Phys. Lett.* 224:529–32
119. Brucat PJ, Zhang LS, Pettiette CL, Yang S, Smalley RE. 1986. *J. Chem. Phys.* 84:3078–88
120. Hettich RL, Freiser BS. 1985. *J. Am. Chem. Soc.* 107:6222–26
120a. Hettich RL, Freiser BS. 1987. 109:3537–42
121. Lessen DE, Brucat PJ. 1988. *Chem. Phys. Lett.* 149:473–76
122. Jarrold MF, Creegan KM. 1990. *Int. J. Mass Spectrom. Ion Proc.* 102:161–81
123. Hild U, Dietrich G, Krückeberg S, Lindinger M, Lützenkirchen K, et al. 1998. *Phys. Rev. A* 57:2786–93
124. Shi Y, Spasov VA, Ervin KM. 1999. *J. Chem. Phys.* 111:938–49
125. Walch SP, Bauschlicher CW. 1985. *J. Chem. Phys.* 83:5735–42
126. Armentrout PB, Kickel BL. 1995. In *Organometallic Ion Chemistry*, ed. BS Freiser, pp. 1–45. Dordrecht, Netherlands: Kluwer Academic
127. Gronbeck H, Rosen A. 1997. *J. Chem. Phys.* 107:10620–25
128. Wu X, Ray AK. 1999. *J. Chem. Phys.* 110:2437–45
129. Bonačić-Koutecký V, Češpiva L, Fantucci P, Koutecký J. 1993. *J. Chem. Phys.* 98:7981–94
130. Sappey AD, Harrington JE, Weisshaar JC. 1989. *J. Chem. Phys.* 91:3854–68
131. Jarrold MF, Illies AJ, Bowers MT. 1985. *J. Am. Chem. Soc.* 107:7339–44
132. Spain EM, Morse MD. 1990. *Int. J. Mass Spectrom. Ion Proc.* 102:183–97
133. Morse MD, Hansen GP, Langridge-Smith PRR, Zheng LS, Geusic ME, et al. 1984. *J. Chem. Phys.* 80:5400–405

134. Stave MS, DePristo AE. 1992. *J. Chem. Phys.* 97:3386–98
135. Knickelbein MB, Yang S. 1990. *J. Chem. Phys.* 93:1476–77
136. Shi Y, Ervin KM. 1998. *J. Chem. Phys.* 108:1757–60
137. Shi Y, Spasov VA, Ervin KM. 2001. *Int. J. Mass Spectrom.* In press
138. Kaplan IG, Santamaria R, Novaro O. 1993. *Int. J. Quantum Chem. Quantum Chem. Symp.* 27:743–68
139. Bonačić-Koutecký V, Češpiva L, Fantucci P, Pittner J, Koutecký J. 1994. *J. Chem. Phys.* 100:490–506
140. DeHeer W. 1993. *Rev. Mod. Phys.* 65:611–76
141. Brack M. 1993. *Rev. Mod. Phys.* 65:677–732
142. Armentrout PB, Hales DA, Lian L. 1994. In *Advances in Metal and Semiconductor Clusters*, Vol. 2, ed. MA Duncan, pp. 1–39. Greenwich, Conn.: JAI
143. Riley SJ. 1996. *J. Non-Cryst. Solids* 205/207:781–87
144. Riley SJ. 1996. In *Clusters and Nanostructured Materials*, ed. P Jena, SN Behera, pp. 77–89. New York: Nova Sci.
145. Raghavan K, Stave MS, DePristo AE. 1989. *J. Chem. Phys.* 91:1904–17
146. Chase MW Jr, Davies CA, Downey JR Jr, Frurip DJ, McDonald RA, Syverud AN. 1985. *J. Phys. Chem. Ref. Data* 14: Suppl. 1
147. Miedema AR. 1978. *Z. Met.kd.* 69:287–92
148. Miedema AR. 1980. *Faraday Symp. R. Soc. Chem.* 14:136–48
149. Whetten RL, Cox DM, Trevor DJ, Kaldor A. 1985. *Phys. Rev. Lett.* 54:1494–97
150. Ho J, Zhu L, Parks EK, Riley SJ. 1993. *J. Chem. Phys.* 99:140–47
151. Zhu L, Ho J, Parks EK, Riley SJ. 1993. *J. Chem. Phys.* 98:2798–804
152. Conceição J, Liyanage R, Armentrout PB. 2000. *Chem. Phys.* 262:115–30
153. Conceição J, Loh SK, Lian L, Armentrout PB. 1996. *J. Chem. Phys.* 104:3976–88
154. Deleted in proof
155. Huber KP, Herzberg G. 1979. *Molecular Spectra and Molecular Structure*, Vol. IV. *Constants of Diatomic Molecules*, p. 263. New York: Van Nostrand-Reinhold
156. Siegbahn P, Wahlgren U. 1991. In *Metal-Surface Reaction Energetics*, ed. E Shustorovich, pp. 1–52. New York: VCH
157. Toyoshima I, Somorjai G. 1979. *Catal. Rev. Sci. Eng.* 19:105–59
158. Frese KW. 1987. *Surf. Sci.* 182:85–97
159. Boszo F, Ertl G, Grunze M, Weiss M. 1977. *Appl. Surf. Sci.* 1:103–19
160. Boszo F, Ertl G, Weiss M. 1977. *J. Catal.* 50:519–29
161. Benziger JB. 1991. In *Metal-Surface Reaction Energetics*, ed. E Shustorovich, pp. 53–107. New York:VCH
162. Liu K, Parks EK, Richtsmeier SC, Pobo LG, Riley SJ. 1985. *J. Chem. Phys.* 83:2882–88
163. Riley SJ, Parks EK, Nieman GC, Pobo LG, Wexler S. 1984. *J. Chem. Phys.* 80:1360–62
164. Nieman GC, Parks EK, Richtsmeier SC, Liu K, Pobo LG, Riley SJ. 1986. *High Temp. Sci.* 22:115–38
165. Xu J, Rodgers MT, Griffin JB, Armentrout PB. 1998. *J. Chem. Phys.* 108:9339–50
166. Griffin JB, Armentrout PB. 1998. *J. Chem. Phys.* 108:8062–74
167. Balducci G, Gigli G, Guido M. 1983. *J. Chem. Phys.* 79:5616–22
168. Fisher ER, Elkind JL, Clemmer DE, Georgiadis R, Loh SK, et al. 1990. *J. Chem. Phys.* 93:2676–91
169. Griffin JB, Armentrout PB. 1998. *J. Chem. Phys.* 108:8075–83
170. Jellinek J, Güvenç ZB. 1991. *Z. Phys. D* 19:371–73
171. Jellinek J. 1996. In *Metal-Ligand Interactions: Structure and Reactivity*, ed. N Russo, DR Salahub, pp. 325–60. Dordrecht, Netherlands: Kluwer Academic
172. Kelly D, Weinberg WH. 1996. *J. Vac. Sci. Technol. A* 14:1588–92

173. Hamza AV, Steinruck HP, Madix RJ. 1987. *Surf. Sci.* 173:L571–75
174. Rettner CT, Stein H, Schweizer EK. 1988. *J. Chem. Phys.* 89:3337–41
175. Arbab M, Hudson JB. 1988. *Surf. Sci.* 206:317–37
176. Brennan D, Hayward DO, Tradnell BMW. 1960. *Proc. R. Soc. London Ser. A* 256:81–105
177. Bragg J, Tomkins FC. 1955. *Trans. Faraday Soc.* 51:1071–80
178. Wedler G. 1961. *Z. Phys. Chem.* 27:388–401
179. Grushow A, Ervin KM. 1995. *J. Am. Chem. Soc.* 117:11612–13
180. Biberian JP, Van Hove MA. 1982. *Surf. Sci.* 118:443–64
181. Froitzheim H, Schulze M. 1989. *Surf. Sci.* 211/212:837–43
182. Deng HT, Kerns KP, Castleman AW. 1996. *J. Phys. Chem.* 100:13386–92
183. Kerns KP, Guo BC, Deng HT, Castleman AW. 1996. *J. Phys. Chem.* 100:16817–21
184. Guo BC, Kerns KP, Castleman AW. 1992. *Science* 255:1411–13
185. Weiller BH, Bechthold PS, Parks EK, Pobo LG, Riley SJ. 1989. *J. Chem. Phys.* 91:4714–27
186. Parks EK, Riley SJ. 1993. *J. Chem. Phys.* 99:5898–904
187. Winter BJ, Klots TD, Parks EK, Riley SJ. 1991. *Z. Phys. D* 19:375–80
188. Deleted in proof
189. Grunze M, Bozso F, Ertl G, Weiss M. 1978. *Appl. Surf. Sci.* 1:241–65
190. Shustorovich E, ed. 1991. *Metal-Surface Reaction Energetics.* New York: VCH

Spin-$\frac{1}{2}$ and Beyond: A Perspective in Solid State NMR Spectroscopy

Lucio Frydman*

Department of Chemistry, University of Illinois at Chicago, 845 W. Taylor St., Rm 4500, Chicago, Illinois 60607; e-mail: lucio@uic.edu

Key Words spin interactions, high resolution NMR, quadrupolar nuclei, structural determinations

■ **Abstract** Novel applications of solid state nuclear magnetic resonance (NMR) to the study of small molecules, synthetic polymers, biological systems, and inorganic materials continue at an accelerated rate. Instrumental to this uninterrupted expansion has been an improved understanding of the chemical physics underlying NMR. Such deeper understanding has led to novel forms of controlling the various components that make up the spin interactions, which have in turn redefined the analytical capabilities of solid state NMR measurements. This review presents a perspective on the basic phenomena and manipulations that have made this progress possible and describes the new opportunities and challenges that are being opened in the realms of spin-$\frac{1}{2}$ and quadrupole nuclei spectroscopies.

INTRODUCTION

Although the foundations of nuclear magnetic resonance (NMR) were laid long ago (1), its scope and range of applications have remained in constant change through the decades (2–14). This progress has resulted from a better understanding of NMR's quantum principles, from new technical developments, and perhaps most importantly, from the unique opportunities provided by NMR itself. Indeed, NMR is in many ways any spectroscopist's dream, enabling nearly arbitrary manipulations of the interactions and the generation of unusually long-lived coherent states. It is thus not surprising that, when it comes to advancing the frontiers of spectroscopy, much new ground is broken first in NMR. In terms of applications, it is also not surprising that NMR could reach so deeply into such diverse realms as medical imaging, structural biology, analytical chemistry, and material sciences. In fact, the spectroscopic principles involved in the application of NMR to such disimilar disciplines are related to one another and in many instances find their

---

*Present address: Department of Chemical Physics, Weizmann Institute of Sciences, 76100 Rehovot, Israel

most challenging test ground in the topic treated by this article: the NMR of solids. A general discussion on solid state NMR seems justified by interesting physical ideas that have recently emerged in the area, by the new challenges and horizons that these new principles have revealed, and by the promising applications that these have opened up towards the characterization of a wide variety of solid materials. Because even modest coverage of all such recent developments would exceed the scope of an article (see 12–16 for excellent recent treatments of these topics), the objectives of this review are limited: to present a contemporary "stand-alone" perspective on the principles of solids NMR, and to exploit this background for introducing some of the latest developments in this area. The latter are described in mostly physical rather than mathematical terms in the hope of stressing their rationale, applications, and potential limitations.

SOLIDS NMR: Interactions and Spectra

NMR Interactions as Scalar Products Between Spin and Spatial Tensors

NMR is based on observing the oscillating signals that arise when an ensemble of nuclear spins is placed inside a strong static magnetic field B_o, and then taken away from equilibrium by the action of radiofrequency (rf) pulses. All NMR-active nuclides are characterized by a magnetic dipole moment μ, and therefore these time-dependent signals will be mostly governed by the spins' magnetic coupling to either external or internal fields. By virtue of the spins' quantum nature, these couplings are best represented by Hamiltonians, defining both the allowed energy levels and the spins' evolution in time (2, 3, 5). Under the sound assumption that fields can be represented by classical continuous functions, these operators take the general form $\mathcal{H}_\lambda = -\mu \cdot B_\lambda = -\gamma S \cdot B_\lambda$, where B_λ is a generic field, S is a spin's angular momentum, and γ is the nuclear magnetogyric constant.

Dominating NMR is the Zeeman interaction between spins and an external magnetic field B_o

$$\mathcal{H}_Z = -\gamma(S_x, S_y, S_z) \cdot (0, 0, B_o) = -\gamma B_o S_z = -\omega_o S_z, \qquad 1.$$

with a formally similar interaction representing the spins' coupling with the magnetic components of a transverse time-dependent field $B_{rf}(t)$. Although such Zeeman and rf couplings are essential for carrying out the NMR experiment, molecular information becomes available through the coupling of spins to locally generated magnetic fields. For instance, separation between inequivalent sites is promoted by the chemical shielding, reflecting the fields B_{ind} that are induced by electrons when a molecule is immersed inside B_o (1, 11). Owing to the anisotropic ease with which B_o can induce electronic currents, these fields are proportional to B_o in magnitude ($|B_{ind}| \approx 10^{-5} \cdot |B_o|$) but not necessarily in spatial direction;

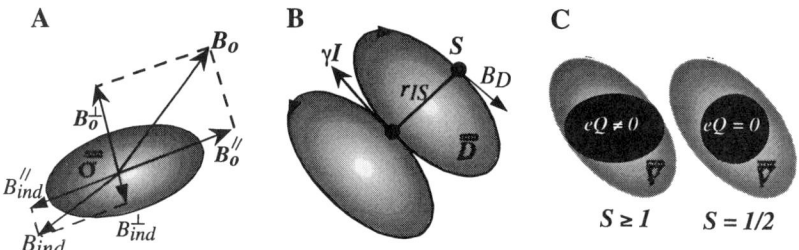

Figure 1 Tensorial nature of coupling tensors in solid state NMR. (A) Chemical shielding $\bar{\bar{\sigma}}$ and the origin of fields B_{ind}, deviating from B_o unless the latter is oriented along one of the spheroid's principal axes. (B) Idem for the I-S dipolar interaction. (C) Field gradient $\bar{\bar{V}}$ tensors and their ensuing electrostatic coupling with charged, nonspherical ($S \geq 1$) nuclei.

therefore, they require 3×3 $\bar{\bar{\sigma}}$ tensors for their complete description (Figure 1A), and lead to a Hamiltonian $\mathcal{H}_{CS} = -\mu \cdot B_{ind} = \gamma S \cdot \bar{\bar{\sigma}} \cdot B_o$.[1]

Because nuclear magnetic dipoles not only couple to fields but also generate them, the NMR evolution of a spin S may be influenced by the fields arising from neighboring nuclei I. Once again these effects will be proportional to the magnetic moments of the spins but only in a tensorial sense (Figure 1B), leading to an interaction Hamiltonian, $\mathcal{H}_D = \gamma_I \gamma_S S \cdot \bar{\bar{D}} \cdot I$. Formally indistinguishable from this dipolar coupling but mediated by a different mechanism is the indirect $\bar{\bar{J}}$ interaction, which though essential in solution-state NMR, can usually be neglected in solid studies thanks to its small size. This is fortunate in view of the direct relation that then remains between the I-S coupling constant and the internuclear distance (r_{IS}^{-3}).

In addition to these magnetic effects, there is an important electric coupling that affects all nuclei with spin $S \geq 1$: the quadrupole interaction (Figure 1C) (11, 17). This arises from the classical energy E_Q between a nuclear charge distribution ρ and its surrounding electrostatic potential V:

$$E_Q = \int \rho(r) V(r) dr \approx \underset{\text{electrostatic energy}}{\text{point}} + \frac{1}{2} \sum_{i,j=x,y,z} \frac{\partial^2 V}{\partial i \partial j} \cdot \int ij\rho(r)dr \qquad 2.$$

$+ \text{ higher-order terms} \ldots$

Here $\frac{\partial^2 V}{\partial i \partial j} = V_{ij}$ denotes the electric field gradient at the nucleus, and its elements constitute an additional 3×3 $\bar{\bar{V}}$ tensor; $Q_{ij} = \int ij\rho(r)dr$ is the classical

---

[1]A 3×3 tensor generalizes the concepts of scalar (an orientation-independent number) and vector (a 3-element array possessing a magnitude and a well-defined rotational transformation $V_i \rightarrow \sum R_{ij} V_j$) into one additional dimension. One way of building such Cartesian tensors is by arranging dyadic products between two vectors V, U into a 3×3 2-dimensional matrix: $A_{ij} = U_i V_j$.

description of the quadrupolar nuclear moment. E_Q is apparently unrelated to NMR, yet it can be shown that on deriving a quantum mechanical Hamiltonian $\mathcal{H}_Q$ for it, the elements Q_{ij} end up expressed in terms of spin operators.[2] The quadrupolar electrostatic energy terms in Equation 2 thus become the NMR-relevant spin Hamiltonian, $\mathcal{H}_Q = \frac{eQ}{4S(2S-1)} S \cdot \bar{\bar{V}} \cdot S$.

All these expressions for the nuclear spin Hamiltonians look similar: They involve products between a characteristic constant C_λ, a spin vector S, a coupling matrix, and another spin or B vector. Their generalized form is therefore

$$\mathcal{H}_\lambda = C_\lambda \overline{U} \cdot \bar{\bar{R}} \cdot \overline{V} = C_\lambda \sum_{i=1}^{3} U_i \sum_{j=1}^{3} R_{ij} V_j. \qquad 3.$$

The $\{R_{ij}\}_{i,j=1-3}$ matrices in these $\mathcal{H}_\lambda$ correspond to the shielding, dipolar or quadrupolar couplings ($\bar{\bar{\sigma}}$, $\bar{\bar{D}}$, $\bar{\bar{V}}$), interactions that depend on the chemical system under observation but not on the spin operators themselves.[3] Because physical rotations of the chemical system will change the individual R_{ij} values, these are collectively denoted as the *spatial* parts of the coupling Hamiltonian. The double sums in Equation 3, however, will end up generating other 3×3 tensors with matrix elements $\{T_{ij} = V_j U_i\}_{i,j=1-3}$ that do not involve any structural coupling parameters; they contain all of the $\mathcal{H}_\lambda$'s dependencies on the spin states, and are the quantum mechanical portions of the Hamiltonian operators. Thanks to this separation between spatial and spin terms, it becomes possible to express all local coupling Hamiltonians as $\mathcal{H}_\lambda = C_\lambda \bar{\bar{R}} \cdot \bar{\bar{T}}$, which is an extension of the scalar product between two vectors to the case of 3×3 tensors. This implies that when considering each individual interaction, its R and T components may change depending on the reference frame used for their description, but their resulting Hamiltonian will not: It is a scalar. Indeed, insensitivity to orientation is one of the most useful characteristics of zero-field magnetic resonance and pure quadrupole resonance, leading to sharp lines even when dealing with polycrystalline samples (1, 19).

There is actually more to the nature of the $\bar{\bar{R}}$ and $\bar{\bar{T}}$ tensors than just a 3×3 matrix character, particularly with regard to describing their changes upon rotating either the spatial or spin coordinates.[4] Indeed, as different reference frames are chosen all nine elements defining a tensor may change, yet certain key features are not really dependent on this choice and will remain constant. (For instance a

---

[2] A fundamental step in this derivation is the Wigner-Eckart theorem (18), stating that in systems with well-defined angular momentum the quantum mechanical expressions of all vector operators (e.g. x, y, z in Q_{ij}) are in fact proportional to one another (that is, to S_x, S_y, S_z).
[3] When external Zeeman or rf couplings are involved R_{ij} becomes δ_{ij}, the identity matrix.
[4] For the case of $\bar{\bar{R}}$ such rotations may be done for the convenience of expressing couplings in a new reference frame, or when accounting for a coherent mechanical motion like magic-angle spinning. $\bar{\bar{T}}$ rotations are usually used to describe the effects of rf pulses or nutations around magnetic fields rather than "physical" rotations, and correspond to stepping into alternative interaction representations.

shielding tensor $\bar{\bar{\sigma}}$ may vary upon rotating a molecule's frame, yet there is a certain isotropic chemical shift component, the one usually observed in solution phase experiments, that is invariant to reorientations.) This reflects the "reducibility" problem of Cartesian tensors, whose resolution requires rearranging the matrix elements into a series of objects that behave differently with respect to rotations. The resulting "irreducible" ranks make up mathematical groups, meaning that a rotation $\varepsilon(\alpha, \beta, \gamma)$ will not transform components $A_{ij}^{(k)}$ of rank k into elements $A_{ij}^{(k')}$ of a different rank k'. One such possible rearrangement, applicable to either the spin or spatial NMR tensors, is (18)

$$A^{(0)} = (A_{11} + A_{22} + A_{33})/3: \text{orientation-independent}$$
(scalar, zero-rank) component 4a.

$$A_{ij}^{(1)} = \frac{1}{2}(A_{ij} - A_{ji}) \quad i = 1\text{-}2, j = i+1\text{-}3: \text{three first-rank}$$
components transforming as a vector 4b.

$$A_{ij}^{(2)} = \frac{1}{2}(A_{ij} + A_{ji}) - A^{(0)} i = 1\text{-}3, j = 1\text{-}3: \text{five second-rank}$$
components transforming as a 3×3 traceless symmetric matrix 4c.

Rather than using these definitions based on Cartesian coordinates, it is customary to take linear combinations within each rank to obtain a tensor's description in spherical coordinates, better behaved with respect to uniaxial rotations.[5] Even when written in this manner, the various elements of a particular tensor are represented as $\{A_{km}\}$, but k now refers to an element's rank and $m = -k, \ldots, k$ indicates its order. Such tensor elements transform under coordinate rotations according to

$$A_{km'} = \varepsilon(\alpha, \beta, \gamma) A_{km} \varepsilon^{-1}(\alpha, \beta, \gamma) = \sum_{m'=-k}^{+k} \mathcal{D}_{m'm}^{(k)}(\alpha, \beta, \gamma) A_{km'}, \quad 5.$$

where the $\mathcal{D}_{m'm}^{(k)}$ define Wigner rotation matrix elements $e^{-im'\alpha} d_{m'm}^{(k)}(\beta) e^{-im\gamma}$ describing how elements within a rank k transform into one another (2, 20).

Truncation by B_o: First- and Second-Order Anisotropies

It follows from these arguments that the various internal NMR couplings can be expressed as products between irreducible spin and spatial spherical tensors possessing ranks $k \leq 2$. These products are orientation-independent scalars, but the dominating Zeeman coupling will break this symmetry and endow each spin interaction with an anisotropic character. This truncation imposed by $\mathcal{H}_Z$ on the

---

[5]This choice is not unlike the one taken for describing atomic orbitals, which can be expressed either as easily visualized Cartesian functions (e.g. p_x, p_y, p_z) or as better behaved spherical harmonics [$p_o = p_z, p_{+1} = -(p_x + ip_y)/\sqrt{2}, p_{-1} = (p_x - ip_y)/\sqrt{2}$]. It also reflects the fact that s orbitals transform as $A^{(0)}$, p as $A^{(1)}$, d as $A^{(2)}$, etc.

smaller $\mathcal{H}_\lambda$ can be appreciated in a number of ways: One is by using standard time-independent perturbation theory; another is by viewing the truncation as resulting from the fast time dependence that $\mathcal{H}_Z$ imposes on the smaller interactions. The latter derivation involves transforming the spin-space components of the $\mathcal{H}_\lambda$'s into an interaction representation, akin to the rotating frame usually employed in the classical description of NMR (1, 9, 11). This is defined quantum mechanically by the time propagator $U_o(t) = exp(-i\mathcal{H}_Z t) = exp(i\omega_o S_z t)$, representing a continuous rotation at a rate ω_o around S_z, the spin-space's z-axis. Thanks to the well-behaved nature of the spherical tensor operators with respect to z-rotations ($e^{iS_z\phi} T_{km} e^{-iS_z\phi} = T_{km} e^{-im\phi}$), this can be simply accounted for as

$$\tilde{\mathcal{H}}_\lambda(t) = U_0(t)^{-1} \mathcal{H}_\lambda U_0(t) = C_\lambda \sum_{k=0}^{2} \sum_{m=-k}^{k} R_{k-m} T_{km} e^{-im\omega_0 t}. \qquad 6.$$

At first sight this transformation seems to have worsened matters by making the $\mathcal{H}_\lambda$ time dependent, but this complication can be dealt with using a versatile approximation known as average Hamiltonian theory (AHT) (2, 21). According to AHT, the effective evolution introduced on all $\tilde{\mathcal{H}}_\lambda$'s at the end of each periodic Larmor cycle $\tau_c = 2\pi/\omega_o$ can be approximated as the time-independent series

$$\overline{\mathcal{H}_{\text{total}}}(\tau_c) = \sum_{\lambda \neq Z} \mathcal{H}_\lambda^{(1)} + \sum_{\lambda,\lambda' \neq Z} \mathcal{H}_{\lambda,\lambda'}^{(2)} + \cdots, \qquad 7.$$

where the leading terms are

$$\mathcal{H}_\lambda^{(1)} = \tau_c^{-1} \int_0^{\tau_c} \tilde{\mathcal{H}}_\lambda(t) dt, \quad \mathcal{H}_{\lambda,\lambda'}^{(2)} = \frac{-i}{2} \tau_c^{-1} \int_0^{\tau_c} dt \int_0^{t} [\tilde{\mathcal{H}}_\lambda(t), \tilde{\mathcal{H}}_{\lambda'}(t')] dt'. \qquad 8.$$

It follows from Equation 6 that the first of these terms will only preserve the time-independent $m = 0$ elements. For the shielding, dipolar, and quadrupolar interactions this leads to the dominant first-order Hamiltonians (2–6)

$$\mathcal{H}_{CS}^{(1)} = -\gamma \left(R_{00}^{CS} T_{00}^{S} + R_{20}^{CS} T_{20}^{S} \right) = -\gamma B_0 \left(R_{00}^{CS} + R_{20}^{CS} \right) S_z = \left(\omega_{CS}^{iso} + \omega_{CS}^{aniso} \right) S_z \qquad 9a.$$

$$\mathcal{H}_D^{(1)} = \gamma_I \gamma_S R_{20}^D \cdot \begin{cases} T_{20}^{IS} = \omega_D (3 I_z S_z - I \cdot S)/2 & \text{if } \omega_0^I = \omega_0^S \\ T_{10}^I T_{10}^S = \omega_D I_z S_z & \text{if } \omega_0^I \neq \omega_0^S \end{cases} \qquad 9b.$$

$$\mathcal{H}_Q^{(1)} = \frac{eQ}{4S(2S-1)} \cdot R_{20}^Q T_{20}^S = \omega_Q [3 S_z^2 - S(S+1)]. \qquad 9c.$$

Thus, the only isotropic term arising from these couplings comes from the chemical shift ω_{CS}^{iso}, with all remaining ones leading to spatial anisotropies that transform as second-rank R_{20}^λ tensors.

In most cases these first-order Hamiltonians, proportional to coupling constants C_λ, are excellent descriptions of H_λ's complete effects. The following terms in the expansion are proportional to $C_\lambda C_{\lambda'}/\omega_0$ and therefore inconsequential, except when dealing with $S \geq 1$ nuclei subject to large quadrupole effects. Indeed, quadrupole coupling constants can often lie in the MHz range (17, 22), and thereby lead to cross terms $\mathcal{H}_{\lambda,Q}^{(2)}$ that are easily detectable by NMR. Most notable among these is the second-order quadrupole effect

$$\mathcal{H}_{Q,Q}^{(2)} = \frac{C_Q^2}{\omega_0} \sum_{m \neq 0} \frac{R_{2m} R_{2-m}[T_{2m}, T_{2-m}]}{2m}, \qquad 10.$$

which like all remaining second-order correlations, brings out new products of both spatial ($R_{2m}R_{2-m}$) and spin ($T_{2m}T_{2-m}$) spherical tensor components.[6] In the same manner that dyadic multiplications of rank-1 vectors lead to second-rank tensors, such products of rank-2 terms will lead to tensors with $k \leq 4$ (23, 24). Symmetry considerations force the order of all elements in this multirank expansion to $m = 0$; further calculations indicate that the products of the spatial tensor components will result in a zero-rank (R_{00}^Q) term analogous to ω_{CS}^{iso} but of quadrupole origin, as well as to second-rank (R_{20}^Q) and fourth-rank (R_{40}^Q) anisotropies. Higher-rank spin-space components will also arise, with the commutators in Equation 10 leading only to odd (T_{10}, T_{30}) terms. When dealing with the central $-\frac{1}{2} \leftrightarrow +\frac{1}{2}$ transition of a half-integer quadrupolar spin ($S = \frac{3}{2}, \frac{5}{2}, \ldots$), which is the only single-quantum transition in these systems that is not affected by the otherwise dominating $\mathcal{H}_Q^{(1)}$ term, both of these operators are proportional to the longitudinal central-transition angular momentum C_z. Therefore, from a spin-space perspective, the type of precession that $\mathcal{H}_{Q,Q}^{(2)}$ imparts on the central transition of these nuclei is akin to that of a chemical shift.

Spin Evolution and the Calculation of NMR Spectra

To calculate the spins' NMR signal after they have been taken away from equilibrium, it is convenient to represent their ensemble by a density matrix ρ that accounts for both the quantum-mechanical nature of the spins and their incoherent statistical superposition (1–13). The spin evolution can then be obtained from integrating Schröedinger's equation as

$$\rho(t) = U(t)\rho_o U(t)^{-1}; \quad U(t) = \exp\left[-i \int_0^t \mathcal{H}(t') dt'\right], \qquad 11.$$

where the operator $U(t)$ describes the dynamics imposed by a rotating frame Hamiltonian like the one in Equation 7 on spins assumed in an initial state ρ_o. The small voltage induced by the spins along a transverse coil can then be derived as $S(t) = Tr[\rho(t)S_+]$.

The density matrix ρ is itself an operator, and can thus be expressed as a linear combination of various spin-space terms (S_z, S_x, T_{20}^S, etc). For instance, an initial

---

[6]Additional terms $R_{2m}R_{20}[T_{2m}, T_{20}]$ representing a tilting in the axis of quantization survive the double time integration in Equation 8, but can be neglected to this degree of approximation.

thermal equilibrium state dictated by the Zeeman interaction will be

$$\rho_{eq} = e^{-\mathcal{H}_z/kT} \approx 1 + \frac{\omega_0}{kT}S_z \sim \frac{\omega_0}{kT}S_z, \qquad 12.$$

where the "1" represents unpolarized spins that remain indifferent to all NMR manipulations and thereby can be ignored. Such an operator description thus results in a state that is analogous to the z-magnetization that could be expected from a classical perspective. Furthermore, the action of single-spin $\mathcal{H}_{rf} = \omega_{rf} S_x$ or $\mathcal{H}_{CS} = \omega_{CS} S_z$ Hamiltonians can be rigorously described according to

$$S_z \xrightarrow{\omega_{rf} S_x t} S_z \cos(\omega_{rf} t) - S_y \sin(\omega_{rf} t);$$
$$S_x \xrightarrow{\omega_{CS} S_z t} S_x \cos(\omega_{CS} t) + S_y \sin(\omega_{CS} t), \qquad 13.$$

where the left-hand operators denote prototypical ρ_o states, the arrows are shorthand for the evolution operators, and the right-hand sides show the $\rho(t)$. Again there is a one-to-one correspondence between these equations and the expectations that result from classical predictions. This parallelism is maintained for as long as linear single-spin interactions are involved[7] but ceases to be complete in more complex cases containing either quadrupolar or spin-spin couplings, for which states not describable by single-spin operators appear. In an effort to preserve even for these cases the simplicity of the spin-½ notation a formalism was developed, in which the new states are described as direct multiplications of single-spin operators. Hence, the effects of heteronuclear dipolar or $S=1$ quadrupolar couplings can be described as (9, 10)

$$S_x \xrightarrow{\omega_D I_z S_z t} S_x \cos \frac{\omega_D t}{2} + 2S_y I_z \sin \frac{\omega_D t}{2};$$
$$S_x \xrightarrow{\omega_Q [3S_z^2 - S(S+1)]t} S_x \cos \omega_Q t + 2(S_y S_z + S_z S_y) \sin \omega_Q t. \qquad 14.$$

The contributions made by these spin-product states to the NMR signal is easy to visualize: Functions that are multiplying single-quantum, in-phase operators (S_x, S_y) are directly detectable by the NMR coil; single-quantum antiphase coherences containing only one transverse operator ($I_z S_y$, $S_z S_y$) are not directly observable but can lead to in-phase signals if acted upon by suitable couplings; spin-order or multiple-quantum states possessing only longitudinal or several transverse operators are not observable unless acted upon by further pulses.

This elegant formalism is particularly suitable for describing experiments defined by commuting interactions, such as chemical shifts and/or weak spin-spin couplings. Such conditions are widespread in solution but not always met in solid state NMR; here interactions may be time-dependent and not mutually commuting, and couplings comparable if not larger than the rf fields. Analytical descriptions

---

[7]This "coincidence" actually reflects a local isomorphism between the elements of the SO(3) space group defining vector rotations in an orthogonal three-dimensional space, and the SU(2) group defining unitary transformations $U(t)$ for 2×2 matrix operators such as those describing isolated spin-1/2 ensembles.

of the spins' evolution may then be difficult to come by, and alternatives are needed for evaluating the experimental results. A common approximation is AHT (Equation 7), which can provide a hierarchical expansion of the effects introduced by periodically time-dependent interactions if the system is probed at proper integer multiples of the modulation period (25). A conceptual and practical alternative for dealing with periodic manipulations is Floquet theory (26, 27), which bypasses the problems associated with finding the evolution imposed by a time-dependent interaction by deriving an alternative Hamiltonian that is time-independent but possesses an infinite dimension.[8] Finally, a general route to the calculation of arbitrary spin evolutions consists of propagating density matrices throughout an interval of interest t by subdividing the time axis into short enough periods Δt. Computations of the evolution operator can then proceed on the assumption of piecewise constant Hamiltonians as $U(0, t) \approx \ldots e^{-i\mathcal{H}(\Delta t)\Delta t} e^{-i\mathcal{H}(0)\Delta t}$. Such a procedure can be highly time consuming, particularly when dealing with powdered samples containing multiply coupled or high-spin nuclei and subject to arbitrary time dependencies; alternatively, certain simplifying assumptions (tensor symmetries, stroboscopic observation) can be exploited, and numerous useful algorithms have been proposed for facilitating solid state spectral simulations under a variety of conditions (28–32).

HIGH RESOLUTION IN SOLIDS NMR

Averaging via Spin-Space Manipulations

Given the different information conveyed by the spin interactions and the orientation-dependence brought upon them by the high field Zeeman truncation, the selective removal of couplings and/or of their anisotropic components becomes an important topic in solid state NMR. The complete elimination of anisotropies becomes particularly relevant when dealing with randomly powdered samples and trying to resolve the broadened signals arising from chemically inequivalent sites. Because NMR Hamiltonians are given by products of spin (T_{k0}^λ) and spatial (R_{k0}^λ) terms, such selective eliminations can generally involve imposing a time dependence on the spin components of $\mathcal{H}_\lambda$ via rf irradiations, on their spatial components via mechanical sample reorientations, or sometimes on both spin and spatial components.

Perhaps the simplest relevant example of selective spin-space averaging is heteronuclear decoupling, which removes the effects of $\mathcal{H}_D^{IS}$ from an S spectrum by continuously irradiating I close to resonance (2). The application of such an rf field can be accounted for by an evolution operator $U_{rf}(t) = exp[i\omega_{rf}I_x t]$, which

---

[8]The resulting Floquet Hamiltonian is no longer defined on the conventional spin manifold $\{|\alpha\rangle, |\beta\rangle\}$ but on a "dressed" basis set $\{|\alpha m\rangle, |\beta m\rangle\}$ associated with a spin state as well as with a multiple mode of the basic modulating frequency; practical calculations involve diagonalizing this Hamiltonian after it has been truncated to a sufficiently high order and then exploiting it to compute the spins' evolution at arbitrary times.

leaves S unaffected but imparts on the I-containing terms in the Hamiltonian a time-modulation

$$\mathcal{H}_D^{IS} + \mathcal{H}_{CS}^I \xrightarrow{U_{rf}} \tilde{\mathcal{H}}(t) = \omega_D S_z (I_z \cos \omega_{rf} t - I_y \sin \omega_{rf} t) \\ - \Delta \omega_I (I_z \cos \omega_{rf} t - I_y \sin \omega_{rf} t),$$ 15.

where $\Delta \omega_I$ is the rf irradiation offset. This time evolution is akin to the one expected from classical nutation arguments, and it clearly makes $\mathcal{H}_D = 0$, at least to first-order in AHT and when $\Delta \omega_I = 0$. Continuous irradiation is consequently a method of choice for achieving heteronuclear (e.g. {^{1}H}^{13}C) decoupling in the solid state (14, 33, 34). When the rates of nutation ω_{rf} are not fast enough, however, second-order terms $\mathcal{H}_{CS,D}^{(2)} \propto (\Delta \omega_I \cdot \omega_D / \omega_{rf}) S_z I_x$ arising from offset/dipolar cross correlations may become relevant (35). These residuals are generally present in the case of ^{1}H-decoupling in organic solids due to site inequivalencies and/or shielding anisotropies (36); they are not susceptible to complete elimination by concurrent sample spinning, and are consequently important factors in broadening the S-spin resonances.[9] In such cases it is possible to improve the decoupling performance by imposing a second time dependence on the spins that, acting orthogonally to I_x, helps quench the $\mathcal{H}_{CS,D}^{(2)}$ residual (37, 38). This is most often implemented with a simple "two-pulse phase modulated" (TPPM) scheme, although more sophisticated alternatives have also been described (39, 40).

An equivalent way of visualizing heteronuclear decoupling is by considering the rotations induced by $U_{rf}(t)$ on the first-rank spin-space elements (Equation 5):

$$T_{10}^I \xrightarrow{U_{rf}} \tilde{T}_{10}^I(t) = d_{00}^{(1)}(\beta) T_{10}^I + d_{10}^{(1)}(\beta) T_{11}^I e^{-i\omega_{rf} t} + d_{-10}^{(1)}(\beta) T_{1-1}^I e^{i\omega_{rf} t};$$ 16.

fast oscillations will then average out the $\{T_{1\pm1}^I\}$, whereas the choice of transverse rf ($\beta = 90°$) eliminates the first-order spherical harmonic $d_{00}^{(1)}(\beta) = \cos \beta$. By contrast, spin terms in the homonuclear $\mathcal{H}_D^{IS} = \omega_D R_{20}^{IS} T_{20}^{IS}$ couplings transform as second-rank tensors and therefore will fail to average out under these conditions. Instead, removing second-rank components requires fast nutations around an axis inclined at the root of $d_{00}^{(2)}(\beta) = (3 \cos^2 \beta - 1)/2$, the magic angle $\beta_m = 54.7°$. A continuous version of this averaging is achieved in the Lee-Goldburg (LG) experiment (41), which applies an rf field that is offset from resonance by $\Delta \omega_{LG} = 0.71 \omega_{rf}$ (Figure 2). The ensuing spin-space rotation does not occur at a root of T_{10}; first-rank tensors such as the chemical shift will then be scaled but not eliminated, thereby enabling the acquisition of shift-based NMR spectra from strongly coupled networks such as protons in organic solids.

Although important as a conceptual starting point, LG experiments are rarely employed in high-resolution acquisitions owing to a number of limitations, including a lack of observation windows, difficulties in strictly fulfilling the first-order

---

[9]Residual couplings also arise when magic-angle spinning (MAS) rates or small integer multiples thereof approach the rates of nutation ω_{rf}, as manifestations of rotary resonance recoupling (see "Chemical Shift/Heteronuclear Coupling Correlations" below).

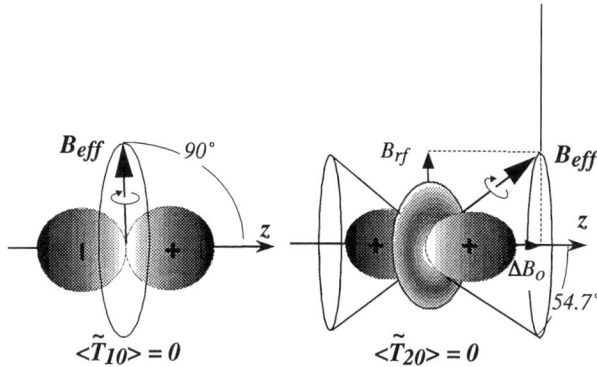

Figure 2 Averaging of first (T_{10})- and second (T_{20})-rank spin-space tensors by continuous rotations around effective fields inclined at $d_{00}^{(\ell)}(\beta_m) = 0$. Prior to irradiation only the z-direction of spin-space is defined (by B_o); hence the axial symmetry.

averaging regime, and high sensitivity to inhomogeneities in ω_{rf}. Multiple-pulse sequences in which the continuous T_{20}^{IS} rotation introduced by LG irradiation is replaced with compensated reorientations separated by windows of free evolution can alleviate all these limitations (21). As in the continuous LG version, the "toggling" motion imposed by these discrete spin-space manipulations on the interaction-frame Hamiltonians needs to fulfill an effective tetrahedral symmetry. This will then average out second- (but not necessarily first-) rank couplings, while overcoming nonidealities stemming from pulse imperfections and higher-order ($\mathcal{H}_{D,CS}^{(2)}, \mathcal{H}_{D,rf}^{(2)}$) interferences. Over the years numerous principles have been developed to meet these ends, and many of them serve as useful general guidelines in the development of solid- and liquid-state pulse sequences (2, 5, 6). One such principle relates to the fact that all even-numbered imperfections in the AHT series may be eliminated by "symmetrizing" the interaction Hamiltonian over the decoupling cycle (42); this in turn entails concatenating into a single "supercycle" period $2\tau_c$ two interaction Hamiltonians fulfilling $\{\tilde{\mathcal{H}}(t)\}_{0 \leq t \leq \tau_c} = \{\tilde{\mathcal{H}}(\tau_c - t)\}_{\tau_c \leq t \leq 2\tau_c}$.[10] Another compensating principle relies on the fact that regardless of the complexity that may characterize higher-order multipulse imperfections, the spin parts of their effective Hamiltonians can still be written as linear combinations of T_{km}'s (44). Given the well-defined rotational properties exhibited by these residual terms with respect to z-axis rotations, repeating the complete decoupling cycle with all pulses shifted by phase $\Delta\phi_j = 2\pi j/N$ will impose phase shifts $T_{km} e^{-im\Delta\phi_j}$ on these imperfections and eventually remove

---

[10]In the LG case τ_c corresponds to $(\omega_{rf}^2 + \Delta\omega_{LG}^2)^{-1/2}$, and the time reversal can be carried out by simultaneous frequency shifts $\Delta\omega_{LG} \leftrightarrow -\Delta\omega_{LG}$ in coordination with 180° phase inversions of the rf. This is the principle of the much more efficient "frequency-shifted LG" (FSLG) experiment (43).

them when summed over a sufficiently large number of cycles N.[11] All these guidelines need to be exercised with care lest they eliminate the desired chemical shift observables together with the imperfections, or end up lasting too long for their AHT premises to remain valid; still, it has been shown that the gains resulting from following them amply overcome their drawbacks.

Averaging First-Order Couplings via Spatial Space Manipulations

Whereas spin-space components vary from coupling to coupling, all first-order spatial anisotropies transform as spherical harmonics of rank-2. In analogy with the LG experiment these can be modulated by imposing on the R_{20}^λ the time dependence that arises upon rotating the sample—now mechanically, in real coordinate space. Spinning at an angle β with respect to B_o then results in

$$R_{20}^\lambda \xrightarrow{\text{spinning}} \tilde{R}_{20}^\lambda(t) = \sum_{m=-2}^{2} d_{m0}^{(2)}(\beta) e^{-im\omega_r t} R_{2m}^\lambda(\Omega), \qquad 17.$$

where Ω is a set of angles transforming the spatial coupling tensor into a reference frame fixed on the spinning rotor. $\tilde{R}_{20}^\lambda(t)$ thus includes four terms oscillating at frequencies $\pm\omega_r, \pm 2\omega_r$ (also expressible as cosines and sines of $\omega_r t, 2\omega_r t$), plus a constant term proportional to $d_{00}^{(2)}(\beta)$. In the $\omega_r \gg \omega_\lambda$ fast-spinning regime the $\pm m\omega_r$ oscillations occur so rapidly that the $\{R_{2m}^\lambda\}_{m\neq 0}$ terms cannot impose a substantial net evolution; the residual is then a constant $R_{20}^\lambda(\Omega)$ identical in form to the static interaction, except for the $d_{00}^{(2)}(\beta)$ scaling. As a function of spinning angle $0° \leq \beta \leq 90°$ this scaling factor sweeps monotonically the $[1, -0.5]$ interval, and for the sake of high resolution its key value is $\beta_m = 54.7°$ magic angle spinning (MAS), for which $d_{00}^{(2)}(\beta_m) = 0$ (45, 46).

Equation 17 entirely describes the modulation imposed by sample spinning on the spatial components of the $\mathcal{H}_\lambda$ interactions, but the actual fate of the spin coherences will depend as well on the type of interaction being averaged (13, 23). Most important is whether the various spin parts of the $\mathcal{H}_\lambda$'s rendered time dependent by the sample rotation commute with one another or not. They do, for instance, when considering shielding, heteronuclear dipolar, or first-order quadrupolar interactions. In these cases the time evolution can be accurately described as

$$U(t) = T \exp\left[-i \int_0^t \mathcal{H}_\lambda(t') dt'\right] = e^{-i\omega_\lambda(t) S_z}, \qquad 18.$$

---

[11] This is an example of second averaging, in which residuals of a partly averaged interaction are further reduced by imposing an additional (slower) time dependence. Decoupling itself, for example, can be viewed as second averaging of the secular residuals left by the B_o truncation.

and then the spins' evolution is as in Equations 13 and 14 except for the fact that couplings ω_λ are no longer constant but have time- and orientation-dependent expressions.[12] The free precession of spins under MAS can then be represented by an ensemble of magnetizations, each one associated with a different single crystallite in the sample and possessing an evolution phase (23, 47)

$$\phi(t) = \omega_{CS}^{iso} t + \sum_\lambda \sum_{m=-2}^{2} \omega_{2m}^\lambda (\Omega) \left[e^{im\omega_r t} - 1 \right] / m\omega_r. \qquad 19.$$

All such "spin packets" in a powder thus begin their evolution in the perfect state of alignment that follows excitation but dephase throughout a rotor period as they become affected by different anisotropic evolution frequencies $\omega_{2m}^\lambda (\Omega)$. At the end of each period T_R, however, when $e^{im\omega_r T_R} = 1$, the cumulative effects of these anisotropic evolutions vanish regardless of crystallite orientation, and all packets meet again at a phase dictated solely by the isotropic shift (Figure 3A). Because of the intervening dephasing between 0 and T_R, MAS signals from inhomogeneously broadened systems such as these usually adopt the form of rotational echo trains, whose spacing and extent of dephasing scales as ω_r^{-1} (Figure 3B). This is reflected in the frequency-domain spectra as sets of sharp spinning sidebands flanking the isotropic centerbands at multiples of $N\omega_r$ [where $N \approx O(\omega_\lambda/\omega_r)$], which can convey valuable information on the anisotropic coupling parameters (48). This unassisted MAS technology finds its widest use in the averaging of dilute spin-½ shielding anisotropies (^{13}C, ^{15}N, ^{31}P) and in the line-narrowing of moderate first-order quadrupole broadening (^{2}H) (4, 34). For moderately symmetric environments and at relatively high fields MAS can also yield considerably sharp ^{14}N resonances and enable the resolution of chemically inequivalent nitrogen sites, though this requires an inordinately high accuracy ($\leq 0.01°$) in the setting of β_m (49, 50).

Different considerations may arise when MAS is used for averaging out couplings that include the homonuclear dipole interaction (6, 23, 51). Actually, a pure $\mathcal{H}_D^{IS}(t) = T_{20}^{IS} \cdot \tilde{R}_{20}^{IS}(t)$ two-spin interaction is analogous to a first-order quadrupole coupling, and as in the latter case spectra will break up into sharp MAS sideband manifolds even when $\omega_r \ll \omega_D$. Isolated pairs of equivalent spins, however, are hardly typical when considering systems such as protons in organic solids, and realistic analyses need to account for the presence of multiple (I-S, I-J, ...) dipole couplings as well as for isotropic and anisotropic shieldings. The spin-space components of these various interactions do not commute among themselves, thereby rendering the overall coupling homogeneous and quenching MAS's averaging effects unless fast ($\gg \omega_D$) spinning rates are employed. Further insight into the effects of sample spinning can be gathered from an AHT expansion of the time-dependent MAS Hamiltonian in powers of ω_r^{-1}, which yields an $\mathcal{H}_D^{(1)}$ that is identically zero and centerband residuals that at high speeds will be dominated

---

[12] This self-commutation corresponds to cases of inhomogeneous broadenings, which in time-independent systems can be distinguished by their susceptibility to spectral hole burning.

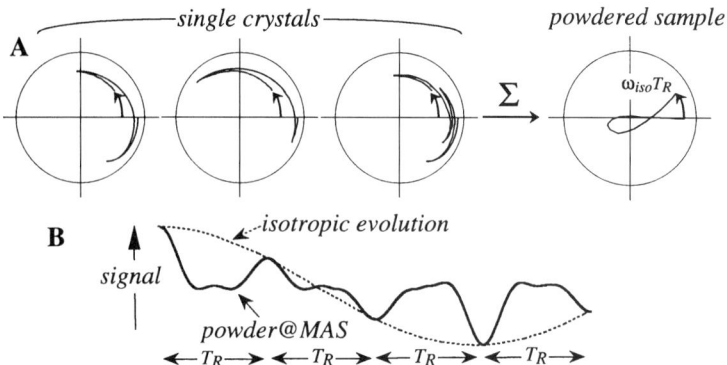

Figure 3 (A) x-y trajectories executed throughout the course of a rotor period by magnetization vectors subject to mutually self-commuting interactions [single-crystal trajectories were progressively contracted for presentation purposes (47)]. (B) Comparison of the powder MAS signals expected in these cases and their purely isotropic counterpart for multiple rotor periods.

by $\mathcal{H}_{D,D}^{(2)} \propto [\mathcal{H}_D^{IS}, \mathcal{H}_D^{IJ}]/\omega_r$ terms. As spinning rates increase, a progressive $\omega_r^{-\alpha}$ ($\alpha = 1$–1.5) scaling is indeed observed experimentally (51, 52), yet this is a fairly shallow ω_r dependence, which suggests that unassisted MAS will only become competitive vis-a-vis multiple pulse at very high (≈ 50–100 kHz) spinning rates (53). There are, however, a number of aids that can endow unassisted MAS with a positive role in the high-resolution solid state NMR of abundant nuclei. One is isotopic dilution, which in combination with currently attainable spinning rates yields narrow ^{1}H lines even at modest levels (54); another may be increasing the magnetic field strength, and spreading the chemical shifts of inequivalent coupled sites until the homogeneous character of their couplings is alleviated.[13]

Manipulating Second-Order Quadrupolar Interactions

As alluded to earlier, second-order effects become particularly relevant when focusing on the central $-\frac{1}{2} \leftrightarrow +\frac{1}{2}$ transitions of half-integer quadrupolar nuclei, unaffected to first-order by quadrupolar couplings thanks to their S_z^2 dependence (Equation 9c) (17, 22). On attempting to remove the residual anisotropic components of these second-order $\mathcal{H}_{Q,Q}^{(2)}$ interactions via spatial manipulations, one is confronted with a factor $d_{00}^{(2)}(\beta)$ that will scale the second-rank broadening R_{20}^Q, as well as with a factor $d_{00}^{(4)}(\beta) = (35\cos^4\beta - 30\cos^2\beta + 3)/8$ that will scale R_{40}^Q. Either of these polynomials can be set to zero at certain spinning angles, yet their roots are not coincident, and therefore no single "magic" axis of rotation will

---

[13] Such a regime has already materialized for the case of ^{19}F NMR, which though abundant, possesses a much wider chemical shift scale than ^{1}H and for which available MAS rates (25–50 kHz) can provide good spectral resolution (55).

simultaneously remove all their associated second-order broadenings. This single-axis spinning deficiency can be overcome by introducing more complex forms of mechanical reorientation: multiple-axes spinning strategies (24, 56, 57). Among these, perhaps conceptually closest to MAS is double-rotation (DOR), in which the quadrupole-containing powdered sample is simultaneously spun around two axes, β_1 and β_2 (58). An extension of the formalism described earlier for MAS (Equation 17) reveals that nonoscillating k-rank anisotropies will be scaled in this case by $d_{00}^{(k)}(\beta_1)d_{00}^{(k)}(\beta_2)$, and therefore all broadenings can be removed if the noncoincident spinning axes are set at the magic angles of the second- and fourth-rank spherical harmonics. An alternative that narrows the central transitions in a technically easier manner consists of consecutively spinning the sample around two different β_1, β_2 axes, each associated with their own evolution times t_1, t_2 (59). The choice of spinning angles in such dynamic-angle-spinning (DAS) experiments is more flexible than in DOR, as all that is demanded is the pointwise cancellation of anisotropies according to

$$d_{00}^{(2)}(\beta_1)t_1 = -d_{00}^{(2)}(\beta_2)t_2; \quad d_{00}^{(4)}(\beta_1)t_1 = -d_{00}^{(4)}(\beta_2)t_2. \qquad 20.$$

At the conclusion of these evolution times a purely isotropic echo forms, and by synchronously increasing the duration of (t_1, t_2) a high resolution signal becomes available. The stepwise nature of this refocusing implies that anisotropies are not instantly removed as in other averaging methods discussed so far but appear, after a two-dimensional (2D) Fourier transformation of $S(t_1, t_2)$, correlated along a sharp ridge for every single-crystallite in the sample. Therefore, unlike MAS, DAS does not bring with its higher resolution an effective increase in signal-to-noise; in fact, signal is lost by virtue of the need for "storing" the evolving coherences along B_o while the spinning axis is reoriented from β_1 to β_2.[14]

A refocusing similar to that carried out by DAS but involving a single axis of sample rotation is feasible if the restriction to central transition observations is lifted (61, 62). Indeed, it follows from the spin energy diagram for half-integer quadrupole nuclei (Figure 4) that not only the central but in fact any $-m \leftrightarrow +m$ multiple-quantum (MQ) transition will be free from the dominant first-order quadrupole broadenings. Yet second-order effects will still influence these transitions. This opens up the possibility of compensating the residual $\mathcal{H}_{Q,Q}^{(2)}$ broadenings affecting the $-\frac{1}{2} \leftrightarrow +\frac{1}{2}$ evolution, with the $\mathcal{H}_{Q,Q}^{(2)}$ anisotropies affecting other symmetric MQ transitions. To evaluate such a possibility it is pertinent to include the transition order m in the description of the second-order frequencies, whose

---

[14] "Storage" is a widespread way of protecting the phase encoded by evolving magnetizations while a relatively slow process like a sample hop is taking place. It involves rotating (with a pulse) spin coherences away from the x-y plane and into the B_o axis, where they can reside for times in the order of T_1 without evolving or losing their original encoding (4, 60). Owing to their low symmetry, storage pulses can only conserve an axial projection of the transverse x-y magnetization, thus decreasing the signal observed upon recall by an average factor of two (plus losses to T_1 relaxation and/or spin diffusion).

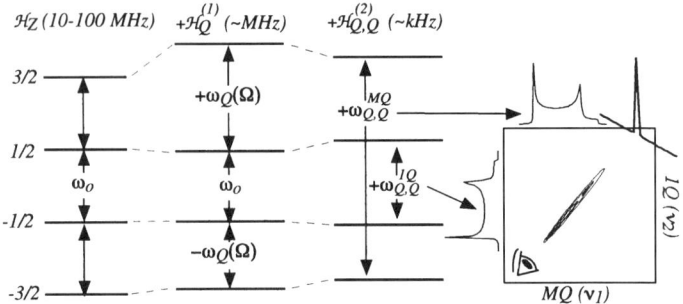

Figure 4 Hierarchical description of Zeeman plus quadrupolar effects on an $S = 3/2$ energy diagram, illustrating how all anisotropies can be removed by correlating under MAS multiple-quantum and central single-quantum transitions within a 2D NMR experiment.

average under rapid sample spinning becomes

$$\omega_{Q,Q}^{-m \leftrightarrow +m}(m, \beta) = C_S^{(0)}(m)\omega_Q^{(0)} + C_S^{(2)}(m)d_{00}^{(2)}(\beta)\omega_Q^{(2)}(\Omega) \\ + C_S^{(4)}(m)d_{00}^{(4)}(\beta)\omega_Q^{(4)}(\Omega),$$

21.

where the $\{\omega_Q^{(k)}\}_{k=0-4}$ denote the zero-, second-, and fourth-rank frequency contributions, and the $\{C_S^{(k)}(m)\}_{k=0-4}$ are polynomials that depend on the spin S and transition order m involved. According to this expression, m imparts on the spin evolution, via the $C_S^{(k)}$-polynomials, an effect similar to that played by β through the $\{d_{00}^{(k)}\}$. Therefore, two analogous routes open up for averaging out second-order $\omega_{Q,Q}$ anisotropies: to keep $m = \frac{1}{2}$ constant and make β time dependent (DAS), or to keep β constant at MAS while making m time dependent through MQ $\leftrightarrow$ 1Q 2D correlations. Experimentally, the latter is a simpler route, whereas from a practical standpoint it has the advantage of concurrently averaging out all remaining shielding and dipolar anisotropies. An important issue in these 2D MQMAS experiments is the optimized manipulation of the MQ excitation and conversion processes; intensive research in this area is being performed (63–72), and sequences that in favorable cases achieve a DAS-like sensitivity have been developed. Also promising is the recent realization that 2D MAS correlations between central and satellite transitions can achieve a similar type of refocusing, even though they involve only single-quantum correlations (73).

CROSS POLARIZATION

Cross Polarization Transfers Between Spin-$\frac{1}{2}$

Collecting solid NMR spectra can be particularly challenging for dilute low-γ nuclei with inherently low sensitivities and long relaxation times (e.g. ^{13}C, ^{15}N, ^{17}O, ^{25}Mg, ^{67}Zn). Throughout the years one technique has proven instrumental for bypassing this signal to noise (S/N) limitation: double-resonance cross polarization (CP) (33, 74). New perspectives have emerged on the effects that fast MAS and

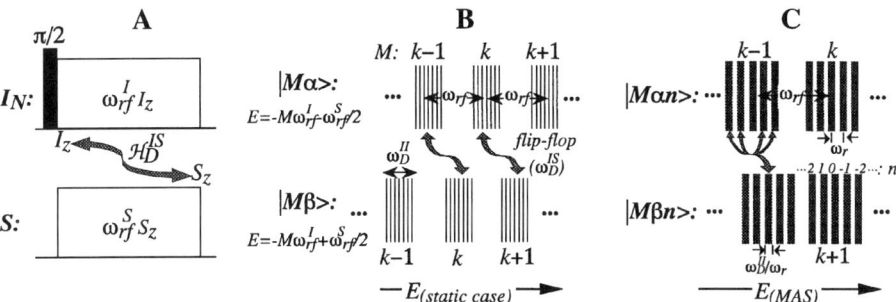

Figure 5 (A) Cross polarization pulse sequence and rf operators in the doubly-tilted rotating frame. (B,C) Ensuing energy-level diagrams for static and fast magic-angle spinning cases. An efficient cross polarization requires the ω_D^{IS}-driven flip-flop effect to be secular, i.e. to have its interconnected energy levels matched up.

quadrupole couplings may have on this sequence, which are worth bringing up in this discussion.

CP transfers polarization from an abundant I- (usually ^{1}H) to a rare S-spin reservoir, thereby increasing the latter's signal by $\sim \gamma_I/\gamma_S$ while making the repetition time of the experiment dependent on the usually much shorter T_1^I time. Two conditions make this $I \rightarrow S$ transfer possible: the generation of a perturbed I state seeking a return to equilibrium via the discharge of its excess polarization, and the establishment of an I-S coupling Hamiltonian that enables this excess to relax primarily into observable S magnetization. In pulsed CP (Figure 5A) the first of these conditions is achieved via spin-locking, a $\pi/2$ pulse followed by a rapid phase shift that places the B_o-equilibrated I magnetization into the x-y plane and parallel to a transverse rf field $B_{rf}^I \ll B_o$. Such an rf field would normally result in the decoupling of I and S reservoirs, but during CP this is prevented by the simultaneous application of a B_{rf}^S field tuned at the same nutation frequency, $\omega_{rf}^S = \omega_{rf}^I$. This imparts on neighboring I- and S-nuclei identical longitudinal oscillation frequencies, making them look, in a suitable interaction frame, like a homonuclear spin pair capable of undergoing back-and-forth transfers of magnetization. Reaching this frame requires a 90° tilt of the I, S quantization directions that places both rf fields along redefined z-axes, followed by a rotating-frame transformation $\exp[-i(\omega_{rf}^I I_z + \omega_{rf}^S S_z)t]$ which truncates chemical shifts, scales homonuclear I-I couplings by $-\frac{1}{2}$, and leaves a heteronuclear dipole Hamiltonian $\mathcal{H}_D^{IS} = \omega_D(I_x S_x + I_y S_y)$ containing a flip-flop exchange character. An I-spin state initially prepared parallel to B_{rf}^I will then transform as (13)

$$\rho_0 = \gamma_I I_z \xrightarrow[\mathcal{H}_D^{IS}]{\omega_{rf}^I = \omega_{rf}^S} \gamma_I \left[I_z \left(\frac{1+\cos \omega_D t}{2} \right) + S_z \left(\frac{1-\cos \omega_D t}{2} \right) + (I_x S_y - I_y S_x) \sin \omega_D t \right]. \qquad 22.$$

These transfer functions evidence that an S_z polarization will grow along B_{rf}^S in an oscillatory fashion that is closely related to the cos $(\omega_D t)$ dipolar signal, and potentially result in a net γ_I/γ_S enhancement. The fine structure of this $I_z \to S_z$ transfer is usually blurred by homonuclear I-I couplings, but it can be observed for certain systems and under suitable conditions (75) (see "Chemical Shift/Heteronuclear Coupling Correlations" below).

CP to dilute $S = \frac{1}{2}$ nuclei is usually carried out in combination with MAS for the sake of line narrowing. These could appear as conflicting procedures because the former is mediated by a heteronuclear coupling that the latter averages out (76). To appreciate why and how polarization can be transferred even under fast MAS it is illustrative to revisit the energy diagram originated by CP in the interaction frame introduced above (Figure 5B) (77): Spacings between the various $\{|IS\rangle = |M\alpha\rangle, |M\beta\rangle\}$ states are here defined by the $\omega_{rf}^I, \omega_{rf}^S$ fields, homonuclear $\{\omega_D^{II}\} < \omega_{rf}^I$ couplings are responsible for a "spread" in these bands, and heteronuclear flip-flop terms enable an exchange and equilibriation of populations between $|M\alpha\rangle \leftrightarrow |(M+1)\beta\rangle$ manifolds.[15] Upon subjecting the sample to MAS—particularly to moderately fast spinning conditions—two distinctive changes will be introduced: $\pm\omega_r$ and $\pm 2\omega_r$ dependencies will be imparted on the heteronuclear couplings, and the width of homonuclear interactions will start scaling as $|\omega_D^{II}|/\omega_r$ (Figure 5C). As has been experimentally observed, the first of these changes modifies the secular transfer condition to $|\omega_{rf}^S - \omega_{rf}^I| = m\omega_r (m = \pm 1, \pm 2)$, whereas the second decreases the energy width of the $|IS\rangle$ manifolds and thus increases the accuracy with which these matching conditions need to be met (76, 78, 79). To deal with these complications a number of simple CP improvements have been proposed, including (a) changes in the rf levels of either I or S irradiation fields to enhance the chances of achieving an effective matching (80), (b) amplitude modulations of the ω_{rf} fields that involve adiabatic passages of the $\{|M\alpha\rangle, |(M+1)\beta\rangle$ manifolds and therefore a more effective exchange of their relative population (81, 82), and (c) repeated inversions in rf field phases in synchrony with reversals in the dipolar couplings (i.e. with the MAS process) (83).[16]

Cross Polarization to Half-Integer Quadrupoles

CP could also be potentially important for enhancing signals in quadrupolar NMR. Features that are relevant in an analytical context are the nature of CP to the sharper

---

[15]$\mathcal{H}_D^{IS}$ also contains double-quantum components that enable $|M\alpha\rangle \leftrightarrow |(M-1)\beta\rangle$ transitions. These terms govern the transfer when magnetization is spin-locked antiparallel to its rf field: S polarization is then generated antiparallel to B_{rf}^S.

[16]Driven by the emergence of sequences that will perform only suitably in the presence of very efficient proton decoupling, interest has also been spurred into finding the condition that minimizes the I-S CP transfer at the conclusion of a π_S pulse: $\omega_{rf}^I/\omega_{rf}^S = (2m+1)(m = 1, 2, \ldots)$ (84). This can place stringent decoupling conditions ($\omega_{rf}^I \geq 150$ kHz), particularly when S pulses much shorter than a rotor period and high ($\geq$8–10 kHz) MAS rates are desired. No similar anti-CP conditions have been found for $\pi/2_S$ pulses.

and easier to observe central transitions of half-integer $S \geq 3/2$ nuclei, and the possibility of combining CP with line-narrowing methodologies such as MAS. Because of the potentially large size of the quadrupolar interaction, its orientation dependence, and the presence of additional energy levels, the $S \geq 3/2$ scenario ends up being quite different from its spin-$1/2$ counterpart (85–87). The most evident difference concerns the nutation frequencies of the various transition: In the common $\omega_Q \gg \omega_{rf}^S$ limit these are $(S + 1/2)$ for the central transitions and $M\omega_{rf}^S(\omega_{rf}^S/\omega_Q)^{m-1}$ for other MQ transitions (with M a coefficient depending on the S and m numbers and ω_Q the Ω-dependent first-order effect).[17] It is this particular set of nutation rates that needs to be matched by ω_{rf}^I in order for the polarization of a particular transition to build up. For instance, when focusing on the central transition of a static powder, CP from a spin-$1/2$ nucleus will occur between tilted spin-locked states

$$\gamma_I I_z \xrightarrow[\mathcal{H}_{IS}]{\omega_{rf}^I = (S+1/2)\omega_{rf}^S} \gamma_I(C_z/2)[1 - \cos(\omega_D t)]/(S + 1/2) + \cdots, \qquad 23.$$

where C denotes the central-transition fictitious spin-$1/2$ operator. This transfer looks similar to the one in Equation 22 except for the fact that only a fraction of I polarization, the portion associated with S's central transition, is actually getting transferred.

When executing MAS, this scenario changes owing to the periodic vanishing of the first-order quadrupole couplings for all crystallites in the powder. Indeed, large $\omega_Q \gg \omega_{rf}^S$ couplings justified neglecting the presence of satellite $m \leftrightarrow m-1$ transitions in the static case, but MAS will now force $\omega_Q(t)$ to vanish either two or four times per rotation period (depending on a single crystallite's orientation). At these zero-crossings ω_{rf}^S brings into contact all the states within the S-spin manifold (Figure 6), and polarization that had been transferred from I_z into C_z may redistribute into other spin populations and/or coherences. The actual fate of the spin-locked C_z will depend on a competition between the strengths of the first-order effect ω_Q separating central from satellite transition peaks, the ω_{rf}^S field recoupling these transitions, and the spinning rate ω_r controlling how long central and satellite transitions stay in contact during the zero-crossings. In fact, the outcome of these MAS-driven ω_Q modulations can be estimated from an analogy to the case of an rf field sweeping through the on-resonance condition of a spin-$1/2$ manifold (Figure 6B); in this scenario an S_z magnetization that was initially along B_o may end up parallel to $-z$ if the rf is swept slowly enough (adiabatic inversion), remain unchanged if the rf sweep is sudden, or begin to follow the effective field but end up in a non-spin-locked state (as coherences rather than populations) if the sweep rate is intermediate. The condition required from the

---

[17]In the opposite and rarely achievable limit $\omega_Q \ll \omega_{rf}^S$ the nutation frequencies for all 1Q transitions are equal and as in a spin-$1/2$ case, whereas for intermediate ranges several frequencies may arise simultaneously. Such complex behavior is the basis for 2D nutation NMR spectroscopy (88).

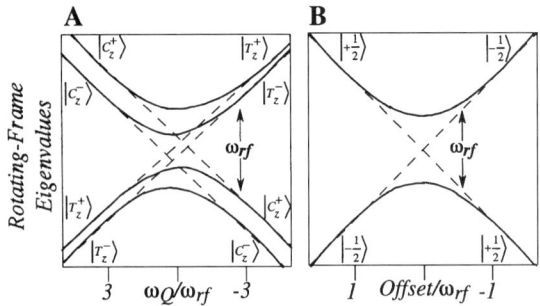

Figure 6 (A) Changes in the rotating-frame eigenvalues of a spin-$3/2$ arising from the oscillations imposed by magic-angle spinning on $\mathcal{H}_Q^{(1)}$. Two or four horizontal sweeps occur every T_R; depending on the rate of these changes states may interconvert (—, adiabatic passages), remain unchanged (---, sudden passages) or end up in non-spin-locked states (i.e. coherences that are not describable by this diagram). (B) Spin-$1/2$ analog of these effects, assumed to be driven by the sweep of an *rf* field through resonance.

changing $\gamma B_o(t)$ field for achieving adiabaticity during such a sweep is derived from the early literature (1): $\gamma \dot{B}_o \ll \omega_{rf}^2$. In the quadrupolar instance the first order $\omega_Q(t)$ takes the role of the $\gamma B_o(t)$ while the MAS-driven $e^{im\omega_r t}$ modulation defines the mechanism of the sweep; the condition for adiabatic transfer thus becomes $\omega_Q(t) \approx \omega_Q \omega_r \ll \omega_{rf}^2$. When this inequality is met each $\omega_Q \to 0$ zero-crossing will be associated with mutual exchanges between the spin-locked C_z populations and outer (e.g. $|3/2\rangle, |-3/2\rangle$) states; if this occurs while polarization is being transferred from I_z to C_z via CP, the net result is an enhancement of both central and MQ populations. On the other, sudden extreme ($\omega_{rf}^2 \ll \omega_Q \omega_r$), polarization transferred to the central transition remains unchanged during the crossing and CP thus proceeds as in the $S = 1/2$ case, except for the MAS- and quadrupole-modified $\omega_{rf}^I = (S + 1/2)\omega_{rf}^S \pm m\omega_r$ matching condition. This sudden-passage condition is easier to satisfy, yet to be truly valid for a majority of crystallites it may demand the use of very weak *rf* fields associated with short relaxation times and inefficient transfers. In many practical cases it is therefore the intermediate $\omega_Q \omega_r \approx \omega_{rf}^2$ regime that is satisfied by a majority of crystallites in a sample; each zero crossing then helps transform the spin-locked populations into single- and multiquantum coherences that rapidly decay and fail to contribute to observable signal, making CP particularly ineffective as a signal-enhancement technique.[18]

---

[18]Because of similar complications, many sophisticated multiple-pulse sequences developed for $S = 1/2$ cases may not be directly applied on half-integer quadrupoles. Alternatively, this rotationally induced dissipation serves as an efficient drain of the *I*-spin reservoir and can be used as an efficient spectral editing mechanism (see "Chemical Shift/Heteronuclear Coupling Correlations" below).

THE SELECTIVE REINTRODUCTION OF SPIN ANISOTROPIES

Avoiding the Penalties of High Resolution

The various manipulations described above may enable the acquisition of high resolution spectra endowed with good S/N, yet they do so at the expense of eliminating an orientation dependence that may otherwise have proven valuable. A practical solution to this information/resolution dichotomy is provided by 2D NMR, which can separate along a high-resolution spectral axis the rich but poorly resolved anisotropic information (4, 9, 13). In fact, some of the experiments described above (DAS, MQMAS) yield, by their very nature, isotropic/anisotropic 2D correlation spectra. Other such experiments that have been realized include pairwise correlations of isotropic and anisotropic shieldings (60, 89, 90), of shifts and dipolar couplings (91, 92), and of shifts and first-order $S=1$ quadrupolar anisotropies (93, 94), as well as higher-dimensional correlations involving various triads of these interactions (95–97). For the sake of maximizing resolution a majority of these experiments encodes the isotropic evolution along the directly detected dimension; the anisotropic evolution that modulates individual peaks can then be extracted either by analyzing $S(t_1, \omega_2)$ time-domain functions or via a second transformation along the anisotropic domain. The two procedures are obviously related, but the former is usually preferred when trying to assess relatively small interactions such as the dipolar coupling between distant spins.

By virtue of the similar $\tilde{R}_{20}^{\lambda}(t)T_{k0}^{\lambda}$ dependencies that characterize all first-order anisotropies subject to sample spinning, there are certain common ways for reintroducing these couplings along an indirect t_1 domain. Conceptually the simplest is probably to spin the sample off-MAS (60), as then anisotropies are reintroduced with a scaling $d_{00}^{(2)}(\beta) \neq 0$.[19] This has been exploited in various applications, even if practical and S/N complications arise from the need to introduce a storage period in between t_1 and t_2 for the sake of rapid reorientation to (and subsequently during the relaxation delay, from) the magic angle. Such demands can be alleviated by replacing the correlated dynamic-angle evolution with a set of conventional variable-angle sample spinning acquisitions, followed by a simple interpolation of the data (99). An alternative to these fast-spinning spatial modulations is imposed by the magic-angle hopping and turning experiments (90, 100), which on the basis of repetitive storages on a slowly reorienting sample make $\langle \tilde{R}_{20}^{\lambda} \rangle = 0$ during the t_1 domain and correlate the ensuing isotropic evolution with a static-like anisotropic lineshape along ω_2.

The demands of these experiments for specialized instrumentation have stimulated the search for anisotropy-recoupling protocols that employ constant MAS

---

[19]Because the static spins' evolution is equivalent to that of a powder rotating at $\beta = 0°$, a related procedure is to stop the sample MAS altogether during t_1; this is the basis for the "stop-and-go" 2D NMR experiment (98).

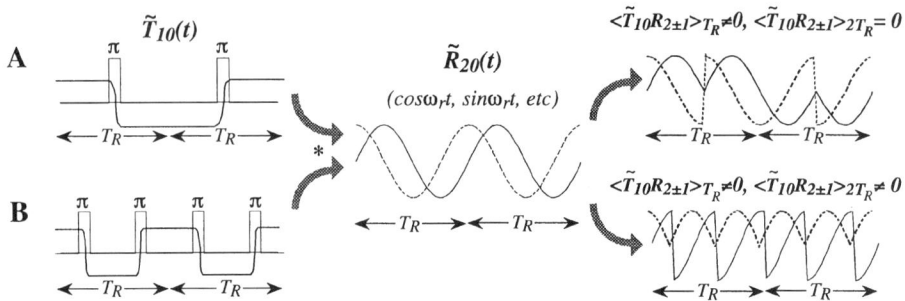

Figure 7 Reintroduction of the anisotropic interactions resulting from the synchronous modulation of spin ($\tilde{T}$) and spatial ($\tilde{R}$) terms in the coupling Hamiltonian. The cumulative behavior expected when using (A) one π-pulse/T_R and (B) two π-pulses/T_R is shown. Similar arguments can be extended to other coupling ranks and orders.

but spoil this averaging during t_1 via rotor-synchronized spin-space manipulations (89, 101). Making $\langle \tilde{R}_{20}^\lambda(t_1)\tilde{T}_{k0}^\lambda(t_1)\rangle_{\text{MAS}} \neq 0$ is best visualized for the simplest case of an isolated spin subject to its local ω_{CS}^{aniso}. As described in Figure 3, MAS will refocus this anisotropy at every multiple of T_R, yet this averaging can be interrupted by the application of a π-pulse during the course of the rotor period (47). Magnetizations from different crystallites will then be taken away from their MAS trajectories, failing to refocus at $t = T_R$ and bringing about a signal decay that depends on the site's anisotropy.[20] Such dephasing can be understood as arising from a destructive interference between the spatial $\tilde{R}_{20}(t)$ and spin $\tilde{T}_{10}(t)$ terms of Hamiltonian (Figure 7), with the latter becoming time-dependent ($S_z \rightarrow -S_z$) in an interaction representation imposed by the π-pulse. A simple single-pulse/T_R approach cannot serve as the basis of pulse sequences that accumulate an anisotropic dephasing over evolution times $t_1 > T_R$, as the application of a second π-pulse will undo the first pulse's effect (Figure 7A). Two π-pulses per T_R, on the other hand, can serve as useful rotor-synchronized anisotropic dephasing blocks (Figure 7B), even if the nature of the dephasing depends on the exact location of the pulses (103). Maximum dephasing will occur if these are spaced $T_R/2$ intervals apart, and the ideality of these experiments usually benefits from short ($\ll T_R$) pulse widths and from an *XY*-type of phase cycling (104). The lineshapes that result from Fourier analyzing these decaying signals carry anisotropic information but do not resemble static-like powder patterns; at least four π-pulses per rotor cycle are needed for obtaining such patterns within a measurable scaling factor (105).

---

[20] Effects similar to those introduced by π-pulses can be achieved by "freezing" the evolution of the spins during a fraction of T_R, using for instance, pairs of rotor-synchronized back-to-back $(2m\pi)_\phi(2m\pi)_{-\phi}$ nutation pulses (102).

Chemical Shift/Heteronuclear Coupling Correlations

As mentioned earlier, dipole-dipole anisotropies provide a convenient means for measuring internuclear distances. This task can be facilitated by the concurrent elimination of complicating local effects such as chemical-shift anisotropies, calling again for pulse sequences that will selectively preserve only one kind of interaction. A selective reintroduction of dipolar couplings is particularly straightforward for isolated heteronuclear spin-$\frac{1}{2}$ pairs via the spin-echo double resonance (SEDOR) experiment (11, 106)

$$\pi/2_S - t_1/2 - (\pi_I, \pi_S) - t_1/2 - \text{observe } S(t_1). \quad 24.$$

When ideal π-pulses are involved the resulting $S(t_1)$ signals are solely a function of spin-relaxation and of the I-S dipolar coupling; normalization by $S_o(t_1)$ signals acquired in the absence of π_I pulses then leads to "universal" $\langle S/S_o \rangle(t_1)$ dephasing curves, depending solely on the dipolar couplings and from which I-S distances can be extracted.[21] Alternatively, SEDOR can be implemented as a full-fledged 2D experiment (in unison with homonuclear I decoupling if I-I couplings are a complication), resulting in "separate-local-field" signals $S(t_1, t_2)$ whose transforms correlate S's dipolar and shielding anisotropies (107, 108). The nature of multiple I-S couplings (e.g. in I_mS systems) can also be analyzed via SEDOR if the π_I-pulse is replaced by a variable θ_I irradiation (109), and changes to the basic sequence may also enable its extension to quadrupolar nuclei [e.g. replacing the π_S-refocusing pulse by $\pi/2_S$ when $S = 1$ (110)]. Yet the most widespread SEDOR modifications are probably those introduced in an effort to merge this dipolar protocol with MAS as a means of enhancing the resolution and sensitivity of the S spectrum (111–114); we briefly turn to describing these experiments.

Perhaps the simplest dipolar/MAS conflict to resolve arises when trying to determine large spin-spin couplings such as those occurring in directly bonded ^{1}H-S systems; enough dipolar dephasing then occurs within one rotor period (or equivalently, enough intensity remains in the spinning sidebands of the separate-local-field spectrum) to require only minor rotor-synchronization modifications on the SEDOR protocol (115–117). The situation is different when trying to quantify I-S couplings between nonbonded and/or low-γ nuclei; here it may still be possible to rely on SEDOR-like sequences, provided that $\omega_D^{IS}(t)$ is actively reintroduced during t_1 periods extending beyond a single T_R. Because heteronuclear dipolar couplings transform as shielding anisotropies, any of the spatial and spin manipulation strategies discussed in the previous subsection of this article could be exploited for such ends. Most widespread among these are those variants relying on the application of two synchronized π-pulses per T_R, collectively known as rotational-echo double-resonance (REDOR; Figure 8) (111, 118). REDOR variants

---

[21]Carrying out such normalization has the important consequence of canceling to a large extent both residual local- as well as T_2-dephasings, enabling the estimation of dipolar couplings even when comparable or smaller than these effects.

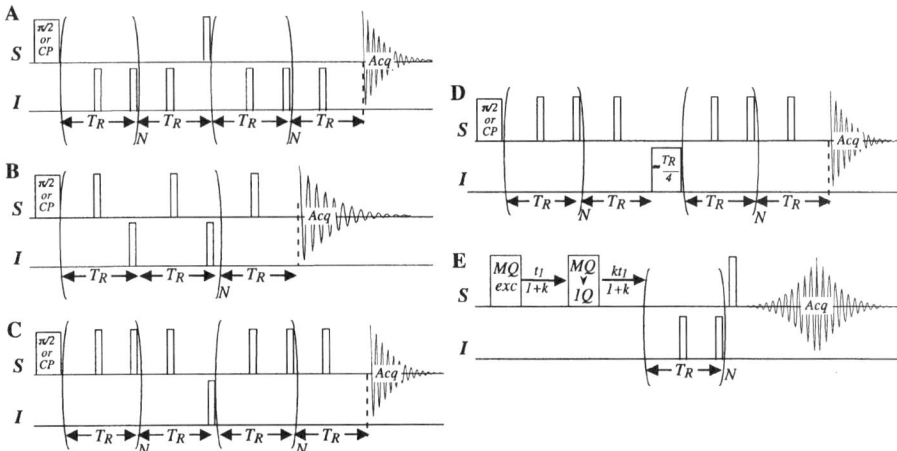

Figure 8 Recoupling alternatives for heteronuclear spin-$1/2$/spin-$1/2$ (A–C) or spin-$1/2$/quadrupole (D, E) distance measurements. (A–C) rotational-echo double-resonance (REDOR) variants; (D) rotational echo by adiabatic passage double resonance ($I \geq 1$); (E) MQMAS REDOR combination ($S \geq 3/2$). Unlabeled rectangles denote π-pulses whose phases are usually cycled to remove nonidealities.

arise from the fact that dipolar couplings transform as I_zS_z, and therefore dephasing can be achieved by placing all pulses at the I frequency, the S frequency, or alternating between the two. Placing all dephasing pulses on the I spins (Figure 8A) has the bonus of never incurring in a ω_{CS}^{aniso} dephasing of the S signal, yet it has been observed that for $I = S = 1/2$ the highest distance accuracies are generally achieved when alternating the dephasing pulses over the two channels (Figure 8B) (119). Because the effective S irradiation involved in this case is only one π-pulse per T_R, both isotropic and anisotropic S shieldings refocus after every other rotor period, making this the basic unit of t_1-incrementation.[22] Extracting long-range distances between heteronuclear spin-$1/2$ pairs also requires collecting a reference set S_o in the absence of π_I pulses, which is then employed for calculating $\langle S/S_o \rangle$ dephasing or $\langle (S_o - S)/S_o \rangle$ build-up curves. In principle, these depend solely on the dimensionless parameter $\lambda_D = \omega_D t_1 = m\omega_D T_R$, and therefore a single λ_D measurement could yield the desired distances. In practice, however, particularly if incomplete isotopic labeling might be involved, accuracy is increased by fitting whole portions of these curves. Direct numerical transforms are also available for extracting one or multiple ω_D values when these are sizable and the quality of the dephasing data is good (120).

A quantitative analysis on the effects of REDOR multiple-pulse trains becomes more challenging if one of the coupled species (I) is quadrupolar. Simplifying

---

[22] At least conceptually, practical increments are usually larger owing to the benefits resulting from XY-type phase cyclings of the I, S pulse trains.

conditions may arise when $\omega_{rf}^I > \omega_Q^I$ (potentially achievable for $I = {}^2$H) (121) or when it can be assumed that only the I central transition has been manipulated, yet even in these cases it is advisable to apply the smallest possible number of pulses on the quadrupolar nucleus (Figure 8C). When the single $I_z \to -I_z$ reversal that then remains cannot be ensured by a π_I pulse, it may be preferable to forgo this scheme altogether and achieve a redistribution of I_z populations by exploiting the zero-crossing phenomena discussed above in relation to quadrupolar cross polarization MAS (122, 123). The rotational echo by adiabatic passage double resonance (REAPDOR) sequence has been derived on these principles (124); it employs an I-irradiation period placed at the center of the sequence and is timed so that a majority of crystallites in the sample experience a zero-crossing through resonance but are unlikely to have undergone two such exchanges (Figure 8D). A quantitative analysis of the ensuing S-decay still requires explicit spin propagations as well as knowledge of I's quadrupolar tensor parameters, even if general "universal-like" curves may be proposed (125). S/N permitting, an alternative that enables a reversal to the simple REDOR-like dephasing analysis arises if the quadrupole nucleus is made the target of observation (Figure 8E). Experiments of this kind involving a MQMAS-driven refocusing of the quadrupolar anisotropies in combination with $I = \frac{1}{2}$ dephasing pulses have been demonstrated and shown to be amenable to interpretations involving solely the I-S dipolar interactions (126, 127).

The dipolar recoupling principles underlying REDOR can also find a role in the spectral assignment of complex systems. For instance, transferred-echo double-resonance (TEDOR), a dipole-based coherence transfer experiment applicable to both organic and inorganic systems (128, 129), can be used for simplifying spectra or assigning their resonances. Solution-like 2D heteronuclear correlations between ^{13}C/^{15}N and ^{13}C/^{2}H in isotope-labeled polycrystalline proteins have also been implemented on the basis of this protocol (Figure 9) (130, 131).

Though simple and efficient, rotor synchronized π-pulses are but one way of precluding the MAS averaging of heteronuclear dipolar couplings: given the $\omega_D(t)I_zS_z = \tilde{R}_{20}^D(t)T_{10}^S T_{10}^I$ form of this interaction, any manipulation that makes T_{10}^λ the periodically time dependent and can interact destructively with the $R_{2m}e^{im\omega_r t}$ terms in $\tilde{R}_{20}^D(t)$ will result in a recoupling. An example of this is rotary resonance recoupling (R^3) (132), which continuously irradiates I spins with a nutation frequency $\omega_{rf}^I = \omega_r$ or $2\omega_r$ and thus achieves $\langle\mathcal{H}_D(t)\rangle_{\text{MAS}} \neq 0$ (see Equation 16).[23] Conceptually similar but more general forms of heteronuclear recoupling involving simultaneous frequency and amplitude modulations (SFAM) of the double-resonance (134), phase-cycled rotor-synchronized cycles (135), and nearly arbitrary forms of phase- and amplitude-modulated irradiation (136), have also been recently demonstrated.

---

[23] The effective $\mathcal{H}^{(1)}$ Hamiltonian also ends up dependent on I's shielding anisotropy parameters, a complication that to some extent can be compensated by alternating the phase of the rf irradiation (133). R^3 also occurs in an increasingly weaker fashion if $(\omega_{rf}^I = m\omega_r)_{m \geq 3}$, by virtue of higher order AHT terms.

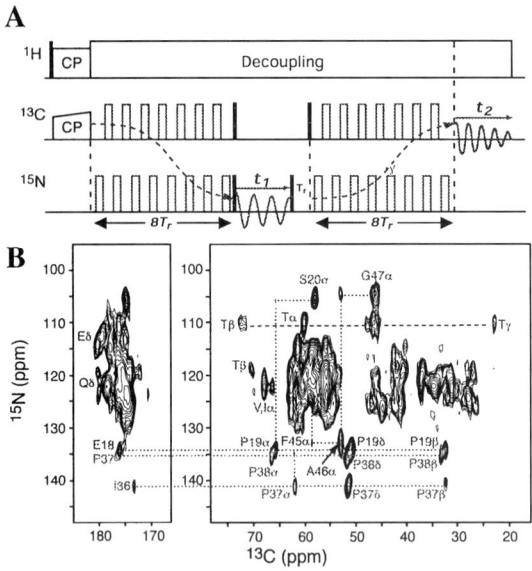

Figure 9 2D ^{15}N-^{13}C heteronuclear correlation experiment on ^{13}C-selectively/^{15}N-uniformly enriched ubiquitin. (A) Pulse sequence based on back-and-forth transferred-echo double-resonance coherence transfers, with thick lines and rectangles denoting $\pi/2$ and π pulses respectively. (B) Partial assignment of the resolved $^{13}C_\alpha$-^{15}N and ^{13}CO-^{15}N resonances. (Adapted from 169)

SEDOR is not the sole starting point for investigating heteronuclear couplings; the CP dynamics in Equation 22 in combination with homonuclear (frequency-shifted LG) decoupling and repetitive inversions of the I and S spin-temperature, also open up opportunities for accurate measurements of large coupling constants and serve as basis for the polarization-inversion with spin-exchange at the magic-angle (PISEMA) sequence (Figure 10) (137, 138). An alternative to these coherent I-S forms of recoupling is also offered by variants of the 2D exchange NMR technique in which the S magnetization is allowed to dephase under I's dipolar field, stored over times $\tau \gg T_1^I$, and subsequently recalled and refocused into a stimulated echo (139). Random fluctuations in I's spin state will lead to a dipole-encoding echo attenuation without having to irradiate the I spins, which for $I > \frac{1}{2}$ cases can be made independent of quadrupole parameters (97).

Chemical Shift/Homonuclear Coupling Correlations

Equally important to the determination of molecular structures can be the measurement of distances between homonuclear I-S pairs under the presence of MAS. Yet the overall SEDOR strategy discussed in the preceding paragraph is complicated in these systems by at least two factors: the more complex nature of the dipolar

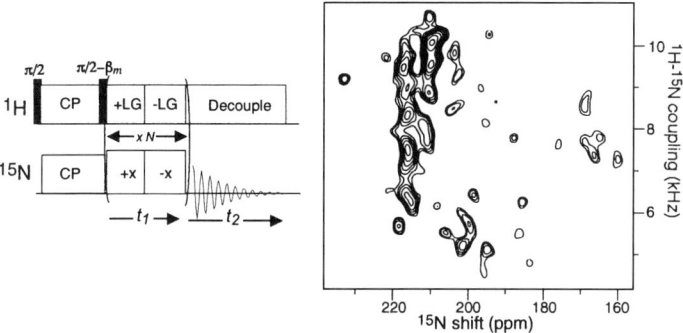

Figure 10 2D local-field polarization-inversion with spin-exchange at the magic-angle (PISEMA) spectrum of static ^{15}N-labeled fd bacteriophage viruses ($M_\omega \approx 16 \cdot 10^3$ kD) magnetically oriented in B_o at 60°C; dipolar couplings are encoded by the combined frequency-shifted LG/CP dephasing shown in the sequence. (Adapted from 170)

Hamiltonian (containing flip-flop terms) and the usual impossibility of manipulating the various coupled sites in the system independently from one another. These factors combine to make the net dipolar effects dependent on the spins' chemical shielding parameters, tensors that are interesting themselves but not necessarily known or being sought when looking for structural information. These factors also complicate the quantification of SEDOR-type S/S_o curves reflecting the spins' decay owing exclusively to dipolar effects, thus restricting the accuracy with which homonuclear distances can be evaluated. In view of these challenges, it is not surprising to encounter a more fluid scenario here than in the heteronuclear recoupling case (112–114, 140). This section provides a brief overview of some of its avenues.

The selective dephasing and rephasing π-pulses on which SEDOR and its daughter techniques rely are not directly applicable to homonuclear systems, as these will simultaneously affect the I and S spins and thereby have no effect on their mutual coupling. Still, a complete refocusing of the homonuclear evolution is possible provided that the heteronuclear π_I, π_S combination is replaced by a $\pi/2$ rotation; this is the principle of the solid-echo sequence (11, 141)

$$(\pi/2)_x - t_1/2 - (\pi/2)_y - t_1/2 - \text{observe } I + S \qquad 25.$$

which is strictly valid only when the two coupled spins are magnetically equivalent (no chemical shift differences). In analogy with the $T_{10} \rightarrow -T_{10}$ effects ascribed to π-pulses (Figure 7), the central $\pi/2$ pulse in this sequence can be thought of as having reversed the net effect of the homonuclear dipolar evolution. Consequently, replacing the π-driven modulation of REDOR with a similar train of rotor-synchronized $\pi/2$ pulses introduces a time dependence of the homonuclear dipolar coupling that prevents its spatial refocusing by MAS. This constitutes the basis for dipolar recovery at the magic angle (DRAMA; Figure 11A) (142, 143), a

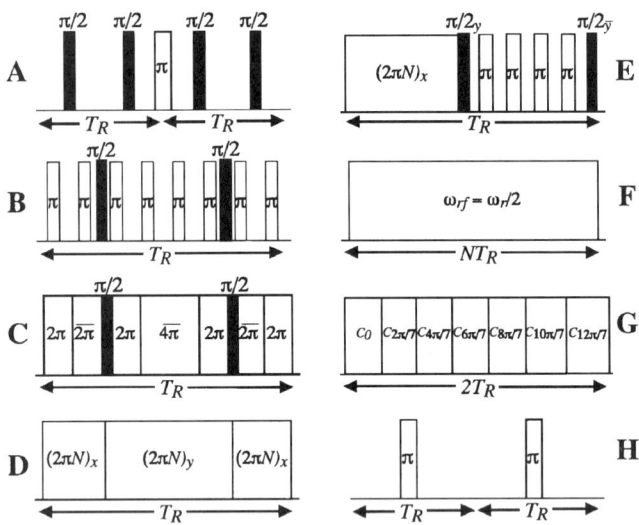

Figure 11 Alternatives for the rotor-synchronized, *rf*-driven recoupling of homonuclear spin-pairs under magic-angle spinning. In practical experiments these rotor-synchronized building blocks are usually further cycled to remove potential shielding/*rf* nonidealities.

sequence whose operation can best be visualized in a toggling frame in which the dipolar Hamiltonian oscillates between $\widetilde{\mathcal{H}}_{zz}^{IS}$ and $\widetilde{\mathcal{H}}_{yy}^{IS}$ every half rotor period.[24] DRAMA exhibits a high recoupling (i.e. fast dephasing) efficiency (140), yet its reliance on a purely dipolar scenario makes it sensitive to the I and S chemical shielding parameters. A number of sequences thus employ DRAMA as a basic recoupling scheme but tailor it to attenuate its dependence on chemical shifts. One such example is XY8-DRAMA, which introduces extensively phase-cycled π-pulses in between the $\pi/2$ nutations in order to refocus the chemical shielding effectively (Figure 11*B*) (144). Another alternative is the dipolar recovery with a windowless sequence (DRAWS) (Figure 11*C*) (145, 146), which replaces DRAMA's periods of free evolution with intervals of forced *rf* spin precessions built around 2π-pulses. Melding of spin-locking and DRAMA (MELODRAMA) (Figure 11*D*) is another offset-compensated variant that bypasses DRAMA's $\pi/2$ pulses altogether, preventing instead the MAS averaging by toggling $\widetilde{\mathcal{H}}_D^{IS}(t)$ between the two (*x*-*y*) rotating-frame transverse axes (147). Yet additional hybrid variations include switching the homonuclear evolution from the rotating to laboratory (R/L) frames every half rotor period, with offsets being compensated during

---

[24] A time dependence of $\widetilde{\mathcal{H}}_{\alpha\alpha}^{IS}(t) = R_{20}^{IS} T_{20}^{IS}(\alpha)$, $T_{20}^{IS}(\alpha) = 3I_\alpha S_\alpha - I \cdot S$, is exploited in the dipolar-averaging condition of all static-sample multiple-pulse sequences: $T_{20}^{IS}(x) + T_{20}^{IS}(y) + T_{20}^{IS}(z) = 0$ (2, 5). In DRAMA, however, these changes are made synchronous with MAS and therefore $\langle R_{20}^{IS}(t) T_{20}^{IS}(t) \rangle_{\text{MAS}} \neq 0$.

the first of these intervals by continuous irradiation and during the second via trains of π-pulses (Figure 11E) (148, 149).

Heteronuclear recoupling alternatives other than REDOR can also be extended to the case of homonuclear spin pairs. One such opportunity is opened by R^3, whose homonuclear rotary resonance (HORROR) variant exploits the fact that both nuclei are now being irradiated in order to modify the recoupling condition to $\omega_{rf} = \omega_r/2$ (Figure 11F) (150). Because all spatial $\tilde{R}_{20}(t)$ components oscillate at $(\pm\omega_r, \pm 2\omega_r)$, such rf-driven nutation rates are in principle too slow for recoupling single-spin interactions like ω_{CS}^{aniso}, but fast enough for making $\langle \tilde{R}_{20}^D(t)\tilde{T}_{20}^{IS}(t)\rangle_{MAS} \neq 0$.[25] An attractive feature of this approach is its dependence on the powder angles (β, γ) defining the orientation of the internuclear I-S vector in the rotor frame: Whereas in DRAMA derivatives these angles scale the effective recoupling as products of trigonometric functions (e.g. $\sin 2\beta \cos \gamma$), a phase-encoded dependence on the γ-angle ($\sin 2\beta e^{i\gamma}$) appears in HORROR. Consequently, when considered over a powdered sample, HORROR gives a more readily detectable decay of the recoupled spins' coherences.

Related to this continuous rf-driven homonuclear MAS recoupling but more immune to rf imperfections is the CN_n family of sequences (Figure 11G), which recouples the MAS-modulated Hamiltonian by concatenating N phase–shifted rf pulses throughout n consecutive rotor periods (151). This pulsing imparts a controlled time dependence on the spin-space components of the coupling and allows one to select, at least to first-order in AHT, particular $R_{2m}^D T_{2\mu}^{IS}$ combinations that are unique to the homonuclear I-S coupling. One of the shortest such solutions involves $N = 7$ pulses, $n = 2$ rotor periods, and rf phases ϕ_{rf} incremented by $2\pi/N$; only purely dipolar $R_{2\pm 1}T_{2\pm 2}$ terms survive such incrementation, and an effective zero-order Hamiltonian results that is analogous to the HORROR one except for a smaller scaling factor.[26] Further improvements on C7's shielding independence have been demonstrated by refining the actual pulses used to define each phase-shifted propagator (POST-C7) (152) or via supercycling combinations (CMR7) (153).

Either the DRAMA or the HORROR/C7 derivations can be thought of as having analogues in the heteronuclear scenario. Contrasting to this is rotational resonance (R^2), which prevents the MAS averaging of the I-S dipolar interaction simply by setting the spinning rate at an integer fraction of the isotropic chemical shift difference between the sites: $|\omega_{CS}^I - \omega_{CS}^S| = m \cdot \omega_r$ (154–156). How and why this condition achieves recoupling can be appreciated when factorizing the total two-spin Hamiltonian into a double-quantum contribution acting on the $\{|\alpha\alpha\rangle, |\beta\beta\rangle\}$ subspace of $\{|IS\rangle\}$, and a commuting zero-quantum Hamiltonian acting on $\{|\alpha\beta\rangle, |\beta\alpha\rangle\}$

---

[25] In practice this selective dipolar recoupling may be significantly affected by the chemical-shift offsets of the coupled sites and, as in many windowless sequences, by rf inhomogeneity.

[26] This time dependence $T_{20} \xrightarrow{C7} \Sigma_{\mu=-2}^{2} d_{\mu 0}^{(2)}(\beta_{rf})e^{-i\mu\phi_{rf}}T_{2\mu}$ is to be compared with $T_{20} \xrightarrow{HORROR} \Sigma_{\mu=-2}^{2} d_{\mu 0}^{(2)}(\frac{\pi}{2})e^{-i\mu\omega_{rf}t}T_{2\mu}$; in either case recoupling occurs by interference with the $\tilde{R}_{20}^D(t)$ in Equation 17, but the former scheme is more robust.

(112, 156, 157). A definition of operators

$$I_z^\Delta = \frac{1}{2}(|\alpha\beta\rangle\langle\alpha\beta| - |\beta\alpha\rangle\langle\beta\alpha|) = \frac{1}{2}(I_z - S_z);$$

$$I_x^\Delta = \frac{1}{2}(|\alpha\beta\rangle\langle\alpha\beta| - |\beta\alpha\rangle\langle\alpha\beta|) = \frac{1}{2}(I_+S_- + I_-S_+)$$

26.

enables one to express the latter contribution as a 2 × 2 irradiation-like Hamiltonian $\mathcal{H}_\Delta = \omega_\Delta I_z^\Delta + \omega_1 I_x^\Delta$, where the offset $\omega_\Delta(t)$ reflects the instantaneous difference between I and S chemical shifts, and $\omega_1 = \omega_D(t)$ is the time-dependent dipolar coupling. Upon MAS these longitudinal and transverse components average to the isotropic difference ω_Δ^{iso} and zero, respectively, thereby leading to a time evolution that is free from dipolar effects. Yet when any of the time dependencies modulating $\omega_D(t)$ match the average chemical shift difference ($\omega_\Delta^{iso} = \omega_r$ or $2\omega_r$) a resonance condition occurs within this 2 × 2 subspace, not unlike the one observed when the laboratory frame irradiation frequency of an *rf* field matches the spins' Larmor frequency. The evolution of an I^Δ vector under this rotational resonance condition will then reflect the strength of *I-S* dipolar couplings and enable their measurement.[27] Still, accurate long-range distance determinations by R^2 require additional a priori knowledge of the remaining parameters that affect the subspace evolution, including the I and S shielding tensors [which influence the instantaneous value of $\omega_\Delta(t)$] and the relaxation times of the zero-quantum vector. Methods have been proposed for independently measuring these quantities (158) as well as for alleviating the narrowness of the R^2 matching condition (159, 160). Furthermore, thanks to its relative simplicity, R^2 is one of the few recoupling mechanisms directly applicable to quadrupole nuclei (161).

By its very nature R^2 is a highly selective method for measuring homonuclear distances. An *rf*-driven, broadband alternative to the R^2 effect is offered by the simple excitation for the dephasing of rotational-echo amplitudes (SEDRA) protocol (162, 163), which achieves a broadband dipolar recoupling by applying one π-pulse per T_R (Figure 11H). SEDRA's dephasing principles can be gathered from considering the effects that its pulse train will have on the various *I,S* interactions within their zero-quantum subspace: The spin part of the homonuclear coupling will remain unaffected, the shielding anisotropy will dephase every odd but refocus every even rotor period (and thus it is to even multiples of T_R that dephasing increments will end up circumscribed), while isotropic shieldings will undergo periodic spin-echo time reversals. Such square-wave modulation of $\omega_\Delta I_z^\Delta$ every T_R can be viewed as effectively splitting the shift difference spectrum into a series of harmonics positioned at multiples of ω_r, and thus being susceptible to undergo R^2-type recoupling with $\omega_D(t)$ regardless of the sites' isotropic shift values. It also follows that in contrast to HORROR and DRAMA, SEDRA's dephasing is biased towards systems with sizeable ω_Δ^{iso} values, on the order of ω_r (140).

---

[27]ω_D will also influence I^Δ's evolution when both sites have identical isotropic shifts ($m = 0$ R^2) or for $m \geq 3$ conditions, via weaker higher-order effects.

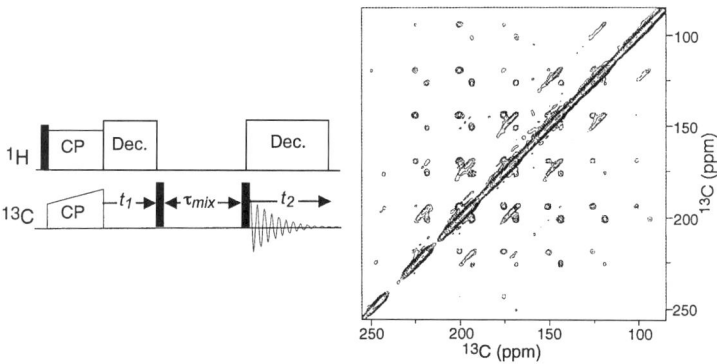

Figure 12 2D homonuclear ^{13}C correlation spectrum of a 24-residue HIV peptide, enriched at adjacent positions P320 and Gly321 and bound to a monoclonal antibody ($M_\omega >$ 50 kD); cross-peak intensities reveal the peptide's local conformation upon binding. Data were acquired using the spin-diffusion driven sequence on the left while executing MAS of a frozen solution at $-120°$C. (Adapted from 171)

As in the heteronuclear case, all these homonuclear recoupling sequences can be used to either introduce a dephasing that is directly monitored along a t_1 dimension or to activate couplings during the mixing periods of 2D chemical shift/chemical shift correlation experiment. Thanks to its simplicity and broad-bandedness, a SEDRA-derived scheme dubbed *rf*-driven recoupling (RFDR) has found the widest use for establishing this type of connectivity (113, 164), even if quantitative RFDR interpretations may be far from trivial when dealing with multiply-coupled networks (165). A peculiar feature of solid state homonuclear 2D correlations is that, given sufficiently long mixing times and often with the aid of a coupled proton network, dipolar-driven cross peaks between inequivalent sites may arise even in the absence of active *I-S* recoupling. These correlations are generated by spin-diffusion (166) and reflect the activity of flip-flop terms that have not been entirely truncated by MAS. The dynamics of these processes are slow (~Hz) and not always amenable to quantitative kinetic analyses, yet in the complete exchange regime their resulting 2D lineshapes are featured and convey a clear picture of the relative geometry between the coupled sites (Figure 12) (167).

CLOSING REMARKS

Although limited in scope, it is hoped that the material summarized above conveys some of the progress that during recent years has characterized solid state NMR. Particularly encouraging has been the gradual inclusion of quadrupolar nuclei and of multiple-resonance distance determination techniques into the mainstream of experiments, as these have helped extend the frontiers of NMR as a spectroscopy while greatly improving its potential for the analysis of complex solids. Notwithstanding this progress, important issues remain to be addressed in the field, including the development of new and simpler high resolution ^{1}H NMR protocols that can be incorporated into routine multidimensional experiments [an area that has already witnessed numerous advances (168)], additional heteronuclear decoupling improvements for furthering resolution, new signal enhancement approaches applicable to quadrupolar nuclei, and the reliable quantification of internuclear distances and angular constraints in multiply-labeled $I, S \geq \frac{1}{2}$ systems. In view of the proven track record of breakthroughs and achievements in solids NMR, it is not so much a matter of if but of when and of how such targets will be achieved.

ACKNOWLEDGMENTS

It is a pleasure to acknowledge my many UIC-based co-workers for their insightful lessons and comments, as well as Ms. Rhonda Staudohar for her patient typing of manuscripts. This work was supported by the US National Science Foundation and Department of Energy, as well as by the Beckman, Dreyfus, University of Illinois, and Sloan Foundations.

Visit the Annual Reviews home page at www.AnnualReviews.org

LITERATURE CITED

1. Abragam A. 1961. *Principles of Nuclear Magnetism*. New York: Oxford Univ. Press
2. Haeberlen U. 1976. In *Advances in Magnetic Resonance*, ed. JS. Waugh, Suppl 1. New York: Academic
3. Spiess HW. 1978. In *NMR Basic Principles and Progress*, ed. P Diehl, E Fluck, R Kosfeld, 15:55–214. New York: Springer-Verlag
4. Fyfe CA. 1983. *Solid State NMR for Chemists*. Ontario: CFC Press
5. Mehring M. 1983. *High Resolution NMR in Solids*. Berlin: Springer-Verlag
6. Gerstein BC, Dybowski C. 1985. *Transient Techniques in NMR of Solids: An Introduction to Theory and Practice*. Orlando, FL: Academic
7. Derome AE. 1987. *Modern NMR Techniques for Chemistry Research*. Oxford: Pergamon
8. Munowitz MG. 1987. *Coherence and NMR*. New York: Wiley & Sons
9. Ernst RR, Bodenhausen G, Wokaun A. 1987. *Principles of Nuclear Magnetic Resonance in One and Two Dimensions*. Oxford: Clarendon
10. Goldman M. 1988. *Quantum Description of High Resolution NMR in Liquids*. New York: Oxford Univ. Press
11. Slichter CP. 1990. *Principles of Nuclear Magnetic Resonance*. New York: Springer-Verlag
12. McBrierty VJ, Packer KJ. 1993. *Nuclear Magnetic Resonance in Solid Polymers*. Cambridge, UK: Cambridge Univ. Press
13. Schmidt-Rohr K, Spiess HW. 1994. *Multidimensional Solid-State NMR and Polymers*. London: Academic
14. Stejskal EO, Memory JD. 1994. *High Resolution NMR in the Solid State: Fundamentals of CPMAS*. New York: Oxford Univ. Press
15. Blümich B, Kosfeld R, eds. 1994. *NMR:

Basic Principles and Progress, Vols. 30–33. New York: Springer-Verlag
16. Grant DM, Harris RK, eds. 1996. *Encyclopedia of NMR*. Chichester, UK: Wiley & Sons
17. Cohen MH, Reif F. 1957. *Solid State Phys.* 5:321–48
18. Sakurai JJ. 1995. *Modern Quantum Mechanics*. Reading, MA: Addison-Wesley
19. Thayer AM, Pines A. 1987. *Acc. Chem. Res.* 20:47–56
20. Rose ME. 1995. *Elementary Theory of Angular Momentum*. New York: Dover
21. Haeberlen U, Waugh JS. 1968. *Phys. Rev.* 175:453–67
22. Freude D, Haase J. 1993. *NMR: Basic Princ. Progr.* 29:1–90
23. Maricq MM, Waugh JS. 1979. *J. Chem. Phys.* 70:3300–16
24. Chmelka BF, Zwanziger JW. 1994. *NMR: Basic Princ. Progr.* 33:79–123
25. Goldman M, Grandinetti PJ, Llor A, Olenniczak Z, Sachleben JR, Zwanziger JW. 1992. *J. Chem. Phys.* 97:8947–60
26. Shirley JH. 1965. *Phys. Rev. B* 138:979–1000
27. Vega S, Olejniczak ET, Griffin RG. 1984. *J. Chem. Phys.* 80:4832–40
28. Alderman DW, Solum MS, Grant DM. 1986. *J. Chem. Phys.* 84:3717–25
29. Smith SA, Levante TO, Meier BH, Ernst RR. 1994. *J. Magn. Reson. A.* 106:75–105. http://gamma.magnet.fsu.edu
30. Eden M, Lee YK, Levitt MH. 1996. *J. Magn. Reson. A.* 120:56–71
31. Hodgkinson P, Emsley L. 2000. *Prog. Nucl. Magnn. Reson. Spectrosc.* 36:201–39. http://www.durham.ac.uk/~dch0ph/pubs
32. Bak M, Nielsen NC. 1997. *J. Magn. Reson.* 125:132–39. http://nmr.imsb.au.dk/simpson
33. Pines A, Gibby MG, Waugh JS. 1973. *J. Chem. Phys.* 59:569–90
34. Schaefer J, Stejskal EO. 1976. *J. Am. Chem. Soc.* 98:1031–32
35. Ernst M, Bush S, Kolbert AC, Pines A. 1996. *J. Chem. Phys.* 105:3387–97
36. vanderHart DL, Campbell GC. 1998. *J. Magn. Reson.* 134:88–112
37. Tekely P, Palmas P, Canet D. 1994. *J. Magn. Reson. A.* 107:129–33
38. Bennett AE, Rienstra CM, Auger M, Lakshmi KV, Griffin RG. 1995. *J. Chem. Phys.* 103:6951–58
39. Eden M, Levitt MH. 1999. *J. Chem. Phys.* 111:1511–19
40. Khitrin A, Fung BM. 2000. *J. Chem. Phys.* 112:2392–98
41. Lee M, Goldburg WI. 1965. *Phys. Rev.* 140:1261–65
42. Wang CH, Ramshaw JD. 1972. *Phys. Rev. B* 6:3253–60
43. Bielecki A, Kolbert AC, Levitt MH. 1989. *Chem. Phys. Lett.* 155:341–46
44. Hohwy M, Nielsen NC. 1997. *J. Chem. Phys.* 106:7571–86
45. Andrew ER, Bradbury A, Eades RG. 1958. *Nature* 182:1659
46. Lowe IJ. 1959. *Phys. Rev. Lett.* 2:285–87
47. Olejnizak ET, Vega S, Griffin RG. 1984. *J. Chem. Phys.* 81:4804–17
48. Herzfeld J, Berger AE. 1980. *J. Chem. Phys.* 73:6021–30
49. Jeschke G, Jansen M. 1998. *Angew. Chem. Int Ed. Engl.* 37:1282–83
50. Ermolaev K, Fung BM. 1999. *J. Chem. Phys.* 110:7977–82
51. Brunner E, Freude D, Gerstein BC, Pfeifer H. 1990. *J. Magn. Reson.* 90:90–99
52. Filip C, Hafner S, Schnell I, Demco DE, Spiess HW. 1999. *J. Chem. Phys.* 110:423–40
53. Ray S, Ladizhansky V, Vega S. 1998. *J. Magn. Reson.* 135:427–34
54. Zheng L, Fishbein KW, Griffin RG, Herzfeld J. 1993. *J. Am. Chem. Soc.* 115:6254–61
55. Isbester PK, Brandt JL, Kestner TA, Manson EJ. 1998. *Macromolecules* 31:8192–8200
56. Llor A, Virlet J. 1988. *Chem. Phys. Lett.* 152:248–53
57. Wooten EW, Mueller KT, Pines A. 1992. *Acc. Chem. Res.* 25:209–13

58. Samoson A, Lippmaa E, Pines A. 1988. *Mol. Phys.* 65:1013–18
59. Mueller KT, Sun BQ, Chingas GC, Zwanziger JW, Terao T, Pines A. 1990. *J. Magn. Reson.* 86:470–87
60. Bax A, Szeverenyi NM, Maciel GE. 1983. *J. Magn. Reson.* 55:494–97
61. Frydman L, Harwood JS. 1995. *J. Am. Chem. Soc.* 117:5367–68
62. Medek A, Harwood JS, Frydman L. 1995. *J. Am. Chem. Soc.* 117:12779–787
63. Amoureux JP, Fernandez C, Steuernagel S. 1996. *J. Magn. Reson.* A123:116–18
64. Amoureux JP, Fernandez C, Frydman L. 1996. *Chem. Phys. Lett.* 259:347–55
65. Brown SP, Wimperis S. 1997. *J. Magn. Reson.* 128:42–61
66. Wu G, Rovnyak D, Griffin RG. 1996. *J. Am. Chem. Soc.* 118:9326–32
67. Duer MJ, Stourton C. 1997. *J. Magn. Reson.* 124:189–99
68. Marinelli L, Medek A, Frydman L. 1998. *J. Magn. Reson.* 132:88–95
69. Kentgens APM, Verhagen R. 1999. *Chem. Phys. Lett.* 300:435–43
70. Larsen FH, Nielsen NC. 1999. *J. Phys. Chem. A* 103:10825–32
71. Madhu PK, Goldbourt A, Frydman L, Vega S. 2000. *J. Chem. Phys.* 112:2377–91
72. Vosegaard T, Florian P, Grandinetti PJ, Massiot D. 2000. *J. Magn. Reson.* 143:217–22
73. Gan Z. 2000. *J. Am. Chem. Soc.* 122:3242–43
74. Hartmann SR, Hahn E. 1962. *Phys. Rev.* 128:2042–53
75. Müller L, Kumar A, Baumann T, Ernst RR. 1974. *Phys. Rev. Lett.* 32:1402–6
76. Stejskal EO, Scheafer J, Waugh JS. 1977. *J. Magn. Reson.* 28:105–12
77. Marks D, Vega S. 1996. *J. Magn. Reson A* 118:157–72
78. Sardashti M, Maciel GE. 1987. *J. Magn. Reson.* 72:467–74
79. Meier BH. 1992. *Chem. Phys. Lett.* 188:201–7
80. Peersen OB, Wu X, Kustanovich I, Smith SO. 1993. *J. Magn. Reson.* A104:334–39
81. Hediger S, Meier BH, Ernst RR. 1995. *Chem. Phys. Lett.* 140:449–56
82. Baldus M, Geurts DG, Hediger S, Meier BH. 1996. *J. Magn. Reson. A* 118:140–44
83. Wu X, Zilm KW. 1993. *J. Magn. Reson. A* 104:154–65
84. Ishii Y, Ashida J, Terao T. 1995. *Chem. Phys. Lett.* 246:439–45
85. Vega AJ. 1992. *J. Magn. Reson.* 96:50–68
86. Vega AJ. 1992. *Solid State NMR* 1:17–32
87. Baltisberger JH, Gann SL, Grandinetti PJ, Pines A. 1994. *Mol. Phys.* 81:1109–24
88. Jannsen R, Veeman WS. 1988. *J. Chem. Soc. Faraday Trans.* 84:3747–69
89. Bax A, Szeverenyi NM, Maciel GE. 1983. *J. Magn. Reson.* 51:400–8
90. Bax A, Szeverenyi NM, Maciel GE. 1983. *J. Magn. Reson.* 52:147–52
91. Stoll ME, Vega AJ, Vaughan RW. 1976. *J. Chem. Phys.* 65:4093–98
92. Waugh JS. 1976. *Proc. Natl. Acad. Sci. USA.* 73:1394–98
93. Blumler P, Jansen J, Blümich B. 1994. *Solid State NMR.* 3:237–40
94. Spaniol T, Kubo A, Terao T. 1997. *J. Chem. Phys.* 106:5303–16
95. Medek A, Sachleben JR, Beverwyk P, Frydman L. 1996. *J. Chem. Phys.* 104:5374–83
96. Hu JZ, Alderman DW, Pugmire RJ, Grant DM. 1997. *J. Magn. Reson.* 126:120–26
97. Sachleben JR, Beverwyk P, Frydman L. 2000. *J. Magn. Reson.* 144:330–42
98. Zeigler RC, Wind RA, Maciel GE. 1988. *J. Magn. Reson.* 79:299–306
99. Frydman L, Chingas GC, Lee YK, Grandinetti PJ, Eastman MA, et al. 1992. *J. Chem. Phys.* 97:4800–8
100. Gan Z. 1992. *J. Am. Chem. Soc.* 114:8307–9
101. Yarim-Agaev Y, Tutunjian PN, Waugh JS. 1982. *J. Magn. Reson.* 47:51–56
102. Hong J, Harbison GS. 1993. *J. Magn. Reson. A* 105:128–36

103. Gullion T. 1989. *J. Magn. Reson.* 85:614–19
104. Gullion T, Baker DB, Conradi MS. 1990. *J. Magn. Reson.* 89:479–83
105. Tycko R, Dabbagh G, Mirau PA. 1989. *J. Magn. Reson.* 85:265–74
106. Emshwiller M, Hahn EL, Kaplan DE. 1960. *Phys. Rev.* 118:414–24
107. Linder M, Hohener A, Ernst RR. 1980. *J. Chem. Phys.* 73:4959–70
108. Terao T, Miura H, Saika A. 1986. *J. Chem. Phys.* 85:3816–22
109. Gullion T, Pennington CH. 1998. *Chem. Phys. Lett.* 290:88–93
110. Ba Y, Ratcliffe CI, Ripmeester JA. 1999. *Chem. Phys. Lett.* 299:201–6
111. Gullion T, Schaefer J. 1989. *Adv. Magn. Reson.* 13:57–83
112. Griffiths JM, Griffin RG. 1993. *Anal. Chim. Acta.* 283:1081–1101
113. Bennett AE, Griffin RG, Vega S. 1994. See Ref. 15, 33:3–77
114. Dusold S, Sebald A. 2000. *Ann. Rep. NMR Spectrosc.* 41:185–264
115. Munowitz MG, Griffin RG, Bodenhausen G, Wang TH. 1981. *J. Am. Chem. Soc.* 103:2529–33
116. Munowitz MG, Griffin RG. 1982. *J. Chem. Phys.* 76:2848–58
117. Schaefer J, McKay RA, Stejskal EO, Dixon WT. 1983. *J. Magn. Reson.* 52:123–29
118. Gullion T, Schaefer J. 1989. *J. Magn. Reson.* 81:57–83
119. Garbow JR, Gullion T. 1992. *Chem. Phys. Lett.* 192:71–76
120. Mueller KT. 1995. *J. Magn. Reson. A* 113:81–93
121. Schmidt A, McKay RA, Schaefer J. 1992. *J. Magn. Reson.* 96:644–50
122. Grey CP, Veeman WS. 1992. *Chem. Phys. Lett.* 192:379–85
123. Grey CP, Veeman WS, Vega AJ. 1993. *J. Chem. Phys.* 98:7711–24
124. Gullion T. 1995. *Chem. Phys. Lett.* 246:325–30
125. Ba Y, Kao HM, Grey CP, Chopin L, Gullion T. 1998. *J. Magn. Reson.* 133:104–14
126. Fernandez C, Lang DP, Amoureux JP, Pruski M. 1998. *J. Am. Chem. Soc.* 120:2672–73
127. Pruski M, Bailly A, Lang DP, Amoureux JP, Fernandez C. 1999. *Chem. Phys. Lett.* 307:35–40
128. Hing AW, Vega S, Schaefer J. 1993. *J. Magn. Reson. A* 103:151–62
129. Fyfe CA, Wong-Mong KC, Huang Y, Grondey H, Mueller KT. 1995. *J. Phys. Chem.* 99:8707–16
130. Hong M, Griffin RG. 1998. *J. Am. Chem. Soc.* 120:7113–14
131. Sandström D, Hong M, Schmidt-Rohr K. 1999. *Chem. Phys. Lett.* 300:213–20
132. Oas TG, Levitt MH, Griffin RG. 1988. *J. Chem. Phys.* 89:692–95
133. Costa PR, Gross JD, Hong M, Griffin RG. 1997. *Chem. Phys. Lett.* 280:95–103
134. Fu R, Smith SA, Bodenhausen G. 1997. *Chem. Phys. Lett.* 272:361–69
135. Gross JD, Costa PR, Griffin RG. 1998. *J. Chem. Phys.* 108:7286–93
136. Ishii Y, Terao T. 1998. *J. Chem. Phys.* 109:1366–74
137. Wu CH, Ramamoorthy A, Opella SJ. 1994. *J. Magn. Reson. A* 109:270–72
138. Ramamoorthy A, Gierasch LM, Opella SJ. 1996. *J. Magn. Reson. B* 110:102–6
139. Sachleben JR, Frydman V, Frydman L. 1996. *J. Am. Chem. Soc.* 118:9786–87
140. Baldus M, Geurts DG, Meier BH. 1998. *Solid State NMR* 11:157–68
141. Powles JG, Mansfield P. 1962. *Phys. Rev. Lett.* 2:58–61
142. Tycko R, Dabbagh G. 1990. *Chem. Phys. Lett.* 173:461–65
143. Tycko R, Smith SO. 1993. *J. Chem. Phys.* 98:932–43
144. Klug CA, Zhu W, Merritt ME, Schaefer J. 1994. *J. Magn. Reson. A* 109:134–36
145. Gregory DM, Mitchell DJ, Stringer JA, Kiihne S, Shiels JC, et al. 1995. *Chem. Phys. Lett.* 246:654–63

146. Kiihne S, Mehta MA, Stringer JA, Gregory DM, Shiels JC, Drobny GP. 1998. *J. Phys. Chem. A* 102:2274–82
147. Sun BQ, Costa PR, Kosisko D Jr, Lansbury PT, Griffin RG. 1995. *J. Chem. Phys.* 102:702–7
148. Fujiwara T, Ramamoorthy A, Nagayama K, Hioka K, Fujito T. 1993. *Chem. Phys. Lett.* 212:81–84
149. Baldus M, Tomaselli M, Meier BH, Ernst RR. 1994. *Chem. Phys. Lett.* 230:329–36
150. Nielsen NC, Bildsoe H, Jakobsen HJ, Levitt MH. 1994. *J. Chem. Phys.* 101:1805–2
151. Lee YK, Kurur ND, Helmle M, Johannessen OG, Nielsen NC, Levitt MH. 1995. *Chem. Phys. Lett.* 242:304–9
152. Hohwy M, Jakobsen HJ, Eden M, Levitt MH, Nielsen NC. 1998. *J. Chem. Phys.* 108:2686–94
153. Rienstra CM, Hatcher ME, Mueller LJ, Sun BQ, Fesik SW, Griffin RG. 1998. *J. Am. Chem. Soc.* 120:10602–612
154. Colombo MG, Meier BH, Ernst RR. 1988. *Chem. Phys. Lett.* 146:189–96
155. Raleigh DP, Levitt MH, Griffin RG. 1988. *Chem. Phys. Lett.* 146:71–76
156. Levitt MH, Raleigh DP, Creuzet F, Griffin RG. 1990. *J. Chem. Phys.* 92:6347–64
157. Gan ZH, Grant DM. 1990. *Mol. Phys.* 67:1419–30
158. Karlsson T, Levitt MH. 1998. *J. Chem. Phys.* 109:5493–5507
159. Costa PR, Sun B, Griffin RG. 1997. *J. Am. Chem. Soc.* 119:10821–30
160. Verel R, Baldus M, Nijman M, vanOs JWM, Meier BH. 1997. *Chem. Phys. Lett.* 280:31–39
161. Nijman M, Ernst M, Kentgens APM, Meier BH. 2000. *Mol. Phys.* 98:161–78
162. Gullion T, Vega S. 1992. *Chem. Phys. Lett.* 194:423–28
163. Weintraub O, Vega S, Hoelger C, Limbach HH. 1994. *J. Magn. Reson A* 109:14–25
164. Bennett AE, Ok JH, Griffin RG, Vega S. 1992. *J. Chem. Phys.* 96:8624–27
165. Hodgkinson P, Emsley L. 1999. *J. Magn. Reson.* 139:46–59
166. Meier BH. 1994. In *Advances in Magnetic and Optical Resonance*, ed. WS Warren, 18:1–116. New York: Academic
167. Weliky DP, Tycko R. 1996. *J. Am. Chem. Soc.* 118:8487–88
168. Hafner S, Spiess HW. 1998. *Concepts Magn. Reson.* 10:99–128
169. Hong M. 1999. *J. Magn. Reson.* 139:389–401
170. Tan WM, Zelink R, Opella SJ, Malik P, Terry TD, Perham RN. 1999. *J. Mol. Biol.* 286:787–96
171. Weliky DP, Bennett AE, Zvi A, Anglister J, Steinbach PJ, Tycko R. 1999. *Nat. Struct. Biol.* 6:141–45

FROM FOLDING THEORIES TO FOLDING PROTEINS: A Review and Assessment of Simulation Studies of Protein Folding and Unfolding

Joan-Emma Shea and Charles L Brooks III
Department of Molecular Biology, TPC6 The Scripps Research Institute La Jolla, California 92037; e-mail: Shea@midway.uchicago.edu; brooks@scripps.edu

Key Words protein folding/unfolding, molecular dynamics, folding mechanism, free energy surfaces, atomic models

■ **Abstract** Beginning with simplified lattice and continuum "minimalist" models and progressing to detailed atomic models, simulation studies have augmented and directed development of the modern landscape perspective of protein folding. In this review we discuss aspects of detailed atomic simulation methods applied to studies of protein folding free energy surfaces, using biased-sampling free energy methods and temperature-induced protein unfolding. We review studies from each on systems of particular experimental interest and assess the strengths and weaknesses of each approach in the context of "exact" results for both free energies and kinetics of a minimalist model for a beta-barrel protein. We illustrate in detail how each approach is implemented and discuss analysis methods that have been developed as components of these studies. We describe key insights into the relationship between protein topology and the folding mechanism emerging from folding free energy surface calculations. We further describe the determination of detailed "pathways" and models of folding transition states that have resulted from unfolding studies. Our assessment of the two methods suggests that both can provide, often complementary, details of folding mechanism and thermodynamics, but this success relies on (*a*) adequate sampling of diverse conformational regions for the biased-sampling free energy approach and (*b*) many trajectories at multiple temperatures for unfolding studies. Furthermore, we find that temperature-induced unfolding provides representatives of folding trajectories only when the topology and sequence (energy) provide a relatively funneled landscape and "off-pathway" intermediates do not exist.

INTRODUCTION

The challenges provided by the development of a fundamental framework for understanding the processes of protein and peptide folding have been a cornerstone of theoretical efforts across chemistry, physics, and biology for nearly four decades (1, 2). As we move into the twenty-first century, these challenges continue to motivate much theoretical and experimental work. However, a paradigm shift has directed our focus from a "chemical reaction-like" pathway perspective (3–5), as posited by Levinthal (6), to a statistical view of the folding landscape (7–13). Levinthal proposed the pathway scenario as a possible solution to the paradoxical question of how a polypeptide chain, with an astronomically sized conformational space, could "find" its well-defined (functional) native state so rapidly. The new theoretical perspective encompasses the pathway-based ideas but moves beyond these specific-scenario approaches to identify the statistical characteristics that differentiate the energetic landscapes of heteropolymeric sequences of polypeptides that rapidly form well-defined native conformations from those that do not. Key to the description of foldable protein-like landscapes is their funneled nature (7, 8, 14); the relative values of characteristic energy scales for folding and glass formation, ideas that have been borrowed from the physics of spin glasses and related statistical materials (15, 16); and the concept of minimal frustration of local interactions (8, 17).

The "new-view" perspective is a theoretical language intended to explore and extend our understanding of folding that emphasizes the many-pathway nature of the folding process and the interplay of local and global driving forces manifest in folding. How these features reveal themselves in the folding of specific sequences, or within the context of specific models, has been elucidated through simulation studies employing models of varying complexity. These include simplified models of the polypeptide chain and its environment, such as lattice-based hydrophobic/hydrophilic models (18) and minimalist interaction scheme models (19–26). Detailed atomic models have also figured significantly into this development (9, 20, 27–30).

Minimalist simulation models played an early role in honing the landscape framework and illustrating the varied nature of folding processes realized through alterations of the underlying (sequence-space) parameters (7, 11, 31). For example, studies of these models demonstrated that suitable folding coordinates are often those that are correlated with global characteristics of the folding process, e.g. the fraction of native contacts specifying the degree of similarity to the native state, or the radius of gyration, R_g, a global measure of the "size" of the folding protein globule (32–34). They also taught us that caution must be exercised in the choice of "reaction coordinates" because in some instances, particularly for less-optimal folders (25), very specific interactions determine the dominant kinetic pathways of folding; e.g. when kinetic partitioning between a "fast track" and a kinetic trap occur (19), when specific nuclei are involved (35–37), or when folding deviates from a two-state-like cooperative process (38, 39). In the former cases, one can learn a great deal about folding kinetics and mechanism from a thorough investigation of the folding free-energy surface projected onto the aforementioned reaction coordinates. In the latter cases, deducing information about the folding mechanism and kinetics from "simple" free-energy surface projections is not possible. More complicated coordinates, which are most likely system specific, must first be identified (40). In either case, it should be clear that simulations of such minimalist models provide insights into detailed aspects of folding not easily accessible from the theoretical framework of the folding landscape theory (24, 25, 41–43). Furthermore, they provide a computational paradigm for exploring algorithmic questions, sharpening conceptual issues, and exploring how best to "ask" questions using more complicated and detailed models. This theme is stressed in this review as we illustrate how minimalist models can be constructed and utilized to examine fundamental technical questions that relate to the use and interpretation of detailed simulation models of protein folding thermodynamics, mechanism, and kinetics.

The primary focus of this review centers on a presentation and assessment of a limited number of the current methods utilized to explore the folding process with atomic detail models. Such models, in which both the protein and its solvent environment are represented at the level of explicit atoms and in which motions are sampled using molecular dynamics or Monte Carlo methodologies, provide the most complete theoretical description of these systems presently available (44–46). The complexity of this detailed representation coupled with the complexity of the folding process, however, make a direct "attack" on folding using brute force, an inviable approach for studies of all but the simplest peptides (47–50). We note this even in light of the recent tour de force demonstration of microsecond time-scale molecular dynamics on the villin headpiece by Kollman and coworkers (51, 52). This demonstration, although impressively pushing the envelope of simulation techniques, falls short of the mark needed to study the folding process by 4–6 orders of magnitude (arising both from the inherent time scale of folding and the necessity of averaging over multiple trajectories, as discussed below). Thus, alternative techniques are being employed to examine folding, which include biased-sampling free-energy surface calculations to elucidate folding free-energy landscapes and nonequilibrium unfolding simulations that permit unfolding trajectories to be followed over time courses of sufficient duration to observe unfolding events. We explore aspects of both of these methodologies below, with an objective of laying out in detail many of their technical features. Particular emphasis is placed upon their strengths and weaknesses throughout our discussions.

Areas we do not aim to address directly include (*a*) the use of implicit solvation models to study peptide and protein folding (53–59); (*b*) generalized biased-sampling techniques that employ multicanonical or entropy-sampling umbrella potentials (though these are, in principle, analogous to the biased-sampling methods we discuss) (54–56, 60–64); and (*c*) generalized "time-stepping" approaches that utilize action-based formulations of Hamilton's equations, the Langevin equation, or Brownian dynamics (e.g. the Onsager-Machlup approach) to achieve very large time steps, thereby potentially making accessible direct simulation of folding processes (65–67). Many of the observations we make apply to these approaches to exploring folding, and some of these techniques will undoubtedly prove (as some already have) useful in studying folding processes. We make special mention in passing of the development of implicit solvent models, based either on the generalized Born approach (68, 69) or other empirical solvation methods (57, 58) because these approaches provide (especially coupled with the sampling and time-stepping methods just noted) the potential to explore directly the folding process while maintaining a detailed representation of the protein.

HISTORICAL PERSPECTIVES: Early Characterizations of Peptide and Protein Folding Using Simulations

Experimental studies in the mid to late 1980s were motivated by the suggestion that secondary structural elements may play an important role in determining protein folding pathways and also by an interest in elucidating the fundamental interactions that stabilize such structures (70–76). These studies focused on "deconstructing" the folding process to that of (*a*) formation of native-like secondary structure in solution and (*b*) coalescence of these nascent secondary structural elements into folded proteins. Simulation studies, in turn, began to focus on peptide-folding kinetics and thermodynamics to complement experiment and provide detailed theoretical models for secondary structure formation.

An early attempt to study the equilibrium folding kinetics of a short turn-forming penta-peptide, known experimentally to adopt a specific β-turn conformation in isolation and in solution (71, 72), is representative of these calculations. It was performed by Brooks and his colleagues and employed detailed representations of peptide and solvent, utilizing the day's most modern computer architectures (77, 78). Even though the time scale of this simulation study was only a few nanoseconds (one of the longest of a biological molecule in solution at the time), it was one of the first studies to illustrate that simulations could be used to explore folding kinetics under equilibrium conditions. It served to demonstrate that turn formation in such peptides could occur on nanosecond time scales. Subsequent analyses from this group demonstrated the first use of clustering techniques to quantify the nature of the landscape for this small peptide (78).

Other simulations of peptides in solution were also published during this time. These calculations were aimed at investigating turn formation and the dissolution of helices in artificial (79–81) and natural sequences (82–84). Most of these simulations focused on nonequilibrium unfolding events because the time scale for helix formation was too long to be simulated directly with today's computers. In these calculations temperature was introduced as a "perturbant" to drive unfolding and thereby expedite the process. It was argued that the increased temperature only perturbed the velocity (or mean speed) of atoms by a small factor and thus rendered the processes under investigation relevant to those under equilibrium, or native, conditions (79).

In parallel with the earliest applications of simulations to study peptide unfolding kinetics, Brooks and his colleagues developed methods to permit the folding thermodynamics for small peptides in aqueous solution to be investigated with detailed simulation models (77, 85–91). These studies aimed to delineate the specific role of hydrogen-bonding interactions in stabilizing secondary structure in water (77). From systematic studies of small peptides or model fragments, the role of hydrogen bonding in "bare" peptide hydrogen bonds (87), turns (88), helices (77, 89, 92, 93), and model β-sheets (90) was elucidated. These studies demonstrated that relatively exposed peptide-hydrogen bonds, as occur in helices and turns, were weakened significantly by competition from water-hydrogen bond donor and acceptor groups and formed/dissolved rapidly, in basic agreement with results from the direct simulation studies noted above. However, in β-sheet hydrogen bonds, even in the simplest model β-sheet comprised of an antiparallel alanine "dipeptide," the alanine side chain provided significant shielding of the peptide interactions. This led to relatively strong interactions between the peptide chains, approximately 4–5 $k_B T$ per hydrogen bond at room temperature compared with only marginal stabilization for solvent-exposed interactions in helices and turns (77). These strong interactions were hypothesized to lead to slow dissolution of β-sheet structure once formed, and to be a possible source for kinetic trapping of misfolded aggregates during peptide and protein folding (77, 90). Also elucidated in these studies were the basic thermodynamic parameters and mechanism for helix initiation and propagation in water (77, 89, 93, 94) and alcohol/water mixtures (92) and the attendant time scales for folding of these structures (77, 95). Studies like these, and related work from other laboratories (96–99), focused directly on a thermodynamic description of peptide folding. They aimed to deduce the mechanism and kinetics of folding by inference from the presence of metastable intermediates [e.g. 3_{10} helix formation on the pathway to α-helix folding (77, 83, 89, 93, 94, 100–105)] and the barriers between free energy minima on the conformational free-energy surface (95). These free-energy methods explore processes under equilibrium conditions and require prior information about potential reaction coordinates to provide the variables onto which the free-energy changes may be mapped. Temperature-induced unfolding studies, on the other hand, employ nonequilibrium simulation methods and explore the response of the system to the perturbation (unfolding as a result of increased temperature).

Early in the past decade, detailed simulation studies were first employed to characterize protein folding intermediates, molten globules, and unfolding pathways (28, 30, 106). Nonequilibrium, temperature-induced unfolding studies were used to explore the "molten-globule" state of the small trypsin inhibitor, PTI (107), and the unfolding pathways of lysozyme (108–110), apomyoglobin (111), and barnase (112, 113). Equilibrium simulation approaches were used in the first of these studies to explore the native state of apomyoglobin (114). Also during this period, pressure (115) and denaturants (116–118) were used in simulations to attempt to "drive" proteins from folded to unfolded states. Glimpses of the behavior of proteins as they unfolded or adopted their new equilibrium conformational distributions were garnered from each of these early studies, and new questions were formulated to shape future investigations of folding and unfolding. However, none of these early studies was really sufficiently long or characterized, and in most cases the systems were too big for thorough study to permit deduction of detailed insights into the folding landscape. They, nevertheless, paved the way for the generation of work that followed by turning the focus of simulation studies to a family of small, single-domain, cooperative, fast-folding proteins. These included fragment B of staphylococcal protein A, chymotrypsin inhibitor-2, segment B1 of streptococcal protein G (GB1), cold shock protein A (CspA), src-SH3 domain, and the designed protein Betanova, a subset of a host of proteins being characterized in detail by experiment. We discuss a number of these below.

The studies of peptides and proteins noted above served to bring to the surface two basic approaches to characterizing protein folding. One of these, which was developed primarily within the Brooks group and applied to proteins first by Boczko & Brooks (27), held as an objective the characterization of equilibrium aspects of peptide and protein free-energy surfaces or thermodynamic states. The other, pioneered by among others, Levitt and Daggett (30, 119), Jorgensen (111), and van Gunsteren (108), uses nonequilibrium methods of protein unfolding (primarily with temperature as the driving force) and focuses on identifying "kinetic" features of the unfolding process. Both efforts aim at elucidating mechanism, kinetics, and thermodynamic forces responsible for folding and have evolved to work around the computational obstacles presented by all-atom simulations. The first approach (method A) consists of deducing the folding mechanism from the free-energy surface of the protein under native-like conditions. This free-energy surface is constructed from biased sampling of structures generated initially from high-temperature unfolding runs. The second (method B) seeks to infer the folding mechanism of a protein by studying its unfolding behavior at high temperatures. Both methods rely on the use of extreme conditions (either high temperature or denaturants) to accelerate the simulation events. Method A, however, only uses the high temperature simulations to generate various conformations. The biased (conformational space) sampling used to generate the free energy profile is performed under native-like conditions. Method B relies heavily on the underlying assumption that unfolding proceeds by the reverse route as folding.

In the pages that follow we discuss both methodologies and their applications to different proteins. We begin by briefly describing the approaches of both and reviewing results that have emerged from each. We next demonstrate the strengths and potential shortcomings of each method through an illustrative application employing a minimalist off-lattice model. Minimalist protein models contain on average two orders of magnitude fewer atoms than fully atomic solvated models, and the complete folding process from an extended chain to the native state can hence be studied. These simplified models do not reproduce the full atomic details of folding but do successfully capture the underlying physics of the process (22, 24, 25, 41). We characterize the folding thermodynamics and kinetics of such a model. Following this, we analyze the same system using both method A and method B. We stress the differences between these two approaches and assess their relevance to studying protein folding. By comparing the unfolding and folding behavior of the model, we investigate the limits of studying folding by means of unfolding simulations. In particular, we seek to establish whether the principle of microscopic reversibility (or detailed balance) holds under the extreme conditions used in such simulations.[1] The principle of microscopic reversibility is a statement made for equilibrium systems and there is no reason, a priori, to assume that it is valid under the strongly nonequilibrium conditions of unfolding simulations. We also compare the potential of mean force generated from folding runs with the one obtained by method A, utilizing high-temperature unfolding runs to generate initial conditions, and biased sampling to provide population estimates of different conformational regions. This enables us to assess the extent to which this method can suitably be used to elucidate the correct free-energy minima (corresponding to the states of the protein) on the potential of mean force surface.

METHOD A: Free Energy Surfaces for Protein Folding

Potentials of Mean Force Generated from Biased Sampling Initiated from Structures Generated at High Temperatures

It is currently impossible to fold a fully atomic protein model in explicit water in a computationally reasonable time frame. Thus, alternative approaches are required. An elegant way of gaining insight into the folding of a protein is through the generation and study of the potentials of mean force (PMF) (or free energy as a function of reaction coordinates describing the protein's folding progress). This approach has been pioneered by Brooks and co-workers for peptides (85, 120) and proteins (27), and has been applied to a number of small, fast-folding proteins. In the following sections, we review the methodology used to generate these PMF and its application to protein models of different native folds.

Methodology for Generating the Potentials of Mean Force The method can be briefly summarized as follows. Denaturing conditions are employed to generate a database of protein conformations spanning the unfolded to the folded state. Specific structures from this database are selected as initial conditions for biased molecular dynamics sampling under a set of desired thermodynamic conditions (e.g. native conditions) using a clustering algorithm. The sampling is then combined to generate a density of states from which the free energy projected onto any number of folding-progress coordinates can be calculated. A number of important considerations arise, namely the need for suitable reaction coordinates to describe the folding progress of the protein, the generation and selection of the initial conditions for the biased thermodynamic sampling and the method and extent of sampling necessary to create a "connected net" of conformations linking the initial conditions, as well as a suitable method of combining the sampling.

---

[1] We use *principle of microscopic reversibility* loosely, in a manner analogous to that employed in arguments in the literature, to mean that folding and unfolding follow similar (the same) pathways.

This method is computationally tractable for two reasons. First, the use of denaturing conditions to generate initial conformations throughout the folding space of the system accelerates the unfolding events by several orders of magnitude. Second, the biased sampling around the initial condition-conformational regions can be performed simultaneously (independently), hence enhancing the parallelism inherent in such calculations and drastically cutting down on the number of "real time" hours required.

In the following paragraphs, we present a "generic" method for generating the PMF. This description represents a composite of the approaches that have evolved from the Brooks group since their first folding studies. The reader can easily modify the protocol to suit the system being considered.

Characterization of the Native State A protein crystal or NMR structure is solvated in a volume of water (typically in a truncated octahedron volume to minimize the amount of solvent). An equilibration molecular-dynamics simulation is then performed at constant temperature (298°K) and constant pressure (1 atm) for several hundreds of picoseconds. Decreasing harmonic restraints are applied to the protein backbone. The heavy atom–hydrogen bonds are usually kept fixed using the SHAKE algorithm (121), and the simulations are propagated with a 2 fs time step. The long range interactions are treated using particle mesh Ewald (122–125), though earlier calculations used long cutoff-based techniques. After the system has equilibrated (as verified by a stabilization of the energies and protein conformations), several constant temperature (298°K), constant volume (or pressure) molecular dynamics simulations of a 2–3 ns duration are performed to characterize the native state. From the native state simulations, properties characterizing this state of the protein, such as native contacts, radius of gyration, native hydrogen bonds, and solvent exposure, are computed. These serve as a reference for potential folding reaction coordinates. It is essential to characterize the protein's native state under the native conditions for one's model (thermodynamic variables, force field, and simulation protocol) to delineate differences between the model and the experimental structure. With current force fields and simulation methods, such models typically agree very well with experimental properties of the native protein.

Reaction Coordinates A number of reaction coordinates may be used to describe the progress of folding. The main criterion is that the reaction coordinate unambiguously distinguishes between the folded and the unfolded/partially folded conformations.

In the Brooks protocol the number of native contacts (C) and the radius of gyration (R_g) are typically used as reaction coordinates. The number of native hydrogen bonds (H) may also be a very useful progress variable, especially for helical systems. Other coordinates, such as the amount of solvent in the protein core (126, 127) and specific secondary structure (20, 27), are often useful in assessing specific details of folding progress.

One possible way of defining the native contacts is to consider all residue pairs with side chain centers of geometry within some separation as being in contact. In what we describe below, any pair with a separation of less than 6.5 Å in simulations of the native state is said to form a native contact. Similarly, hydrogen-oxygen pairs less than 3.5 Å apart in the native simulations can be taken to be native hydrogen bonds.

Generation of the Initial Conditions for Biased Sampling Generation of the database: At least 3 to 4 (and often up to 6) high-temperature (350–500°K) molecular dynamics unfolding runs are performed to generate configurations, spanning conformations from the folded to the unfolded state. The simulations can be initiated from different structures taken from the native runs at 298°K. It is typically observed that small (fewer than 80 residues) single-domain proteins are significantly unfolded after 1–2 ns of temperature-induced unfolding. It is possible to unfold the protein through different means [for example, R_g or root-mean-square deviation (rmsd) restrained dynamics (109, 128)]; however, it is essential to generate conformations that are representative (i.e. that are not so high in energy that they will not equilibrate in a computationally accessible time frame). With this in mind, our studies have shown that the resulting PMF is not significantly affected by the method employed to generate the initial conditions, provided sufficient reequilibration is permitted. The trajectories from these simulations are combined to form a database. This database of potential initial conditions is partitioned among "bins" in a relevant property space of reaction coordinates to ensure coverage of the folding space. Typically, the unfolding trajectories are partitioned on a grid of C and R_g bins (e.g. 20 × 5), and the population of each bin is examined to see that sufficient samples are present to provide meaningful clustering. When this is not the case, further sampling is carried out to generate conformations that reside in the low-population bins.

Clustering of the database: The conformations used as initial conditions for the biased sampling are chosen from the partitions of the initial condition database through a hierarchical clustering algorithm, which is based on a measure of structural dissimilarity combined with a stopping rule (129). The structures in each partition are hierarchically clustered, and the particular level in the hierarchy providing initial conditions for that bin is chosen based on a "stopping" rule that optimizes the compactness and separation of clusters. This choice is identified as the maximum of the function E(i) (27),

$$E(i) = \frac{M(i) - M(i+1)}{\sigma(i) - \sigma(i+1)}, \quad i = n-1, n-2, \ldots, 3, 2 \qquad 1.$$

at the level i in the hierarchy. M(i) is the minimal between cluster distance at level i; n is the number of data points to be clustered; $\sigma^2(i) = \Sigma(d^i_{\alpha\beta})^2$; $d^i_{\alpha\beta}$ is the distance between clusters α and β at level i. The distances between the newly formed clusters are updated with the minimum variance (Ward's) method (130).

A number of descriptors of the protein structure can be employed to define the dissimilarity function used in clustering the conformations. An appropriate set could include the number of native contacts C (a measure of tertiary structure), the number of hydrogen bonds H (a measure of secondary structure), and the solvation energy (131). The dissimilarity $D(\alpha, \beta)$ between two structures α and β in the space of these descriptors is given by (27)

$$D(\alpha, \beta) = \frac{1}{\text{var}(d_c)} \cdot \sum_{i=1}^{\#C} (d_{Ci}(\alpha) - d_{Ci}(\beta))^2$$

$$+ \frac{1}{\text{var}(d_H)} \cdot \sum_{i=1}^{\#H} (d_{Hi}(\alpha) - d_{Hi}(\beta))^2$$

$$+ \frac{1}{\text{var}(SE)} \cdot (SE(\alpha) - SE(\beta))^2, \qquad 2.$$

where $d_{Ci}(\nu)$ is the distance of the ith contact in the structure ν, $d_{Hj}(\nu)$ is the distance of the jth hydrogen bond in the structure ν, and $SE(\nu)$ is the solvation energy of the structure ν. From the clustering, it is typical that 2–5 clusters for each partition emerge. The protein conformation nearest to each cluster center in each partition is used as an initial condition in sampling the folding free-energy surface.

Biased Sampling of the Initial Conditions Once the initial conditions have been generated, biased molecular dynamics sampling is performed around these starting structures. The goal of the sampling is to link the different initial conditions through a distribution of conformations (20). It is imperative to create such a connected network of conformational sampling if one wishes to extract a meaningful density of states and consequently free energy for the system.

Biased sampling is performed along points on a suitable reaction coordinate so as to "concentrate" the sampling to specific regions of conformational space as determined from the clustering. A number of coordinates, such as the number of native contacts, C, or the radius of gyration, R_g, are possible candidates. Recall that the reaction coordinate must follow the progress from the unfolded to the folded state and discriminate between the native and the nonnative structures. Although many reaction coordinates may be chosen, we illustrate the biasing potential with a reaction coordinate related to the fraction of native contacts, following the work of Sheinerman & Brooks (126, 127). We define the native contacts from the native state simulations as consisting of pairs of residues with side chain centers of geometry less than a certain cutoff K (e.g. 6.5 Å). The state of each contact x(i) is set to 1 if the distance d(i) between the centers of geometry of the side chains is less than the cutoff K and to 0 if the distance is larger than the cutoff plus a given tolerance ("tol") using the continuous function

$$x(i) = \frac{1}{\left(1 + \exp(N_{bins} \cdot \left(d(i) - \left(K + \frac{tol}{2}\right)\right)\right)}. \qquad 3.$$

The reaction coordinate in this case has been separated into N_{bins} bins of width y.

The reaction coordinate ρ that will be used in the biasing simulations is related to the overall fraction of native contacts. It is defined in terms of the state x(i) of a native contact as well as its weight p(i). The weight p(i) may be determined from the fraction of time each contact was present in the native state simulations or set to unity,

$$1 - \rho = \frac{\sum p(i) \cdot (1 - x(i))}{\sum p(i)}. \qquad 4.$$

The biased sampling along the reaction coordinate ρ is performed following the standard umbrella sampling protocol by adding the quadratic biasing potential (V_j) to the energy function,

$$V_j = \frac{1}{2} k \cdot (\rho - \rho(j))^2. \qquad 5.$$

For each window j, the center of the biasing potential is set to $\rho(j) = y \cdot (j - 0.5)$, where y is the width of the reaction-coordinate bin. The force constant k is chosen so that the value of the biasing potential is on the order of $k_B T$ at the edge of the bin. In typical calculations from our group force, constants ranging from 500 kcal/mol, for the 20-residue Betanova protein (59), to 2000 kcal/mol, for the 69-residue cold shock protein A (132), were used. This bias is only felt by the protein during the net formation and loss of a contact; exchange of contacts yields no net forces on the protein.

The protocol used in the biased molecular-dynamics simulations is as follows. Each initial condition is resolvated in a volume of water and reequilibrated for several hundreds of picoseconds. Care is taken to ensure that the same number of water molecules is used in solvating each initial condition (this makes subsequent analysis simpler). The equilibration period is performed with restraints on the polypeptide backbone and under conditions of constant temperature and pressure. It is useful to examine the distributions of energies for the protein-protein, protein-solvent, and solvent-solvent components to ensure that they have reached a quasi-stable state. Biased molecular-dynamics simulations (as described above) are then performed for 400–600 ps, longer if the structures are observed to sample new regions of configurational space, for each of the 75–200 initial conditions.

Weighted Histogram Analysis The sampling obtained from the biased molecular-dynamics simulations is then combined by using a constant temperature variant of the weighted histogram analysis method (WHAM) (133, 134). This method provides an optimal estimate of the density of states from which one can obtain the free energy (or PMF) as a function of a number of reaction coordinates. The WHAM equations are described in general form in a subsequent section of this chapter (see Equations 6–8 below). However, we note that typically, sampling precludes more than three variables to be used in constructing the histograms and corresponding free-energy surfaces.

TABLE 1 Thermodynamic stability computed from calculated potential of mean force surfaces[a]

Protein	Stability (kcal/mol)	Temperature (K)	Cutoff
Protein A	−1.03	298	$R_g \leq 10.75$ Å
Protein G	−1.77	298	$\rho \geq 0.7$
Protein G	−0.31	343	$\rho \geq 0.7$
src-SH3	−3.10	298	$\rho \geq 0.7$
CspA	−3.75	298	$\rho \geq 0.7$

[a]The stability computed as $-k_B T \cdot \ln(K_{eq})$, where K_{eq} is the calculated equilibrium population of the native basin defined by the measure denoted as cutoff.

All-Atom Simulations: Potentials of Mean Force: Shape and Folding

The methodology described in the previous section to generate the free-energy surface was applied by the Brooks group to fully atomic models of proteins in explicit water. The PMFs of the α-helical protein A (20, 27), the mixed α-β protein GB1 (126, 127), the mostly β (with a short 3–10 helix) protein src-SH3 (135), and the all-β protein CspA (132) were calculated and are displayed in Figure 1 (see color insert). Table 1 shows the stability of each protein under the specified conditions. We note that all proteins have a stability that is consistent with experimental measurements under similar conditions. This suggests that, in general, the energy functions used in these calculations provide a reasonable thermodynamic representation of the native state relative to unfolded states. Finally we note that, although we do not review this work here because our focus is restricted to protein sequences derived from natural proteins, Bursulaya & Brooks successfully applied these methods to the artificial 22-residue, 3-stranded β-sheet peptide Betanova, using the same techniques (59).

A relationship between the shape of the PMF projected onto these coordinates, the folding mechanism, and the topology of the protein has emerged from these studies (28). Proteins with mostly local contacts, such as the all-α protein fragment B of staphylococcal protein A, tend to have a "diagonal" landscape in (R_g, ρ) space, reflective of a concomitant collapse and formation of native structure. Proteins in which longer-range interactions dominate (such as the mostly β sheet proteins segment B1 of streptococcal protein G, src-SH3, and CspA) show a more "L-shaped" profile, corresponding to a folding mechanism characterized by an early collapse followed by a rearrangement of the protein to the native state. This observation, from the calculations we describe briefly below, has found subsequent confirmation in an analysis of experimental data by Baker and colleagues (136).

Fragment B of Staphylococcal Protein A This small helical protein fragment (see Figure 1a; see color insert) was the first protein to be studied in detail with atomic-level simulation methods (20, 27). It represents a 46-residue fragment from the IgG binding domain of protein A. The 60-residue parent fragment has been characterized both structurally (137–140) and thermodynamically (141). It is known to be a relatively fast, two-state folder. The truncated version was synthesized and shown to be marginally stable (142), consistent with the thermodynamic data presented in Table 1 and Figure 1a (see color insert).

The free-energy surface (PMF) versus R_g and ρ (radius of gyration and fraction of native contacts) illustrates the "mobility" of the native basin for this marginally stable system, indicated by the multiple minima that exist for $R_g < 10.5$ Å and $\rho > 0.4$. Also evident in the PMF is an "entropic" bottleneck near $R_g \approx 11.5$ Å, $\rho \approx 0.35$, which corresponds to a "transition state barrier" for folding (20, 27). This barrier is small, consistent with the rapid, two-state-like folding observed experimentally.

Guo et al (20) examined the free energy surface for protein A projected onto alternative coordinates, such as the fraction of native hydrogen bonds and specific hydrogen bonds in each helix, to explore the mechanism for folding. Their observations are consistent with early formation of structure around the helix I–helix II interface (a proline-containing turn with hydrophobic residues flanking the proline) and later formation of the helix II–helix III interface, with concomitant rearrangement of the helix I–helix II interface. This finding is of interest because, although consistent with observations from other simulations of the folding of this system that utilized simplified models of the protein (143–145), it appears to disagree with the interpretation of H/D exchange protection factor measurements on the 60-residue protein (141). The experimental findings report the protection of amide hydrogen atoms from exchange with solvent and observe that helix II and helix III show greater protection than helix I, protection that appears to be consistent with the overall stability of the protein. From these results the authors argue that because helix II and helix III show a greater stability than helix I (with respect to H/D exchange protection) under native conditions, helix I must unfold first and therefore fold last (146–149). This model assumes that rearrangement after the initial formation of partial structure in the protein does not occur. In the calculations, one sees that rearrangement can occur late in folding and also lead to a consistent picture of the "stability" of individual peptide regions in the folded protein, i.e. the high mobility of helix I in the native folded basin is consistent with the lower H/D protection measurements, whereas the folding process still seems to involve early formation of structure around the helix I–helix II turn interface. We note, however, that the differences may be due to the fact that the experimental measurements are on the longer polypeptide, and the two systems may be inherently different.

Also of note in this system is the nature of the "pathways" sampled during folding. The free-energy surface shown in the figure suggests that regions near the R_g/ρ diagonal are heavily populated as this protein folds. This implies that there is a concomitant formation of secondary (local) and tertiary (long-range) structure during folding. Such would be the case for proteins that fold on "near-ideal" funneled landscapes with fully delocalized transition states (34). One measure of the distributed nature of the transition state is the similarity of structures within the folding ensemble to each other as folding progresses. Boczko & Brooks (27) illustrated that indeed conformations of the protein sampled prior to the transition-state region are broadly, and nearly uniformly, distributed in their dissimilarity, whereas conformations adopted after this point are significantly more similar to each other. This too was indicated by examining the distribution of native contacts as one approached the transition region (20), which was shown to be nearly uniformly distributed, consistent with partial participation of most of the native interactions in transition from the unfolded to the folded state—a hallmark of folding on a smooth funneled landscape (24).

Segment B1 of Streptococcal Protein G   Segment B1 of streptococcal protein G (GB1) is a small mixed α/β protein comprising 56 residues (see Figure 1b; see color insert). Its structure has been elucidated via X ray (150) and NMR (151–153), as well as being characterized by molecular simulation (154). The folding kinetics of GB1 have been classified as two-state, although experiments have indicated that some structure forms early in folding. Specifically, rapid mixing measurements of fluorescence quenching suggest early formation of a relatively collapsed metastable state that buries a tryptophan residue in "rapid" equilibrium with the unfolded state preceding the rate-limiting barrier crossing to the native state (155). This study supports earlier work based on quenched-flow H/D exchange protection and characterization of a denaturant-induced unfolded state, which indicates that early collapse involves formation of structure around the C-terminal β-hairpin, separating β-strands 3 and 4, and in the middle and near the N-terminus of the central α-helix (156, 157).

This small IgG-binding protein provided a target for folding studies using the biased-sampling, free-energy surface approach described above. It differs in topology from the earlier studied helical protein, it folds rapidly, and it has been examined in some detail by experiment. In a series of papers, Sheinerman & Brooks explored the folding free-energy surface and a range of properties characterizing the folding mechanism for this protein by using detailed atomic models (126, 127, 154). The L-shaped free energy surfaces projected onto ρ and R_g are shown in the panels of Figure 1b (see color insert) for folding under native conditions (298°K) and near the folding temperature (343°K). What is clearly evident in these plots is the presence of a meta-stable state at relatively low R_g and about 50% of the native contacts ($\rho \approx 0.5$). This state is only marginally populated under the native conditions of the first calculations but significantly more so at 343°K, as indicated by both the free-energy contouring and coloring in the PMF surface as well as in the change in stability indicated in Table 1. The nature of this meta-stable state was identified to correspond well to that suggested by experiment, for both tryptophan burial and the presence of highly populated native interactions around the C-terminal β-turn and in regions of the α-helix. Thus, the nature of the folding landscape, the general stability of the folded protein, and the mechanism of folding are in excellent agreement with experimental observations for this system.

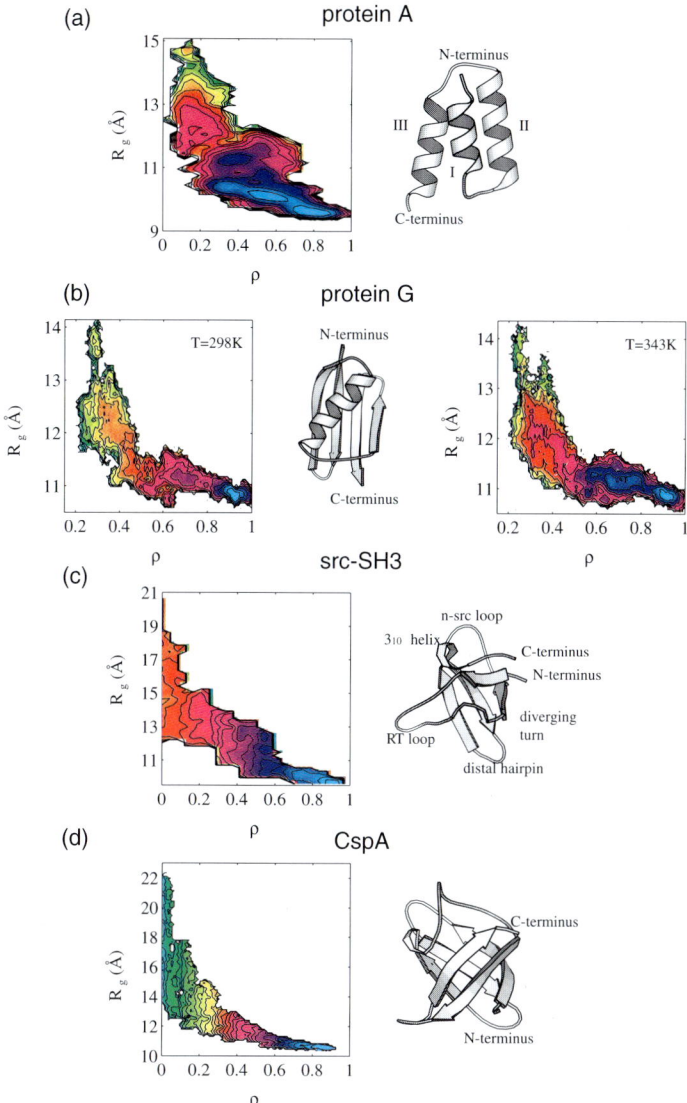

Figure 1 Free energy surfaces as a function of the fraction of native contacts (ρ) and the radius of gyration (R_g) for (*a*) fragment B of staphylococcal protein A (at 298°K, free-energy contours every 0.5 kcal/mol), (*b*) segment B1 of streptococcal protein G (at 298°K and 343°K, free energy contours every 0.5 kcal/mol), (*c*) *src*-SH3 protein (at 298°K, free-energy contours every 2 kcal/mol), and (*d*) cold shock protein A (at 298°K, free-energy contours every 2 kcal/mol). These surfaces were computed as described in text using the biased sampling of conformations generated from fully atomic denaturing simulations. The free-energy contours are colored from light blue (low energy) through to the highest energy contours which are red to green.

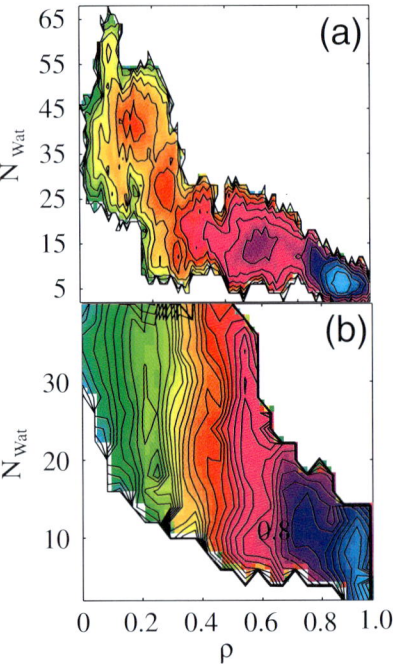

Figure 2 Free-energy surfaces at 298°K as a function of the fraction of native contacts (ρ) and the number of core water molecules (N_{wat}) for the fully atomic models of (*a*) segment B1 of streptococcal protein G (298°K) and (*b*) src-SH3 protein. The water is observed to serve as a "lubricant," facilitating the final stages of folding. Surfaces computed at 298°K and free-energy contours every 0.5 kcal/mol.

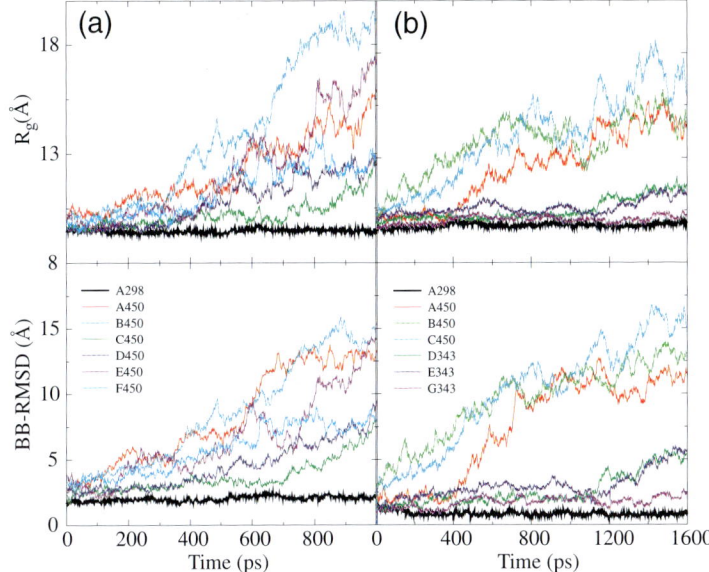

Figure 3 (*a*) Fragment B of staphylococcal protein A: time development of the radius of gyration R_g and the root-mean square deviation (rmsd) from the NMR structure for six high-temperature fully atomic unfolding runs (450°K) and one native state simulation (298°K). (*b*) *src*-SH3: time evolution of R_g and rmsd for high-temperature fully atomic unfolding (350°K to 450°K) and native state simulations (298°K). The unfolding runs for both proteins are observed to follow different pathways from the native to the denatured states.

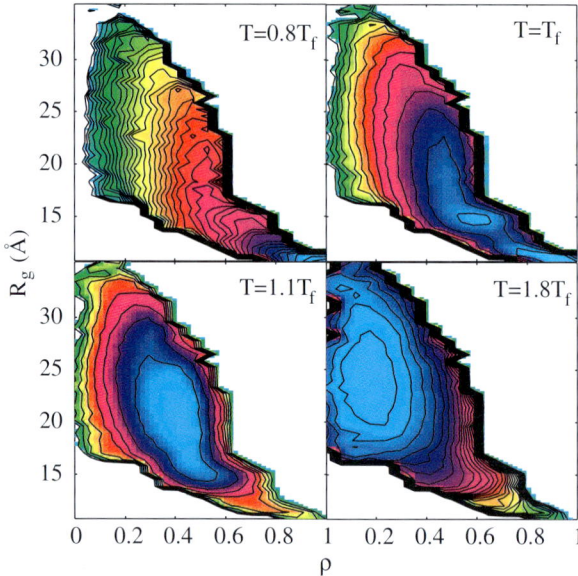

Figure 6 Free energy surface as a function of the radius of gyration R_g and the fraction of native contacts (ρ) for the minimalist model of cold shock protein A below (*a*), at (*b*) and above (*c, d*) the folding transition temperature (0.8 T_f, T_f, 1.1 T_f, and 1.8 T_f, respectively). The native and nonnative basins are equally populated at the folding transition temperature. Only the folded (unfolded) basins are significantly populated below (above) the folding temperature. The metastability of the unfolded (folded) state is lost under strongly native (nonnative) conditions, making the folding (unfolding) appear to be downhill (uphill) in nature. Surface free-energy contours are shown every 2 kcal/mol.

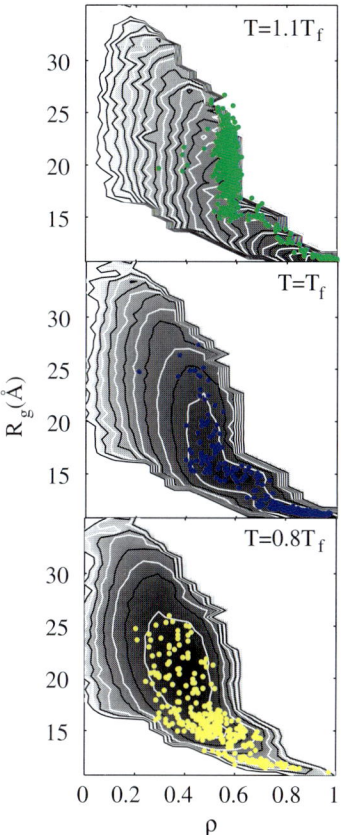

Figure 7 Typical simulation runs for the minimalist model of cold shock protein A below (*a*), at (*b*), and above (*c*) the folding transition temperature (0.8 T_f, T_f, and 1.1T_f, respectively) superimposed onto their respective free energy surfaces as a function of R_g and ρ. The trajectories closely follow the free energy surfaces. Surface free energy contours are shown every 2 kcal/mol.

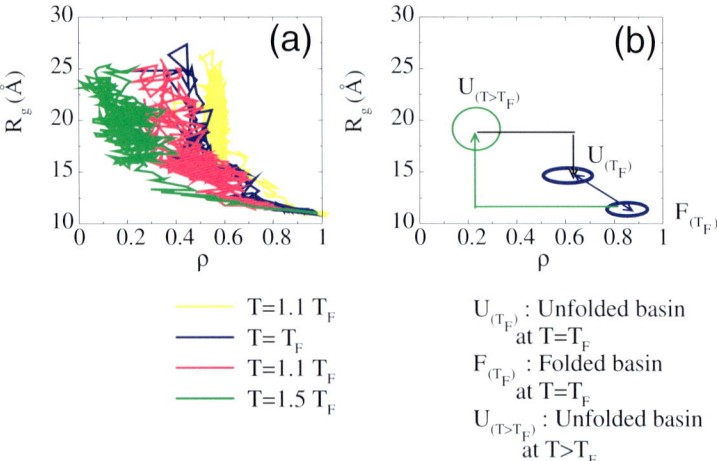

Figure 8 Typical trajectories for the minimalist model of cold shock protein A at different temperatures plotted as a function of R_g and ρ. Folding proceeds by the reverse route as unfolding under equilibrium conditions at T_f. The unfolding trajectories, however, start to deviate from the pathways taken in folding trajectories as the temperature is increased. Whereas folding at T_f occurs through an initial collapse followed by concomitant formation of secondary and tertiary structure, the unfolding under high temperature proceeds through a "swelling" of the protein. Secondary structure is lost first, followed by tertiary structure. (*b*) A schematic representation of this hysterisis.

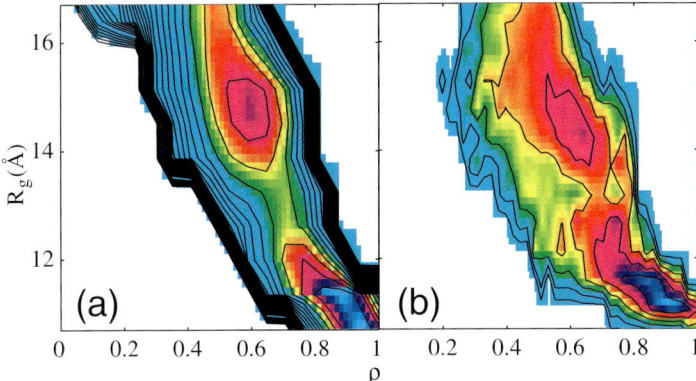

Figure 9 Free-energy surfaces just below the folding transition temperature for the minimalist model of cold shock protein A computed by (*a*) combining simulations at different temperatures ("exact method") and (*b*) using the biased-sampling method described in the section on minimalist models. The biased-sampling method successfully reproduces the native and nonnative basins as well as the location of the transition state with significantly less sampling time than the folding time scale. Surface free-energy contours are shown every 1 kcal/mol.

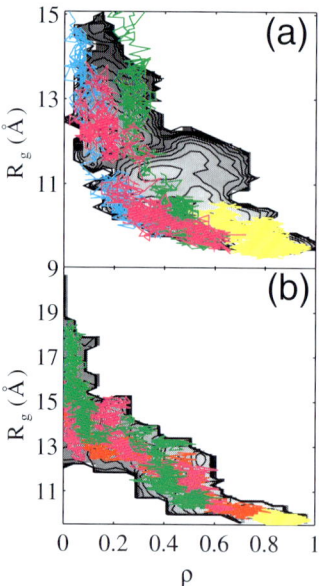

Figure 10 Unfolding simulations at different temperatures (350°K to 450°K) projected onto the free-energy profile at 298°K for (*a*) fragment B of staphylococcal protein A and (*b*) *src*-SH3. The unfolding runs parallel the free-energy profile in the case of SH3 but not in the case of protein A. The consistency between the folding and unfolding routes are linked to the nature of the folding mechanism. Surfaces contoured as in Figure 1.

The quality of this agreement gives one confidence to explore aspects of folding for this protein not directly amenable to experiment at this time and to learn more about how folding proceeds. Sheinerman & Brooks pursued a number of such questions (126, 127), of which we highlight only two here. The first is directed toward characterizing the nature of the folding transition state in this system. As noted above, this study, and others from the Brooks group, suggested the link between folded topology and the general mechanism for folding, later reinforced by experimental correlation (136). Furthermore, we described above the highly delocalized, but approximately equal, population of native interactions in the transition region for folding of the simple helical topology of protein A (20, 27, 34). Does GB1, another representative of the family of "simple," "fast-folding" proteins, show similar characteristics as it moves from unfolded to folded states? The answer is no, GB1 has a more polarized transition state involving greater probability of structure formation in specific regions involving the β-turn and helix noted above. This suggests that modification of residues in these regions of the protein's sequence should show the greatest influence on folding rates. Also, this observation suggests that the folding funnel for protein G is less smooth and less ideal than that of the all-helical protein A.

The second point we highlight involves a proposal for the microscopic origin of the barrier to folding for GB1. Sheinerman & Brooks (126, 127) examined the free-energy surface for folding of this protein by using a unique set of reaction coordinates that probed the role of protein interior water molecules in the late stages of folding. By projecting the free-energy surface onto coordinates that describe the loss of core water molecules (N_{wat}) and the fraction of native contacts (ρ), they discovered that the meta-stable state noted above was also a minimum on this free-energy surface and was stabilized by an excess of 4–6 buried core water molecules (see the free energy surface in Figure 2*b*; see color insert). Detailed examination of the nature of these core water interactions with the polypeptide chain revealed that they were generally bridging nascent β-sheet hydrogen bonds, often between mis-registered donor/acceptor partners. From the observations based on their calculations, Sheinerman & Brooks postulated that water was serving as a "lubricant," facilitating the "search" for correct hydrogen-bonding partners in this collapsed, near-native state, and that the barrier was most likely due to the final expulsion of these water molecules. This novel proposal points to the potential strength of complementing both experiment and theory by simulation studies. While this suggestion awaits confirmation for GB1, the complex folding kinetics—associated with formation of water-mediated hydrogen-bonding iterations later exchanging to native-like hydrogen bonds—that can occur in such systems, i.e. in proteins with overall topologies dominated by β-sheet interactions, may have been observed in ultrafast laser T-jump experiments performed on the all-β protein cold shock protein A (158).

src-**SH3 Domain**   The *src*-SH3 protein, represented in Figure 1*c* (see color insert), possesses 5 β-strands and 1 3–10 helix. Its hydrophobic core comprises two orthogonal β-sheets consisting of the terminal β strands, the RT loop (sheet 1), the nrc loop, and the distal hairpin (sheet 2). The protein has been studied experimentally by Baker and co-workers (136, 159–162). SH3 was found to fold rapidly without forming any detectable intermediates. Mutagenic studies (ϕ-value analysis) revealed that the distal loop beta hairpin and the diverging turn interact in the transition state, whereas the rest of the protein remains fairly unstructured.

The PMF for SH3 under native conditions appears to be almost downhill in free energy, implying a strong driving force for folding, consistent with the rapid experimental folding rate. Our analysis indicates that the protein samples a large number of different structures en route to the folded state. The probability of forming particular contacts along the reaction coordinates ρ and R_g, however, clearly show consistent early structure formation in the distal loop and the diverging turn, the very regions that are observed to be formed in experimental studies of the transition state.

Additional examination of the contact probabilities at various stages of folding reveals that the contacts between the two sheets of the hydrophobic core form very late in the folding process. This observation is consistent with both our unfolding simulations and those of Tsai et al (see next section) (163), in which the hydrophobic core is observed to be one of the first structural elements to disappear.

In order to explore the formation of the hydrophobic core further, we computed the free energy surface of the protein as a function of the fraction of native contacts (ρ) and the number of core water molecules (N_{wat}). The (ρ-N_{wat})-PMF resembles the (ρ-R_g)-PMF by its downhill nature and minimum at the native state. The number of core water molecules decreases very rapidly (from nearly 15 water molecules at $\rho = 0.85$ to only 4 at $\rho = 0.95$), implying that the final stages of folding involve the expulsion of water molecules from the hydrophobic core. The water molecules are observed to act as hydrogen bond "bridges" between the two core sheets. These water-sheet bonds are lost in the last stages of folding, allowing the native intersheet hydrogen bonds to form. Thus, it appears that in analogy with GB1, src-SH3 also uses water molecules as "lubricants" to assist the final stages of folding.

Although the topology of the protein determines the shape of the PMF (L-shaped versus diagonal), the detailed folding mechanism can influence the extent to which the PMF adopts a specific shape. To explore this relationship, Shea et al (22) compared a series of protein models with the same native fold, but with folding mechanisms spanning a nucleation-collapse type mechanism (involving concurrent collapse and formation of native contacts) to a mechanism in which the protein can get trapped in a collapsed, nonnative state en route to the folded state. This model built upon the all-β four-stranded bundle originally developed by Honeycutt & Thirumalai (23). By varying the strength of the nonnative hydrophobic interactions, model proteins with the range of folding behaviors described above could be created and their folding studied. The details of this study can be found in the paper of Shea et al (22). The results are summarized as follows: (*a*) The β-topology led to an overall L-shaped PMF. (*b*) The β-protein that folded with concurrent formation of secondary and tertiary elements had a more "diagonal L-shape" than the protein that went through a trapped state. Thus, the shape of the free energy surface and the mechanism of folding are determined by the topology to which the protein is folding and the sequence-specific interactions in the protein chain.

Cold Shock Protein A Cold shock protein A (CspA) and its relative CspB represent simple β-barrel folded topologies (Figure 1d; see color insert) (164–166) that fold rapidly and without detectable intermediates on a millisecond time scale (167–169). CspA has been the subject of folding free-energy surface calculations in which the biased sampling methods described above by Brooks and colleagues was used (132). The free-energy surface for this protein as computed in these studies is shown in Figure 1d (see color insert) and the stability appears in Table 1. What is most evident from the L-shaped landscape is the steep slope and strong driving force toward the native state under salt-free conditions at 298°K. The steep nature of this surface suggests that folding may be quite fast, a conclusion that is confirmed by experiment (168). Also of interest in this system is the complexity of folding dynamics on quite short time scales, as recently noted by Leeson et al (158). It was suggested above that these complications may be due to water molecules in the protein core late in the folding process. However, this speculation remains to be investigated by Brooks and his coworkers.

METHOD B: Probing Protein Folding Through Unfolding Simulations

An alternate means of studying protein folding in a fully atomic representation of the protein is through temperature (or denaturant)-induced unfolding simulations. A series of such simulations are used to infer the folding mechanism of a protein. This method was pioneered, and most extensively exercised, by Levitt & Daggett (30, 119). The underlying assumption in this approach is that unfolding proceeds by the reverse route as folding. In the following sections, we discuss the results of recent unfolding simulations utilizing this protocol. Following that, the folding and unfolding of a simplified protein model is examined to investigate conditions under which protein folding is the reverse of protein unfolding.

The Protocol of Temperature-Induced Unfolding Simulations

The methodology of high-temperature protein folding is quite straightforward. A standard protocol involving equilibration of the solvated protein at constant pressure and the temperature of interest (usually over 400°K) is carried out for several trials, possibly at differing temperatures. The constant pressure conditions allow the simulation volume to change and thus achieve the correct density. Constant-volume simulations have also been employed at reduced water densities (170).

Production runs are performed for up to several nanoseconds. The conditions used to unfold the protein can accelerate the simulation events by approximately six orders of magnitude (171), allowing the unfolding barrier to be overcome in a matter of nanoseconds.

Unfolding Pathways from Temperature-Induced Unfolding

Among the more extensive unfolding simulations are those performed by Daggett and colleagues (30, 118, 119, 170, 172–174). Over the past few years, the Daggett group has studied the unfolding of a number of proteins, including chymotrypsin inhibitor 2 (CI2), barnase, fragment B of staphylococcal protein A (and related variants), and lysozyme. They have worked in close collaboration with experimental groups, often obtaining results consistent with experiment. In the cases of barnase and lysozyme, they believe they have identified the major kinetic unfolding intermediates. Studies have also been performed on systems related to those for which free energy surfaces have been computed. We focus our discussion below on results from these calculations as well as others that show particular correspondence to experimental studies.

Fragment B of Staphylococcal Protein A   The folding free-energy surface and properties relating to the folding mechanism of protein A, computed using method A, are described in the previous section. In this section we review recent simulations that have used temperature-induced unfolding to explore simulated unfolding pathways and to infer information about relative stability of helical fragments and pathways for folding of this system. Our discussions are centered around work from Alonso & Daggett (175), as well recent unpublished results from Brooks' group (176).

Alonso & Daggett studied the unfolding pathways of a number of 3-helical domains related to, and including, the 46-residue truncated fragment of protein A studied by Brooks and coworkers (20, 27). These comprised the 57–60-residue fragments of the E and B domains, the so-called Z domain, and the 46-residue B fragment. For the E, B, and truncated B domains, two unfolding simulations at 498°K (with reduced water density) were carried out for each beginning structure. In all cases, unfolding proceeded rapidly at 498°K, with the structures reaching >90% of their unfolded R_g values in 1–2 ns. Alonso & Daggett's conclusions based on these trajectories were that (*a*) helix III appears to persist longest, and hence is "most stable" in all structures; (*b*) the relative "stability" of helix I and helix II depend on the particular realization of the unfolding simulation as well as the sequence, e.g. helix II appears to be more persistent than helix I in the B domain unfolding simulations but less so in the E domain, and all helices seem to be more persistent in one of the B domain simulations than the other; and (*c*) the B domain appears to adopt a more compact unfolded state than the E domain. The basic findings regarding helix stability appear to coincide with experimental findings based on H/D exchange and fragment studies, as noted above.

It is interesting to compare the conclusions from these studies to those described earlier in the work of Boczko and Brooks & Guo et al (20, 27). There are clear differences in the conclusions from these two studies regarding the order of events in folding. The studies of Brooks and coworkers, inferred from an equilibrium population analysis of conformations from the biased-sampling PMF calculations, support the idea that the proline-induced turn around the helix I–helix II interface forms earliest and that helix III has transient structure until late in folding. According to the nonequilibrium, temperature-induced unfolding studies of Alonso & Daggett, helix III forms a "scaffold" for structure formation in this system, and folding is centered around the stability of this helical segment. Guo et al also noted that they saw potential pathways for folding involving almost complete formation of helices prior to significant collapse. However, because of sampling issues, they were unable to quantify with high confidence the fraction of such "pathways" that may exist. They also noted the relative mobility of helix I once the protein enters the native basin, and this is consistent with the lower H/D protection for this helix under native conditions. Finally, both the free-energy surface calculations and the unfolding simulations support the idea that the unfolded state of the B domain is relatively compact. Brooks and coworkers saw a sharp rise in the free energy as a function of R_g for radii of gyration around 12.5 Å, whereas Alonso & Daggett observed that the average R_g value obtained in their unfolding simulations of the B domain is around 12–13 Å, compared with the more extended unfolded state of the E domain.

We also briefly review recent unfolding studies by Brooks and colleagues on the truncated protein A fragment (176). The results from 6 independent unfolding runs at 450°K and 2 at 298°K starting from the native NMR structure of fragment B of staphylococcal protein A are shown for the time development of the radius of gyration, R_g, and the root-mean-square deviation (rmsd) of the evolving structure from the initial NMR structure (Figure 3a; see color insert). What is quite evident from the data presented in the figure is that each different unfolding realization takes a very different unfolding "pathway." This is also evident from the work of Alonso & Daggett (see Figure 2, see color insert; 175). This raises two critical questions regarding the potential utility of such calculations in studying unfolding/folding pathways. (a) What statistical weight should one apply to these trajectories in analyzing them for consensus unfolding pathways and relating these to observable properties from folding experiments? (b) If only a small number of such trajectories were going to be carried out in studying the unfolding behavior of this system, which should one choose? Again, this is an issue of which, if any, of these realizations is representative of what happens with greatest frequency. These problems are inherent to the utilization of nonequilibrium methods, and the solution is to utilize many, many trials, allowing the appropriate statistical weight to emerge through the frequency of occurrence of events in the time-course of the "relaxation process" being measured. In the sections below we attempt to address this issue directly by examining behavior of a model system.

Chymotrypsin Inhibitor 2 CI2 is a small mixed α-β protein that has been studied extensively experimentally by Fersht's group (177). It is a two-state folder and its transition state has been characterized through mutagenic studies. Daggett et al (170, 173) examined CI2 computationally with high-temperature unfolding simulations. They identified the transition state as consisting of those conformations formed prior to a large structural change based on clustering in a space of root-mean-square deviation between conformations. The idea is that the transition state structures are highly unstable and that the protein will undergo a significant structural rearrangement once it has passed the transition state. This is clearly far from an ideal way of determining the transition state. However, both experiment and simulation are in qualitative agreement, suggesting that the transition state is close to the native state, with a disrupted β-sheet and a mostly intact helix. The unfolding simulations were subsequently used to design faster folding mutants of CI2, based on the observation that the transition state structures identified from the molecular dynamics study appeared to have a number of unfavorable contacts. Experiments performed on mutants in which these contacts were removed or replaced by stabilizing interactions showed an increase in the rates of folding (178).

src-SH3 Domain Levitt and co-workers (163) recently performed a series of unfolding simulations on the *src*-SH3 protein (see Figure 1*c*; see color insert). This protein is experimentally known to have a very polarized transition state in which the distal loop and part of the diverging turn are structured, whereas the rest of the protein is largely unstructured. The unfolding simulations of SH3 show that the regions that form the experimental transition state are persistent, i.e. they are the last structural elements to disappear as the protein unfolds, in agreement both with experiment and the more thorough analysis of the folding PMF presented in the preceding section. ϕ-values are a measure of the extent of structure in the transition state (179) and molecular dynamics ϕ-values calculated from the unfolding runs are also in reasonable agreement with the experimental observations. Experimentally, this quantity is related to the ratio of folding rates for native and mutant proteins relative to the overall change in protein stability upon mutation. The molecular dynamics ϕ-values are a rough estimate of the experimental ϕ-values. They were calculated by summing the native side chain hydrophobic contacts of each residue for structures possessing between 50% and 60% of the native contacts (the estimated transition state from simulations). More detailed analysis of the nature of such descriptions of this quantity, and the breakdown of their simple interpretation has been performed by Shea et al for simplified models (24, 25).

As a component of calculating the folding free-energy surface described above for the *src*-SH3 domain (135), unfolding studies were also performed at temperatures varying between 350°K and 450°K. Figure 3*b* (see color insert) shows typical profiles of unfolding for rmsd and R_g versus time. What is most clear from these calculations is that differing, independent realizations of the unfolding process yield different unfolding pathways. Whether consensus regarding the nature of those pathways emerges from clustering and related analysis, and hence whether one can confidently draw conclusions about either unfolding or folding pathways, is clearly dependent on whether sufficient sampling of the nonequilibrium unfolding events has occurred. Although the analyses just described for *src*-SH3 and CI2 appear to be in good agreement with experiment, caution must be exercised in assessing the robustness of this agreement. As we discuss below, many tens of trajectories (58) at differing temperatures may be required to achieve consensus regarding the unfolding "pathway" and then only for proteins that fold via specific mechanisms (to specific topologies).

MINIMALIST MODELS TO ASSESS BIASED SAMPLING POTENTIALS OF MEAN FORCE AND TEMPERATURE INDUCED UNFOLDING APPROACHES FOR CHARACTERIZING FOLDING LANDSCAPES

In the following sections we explore the strengths and shortcomings of both the equilibrium biased–sampling PMF method (method A) and the nonequilibrium temperature-induced unfolding method (method B) by applying them to a minimalist off-lattice model for which the complete folding process can be studied. We begin with a complete characterization of the thermodynamics and kinetics of folding of this model to provide a reference for comparison. We then analyze the model using both methodologies. Comparing the unfolding and folding behavior of the models allows us to investigate the validity of studying folding by means of unfolding simulations. In particular, we seek to establish whether the principle of microscopic reversibility (or detailed balance) holds under the extreme conditions used in such simulations. We also compare the PMF generated from complete knowledge of the density of states for the model system with the one obtained by the Brooks method, which combines temperature induced unfolding dynamics with biased sampling. This enables us to determine whether this method can suitably locate the correct free-energy minima (corresponding to the states of the protein) on the PMF surface as well as lend insights into the folding mechanism.

Protein Model and Methods

The model protein we study is an α-carbon (C_α)–based off-lattice minimalist representation of the 5-stranded, 69-residue β-barrel cold shock protein CspA (see Figure 1*d*; see color insert). CspA and its homologues of the CspB and CspC families have been extensively studied by the experimental groups of Gregoret (180, 181) and Schindler et al (167, 168). These proteins fold extremely rapidly in a two-state manner without forming any detectable intermediates.

We have opted to model CspA in the simplest way possible by using a Gō-type representation (17). Attractive interactions are assigned only between residue pairs that are in contact in the native state (favorable pairs); all other pairs interact through hard-sphere potentials. Gō-models adequately represent proteins, such as CspA, that have little energetic frustration and for which the folding is dominated by the topology of the protein (24, 41). They provide a simple representation of the protein and yield the desired thermodynamic and kinetic two-state folding behavior. An additional advantage of using Gō-models is that for such systems the location of the transition state for the folding "reaction" can be simply related to the barrier in the free energy projected onto a reaction coordinate (such as the number of native contacts) (24, 42). This is an extremely useful property that we exploit in the following analysis. The model is similar in nature to the one used by Shea et al (24) for the helical bundle protein A. We refer the reader to this work for further details.

Thermodynamic Simulations

The protein model was incorporated into the molecular dynamics program CHARMM (182), and conformational populations and properties were sampled over a broad range of temperatures from 900°K to 200°K, in 25°K increments. This range of temperatures allowed us to probe high-energy extended conformations as well as the low-energy compact states. The time was measured in units of $\tau = (m/\varepsilon)^{1/2} r_0$, where m, r_o, and ε are 50 atomic mass units, 3.78 Å, and 2 kcal/mol, respectively. The model was allowed to evolve in a continuum manner following Langevin dynamics, with a friction coefficient of $0.2\tau^{-1}$ for times of $\sim 30,000\tau$ at each temperature. Snapshots were saved only every 3.75τ, sufficiently infrequently that each data point can be considered to be independent. The sampling from the simulations at each temperature was combined by using the weighted histogram analysis method (WHAM) (134). This method yields an optimal estimate of the density of states of the system and hence allows the calculation of relevant thermodynamic quantities.

The WHAM equations used here differ slightly from the constant-temperature version mentioned above, we present a brief overview of the WHAM equations tailored to our system. We describe the evolution of the model from a random coil to the native state with two progress variables, the fraction of native contacts (ρ) and the radius of gyration R_g. The density of states $\Omega(V, R_g, \rho)$ for our system (where V is the potential energy) is given by

$$\Omega(V, R_g, \rho) = \frac{\sum_{k=1}^{R} N_k(V, R_g, \rho)}{\sum_{m=1}^{R} n_m \exp(+f_m - \beta_m V)}, \qquad 6.$$

where n_m is the number of data points in the mth simulation, $\beta_m = 1/k_B T_m$, and k_B and T_m are Boltzmann's constant and the temperature at which the mth simulation was performed, respectively. $N_k(V, R_g, \rho)$ is the value taken by the histogram, and f_m is the scaled free energy obtained by solving the following equations self-consistently.

$$P_\beta(V, R_g, \rho) = \frac{\sum_{k=1}^{R} N_k(V, R_g, \rho) \exp(-\beta V)}{\sum_{m=1}^{R} n_m \exp(+f_m - \beta_m V)} \qquad 7a.$$

and

$$\exp(-f_j) = \sum_{V, R_g, \rho} \Omega(V, R_g, \rho) \exp(-\beta_j \cdot V). \qquad 7b.$$

The potential of mean force (PMF) W_β at temperature T, or free energy projected onto the reaction coordinates (ρ and R_g), is obtained from the probability density P_β

$$W_\beta(R_g, \rho) = -k_B T \ln[P_\beta(R_g, \rho)] \qquad 8.$$

The thermodynamic properties for the model can be calculated from the probability density (22).

Two-State Thermodynamics of Folding The Gō-model for CspA presents the kinetic and thermodynamic signatures of a two-state folder. The collapse (T_c) and folding transition (T_f) temperatures, determined independently from the behavior of the specific heat and the average fraction of native contacts $\langle \rho \rangle$ as a function of temperature (Figure 4a and b), illustrate this point. The number of native contacts was redefined from the average of the low temperature (200°K) conformations as all i, i + 4, and greater contacts with separations less than 8 Å. Folding and collapse are seen to occur concurrently at $T_f = T_c$, indicating that the model is a good folder. The heat-capacity curve is narrow and the average number of native

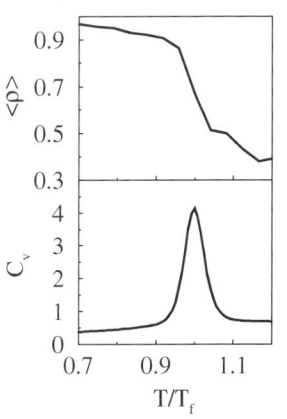

Figure 4 Collapse and folding transition temperatures for the minimalist model of cold shock protein A, determined from the behavior of (a) the specific heat C_v as a function of temperature and (b) the average fraction of native contacts $\langle \rho \rangle$ as a function of temperature. Folding and collapse are observed to occur concurrently at the folding temperature T_f.

contacts rises sharply at the transition temperature. These are indications that the transition is first-order-like.

Free Energy Profiles Above, Below, and at the Folding Transition Temperature T_f
The PMF surfaces (free energy as a function of a reaction coordinate) for the CspA model were calculated at different temperatures from the probability histograms determined through the WHAM equations.

The free energy projected onto the reaction coordinate ρ (fraction of native contacts) (Figure 5b) at the folding transition temperature T_f presents two equal free-energy minima corresponding to the native and nonnative basins around $0.55 < \rho < 0.65$ and $0.85 < \rho < 0.95$, respectively. These minima are separated by a barrier of 1.5 $k_B T_f$ at $\rho = 0.7$. At temperatures above the folding transition, the curve shifts to the unfolded basin, whereas below the transition temperature the curve shifts to the native basin (Figure 5a, c). This, too, is a signal of a two-state, first-order-like transition. The folding transition in this model is energetically and entropically driven, with the energy and entropy varying almost linearly with ρ. The almost complete cancellation of these two terms leads to the small free-energy barrier at the folding temperature, as observed in Figure 5b.

The two-dimensional PMF projected onto the fraction of native contacts ρ and the radius of gyration R_g offers a more detailed picture of the thermodynamics of

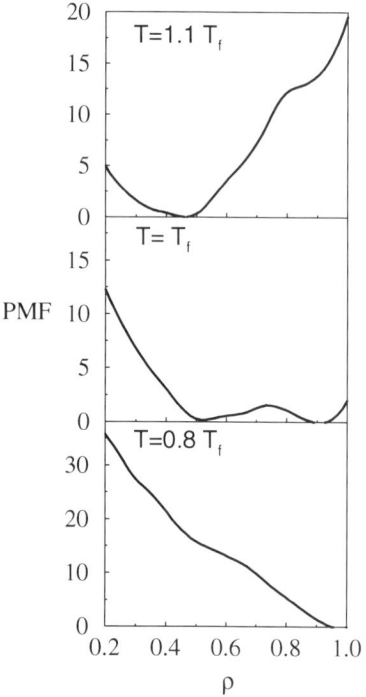

Figure 5 Free energy as a function of the fraction of native contacts (ρ) for the minimalist model of cold shock protein A at 1.1 T_f, T_f, and 0.8 T_f. The free-energy profile at the folding temperature T_f displays two equally populated basins corresponding to the native and nonnative states. The free energy profile is observed to shift to the unfolded (folded) basin with increasing (decreasing) temperature.

folding. At the folding temperature, the native ($0.85 < \rho \approx 0.95$, 10.7 Å $< R_g < 11.2$ Å) and nonnative ($0.55 < \rho < 0.65$, 14 Å $< R_g \leq 15$ Å) basins are separated by a barrier located near ($\rho = 0.7$, $R_g = 13$ Å) (Figure 6b; see color insert). The population of the basin shifts from the native (nonnative) to the nonnative (native) as the temperature is raised (lowered) (Figures 6a, c, d; see color insert).

The folded and unfolded states are metastable states and their metastability is lost under extreme conditions, as evidenced by the free-energy surfaces in Figure 5a, c and Figure 6a, c–d. Under strongly native or nonnative conditions, the free energy barrier at the transition state disappears and folding becomes apparently downhill below T_f and uphill above T_f.

Comparison of Folding and Unfolding: Can Unfolding Simulations be Used to Study Folding?

Folding Under Native Conditions   To explore the kinetics of folding for this model, several hundred independent folding runs, initiated from different high-temperature random coil structures, were performed under native conditions (T < T_f). A typical folding run below the folding temperature (0.8 T_f) is superimposed onto the corresponding PMF in Figure 7a (see color insert). At temperatures below the folding temperature, only the native basin is significantly populated, and the free-energy surface slopes in a downhill manner. The kinetics closely parallel the thermodynamics of folding, with the structures following the free-energy surface from the high-energy (unfolded) regions to the low-energy folded basin.

Folding and Unfolding at the Transition Temperature   The behavior of the protein at the folding transition temperature was investigated by initiating folding from high-temperature unfolded conformations and allowing the system to equilibrate at T_f. At the folding temperature, the conformations populate equally the native and nonnative basins, alternating between the folded and unfolded state as the kinetic runs proceed. The first passage times for folding were obtained by recording the first time at which a conformation reaches the native basin. The distribution of first passage times is single exponential, indicating a first-order-like transition without the formation of kinetic intermediates. Typical mean first passage times are around $187,500\tau$. The kinetic behavior, as with the thermodynamic characteristics, is reflective of a two-state system. The superposition of a run at T_f over the corresponding PMF reinforces the parallel between the kinetics and thermodynamics of the model, Figure 7b (see color insert). The population shifts from the unfolded to the folded basin without occupying an intermediate state.

Clearly, folding is the reverse of unfolding at this temperature, as it corresponds to the temperature at which the system is in equilibrium between the folded and unfolded states. Whereas it is clear that the protein samples a large number of different trajectories, our detailed analysis reveals that the protein folds and unfolds on average in the same manner, repeatedly sampling the unfolded and folded basins, separated by the transition state at ($\rho = 0.7$, $R_g = 13$ Å).

Unfolding at High Temperatures The unfolding of the CspA model was studied by heating structures belonging to the native basin to temperatures of 1.1 T_f and 1.5 T_f. An unfolding run at 1.1 T_f is plotted with its corresponding PMF in Figure 7c (see color insert). The kinetics parallel the thermodynamics, following the free-energy surface to the unfolded basin at ($0.3 < \rho < 0.5$, $17 \text{ Å} < R_g < 22 \text{ Å}$).

Figure 8a (see color insert) shows typical folding and unfolding trajectories at temperatures below, at, and above T_f. The trajectories are similar for similar temperatures but start to significantly deviate as the temperature difference increases. The trajectories at 1.5 T_f clearly do not sample the same regions of conformational space as the folding/unfolding runs at T_f. The unfolding basin observed at T_f is completely bypassed in the high-temperature unfolding runs. In these cases, the protein initially "swells" (i.e. gradually losing secondary structure in the beta sheets) at a low radius of gyration followed by a loss of tertiary structure. The unfolding runs at T_f show a more concomitant loss of secondary and tertiary structure.

Figure 8b (see color insert) presents a pictorial summary of the trajectories at T_f and 1.5T_f. As discussed above, once the system has equilibrated at T_f, folding is the reverse of unfolding at this temperature. The system samples the unfolded and folded basins shown in Figure 8b (see color insert) as the protein unfolds and refolds at T_f. A clear hysterisis can be observed in Figure 8b (see color insert) for refolding initiated from the high-temperature unfolded basin (high R_g, low ρ) and run at T_f. These trajectories rapidly form a significant amount of native structure (namely in the terminal beta-sheet regions) and collapse to the unfolded T_f basin (mid R_g, mid ρ) after passing through a pre-equilibrium (high R_g, mid ρ) state. Folding then proceeds to the native (low R_g, high ρ) basin with a concurrent formation of secondary and tertiary structure. Unfolding at T_f proceeds in the reverse manner. Unfolding at 1.5 T_f, alternatively, proceeds through a swelling of the protein: The protein starts losing its native contacts while still in a relatively compact state (decreasing ρ at low R_g). Once a significant number of native contacts has been lost (nearly 70%), the protein expands to the high-temperature unfolded basin (low ρ, high R_g). Unfolding at high temperatures (1.5 T_f) hence proceeds through a loss of secondary structure followed by a loss of tertiary structure. The unfolding at T_f shows a concurrent loss of secondary and tertiary structure.

It is apparent from our studies that the principle of microscopic reversibility does not hold under strongly nonequilibrium conditions. It should be noted, however, that the "low-resolution" differences between the unfolding and folding runs are not dramatic. On the whole, the terminal beta sheets are formed first and lost last. In a very crude manner, unfolding runs can be used to gain insight into the overall folding mechanism of the protein. What the unfolding runs fail to do is pass through the transition state for folding. It is apparent from Figure 5 that the transition state (as given by the maximum of the free energy projected onto ρ) shifts toward the native state as the temperature increases, in agreement with Hammond's principle (183).

As mentioned above, whereas the trajectories at and above T_f can differ quite significantly, the overall folding/unfolding does seem to proceed grossly in a similar manner. This agreement, not in the details, but in the overall picture, accounts for the surprising success of high-temperature unfolding simulations to describe the folding of proteins (30, 58, 118, 119, 170, 173, 174). Although this is good news, a word of caution is necessary. This general agreement seems to hold for simple two-state folders, i.e. proteins that do not present long-lived traps or intermediates. We have investigated the unfolding/folding behavior of a more complex model, one that folds following two pathways, a fast track directly from the unfolded state to the folded state and a slower pathway in which the protein gets trapped in an intermediate step. In this system, high-temperature unfolding runs completely bypass the intermediate state and proceed directly from the folded state to the unfolded state. Our work indicates that high-temperature unfolding probes of such multitrack folders (for example, lysozyme or barnase) will be unsuccessful in providing a complete picture of folding. High-temperature unfolding runs are best reserved for the study of proteins that follow a nucleation-collapse mechanism [such as chymotrypsin inhibitor 2 (CI2)] (58). Recent studies by Dinner et al utilizing lattice-based models have reached similar conclusions (184).

Comparison with the Potentials of Mean Force Generated from Biased Sampling

In the preceding section, potentials of mean force (PMFs) were computed by combining multiple simulation runs at different temperatures. We will refer to these PMFs as exact PMFs because the extensive sampling required for their generation far exceeded the mean folding time of the protein model. This method is very useful for generating the free-energy surfaces of proteins that can be folded from a random coil on a reasonable time scale. For solvated protein models described in atomic detail, this approach is clearly not possible. For such cases it is necessary to turn to the methodology of biased sampling introduced in the section describing method A. In order to verify the validity of the biased-sampling PMF approach, we use this methodology to compute the free-energy surface for the minimalist CspA model and compare the PMFs obtained by both methods.

The PMFs at a temperature just below T_f obtained by each of the methods described above (i.e. from the combination of sampling at different temperatures and from the sampling at one temperature with initial conditions obtained from temperature-induced unfolding runs) are given in Figure 9a, b (see color insert). The PMFs from both methods are very similar. Method A, the biased-sampling PMF calculation, successfully reproduces the location of the unfolded and folded basins as well as the transition state. Some sampling deficiencies are evident, especially along the "edges" of the free-energy surface. However, this calculation, which considered 50 initial conditions and sampled each for a period of $1,500\tau$, represents significantly less computation than what was employed in the "exact" case (total simulation time of $750,000\tau$). (The minima in the free-energy surface are reproduced even with sampling as small as 100τ per initial condition.) We note that increasing the number of initial conditions further permits one to decrease sampling per initial condition while maintaining semiquantitative agreement with the "exact" surface.

Our objective was to assess the ability of the biased-sampling approach (method A to reproduce the features and statistical weights (free-energy differences) of the folding free-energy surface under conditions that favor folding. We illustrated that, even with sampling far smaller than the actual folding time of the protein, this approach provides an accurate representation of the PMF. Clearly, there will be exceptions to these observations, and particular care must be given to inferring kinetic mechanisms from these surfaces. However, we suggest that if the protocol outlined in the section describing method A above is followed, and if characteristics of the convergence of the calculated PMF are monitored as noted, the biased-sampling methods provide a robust technique to obtain thermodynamic properties for folding under specific thermodynamic conditions.

Comparison of Detailed Model Unfolding Trajectories and Folding Potentials of Mean Force

In the above sections, we have shown that under suitable conditions of temperature and sampling both nonequilibrium unfolding trajectories and biased-sampling free-energy surfaces agree with unbiased, equilibrium calculations of analogous properties ("exact" results for the same model). In this section we examine whether temperature-induced unfolding trajectories, run on systems with full atomic detail for a range of temperatures, coincide with the free-energy surfaces computed for the same systems under native conditions (298°K). We focus on two specific systems for which both high-temperature unfolding runs and biased-sampling free-energy surfaces, both in explicit solvent, have been computed. The systems we describe are the 46-residue fragment B from staphylococcal protein A and the *src*-SH3 domain. In Figure 10 (see color insert) we illustrate the "projection" of the ρ-R_g trajectories during these unfolding runs onto the ρ-R_g PMF surface.

Fragment B of Staphylococcal Protein A The simple three-helical bundle structure of protein A provides one particular system for exploring, with all-atom models, whether unfolding trajectories computed at elevated temperatures coincide with the folding free-energy surface evaluated from biased sampling under native conditions at 298°K. In Figure 10*a* (see color insert), we illustrate this by plotting several representative unfolding trajectories of this protein computed at 450°K onto the free-energy surface at 298°K. Additionally, we illustrate the behavior of a "native" trajectory at 298°K projected onto this surface. The native trajectory moves within the low-lying free-energy states. The 450°K trajectories all begin near the native basin but unfold via different pathways depending on the specific realization. One trajectory can be seen to move from the native basin through regions highly populated at 298°K and into the unfolded conformational space while others sample only the high free energy, low population, edges of the free-energy surface. The former trajectory may provide information characteristic of folding under native conditions, but it is unclear whether the other trajectories are at all representative.

These observations point to one potential problem in using thermally driven unfolding runs to characterize the folding process: Such trajectories do not necessarily coincide with regions of high population under native conditions, and hence are at best anecdotal of folding. One cannot know a priori whether this will be the case or not, though we are gathering evidence to suggest that coincidence of unfolding and refolding under different thermodynamic conditions depends on the protein topology and the degree to which the folding transition state is "polarized." The examples we illustrated with the minimalist model discussed in the previous section are further reinforced in the context of detailed molecular-simulation studies with this example.

src-SH3 We also illustrate the behavior of unfolding trajectories for SH3. As is obvious from the trajectories shown in Figure 10*b* (see color insert), the unfolding runs for SH3 all fall reasonably well onto low free-energy regions of the free-energy surface. This suggests that unfolding over the range of temperatures examined here closely parallels the folding of the protein (at least in ρ and R_g space), or alternatively that our sampling with biased molecular dynamics is insufficient. We are confident that the sampling is adequate, as extending the sampling, as well as adding biased simulations from different initial conditions, did not significantly change the PMF. We believe that in the case of SH3, the folding and unfolding of the protein proceed in similar manners. This is due to the fact that SH3 has a very polarized transition state as a component of its folding mechanism, governed in large part by the topology of the protein. Early formation of a structurally necessary portion of the protein involving the distal hairpin and the diverging turn is required for folding.

It is not possible to predict with confidence whether a protein will unfold and fold in a similar manner under differing thermodynamic conditions, e.g. high temperature versus native-favoring temperatures. Hence, we emphasize the use of free-energy surface calculations to characterize the statistical weights of "trajectories" under native conditions. The importance of this step is particularly clear in the cases of the minimalist model of CspA and the atomic model of protein A, in which certain unfolding runs do not overlap with the free-energy profiles. Attempting to infer the folding mechanism from these trajectories may lead to an erroneous understanding of how these proteins fold.

CONCLUSION

Fully atomic simulations detailing the time course of protein folding remain a great challenge, despite significant advances in computational methodologies and the ever increasing computational power available. Whereas small proteins fold on the order of milliseconds to minutes, the longest folding simulation using conventional molecular dynamics has so far not exceeded one microsecond (51, 52). Undoubtedly, we will continue to push the envelop of real-time at which one can simulate biological systems at atomic detail as we move into the twenty-first century. However, Herculean computational feats such as the simulation by Duan & Kollman (51, 52) require months of simulation on highly parallel computers still remain several orders of magnitude short of the actual folding time of a small protein. Another issue concerns the fact that proteins fold following a large number of different routes to the folded state. A single simulation only follows one trajectory initiated from a single conformation. In order to gain a reasonable insight into the folding of a protein, it is hence necessary to perform not one, but several hundred folding simulations. This serves to move the target of direct observation of protein folding by molecular simulation using explicit models (of protein and solvent) even further into the future.

Why then invest such effort into a seemingly intractable problem? Aside from the intellectual challenges this problem presents, simulations are invaluable as a complement to experimental and theoretical studies. They provide an atomically detailed picture of the protein that rivals, if not exceeds, the one obtained from experiment. A simulation can track a single conformation from the unfolded to the folded state, much as in single-molecule experiments. (Simulations are by their very nature the equivalent of single-molecule experiments, but have the advantage of being able to produce detailed visual "snapshots" of the protein as it folds.) One should note that, although single-molecule experiments hold great promise, they are still in their infancy, and the majority of folding experiments must still be performed as bulk measurements.

In this review we have focused on two schools of atomic modeling that have evolved to work around the computational obstacles presented by all-atom simulations. We have summarized the findings from this body of work and assessed the strengths and weaknesses of each method using a simplified protein model for which the complete folding from a random coil to the native state could be performed in a reasonable time frame.

The first method (method A), developed by Brooks and his colleagues, (20, 27, 59, 126, 127, 132, 135) attempts to construct the free-energy surface of the protein under native conditions (or more generally, under any conditions of interest). This is achieved by performing equilibrium-biased sampling under the desired thermodynamic conditions (temperature, pressure, etc) of initial conditions chosen from conformations generated via denaturing simulations (usually high-temperature unfolding). The second method (method B), developed in part by Levitt, Daggett, Jorgensen, & van Gunsteren, studies the folding of the protein by means of nonequilibrium denaturation simulations alone (30, 108, 111, 119). The fundamental difference between the two methods is that the first is an equilibrium "thermodynamic" study, whereas the second is a nonequilibrium "kinetic" investigation. Both methods have their place in protein folding studies, as has been illustrated by the successes from each in applications to specific protein systems.

However, we argue that method A is clearly the more complete of the two. Method B is a subset of the free-energy methodology.

From a theoretical perspective, method B may appear to be quite unsuitable for the study of protein folding. This method is implicitly based on the principle of "microscopic reversibility" (or detailed balance), a concept that is only valid under equilibrium conditions. In order to accelerate the unfolding of the protein, method B must, however, rely on the use of extreme conditions (i.e. a large perturbation to the system), such as high temperatures or other strong denaturants. Our investigations in which a simplified protein model was used (section on method B) clearly show that a protein need not unfold and fold in the same manner under very different conditions. Moreover, there is not one, but several different pathways the protein can adopt upon folding and conversely unfolding, so deducing the folding mechanism from a few unfolding simulations may be quite difficult. Despite these very valid concerns, method B has been observed, in practice, to provide meaningful results in a number of studies of the folding of small proteins. Although it cannot be used to locate the transition states for folding (the transition states shift significantly with denaturant and temperature changes), this approach often reproduces the main mechanistic elements of folding. In a very gross manner, the structural features that are the last to fold are usually the first to unfold. High-temperature simulations have been applied by a number of groups (30, 108, 111, 118, 119, 170, 172–174) to study the folding and unfolding mechanisms of, among other proteins, CI2, fragment B of protein A, lysozyme, barnase, and SH3. The method appears best suited for proteins that fold following a nucleation-collapse mechanism, explaining the remarkable agreement between experimental folding data and the unfolding simulations of CI2. It is much less reliable for proteins that display off-pathway intermediates, as these structures will most likely not be populated under extreme conditions.

Method A is in many ways intellectually more pleasing, as it relies on the unfolding simulations merely to generate a sampling of conformations from folded to unfolded states for the initial conditions used in the biased sampling under a chosen set of thermodynamic conditions. This sampling is combined by using the weighted histogram analysis method (WHAM), which provides the correct statistical weighting to generate the density of states and hence the free energy as a function of a given set of reaction coordinates. Key concerns addressed in this review regarding method A are an appropriate means of generating the initial conditions and the extent of simulation required to obtain an accurate PMF, as well as the need for suitable reaction coordinates to describe the folding of the protein. Free-energy profiles have been generated by the Brooks group (20, 27, 126, 127, 132, 135) for fragment B of protein A, segment B1 of protein G, CspA, betanova, and the *src*-SH3 protein domain. These investigations have provided considerable insight into both the folding and unfolding mechanisms of these proteins. Interestingly, not all unfolding trajectories followed the folding free-energy profiles, implying that the unfolding trajectories can differ quite significantly from the folding ones. This was the case for fragment B of protein A. In the case of SH3, on the other hand, the unfolding trajectories lay directly on the folding free-energy surface. The polarized nature of the folding mechanism of SH3 explains the consistency between the unfolding and folding mechanisms. As one cannot know before hand whether the unfolding will parallel the folding behavior, it is imperative to compute the PMF for each protein.

Other features have emerged from these studies. Of particular interest is the relationship between the topology of the protein, its folding mechanism, and the shape of the PMF. Proteins in which short range contacts are predominant show a diagonal free-energy profile in R_g/ρ space, indicative of a concomitant collapse and formation of native structure. Proteins that possess a greater number of long-range contacts have a more "L-shaped" profile, consistent with an early collapse followed by a rearrangement to the native state. Also noteworthy is the proposal that water may play a key role in facilitating folding for proteins involving long-range β-sheet hydrogen bonding by mediating interactions in the protein core late into the folding process and thereby lubricating the conformational search for the correct native pairings of hydrogen-bonding partners across β-strands.

It is our intention to provide a critical evaluation of the role of detailed atomic simulations in investigations of protein folding. We believe these techniques are essential in moving forward our understanding of the interactions that control folding. Detailed models from simulations are already used to provide insights for the design of new proteins with altered folding characteristics, e.g. faster folding. As the technical aspects of these approaches improve (in great part through the utilization of minimalist models as a "staging ground" for methods development) and as computational power increases, such models will be important in understanding medically relevant processes that involve folding and misfolding; examples in which simulations are likely to have an impact include Alzheimer's and related "protein aggregation" maladies. Furthermore, components of the methods we describe in this review will be important tools in assessing the validity and, potentially, the role of "newly discovered" protein structures from proteomics efforts.

ACKNOWLEDGMENTS

This work is based on important contributions from many members of the Brooks group over nearly a decade of effort. We thank all of these individuals for their contributions, with a special note of thanks to Erik Boczko, Zhuyan Guo, Felix Sheinerman, and Badry Bursulaya, on whose work the current developments rest. We also appreciate our interactions with Professor José Onuchic and the Onuchic group. Financial support from the National Institutes of Health (GM48807, GM57513, RR06009, RR12255) and fellowship support from a Canadian NSERC and the La Jolla Interfaces in Science Interdisciplinary Training Program (LJIS) are acknowledged. Finally, a major portion of the computational resources utilized in performing the studies presented here came from the Pittsburgh Supercomputing Center; we acknowledge them and credit their strong conviction to support computational biology.

Visit the Annual Reviews home page at www.AnnualReviews.org

LITERATURE CITED

1. Anfinsen CB. 1973. *Science* 181:223–30
2. Anfinsen CB, Haber E, Sela M, White FH. 1961. *Proc. Natl. Acad. Sci. USA* 47:1309–14
3. Kim PS, Baldwin RL. 1990. *Annu. Rev. Biochem.* 59:631–60
4. Karplus M, Weaver DL. 1994. *Protein Sci.* 3:650–68
5. Cohen FE, Richmond TJ, Richard FM. 1979. *J. Mol. Biol.* 132:275–88
6. Levinthal C. 1968. *J. Chem. Phys.* 65:44–45
7. Onuchic JN, Luthey-Schulten Z, Wolynes PG. 1997. *Annu. Rev. Phys. Chem.* 48:545–600
8. Bryngelson JD, Onuchic JN, Wolynes PG. 1995. *Proteins: Struct. Funct. Genet.* 21:167–95
9. Dobson CM, Karplus M. 1999. *Curr. Opin. Struct. Biol.* 9:92–101
10. Dobson CM, Sali A, Karplus M. 1998. *Angew. Chem. Int. Ed.* 37:868–93
11. Shakhnovich EI. 1997. *Curr. Opin. Struct. Biol.* 7:29–40
12. Chan HS, Dill KA. 1998. *Proteins: Struct. Funct. Genet.* 30:2–33
13. Thirumalai D, Klimov D. 1999. *Curr. Opin. Struct. Biol.* 9:197–207
14. Leopold PE, Montal M, Onuchic JN. 1992. *Proc. Natl. Acad. Sci. USA* 89:8721–25
15. Bryngelson JD, Wolynes PG. 1987. *Proc. Natl. Acad. Sci. USA* 84:7524–28
16. Bryngelson JD, Wolynes PG. 1990. *Biopolymers* 30:177–88
17. Gō N. 1983. *Annu. Rev. Biophys. Bioeng.* 12:183–210
18. Chan HS, Dill KA. 1997. *Proteins: Struct. Funct. Genet.* 8:2–33
19. Guo Z, Thirumalai D. 1995. *Biopolymers* 36:745–57
20. Guo Z, Brooks CL III, Boczko EM. 1997. *Proc. Natl. Acad. Sci. USA* 94:10161–66
21. Guo Z, Brooks CL III. 1997. *Biopolymers* 42:745–57
22. Shea J-E, Nochomovitz YD, Guo Z, Brooks CL III. 1998. *J. Chem. Phys.* 109:2895–903
23. Honeycutt JN, Thirumalai D. 1992. *Biopolymers* 32:695–709
24. Shea J-E, Onuchic JN, Brooks CL III. 1999. *Proc. Natl. Acad. Sci. USA* 96:12512–17
25. Shea J-E, Onuchic JN, Brooks CL III. 2000. *J. Chem. Phys.* 113:7663–71
26. Rey A, Skolnick J. 1991. *Chem. Phys.* 158:199–219
27. Boczko EM, Brooks CL III. 1995. *Science* 269:393–96
28. Brooks CL III. 1998. *Curr. Opin. Struct. Biol.* 8:222–26
29. Daggett V. 2000. *Curr. Opin. Struct. Biol.* 10:160–64
30. Daggett V, Levitt M. 1994. *Curr. Opin. Struct. Biol.* 4:291–95
31. Klimov D, Thirumalai D. 1996. *Proteins: Struct. Funct. Genet.* 26:411–41
32. Socci ND, Onuchic JN. 1995. *J. Chem. Phys.* 103:4732–44
33. Socci ND, Onuchic JN, Wolynes PG. 1996. *J. Chem. Phys.* 104:5860–68
34. Onuchic JN, Socci ND, Luthey-Schulten Z, Wolynes PG. 1996. *Fold. Des.* 1:441–50
35. Shakhnovich E, Abkevich V, Ptitsyn O. 1996. *Nature* 379:96–98
36. Gutin AM, Abkevich VI, Shakhnovich EI. 1998. *Fold. Des.* 3:183–94
37. Li L, Mirny LA, Shakhnovich EI. 2000. *Nat. Struct. Biol.* 7:336–42
38. Pande VS, Grosberg AY, Tanaka T, Rokhsar DS. 1998. *Curr. Opin. Struct. Biol.* 8:68–79
39. Dellago C, Bolhuis PG, Chandler D. 1999. *J. Chem. Phys.* 110:6617–25
40. Bryant Z, Pande VS, Rokhsar DS. 2000. *Biophys. J.* 78:584–89

41. Clementi C, Nymeyer H, Onuchic JN. 2000. *J. Mol. Biol.* 298:937
42. Nymeyer H, Socci ND, Onuchic JN. 2000. *Proc. Natl. Acad. Sci. USA* 97:634–39
43. Gutin AM, Abkevich VI, Shakhnovich EI. 1998. *Fold. Des.* 3:183–94
44. Brooks CL III, Karplus M, Pettitt BM. 1988. *Proteins: A Theoretical Perspective of Dynamics, Structure and Thermodynamics*. New York: Wiley & Sons
45. Allen MP, Tidesley DJ. 1989. *Computer Simulation of Liquids*. Oxford :Clarendon
46. McCammon JA, Harvey S. 1987. *Dynamics of Proteins and Nucleic Acids*. Cambridge Univ. Press
47. Demchuk E, Bashford D, Gippert GP, Case DA. 1997. *J. Mol. Biol.* 270:305–17
48. Demchuk E, Bashford D, Case DA. 1997. *Fold. Des.* 2:35–46
49. Daura X, Jaun B, Seebach D, van Gunsteren WF, Mark AE. 1998. *J. Mol. Biol.* 280:925–32
50. Daura X, van Gunsteren WF, Mark AE. 1999. *Proteins: Struct. Funct. Genet.* 34:269–80
51. Duan L, Wang L, Kollman PA. 1998. *Proc. Natl. Acad. Sci. USA* 95:9897–902
52. Duan L, Kollman PA. 1998. *Science* 282:740–44
53. Bashford D, Case DA. 2000. *Annu. Rev. Phys. Chem.* 51:129
54. Sugita Y, Okamoto Y. 2000. *Prog. Theor. Phys. Suppl.* 138:402–3
55. Okamoto Y. 2000. *Prog. Theor. Phys. Suppl.* 138:301–10
56. Hansmann UHE, Okamoto Y. 1999. *J. Chem. Phys.* 111:1339
57. Ferrara P, Apostolakis J, Caflisch A. 2000. *Proteins: Struct. Funct. Genet.* 39:252–60
58. Lazaridis T, Karplus M. 1997. *Science* 278:1928–31
59. Bursulaya B, Brooks CL III. 1999. *J. Am. Chem. Soc.* 121:9927
60. Bartels C, Schaefer M, Karplus M. 1999. *J. Chem. Phys.* 111:8048–67
61. Bartels C, Schaefer M, Karplus M. 1999. *Theor. Chem. Acc.* 101:62–66
62. Bartels C, Karplus M. 1998. *J. Phys. Chem. B* 102:865–80
63. Bartels C, Karplus M. 1997. *J. Comput. Chem.* 18:1450–62
64. Scheraga HA, Hao M-H. 1999. *Adv. Chem. Phys.* 105:243–72
65. Zaloj V, Elber R. 2000. *Comput. Phys. Commun.* 128:118–27
66. Elber R, Meller J, Olender R. 1999. *J. Phys. Chem. B* 103:899–911
67. Olender R, Elber R. 1996. *J. Chem. Phys.* 105:9299–315
68. Lazaridis T, Karplus M. 2000. *Curr. Opin. Struct. Biol.* 10:139–45
69. Bursulaya BD, Brooks CL III. 2000. *J. Phys. Chem. B* 104: In press
70. Kim PS, Baldwin RL. 1982. *Annu. Rev. Biochem.* 51:459–89
71. Dyson HJ, Rance M, Houghten RA, Wright PE, Lerner RA. 1988. *J. Mol. Biol.* 201:201–17
72. Dyson HJ, Rance M, Houghten RA, Lerner RA, Wright PE. 1988. *J. Mol. Biol.* 201:161–200
73. Wright PE, Dyson HJ, Lerner RA. 1988. *Biochemistry* 27:7167–75
74. Wright PE, Lerner RA, Dyson HJ. 1989. *GBF Monogr. (Adv. Prot. Des.)* 12:13–19
75. Wright PE, Dyson HJ, Feher VA, Tennant LL, Waltho JP, et al. 1990. *UCLA Symp. Mol. Cell. Biol., New Ser.* 109:1–13
76. Waltho JP, Feher VA, Wright PE. 1990. In *Current Research in Protein Chemistry*, pp. 283–93. New York: Academic
77. Tobias DJ, Mertz E, Brooks CL III. 1991. *Biochemistry* 30:6054–58
78. Karpen ME, Tobias DJ, Brooks CL III. 1993. *Biochemistry* 32:412–20
79. Daggett V, Levitt M. 1992. *J. Mol. Biol.* 223:1121–38
80. DiCapua FM, Swaminathan S, Beveridge DL. 1990. *J. Am. Chem. Soc.* 112:6768–71
81. DiCapua FM, Swaminathan S, Beveridge DL. 1991. *J. Am. Chem. Soc.* 113:6145–55
82. Tirado-Rives J, Jorgensen WL. 1991. *Biochemistry* 30:3864–71

83. Soman KV, Karimi A, Case DA. 1991. *Biopolymers* 31:1351–61
84. Hirst JD, Brooks CL III. 1995. *Biochemistry* 34:7614–21
85. Tobias DJ, Brooks CL III. 1987. *Chem. Phys. Lett.* 142:472–76
86. Tobias DJ, Brooks CL III. 1988. *J. Chem. Phys.* 89:5115–27
87. Sneddon SF, Tobias DJ, Brooks CL III. 1989. *J. Mol. Biol.* 209:817–20
88. Tobias DJ, Sneddon SF, Brooks CL III. 1990. *J. Mol. Biol.* 216:783–96
89. Tobias DJ, Brooks CL III. 1991. *Biochemistry* 30:6059–70
90. Tobias DJ, Sneddon SF, Brooks CL III. 1992. *J. Mol. Biol.* 227:1244–52
91. Tobias DJ, Brooks CL III. 1992. *J. Phys. Chem.* 96:3864–70
92. Brooks CL III, Nilsson L. 1993. *J. Am. Chem. Soc.* 115:11034–35
93. Young WS, Brooks CL III. 1996. *J. Mol. Biol.* 259:560–72
94. Sheinerman FB, Brooks CL III. 1995. *J. Am. Chem. Soc.* 117:10098–103
95. Brooks CL III. 1996. *J. Phys. Chem.* 100:2546–49
96. Tropsha A, O'Connell T, Wang L, Hermans J. 1996. *Book of Abstracts, 212th ACS Natl. Meet., Orlando, FL, August 25–29.* ACS Abstr.#COMP-044
97. Wang L, O'Connell T, Tropsha A, Hermans J. 1995. *Proc. Natl. Acad. Sci. USA* 92:10924–28
98. Zhang L, Hermans J. 1994. *J. Am. Chem. Soc.* 116:11915–21
99. Yan Y, Tropsha A, Hermans J, Erickson BW. 1993. *Proc. Natl. Acad. Sci. USA* 90:7898–902
100. Sundaralingam M, Sekharudu YC. 1989. *Science* 244:1333–37
101. Sundaralingam M, Sekharudu YC. 1990. *Struct. Methods, Proc. Conversation Discip. Biomol. Stereodyn., 6th,* 2:115–27
102. Miick SM, Martinez GV, Fiori WR, Todd AP, Millhauser GL. 1992. *Nature* 359:653–55
103. Fiori WR, Miick SM, Millhauser GL. 1993. *Biochemistry* 32:11957–62
104. Millhauser GL. 1995. *Biochemistry* 34:10318
105. Bolin KA, Millhauser GL. 1999. *Acc. Chem. Res.* 32:1027–33
106. Brooks CL III. 1993. *Curr. Opin. Struct. Biol.* 3:92–98
107. Daggett V, Levitt M. 1992. *Proc. Natl. Acad. Sci. USA* 89:5142–46
108. Mark AE, van Gunsteren WF. 1992. *Biochemistry* 31:7745–48
109. van Gunsteren WF, Huenenberger PH, Kovacs H, Mark AE, Schiffer CA. 1995. *Philos. Trans. R. Soc. London Ser. B* 348:49–59
110. Smith LJ, Mark AE, Dobson CM, van Gunsteren WF. 1995. *Biochemistry* 34:10918–31
111. Tirado-Rives J, Jorgensen WL. 1993. *Biochemistry* 32:4175–84
112. Caflisch A, Karplus M. 1994. *Proc. Natl. Acad. Sci. USA* 91:1746–50
113. Caflisch A, Karplus M. 1995. *J. Mol. Biol.* 252:672–708
114. Brooks CL III. 1992. *J. Mol. Biol.* 227:375–80
115. Kitchen DB, Reed LH, Levy RM. 1992. *Biochemistry* 31:10083–93
116. Tirado-Rives J, Orozco M, Jorgensen WL. 1997. *Biochemistry* 36:7313–29
117. Alonso DOV, Daggett V. 1995. *J. Mol. Biol.* 247:501–20
118. Alonso DOV, Daggett V. 1998. *Protein Sci.* 7:860–74
119. Daggett V, Li A, Itzhaki LS, Otzen DE, Fersht AR. 1996. *J. Mol. Biol.* 257:430–40
120. Tobias DJ, Sneddon SF, Brooks CL III. 1991. In *AIP Conf. Proc.*, pp. 174–99. Obernai, France
121. Ryckaert JP, Ciccotti G, Berendsen HJC. 1977. *J. Comput. Phys.* 23:327–41
122. York DM, Darden TA, Pedersen LG. 1993. *J. Chem. Phys.* 99:8345–48
123. Darden T, York D, Pedersen L. 1993. *J. Chem. Phys.* 98:10089–92

124. Essmann U, Perera L, Berkowitz ML, Darden T, Lee H, Pedersen LG. 1995. *J. Chem. Phys.* 103:8577–93
125. Darden TA, Toukmaji A, Pedersen LG. 1997. *J. Chim. Phys. Phys.-Chim. Biol.* 94:1346–64
126. Sheinerman FB, Brooks CL III. 1998. *Proc. Natl. Acad. Sci. USA* 95:1562–67
127. Sheinerman FB, Brooks CL III. 1998. *J. Mol. Biol.* 278:439–55
128. Hunenberger PH, Mark AE, van Gunsteren WF. 1995. *Proteins: Struct. Funct. Genet.* 21:196–213
129. Xu S, Kamath MV, Capson DW. 1993. *Patt. Recog. Lett.* 14:7–15
130. Ward JH Jr. 1963. *J. Am. Stat. Assoc.* 58:236–44
131. Wesson L, Eisenberg D. 1992. *Protein Sci.* 1:227–35
132. Brooks CL III. 2000. Work in progress
133. Boczko EM, Brooks CL III. 1993. *J. Phys. Chem.* 97:4509–13
134. Ferrenberg AM, Swendsen RH. 1989. *Phys. Rev. Lett.* 63:1195–98
135. Shea J-E, Onuchic JN, Brooks CL III. 2000. *J. Mol. Biol.* In preparation
136. Plaxco KW, Simons KT, Baker D. 1998. *J. Mol. Biol.* 277:985–94
137. Tashiro M, Tejero R, Zimmerman DE, Celda B, Nilsson B, Montelione GT. 1997. *J. Mol. Biol.* 272:573–90
138. Jendeberg L, Tashiro M, Tejero R, Lyons BA, Uhlen M, et al. 1996. *Biochemistry* 35:22–31
139. Gouda H, Torigoe H, Saito A, Sato M, Arata Y, Shimada I. 1992. *Biochemistry* 31:9665–72
140. Gouda H, Shiraishi M, Takahashi H, Kato K, Torigoe H, et al. 1998. *Biochemistry* 37:129–36
141. Bai Y, Karimi A, Dyson HJ, Wright PE. 1997. *Protein Sci.* 6:1449–57
142. Witte K, Skolnick J, Wong C-H. 1998. *J. Am. Chem. Soc.* 120:13042–45
143. Zhou Y, Karplus M. 1999. *J. Mol. Biol.* 293:917–51
144. Zhou Y, Karplus M. 1999. *Nature* 401:400–3
145. Zhou Y, Karplus M. 1997. *Proc. Natl. Acad. Sci. USA* 94:14429–32
146. Li R, Woodward C. 1999. *Protein Sci.* 8:1571–91
147. Woodward C, Li R. 1998. *Trends Biochem. Sci.* 23:379
148. Barbar E, LiCata VJ, Barany G, Woodward C. 1997. *Biophys. Chem.* 64:45–57
149. Woodward C. 1993. *Trends Biochem. Sci.* 18:359–60
150. Gallagher T, Alexander P, Bryan P, Gilliland GL. 1994. *Biochemistry* 33:4721–29
151. Clore GM, Gronenborn AM. 1992. *J. Mol. Biol.* 223:853–56
152. Gronenborn AM, Filpula DR, Essig NZ, Achari A, Whitlow M, et al. 1991. *Science* 253:657–61
153. Barchi JJ Jr, Grasberger B, Gronenborn AM, Clore GM. 1994. *Protein Sci.* 3:15–21
154. Sheinerman FB, Brooks CL III. 1997. *Proteins: Struct. Funct. Genet.* 29:193–202
155. Park SH, O'Neil KT, Roder H. 1997. *Biochemistry* 36:14277–83
156. Frank MK, Clore GM, Gronenborn AM. 1995. *Protein Sci.* 4:2605–15
157. Kuszewski J, Clore GM, Gronenborn AM. 1994. *Protein Sci.* 3:1945–52
158. Leeson DT, Gai F, Rodriguez HM, Gregoret LM, Dyer RB. 2000. *Proc. Natl. Acad. Sci. USA* 97:2527–32
159. Baker D. 2000. *Nature* 405:39–42
160. Grantcharova VP, Baker D. 1997. *Biochemistry* 36:15685–92
161. Grantcharova VP, Riddle DS, Santiago JV, Baker D. 1998. *Nat. Struct. Biol.* 5:714–20
162. Riddle DS, Grantcharova VP, Santiago JV, Alm E, Ruczinski I, Baker D. 1999. *Nat. Struct. Biol.* 6:1016–24
163. Tsai J, Levitt M, Baker D. 1999. *J. Mol. Biol.* 291:215–25

164. Mueller U, Perl D, Schmid FX, Heinemann U. 2000. *J. Mol. Biol.* 297:975–88
165. Feng W, Tejero R, Zimmerman DE, Inouye M, Montelione GT. 1998. *Biochemistry* 37:10881–96
166. Schindelin H, Jiang W, Inouye M, Heinemann U. 1994. *Proc. Natl. Acad. Sci. USA* 91:5119–23
167. Schindler T, Graumann PL, Perl D, Ma S, Schmid FX, Marahiel MA. 1999. *J. Biol. Chem.* 274:3407–13
168. Schindler T, Herrler M, Marahiel MA, Schmid FX. 1995. *Nat. Struct. Biol.* 2:663–73
169. Schindler T, Schmid FX. 1996. *Biochemistry* 35:16833–42
170. Li A, Daggett V. 1994. *Proc. Natl. Acad. Sci. USA* 9:10430–34
171. Karplus M, Sali A. 1995. *Curr. Opin. Struct. Biol.* 5:58–73
172. Daggett V, Levitt M. 1992. *J. Mol. Biol.* 223:1121–38
173. Li A, Daggett V. 1996. *J. Mol. Biol.* 257:412–29
174. Li A, Daggett V. 1998. *J. Mol. Biol.* 275:677–94
175. Alonso DO, Daggett V. 2000. *Proc. Natl. Acad. Sci. USA* 97:133–38
176. Brooks CL III. 2000. Work in progress
177. Itzhaki LS, Otzen DE, Fersht AR. 1995. *J. Mol. Biol.* 254:260–88
178. Ladurner AG, Itzhaki LS, Daggett V, Fersht AR. 1998. *Proc. Natl. Acad. Sci. USA* 95:8473–78
179. Fersht AR. 1995. *Curr. Opin. Struct. Biol.* 5:79–84
180. Reid K, Rodriguez HM, Hillier BJ, Gregoret LM. 1998. *Protein Sci.* 7:470–79
181. Hillier BJ, Rodriguez HM, Gregoret LM. 1998. *Fold. Des.* 3:87–93
182. Brooks BR, Bruccoleri RE, Olafson BD, States DJ, Swaminathan DJ, Karplus M. 1982. *J. Comp. Chem.* 4:187–217
183. Hammond GS. 1955. *J. Am. Chem. Soc.* 77:334
184. Dinner AR, Karplus M. 1999. *J. Mol. Biol.* 292:403–19

# POLYMER ADSORPTION–DRIVEN SELF-ASSEMBLY OF NANOSTRUCTURES

Arup K Chakraborty[1,2,3] and Aaron J Golumbfskie[1,3]

Department of Chemical Engineering[1] and Department of Chemistry[2], University of California, Berkeley and Materials Science Division, Lawrence Berkeley National Laboratory[3] Berkeley, California 94720; e-mail: arup@lolita.cchem.berkeley.edu

Key Words polymer adsorption, frustrated systems, disorder, nanostructures

■ **Abstract** Driven by prospective applications, there is much interest in developing materials that can perform specific functions in response to external conditions. One way to design such materials is to create systems which, in response to external inputs, can self-assemble to form structures that are functionally useful. This review focuses on the principles that can be employed to design macromolecules that when presented with an appropriate two-dimensional surface, will self-assemble to form nanostructures that may be functionally useful. We discuss three specific examples: (*a*) biomimetic recognition between polymers and patterned surfaces. (*b*) control and manipulation of nanomechanical motion generated by biopolymer adsorption and binding, and (*c*) creation of patterned nanostructuctures by exposing molten diblock copolymers to patterned surfaces. The discussion serves to illustrate how polymer sequence can be manipulated to affect self-assembly characteristics near adsorbing surfaces. The focus of this review is on theoretical and computational work aimed toward elucidating the principles underlying the phenomena pertinent to the three topics noted above. However, synergistic experiments are also described in the appropriate context.

I. INTRODUCTION

The adsorption of macromolecules onto two-dimensional surfaces has been considered since ancient times. For example, inks have included naturally occurring substances that we now understand to be polymers that adsorb onto the dye molecules and prevent their aggregation and precipitation from solution. In modern times, polymer adsorption has been the focus of much attention because of its importance in a host of practical applications and because developing a proper understanding of the pertinent phenomena presents some fundamental challenges in statistical physics.

Examples of traditional technological applications wherein polymer adsorption plays a crucial role include paints, coatings, boundary lubricants, ceramics processing, and adhesives. The fundamental interest in polymer adsorption stems from the fact that in addition to a short-range primary structure defined by the chemical identity of the monomers polymers are long flexible objects over larger length scales. Long flexible molecules can adopt a large number of conformations, and there is a significant entropy associated with sampling this multitude of conformations. Many macroscopic properties of polymers are dominated by entropic effects determined by their many possible molecular conformations. Adsorption of a flexible macromolecule onto an impenetrable surface reduces the number of accessible conformations that a macromolecule can adopt. The balance between the reduction in entropy due to adsorption and the energetic advantage of binding determines the molecular conformations (and hence, properties) of adsorbed polymers. Understanding how this balance is influenced by the nature of the polymer and the surface from a fundamental standpoint has been, and continues to be, a challenge. It is essential to develop such an understanding if we are to learn how to manipulate the properties of adsorbed polymers by appropriate choice of polymer and surface.

The adsorption of homopolymers (i.e. polymers with only one kind of monomer unit) onto chemically homogeneous surfaces has been studied extensively for over three decades. Experiments, theory, and computer simulations have elucidated the conformational statistics of adsorbed homopolymers and their implications for properties. These important advances have been reviewed many times (e.g. 1–3). In the preceding two decades, the adsorption of diblock copolymers onto homogeneous surfaces has also received much attention. These are amphiphilic polymers wherein two contiguous blocks of chemically distinct monomers are connected at a well-defined junction point. The adsorption of diblock copolymers with one block being well-solvated in the solvent adjacent to the solid surface and the other exhibiting a preferential affinity for the surface has been studied extensively using a battery of experimental tools, theory, and computer simulations (e.g. 4–6). In this instance, the poorly solvated block binds strongly to the surface, while the well-solvated block extends out into the solvent. The number of adsorbed polymer chains per unit area is called the grafting density. For sufficiently high grafting densities, adsorbed diblock copolymer layers are referred to as polymer brushes because the strong stretching of the individual chains makes them resemble this object in everyday use. Studies of the adsorption of homopolymers and diblock copolymers have advanced our understanding of polymer conformational statistics while assisting the design of effective polymers for applications (e.g. colloid stabilization). With the exception of systems where the monomers carry an electric charge (7, 8), however, it is fair to say that our understanding of such systems is fairly mature.

Many current and prospective technologies demand materials that can function with precision over length scales ranging from nanometers to microns. Functional materials of this type could be realized if the response of the material's molecular constituents to a particular set of conditions is to self-assemble into a structural motif that is functionally interesting. Such self-assembly into nano- and micro-scale structures that perform important functions is ubiquitous in biology. In many instances, evolution has led nature to use macromolecules as the molecular building blocks that self-assemble into functionally useful nanostructures. Thus, exploring whether the sequence and architecture of synthetic polymers can be manipulated so that we can control their self-assembly characteristics appears to be a fruitful avenue to pursue in taking steps toward the design of functional nanostructures.

Research aimed toward designing polymeric materials that can self-assemble into functional nanostructures in response to external stimuli is just beginning to emerge. A number of different environments and target applications are being employed. In this article, we restrict attention to situations involving polymers interacting with surfaces. Specifically, we discuss three examples where polymeric molecules form self-assembled nanostructures upon being exposed to certain types of surfaces. The first example concerns the development of strategies that may enable polymeric molecules to recognize target patterns of binding sites on a surface (Section II). The second example illustrates how interactions of an adsorbed polymer layer with biomolecules can affect nanomechanical motion, which might prove useful in sensing applications (Section III). The third example considers how nanometer-scale periodic patterns can be induced by the self-assembly of certain molten polymers due to interactions with a similarly patterned surface (Section IV).

Discussion of these examples also illuminates interesting aspects of the effects of sequence on the statistical physics of long polymer chains. While the focus of this review is on theoretical and computational research, synergistic experiments are described in the appropriate context.

II. BIOMIMETIC RECOGNITION BETWEEN POLYMERS AND SURFACES

Many important biological processes such as intracellular signaling are initiated by a protein searching for and recognizing a particular pattern of binding sites (a receptor) on a cell surface. Three important hallmarks of such recognition events are that (a) a sharp discrimination exists between patterns of binding sites to which the macromolecule binds strongly and those to which it binds weakly; (b) the biopolymer is not kinetically trapped in metastable states for long periods by adsorbing to the nontarget patterns of binding sites on the surface; and (c) adsorption on the target pattern occurs in a particular conformation (note that the pattern has a length scale commensurate with chain dimensions). Adsorption of synthetic macromolecules with ordered sequence distributions from solution does not exhibit all the features that characterize recognition. The following question presents itself: Since nature does it all the time, why can we not?

One way to mimic recognition in synthetic systems is to attempt to reproduce the detailed chemistries to which nature has been led via evolution. This strategy would be very difficult to implement for a synthetic system even if the mechanisms that lead to recognition were known. This prompts us to ask whether there are any essential (and universal) ingredients that would allow synthetic systems to mimic recognition.

It is important to remark that the work reviewed here does not aim to describe recognition in biological systems. However, coarse-grained observations concerning biological systems provided the inspiration for deducing candidate strategies that could be used to mimic recognition. In the context of protein folding, for example, many coarse-grained representations of protein sequences have been developed. An example is the HP model (9), wherein amino acids are considered to belong to two classes, hydrophobic (H) and polar (P). When protein sequences are represented using this or other coarse-grained models, it has been observed that in most instances the sequence of Hs and Ps is not periodically repeating (or ordered) (10–12). Similarly, while the pattern of binding sites that constitute a receptor are very specifically arranged, there is usually no periodically repeating pattern. These observations led to the following speculation: Is it possible that the disorder (in polymer sequence and pattern of binding sites) and competing interactions due to multiple types of segment-site interactions are the essential ingredients required to mimic recognition in synthetic polymeric systems? It is worth remarking that while coarse-grained observations concerning biological systems inspired this question, one can ask it independent of such observations.

Investigating the interactions of disordered heteropolymers (DHPs) with surfaces bearing binding sites of more than one type provides a good model system to explore the question noted above. Figure 1 is a schematic depiction of a DHP bearing two kinds of segments interacting with a surface that has two kinds of sites. The white (gray) segments interact favorably with the white (gray) sites on the surface. This introduces competing interactions while disorder is embodied in the sequence and surface site distribution.

DHP sequences are described statistically, and hence they encode a statistical rather than specific pattern. Similarly, binding sites on a surface can be arranged in a pattern that is also described statistically. For example, the average density of sites, the fraction of each type of site, and a two-point correlation function that measures the probability of finding sites of the same type a certain distance

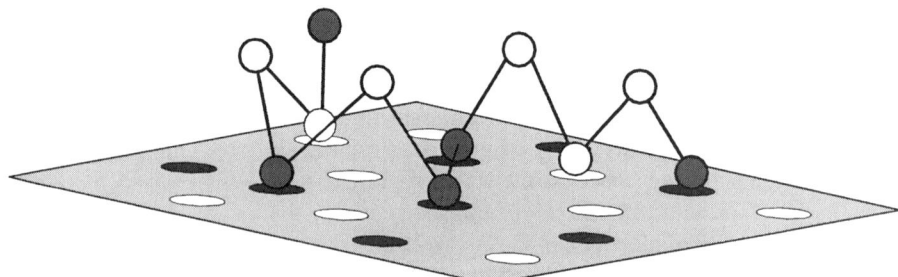

Figure 1 Cartoon representation of DHP interacting with a surface onto which two types of sites have been distributed. A pattern matched conformation is shown.

away could be specified. In nature, when the specific pattern encoded in the biopolymer sequence and the pattern of binding sites are related in a special way (i.e. matched), recognition occurs. The question that has been asked (13–19) for the class of synthetic systems depicted in Figure 1 is, If the statistics characterizing the DHP sequence and the surface site distribution are related in a special way, will the ensemble of DHP sequences in question be able to recognize the statistical pattern of binding sites? In other words, is statistical pattern matching sufficient to mimic recognition?

Answering these questions requires that both thermodynamic and kinetic issues be analyzed. Theoretical studies, computer simulations, and recent experiments have provided partial answers to some of the pertinent questions. In this section, we discuss some of this progress.

Thermodynamic issues were first considered using a toy model (13). While these were primitive, later studies have shown that this model contains much of the essential physics. Imagine that the segment-surface interactions are so much stronger than the intersegment interactions that the latter can be ignored. In this circumstance, the Edwards Hamiltonian for the situation depicted in Figure 1 can be written as:

$$-\beta H = -\frac{3}{2l} \int_0^N dn \left(\frac{d\mathbf{r}}{dn}\right)^2 - \int_0^N dn \int d\mathbf{r} \times k(\mathbf{r})\delta(\mathbf{r}(n) - \mathbf{r})\theta(n)\delta(z), \quad 1.$$

where $\mathbf{r}(n)$ is the spatial position of the nth polymer segment in three-dimensional space. $\theta(n)$ equals $+1$ if the nth segment is of type A and is -1 if it is a B-type segment. $k(\mathbf{r})$ is the interaction strength with a surface site located at $\mathbf{r}$. The Hamiltonian is written such that if a surface site is attractive to one type of DHP segment, it is equally repulsive for the other kind of segment. This is for simplicity only; the qualitative physics does not change as long as one type of segment interacts preferentially with one type of site while the other type of segment prefers the other kind of site. The factor of $\delta(z)$ in the Hamiltonian restricts the sites to a two-dimensional manifold.

Since the sites on the surface and the DHP sequence are disordered, $k(\mathbf{r})$ and $\theta(n)$ are fluctuating variables. Their fluctuations are described statistically, and the statistics encode the pattern carried by the surface sites and the DHP sequence. In the simplest scenario, $k(\mathbf{r})$ and $\theta(n)$ are taken to be random variables described by Gaussian statistics. The statistics are therefore described by the variance, which is trivially related to the average composition of the DHP or the average total density of the two types of sites on the surface. In the high temperature limit, this description is equivalent to an uncorrelated discrete distribution characterized by a mean value. For DHP sequences, using this representation also generalizes the results to sequences with more than two types of segments.

Let us employ this simplest version of statistical patterns to examine some elementary ideas concerning pattern recognition between polymers and surfaces. We will soon examine more complicated statistical patterns. The probability distributions for k(**r**) and $\theta(n)$ are:

$$P[k(\mathbf{r})] \propto \exp\left[-\frac{1}{2\sigma_1^2}\int d\mathbf{r}\, k^2(\mathbf{r})\right]$$

$$P[\theta(n)] \propto \exp\left[-\frac{1}{2\sigma_2^2}\int dn[\theta(n) - (2f-1)^2]\right], \qquad 2.$$

where σ_1^2 and σ_2^2 are the variances for the distributions. σ_1^2 is proportional to the total number density of surface sites; σ_2^2 equals $4f(1-f)$, where f is the average composition characterizing the DHP sequence.

Depending on the preparation conditions, the distribution of surface sites could either change in response to interactions with the DHP chains (annealed) or not (quenched). In the former case, an interesting question is how the final pattern adopted by the surface sites is related to the sequence of the DHP. This would tell us whether it is possible to transfer information from the DHP to the surface. How much entropic loss of information would occur? In the latter case, the issue is whether DHPs with a particular statistical class of sequences can recognize a prescribed pattern of binding sites.

For disorders external to the fluid of interest (the surface sites, in our case), in the thermodynamic limit, treating this disorder as quenched or annealed leads to the same equilibrium properties. This has been argued in many contexts (e.g. 20–22). However, it is important to note that quenched external disorders exhibit the same thermodynamics if the sample is sufficiently large and the observation time is sufficiently long. Quantifying this statement for a given problem is difficult, and the dynamics are certainly different for quenched and annealed disorders external to the fluid. For the moment, we restrict attention to thermodynamic properties and imagine that the surface is sufficiently large and the DHP has sufficient time to sample the surface sites. In this case, we can integrate out the external disorder to obtain the following influence functional (23, 24):

$$\exp[-\beta H_{eff}] = \iint Dk(\mathbf{r})\exp\left[-\frac{1}{2\sigma_1^2}\int d\mathbf{r}\int d\mathbf{r}'\, k(\mathbf{r})\delta(\mathbf{r}-\mathbf{r}')k(\mathbf{r}')\delta(z)\delta(z')\right]$$

$$\times \exp\left[-\frac{3}{2l}\int_0^N dn\left(\frac{d\mathbf{r}}{dn}\right)^2 - \int dn\int d\mathbf{r}\, k(\mathbf{r})\delta(\mathbf{r}(n)-\mathbf{r})\theta(n)\delta(z)\right]. \quad 3.$$

The quenched disorder embodied in the DHP sequence, however, is carried by the fluid of interest. The arguments made above for the external disorder no longer apply, and we must explicitly treat the sequence as quenched even for thermodynamic properties. This is to say that the partition function is not self-averaging with respect to $\theta(n)$; the free energy is. One way to carry out this quenched average is to use the replica trick (25, 26). Replicating the effective Hamiltonian, H_{dis}, in Equation 3 and carrying out the functional integral over the distribution of $\theta(n)$ leads to the following m-replica partition function:

$$\langle G^m \rangle = \prod_{\alpha=1}^{m} \int\!\!\int D\mathbf{r}_\alpha(n) \int\!\!\int Dk_\alpha(\mathbf{r}) \exp\left[-\frac{3}{2l}\sum_{\alpha=1}^{m}\int dn \left(\frac{d\mathbf{r}_\alpha}{dn}\right)^2\right]$$

$$\times \exp\left[-\frac{1}{2\sigma_1^2}\sum_{\alpha,\beta}\int d\mathbf{r}\int d\mathbf{r}'\, k_\alpha(\mathbf{r})\delta_{\alpha\beta}\delta(\mathbf{r}-\mathbf{r}')k_\beta(\mathbf{r}')\delta(z)\delta(z')\right]$$

$$\times \exp\left[\frac{\sigma_2^2}{2}\sum_{\alpha,\beta}\int dn\int d\mathbf{r}\int d\mathbf{r}'\, k_\alpha(\mathbf{r})k_\beta(\mathbf{r}')\right.$$

$$\left.\times \delta(\mathbf{r}_\alpha(n)-\mathbf{r})\delta(\mathbf{r}_\beta(n)-\mathbf{r}')\delta(z)\delta(z')\right]. \qquad 4.$$

Let us define the following order parameter that measures the conformational overlap in the plane of the surface between the various replicas:

$$Q_{\alpha\beta}(\mathbf{r},\mathbf{r}') = \int dn\, \delta(\mathbf{r}_\alpha(n)-\mathbf{r}) \times \delta(\mathbf{r}_\beta(n)-\mathbf{r}')\delta(z)\delta(z') \qquad 5.$$

In terms of $Q_{\alpha\beta}$, the replicated partition function can be rewritten as:

$$\langle G^m \rangle = \int\!\!\int DQ_{\alpha\beta}\exp(-E[Q_{\alpha\beta}] + S[Q_{\alpha\beta}])$$

$$E = -\ln\prod_{\alpha=1}^{m}\int\!\!\int Dk_\alpha(\mathbf{r})\exp\left[-\frac{1}{2}\int d\mathbf{r}\int d\mathbf{r}'\, k_\alpha(\mathbf{r})P_{\alpha\beta}(\mathbf{r},\mathbf{r}')k_\beta(\mathbf{r}')\right]$$

$$S = \ln\prod_{\alpha=1}^{m}\int\!\!\int D\mathbf{r}_\alpha(n)\exp\left[-\frac{3}{2l}\sum_{\alpha=1}^{m}\int dn\left(\frac{d\mathbf{r}_\alpha}{dn}\right)^2\right] \qquad 6.$$

$$\times \delta\left[Q_{\alpha\beta}(\mathbf{r},\mathbf{r}') - \int dn\,\delta(\mathbf{r}_\alpha(n)-\mathbf{r})\delta(\mathbf{r}_\beta(n)-\mathbf{r}')\right],$$

where

$$P_{\alpha\beta}(\mathbf{r},\mathbf{r}') = \frac{\delta(\mathbf{r}-\mathbf{r}')\delta_{\alpha\beta}}{\sigma_1^2}\delta(z)\delta(z') - \sigma_2^2 Q_{\alpha\beta}(\mathbf{r},\mathbf{r}'). \qquad 7.$$

The expression above shows that S equals the logarithm of the number of chain conformations that have a particular surface overlap function while obeying the connectivity constraint. Thus, it is the entropy corresponding to a particular $Q_{\alpha\beta}$. Then, E must be the energy.

Physical systems characterized by competing interactions and quenched disorder can, under certain circumstances, have thermodynamic properties that are determined by a few dominant conformations. In other words, phase space can be partitioned into different parts with very large free energy barriers separating the different regions. DHPs exhibit this kind of behavior, wherein a few conformations are thermodynamically important, in a number of different contexts

(14, 15, 27–34). We admit the possibility of such thermodynamic states in the mathematical analysis by allowing for broken replica symmetry. Parisi pioneered the way to think about broken replica symmetry in spin glass physics (35). We follow this line of thinking and its adaptation to DHPs by other workers (27, 28). It has been argued that DHPs can exhibit characteristics akin to the random energy model (REM) or p-spin models with p > 2 (e.g. 36). This has been the basis for employing a one-step replica symmetry breaking (RSB) scheme (26) to examine thermodynamic states with only a few dominant conformations. We use such an analysis here.

In a one-step RSB scheme, replicas are divided into groups. Replicas within a group have perfect conformational overlap (in our case, in the plane of the surface), while replicas in different groups have no conformational overlap. The energy in Equation 7 is obtained by evaluating the Gausian integral, and it equals the logarithm of the determinant of the matrix **P**. The computation of this quantity, while allowing for broken replica symmetry, has been detailed by Mezard & Parisi (37). Following their method, for a one step RSB scheme, we find that

$$E = \frac{1}{2}\left[-\ln \sigma_1^2 + \frac{1}{x_0}\ln(1 - C_1 \overline{p} x_0)\right]$$

$$\overline{p} = p/N$$

$$C_1 = \sigma_1^2 \sigma_2^2 / A,$$

8.

where x_0 is the number of replicas in a group, A is the surface area, and p is the number of segments in contact with the surface. The derivation of Equation 8 involves approximating the total segmental density on the surface to be uniform.

The number of ways in which replicas in a group can be arranged such that they have perfect conformational overlap on the surface and p-adsorbed segments can be calculated as follows. The conformations of adsorbed polymers are comprised of loops, trains, and tails (see Figure 2). Trains are a contiguous set of adsorbed segments. Loops are made up of a contiguous set of unbound segments between two bound segments. Tails extend away from the surface. In the long chain limit, tails can be ignored. Furthermore, trains can be considered to be composed of a number of consecutive loops, each one segment long. Let the probability for a loop of length n_i to begin at a position $\mathbf{r}_i''$ and end at $\mathbf{r}_{i+1}''$ be $f_{ni}(\mathbf{r}_{i+1}'' - \mathbf{r}_i'')$. Then, the number of ways to arrange x_0 replicas that overlap perfectly on the surface and have the first segment adsorbed at $\mathbf{r}_1''$ is:

$$Z(\mathbf{r}_1'') = \sum_{n_1,n_2,\dots,n_p} \int d\mathbf{r}_2'' \cdots \int d\mathbf{r}_p'' f_{n_1}^{x_0}(\mathbf{r}_2'' - \mathbf{r}_1'') \cdots f_{n_p}^{x_0}(\mathbf{r}_p'' - \mathbf{r}_{p-1}'')$$

$$\times \delta(n_1 + n_2 + \cdots + n_p - N).$$

9.

The entropy that we seek is obtained by integrating this restricted partition function,

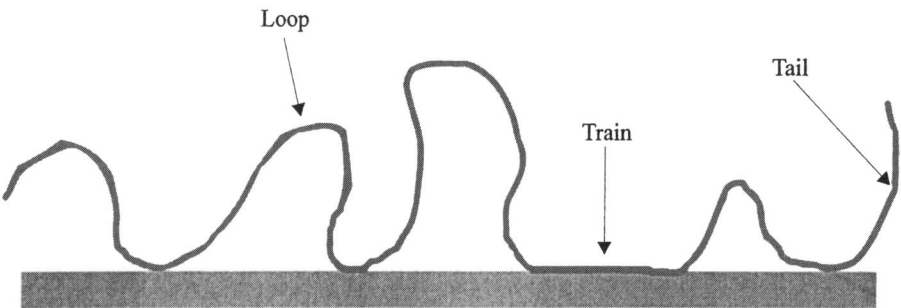

Figure 2 Cartoon depicting adsorbed polymer conformation. An example loop, train, and tail are labeled.

Z, over r_1'' and taking the negative logarithm. The Dirac delta function in Equation 9 enforces a constant chain length equal to N.

This computation is done most easily by defining the Fourier-Laplace transform of Equation 9. The Laplace transform is conjugate to chain length and the Fourier transform is defined with respect to the two-dimensional spatial coordinates in Equation 9. Taking advantage of the convolution structure of Equation 9 yields

$$Z(k, \lambda) = \left[\sum_{n=1}^{N} f_n^{x_0}(k) e^{-\lambda n} \right]^p, \qquad 10.$$

where k and λ are Fourier and Laplace variables. Different functions are used for f_1 and f_i for all $i \neq 1$. For $i = 1$, we use the well-established expression due to Hoeve et al. (38); i.e.

$$f_1(r) = \omega \delta(r - l). \qquad 11.$$

For larger values of i, we imagine that the loops extending away from the surface are Gaussian (38, 39). Using these expressions for f_i, and following the steps noted earlier, we can obtain the entropy as a function of $\bar{p}$ and x_0. Combining this expression with Equation 2 yields the free energy F.

$$F = \frac{1}{2} \left[-\ln \sigma_1^2 + \frac{1}{x_0} \ln(1 - C_1 \bar{p} x_0) \right] - \frac{1}{x_0 N}$$

$$\times \ln \left[\sum_{q=0}^{\bar{p}N} \binom{\bar{p}N}{q} \left(\frac{2\pi l^2}{3x_0} \right)^q C^{q x_0} \omega^{(\bar{p}N - q) x_0} \frac{\Gamma^q((4 - 3x_0)/2)}{\Gamma(q(4 - 3x_0)/2)} \right.$$

$$\left. \times [N - (\bar{p}N - q)]^{[(4 - 3x_0)/2]q - 1} \right]. \qquad 12.$$

In deriving this expression, the discrete sum over loop lengths was replaced by a continuous integral.

Mean field predictions for $\bar{p}$ and x_0 are obtained by extremizing the free energy expression with respect to these order parameters. The quantity $\bar{p}$ is just the adsorbed fraction and measures the extent of adsorption. The parameter x_0 has been interpreted to be $1 - \sum_i P_i^2$, where P_i is the probability of occurrence of conformation i. When a multitude of conformations are sampled, x_0 is, therefore, unity. Values of x_0 smaller than unity signal a thermodynamic state with a few dominant conformations.

Let us study the behavior of $\bar{p}$ and x_0 for a given statistical ensemble of DHP sequences interacting with different surfaces. In this simple model, this implies varying the total number density of sites (loading, measured by the parameter C). Figure 3 shows the variation of p and x_0 with C; note that every point on the abcissa of this figure corresponds to a different surface. There is no adsorption on surfaces that have small loadings. This is simply because the energetic advantage of binding to a few sites is insufficient to overcome the entropic penalties associated with adsorption. When the loading is sufficiently high, Figure 3 shows a continuous transition to adsorbed states. At this stage, x_0 remains unity. This is similar to the phenomenon that is observed for adsorption of chains with ordered sequence distributions. Figure 3 makes a much more interesting prediction when the surface

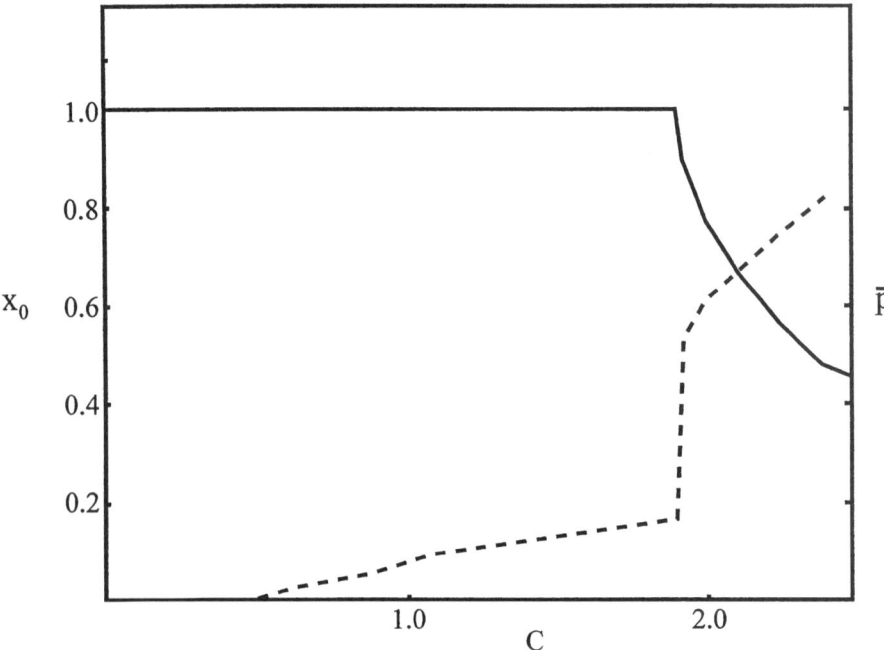

Figure 3 The order parameters $\bar{p}$ (dashed line) and x_0 (solid line) plotted as a function of C. For the calculation described in the text, each point on the abcissa represents a different statistically patterned surface.

loading exceeds a higher threshold value. The adsorbed fraction exhibits a sharp transition to strongly adsorbed states. This sharp transition from weak to strong adsorption is accompanied by x_0 acquiring values smaller than unity, signaling that only a few adsorbed conformations are thermodynamically important.

Consider the situation when the loading is sufficiently large to induce adsorption, but not large enough to cause the sharp transition from weak to strong adsorption. In this scenario, since the loading is not large, it is relatively easy for adsorbed chains to avoid unfavorable segment-surface contacts while still making favorable ones. Furthermore, there are many adsorbed conformations that can acquire the same energy by having different combinations of the same number of favorable segment-surface interactions. Thus, the system minimizes free energy by sampling these adsorbed conformations and x_0 equals unity. As the loading increases, the adsorbed fraction increases. However, larger values of the adsorbed fraction and the loading imply that it becomes increasingly difficult to avoid unfavorable interactions. Imagine generating arbitrary adsorbed conformations for a particular adsorbed fraction when the loading is high. Intuitively it is clear that most of these arbitrary conformations will have a rather high energy because of many unfavorable segment-surface contacts. However, there will be a few conformations that carefully avoid these unfavorable contacts. These few conformations are much lower in energy than the other adsorbed conformations. Sampling only these states is entropically not favored, however. As the loading increases, the energy gap between the few pattern-matched conformations and the others increases. (We use the term pattern-matched conformations to refer to those conformations that exhibit a high degree of registry between complementary segments and surface sites.) When this energy gap becomes much larger than the thermal energy, the system prefers to sacrifice the entropic advantage of sampling a multitude of conformations and so adopts the few energetically favorable ones. Of course, these pattern-matched conformations are strongly adsorbed. This is the physical origin of the transition from weak to strong adsorption leading to a thermodynamic state with a few dominant conformations.

The arguments above are very similar to those that describe the behavior of the random energy model (REM) in spin glass physics. It is worth remarking that the system we have been studying is more frustrated than the two-letter DHP in solution by itself (39a). This is because the competing interactions are confined to a two-dimensional manifold. It is important to note that this does not mean that the dimensionality of the system that we have been describing is two. This is because the polymer chains can escape into a third dimension by forming loops. The importance of these loops and their fluctuations will be made vivid later.

Figure 3 shows that the transition from weak to strong adsorption is sharp. In fact, replica mean-field theory predicts the transition to be first order (13). In such a theory, x_0 acquiring values less than unity corresponds to broken replica symmetry of the underlying Hamiltonian. The mean-field theory shows that there are two free energy minima, one that is replica symmetric and the other with broken replica symmetry. When the loading acquires the value required for a transition from weak to strong adsorption, the global free energy minimum switches from being the replica symmetric one to that with broken replica symmetry. The mean-field prediction of the order of the transition could be incorrect because fluctuations could change the order of the transition. It is worth noting, however, that Monte-Carlo simulations are also suggestive of a first order transition (16–18).

From a practical standpoint, what is important is that the transition is sharp. This implies a sharp discrimination between weakly and strongly adsorbing surfaces—one of the previously mentioned hallmarks of recognition. The arguments noted above do not provide a reason for the sharpness of the transition or its order. An intuitive argument, and associated phenomenological model, suggests that the physical origin of the sharp transition is the strong suppression of loop fluctuations when adsorption occurs in pattern-matched conformations (40). The basic idea is this: Consider homopolymer adsorption on a chemically homogeneous surface. All segment-surface contacts are favorable, so the loop lengths and distance between loop ends on the surface can change without changing the nature of the segment-surface interactions. Thus, strong loop fluctuations occur leading to a lower free energy. Now consider the problem we are considering—a DHP interacting with a heterogeneous surface with a disordered distribution of sites. Now, there are two types of contacts with the surface: favorable and unfavorable. Concomitantly, there are two kinds of loops: those associated with good contacts at both ends, and the others. For loops associated with good contacts, only certain values of loop lengths and distances between loop ends on the surface are allowed. Loops associated with good contacts require favorable segment-surface contacts at both ends. This indicates that the fluctuations of this type of loop are suppressed due to competing interactions and quenched disorder. These loops are thus fundamentally different in character from the other type of loops or the strongly fluctuating loops associated with homopolymer adsorption on homogeneous surfaces. The allowed values of loop lengths and distance between loop ends for loops associated with good contacts are intimately related to the probabilities of finding certain types of sites and segments at different locations on the surface and along the chain because favorable segment-surface contacts must be established. When the statistics of the DHP sequence and the surface site distribution are related in such a way that there is a high probability of forming good contacts, it is reasonable to suppose that loop fluctuations are strongly suppressed. Suppression of loop fluctuations is tantamount to making the chain stiffer. Sufficiently stiff chains are known to undergo first-order adsorption transitions (39), and thus, suppression of loop fluctuations is argued to be the origin of the sharp adsorption transition depicted in Figure 3.

The analysis described above predicts that when the statistics characterizing the distribution of surface sites and those characterizing the DHP sequence are related in a special way, a sharp adsorption transition accompanied by the adoption of a few thermodynamically important conformations occurs. A sharp transition implies sharp discrimination between surfaces, one hallmark of recognition. Thus, one hallmark of recognition is realized when the statistics of the DHP sequence and surface site distribution are matched. This has been shown for a very simple model where the statistics for uncorrelated surface site and sequence distributions are measured by total loading and average composition, respectively. Monte-Carlo simulations have shown that the same phenomenon occurs for more complicated and interesting statistical patterns (16). The thermodynamic principles described above cause DHPs bearing a particular statistical pattern encoded in their sequence to sharply distinguish between surfaces bearing complementary statistical patterns and those that do not. The complementarity condition, based upon thermodynamic considerations, is found to be (16):

$$\sum_{m=1}^{N} P_s(m) P_c(m) = \Gamma, \qquad 13.$$

where $P_s(m)$ is the probability of finding a patch of size m of like sites at an arbitrary point on the surface, $P_c(m)$ is the probability of finding a contiguous "run" of like segments of length m at arbitrary locations along the chain, and Γ is a threshold number that depends strongly on chain flexibility and the strength of the interactions. $P_s(m)$ and $P_c(m)$ can be determined from the statistics of the DHP sequence and surface site distribution.

Our discussion so far has been concerned with thermodynamics. If recognition is to occur due to statistical pattern matching, a DHP bearing a particular statistical pattern encoded in the sequence, when exposed to a surface bearing different statistical patterns in different regions, should be able to find the target pattern without getting kinetically trapped in the wrong parts of the surface. Whether or not this can occur, and if so, what the pertinent principles for designing DHP sequences and surface site distributions are, can only be understood by studying the dynamics of DHP chains near surfaces bearing multiple types of statistical patterns. This has recently been done using kinetic Monte-Carlo simulations (19).

The simplest way to create statistical patterns that go beyond the completely uncorrelated distributions we have considered so far is to introduce two point correlations. The statistical patterns carried by chemically synthesized DHPs can be characterized naturally in terms of the chemical reactivities of the segments. If only two point correlations are included, the statistics of two-letter DHP sequences can be characterized by the average composition and a 2×2 matrix of reaction probabilities. An element of this matrix, P_{ij}, is the probability of finding a segment of type j immediately following a segment of type i. This matrix has one nontrivial eigenvalue (41), $\lambda = P_{AA} + P_{BB} - 1$. Positive values of λ imply that, below a certain correlation length q, contiguous runs of the same type of segment occur with high probability. We shall refer to this type of statistical pattern as statistically blocky. Negative values of λ correspond to sequences where, for lengths smaller than q, there is a propensity to form blocks of alternating segments. We shall refer to this type of sequence as statistically alternating. The value of q is set by the absolute magnitude of λ. Examples of statistically blocky and alternating DHP sequences are depicted schematically in Figure 4a (see color insert).

Figure 4b (see color insert) depicts statistically alternating and blocky distributions of A- and B-type surface sites. The simplest statistical measures of such site distributions are the correlation length, the total number density of A- and B-type sites, and the fraction of sites of each type (always equal to half in the following discussion). Such surface patterns can be generated by simulating an Ising-like Hamiltonian using a Monte-Carlo algorithm. Details are provided in (19). These types of patterned surfaces can be prepared in practice by self-assembly of mixed molecular adsorbents or molten DHPs on solid surfaces (42–44).

Figure 5 (see color insert) shows a surface divided into four quarters. The pattern of sites in the top right corner is statistically alternating, while that in the bottom left corner is statistically blocky. In the other two quarters the distribution of sites is completely random (i.e. uncorrelated). If a solution with a mixture of DHPs carrying statistically alternating and statistically blocky sequences is exposed to this surface, will the statistically alternating (blocky) chains rapidly find and bind strongly to the complementary statistically alternating (blocky) patch of the surface? An affirmative answer to this question would indicate that a phenomenology that mimics recognition can occur due to statistical pattern matching.

Figure 6 (see color insert) shows typical trajectories of the center of mass for a statistically alternating and a statistically blocky DHP generated by kinetic Monte Carlo simulations carried out on a lattice. The parameters characterizing the DHP and the surface site patterns are detailed in the figure caption. A standard Verdier-Stockmayer algorithm (45) is used to carry out the simulations. The interaction energy between complementary segments and surface sites is -1 kT_{ref}, and the unfavorable interactions are equal but of opposite sign. Only nearest neighbor interactions are allowed. As is clear, both trajectories start out in parts of the surface characterized by random site distributions. Each trajectory first samples wrong parts of the surface, but ultimately finds the target pattern and binds strongly. For $T/T_{ref} = 0.6$, separation due to statistical pattern matching is over 90% efficient (1000 trials) for a simulation time corresponding to the order of one second of real time. (Local motion on the scale of monomers occurs in $10^{-10}-10^{-11}$ sec depending on solvent conditions and monomer size.) These phenomenological results demonstrate that recognition due to statistical pattern matching can be successful, and it may be possible to exploit this notion in applications where biomimetic specificity is desired.

The DHPs do adsorb onto the wrong parts of the surface. These adsorbed states are, however, short lived compared to the length of the trajectories. In contrast, once strong binding occurs in the target region of the surface, the chains do not desorb in simulation time scales [or, it is estimated (19), on any time scale of experimental interest]. Figure 7 shows the free energy landscape that a statistically alternating DHP negotiates when interacting with the surface shown in Figure 6. This has been computed by extensive Monte Carlo simulations. As is evident, the free energy barriers separating local minima in the wrong parts of the surface are relatively small. The target region of the surface is, however, characterized by a few deep minima. Each such minima exhibits a rugged topography. This free

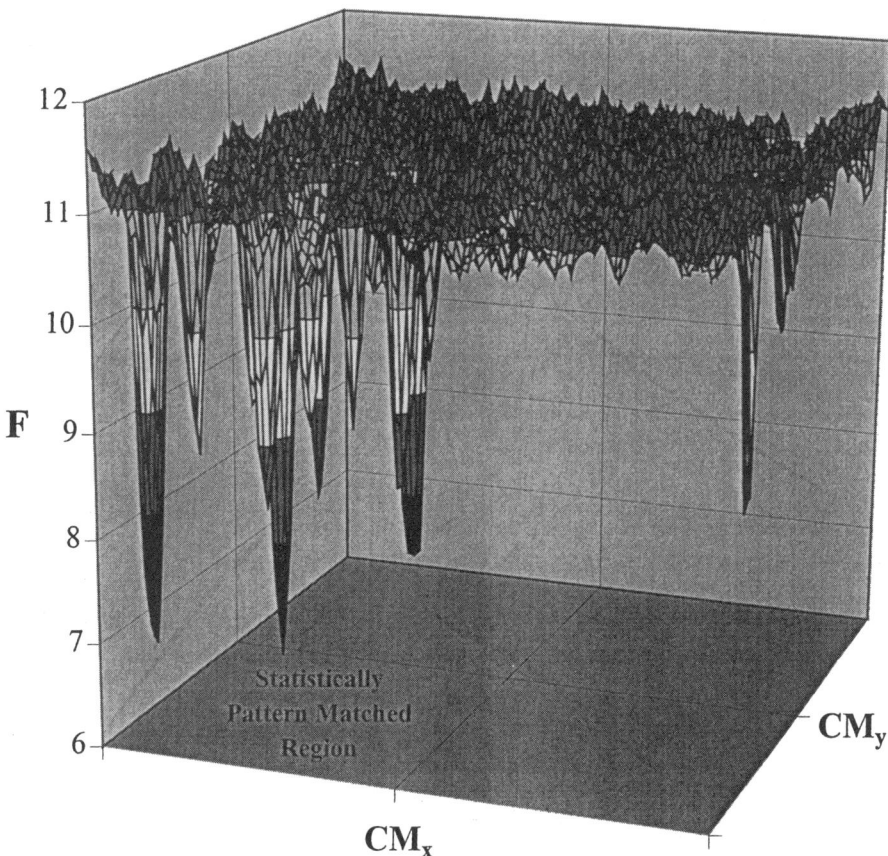

Figure 7 Free energy landscape (in arbitrary units) as a function of DHP center of mass position generated for an alternating DHP interacting with the surface shown in Figure 5. The deep free energy minima on the right side of the figure are artifacts of periodic boundary conditions.

energy landscape makes clear why the chain traverses the wrong parts of the surface relatively fast and ultimately finds and binds strongly to a region corresponding to one of the deep free energy minima in the target region of the surface.

The strength of the segment-surface interactions is set by the chemical identities of the chain segments and the surface sites. For our model, this chemistry sets the value of T_{ref}. The temperature of the system relative to the strength of the segment-surface energetics (T/T_{ref}) plays an important role in determining the behavior of the system. If the temperature is too high, entropic factors will prevent appreciable adsorption to all parts of the surface due to thermodynamic reasons. If the temperature is too low, then the free energy minima in the wrong parts of the surface will serve as long-lived kinetic traps. The existence of an optimal value of T/T_{ref} suggests that, given the chemical identity and statistics characterizing the surface site distribution, the chemical identity of the monomers that make up the statistically pattern-matched DHP sequences can be chosen in an optimal way. This could provide guidelines for design and synthesis of devices exploiting the notion of recognition due to statistical pattern matching.

One example of such an application is work being done by Whitesides and coworkers aimed toward designing copolymeric viral inhibition agents (47–49). The idea is to choose the chemistry and sequence of synthetic copolymers that could recognize and bind strongly to the pattern of receptors on a specific virus, thereby inhibiting its ability to interact with cells. Using copolymers made of sugars and acrylic acid, experiments suggest that a range of compositions (i.e. sequence) leads to optimal inhibition of the influenza virus (49). We briefly mention current efforts to develop computational tools that may prove useful in designing optimal DHP sequences that can recognize specific patterns of binding sites (e.g. a receptor).

The optimum in T/T_{ref} results from the importance of both kinetics and thermodynamics in determining the efficacy of recognition due to statistical pattern matching. This can also be demonstrated by comparing two situations: (a) the behavior of a statistically alternating DHP interacting with the surface shown in Figure 5 (see color insert) and (b) a surface with two regions, both bearing random site distributions with loadings equal to 20% and 40% exposed to DHP sequences that strongly favor the latter region on thermodynamic grounds. Recall that our replica field theory showed that, on thermodynamic grounds, DHPs with random sequences could differentiate sharply between surfaces that have different loadings of randomly distributed sites (13).

Monte Carlo simulations show that the window of T/T_{ref} over which successful recognition occurs (defined as >80% efficiency) is 50% wider for scenario (a) compared to scenario (b). This is so even though the equilibrium adsorbed fraction in the target region is the same for both cases. The reason is that, for uncorrelated sequence and surface site distributions, the free energy landscape that the chains negotiate is uncorrelated and rife with optima that serve as kinetic traps. (It is amusing to remark that similar arguments have been made in the context of species evolution on fitness landscapes (50).)

The free energy landscape shown in Figure 7 suggests that the dynamics of DHP chains traversing it should exhibit interesting features on different scales of length and time. To begin, consider the distribution of first passage times for a chain starting in the random part of the surface to find the target region and adsorb strongly. Strong adsorption is quantified by an energy cut off (-20 kT in the discussion to follow). Figure 8 shows the first passage time distribution for a 100-segment statistically alternating DHP interacting with the surface shown in Figure 5 (see color insert). After the usual turn-on time, the distribution of first passage times is decidedly nonexponential. In fact, a stretched exponential function with a stretching parameter equal to 0.43 best fits the simulation results. This indicates that highly cooperative events occur during the phenomenon that we are simulating.

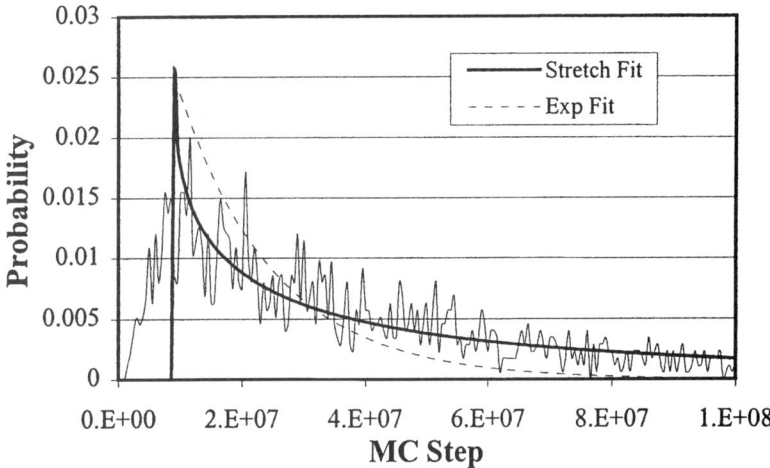

Figure 8 First passage time distribution for alternating DHP interacting with the surface shown in Figure 5. An exponent of $\beta = 0.43$ was used to generate the curve shown.

The event in question can be naturally divided into two events that occur in succession: center of mass motion from the random part of the surface to the edge of the target region followed by motion leading to strong adsorption. Simulation results show that only the first passage time distribution for the second event is nonexponential. It is also important to note that single-particle Kramers diffusion on the free energy landscape shown in Figure 7 does not lead to nonexponential first passage time distributions. This last observation suggests that the DHP center of mass is not the slowest dynamic mode through the duration of the entire event.

Further insight into the origin of the cooperative dynamics (data in Figure 8) is obtained by examining the distribution of loops during the course of a typical trajectory. This is shown in Figure 9 for a typical trajectory of a statistically alternating chain interacting with the surface shown in Figure 5 (see color insert). The distribution of loops is quantified by a quantity, P_n, which is the probability of a randomly chosen segment belonging to a loop of length n. This probability distribution is shown for different time periods in the trajectory. During the early stage of the trajectory, when the DHP center of mass lies above the wrong parts of the surface (first panel), the probability distribution is essentially structureless and the values of P_n are uniformly small. This is because different adsorbed conformations are energetically similar, and so the DHP minimizes free energy by sampling a multitude of conformations (and thus, loop lengths). Thus, no particular adsorbed conformations are greatly favored, leading to an essentially structureless distribution for P_n. As the trajectory enters the statistically pattern-matched region (second panel), we find that the distribution P_n begins to acquire structure. By the time the DHP is strongly adsorbed in a deep free energy minimum in the target region of the surface (last panel) the distribution resembles a spectrum, indicating that the DHP

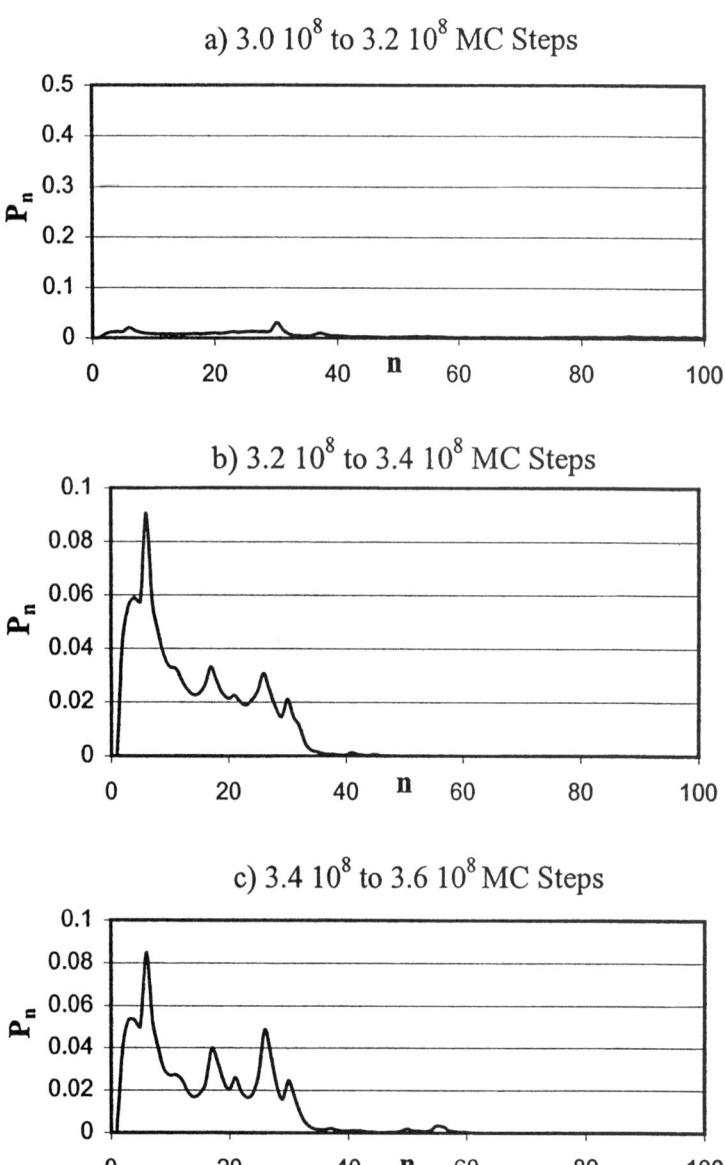

Figure 9 Loop length distributions, P_n, for different time windows along a typical trajectory: (*a*) DHP in the wrong region of the surface, (*b*) DHP center of mass entering the pattern-matched region, and (*c*) DHP strongly adsorbed in the pattern-matched region.

is bound in a class of conformations (or shapes). This is because in the statistically pattern-matched region, with high probability, there exist a few dominant conformations that can bind very strongly by carefully avoiding unfavorable interactions. However, most arbitrary conformations correspond to many unfavorable interactions because of the disordered sequence and competing interactions. Thus, the DHP sacrifices the entropic advantage of sampling a multitude of conformations, and adsorbs in a class of shapes. This is reminiscent of biopolymers binding to specific receptors in particular shapes.

These findings suggest that the time scales associated with loop fluctuations may be very different in different parts of the trajectory. This is seen clearly in the simulation results shown in Figure 10 for the time correlation function $\langle P_n(t)P_n(t+\tau)\rangle$. Specifically, we plot the time in which these time correlations decay along the trajectory. Only one value of n (corresponding to a peak in Figure 9) is considered in Figure 10, but the results are similar for other values of n corresponding to peaks in the distribution of P_n. In the wrong parts of the surface, these correlations decay very fast, and on the scale of the plot in Figure 10, the decay time is essentially zero (data not shown). As we enter the statistically pattern-matched region, the decay time begins to rise and ultimately oscillates about a large value as in a crystal. This implies that once adsorption occurs in a particular class of shapes, $\langle P_n(t)P_n(t+\tau)\rangle$ decays very slowly as these adsorbed shapes persist for long times.

Chain dynamics are thus very different in the wrong and target regions of the surface. In the wrong regions, the loop fluctuations are very fast, and the center of mass motion is the slowest dynamic mode. In the target region of the surface, once

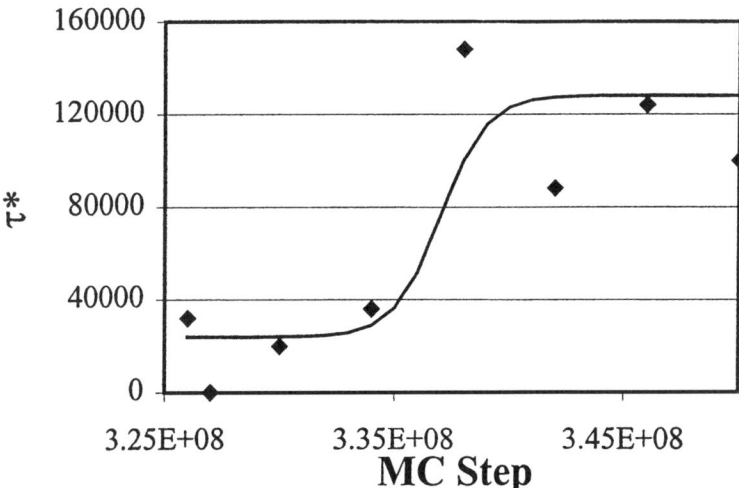

Figure 10 Relaxation time, τ^* (time for $\langle P_n(t)P_n(t+\tau)\rangle$ to relax to within 5% of its equilibrium value) for loops of length 26, calculated for same trajectory as Figure 9. During the time window shown, the polymer is in the statistically pattern-matched region.

the chain center of mass localizes over a region corresponding to a deep free energy minimum, the center of mass essentially stops. On the time scale of observation, it is effectively equilibrated. Now, the loops rearrange to acquire the conformation that corresponds to the bottom of the free energy minimum. This rearrangement takes place over long time scales, as there are many important entropic barriers that must be surmounted. These entropic barriers are made vivid in a movie that shows an animation of a trajectory for a DHP chain searching and finding a target pattern. These entropic traps have also been considered by Muthukumar (51, 52) in the context of a polyelectrolyte binding to a patterned array of charges of the opposite sign, and he has called a related phenomenon topological dereliction. Figure 11 (see color insert) shows snapshots from the movie mentioned above, and the fact that many adsorbed shapes are sampled in the wrong parts of the surface and only a few in the target region is made vivid.

After the chain center of mass has localized over a deep free energy minimum, the chain center of mass is the fastest degree of freedom, and the loop fluctuations are the slow dynamic modes. This explains why Kramer's dynamics for a particle on the free energy landscape parametrized by the center of mass coordinates does not reproduce the first passage time distribution observed in the simulations. The center of mass is not the slowest dynamic mode during the course of the trajectory.

Careful experimental studies of these issues pertinent to dynamics in frustrated systems and the phenomenon of statistical pattern matching are under way. One study aims to synthesize polymers made up of two naturally occurring amino acids and disordered sequences (SJ Muller, JD Keasling, personal communication, 2000). Following the methods pioneered by Tirrell et al (54), carefully controlled sequences can be synthesized. Studying the adsorption characteristics of such polypeptides on patterned surfaces created by mixed self-assembled monolayers is expected to address issues similar to those described in the text. Another study focuses on the characteristics of synthetic DHPs adsorbing onto patterned surfaces prepared by casting molten films of the same DHP onto silicon wafers. Neutron and X-ray reflectivity studies of the adsorption characteristics are planned (TP Russell, personal communication, 2000).

While these careful experiments have been initiated, three classes of experiments that provide coarse-grained information are worth mentioning. We have already mentioned the viral inhibition experiments carried out by Whitesides and coworkers (47–49). Recently, Gunning et al (56) have used time-resolved AFM to study the dynamics of a single plant polysaccharide chain interacting with a mica surface with a random distribution of charges. The dynamics they observe is reminiscent of our simulation results. In fact, movies of their AFM observations (56) and animations of our simulation results resemble each other in important ways. Further work comparing this class of experiments and theoretical results are suggested. Finally, some experimental studies of horse cytochrome c adsorption on surfaces bearing a disordered distribution of copper (which favorably interacts with the hystidine residues of the protein) exhibits adsorption characteristics that are similar to expected from statistical pattern matching. For example, it is found (57, 58) that a plot of adsorbed fraction versus loading of copper sites is strikingly similar to the prediction of the replica field theory (Figure 3).

In this section we have described how to roughly design ensembles of sequences that can discriminately bind to target patterns of binding sites by statistical pattern matching. For this notion to be useful in applications, however, the following question needs to be addressed: Can the optimal DHP sequence for binding to a target pattern be determined? Or conversely, can the optimal pattern of binding sites to elicit adsorption of a DHP with a specific sequence be determined? Consider the first question. The design of such DHPs is complicated by the fact that the adsorbed conformation is a priori unknown.

In natural systems, this optimization is continuously being carried out by molecular evolution. Currently, calculations are being performed that utilize an algorithm designed to mimic directed molecular evolution (59). In these simulations, trajectories are run that evolve not only in real space, but also in DHP sequence space. This algorithm generates DHP sequences optimal for binding to a target pattern. Such a computational tool may find use in applications such as viral inhibition (47–49).

III. NANOMECHANICAL MOTION DUE TO BIOPOLYMER BINDING AND ADSORPTION

Interactions between biomolecules are often translated into motion that is crucial for biological function. Understanding the mechanisms by which such molecular motors work in physiological processes is a subject of great current interest (60–68). Recent experiments have also demonstrated how interfacing molecular biology with microcantilever technology and biopolymer adsorption can result in controlled nanomechanical motion (69, 70). Quantitative differences in the generated motion can be related to differences in the biochemistry of the involved macromolecules. This observation, and its origin, suggests ways in which this class of phenomena can be exploited to devise diagnostic tools that can accurately detect the presence of specific biomolecules in solution. Such devices may prove useful for rapid and accurate detection of pathogens and proteins implicated as causes of certain diseases.

Here, we describe only one set of experiments (69) that illustrates the basic phenomenon, its origins, and its potential for applications. The experimental setup consists of a transparent fluid cell within which a gold-coated silicon nitride cantilever is mounted. Only one side of the cantilever is coated with gold. The cantilevers are typically 200 μm long and 0.5 μm thick, with each leg being 20 μm wide. A laser reflected off the cantilever and focused onto a position-sensitive detector is used to measure cantilever deflection. This methodology for measuring cantilever deflection is commonly used in atomic force microscopy (e.g. 71, 72).

Phosphate buffer, which maintains a desired pH, is first injected into the cell. Figure 12 (69) shows the measured deflection as a function of time upon injecting 50-nucleotide-long single stranded (ss) DNA molecules that are end-thiolated with HS-$(CH_2)_6$. These molecules form a self-assembled monolayer on the gold-coated side of the microcantilever. As Figure 12 shows, adsorption (or end-grafting) of the ssDNA molecules leads to a downward deflection of the microcantilever (see Figure 13). The inset in Figure 13 makes clear that the steady state deflection increases as the number of nucleotides (length) of the adsorbing DNA molecules increases.

Figure 14 shows the deflection that results upon subsequently injecting a solution containing ssDNA molecules, which are complementary to the end-grafted molecules and hence bind to them with great specificity. This process of complementary ssDNA molecules binding to form double-stranded (ds) DNA is called

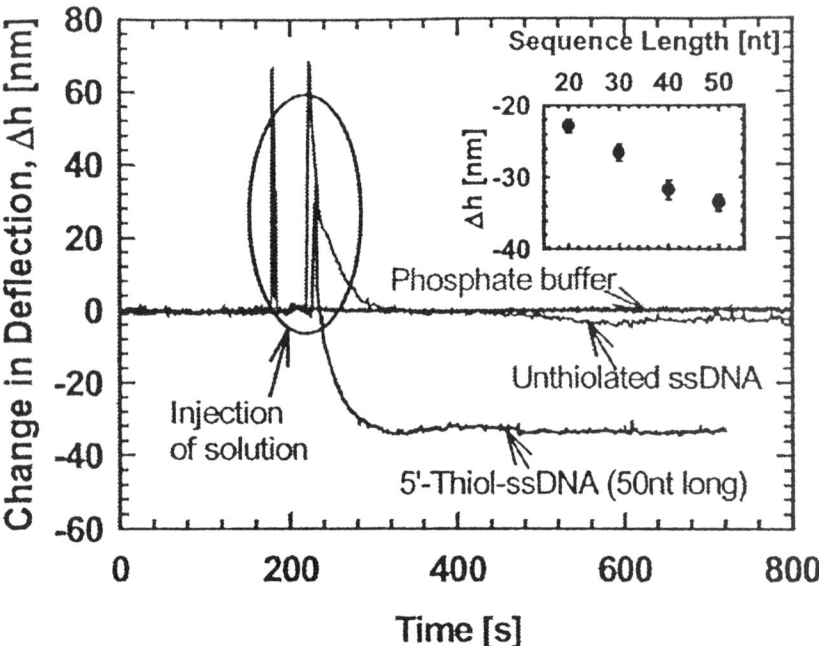

Figure 12 Change in cantilever deflection as a function of time. The data suggest exposure to either 0.1 M phosphate buffer or unthiolated probe ssDNA causes very little change in cantilever deflection while exposure to probe ssDNA thiolated at the 5' end does generate deflection (all ssDNA concentrations approximately 3.2 μM). The effect of the length of the probe molecule on the steady state deflection is shown in the inset.

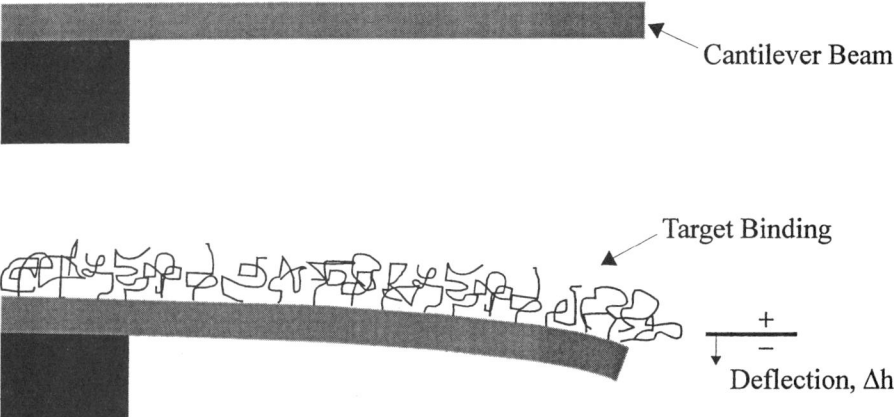

Figure 13 Interactions generated by adsorption of molecules to the cantilever surface can be designed to result in repulsive forces between adsorbed species. These forces may be large enough to generate nanomechanical motion in the form of the deflection of the cantilever as shown.

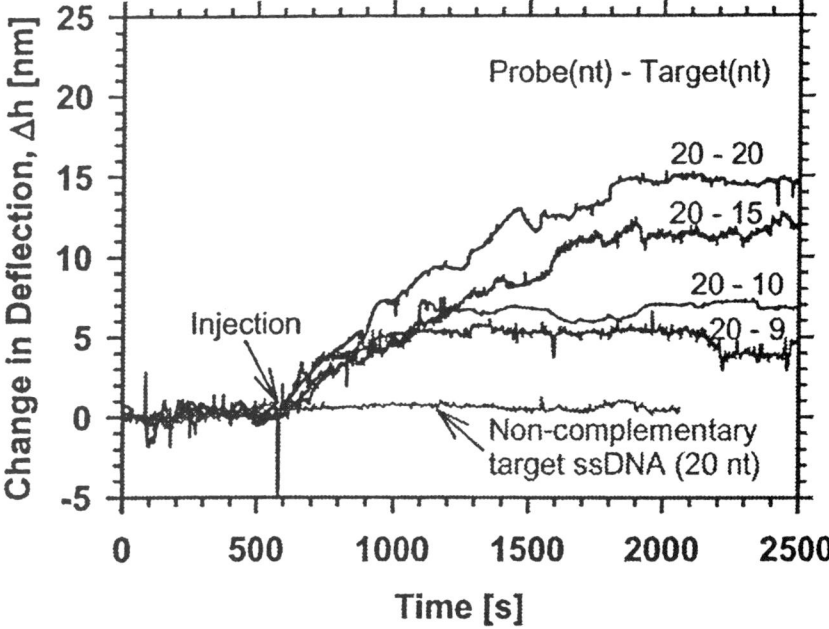

Figure 14 Change in cantilever deflection due to hybridization of a complementary target fragment of ssDNA with the probe ssDNA at its distal end. Results are shown for several target fragment lengths—20, 15, 10, and 9 nucleotides (nt). The degree of deflection increases with the length of the complementary fragment. Results for a typical noncomplementary fragment show little impact on cantilever deflection.

hybridization. While results are shown for only 20-nucleotide-long adsorbed DNA molecules, the same behavior is observed for adsorbed DNA molecules containing 30–50 nucleotides. The experimental data shows that hybridization leads to upward motion of the microcantilever. The upward deflection at steady state is directly related to the length of the complementary ssDNA that is injected into the solution after creating the original adsorbed layer. The longer the length of these target DNA molecules, the greater the extent of upward deflection. Thus, measurement of the deflection can be used to distinguish between small (the data indicates 1 nucleotide) changes in the number of nucleotides in the target DNA. The data shown in Figure 14 (69) correspond to situations where the target DNA molecules are all distally complementary. However, recent experiments (A Majumdar, T Thundat, personal communication, 2000) show that the deflection is also sensitive to the sequence and the number of complementary nucleotides on target DNA molecules of the same overall length. This again is suggestive that such observations could be exploited in diagnostic applications.

The origin of the downward deflection of the microcantilever upon end-grafting biopolymers onto one surface has been argued to be the compressive stress created by the intermolecular repulsion between adsorbed polymer chains. Since the polymers are end-grafted only to one side of the microcantilever, the repulsive interactions and concomitant compressive stress are alleviated by bending the cantilever at the expense of the strain energy required for bending the microcantilever by a certain amount.

Quantitative cantilever deflections for both the end-grafting step and the hybridization step are found to depend on the concentration of the buffer (the ionic strength) in the solution. This suggests that repulsive electrostatic forces and their screening play an important role. Since each nucleotide carries a net negative charge, one would expect hybridization to further increase the repulsive forces between the adsorbed macromolecules, and hence cause the cantilever to bend further. Clearly, the opposite is observed in the experiments. Thus, steric and electrostatic repulsions between the adsorbed macromolecules cannot be the sole reason underlying the observed phenomenon.

At the ionic strengths used in the experiments (0.05–1 M), ssDNA is known to have a persistence length on the order of 0.75 nm (corresponding to two nucleotides) (74). When these chains are end-grafted onto a surface, if the grafting density is sufficiently high, each chain will occupy a region of space that is smaller than its natural size due to the intermolecular repulsions. This reduction in the configurational entropy can be alleviated by adsorption onto a convex surface. This is because the curvature allows each chain to occupy a larger region of space as distance from the surface increases. Thus, in addition to intersegment repulsions, there is an entropic driving force that balances the strain energy of the cantilever and enhances the extent of bending. In contrast to ssDNA, at the experimental conditions, dsDNA has a persistence length of 50–80 nm (74), which is approximately 150 base pairs. Thus, the dsDNA is effectively a rod at the

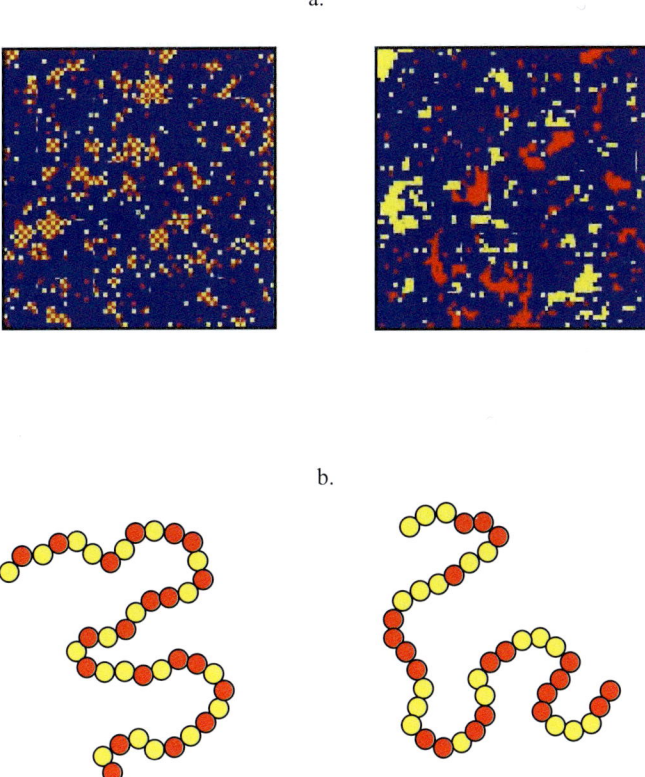

Figure 4 These panels show examples of statistical patterns. Panel a shows alternating and blocky surface patterns while panel b shows alternating and blocky DHP sequences.

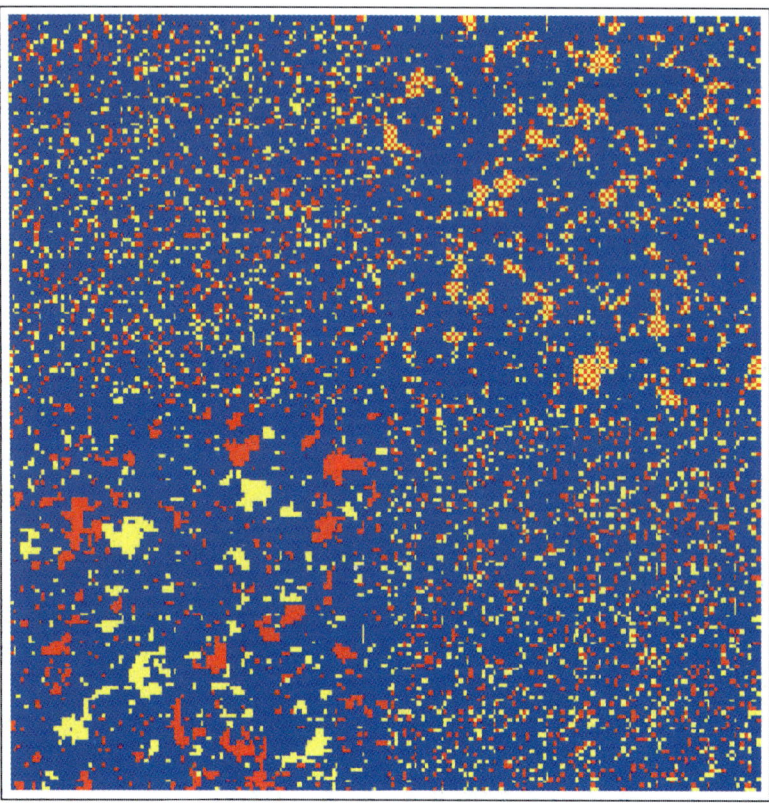

Figure 5 Surface onto which four patterns with two types of sites have been put. The top right (bottom left) quadrant bears an alternating (blocky) pattern (within a correlation length of approximately 1.6 lattice units) while remaining two quadrants have a statistically uncorrelated distribution of sites (zero correlation length).

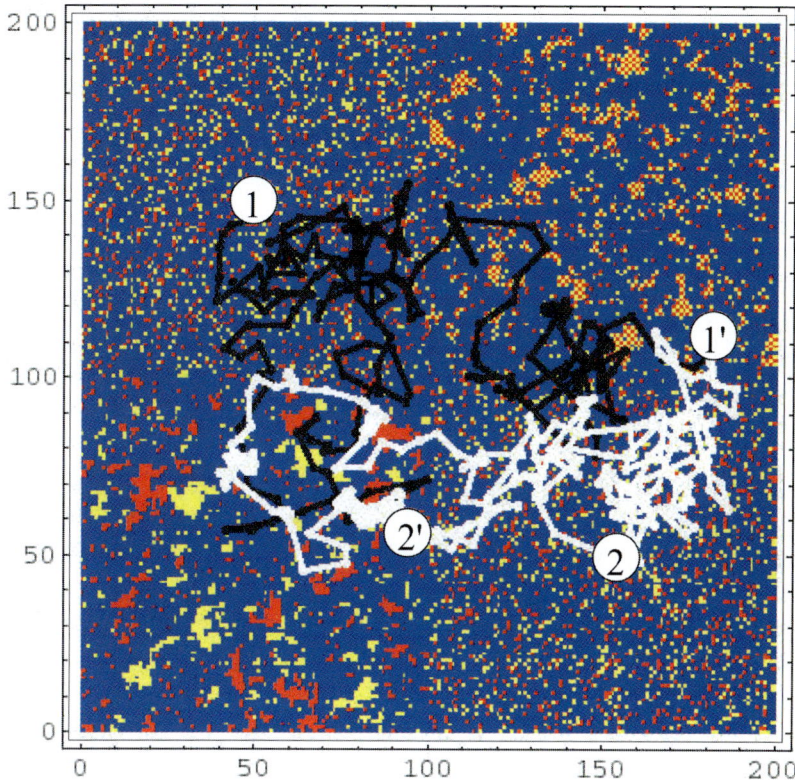

Figure 6 Surface from Figure 5 with typical polymer center of mass trajectories shown. The DHPs are characterized by f = 0.5 and λ = 0.4 or λ = -0.4. The black (white) trajectory is for a statistically alternating (blocky) DHP. Starting and ending positions are denoted by 1(1') and 2(2') respectively.

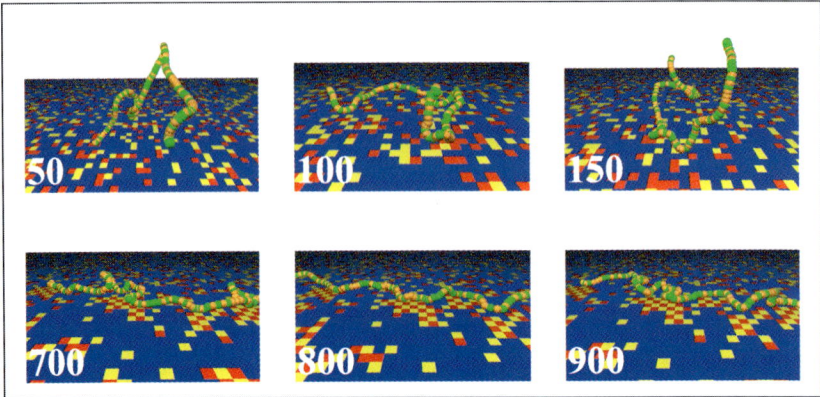

Figure 11 Six sequential snapshots taken from a typical trajectory. The top three panels depict the DHP in the wrong region of the surface, while the bottom three panels show the DHP strongly adsorbed in the pattern matched region.

experimental conditions, and the configurational entropy gain by forming a curved surface is inconsequential compared to the adsorbed ssDNA molecules. Therefore, the cantilever strain energy and intersegment repulsions are balanced at a lower cantilever deflection (or curvature). This explains the upward deflection upon hybridization.

The qualitative energy-configurational entropy balance described above is supported by the following experiment (69). ssDNA molecules were end-grafted with a buffer concentration of 1 M. The ionic strength was then reduced by an order of magnitude reduction in the buffer concentration. The grafting density resulting from adsorption at the higher buffer concentration is too high at the lower ionic strength because now the range of the intersegment repulsions is larger. This can cause the DNA chains to reduce the configurations sampled and to maximize the distance between neighboring grafted molecules by adopting the stretched configurations. In this instance, subsequent hybridization should lead to minimal changes in configurational entropy. The increased intersegment repulsions between the dsDNA molecules should be dominant, leading to a downward deflection upon hybridization. This is exactly what is observed in experiments (69). In addition to supporting arguments pointing to the importance of configurational entropy of adsorbed macromolecules in generating nanomechanical motion, this experiment also indicates that the direction of such motion can be controlled by tuning the conditions.

A quantitative analysis of the entropic and energetic contributions that determine cantilever bending can be easily carried out for sufficiently long chains using methods described in the polymer adsorption literature (e.g. 75–78). However, the DNA chains used in the experiments are short, and so a quantitative analysis is more delicate. It can be carried out, however, and should lead to a quantitative method for predicting and analyzing nanomechanical motion generated by biomolecular binding to polymers adsorbed on microcantilevers. Such an analysis, in conjunction with more extensive experiments, can prove useful in understanding how to control and manipulate nanomechanical motion generated by biomolecule binding to adsorbed layers.

Microcantilever deflections upon immobilizing proteins and then binding complementary ligands to immobilized proteins as well as that due to antibody-antigen interactions have also been observed and quantified for a few systems (68, 69). These results and those described in more detail in this section suggest that the binding of biomolecules to adsorbed polymer layers, and an understanding of the principles that govern the chemomechanical phenomena, could be exploited to design medical diagnostic devices. For example, an array of microcantilevers bearing different adsorbed macromolecules could be fabricated. Serum containing a known pathogen could then be injected into the solution in which the array is immersed. The deflection of the specific cantilever bearing the adsorbed macromolecule complementary to the target pathogen could then be used as a way to accurately and simply detect the presence of this pathogen. Much progress in measurement methods, fabrication, and theoretical analyses is required before such devices can be used in routine application. However, research aimed in this direction is fruitful, as the payoff could be substantial. It is also an arena where knowledge of polymer adsorption and molecular biology, acquired in the last three decades, can be used fruitfully toward creating useful medical diagnostic tools.

IV. ORDERING OF DIBLOCK MELTS NEAR PATTERNED SURFACES

As noted earlier, diblock copolymers are macromolecules consisting of two chemically distinct polymer chains covalently bonded together at one junction point. It is well known that a molten collection of these molecules microphase segregates below an order-disorder transition temperature (ODT) to form a myriad of interesting nano/microstructures (e.g. 79, 80). Efforts to exploit this ability to form regular nanoscale features in applications is an active research area. In this section we briefly review research aimed toward the creation of regular morphologies in thin films of diblock copolymers by utilizing interactions with patterned surfaces.

Molten diblock copolymers with symmetric compositions will spontaneously form lamellae with the stacking pattern AB–BA, and natural lamellar period L_b below the ODT (e.g. 79–82). In the case of a thin film (or slit), the lamellae may orient either parallel to the film or perpendicular to it (83–91) (see Figure 15). The case in which the lamellae orient perpendicularly is of particular interest, as the resulting exposed surface will bear a well-ordered repeating pattern on the scale of nanometers. This surface is potentially useful as a template for nanolithography.

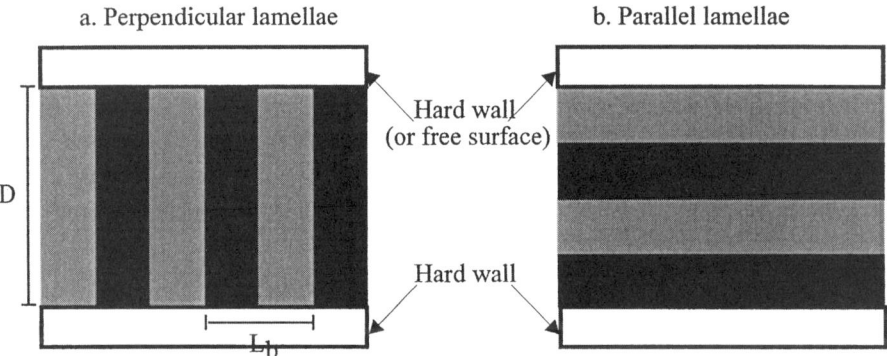

Figure 15 Schematic depiction of symmetric DCP morphologies in thin slits. Panel *a* shows a perpendicular lamellar structure while panel *b* shows parallel lamellae. D is the slit thickness and L_b is the natural lamellar period.

Much work has been done in predicting the orientation of these lamellae, which is strongly dependent upon the nature of the confining surface(s) and the film thickness D (83–85).

Surfaces that interact preferentially with one block of the diblock copolymer induce wetting of the surface by that component, resulting in the formation of parallel lamellae. Depending on the nature of each of the surfaces (where the air-polymer interface of a film may be considered as a surface), either symmetric or asymmetric wetting will be favored, which have preferred film thicknesses of either nL_b or $(n + 1/2)L_b$, respectively. Molten diblock copolymers constrained to different film thicknesses will either (for slits) form lamellae with periods streched or compressed from their unconstrained values or (for films) form islands or holes at the free surface. The formation of both parallel and perpendicular lamellae, as well as the formation of islands and holes at the free surface, has been observed experimentally by a number of researchers (92–98).

It has been suggested that patterning one of the surfaces with alternating stripes that interact preferentially with the two types of diblock segments can lead to stable perpendicular lamellar morphologies with a greater degree of lateral homogeneity than those formed on homogeneous surfaces (99–107). The remainder of this section is devoted to understanding how factors such as the natural bulk lamellar period (L_b), film thickness, surface interfacial energy, and surface patterning period (L_s) influence the morphology of the confined melt.

This problem has been studied theoretically by many researchers. Petera & Muthukumar employed both mean-field and self-consistent field theories (99, 100) to examine the problem. Another coarse-grained model was proposed by Chen & Chakrabarti (101), who employed a Landau-Ginzburg free energy functional combined with the Cahn-Hilliard formalism to study a diblock melt confined between striped and neutral walls. Nath, Nealey & de Pablo (106) utilized density functional theory to provide information on the length scale of the molecule. Wang and coworkers (108) attacked the problem using exhaustive lattice Monte Carlo simulations. The most useful strategy appears to use phenomenological models, first employed by Pereira & Williams (102–105, 107) and elaborated by Wang et al (109).

The essence of work in the area can be captured by briefly discussing the SCF theory, the phenomenological model, and their connections with experiments.

Petera & Muthukumar (100) use a SCF theory to generate possible equilibrium structures for the molten diblock copolymers below the ODT between striped and neutral or (preferentially) attractive walls. They do this by slowly lowering the temperature of an initially disordered state through the ODT while allowing the system to relax to a free energy minimum. They find the ratio of parameters L_s (the surface pattern period) and L_b (the natural bulk lamellar period) to be important, as well as the film thickness D. In addition to these geometrical parameters, the degree of chemical inhomogeneity χ (where χn is the Flory interaction parameter between segment types and n is the number of segments constituting each diblock) and the nature of the second wall (neutral or repulsive) also play pivotal roles in determining morphologies.

We begin a brief discussion of their theory by writing their Hamiltonian (H) for molten diblocks of length N (with n Kuhn segments of length l) composed of fn segments of type A.

$$H = \sum_{i=1}^{N} \frac{1}{2l^2} \int_0^n d\tau \left(\frac{\partial \mathbf{R}_i(\tau)}{\partial \tau}\right)^2$$

$$+ \sum_{i,j=1}^{N} \begin{bmatrix} \frac{\omega_{AA}}{2} \int_0^{fn} d\tau \int_0^{fn} d\tau' \delta(\mathbf{R}_j(\tau) - \mathbf{R}_i(\tau')) + \\ \frac{\omega_{BB}}{2} \int_{fn}^{n} d\tau \int_{fn}^{n} d\tau' \delta(\mathbf{R}_j(\tau) - \mathbf{R}_i(\tau')) + \\ \frac{\omega_{AB}}{2} \int_0^{fn} d\tau \int_{fn}^{n} d\tau' \delta(\mathbf{R}_j(\tau) - \mathbf{R}_i(\tau')) + \end{bmatrix} + \sum_{i=1}^{N} \begin{bmatrix} \int_0^n d\tau V_{SA}(\mathbf{R}_i(\tau)) + \\ \int_{fn}^n d\tau V_{SB}(\mathbf{R}_i(\tau)) \end{bmatrix},$$

14.

where ω_{ij} is the segment-segment interaction energy and V_{si} is the segment-surface interaction energy.

Providing for weak compressibility with a quadratic term and introducing coarse grained density fields

$$\rho_A(\mathbf{x}) = \sum_{i=1}^{N} \int_0^{fn} d\tau \delta(\mathbf{x} - \mathbf{R}_i(\tau)) \quad \rho_B(\mathbf{x}) = \sum_{i=1}^{N} \int_{fn}^{n} d\tau \delta(\mathbf{x} - \mathbf{R}_i(\tau)) \quad 15.$$

the Hamiltonian becomes

$$H' = N \ln(N-1) - N \ln S[\Phi_A, \Phi_B]$$

$$+ \int d^d x \begin{pmatrix} \frac{\omega_{AA}}{2} \rho_A^2(\mathbf{x}) + \frac{\omega_{BB}}{2} \rho_B^2(\mathbf{x}) + \omega_{AB}\rho_A(\mathbf{x})\rho_B(\mathbf{x}) \\ -i\Phi_A(\mathbf{x})\rho_A(\mathbf{x}) - i\Phi_B(\mathbf{x})\rho_B(\mathbf{x}) \\ +\frac{\lambda}{2}(\rho_A(\mathbf{x}) + \rho_B(\mathbf{x}) - \rho_o) \end{pmatrix}, \quad 16.$$

where

$$S[\Phi_A, \Phi_B] =$$

$$\int d\mathbf{R}(\tau) \exp \begin{bmatrix} -\frac{1}{2l^2} \int_0^n d\tau \left(\frac{\partial \mathbf{R}(\tau)}{\partial \tau}\right)^2 - \int_0^{fn} d\tau (i\Phi_A(\mathbf{R}(\tau)) + V_{SA}(\mathbf{R}(\tau))) \\ - \int_{fn}^n d\tau (i\Phi_B(\mathbf{R}(\tau)) + V_{SB}(\mathbf{R}(\tau))) \end{bmatrix}$$

$$\rho_o = \frac{nN}{V}. \quad 17.$$

The functional integrals over the fields $\rho_A(\mathbf{x})$, $\rho_B(\mathbf{x})$, $\phi_A(\mathbf{x})$, and $\phi_B(\mathbf{x})$ are evaluated using the saddle point approximation. With the assumption $\omega_{AA} = \omega_{BB} = \omega$, this leads to the following SCF equations:

$$i\Phi_A(\mathbf{x}) = \chi\rho_B(\mathbf{x}) + (\lambda + \omega)(\rho_A(\mathbf{x}) + \rho_B(\mathbf{x}) + \rho_o)$$
$$i\Phi_B(\mathbf{x}) = \chi\rho_A(\mathbf{x}) + (\lambda + \omega)(\rho_A(\mathbf{x}) + \rho_B(\mathbf{x}) + \rho_o)$$

18.

$$\rho_A(\mathbf{x}) = \frac{N \int_0^{fn} d\tau q_1(\mathbf{x}, \tau) q_2(\mathbf{x}, n - \tau)}{\int d^d x q_1(\mathbf{x}, n)}$$

$$\rho_B(\mathbf{x}) = \frac{N \int_0^{(1-f)n} d\tau q_2(\mathbf{x}, \tau) q_1(\mathbf{x}, n - \tau)}{\int d^d x q_2(\mathbf{x}, n)}$$

19.

$$\left(\frac{\partial}{\partial t} - \frac{l^2}{2}\nabla^2 + \Theta(fn - \tau)(i\Phi_A(\mathbf{x}) + V_{SA}(\mathbf{x}))\right.$$
$$\left. + \Theta(\tau - fn)(i\Phi_B(\mathbf{x}) + V_{SB}(\mathbf{x}))\right) q_1(x, t) = 0$$

$$\left(\frac{\partial}{\partial t} - \frac{l^2}{2}\nabla^2 + \Theta((1 - f)n - \tau)(i\Phi_B(\mathbf{x}) + V_{SB}(\mathbf{x}))\right.$$
$$\left. + \Theta(\tau - (1 - f)n)(i\Phi_A(\mathbf{x}) + V_{SA}(\mathbf{x}))\right) q_2(x, t) = 0$$

20.

Beginning with a homogeneously disordered density field, these equations are iterated until the density stops changing. To simulate the temperature decrease, the final state of a higher temperature calculation is used as the initial state of a lower temperature calculation. Resultant density fields for several different sets of parameters are displayed in Figure 16.

In the case of commensurate values of L_s and L_b (Figure 16a and b), perpendicular lamellae are only supported provided that the top wall is neutral. The presence of a preferentially attractive top wall was found to promote parallel lamellae in that region. This arrangement minimizes interfacial energy, as well as that through most of the film; very near the patterned surface, however, the diblocks align with the other surface to minimize its interfacial free energy. The authors also consider cases of incommensurability. When $L_s > L_b$ (Figure 16c and d) they find the lamellae tilt in order to minimize both the interfacial free energy and stretching free energy of the diblock chains. For $L_b > L_s$ (Figure 16e and f) they find that the chains undergo compression near the surface to match their period with that of the surface. This effect dissipates further into the film and parallel lamellae emerge unless the film is very thin (D < 2 L_b, as shown in f), whereupon the film cannot

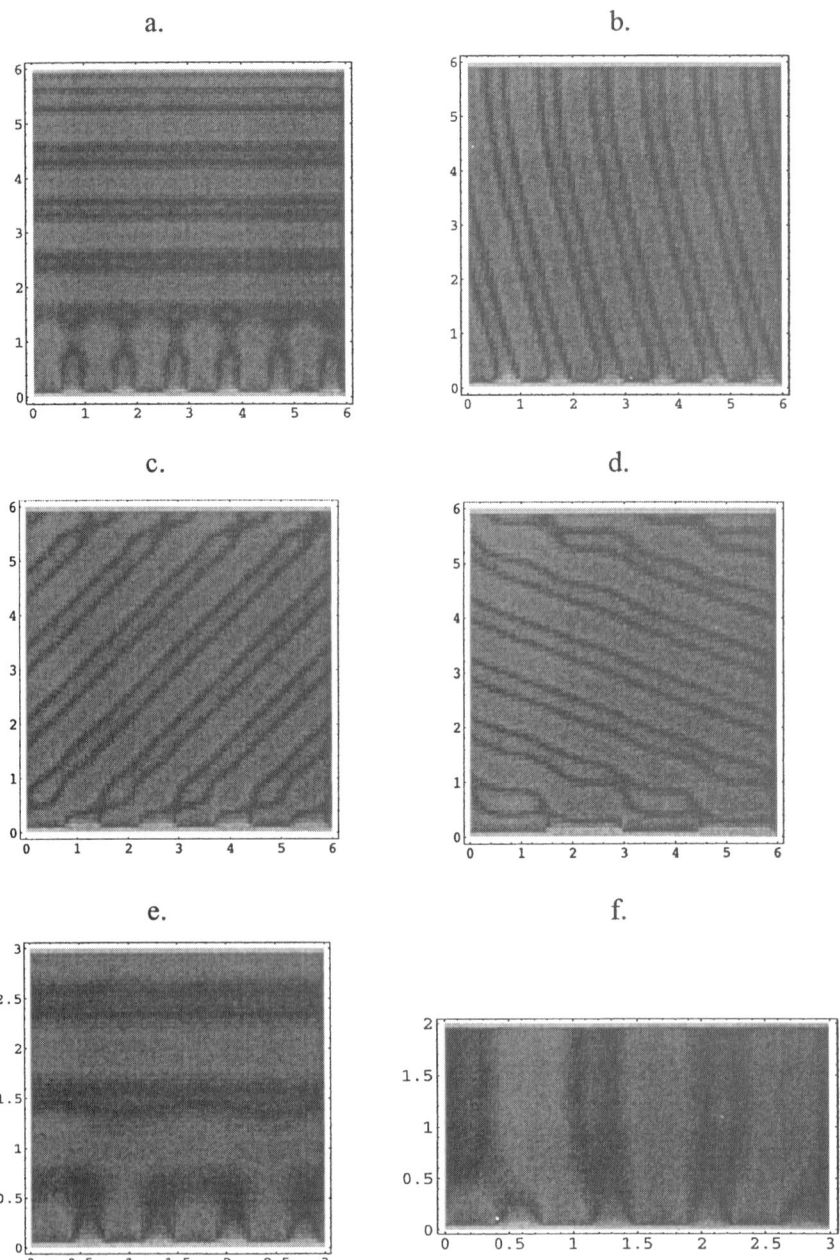

Figure 16 Graphical depiction of self-consistent field theory results. Adapted from (100). Shading indicates density variations. (*a*) Commensurate case with neutral top surface. (*b*) Commensurate case with preferential top surface. (*c*) Neutral top surface with ratio $L_s/L_b = 1.5$. (*d*) Neutral top surface with ratio $L_s/L_b = 3$. (*e*) Neutral top surface with ratio $L_s/L_b = 0.75$. (*f*) Thinner slit with neutral top surface and ratio $L_s/L_b = 0.75$.

support the high energy transition region between morphologies, and purely perpendicular lamellae persist throughout the film.

Some of these predictions are verified by experiments conducted by Rockford et al (110) for symmetric PS-PMMA diblock copolymer films on striped SiO_2-Au surfaces. The films are annealed under vacuum, corresponding to a neutral free surface. This study examines the regime in which $D < 2L_b$ for varying ratios of L_b/L_s. Only for the commensurate case are perpendicular lamellae formed that align with the surface pattern. For near commensurate cases, perpendicular lamellae are formed, but with significant structural defects. Cases in which L_b and L_s differ greatly show very little structure. Unfortunately, cases with $D > 2L_b$ are not studied, so the effects of the second surface and the possible formation of mixed lamellar structures have not been observed.

Forgoing a treatment that starts from a microscopic Hamiltonian (vide supra), many researchers have attempted to understand the morphologies present in thin diblock copolymer films by employing variants of essentially the same phenomenological theory. In this method, the free energy of the molten diblock copolymer film (in the strong segregation limit) is calculated as a sum of several contributions: a copolymer-copolymer interfacial contribution, copolymer-surface interfacial contribution, an elastic contribution and, if undulations in the lamellae are being considered, a bending contribution. In this method, a set of possible morphologies is proposed, and the free energy is calculated by summing the various contributions listed above. The theory predicts the melt to adopt the morphology that minimizes the free energy. With the variation of system parameters such as interfacial energies (segment-segment or segment-surface) and surface pattern periods, for example, a phase diagram for the melt may be constructed.

Pereira & Williams (102–105, 107) first applied this type of theory to the problem of a molten diblock copolymer confined in a thin slit with a heterogeneous surface. [This type of theory has been previously applied to molten diblock polymers (89, 111–115).] In their most inclusive study (105) they consider five different potential morphologies: perpendicular lamellae with period matching the surface pattern, perpendicular lamellae with the natural bulk period, parallel lamellae, and two mixed morphologies consisting of parallel lamellae over each of the two perpendicular lamellae (see Figure 17). Wang and co-workers (103) later applied the same type of theory to a more general set of eight morphologies that were observed in Monte Carlo simulations (108). Their set of morphologies is shown in Figure 18 and contains several that Pereira & Williams did not include, namely a checkerboard morphology, a mixed perpendicular lamellar morphology, and mixed checkerboard/lamellar structures.

In this section we have briefly discussed work that seeks to identify the regions of parameter space in which upon placing a diblock copolymer melt near a patterned surface, a regular periodic domain structure is induced at the exposed polymer surface. This work has been illustrated through both a SCF theory and a phenomenological model. The results give some general guidelines for the fabrication

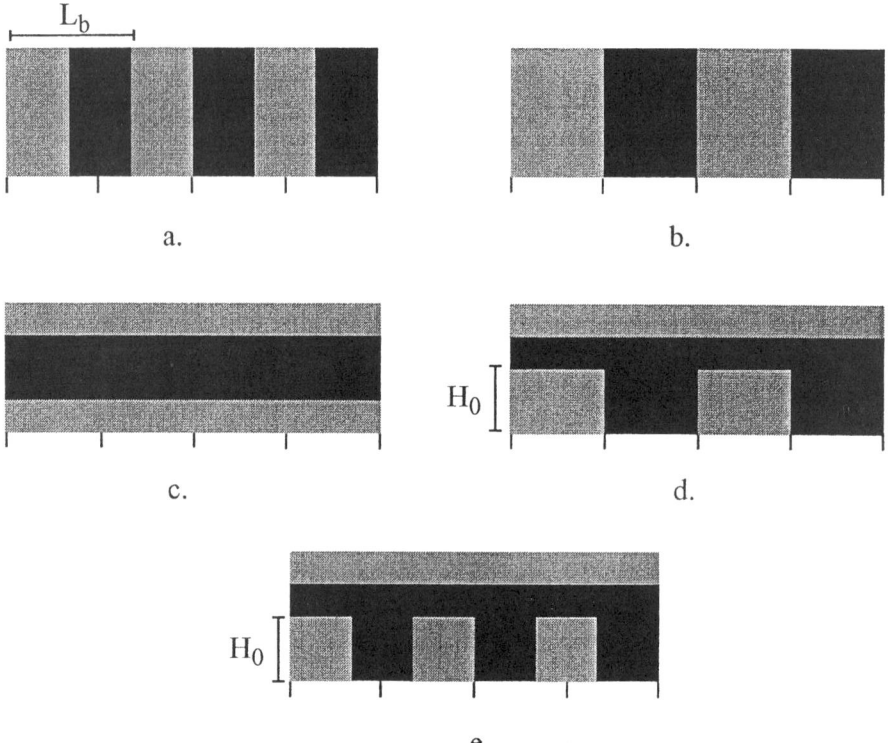

Figure 17 Potential morphologies of Pereira & Williams (adapted from 104). They are: (*a*) Perpendicular lamellae with surface pattern period. (*b*) Perpendicular lamellae with natural period. (*c*) Parallel lamellae. (*d*) Mixed morphology with parallel lamellae near top surface and perpendicular surface–directed lamellae penetrating a distance H_0 from the patterned surface. (*e*) Mixed morphology with parallel lamellae near top surface and perpendicular natural lamellae penetrating a distance H_0 from the patterned surface. Periodic lines below bottom surface indicate surface patterning.

of such nanostructures. Efforts to fabricate such structures should concentrate on thin films between patterned-neutral surfaces with surface pattern periods commensurate with the bulk lamellar period of the diblock melt. Future efforts will illuminate whether or not the creation of defect-free nanostructures of this type is possible, and whether the fabrication of such nanoscale features is useful in applications.

CONCLUDING REMARKS

In the past three decades, much progress has been made toward understanding the fundamental physical chemistry of polymer adsorption processes. This understanding has also been harnessed for the creation of better commodity products where adsorbed polymer layers play a crucial role. It is expected that future

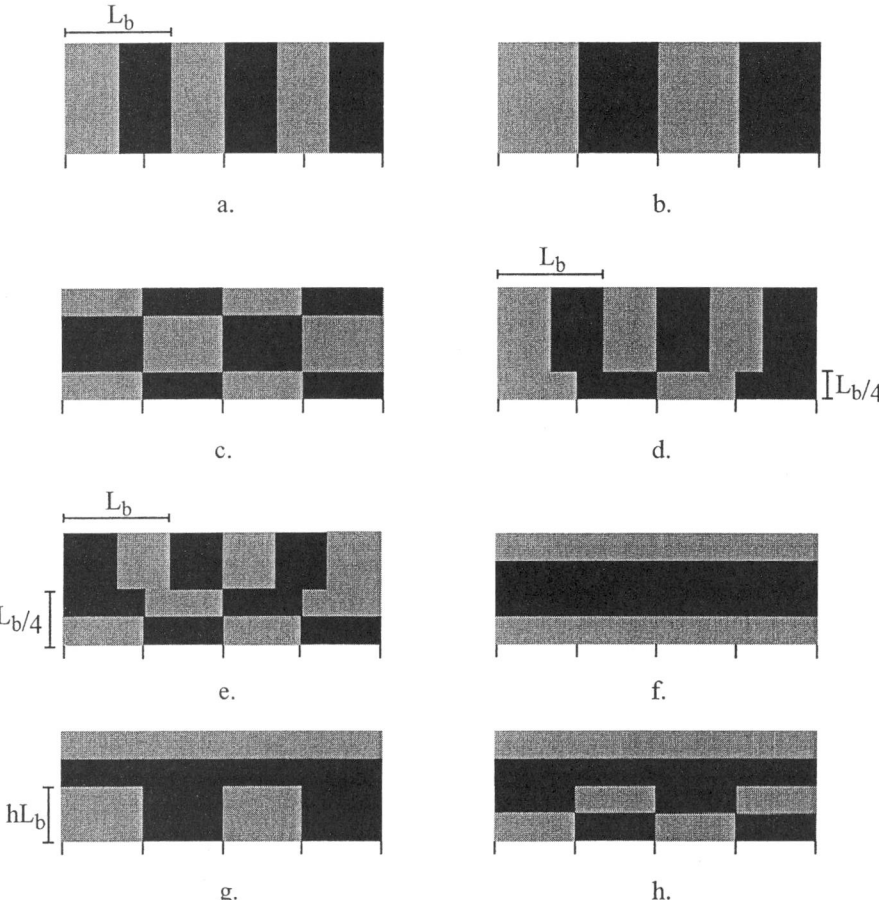

Figure 18 Potential morphologies of Wang et al (adapted from 109). They are (*a*) Natural perpendicular lamellae. (*b*) Surface directed perpendicular lamellae. (*c*) Checkerboard morphology. (*d*) Mixed morphology of surface lamellae near the patterned surface and bulk lamellae near the top surface. (*e*) Mixed morphology of checkerboard near the patterned surface combined with perpendicular bulk lamellae. (*f*) Parallel lamellae. (*g*) Mixed morphology of perpendicular surface lamellae near the patterned surface combined with parallel lamellae. (*h*) Mixed morphology of checkerboard morphology near the patterned surface combined with parallel lamellae. Periodic lines below bottom surface indicate surface patterning.

applications will demand the creation and processing of information on short length scales (tens of nanometers). In response to external input (stimuli), nature often employs self-assembly processes to create specific structural motifs that are useful for a specific function. In other words, given certain inputs, self-assembly is used to process this information on short length scales and generate a specific function. In many instances, macromolecules are used as the molecular building blocks that self-assemble into specific structures. A question that suggests itself is: Can we design synthetic macromolecular systems that can self-assemble into functionally useful nanoscale structures when stimulated by certain external conditions? Such processes may be found useful in prospective technologies where creation and processing of information on short length scales are desired.

In this article, we have briefly reviewed three examples where polymer adsorption processes can mediate self-assembly into specific nanostructures when particular external conditions are presented. In the first example, the external stimulus is a surface bearing a target pattern of binding sites. We have considered how the sequences of polymers can be designed such that they can rapidly search the patterns of binding sites on a surface, find the target pattern, and then bind to it in a class of conformations. In other words, we have considered the minimal requirements for biomimetic recognition between polymers and surfaces. In the second example, the external stimulus is provided by biological molecules (DNA) in solution that can bind with specificity to an adsorbed layer of complementary ssDNA immobilized on a microcantilever. The physical origins of strategies that allow the manipulation and control of nanomechanical motion of these cantilevers due to such specific biomolecular interactions were described. These strategies may prove useful in future applications requiring sensing and actuation on short length scales (e.g. medical diagnostic tools). The third example considered how surfaces bearing periodic patterns on the nanometer scale could be created by self-assembly of diblock copolymers adjacent to surfaces bearing periodic patterns on the nanometer scale.

The examples described in this review, and many others, constitute early attempts to develop functional materials that self-assemble in response to external inputs. Future research at the crossroads of physical chemistry, materials science, and biology, using theory, computer simulation, physical characterization experiments, and synthesis, should allow progress toward discovering strategies and implementing them in the creation of such responsive nanometer scale objects. Perhaps, some day, such studies will enable the design of synthetic systems that can mimic processes such as the assembly of micron-sized patterns of nanometer-size proteins that constitute synapses formed during exquisitely specific cell-cell recognition processes necessary for signal transduction and concomitant biological responses (116, 117).

Visit the Annual Reviews home page at www.AnnualReviews.org

LITERATURE CITED

1. Chakraborty AK, Tirrell M. 1996. *MRS Bull.* 21:28–32
2. Stuart MAC, Cosgrove T, Vincent B. 1986. *Adv. Coll. Int. Sci.* 24:143–239
3. Takahashi A, Kawaguchi M. 1982. *Adv. Polym. Sci.* 46:1
4. Patel SS, Tirrell M. 1989. *Annu. Rev. Phys. Chem.* 40:597–635
5. Milner ST, Witten TA, Cates ME. 1988. *Macromolecules* 21:2610–19
6. Potanin AA, Russel WB. 1995. *Phys. Rev. E* 52:730–37

7. Borisov OV, Birshtein TM, Zhulina EB. 1991. *J. Phys. II* 1:521–26
8. Haraharan R, Biver C, Russel WB. 1998. *Macromolecules* 31:7514–18
9. Lau KF, Dill KA. 1989. *Macromolecules* 22:3986–97
10. Irback A, Peterson C, Potthast F. 1996. *Proc. Natl. Acad. Sci. USA* 93:9533–38
11. Pande VS, Grosberg AY, Tanaka T. 1994. *Proc. Natl. Acad. Sci. USA* 91:12972–75
12. Dewey G. 1997. *Fractals in Molecular Biophysics*. Oxford: Oxford Univ. Press
13. Srebnik S, Chakraborty AK, Shakhnovich EI. 1996. *Phys. Rev. Lett.* 77:3157–60
14. Chakraborty AK, Shakhnovich EI. 1995. *J. Chem. Phys.* 103:10751–63
15. Bratko D, Chakraborty AK, Shakhnovich EI. 1996. *Phys. Rev. Lett.* 76:1844–47
16. Bratko D, Chakraborty AK, Shakhnovich EI. 1997. *Chem. Phys. Lett.* 280:46–52
17. Bratko D, Chakraborty AK, Shakhnovich EI. 1998. *Comp. Theor. Polym. Sci.* 8:113–26
18. Srebnik S, Bratko D, Chakraborty AK. 1998. *J. Chem. Phys.* 109:6514–19
19. Golumbfskie AJ, Pande VS, Chakraborty AK. 1999. *Proc. Natl. Acad. Sci. USA* 96:11707–12
20. Cates ME, Ball RC. 1988. *J. Phys.* 49:2009–18
21. Bouchaud JP, Georges A. 1990. *Phys. Rep. C* 195:127–293
22. Chandler D. 1991. In *Liquids, Freezing and the Glass Transition, Proceedings of the Les Houches Summer School* ed. D Levesque, et al. New York: Elsevier
23. Feynman RP, Vernon JL Jr. 1963. *Annu. Phys. NY* 24:118–73
24. Chandler D, Singh Y, Richardson DM. 1984. *J. Chem. Phys.* 81:1975–82
25. Edwards SF, Anderson PA. 1975. *J. Phys. F* 5:965
26. Fischer KH, Hertz JA. 1993. *Spin Glasses*. Cambridge: Cambridge Univ. Press
27. Chan HS, Dill KA. 1991. *Annu. Rev. Biophys. Chem.* 20:447–90
28. Bryngelson J, Wolynes PG. 1987. *Proc. Natl. Acad. Sci. USA* 84:7524–28
29. Shakhnovich EI, Gutin AM. 1990. *Nature* 346:773–75
30. Bryngelson J, Onuchic JN, Socci ND, Wolynes PG. 1995. *Protein.-Struct. Funct. Gen.* 21:167–95
31. Shakhnovich EI. 1996. *Folding Design* 1:50–54
32. Chan HS, Dill KA. 1997. *Nat. Struct. Biol.* 4:10–19
33. Karplus M, Shakhnovich EI. 1995. *Protein Folding*, ed. TE Creighton. New York: Freeman
34. Pande VS, Grossberg AY, Tanaka T. 1997. *Biophys. J.* 73:3192–210
35. Parisi G. 1979. *Phys. Lett.* 73:203–5
36. Shakhnovich EI, Gutin A. 1989. *J. Phys.* 50:1843–50
37. Mezard M, Parisi G. 1991. *J. Phys. I* 1:809–36
38. Hoeve CAJ, diMarzio EA, Peyser P. 1965. *J. Chem. Phys.* 42:2558–63
39. Grossberg AY, Khokhlov AK. 1994. *Statistical Physics of Macromolecules*. New York: Am. Inst. Phys.
39a. Chakraborty AK. 2001. *Phys. Rep.* 342:2–61
40. Bratko D, Chakraborty AK. 1998. *J. Chem. Phys.* 108:1676–82
41. Odian GG. 1991. *Principles of Polymerization*. New York: Wiley
42. Berndt P, Fields GB, Tirrell M. 1995. *J. Am. Chem. Soc.* 117:9515–22
43. Stranick SJ, Parikh AN, Tao YT, Allara DL, Weiss PS. 1994. *J. Phys. Chem.* 98:7636–46
44. Leduc MR, Hayes W, Frechet JMJ. 1998. *J. Polym. A* 36:1–10
45. Verdier PH, Stockmayer WH. 1962. *J. Chem. Phys.* 36:227–35
46. Deleted in proof
47. Spaltenstein A, Whitesides GM. 1991. *J. Am. Chem. Soc.* 113:686–87
48. Mammen M, Dahmann G, Whitesides GM. 1995. *J. Med. Chem.* 38:4179–90
49. Sigal G, Mammen M, Dahmann G,

Whitesides GM. 1996. *J. Am. Chem. Soc.* 118:3789–800
50. Kauffman SA. 1993. *The Origins of Order: Self Organization and Selection in Evolution.* New York: Oxford Univ. Press
51. Muthukumar M. 1995. *J. Chem. Phys.* 103:4723–31
52. Kong CY, Muthukumar M. 1998. *J. Chem. Phys.* 109:1522–27
53. Deleted in proof
54. Tirrell DA, Fournier MJ, Mason TL. 1991. *Curr. Opin. Struct. Biol.* 1:638
55. Deleted in proof
56. Gunning AP, Mackie AR, Kirby AR, Kroon P, Williamson G, Morris VJ. 2000. *Macromolecules* 33:5680–85
57. Todd RJ, Johnson RD, Arnold FH. 1994. *J. Chromatogr. A* 662:13–26
58. Johnson RD, Wang Z-G, Arnold FH. 1996. *J. Phys. Chem.* 100:5134–39
59. Golumbfskie AJ, Chakraborty AK. 2000. In press
60. Rastogi VK, Girvin ME. 1991. *Nature* 402:263–68
61. Vale RD, Milligan RA. 2000. *Science* 288:88–95
62. Mahadevan L, Matsudaira P. 2000. *Science* 288:95–99
63. Astumian RD, Bier M. 1994. *Phys. Rev. Lett.* 72:1766–69
64. Astumian RD. 1997. *Science* 276:917–22
65. Keller D, Bustamante C. 2000. *Biophys. J.* 78:541–56
66. Kuo SC, Sheetz MP. 1993. *Science* 260:32–34
67. Svoboda K, Schmidt CF, Schnapp BJ, Block SM. 1993. *Nature* 365:721–27
68. Finer JT, Simmons RB, Spuldich JA. 1994. *Nature* 368:113–19
69. Wu G, Ji H, Hansen K, Thundat T, Datar R, et al. 2001. *Proc. Natl. Acad. Sci. USA* 98:1560–64
70. Fritz J, Baller M, Lang HP, Rothuizen H, Vettiger P, et al. 2000. *Science* 288:316–18
71. Madou MJ. 1997. *Fundamentals of Microfabrication.* Boca Raton, FL: CRC
72. Perazzo T, Mao M, Kwon O, Majumdar A. 1999. *Appl. Phys. Lett.* 74:3576–69
73. Deleted in proof
74. Baumann CG, Smith SB, Bloomfield VA, Bustamante C. 1997. *Proc. Natl. Acad. Sci. USA* 94:6185–90
75. Ross RS, Pincus P. 1992. *Macromolecules* 25:2177–83
76. Zhulina EB, Borisov OV. 1996. *Macromolecules* 29:2618–26
77. Hariharan R, Biver C, Mays J, Russel WB. 1998. *Macromolecules* 31:7506–13
78. Biver C, Hariharan R, Mays J, Russel WB. 1997. *Macromolecules* 30:1787–92
79. Hamley IW. 1998. *The Physics of Diblock Copolymers.* New York: Oxford Univ. Press
80. Bates FS, Fredrickson GH. 1999. *Phys. Today* 52:32–38
81. Leibler L. 1980. *Macromolecules* 13:1602–7
82. Bates FS, Fredrickson GH. 1990. *Annu. Rev. Phys. Chem.* 41:525–57
83. Fredrickson GH. 1987. *Macromolecules* 20:2535–42
84. Anastasiodis SH, Russell TP. 1989. *Phys. Rev. Lett.* 62:1852–55
85. Shull KR. 1992. *Macromolecules* 25:2122–33
86. Tang H, Freed KF. 1992. *J. Chem. Phys.* 97:4496–504
87. Brown G, Chakrabarti A. 1994. *J. Chem. Phys.* 101:3310–17
88. Brown G, Chakrabarti A. 1995. *J. Chem. Phys.* 102:1440–48
89. Matsen MW. 1997. *J. Chem. Phys.* 106:7781–91
90. Pickett GT, Balazs AC. 1997. *Macromolecules* 30:3097–103
91. Wang Q, Yan Q, Nealey PF, de Pablo JJ. 2000. *J. Chem. Phys.* 112:450–63
92. Kellogg GJ, Walton DG, Mayes AM, Lambooy P, Russell TP, et al. 1996. *Phys. Rev. Lett.* 76:2503–6
93. Mansky P, Russell TP, Hawker CJ, Mays J, Cook DC, Satija SK. 1997. *Phys. Rev. Lett.* 79:237–40

94. Mansky P, Russell TP, Hawker CJ, Pitsikalis M, Mays J. 1997. *Macromolecules* 30:6810–13
95. Huang E, Russell TP, Harrison C, Chaikin PM, Register RA, et al. 1998. *Macromolecules* 31:7641–50
96. Mansky P, Tsui OKC, Russell TP, Gallot Y. 1999. *Macromolecules* 32:4832–37
97. Peters RD, Yang XM, Kim TK, Sohn BH, Nealey PF. 2000. *Langmuir* 16:4625–31
98. Huang E, Mansky P, Russell TP, Harrison C, Chaikin PM, et al. 2000. *Macromolecules* 33:80–88
99. Petera D, Muthukumar M. 1997. *J. Chem. Phys.* 107:9640–44
100. Petera D, Muthukumar M. 1998. *J. Chem. Phys.* 109:5101–7
101. Chen H, Chakrabarti A. 1998. *J. Chem. Phys.* 108:6897–905
102. Pereira GG, Williams DRM. 1998. *Phys. Rev. Lett.* 80:2849–52
103. Pereira GG, Williams DRM. 1998. *Macromolecules* 31:5904–15
104. Pereira GG, Williams DRM. 1999. *Macromolecules* 32:758–64
105. Pereira GG, Williams DRM. 1999. *Langmuir* 15:2125–29
106. Nath SK, Nealey PF, de Pablo JJ. 1999. *J. Chem. Phys.* 110:7483–90
107. Pereira GG, Williams DRM. 1999. *Phys. Rev. E.* 60:5841–47
108. Wang Q, Yan Q, Nealey PF, de Pablo JJ. 2000. *Macromolecules* 33:4512–25
109. Wang Q, Nath SK, Graham MD, Nealey PF, de Pablo JJ. 2000. *J. Chem. Phys.* 112:9996–10010
110. Rockford L, Liu Y, Mansky P, Russell TP, Yoon M, Mochrie SGJ. 1999. *Phys. Rev. Lett.* 82:2602–5
111. Turner MS. 1992. *Phys. Rev. Lett.* 69:1788–91
112. Kikuchi M, Binder K. 1994. *J. Chem. Phys.* 101:3367–77
113. Semenov AN. 1985. *Sov. Phys. JETP* 61:733–41
114. Ohta T, Kawasaki K. 1986. *Macromolecules* 19:2621–32
115. Wang ZG, Safran SA. 1991. *J. Chem. Phys.* 94:679–87
116. Grakoui A, Bromley SK, Sumen C, Davis MM, Shaw AS, et al. 1999. *Science* 285:221–27
117. Monks CR, Freiberg H, Kupfer H, Sciaky N, Kupfer A. 1998. *Nature* 395:82–86

BIOMOLECULAR SOLID STATE NMR: Advances in Structural Methodology and Applications to Peptide and Protein Fibrils[1]

Robert Tycko

Laboratory of Chemical Physics, National Institute of Diabetes and Digestive and Kidney Diseases, National Institutes of Health, Bethesda, Maryland 20892-0520; e-mail: tycko@helix.nih.gov

Key Words magnetic resonance, structural biology, amyloid

■ **Abstract** Solid state nuclear magnetic resonance (NMR) methods can provide atomic-level structural constraints on peptides and proteins in forms that are not amenable to characterization by other high-resolution structural techniques, owing to insolubility, high molecular weight, noncrystallinity, or other characteristics. Important examples include peptide and protein fibrils and membrane-bound peptides and proteins. Recent advances in solid state NMR methodology aimed at structural problems in biological systems are reviewed. The power of these methods is illustrated by experimental results on amyloid fibrils and other protein fibrils.

INTRODUCTION

Solid state nuclear magnetic resonance (NMR) is the application of NMR spectroscopy to systems that are solids, nearly solid, or strongly anisotropic. This chapter describes recent developments in the use of solid state NMR as a structural probe of peptides and proteins, an area of research that has experienced accelerated progress and growth in the past five years. This chapter is not intended to be a comprehensive review. Instead, selected aspects of solid state NMR techniques that are of general interest and importance and recent applications of these techniques in structural studies of peptide and protein fibrils are described. These applications illustrate the potential of solid state NMR to address real structural issues in complex systems of biological origin and relevance. The experimental examples are taken primarily from work in the author's laboratory, but of course progress in this field is the result of the combined efforts of many research groups.

---

[1]The US Government has the right to retain a nonexclusive, royalty-free license in and to any copyright covering this paper.

Aspects of biomolecular solid state NMR not covered in this chapter have been reviewed elsewhere (1–9).

The importance of peptides and proteins in biology, as structural materials, catalysts, mechanical force-generating elements, regulators of gene expression, carriers and transducers of chemical signals between cells, energy transducers, components of pathogens and the defense against pathogens, and in other capacities, is obvious. The importance of structural information about peptides and proteins, as a key to understanding biological function and intermolecular interactions, as a means of classifying, organizing, and understanding relationships among proteins, and as a guide to the development of therapeutic agents, is also obvious. What may be less obvious to the nonexpert is the role that solid state NMR measurements can play in structural studies of peptides and proteins. The vast majority of atomic-resolution structural data about peptides and proteins comes from two techniques, namely X-ray crystallography and liquid state NMR, that are extremely well established and widely practiced. So why use solid state NMR? There are many answers to this question: (*a*) X-ray crystallography depends on the availability of high-quality crystals. Liquid state NMR generally requires solubility at concentrations in excess of 100 μM and is currently limited to proteins with molecular weights less than approximately 40 kD. Solid state NMR techniques do not require crystallinity or solubility and can be applied to proteins and complexes with molecular weights that greatly exceed 100 kD. Thus, systems that prove resistant to crystallization (e.g. because of inherent flexibility), have insufficient solubility, or are too large for liquid state NMR may be amenable to solid state NMR methods. (*b*) Solid state NMR techniques have the unique capability of providing atomic-level constraints on the molecular structures of complex, noncrystalline solids. Peptide and protein fibrils (e.g. amyloid fibrils) are an important class of systems of this type. (*c*) Approximately 30% of proteins are believed to be integral membrane proteins. Although an increasing number of membrane proteins have been crystallized and studied by diffraction, this number remains small. Solid state NMR techniques are applicable to membrane proteins. (*d*) Solid state NMR techniques can provide structural information that is complementary to information from X-ray crystallography or liquid state NMR. For example, large proteins are often constructed from smaller domains with independent structural integrity. X-ray crystallography or liquid state NMR may be used to determine separate structures of these domains, but the mode of assembly of the domains into the full structure often remains unclear. Solid state NMR measurements can be designed to establish the contacts between and relative orientations of the domains in the intact state. (*e*) Solid state NMR data can be used in principle to determine the distribution of conformations in conformationally disordered peptides and proteins (e.g. protein folding intermediates). Quantitative characterization of a conformational distribution is difficult to obtain by other means. (*f*) Quite frequently, a specific protein of interest cannot be crystallized or stably solubilized because of its own idiosyncratic properties, even though it may belong to a class of proteins that is usually amenable to

X-ray crystallography or liquid state NMR. Solid state NMR methods may then be applied.

SPECTROSCOPIC BACKGROUND

Nuclear Spin Interactions

Solid state NMR spectra in the examples described below are determined by the nuclear spin Zeeman interaction H_Z, the chemical shift H_{CS}, homonuclear and heteronuclear magnetic dipole-dipole interactions H_{II} and H_{IS}, and interactions with applied radio-frequency (rf) fields H_{RF}. NMR experiments are analyzed in the rotating frame, where the Hamiltonian terms (units of rad/s) representing these interactions have the following forms:

$$H_Z = \Delta\omega I_z \qquad \text{1a.}$$

$$H_{CS} = \omega_0[\delta_{iso} + \delta_{11} \sin^2\theta \cos^2\phi + \delta_{22} \sin^2\theta \sin^2\phi + \delta_{33} \cos^2\theta] \qquad \text{1b.}$$

$$H_{II} = \frac{-\gamma_I^2 \hbar}{r_{12}^3} \frac{(3\cos^2\theta' - 1)}{2}(3I_{z1}I_{z2} - I_1 \cdot I_2) \qquad \text{1c.}$$

$$H_{IS} = \frac{-2\gamma_I\gamma_S\hbar}{r_{IS}^3} \frac{(3\cos^2\theta'' - 1)}{2} I_z S_z \qquad \text{1d.}$$

$$H_{RF} = \omega_1(t)[I_x \cos\chi(t) + I_y \sin\chi(t)] \qquad \text{1e.}$$

I and S are spin angular momentum vector operators for two different spin species. $\Delta\omega$ in Equation 1a is the resonance offset, i.e. the difference between the nuclear Larmor frequency ω_0 and the rf carrier frequency ω. δ_{iso}, δ_{11}, δ_{22}, and δ_{33} in Equation 1b are the isotropic chemical shift and the three principal values of the chemical shift anisotropy (CSA) tensor. θ and ϕ specify the direction of the magnetic field in the CSA principal axis system. γ_I, r_{12}, and θ' in Equation 1c are the I spin magnetogyric ratio, the distance between the two coupled I spins, and the angle between the internuclear displacement vector and the externally applied magnetic field (directed along the laboratory z axis). Similar definitions hold in Equation 1d. $\omega_1(t)$ and $\chi(t)$ are the time-dependent rf field amplitude and phase, with respect to a constant-phase carrier signal at frequency ω. In systems of many spins, H_{CS}, H_{II}, and H_{IS} would contain many terms with different values of the various parameters for each type of spin or spin pair. In heteronuclear spin systems, H_Z and H_{RF} would contain one term for each spin species.

H_{II} and H_{IS} obviously contain structural information through the dependence on internuclear distances and directions. Thus, information in dipole-dipole interactions has the advantage of being independent of electronic structure and therefore independent of quantum chemical calculations or empirical calibration from model compound studies. The angular terms in H_{CS} also contain structural information, but δ_{iso}, δ_{11}, δ_{22}, and δ_{33} (as well as the CSA principal axis directions) depend on electronic structure. Fortunately, these chemical shift parameters are often well established from measurements on model compounds (10–21) or can be calculated with increasing accuracy (22–28). Rf pulses, pulse sequences, or pulse shapes, represented by H_{RF}, are designed to excite NMR signals and to manipulate the other Hamiltonian terms in order to permit the extraction of structural information in as direct and robust a manner as possible.

Powder Patterns, Magic Angle Spinning, and Recoupling

The samples of interest in most of the examples described below are polycrystalline or noncrystalline solids, so the molecules in the sample take on all possible orientations with an isotropic orientational distribution. The orientation dependences of H_{CS}, H_{II}, and H_{IS} imply a strong orientation dependence of the NMR frequencies. NMR spectra of polycrystalline or noncrystalline solids therefore exhibit inhomogeneously broadened lines, commonly called powder patterns (29–31). Particularly in the case of spectra determined by H_{CS}, powder patterns have characteristic lineshapes dictated by the functional form of the orientation dependence and by the values of the spin interaction parameters. Figure 1a shows a simple example of a powder pattern lineshape for the case of a ^{13}C-labeled compound in polycrystalline form. Note that the inhomogeneous broadening in such cases is a consequence of the static, random orientations of molecules in the sample with respect to the external magnetic field, not a consequence of structural disorder or defects.

Although powder pattern lineshapes (in one or more dimensions) can contain useful structural information (32–43), broad lineshapes result in lower resolution and sensitivity. In solid state NMR experiments on biological materials, sensitivity is usually a dominant consideration. Resolution is also important when biopolymers are isotopically labeled at multiple sites. For these reasons, the majority of biological solid state NMR experiments on unoriented samples are carried out with magic-angle spinning (MAS), a technique in which samples are rotated rapidly around an axis at the magic angle θ_m to the external magnetic field (31, 44, 45). The magic angle is defined by $3\cos^2\theta_m - 1 = 0$ and is approximately $54.736°$. The dramatic effect of MAS on solid state NMR spectra is shown in Figures 1b–d. At the magic angle the anisotropic terms of H_{CS}, H_{II}, and H_{IS} average to zero over one rotation period τ_R. When the MAS rotation frequency ν_R greatly exceeds the inhomogeneous linewidth the powder pattern collapses into a single, much

Figure 1 (a) ^{13}C NMR spectrum of static, polycrystalline glycine with approximately 5% ^{13}C enrichment at the carboxyl site. A chemical shift anisotropy powder pattern lineshape is observed for the carboxyl carbons, spanning the 100–250 ppm range. (b–d) Spectra obtained with magic-angle spinning at the indicated spinning frequencies ν_R. Lines at 177 ppm and 43 ppm are centerbands at the isotropic chemical shifts of the labeled carboxyl and natural-abundance α-carbon sites. Lines whose frequencies vary with ν_R are spinning sidebands. Spectra obtained at 100.4 MHz ^{13}C NMR frequency.

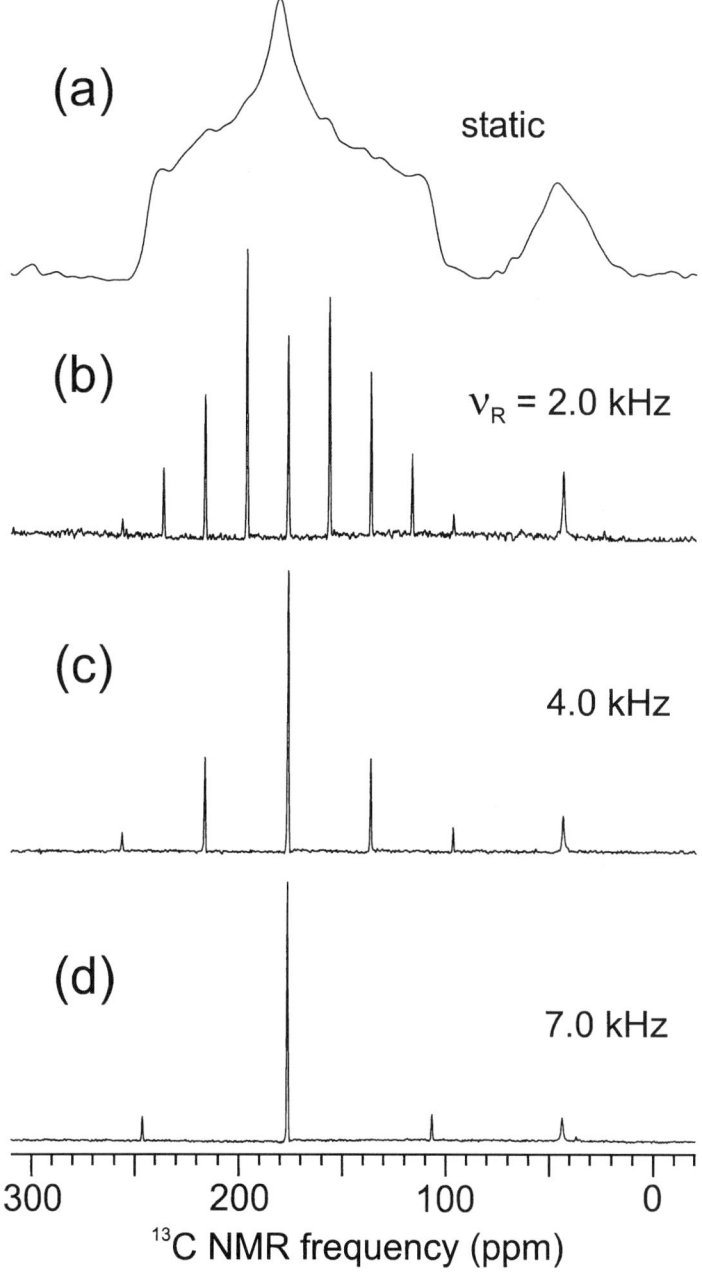

sharper line at the isotropic chemical shift frequency. The residual linewidth is then often determined by true structural disorder in noncrystalline solids, or possibly by residual dipole-dipole couplings, magnetic susceptibility effects, transverse nuclear spin relaxation, or other linebroadening mechanisms. The line-narrowing brought about by MAS results in large increases in signal-to-noise ratio (because the total area of the lineshape is unchanged) and resolution.

MAS also destroys the structural information in solid state NMR spectra. A number of clever tricks have been devised to recover this information. One such trick is simply to spin slowly. When ν_R does not exceed the powder pattern linewidth, spinning sideband lines appear in the spectrum at frequencies separated from the isotropic chemical shift frequency by multiples of ν_R, owing to periodic modulation of the NMR frequencies under MAS. As shown in Figure 1, the spinning sidebands become more numerous and intense as ν_R decreases. The amplitudes of the spinning sidebands can be calculated for any given set of spin interaction parameters. Conversely, the spin interaction parameters can be determined experimentally from the spinning sideband amplitudes (46). Most of the structural information contained in powder pattern lineshapes is also contained in spinning sidebands under slow MAS (i.e. $2\pi\nu_R < \omega_0|\delta_{11} - \delta_{33}|$), permitting structural techniques based on analyses of sideband amplitudes under slow MAS (47–54).

A second trick is to combine MAS with rotation-synchronized pulse sequences. Interference between the time dependence induced by MAS and the time dependence induced by pulse sequences can prevent anisotropic spin interactions from being averaged to zero, even under high-speed MAS (55–90). This phenomenon has come to be called "recoupling." In effect, recoupling techniques allow anisotropic nuclear spin interactions in solids to be turned on and off at will, so that the advantages of MAS and the structural information in anisotropic nuclear spin interactions can be preserved simultaneously.

As a simple illustration of recoupling of homonuclear dipole-dipole couplings, consider the two-pulse sequence in Figure 2a, which is the basic element of a recoupling technique called "dipolar recovery at the magic angle" (DRAMA) (81). MAS alone makes the $(3\cos^2\theta - 1)$ term in Equation 1c time dependent, with terms that oscillate at ν_R and $2\nu_R$ and therefore average to zero. The rf pulses introduce an additional time dependence of the spin operator term $(3I_{z1}I_{z2} - I_1 \cdot I_2)$, effectively rotating this term about x by 90° in the period between the two pulses. H_{II} then assumes the form

$$\widetilde{H}_{II}(t) = \frac{\gamma^2\hbar}{r_{12}^3}(C_1 \cos\omega_R t + S_1 \sin\omega_R t + C_2 \cos 2\omega_R t + S_2 \sin 2\omega_R t)\widetilde{T}(t) \qquad 2a.$$

$$\widetilde{T}(t) = \begin{cases} 3I_{z1}I_{z2} - I_1 \cdot I_2, & 0 < t < \tau' \\ 3I_{y1}I_{y2} - I_1 \cdot I_2, & \tau' < t < \tau' + \tau. \\ 3I_{z1}I_{z2} - I_1 \cdot I_2, & \tau' + \tau < t < \tau_R \end{cases} \qquad 2b.$$

The coefficients C_1, S_1, C_2, and S_2 are functions of the internuclear direction in an axis system fixed in the MAS rotor (i.e. the sample container) (31, 48, 50).

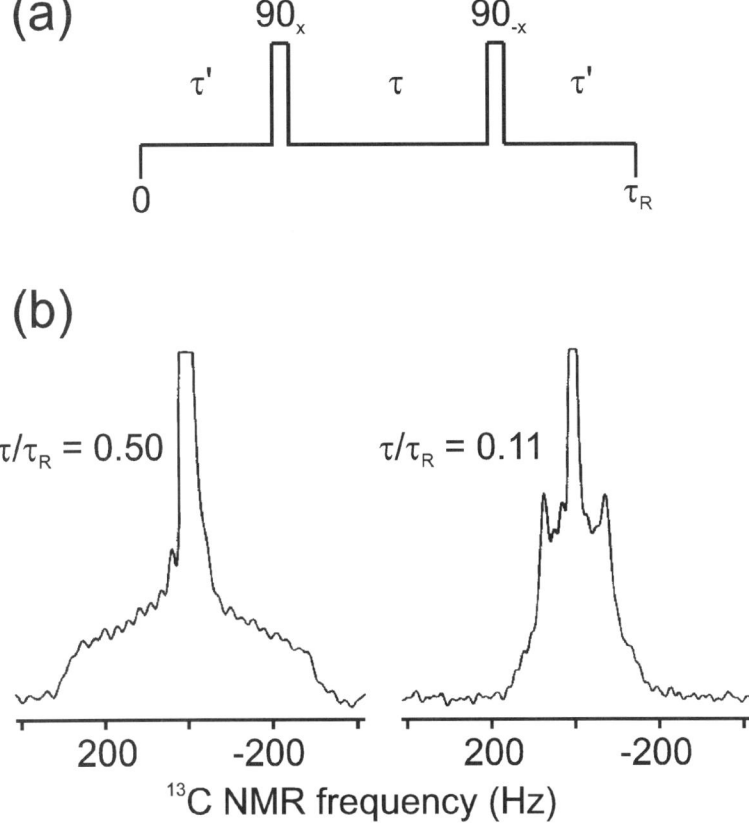

Figure 2 (*a*) Radio-frequency pulse sequence for dipolar recovery at the magic angle (DRAMA), consisting of two $\pi/2$ pulses with phases X and $-$X per MAS rotation period τ_R. (*b*) ^{13}C NMR spectra of a doubly ^{13}C-labeled model compound (bisulfite adduct of acetone) under the DRAMA sequence. The powder pattern lineshapes illustrate the recoupling effect, i.e. the recovery of ^{13}C-^{13}C dipole-dipole couplings that would be averaged to zero by MAS in the absence of the rotation-synchronized pulses. Spectra obtained at 25.3 MHz ^{13}C NMR frequency, with $\nu_R = 3.33$ kHz (Reprinted with permission from Tycko R, Dabbagh G. 1990. *Chem. Phys. Lett.* 173:461–465.)

Because of the time dependence of the operator $\widetilde{T}(t)$, the time-average of $\widetilde{H}_{II}(t)$, which determines the effect of $\widetilde{H}_{II}(t)$ to lowest order when nuclear spin evolution is observed stroboscopically at multiples of τ_R (a common experimental situation), becomes

$$\langle \widetilde{H}_{II} \rangle = \frac{\gamma^2 \hbar}{r_{12}^3} \left[C_1 \frac{\sin \omega_R (\tau' + \tau) - \sin \omega_R \tau'}{2\pi} \right.$$

$$\left. + C_2 \frac{\sin 2\omega_R (\tau' + \tau) - \sin 2\omega_R \tau'}{4\pi} \right] (3I_{y1}I_{y2} - 3I_{z1}I_{z2}), \qquad 3.$$

which is generally nonzero. Because C_1 and C_2 are orientation dependent, spectra acquired with DRAMA (or other recoupling techniques) exhibit powder pattern lineshapes that are dependent on the details of the pulse sequence, as shown in Figure 2*b*. It is interesting to note that the form and rotational symmetry properties of the spin operator term in Equation 3 differ from those in Equation 1c. The symmetry properties of recoupled spin interactions can lead to useful features in applications such as multidimensional spectroscopy and multiple quantum spectroscopy (81, 89, 91, 92).

A wide variety of recoupling techniques for heteronuclear dipole-dipole couplings (58, 61–66, 74, 93), homonuclear dipole-dipole couplings (59, 60, 67–73, 75–79, 81, 83–90), and CSA interactions (55–58, 74) have been devised. Important practical considerations in the design of useful recoupling techniques include effectiveness over large chemical shift ranges and in the presence of large CSA, stringency of requirements on the amplitude and homogeneity of rf fields, the rf duty factor, and the magnitude (i.e. scaling factor) of the recoupled interaction. Frequency-selective recoupling sequences have recently been developed for the purpose of turning on dipole-dipole couplings (and measuring distances) between selected pairs of labels in multiply isotopically labeled compounds (66, 94–96).

A conceptually distinct class of recoupling techniques exists in the case of dipole-coupled homonuclear spin systems with large chemical shift or CSA differences, exemplified by the "rotational resonance" (R^2) (67, 68, 97–100) and "radio-frequency-driven recoupling" (RFDR) (62, 70, 73) techniques. In this class of techniques recoupling occurs owing to interference between time dependences induced by MAS and chemical shift differences, rather than by rf pulse sequences. In the case of R^2, the time-dependent spin operator $\widetilde{T}(t)$ (viewed in an interaction representation with respect to the chemical shift interactions, rather than in an interaction representation with respect to H_{RF} as was assumed implicitly in the discussion of DRAMA above) becomes

$$\widetilde{T}(t) = 2I_{z1}I_{z2} - \frac{1}{2}R_z[-\xi_{12}(t)](I_{+,1}I_{-,2} + I_{-,1}I_{+,2})R_z[\xi_{12}(t)], \qquad 4.$$

where $\xi_{12}(t)$ is the difference in the net precession angles of the two coupled spins up to time t, and R_z represents a rotation of spin angular momenta around z. As usual, $I_\pm = I_x \pm iI_y$. When the isotropic chemical shift difference is a multiple of ω_R, $\xi_{12}(t)$ has the form

$$\xi_{12}(t) = n\omega_R t + A_1 \cos \omega_R t + B_1 \sin \omega_R t + A_2 \cos 2\omega_R t + B_2 \sin 2\omega_R t, \qquad 5.$$

where the oscillatory terms arise from the CSA difference. Then the time dependence of $\widetilde{T}(t)$ is a Fourier series in $\omega_R t$, and $\langle \widetilde{H}_{II} \rangle = \frac{\gamma^2 \hbar}{r_{12}^3}(gI_{+,1}I_{-,2} + g^*I_{-,1}I_{+,2})$ with g being an orientation-dependent, nonzero number. Note that the average Hamiltonian contains only the "flip-flop" part of the dipole-dipole interaction, unlike the average Hamiltonian under the DRAMA sequence. The recoupling mechanism and average Hamiltonian under RFDR are essentially the same as under R^2, but a train of rotor-synchronized π pulses is applied that forces $\widetilde{T}(t)$ to be periodic with period equal to a multiple of τ_R even when the isotropic chemical shift difference is not a multiple of ω_R.

Structural Techniques

Distance Measurements Perhaps the most conceptually straightforward, but not always experimentally robust or useful, approach to structural studies in biomolecular solid state NMR is to introduce a pair of isotopic labels, most commonly ^{13}C and ^{15}N, and use dipolar recoupling techniques to create a nonzero heteronuclear or homonuclear dipole-dipole coupling under MAS. Measurement of the dipole-dipole coupling then permits extraction of the internuclear distance. As one concrete and important example of a structural problem that can be addressed by distance measurements, Figure 3 shows a segment of a polypeptide backbone that includes two amino acid residues. The conformation of a polypeptide backbone is defined by one pair of variable dihedral angles, conventionally termed ϕ and ψ, for each residue. A third angle, representing rotations about the peptide bonds between backbone carbonyl carbons and amide nitrogens, is generally limited to $180° \pm 5°$ for *trans* peptide bonds and can therefore be considered fixed. To a good approximation, the distance between amide nitrogens of residues $i - 1$ and i depends only on ψ of residue $i - 1$. The distance between carbonyl carbons of residues $i - 1$ and i depends only on ϕ of residue i. Thus, peptide backbone conformations could be determined by measurements of homonuclear carbonyl ^{13}C-^{13}C distances and amide ^{15}N-^{15}N distances. Studies of this sort have been carried out (101–104). Heteronuclear ^{15}N-^{13}C distances can also be used as constraints on polypeptide backbone conformations, although each distance then depends on more than one dihedral angle (105–107).

Internuclear distances of interest typically exceed 2.5 Å, so that the magnitude of the recoupled interaction is of order 200 Hz or less. Such small couplings do not usually produce measurable splittings or other effects in simple one-dimensional solid state NMR spectra. Consequently, the recoupled interactions are measured in a two-dimensional (2D) manner, using rf pulse sequences of the form PREP-τ_d-MIX-FID, where PREP represents the preparation of nuclear spin polarization

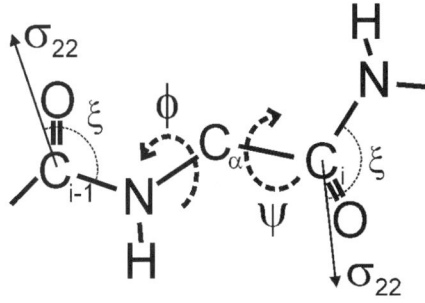

Figure 3 Polypeptide backbone, from carbonyl carbon of residue $i - 1$ to amide nitrogen of residue $i + 1$. Dihedral angles ϕ and ψ of residue i are indicated. The carbonyl CSA tensor has its σ_{11} and σ_{22} principal axes approximately in the carbonyl plane, with $\xi \approx 130°$. The σ_{33} principal axis is perpendicular to this plane. This CSA orientation is assumed in analyses of solid state NMR experiments on doubly carbonyl-labeled peptides and proteins designed to determine ϕ and ψ.

or coherence, τ_d represents evolution of the polarization or coherence under the recoupled interactions for a time τ_d, MIX represents a conversion of the magnetization or coherence to transverse magnetization of observable nuclei, and FID represents the detection of nuclear free-induction-decay signals. FIDs are recorded for a series of τ_d values. Experimental dipolar evolution curves can be compared with numerical simulations to extract the desired structural parameters.

A significant practical problem with the measurement of weak recoupled interactions is that FID signals decay with increasing τ_d owing to factors other than the dipole-dipole couplings of interest, in particular pulse imperfections, spin relaxation processes, and residual couplings to abundant proton spins due to insufficient proton decoupling fields. It is then necessary to calibrate or estimate extraneous sources of signal decay before experimental dipolar evolution curves can be compared with simulations (73, 101, 108). Two solutions to this problem have been developed: (*a*) In the case of heteronuclear recoupling techniques such as REDOR (61, 63, 64), signal decay is calibrated accurately by a direct comparison of signals obtained with recoupling pulses on both nuclear species and signals obtained with pulses on the observed species only; (*b*) In the case of homonuclear recoupling techniques, it is sometimes possible to record dipolar evolution curves in a constant-time manner, i.e. with a fixed total period τ_d within which certain rf pulses are shifted in time in order to generate the dipolar evolution (109–111).

A second significant practical problem, particularly in solid state ^{13}C NMR measurements on high molecular weight systems, is the presence of natural-abundance background signals. The natural abundance of ^{13}C is approximately 1.1%, so that the number of natural-abundance ^{13}C nuclei is ten times greater than the number of ^{13}C labels in a doubly labeled protein with a molecular weight of approximately 50 kD. Natural-abundance background signals can be suppressed to a large extent by taking the difference between signals from labeled and unlabeled samples acquired under otherwise identical conditions (112), by using 2D spectroscopic techniques in which only the labels contribute to crosspeaks (47–49, 113, 114), or by employing double quantum (DQ) filtering (60, 72, 75, 76, 83, 84, 115). In a DQ-filtered measurement DQ coherences (i.e. superpositions of the $|++>$ and $|--->$ spin states of a two-spin system) are prepared during the PREP period and are allowed to evolve during τ_d. Contributions to the FID signals from the DQ coherences are selected by applying phase shifts ϖ to the rf pulses in the PREP period and taking advantage of the fact that such phase shifts have the effect of multiplying these signals by $\exp(-2i\varpi)$. Excitation of DQ coherences requires a coupling between the two spins involved in the coherence. Consequently, recoupling sequences must be applied during the PREP period, and the majority of natural-abundance ^{13}C nuclei do not contribute to the DQ coherences.

As an illustrative example of the ideas discussed above, Figure 4 shows the application of a solid state ^{13}C NMR technique called "constant-time double-quantum-filtered dipolar evolution" (CTDQFD) to a doubly labeled model system, the tripeptide L-alanylglycylglycine (AGG), in polycrystalline form.

Angular Measurements An alternative approach to placing experimental constraints on the conformation of a biopolymer is to make measurements that are primarily angularly dependent, rather than distance dependent. A wide variety of angularly dependent techniques have been developed (32, 33, 35, 36, 38, 39, 41, 42, 47–49, 51–54, 116–125), all of which lead to solid state NMR data that depend on the relative orientation of spin interaction tensors at two separate labeled sites. Comparison of the experimental data with numerical simulations allows the determination of dihedral or torsion angles that link the two sites. The data can take several forms. The earliest angularly dependent structural techniques were simple 2D exchange experiments, performed on static samples (32, 33, 35, 36, 38, 39). In a 2D exchange experiment one essentially measures NMR frequencies associated with one labeled site (site A) during the evolution period t_1, waits an exchange time τ_e during which nuclear spin polarization transfers from site A to the second labeled site (site B), and then measures NMR frequencies associated with site B. Polarization transfer depends on dipole-dipole couplings, but in the limit of full exchange (sufficiently large τ_e) the data are not distance dependent. When the NMR frequencies in the evolution and detection periods depend on the absolute orientation of the labeled sites, through dependences on the spin interaction terms in Equations 1b–d, the correlations of frequencies of sites A and B revealed in the 2D exchange spectrum depend on the relative orientations of the two sites, allowing the relative orientations to be extracted from the data.

In static samples the 2D exchange spectra are 2D powder patterns with shapes that depend strongly on the molecular conformation (32, 33, 35, 36, 38, 39). The relatively low sensitivity of static 2D exchange measurements makes such measurements impractical in many biological applications. However, MAS versions of 2D exchange measurements have been demonstrated, originally for studies of slow molecular motions (126, 127) and subsequently for structural studies (47–49). 2D MAS exchange measurements depend on relatively slow MAS to retain the structural information in the anisotropic spin interactions and on proper synchronization of rf pulses with the sample rotation (49, 126, 127). In 2D MAS exchange measurements the 2D spectra consist of crosspeaks that connect spinning sideband lines of site A with those of site B. Although the positions of these crosspeaks are independent of conformation, their relative intensities are dependent on conformation and can be simulated numerically for any assumed relative orientation of the two sites (48). Figure 5 shows an example of the application of 2D MAS exchange spectroscopy to doubly labeled AGG.

A second approach to angular measurements is to prepare a DQ coherence involving labeled sites A and B and to record the evolution of this coherence under the sum of anisotropic nuclear spin interactions at the two sites. Both static (41, 42) and MAS versions (52, 54, 116, 117) of this approach have been demonstrated. As an example, constraints on the ϕ and ψ dihedral angles of a polypeptide backbone can be derived from the time-domain DQ evolution (54) or frequency-domain DQ spectrum (52) of a sample labeled with ^{13}C at two sequential carbonyl sites (see Figure 3), because the DQ coherence evolves under the sum

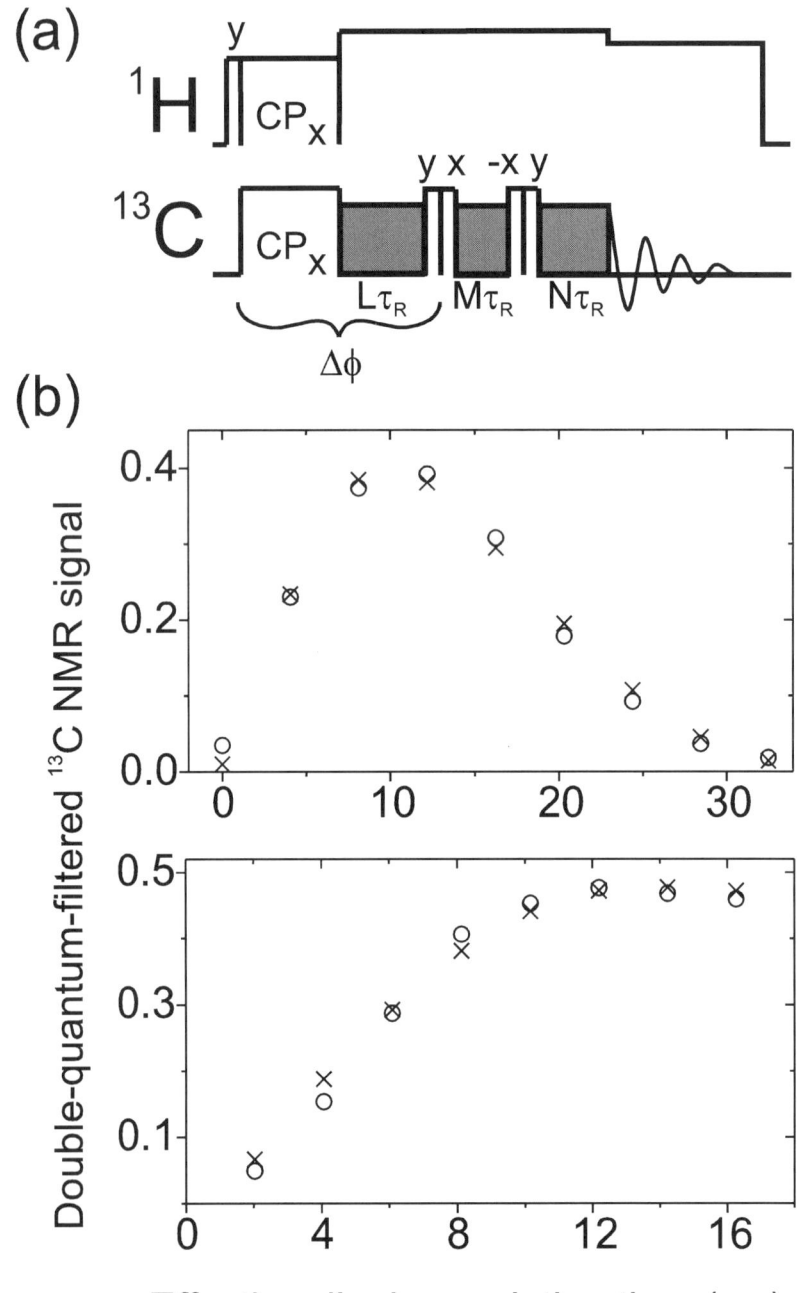

of the two carbonyl CSA tensors and the "sum tensor" depends on the relative orientation of the two carbonyl groups. Techniques of this type applicable to doubly carbonyl-labeled samples include DOQSY for static (41, 118) and DQ-DRAWS (51, 52) and DQCSA (54) for MAS experiments.

A third approach to angular measurements is to correlate the evolution of nuclear spin polarization (or coherence) at ^{13}C-labeled site A with evolution of the same polarization (or coherence) at ^{15}N-labeled site B, after polarization transfer (or excitation of coherence) between ^{13}C and ^{15}N nuclei (120, 122, 124, 125). As an example, ϕ dihedral angles can be determined with this approach in polypeptides that are labeled with ^{15}N at the backbone amide and ^{13}C at the α-carbon of a single residue.

A fourth approach is to take advantage of the dependence of R^2 recoupling on the CSA tensors of recoupled, labeled sites, which makes dipolar evolution curves under R^2 recoupling dependent on relative CSA tensor orientations as well as internuclear distances (119). Similarly, dipolar evolution curves obtained with RFDR recoupling (as in Figure 4) depend on CSA tensor orientations as well as internuclear distances (73, 109).

Multiply Labeled Samples   Recently, several groups have investigated the feasibility of solid state NMR spectroscopy under MAS of biopolymers that are multiply or uniformly labeled with ^{13}C and/or ^{15}N (69, 72, 91, 128–140), with the goal of obtaining many more structural constraints than can be obtained from measurements on selectively labeled samples of the types described above. A persistent problem is that ^{13}C and ^{15}N NMR lines are inhomogeneously broadened by the structural disorder inherent in noncrystalline solids, so that spectral resolution limits the number of sites that can be usefully labeled (141). In the case of highly ordered (e.g. microcrystalline), relatively small proteins, however, well-resolved multidimensional spectra can be obtained and the majority of resonances can be assigned (134, 137). Protein expression techniques that limit the number of labels

―――

Figure 4   (*a*) Radio-frequency pulse sequence for constant-time double-quantum-filtered dipolar (CTDQFD) evolution measurements. Following cross-polarization (CP) from ^{1}H to ^{13}C spins, three RFDR recoupling blocks are applied with durations Lτ_R, Mτ_R, and Nτ_R, separated by pairs of $\pi/2$ pulses with indicated phases, before acquisition of ^{13}C NMR signals. Pulses are synchronized with magic-angle spinning, with rotation period τ_R. Signals arising from double-quantum coherence are selected by alternate addition and subtraction of signals acquired with overall phase shifts $\Delta\phi = 0$, $\pi/2$, π, and $3\pi/2$. (*b*) Results of CTDQFD experiments on doubly-carbonyl-labeled, polycrystalline L-alanylglycylglycine (^{13}C$_2$-AGG), at 100.8 MHz ^{13}C NMR frequency and $\tau_R = 254$ μs. Top curve is acquired with L = 32, M + N = 128, and effective evolution time defined to be (M − N)τ_R. Bottom curve is acquired with L + M + N = 128, L = M − N, and effective evolution time defined to be Lτ_R. Circles and crosses are experiments and simulations, based on the known AGG crystal structure. (Adapted from 109.)

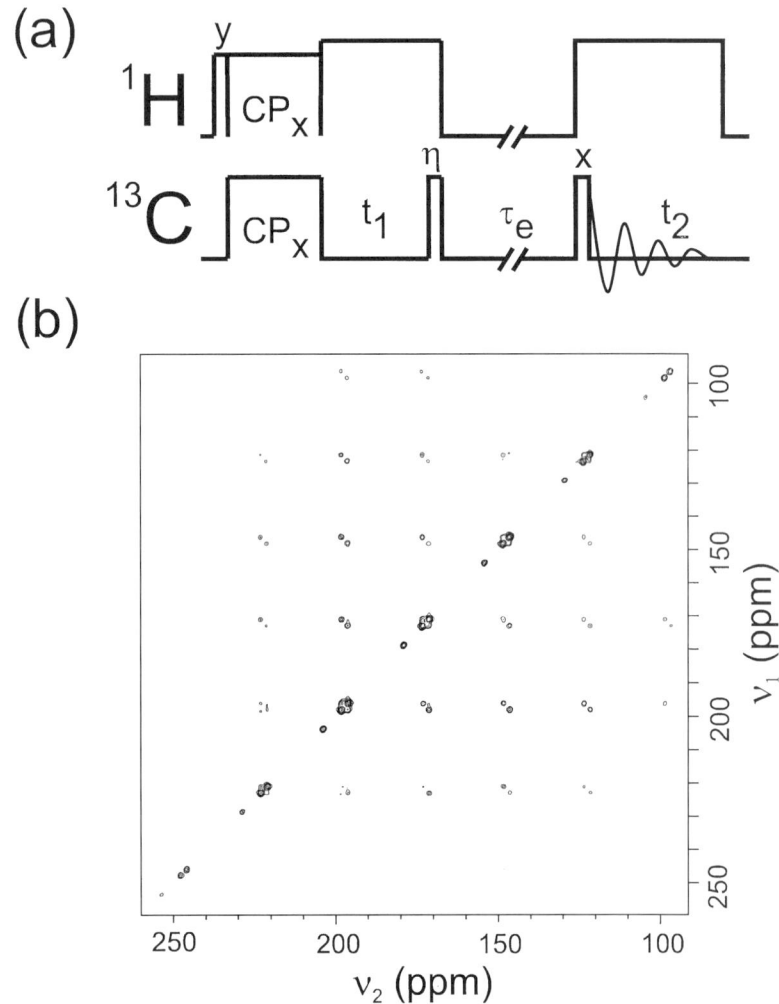

Figure 5 (*a*) Radio-frequency pulse sequence for two-dimensional magic-angle spinning (2D MAS) exchange measurements. 2D data sets are acquired with η = x and y, and with rotor synchronization such that τ_e is a multiple of τ_R, and $t_1 + \tau_e$ is a multiple of τ_R. These four data sets are combined to yield a phase-sensitive 2D spectrum in which crosspeaks between spinning sidebands arise from exchange processes. (*b*) 2D MAS exchange spectrum of $^{13}C_2$-AGG, obtained at 188.6 MHz ^{13}C NMR frequency; ν_R = 4.695 kHz and τ_e = 500 ms. Crosspeaks connect spinning sidebands of the two labeled carbonyl carbons. Comparison of experimental crosspeak amplitudes with simulations permits the determination of ϕ and ψ dihedral angles. (Adapted from 49.)

(133, 142, 143) and solid state NMR techniques that select particular classes of resonances (144) have been proposed to alleviate spectral congestion.

Assuming that ^{13}C and/or ^{15}N signals from single sites in multiply-labeled samples can be resolved and assigned, the task of obtaining structural information from the solid state NMR spectra remains. Low-resolution information, at the level of secondary structure, can be extracted from the isotropic chemical shifts of assigned resonances, as discussed below. When resonances are sufficiently well resolved, selective dipolar recoupling techniques that depend on matching differences in the effective rotating-frame fields at particular pairs of labeled sites to multiples of ν_R can be employed to measure particular internuclear distances (94–96, 139, 140). Alternatively, certain angularly dependent techniques have been shown to be compatible with uniform labeling (120, 123, 142). At present, the determination of biopolymer structures from solid state NMR measurements on multiply or uniformly labeled samples remains a challenging area for methodological development.

Multiple Quantum Spectroscopy Techniques for multiple quantum (MQ) solid state NMR spectroscopy were developed principally by Pines and coworkers (145–148), originally to elucidate fundamental aspects of spectroscopy and spin physics (147–150) and as a tool for materials science (151–153). In an MQ NMR spectrum, one observes signals corresponding to transitions in which groups of N spins flip in the magnetic field. For excitation and detection of MQ coherences to be possible in a solid, the spins must be linked by dipole-dipole couplings. Thus, MQ NMR spectra contain structural information, which can be analyzed either qualitatively in terms of the numbers of spins in a coupled group or quantitatively in terms of the precise internuclear distances and geometry by comparison of experimental MQ signal amplitudes with simulated amplitudes. MQ solid state NMR techniques have recently been adapted for structural studies of ^{13}C-labeled organic and biological systems (92, 154–156). The utility of these techniques in structural studies of biological systems has not yet been fully explored, but MQ NMR spectroscopy has been used to establish the supramolecular organization of amyloid fibrils (see below). In principle, MQ NMR spectra of multiply labeled biopolymers could be used as constraints on their global folds.

Spectroscopy of Oriented Samples

The techniques described above are intended for applications to unoriented samples, whose NMR spectra are powder patterns unless MAS is employed. A qualitatively different set of techniques has been developed for samples that are highly oriented with respect to the magnetic field of the NMR spectrometer. The principal areas of application for these techniques are peptides and proteins that are bound to or embedded in phospholipid bilayer membranes, which can be oriented either by deposition on planar substrates (5, 157, 158) or by magnetic alignment (159–163), and to oriented protein fibrils (see below). Techniques for oriented samples are largely outside the scope of this chapter, but have been reviewed elsewhere (3–6, 164). Briefly, in the case of selectively ^{13}C-, ^{15}N-, or ^{2}H-labeled samples the anisotropic nuclear spin interactions lead to spectra in which the NMR frequencies or splittings are direct constraints on the orientation of labeled chemical groups or bonds with respect to the magnetic field. Measurements on a series of selectively labeled, oriented samples can be used to develop a set of orientational constraints that may be sufficient to determine or refine a full molecular structure. An important recent development has been the demonstration of 2D and 3D spectroscopic techniques that are applicable to uniformly labeled peptides or proteins and yield multiple high-resolution structural constraints (165–168) or constraints on orientations of helical segments (169, 170). In the case of uniformly ^{15}N- and ^{13}C-labeled, oriented polypeptides, homonuclear ^{13}C-^{13}C decoupling techniques and ^{15}N-^{13}C polarization transfer techniques have been proposed that permit simultaneous single-site resolution, sequential assignment, and extraction of structural constraints from 2D or 3D spectra (171–173).

STRUCTURAL STUDIES OF PEPTIDE AND PROTEIN FIBRILS

General Remarks

As discussed in the introduction, one of the major motivations for developing solid state NMR techniques for structural studies of biopolymers is that these techniques can be applied to classes of biochemical systems that are not generally amenable to study by more well-established techniques such as X-ray crystallography and multidimensional liquid state NMR. Two classes that are of particular importance are polypeptides that bind to or embed in biological membranes and polypeptides that form fibrils (i.e. fibers with nanoscopic dimensions). Membrane proteins have been the primary focus of many solid state NMR laboratories (133, 158, 162, 168, 174–202) because of the importance of membrane proteins as receptors, channels, components of signal transduction pathways, and effectors of energy transduction, among other roles. Quite remarkable progress has been made in solid state NMR studies of membrane proteins despite the challenges, both spectroscopic and biochemical, associated with the high molecular weight of many membrane proteins, the difficulty of expressing the requisite quantities of protein, the difficulty of incorporating isotopic labels (particularly selective labels), and uncertainties regarding the sample conditions that optimize protein stability and spectral quality. A promising direction for experiments on membrane systems is the investigation of conformations of peptide ligands. The feasibility of experiments on peptide/protein complexes (e.g. peptide hormone/receptor complexes) has been demonstrated in conformational studies of an antibody-bound peptide representing the V3 loop of the HIV-1 gp120 protein (115, 203). Recent work in the author's laboratory (138, 204), as well as in other laboratories (108, 118, 205–219), has been directed at peptide and protein fibrils. Several intrinsic features of fibrils make them attractive targets for solid state NMR. First, fibrils are often composed of peptides or proteins with relatively low molecular weight or with repeated sequence (and structural) motifs, facilitating isotopic labeling by synthetic or biosynthetic means. Second, fibril samples can be nearly pure peptide or protein, maximizing NMR sensitivity and allowing for relatively small sample volumes that permit the high MAS frequencies and high rf fields required by some sophisticated techniques. Third, peptide and protein fibrils can yield high-quality spectra, as shown below, making the structural measurements reliable and robust. Finally, peptide and protein fibrils are of considerable current biophysical interest and biomedical importance and are inherently insoluble and noncrystalline. Site-specific, atomic-level structural information obtained in solid state NMR experiments is therefore unique and is likely to have substantial scientific impact.

Amyloid Fibrils

Structural Issues Amyloid fibrils are filamentous structures, with diameters of order 10 nm and typical lengths of order 1 μm or more, that are formed by a wide variety of peptides and proteins with disparate sequences and molecular weights (220). Figure 6 shows an electron micrograph of typical amyloid fibrils. Amyloid fibrils are associated with amyloid diseases, which include Alzheimer's disease, Parkinson's disease, type2 diabetes, Huntington's disease, and spongiform encephalopathies (i.e. prion diseases). Amyloid fibrils can also be formed by peptides and proteins that are not associated with diseases, including proteins with soluble monomeric structures that have been destabilized by appropriate solvent conditions (221, 222). It appears that the amyloid fibril may be a stable form for many, perhaps all, polypeptides. The nature of the intermolecular and intramolecular interactions that stabilize amyloid fibrils is not well understood. Thus, amyloid fibrils are currently a subject of both biophysical interest and biomedical importance.

High-resolution structural information is likely to facilitate an understanding of amyloid fibril formation, but relatively little is known about the molecular structure of amyloid fibrils. X-ray fiber diffraction measurements have established that amyloid fibrils contain extended, ribbon-like β-sheets, oriented such that the polypeptide chains in the β-sheets run approximately perpendicular to the long axis of the fibril and the interchain hydrogen bonds are approximately parallel to the long axis (220, 223, 224). Such a structure gives rise to the "cross-β" diffraction pattern that is a defining characteristic of amyloid fibrils. This β-sheet structure has received additional support from cryo-electron microscopy (225). Most other aspects of amyloid fibril structure had not been established definitively prior to the solid state NMR studies discussed below. Questions that have been or may be addressed by solid state NMR include: (*a*) What is the supramolecular organization of the β-sheets? That is, are amyloid fibrils comprised of antiparallel β-sheets [as often assumed (226–229)], parallel β-sheets, or β-sheets with a more complex

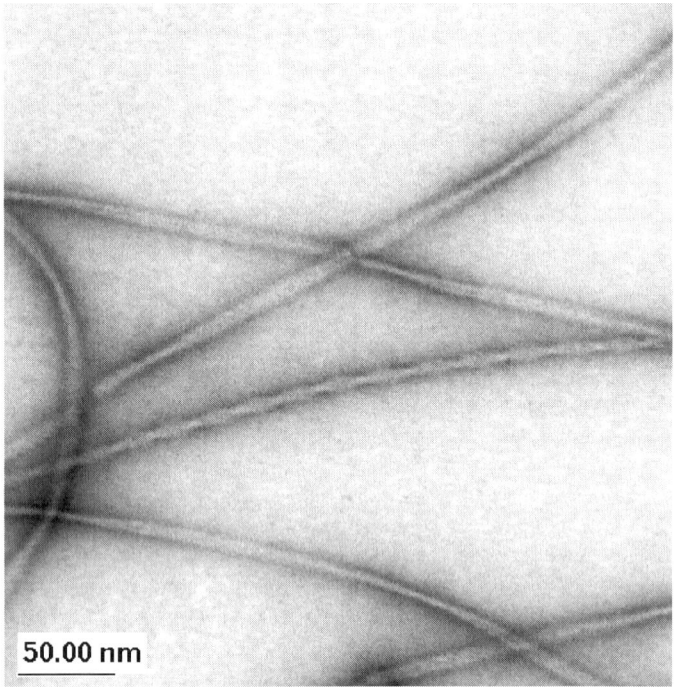

Figure 6 Negatively stained electron micrograph of amyloid fibrils formed by the 40-residue Alzheimer's β-amyloid peptide, in this case with the rodent, rather than human, amino acid sequence. Morphologically similar unbranched fibrils, with diameters of approximately 10 nm and lengths on the order of microns, are formed by a wide variety of disparate peptides and proteins.

organization? (*b*) What segments of the polypeptide chain contribute to the β-sheets? Are other secondary structures present? (*c*) What is the degree of structural order at the atomic level in amyloid fibrils? Do amino acid sidechains have well-defined conformations? Are there regions of disorder? (*d*) The β-sheets in amyloid fibrils are often proposed to form laminated structures (223). Can a laminated structure be confirmed experimentally? What is the nature of contacts between β-sheet laminae?

Supramolecular Organization of β-Sheets Most, but not all (103), solid state NMR experiments on amyloid fibrils have been carried out on the β-amyloid peptide (Aβ) of Alzheimer's disease, which ranges from 39 to 43 residues in length in vivo, or on specific shorter fragments of Aβ that serve as model systems. The issue of the supramolecular organization of β-sheets in amyloid fibrils was first addressed by Griffin, Lansbury, and coworkers (104) in the case of fibrils formed by a hydrophobic nine-residue peptide representing residues 34 through 42 of the 42-residue version of Aβ (Aβ_{34-42}). They used R^2 recoupling measurements on a series of doubly ^{13}C-labeled, fibrillized Aβ_{34-42} samples, in particular the observation of effects caused by intermolecular dipole-dipole couplings on the R^2 dipolar evolution curves, to show that β-sheets in Aβ_{34-42} fibrils were antiparallel (104). This observation of antiparallel β-sheets is in agreement with expectations based on infrared spectroscopy (230, 231).

Lynn, Meredith, Botto, and coworkers subsequently used dipolar recoupling measurements to investigate the organization of β-sheets in fibrils formed by a 26-residue peptide representing residues 10 through 35 of Aβ (Aβ_{10-35}), which contains both hydrophobic and nonhydrophobic segments (108, 212, 213). They used the DRAWS (dipolar recoupling with a windowless sequence) technique, a modification of DRAMA developed by Drobny and coworkers for systems with large CSA (88), to record dipolar evolution curves for a series of fibrillized Aβ_{10-35} samples with single ^{13}C labels at backbone carbonyl sites. Strikingly, they found nearly identical dipolar evolution curves for all samples, regardless of the position of the labeled site. This result provided strong evidence for a parallel, rather than antiparallel, β-sheet structure. Moreover, detailed comparison of experimental and simulated curves revealed the best agreement for an in-register, parallel structure, i.e. a structure in which hydrogen bonds link residue i of one peptide molecule with residues i $-$ 1 and i $+$ 1 of neighboring molecules in a single β-sheet layer.

More recently, Antzutkin et al (204) have used MQ NMR measurements to investigate the organization of β-sheets in fibrils formed by the full-length, 40-residue form of Aβ (Aβ_{1-40}). ^{13}C MQ NMR spectra of two fibrillized Aβ_{1-40} samples, labeled at methyl carbons of either Ala21 or Ala30, are shown in Figure 7a. Strong two-, three-, and four-quantum signals are observed in both samples, with nearly identical relative intensities. The observation of four-quantum signals indicates the existence of dipole-coupled groups of at least four ^{13}C nuclei. Qualitatively, this result favors a fully parallel β-sheet structure over the antiparallel or mixed structures shown in Figure 7b. Quantitative analysis by comparison with numerical simulations confirms that an in-register, parallel β-sheet structure is in best agreement with the experimental MQ NMR data (204). Thus, amyloid fibrils formed by full-length Aβ are composed of parallel, not antiparallel, β-sheets. Structural models that invoke antiparallel β-sheets (204) are excluded by these data. MQ NMR spectra of Aβ_{1-40} fibrils labeled at the methyl carbon of Ala2 exhibit weak three-quantum and no four-quantum signals, indicating that the N-terminus of Aβ_{1-40} does not participate in the parallel β-sheets. This result and the overall parallel structure are consistent with biochemical experiments that show limited proteolysis of Aβ_{1-40} fibrils in vivo exclusively at the N-terminus.

MQ NMR and REDOR recoupling experiments by Balbach et al on fibrils formed by the seven-residue peptide N-acetyl-Lys-Leu-Val-Phe-Phe-Ala-Glu-amide, representing residues 16 through 22 of full-length Aβ (Aβ_{16-22}), indicate an in-register, antiparallel β-sheet organization (138). Taken together, solid state NMR data on the organization of β-sheets in amyloid fibrils show that both parallel and antiparallel organizations are possible, depending on the details of the

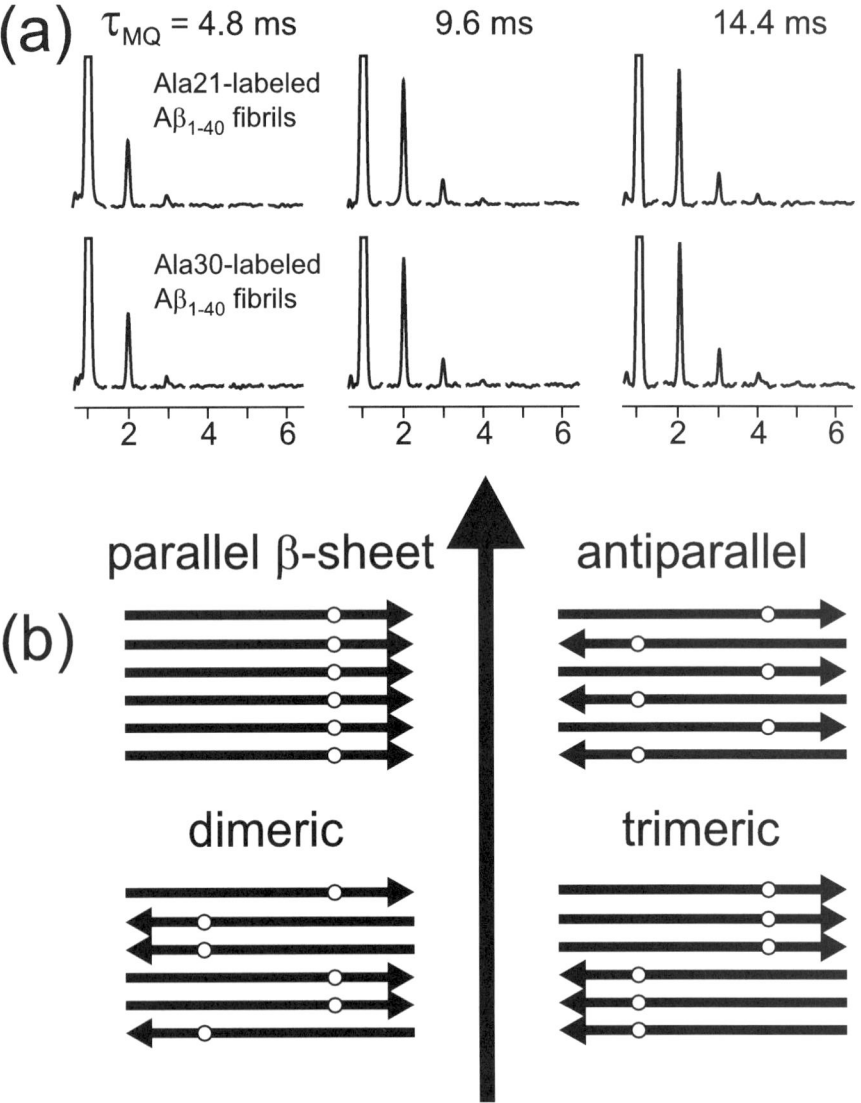

Figure 7 (*a*) Multiple quantum ^{13}C NMR spectra of fibrillized, 40-residue Alzheimer's β-amyloid (Aβ$_{1-40}$), showing MQ orders from one through six for each value of the multiple quantum excitation time τ_{MQ}. Data for two samples are shown, with ^{13}C labels at methyl carbons of Ala21 or Ala30. In both samples, significant two-, three-, and four-quantum peaks are observed, with amplitudes that increase with increasing τ_{MQ}. (*b*) Depiction of possible supramolecular organizations of β-sheets in amyloid fibrils. Large vertical arrow represents the long axis of the fibril. Horizontal arrows represent peptide chains in β-sheets. In a parallel β-sheet, ^{13}C labels would form linear chains with approximately 4.8 Å internuclear distances. In an antiparallel β-sheet, internuclear distances would be significantly larger and would depend on the position of the labeled site. Of the four models shown, only the fully parallel model agrees with the experimental data, as shown by numerical simulations of MQ signal amplitudes. (Adapted from 204.)

peptide sequence, and suggest that juxtaposition of hydrophobic segments may be the decisive factor governing β-sheet organization (138, 204).

Peptide Conformations The conformations of peptides in amyloid fibrils have been investigated by distance measurements, using R^2 and DRAWS recoupling techniques (102–104, 108, 213) and by angular measurements (138). Figure 8 illustrates the application of the CTDQFD and 2D MAS exchange techniques to $A\beta_{16-22}$ fibrils. High-quality data can be obtained from approximately 0.8 μmol of ^{13}C-labeled, fibrillized peptide with several days of signal acquisition at room temperature and a 9.4 T field strength. A combined analysis of the two data sets defines a single region of ϕ and ψ dihedral angles that fits the data adequately. This region is appropriate for a β-strand conformation at the labeled site. It will be particularly interesting to identify non-β-strand conformations in amyloid fibrils. In the case of $A\beta_{1-40}$ fibrils, biochemical data and circular dichroism data (230) and molecular modeling (227, 228) suggest the existence of a β-hairpin or turn conformation in the segment from residue 23 to residue 26. Preliminary solid state NMR data on samples that are doubly-^{13}C-labeled at carbonyl positions in this putative turn segment support a non-β-strand conformation in this region (ON Antzutkin, JJ Balbach, and R Tycko, unpublished results).

Figure 9 shows a 2D ^{13}C-^{13}C chemical shift correlation spectrum of $A\beta_{16-22}$ fibrils that are uniformly ^{13}C-labeled in the central five hydrophobic residues (Leu17 through Ala21). A striking feature of this spectrum is the relatively narrow ^{13}C lines (less than 2 ppm linewidths) at all backbone and sidechain carbon sites, which permits resolution and assignment of the majority of crosspeaks (138).

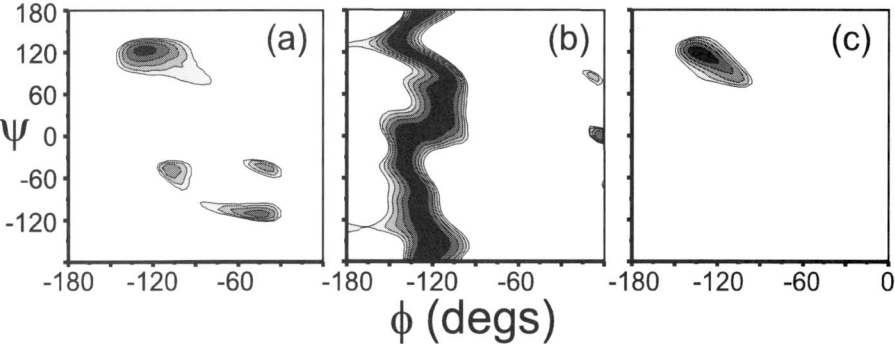

Figure 8 (*a*) Analysis of 2D MAS exchange measurements on amyloid fibrils formed by $A\beta_{16-22}$, a seven-residue fragment of the full-length Alzheimer's β-amyloid peptide. The sample is ^{13}C-labeled at carbonyl sites of Val18 and Phe19. The contour plot shows the χ^2 deviation between experimental data and simulations as a function of the ϕ and ψ dihedral angles of Phe19. Black region is the best fit. (*b*) Analysis of CTDQFD measurements. (*c*) χ^2 contour plot for the combined measurements. A single region of agreement between experiments and simulations is found, with best fit at $\phi \approx -130°$ and $\psi \approx 115°$ indicating a β-strand conformation. (Reprinted from 138.)

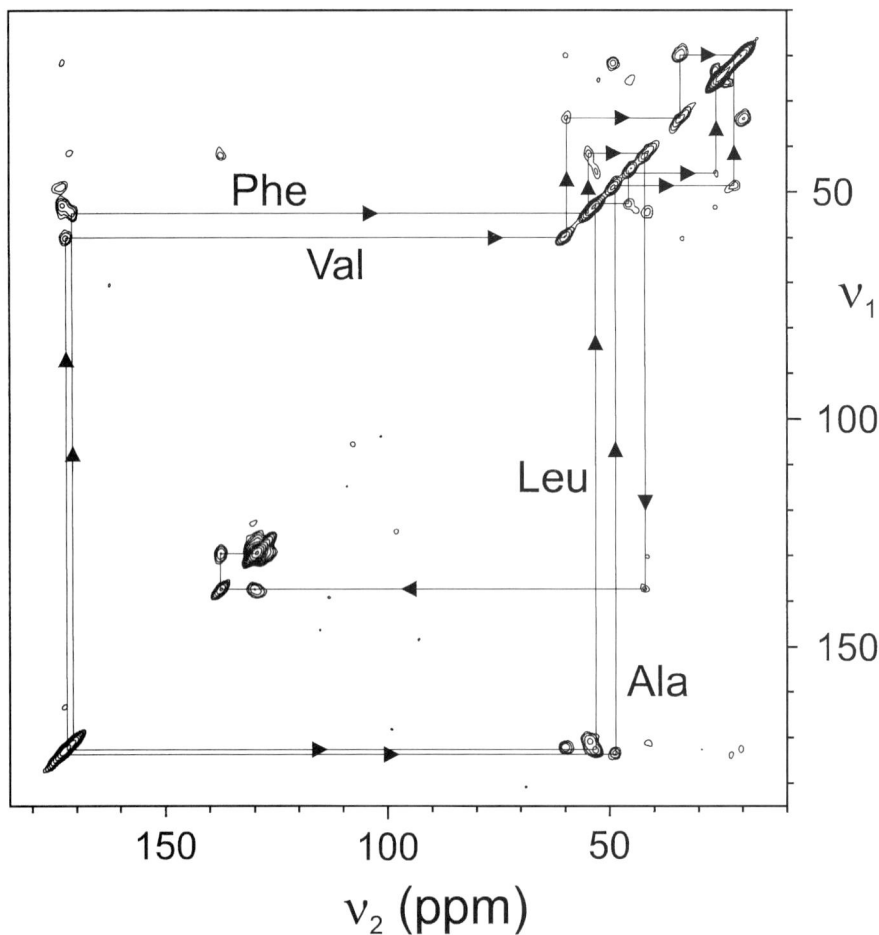

Figure 9 Two-dimensional ^{13}C-^{13}C chemical shift correlation spectrum of amyloid fibrils formed by Aβ_{16-22}, with uniform ^{13}C labeling of the central Leu17-Val18-Phe19-Phe20-Ala21 segment. Remarkably sharp ^{13}C NMR lines are observed, indicating a high degree of structural order in the amyloid fibrils and permitting assignments of backbone and sidechain carbon signals by connectivities shown by arrows. Spectrum obtained at 100.8 MHz ^{13}C NMR frequency, with $\nu_R =$ 24 kHz. (Reprinted from 138.)

These narrow lines establish that both the backbone and the sidechain conformations of peptide molecules are well ordered in the amyloid fibrils, a fact that cannot be established from fiber diffraction or other measurements. Once the signals from carbonyl, α-carbon, and β-carbon sites are resolved and assigned, the isotropic ^{13}C chemical shifts can be used to determine the backbone conformation. In the case of Aβ_{16-22} fibrils the ^{13}C chemical shifts indicate a β-strand conformation for the entire central hydrophobic segment (138).

Other Protein Fibrils

Solid state NMR methods have been used to investigate the molecular conformation and supramolecular structure of high-molecular-weight fibrous proteins that comprise silks produced by spiders (205–211) and silkworms (118, 214–219). Silk fibroins from *Bombyx mori* and *Samia cynthia ricini* silkworms have qualitatively different primary sequences, dominated by (Gly-Ala-Gly-Ala-Gly-Ser)$_n$ segments in the case of *B. mori* and by poly-Ala and Gly-rich segments in the case of *S. cynthia ricini*. Asakura and coworkers have carried out a series of measurements on selectively ^{13}C-labeled and ^{15}N-labeled silk samples from both species (214–219). Isotropic ^{15}N and ^{13}C chemical shifts determined from MAS spectra of unoriented samples support β-strand conformations (214). More detailed structural information is obtained from analyses of spectra of oriented silk fibroin samples, which are strongly dependent on the angle between the external magnetic field and the axis of fiber orientation. Experiments on samples labeled at both Ala and Gly residues lead to ϕ and ψ dihedral angles, indicative of β-strand conformations in silk fibroins of both species (215, 216, 219).

van Beek et al (118) have recently described investigations of the secondary structure at ^{13}C-labeled Ala residues in *S. cynthia ricini* silk fibroin using 2D DOQSY spectroscopy, a technique for determining the relative orientations of labeled backbone carbonyl groups at sequential sites (41). An interesting aspect of this work is the derivation of a complete ϕ, ψ probability distribution, rather than a single ϕ, ψ pair, from the solid state NMR data. Because each solid state NMR measurement yields a set of data points that can be considered independent constraints on a static conformational distribution, it is possible in principle to determine complete conformational distributions from sufficiently large data sets (233, 234). In practice, the number of truly independent data points may be limited and experimental signal-to-noise may not be high, so that algorithms that are capable of handling under-determined problems and tolerating noise are required. van Beek et al used a Tikhonov regularization algorithm (118) to treat their DOQSY data. They found strongly peaked ϕ, ψ distributions indicative of a β-strand conformation in poly-Ala segments of *S. cynthia ricini* silk fibers and an α-helical conformation in films of the same protein.

Meier and coworkers have used 2D exchange spectroscopy (without MAS) to investigate the secondary structure in spider dragline silk (205, 206). They found strikingly different 2D exchange spectra for samples labeled with ^{13}C at Ala carbonyl positions and at Gly carbonyl positions. Their analyses support a β-sheet structure for poly-Ala segments and suggest a 3_1-helical structure in (Gly-Gly-X)$_n$ segments of the dragline silk protein (206). Michal & Jelinski have investigated putative turn regions, composed of Leu-Gly-Ser-Gln and Leu-Gly-Asn-Gln motifs in dragline silk, using REDOR measurements on selectively ^{13}C- and ^{15}N-labeled samples. Distances measured between carbonyl or α-carbon sites of Leu residues and amide ^{15}N sites of Ser or Gln residues are in best agreement with a type I β-turn (211). An important aspect of this work is the demonstration that useful structural constraints can be obtained even in highly complex solid proteins with repeated but inequivalent sequence motifs and with substantial background signals.

As a final example, Blanco & Tycko have recently applied solid state NMR techniques to the 116-residue Rev protein encoded by the genome of HIV-1, which plays an essential role in the viral life cycle by regulating transport of unspliced viral mRNA from the nuclei of infected cells. Although Rev is a relatively small protein, it fibrillizes at concentrations required for crystal growth or liquid state NMR (236) and has therefore eluded structure determination by those techniques. Based on sequence analysis, circular dichroism spectra, and liquid state NMR experiments on complexes of Rev-derived peptides with RNA (237), models of the structure of Rev have been proposed that include two helical segments (238), one of which is the RNA-binding region spanning residues 34 through 55. This putative helical segment contains 10 arginine residues, including the sequential residues 38–39 and 41–44. To examine the secondary structure in this segment, samples were prepared in which all arginines were labeled with ^{13}C at carbonyl positions. As shown in Figure 10, solid state NMR data obtained with the DQCSA technique support a helical conformation in the RNA-binding region of the intact, fibrillized protein (FJ Blanco and R Tycko, manuscript in preparation).

FUTURE DIRECTIONS

As illustrated by the experimental examples cited above and displayed in the figures, recently developed solid state NMR methodology can now provide accurate and robust structural constraints on peptides and proteins of substantial complexity in noncrystalline solids. In systems such as amyloid fibrils, other protein fibrils, and membrane-bound systems, atomic-level structural information from solid state NMR can be crucial. The next several years are likely to see significant advances in the following areas:

1. Exploration of new systems. As the number of research laboratories pursuing biomolecular solid state NMR grows, an increasing number of applications of the techniques described above will be attempted. The range of applicability of these techniques will become better defined, and the power of these techniques will become more evident to biophysical chemists and structural biologists who are not solid state NMR experts.

2. Development of techniques with increased information content. An important goal of methodological development is that of determining complete molecular structures, or complete structures of regions within macromolecules, without dependence on structural models. Techniques that yield multiple structural constraints from experiments on uniformly or multiply labeled, noncrystalline samples will receive increased attention.

3. Sensitivity enhancement. Applications of solid state NMR are often limited by the current requirement for greater than 0.1 μmol of labeled protein. Techniques for sensitivity enhancement by indirect detection

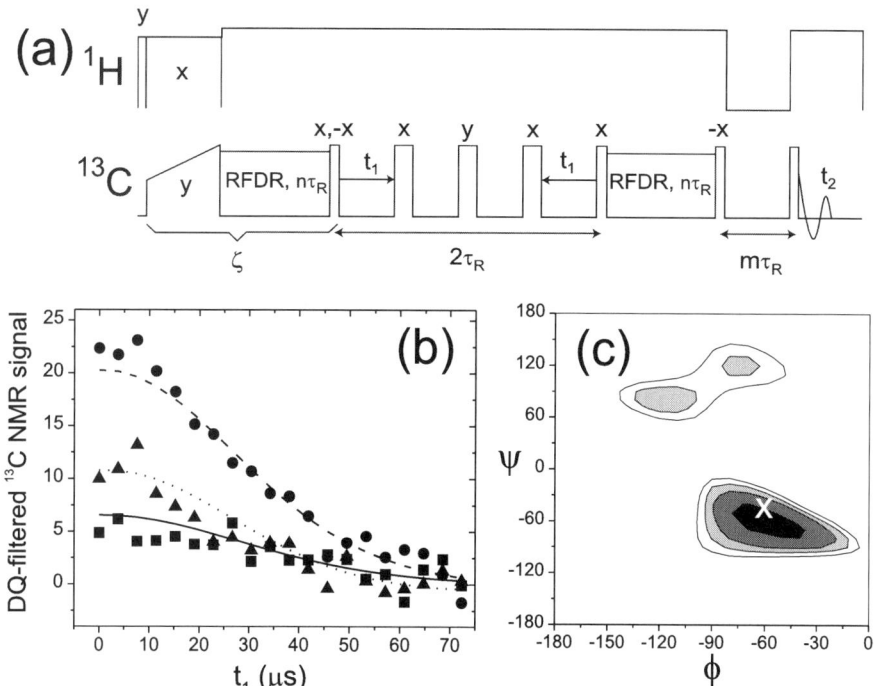

Figure 10 (*a*) Radio-frequency pulse sequence for double-quantum chemical shift anisotropy (DQCSA) measurements. In a doubly carbonyl-labeled sample, DQ coherence is excited in an RFDR recoupling period, evolves under the sum of the two carbonyl CSA interactions during a symmetrized t_1 period, and is converted to observable magnetization in a second recoupling period. If the recoupling period is kept short, data from a multiply carbonyl-labeled sample can be analyzed in terms of two-spin systems. (*b*) Experimental DQCSA data from HIV-1 Rev fibrils labeled with ^{13}C at the carbonyl sites of the 15 arginine residues in the Rev sequence, obtained at 100.8 MHz ^{13}C NMR frequency and $\tau_R = 250$ μs. Circles, triangles, and squares indicate the DQ-filtered amplitudes of the carbonyl centerband, +1 sideband, and −1 sideband, respectively. Dashed, dotted, and solid lines are the corresponding best-fit simulations. (*c*) χ^2 contour plot, comparing experimental and simulated DQCSA curves. Black region is the best fit. White X indicates standard α-helical ϕ and ψ values.

(239), dynamic nuclear polarization (240, 241), and optical pumping (242–247) are under active development.

4. Sample preparation. Applications of solid state NMR are also limited by issues that are the province of molecular biologists, including high-level expression of membrane proteins (248) and selective or segmental labeling of large proteins (249–251). Progress in techniques for protein expression and labeling, driven also by the needs of crystallographers and liquid state NMR spectroscopists, will be of great benefit to biomolecular solid state NMR.

ACKNOWLEDGMENTS

Methodological development and applications to HIV-related systems were supported by grants from the NIH Intramural AIDS Targeted Antiviral Program. Data shown in the figures were obtained by ON Antzutkin, RD Leapman, and NW Rizzo (Figure 6), ON Antzutkin and JJ Balbach (Figure 7), Y Ishii (Figure 9), and FJ Blanco (Figure 10). Contributions to work described above by DP Weliky, AE Bennett, HW Long, CA Michal, AT Petkova, J Reed, and JP Yesinowski are also gratefully acknowledged.

Visit the Annual Reviews home page at www.AnnualReviews.org

LITERATURE CITED

1. McDowell LM, Schaefer J. 1996. *Curr. Opin. Struct. Biol.* 6:624–29
2. Griffin RG. 1998. *Nat. Struct. Biol.* 5:508–12
3. Marassi FM, Opella SJ. 1998. *Curr. Opin. Struct. Biol.* 8:640–48
4. Opella SJ. 1997. *Nat. Struct. Biol.* 4:845–48
5. Cross TA, Opella SJ. 1994. *Curr. Opin. Struct. Biol.* 4:574–81
6. Opella SJ. 1994. *Annu. Rev. Phys. Chem.* 45:659–83
7. Bechinger B, Kinder R, Helmle M, Vogt TCB, Harzer U, Schinzel S. 1999. *Biopolymers* 51:174–90
8. Smith SO, Aschheim K, Groesbeek M. 1996. *Q. Rev. Biophys.* 29:395–449
9. Watts A. 1999. *Curr. Opin. Biotechnol.* 10:48–53
10. Teng Q, Iqbal M, Cross TA. 1992. *J. Am. Chem. Soc.* 114:5312–21
11. Teng Q, Cross TA. 1989. *J. Magn. Reson.* 85:439–47
12. Oas TG, Hartzell CJ, McMahon TJ, Drobny GP, Dahlquist FW. 1987. *J. Am. Chem. Soc.* 109:5956–62
13. Oas TG, Hartzell CJ, Dahlquist FW, Drobny GP. 1987. *J. Am. Chem. Soc.* 109:5962–66
14. Hartzell CJ, Whitfield M, Oas TG, Drobny GP. 1987. *J. Am. Chem. Soc.* 109:5966–69
15. Naito A, McDowell CA. 1984. *J. Chem. Phys.* 81:4795–803
16. Naito A, Ganapathy S, Raghunathan P, McDowell CA. 1983. *J. Chem. Phys.* 79:4173–82
17. Naito A, Ganapathy S, Akasaka K, McDowell CA. 1981. *J. Chem. Phys.* 74:3190–97
18. Harbison GS, Jelinski LW, Stark RE, Torchia DA, Herzfeld J, Griffin RG. 1984. *J. Magn. Reson.* 60:79–82
19. Stark RE, Jelinski LW, Ruben DJ, Torchia DA, Griffin RG. 1983. *J. Magn. Reson.* 55:266–73
20. Gu ZT, Zambrano R, McDermott A. 1994. *J. Am. Chem. Soc.* 116:6368–72
21. Gu ZT, McDermott A. 1993. *J. Am. Chem. Soc.* 115:4282–85
22. Facelli JC, Gu ZT, McDermott A. 1995. *Mol. Phys.* 86:865–72
23. Dedios AC, Oldfield E. 1994. *J. Am. Chem. Soc.* 116:11485–88
24. Wei YF, deDios AC, McDermott AE. 1999. *J. Am. Chem. Soc.* 121:10389–94
25. Walling AE, Pargas RE, deDios AC. 1997. *J. Phys. Chem. A* 101:7299–303
26. deDios AC. 1996. *Prog. Nucl. Magn. Reson. Spectrosc.* 29:229–78
27. Yamanobe T, Ando I, Saito H, Tabeta R, Shoji A, Ozaki T. 1985. *Chem. Phys.* 99:259–64
28. Jiao D, Barfield M, Hruby VJ. 1993. *J. Am. Chem. Soc.* 115:10883–87
29. Pake GE. 1948. *J. Chem. Phys.* 16:327–35
30. Haeberlen U. 1976. *High Resolution NMR*

in Solids: Selective Averaging. New York: Academic
31. Mehring M. 1983. *Principles of High Resolution NMR in Solids.* Berlin: Springer-Verlag
32. Henrichs PM, Linder M. 1984. *J. Magn. Reson.* 58:458–61
33. Edzes HT, Bernards JPC. 1984. *J. Am. Chem. Soc.* 106:1515–17
34. Robyr P, Meier BH, Ernst RR. 1991. *Chem. Phys. Lett.* 187:471–78
35. Robyr P, Meier BH, Fischer P, Ernst RR. 1994. *J. Am. Chem. Soc.* 116:5315–23
36. Tomaselli M, Robyr P, Meier BH, GrobPisano C, Ernst RR, Suter UW. 1996. *Mol. Phys.* 89:1663–94
37. Tycko R, Opella SJ. 1987. *J. Chem. Phys.* 86:1761–74
38. Tycko R, Dabbagh G. 1991. *J. Am. Chem. Soc.* 113:3592–93
39. Weliky DP, Dabbagh G, Tycko R. 1993. *J. Magn. Reson. Ser. A* 104:10–16
40. Dabbagh G, Weliky DP, Tycko R. 1994. *Macromolecule* 27:6183–91
41. Schmidt-Rohr K. 1996. *Macromolecule* 29:3975–81
42. Schmidt-Rohr K. 1996. *J. Am. Chem. Soc.* 118:7601–3
43. Schmidt-Rohr K, Hu W, Zumbulyadis N. 1998. *Science* 280:714–17
44. Andrew ER, Bradbury A, Eades RG. 1958. *Nature* 182:1659
45. Lowe IJ. 1959. *Phys. Rev. Lett.* 2:285–86
46. Herzfeld J, Berger AE. 1980. *J. Chem. Phys.* 73:6021–30
47. Weliky DP, Tycko R. 1996. *J. Am. Chem. Soc.* 118:8487–88
48. Tycko R, Weliky DP, Berger AE. 1996. *J. Chem. Phys.* 105:7915–30
49. Tycko R, Berger AE. 1999. *J. Magn. Reson.* 141:141–47
50. Munowitz MG, Griffin RG. 1982. *J. Chem. Phys.* 76:2848–58
51. Gregory DM, Mehta MA, Shiels JC, Drobny GP. 1997. *J. Chem. Phys.* 107:28–42
52. Bower PV, Oyler N, Mehta MA, Long JR, Stayton PS, Drobny GP. 1999. *J. Am. Chem. Soc.* 121:8373–75
53. Goetz JM, Schaefer J. 1997. *J. Magn. Reson.* 129:222–23
54. Blanco FJ, Tycko R. 2001. *J. Magn. Reson.* In press
55. Alla M, Kundla EI, Lippmaa ET. 1978. *JETP Lett.* 27:194
56. Yarimagaev Y, Tutunjian PN, Waugh JS. 1982. *J. Magn. Reson.* 47:51–60
57. Tycko R, Dabbagh G, Mirau PA. 1989. *J. Magn. Reson.* 85:265–74
58. Ishii Y, Terao T. 1998. *J. Chem. Phys.* 109:1366–74
59. Meier BH, Earl WL. 1986. *J. Chem. Phys.* 85:4905–11
60. Meier BH, Earl WL. 1987. *J. Am. Chem. Soc.* 109:7937–42
61. Gullion T, Schaefer J. 1989. *J. Magn. Reson.* 81:196–200
62. Gullion T, Vega S. 1992. *Chem. Phys. Lett.* 194:423–28
63. Gullion T. 1998. *Concepts Magn. Reson.* 10:277–89
64. Gullion T, Pennington CH. 1998. *Chem. Phys. Lett.* 290:88–93
65. Oas TG, Griffin RG, Levitt MH. 1988. *J. Chem. Phys.* 89:692–95
66. Bennett AE, Rienstra CM, Lansbury PT, Griffin RG. 1996. *J. Chem. Phys.* 105:10289–99
67. Raleigh DP, Levitt MH, Griffin RG. 1988. *Chem. Phys. Lett.* 146:71–76
68. Raleigh DP, Creuzet F, Gupta SKD, Levitt MH, Griffin RG. 1989. *J. Am. Chem. Soc.* 111:4502–3
69. Ok JH, Spencer RGS, Bennett AE, Griffin RG. 1992. *Chem. Phys. Lett.* 197:389–95
70. Bennett AE, Ok JH, Griffin RG, Vega S. 1992. *J. Chem. Phys.* 96:8624–27
71. Sun BQ, Costa PR, Kocisko D, Lansbury PT, Griffin RG. 1995. *J. Chem. Phys.* 102:702–7
72. Rienstra CM, Hatcher ME, Mueller LJ, Sun BQ, Fesik SW, Griffin RG. 1998. *J. Am. Chem. Soc.* 120:10602–12
73. Bennett AE, Rienstra CM, Griffiths JM,

Zhen WG, Lansbury PT, Griffin RG. 1998. *J. Chem. Phys.* 108:9463–79
74. Gross JD, Costa PR, Griffin RG. 1998. *J. Chem. Phys.* 108:7286–93
75. Hohwy M, Rienstra CM, Jaroniec CP, Griffin RG. 1999. *J. Chem. Phys.* 110:7983–92
76. Nielsen NC, Bildsoe H, Jakobsen HJ, Levitt MH. 1994. *J. Chem. Phys.* 101:1805–12
77. Lee YK, Kurur ND, Helmle M, Johannessen OG, Nielsen NC, Levitt MH. 1995. *Chem. Phys. Lett.* 242:304–9
78. Hohwy M, Jakobsen HJ, Eden M, Levitt MH, Nielsen NC. 1998. *J. Chem. Phys.* 108:2686–94
79. Carravetta M, Eden M, Zhao X, Brinkmann A, Levitt MH. 2000. *Chem. Phys. Lett.* 321:205–15
80. Tycko R. 1988. *Phys. Rev. Lett.* 60:2734–37
81. Tycko R, Dabbagh G. 1990. *Chem. Phys. Lett.* 173:461–65
82. Tycko R. 1990. *J. Chem. Phys.* 92:5776–93
83. Tycko R, Dabbagh G. 1991. *J. Am. Chem. Soc.* 113:9444–48
84. Tycko R, Smith SO. 1993. *J. Chem. Phys.* 98:932–43
85. Tycko R. 1994. *J. Am. Chem. Soc.* 116:2217–18
86. Fujiwara T, Ramamoorthy A, Nagayama K, Hioka K, Fujito T. 1993. *Chem. Phys. Lett.* 212:81–84
87. Fujiwara T, Khandelwal P, Akutsu H. 2000. *J. Magn. Reson.* 145:73–83
88. Gregory DM, Mitchell DJ, Stringer JA, Kiihne S, Shiels JC. et al. 1995. *Chem. Phys. Lett.* 246:654–63
89. Baldus M, Tomaselli M, Meier BH, Ernst RR. 1994. *Chem. Phys. Lett.* 230:329–36
90. Baldus M, Geurts DG, Meier BH. 1998. *Solid State Nucl. Magn. Reson.* 11:157–68
91. Baldus M, Meier BH. 1997. *J. Magn. Reson.* 128:172–93
92. Tycko R. 1999. *J. Magn. Reson.* 139:302–7
93. Balazs YS, Thompson LK. 1999. *J. Magn. Reson.* 139:371–76
94. Takegoshi K, Nomura K, Terao T. 1995. *Chem. Phys. Lett.* 232:424–28
95. Takegoshi K, Nomura K, Terao T. 1997. *J. Magn. Reson.* 127:206–16
96. Baldus M, Petkova AT, Herzfeld J, Griffin RG. 1998. *Mol. Phys.* 95:1197–207
97. Andrew ER, Bradbury A, Eades RG, Wynn VT. 1963. *Phys. Lett.* 21:99
98. Colombo MG, Meier BH, Ernst RR. 1988. *Chem. Phys. Lett.* 146:189–96
99. Maas W, Veeman WS. 1988. *Chem. Phys. Lett.* 149:170–74
100. Kubo A, McDowell CA. 1988. *J. Chem. Soc. Faraday Trans. I* 84:3713–30
101. Shaw WJ, Long JR, Dindot JL, Campbell AA, Stayton PS, Drobny GP. 2000. *J. Am. Chem. Soc.* 122:1709–16
102. Spencer RGS, Halverson KJ, Auger M, McDermott AE, Griffin RG, Lansbury PT. 1991. *Biochemistry* 30:10382–87
103. Griffiths JM, Ashburn TT, Auger M, Costa PR, Griffin RG, Lansbury PT. 1995. *J. Am. Chem. Soc.* 117:3539–46
104. Lansbury PT, Costa PR, Griffiths JM, Simon EJ, Auger M. et al. 1995. *Nat. Struct. Biol.* 2:990–98
105. Garbow JR, Breslav M, Antohi O, Naider F. 1994. *Biochemistry* 33:10094–99
106. Arshava B, Breslav M, Antohi O, Stark RE, Garbow JR. et al. 1999. *Solid State Nucl. Magn. Reson.* 14:117–36
107. Anderson RC, Gullion T, Joers JM, Shapiro M, Villhauer EB, Weber HP. 1995. *J. Am. Chem. Soc.* 117:10546–50
108. Gregory DM, Benzinger TLS, Burkoth TS, Miller-Auer H, Lynn DG. et al. 1998. *Solid State Nucl. Magn. Reson.* 13:149–66
109. Bennett AE, Weliky DP, Tycko R. 1998. *J. Am. Chem. Soc.* 120:4897–98
110. Ishiiy Y. 2000. *J. Chem. Phys.* In press
111. Tycko R, Balbach JJ, Ishii Y. 2000. *Chem. Phys.* In press
112. De Groot HJM, Copie V, Smith SO, Allen PJ, Winkel C. et al. 1988. *J. Magn. Reson.* 77:251–57

113. deAzevedo ER, Bonagamba TJ, Schmidt-Rohr K. 2000. *J. Magn. Reson.* 142:86–96
114. Ernst M, Kentgens APM, Meier BH. 1999. *J. Magn. Reson.* 138:66–73
115. Weliky DP, Bennett AE, Zvi A, Anglister J, Steinbach PJ, Tycko R. 1999. *Nat. Struct. Biol.* 6:141–45
116. Feng X, Lee YK, Sandstrom D, Eden M, Maisel H. et al. 1996. *Chem. Phys. Lett.* 257:314–20
117. Feng X, Eden M, Brinkmann A, Luthman H, Eriksson L. et al. 1997. *J. Am. Chem. Soc.* 119:12006–7
118. van Beek JD, Beaulieu L, Schafer H, Demura M, Asakura T, Meier BH. 2000. *Nature* 405:1077–79
119. Tomita Y, O'Connor EJ, McDermott A. 1994. *J. Am. Chem. Soc.* 116:8766–71
120. Hong M, Gross JD, Griffin RG. 1997. *J. Phys. Chem. B* 101:5869–74
121. Costa PR, Gross JD, Hong M, Griffin RG. 1997. *Chem. Phys. Lett.* 280:95–103
122. Hong M, Gross JD, Hu W, Griffin RG. 1998. *J. Magn. Reson.* 135:169–77
123. Reif B, Hohwy M, Jaroniec CP, Rienstra CM, Griffin RG. 2000. *J. Magn. Reson.* 145:132–41
124. Ishii Y, Terao T, Kainosho M. 1996. *Chem. Phys. Lett.* 256:133–40
125. Ishii Y, Hirao K, Terao T, Terauchi T, Oba M. et al. 1998. *Solid State Nucl. Magn. Reson.* 11:169–75
126. Dejong AF, Kentgens APM, Veeman WS. 1984. *Chem. Phys. Lett.* 109:337–42
127. Luz Z, Spiess HW, Titman JJ. 1992. *Isr. J. Chem.* 32:145–60
128. Boender GJ, Raap J, Prytulla S, Oschkinat H, de Groot HJM. 1995. *Chem. Phys. Lett.* 237:502–8
129. Fujiwara T, Sugase K, Kainosho M, Ono A, Akutsu H. 1995. *J. Am. Chem. Soc.* 117:11351–52
130. Sun BQ, Rienstra CM, Costa PR, Williamson JR, Griffin RG. 1997. *J. Am. Chem. Soc.* 119:8540–46
131. Hong M, Griffin RG. 1998. *J. Am. Chem. Soc.* 120:7113–14
132. Hong M. 1999. *J. Biomol. NMR.* 15:1–14
133. Hong M, Jakes K. 1999. *J. Biomol. NMR.* 14:71–74
134. Pauli J, van Rossum B, Forster H, de Groot HJM, Oschkinat H. 2000. *J. Magn. Reson.* 143:411–16
135. Straus SK, Bremi T, Ernst RR. 1996. *Chem. Phys. Lett.* 262:709–15
136. Straus SK, Bremi T, Ernst RR. 1998. *J. Biomol. NMR* 12:39–50
137. McDermott A, Polenova T, Bockmann A, Zilm KW, Paulsen EK. et al. 2000. *J. Biomol. NMR* 16:209–19
138. Balbach JJ, Ishii Y, Antzutkin ON, Leapman RD, Rizzo NW. et al. 2000. *Biochemistry.* 39:13748–59
139. Nomura K, Takegoshi K, Terao T, Uchida K, Kainosho M. 1999. *J. Am. Chem. Soc.* 121:4064–65
140. Nomura K, Takegoshi K, Terao T, Uchida K, Kainosho M. 2000. *J. Biomol. NMR* 17:111–23
141. Tycko R. 1996. *J. Biomol. NMR* 8:239–51
142. Hong M. 1999. *J. Magn. Reson.* 139:389–401
143. Coughlin PE, Anderson FE, Oliver EJ, Brown JM, Homans SW. et al. 1999. *J. Am. Chem. Soc.* 121:11871–74
144. Hong M. 2000. *J. Am. Chem. Soc.* 122:3762–70
145. Yen YS, Pines A. 1983. *J. Chem. Phys.* 78:3579–82
146. Warren WS, Weitekamp DP, Pines A. 1980. *J. Chem. Phys.* 73:2084–99
147. Baum J, Munowitz M, Garroway AN, Pines A. 1985. *J. Chem. Phys.* 83:2015–25
148. Suter D, Liu SB, Baum J, Pines A. 1987. *Chem. Phys.* 114:103–9
149. Cho G, Yesinowski JP. 1996. *J. Phys. Chem.* 100:15716–25
150. Feldman EB, Lacelle S. 1997. *J. Chem. Phys.* 107:7067–84
151. Baum J, Gleason KK, Pines A, Garroway AN, Reimer JA. 1986. *Phys. Rev. Lett.* 56:1377–80

152. Levy DH, Gleason KK. 1992. *J. Phys. Chem.* 96:8125–31
153. Scruggs BE, Gleason KK. 1993. *J. Phys. Chem.* 9:9187–95
154. Antzutkin ON, Tycko R. 1999. *J. Chem. Phys.* 110:2749–52
155. Eden M, Brinkmann A, Luthman H, Eriksson L, Levitt MH. 2000. *J. Magn. Reson.* 1:4:266–79
156. Eden M, Levitt MH. 1998. *Chem. Phys. Lett.* 293:173–79
157. Ketchem RR, Hu W, Cross TA. 1993. *Science* 261:1457–60
158. Bechinger B, Opella SJ. 1991. *J. Magn. Reson.* 95:585–88
159. Sanders CR, Prestegard JH. 1990. *Biophys. J.* 58:447–60
160. Vold RR, Prosser RS. 1996. *J. Magn. Reson. Ser. B* 113:267–71
161. Prosser RS, Hunt SA, DiNatale JA, Vold RR. 1996. *J. Am. Chem. Soc.* 118:269–70
162. Sanders CR, Landis GC. 1995. *Biochemistry* 34:4030–40
163. Sanders CR, Schwonek JP. 1992. *Biochemistry* 31:8898–905
164. Ketchem RR, Lee KC, Huo S, Cross TA. 1996. *J. Biomol. NMR* 8:1–14
165. Ramamoorthy A, Wu CH, Opella SJ. 1995. *J. Magn. Reson. Ser. B* 107:88–90
166. Ramamoorthy A, Gierasch LM, Opella SJ. 1995. *J. Magn. Reson. Ser. B* 109:112–16
167. Ramamoorthy A, Gierasch LM, Opella SJ. 1996. *J. Magn. Reson. Ser. B* 111:81–84
168. Marassi FM, Ramamoorthy A, Opella SJ. 1997. *Proc. Natl. Acad. Sci. USA* 94:8551–56
169. Wang J, Denny J, Tian C, Kim S, Mo Y. et al. 2000. *J. Magn. Reson.* 144:162–67
170. Marassi FM, Opella SJ. 2000. *J. Magn. Reson.* 144:150–55
171. Gu ZT, Opella SJ. 1999. *J. Magn. Reson.* 138:193–98
172. Tan WM, Gu ZT, Zeri AC, Opella SJ. 1999. *J. Biomol. NMR* 13:337–42
173. Ishii Y, Tycko R. 2000. *J. Am. Chem. Soc.* 122:1443–55
174. Creuzet F, McDermott A, Gebhard R, Vanderhoef K, Spijkerassink MB. et al. 1991. *Science* 251:783–86
175. Thompson LK, McDermott AE, Raap J, Vanderwielen CM, Lugtenburg J. et al. 1992. *Biochemistry* 31:7931–38
176. Ulrich AS, Watts A. 1993. *Solid State Nucl. Magn. Reson.* 2:21–36
177. Prosser RS, Daleman SI, Davis JH. 1994. *Biophys. J.* 66:1415–28
178. Griffiths JM, Lakshmi KV, Bennett AE, Raap J, Vanderwielen CM. et al. 1994. *J. Am. Chem. Soc.* 116:10178–81
179. Smith SO, Bormann BJ. 1995. *Proc. Natl. Acad. Sci. USA* 92:488–91
180. Ramamoorthy A, Marassi FM, Zasloff M, Opella SJ. 1995. *J. Biomol. NMR* 6:329–34
181. Ketchem RR, Roux B, Cross TA. 1997. *Structure* 5:1655–69
182. Wang JX, Balazs YS, Thompson LK. 1997. *Biochemistry* 36:1699–703
183. Middleton DA, Robins R, Feng XL, Levitt MH, Spiers ID. et al. 1997. *FEBS Lett.* 410:269–74
184. Smith SO, Smith CS, Bormann BJ. 1996. *Nat. Struct. Biol.* 3:252–58
185. Feng X, Verdegem PJE, Lee YK, Sandstrom D, Eden M. et al. 1997. *J. Am. Chem. Soc.* 119:6853–57
186. Kovacs FA, Cross TA. 1997. *Biophys. J.* 73:2511–17
187. Kumashiro KK, Schmidt-Rohr K, Murphy OJ, Ouellette KL, Cramer WA, Thompson LK. 1998. *J. Am. Chem. Soc.* 120:5043–51
188. Kim Y, Valentine K, Opella SJ, Schendel SL, Cramer WA. 1998. *Protein Sci.* 7:342–48
189. Struppe J, Komives EA, Taylor SS, Vold RR. 1998. *Biochemistry* 37:15523–27
190. Opella SJ, Marassi FM, Gesell JJ, Valente AP, Kim Y. et al. 1999. *Nat. Struct. Biol.* 6:374–79

191. Marassi FM, Opella SJ, Juvvadi P, Merrifield RB. 1999. *Biophys. J.* 77:3152–55
192. Marassi FM, Ma C, Gratkowski H, Straus SK, Strebel K. et al. 1999. *Proc. Natl. Acad. Sci. USA* 96:14336–41
193. Grobner G, Glaubitz C, Watts A. 1999. *J. Magn. Reson.* 141:335–39
194. Spooner PJR, Veenhoff LM, Watts A, Poolman B. 1999. *Biochemistry* 38:9634–39
195. Gorzelle BM, Nagy JK, Oxenoid K, Lonzer WL, Cafiso DS, Sanders CR. 1999. *Biochemistry* 38:16373–82
196. Glaubitz C, Burnett IJ, Grobner G, Mason AJ, Watts A. 1999. *J. Am. Chem. Soc.* 121:5787–94
197. Feng X, Verdegem PJE, Eden M, Sandstrom D, Lee YK. et al. 2000. *J. Biomol. NMR* 16:1–8
198. Griffiths JM, Bennett AE, Engelhard M, Siebert F, Raap J. et al. 2000. *Biochemistry* 39:362–71
199. Grobner G, Burnett IJ, Glaubitz C, Choi G, Mason AJ, Watts A. 2000. *Nature* 405:810–13
200. Czerski L, Sanders CR. 2000. *Anal. Biochem.* 284:327–33
201. Glaubitz C, Grobner G, Watts A. 2000. *Biochim. Biophys. Acta Biomembr.* 1463:151–61
202. Hing AW, Schaefer J, Kobayashi GS. 2000. *Biochim. Biophys. Acta Biomembr.* 1463:323–32
203. Balbach JJ, Yang J, Weliky DP, Steinbach PJ, Tugarinov V, Anglister J, Tycko R. et al. 2000. *J. Biomol. NMR* 16:313–27
204. Antzutkin ON, Balbach JJ, Leapman RD, Rizzo NW, Reed J, Tycko R. 2000. *Proc. Natl. Acad. Sci. USA*. 204:13045–50
205. van Beek JD, Kummerlen J, Vollrath F, Meier BH. 1999. *Int. J. Biol. Macromol.* 24:173–78
206. Kummerlen J, van Beek JD, Vollrath F, Meier BH. 1996. *Macromolecules* 29:2920–28
207. Jelinski LW, Blye A, Liivak O, Michal C, LaVerde G. et al. 1999. *Int. J. Biol. Macromol.* 24:197–201
208. Liivak O, Flores A, Lewis R, Jelinski LW. 1997. *Macromolecules* 30:7127–30
209. Hijirida DH, Do KG, Michal C, Wong S, Zax D, Jelinski LW. 1996. *Biophys. J.* 71:3442–47
210. Simmons A, Ray E, Jelinski LW. 1994. *Macromolecules* 27:5235–37
211. Michal CA, Jelinski LW. 1998. *J. Biomol. NMR* 12:231–41
212. Benzinger TLS, Gregory DM, Burkoth TS, Miller-Auer H, Lynn DG. et al. 1998. *Proc. Natl. Acad. Sci. USA* 95:13407–12
213. Benzinger TLS, Gregory DM, Burkoth TS, Miller-Auer H, Lynn DG. et al. 2000. *Biochemistry* 39:3491–99
214. Asakura T, Ito T, Okudaira M, Kameda T. 1999. *Macromolecules* 32:4940–46
215. Demura M, Yamazaki Y, Asakura T, Ogawa K. 1998. *J. Mol. Struct.* 441:155–63
216. Demura M, Minami M, Asakura T, Cross TA. 1998. *J. Am. Chem. Soc.* 120:1300–8
217. Asakura T, Demura M, Hiraishi Y, Ogawa K, Uyama A. 1994. *Chem. Lett.,* (12):2249–52
218. Asakura T, Demura M, Date T, Miyashita N, Ogawa K, Williamson MP. 1997. *Biopolymers* 41:193–203
219. Nicholson LK, Asakura T, Demura M, Cross TA. 1993. *Biopolymers* 33:847–61
220. Sunde M, Blake CCF. 1998. *Q. Rev. Biophys.* 31:1–39
221. Guijarro JI, Sunde M, Jones JA, Campbell ID, Dobson CM. 1998. *Proc. Natl. Acad. Sci. USA* 95:4224–28
222. Chiti F, Webster P, Taddei N, Clark A, Stefani M, Ramponi G, Dobson CM et al. 1999. *Proc. Natl. Acad. Sci. USA* 96:3590–94
223. Blake C, Serpell L. 1996. *Structure* 4:989–98
224. Sunde M, Serpell LC, Bartlam M, Fraser PE, Pepys MB, Blake CCF. 1997. *J. Mol. Biol.* 273:729–39
225. Serpell LC, Smith JM. 2000. *J. Mol. Biol.* 299:225–31

226. George AR, Howlett DR. 1999. *Biopolymers* 50:733–41
227. Li LP, Darden TA, Bartolotti L, Kominos D, Pedersen LG. 1999. *Biophys. J.* 76:2871–78
228. Tjernberg LO, Callaway DJE, Tjernberg A, Hahne S, Lilliehook C. et al. 1999. *J. Biol. Chem.* 274:12619–25
229. Chaney MO, Webster SD, Kuo YM, Roher AE. 1998. *Protein Eng.* 11:761–67
230. Hilbich C, Kisterswoike B, Reed J, Masters CL, Beyreuther K. 1991. *J. Mol. Biol.* 218:149–63
231. Fraser PE, Nguyen JT, Inouye H, Surewicz WK, Selkoe DJ. et al. 1992. *Biochemistry* 31:10716–23
232. Deleted in proof
233. Utz M. 1998. *J. Chem. Phys.* 109:6110–24
234. Long HW, Tycko R. 1998. *J. Am. Chem. Soc.* 120:7039–48
235. Deleted in proof
236. Watts NR, Misra M, Wingfield PT, Stahl SJ, Cheng NQ. et al. 1998. *J. Struct. Biol.* 121:41–52
237. Battiste JL, Mao HY, Rao NS, Tan RY, Muhandiram DR. et al. 1996. *Science* 273:1547–51
238. Auer M, Gremlich HU, Seifert JM, Daly TJ, Parslow TG. et al. 1994. *Biochemistry* 33:2988–96
239. Ishii Y, Tycko R. 2000. *J. Magn. Reson.* 142:199–204
240. Gerfen GJ, Becerra LR, Hall DA, Griffin RG, Temkin RJ, Singel DJ. 1995. *J. Chem. Phys.* 102:9494–97
241. Hall DA, Maus DC, Gerfen GJ, Inati SJ, Becerra LR. et al. 1997. *Science* 276:930–32
242. Tycko R, Reimer JA. 1996. *J. Phys. Chem.* 100:13240–50
243. Tycko R. 1998. *Solid State Nucl. Magn. Reson.* 11:1–9
244. Michal CA, Tycko R. 1999. *Phys. Rev. B Condens. Matter* 60:8672–79
245. Zysmilich MG, McDermott A. 1996. *Proc. Natl. Acad. Sci. USA* 93:6857–60
246. Zysmilich MG, McDermott A. 1996. *J. Am. Chem. Soc.* 118:5867–73
247. Polenova T, McDermott AE. 1999. *J. Phys. Chem. B* 103:535–48
248. Eilers M, Reeves PJ, Ying WW, Khorana HG, Smith SO. 1999. *Proc. Natl. Acad. Sci. USA* 96:487–92
249. Xu R, Ayers B, Cowburn D, Muir TW. 1999. *Proc. Natl. Acad. Sci. USA* 96:388–93
250. Ellman JA, Volkman BF, Mendel D, Schultz PG, Wemmer DE. 1992. *J. Am. Chem. Soc.* 114:7959–61
251. Otomo T, Ito N, Kyogoku Y, Yamazaki T. 1999. *Biochemistry* 38:16040–44

PHOTOFRAGMENT TRANSLATIONAL SPECTROSCOPY OF WEAKLY BOUND COMPLEXES: Probing The Interfragment Correlated Final State Distributions

L Oudejans and RE Miller
Department of Chemistry, University of North Carolina, Chapel Hill, North Carolina 27599; e-mail: remiller@unc.edu

Key Words vibrational predissociation dynamics, dissociation energy

■ **Abstract** The vibrational predissociation dynamics of weakly bound complexes is well known to be highly nonstatistical. In particular, the associated photofragment final state distributions are often far from statistical, consequently reflecting the nature of the dissociation process. For binary complexes consisting of two molecules, a complete description of the final state of the system must include the associated interfragment correlations, specifically between their internal states. Information of this type is imprinted in the translational energies of the fragments, which can be measured using a number of recently developed translational spectroscopy methods. These data can provide detailed insights into the nature of the bond rupture process, as well as accurate values for the dissociation energy of the complexes. The focus of the present review is on experiments that provide correlated final state distributions for weakly bound binary complexes.Where possible, comparisons with theoretical calculations are made.

INTRODUCTION

The processes associated with the making and breaking of bonds between atoms and molecules are of fundamental importance in chemistry. By their very nature, these processes are highly anharmonic, making their detailed characterization a formidable challenge. In recent years attempts have been made to address this challenge by spectroscopic methods that characterize the quantum states of a molecule at high vibrational energies, corresponding to the chemically interesting regime (1, 2). At these energies the density of states becomes extremely high and the coupling between the states very strong, the result being that the vibrations can no longer be characterized in terms of simple isolated local or normal modes. In the extreme limit, where Rice, Ramsperger, Kassel, and Marcus (RRKM) theory (3) applies, there is rapid energy redistribution that, at least approximately, samples the available states statistically, allowing us to overlook many of the fine details. Although we are still far from having a complete understanding of the quantum state dynamics of systems in this regime, the recent progress in both experiment (4–8) and theory (9, 10) is helping to better define the important processes. Ultimately, the detailed characterization of all of the intramolecular couplings in a molecule would provide us with a basis for understanding the chemistry at a fundamental level, in both the statistical and nonstatistical regimes.

Weakly bound complexes represent an interesting class of systems, given that dissociation can occur at low energies, where nonstatistical behavior is often observed. Vibrational predissociation is common in these complexes (11, 12), given that the intermolecular binding energies are often less than the vibrational energies associated with the constituent molecules. In these cases, intramolecular vibrational excitation produces metastable states of the complex. Owing to the large frequency difference between the intramolecular and intermolecular degrees of freedom, relaxation from these states is often quite slow (13), corresponding to thousands of intramolecular vibrational periods. Because of this, the corresponding infrared spectra are often highly resolved, showing fine structure associated with the rotational and tunneling motions. In many cases, it is possible to observe combination bands involving the intermolecular vibrations (14, 15), which provides further information on the corresponding intermolecular potential surface. The theoretical methods needed to relate this vibration/rotation/tunneling (VRT) spectroscopy to the corresponding intermolecular potential surface have also undergone considerable development (16). The combination of these experimental and theoretical methods can provide high-quality empirical potential surfaces, in cases where the spectroscopy samples a significant fraction of the potential well. A number of reviews have already been published on this VRT approach for determining potential energy surfaces (17, 18). In light of this, the focus of the present article is on what happens after the complex is vibrationally excited. In particular, we are interested in the final outcome of the vibrational predissociation process, that is, the characterization of the final state distributions of the fragments. As we will see, these final state distributions can provide detailed insights into the nature of the dynamical processes that lead to bond rupture.

Much of what is known about the vibrational predissociation dynamics of complexes in the ground electronic state has come from the interpretation of the homogeneous line broadening observed in these spectra in terms of the vibrational predissociation lifetimes or rates (19–23). Unfortunately, experiments of this type leave one to speculate about (*a*) the final outcome of the photochemical event and (*b*) the detailed dynamical events leading up to dissociation. To address these questions we clearly need more sophisticated experiments that allow us to determine the final states of the photofragments. The ideal for such an experiment also provides detailed information on the initial parent state. It is also important to determine the various scalar and vector correlations that may exist (24–26). Although these correlations complicate the problem, they also provide rich detail that can be used to obtain important insights into the nature of the associated dynamical processes, particularly if they are nonstatistical. In recent years, a vast literature has evolved that deals with the characterization of final state distributions, which includes both vector and scalar correlations (25).

For chemically bound molecules, UV excitation is required to dissociate the intramolecular bonds, and the time scales are often so short that the spectra become strongly lifetime broadened. For this reason, the measured final state distributions often correspond to the excitation of a range of initial states.

As noted above, the situation is quite different for the metastable states of binary complexes, where the lifetimes are often long enough to permit the excitation of individual initial states, allowing us to explore the initial state dependence of the final state distributions, often resolved at the rotational level. In this review, we focus on systems for which the interfragment correlations have been determined experimentally by some form of translational spectroscopy. Where possible we compare these with the results from theory.

TRANSLATIONAL SPECTROSCOPY AS A PROBE OF VIBRATIONAL PREDISSOCIATION

In the following sections we discuss a number of experimental methods that have been applied to the study of weakly bound complexes, with the goal of providing correlated final state distributions. The basic approach is to use translational spectroscopy, which encompasses a family of techniques that provide internal state distributions for photofragments by measuring the translational recoil distributions. This approach is perhaps best illustrated by considering the case of hydrogen Rydberg atom photofragment translational spectroscopy (27), a widely used technique in UV photochemistry (28). The translational energy of an ejected hydrogen atom is clearly correlated with the internal energy of the molecular cofragment, so that the internal state distribution of the latter can be determined from the velocity distribution of the hydrogen atom. In this case, the velocity distribution is measured by time of flight (TOF) of hydrogen atoms. Detection is made possible by laser excitation of the atoms to long-lived Rydberg states, which are easily detected by field ionization just before they impinge on an ion detector. The attractiveness of this method comes from the fact that a wide range of systems can be studied using the same laser detection scheme (i.e. it does not depend upon the identity of the molecular cofragment). In principle, of course, the same information could be obtained by directly probing the internal state distribution of the molecular fragment using an appropriate spectroscopic method. In general, however, the spectroscopy of the molecule of interest may not be accessible.

Although the above example demonstrates one use of translational spectroscopy, its unique capabilities become apparent when we consider photodissociation processes that give rise to two molecular fragments, formed in a range of internal states. A standard approach for characterizing such a system would be to probe the internal states of the two fragments separately, using spectroscopic probes.

Consider, for example, the photodissociation of HONO to produce OH and NO (29). Although the final state distributions for the two fragments can be probed by separate spectroscopy studies, such experiments would not provide the correlations between their internal states. Namely, which NO rotational vibrational states are produced in coincidence with which rotational vibrational states of the OH. This is precisely the kind of information that translational spectroscopy can provide (24, 30). If the translational energy distribution can be measured for each internal state of one (or both) of the fragments, conservation of energy can be used to determine the state distribution of the cofragment, assuming of course that the method provides sufficient velocity resolution.

Several approaches have been developed for determining these state specific translational distributions. For the present purposes we focus our attention on those that have been applied to study the photodissociation of weakly bound binary complexes. In particular, we consider ground electronic state vibrational predissociation, initiated by infrared laser excitation of the complex. In the frame of reference of the complex, dissociation results in a plume of photofragments. The associated angular distribution depends upon many different factors and need not be spherical (31), particularly when using a polarized dissociation laser. Note that fragments in different internal states have different recoil velocities and are thus expanding at different rates.

One approach for determining the state-specific velocity distributions is to use a high-resolution laser to measure the Doppler line shapes for transitions to the various states of the fragments (hereafter, this technique is referred to as the Doppler method) (30). We give examples of experiments that have been conducted in this way. Another approach is to take advantage of the fact that the various internal states of the molecular fragments will become spatially separated as time progresses since they have different translational energies. This suggests the use of imaging methods for determining the recoil velocities and hence energies. The first experiments to reveal this structure were carried out in our laboratory (32), where the spatial resolution was obtained by scanning the photofragment angular distributions (PHOFAD) in a molecular beam, as shown in Figure 1. In these experiments, a bolometer detector was used to detect the photofragment intensity as a function of the recoil angle. Thus, the velocity distribution is reflected in the angular distribution of the photofragments. Since the bolometer itself is not a state-specific detector, a probe laser may be necessary to record the angular distributions (velocity distributions) of the individual fragment states (33). In addition, a large electric field has been used in some cases to orient the parent molecules by brute force (pendular state method) (34, 35), so that the two fragments recoil in opposite directions in the laboratory frame and thus can be detected separately (36). Valentini and coworkers (37) have obtained similar slices through the velocity distributions by using a method they call position-sensitive translational spectroscopy (POSTS). In this case, a pulsed laser is used to pump the complex into the vibrationally excited state (in this case using stimulated Raman excitation), and a second laser, delayed in time, provides state-specific ionization [by use of

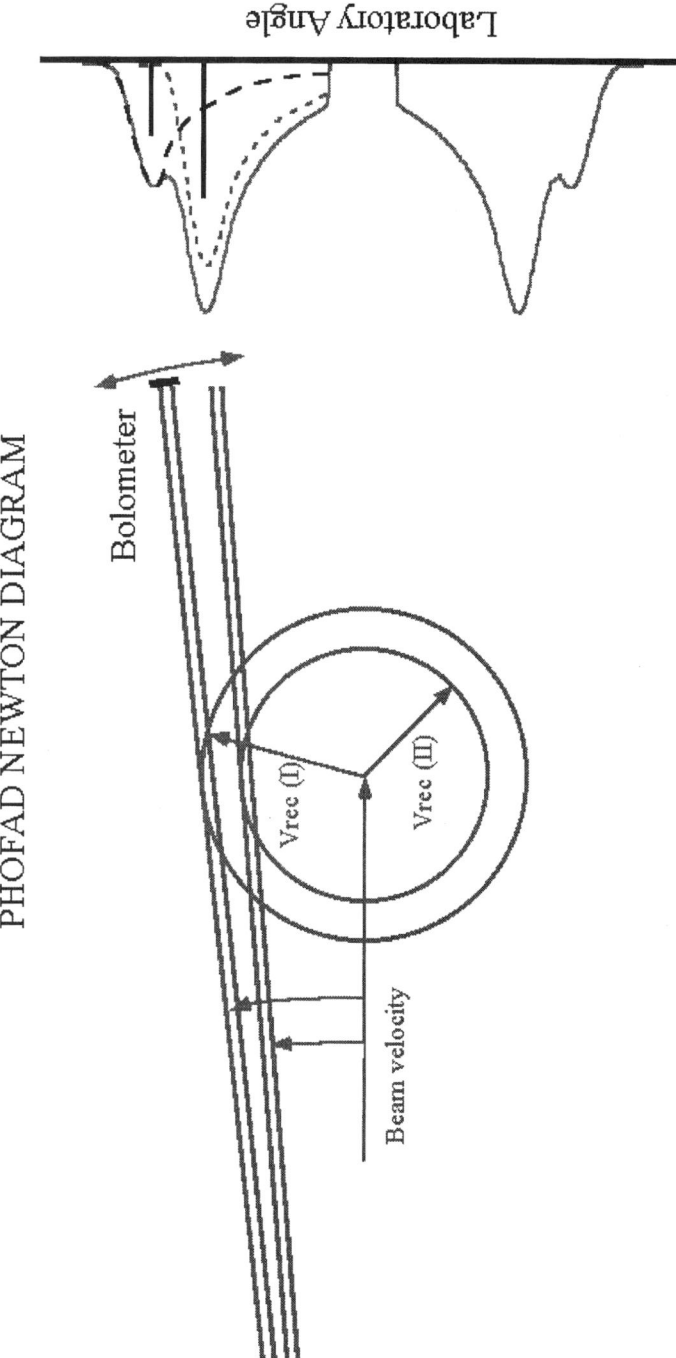

Figure 1 A schematic diagram showing a slice through the photofragment distributions. Data of this type can be measured using either the PHOFAD method, by measuring the photofragment angular distribution downstream of the photolysis volume, or using the POSTS methods, based on scanning the spatial overlap between time-delayed pump and probe lasers. Similar results can be obtained from measuring the Doppler profiles associated with the various photofragment states. These are all variants of translational spectroscopy used to correlate the internal states of one fragment with its cofragment.

resonant enhanced multi photon ionization (REMPI)] of the fragments. Since the lasers used for these two purposes are counter propagating, the probe laser can be spatially separated from the pump. Scanning the distance between the two lasers at a fixed time delay provides the necessary spatial resolution. In all of these techniques the dependence of the data on laser polarization direction can provide information on the vector correlations that might also exist (31, 38). This is often true even for long lived excited states because the excitation process is often rotational-state selective, so that the polarization information imprinted by the pump laser is not lost because of molecular rotation (31).

HYDROGEN FLUORIDE (HF) DIMER—A PROTOTYPICAL SYSTEM

Hydrogen fluoride-dimer has become a prototype in the study of hydrogen-bonded molecular complexes, as illustrated by the extensive experimental (39–45) and theoretical (46–52) work. Klemperer and coworkers carried out the first gas phase study of this complex using molecular beam electric resonance spectroscopy (53), from which they were able to obtain detailed structural information, including information on the associated tunneling dynamics. Subsequent microwave, near and far infrared, and overtone studies of several isotopomers of the dimer have mapped out much of the spectroscopy and have shown that the vibrational predissociation lifetime is strongly dependent upon the nature of the initial intramolecular vibrational mode that is excited (21, 42). In particular, excitation of the fundamental free HF vibration results in a vibrational predissociation lifetime (estimated from the line width), which is approximately 20 times longer than that resulting from excitation of the hydrogen bonded HF stretch. This behavior can be rationalized if one considers that the hydrogen-bonded HF stretch is much more strongly coupled to the intermolecular hydrogen-bonding coordinate than is the free HF stretch. The theoretical work includes studies of the potential energy surface (46–52), the lifetimes, and the state-to-state photodissociation dynamics (48, 54). Particularly noteworthy are the early calculations of the final state distributions (48) based upon a pseudo atom-diatom approximation, which predicted that the proton donor HF would be produced in the highest energetically accessible rotational state, consistent with an impulsive dissociation in which the proton donor "pushes off" its partner (in this, represented by a spherical atom). The result is a large torque on the proton donor molecule and hence a large amount of rotational excitation.

The first direct experimental measurement of the final state distributions for HF dimer were carried out in our laboratory using the PHOFAD method discussed above (31, 32, 55, 56). Angular distributions are shown in Figure 2, corresponding to excitation of different initial states. The corresponding fits to the data are shown as the solid lines, from which the populations of the various correlated final states are determined. These are shown in the figure as vertical bars. The shapes of the individual peaks in these distributions depend upon the experimental geometry and the anisotropy parameter (β), all of which are included in the fitting procedure.

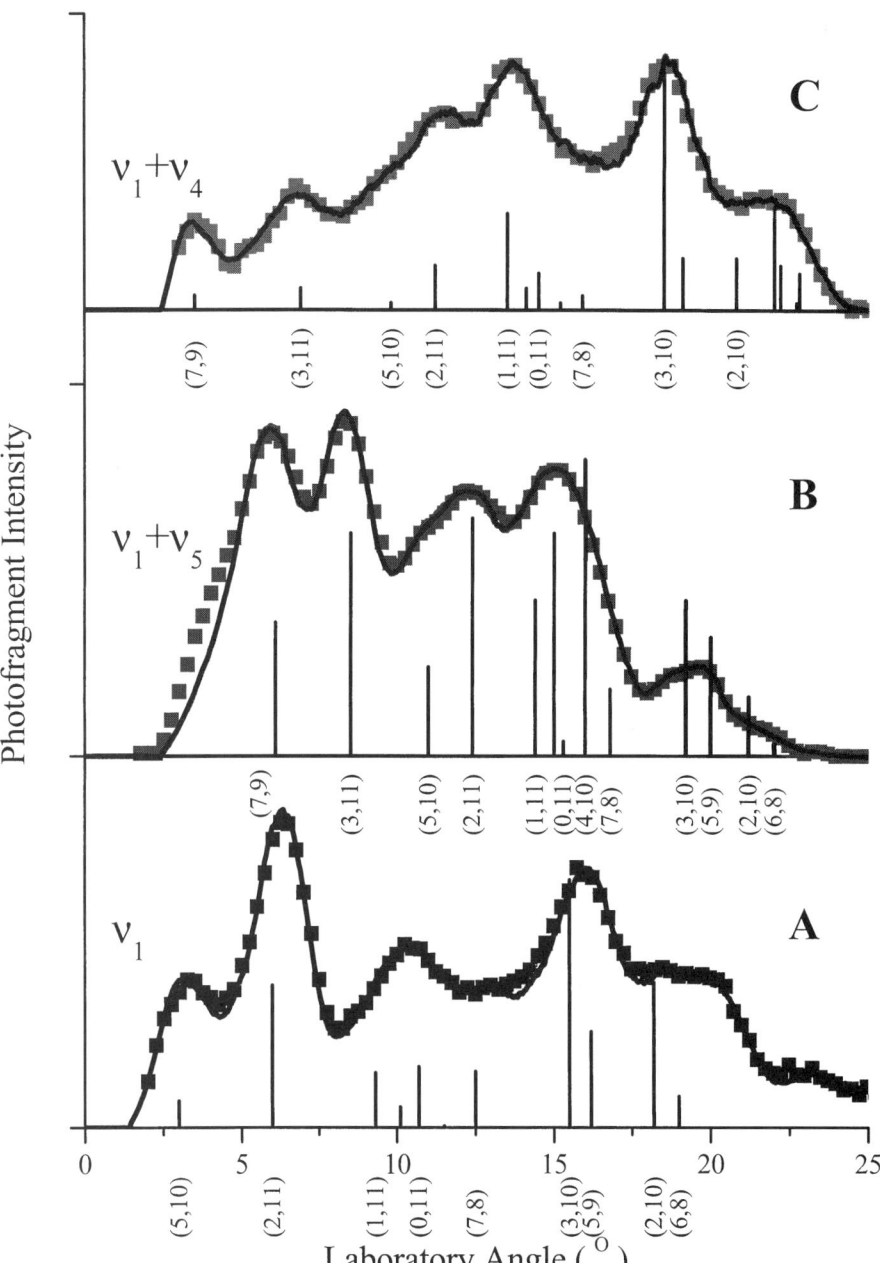

Figure 2 Three sets of PHOFAD data for HF dimer, corresponding to (A) excitation of the fundamental (free) H-F stretch, (B) a combination band that results in excitation of the bending vibration of the dimer in combination with the H-F stretch, and (C) a combination band involving the F-F stretch. Detailed analysis of the data shows that bending tends to reduce the high j–low j propensity, while the additional energy associated with the F-F stretch evolves directly into the translational energy of the recoiling fragments.

The anisotropy parameters obtained from this fitting procedure are generally consistent with an axial recoil process (31). As pointed out above, angular positions of the intensity maxima provide a measure of the translational recoil energy associated with a particular dissociation channel.

For HF dimer, the probabilities determined from these fits reveal a preference for the production of a highly rotationally excited fragment in coincidence with one in a low j state. In addition, correlated states having large translational energies (large scattering angles) are improbable, in agreement with energy (or momentum) gap arguments (57). This is in qualitative agreement with the simple triatomic theoretical model discussed above, the high j fragment resulting from the proton donor and the low j fragment from the acceptor. A more quantitative comparison is possible only if the rotational degrees of freedom of both fragments are accounted for in the theory. Zhang & Zhang (54) carried out such calculations on two potential surfaces (54, 58), the results being shown in Figure 3. The agreement with experiment is still only qualitative, suggesting that further improvements to the potential energy surface will be necessary, particularly in the regions that control the dissociation dynamics. In contrast with the spectroscopic data, which probes the region of the potential near the minimum, the dissociation dynamics tends to be more sensitive to the repulsive wall, where the anisotropy is greatest. This is important in this system since high rotational states of the proton donor fragment are populated. It is interesting to note that further refinements to the dimer surface have been made subsequent to these theoretical calculations (59), demonstrating the continued interest in this system. In fact, the newest potential has been adjusted to reproduce the dissociation of the HF dimer [1062 cm^{-1} (55)] determined from these PHOFAD experiments.

As noted above, weakly bound complexes afford us the opportunity of studying the initial state dependence of the dissociation dynamics. This point is illustrated by the three angular distributions shown in Figure 2, corresponding to excitation of the fundamental H-F stretch and two of the associated intermolecular combination bands. The upper distribution corresponds to excitation of the F-F intermolecular stretching mode in combination with the H-F stretch. It is clear from the figure that the maximum intensity in the angular distribution occurs at larger angles than in the angular distribution corresponding to the fundamental (bottom distribution in Figure 3), the implication being that the energy initially deposited in the F-F stretching coordinate results in higher average translational energy in the fragments. When the trans-bending vibration is excited, the states with lower translational energy are populated. Comparing this distribution with that of the fundamental, it is evident that the bending excitation shifts the population distribution more toward channels where both HF fragments are rotationally excited. The implication is that bending of the dimer gives a less asymmetric vibrational averaged geometry (the two HF molecules are somewhat more equivalent), such that both HF molecules can be rotationally excited. It is interesting to note that theoretical calculations for these excited intermolecular vibrations give results that are in qualitative agreement with these experiments (60).

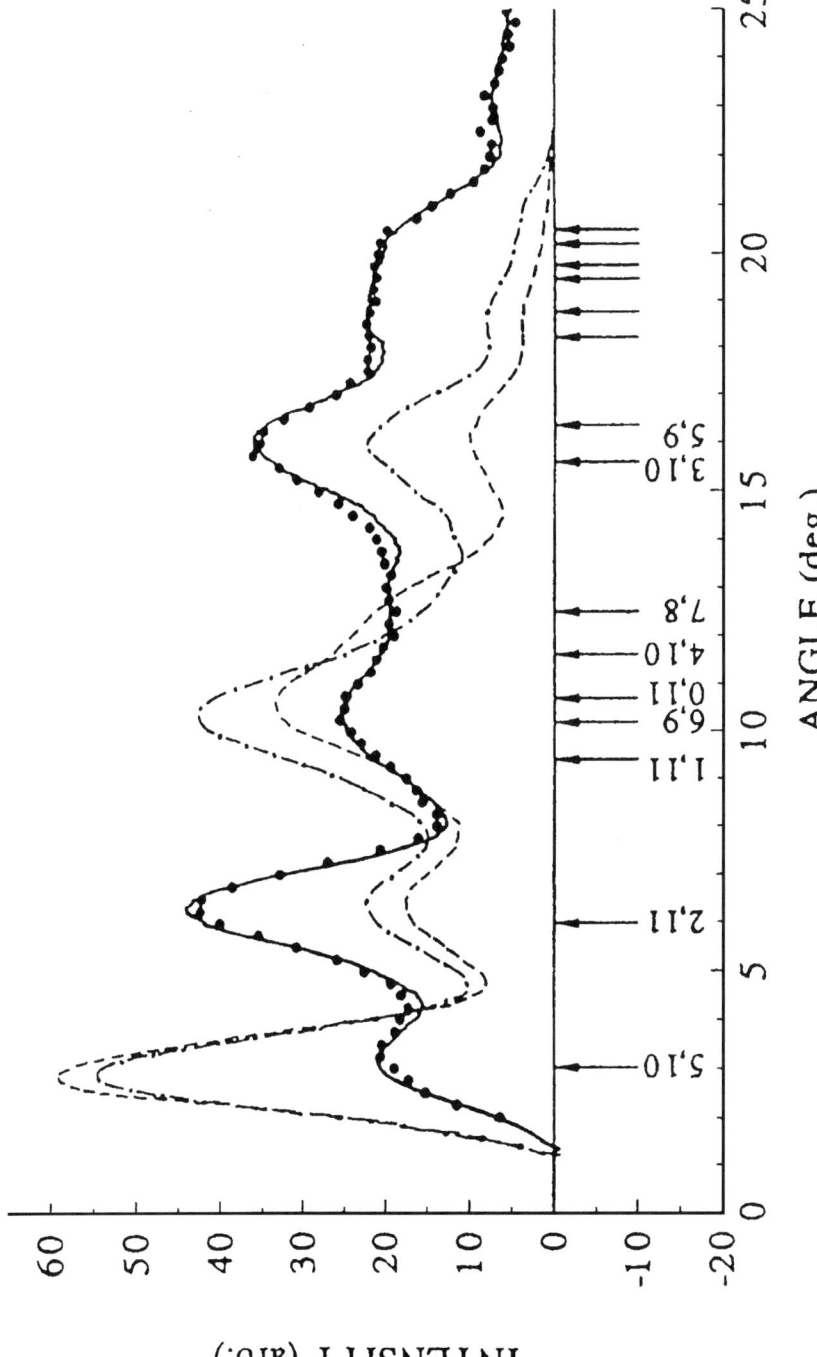

Figure 3 Comparisons between the experimental PHOFAD data and the theoretical calculations of Zhang & Zhang (100) for HF dimer.

Isotopic substitution can be used to distinguish between the two fragments, making the DF-HF and HF-DF systems of interest. The spectroscopy of both of these isomers has been reported (61, 62), consistent with the large barrier between them. In this case, the PHOFAD method has been used in combination with the pendular state orientation method to separately detect the two fragments (63), given that impulsive dissociation leads to the HF and DF fragments recoiling in opposite directions in the laboratory frame. The resulting angular distributions for both isomers are shown in Figure 4. For DF-HF the dissociation results in the preferential population of highly rotationally excited states of the HF fragment, the dominant channel being $j_{DF}, j_{HF} = 2, 11$. This is clearly in good agreement with the impulsive, nonstatistical dynamics discussed above, which favors rotational excitation of the proton donor molecule. Note that the signals on the HF side of the angular distribution are larger than on the DF side, owing to the fact that the bolometer is an energy detector and is thus more sensitive to the highly rotationally excited HF.

The surprise comes with the HF-DF angular distribution shown in Figure 4b. From the impulsive model we would predict that the DF fragment would be highly rotationally excited in this case. However, we again see that the intensity of the angular distribution is higher on the HF side, indicating that the HF carries away more internal energy. In considering possible reasons why the impulsive model fails in this case, we note that disposing of the excess energy in the rotational degree of freedom of the DF fragment is more difficult than in HF, owing to the relatively smaller rotational constant of DF. For this reason, higher j states of the DF fragment must be excited in order to carry away the same amount of energy. We should therefore consider mechanisms that would allow this to occur. An interesting possibility is illustrated schematically in Figure 5. The complex is excited to a state $HF(v = 1)$-$DF(v = 0)$ by the infrared laser. The corresponding ground intermolecular vibrational state is seen to be embedded in the well of the $HF(v = 0)$-$DF(v = 1)$ state. Although intramolecular vibrational redistribution (IVR) into this state cannot lead to dissociation directly (the dashed line shows that this channel is closed), the high degree of intermolecular vibrational excitation from such a process would result in the complex no longer having a vibrationally averaged structure close to that of the equilibrium geometry. Therefore, dissociation from this state, which would involve quenching the $DF(v = 1)$ level, might very well occur from a geometry that would favor the production of rotationally excited HF. This indirect dissociation mechanism has been seen previously in other systems (64) but remains to be confirmed for HF-DF.

Figure 4 PHOFAD data for HF-DF and DF-HF. The latter dissociates with a high j–low j propensity that is consistent with the proton donor–proton acceptor model discussed in the text. The final state distribution of the HF-DF isomer suggests that the dissociation is indirect in this case. A large DC electric field was used in this case to orient the complexes, resulting in the two fragments recoiling in opposite directions in the laboratory frame, shown here as positive and negative angles.

PHOTOFRAGMENT TRANSLATIONAL SPECTROSCOPY 617

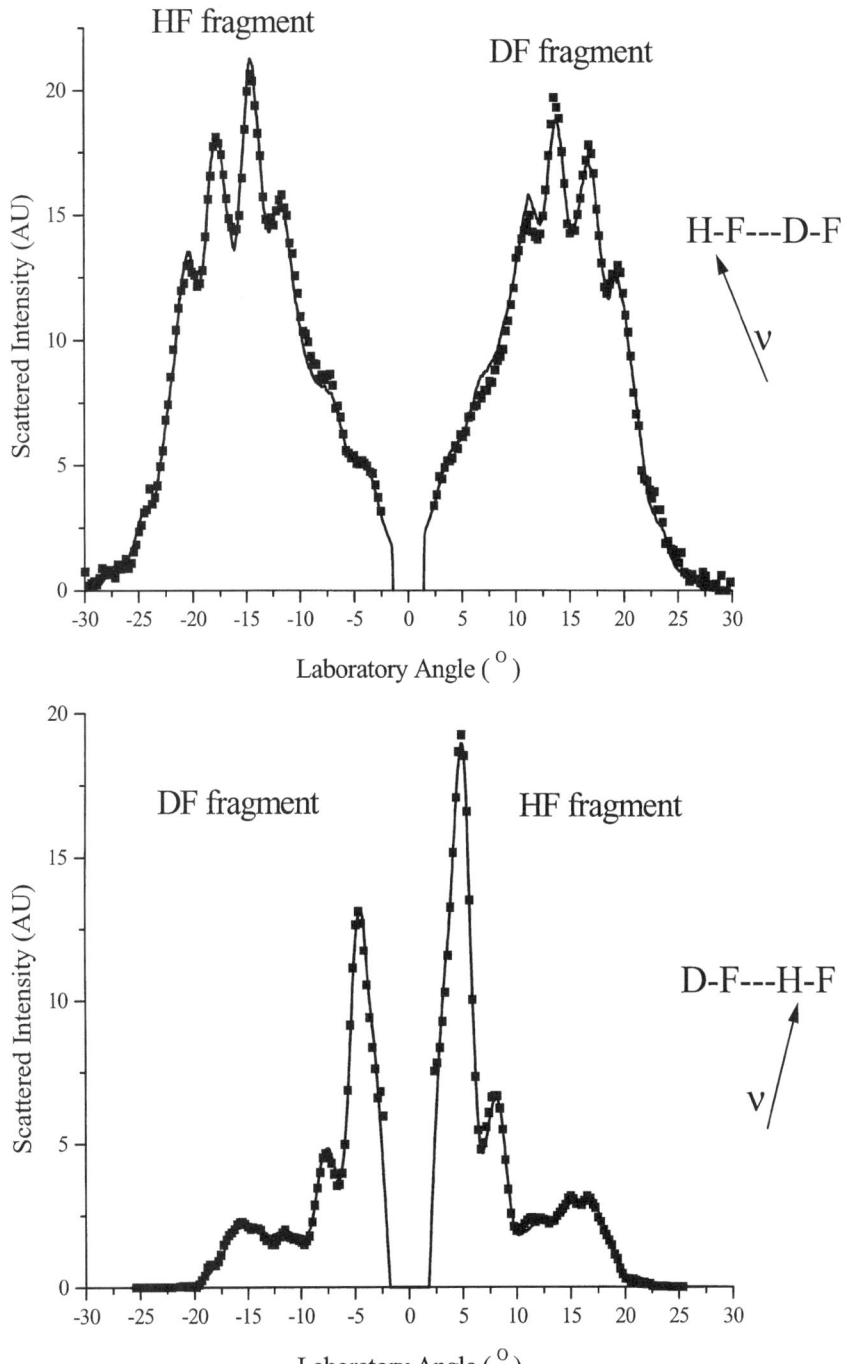

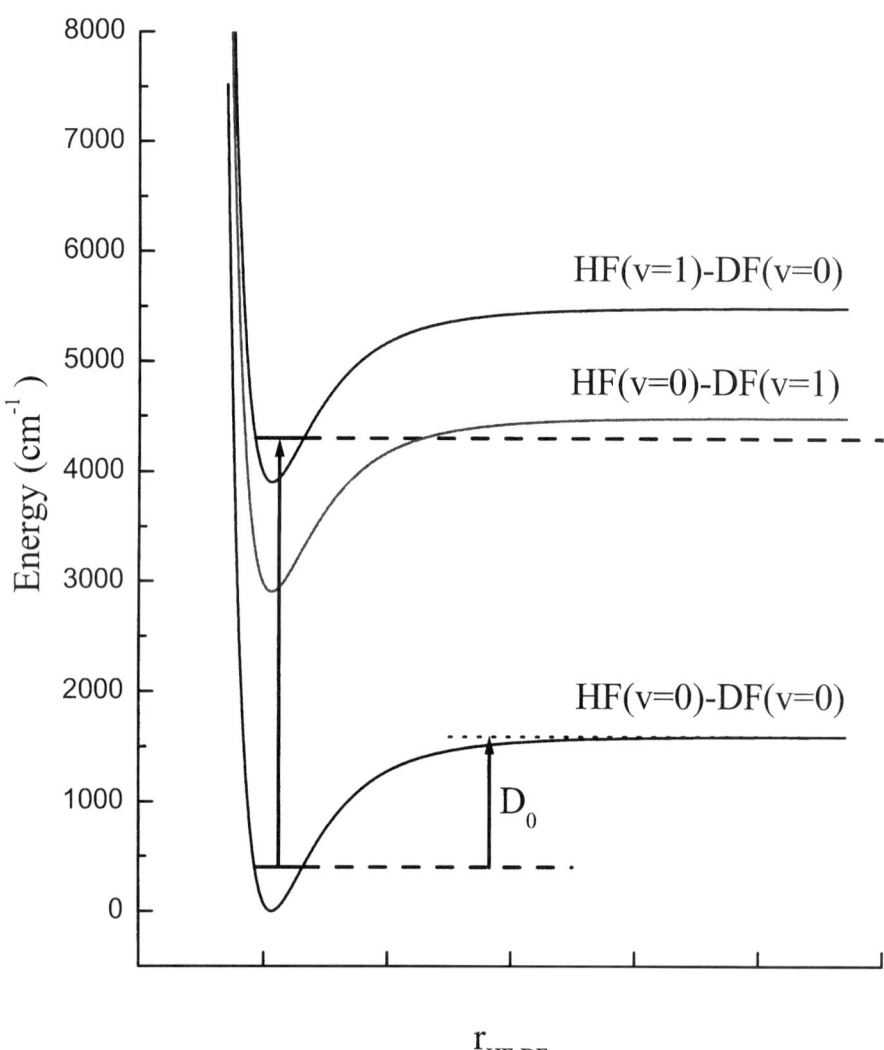

Figure 5 A schematic diagram showing a possible indirect dissociation pathway for the HF-DF complex that could account for the fact that the DF fragment is not preferentially populated in high j states.

In summary, the HF dimer displays highly nonstatistical dynamics, the final state distributions reflecting the impulsive nature of the photodissociation process and the structure of the parent complex. The main features of these final state distributions are attributed to the asymmetric equilibrium structure, corresponding to one fragment correlated with the proton donor in the complex and the other to the proton acceptor. In HF-DF there is evidence of a more complex, IVR-mediated dissociation process.

HCL DIMER—A CONTRASTING CASE

The HCl dimer has also been studied extensively, using a number of experimental (65–70) and theoretical (71–74) approaches. Although the equilibrium structure is similar to that of the HF dimer, corresponding to a proton donor/proton acceptor geometry, this is a floppier system, with the HCl monomer units undergoing much wider amplitude-bending motion in the ground vibrational state. In addition, the vibrational dependence of the predissociation rate is weaker than in HF dimer, and the lifetimes are approximately two orders of magnitude longer. Presumably this should also be reflected in the correlated rotational distributions, which have been measured using POSTS (75). Figure 6 shows a POSTS scan corresponding to probing the $j = 14$ HCl state. The position of the peaks in this distribution give the translational energies of the corresponding dissociation channels, which can be assigned using the energy level diagram given in the figure. Data of this type has been obtained for both the free and hydrogen bonded stretches of the dimer. For this system they are essentially the same (75). As with the HF dimer, there is a strong propensity in HCl dimer for the production of channels with small translational energies. However, in this case the data is quite accurately represented by a linear surprisal function (75), suggesting that this single dynamical constraint of low translational energies is sufficient to describe the dynamics of the system. This is consistent with the fact that the HCl dimer is less rigid, so that it samples a wider range of geometries than does the HF dimer. The situation is not unlike that encountered in the HF-DF system discussed above, where dissociation is also thought to occur from a state that is much less rigid than the ground state.

D_2-HF AND H_2-HF—INTERMOLECULAR ENERGY TRANSFER

In all of the systems discussed so far, the only open vibrational channel corresponds to both molecular fragments being in the ground vibrational state. We now consider a case where the amount of excess energy in the vibrational predissociation process is sufficient to populate at least one vibrationally excited state. For this we consider the D_2-HF system, where estimates of the dissociation energy from theory (76, 77) indicated that both the D_2 $v = 0$ and $v = 1$ channels are energetically accessible. The first evidence that the vibrational excited state channel was important came from the spectroscopic work of Lovejoy et al (78) (exciting the H-F stretching vibration). This study showed that the vibrational predissociation lifetime of this system is considerably shorter than for H_2-HF (79), which was interpreted as

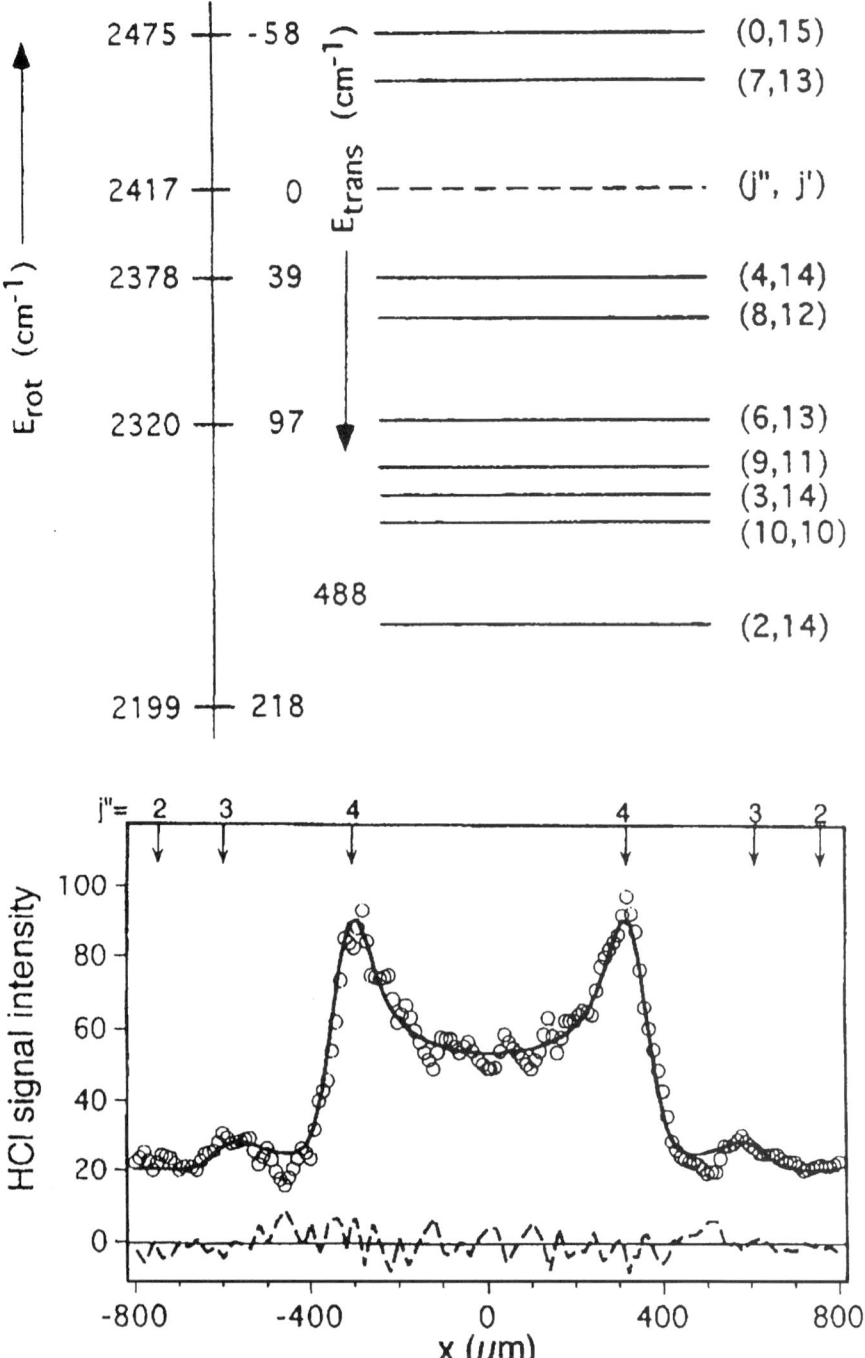

Figure 6 A POSTS scan of HCl (v = 0, j = 14) from vibrational predissociation of HCl dimer. The energy level diagram shows the channels that are responsible for the structure in the data.

resulting from the fact that it is easier to populate the v = 1 state of the D_2 fragment than to produce highly rotationally excited fragments.

The D_2-HF is also of interest since it is one of the few systems for which final state distributions have been calculated from first principles (77). The lifetimes obtained from these calculations, which assume that the V-V channel dominates the dissociation dynamics, are in excellent agreement with experiment (77, 78). We have obtained PHOFAD results for this system (80) so that direct comparisons can be made between experiment and theory (81). Although some minor quantitative differences suggest that the potential energy surface and/or the V-V coupling may still need some modification, the agreement is remarkably good and conclusively shows that the V-V mechanism is the most important. Indeed, the electrostatic quadrupole-dipole vibrational coupling used in the theory (76, 77) must be primarily responsible for the V-V process. Part of the reason for the good agreement in this system is undoubtedly the fact that the available energy, following excitation of the D_2 vibration, is rather small. The correspondingly lower rotational states populated in this system are easily accessible because they do not require such high-order terms in the anisotropy of the potential surface (as in the case of HF dimer). Therefore, dissociation may very well occur from regions of the potential that were rather well sampled by the ab initio calculations (76, 77).

The situation is qualitatively different for the H_2-HF complex. Here the experimental results are in stark disagreement with the calculations of Zhang et al (82), based on the same potential surface as used for D_2-HF. In particular, the calculations show significant probability for populated low j states of the fragments, with correspondingly high kinetic energy. In contrast, the experimental results indicate that the average recoil kinetic energy is once again rather small (80). Interestingly, the calculated lifetime is much longer than that observed experimentally (83). The explanation for these discrepancies, in light of the good agreement for D_2-HF, involves the fact that the V-V channel is no longer open for H_2-HF, as shown in Figure 7. As a result, dissociation must proceed via a V-R,T process. Since it is difficult to dispose of the energy in translation (energy gap considerations), the system will clearly prefer to populate high rotational states. However, this requires the appropriate terms in the anisotropy of the potential, which may very well be missing in the ab initio calculations used in these calculations, given that the repulsive region was only an extrapolation from the points in the attractive region (76). These systems clearly demonstrate how the combination of PHOFAD data and dynamical calculations can help to pinpoint regions of the potential energy surfaces that still need attention.

HF-HCL AND HCL-HF

The mixed HCl and HF binary complexes are also capable of dissociating via an intermolecular V-V energy transfer process. The question here is whether both isomers, namely the HF acceptor and HF donor complexes, behave in the same

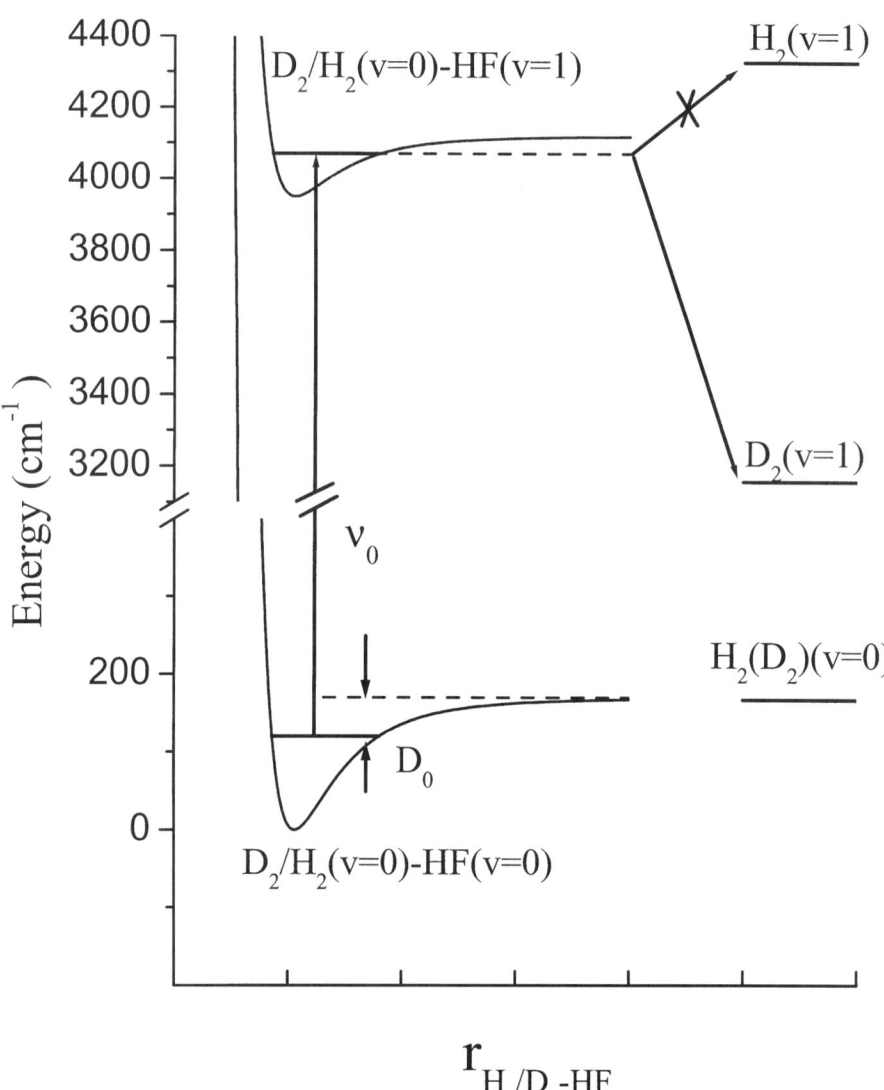

Figure 7 A schematic energy level diagram showing that the V-V channel is open in D_2-HF (data shows this channel dominates the dynamics) and closed for H_2-HF (where V-R channels are important).

way. The spectroscopy of both isomers was first observed by Pine & Fraser (84). PHOFAD data was since obtained in our laboratory, making use of both the pendular state method to separate the two fragments and a second laser to probe the internal states of the HF fragment (85). These results show conclusively that vibrational excitation of the H-F stretch in the HF-HCl complex resulted in the production of the HCl fragment in the v = 1 state. This leaves very little excess energy to be disposed of in the rotational and translational degrees of freedom, which results in a fairly wide range of low j states of both fragments being populated. Although a complete analysis of the HCl-HF isomer was not possible, the data clearly showed that in this case the HCl was produced in the ground vibrational state. In light of the results for the other complexes we have discussed so far, this behavior can be understood. In particular, when the HF is in the proton acceptor position it is difficult to torque and thus, high j states needed for dissociation into the ground vibrational state HCl are improbable. Of course it is in principle possible to dispose of all of this excess energy in the HCl rotational degree of freedom, given that it is in the donor position. However, its relatively small rotational constant, compared to HF, means that much higher j states would have to be populated. Presumably, since we do not observe such channels, this is more difficult than transferring the majority of the excess energy to the HCl vibrational degree of freedom. When the HF is in the donor position, it is easier to excite the rotations of this fragment, and dissociation proceeds via the v = 0 HCl channel. Here again, therefore, the dissociation dynamics is consistent with our intuitive ideas concerning how the molecules push apart. Of course, this is only a qualitative explanation; a full quantitative understanding of these processes will require a detailed theoretical study of the associated potential surfaces, including the coupling between the various intramolecular degrees of freedom of the system.

N_2-HF, OC-HF, AND NO-HF–V-V ENERGY TRANSFER

A number of other diatomic-diatomic systems have been studied, where more than one photofragment vibrational channel is open. For example, consider the NO-HF complex, which has been studied using translational spectroscopy, based upon the measurement of the Doppler profiles associated with the NO fragment (38). In this case, pulsed infrared and UV lasers were used for the pump and probe (LIF) steps, respectively. Figure 8 shows Doppler profiles for two NO final states, the fit to the data corresponding to a delta function translational recoil distribution. No structure is observed in the Doppler profile since only a single HF rotational state is produced in coincidence with these NO states. To the authors' surprise, the results showed that both the v = 0 and v = 1 states of the NO fragment were populated, following vibrational predissociation from the H-F stretch excited state. The reason this is so surprising is that a zero order model for the dissociation, based upon the initial state coupling directly to these two independent final states [namely NO(v = 0)

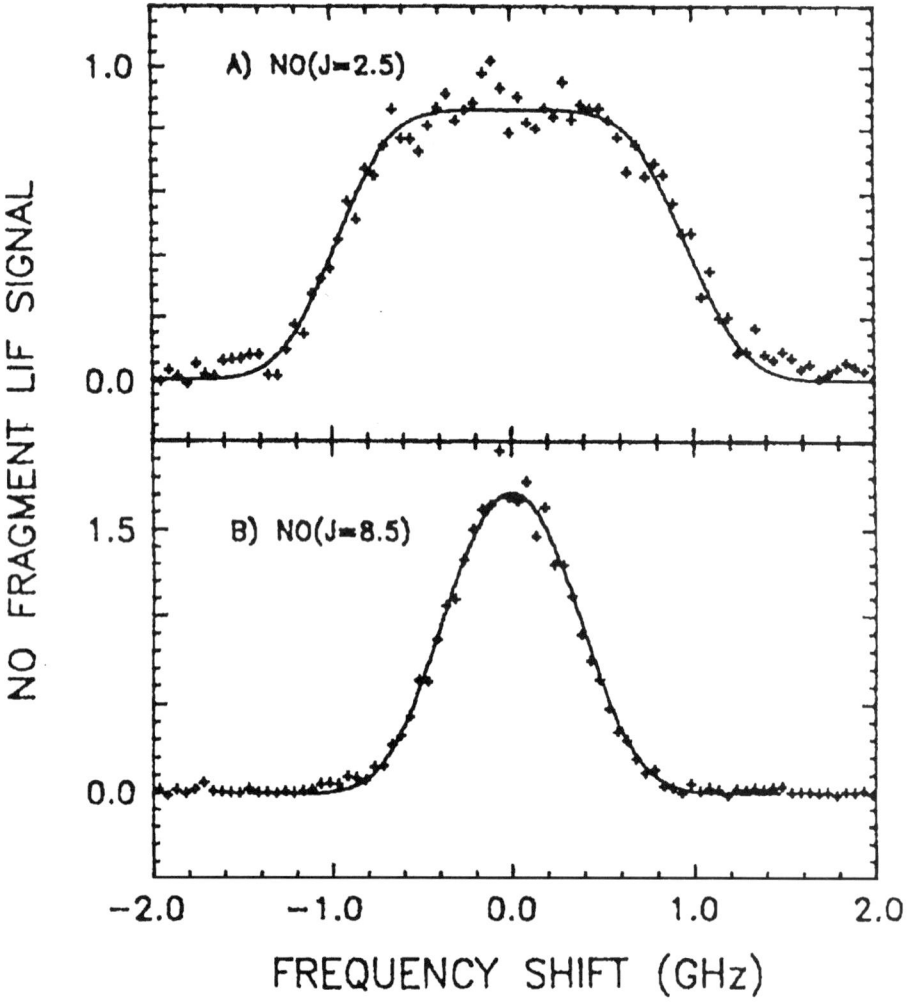

Figure 8 Doppler profiles for two rotational states of the NO fragment produced upon vibrational predissociation of the NO-HF complex (excitation of the H-F stretch). Multiple peaks are not observed in these spectra since only one HF rotational state is produced in coincidence with these NO states.

and NO(v = 1)], gives very different rates for the two channels. Indeed, Shorter et al (38) have applied the ideas of Ewing (namely that the rate is determined by the sum of the quantum number changes in the dissociation process) to show that the rates should be very different if the coupling is direct (85a). Conservation of energy insists that the NO(v = 0) fragment be produced in coincidence with a high j state of the HF fragment. Thus, the HF must experience a large torque in this case, making this NO(v = 0) channel very sensitive to the anisotropy of the potential (similar to HF dimer H_2-HF). On the other hand, the V-V channel, which populates the NO(v = 1) state, depends upon the coupling between the two intramolecular vibrational coordinates (as in D_2-HF). Again, it seems unlikely that such qualitatively different coupling mechanisms would give rise to the same rates. One possibility is that the coupling is not direct, but rather is mediated by some form of anharmonic resonance involving the low frequency vibrational modes of the complex. If this were the case, however, one might expect that the behavior would be specific to this complex and that similar systems would not have this resonance and thus would behave very differently. The Doppler results are unable to differentiate between two possible assignments of the HF rotational states produced in coincidence with the NO. One possibility is NO (v = 0) + HF(j = 12) and NO(v = 1) + HF(j = 8) and the other is NO(v = 0) + HF(j = 9) and NO(v = 1) + HF(j = 2). The corresponding dissociation energies for these two assignments are 448 cm^{-1} and 1769 cm^{-1}, respectively (38).

For comparison, we now consider the N_2-HF complex, for which PHOFAD experiments have been successfully performed (36, 86). Compared with the lighter systems discussed previously, the photofragment shells are much closer together, making them more difficult to resolve. This problem can be overcome, however, by extending the PHOFAD method to include a probe laser, so that the angular distributions for the various HF rotational states can be recorded separately (33). When combined with the pendular state method (35), used to orient the parent complex in order to separate the two fragments, it is possible to get a complete assignment of the correlated final state distribution. In fact, the N_2-HF complex was the first system for which this new experimental approach was applied. In this case, the rotational states of the HF fragment were probed by a second infrared laser system, which excited the fragments between the photolysis volume and the bolometer detector (42). Here again, the experiments revealed that the N_2(v = 0) and N_2(v = 1) were populated nearly equally. The vibrational state of the N_2 fragment was determined by conservation of energy, and only the HF j = 12 and j = 7 were populated, which could only occur in coincidence with N_2(v = 0) and N_2(v = 1), respectively (33). The structure in the angular distributions could then be used to determine the rotational state distribution of the N_2 fragment (36). The dissociation energy for this complex was determined to be 398 cm^{-1} (36).

PHOFAD experiments have also been carried out for the OC-HF complex (87), where again the two open CO vibrational channels are observed with nearly equal probability. The higher binding energy of this complex reduces the available energy, such that the CO(v = 0) channel is produced in coincidence with HF(j = 11), while for the CO(v = 1) channel the HF is produced in j = 6, 5, and 4. The fact that the latter is not as selective in the HF rotational state is thought to result simply from the fact that the separation between these HF rotational states is smaller than in the N_2(v = 1)/HF(j = 7) case. As a result, the translational degree of freedom is able to take up the energy difference resulting from population of the lower rotational states. The dissociation energy determined from these PHOFAD results is 732 cm^{-1}.

In light of the N_2-HF and OC-HF results, we return to the fact that the Doppler experiments on NO-HF were consistent with two different dissociation energies. In one case (448 cm^{-1}), there is nothing particularly special about the open shell system (a dissociation energy midway between the other two), whereas for the other (1769 cm^{-1}), the NO-HF is much more strongly bound than the other two. Although the latter would be very interesting, suggesting that the open shell nature of the complex results in a much stronger bond, we feel that this assignment is unlikely. Indeed, with such a large binding energy, the HF would be produced in low j states. The OC-HF results show what we expect in this case, namely that when the rotational states get close together the dissociation is much less j state selective. The NO Doppler profiles clearly show that only one HF rotational state is populated. We thus conclude that there is nothing particularly special about the bonding in the NO-HF system and that it is intermediate in strength between N_2-HF and OC-HF. The fact that both of the open vibrational channels are observed in all three of these systems must mean that the mechanism responsible for this behavior is not dependent upon chance resonances that would certainly change from one system to another. Unfortunately, detailed theoretical studies of the final state distributions in these systems are still lacking. Given the advances that have occurred in the theory of multidimensional vibrational dynamics, these systems are certainly within the reach of current methods and would be worthy of such study.

NO DIMER

Doppler methods have been used to study the correlated final state distributions for vibrational predissociation of fundamental and overtone excited NO dimer (88, 89). In both cases one observes that the vast majority of the available energy appears in translation of the two NO fragments, suggesting that the vibrational predissociation dynamics of this system is very different from the ones discussed so far. The explanation for this comes from the fact that there are a total of 16 potential energy surfaces for the $^2\Pi$ configurations of the NO fragments, combining in the C_{2v} symmetry of the dimer. It therefore seems reasonable to turn to nonadiabatic processes to explain these results (89). Further evidence for nonadiabatic processes comes from the observation of a nonstatistical distribution in the spin-orbit channels (89). A barrier for the reverse association reaction could therefore account for the large kinetic energy release on this system. It is important to point out that the small rotational constant associated with NO (relative to HF) makes it much more difficult to dispose of large amounts of excess energy in rotation. Thus, it is perhaps not surprising to find that these nonadiabatic effects are important in channeling large amounts of excess energy into translation.

CO_2-HF—SO MANY CHOICES

We now turn our attention to the more complex vibrational dynamics associated with binary complexes containing at least one polyatomic molecule and having several vibrational degrees of freedom. A simple extrapolation of the results discussed above might lead to the conclusion that all open vibrational channels would be important in the vibrational predissociation dynamics of these systems. As shown schematically in Figure 9, the available energy resulting from H-F stretch excitation in CO_2-HF is sufficient to give rise to 18 open vibrational channels (90). If all of these states were populated with appreciable probabilities, the corresponding angular distributions in a PHOFAD experiment would show no discernable structure, particularly once the numerous rotational channels are included. The fact that the PHOFAD results are actually quite structured (33, 90) indicates that the dissociation process is much more selective than this. By orienting the complexes using the pendular state method, the rotational and vibrational states of the CO_2

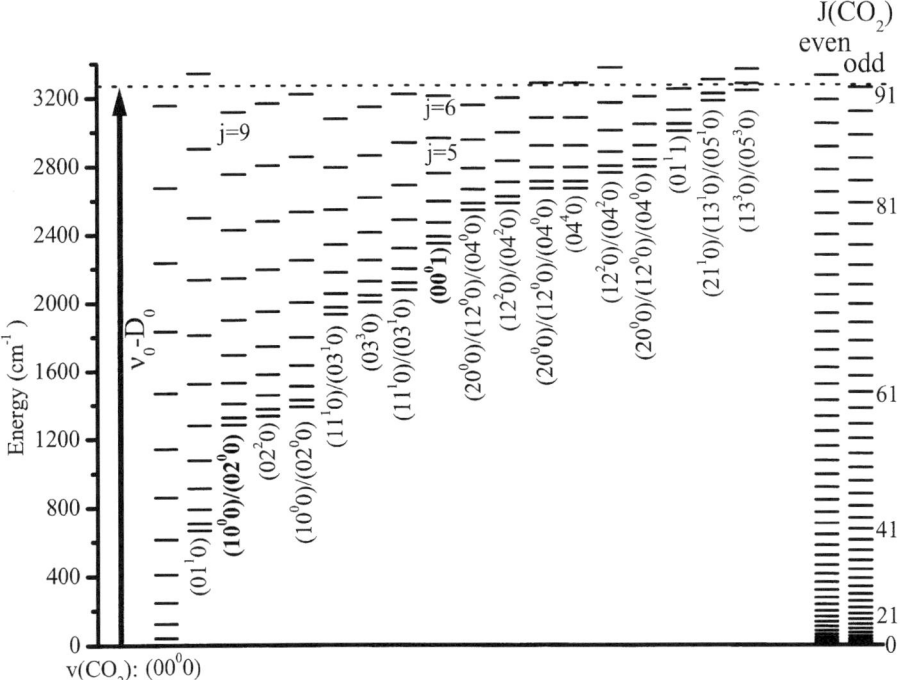

Figure 9 A schematic energy level diagram showing all of the open rotational-vibrational channels in the CO_2-HF system (corresponding to excitation of the H-F stretch). Vibrational predissociation is very selective in this case, populating primarily the asymmetric stretch of the CO_2 fragment, with a smaller amount in the symmetric stretch.

cofragment were determined using the probe laser. These results show that the dominant feature in the angular distribution comes from a channel involving the population of the j = 6 state of the HF fragment, produced in coincidence with the (00°1) vibrational state (the asymmetric stretch) of CO_2. Note that the small rotational constant of the CO_2 fragment ensures that the amount of energy that can be accommodated in this rotation degree of freedom is small. A second channel is also observed, corresponding to populating the (10°0) (the symmetric stretch, or more rigorously the lower diad of the corresponding Fermi diad) (90). Thus, instead of seeing all of the open vibrational channels in this system, we observe only two. Although both of these involve transferring of vibrational energy from one monomer unit (HF) to the other (CO_2) (intermolecular V-V energy transfer), the coupling mechanisms are presumably very different for these channels. Indeed, the asymmetric stretch might be accessed by strong transition dipole–transition dipole coupling, while this term is clearly unimportant in the case of the symmetric stretch. This further illustrates that a general and quantitative understanding of the coupling mechanisms in these systems is still lacking.

ACETYLENE-HF AND HCN-HF—MODE SPECIFIC EFFECTS

In this section we discuss binary complexes for which it has been possible to excite initial vibrational states associated with both molecules and to determine the corresponding final state distributions. Consider for example the HCN-HF complex, a system that has been studied extensively by infrared spectroscopy (91). The C-H and H-F stretches of this complex are easily accessible with existing lasers. PHOFAD experiments have been carried out for both of these excited states (92), providing an accurate value for the dissociation energy of the complex, namely 1970 cm^{-1}. This large dissociation energy is generally consistent with the available ab initio calculations for this system. The large binding energy in this system greatly reduces the available energy, compared to the previous systems. This alone has an important effect on the dissociation dynamics. The most surprising finding in this system is that the final state distributions resulting from excitation of the C-H and H-F stretches of the complex are very similar. In both cases, the majority of the excess energy appears in HF rotation, while a significant amount is also observed in translation (40%). The only real difference between the two distributions is that, because of the lower excess energy associated with the C-H stretch, the corresponding HF rotational distribution is shifted one unit lower than upon H-F stretch excitation. It is also surprising that the bending excited states of the HCN fragment are not populated, even though these channels are open. As seen above for CO_2-HF, the HF rotational distribution is not peaked in a single state, presumably because the available energy is lower than for some of the other systems discussed. In addition, the highly anisotropic nature of the intermolecular potential surface in HCN-HF presumably makes it easier for this system to generate angular momentum in the HF fragment. This may be the reason that the excited vibrational states of the HCN fragment are not populated.

Acetylene-HF is another system for which PHOFAD results have been obtained corresponding to excitation of both molecules (93). In this case both the asymmetric stretch of the acetylene and the HF stretch can be pumped with the existing lasers. In addition, a probe laser can be used to obtain final state distributions for both fragments, as illustrated in Figure 10. This complex is more weakly bound (1088 cm^{-1}) than HCN-HF, making the available energy considerably larger. For this system, excitation of the H-F stretch results primarily in the production of the HF fragment in $j = 11$, with a small amount in $j = 10$. These states obviously correlate with the ground vibrational state of the acetylene fragment, as confirmed from the corresponding probe laser experiments. However, in this system the channel corresponding to binding excitation of the acetylene fragment is also observed. This appears to be consistent with the fact that an impulsive dissociation could naturally result in coupling to the bend in acetylene, since the complex is T-shaped, while for the linear HCN-HF complex bending would be unlikely. Here again, we find that the (heavy) acetylene fragment is only in low j states, due to its small rotational constant (93).

Excitation of the asymmetric stretch of acetylene leads to a very different outcome. Now most of the excess energy remains with the acetylene fragment, corresponding to excitation of the intramolecular C-C stretching vibration. The corresponding HF fragment is produced primarily in $j = 2$. Despite the fact that the density of photofragment channels is quite high in this system, the dissociation dynamics is still highly nonstatistical and mode dependent. Given the weakness of the coupling in these systems, this behavior could very well persist to much larger binary complexes.

Spectroscopic (20) and laser probing (94) experiments have been reported for C_2H_2-HCl, which suggest that intermolecular V-V energy transfer is important in this system. In particular, it was suggested that excitation of the asymmetric stretch of the acetylene resulted in a near resonant energy transfer to the HCl fragment during the predissociation process. More recent PHOFAD experiments have nevertheless shown that the dissociation of this system is quite similar to that of acetylene-HF (95), resulting in the excitation of the C-C stretch in the acetylene fragment. In fact, the dissociation energy determined from this study suggests that the intermolecular V-V channel, corresponding to the HCl being vibrationally excited, is actually closed.

PHOTODISSOCIATION OF CYCLIC $(HF)_n$

All of the above discussions have focused on the vibrational predissociation dynamics of binary complexes, where the available methods are clearly capable of providing a rather complete characterization of the final state distributions. In

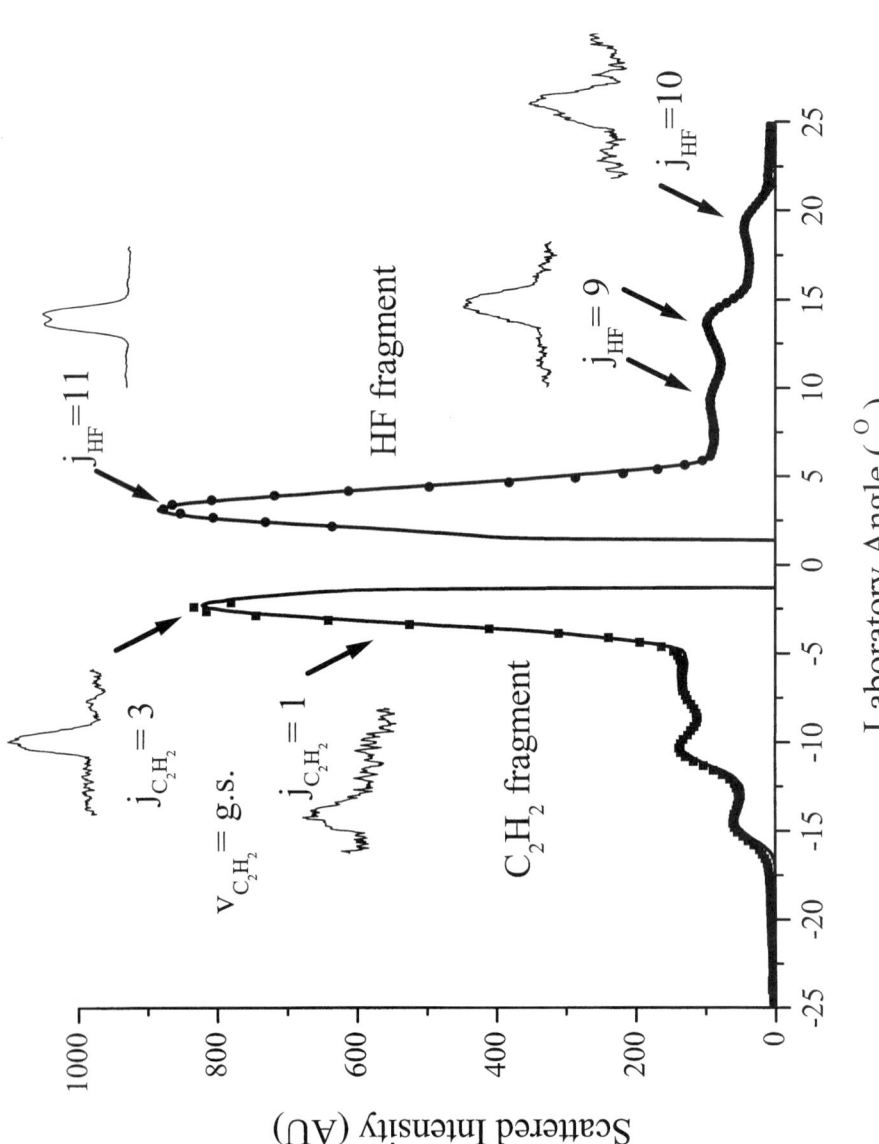

contrast, the study of vibrationally induced dissociation in larger clusters is still in its infancy. In the large cluster limit, it is generally assumed that the excitation energy will be quickly thermalized, leading to a process that is best described as laser-induced evaporation (96). Given the highly nonstatistical nature of the dynamics in binary complexes, it is interesting to consider at what cluster size the transition to thermal behavior occurs. Higher order clusters clearly present a formidable challenge for studies at the state-to-state level, given that the density of photofragment channels is much higher than in the systems we have discussed thus far. We might therefore anticipate that somewhat less refined data will be available for these systems.

Consider for a moment the vibrational excitation of a molecule within a cluster of moderate size. Assuming the energy is quickly randomized in the thermal bath of the cluster, the net result of this specific vibrational excitation is simply to increase the temperature of the cluster. If this temperature jump is sufficient to make the cluster metastable, monomers will begin to evaporate. This process will cool the cluster, thus decreasing the subsequent evaporation rate (96). Since the photon energy is quickly redistributed among the many degrees of freedom of the complex, one expects that the kinetic energy of the evaporating molecules will be quite low, in comparison with the corresponding binary complex. Figure 11 shows an example of a PHOFAD data set for the HF pentamer (97). In contrast with the previous data sets, this one shows no discernable structure, presumably because of the large density of states of the cofragment. It is important to point out that detailed theoretical calculations on these higher order complexes show that the energy associated with excitation of the H-F stretch is sufficient to evaporate only a single HF monomer unit (98). As a result, the cofragment in this case is an HF tetramer, which clearly has a very high density of vibrational states. The resulting smooth angular distribution provides only an indication that the recoil energy of the evaporating HF monomer is quite small.

Further progress is possible, however, by making use of a probe laser in the PHOFAD experiments to measure the rotational distribution of the HF fragment. Measurements of this type have been reported for the pentamer, hexamer, and heptamer (97), revealing that the rotational state distribution of the HF monomer is essentially Boltzmann, corresponding to a rotational temperature of approximately 180 K. Fits to the angular distribution suggest that the translational distribution is characteristic of a similar temperature. These results are clearly consistent with the vibrational dynamics of these larger clusters being described as a laser-induced evaporation process.

Figure 10 PHOFAD data for the acetylene-HF complex, corresponding to excitation of the H-F stretch. The insets show probe laser spectra used to determine the internal state distributions of the two fragments. In this case, a large electric field was used to orient the complex in order to separate the two fragments as shown.

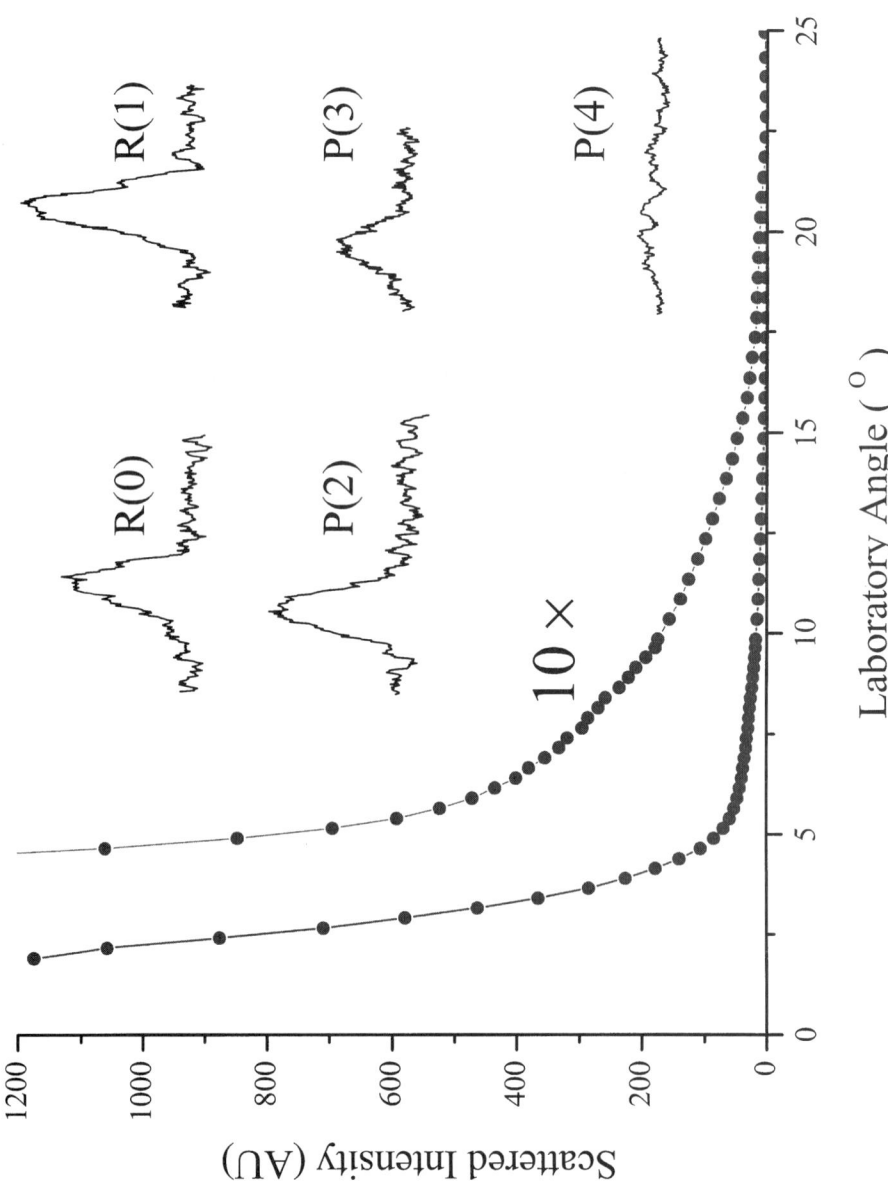

TABLE 1 A summary of the dissociation energies determined from translational spectroscopy

Complex	Experimental D_0 (cm^{-1})
HF Dimer	1062(2) (55)
NO Dimer	710(40) (99)
HF-DF	1157(2) (63)
DF-HF	1082(2) (63)
HF-HCl	642(2) (85)
C_2H_2-HF	1088(2) (93)
C_2H_2-HCl	830(6) (95)
(ortho)D_2-HF	51(2) (80)
(para)D_2-HF	68(2) (80)
N_2-HF	398(2) (36)
OCO-HF	672(4) (90)
HCN-HF	1970(10) (92)
HCl Dimer	439(2) (75)
NO-HF	448 (38)
OC-HF	732(2) (87)
(HF)$_5$	<2941 (97)
(HF)$_6$	<2854 (97)

DISSOCIATION ENERGIES FOR BINARY COMPLEXES

As noted previously, dissociation energies for weakly bound complexes are difficult to obtain directly from the corresponding spectroscopic results. At best, such estimates depend upon extrapolating from the bond states of the system using some form of model. Fortunately, the correlated photofragment studies reviewed here provide this information, directly from conservation of energy, requiring no modeling of the potential energy surface. Table 1 summarizes the dissociation energies (D_0) obtained from the data discussed above. The list is now quite extensive,

Figure 11 A PHOFAD data set for the HF pentamer, which dissociates by evaporating a single HF monomer, leaving an HF tetramer cofragment. The high density of states associated with the latter prevents (us) from resolving structure in the angular distributions. The insets show probe laser spectra of the HF fragment, which is produced in low j states with a near Boltzmann distribution in rotational temperature.

providing a database for testing available and future potential energy surfaces and ab initio calculations. To date, with the exception of the HF dimer and D_2/H_2-HF, the comparisons with the latter have been qualitative at best. In general, D_0 calculations are difficult, not only because they require large basis sets and high-level methods, but because they also depend upon the vibrational frequency calculations in order to include the zero point energy effects. Since in most cases these are done at the harmonic level, the comparisons are always qualitative at best. Once again, the exceptions are HF dimer and H_2/D_2-HF, where the vibrations have been treated in the full dimensionality of the problem on a realistic potential energy surface, thus providing a fully anharmonic estimate of the zero point energy. The data in Table 1 clearly provides benchmarks that theories can be tested against.

SUMMARY

We have reviewed the state-to-state dissociation dynamics of a wide range of binary and higher order complexes. The focus of this paper has been on systems that dissociate into two molecular fragments, where the correlations between the internal states of the two fragments are needed to fully describe the final state of the system. In most cases, a qualitative understanding of the rotational state distributions can be obtained from classical arguments, based upon how the two molecules impulsively push apart. The diversity of nonstatistical behavior observed in these systems illustrates that the interactions responsible for the associated energy transfer processes are also quite different from system to system. In particular, we find dissociation channels in several systems that arise from fundamentally different types of couplings between the modes of the system, which are often competitive with one another, while others are not observed at all. In only a few cases do we have detailed theoretical work to compare with the experimental results. Examples have been given for essentially every possible dissociation process where vibrational energy is converted to rotation in the fragments and where it is transferred both within the excited molecule and to the cofragment. A quantitative theoretical treatment of these systems would clearly provide us with many more insights concerning the nature of the terms in the potential surface that are responsible for the observed distributions. Understanding interactions of this type in these prototype systems will greatly aid in our understanding of the dynamics of molecules at much higher energies.

ACKNOWLEDGMENTS

Support for this research is gratefully acknowledged from the National Science Foundation (Grant No. CHE-99-87740) and the Donors of the Petroleum Research Fund (administered by the ACS).

Visit the Annual Reviews home page at www.AnnualReviews.org

LITERATURE CITED

1. Jacobson MP, Field RW. 2000. *J. Phys. Chem. A* 104:3073–86
2. Nesbitt DJ, Field RW. 1996. *J. Phys. Chem.* 100:12735–56
3. Wardlaw DM, Marcus RA. 1988. *Adv. Chem. Phys.* 70:231–63
4. Parmenter CS. 1983. *Faraday Discuss. Chem. Soc.* 75:7–22
5. Parmenter CS. 1982. *J. Phys. Chem.* 86:1735–50
6. Smalley RE. 1982. *J. Phys. Chem.* 86:3504–52
7. Felker PM, Zewail AH. 1985. *J. Chem. Phys.* 82:2961–74
8. Go JS, Bethardy GA, Perry DS. 1990. *J. Phys. Chem.* 94:6153–56
9. Uzer T. 1991. *Phys. Rep.* 199:73–146
10. Stuchebrukhov AA, Marcus RA. 1993. *J. Chem. Phys.* 98:6044–61
11. Miller RE. 1988. *Science* 240:447–53
12. Nesbitt DJ. 1988. *Chem. Rev.* 88:843–70
13. Ewing GE. 1980. *J. Chem. Phys.* 72:2096–107
14. Jucks KW, Miller RE. 1988. *Chem. Phys. Lett.* 147:137–41
15. Nesbitt DJ, Lovejoy CM. 1990. *J. Chem. Phys.* 93:7716–30
16. Hutson JM. 1990. *Annu. Rev. Phys. Chem.* 41:123–54
17. Saykally RJ. 1989. *Acc. Chem. Res.* 22:295–300
18. Cohen RC, Saykally RJ. 1992. *J. Phys. Chem.* 96:1024–40
19. Dayton DC, Miller RE. 1988. *Chem. Phys. Lett.* 143:181–85
20. Dayton DC, Block PA, Miller RE. 1991. *J. Phys. Chem.* 95:2881–88
21. Pine AS, Fraser GT. 1988. *J. Chem. Phys.* 89:6636–43
22. Jucks KW, Miller RE. 1987. *J. Chem. Phys.* 86:6637–45
23. Huang ZS, Jucks KW, Miller RE. 1986. *J. Chem. Phys.* 85:6905–9
24. Vasudev R, Zare RN, Dixon RN. 1984. *J. Chem. Phys.* 80:4863–78
25. Hall GE, Houston PL. 1989. *Annu. Rev. Phys. Chem.* 40:375–405
26. Hall GE, Sivakumar N, Chawla D, Houston PL, Burak I. 1988. *J. Chem. Phys.* 88:3682–91
27. Schnieder L, Seekamp-Rahn K, Wrede E, Welge KH. 1997. *J. Chem. Phys.* 107:6175–95
28. Zhang J, Dulligan M, Segall J, Wen Y, Wittig C. 1995. *J. Phys.Chem.* 99:13680–90
29. Dixon RN, Rieley H. 1989. *J. Chem. Phys.* 91:2308–20
30. Novicki SW, Vasudev R. 1991. *J. Chem. Phys.* 95:7269–74
31. Marshall MD, Bohac EJ, Miller RE. 1992. *J. Chem. Phys.* 97:3307–17
32. Dayton DC, Jucks KW, Miller RE. 1989. *J. Chem. Phys.* 90:2631–38
33. Bohac EJ, Miller RE. 1993. *Phys. Rev. Lett.* 71:54–57
34. Rost JM, Griffin JC, Friedrich B, Herschbach DR. 1992. *Phys. Rev. Lett.* 68:1299–301
35. Block PA, Bohac EJ, Miller RE. 1992. *Phys. Rev. Lett.* 68:1303–6
36. Bemish RJ, Bohac EJ, Wu M, Miller RE. 1994. *J. Chem. Phys.* 101:9457–68
37. Ni H, Serafin JM, Valentini JJ. 1996. *J. Chem. Phys.* 104:2259–70
38. Shorter JH, Casassa MP, King DS. 1992. *J. Chem. Phys.* 97:1824–31
39. Howard BJ, Dyke TR, Klemperer W. 1984. *J. Chem. Phys.* 81:5417–25
40. Pine AS, Lafferty WJ. 1983. *J. Chem. Phys.* 78:2154–62
41. Pine AS, Lafferty WJ, Howard BJ. 1984. *J. Chem. Phys.* 81:2939–50
42. Huang ZS, Jucks KW, Miller RE. 1986. *J. Chem. Phys.* 85:3338–41
43. von Puttkamer K, Quack M. 1989. *Chem. Phys.* 139:31–53

44. von Puttkamer K, Quack M, Suhm MA. 1988. *Mol. Phys.* 65:1025–45
45. Suhm MA, Farrell JT Jr, McIlroy A, Nesbitt DJ. 1992. *J.Chem. Phys.* 97:5341–54
46. Truhlar DG. 1990. In *Dynamics of Polyatomic van der Waals Complexes, NATO ASI Ser. B*, ed. N Halberstadt, KC Janda, pp. 159–85. New York: Plenum
47. Jensen P, Bunker PR, Karpfen A, Kofranek M, Lischka H. 1990. *J. Chem. Phys.* 93:6266–80
48. Halberstadt N, Bréchignac Ph, Beswick JA, Shapiro M. 1986. *J. Chem. Phys.* 84:170–75
49. Bunker PR, Carrington T, Gomez PC, Marshall MD, Kofranek M, et al. 1989. *J. Chem. Phys.* 91:5154–59
50. Hancock GC, Truhlar DG, Dykstra CE. 1988. *J. Chem. Phys.* 88:1786–96
51. Barton AE, Howard BJ. 1982. *Faraday Discuss. Chem. Soc.* 73:45–62
52. Sun H, Watts RO. 1990. *J. Chem. Phys.* 92:603–16
53. Dyke TR, Howard BJ, Klemperer W. 1972. *J. Chem. Phys.* 56:2442–54
54. Zhang DH, Zhang JZH. 1993. *J. Chem. Phys.* 99:6624–33
55. Bohac EJ, Marshall MD, Miller RE. 1992. *J. Chem. Phys.* 96:6681–95
56. Bohac EJ, Miller RE. 1993. *J. Chem. Phys.* 99:1537–44
57. Ewing GE. 1979. *J. Chem. Phys.* 71:3143–44
58. Bemish RJ, Wu M, Miller RE. 1994. *Faraday Discuss. Chem. Soc.* 97:57–68
59. Klopper W, Quack M, Suhm MA. 1998. *J. Chem. Phys.* 108:10096–115
60. von Dirke M, Bačić Z, Zhang DH, Zhang JZH. 1995. *J. Chem. Phys.* 102:4382–89
61. Fraser GT, Pine AS. 1989. *J. Chem. Phys.* 91:633–36
62. Farrell JT Jr, Suhm MA, Nesbitt DJ. 1996. *J. Chem. Phys.* 104:9313–31
63. Oudejans L, Miller RE. 1997. *J. Phys. Chem. A* 101:7582–92
64. Halberstadt N, Serna S, Roncero O, Janda KC. 1992. *J. Chem. Phys.* 97:341–54
65. Blake GA, Busarow KL, Cohen RC, Laughlin KB, Lee YT, Saykally RJ. 1988. *J. Chem. Phys.* 89:6577–87
66. Blake GA, Bumgarner RE. 1989. *J. Chem. Phys.* 91:7300–1
67. Elrod MJ, Saykally RJ. 1995. *J. Chem. Phys.* 103:933–49
68. Schuder MD, Lovejoy CM, Lascola R, Nesbitt DJ. 1993. *J. Chem. Phys.* 99:4346–62
69. Meads RF, McIntosh AL, Arno JI, Hartz CL, Lucchese RR, Bevan JW. 1994. *J. Chem. Phys.* 101:4593–98
70. Moazzen-Ahmadi N, McKellar ARW, Johns JWC. 1989. *J. Mol. Spectrosc.* 138:282–301
71. Bunker PR, Epa VC, Jensen P, Karpfen A. 1991. *J. Mol. Spectrosc.* 146:200–19
72. Bačić Z, Qiu Y. 1998. In *Advances in Molecular Vibrations and Collision Dynamics*, Vol. 3 pp. 183–204. Greenwich, Conn.: JAI
73. Qiu Y, Bačić Z. 1997. *J. Chem. Phys.* 106:2158–70
74. Qiu Y, Zhang JZH, Bačić Z. 1998. *J. Chem. Phys.* 108:4804–16
75. Ni H, Serafin JM, Valentini JJ. 2000. *J. Chem. Phys.* 113:3055–66
76. Clary DC, Knowles PJ. 1990. *J. Chem. Phys.* 93:6334–49
77. Clary DC. 1992. *J. Chem. Phys.* 96:90–97
78. Lovejoy CM, Nelson DD Jr, Nesbitt DJ. 1988. *J. Chem. Phys.* 89:7180–88
79. Lovejoy CM, Nelson DD Jr, Nesbitt DJ. 1987. *J. Chem. Phys.* 87:5621–28
80. Bohac EJ, Miller RE. 1993. *J. Chem. Phys.* 98:2604–13
81. Zhang DH, Zhang JZH, Bačić Z. 1992. *J. Chem. Phys.* 97:927–34
82. Zhang DH, Zhang JZH, Bačić Z. 1992. *J. Chem. Phys.* 97:3149–56
83. Jucks KW, Miller RE. 1987. *J. Chem. Phys.* 87:5629–33
84. Fraser GT, Pine AS. 1989. *J. Chem. Phys.* 91:637–45

85. Oudejans L, Miller RE. 1995. *J. Phys. Chem.* 99:13670–79
85a. Ewing GE. 1987. *J. Chem. Phys.* 91:4662–71
86. Wu M, Bemish RJ, Miller RE. 1994. *J. Chem. Phys.* 101:9447–56
87. Oudejans L, Miller RE. 2000. *J. Chem. Phys.* 113:4581–87
88. Casassa MP, Stephenson JC, King DS. 1986. *J. Chem. Phys.* 85:2333–34
89. Hetzler JR, Casassa MP, King DS. 1991. *J. Phys. Chem.* 95:8086–95
90. Oudejans L, Miller RE. 1998. *J. Chem. Phys.* 109:3474–84
91. Quiñones A, Bandarage G, Bevan JW, Lucchese RR. 1992. *J. Chem. Phys.* 97:2209–23
92. Oudejans L, Miller RE. 1998. *J. Chem. Phys.* 239:345–56
93. Oudejans L, Moore DT, Miller RE. 1999. *J. Chem. Phys.* 110:209–19
94. Rudich Y, Naaman R. 1992. *J. Chem. Phys.* 96:8616–17
95. Oudejans L, Miller RE. 1999. *J. Phys. Chem.* 103:4791–97
96. Insepov ZA, Karataev EM. 1991. *Pisma. Zh. Tekh. Fiz.* 17:36–39
97. Oudejans L, Miller RE. 2000. *J. Chem. Phys.* 113:971–78
98. Quack M, Stohner J, Suhm MA. 1993. *J. Mol. Struct.* 294:33–36
99. Hetzler JR, Casassa MP, King DS. 1991. *J. Phys. Chem.* 95:8086–95
100. Zhang DH, Zhang JZH. 1993. *J. Chem. Phys.* 98:5978–81

COHERENT NONLINEAR SPECTROSCOPY: From Femtosecond Dynamics to Control

Marcos Dantus

Department of Chemistry and Center for Fundamental Materials Research, Michigan State University, East Lansing, Michigan 48824; e-mail: dantus@msu.edu

Key Words ultrafast, four-wave mixing, photon echo, transient-grating, wavepacket

■ **Abstract** This review focuses on the study of the dynamics of isolated molecules and their control using coherent nonlinear spectroscopic methods. Emphasis is placed on topics such as bound-to-free excitation and the study of concerted elimination reactions, free-to-bound excitation and the study of bimolecular reactions, and bound-to-bound excitation and the study of intramolecular rovibrational dynamics and coherence relaxation. For each case the detailed time-resolved information reveals possible strategies to control the outcome. Experimental results are shown for each of the reactions discussed. The methods discussed include pump-probe and four-wave mixing processes such as transient grating and photon echo spectroscopy. Off-resonance transient-grating experiments are shown to be ideal for the study of ground state dynamics, molecular structure, and the molecular response to strong field excitation.

INTRODUCTION

The early calculations carried out by Hirschfelder & Eyring on the dynamics of the $H + H_2$ reaction in the 1930s revealed the femtosecond time scale of fundamental chemical processes such as bond breakage and bond formation (1). Scientists have been pushing the technological edge to be able to reach the time scale at which these chemical events occur. In the mid-1950s the resolution was in the microsecond time scale with the pump-probe work of Porter (2, 3). The invention of the laser in the 1960s facilitated the rapid progress to the nanosecond and then to the picosecond time scale (4). By 1984, scientists had arrived at the femtosecond time scale with measurements of fundamental processes such as the photochemistry of bacteriorhodopsin and the intramolecular relaxation time of large organic molecules in the gas phase and in solution (5–10). In 1985, Zewail began to probe the ultrafast dynamics of isolated molecules with subpicosecond time resolution (11), and by 1987, published the first time-resolved observation of transition states in a chemical reaction (12). The significance of these measurements, recognized by the Nobel prize in 1999, was that Zewail had captured the essence of a chemical reaction in a system that had a well-defined reaction coordinate and was not complicated by the solvent response.

Given that the generation of ultrafast pulses has outpaced the development of ultrafast detectors, the study of ultrafast chemical dynamics requires the use of methods that involve multiple laser pulses. One or more pulses initiate the chemical reaction and one or more probe its progress. The involvement of multiple laser pulses implies that these measurements belong to the realm of nonlinear optical spectroscopy. This distinction is of importance when seeking a thorough understanding of the signals. In linear spectroscopy, one is concerned with absorption or emission. In nonlinear optics, the processes involve the coherent interaction between the sample and one or more of the laser pulses (13–15). Coherent nonlinear spectroscopic methods provide the means to study molecular dynamics and to explore laser control of chemical reactivity.

The goal for active laser control is to devise electromagnetic fields that drive the outcome of a chemical reaction in the desired direction (16–20). There are two main approaches to this problem. The frequency-resolved scheme (also known as coherent control), proposed by Brumer & Shapiro (21, 22) utilizes quantum interference between different pathways to a final state to exert control over the outcome. One of the most striking demonstrations of this scheme is found in the work of Gordon and co-workers, controlling autoionization versus predissociation in HI and DI molecules (23, 24). The time-resolved scheme (also known as pump-dump), proposed by Tannor et al (25, 26), exploits the time-dependent motion of wave packets created by ultrafast (usually femtosecond) laser pulses to manipulate the outcome of the reaction. Experimental demonstrations of this control scheme are found in many pump-probe time-resolved experiments, for example, the excitation of I_2 to produce either the D ($^1\Sigma_u^+$) or the F ($^3\Pi_u$) states (27), or the production of Na^+ or Na^{2+} as a function of time delay between pump and probe pulses (28–30) and the isotopic separation of bromine (31), and more recently, the preparation of groundstate wave packets of K_2 (32). Wilson and coworkers generalized this approach to obtain a formalism that is more amenable for the study of thermal ensembles of molecules (33, 34).

The search for an optimal electromagnetic field in terms of spectral and temporal composition to control the outcome of a chemical reaction was formalized by Rabitz and coworkers (35, 36). Optimization of the Tannor-Rice pump-dump scheme for controlling the selectivity of product formation was considered by Kosloff et al (37). The application of chirped pulses to shape nuclear wave packets and enhance vibrational coherence was proposed by Ruhman & Kosloff (38). Broers et al demonstrated the use of chirped pulses to enhance the population transfer in the three-state ladder of the rubidium atom (39). Experimental (40) and theoretical (41, 42) studies on the effect of chirped pulses on the multiphoton excitation of molecules showed that the traditional saturation limits can be exceeded, thereby facilitating population inversion. The groups of Shank, Wilson, Leone, and others have shown experimental evidence that tailored femtosecond pulses can be used to modify the initial wave packet formed by the excitation laser (43–53). In some cases, tailored pulses (chirp) can be used to enhance single and multi-photon transitions (54–57) as well as excitation of high vibrational states (58, 59). Experimental demonstration of optical control using shaped laser pulses on multi-dimensional systems has been shown by Gerber's group, who optimized the yield of different product channels (60). Bucksbaum's group showed selective Raman excitation of the symmetric and antisymmetric OH stretch in liquid methanol (61).

In the following sections we focus on the application of coherent nonlinear spectroscopic techniques to study isolated chemical reactions and their control, in which all laser interactions take place on a time scale that is short compared with the coherence relaxation. The protocol we follow is first to understand the ultra-fast dynamics involved, selecting systems with a well-defined reaction coordinate, and then to consider a rational method for their control using lasers. We explore bound-free concerted elimination reactions and free-bound bimolecular reactions, both of which are best studied in the gas phase, in which there is no solvent cage to dictate the progress of the reaction. The section on bound-bound molecular transitions focuses on inter- and intramolecular dynamics and relaxation. Experiments using four-wave mixing methods reveal mechanisms for controlling the dynamics and energy flow. We have attempted to illustrate each section with a number of experimental results from our group, rather than attempt a comprehensive review of each topic.

METHODS

Advances in ultrafast laser technology have permitted the development of commercial units capable of carrying out most ultrafast experiments. For this reason, we do not describe these laser systems in detail. Instead we concentrate on describing the setups typically used for nonlinear optical measurements. The experiments described here were carried out with two types of laser systems (see Figure 1). The first is based on a home-built colliding pulse mode–locked laser (62–64) amplified by a four-stage dye amplifier (65). After amplification the system produces pulses centered at 620 nm with 0.5 mJ in energy at a repetition rate of 30 Hz. The second system is based on a Kapteyn-Murnane oscillator capable of producing 13 fs pulses when compressed. This laser is regeneratively amplified by an Evolution X pumped Spitfire (Spectra Physics). The output, centered at 810 nm consists of pulses with 0.8 mJ in energy at a repetition rate of 1 kHz. Both laser systems produce transform-limited pulses of 50-fs duration, as measured by frequency-resolved optical gating (66).

The two techniques used most frequently in our laboratory are pump-probe and degenerate four-wave mixing (FWM). For pump-probe measurements the laser, after the compression, is split into two arms of a Mach-Zhender interferometer. Typically, one beam is frequency doubled. The two pulses are then recombined and

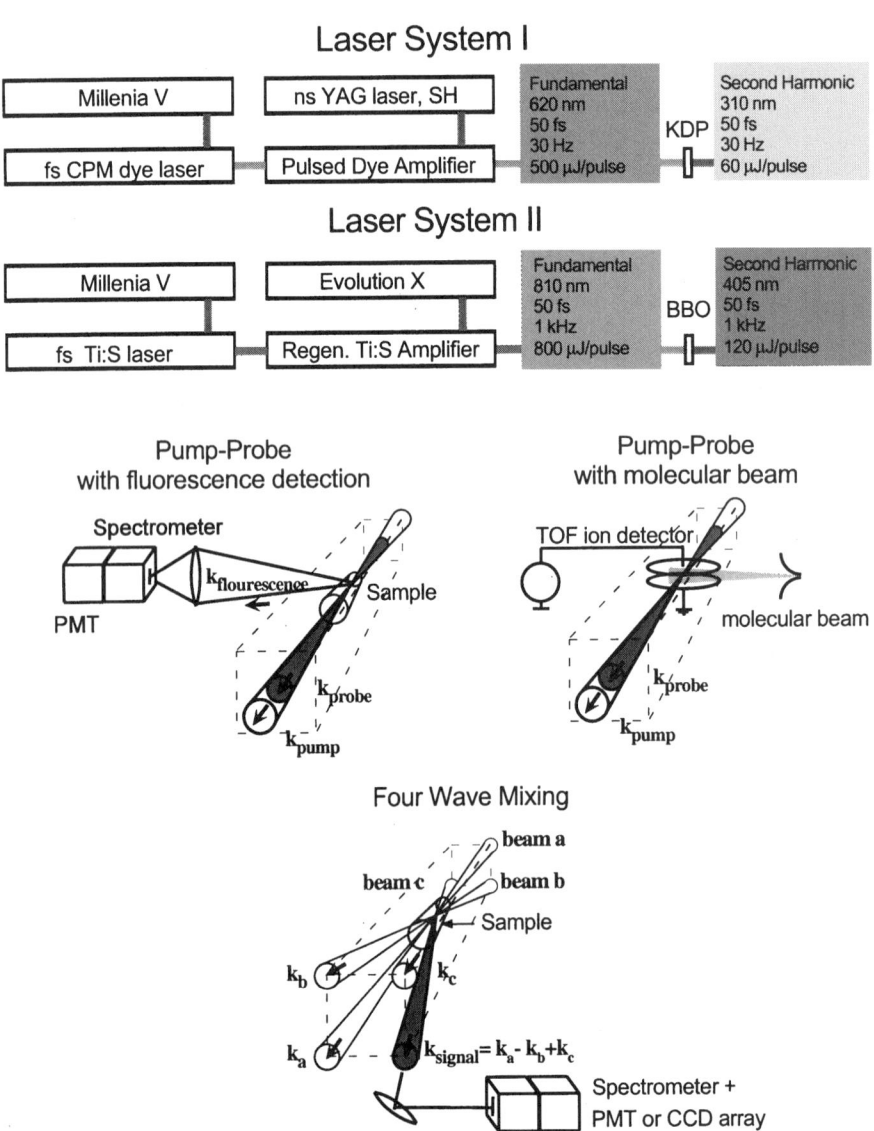

Figure 1 (*Top*) Schematic of two experimental set-ups used in our laboratory with their main characteristics. (*Bottom*) Three most commonly used nonlinear optical techniques in time resolved spectroscopy.

focused into the sample cell and fluorescence is detected at right angles (67). When required, tunability is achieved with two-stage noncollinear optical parametric amplifier system, generating tunable ~20-fs pulses in the 450- to 1600-nm range (68). Gas-phase measurements can be carried out in gas bulbs or in molecular beams. The signal detected can be laser-induced fluorescence, mass selected ions, or photoelectrons.

For FWM measurements, the laser beam is split by two successive beam splitters into three beams of approximately equal intensity and attenuated down to ~20 μJ each. Pulses are delayed with respect to each other by a computer-controlled actuator. The three beams are combined in a specific phase-matching geometry (69, 70) and focused by a 0.5 m lens in a quartz cell containing the gas-phase sample (see Figure 1). In this configuration the FWM signals are detected in the direction $\mathbf{k}_S = \mathbf{k}_a - \mathbf{k}_b + \mathbf{k}_c$. The subindices a, b, and c are used to identify the beams in space, but the pulses can take any time order 1, 2, or 3; therefore, for the same phase-matching geometry, different phenomena such as stimulated photon echo, virtual echo, transient grating (TG), and reverse TG (RTG) can be detected (71). The signal can be time integrated (homodyned) or time gated (heterodyned); similarly, the signal can be spectrally integrated or spectrally dispersed (71–75).

BOUND-FREE TRANSITIONS: CONCERTED-ELIMINATION CHEMICAL REACTIONS

Concerted chemical and biochemical processes have been of great interest, particularly since the publication of Woodward & Hoffmann's work on pericyclic reactions (76). A concerted reaction is defined as one for which multiple fundamental changes (such as bond formation, charge transfer, etc) occur in a single kinetic step (76–79). In practice, this means that a reaction is considered to be concerted if there is no evidence of intermediate stages. Therefore, the classification depends on the sensitivity of the chosen method to detect the short-lived intermediates. A reaction that is rapid compared with the detection method could be erroneously considered concerted. Molecular beam techniques have been used to determine gas-phase reaction mechanisms (80). These methods have a temporal resolution comparable with the rotational period of the molecules ($\approx 10^{-12}$ s) and have been useful in determining the concertedness of chemical reactions by analysis of the velocity and angular distribution of the products (81–83).

The use of femtosecond transition state spectroscopy to detect the presence of reaction intermediates was introduced by Zewail and coworkers (84). Their time-resolved experiments on the α-cleavage reaction of acetone and on the decarbonylation of cyclopentanone have shown that both reactions proceed by a stepwise (nonconcerted) mechanism (85, 86). In the condensed phase, the solvent provides a "cage" that keeps the reagents and products of a chemical reaction in close proximity, making it difficult to determine if a reaction proceeds by a concerted mechanism. This determination, therefore, is ideally carried out on isolated molecules. One example that has received a great deal of attention recently is the tautomerization reaction of 7-aza-indole (87–90). Time-resolved gas-phase measurements indicate that the two hydrogen bonds are traded in a nonconcerted fashion (87, 88).

Recently, several groups have addressed the ultrafast dynamics involved in the concerted elimination of halogen molecules following the high-energy excitation of halogenated alkanes (91–104). The reaction channel $CH_2I_2 \rightarrow CH_2 + I_2(D')$ was first investigated by Style (105, 106) and Okabe (107). Our group has investigated the femtosecond time scale dynamics of this reaction (95, 97, 98). Huber and coworkers have studied the high-energy dissociation pathways of CF_2I_2 with excitation between 248 and 351 nm. Their findings indicate that production of I and I* constitute the major dissociation channels for the reaction at these wavelengths (83, 91, 101, 102). As in diiodomethane, difluorodiiodomethane also produces I_2 following high-energy excitation; in this case, the reaction is initiated with absorption of two photons of 267 nm (101). This study did not identify the electronic state of the molecular halogen or its vibrational or rotational energetics. Schwartz et al have studied the photodissociation dynamics of CH_2I_2 with femtosecond lasers in different solvents (92). Their work focused on the reaction pathway initiated by the 310-nm pump that leads to the production of $CH_2I + I$. They followed the geminate recombination reforming the parent molecule by measuring transient absorption at 620 nm. The initial dynamics that take place upon excitation of CH_2I_2 have been studied by Duschek et al in the gas phase (108) and in solution by Kwok & Phillips using resonance Raman scattering between 342 and 369 nm (93, 94), and by Sundström and coworkers using pump-probe spectroscopy (103). Their findings indicate the involvement of the I-C-I symmetric stretch, antisymmetric stretch, and bending vibrational modes.

Femtosecond pump-probe measurements from our group showed that high-energy excitation produces molecular iodine by a concerted process (56, 95–100). Coherent vibrational motion in the I_2 product was observed (95, 98, 100). Analysis of the transition state dynamics shows that the two-carbon halogen bonds are broken and the new interhalogen bond is formed within 50 fs. When the dissociation dynamics of CH_2I_2 and $CH_3(CH_2)_2CHI_2$ were compared, the transition state lifetime for $CH_3(CH_2)_2CHI_2$ was found to be approximately two times longer than the lifetime of CH_2I_2 (97, 98). Analysis indicated that the difference in lifetime could be attributed to the change in the reduced mass of the carbene fragment, therefore, the reaction occurs faster than intramolecular vibrational relaxation (IVR) to the alkane chain.

Figure 2a (see color insert) depicts a cut of the potential energy surfaces for this system. The pathways that lead to the production of I and I* by stepwise photodissociation processes are shown on the left. The pathways shown on the right-hand side of the figure lead to dihalogen molecules produced through concerted elimination mechanisms, the $CH_2 + I_2(D')$ channel, and a less probable channel leading to $CH_2 + I_2(f)$ (98, 100). The former showed very small differences for time delay greater than zero when probed with parallel or perpendicular polarized light while the latter showed very clear rotational anisotropy (99, 100) (see Figure 2b and 2c). This rotational component of the data allowed us to have a much clearer picture of the mechanism involved in this dissociation process. Analysis of the rotational anisotropy indicated that $I_2(f)$ is produced with a very hot rotational distribution (100). Using classical mechanical modeling of the dynamics, we confirmed that the symmetry of the molecule must be broken in the dissociation to achieve such a hot rotational distribution. An asynchronous concerted process does not conserve the C_{2v} symmetry and is consistent with the observed dynamics. A synchronous concerted mechanism in which the C_{2v} symmetry of the parent is conserved would produce rotationally cold products. In conclusion, the $I_2(D')$ pathway is consistent with a synchronous concerted mechanism, whereas the $I_2(f)$ pathway is consistent with an asynchronous concerted mechanism.

The concerted reaction occurs following 12 eV excitation. Electronic structure calculations are not capable of providing an accurate potential energy surface for molecules containing two heavy atoms at these high energies. In the absence of a potential energy surface, we assume, as a first-order approximation, that the parent molecule behaves as a pseudodiatomic, breaking into the carbene radical and the halogen molecule, and that the fragments reach terminal velocity immediately following the dissociation. We have studied the concerted elimination of halogen molecules from a family of compounds, CX_2YZ (where X = H, F, or

TABLE 1 Energetics and dynamics for concerted molecular elimination[a]

Reaction	Reaction enthalpy (eV)	Energy available[a] (eV)	Reduced mass[b] (a.m.u.)	Experimental time (fs)	E_{kin}/E_{avail}[c]
$CH_2I_2 \longrightarrow CH_2 + I_2(D')$	8.38	3.62	13.3	47 ± 3	0.34
$CH_2 + I_2(f)$	9.20	2.80	13.3	—	
$CD_2I_2 \longrightarrow CD_2 + I_2(D')$	8.38	3.62	15.1	47 ± 3	0.50
$BuI_2 \longrightarrow Bu: + I_2(D')$	8.38	3.62	46.0	87 ± 5	0.35
$CH_2Br_2 \longrightarrow CH_2 + Br_2(D')$	10.50	1.50	12.9	59 ± 1	0.52
$CF_2Br_2 \longrightarrow CF_2 + Br_2(D')$	8.50	3.50	38.1	—	
$CCl_2Br_2 \longrightarrow CCl_2 + Br_2(D')$	8.80	3.20	54.6	81 ± 4	0.54
$CH_2ICl \longrightarrow CH_2 + ICl(D')$	8.40	3.60	12.9	48 ± 1	0.32
$CH_2 + ICl(G)$	9.20	2.80	12.9	71 ± 4	0.19

[a](95–100, 104).
[b]Energy available is calculated by subtracting the reaction enthalpy from the photon energy (12 eV).
[c]Reduced mass is calculated assuming a two-body, carbene-halogen, molecular dissociation. The lifetimes are the results of femtosecond time-resolved measurements from our group (95–100, 104).
[d]Kinetic energy is calculated from the experimental dissociation time (see text), and the ratio between the kinetic energy and the available energy can be used to compare energy partitioning among the different reactions.

Cl and Y, Z = Cl, Br, or I) (see Table 1). We have assumed that the potentials are similar within this family, which allows us to compare the transition state lifetimes among these compounds. If we consider that the time it takes the product to achieve terminal velocity is negligible, we can use the expression for the kinetic energy, $E_{kin} = (1/2)\mu v^2$, and substitute the velocity v by L/τ_{exp}. Where L is the distance required for bond breaking, τ_{exp} is the experimental dissociation time (i.e. transition state lifetime) and μ is the reduced mass calculated for the psuedodiatomic (carbene-dihalogen) molecule. The kinetic energy E_{kin} represents the energy available for recoil after the enthalpy of reaction and internal energy (vibrational and rotational energy) of the products have been subtracted from the photon energy.

Based on this model we can compare the dissociation dynamics of nine related concerted elimination reactions. Table 1 compares the transition state lifetimes of CH_2I_2, CD_2I_2, $CH_3(CH_2)_2CHI_2$, CH_2Br_2, CF_2Br_2, CCl_2Br_2, and CH_2ICl. The data were collected consecutively, with the laser intensity kept constant to avoid apparent differences in dissociation times resulting from saturation of the transitions. For this comparison, we take the total photon energy and subtract the reaction enthalpy to yield the available energy for the reaction, $E_{avail} = h\nu - \Delta H$. The ratio E_{kin}/E_{avail}, where E_{kin} is estimated from τ_{exp} as discussed above, gives us a parameter to compare the energy partitioning in the reactions and is given in the sixth column of Table 1. When $E_{kin}/E_{avail} = 0.34$, the energy partitioning is similar to the reaction for CH_2I_2. When the ratio is greater, more energy is partitioned into translation. Finally, when the ratio is smaller, the fragments acquire more internal energy in the form of vibrations and rotations.

The data in Table 1 can be used to estimate dissociation times. The ratio of the dissociation time is given by

$$\frac{\tau_1}{\tau_2} = \left[\frac{(E_{photon} - \Delta H_2)\mu_1}{(E_{photon} - \Delta H_1)\mu_2}\right]^{1/2}, \qquad 1.$$

where we assume that $L_1 \approx L_2$ and $E_{kin} \approx E_{avail}$ as a first approximation. We first compare the dissociation times of reactions with equal reaction enthalpy, such as CH_2I_2 (reagent 1) and $CH_3(CH_2)_2CHI_2$ (reagent 2), producing $I_2(D')$. The ratio between the experimentally determined dissociation times is $\tau_2/\tau_1 = 1.85$ (see Table 1); the ratio of the estimated dissociation times based on Equation 1 is $\tau_2/\tau_1 = 1.85$. The agreement in this case is remarkable, indicating that the assumptions $L_1 \approx L_2$ and a similar kinetic energy partition are valid. Second, we compare the experimental dissociation times of reactions having different enthalpy, such as CH_2Br_2 (reagent 1) and CCl_2Br_2 (reagent 2). The ratio of the experimental dissociation times is $\tau_2/\tau_1 = 0.73$, where the ratio of the estimated times using Equation 1 is $\tau_2/\tau_1 = 0.71$. The small difference between the experimental ratio and the value obtained using Equation 1 might indicate small differences in the energy partitioning or reflects differences in the potential energy surface. From these results it is clear that the concerted elimination of halogen molecules is a direct process that takes place on a time scale faster than IVR. These observations indicate that there is a time window of ~100 fs to control the yield of these chemical reactions using a tailored laser pulse.

Inspired by the work of Shank and coworkers and Wilson and coworkers on the enhancement of three-photon excitation using chirp (55, 109, 110), we have explored the effect of chirp on the multiphoton excitation of CH_2I_2 with 312-nm laser pulses (56). The product-state distribution (electronic, vibrational, rotational, and translational) resulting from the photodissociation of polyatomic molecules depends upon the potential energy surfaces participating in the fragmentation process, along with their couplings and the characteristics of the incident electromagnetic field (111). The field is characterized by its frequency, duration, intensity, and chirp. Chirp is caused by the propagation of laser pulses through matter that leads to group velocity variations as a function of frequency within the pulse, thus causing a frequency sweep (112, 113). In most cases, the group velocity variation causes a positive chirp in which the leading edge of the pulse is red-shifted and the trailing edge is blue-shifted with respect to the central frequency of the pulse. Negative chirp corresponds to the opposite effect. Increases in absolute chirp lead to a temporal broadening of the pulse and are usually considered detrimental for the study of ultrafast phenomena, in which the best time resolution is required.

The shape and time evolution of a wave packet $\psi(t)$ produced through absorption of an ultrafast laser pulse are determined by the phase factors in the following expression:

$$\psi(t) = \sum_n a_n e^{-iE_n t/\hbar} \varphi_n, \qquad 2.$$

where E_n and φ_n denote the eigenvalue and eigenfunctions of each eigenstate n. The coefficient a_n is given by $a_n = b_n e^{i\phi_n}$, where b_n depends on the Franck-Condon overlap between the initial state and each final state n. The phase ϕ_n depends on the linear chirp of the pulse, ϕ'', where $\phi_n = \frac{1}{2}\phi''(\omega_n - \omega)^2$, $\omega_n = (E_n - E_o)/\hbar$ is the transition frequency of level n, and ω is the carrier frequency of the laser pulse. The initial phase factor of each coefficient is equal for all states when excitation takes place with transform limited pulses, that is, $\phi'' = 0$. Therefore, the shape and dynamics of the wave packet can be controlled with the goal of affecting the outcome of a chemical process using chirped pulses.

Figure 3 presents the yield of the molecular pathway determined by detection of I_2 D′ → A′ fluorescence intensity as a function of chirp. Increasing the chirp enhances the photodissociation yield significantly (56). The molecular pathway enhancement is found to be nonsymmetric for high fields, favoring positive over negative chirps. For example, the enhancement for $\phi'' = -1500 \text{ fs}^2$ is 1.2, whereas for $+1500 \text{ fs}^2$ it is 3.2 (see black circles in Figure 3). This observation implies that the observed enhancements are not due to pulse width effects, but they depend on the magnitude and sign of the linear chirp. The two scans shown in Figure 3a were obtained under identical conditions except for the intentional changes in the pulse intensity from 0.8×10^{12} to 1.6×10^{12} W/cm^2 (calculated for zero chirp pulses). The data are shown normalized to laser pulse energy (such that the yield

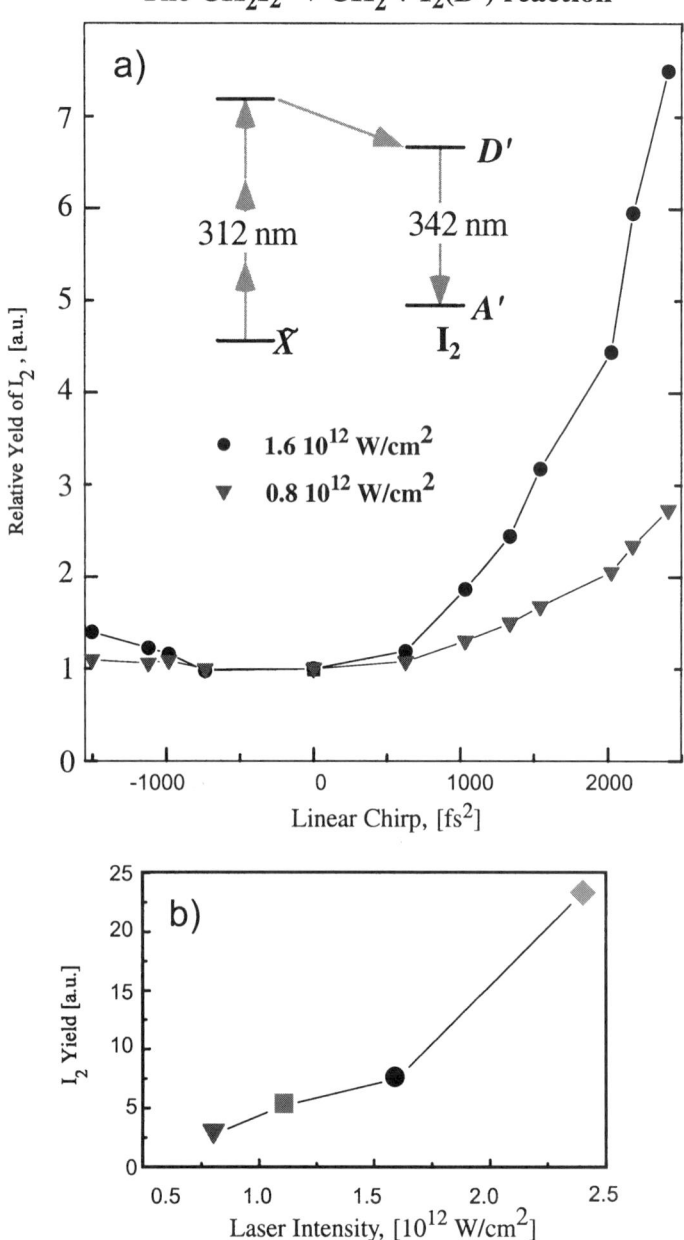

for the transform-limited pulses equals unity) but are not corrected for the change in peak intensity caused by pulse broadening as a function of chirp. The effect of laser intensity on these control experiments is shown in Figure 3b, where the molecular pathway yield is shown to increase for positive 2400 fs^2 chirp by factors of 3 to 24 as the intensity of the laser pulses is increased from 0.8×10^{12} to 2.4×10^{12} W/cm^2.

Considerable differences were observed between the dependence of the molecular pathway yield on laser pulse chirp depending on the central wavelength of the excitation pulse (56). For 624 nm, the yield of $I_2(D')$ was found to decrease with absolute chirp, whereas for 312 nm it increased. The probability of multiphoton transitions, in general, is proportional to the peak intensity raised to the nth power (I^n), where n is the number of photons. Therefore, multiphoton excitation is expected to be maximized for transform-limited pulses. For 624 nm excitation, the maximum yield was found for a chirp of -500 fs^2; however, for 312 nm the maximum yield was found for a chirp of 2400 fs^2 (56). Based on a peak intensity argument, the transition probability for the three-photon transition and hence the yield would have been expected to decrease by one order of magnitude because the pulse width triples at this chirp value.

The effects caused by chirp in the excitation pulses reflect characteristics of the potential energy surfaces and the nascent wave packet dynamics. For diatomic I_2, Cao et al explained their observed chirp effects based on quantum mechanical calculations that show a "wave-packet following" effect for positive chirp (110). The potential energy surfaces of CH_2I_2 are not known, which prevents us from giving an accurate quantum mechanical description of the effect. In principle, a similar wave-packet following effect could be responsible, given that the first photon transition is resonant, as in the I_2 experiment from Wilson's group (55).

FREE-BOUND BIMOLECULAR REACTIONS: PHOTOASSOCIATION

Bimolecular reactions have proven to be extremely difficult to study using time-resolved methods because an encounter between two reagents is required. In the gas phase these encounters occur at random times, with random configurations and

---

←

Figure 3 (a) Experimental measurement of the yield of the concerted elimination pathway producing $I_2(D')$ from the multiphoton dissociation of CH_2I_2 using 312 nm laser pulses as a function of chirp. Results at two different intensities are shown. The insert shows the relevant energetics of the reaction. (b) Maximum $I_2(D')$ yield enhancement recorded at 2400 fs^2 chirp as a function of laser peak intensity. In all cases, the peak intensity was measured for zero chirp. Yields were normalized to unity for each laser intensity when the pulses were not chirped.

random energies. The experimental challenge is to devise ways to determine or restrict the initial collision conditions, such as impact parameter, orientation, collision energy, and time of collision. Traditional methods for studying bimolecular reactions include the use of molecular beams where the energy of the reactions can be regulated and special detectors to track the energy and position of the products (80, 81, 114, 115).

A number of methods have been proposed to obtain a detailed understanding of a bimolecular encounter. The interpretation of unimolecular dissociation reactions as the "half collision" is one such effort based on the principle of microscopic reversibility (116). According to this interpretation, a unimolecular photodissociation is equivalent to the second half of a full collision. The first half would involve the collision of the fragments. Clearly, only very specific initial conditions, such as impact parameter and reagent energies, would reproduce the observed dissociation dynamics. Therefore, unimolecular dissociation provides detailed information that is relevant to a very small subset of the possible bimolecular pathways.

Brooks et al used laser excitation during reactive collisions to study the transition states of chemical reactions (117). In their study, a nanosecond laser, not resonant with the asymptotic transitions of reactants or products, was used to open a chemiluminescent product channel; this was perhaps one of the first attempts to change the course of a bimolecular reaction by excitation of the transient collision complex. A different approach to the study of bimolecular reactions with ultrafast pulses takes advantage of van der Waals clusters involving the two precursors (118–122). These clusters, formed in a supersonic jet expansion, are cooled into the most energetically stable configuration, thereby reducing the range of initial reaction parameters. Essentially, the bimolecular encounter is converted into a unimolecular dissociation; therefore, the collision geometry is restricted, and with it the impact parameter. Laser excitation liberates one of the reagents or excites one of the reagents to a reactive state in order to initiate the reaction. The laser determines the available energy and the time of collision. These studies have been carried out using frequency (119) as well as time-resolved methods (118, 120–122). The computation of time-dependent dynamics in excimer molecules formed from van der Waals clusters has been considered by Petsalakis et al (123).

A method that has emerged as a new possibility in the study of bimolecular reactions is femtosecond photoassociation spectroscopy (99, 124–126). The photoassociation process, involving cooperative absorption of a photon by a pair of unbound atoms or molecules, causes bond formation between them in a free-to-bound photonic transition. Although photoassociation has been known at least since 1937 (127), the chemical implications of this process were described conceptually much later by Dubov et al (128). Photoassociation has been used for the spectroscopic study of excimer and exciplex molecules that have a repulsive ground state (129–138) and more recently has gained interest because of its role in the generation of ultracold molecules (139–153), the real time observation of bimolecular reactions (124, 154), and control of bimolecular encounters (99, 126, 155).

Even though most of the work in the area of laser control of chemical reactions has been dedicated to unimolecular processes (20, 156), some groups have begun to investigate how to control bimolecular reactions (157–160). The yield of a bimolecular reaction is determined by the energy of the collision, relative orientation of the reactants, and impact parameter of the encounter. The photoassociation process has been demonstrated to achieve control of these three key parameters (99, 126). Short-pulse photoassociation provides a well-determined initiation time for the reaction as well as an alignment with respect to the laboratory frame. These additional parameters allow for very detailed studies of bimolecular chemical reactions. Studies from our group have established that photoassociation is possible with femtosecond laser pulses, bringing the technique to the time scale of vibrational motion (10^{-14}–10^{-12} s) (124, 126). The goal of our work has been to perform time-resolved measurements of transition state dynamics during reactive bimolecular collisions, to demonstrate control over the impact parameter of bimolecular reactions, and to establish the dependence of the photoassociation process on different laser characteristics (duration, frequency, and chirp) in order to explore the optimum balance between wavelength selectivity and temporal resolution. In Figure 4a (see color insert) a sketch of the photoassociation process $Hg + Hg \rightarrow Hg_2^*$ is presented. The gray region represents the thermal population of free continuum states. Notice that resonance occurs only in a narrow range of internuclear distances, primarily at the "repulsive wall." The binding wavelength is not absorbed by van der Waals clusters near the equilibrium distance or by the free reagents. The selectivity of the photoassociative process arises from the Franck-Condon overlap between the continuum wave functions in the ground electronic state and the bound wave functions of the upper state.

We have performed pump-probe experiments on the reaction $Hg + Hg \rightarrow Hg_2^*$, where a 312-nm pump pulse photoassociates a pair of ground state Hg atoms into the bound excited state $D1_u$. The fluorescence of the Hg_2 $D \rightarrow X$ is collected as a function of the delay time between the pump and probe pulses as shown in Figure 4b for pulses that are polarized parallel and perpendicular to each other. For positive time delays, depletion of the $D1_u$ state takes place as the molecules are excited to the 1_g state by the probing pulse. The difference between both parallel and perpendicular transients indicates that the photoassociation process of this reaction is anisotropic with respect to the collision pair alignment. Because the pump laser is polarized, the nascent product molecules are aligned. This implies that a maximum in the depletion probability is expected for parallel pump-probe relative polarization (99). The fast onset of the depletion indicates that the photoassociation of free mercury atoms occurs within the laser pulse duration time. The large rotational distribution of the excimers leads to the dephasing of the rotational coherence. In that way, the rotational anisotropy decay depends on the rotational distribution parameters: the central quantum rotational level, j_{max}, and the range Δj. When a rotational distribution model is adopted, j_{max} and Δj can be obtained from a numerical fitting of the experimental rotational anisotropy, indicating that the photoassociation products have a narrow rotational distribution. The rotational level distribution of the products reflects the range of collisional impact parameters that contribute to the photoassociation process, $b = j_{max} \hbar/(\mu v)$. Control of the range of the collisional impact parameter can be achieved with the wavelength of the binding pulse through the Franck-Condon dependence of the photoassociation process (99, 126), which in turn determines the rotational excitation of the products. The results obtained for photoassociation at 312 nm (shown in Figure 4c) yield a rotational distribution with $j_{max} \approx 30$ and $\Delta j = 90$ based on a Gaussian distribution model. From this we can estimate the most probable impact parameter to be $b \approx 0.6$ Å. The experimental results are in agreement with the quantum dynamics calculations of Backhaus & Schmidt (154). The most probable impact parameter for hard-sphere collisions in the absence of photoassociation is $b \approx 3.2$ Å. The difference in the impact parameter for the photoassociation at 310 nm and hardsphere collisions indicates how the photoassociation process can be used to control the collision geometry and to limit the range of impact parameters.

Calculations of the photoassociation yield for a free-to-bound transition as a function of the pulse duration are shown in Figure 4d for different binding pulse wavelengths (155). In all cases the number of photons per laser pulse is kept constant, and a constant initial kinetic energy of the colliding atoms is assumed. The association yield features the following two interesting aspects. Starting from 100 fs and shorter pulse lengths, the yield shows a local maximum at approximately 10 fs for each laser wavelength. For increasing pulse length ($\tau_{pulse} > 100$ fs), the yield exhibits an additional maximum in the range from 10^4 to 10^6 fs for some laser frequencies. The enhancement of the association yield in the regime of ultrashort pulses can be readily explained by the following argument. For $\tau_{pulse} \approx 10$ fs, the pulse spectrum overlaps almost the entire bond potential, resulting in an increase of the association yield. For even shorter pulse lengths, the energetic width exceeds the range of bound states, and the association yield decreases again. These competing effects lead to the maximum in the association yield at about 10 fs. For picosecond pulses (10^4–10^5 fs), resonance with a free-to-bound transition for a certain final rovibrational level of the excited state leads to the enhancement of the photoassociation probability.

BOUND-BOUND MOLECULAR TRANSITIONS: VIBRATIONAL DYNAMICS AND COHERENCE

Molecular dynamics are critically dependent on the inter- and intramolecular flow of energy. In the past decades there has been a considerable effort to measure and understand the flow of energy in the gas and condensed phases. The initial concepts of laser control of chemical reactions assumed that energy would remain localized in certain chemical bonds long enough to control reactivity (161–168). It was soon discovered that even for isolated molecules the energy dispersed among all accessible degrees of freedom in the picosecond time scale. Extensive studies carried out in the 1980s and 1990s on IVR processes and advances in ultrafast lasers are combining to obtain a better understanding of these processes from small isolated molecules to large proteins in solution (169–171). Conceptually, the laser-sample interactions must take place in a time that is short compared to IVR (170). This concept is embodied in the pump-dump theory (16, 25). On the experimental front, the development of lasers with femtosecond pulse duration made the pursuit of this work possible (10, 62, 63, 65). Zewail's group quickly incorporated these techniques and dedicated their work to study the ultrafast dynamics of chemical reactions in the gas phase (172–174). The experimental observation of vibrational dynamics caused by impulsive excitation of multiple vibrational levels using femtosecond pulses in a pump-probe or in a TG arrangement provided experimental evidence of wave packet localization (9, 175–182). The ability to determine the position of a wave packet in time and space (within the constraint of the uncertainty principle) indicated that the pump-probe method is equivalent to the pump-dump technique (16, 25, 26) and could be used to control chemical processes (27, 28).

There are a number of successful laser techniques that have shown promise in schemes aimed at controlling chemical reactivity with lasers (e.g. pump-dump, interference of two or more pathways, and multiphoton excitation). All of these methods can be combined to achieve more general schemes for controlling chemical reactivity. Four-wave mixing with phase-matching detection is an ideal platform for the coherent combination of degenerate laser pulses. Phase-matching detection ensures that the signal arises from the coherent contribution of the laser beams without requiring active phase control. Our work on degenerate four-wave mixing (FWM) has shown that pulse sequences can be used to probe the vibrational dynamics of molecules in a specific electronic state and to control coherence and population transfer between different states (71, 73, 74, 183, 184). This work has been a combination of experimental observation and theoretical interpretation based on density matrix and wave packet simulations (73, 184, 185).

The FWM signal results from the polarization of the sample following three consecutive electric field interactions. The lasers and detector are arranged in a phase matching configuration that conserves energy and momentum. This ensures coherent interactions among the beams and determines the sign of the electric field interactions with the sample. Each electric field can be described by $\widetilde{E}(t) = E(t)e^{i(\mathbf{k}\cdot\mathbf{x}-\omega t)} + \widetilde{E}(t)^* e^{-i(\mathbf{k}\cdot\mathbf{x}-\omega t)}$, where $\widetilde{E}(t)$ is the time-dependent amplitude of the field, $\mathbf{k}$ is the wave vector, $\mathbf{x}$ is the space coordinate of the sample and ω is the carrier frequency of the laser. Because of the geometrical arrangement, the contribution of beams E_a and E_c always carry a positive wave vector, and beam E_b a negative one (see Figure 1). The pulse sequence can be chosen to obtain different nonlinear optical processes. For example, the pulse sequence with the temporal order $\exp[i(\mathbf{k}_1\mathbf{x} - \omega t_1)]$, $\exp[-i(\mathbf{k}_2\mathbf{x} - \omega t_2)]$, and $\exp[i(\mathbf{k}_3\mathbf{x} - \omega t_3)]$, is known as virtual echo (186), whereas the pulse sequence with the temporal order $\exp[-i(\mathbf{k}_1\mathbf{x} - \omega t_1)]$, $\exp[i(\mathbf{k}_2\mathbf{x} - \omega t_2)]$, and $\exp[i(\mathbf{k}_3\mathbf{x} - \omega t_3)]$, is known as stimulated photon echo (15). Notice that each beam interacts only once with the molecules and that one is able to determine the signs for the electric field interactions within the constraints of the phase-matching detection geometry. This flexibility is not usually available with collinear pulses.

Here we illustrate how the time delay between the first two pulses can be used to control the FWM signal. The signal is proportional to the square of the average third-order polarization given by $P^{(3)} = Tr[\hat{\rho}^{(3)}\hat{\mu}]$, where $\hat{\rho}^{(3)}$ is the third-order density matrix and $\hat{\mu}$ the dipole moment operator. We derive simplified expressions for the density matrix elements with first- second-, and third-order dependence in the electric field interaction. This can be demonstrated with a simple model that includes two vibrational levels in the ground state and two in the electronically excited state. The vibrational levels are separated by $\hbar\omega_g$ and $\hbar\omega_e$ in the ground and excited states, respectively. The laser pulses are considered very short, such that their bandwidth is larger than ω_g or ω_e. The three pulses are degenerate and resonant with the electronic transition. The system is assumed to be at a temperature that allows the two ground state vibrational levels to be equally populated. After the first laser interaction the first-order density-matrix elements, $\rho_{ij}^{(1)}(t)$, depend on the sign of the interaction with the electric field. Interaction with $e^{-i\omega t}$ yields a first-order electronic coherence (off-diagonal elements)

$$\rho_{eg}^{(1)}(t) \propto e^{-i\omega_{eg}t}, \qquad \text{3a.}$$

where e, g denotes vibrational levels of the excited and ground state, respectively. Interaction with $e^{i\omega t}$ yields

$$\rho_{ge}^{(1)}(t) \propto e^{i\omega_{eg}t}. \qquad \text{3b.}$$

Population transfer occurs upon interaction with the second laser pulse. The resulting expression contains the population of the four levels. Notice that populations (diagonal elements) are time independent when no relaxation is included (as in this model). The simplified second-order density-matrix elements are given by the populations

$$\rho_{gg}^{(2)}(t) \propto -A\cos(\omega_e \tau_{12}/2), \qquad \text{4a.}$$

$$\rho_{ee}^{(2)}(t) \propto A\cos(\omega_g \tau_{12}/2), \qquad \text{4b.}$$

and the vibrational coherences

$$\rho_{g'g}^{(2)}(t) \propto -A\cos(\omega_e \tau_{12}/2)e^{-i\omega_g t}, \qquad \text{4c.}$$

$$\rho_{ee'}^{(2)}(t) \propto A\cos(\omega_g \tau_{12}/2)e^{i\omega_e t}, \qquad \text{4d.}$$

where the time delay between the first two laser pulses is τ_{12}, and A is a constant that depends on the laser intensity and the transition dipole moment. The primes indicate different vibrational levels in each electronic state. The dependence on

τ_{12} is manifested as well in the third-order density matrix expression for a virtual echo pulse sequence,

$$\rho_{eg}^{(3VE)}(t) \propto e^{-i\omega(t-(\tau_{13}-\tau_{12}))}[\cos(\omega_g\tau_{12}/2)\cos(\omega_e\tau_{23}/2)$$
$$+ \cos(\omega_e\tau_{12}/2)\cos(\omega_g\tau_{23}/2)e^{i\varphi_{eg}}], \qquad 5.$$

where the phase is given by $\varphi_{eg} = (1 + (-1)^{(e+g)})\omega_g\tau_{13}/2$. Notice that the third-order matrix elements achieve their maximum before the third pulse is applied, when $t = \tau_{13} - \tau_{12}$, hence the name virtual echo (186). From Equation 5 we can distinguish two oscillatory components as a function of the scanning time τ_{23}; one oscillates with ω_e, reflecting excited-state dynamics and the other with ω_g, reflecting ground-state dynamics. The molecular response for a virtual echo process can be expressed diagrammatically by using two double-sided Feynman diagrams (see Figure 5a), R_1 and R_4, corresponding to excited-state and ground-state dynamics, respectively (13–15, 185). The amplitude of each response function is controlled by the fixed time delay τ_{12}. Note that the parameter $\tau_{12} = 2\pi(n + 1/2)/\omega_e$, where n is an integer, can be used to cancel the R_4 response (see Equation 5). These values for τ_{12} would cancel the observation of ground-state dynamics as a function of τ_{23}. Notice that a similar condition exists for $\tau_{12} = 2\pi(n + 1/2)/\omega_g$ for canceling the excited-state contribution. The time delay between the first two pulses can be set such that two wave packets formed from the ground state vibrational levels interfere destructively, leading to the cancellation of the excited-state contribution (185). The control mechanism in a stimulated photon echo pulse sequence is apparent from the third-order density matrix elements,

$$\rho_{eg}^{(3SPE)}(t) \propto e^{-i\omega(t-(\tau_{13}+\tau_{12}))}[\cos(\omega_g\tau_{12}/2)\cos(\omega_e\tau_{13}/2)$$
$$+ \cos(\omega_e\tau_{12}/2)\cos(\omega_g\tau_{13}/2)e^{i\varphi_{eg}}]. \qquad 6.$$

Notice that the matrix element achieves its maximum at $t = \tau_{13} + \tau_{12}$. The stimulated photon echo technique offers two possibilities for control of the intramolecular dynamics. In one case, the time delay τ_{12} is fixed (scanning time τ_{23}); in the other case, the time delay τ_{13} is fixed (scanning time τ_{12}). The latter has been called mode suppression (187, 188). For both cases we can distinguish two oscillatory components as a function of the scanning time; one oscillates with ω_e, and the other with ω_g. The contribution of each component is controlled by the fixed time delay, τ_{13} or τ_{12}, allowing the control of molecular responses R_2 and R_3, respectively (see Figure 6a).

The coherent nature of FWM experiments provides the opportunity to harness the coherent properties of lasers for controlling intra- and intermolecular degrees of freedom. We have been exploring FWM methods in our group to achieve coherent control of molecular dynamics (71, 74, 183, 184) and illustrate this work with the following example. The data in Figure 5b were obtained with molecular iodine. The time delay between the first two pulses was fixed to be 460 fs (left transient)

Coherent Control of Excited and Ground Vibrational Wave-packets

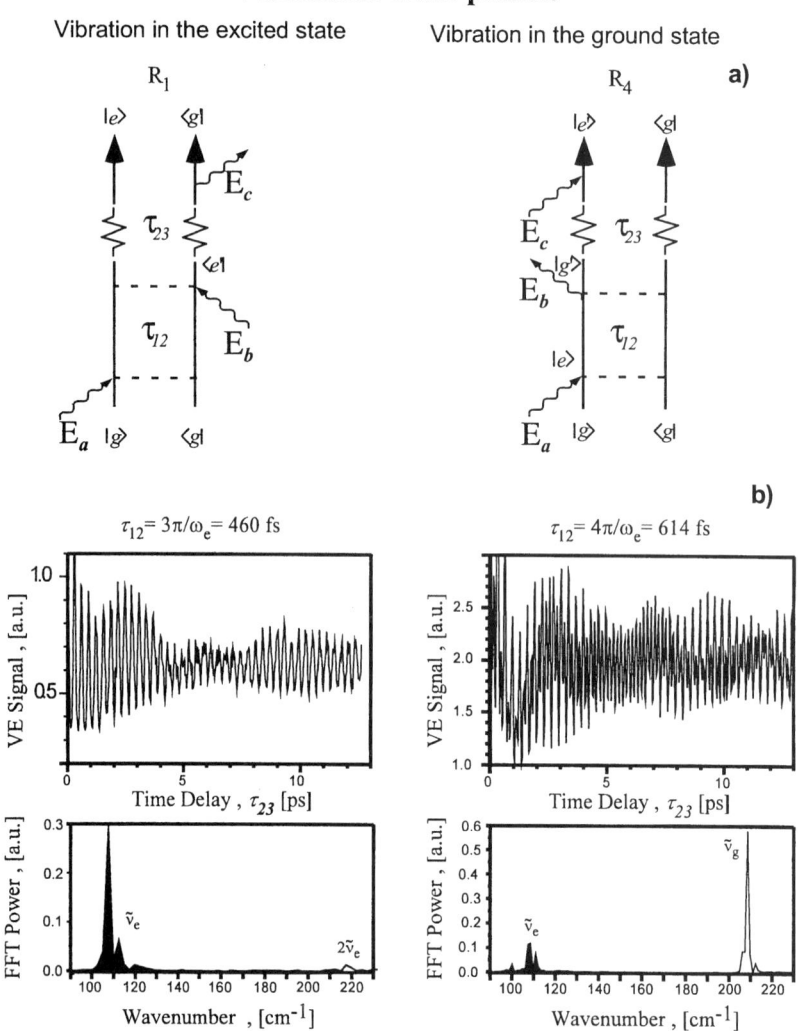

Figure 5 (*a*) Double-sided Feynman diagrams of the experimental processes observed. The response functions R_1 and R_4 are responsible for the observation of excited- and ground-state intramolecular dynamics, respectively. (*b*) Experimental FWM data on molecular iodine obtained for pulse-sequences with a fixed time-delay between the first two pulses. The signal corresponds to a virtual photon echo pulse sequence. For the left-side transient, the delay time τ_{12} was fixed at 460 fs, one and a half vibrational periods of the excited state of iodine. The observed vibrations have a period of 307 fs, corresponding to the excited-state vibrational motion. The power FFT (*bottom left*) shows that the main contribution is at 108 cm^{-1} corresponding to excited-state vibrational motion. For the right-side transient the delay time τ_{12} was fixed at 614 fs, two vibrational periods in the excited state of iodine. The observed dynamics have a period of 160 fs, corresponding to ground state vibrations. The power FFT (*bottom right*) shows a main contribution at 208 cm^{-1} corresponding to ground-state vibrations and a smaller component at 110 cm^{-1} that corresponds to excited-state vibrations.

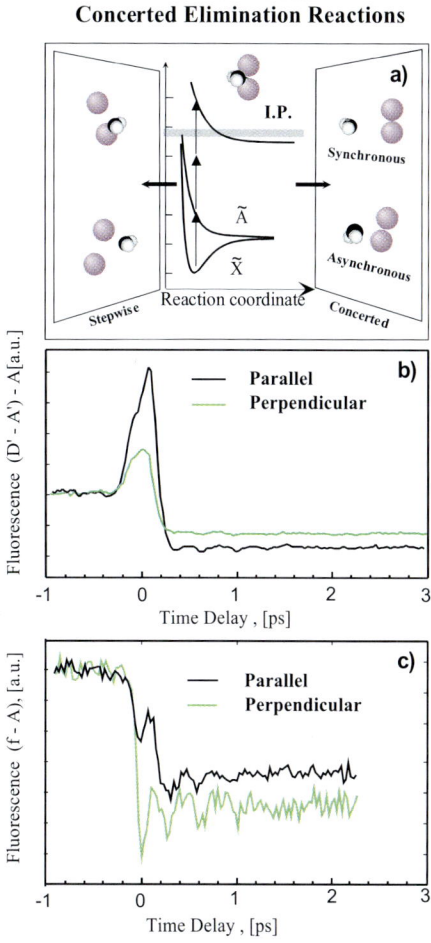

Figure 2 (*a*) Potential energy surfaces for CH_2I_2 in the gas phase showing stepwise (*left*) and concerted (*right*) photodissociation pathways following excitation by the pump pulse (312 nm). The concerted mechanisms are responsible for the production of I_2 in the D′ and f excited states as well as carbene. The concerted synchronous mechanism involves the simultaneous breakage of two C-I bonds and the formation of the I-I bond, while the concerted asynchronous mechanism requires the asymmetric progress of the reaction such that one C-I bond starts breaking before the other, but all bond rearrangement takes place in a single kinetic step. The later mechanism results in large rotational excitation of the product. (*b*) Transients obtained by depletion probing of the D′→A′ fluorescence collected at 340 nm of the nascent $I_2(D')$ product, for pulses polarized parallel (*black*) and perpendicular (*green*) to each other. (*c*) Transients obtained by depletion probing of the f → A fluorescence collected at 272 nm of the nascent $I_2(f)$ product, for pulses polarized parallel (*black*) and perpendicular (*green*) to each other.

Free → Bound Bimolecular Reactions
Femtosecond Photoassociation Spectroscopy

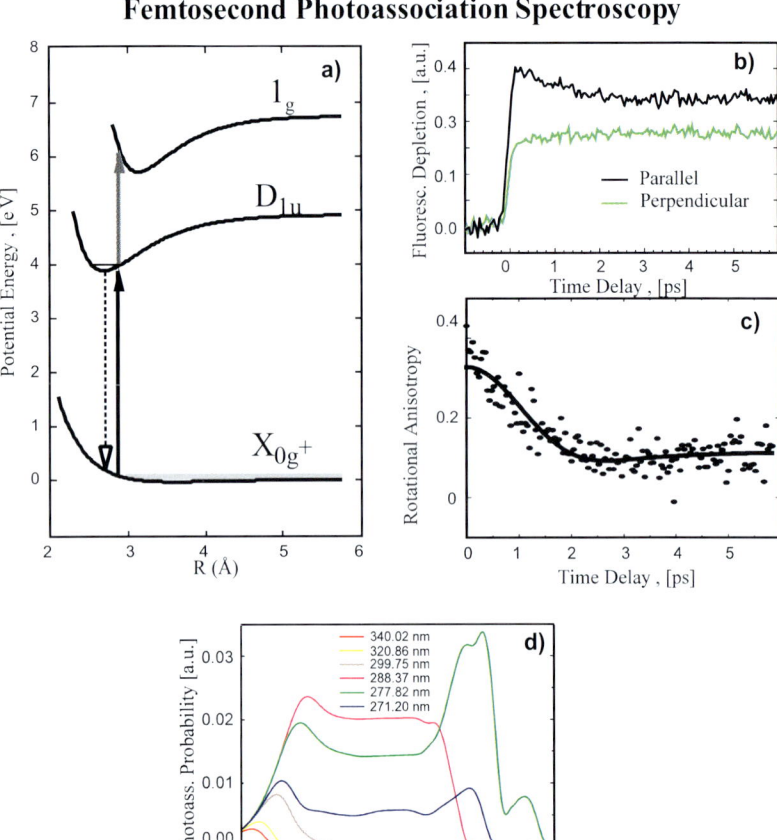

Figure 4 (*a*) Potential energy surfaces involved in the photoassociation reaction of Hg atoms. Hg_2 is formed in the $D1_u$ state after the pump pulse is applied. The probe pulse causes a depletion of the $D1_u$ state population by exciting the molecules into the 1_g state. (*b*) Fluorescence signal of Hg_2 $D\,1_u \rightarrow X\,O^+_g$ as a function of the delay between the pump and the probe pulses, for pulses polarized parallel and perpendicular to each other. At time zero, a sudden onset of the fluorescence takes place indicating that the photoassociated molecules are formed within the pulse duration. The maximum depletion probability corresponds to collision pairs that are aligned parallel with respect to the probing pulse. (*c*) Rotational anisotropy obtained from the experimental data (*dots*) and from a numerical fitting (*line*). The rotational dephasing of the initial anisotropy results from the rotational distribution of the excimers respect to a central rotational level. (*d*) Calculated photoassociation probability assuming a constant initial colliding energy, as a function of pulse duration for different binding pulse wavelengths.

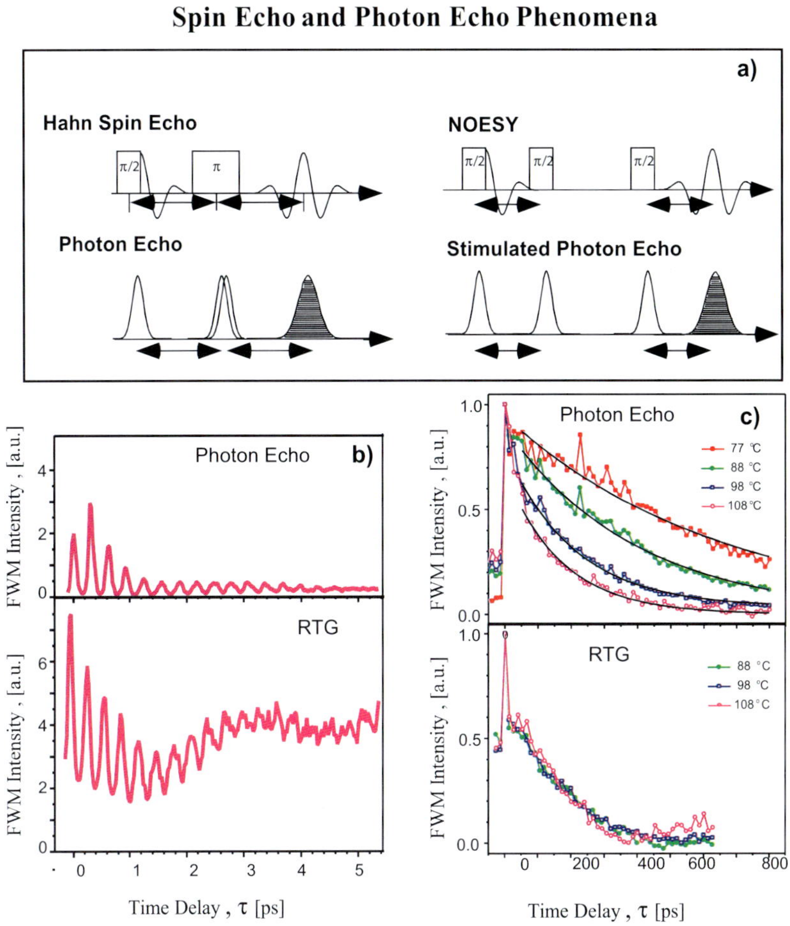

Spin Echo and Photon Echo Phenomena

Figure 7 (*a*) Schematic representation of the Hahn spin echo and of the Nuclear Overhauser Effect Spectroscopy (NOESY) pulse sequences used in nuclear magnetic resonance (*top*) and their optical counterparts, photon echo (PE) and stimulated photon echo (*bottom*). (*b*) Experimental time-integrated transients for photon echo and reverse transient grating (RTG) pulse sequences as a function of time delay between the scanned and the overlapped pulses. The PE signal is observed to be background free while the RTG data is strongly modulated by a dephasing process with a dip at 1.4 ps. (*c*) Homogenous vibronic coherence relaxation measurements of molecular iodine as a function of temperature using the PE (*top*) pulse sequence. When the RTG pulse sequence is used (*bottom*), we observe a temperature independent decay suggesting the large inhomogeneous relaxation process overwhelms the homogeneous rate of relaxation.

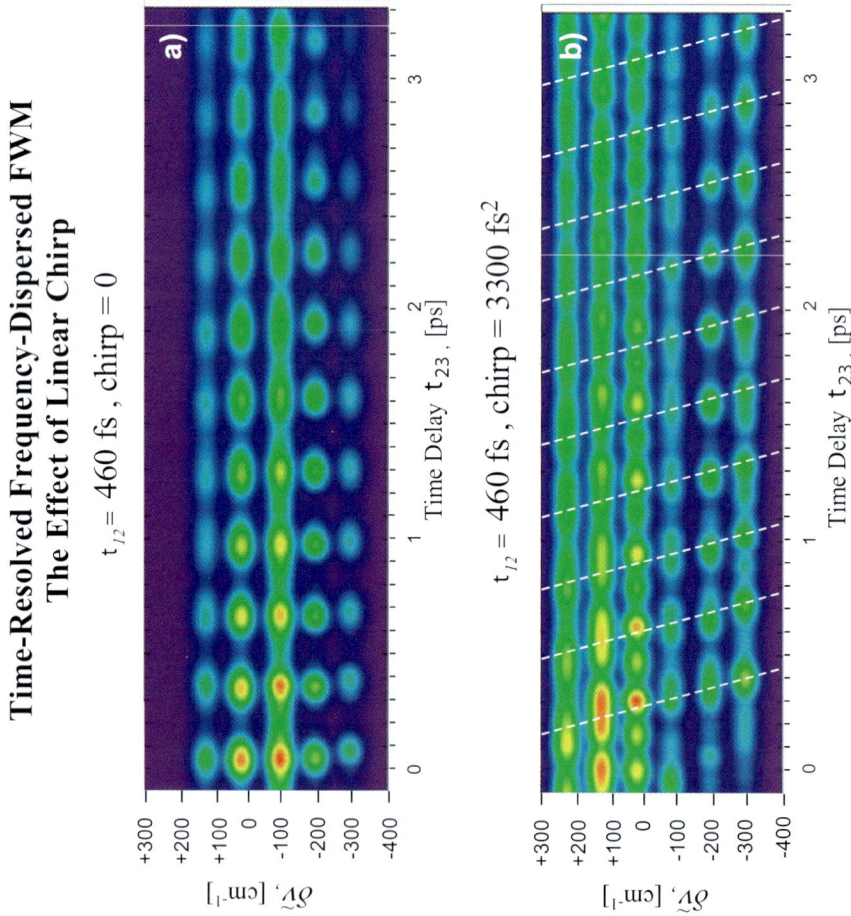

Figure 9 (*a*) Experimental time-resolved frequency dispersed FWM data for iodine vapor with $\tau_{12} = 460$ fs and transform limited pulses. This two dimensional contour plot shows only 307 fs oscillations from the excited state dynamics. There is no evidence of ground state dynamics contribution and all spectral components oscillate in phase. (*b*) Experimental time-resolved frequency dispersed data when $\tau_{12} = 460$ fs and $\phi'' = 3300$ fs^2. The chirped pulses lead to the formation of a chirped vibrational wave packet in the excited state. The dotted lines correspond to the chirp magnitude of the lasers. Note that the signal on the anti-Stokes side $\delta\tilde{v} > 0$ contains some contribution from ground and excited state dynamics.

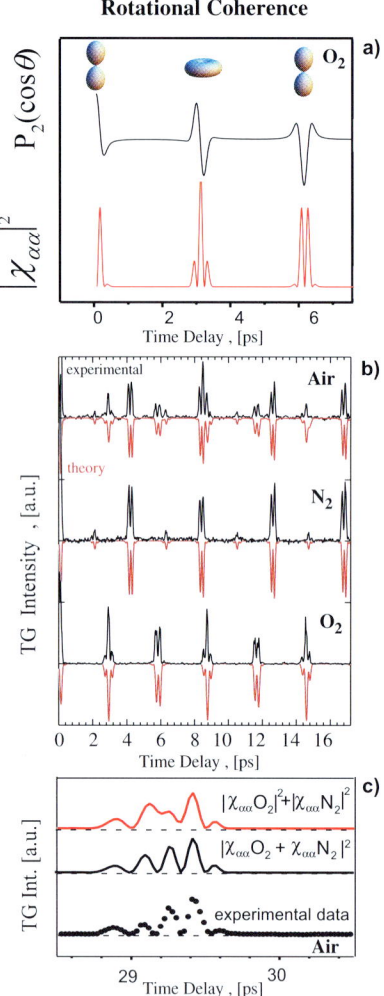

Figure 10 (*a*) Simulation of the time-dependent alignment $P_2(\cos\theta(\tau))$, and the impulsive coherent Raman scattering signal, proportional to $|\chi_{aa}(\tau)|^2$, for O_2. (*b*) Experimental (*black*) and theoretical (*red*) transient grating (TG) transients of air, nitrogen, and oxygen. Full rotational recurrences are observed at 4.15 ps for N_2 and at 5.77 ps for O_2. Half recurrences are also observed. The peaks in the air transient directly correspond to recurrences in the N_2 and O_2 scans. Simulated TG signals for these samples were calculated using Equation (9). Note that in the simulations, the full and half recurrences are reproduced at the same recurrence times and with the same intensity and shape as in the experimental signal. (*c*) Experimental TG signal from air for time delay in the range of 29 to 30 ps (*dots*). Simulations of the data using two models, the sum of O_2 and N_2 signals (*red*) and of the square of the sum of the O_2 and N_2 susceptibilities (*black*). The later simulation gives the best fit of the experimental data.

Off-resonance Transient Grating Measurements

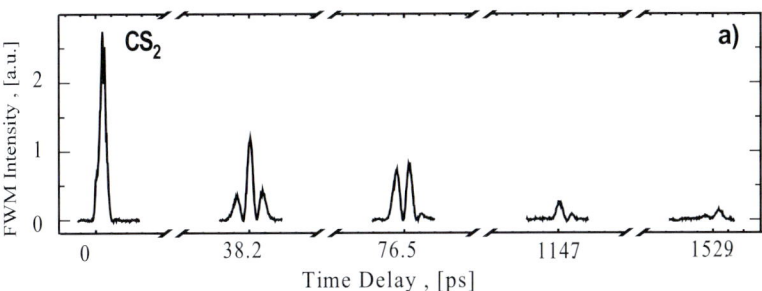

Time-Resolved and Spectrally-Dispersed TG on Benzene

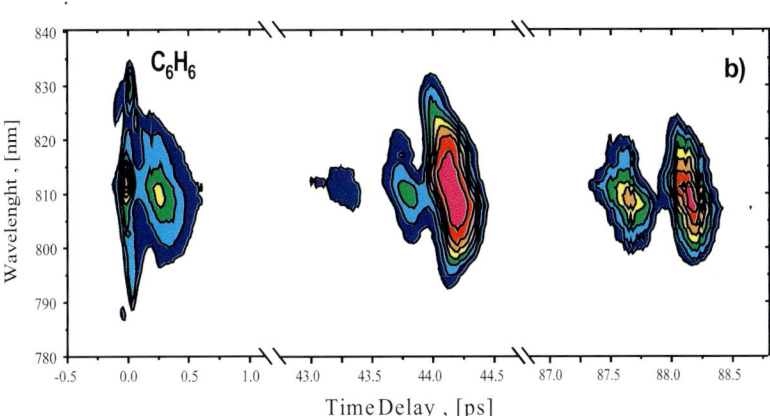

Figure 11 (*a*) Off-resonance transient grating experimental data obtained from carbon disulfide vapor. Half rotational recurrences are observed at 38.2 and 114.7 ps while full rotational recurrences are observed at 76.5 and 152.9 ps. (*b*) Off-resonance transient grating experimental data obtained from benzene vapor. The time-resolved spectrally dispersed signal is detected with a spectrometer and CCD detector. The initial feature corresponds to time-zero observed because of the instantaneous polarizability of the sample. The feature near 2.5 ps is caused by the initial rotational dephasing. The subsequent features correspond to the first and second rotational recurrences. Notice that the time axis is not continuous.

or 614 fs (right transient). The signal for $\tau_{12} = 460$ fs can be assigned to the dynamics of the excited state with an oscillation period of $\tau_e = 307$ fs corresponding to vibrational levels $v' = 6-11$. No evidence of ground-state dynamics is apparent in this transient or in the corresponding fast Fourier transform (FFT). For $\tau_{12} = 614$ fs, the signal oscillates with $\tau_g = 160$ fs, corresponding to the molecular dynamics in the ground state for vibrational levels $v'' = 2-4$. The FFT of these data confirms the predominant ground-state contribution.

The mechanism for signal emission corresponding to the ground-state transient can be understood in terms of a coherent Raman scattering process. The second pulse creates a coherent superposition of vibrational states in the ground state. This process is enhanced when the time between the first and second pulses matches the vibrational period of the excited state (71, 183, 184, 189, 190). The third laser pulse probes the resulting ground-state vibrational coherence. Coherent control over the ground- and excited-state vibrational wave packets is achieved by a combination of pump-dump and interference methods. When the time delay is set at $\tau_{12} = 460$ fs, the transfer of the excited state superposition of states to the ground state cannot take place. This cancels the contribution from R_4, leaving only R_1. Probing with pulse E_3 results in the observation of an excited-state vibrational coherence. The wave-packet representation of this process has been given by Pastirk et al (73, 204). Chen et al have shown that by using different color lasers they are able to detect the Stokes and anti-Stokes coherent Raman scattering and hence collect ground- and excited-state dynamics (191). Control over ground- or excited-state dynamics has been recently shown by Motzkus and coworkers using an adaptive pulse shaper with a learning algorithm (192). By detecting the photoelectron signal, Zanni et al have developed a very sensitive method to detect the ground-state dynamics observed after the second pulse (193).

Shank's group introduced a method aimed at suppressing the contribution of excited-state vibrational dynamics in order to improve relaxation rate measurements in liquids (187, 188, 194). They observed that when τ_{13} is in phase with the excited-state dynamics, $\tau_{13} = 2\pi n/\omega_e$ (mode suppression is on), the amplitude of the excited-state vibrations was greatly reduced. When τ_{13} was out of phase (mode suppression is off), the excited-state vibrations were very prominent (187, 188).

We have explored control of the response functions responsible for stimulated photon echo signals (68). For these measurements the time delay τ_{13} was kept fixed while the time τ_{12} was scanned. Based on the literature (187, 188), mode suppression should take place when τ_{13} is in phase with the vibrational motion of the excited state, 614 fs for gas-phase iodine, and mode suppression should not take place when the time delay τ_{13} is out of phase, here 460 fs. In Figure 6b we present measurements for these cases on gas-phase molecular iodine. When $\tau_{13} = 460$ fs, mode suppression is off and the data show pronounced 307 fs vibrations corresponding to the excited state without background or ground-state contributions. This is consistent with liquid-phase observations (188). The data obtained for $\tau_{13} = 614$ fs, when mode suppression is on, show a considerable background signal as well as ground- and excited-state vibrational dynamics. The two transients in

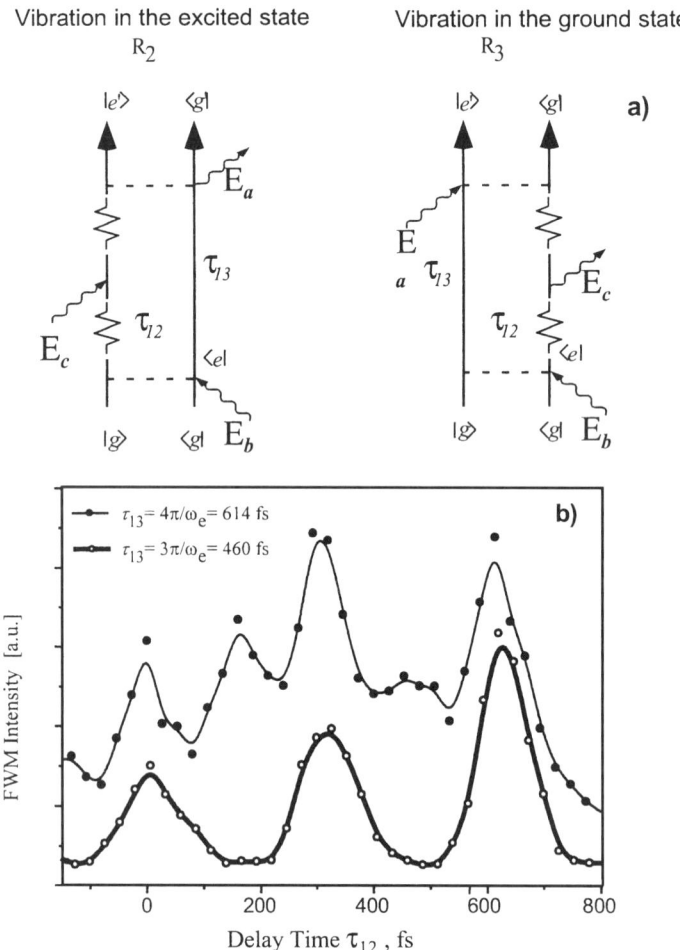

Figure 6 Stimulated photon echo measurements on gas-phase molecular iodine when the delay time between the first and last pulses is fixed and the second pulse is scanned between them. (*a*) Double-sided Feynman diagrams describing the Liouville pathways that contribute to the signal arising from this method. Molecular responses R_2 and R_3 correspond to the excited- and ground-state vibrational motion, respectively. (*b*) Four-wave mixing (FWM) signal for two values of the fixed time τ_{13}. When $\tau_{13} = 460$ fs, the signal is modulated with a period of 307 fs (open circles), corresponding to the vibrational period of the excited state. When $\tau_{13} = 614$ fs, the signal is modulated by 160 and 307 fs oscillations (filled circles), corresponding to a mixture of excited- and ground-state dynamics.

Figure 6b indicate that the background of the "mode suppressed" signal corresponds to a contribution that is independent of vibrational motion. However, excited- and ground-state vibrations are not suppressed in these measurements. When mode suppression is off, the molecular response known as R_2 is canceled. When mode suppression is on, both R_2 and R_3 contribute to the signal. Mode suppression is useful in liquid phase studies because when mode suppression is on, R_3 contributes a large signal that overwhelms the excited-state vibrational coherence.

We can summarize our work as follows: By selecting the fixed time delay between the pulses of different sequences we can control the different molecular responses. In this way, a molecular response can be canceled with specific time delays, as shown in Table 2 (71). Control over these response functions for a setup involving two pairs of collinear-phase locked pulses has recently been considered by Cina (195).

The previous example illustrates how the timing between the first two pulses can be used for control of the intramolecular dynamics (183, 184, 189). In a sense, the different pulse sequences can be thought of as optical analogues to multiple pulse NMR sequences (196, 197). The photon echo (PE) sequence is similar to the Hahn spin echo in NMR (198), and the stimulated photon echo is similar to the nuclear Overhauser effect spectroscopy method in NMR (199) (see Figure 7a, color insert). The cancellation of inhomogeneous broadening in PE measurements has been recognized since the first photon echo measurement in 1964 (200, 201). This advantage has been exploited to measure the homogeneous lifetime of complex systems such as large organic molecules in solution (72, 194, 202). Here we illustrate how this method works for molecular iodine.

TABLE 2 Pulse sequence control of third-order response functions[a]

R_1	R_2	R_3	R_4
$\tau_{12} = 2\pi\left(n + \frac{1}{2}\right)/\omega_g$	$\tau_{12} = 2\pi\left(n + \frac{1}{2}\right)/\omega_g$	$\tau_{12} = 2\pi\left(n + \frac{1}{2}\right)/\omega_e$	$\tau_{12} = 2\pi\left(n + \frac{1}{2}\right)/\omega_e$
$\tau_{23} = 2\pi\left(n + \frac{1}{2}\right)/\omega_e$	$\tau_{13} = 2\pi\left(n + \frac{1}{2}\right)/\omega_e$	$\tau_{13} = 2\pi\left(n + \frac{1}{2}\right)/\omega_g$	$\tau_{23} = 2\pi\left(n + \frac{1}{2}\right)/\omega_g$

[a]The third-order response functions relevant to three-pulse four-wave mixing are given using double-sided Feynman diagrams (13–15). The time delays given correspond to the values that minimize the particular response function. The principle we have used is to make the pulses arrive out of phase with respect to the molecular dynamics. The maximum response for each response function can be achieved when the pulses arrive in phase with the molecular dynamics (when the factor of $\frac{1}{2}$ is omitted).

Optical PE and spin echo processes are quite different phenomena. However, they have a large number of similarities. In the Hahn spin echo a coherent superposition of spins, originally pointed in the Z-axis, is rotated by 90 degrees into the XY plane. Inhomogeneous broadening in the sample causes dephasing of the coherent superposition as a function of time. Application of a π pulse causes an inversion in space, and hence the spreading motion becomes a focusing motion that leads to a rephasing of the original superposition. This generates the spin echo. The process can also be carried out by separating the π pulse into two $\pi/2$ pulses, known as nuclear Overhauser effect spectroscopy. In the optical PE process the first electric field creates a coherence on the bra (see Equation 3b). After interaction with $e^{i\omega t}$, all the electronic coherences evolve with a positive sign while the relaxation process takes place. Subsequent $e^{-i\omega t}$ interaction with two electric fields inverts the sign of the evolution, and the initial dephasing dynamics rephase to form an echo signal at $\tau = 2\tau_{12}$ (see Equation 6). The main differences between spin echo and optical PE signals are the following: In spin echo the signal is proportional to the entire polarization of the system, whereas in PE the signal is proportional to the third-order polarization (which can be quite small). Therefore, π or $\pi/2$ pulses that transform the entire population are not necessary to observe PE phenomena. In spin echo one usually works with a small number of levels, whereas in PE many more levels are available.

Experimental data for PE and reverse transient grating (RTG) measurements are shown in Figure 7b. In the pulse sequence for PE beam, E_b is followed by beams E_a and E_c overlapped in time ($\mathbf{k}_S = -\mathbf{k}_1 + \mathbf{k}_2 + \mathbf{k}'_2$), whereas for RTG beam, E_c is followed by E_b and E_a overlapped in time ($\mathbf{k}_S = \mathbf{k}_1 + \mathbf{k}_2 - \mathbf{k}'_2$). The differences observed in the background, undulation, and apparent signal-to-noise ratio in these data result from the difference in the first pulse interaction. When the first interaction is with $e^{i\omega t}$, action on the bra, the subsequent laser interactions lead to a cancellation of the inhomogeneous broadening in the sample and to the observation of the photon echo. When the first interaction is with $e^{-i\omega t}$, action on the ket, there is no mechanism to cancel the inhomogeneous contributions to the signal. The RTG data show a strong background and a slow undulation that results from the inhomogeneous rotational dephasing of the sample molecules. After the first 3 ps the RTG data show a mixture of ground- and excited-state dynamics. The observation of ground-state dynamics results from the initial thermal population of different vibrational modes. In the PE data only excited-state vibrational dynamics are observed.

In Figure 7c we present RTG and PE measurements for molecular iodine taken with long time delays. The measurements are taken as a function of temperature to illustrate the different mechanisms for coherence relaxation. Notice that the RTG measurements appear not to be temperature dependent in this temperature range (see Figure 7c). The reason for this observation is that the inhomogeneous contributions are caused by Dopler broadening, having a $T^{1/2}$ dependence. In the PE measurements we can see that the homogeneous relaxation times are much longer and are found to decrease with temperature. The transients are fit by a single exponential decay. The single exponential behavior is consistent with the random nature of dephasing collisions, and hence the Poisson statistics. The cause for the decreased coherence lifetime as a function of the temperature is the increase in the number density and hence an increase in the collision frequency. From these measurements a cross section for homogeneous vibronic relaxation, $\sigma = 1150 \pm 150$ Å^2, is obtained for iodine-iodine collisions (203). When the PE signal is recorded with higher temporal resolution vibrational and rotational features are revealed. The data in Figure 8a show that for PE the rotational coherence is maintained for hundreds of picoseconds. The transient shows the excited-state vibrational dynamics superimposed on the ground-state rotational coherence (Figure 8b). This combination is unexpected and may be the result of a macroscopic coherence (15, 204). The RTG data do not show the rotational and vibrational recurrences for time delays longer than 100 ps. Analysis of the Fourier transformed PE data (see Figure 8c) confirms that the vibrational coherence is due to the excited state.

The FWM signal contains valuable spectroscopic information, which can be extracted by detection with a spectrometer to obtain a signal as a function of three parameters, τ_{12}, τ_{23}, and ω_{eg} (73, 74, 184). Materny and coworkers have studied coherent anti-Stokes Raman scattering and degenerate FWM on iodine vapor (191, 205–208). Their data obtained using TG (positive time) and RTG (negative time) showed ground- or excited-state dynamics, depending on the detection wavelength (205, 209). In Figure 9a (see color insert), we show a time-resolved, spectrally dispersed virtual echo transient obtained with transform-limited pulses for $\tau_{12} = 460$ fs. Notice that all frequency components oscillate in phase with a period of 307 fs, corresponding to the excited-state vibrations. In order to explore the role of the pulse chirp in the control of the molecular dynamics, experimental data with $\tau_{12} = 460$ fs were obtained when beams E_1, E_2, and E_3 were equally chirped, $\phi'' = +3300$ fs^2. The bottom of Figure 9b shows the spectrally dispersed data for the above conditions. These data make it clear that chirped pulses lead to the formation of a chirped wave packet. The dotted lines correspond to the chirp of the laser and, as expected, the phase difference as a function of wavelength is imprinted on the wave packet (see Equation 2). The transient shows an increased contribution from the ground state (compare with upper plot). The mixing of both states' dynamics is evident in the anti-Stokes-shifted frequencies ($\delta\tilde{\nu} > 0$), whereas the excited-state dynamics prevail in the Stokes frequencies ($\delta\tilde{\nu} < 0$). Notice that the observed vibrations shift $\approx$40 fs for every 100 cm^{-1}, as expected from the introduced linear chirp. The complex dynamics observed following chirped pulse excitation could not be obtained from spectrally integrated transients.

OFF-RESONANCE PROBING OF GROUND STATE DYNAMICS

When the bandwidth of a laser overlaps several rotational and/or vibrational states the impulsive limit, a coherent superposition of states, can be formed by off-resonance excitation and its time evolution can be probed. The transient grating (TG) method is based on this principle and has been used extensively by the groups

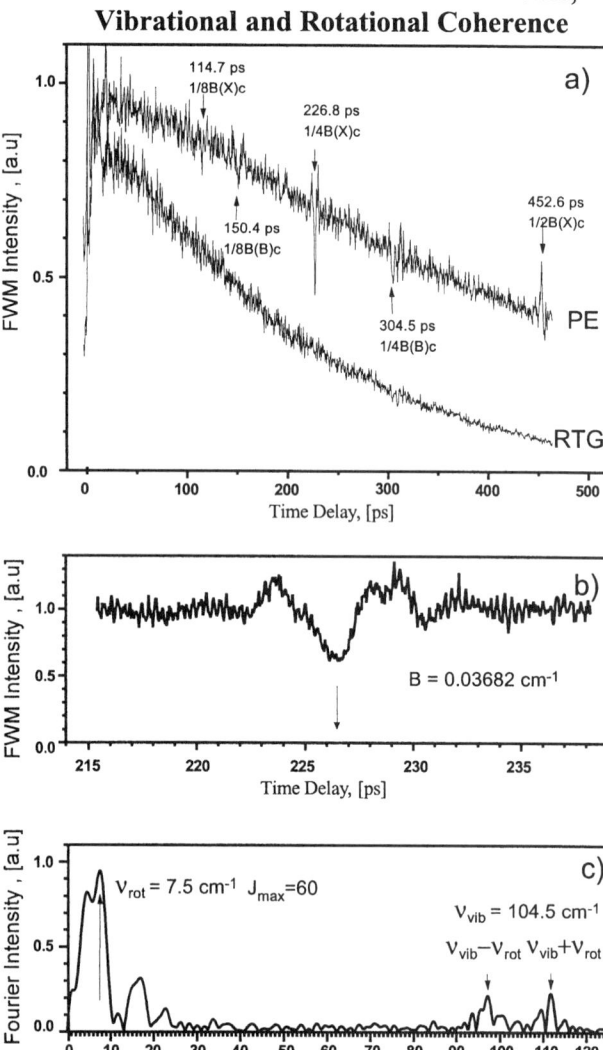

Figure 8 (*a*) Experimental measurements of reverse transient grating (RTG) signal and photon echo (PE) signal obtained with high temporal resolution. The PE signal contains rotational recurrences, which can be assigned to the ground (X) and excited (B) states. (*b*) Higher temporal resolution of a section of the PE signal around the largest rotational recurrence at 226.8 ps. The oscillatory features with a period of 307 fs correspond to excited-state vibrations, whereas the rotational revival corresponds to the ground state. (*c*) Fourier transform of the PE signal, showing the rotational component at low frequencies and the rovibrational components at higher frequencies. No other features were observed at higher frequencies.

of Fayer and Nelson to explore molecular dynamics in gas and condensed phases (175, 210–212). The crossing of the two plane-wave beams forms a grating in the sample with regions of high and low polarization. The third laser Bragg-diffracts from the grating to generate the observed signal. The diffraction process is very similar to the diffraction of X-rays from crystalline systems because the spatial arrangement of the lasers leads to the spatial coherence in the sample. More recent work on off-resonance probing of gas phase samples includes the work of Chen and coworkers using Raman-induced polarization spectroscopy (213), as well as others (214, 215).

The homodyne detected off-resonance FWM signal can be classified as a coherent Raman scattering process. When the three incident lasers pulses are ultrafast, the impulsive limit, the impulsive coherent Raman scattering (ICRS) signal can be evaluated using the following expression (15)

$$S_{ICRS}(\tau) = |\chi_{\alpha\alpha}(\tau)|^2 \qquad 7.$$

where

$$\chi_{\alpha\alpha}(\tau) \equiv -\frac{i}{\hbar}\langle[\alpha(\tau),\alpha(0)]\rho_g\rangle. \qquad 8.$$

In Equation 8, α is the electronic polarizability, and ρ_g represents the equilibrium ground-state density operator. Notice that the time-domain ICRS signal depends on the purely imaginary $\chi_{\alpha\alpha}(\tau)$, a quantity that depends on the response function associated with the electronic polarizability. For a system close to the classical limit, the imaginary part of any operator is proportional to time derivative of the full operator, therefore

$$\chi_{\alpha\alpha}(\tau) \cong \frac{1}{k_BT}\frac{d}{d\tau}\langle\alpha(\tau)\alpha(0)\rho_g\rangle. \qquad 9.$$

This result comes from the fluctuation-dissipation theory (15). Where the time correlation function, the expression in the brackets in Equation 9, depends on two factors. First, only Raman active modes, those for which $\partial\alpha/\partial q_j \neq 0$, where q_j represents a given normal mode, can be observed. Second, only the Raman active modes whose frequencies lie within the spectral window of the laser pulses participate. The bandwidth restriction is not expressly written in Equation 9 because we have assumed the impulsive limit, hence infinite bandwidth. In the gas phase, changes in the polarizability are caused by rotational and vibrational motion of the molecules. Raman transitions with linearly polarized light depend on the polarizability operator,

$$\hat{\alpha} = \hat{\alpha}_0 + \hat{\gamma}_0\frac{2}{3}P_2(\cos\theta), \qquad 10.$$

where $\hat{\alpha}_0$ and $\hat{\gamma}_0$ are the isotropic and anisotropic components of the polarizability operator and P_2 is the Legendre polynomial. The isotropic component gives the signal near time-zero delay of the scanning pulse; the anisotropic term gives the signal for later times. The selection rule for linear molecules is $\Delta J = 0, \pm 2$. After interaction with two overlapped pulses the rotational wave packet consists of three types of coherent states with amplitudes that depend on ΔJ. For $\Delta J = 0$ transitions, the alignment is time independent; therefore, these transitions do not contribute to the signal Keeping only the $\Delta J = \pm 2$ terms, the beating frequencies are $\Omega_J^{+2} = (\varepsilon_{J+2} - \varepsilon_J)/\hbar$, $\Omega_J^{-2} = (\varepsilon_J - \varepsilon_{J-2})/\hbar$. Where the rotational energy levels are given by $\varepsilon_J = [BJ(J+1) - DJ^2(J+1)^2]ch$, with B the rotational constant and D the centrifugal distortion in wave numbers. If we assume $\Omega_J^{+2} \approx \Omega_J^{-2} = \omega_J$ with $\omega_J = 2\pi c[(4B - 6D^3)(J+3/2) - 8D(J+3/2)^3]$, a situation that is true for large J, we obtain the following expression for the off-resonant signal:

$$S_{ICRS}(\tau) \propto \left| \sum_J n_J \omega_J \sin(\omega_J \tau) \right|^2, \qquad 11.$$

where n_J represents the initial population of each rotational state defined by the Boltzmann distribution. Figure 10a (see color insert) depicts the initial alignment of the molecules at time zero, represented by a cosine-squared distribution from $P_2(\cos\theta)$ in Equation 10. The broad rotational distribution causes fast rotational dephasing of the wave packet. Notice that the alignment is re-established after a time equivalent to $(1/4Bc)$, giving rise to a rotational recurrence or revival. Notice that a half recurrence is also depicted. Half recurrences can be observed whenever the contributions of odd and even J levels are unequal. The relative contributions are determined by the nuclear spin statistics. The signal is proportional to the square of the derivative of the time-dependent alignment; this function is depicted in the bottom of Figure 10a.

Figure 10b shows data for air, nitrogen, and oxygen. The experimental data are shown in black, and the simulations, using Equation 11, are shown as mirror images in red. From the data, it is clear that the rotational recurrence time for nitrogen is 4.15 ps and the rotational recurrence time for oxygen is 5.77 ps. Figure 10b shows that off-resonance probing is ideal to study mixtures, no wavelength tunability is required, and the signals from all components are separated by the differences in their rotational constants. Our results are in very good agreement with those of Chen and coworkers obtained using Raman-induced polarization spectroscopy (213, 217). It is particularly interesting when signals from two different species overlap. This situation is shown in Figure 10c. The sample is air, at a time delay at which oxygen and nitrogen recurrences overlap. Two simulations are given for these data. The first contains a sum of both signals squared, and the second contains the square of the sum of both contributions. The latter simulation gives the best fit to the data. Because the signal arises from the macroscopic polarization of the sample, it is proportional to the square of the sum of all anisotropic contributions; therefore, cross terms are expected. These data are valuable because they show that overlapping recurrences can be used to calibrate a time-domain spectrometer and to amplify the signal of a weak sample by mixing it with a strongly scattering sample. Hayden & Chandler used a pair of femtosecond pulses with different wavelengths to coherently excite high vibrational overtones by a CARS process (225). A third laser pulse probed the grating generated by the first two pulses. These gas-phase measurements showed the early rotational dephasing of the molecular ensemble.

In the 1970s, Heritage et al used a picosecond transient birefringence method to observe the rotational recurrences of CS_2, showing that time-resolved data can provide accurate rotational constants (218). Later, measuring polarized fluorescence from jet-cooled molecules, Zewail, Felker, and others have explored the molecular structure of large organic molecules and clusters using time-resolved rotational coherence spectroscopy (219–224).

In Figure 11 (see color insert), we show time-resolved, off-resonance TG measurements on CS_2 and benzene vapors. The CS_2 data (shown in Figure 11a) contain the initial rotational dephasing near time zero, two half-rotational recurrences at 38.2 and at 114.7 ps, and two full rotational recurrences at 76.5 and 152.4 ps. From these recurrences, a rotational constant of $B = 0.10912 \pm 0.00002$ cm^{-1} and a centrifugal distortion of $D = (6.4 \pm 0.2)10^{-9}$ cm^{-1} can be determined. These constants are in very good agreement with the literature (226). The signal is found to decay owing to collisional dephasing. Figure 11b shows off-resonance, time-resolved, and spectrally dispersed FWM data for benzene. The signal, in this case, is detected with a spectrometer and a CCD detector, allowing frequency resolution. The initial features are the time-zero instantaneous polarizability response and the initial rotational dephasing. The following features correspond to the rotational recurrences. From these data, the rotational constant for benzene can be determined to be 0.1897 ± 0.0002 cm^{-1}. This value is in excellent agreement with the literature value of 0.1896 cm^{-1} (227). The spectral information can be used to separate imperfections in the laser pulse, such as chirp from the molecular dynamics.

Experimental data on gas phase HgI_2 together with the theoretical simulation are shown in Figure 12. The transient consists of three contributions. At time zero there is a sharp feature corresponding to the isotropic instantaneous polarizability. This feature has no dependence on the intramolecular degrees of freedom and it is observed for all media, even for isolated atoms (13, 216). The data show a fast vibration that is modulated by a very low frequency envelope. The vibrations with a 211-fs period correspond to the symmetric stretch, the only Raman-active mode in this linear molecule (228). The slow modulation belongs to the anisotropic contribution to the signal that depends on the molecular orientation.

To model the data in Figure 12, we separate the isotropic and anisotropic contributions to the susceptibility (229). The vibrational motion makes the major contribution to the isotropic susceptibility. Based on the bandwidth of our laser pulse, only a few vibrational overtones are excited coherently. The anisotropic part of the susceptibility depends on the changes in orientation of the molecules caused by rotational motion, with some contribution from vibrational motion. There is an additional zero-time feature with amplitude A_z that arises from the equilibrium isotropic polarizability α_0. The off-resonance transient-grating (TG)

Ground State Rotational and Vibrational Dynamics

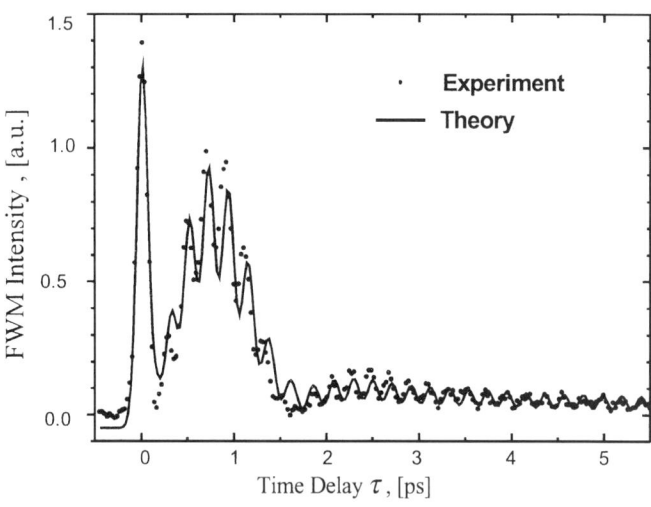

Figure 12 Off-resonance transient-grating signal for HgI$_2$. The experimental transient (*circles*) is modeled using Equation 12 to obtain the theoretical simulation (*solid line*). The time-zero feature corresponds to the instantaneous polarizability. The fast vibrations, with a 211-fs period, correspond to the symmetric stretch, and the slow modulation corresponds to the rotational anisotropy.

signal for a molecular system in the gas phase with an active Raman vibrational mode is

$$S_v(\tau) = \left| A_Z \delta(\tau) + A_v e^{-(\tau/T_{2v})^2} \cos\left(\frac{1}{2}\omega_v \tau + \phi_{iso}\right) \right.$$
$$\left. + A_{rot} \cos\left(\frac{1}{2}\omega_v \tau + \phi_{aniso}\right) \sum_J n_J \omega_J \sin(\omega_J \tau) \right|^2, \qquad 12.$$

where the constants A_v and A_{rot} are the amplitudes of the time-zero vibrational and rotational components, respectively; ϕ_{iso} and ϕ_{aniso} are vibrational phases, and ω_v is the vibrational frequency of the mode involved. T_{2v} represents the vibrational relaxation time. The first term represents the contribution of the instantaneous polarizability, and the second term represents the vibrational contribution, both caused by the isotropic component of the polarizability. Finally, the last term is related to the anisotropic polarizability, depending on the vibrational and rotational motions. Convolution of the simulation by the finite temporal width of our laser pulses yields the final result.

The simulation of the HgI$_2$ data is shown in Figure 12 (circles) together with the experimental data (line). The data were simulated using Equation 12. In our model the three amplitude parameters, the phases, and the relaxation time were adjusted; all the other values were obtained from spectroscopic parameters (228, 230). The relaxation time obtained from the simulation is $T_{2v} = 20$ ps. It is clear that the model reproduces the most salient characteristics of the data. The small differences between model and data could be reduced by using a nonlinear least-squares fitting routine.

We have used the TG technique to explore the behavior of molecules in strong laser fields. Strong nonresonant laser fields from ultrafast laser pulses can cause very large electric-field gradients. Molecules experience a large torque along the polarization vector of the field owing to their anisotropic polarizability. With long laser pulses the torque is enough to cause adiabatic alignment (231, 232). However, with ultrafast pulses the torque provides an "instantaneous" kick towards alignment (233). For strong enough fields electronic state mixing can occur. When the electronic states have different geometry, this process leads to molecular deformation. Corkum and coworkers have combined intense off-resonance, chirped, circularly polarized fields to induce rotational acceleration and thus constructed a molecular centrifuge (234). For our measurements we use two high-intensity laser pulses followed by a weak probe pulse. The signal corresponds to the diffraction of the probe laser from the TG formed by the two intense pulses in the sample (216). The degree of molecular alignment and deformation are measured from changes in the rotational recurrences that are observed after field-free evolution of the molecular sample.

The full rotational recurrence of CS_2 occurs at 76.5 ps, as seen in Figure 11a (216, 218). When the laser pulses are weak this feature can be observed, and an accurate rotational constant can be determined for the linear molecule (216, 218). In Figure 13a we show experimental data for the cases in which the first two laser pulses are one or two orders of magnitude stronger than the probe pulse. For low intensities the rotational recurrence can be modeled accurately by Equation 11. When the laser intensity of the first two pulses is increased by an order of magnitude the full rotational recurrence increases in intensity. The data shown in Figure 13a, when the laser intensity is 10^{11} W/cm^2 (open squares), can be fitted by a room-temperature distribution of linear CS_2 molecules. When the intensity is increased above 10^{12} W/cm^2 the full rotational recurrence increases in intensity and changes shape [see Figure 13a (filled squares)]. Most markedly, the first feature at 76.1 ps becomes more intense than the second feature at 76.8 ps. The features observed at long time delays, >77.5 ps, increase in intensity, and new oscillations are observed. This region of the transient is amplified by a factor of five in the figure (green trace). The high intensity data in Figure 13a are consistent with bending and stretching induced by the electric field. Roberts and coworkers (235) have recently studied electronic state mixing caused by high-intensity femtosecond laser excitation.

We have explored the use of off-resonance probing to study transients in flames. Figure 13b shows the changes caused by increasing the temperature of a sample. The data shown here were obtained for CO_2 at room temperature and at 400°C. Notice that at higher temperatures the rotational coherence signal sharpens up and shifts slightly toward longer values. The narrowing is caused by the broader

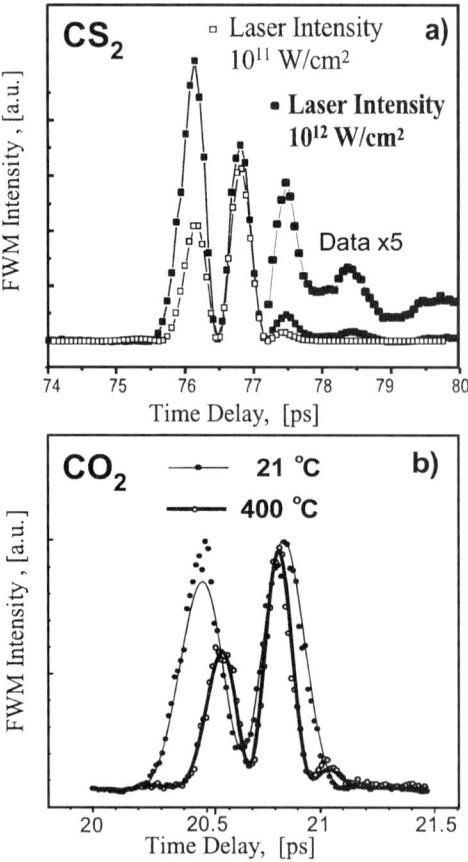

Figure 13 (a) Effect of strong field intensity on the first full rotational recurrence of CS_2. Low laser intensity experimental data (open squares) are compared with high laser intensity data (filled squares) at room temperature. The transients were normalized to the laser intensity. A portion of the strong field transient is shown after a 5x amplification to highlight the additional features observed after 77 ps. (b) The effect of the temperature on the first full recurrence of CO_2. Experimental data are shown as dots and the simulations using Equation 11 are shown as lines. The room temperature (21°C) transient (filled circles) is compared with the high temperature (400°C) transient (open circles).

rotational distribution, and the shift toward longer times is caused by the centrifugal distortion. The main limitation of the TG technique as discussed here to study molecular species in flames is the faster collisional-dephasing time caused by the very fast translational motion.

Strong field laser excitation of atoms and molecules is an active area of research. Among the current topics of interest are generation of high harmonics, (236–242), above-threshold ionization (243–246), and plasma generation (247–249). The goal of these studies is usually to detect the high-order outcome of the interaction. Alternatively, one could explore the effects of strong fields on the sample. We have reviewed work from our group on a four-wave mixing (FWM) method to probe the sample after high-intensity laser interaction. This method provides high-resolution structural information and reports on changes induced by strong fields.

CONCLUDING REMARKS

The quest to observe and control chemical reaction dynamics has been one of the most formidable scientific endeavors in physical chemistry since the nature of the chemical bond was uncovered (250). Progress over the past four decades has been astonishing: from the microsecond experiments of Porter (2, 3), to the indirect methods based on molecular beams (80, 81, 115, 251), and more recently, to the direct probing of the transition states of chemical reactions (252); we have witnessed a nine-order-of-magnitude improvement in time resolution. Many important lessons have been learned along the path of discovery. This review focuses on fundamental questions that are best studied in isolated systems, including concerted elimination chemical processes, bimolecular chemical reactions, wave packet dynamics, and molecules in strong fields. Regarding laser control, a number of schemes have been proven effective. Our approach has been to show that femtosecond time-resolved studies of isolated molecules provide crucial information regarding the best strategies for control.

Here we concentrate on the elimination of halogen molecules from dihalogenated alkanes. The experiments have shown that two bonds are broken and one is formed within 50 fs. The process has been shown to be faster than intramolecular vibrational relaxation and therefore it lends itself to relatively simple control schemes such as chirped laser excitation. There are a large number of chemical reactions thought to occur by concerted mechanisms. Improvements in time-resolved techniques would allow us to determine the precise chemical mechanism. The issue of concertedness is of importance in solution as well as in biological molecules. For example, the release of oxygen during the photosynthetic process is thought to proceed by a concerted process that avoids the formation of oxygen radicals (253–255).

The study of bimolecular reactions with the same level of detail and time resolution as unimolecular reactions remains a major challenge. We review work on studies based on van der Waals clusters that serve as precursors for bimolecular reactions. More attention has been given to studies based on femtosecond photo-association. These studies promise to bring the powerful techniques being used to study unimolecular processes to bear on the rich dynamics involved in bimolecular reactions. These studies are challenging because they involve a larger number of chemical species that can interfere with the signals, and the concentration of collision partners is very small. The studies presented on the photoassociation of mercury atoms to form the excimer molecule are an important step in this direction (96, 124, 126, 154, 155, 256, 257).

We have reviewed the use of nonlinear optical methods like three-pulse four-wave mixing to measure dephasing times, to observe and to control rovibrational wave packets in the ground and excited states. The observation of vibrational and rotational wave packet dynamics by pump-probe methods has been reviewed previously. The pulse sequences in our FWM experiments combine the pump-probe concept and interference between quantum mechanical pathways to select the dynamics between the different Liouville pathways that yield signal. Therefore, with certain pulse sequences we are able to follow exclusively ground- or excited-state dynamics. These multiple pulse methods have some analogies with multiple pulse NMR. Extension of these third-order optical methods to fifth and higher orders is already under way following the theoretical work of Tanimura & Mukamel (258) and Ivanecky & Wright (259). Experimental results have already been obtained in some laboratories (260–272). The signal from some of the early measurements included cascading effects between third- and higher-order processes. In our work we have observed the cascading of a first-order process with a third-order process (273). This signal can be isolated with a specific pulse sequence that does not permit FWM to occur. The future goal of our studies is to control the wave-packet motion in molecules with multiple vibrational normal modes. These studies will seek to control the pathways for intramolecular vibrational energy redistribution with the goal of controlling reactivity. These methods can be applied to both ground and excited states.

Recent advances in laser technology allow the creation of extremely strong fields. Here we introduce off-resonance transient-grating techniques to study the effects of these high fields on molecules. Our experiments are relevant to the goals of controlling the external degrees of freedom of molecules with the purpose of focusing and aligning molecular beams using intense off-resonance fields (274, 275). Different regimes are observed, from perturbation of spectroscopic lines, to electronic state mixing, and finally to ionization.

In Figure 14 we seek to find some unifying concepts in the quest for laser observation and control of molecular dynamics. The roots of the tree identify some concepts and methods that have been introduced over the years based on the fundamental interactions of lasers and molecules. The trunk depicts the double-sided Feynman diagrams of the four Liouville paths that govern laser interactions with matter (up to fourth order). Here we find brave scientists taking these concepts and methods to apply them to the fundamental problems in physics, chemistry, and biology, as well as more applied disciplines such as computer science and medicine

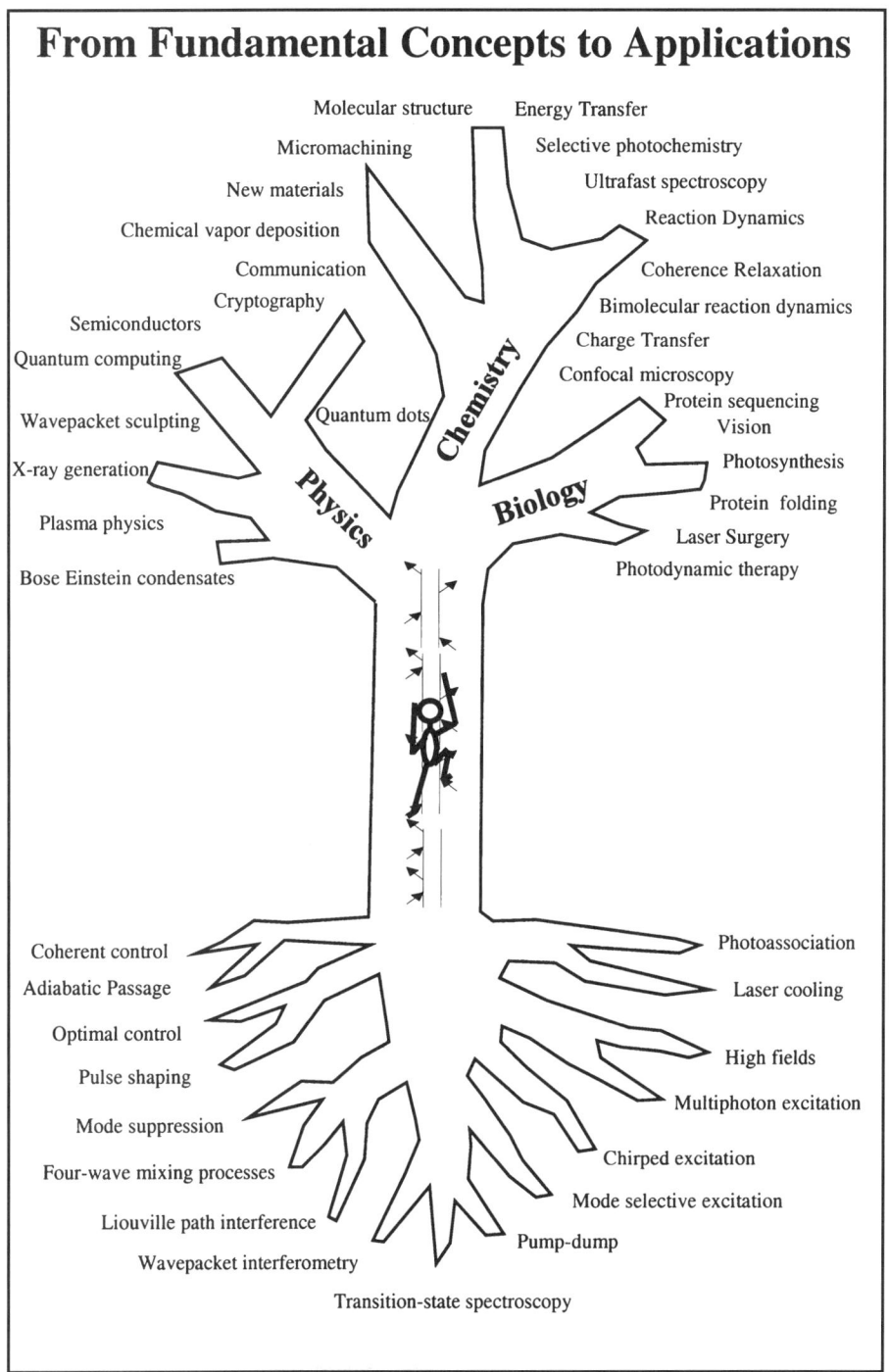

Figure 14 Conceptual representation of the progress from fundamental ideas to applications in the major sciences and industry. See text for details.

and in industry. Many of the concepts in this tree are interrelated and can be associated with more than one discipline. Throughout the review, we reference the methods that we have used in our group's quest to take fundamental dynamics observations and use them for controlling molecular dynamics.

Advances in the past decade give definite hope for laser-controlled chemical processes. Progress will likely proceed along two lines. The first is the continued fundamental work on understanding the interaction of photons and molecules. The second is an industrial type of approach in which the main goal will be to find the optimum field to obtain a product. Both will continue to expand our knowledge and will allow us to deliver laser control to complex systems in which the Hamiltonian cannot be calculated. We hope this review stimulates a number of scientists and justifies continued funding of fundamental studies that will eventually lead to future applications.

ACKNOWLEDGMENTS

It is a pleasure to acknowledge Shaul Mukamel, Jeff Cina, YJ Yan, and Gary Blanchard, Robert Cukier, James Harrison, and David Weliky for many stimulating discussions. The research presented here has been carried out by the following very talented students and postdocs: EJ Brown, M Comstock, BI Grimberg, P Gross, VV Lozovoy, U Marvet, I Pastirk, K Walowicz, and Q Zhang. This research was partially funded by a grant from the National Science Foundation (CHE-9812584). Major funding for our work came from a Lucille and David Packard Science and Engineering fellowship and an Arnold and Mabel Beckman Young Investigator award. The investment of these foundations on fundamental scientific discovery is gratefully acknowledged. MD is a Camille Dreyfus Teacher-Scholar and an Alfred P. Sloan Fellow. During the course of our research, EJ Brown was supported by a National Science Foundation Graduate Fellowship and I Pastirk by a James L Dye Endowment Fellowship.

Visit the Annual Reviews home page at www.AnnualReviews.org

LITERATURE CITED

1. Hirschfelder JO, Eyring H, Topley B. 1936. *J. Chem. Phys.* 4:170–77
2. Porter G. 1950. *Discuss. Faraday Soc.* 9:60–82
3. Porter G, Ward B. 1965. *Proc. R. Soc. London* 287:457
4. Hochstrasser R, Kaiser W, Shank CV. 1980. In *Chemical Physics*, Vol. 14. Berlin: Springer-Verlag
5. Wiesenfeld JM, Greene BI. 1983. *Phys. Rev. Lett.* 51:1745–48
6. Taylor AJ, Erskine DJ, Tang CL. 1984. *Chem. Phys. Lett.* 103:430–35
7. Nuss MC, Zinth W, Kaiser W, Kolling E, Oesterhelt D. 1985. *Chem. Phys. Lett.* 117:1–7
8. Doany FE, Hochstrasser RM, Greene BI, Millard RR. 1985. *Chem. Phys. Lett.* 118:1–5
9. Rosker MJ, Wise FW, Tang CL. 1986. *Phys. Rev. Lett.* 57:321–24
10. Shank CV. 1986. *Science* 233:1276–80

11. Scherer NF, Knee JL, Smith DD, Zewail AH. 1985. *J. Phys. Chem.* 89:5141–43
12. Dantus M, Rosker MJ, Zewail AH. 1987. *J. Chem. Phys.* 87:2395–97
13. Shen YR. 1984. *The Principle of Nonlinear Optics.* New York: Wiley
14. Boyd RW. 1992. *Nonlinear Optics.* San Diego: Academic
15. Mukamel S. 1995. *Principles of Nonlinear Optical Spectroscopy.* New York: Oxford Univ. Press
16. Tannor DJ, Rice SA. 1985. *J. Chem. Phys.* 83:5013–18
17. Brumer P, Shapiro M. 1989. *Acc. Chem. Res.* 22:407–13
18. Shi S, Rabitz H. 1990. *J. Chem. Phys.* 92:364–76
19. Shapiro M, Brumer P. 1993. *J. Chem. Phys.* 98:201–5
20. Gordon RJ, Rice SA. 1997. *Annu. Rev. Phys. Chem.* 48:601–41
21. Shapiro M, Brumer P. 1986. *J. Chem. Phys.* 84:4103–4
22. Brumer P, Shapiro M. 1992. *Annu. Rev. Phys. Chem.* 43:257–82
23. Zhu LC, Kleiman V, Li XN, Lu SP, Trentelman K, Gordon RJ. 1995. *Science* 270:77–80
24. Zhu LC, Suto K, Fiss JA, Wada R, Seideman T, Gordon RJ. 1997. *Phys. Rev. Lett.* 79:4108–11
25. Tannor DJ, Kosloff R, Rice SA. 1986. *J. Chem. Phys.* 85:5805–20
26. Tannor DJ, Rice SA. 1988. *Adv. Chem. Phys.* 70:441–523
27. Bowman RM, Dantus M, Zewail AH. 1990. *Chem. Phys. Lett.* 174:546–52
28. Baumert T, Gerber G. 1994. *Isr. J. Chem.* 34:103–14
29. Baumert T, Thalweiser R, Weiss V, Gerber G. 1995. In *Femtosecond Chemistry*, ed. J Manz, L Wöste, pp. 397–432. Weinheim: VCH
30. Assion A, Baumert T, Helbing J, Seyfried V, Gerber G. 1996. *Chem. Phys. Lett.* 259:488–94
31. Averbukh IS, Vrakking MJJ, Villeneuve DM, Stolow A. 1996. *Phys. Rev. Lett.* 77:3518–21
32. Pausch R, Heid M, Chen T, Kiefer W, Schwoerer H. 1999. *J. Chem. Phys.* 110:9560–67
33. Yan YJ, Gillilan RE, Whitnell RM, Wilson KR, Mukamel S. 1993. *J. Phys. Chem.* 97:2320–33
34. Krause JL, Whitnell RM, Wilson KR, Yan Y, Mukamel S. 1993. *J. Chem. Phys.* 99:6562–78
35. Shi S, Woody A, Rabitz H. 1988. *J. Chem. Phys.* 88:6870–83
36. Judson RS, Rabitz H. 1992. *Phys. Rev. Lett.* 68:1500–3
37. Kosloff R, Rice SA, Gaspard P, Tersigni S, Tannor DJ. 1989. *Chem. Phys.* 139:201–20
38. Ruhman S, Kosloff R. 1990. *J. Opt. Soc. Am. B* 7:1748–52
39. Broers B, van den Heuvell HBV, Noordam LD. 1992. *Phys. Rev. Lett.* 69:2062–65
40. Melinger JS, Hariharan A, Gandhi SR, Warren WS. 1991. *J. Chem. Phys.* 95:2210–13
41. Band YB, Julienne PS. 1992. *J. Chem. Phys.* 96:3339–41
42. Band YB. 1994. *Phys. Rev. A* 50:5046–50
43. Cerullo G, Bardeen CJ, Wang Q, Shank CV. 1996. *Chem. Phys. Lett.* 262:362–68
44. Bardeen CJ, Wang Q, Shank CV. 1998. *J. Phys. Chem. A* 102:2759–66
45. Kohler B, Yakovlev VV, Che JW, Krause JL, Messina M, et al. 1995. *Phys. Rev. Lett.* 74:3360–63
46. Bardeen CJ, Che JW, Wilson KR, Yakovlev VV, Cong PJ, et al. 1997. *J. Phys. Chem. A* 101:3815–22
47. Papanikolas JM, Williams RM, Leone SR. 1997. *J. Chem. Phys.* 107:4172–78
48. Bardeen CJ, Che JW, Wilson KR, Yakovlev VV, Apkarian VA, et al. 1997. *J. Chem. Phys.* 106:8486–503
49. Zadoyan R, Schwenter N, Apkarian VA. 1998. *Chem. Phys.* 233:353–63
50. Weinacht TC, Ahn J, Bucksbaum PH. 1998. *Phys. Rev. Lett.* 80:5508–11
51. Lozovoy VV, Antipin SA, Gostev FE, Titov

AA, Tovbin DG, et al. 1998. *Chem. Phys. Lett.* 284:221–29
52. Lozovoy VV, Sarkisov OM, Vetchinkin AS, Umanskii SY. 1999. *Chem. Phys.* 243:97–114
53. Sarkisov OM, Tovbin DG, Lozovoy VV, Gostev FE, Antipin AA, Umanskii SY. 1999. *Chem. Phys. Lett.* 303:458–66
54. Bardeen CJ, Yakovlev VV, Wilson KR, Carpenter SD, Weber PM, Warren WS. 1997. *Chem. Phys. Lett.* 280:151–58
55. Yakovlev VV, Bardeen CJ, Che JW, Cao JS, Wilson KR. 1998. *J. Chem. Phys.* 108:2309–13
56. Pastirk I, Brown EJ, Zhang Q, Dantus M. 1998. *J. Chem. Phys.* 108:4375–78
57. Kleiman VD, Arrivo SM, Melinger JS, Heilweil EJ. 1998. *Chem. Phys.* 233:207–16
58. Legare F, Chelkowski S, Bandrauk AD. 2000. *J. Raman Spectrosc.* 31:15–23
59. Chelkowski S, Bandrauk AD. 1997. *J. Raman Spectrosc.* 28:459–66
60. Assion A, Baumert T, Bergt M, Brixner T, Kiefer B, et al. 1998. *Science* 282:919–22
61. Weinacht TC, White JL, Bucksbaum PH. 1999. *J. Phys. Chem. A* 103:10166–68
62. Fork RL, Greene BI, Shank CV. 1981. *Appl. Phys. Lett.* 38:671–72
63. Valdmanis JA, Fork RL, Gordon JP. 1985. *Opt. Lett.* 10:131–33
64. Valdmanis JA, Fork RL. 1986. *IEEE J. Quantum Electron.* QE22:112–18
65. Fork RL, Shank CV, Yen RT. 1982. *Appl. Phys. Lett.* 41:223–25
66. Kane DJ, Trebino R. 1993. *Opt. Lett.* 18:823–2567
67. Dantus M, Rosker MJ, Zewail AH. 1988. *J. Chem. Phys.* 89:6128–40
68. Wilhelm T, Piel J, Riedle E. 1997. *Opt. Lett.* 22:1494–96
69. Prior Y. 1980. *Appl. Opt.* 19:1741–43
70. Shirley JA, Hall RJ, Eckbreth AC. 1980. *Opt. Lett.* 5:380–82
71. Lozovoy VV, Pastirk I, Brown EJ, Grimberg BI, Dantus M. 2000. *Int. Rev. Phys. Chem.* 19:531–52
72. Vohringer P, Arnett DC, Yang TS, Scherer NF. 1995. *Chem. Phys. Lett.* 237:387
73. Pastirk I, Lozovoy VV, Grimberg BI, Brown EJ, Dantus M. 1999. *J. Phys. Chem. A* 103:10226–36
74. Lozovoy VV, Grimberg BI, Brown EJ, Pastirk I, Dantus M. 2000. *J. Raman Spectrosc.* 31:41–49
75. Zadoyan R, Apkarian VA. 2000. *Chem. Phys. Lett.* 326:1–10
76. Woodward RB, Hoffmann R. 1969. *Angew. Chem. Int. Ed. Engl.* 8:781–853
77. Dewar MJS. 1984. *J. Am. Chem. Soc.* 106:209–19
78. Borden WT, Loncharich RJ, Houk KN. 1988. *Annu. Rev. Phys. Chem.* 39:213–36
79. Strauss CEM, Houston PL. 1990. *J. Phys. Chem.* 94:8751–62
80. Bernstein RB. 1982. *Chemical Dynamics Via Molecular Beam and Laser Techniques.* New York: Oxford Univ. Press
81. Lee YT. 1987. *Agnew. Chem. Int. Ed. Engl.* 26:936–51
82. Butler LJ, Hintsa EJ, Shane SF, Lee YT. 1987. *J. Chem. Phys.* 86:2051–74
83. Wannenmacher EAJ, Felder P, Huber JR. 1991. *J. Chem. Phys.* 95:986–97
84. Zewail AH. 1994. *Femtochemistry: Ultrafast Dynamics of the Chemical Bond.* Singapore: World Sci.
85. Pedersen S, Herek JL, Zewail AH. 1994. *Science* 266:1359–64
86. Kim SK, Pedersen S, Zewail AH. 1995. *J. Chem. Phys.* 103:477–80
87. Chachisvilis M, Fiebig T, Douhal A, Zewail AH. 1998. *J. Chem. Phys. A* 102:669–73
88. Fiebig T, Chachisvilis M, Manger M, Zewail AH, Douhal A, et al. 1999. *J. Phys. Chem. A* 103:7419–31
89. Folmer DE, Wisniewski ES, Castleman AW. 2000. *Chem. Phys. Lett.* 318:637–43
90. Catalan J, del Valle JC, Kasha M. 2000. *Chem. Phys. Lett.* 318:629–36
91. Baum G, Felder P, Huber JR. 1993. *J. Chem. Phys.* 98:1999–2010

92. Schwartz BJ, King JC, Zhang JZ, Harris CB. 1993. *Chem. Phys. Lett.* 203:503–8
93. Kwok WM, Phillips DL. 1995. *Chem. Phys. Lett.* 235:260–67
94. Kwok WM, Phillips DL. 1996. *J. Chem. Phys.* 104:2529–40
95. Marvet U, Dantus M. 1996. *Chem. Phys. Lett.* 256:57–62
96. Marvet U, Dantus M. 1996. See Ref. 173, pp. 134–37
97. Zhang Q, Marvet U, Dantus M. 1997. *J. Chem. Soc. Faraday Discuss.* 108:63–80
98. Marvet U, Zhang Q, Brown EJ, Dantus M. 1998. *J. Phys. Chem.* 109:4415–27
99. Marvet U, Zhang Q, Dantus M. 1998. *J. Phys. Chem. A* 102:4111–17
100. Zhang Q, Marvet U, Dantus M. 1998. *J. Chem. Phys.* 109:4428–42
101. Radloff W, Farmanara P, Stert V, Schreiber E, Huber JR. 1998. *Chem. Phys. Lett.* 291:173–78
102. Bergmann K, Carter RT, Hall GE, Huber JR. 1998. *J. Chem. Phys.* 109:474–83
103. Tarnovsky AN, Alvarez JL, Yartsev AP, Sundström V, Akesson E. 1999. *Chem. Phys. Lett.* 312:121–30
104. Marvet U, Brown EJ, Dantus M. 2000. *Phys. Chem. Chem. Phys.* 2:885–91
105. Dyne PJ, Style DWG. 1952. *J. Chem. Soc. 1952:* 2122–24
106. Style DWG, Ward JC. 1952. *J. Chem. Soc. 1952:* 2125–27
107. Okabe H, Kawasaki M, Tanaka Y. 1980. *J. Chem. Phys.* 12:6162–66
108. Duschek F, Schmitt M, Vogt P, Materny A, Kiefer W. 1997. *J. Raman Spectrosc.* 28:445–53
109. Bardeen CJ, Wang Q, Shank CV. 1995. *Phys. Rev. Lett.* 75:3410
110. Cao JS, Che JW, Wilson KR. 1998. *J. Phys. Chem. A* 102:4284–90
111. Hammerich AD, Kosloff R, Ratner MA. 1992. *J. Chem. Phys.* 97:6410–31
112. Treacy EB. 1969. *IEEE J. Quantum Electron.* 5:454
113. de Silvestri S, Laporta P, Svetlo O. 1984. *J. Quantum Electron.* QE20:533–30
114. Lee YT. 1982. In *Atomic and Molecular Beam Method*, ed. G Scoles, pp. 553–68. New York: Oxford Univ. Press
115. Herschbach DR. 1987. *Agnew. Chem. Int. Ed. Engl.* 26:1221–43
116. Atabek O, Lefebre R. 1977. *J. Chem. Phys.* 67:4983–89
117. Brooks PR, Curl RF, Maguire TC. 1982. *Ber. Bunsenges Phys. Chem.* 86:401–7
118. Scherer NF, Khundkar LR, Bernstein RB, Zewail AH. 1987. *Chem. Phys.* 87:1451–53
119. Visticot JP, Soep B, Whitham CJ. 1988. *J. Phys. Chem.* 92:4574–76
120. Gruebele M, Roberts G, Dantus M, Bowman RM, Zewail AH. 1990. *Chem. Phys. Lett.* 166:459–69
121. Gruebele M, Sims IR, Potter ED, Zewail AH. 1991. *J. Chem. Phys.* 95:7763–66
122. Ionov SI, Brucker GA, Jaques C, Valachovic L, Witting C. 1993. *J. Chem. Phys.* 99:6553–61
123. Petsalakis ID, Mercouris T, Nicolaides CA. 1994. *Chem. Phys.* 189:615–28
124. Marvet U, Dantus M. 1995. *Chem. Phys. Lett.* 245:393–99
125. Marvet U, Dantus M. 1996. See Ref. 173, pp. 138–42
126. Gross P, Dantus M. 1997. *J. Chem. Phys.* 106:8013–21
127. Mrozowski S. 1937. *Z. Physik.* 106:458–62
128. Dubov VS, Gudzenko LI, Gurvich LV, Iakovlenko SI. 1977. *Chem. Phys. Lett.* 45:330–33
129. Bergeman T, Liao PF. 1980. *J. Chem. Phys.* 72:886–98
130. Inoue G, Ku JK, Setser DW. 1984. *J. Chem. Phys.* 80:6006–19
131. Ku JK, Setser DW, Oba D. 1984. *Chem. Phys. Lett.* 109:429–35
132. Rodriguez G, Eden JG. 1991. *J. Chem. Phys.* 95:5539–52
133. Schloss JH, Jones RB, Eden JG. 1992. *Chem. Phys. Lett.:* 195–202
134. Jones RB, Schloss JH, Eden JG. 1993. *J. Chem. Phys.* 98:4317–34

135. Cline RA, Miller JD, Heinzen DJ. 1994. *Phys. Rev. Lett.* 73:632–35
136. Pavlenko VS, Nalivaiko SE, Egorov VG, Rzhevskii OS, Gordon EB. 1994. *J. Quantum Electron.* 24:199–206
137. Azinovic D, Li X, Milosevic S, Pichler G. 1996. *Phys. Rev. A* 53:1323–29
138. Gruber D, Li X, Windholz L, Gleichmann MM, Hess BA, et al. 1996. *J. Phys. Chem.* 100:10062–69
139. Thorsheim HR, Weiner J, Julienne PS. 1987. *Phys. Rev. Lett.* 58:2420–23
140. Miller JD, Cline RA, Heinzen DJ. 1993. *Phys. Rev. Lett.* 71:2204–7
141. Napolitano J, Weiner J, Williams CJ, Julienne PS. 1994. *Phys. Rev. Lett.* 73:1352–55
142. Williams CJ, Julienne PS. 1994. *J. Chem. Phys.* 101:2634–37
143. Moerdijk AJ, Verhaar BJ. 1995. *Phys. Rev. A* 51:R4333–36
144. Chapman MS, Ekstrom CR, Hammond TD, Rubenstein RA, Schmiedmayer J, et al. 1995. *Phys. Rev. Lett.* 74:4783–86
145. Band YB, Julienne PS. 1995. *Phys. Rev. A* 51:R4317–20
146. Lett PD, Julienne PS, Phillips WD. 1995. *Annu. Rev. Phys. Chem.* 46:423–52
147. Bahns JT, Stwaley WC, Gould PL. 1996. *J. Chem. Phys.* 104:9689–97
148. Vardi A, Abrashkevich D, Frishman E, Shapiro M. 1997. *J. Chem. Phys.* 107:6166–74
149. Wang H, Wang XT, Gould PL, Stwaley WC. 1997. *Phys. Rev. Lett.* 78:4173–76
150. Wang H, Gould PL, Stwaley WC. 1997. *J. Chem. Phys.* 106:7899–7912
151. Wang H, Stwalley WC. 1998. *J. Chem. Phys.* 108:5767–71
152. Gensemer SD, Gould PL. 1998. *Phys. Rev. Lett.* 80:936–39
153. Fioretti A, Coomparat D, Crubellier A, Dulieu O, Masnou-Seeuws F, Pillet P. 1998. *Phys. Rev. Lett.* 80:4402–5
154. Backhaus P, Schmidt B. 1997. *Chem. Phys.* 217:131–43
155. Backhaus P, Schmidt B, Dantus M. 1999. *Chem. Phys. Lett.* 306:18–24
156. Gaspard P, Burghardt I. 1997. *Adv. Chem. Phys.* 101:491–581
157. Krause JL, Shapiro M, Brumer P. 1990. *J. Chem. Phys.* 92:1126–31
158. Potter ED, Herek JL, Pedersen S, Liu Q, Zewail AH. 1992. *Nature* 355:66–68
159. Brumer P, Shapiro M. 1997. *Adv. Chem. Phys.* 101:295–300
160. Apkarian VA. 1997. *J. Chem. Phys.* 106:5298–99
161. Letokhov VS. 1973. *Science* 180:451–58
162. Stone J, Goodman MF, Dows DA. 1976. *Chem. Phys. Lett.* 44:411–14
163. Mukamel S, Jortner J. 1976. *Chem. Phys. Lett.* 40:150–56
164. Letokhov VS, Makarov AA. 1978. *Appl. Phys.* 16:47–57
165. Quack M. 1978. *J. Chem. Phys.* 69:1282–1307
166. Quack M. 1979. In *Laser Induced Processes in Molecules*, ed. K Kompa, SD Smith, 6:142–44. Berlin: Springer-Verlag
167. Vander Wal RL, Scott JL, Crim FF. 1990. *J. Chem. Phys.* 92:803–5
168. Bar I, Cohen Y, David D, Rosenwaks S, Valentini JJ. 1990. *J. Chem. Phys.* 93:2146–48
169. Robinson PJ, Holbrook KA. 1972. *Unimolecular Reactions*. New York: Wiley
170. Zewail AH. 1980. *Phys. Today* 33:27–33
171. Manz J, Parmenter CS. 1989. *Chem. Phys.* 139:U1–U4
172. Manz J, Wöste L. 1995. *Femtosecond Chemistry*. Weinheim:VCH
173. Chergui M. 1996. *Femtochemistry: Ultrafast Chemical and Physical Processes in Molecular Systems*. Singapore: World Sci.
174. Sundström V. 1998. *Femtochemistry and Femtobiology*. Singapore: World Sci.
175. Yan Y-X, Cheng L-T, Nelson KA. 1987. In *Advances in Non-Linear Spectroscopy*, ed. RJH Clark, RE Hester, pp. 299–355. Chichester, UK: Wiley

176. Ruhman S, Williams LR, Joly AG, Kholer B, Nelson KA. 1987. *J. Phys. Chem.* 91:2237–40
177. Ruhman S, Joly AG, Nelson KA. 1987. *J. Chem. Phys.* 86:6563–65
178. Engel V, Metiu H, Almeida R, Marcus RA, Zewail AH. 1988. *Chem. Phys. Lett.* 152:1–7
179. Bowman M, Dantus M, Zewail AH. 1989. *Chem. Phys. Lett.* 161:297–302
180. Dantus M, Bowman RM, Zewail AH. 1990. *Nature* 343:737–39
181. Rosker MJ, Rose TS, Zewail AH. 1988. *Chem. Phys. Lett.* 146:175–79
182. Rose TS, Rosker MJ, Zewail AH. 1988. *J. Chem. Phys.* 88:6672–73
183. Brown EJ, Pastirk I, Grimberg BI, Lozovoy VV, Dantus M. 1999. *J. Chem. Phys.* 111:3779–82
184. Pastirk I, Brown EJ, Grimberg BI, Lozovoy VV, Dantus M. 1999. *Faraday Discuss.* 113:401–24
185. Grimberg BI, Lozovoy VV, Dantus M, Mukamel S. 2001. *J. Phys. Chem.* Submitted
186. Pshenichnikov MS, de Boeij WP, Wiersma DA. 1996. *Phys. Rev. Lett.* 76:4701–4
187. Schoenlein RW, Mittleman DM, Shiang JJ, Alivisatos AP, Shank CV. 1993. *Phys. Rev. Lett.* 70:1014–17
188. Bardeen CJ, Shank CV. 1993. *Chem. Phys. Lett.* 203:535–39
189. Knopp G, Pinkas I, Prior Y. 2000. *J. Raman Spectrosc.* 31:51–58
190. Materny A, Chen T, Schmitt M, Siebert T, Vierheilig A, et al. 2000. *Appl. Phys. B* 71:299–317
191. Chen T, Engel V, Heid M, Kiefer W, Knopp G, et al. 1999. *J. Mol. Struct.* 481:33–43
192. Hornung T, Meier R, Motzkus M. 2000. *Chem. Phys. Lett.* 326:445–53
193. Zanni MT, Davis AV, Frischkorn C, Elhanine M, Neumark DM. 2000. *J. Chem. Phys.* 112:8847–54
194. de Boeij WP, Pshenichnikov MS, Wiersma DA. 1998. *Annu. Rev. Phys. Chem.* 49:99–123
195. Cina JA. 2000. *J. Chem. Phys.* 113:9488–96
196. Hybl JD, Albrecht AW, Faeder SMG, Jonas DM. 1998. *Chem. Phys. Lett.* 297:307–313
197. Keusters D, Tan HS, Warren WS. 1999. *J. Phys. Chem.* 103:10369–80
198. Hahn EL. 1950. *Phys. Rev.* 80:580–94
199. Ernst RR, Bodenhausen G, Wokaun A. 1987. *Principles of Nuclear Magnetic Resonance in One and Two Dimensions.* New York: Oxford Univ. Press
200. Kurnit NA, Abella ID, Hartmann SR. 1964. *Phys. Rev. Lett.* 13:567–70
201. Patel CKN, Slusher RE. 1968. *Phys. Rev. Lett.* 20:1087–89
202. Fleming GR, Cho MH. 1996. *Annu. Rev. Phys. Chem.* 47:109–34
203. Pastirk I, Lozovoy VV, Dantus M. 2001. *Chem. Phys. Lett.* In press
204. Lozovoy VV, Grimberg BI, Pastirk I, Dantus M. 2001. *Chem. Phys.* (*Special issue Laser Control of Quantum Dynamics*). In press
205. Schmitt M, Knopp G, Materny A, Kiefer W. 1997. *Chem. Phys. Lett.* 280:339–47
206. Schmitt M, Knopp G, Materny A, Kiefer W. 1997. *Chem. Phys. Lett.* 270:9–15
207. Meyer S, Schmitt M, Materny A, Kiefer W, Engel V. 1997. *Chem. Phys. Lett.* 281:332–36
208. Meyer S, Schmitt M, Materny A, Kiefer W, Engel V. 1998. *Chem. Phys. Lett.* 287:753–54
209. Siebert T, Schmitt M, Michelis T, Materny A, Kiefer W. 1999. *J. Raman Spectrosc.* 30:807–13
210. Fayer MD. 1982. *Annu. Rev. Phys. Chem.* 33:63–87
211. Rose TS, Wilson WL, Wäckerle G, Fayer MD. 1987. *J. Chem. Phys.* 86:5370–91
212. Dhar L, Rogers JA, Nelson KA. 1994. *Chem. Rev.* 94:157–93
213. Morgen M, Price W, Ludowise P, Chen Y. 1995. *J. Chem. Phys.* 102:8780–89

214. Lavorel B, Faucher O, Morgen M, Chaux R. 2000. *J. Raman Spectrosc.* 31:77–83
215. Frey HM, Beaud P, Gerber T, Mischler B, Radi PP, Tzannis AP. 2000. *J. Raman Spectrosc.* 31:71–76
216. Brown EJ, Zhang Q, Dantus M. 1999. *J. Phys. Chem.* 110:5772–88
217. Morgen M, Price W, Hunziker L, Ludowise P, Blackwell M, Chen Y. 1993. *Chem. Phys. Lett.* 209:1–9
218. Heritage JP, Gustafson TK, Lin CH. 1975. *Phys. Rev. Lett.* 34:1299–1302
219. Felker PM, Baskin JS, Zewail AH. 1986. *J. Chem. Phys.* 90:724–28
220. Baskin JS, Fekler PM, Zewail AH. 1986. *J. Chem. Phys.* 84:4708–10
221. Dantus M, Bowman RM, Baskin JS, Zewail AH. 1989. *Chem. Phys. Lett.* 159:406–12
222. Baskin JS, Zewail AH. 1994. *J. Phys. Chem.* 98:3337–51
223. Connell LL, Corcoran TC, Joireman PW, Felker PM. 1990. *J. Phys. Chem.* 94:1229–32
224. Felker PM. 1992. *J. Phys. Chem.* 96:7844–57
225. Hayden CC, Chandler DW. 1995. *J. Chem. Phys.* 103:10465–72
226. Herzberg G. 1966. *Molecular Spectra and Molecular Structure. III. Electronic Spectra and Electronic Structure of Polyatomic Molecules.* New York: van Nostrand
227. Herzberg G. 1991. *Molecular Spectra and Molecular Structure. I. Diatomic Molecules.* New York: van Nostrand
228. Clark RJH, Rippon DM. 1973. *J. Chem. Soc. Faraday Trans. 2.* 69:1496–1501
229. Cho M, Du M, Scherer NF, Fleming GR. 1993. *J. Chem. Phys.* 99:2410–28
230. Spiridonov VP, Gershikov AG, Butayev BS. 1979. *J. Mol. Struct.* 52:53–62
231. Friedrich B, Herschbach D. 1995. *Phys. Rev. Lett.* 74:4623–26
232. Kim W, Felker PM. 1996. *J. Chem. Phys.* 104:1147–50
233. Dion CM, Keller A, Atabek O, Bandrauk AD. 1999. *Phys. Rev. A* 59:1382–91
234. Villeneuve DM, Aseyev SA, Dietrich P, Spanner M, Ivanov MY, Corkum PB. 2000. *Phys. Rev. Lett.* 85:545–42
235. Lopez-Martens RB, Schmidt TW, Roberts G. 2000. *Phys. Rev. A* 62:13414–19
236. Lewenstein M, Balcoup P, Ivanov MY, Lhuillier A, Corkum PB. 1994. *Phys. Rev. A* 49:2117–32
237. Dietrich P, Burnett NH, Ivanov M, Corkum PB. 1994. *Phys. Rev. A* 50:R3585–88
238. Zuo T, Bandrauk AD. 1995. *J. Nonlinear Opt. Phys. Mater.* 4:533–46
239. Constant E, Taranukhin VD, Stolow A, Corkum PB. 1997. *Phys. Rev. A* 56:3870–78
240. Sheehy B, Martin JDD, DiMauro LF, Agostini P, Schafer KJ, et al. 1999. *Phys. Rev. Lett.* 83:5270–73
241. Bandrauk AD, Yu HT. 1999. *Phys. Rev. A* 59:539–48
242. Bartels R, Backus S, Zeek E, Misoguti L, Vdovin G, et al. 2000. *Nature* 406:164–66
243. Schumacher DW, Bucksbaum PH. 1996. *Phys. Rev. A* 54:4271–78
244. Hertlein MP, Bucksbaum PH, Muller HG. 1997. *J. Phys. B* 30:L197–205
245. Bensky TJ, Haeffler G, Jones RR. 1997. *Phys. Rev. Lett.* 79:2018–21
246. van Leeuwen R, Bajema ML, Jones RR. 2000. *Phys. Rev. A* 6102:2716–17
247. Villeneuve DM, Ivanov MY, Corkum PB. 1996. *Phys. Rev. A* 54:736–41
248. Yoneda H, Hasegawa N, Kawana S, Ueda K. 1999. *Fusion Eng. Design* 44:141–46
249. Riley D, Woolsey NC, McSherry D, Weaver I, Djaoui A, Nardi E. 2000. *Phys. Rev. Lett.* 84:1704–7
250. Pauling L. 1955. *Angew. Chem.* 67:241–60
251. Polanyi JC. 1987. *Angew. Chem. Int. Ed. Engl.* 26:952–71
252. Zewail AH. 2000. *J. Phys. Chem. A* 104:5660–94

253. Hoganson CW, Babcock GT. 1997. *Science* 277:1453–56
254. Tommos C, Babcock GT. 1998. *Acc. Chem. Res.* 31:18–25
255. Schelvis JPM, Varotsis C, Deinum G, Babcock GT. 1999. *Laser Chem.* 19:223–25
256. Okunishi M, Hashimoto J, Chiba H, Ohmori K, Ueda K, Sato Y. 1999. *J. Phys. Chem. A* 103:1734–41
257. Ohmori K, Kurosawa T, Chiba H, Okunishi M, Sato Y, et al. 1999. *Chem. Phys. Lett.* 315:411–15
258. Tanimura Y, Mukamel S. 1993. *J. Chem. Phys.* 99:9496–511
259. Ivanecky JE, Wright JC. 1993. *Chem. Phys. Lett.* 206:437
260. Tominaga K, Keogh GP, Naitoh Y, Yoshihara K. 1995. *J. Raman Spectrosc.* 26:495–501
261. Steffen T, Duppen K. 1996. *Phys. Rev. Lett.* 76:1224–27
262. Tominaga K, Yoshihara K. 1996. *Phys. Rev. Lett.* 76:987–90
263. Tominaga K, Yoshihara K. 1996. *J. Chem. Phys.* 104:4419–26
264. Tominaga K, Yoshihara K. 1997. *Phys. Rev. A* 55:831–34
265. Tokmakoff A, Lang MJ, Larsen DS, Fleming GR. 1997. *Chem. Phys. Lett.* 272:48–54
266. Tokmakoff A, Fleming GR. 1997. *J. Chem. Phys.* 106:2569–82
267. Steffen T, Duppen K. 1997. *Chem. Phys. Lett.* 273:47–54
268. Ulness DJ, Kirkwood JC, Albrecht AC. 1998. *J. Chem. Phys.* 108:3897–902
269. Blank DA, Kaufman LJ, Fleming GR. 1999. *J. Chem. Phys.* 111:3105–14
270. Kirkwood JC, Albrecht AC. 2000. *J. Raman Spectrosc.* 31:107–24
271. Astinov V, Kubarych KJ, Milne CJ, Miller RJD. 2000. *Chem. Phys. Lett.* 327:334–42
272. Astinov V, Kubarych KJ, Milne CJ, Miller RJD. 2000. *Opt. Lett.* 25:853–55
273. Lozovoy VV, Pastirk I, Comstock M, Dantus M. 2001. *Chem. Phys.* In press
274. Yan ZC, Seideman T. 1999. *J. Chem. Phys.* 111:4113–20
275. Seideman T. 1999. *J. Chem. Phys.* 111:4397–405

ELECTRON TRANSMISSION THROUGH MOLECULES AND MOLECULAR INTERFACES

Abraham Nitzan

School of Chemistry, The Sackler Faculty of Science, Tel Aviv University, Tel Aviv, 69978, Israel; e-mail: nitzan@post.tau.ac.il

Key Words electron transfer, molecular conduction, molecular electronics, tunneling

■ **Abstract** Electron transmission through molecules and molecular interfaces has been a subject of intensive research due to recent interest in electron-transfer phenomena underlying the operation of the scanning-tunneling microscope on one hand, and in the transmission properties of molecular bridges between conducting leads on the other. In these processes, the traditional molecular view of electron transfer between donor and acceptor species gives rise to a novel view of the molecule as a current-carrying conductor, and observables such as electron-transfer rates and yields are replaced by the conductivities, or more generally by current-voltage relationships, in molecular junctions. Such investigations of electrical junctions, in which single molecules or small molecular assemblies operate as conductors, constitute a major part of the active field of molecular electronics. In this article I review the current knowledge and understanding of this field, with particular emphasis on theoretical issues. Different approaches to computing the conduction properties of molecules and molecular assemblies are reviewed, and the relationships between them are discussed. Following a detailed discussion of static-junctions models, a review of our current understanding of the role played by inelastic processes, dephasing and thermal-relaxation effects is provided. The most important molecular environment for electron transfer and transmission is water, and our current theoretical understanding of electron transmission through water layers is reviewed. Finally, a brief discussion of overbarrier transmission, exemplified by photoemission through adsorbed molecular layers or low-energy electron transmission through such layers, is provided. Similarities and differences between the different systems studied are discussed.

1. INTRODUCTION

Electron transfer, a fundamental chemical process underlying all redox reactions, has been under experimental and theoretical study for many years (1, 2). Theoretical studies of such processes seek to understand the ways in which their rates depend on donor and acceptor properties, on the solvent, and on the electronic coupling between the states involved. The different roles played by these factors and the way they affect qualitative and quantitative aspects of the electron-transfer process have been thoroughly discussed in the past half-century. These kinds of processes, which dominate electron transitions in molecular systems, are to be contrasted with electron transport in the solid state, i.e. in metals and semiconductors. Electrochemical reactions, which involve both molecular and solid state donor/acceptor systems, bridge the gap between these phenomena (2). Here electron transfer takes place between quasi-free electronic states on one side and bound molecular electronic states on the other.

The focus of this discussion is another class of electron-transfer phenomena: electron transmission between two regions of free or quasifree electrons through molecules and molecular layers. Examples of such processes are photoemission (PE) through molecular overlayers, the inverse process of low-energy electron transmission (LEET) into metals through adsorbed molecular layers, and electron transfer between metal and/or semiconductor contacts through molecular spacers. Figure 1 depicts a schematic view of such systems. The "standard" electron-transfer model in Figure 1a shows donor and acceptor sites, with their corresponding polarization wells connected by a molecular bridge. In Figure 1b the donor and the acceptor are replaced by a continuum of electronic states representing free space or metal electrodes. (This replacement can occur on one side only, representing electron transfer between a molecular site and an electrode.) In Figure 1c the molecular bridge is replaced by a molecular layer. In addition, coupling to the thermal environment may affect transmission through the bridge.

The first two of the examples given above, PE and LEET, involve electrons of positive energy (relative to zero kinetic energy in vacuum) and as such are related to normal scattering processes. The third example, transmission between two conductors through a molecular layer, involves negative energy electrons and as such is closely related to regular electron-transfer phenomena. The latter type of process has drawn particular attention in recent years because of the growing interest in conduction properties of individual molecules and of molecular assemblies. Such processes have become subjects of intensive research because of recent interest in electron-transfer phenomena underlying the operation of the scanning tunneling microscope (STM), and in the transmission properties of molecular bridges between conducting leads. In the latter case, the traditional molecular view of electron transfer between donor and acceptor species gives rise to a novel view of the molecule as a current-carrying conductor, and observables such as electron-transfer rates and yields are replaced by the conductivities, or more generally by current-voltage relationships, in molecular junctions. Of primary importance is the need to understand the interrelationship between the molecular structure of such junctions and their function, i.e. their transmission and conduction properties. Such investigations of electrical junctions, in which single molecules or small molecular assemblies operate as conductors connecting "traditional" electrical components such as metal or semiconductor contacts, constitute a major part of what has become the active field of molecular electronics (3, 4). Their diversity, versatility, and amenability to control and manipulation make molecules and molecular

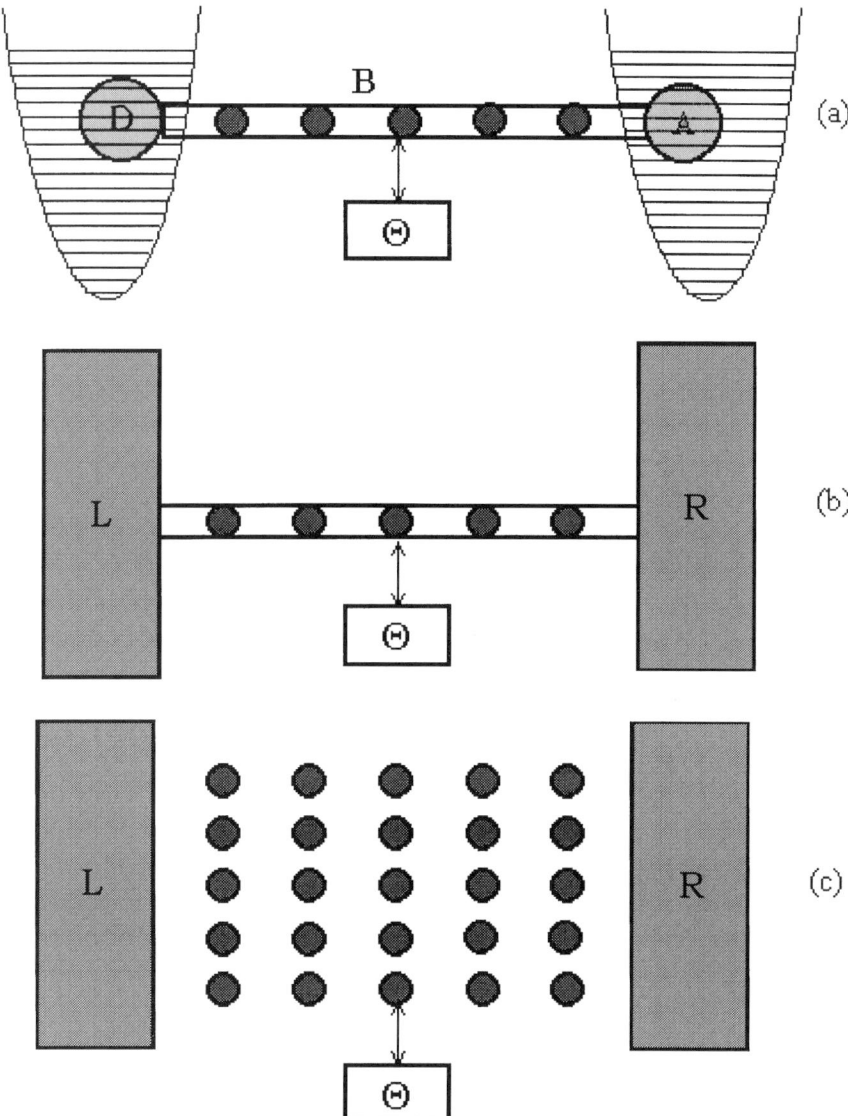

Figure 1 Schematic views of typical electron transmission systems. (*a*) A standard electron-transfer system containing a donor, an acceptor, and a molecular bridge connecting them (not shown are nuclear motion baths that must be coupled to the donor and acceptor species). (*b*) A molecular bridge connecting two electronic continua, L and R, representing e.g. two metal electrodes. (*c*) Same as panel *b* with the bridge replaced by a molecular layer. The Θ blocks represent the thermal environment.

assemblies potentially important components in nano-electronic devices. Indeed, basic properties pertaining to single electron transistor behavior and to current rectification have already been demonstrated. At the same time, major difficulties lie in the way of discovering real technological applications. These difficulties stem from problems associated with the need to construct, characterize, control, and manipulate small molecular structures at confined interfaces with a high degree of reliability and reproducibility, and from issues of stability of such small junctions.

It should be obvious that while the different processes outlined above correspond to different experimental setups, fundamentally they are controlled by similar physical factors. Broadly speaking, we may distinguish between processes for which lifetimes or rates (more generally the time evolution) are the main observables and those that monitor fluxes or currents. In this review, we focus on the second class, which may be further divided into processes that measure current-voltage relationships, mostly near equilibrium, and those that monitor the nonequilibrium electron flux, e.g. in photoemission experiments.

1.0.1 Notations A problem characteristic of an interdisciplinary field such as the one we are covering is that notations that became standard in particular disciplines overlap similarly standard notations of other disciplines. The T operator of scattering theory and the temperature constitute one example; the β parameter of bridge-mediated electron-transfer theory and the inverse (temperature × Boltzmann constant) is another. I have therefore used nonstandard notations for some variables in order to avoid confusion. Table 1 lists of the main notations used in this article.

2. THEORETICAL APPROACHES TO MOLECULAR CONDUCTION

The focus of this chapter is electron transfer between two conducting electrodes through a molecular medium. Such processes bear strong similarity to the more conventional systems that involve at least one molecular species in the donor/acceptor pair. Still, important conceptual issues arise from the fact that such systems can be studied as part of complete electrical circuits, providing current-voltage characteristics that can be analyzed in terms of molecular resistance, conductance, and capacitance.

2.1 Standard Electron-Transfer Theory

To set the stage for our later discussion, we first briefly review the rate expressions for standard electron-transfer processes (Figures 1a, 2a). We focus on the particular limit of nonadiabatic electron transfer, where the electron-transfer rate is given

TABLE 1 Notations used

Notation	Variable
T	Scattering operator
$\mathcal{T}$	Transmission coefficient
Θ	Temperature
β	$(k_B \Theta)^{-1}$
β'	Range parameter in electron-transfer rate theory
G	Conduction
σ	Used in different contexts for conductivity and for the reduced system's density operator
I	Current
J	Flux
ρ	Used in different contexts for charge density and for the density operator of the total system
E_F	Fermi energy (E_{FL} and E_{FR} sometimes used for left and right electrodes)
M	Electron electrochemical potential (μ_L and μ_R sometimes used for left and right electrodes)
$\mathcal{F}$	The thermally averaged and Franck-Condon (FC) weighted density of nuclear states
F	System-thermal bath interaction. In specific cases we also use $H_{el\text{-}ph}$
V	Electronic coupling between zero order molecular states
H	System's Hamiltonian
H_B	Bridge Hamiltonian
H_Θ	Hamiltonian of the thermal bath. In some specific cases we also use H_{ph}
Z	Overlap Matrix: $Z_{i,j} = \langle i | j \rangle$
$\mathbb{H}$	EZ–H
$\mathcal{H}$	Combined system+thermal bath Hamiltonian
S	S matrix
v	Speed
Φ	Potential or potential difference
Σ	Self energy
Γ	Width (decay rate)
Acronyms	
MMM	Metal-Molecule-Metal (junction)
MIM	Metal-Insulator-Metal (junction)
EH	Extended Huckel
HF	Hartree Fock
FC	Franck-Condon
STM	Scanning tunneling microscope
LEET	Low energy electron transmission
PE	Photoemission

(under the Condon approximation) by the golden rule–based expression

$$k_{et} = \frac{2\pi}{\hbar}|V_{DA}|^2 \mathcal{F}, \qquad 1.$$

where V_{DA} is the coupling between the donor (D) and acceptor (A) electronic states and where

$$\mathcal{F} = \mathcal{F}(E_{AD}) = \sum_{v_D}\sum_{v_A} P_{th}(\varepsilon_D(v_D))|\langle v_D|v_A\rangle|^2 \delta(\varepsilon_A(v_A) - \varepsilon_D(v_D) + E_{AD}), \qquad 2.$$

is the thermally averaged and Franck-Condon (FC) weighted density of nuclear states. In Equation 2, v_D and v_A denote donor and acceptor nuclear states, P_{th} is the Boltzmann distribution over donor states, $\varepsilon_D(v_D)$ and $\varepsilon_A(v_A)$ are nuclear energies above the corresponding electronic origins, and $E_{AD} = E_A - E_D$ is the electronic energy gap between the donor and acceptor states. In the classical limit, $\mathcal{F}$ is given by

$$\mathcal{F}(E_{AD}) = \frac{e^{-(\lambda + E_{AD})^2/4\lambda k_B \Theta}}{\sqrt{4\pi \lambda k_B \Theta}}, \qquad 3.$$

where k_B is the Boltzmann constant, Θ is the temperature, and λ is the reorganization energy, a measure of the electronic energy that would be dissipated after a sudden jump from the electronic state describing an electron on the donor to that associated with an electron on the acceptor. If the donor is replaced by an electrode (2, 5, 6), we have to sum over all occupied electrode states:

$$|V_{DA}|^2 \mathcal{F} \Rightarrow \sum_k f(\varepsilon_k)\mathcal{F}(\varepsilon_k - e\Phi)|V_{kA}|^2$$

$$= \int d\varepsilon f(\varepsilon)\mathcal{F}(\varepsilon - e\Phi)\sum_k \delta(\varepsilon - \varepsilon_k)|V_{kA}|^2, \qquad 4.$$

where

$$f(\varepsilon) = \frac{1}{1 + e^{\varepsilon/k_B \Theta}} \qquad 5.$$

is the Fermi-Dirac distribution function, with ε measured relative to the electron chemical potential μ in the electrode and Φ, which determines the position of the acceptor level relative to μ, is the overpotential. Defining

$$\sum_k \delta(\varepsilon - \varepsilon_k)|V_{kA}|^2 \equiv |V(\varepsilon)|^2, \qquad 6.$$

the electron-transfer rate takes the form

$$k_{et} = \frac{2\pi}{\hbar}\int d\varepsilon \frac{e^{-(\lambda - e\Phi + \varepsilon)^2/4\lambda k_B \Theta}}{\sqrt{4\pi \lambda k_B \Theta}}|V(\varepsilon)|^2 f(\varepsilon). \qquad 7.$$

Note that the reorganization energy that appears in Equation 7 is associated with the change in the redox state of the molecular species only. The nominal change in the "oxidation state" of the macroscopic electrode does not affect the polarization state of the surrounding solvent because the transferred electron or hole does not stay localized.

Much of the early work on electron transfer has used expressions like those found in Equations 3 and 7, with the electronic coupling term V_{DA} used as a fitting parameter. More recent work has focused on ways to characterize the dependence of this term on the electronic structure of the donor/acceptor pair and on the environment. In particular, studies of bridge-mediated electron transfer, where the donor and acceptor species are rigidly separated by molecular bridges of well-defined structure and geometry, have been very valuable in characterizing the interrelationship between structure and functionality of the separating environment in electron-transfer processes. As expected for a tunneling process, the rate is found to decrease exponentially with the donor-acceptor distance

$$k_{et} = k_0 e^{-\beta' R_{DA}}, \qquad 8.$$

where β' is the range parameter that characterizes the distance dependence of the electron-transfer rate. The smallest values for β' are found in highly conjugated organic bridges for which β' is in the range 0.2–0.6 Å^{-1} (7, 8). In contrast, for free space, taking a characteristic ionization barrier $U_B = 5 eV$, we find $\beta' = \sqrt{8m U_B/\hbar^2} \approx 2.4$ Å^{-1} (m is the electron mass). Lying between these two regimes are many motifs, both synthetic and natural, including cytochromes and docked proteins (9, 10), DNA (11–13), and saturated organic molecules (14, 15). Each displays its own characteristic range of β' values, and hence its own timescales and distance dependencies of electron transfer. A direct measurement of β' along a single molecular chain was recently demonstrated (16). In addition to the bridge-assisted transfer between donor and acceptor species, electron transfer has been studied in systems where the spacer is a well-characterized Langmuir-Blodgett film (17). The STM provides a natural apparatus for such studies (16, 18–22). Other approaches include break junctions (23) and mercury drop contacts (24, 25).

Simple theoretical modeling of V_{DA} usually relies on a single electron (or hole) picture in which the donor-bridge-acceptor (DBA) system is represented by a set of levels: $|D\rangle, |A\rangle, \{|1\rangle, \ldots |N\rangle\}$, as depicted in Figure 2. In the absence of the coupling of these bridge states to the thermal environment, and when the energies E_n ($n = 1, \ldots, N$) are high relative to the energy of the transmitted electron (the donor/acceptor orbital energies in Figures 1a and 2a or the incident electron energy in Figures 1b–c and 2b), this is the superexchange model for electron transfer (26). Of particular interest are situations where the states $|1\rangle, \ldots |N\rangle$ are localized in space, so that the state index n corresponds to a position in space between the donor and acceptor sites (Figure 2a) or between the two electron reservoirs (Figure 2b). These figures depict generic tight-binding models of this type, where the states $n = 1, \ldots, N$ are the bridge states, here taken degenerate in zero order. Their localized nature makes it possible to assume only nearest-neighbor coupling between them, i.e. $V_{n,n'} = V_{n,n\pm 1} \delta_{n',n\pm 1}$. We recall that the appearance of V_{DA} in Equation 1 is a low-order perturbation theory result. A more

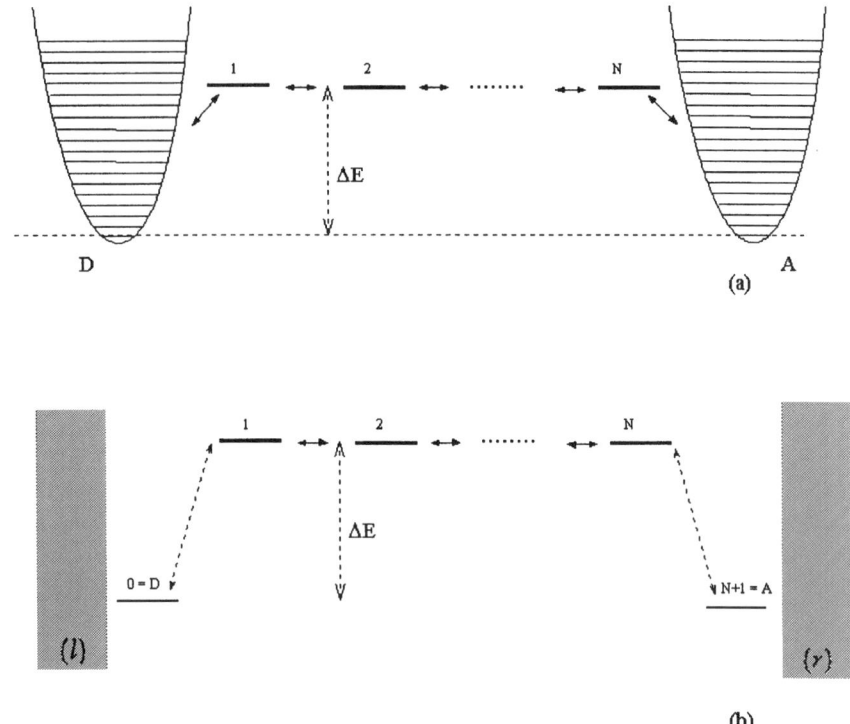

Figure 2 Simple level structure models for molecular electron transfer (*a*) and for electron transmission (*b*). The molecular bridge is represented by a simple set of levels that represents local orbitals of appropriately chosen bridge sites. This set of levels is coupled to the donor and acceptor species (with their corresponding nuclear environments) in panel *a*, and to electronic continua ($\{\ell\}$ for left, $\{r\}$ for right) representing metal leads in panel *b*. In the latter case, the physical meaning of states 0 and N + 1 depends on the particular physical problem: They can denote donor and acceptor states coupled to the continua of environmental states (hence the notation $0 = D, N + 1 = A$), surface localized states in a metal-molecule-metal junction, or belong to the right and left scattering continua.

general expression is obtained by replacing V_{DA} by T_{DA}, where the T operator is defined by $T(E) = V + VG(E)V$, with $G(E) = (E - H + (1/2)i\Gamma)^{-1}$ and where Γ stands for the inverse lifetime matrix of bridge levels. Assuming for simplicity that the donor level $|D\rangle$ is coupled only to bridge state $|1\rangle$ and that the acceptor level $|A\rangle$ is coupled only to bridge level N, the effective coupling between donor and acceptor is given by

$$T_{DA}(E) = V_{DA} + V_{D1}G_{1N}(E)V_{NA}. \qquad 9.$$

This naturally represents the transition amplitude as a sum of a direct contribution, V_{DA}, which is usually disregarded for long bridges, and a bridge-mediated contribution. In using T_{DA} instead of V_{DA} in Equation 1 the energy parameter E in Equation 9 should be taken equal to $E_D = E_A$ at the point where the corresponding potential surfaces cross (or go through an avoided crossing). For the level structure of Figure 2a that corresponds to the DBA system in Figure 1a, making the tight binding approximation and in the weak coupling limit, $\max|V| \ll \min(E_B - E)$,[1] the Green's function element in Equation 9 is given by

$$G_{1N}(E) = \frac{1}{E - E_N} \prod_{n=1}^{N-1} \frac{V_{n,n+1}}{E - E_n}. \qquad 10.$$

For a model with identical bridge segments, E_n and $V_{n,n+1}$ are independent of n and are denoted $E_n = E_B$ and $V_{n,n+1} = V_B$. Using this in Equation 1 leads to

$$k_{et} = \frac{2\pi}{\hbar} \left| \frac{V_{1D} V_{NA}}{V_B} \right|^2 \left(\frac{V_B}{\Delta E_B} \right)^{2N} \mathcal{F}, \qquad 11.$$

where $\Delta E_B = E_B - E$. Similarly, for a bridge-assisted transfer between a molecule and an electrode, Equation 7 applies, with $|V(\varepsilon)|^2$ given by

$$|V(\varepsilon)|^2 = \left(\frac{V_B}{\Delta E_B} \right)^{2N} \sum_k \delta(\varepsilon - \varepsilon_k) \left| \frac{V_{1k} V_{NA}}{V_B} \right|^2. \qquad 12.$$

These results imply a simple form for the distance parameter β' of Equation 8

$$\beta' = \frac{2}{a} \ln\left(\frac{\Delta E_B}{V_B} \right), \qquad 13.$$

where a measures the segment size, so that the bridge length is Na. The exponential dependence on the bridge length is a manifestation of the tunneling character of this process. For typical values, e.g. $\Delta E_B/V_B = 10$ and $a = 5$ Å, Equation 13 gives $\beta' = 0.92$ Å^{-1}. More rigorous estimates of the electronic coupling term in electron-transfer processes involve electronic-structure calculation for the full DBA system. Such calculations, in the context of molecular conduction, are discussed later.

2.2 Transmission Between Conducting Leads

Equations 1, 7, and 11 are expressions for the rate of electron transfer between donor and acceptor molecules or between a molecule and a metal electrode. As already mentioned, for electron transfer in metal-molecule-metal (MMM) junctions, the primary observable is the current-voltage characteristics of the system. Putting it another way, while the primary observable in standard charge-transfer processes involving molecular donors and/or acceptors is a transient quantity,[2] in

---

[1] For a generalization of Equation 10 that does not assume weak coupling see Onipko & Klymenko (27).
[2] In addition to rates, other observables are the yields of different products of the electron-transfer reaction.

metal-molecule-metal junctions we focus on the steady-state current through the junction for a given voltage difference between the two metal ends.

Consider first a simple model for a metal-insulator-metal (MIM) system, where the insulator is represented by a continuum characterized by a dielectric constant ε (28). For specificity assume that the electrode surfaces are infinite parallel planes perpendicular to the x direction. In this case, the transmission problem is essentially one dimensional and depends only on the incident particle velocity in the x direction, $v_x = \sqrt{2E_x/m}$. In the WKB approximation, the transmission probability is given by

$$T(E_x) = \exp\left[-\frac{4\pi}{\hbar}\int_{s_1}^{s_2}[2m(U_B(x) - E_x)]^{1/2}dx\right], \quad 14.$$

where $U_B(x)$ is the barrier potential that determines the turning points s_1 and s_2 and m is the mass of the tunneling particle. The tunneling flux is given by $T(E_x)n(E_x)\sqrt{2E_x/m}$, where $n(E_x)$ is the density per unit volume of electrons of energy E_x in the x direction. $n(E_x)$ is obtained by integrating the Fermi-Dirac function with respect to E_y and E_z. When a potential Φ is applied so that the right electrode is positively biased, the net current density is obtained in the form (28)

$$J = \int_0^\infty dE_x T(E_x)\xi(E_x), \quad 15.$$

where

$$\xi(E_x) = \frac{2m^2e}{(2\pi\hbar)^3}\int_{-\infty}^\infty dv_y \int_{-\infty}^\infty dv_z[f(E) - f(E + e\Phi)]$$

$$= \frac{4\pi me}{(2\pi\hbar)^3}\int_0^\infty dE_r[f(E) - f(E + e\Phi)], \quad 16.$$

and where $E_r = E - E_x = (1/2)m(v_y^2 + v_z^2)$ is the energy in the direction perpendicular to x. In obtaining this result, it is assumed that the electrodes are chemically identical. At zero temperature and when $\Phi \to 0$, $f(E) - f(E + e\Phi) = e\Phi\delta(E - E_F)$. Equations 15 and 16 then lead to an expression for the conduction per unit area, i.e. the conductivity per unit length

$$\sigma_x = \frac{4\pi me^2}{(2\pi\hbar)^3}\int_0^{E_F} dE_x T(E_x). \quad 17.$$

For finite Φ, these expressions provide a framework for predicting the current-voltage characteristics of the junction; explicit approximate expressions were given by Simmons (28). Here we only emphasize (28) that the dependence on Φ arises partly from the structure of Equations 15 and 16, for example, at zero temperature,

$$J = \frac{4\pi me}{(2\pi\hbar)^3}\left[e\Phi\int_0^{E_F-e\Phi}dE_x T(E_x) + \int_{E_F-e\Phi}^{E_F}dE_x(E_F - E_x)T(E_x)\right], \quad 18.$$

but mainly from the voltage dependence of $\mathcal{T}$. The simplest model for a metal-vacuum-metal barrier between identical electrodes without an external field is a rectangular barrier of height above the Fermi energy given by the metal workfunction. When a uniform electric field is imposed between the two metals, a linear potential drop from E_F on one electrode to $E_F - e\Phi$ on the other is often assumed. In addition, the image potential experienced by the electron between the two metals will lower the potential barrier (28). This has been invoked to explain the lower-than-expected barrier observed in STM experiments (29). It should be kept in mind that quantum mechanical and atomic size effects, as well as the dynamic nature of the image response, should be taken into account in estimating this correction.

The planar geometry implied by the assumption that transmission depends only on the energy of the motion parallel to the tunneling direction, as well as the explicit form of Equation 14, are not valid for a typical STM configuration that involves a tip on one side and a structured surface on the other. To account for these structures, Tersoff & Hamman (30) have applied Bardeen's formalism (31), which is a perturbative approach to tunneling in arbitrary geometries. Bardeen's formula for the tunneling current is[3]

$$I = \frac{4\pi e}{\hbar} \sum_{l,r} [f(E_l)(1 - f(E_r + e\Phi))$$

$$- (1 - f(E_l))f(E_r + e\Phi)]|M_{lr}|^2 \delta(E_l - E_r)$$

$$= \frac{2\pi e}{\hbar} \sum_{l,r} [f(E_l) - f(E_r + e\Phi)]|M_{lr}|^2 \delta(E_l - E_r), \qquad 19.$$

where

$$M_{lr} = \frac{\hbar^2}{2m} \int d\vec{S} \cdot (\psi_l^* \nabla \psi_r - \psi_l \nabla \psi_r^*) \qquad 20.$$

is the transition-matrix element for the tunneling process. In these equations, ψ_ℓ and ψ_r are electronic eigenstates of the negatively biased (left), and positively biased (right), electrodes, respectively, Φ is the bias potential, and the integral is over any surface separating the two electrodes and lying entirely in the barrier region. The wavefunctions appearing in Equation 20 are eigenfunctions of Hamiltonians that describe each electrode in the absence of the other, i.e interfaced with an infinite spacer medium. These functions therefore decay exponentially in the space between the two electrodes in a way that reflects the geometry and chemical nature of the electrodes and the spacer. For $\Phi \to 0$, Equation 19 yields the

---

[3]This is just the golden rule rate expression (multiplied by the electron charge e), with M playing the role of coupling. In (30) only the first term in the square brackets of the first line appears. This gives the partial current from the negative to the positive electrode. The net current is obtained by subtracting the reverse current, as shown in Equation 19. Also, compared with (30), Equation 19 contains an additional factor of two that accounts for the spin multiplicity of the electronic states.

conduction

$$g \equiv \frac{I}{\Phi} = \frac{4\pi e^2}{\hbar} \sum_{l,r} |M_{lr}|^2 \delta(E_l - E_F)\delta(E_r - E_F). \qquad 21.$$

Tersoff & Hamman (30) have used substrate wavefunctions that correspond to a corrugated surface of a generic metal while the tip is represented by a spherical s orbital centered about the center $\mathbf{r}_0$ of the tip curvature. In this case they find

$$I \propto \sum_\nu |\psi_\nu(\mathbf{r}_0)|^2 \delta(E_\nu - E_F). \qquad 22.$$

The right hand side of Equation 22 is the local density of states of the metal. While this result is useful for analysis of spatial variation of the tunneling current on a given metal surface, the contributions from the coupling matrix elements in Equation 21 cannot be disregarded when comparing different metals and/or different adsorbates (6).

2.2.1 The Landauer Formula

The results of Equations 14–17 and 19–21 are special cases of a more systematic representation of the conduction and the current-voltage characteristic of a given junction, as shown by Landauer (32). Landauer's original result was obtained for a system of two one-dimensional leads connecting two macroscopic electrodes (electron reservoirs) via a scattering object or a barrier characterized by a transmission function $\mathcal{T}(E)$. The zero temperature conductance, measured as the limit $\Phi \to 0$ of the ratio I/Φ between the current and the voltage drop between the reservoirs, was found to be[4]

$$g = \frac{e^2}{\pi\hbar}\mathcal{T}(E_F). \qquad 23.$$

This result is obtained by computing the total unidirectional current carried in an ideal lead by electrons in the energy range $[(0, E) = (0, \hbar^2 k_E^2/2m)]$. In a one-dimensional system of length L the density of electrons, including spin, with wavevectors in the range between k and $k + dk$, is $n(k)dk = 2(1/L)(L/2\pi)f(E_k) dk = f(E_k)dk/\pi$. The corresponding velocity is $v = \hbar k/m$. Thus

$$I(E) = e\int_0^{k_E} dk v(k)n(k) = e\int_0^{k_E} dk(\hbar k/m)f(E_k)/\pi = \frac{e}{\pi\hbar}\int_0^E dE'f(E'). \qquad 24.$$

At zero temperature, the net current carried under bias Φ is

$$I = \frac{e}{\pi\hbar}\int_0^\infty dE(f(E) - f(E + e\Phi)) \xrightarrow{\Theta \to 0} \frac{e^2}{\pi\hbar}\Phi. \qquad 25.$$

---

[4]The corresponding resistance, g^{-1}, can be represented as a sum of the intrinsic resistance of the scatterer itself, $[(e^2/\pi\hbar)(\mathcal{T}/(1-\mathcal{T}))]^{-1}$, and a contribution $(e^2/\pi\hbar)^{-1}$ from two contact resistances between the leads and the reservoirs. [See Chapter 5 of (33) for a discussion of this point.]

Thus the conductance of an ideal one-dimensional lead is $I/\Phi = e^2/\pi\hbar = (12.9K\Omega)^{-1}$. In the presence of the scatterer this is replaced by

$$I = \frac{e}{\pi\hbar} \int_0^\infty dE T(E)(f(E) - f(E + e\Phi)) \xrightarrow{\Theta \to 0, \Phi \to 0} \frac{e^2}{\pi\hbar} T(E_F)\Phi, \qquad 26.$$

which leads to Equation 23. This result is valid for one-dimensional leads. In cases where the leads have finite size in the direction normal to the propagation so that they support traversal modes, a generalization of Equation 23 yields (33)[5]

$$g = \frac{e^2}{\pi\hbar} \sum_{i,j} T_{ij}(E_F) = \frac{e^2}{\pi\hbar} \text{Tr}(SS^\dagger)_{E_F}, \qquad 27.$$

where $T_{ij} = |S_{ij}|^2$ is the probability that a carrier coming from the left, say, of the scatterer in transversal mode i will be transmitted to the right into transversal mode j (S_{ij}, an element of the S matrix, is the corresponding amplitude). The sum in Equation 27 is over all traversal modes whose energy is smaller than E_F. More generally, the current for a voltage difference Φ between the electrodes is given by

$$I = \int_0^\infty dE[f(E) - f(E + e\Phi)] \frac{g(E)}{e}, \qquad 28.$$

$$g(E) = \frac{e^2}{\pi\hbar} \sum_{i,j} T_{ij}(E). \qquad 29.$$

As an example, consider the case of a simple planar-tunnel junction (see Equations 14–17), where the scattering process does not couple different transversal modes. In this case, the transmission function depends only on the energy in the tunneling direction

$$\sum_{i,j} T_{ij}(E) = \sum_i T_{ii}(E) = \frac{L_y L_z}{(2\pi)^2} \int dk_y \int dk_z T\left[E - (\hbar^2/2m)(k_y^2 + k_z^2)\right]$$

$$= \frac{L_y L_z}{(2\pi)^2} \frac{2\pi m}{\hbar^2} \int_0^E dE_r T(E - E_r). \qquad 30.$$

E_r is defined below Equation 16. Using this in Equation 27 yields the conductivity per unit length

$$\frac{g}{L_y L_z} \equiv \sigma = \frac{4\pi m e^2}{h^3} \int_0^{E_F} dE_x T(E_x), \qquad 31.$$

in agreement with Equation 17.

---

[5]The analog of Equation 27 for the microcanonical chemical-reaction rate was first written by Miller (34). Similarly, Equation 32 was first written in a similar context by Miller et al (35).

Similarly, Equations 19 and 21 are easily seen to be equivalent to Equation 23 or 29 if we identify M_{lr} with T_{lr} in Equation 35 below. An important difference between the results of Equations 27–29 and results based on the Bardeen's formalism, Equations 19–21, is that the former are valid for any set of transmission probabilities, even close to 1, whereas the latter yields a weak coupling result. Another important conceptual difference is the fact that the sums over ℓ and r in Equations 19–21 are over zero-order states defined in the initial and final subspaces, whereas the sums in Equations 27–29 are over scattering states, i.e. eigenstates of the exact system's Hamiltonian. It is the essence of Bardeen's contribution (31) that in the weak coupling limit (i.e. high/wide barrier) it is possible to write the transmission coefficient T_{ij} in terms of a golden rule expression for the transition probability between the zero-order standing-wave states $|l\rangle$ and $|r\rangle$ localized on the left and right electrodes, thus establishing the link between the two representations. [For an alternative formulation of this link, see Galperin et al (36).]

To explore this connection on a more formal basis, we can replace the expression based on transmission coefficients T by an equivalent expression based on scattering amplitudes, or T matrix elements, between zero-order states localized on the electrodes. This can be derived directly from Equations 27 or 29 by using the identity

$$\sum_{i,j} T_{ij}(E) = 4\pi^2 \sum_{l,r} |T_{lr}|^2 \delta(E - E_l)\delta(E - E_r). \qquad 32.$$

On the left side of Equation 32 a pair of indices (i,j) denote an exact scattering state of energy E, characterized by an incoming state i on the left, say, electrode and an outgoing state j on the right electrode. On the right, l and r denote zero-order states confined to the left and right electrodes, respectively. T is the corresponding transition operator whose particular form depends on the details of this confinement. Alternatively, we can start from the golden rule–like expression

$$I = e \frac{4\pi}{\hbar} \sum_{l,r} [f(E_l)(1 - f(E_r + e\Phi))$$
$$- f(E_r + e\Phi)(1 - f(E_l))] |T_{lr}|^2 \delta(E_l - E_r)$$
$$= \frac{4\pi e}{\hbar} \sum_{l,r} [f(E_l) - f(E_r + e\Phi)] |T_{lr}|^2 \delta(E_l - E_r). \qquad 33.$$

(An additional factor of 2 on the right-hand side accounts for the spin degeneracy.) It is convenient to recast this result in the form

$$I = \frac{4\pi e}{\hbar} \int_0^\infty dE [f(E) - f(E + e\Phi)] \sum_{l,r} |T_{lr}|^2 \delta(E - E_l)\delta(E - E_r)$$
$$= \int_0^\infty dE [f(E) - f(E + e\Phi)] \frac{g(E)}{e}, \qquad 34.$$

where

$$g(E) \equiv \frac{4\pi e^2}{\hbar} \sum_{l,r} |T_{lr}|^2 \delta(E - E_l)\delta(E - E_r). \qquad 35.$$

Note that Equations 32 and 35 imply again Equation 29. For $\Phi \to 0$, Equations 34 and 35 lead to $I = g\Phi$, with

$$g = g(E_F). \qquad 36.$$

The analogy of this derivation to the result in Equation 21 is evident.

2.3 Molecular Conduction

Equations 34–36 provide a convenient starting point for most treatments of currents through molecular junctions where the coupling between the two metal electrodes is weak. In this case it is convenient to write the system's Hamiltonian as the sum, $H = H_0 + V$, of a part, H_0, that represents the uncoupled electrodes and spacer and the coupling V between them. In the weak coupling limit the T operator

$$T(E) = V + VG(E)V; \quad G(E) = (E - H + i\varepsilon)^{-1} \qquad 37.$$

is usually replaced by its second term only. The first "direct" term V can be disregarded if we assume that V couples the states ℓ and r only via states of the molecular spacer. In the simple model (analog of the model that leads to Equation 10) where this spacer is an N-site bridge connecting the two electrodes so that site 1 of the bridge is attached to the left electrodes and site N to the right electrode, we have $T_{lr} = V_{l1} G_{1N} V_{Nr}$. At zero temperature this leads to (37)

$$\sum_{i,j} T_{ij}(E) = |G_{1N}(E_F)|^2 \Gamma_1^{(L)}(E_F) \Gamma_N^{(R)}(E_F) \qquad 38.$$

and, using Equations 34 and 35,

$$I(\Phi) = \frac{e}{\pi\hbar} \int_{E_F - e\Phi}^{E_F} dE |G_{1N}(E, \Phi)|^2 \Gamma_1^{(L)}(E) \Gamma_N^{(R)}(E + e\Phi). \qquad 39.$$

Here G_{1N} is an element of the reduced Green's function in the bridge's subspace, obtained by projecting out the metals' degrees of freedom

$$G = (E - H_B - \Sigma_B(E))^{-1}, \qquad 40.$$

where $H_B = H_B^0 + V_B$ is the Hamiltonian of the isolated-bridge entity given in the basis of eigenstates of H_B^0 by

$$H_B^0 = \sum_{n=1}^{N} E_n |n\rangle\langle n|; \quad V_B = \sum_{n=1}^{N} \sum_{n'=1}^{N} V_{n,n'} |n\rangle\langle n'|, \qquad 41.$$

and where

$$\Sigma_{nn'}(E) = \delta_{n,n'}(\delta_{n,1} + \delta_{n,N})[\Lambda_n(E) - (1/2)i\Gamma_n(E)], \qquad 42.$$

$$\Gamma_n(E) = 2\pi \sum_l |V_{ln}|^2 \delta(E_1 - E) + 2\pi \sum_r |V_{rn}|^2 \delta(E_N - E)$$

$$\equiv \Gamma_n^{(L)}(E) + \Gamma_n^{(R)}(E), \qquad 43.$$

$$\Lambda_n(E) = \frac{PP}{2\pi} \int_{-\infty}^{\infty} dE' \frac{\Gamma_n(E')}{(E - E')}. \qquad 44.$$

The transmission problem is thus reduced to evaluating a Green's function matrix element and two width parameters. The first calculation is a simple inversion of a finite (order N) matrix. The width Γ and the associated shift Λ represent the finite lifetime of an electron on a molecule adsorbed on the metal surface and can be estimated, for example (37), using the Newns-Anderson model of chemisorption (38). In the simple, tight-binding model of the bridge and in the weak coupling limit, G_{1N} is given by Equation 10 modified by the inclusions of the self-energy terms

$$G_{1N}(E) = \frac{V_{1,2}}{(E - E_1 - \Sigma_1(E))(E - E_N - \Sigma_N(E))} \prod_{j=2}^{N-1} \frac{V_{j,j+1}}{E - E_j}. \qquad 45.$$

Equations 38–45 thus provide a complete simple model for molecular conduction, equivalent to similar approximations used in theories of molecular electron transfer (e.g. 39 and references therein). Below we discuss more general forms of this formulation.

2.4 Relation to Electron-Transfer Rates

It is interesting to examine the relationship between the conduction of a molecular species and the electron-transfer properties of the same species (40). We should keep in mind that because of tunneling there is always an Ohmic regime near zero bias, with conduction given by the Landauer formula. Obviously this conduction may be extremely low, indicating in practice an insulating behavior. Of particular interest is the estimating of the electron-transfer rate in a given donor-bridge-acceptor (DBA) system that will translate into a measurable conduction of the same system when used as a molecular conductor between two metal leads. To this end consider a DBA system, with a bridge that consists of N identical segments (denoted 1, 2, ..., N), with nearest neighbor coupling V_B. The electron-transfer rate is given by Equation 11, which we rewrite in the form

$$k_{D \to A} = \frac{2\pi}{\hbar} |V_{D1} V_{NA}|^2 |G_{1N}(E_D)|^2 \mathcal{F}, \qquad 46.$$

where, in the weak coupling limit, $|V_B| \ll |E_B - E|$ (see Equation 10)

$$G_{1N}(E) = \frac{|V_B|^{N-1}}{(E_B - E)^N}, \qquad 47.$$

and where $\mathcal{F}$ is the Franck-Condon-weighted density of nuclear states, given in the classical limit by Equation 3. The appearance of $\mathcal{F}$ in Equation 46 indicates that the process is dominated by the change in the nuclear configuration between the two localization states of the electron. Consider now the conduction of a junction where the same DBA complex is used to connect between two metal contacts such that the donor and acceptor species are chemisorbed on the two metals ("left" and "right," respectively). Note that the conduction process does not involve localized states of the electron on the donor or the acceptor, so the factor $\mathcal{F}$ will be absent. Using the model of Section 2.4, we get

$$g(E) = \frac{e^2}{\pi\hbar}|G_{DA}(E)|^2 \Gamma_D^{(L)}(E) \Gamma_A^{(R)}(E), \qquad 48.$$

where, in analogy to Equation 45,

$$G_{DA}(E) = \frac{V_{D1} V_{NA}}{(E - E_D - \Sigma_D(E))(E - E_A - \Sigma_A(E))} G_{1N}(E). \qquad 49.$$

Since the donor and acceptor species are chemisorbed on their corresponding metal contacts, their energies shift closer to the Fermi energies, and we therefore assume that the denominator in Equation 49 is dominated by the imaginary parts of the self-energies Σ. Assuming also that the electronic structure of the bridge is not strongly distorted by the proximity to the metals leads to (40)

$$g = g(E_F) = \frac{16e^2}{\pi\hbar} \frac{|V_{D1}V_{NA}|^2}{\Gamma_D^{(L)}(E_F)\Gamma_A^{(R)}(E_F)} |G_{1N}(E_F)|^2; \quad E_F = E_D = E_A. \qquad 50.$$

Comparing Equations 6 to Equation 50, we get

$$g = \frac{e^2}{\pi\hbar} \frac{k_{D \to A}}{\mathcal{F}} \frac{8\hbar}{\pi \Gamma_D^{(L)} \Gamma_A^{(R)}}. \qquad 51.$$

In the symmetric case, $E_D = E_A$, we have $\mathcal{F} = (\sqrt{4\pi\lambda k_B T})^{-1} \exp(-\lambda/4k_B T)$. For a typical value of the reorganization energy $\lambda \sim 0.5$ eV, and at room temperature, this is ~ 0.02 (eV)$^{-1}$. Taking also $\Gamma_D^{(L)} = \Gamma_A^{(R)} \sim 0.5$ eV leads to $g \sim (e^2/\pi\hbar)$ $(10^{-13} k_{D \to A}(s^{-1})) \cong [10^{-17} k_{D \to A}(s^{-1})]\Omega^{-1}$. This sets a criterion for observing Ohmic behavior for small voltage bias in molecular junctions: With a current detector sensitive to pico-amperes, $k_{D \to A}$ has to exceed $\sim 10^6$ s^{-1} (for the estimates of $\mathcal{F}$ and Γ given above) before measurable current can be observed at 0.1-V voltage across such a junction.

2.5 Quantum Chemical Calculations

The simple models discussed above are useful for qualitative understanding of molecular conductivity; however, the Landauer formula or equivalent formulations can be used as a basis for more rigorous molecular calculations using extended Huckel (EH) (20, 21, 41–45) or Hartree Fock (HF) (46–48) calculations. These approaches follow similar semiempirical and ab initio calculations of electron-transfer rates in molecular systems (49). Such atomic-level calculations usually start from a (nonorthogonal) basis set of atomic orbitals, so the formalism described above has to be generalized for this situation.[6] We also relax the assumption that the molecule-metal contact is represented by coupling to a single molecular orbital. Finally, to account for possible strong coupling between the molecular species and the metals, the bridge is usually defined to include small portions of the metals on its two sides. We refer to such a bridge as a supermolecule. Defining the operator

$$\mathbb{H}(E) = EZ - H \text{ with } Z_{ij} = \langle i|j\rangle, \qquad 52.$$

the Green's function is $G(E) = \mathbb{H}(E)^{-1}$. In Equation 52, i and j denote atomic orbitals that may be assigned to the supermolecule (M), the left metal (L), and the right metal (R) subspaces. Denoting formally the coupling between the subspace M and the subspaces $K = L, R$ by the corresponding submatrices $\mathbb{H}_{MK}$, the Green's function for the supermolecule subspace is

$$G^{(M)}(E) = (\mathbb{H} - \Sigma^{(L)} - \Sigma^{(R)})^{-1}, \qquad 53.$$

with[7]

$$\Sigma^{(K)} = \mathbb{H}_{MK}(\mathbb{H}^{-1})_{KK}\mathbb{H}_{KM}. \qquad 54.$$

Using also

$$T_{lr} = \sum_{n,n'} \mathbb{H}_{ln} G_{nn'} \mathbb{H}_{n'r}, \qquad 55.$$

(l and r in the metal L and R subspaces, respectively; n,n' in the supermolecule subspace) in Equation 35 leads to

$$g(E) = \frac{e^2}{\pi\hbar} Tr\big[G^{(M)}(E)\Gamma^{(R)}(E)G^{(M)\dagger}(E)\Gamma^{(L)}\big], \qquad 56.$$

where, e.g. for the left metal,

$$\Gamma^{(L)}_{n,n'} = 2\pi \sum_{l} \mathbb{H}_{nl}\mathbb{H}_{ln'}\delta(E - E_l) \text{ (n and n' in the molecular subspace)}. \qquad 57.$$

In practice, Σ and $\Gamma = -2\text{Im}(\Sigma)$ can be computed by using closure relations based on the symmetry of the metal lattice (42). The trace in Equation 56 is over all basis

---

[6]Alternatively, it has been shown by Emberly & Kirczenow (50) that one can map the problem into a new Hilbert space in which the basis states are orthogonal.

[7]$\Sigma^{(K)}$ is a matrix in the molecular subspace and Equation 54 is a compact notation for $(\Sigma^{(K)})_{n,n'} = \sum_{k,k'} \mathbb{H}_{nk}(\mathbb{H}^{-1})_{kk'}\mathbb{H}_{k'n'}$, where k and k' are states in the metal K subspace.

states in the (super) molecular subspace. The evaluation of the Green's function matrix elements and of this trace is straightforward in semiempirical single-electron representations such as the extended Huckel approximation, and can be similarly done at the Hartree-Fock level using, after convergence, the Fock rather then the Hamiltonian matrix in Expressions 52–57, and invoking Koopmans' theorem (51). There are obviously considerable weaknesses in these procedures: Both are based essentially on an independent electron (hole) picture, and both are single-electron orbitals associated with the ground-state electronic configuration of the isolated neutral supermolecule. For example, Koopmans' theorem is accurate only for large systems, and the approximation involved in applying it to small systems is one reason why HF is not necessarily superior to EH for calculating the conduction properties of small molecular junctions. [This is true particularly for the lowest unoccupied molecular orbital (LUMO)-dominated conduction, because the HF method is notoriously inadequate for electron affinities while highest occupied molecular orbital (HOMO)-dominated conduction is better represented by this approach; (for further discussion, see 48)].

In spite of these limitations, EH- and HF-based calculations have provided important insight into the conduction properties of molecular junctions. Figure 3 shows a remarkable example. The (EH) calculation is done for a single α,α'-xylyl dithiol molecule adsorbed between two gold contacts. The experiment monitors the current between an STM tip (obtained by cutting a Pt/Ir wire) and a monolayer of such molecules deposited on gold, and it is assumed that lateral interaction between the molecules is unimportant. Two unknown parameters are used for fitting. The first is the position of the metal's Fermi energy in the unbiased junction relative to the molecular energy levels expressed by $E_{FH} \equiv E_F - E_{HOMO}$. The second describes the electrostatic-potential profile along the junction, represented by a parameter η that expresses the distribution of the voltage drop between the two metal leads (see Equation 67 below). As seen in Figure 3, good agreement between theory and experiment is obtained for $E_{FH} = 0.9$ eV and $\eta = 0.5$.

In view of the other unknowns, associated both with the uncertainty about the junction structure and with the simplified computation, the main value of these results is not in the absolute numbers obtained but rather in highlighting the importance of these parameters in determining the junction-conduction behavior. We return to the issue of the junction-potential profile below. Other qualitative issues that were investigated with these types of calculations include the effect of the nature (length and conjugation) of the molecular bridge (41), the effect of the molecule-electrode binding and of the molecular binding site (42), the relation of conductance spectra to molecular electronic structure (52), and the effect of bonding molecular wires in parallel (41, 53).

2.6 Spatial-Grid Based Pseudopotential Approaches

Another way to evaluate the expressions appearing in Equations 32 and 35 as well as related partial sums is closely related to the discrete variable representation

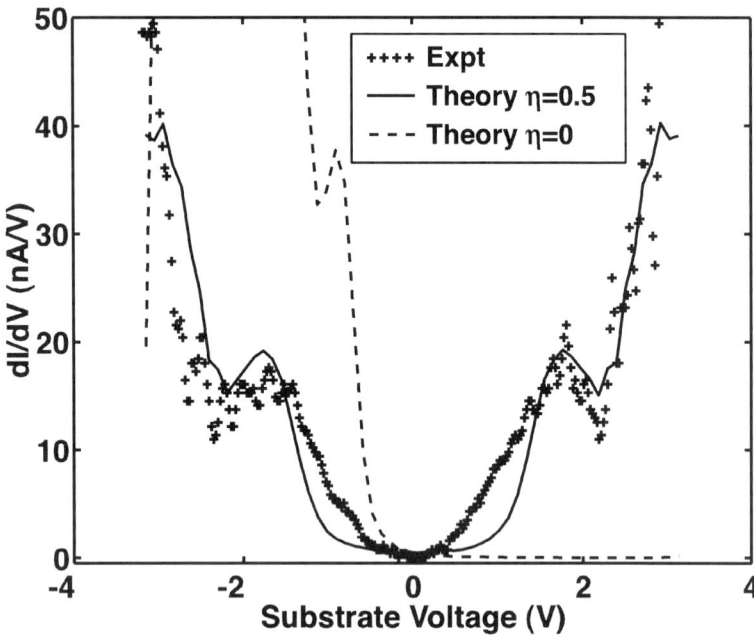

Figure 3 Measured and computed differential conduction of a single α,α'-xylyl dithiol molecule adsorbed between two gold contacts. See text for details. (From Reference 21.)

of reaction probabilities as formulated by Seideman & Miller (54, 54a). We have already seen that the sum

$$s(E) \equiv \sum_{l,r} |T_{lr}|^2 \delta(E - E_l)\delta(E - E_r), \qquad 58.$$

which is related to the conduction by $g(E) = (4\pi e^2/\hbar)s(E)$ (see Equation 35 can be represented by Equation 56).

$$s(E) = \frac{1}{4\pi^2} Tr[G^{(M)}(E)\Gamma^{(R)}(E)G^{(M)*}(E)\Gamma^{(L)}]. \qquad 59.$$

If, instead of considering transitions from left to right electrode, we think of Equation 58 as expressing a sum over transition probabilities from all initial (i) states of energy E in the reactant space to all final (f) states of the same energy in the product space, $s(E)$ is also associated with the so-called cumulative reaction probability (34, 54, 54a), which in terms of the reaction S matrix is defined by $N(E) = \sum_{i,f} |S_{if}(E)|^2 = 4\pi^2 s(E)$, i.e. $N(E) = \sum_{i,f} T_{if}(E)$. Equation 59 now expresses the important observation that the cumulative reaction probability for a reactive scattering process can be expressed as a trace over states, defined in a finite subspace that contains the interaction region, of an expression that depends on the reduced Green's function and the associated self energy defined in that subspace. Following Seideman & Miller, we can use a spatial grid representation for the states in this subspace, so that the trace in Equation 59 becomes a sum over grid points. Also, in this representation the overlap matrix **Z** is zero. In general, any subspace of position space that separates reactants from products (i.e. that encompasses the entire interaction region; the molecular bridge in our application) can be used in Equation 59, provided that the consequences of truncating the "rest of the universe," expressed by the corresponding Σ and Γ, can be computed. The absorbing boundary condition Green's function (ABCGF) method of Seideman & Miller is based on the recognition that if this subspace is shown to be so large that its boundaries are far from the interaction region, the detailed forms of Σ and Γ are not important; the only requirement is that scattered waves that approach these boundaries will be absorbed and not reflected back into the interaction zone. In the ABCGF method, this is accomplished by taking $\Sigma = -(1/2)i\Gamma = -i\varepsilon(\mathbf{r})$, a local function in position space, taken to be zero in the interaction region, and gradually increasing from zero when approaching the subspace boundaries. Its particular form is chosen to affect complete absorption of waves approaching the boundary to a good numerical accuracy. Equation 59 then becomes

$$s(E) = 4Tr\left[G^{ABC}(E)\varepsilon_R G^{ABC*}(E)\varepsilon_L\right], \qquad 60.$$

where $G^{ABC}(E) = (E - H + i\varepsilon)^{-1}$; $\varepsilon = \varepsilon_R + \varepsilon_L$, and where ε_R and ε_L are different from zero only on grid points near the right side (more generally the product side) and the left (reactant) side of the inner subspace, respectively.

A similar development can be seen for the partial sum

$$s_l(E) \equiv \sum_r |T_{lr}|^2 \delta(E - E_r), \qquad 61.$$

which, provided that l is taken as an eigenstate of the Hamiltonian describing the left electrode (or the reactant sunspace), is related the "one-to-all"–rate, $k_l(E)$, to go from an initial state of energy E on the left electrode (or in reactant space) to all possible states on the right one (product space) according to $k_l = (2\pi/\hbar)s_l$.[8] We use the same definition of the coupling V between our subspace (bridge) and the reactant and product (electrode) states. Putting $T = VGV$ into Equation 61, we get

$$s_l(E) = \frac{1}{2\pi} \langle l|VG^{(M)}\Gamma^{(R)}G^{(M)*}V|l\rangle. \qquad 62.$$

Using again a position grid representation of the intermediate states to evaluate this expression, and applying the same methodology as above, Equation 62 can be

---

[8]The "microcanonical rate" is defined by $k(E) = \rho_L^{-1}(E)\sum_l k_l\delta(E - E_l) = (2\pi\hbar\rho_L(E))^{-1}4\pi^2 s(E) = (2\pi\hbar\rho_L(E))^{-1}N(E)$.

recast in the form[9]

$$s_l(E) = \frac{1}{\pi}\langle l|VG^{ABC}(E)\varepsilon_R G^{ABC*}(E)V|l\rangle$$
$$= \frac{1}{\pi}\langle l|\varepsilon_L G^{ABC}(E)\varepsilon_R G^{ABC*}(E)\varepsilon_L|l\rangle. \qquad 63.$$

The results of Equations 60 and 63 are very useful for computations of transmission probabilities in models where the interaction between the transmitted particle and the molecular spacer is given as a position-dependent pseudopotential. Applications to electron transmission through water and other molecular layers are discussed in Section 4.

2.7 Density Functional Calculations

Density functional methods provide a convenient framework for treating metallic interfaces (55). Applications of this methodology to the problem of electron transport through atomic and molecular bridges have been advanced by several workers. In particular, Lang's approach (56, 57) is based on the density-functional formalism (58) in which the single-electron wavefunctions $\psi_0(r)$ and the electron density $n_0(r)$ for two bare metal (jellium) electrodes are computed, then used in the Lippman-Schwinger equation

$$\psi(\mathbf{r}) = \psi_0(\mathbf{r}) + \int d\mathbf{r}'d\mathbf{r}'' G^0(\mathbf{r}, \mathbf{r}')\delta V(\mathbf{r}', \mathbf{r}'')\psi(\mathbf{r}''), \qquad 64.$$

to get the full single-electron scattering wavefunctions $\psi(\mathbf{r})$ in the presence of the additional bridge, where G^0 is the Green's function of the bare electrode system and δV is the difference between the potential of the full system containing an atomic or a molecular spacer and that of the bare electrodes. Equation 64 yields scattering states that can be labeled by their energy E, the momentum $\mathbf{k}_\parallel$ in the direction (yz) parallel to the electrodes, the sign, $\pm$, of k_x, and the spin. Denoting by μ_L and μ_R the electron-electrochemical potential in the left and right electrode, respectively, the zero-temperature electrical current density from left to right (for $\mu_L > \mu_R$) is then

$$J(\mathbf{r}) = -2\int_{\mu_L}^{\mu_R} dE \int d^2k_\parallel \, \text{Im}\{\psi_+^* \nabla \psi_+\}. \qquad 65.$$

The factor 2 accounts for the double occupancy of each orbital. This approach was used recently (59) to calculate current through a molecular species, benzene 1,4-dithiolate molecule [as used in the experiment of Reed et al (19)], between two jellium surfaces and has demonstrated the large sensitivity of the computed current to the microscopic structure of the molecule-metal contacts.

---

[9]The second part of Equation 63 is obtained by using the identity $\varepsilon_r|l\rangle = 0$ to write $\varepsilon_R G^* V|l\rangle = \varepsilon_R(1 + G^*V)|l\rangle = \varepsilon_R G^*(G^{*-1} + V)|l\rangle$, which, together with $G^{*-1} = E - H_0 - V + i\varepsilon$, $(E - H_0)|l\rangle = 0$, and $\varepsilon|l\rangle = \varepsilon_l|l\rangle$, yields the desired result.

The calculations described above were done in the linear response regime. In contrast, the density-functional approach of Hirose & Tsukada (60) calculates the electronic structure of a metal-insulator-metal system under strong applied bias. This is accomplished by computing the effective one-electron potential in a way that accounts for this bias. This potential contains the standard contributions from the Coulomb and the exchange-correlation interactions as well as from the ionic cores. However, the Coulomb (Hartree) contribution is obtained from the solution of a Poisson equation

$$\nabla^2 V_H(\mathbf{r}) = -4\pi[\rho(\mathbf{r}) - \rho_+(\mathbf{r})], \qquad 66.$$

in the presence of the applied potential-boundary conditions. $\rho_+(\mathbf{r})$ is the fixed positive-charge density, and the electron density $\rho(\mathbf{r})$ is constructed by summing the squares of the wavefunctions over the occupied states. The resulting formalism thus approximately accounts for nonequilibrium effects within the density-functional calculation. A simplified version of the same methodology has recently been presented by Mujica et al (61).

To end this brief overview of density functional-based computations of molecular conduction, we should note that this approach suffers in principle from problems similar to those encountered in using the HF-approximation, namely, the inherent inaccuracy of the computed LUMO energy and wavefunctions. The errors are different, for example, HF overestimates the (HOMO)-LUMO gap [since the HF-LUMO energy is too high (62) whereas DFT underestimates it (63)]. Common to both approaches is the observation that processes dominated by the HOMO level will be described considerably better by these approaches than processes controlled by coupling to the LUMO (48, 64).

2.8 Potential Profiles

The theoretical and computational approaches described above are used to compute both the Ohm-law conduction, $g(E_F)$, of a molecular bridge connecting two metals, (Equation 35 or 56) and the current-voltage characteristics of the junction, also beyond the Ohmic regime (Equation 34). We should keep in mind that these calculations usually disregard a potentially important factor—the possible effect of the imposed electrostatic field on the nuclear configuration as well as on the electronic structure of the bridge. A change in nuclear configuration under the imposed electrostatic field is, in fact, not very likely for stable, chemisorbed molecular bridges. On the other hand, the electronic wavefunctions can be distorted by the imposed field, and this in turn may affect the electrostatic-potential distribution along the bridge,[10] the electronic coupling between bridge segments, and the position of the molecular energy levels vis-à-vis the metal's Fermi energies. These

---

[10]In a single electron description this local electrostatic potential will be an input, associated with the underlying many-electron response of the molecular bridge, to the position-dependent energies of the bridge electronic states in the site representation.

effects were, in fact, taken approximately into account by Hirose & Tsukada (60) and by Mujica et al (61) by solving simultaneously the coupled Schrödinger and Poisson equations. The latter yields the electrostatic potential for the given electron density and under the imposed potential boundary conditions.

The importance of the electrostatic-potential profile on the molecular bridge in determining the conduction properties of an MMM junction was recently discussed by Tian et al (21) in conjunction with the current-voltage characteristics of a junction comprised of an STM tip, a gold substrate, and a molecule with two bonding sites (e.g. α,α'-xylyl dithiol) connecting the two. For a given potential bias $\Phi = \mu_L - \mu_R$ between substrate (left electrode, for instance) and tip (right electrode), a common model assumption is that the electrostatic potential on the molecule is pinned to that of the substrate so that all the potential drop occurs between the molecule and the tip. In contrast, for MMM junctions with a strong chemical bonding of the molecule to both metals, one often assumes a linear-potential ramp interpolating between the potentials on the two metal leads. In fact, because the molecule is a polarizable object we may expect that most of the potential drop takes place near the molecule-metal contacts (61). Denoting these drops by Φ_L and Φ_R for the left and right electrodes, respectively (so that $\Phi_L + \Phi_R = \Phi$), the conduction properties of the junction are determined by the position of the molecular-bridge states relative to the equilibrium Fermi energy, and by the voltage division factor η defined by

$$\frac{\Phi_L}{\Phi_R} = \frac{\eta}{1-\eta} \quad \text{or} \quad \eta = \frac{\Phi_L}{\Phi}. \qquad 67.$$

If $\eta = 0$, all the potential drop occurs at the molecule tip (right) interface. In this case, changing the voltage across the junction amounts to changing the energy difference between the molecular levels and the tip electrochemical potential. Enhanced conduction is expected when the latter matches either the HOMO (when the tip is positively biased) or the LUMO (when the tip is negatively biased) energies. However, because the HOMO and the LUMO states are usually coupled differently to the metals (for example, in the aromatic thiols the HOMO is a sulfur-based orbital that couples strongly to the metal whereas the HOMO is a ring-based orbital that couples weakly to it), this implies strong asymmetry, around zero voltage, in the current-voltage dependence, i.e. rectification. In contrast, the observed dependence is essentially symmetric around $\Phi = 0$, a behavior obtained from Equation 34 for a symmetric voltage division factor $\eta = 0.5$ (21).

So far, only a few studies (61, 64) have addressed the computational problem of finding the potential distribution across biased molecular junctions. To what extent the electrostatic potential calculated in these works is relevant for single-electron models of molecular junctions still remains to be clarified. In particular, in other treatments of excess electrons at the insulator side of a metal-insulator interface, the image potential attracting the electron to the interface plays an important role if the insulator dielectric constant is not too large (28, 65). Also, experimental implications of this potential are well known (66). The observation (21) that details of the electrostatic-potential distribution across an MMM junction can significantly affect qualitative aspects of the junction electrical properties makes further theoretical work in this direction highly desirable.

2.9 Rectification

The possibility of constructing molecular junctions with rectifying behavior has been under discussion since Aviram & Ratner (67) suggested that an asymmetric donor-bridge-acceptor system connecting two metal leads can rectify current. The proposed mechanism of operation of such a device is shown in Figure 4. When the left electrode is negatively biased, i.e. the corresponding electrochemical potentials satisfy $\mu_L > \mu_R$, as shown, electrons can move from this electrode to the LUMO of molecular segment A as well as from the HOMO of molecular segment D to the right electrode. Completion of the transfer by moving an electron from A to D is assisted by the intermediate bridge segment B. When the polarity of the bias is reversed, the same channel is blocked. This simple analysis is valid only if the molecular-energy levels do not move together with the metal electrochemical potentials, and if the coupling through the intermediate bridge is weak enough so

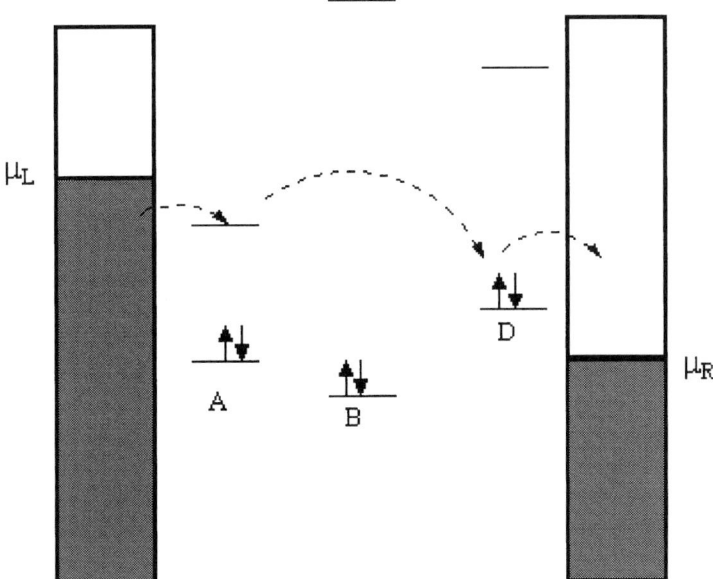

Figure 4 A model for current rectification in a molecular junction. Shown are the chemical potentials μ_L and μ_R in the two electrodes, and the HOMO and LUMO levels of the donor, acceptor, and bridge. When the right electrode is positively biased (as shown), electrons can hop from left to right, as indicated by the dotted arrows. If the opposite bias can be set without affecting the electronic structure of the DBA system too strongly, the reverse current will be blocked.

that the orbitals on the D and A species maintain their local nature. Other models for rectification in molecular junctions have been proposed (68). As discussed above, the expected rectifying behavior can be very sensitive to the actual potential profile in the ABD complex, which in turn depends on the molecular response to the applied bias (21, 69). This explains why rectification is often not observed even in asymmetric molecular junctions (69). Still, rectification has been observed in a number of MMM junctions as well as in several STM experiments involving adsorbed molecules (17, 18, 70, 71).

2.10 Carrier-Carrier Interactions

The models and calculations discussed so far focus on processes for which the probability that a charge carrier occupies the bridge is low so that carrier-carrier interactions can be disregarded. Electron-electron interactions were taken into account only in so far as they affected the single-electron states, either in constructing the molecular spectrum (in the ab initio HF or DFT calculations) or in affecting the junction electrostatic potential through the electronic-polarization response of the molecule or the metal contacts. When the density of carriers in the space between the metal contacts becomes large, Coulomb interactions between them have to be taken into account explicitly. Here we briefly discuss the effect of such interactions.

In classical (hopping) transport of carriers through insulating films separating two metals, intercarrier interactions appear as suppression of current due to film charging (72). In nano-junctions involving double-barrier structures, increased electron population in the intermediate well under resonance transmission should affect the transport process for similar reasons. For example, consider a small metal sphere of radius R in the space between two metal electrodes and assume that both sphere and electrodes are made of the same metal of workfunction W. Neglecting the possible proximity effect of these electrodes, the classical energy for removing an electron from the sphere to infinity is $W + e^2/2R$ and the classical energy for the opposite process is $W - e^2/2R$.[11] Here the sphere plays the role of a molecular bridge in assisting electron tunneling between the two electrodes, and these energies now play the same role as the corresponding HOMO and LUMO energies of the bridge. This implies that a finite voltage difference is needed before current can flow in this sphere-assisted mode between the two metals, a phenomenon known as Coulomb blockade. For a larger potential bias, other conduction channels, corresponding to more highly charged states of the sphere, give rise to the phenomenon of Coulomb steps (74). For experimental manifestations of such and related phenomena (see for instance, 75, 76). The possibility of observing such phenomena in electrochemical systems was discussed by Kuznetsov & Ulstrup (77) and possibly demonstrated by Fan & Bard (78).

---

[11]From experimental and theoretical work on ionization potentials of small metal clusters (73) we know that the actual energies are approximately $W + 0.4e^2/R$ and $W - 0.6e^2/2R$, respectively, with the differences arising from quantum-size effects.

When the junction consists of a molecule or a few molecules connecting two metal leads, such Coulomb blockade phenomena are not expected to appear so clearly. The first Coulomb threshold is replaced, as just described, by the gap associated with the position of the metal's Fermi energies relative to the molecular HOMO and LUMO levels (modified by appropriate electron correlations). However, the discreteness (in the sense that $\Delta E \gg k_B T$) of the molecular spectrum implies that for any given charging state of the molecule, e.g. a molecule with one excess electron or one excess hole, there will be several distinct conduction channels that will appear as steps in the current vs voltage plot. It will be hard to distinguish between this structure and between a genuine Coulomb blockade structure. It should be emphasized that for potential applications, e.g. using the molecular junction in single-electron transistor devices, the distinction between the origins of these conduction structures is, in principle, not important.

Understanding intercarrier interactions, in particular correlated-carrier transport in molecular junctions, continues to be an important experimental and theoretical challenge. Several recent theoretical works have addressed this problem within the Hubbard model with (79, 80) or without (81) the mean-field approximation. Recent work by Gurvitz & Prager (82), using exactly solvable models of electron transport in two- and three-barrier structures, has indicated that new phenomenology may arise from the interplay of inelastic transitions and intercarrier interactions in the barrier. In fact, dephasing transitions in the barrier may prove instrumental in explaining the charge quantization that gives rise to the single-electron transport behavior of such junctions (see Equation 83, Section 6.3).

2.11 Some Open Issues

This section discusses some subtle difficulties that are glossed over in most treatments of electron transmission using the formalisms described above. These should be regarded as open theoretical issues that should be addressed in the future. The source of these problems is our simplified treatment of what is actually a complex many-body open system. In particular, common ways of incorporating many-body effects using single-body effective potentials becomes questionable in particular limits of timescales and interaction strengths.

One such issue, already mentioned, is the use of a static image to account for the effect of metal polarizability (namely the response of the metal electrons) on charge transfer processes at metal surfaces. The timescales estimated in Section 3.1 below are of the same order as metal-plasma frequencies that measure the electronic-response time of metals. Still, static-image theory has been used in the analysis of Section 2.2 and in other treatments of electron injection from metals into insulating phases (84). To what extent dynamic-image effects are important is not known, although theories that incorporate such effects have been developed (55a).

Assuming that image interactions at metal surfaces should be accounted for in the static limit, namely that the metal responds instantaneously to the tunneling charge, opens other questions. Many calculations of electronic processes near metal surfaces [e.g. Equation 28; (See Section 2.2 above)] assume that the metal electrons respond instantaneously to the position of the tunneling electron. Other approaches used in different contexts [e.g. reaction-field (cavity) models in quantum chemistry calculations for solvated molecules] calculate the response to electrons in their atomic or molecular orbitals, or, more generally, electronic-charge distributions, and computing these under the given potential-boundary conditions (e.g. see 61) implies that the corresponding orbitals or charge distributions are well defined on timescales shorter than the metal-response times. These two approaches are not equivalent, because the Schrödinger equations derived from them are nonlinear in the electronic wavefunctions. Examination of the energies and timescales involved suggests that in most situations assuming instantaneous metal response to the electron position is more suitable than assuming instantaneous response to the charge distribution defined by a molecular orbital, but the corresponding timescales are not different enough to make this a definite statement. A similar issue appears in attempts to account for the electronic polarizability of a solvent in treating fast electronic processes involving solute molecules or excess electrons in this solvent. We return to this point in Section 4.

Finally, an interesting point of concern is related to the way the Fermi distribution functions enter into the current equations. For example, Bardeen's transmission formula (19) is based on weak coupling between states localized on the two electrodes. Consequently, unidirectional currents contain a product, $f(1-f)$, i.e. the probability that the initial state is occupied multiplied by the probability that the final state is not. In this viewpoint, the transitions occur between two weakly coupled systems, each of them in internal thermal equilibrium, which are out of equilibrium with each other because of the potential bias. Alternatively, we could work on the basis of exact eigenstates of the whole system comprising the two electrodes and the spacer between them. This system is in an internal nonequilibrium state in which transmission can be described as a scattering problem. The relevant eigenstates correspond to incident (incoming) waves in one electrode and transmitted waves in the other. The flux associated with those scattering states arising from an incident state in the negatively biased electrode is proportional to $f(E)$, while that associated with incoming waves in the positively biased electrode is proportional to $f(E + e\Phi)$. The net flux is therefore found again to be proportional to the difference $f(E) - f(E + e\Phi)$. This argument cannot be made unless the process can be described in terms of coherent scattering states defined over the whole system. When inelastic scattering and dephasing processes take place the description in terms of exact scattering states of the whole system becomes complicated (83, 85), although kinetic equations for electron transport can be derived for relatively simple situations (82). On the other hand, it appears that for weakly coupled contacts the perturbative approach that leads to Equation 19 is valid. This approach describes the transmission in terms of electron states localized on the two electrodes where unidirectional rates appear with $f(1-f)$ factors and can, in principle, be carried over to the inelastic regime (see also Section 3.4). The exact correspondence between these different representation needs further study [see (86) for a recent discussion of experimental implications].

3. DEPHASING AND RELAXATION EFFECTS

The theoretical treatments of electron transmission and conduction through insulating barriers reviewed in the previous section have assumed that the barrier nuclear configuration is static. Consequently, the conduction was computed from the electronic structure of static interfacial configurations. Nuclear reorganization does play a dominant role in the analogous theory of electron transfer in molecular systems; however, here again the electronic coupling itself is computed for static structures, while coupling to nuclear motion is assumed to be associated with the initial and final localized states of the transferred electron. As discussed in Section 2.5, the corresponding nuclear reorganization energies are unimportant in an MMM junction because the transferred electron does not stay localized on the molecular species. Disregarding thermal interactions also during the transmission process therefore leads to a rigid junction model. While we cannot rule out the possible validity of such a model, it is necessary to consider possible scenarios where thermal relaxation on the bridge is important for two reasons. First, dephasing processes associated with electron-phonon coupling are the primary source for converting the transmission process from coherent transfer to incoherent hopping. Therefore ignoring nuclear dynamics disregards a potentially important transfer mechanism. Second, as discussed in the introduction, an important factor in designing molecular conductors is their structural stability; therefore understanding heat generation and dissipation in molecular conductors is an important issue (87, 88). This naturally motivates a study of inelastic effect and thermal relaxation during electron transmission. Indeed, the effect of dephasing and relaxation on carrier transport through molecular junctions (as well as other microscopic charge-transport devices) on its temperature and system-size dependence, and on possible interference effects, has recently attracted much attention.

3.1 Tunneling Traversal Times

The underlying assumption in the treatment of electron transfer and transmission described in Section 2 is that the junction nuclear structure is rigid. The validity of this assumption should be scrutinized. Obviously, whether the barrier appears rigid to the tunneling electron, and to what extent inelastic transitions can occur and affect transmission and conductance, depends on the relative scales of barrier motions and the transmission traversal time, properly defined.

A framework for discussing these issues is the theory of tunneling traversal times. "Straightforward" timescales for tunneling, such as the rate for probability buildup on one side of a barrier following a collision of an incoming particle on the other side, or the time associated with the tunneling splitting in a symmetric, double-well potential, are important measures of the tunneling rate. Following the work of Büttiker & Landauer (89–91) and others (92), it has been recognized that other timescales may be relevant for other observables associated with the tunneling process. The question of how long the tunneling particle actually spends in the classically forbidden region of the potential is of particular interest. This traversal time for tunneling is useful in estimates of the relative importance of processes that may potentially occur while the particle is in the tunneling region. Energy exchange with other degrees of freedom in the barrier and interaction with external fields focused in the barrier region (e.g. deflection of a tunneling electron by an electrostatic field induced by a heavy ion) are important examples.

The Büttiker-Landauer approach to tunneling timescales is based on imposing an internal clock on the tunneling system, for example a sinusoidal modulation of the barrier height (89). At modulation frequencies much smaller than the inverse-tunneling time, the tunneling particle sees a static barrier that is lower or higher than the unperturbed barrier, depending on the phase of the modulation. At frequencies much higher than the inverse-tunneling time, the system sees an average perturbation and thus no effective change in the barrier height, but inelastic tunneling can occur by absorption or emission of modulation quanta. The inverse of the crossover frequency separating these regimes is the estimated traversal time for tunneling. For tunneling through the one-dimensional rectangular barrier,

$$V(x) = \begin{cases} U_B; & x_1 \leq x \leq x_2 \\ 0 & \text{otherwise} \end{cases}, \qquad 68.$$

and provided that $d = x_2 - x_1$ is not too small, and that the tunneling energy E is sufficiently below U_B, this analysis gives

$$\tau = \frac{d}{v_I} = \sqrt{\frac{m}{2(U_B - E_0)}} d \qquad 69.$$

for a particle of mass m and energy $E_0 < U_B$. v_I, defined by Equation 69, is the imaginary velocity for the under-barrier motion. A similar result is obtained by using a clock based on population transfer between two internal states of the tunneling particle induced by a small barrier-localized coupling between them (90). Using the same clock for electron transfer via the superexchange mechanism in the model of Figure 2 (equal donor and acceptor energy levels, $E_A = E_D$, coupled to opposite ends of a molecular bridge described by an N-state tight-binding model with nearest-neighbor coupling V_B, with an energy gap $\Delta E_B = E_B - E_D \gg V_B$) yields (93)

$$\tau = \frac{\hbar N}{\Delta E_B}. \qquad 70.$$

Nitzan et al (93) have shown that the results in Equations 69 and 70 are limiting

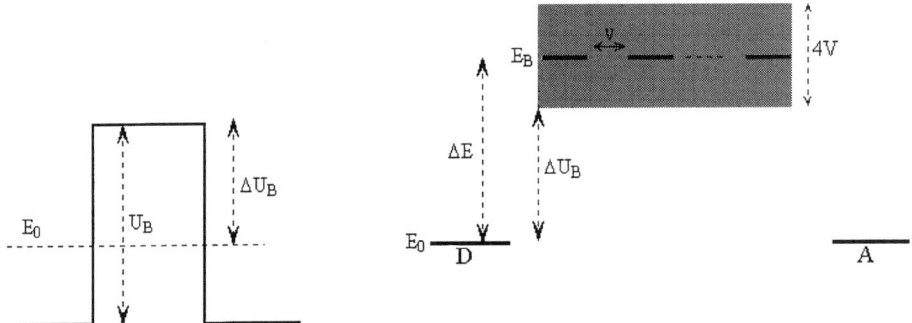

Figure 5 Parameters used in the expressions for tunneling traversal times. *Left*: Tunneling through a rectangular barrier. *Right*: Bridge-mediated transfer, where the gray area denotes the band associated with the tight-binding level structure of the bridge.

cases (wide- and narrow-band limits) of a more general expression

$$\tau = \frac{\hbar N}{2V_B \sqrt{\frac{\Delta U_B}{V_B} + \left(\frac{\Delta U_B}{2V_B}\right)^2}}, \qquad 71.$$

where $\Delta U_B \equiv E_B - 2V_B - E_D$ is the difference between the initial energy E_D and the bottom of the conduction band, $E_B - 2V$ (see Figure 5). When $V_B \to 0$, $\Delta U_B \to \Delta E_B$ and the r.h.s of Equation 71 become that of Equation 70. In the opposite limit, $V_B \to \infty$ with ΔU_B kept constant, Equation 71 becomes

$$\tau = \frac{\hbar N}{2\sqrt{V_B \Delta U_B}}. \qquad 72.$$

Expressing V_B in terms of the effective mass for the band motion, $m = \hbar^2/2V_B a^2$, using $a = d/N$, Equation 72 yields the Büttiker-Landauer result, which is Equation 69.

The interpretation of τ defined above as a characteristic time for the tunneling process should be used with caution. An important observation made by Büttiker (90) is that the tunneling time is not unique but depends on the observable used as a clock. Still, as shown in Büttiker & Landauer (89), for a proper choice of clock the traversal time provides a useful measure for the adiabaticity or nonadiabaticity of the interaction of the tunneling particle with barrier degrees of freedom. The calculation that leads to Equations 70–72 uses a clock based on two internal states, $|1\rangle$ and $|2\rangle$, of the tunneling particle with a small barrier-localized coupling, $\lambda(|1\rangle\langle 2| + |2\rangle\langle 1|)$, between them. The incident particle is in state $|1\rangle$. The population of state $|2\rangle$ in the transmitted wavefunction can be related to the duration of the interstate coupling, i.e. to the traversal time. Writing the transmitted state in the form $c_1|1\rangle + c_2|2\rangle$ this procedure yields

$$\tau = \lim_{\lambda \to 0} \left(\frac{\hbar}{|\lambda|} \left|\frac{c_2}{c_1}\right| \right). \qquad 73.$$

For the one-dimensional rectangular barrier model, Equation 68, and in the limit $\kappa d \gg 1$, this leads again to Equation 69. Galperin et al (94) have applied the same approach to compute traversal times through water layers (see Section 4).

For tunneling through a molecular spacer modeled as a barrier of width ~ 10 Å (N = 2–3) and height $U_B - E \cong \Delta E \sim 1$ eV, Equations 69 and 70 yield $\tau \cong 0.2$ fs and $\tau \cong 2$ fs, respectively, both considerably shorter than the vibrational period of molecular vibrations. When the barrier is lower or when tunneling is affected or dominated by barrier resonances, the traversal time becomes longer, and competing relaxation and dephasing processes in the barrier may become effective. This is expected to be the rule for resonance transmission through molecular bridges, because the bandwidth associated with the bridge states (i.e. the electronic coupling between them; see Figure 5) is considerably smaller than in metals. As a consequence, thermal relaxation and dephasing are expected to dominate electron transport at and near resonance. This issue is discussed next.

3.2 Nuclear Relaxation During Electron Transmission

It has long been recognized that tunneling electrons interact, and may exchange energy, with nuclear degrees of freedom in the tunneling medium. One realization of such processes is inelastic electron-tunneling spectroscopy (84), where the opening of inelastic channels upon increasing the electrostatic potential difference between the source and sink metals is manifested as a peak in the second derivative of the tunneling current with respect to this potential drop. Recent applications of this phenomenon within scanning-tunneling spectroscopy hold great promise for making the STM a molecular analytical tool (95). Inelastic electron tunneling may also cause chemical bond breaking and chemical rearrangement in the tunneling medium, either by electron-induced consecutive excitation or via transient formation of a negative ion[12] (96).

As discussed by Gadzuk (97), the phenomenology of inelastic electron transmission is also closely related to other electronic processes in which transient occupation of an intermediate state drives a phonon field. Intramolecular vibrational excitation in resonant electron scattering (98), phonon excitation in resonant electron tunneling in quantum-well heterostructures (99, 99a), and electron-induced desorption (100) can all be described using similar models. A prototype Hamiltonian describing these models is

$$H = H_{el} + H_{ph} + H_{el-ph}, \qquad 74.$$

---

[12]While our language refers to electron transport and electron tunneling, hole transport and nuclear excitation via transient positive ion formation are equally possible.

where H_{el} is the electronic Hamiltonian

$$H_{el} = \sum_n E_n c_n^\dagger c_n + \sum_{n,n'(n \neq n')} V_{n,n'} c_n^\dagger c_{n'} + \sum_k E_k c_k^\dagger c_k$$
$$+ \sum_k \sum_n \left(V_{k,n} c_k^\dagger c_n + V_{n,k} c_n^\dagger c_k \right), \qquad 75.$$

H_{ph} is the Hamiltonian of the phonon bath

$$H_{ph} = \sum_v \hbar \omega_v b_v^\dagger b_v, \qquad 76.$$

and H_{el-ph} is the electron-phonon interaction, usually written in the form

$$H_{el-ph} = \sum_n \sum_v \lambda_{nv} c_n^\dagger c_n \left(b_v^\dagger + b_v \right). \qquad 77.$$

Here $c_j^\dagger$ and c_j ($j = n, n', k$) create and annihilate an electron in electronic state j, while $b_v^\dagger$ and b_v similarly create and annihilate a phonon of mode v, of frequency ω_v. In Equation 75 the states (k) are taken to be different manifolds of continuous-scattering states, denoted by a continuous index k. Figure 2b shows two such manifolds, $k = \{\ell\}, \{r\}$, with intermediate states that are discrete electronic states of the observed molecular system. The electronic Hamiltonian (Equation 75) can describe a scattering process in which the electron starts in one continuous manifold and ends in another, and the states $\{n\}$ belong to the target that causes the scattering process. These states may be the eigenstates of the target Hamiltonian, in which case $V_{n,m}$ in Equation 75 vanishes, or some zero-order representation in which the basis states are mutually coupled by the exact-target Hamiltonian. Equation 76 represents the thermal environment as a harmonic-phonon bath. The coupling between the electronic system and this bath is assumed in Equation 77 to originate from a target-state dependent shift in the equilibrium position of each phonon mode. An exact solution to this scattering problem can be obtained for the particular case where the target is represented by a single state $n = 1$ and the phonon bath contains one oscillator of frequency ω. In this case, it is convenient to consider the oscillator as part of the target that is therefore represented by a set of states $|m\rangle$ with energies $E_1 + m\hbar\omega$ (the zero-point energy can be set to 0). If the oscillator is initially in the ground state ($m = 0$), the cross-section for electron tunneling (or scattering) from left to right is given by (98–99a)

$$\mathcal{T}(E_i, E_f) \sim \Gamma^{(L)} \Gamma^{(R)} \sum_{m'=0}^{\infty} \delta(E_i - E_f - m'\hbar\omega)$$
$$\times \sum_{\tilde{m}=0}^{\infty} \frac{\langle m' | \tilde{m} \rangle \langle \tilde{m} | 0 \rangle}{E_i - E_{\tilde{m}} - \Lambda_{\tilde{m}}(E_i) + (i/2)\Gamma_{\tilde{m}}(E_i)}, \qquad 78.$$

where $|\tilde{m}\rangle$ are states of the shifted harmonic oscillator that corresponds to the temporary negative ion (electron residing on the target) and $E_{\tilde{m}} = E_1 + m\hbar\omega - \lambda^2/\hbar\omega$. $\Lambda_{\tilde{m}}$ and $\Gamma_{\tilde{m}}$ are the shifts and widths of the dressed-target states associated with their coupling to the continuous manifolds and

$$\Gamma^{(K)}(E) = 2\pi \sum_k |V_{k,1}|^2 \delta(E - E_k); \quad K = L, k = l \text{ or } K = R, k = r. \quad 79.$$

The exact solution shown in Equation 78 can be obtained because of the simplicity of the system, which was characterized by a single-intermediate electronic state and a single-phonon mode. In more realistic situations characterized by many-bridge electronic states and many-phonon modes, one needs to resort to approximations or to numerical simulations. We discuss such systems next.

To get the proper perspective on the nature of this problem, consider again the standard electron-transfer process in a DBA system without metal electrodes. As already emphasized (see Section 2.5), nuclear dynamics and conversion of electronic energy to nuclear motions, resulting from solvent reorganization about the donor and acceptor sites upon changing their charge state, are essential ingredients of this process. The reason for the prominent role of nuclear dynamics in this case is that the transferred charge is localized on the donor/acceptor orbitals, consequently affecting distortion of their nuclear environments (represented by the parabolas in Figures 1a and 2a). Standard electron-transfer theory assumes that nuclear motion is coupled to the donor and acceptor electronic states only, and the electronic coupling itself is taken independent of the nuclear configuration (the Condon approximation). This assumption is sometimes questionable, in particular when intermediate electronic states are involved, as in Figures 1 and 2. The possible role of nuclear motion on such intermediate electronic potential surfaces has been discussed by Stuchebrukhov and coworkers (101). Focusing on bridge-assisted, electron-transfer processes, these authors separate the nuclear degrees of freedom into two groups. The first includes those nuclear modes that are strongly coupled to the donor-acceptor system (solvent polarization modes and vibrational modes of the donor and acceptor species). In the absence of the other modes, this coupling leads to the standard electron-transfer rate expression shown by Marcus (see Equations 1, 3 and 9)

$$k_{et} = \frac{2\pi}{\hbar} |T_{DA}|^2 \frac{e^{-(\lambda + E_{AD})^2/4\lambda k_B \Theta}}{\sqrt{4\pi \lambda k_B \Theta}}, \quad 80.$$

where λ is the reorganization energy, E_{AD} is the free-energy difference between the initial (electron on donor) and final (electron on acceptor) equilibrium configurations, and T_{DA} is the nonadiabatic electronic-coupling matrix element that incorporates the effect of the bridge via, for example, Equations 9 and 10. The other group of degrees of freedom, "bridge modes," are coupled relatively weakly to the electron-transfer process, and it is assumed that their effect can be incorporated using low-order perturbation theory. This is accomplished by considering the modulation of the electronic coupling T_{DA} by these motions, $T_{DA} = T_{DA}(\{x_\nu\})$, where $\{x_\nu\}$ is the set of the corresponding nuclear coordinates. It is important to note that the separation of nuclear modes into those coupled to the donor and acceptor states (schematically represented by the Marcus parabolas in Figures 1a and 2a) and those associated with electronic coupling between them is done for convenience only and is certainly not a rigorous procedure. Within this picture the electron-transfer rate is obtained (101) as a convolution

$$k = \int d\varepsilon \rho_B(\varepsilon) k_0(E_{AD} + \varepsilon),$$ 81.

where

$$\rho_B(\varepsilon) = \int dt e^{i\varepsilon t} \frac{\langle T_{DA}(t) T_{DA}(0) \rangle}{\langle T_{DA}^2 \rangle},$$ 82.

and

$$T_{DA}(t) = e^{iH_B t/\hbar} T_{DA} e^{-iH_B t/\hbar},$$ 83.

where H_B is the bridge Hamiltonian including the thermal environment ($\Theta\Theta$ of Figure 1). Calculations based on this formalism indicate (101) that inelastic contributions to the total electron-transfer flux are substantial for long (>10 segments) bridges.

It should be emphasized that dynamical fluctuations in the bridge can considerably affect also the elastic transmission probability. For example, a substantial effect of the bridge nuclear motion on the electron-transfer rate has been observed in simulations of electron transfer in aqueous azurin carried out by Xie et al (102), in agreement with earlier theoretical predictions (103). There are some experimental indications that electron-transfer rates in proteins are indeed substantially affected by the protein nuclear motion (104).

Equations 81–83 correspond to the lowest-order correction, associated with intermediate-state nuclear relaxation, for bridge-mediated, electron-transfer rates. At the other extreme, we find sequential processes that are best described by two or more consecutive electronic transitions. For this to happen two conditions have to be satisfied. First, the intermediate state(s) energy should be close to that of the donor/acceptor system, so these states are physically populated either directly or by thermal activation. Second, nuclear relaxation and dephasing should be fast enough so that the bridging states can be treated as well-defined, thermally averaged electronic configurations. Obviously, intermediate situations can exist. Bridge-mediated electron transfer can be dominated by two (donor-acceptor) electronic states coupled via intermediate high-lying states that are only virtually populated, by real participation of such intermediate states in a coherent way (when thermal relaxation and dephasing are slow), or by sequential transfer through such states. This issue was extensively discussed (105, 106) for three-state models of electron transfer that were recently used to describe primary charge separation in bacterial photosynthesis. The possibility of observing similar effects in STM studies of molecules adsorbed at electrochemical interfaces was discussed by Schmickler & Tao (107).

Closely related to this phenomenology is the process of light scattering from molecular systems where the donor and acceptor states are replaced by the incoming and outgoing photons. Elastic (Rayleigh) scattering is the analog of the two-state standard electron-transfer process. Inelastic (Raman) scattering is the analog of the inelastic electron-transfer process analyzed above, except that our ability to resolve the energy of the scattered photon makes it possible to separate the total rate (or flux), the analog of Equation 81, into its elastic and different inelastic components (108). Resonance Raman scattering and resonance fluorescence are the processes that take place when excited molecular states are physically, as opposed to virtually, occupied during the light-scattering process. The former is a coherent process that takes place in the absence of dephasing and thermal relaxation while the latter follows thermal relaxation in the excited molecular state. Reemitting the photon after dephasing has occurred, but before full thermal relaxation takes place, is the process known as hot luminescence.

3.3 Thermal Interactions in Molecular Conduction

Coming back to electron transfer and transmission, the importance of dephasing effects in the operation of microscopic junctions has long been recognized (33, 83). The Landauer formula for the conduction of a narrow constriction connecting two macroscopic metals, Equations 23 or 27, is derived by assuming that the transmission is elastic and coherent, i.e. without dephasing and energy-changing interactions taking place in the constriction. If the constriction is small relative to the mean free path of the electron in it, these effects may indeed be disregarded. When the constriction becomes macroscopic, multiple scattering and dephasing are essential to obtain the limiting Ohm's law behavior. A simple demonstration is obtained (83, p. 63) by considering a conductor of length L as a series of N macroscopic scatterers, each of the type that, by itself, would yield Equation 23. At each scatterer, the electron can be transmitted with probability $\mathcal{T}$, or reflected with probability $\mathcal{R} = 1 - \mathcal{T}$. Let the total transmission through N such objects be $\mathcal{T}_N$, so that $\mathcal{T} = \mathcal{T}_1$. Provided the phase of the wavefunction is destroyed after each transmission-reflection event, so that we can add probabilities, the transmission through an N-scatterer system is obtained by considering a connection in a series of an $N - 1$ scatterer system with an additional scatterer, and summing over all multiple scattering paths,

$$\mathcal{T}_N = \mathcal{T}_{N-1}(1 + \mathcal{R}\mathcal{R}_{N-1} + (\mathcal{R}\mathcal{R}_{N-1})^2 + \ldots)\mathcal{T} = \frac{\mathcal{T}\mathcal{T}_{N-1}}{1 - \mathcal{R}\mathcal{R}_{N-1}}, \qquad 84.$$

with $\mathcal{R} = 1 - \mathcal{T}$ and $\mathcal{R}_N = 1 - \mathcal{T}_N$. This implies

$$\frac{1 - \mathcal{T}_N}{\mathcal{T}_N} = \frac{1 - \mathcal{T}_{N-1}}{\mathcal{T}_{N-1}} + \frac{1 - \mathcal{T}}{\mathcal{T}} = N\frac{1 - \mathcal{T}}{\mathcal{T}}, \qquad 85.$$

so that

$$\mathcal{T}_N = \frac{\mathcal{T}}{N(1 - \mathcal{T}) + \mathcal{T}} = \frac{L_0}{L + L_0}, \qquad 86.$$

where $L_0 = \mathcal{T}/\nu(1 - \mathcal{T})$ and $\nu = N/L$ is the scatterer density. Using this in Equation 23 yields

$$g(E) = \frac{e^2}{\pi\hbar} \frac{L_0}{L + L_0}, \qquad 87.$$

which gives the inverse length-dependence characteristic of Ohm's law as $L \to \infty$ (but see 33, p. 107).

A more detailed treatment of the role played by dephasing in quantum charge transport in microscopic junctions was given by Büttiker (109). He has introduced phase destruction processes by conceptually attaching an electron reservoir onto the constriction, under the condition that, while charge carriers are exchanged between the current-carrying system and the reservoir, no net-averaged current is flowing into this reservoir. Büttiker has observed that such a contact, essentially a voltage probe, acts as a phase-breaking scatterer. By adjusting the coupling strength between this device and the system, a controlled amount of incoherent current can be made to be carried through the system. This approach has been very useful in analyzing conduction properties of multigate junctions and connected nano-resistors.

In molecular systems, a very different approach to dephasing was considered by Bixon & Jortner (110), who pointed out that the irregular nature of Franck-Condon overlaps between intramolecular vibrational states associated with different electronic centers can lead to phase erosion in resonant electron transfer. Consequently, bridge-assisted electron transfer, which proceeds via the superexchange mechanism in off-resonance processes, will become sequential in resonance situations. For a finite temperature system with an electronic energy gap between donor and bridge that is not too large relative to $\kappa_B \Theta$, the thermally averaged rate from a canonical distribution of donor states results in a superposition of both superexchange and sequential mechanisms.

While coupling to the thermal environment is implicit in the models described above, using molecular bridges embedded in condensed environments as conductors immediately suggests the need to consider the coupling to intramolecular and environmental nuclear motions explicitly, as in the Hamiltonian shown in Equations 74–77. The models of Figures 1 and 2, where transition between the two electron reservoirs or between the donor and acceptor species is mediated by a bridge represented by the group of states {n}, are again the starting point of our discussion. Several workers have recently addressed the theoretical problem of electron migration in such models, where the electron is coupled to a zero-temperature phonon bath. Bonča & Trugman (111, 111a) have provided an exact numerical solution for such a problem. Their model is similar to that described by Equations 74–77, except that the metal leads connected to the molecular target are represented by one-dimensional, semi-infinite, tight-binding Hamiltonians:

$$H = H_{el} + H_{ph} + H_{el-ph}, \qquad 88.$$

$$H_{el} = \sum_n E_n c_n^\dagger c_n + \sum_k E_k c_k^\dagger c_k + \sum_{n,n'} V_{n,n'} c_n^\dagger c_{n'}$$
$$+ \sum_{k,k'} V_{k,k'} c_k^\dagger c_k + \left(\sum_{n,k} V_{n,k} c_n^\dagger c_k + h.c \right), \qquad 89.$$

$$H_{ph} = \sum_\nu \hbar \omega_\nu b_\nu^\dagger b_\nu, \qquad 90.$$

$$H_{el-ph} = \sum_n \sum_\nu \lambda_{n\nu} c_n^\dagger c_n (b_\nu^\dagger + b_\nu). \qquad 91.$$

Here, H_{el} desribes both the metal leads [represented by the manifold(s) of states {k}] and the molecular target (with states {n}). The coupling to the phonon field is assumed to vanish on the metal sites. The electron transport problem is treated as a one-particle, multichannel scattering problem, where each of the (one incoming, many outgoing) channels corresponds to a given vibrational state of the target. A finite basis is employed by using a finite number of phonon modes and limiting the number of phonons quanta associated with each site, and by projecting out leads that carry only outgoing states; however, the size of this basis can be increased until convergence is achieved. Yu et al (112) have studied the same one-dimensional electronic model with a different electron-phonon interaction: Instead of the Holstein-type interaction, as seen in Equations 77 and 91, they use a model similar to the Su-Schrieffer-Heeger (SSH) Hamiltonian (113), where Equations 89–91 are replaced by

$$H_{el} + H_{el-ph} = \sum_n E_n c_n^\dagger c_n$$
$$+ \sum_n \{ [V_{n,n+1} - \alpha_{n,n+1}(u_{n+1} - u_n)] c_n^\dagger c_{n+1} + h.c. \}, \qquad 92.$$

$$H_{ph} = \frac{1}{2} K \sum_{n=1}^{N-1} (u_{n+1} - u_n)^2 + \frac{1}{2} \sum_{n=1}^{N} m_n \dot{u}_n^2, \qquad 93.$$

where u_n ($n = 1, \ldots, N$) are displacements of the target atoms. The segment of the lattice between $n = 1$ and $n = N$ represents an organic oligomer, connecting between two metals, and the model for the oligomer is the same as that used in the SSH theory of conducting conjugate polymers, with the nuclear degrees of freedom treated classically. The electron-phonon coupling is again assumed to vanish outside the bridge, i.e. in Equation 92, $\alpha_{n,n+1}$ is taken to be zero unless $n = 1, 2, \ldots, N - 1$. A special feature (in the context of this review) of this calculation is that it is done using the exact many-electron ground state of the metal-oligomer-metal system, which takes into account the Peierls distortion that leads to a dimerization in the oligomer's structure (113). The time evolution of an excess-electron wavepacket going through the oligomer segment is computed using the quantum-classical, time-dependent, self-consistent field (TDSCF) approximation, whereupon the electron wavefunction is propagated under the instantaneous nuclear configuration, while the latter is evolved classically using the expectation value of the Hamiltonian with the instantaneous electronic wavefunction.[13] It was found that lattice dynamics can be quite important at an intermediate window of electron energies, where the electronic and nuclear timescales are comparable.

A fully quantum analog of this model was studied by Ness & Fisher (114). Their Hamiltonian is

$$H_{el} = \sum_n E_n c_n^\dagger c_n + \sum_\nu \hbar\omega_\nu b_\nu^\dagger b_\nu + \sum_{\nu,n,m} \gamma_{\nu,n,m}\left(b_\nu^\dagger + b_\nu\right) c_n^\dagger c_m, \quad 94.$$

where, again, the distinction between the metal leads and the molecular system enters through the values of the site energies E_n, and through the fact that coupling to phonons exists only at the oligomer sites. The ground state of the neutral N electron-dimerized chain is the reference system, and the time evolution in the corresponding $N + 1$ or $N - 1$ electron system is studied at zero temperature using the multichannel, time-independent scattering-theory approach of Bonča & Trugman (111, 111a). The result of this calculation is a considerable increase in the tunneling current when the electron-phonon interaction is switched on, in particular for long chains. The origin of this behavior seems to be the existence of a polaron state below the conduction band edge of the molecular segment that effectively lowers the barrier energy experienced by the tunneling electron. Close to resonance, however, the effect of electron-phonon coupling may be reversed, leading to a smaller total overall conduction (115).

The Bonča & Trugman approach (111, 111a) has also been used recently by Emberly & Kirczenow (85), also for a one-dimensional, tight-binding model described by the SSH Hamiltonian. These authors attempt to take into account the Pauli exclusion principle in calculating the inelastic contributions to electron transmission and reflection. While the formalism can, in principle, be applied to finite temperature processes, the implementation is done for a low-temperature system. The result again indicates that inelastic processes can substantially modify electron transport for long molecular chains and large potential drops.

3.4 Reduced Density Matrix Approaches

The research described above uses models for quantum transport that yield practically exact numerical solutions at the cost of model simplicity: one-dimensional, tight-binding transport models; only a few harmonic oscillators; and essentially zero temperature systems. An alternative approach uses the machinery of nonequilibrium statistical mechanics, starting from a Hamiltonian such as in Equation 88 and projecting out the thermal bath part. The resulting reduced equations of motion for the electronic subsystem contain dephasing and energy-relaxation rates that are related explicitly to properties of the thermal-bath and the system-bath coupling.

---

[13]An open issue in this calculation is the validity of the TDSCF approximation. This approximation is known to be problematic in tunneling and scattering calculations where the quantum wavefunction splits to several distinct components.

Such approaches to bridge-mediated electron transport were made by several workers (116–119). For simplicity we limit ourselves to the tight-binding superexchange model for bridge-mediated electron transfer (see Section 2.1). Also, for simplicity of notation we consider N bridge states between the two electrodes, without assigning special status to donor and acceptor states, as in Figure 2b. (It should be obvious that this makes only a notational difference.) The Hamiltonian for the athermal system is

$$H = H_0 + V, \qquad 95.$$

$$H_0 = \sum_{n=1}^{N} E_n |n\rangle\langle n| + \sum_{l} E_l |l\rangle\langle l| + \sum_{r} E_r |r\rangle\langle r|, \qquad 96.$$

$$V = \sum_{l}(V_{l,1}|l\rangle\langle 1| + V_{1,l}|1\rangle\langle l|) + \sum_{n=1}^{N-1}(V_{n,n+1}|n\rangle\langle n+1|$$
$$+ V_{n+1,n}|n+1\rangle\langle n|) + \sum_{r}(V_{r,N}|r\rangle\langle N| + V_{N,r}|N\rangle\langle r|), \qquad 97.$$

where $\{l\}$ and $\{r\}$ are again continuous manifolds corresponding to the left and right metal leads and $\{n\}$ is a set of bridge states connecting these leads in the way specified by the corresponding elements of the coupling V. In the absence of thermal interactions, and when the left and right electrodes are coupled only to levels 1 and N of the bridge, respectively, transport in this system is descibed by the conduction function (see Equations 29 and 38)

$$g(E) = \frac{e^2}{\pi\hbar} |G_{1N}(E)|^2 \Gamma_1^{(L)}(E) \Gamma_N^{(R)}(E), \qquad 98.$$

with

$$\Gamma_1^{(L)}(E) = 2\pi \sum_{l} |V_{l1}|^2 \delta(E_1 - E); \quad \Gamma_N^{(R)}(E) = 2\pi \sum_{r} |V_{Nr}|^2 \delta(E_N - E). \quad 99.$$

In general, $G(E)$ is evaluated numerically by inverting the corresponding Hamiltonian matrix. For $E_n = E_B$ and $V_{n,n+1} = V_B$, identical for all bridge levels and for all mutual couplings, respectively, and in the superexchange limit, $|V_B| \ll |E_B - E|$, the Green's function element is $V_B^{N-1}/\Delta E_B^N$ (see Equation 10), with $\Delta E_B = E - E_B$. In this case, g depends exponentially on the bridge length N according to $g \sim \exp[-\beta' N]$ with $\beta' = 2\ln(|\Delta E_B/V_B|)$ (see Equation 13).

3.4.1 Weak Thermal Coupling

To see how these dynamics are modified by thermal relaxation and dephasing effects, we follow the formulation of Segal et al (118) The Hamiltonian H is supplemented by terms describing a thermal-bath and a system-bath interaction

$$\mathcal{H} = H + H_\Theta + F, \qquad 100.$$

where H_Θ is the Hamiltonian for the thermal environment or bath, and where the system-bath interaction F is assumed weak. In this case, thermal coupling between different bridge levels is neglected relative to the internal coupling V between them, so

$$F = \sum_{n=1}^{N} F_n |n\rangle\langle n|, \qquad 101.$$

where F_n are operators in the bath degrees of freedom that satisfy $\langle F_n \rangle \equiv Tr_\Theta$ $(e^{-\beta H_\Theta} F_n) = 0$ (Tr_Θ is a trace over all thermal bath states). F is characterized by its time-correlation function. As a simple model we postulate

$$\langle F_n(t) F_{n'}(0) \rangle = f(t) \delta_{n,n'}. \qquad 102.$$

The Fourier transform of the remaining correlation functions satisfies the detailed balance condition,

$$\int_{-\infty}^{\infty} dt e^{i\omega t} \langle F_n(t) F_n(0) \rangle = e^{\beta \hbar \omega} \int_{-\infty}^{\infty} dt e^{i\omega t} \langle F_n(0) F_n(t) \rangle; \, \beta = (k_B \Theta)^{-1}, \qquad 103.$$

where Θ is the temperature and β is the Boltzmann constant. For specificity we sometimes use

$$f(t) = \frac{\kappa}{2\tau_c} \exp(-|t|/\tau_c), \qquad 104.$$

which becomes $\kappa \delta(t)$ in the Markovian, $\tau_c \to 0$, limit. Note that Equation 101 is a particular model for the thermal interactions, sufficient to show their general consequences, but by no means adequate for quantitative predictions. In particular, the assumption in Equation 102 will be replaced by a more realistic model below.

Galperin et al (36) have shown that the conduction properties of a system like that described by the Hamiltonian Equations 95–100 can be obtained by studying a steady state in which the amplitude of one state $|0\rangle$ in the initial $\{l\}$ manifold remains constant and the amplitudes of other states evolve under this restriction. Segal et al (118) have generalized this approach to thermal systems of the kind described by the Hamiltonian Equation 100 using, in the weak thermal-coupling limit, the Redfield approximation (116, 120). This approximation combines two steps that rest on the weak-coupling limit: an expansion up to second order in the coupling F and the assumption that the thermal bath is not affected by its coupling to the molecular system. In this approach, one starts from the set of states $|0\rangle, |1\rangle, \ldots, |n\rangle, \{|l\rangle\}, \{|r\rangle\}$, where $|0\rangle$ is the incoming state in the $\{l\}$ manifold, and projects out the continuous manifolds $\{l\}$ (except $|0\rangle$) and $\{r\}$. This amounts to replacing H of Equations 95–100 by an effective Hamiltonian, H^{eff}, in the space spanned by states $|0\rangle, |1\rangle, \ldots, |n\rangle$, in which the energies E_1 and E_N are modified by adding self-energy terms whose imaginary parts are, respectively, $\Gamma_1^{(L)}/2$ and $\Gamma_N^{(R)}/2$. This effective Hamiltonian of order $N+1$ is then diagonalized and the resulting set of $N+1$ states (originating from N bridge states and one incoming state) is used to represent the Liouville equation for the density operator ρ of the overall electrode-bridge-bath system, $\dot{\rho} = -i[\mathcal{H}, \rho]$. This Liouville equation is expanded to second order in F and traced over bath degrees of freedom using the approximation $\rho(t) = \rho_\Theta \sigma(t)$, with $\rho_\Theta = e^{-\beta H_\Theta}$ and $\sigma(t) = Tr_\Theta \rho(t)$. This leads to an equation of motion for the reduced density matrix $\sigma(t)$ for the electrode-bridge system that takes the form

$$\dot{\sigma}_{jk} = -iE_{jk}\sigma_{jk} - \Gamma_{jk}\sigma_{jk}$$

$$-\int_0^t dt' \sum_{lm} \{\langle \tilde{F}_{jl}(t-t')\tilde{F}_{lm}(0)\rangle e^{-iE_{lk}(t-t')}\sigma_{mk}(t')$$

$$-\langle \tilde{F}_{mk}(0)\tilde{F}_{jl}(t-t')\rangle e^{-iE_{kl}(t-t')}\sigma_{lm}(t')$$

$$-\langle \tilde{F}_{mk}(t-t')\tilde{F}_{jl}(0)\rangle e^{-iE_{jm}(t-t')}\sigma_{lm}(t')$$

$$+\langle \tilde{F}_{ml}(0)\tilde{F}_{lk}(t-t')\rangle e^{-iE_{jl}(t-t')}\sigma_{jm}(t')\}, \qquad 105.$$

where $E_{jl} = E_j - E_l$ and $\tilde{F}(t) = e^{iH_\Theta t} F e^{-iH_\Theta t}$. Here the indices j,k,l,m refer to molecular states that diagonalize the effective Hamiltonian H_{eff}. The damping terms Γ originate from the decay of states $|1\rangle$ and $|N\rangle$ distributed into these eigenstates. At steady state all σ elements are constant and Equation 105 becomes

$$0 = -iE_{jk}\sigma_{jk} - \Gamma_{jk}\sigma_{jk}$$

$$+\sum_{lm} \left\{ \sigma_{lm} \int_0^\infty d\tau \left(\langle \tilde{F}_{mk}(0)\tilde{F}_{jl}(\tau)\rangle e^{-iE_{lk}\tau} + \langle \tilde{F}_{mk}(\tau)\tilde{F}_{jl}(0)\rangle e^{-iE_{jm}\tau} \right) \right.$$

$$-\sigma_{mk} \int_0^\infty d\tau \langle \tilde{F}_{jl}(\tau)\tilde{F}_{lm}(0)\rangle e^{-iE_{lk}\tau}$$

$$\left. -\sigma_{jm} \int_0^\infty d\tau \langle \tilde{F}_{ml}(0)\tilde{F}_{lk}(\tau)\rangle e^{-iE_{jl}\tau} \right\}. \qquad 106.$$

Transforming Equation 106 back to the local bridge representation $\{0, n = 1, \ldots, N\}$ leads to a set $(N+1)(N+1)$ equations of the form

$$-iE_{nn'}\sigma_{nn'} - i[V,\sigma]_{nn'} + \sum_{n_1}\sum_{n_2} R_{nn'n_1n_2}\sigma_{n_1n_2} = \frac{1}{2}(\Gamma_n + \Gamma_{n'})\sigma_{nn'};$$

$$n, n' = 0, \ldots, N, \quad 107.$$

where the elements of R are linear combinations of the integrals appearing in Equation 106 and where $\Gamma_n = \Gamma_N^{(R)}\delta_{n,N} + \Gamma_1^{(L)}\delta_{n,1}$. Again, at steady state the first $(n = n' = 0)$ equation is replaced by the boundary condition σ_{00} = constant. The remaining $(N+1)(N+1) - 1$ equations constitute a set of linear nonhomogeneous algebraic equations in which the terms containing σ_{00} constitute source terms.

Thus, all elements $\sigma_{nn'}$, and in particular σ_{NN}, can be obtained in the form $\sigma_{nn'} = U_{nn'}\sigma_{00}$, in terms of the fixed population σ_{00} in the incoming state $|0>$ of the $\{l\}$ manifold, where the coefficients $U_{nn'}$ are related to the inverse of the $(N+1)(N+1)-1$ order matrix of thermal rates. The steady state flux into the $\{r\}$ manifold is $\Gamma_N^{(R)}\sigma_{NN}$, and the corresponding rate is

$$k_{0\to R} = \Gamma_N^{(R)}\sigma_{NN}/\sigma_{00} = \Gamma_N^{(R)}U_{NN}. \qquad 108.$$

While the general expression for U_{NN} is very cumbersome, involving the inverse of an $(N+1)(N+1)-1$ order matrix, numerical evaluation of the resulting rate and its dependence on coupling parameters, bridge length, and temperature is an easy numerical task for reasonable bridge lengths. A final technical point stems from the observation that the resulting $k_{0\to R}$ must be proportional to $|V_{10}|^2$, the squared coupling between the first bridge level and the left continuous manifold. We therefore rewrite Equation 108 in terms of new variables $k'_{0\to R}$ and U'_{NN}, defined by

$$k_{0\to R} = k'_{0\to R}|V_{10}|^2 = \Gamma_N^{(R)}U'_{NN}|V_{10}|^2. \qquad 109.$$

We can make contact with results obtained in the athermal case by writing $|0\rangle = |k_\parallel, k_x\rangle$, where x is the direction of transmission, $k_\parallel$ is the momentum in the yz plane, and $(\hbar^2/2m)(k_\parallel^2 + k_x^2) = E_\parallel + E_x = E_0$. The transmission coefficient $\mathcal{T}(E_0, k_\parallel)$ for electron incident from the left electrode with total energy E_0 in channel $k_\parallel$ is related to $k_{0\to R}$ by

$$k_{0\to R} = \frac{k_x}{mL}\mathcal{T}(E_0, k_\parallel) = (2\pi\rho(E_x))^{-1}\mathcal{T}(E_0, k_\parallel), \qquad 110.$$

where $\rho(E_x)$ is the one-dimensional density of states for the motion in the x direction. Therefore,

$$\mathcal{T}(E_0, k_\parallel) = 2\pi\rho(E_x)k_{0\to R} = \Gamma_{1,k_\parallel}^{(L)}k'_{0\to R}, \qquad 111.$$

and the all-to-all transmission at energy E_0 is the sum over all channels with energy $E_\parallel < E_0$

$$\mathcal{T}(E_0) = \Gamma_1^{(L)}k'_{0\to R} = \Gamma_1^{(L)}\Gamma_N^{(R)}U'_{NN}. \qquad 112.$$

Comparing Equation 112 with Equation 98, we see that Equation 112 is the analog of Equation 38, where, in the thermal case, U'_{NN} has replaced $|G_{1N}|^2$.

In the athermal case, the conduction of a junction characterized by a given transmission coefficient is obtained from the Landauer formula (27). Here the issue is more complex since, while $\mathcal{T}(E_0)$ is the probability that an incident electron with energy E_0 will be transmitted through the molecular barrier, it is obvious that the transmitted electron can carry energy different from E_0. As an example, consider the case where the bridge has only one intermediate state, i.e. $N = 1$. Within the same model and approximations as outlined above, it is possible (121) to obtain the energy-resolved transmission. In the Markovian limit ($\tau_c \to 0$ in Equation 104) the result is

$$T'(E_0, E) = T_0(E_0)\left[\delta(E_0 - E) + \frac{(\kappa/2\pi)e^{-\beta(E_1-E_0)}}{(E_1 - E)^2 + (\Gamma_1/2)^2}\right], \qquad 113.$$

[we use T' to denote the differential (per unit energy range) transmission coefficient], where $\Gamma_1 = \Gamma_1^{(L)} + \Gamma_1^{(R)}$ and T_0 is elastic transmission coefficient

$$T_0(E_0) = \frac{\Gamma_1^{(L)} \Gamma_1^{(R)}}{(E_1 - E_0)^2 + (\Gamma_1/2)^2}.$$

The total transmission coefficient, including inelastic contribution, is given by

$$T(E_0) = \int dE\, T'(E_0, E) = T_0(E_0)\left[1 + \frac{\kappa}{\Gamma_1}e^{-\beta(E_1-E_0)}\right]. \qquad 114.$$

In the absence of thermal interactions ($\kappa = 0$ in Equation 104), T is reduced to T_0, and the electron is transmitted with $E = E_0$. For a finite κ we get an additional, thermally activated, component peaked around the energy E_1 of the bridge level.

How will this affect the conduction? It has been argued (see 83) that simple expressions based on the Pauli principle (e.g. Equations 19 and 33) are not valid in the presence of inelastic processes, including thermal relaxation. It may still be used however in the weak metal-bridge coupling limit (see discussion in Section 2.12). Proceeding along this line, an equation equivalent to Equation 33 can be written

$$I = \frac{e}{\pi\hbar} \int_0^\infty dE_0 \int_0^\infty dE\, T'(E_0, E)$$

$$\times [f(E_0)(1 - f(E + e\Phi)) - f(E_0 + e\Phi)(1 - f(E))]. \qquad 115.$$

For small bias and low enough temperature [so that $f(E + e\Phi) \sim f(E) - e\Phi\delta(E - E_F)$], this leads to (121)

$$g(E_0) = \frac{I}{\Phi} = \frac{e^2}{\pi\hbar}T_0(E_0)\left(1 + (1 - f(E_1))\frac{\kappa}{\Gamma_1}e^{-\beta(E_1-E_0)}\right). \qquad 116.$$

The equivalent result for electron-transfer rates is familiar: At zero-temperature, the rate is determined by a tunneling probability, and at higher temperature an activated component takes over. For an experimental manifestation of this behavior (e.g. see 122, Figure 5).

It is also interesting to examine the bridge-length dependence of the transfer rate and the associated conduction. Here analytical results are cumbersome, but numerical evaluation of the rate (Equation 108), and the transmission coefficients (Equations 111 and 112), in terms of the system parameters (Hamiltonian

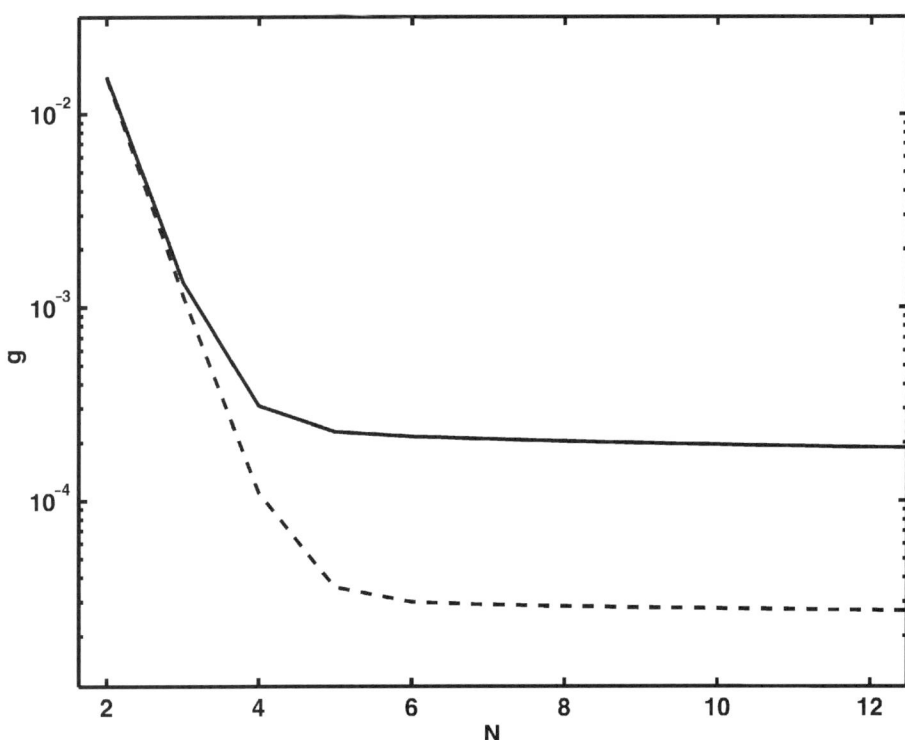

Figure 6 Finite temperature conduction of a simple tight-binding model of a molecular junction as a function of bridge length N. (See text for details.)

couplings and the parameters κ and τ_c of Equation 104) is straightforward (118). Figure 6 shows the conduction (in units of $e^2/\pi\hbar$) obtained from such a model calculation using $V_B = 0.05\,\text{eV}$, $\Delta E_B = E_B - E_F = 0.2\,\text{eV}$, $\Gamma_l^{(L)} = \Gamma_N^{(R)} = 0.1\,\text{eV}$, $\tau_c = 0, \kappa = 0.01$ eV, plotted against the number of bridge segments N for two different temperatures, $T = 300$ K and 500 K. An exponential dependence on N, characteristic of the superexchange model, is seen to give way to a weak bridge-length dependence at some crossover value of N. Further analysis of these results (118, 119) reveals that the dependence on bridge length beyond the crossover may be written in the form $(k_{up}^{-1} + k_{diff}^{-1} N)^{-1}$, where k_{up} is the rate associated with the thermal-activated rate from the Fermi-level into the bridge, whereas k_{diff} corresponds to hopping (diffusion) between bridge sites. As N increases, the conduction behaves as N^{-1}, indicating Ohmic behavior. This inverse length dependence should be contrasted with nondirectional diffusion, where the rate to reach a distance N from the starting position behaves like N^{-2}. Furthermore, if other loss channels exist, carriers may be redirected or absorbed with a rate

TABLE 2 Bridge-length dependence of the transmission rate (118)

Physical Process	Bridge-length (N) dependence	
Super exchange (small N, large $\Delta E_B/V_B$, large $\Delta E_B/\kappa_B\Theta$)	$e^{-\beta' N}$	$\beta' = 2\ln(V_B/\Delta E_B)$
Steady state hopping (large N, small $\Delta E_B/V_B$, small $\Delta E_B/k_B\Theta$)	N^{-1}	
Nondirectional hopping (large N, small $\Delta E_B/V_B$, small $\Delta E_B/k_B\Theta$)	N^{-2}	
Intermediate range (intermediate N, small $\Delta E_B/V_B$)	$(k_{up}^{-1} + k_{diff}^{-1} N)^{-1}$	$k_{up} \sim (V_B^2 \kappa/\Delta E^2)e^{-\Delta E_B/k_B\Theta}$ $k_{diff} \sim (4V_B^2/\kappa)e^{-\Delta E_B/k_B\Theta}$
Steady state hopping + competing loss at every bridge site	$e^{-\alpha N}$	$\alpha = \sqrt{\Gamma_B(\Gamma_B + \kappa)}/2V_B$

Γ_B once they populate the bridge, the bridge-length dependence again becomes exponential and may be written $g \sim (k_{up}^{-1} + k_{diff}^{-1}N)^{-1}e^{-\alpha N}$, where α is related to this loss rate (13, 123). Table 1 (118) summarizes these results for the Markovian limit of the thermal-relaxation process.

Experimental observation of the behaviors indicated in Table 2 is not easy since it is usually not possible to change the length of a molecular bridge without affecting its other properties, e.g. the positions of molecular HOMOs and LUMOs relative to donor and acceptor energies or an electrode Fermi energy (124). A nice example (125) of a crossover behavior observed in a LEET experiment (see Section 6) as a function of thickness of an absorbed molecular layer is seen in Figure 7. Here electrons are injected into N-hexane films adsorbed on a polycrystaline Pt foil at energies below the bottom of the conduction band ($\sim$0.8 eV). The role of bridge states is here assumed by impurity states in the hydrocarbon band gap. Since the energy and localization position of these states is not known, the observed results cannot be quantitatively analyzed with the model described above. However, a crossover from tunneling to hopping behavior is clearly seen.

3.4.2 Strong Thermal Coupling The weak system-thermal, bath-coupling model discussed above rests on two approximations: (*a*) The system-bath interaction can be considered in low order, and (*b*) the bath degrees of freedom are essentially unaffected by the electronic process. Using these assumptions has enabled us to obtain the general characteristics of electron transmission through

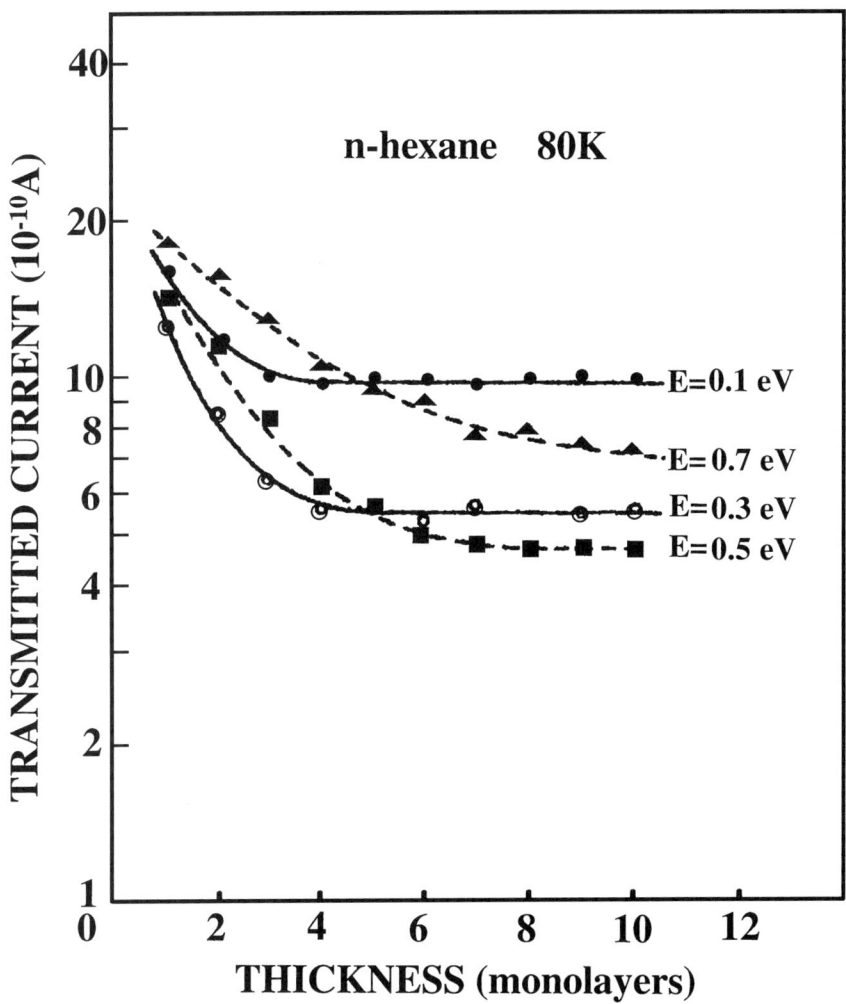

Figure 7 Transmitted current in n-hexane films as a function of thickness for various incident energies, showing the transition from tunneling to activation-induced transport. (Reproduced from Reference 125 and used by permission.)

molecular barriers in the presence of barrier-localized thermal interactions. When the interaction between the electronic system and the underlying bath is stronger, these assumptions break down, and distortions in the bath configuration induced by the electronic process can play an important role. One example is the analysis of Ness & Fisher (114) discussed above, where coupling to phonons increases the overall transmission because of the existence of a polaron state below the conduction band edge of the electronic system. However, because the overall transmission efficiency depends both on energetics (the polaron state lowers the effective barrier height) and coupling strength (small nuclear overlaps between distorted and undistorted nuclear configurations decreases the effective coupling), the issue is more involved, and depending on details of coupling and frequencies, both enhancement or reduction of transmission probabilities can occur. Similarly, at finite temperatures, the relative importance of the two transmission routes, tunneling and activated hopping, is sensitive to these details. Relatively simple results are obtained in the particular limit where the thermal coupling is strong while the bare electronic-coupling V_B is weak. In this case, it may still be assumed that the bath degrees of freedom remain in thermal equilibrium throughout the process. Taking the bath to be a system of harmonic oscillators, $H_B = \sum_\alpha [(p_\alpha^2/2m_\alpha) + (m_\alpha \omega_\alpha^2/2)x_\alpha^2]$, and taking F_n in Equation 101 to be linear in the coordinates x_α,

$$F_n = (1/2) \sum_\alpha C_{n\alpha} x_\alpha \qquad 117.$$

(so that the Hamiltonian used in Equation 100 is similar to the polaron-type Hamiltonian used in Equations 74–77 and 88–91), a small polaron transformation is applied in the form

$$\mathcal{H}' = U\mathcal{H}U^{-1}$$

$$U = U_1 U_2 \ldots U_N$$

$$U_n = \exp(-|n\rangle\langle n|\Omega_n)$$

$$\Omega_n = \sum_\alpha \Omega_{n\alpha}; \quad \Omega_{n\alpha} = \frac{C_{n\alpha} p_\alpha}{2m_\alpha \omega_\alpha^2}, \qquad 118.$$

leading to the transformed Hamiltonian

$$\mathcal{H}' = H + H_B + F' + E_{shift}$$

$$F' = V_B \sum_{n=1}^{N-1} \left(|n\rangle\langle n+1| e^{i(\Omega_{n+1}-\Omega_n)} + |n+1\rangle\langle n| e^{-i(\Omega_{n+1}-\Omega_n)} \right)$$

$$E_{shift} = -\frac{1}{8} \sum_n \sum_\alpha \frac{C_{n\alpha}^2}{m_\alpha \omega_\alpha^2} |n\rangle\langle n|, \qquad 119.$$

where H is given by Equations 95–97. If V_B is small, the procedure based on the Redfield approximation, which leads to Equation 107, can be repeated. Note that keeping terms only up to second order in F' still includes terms of arbitrary order in the system-bath coupling. This procedure leads to (D Segal, A Nitzan, unpublished result)

$$\dot{\sigma}_{jk} = -i\omega_{jk}\sigma_{jk} - iV_B \sum_m \left(\langle F'_{jm}\rangle\sigma_{mk} - \langle F'_{mk}\rangle\sigma_{jm}\right)$$

$$+ V_B^2 \sum_{l,m} \left\{ \sigma_{lm} \int_0^\infty d\tau \left(\langle \tilde{F}_{mk}(0)\tilde{F}_{jl}(\tau)\rangle e^{-iE_{lk}\tau} + \langle \tilde{F}_{mk}(\tau)\tilde{F}_{jl}(0)\rangle e^{-iE_{jm}\tau}\right)\right.$$

$$\left. - \sigma_{mk}\int_0^\infty d\tau \langle \tilde{F}_{jl}(\tau)\tilde{F}_{lm}(0)\rangle e^{-iE_{lk}\tau} - \sigma_{jm}\int_0^\infty d\tau \langle \tilde{F}_{ml}(0)\tilde{F}_{lk}(\tau)\rangle e^{-iE_{jl}\tau},\right.$$

120.

where $\tilde{F} = F' - \langle F'\rangle$. The terms in the first line of Equation 120 account for coherent motion with a modified coupling operator, while the terms proportional to V_B^2 describe incoherent hopping between bridge sites. An important new element in this formulation is the temperature-dependent renormalization of the coupling responsible for the coherent transmission. Using Equation 119 results in

$$\langle F'\rangle = \exp(-S_T)$$
$$S_T = (1/2)\sum_\alpha d_{n\alpha}^2(2\bar{n}_\alpha + 1)$$
$$\bar{n}_\alpha = (\exp(\omega_\alpha/k_B T) - 1)^{-1}$$
$$d_{n\alpha}^2 = \frac{(C_{n\alpha} - C_{n+1,\alpha})^2}{8m_\alpha\omega_\alpha^3},$$

121.

so that coherent transfer becomes less important at higher temperatures. This reduction in the coherent hopping rate is associated with the small overlap between bath degrees of freedom accommodating the electron at different sites. In fact, $\langle F'\rangle$ is recognized as the thermally averaged Franck-Condon factor associated with the electron transfer between two neighboring bridge sites. In terms of the spectral density

$$J(\omega) = \frac{\pi}{2}\sum_\alpha \frac{(C_{n\alpha} - C_{n+1,\alpha})^2}{m_\alpha\omega_\alpha}\delta(\omega - \omega_\alpha)$$

122.

(independent of n if the bridge sites are equivalent), we have

$$S_T = \frac{1}{8\pi}\int_0^\infty \frac{J(\omega)\coth(\omega/2k_B T)}{\omega^2}d\omega \xrightarrow[\text{finite }T]{\omega\to 0} \frac{k_B T}{4\pi}\int_0^\infty \frac{J(\omega)}{\omega^3}d\omega.$$

123.

Depending on the spectral density, this integral may diverge. More specifically, if $J(\omega) \sim \omega^s$ with $s < 2$, S_T diverge at any finite temperature and the coherent route is blocked. In other cases, the coherent route quickly becomes insignificant with increasing temperature.

We have extended this discussion of thermal relaxation and dephasing effects in bridge-assisted electron transport both because these effects are inherently important in determining transport and conduction properties of molecular junctions, and because the issue of heat generation in these current-carrying nanostructures is intimately related to these relaxation phenomena. As we have seen, this problem is far from being solved and more research along these lines can be expected.

4. ELECTRON TUNNELING THROUGH WATER

Electron tunneling through water is obviously an important element in all electron-transfer processes involving hydrated solutes and in many processes that occur in water-based electrochemistry. Still, only a few systematic experimental studies of the effect of the water structure on electron-transfer processes have been done (22, 24, 126–131). Porter & Zinn (24) have found, for a tunnel junction made of a water film confined between two mercury droplets, that at low (<1 nm) film thickness, conduction reflects the discrete nature of the water structure. Nagy (130) has studied STM current through adsorbed water layers and has found that the distance dependence of the tunneling current depends on the nature of the substrate and possibly indicates the existence of resonance states of the excess electron in the water layer. Vaught et al (129) have seen a nonexponential dependence on tip-substrate distances of tunneling in water, again indicating that at small distances water structure and possibly resonance states become important in affecting the junction conductance. Several workers have found that the barrier to tunneling through water is significantly lower than in a vacuum for the same junction geometry (22, 126–128, 130, 131). The observed barrier is considerably lower than the threshold observed in photoemission into water (132) and, in contrast to tunneling in a vacuum, cannot be simply explained by image effects (24).

The present section focuses on attempts (133–138) to correlate these observations with numerical and theoretical studies. In the spirit of most calculations of electron-transfer rates (as in Section 2) and of earlier dielectric continuum modes that neglect water structure altogether, we assume at the outset that in films consisting of a few monolayers, transmission is dominated by elastic processes. The discussion in Section 3 emphasizes the need to justify this assumption. Since we are dealing with negative-energy (tunneling) processes, electronic excitations of water molecules by the transmitting electron can be ruled out. In addition, photoemission through thin water films adsorbed on metals indicates that inelastic processes associated with the water's nuclear motion contributes relatively weakly at such energies (139). Numerical simulations of subexcitation electron transmission through 1–4 water monolayers adsorbed on Pt (1,1,1) (140) are in agreement with this observation.[14] Theoretical calculations of inelastic tunneling (144)

---

[14]It should be kept in mind that energy transfer from the transmitting electron to water, nuclear degrees of freedom, the mechanism responsible for capturing and localizing the electron as a solvated species, must play an important role for thicker layers (141–143).

similarly show that sufficiently far from resonance, the overall transmission is only weakly affected by inelastic processes. In both cases, this can be rationalized by the short interaction times (see 140; see also Section 3.1). In such cases, a static medium assumption appears to provide a reasonable starting point for discussing the overall transmission, i.e. we assume that the transmission event is completed before substantial nuclear motion takes place. The computation of the transmission probability can therefore be done for individual static water configurations sampled from an equilibrium ensemble, and the results averaged over this ensemble. This assumption is critically examined below. It should be emphasized that while solvent nuclear motion is slow relative to the transmission timescale, solvent electronic response (electronic polarizability) is not. We return to this issue below.

In Section 2 we have summarized the theoretical and computational approaches available for studying electron transfer and electron transmission. The following account (see also 137) summarizes recent computational work on electron transmission through water that uses the pseudopotential method (133–136, 138). Here the detailed information about the electronic structure of the molecular spacer is disregarded and replaced by the assumption that the underlying electron scattering or tunneling can be described by a one-electron potential surface. This potential is taken to be a superposition of the vacuum potential experienced by the electron and the interaction potential between an excess electron and the molecular spacer. The latter is written as a sum of terms representing the interaction between the electron and the different atomic (and sometimes other suitably chosen) centers. The applicability of this method depends on our ability to construct reliable pseudopotentials of this type. In the work described below, we use the electron-water pseudopotential derived and tested in studies of electron hydration (145); [for an alternative pseudopotential, see Rossky & Schnitker (145a)], and a modified pseudopotential that includes the many-body interaction associated with the water electronic polarizability.

With such a potential given, the problem is reduced to evaluating the transmission probability of an electron when it is incident on the molecular layer from one side only. In recent years, various time-dependent and time-independent numerical-grid techniques were developed for such calculations. In the time-dependent mode, an electron wavepacket is sent toward the molecular barrier and propagated on the grid using a numerical solver for the time-dependent Schrödinger equation. This propagation continues until such time t_f at which the collision with the barrier has ended, i.e. until the probability that the electron is in the barrier region, $\int_{barrier} |\psi(\mathbf{r}, t)|^2 d\mathbf{r}$, has fallen below a predetermined margin. Since only the result at the end of the time evolution is needed, a propagation method based on the Chebychev polynomial expansion of the time-evolution operator (146) is particularly useful.

In the time-independent mode, Nitzan and coworkers (137, 147) have applied the spatial-grid–based, absorption-boundary condition, Green's-function (ABCGF) technique described in Section 2.7 (Equations 60 and 63). Taking x to be the tunneling direction, periodic boundary conditions are used in the y–z plane parallel to the molecular layer, and the absorption function, $\varepsilon(r) = \varepsilon(x)$, is taken to be different from zero near the grid boundaries in the z direction, far enough from the interaction region (i.e. the tunneling barrier), and gradually diminishing to zero as the interaction region is approached from the outside. The stability of the computed transmission to moderate variations of this function provides one confidence test for this numerical procedure. The cumulative microcanonical transition probability and the one-to-all transition rates are calculated as outlined in Section 2.7. In addition, exact outgoing and incoming wavefunctions Ψ_i^+ and Ψ_f^-, which correspond to initial and final states (eigenfunctions of H_0 with energy E) ϕ_i and ϕ_f, respectively, can be computed from

$$\psi_i^+ = \frac{1}{E - H + i\varepsilon} i\varepsilon \phi_i$$

$$\psi_f^- = \frac{1}{E - H - i\varepsilon}(-i\varepsilon)\phi_f,$$

124.

and provide a route for evaluating state-selected transition probabilities, $S_{if} = \langle \psi_f^- | \psi_i^+ \rangle$. The evaluation of these expressions requires (a) evaluating the Hamiltonian matrix on the grid, and (b) evaluating the operation of the corresponding Green's operator on a known vector. As in most implementations of grid Hamiltonians, the resulting matrix is extremely sparse, which suggests the applicability of Krylov space-based iterative methods (148).

Although considerable sensitivity to the water structure is found in these studies, water layers prepared with different reasonable water-water interaction models have similar transmission properties (134, 135). On the other hand, the results are extremely sensitive to the choice of the electron-water pseudopotential. Most previous studies of electron solvation in water represent the electron-water pseudopotential as a sum of two-body interactions. Studies of electron hydration and hydrated-electron spectroscopy show that the potentials developed for this purpose (145) could account semiquantitatively for the general features of electron-solvation structure and energetics in water and water clusters. Taking into account the many-body aspects of the electronic polarizability contributions to the electron-water pseudopotential (149) has led to improved energy values that were typically different by 10%–20% from the original results. In contrast, including these many-body interactions in the tunneling calculation is found to make a profound effect (see below), an increase of ∼2 orders of magnitudes in the transmission probability of electron through water in the deep tunneling regime. There are two reasons for this. First, as already noted, tunneling processes are fast relative to characteristic nuclear relaxation times. The latter is disregarded, leaving the electronic polarizability as the only solvent response in the present treatment. Second, variations of the interaction potentials enter exponentially into the tunneling probability, making their effects far larger than the corresponding effect on solvation. It should be kept in mind that including the solvent electronic polarizability in simulations of quantum mechanical processes in solution raises some conceptual difficulties. The simulation results described below are based on the approach to this problem described elsewhere (133, 135). In what follows, model B refers to the the corrected electron-water pseudopotential used in these papers whereas model A refers to the original pseudopotential of Barnett et al (145) [see the original publications 133–137 for details of the water-water and water-metal potentials used in these calculations].

The results described below illustrate the principal factors affecting the transmission process: (a) the dimensionality of the process, (b) the effect of layer structure and order, (c) the effect of resonances in the barrier, and (d) the signature of band motion. The simulations consist of first preparing water layer structures on (or between) the desired substrates using classical MD simulations, then setting the Schrödinger equation for the electron-transmission problem on a suitable grid, and finally, computing the transmission probabilities.

Figure 8 shows the results of such calculations for the transmission probability as a function of the incident electron energy. The results for the polarizable model (B) are seen to be in remarkable agreement with the expectation based on the lowering of the effective rectangular barrier by 1.2 eV, whereas those obtained using model A, which does not take into account the many-body nature of the interaction associated with the water electronic polarizability, strongly underestimates the transmission probability. In fact, model A predicts the transmission probability in water to be lower than in a vacuum, in qualitative contrast to our observations.

Next, consider the effect of orientational ordering of water dipoles on the metal walls. Water adsorbs with its oxygen on the metal surface and the hydrogen atoms pointing away from it, leading to net surface dipole density directed away from the wall. Simulations yield $\sim 5 \cdot 10^{-11}$ Coulomb/m for this density. (I. Benjamin, A. Nitzan, unpublished data). This is an important factor in the reduction of the surface-work function of many metals due to water adsorption (132, 150). The sparse-dotted line in Figure 8 shows the transmission probability for a model obtained from model A by eliminating the attractive oxygen/metal-wall interaction, thereby destroying the preferred orientational ordering at the water-metal interface. We see that the existence of a surface dipole in the direction that reduces the work function is associated with a larger transmission probability, as expected.

Traditional approaches to electron transfer are based on a continuum dielectric picture of the solvent, where the issue of the tunneling path rarely arises. Barring other considerations, the exponential dependence of tunneling probabilities on the path length suggests that the tunneling process will be dominated by the shortest possible, i.e. one-dimensional, route. A closer look reveals that electron tunneling through water is inherently three-dimensional (e.g. see 133, Figure 7). An interesting demonstration of the importance of the three-dimensional structure of the water layer in determining the outcome of the tunneling process was given in Benjamin et al (136), where the transmission probability was computed, using the configuration of Figure 8 and model B at room temperature, for water configurations prepared in the presence of a strong electric field pointing along the tunneling (x) axis. In such layers, the water dipoles point on the average along this axis. The

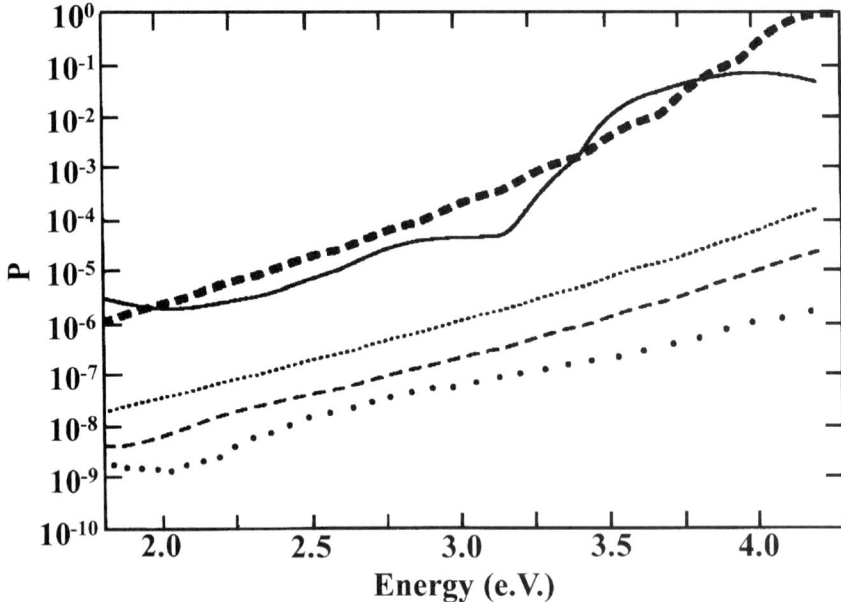

Figure 8 Electron transmission probability as a function of the incident energy. Shown are one-to-all transmission results with the electron incident in the direction normal to the water layer. These results are averaged over six equilibrium water configurations sampled from an equilibrium trajectory for the water system. This system contains 192 water molecules confined between two walls separated by 10 Å, with periodic boundary conditions, with period 23.5 Å in the directions parallel to the walls, at 300 K. These data correspond to three water monolayers between the walls. *Thin dashed line*: Results from model A (see text). *Full line*: Results from model B. Also shown are the corresponding results for tunneling through a vacuum, i.e. through a bare rectangular potential barrier of height 5 eV (*thick-dotted line*), and through a similar barrier of height 3.8 eV (*thick-dashed line*), which corresponds to the expected lowering of the effective barrier for tunneling through water. The *thin-dotted line* is the transmission probability computed for model A modified by eliminating the attractive part of the water-metal interaction, thereby eliminating the preferred orientational ordering of the water dipoles at the water-metal interface. (Reproduced from Reference 135 and used by permission.).

electric field was removed during the subsequent tunneling calculation. The computed one-to-all transmission for electrons incident in the x direction shows several orders of magnitude difference between the probabilities calculated for electron incident in the direction of the induced polarization and against this direction. Microscopic reversibility implies that the corresponding one-dimensional process should not depend on the tunneling direction, positive or negative, along the x axis. The observed behavior is therefore associated with the three-dimensional nature of the process. It shows that the angular distribution associated with the transmission through such layers depends strongly on the transmission direction and suggests that asymmetry in current-voltage dependence of transmission current should exist beyond the linear regime.

Next consider the possibility of resonance-assisted tunneling. Such resonances are found (138) in a range of ~1 eV below the 5 eV vacuum barrier, and their existence correlates with the observation of weakly bound states of an electron in neutral configurations of bulk water. Mosyak et al (135) have found that such states appear in neutral water configurations in both models A and B; however, only model B shows such states at negative energies. Moreover, these states are considerably more extended in systems described by model B compared with the corresponding states of model A (135). The possible effect of bound electron states in water on electron-transmission probability through water was raised in the past (151). Peskin et al (138) have recently identified the source of the resonances seen in our simulations as transient vacancies in the water structure. We emphasize again that because these results were obtained for static water configurations, their actual role in electron transmission through water is yet to be clarified.

The effective barrier to electron tunneling in water has been the subject of many discussions in the STM literature (22, 128, 131, 152). Although the absolute numbers obtained vary considerably depending on the systems studied and on experimental setups and conditions, three observations can be made. (a) Tunneling is observed at large tip-surface distances, sometimes exceeding 20 Å (22, 131, 152). (b) The barrier, estimated using a one-dimensional model from the distance dependence of the observed current, is unusually low, of the order of 1 eV in systems involving metals with work functions of 4–5 eV. (c) The numbers obtained scatter strongly: The estimated barrier height may be stated to be 1 ± 1 eV. (d) The apparent barrier height appears to depend on the polarity of the bias potential.

It should be kept in mind that even in a vacuum STM, the barrier to tunneling is expected to be lower than the work functions of the metals involved because of image effects associated with the fast electronic response of the electrodes (29). Nevertheless, the reduction of barrier height in the aqueous phase seems to be considerably larger. Taking the vacuum barrier as input in our discussion, let us consider the possible roles of the solvent. These can arise from the following factors: (a) the position, on the energy scale, of the "conduction band" of the pure solvent (by "conduction band" we mean extended electronic states of an excess electron in the neutral solvent configuration); (b) the effect of the solvent on the electrode workfunction; (c) the hard cores of the atomic constituents (in the present case the water oxygens, which make a substantial part of the physical space between the electrodes inaccessible to the electron); and (d) the possibility that the tunneling is assisted by resonance states supported by the solvent. Such resonances can be associated with available molecular orbitals—this does not appear to be the case in water—or with particular transient structures in the solvent configurations, as discussed above.

Factors b–d are usually disregarded in theories of electron transfer, whereas a common practice is to account for the first factor by setting the potential barrier height at a value, below the vacuum level, determined by the contribution of the solvent electronic polarizability. This value can be estimated as the Born energy of a point charge in a cavity of intermolecular dimensions, say a radius of ~5 au, in a continuum with the proper dielectric constant, here the optical dielectric constant of water, $\varepsilon_\infty = 1.88$. This yields $e^2(2a)^{-1}[\varepsilon_\infty^{-1} - 1] \sim -1.3$ eV, the same order as the result of a more rigorous calculation by Schmickler & Henderson (153) and in agreement with experimental results on photoemission into water (132). It should be noted that this number was obtained for an infinite bulk of water and should be regarded as an upper limit for the present problem.

The simulations described above shed some light on the roles played by the other factors listed above. First, we find that lowering the metal work function by the orientational ordering of water dipoles at the metal surface does affect the tunneling probability, as discussed above and shown in Figure 8. Second, the occupation of much of the physical space between the electrodes by the impenetrable oxygen cores strongly reduces the tunneling probability. In fact, if these two factors exist alone, the computed tunneling probability is found to be considerably lower than in the corresponding vacuum process (see 133, Figure 7). Even including the effect of the water electronic polarizability (i.e. attractive r^{-4} terms) in the two-body electron-water pseudopotential (model A), it is not sufficient to reverse this trend, as seen in Figure 8. Taking into account the full many-body nature of this interaction is found to be essential for obtaining the correct qualitative effect of water, i.e. barrier lowering relative to vacuum.

The estimate of the magnitude of this lowering effect in our simulations can be done in two ways. One is to fit the absolute magnitude of the computed transmission probability to the result obtained from a one-dimensional rectangular barrier of width given by the distance s between the electrodes (134). This is done in Figure 9 for systems with 1–4 monolayers of water ($s = 3.6, 6.6, 10.0, 13.3$ Å).[15] The following points should be noted:

(1) The effective barrier to tunneling computed with the fully polarizable model B is reduced by at least 0.5 eV (from the bare value of 5 eV used in these simulations) once a bulk has been developed in the water layer, i.e. once the number of monolayers is larger than two.

(2) The equivalent calculation done with model A, in which water polarizability is accounted for only on the two-body level, yields an effective barrier higher than the vacuum barrier.

(3) For the very thin layers studied, the effective barrier height depends on the layer thickness. This behavior [which supports a recent experimental observation by Nagy (130)] is expected to saturate once a well-defined bulk is developed.

Following common practice in STM studies, another way to discuss the effective simulated barrier is to fit the distance dependence of the observed tunneling

---

[15] It should be emphasized that these results were not statistically averaged over many water configurations, so the absolute numbers obtained should be taken only as examples of a general qualitative behavior.

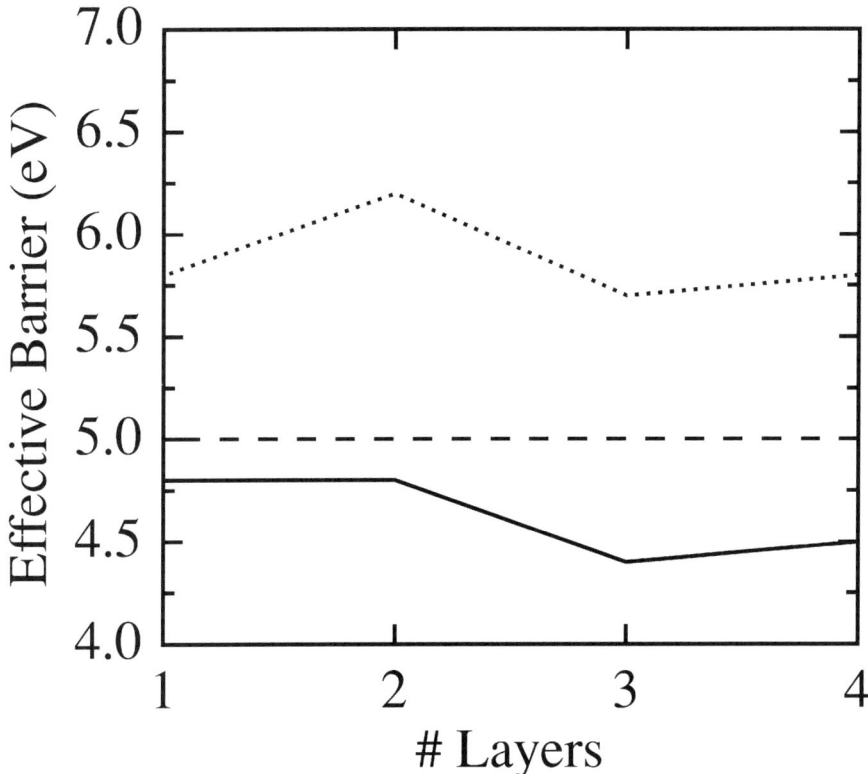

Figure 9 Effective one-dimensional barrier height for electron transmission through water, displayed as a function of number of water layers. *Solid, dotted,* and *dashed lines* correspond to models A and B, and to the bare (5 eV) barrier, respectively. See text for details. (Reproduced from Reference 134 and used by permission.)

probability to the analytical result for a rectangular barrier. This practice can yield very low apparent barriers in cases where tunneling is influenced by resonance structures (137). Moreover, since the existence and energies of these resonances in water depend on local structures that evolve over time, it is possible that the characteristic scatter of data that appears in these measurements (22, 128, 131, 152) may arise not only because of experimental difficulties but also from intrinsic system properties.

The existence in water of transient structures that support excess electron resonances and the possible implications of these resonances in enhancing the tunneling probability, and the apparent barrier height, raises again the issue of timescales. In particular, the lifetimes of these resonance states is of considerable interest, since they determine the duration of the electron capture by the water film and, as a consequence, the possibility that water dynamics and thermal relaxation become important on this timescale. Peskin et al (138) have determined these lifetimes by a direct evaluation of the complex eigenvalues associated with the corresponding resonance structures, using a filter diagonalization method with the imaginary boundary-conditions Hamiltonian. The resulting eigenvalues have imaginary parts of the order ~ 0.05 eV, implying lifetimes of the order ≤ 10 fs. An alternative way to probe the dynamics of electron tunneling in water is by evaluating the corresponding traversal times (see Section 3.1). Here the timescale for possible interaction between the excess electron and barrier motions can be determined both near and away from resonance energies. Galperin et al (94) have applied the internal clock approach of Section 3.1 to this problem, starting from the one-to-all transmission probability, shown in Equation 63, written in the form

$$\sigma = \frac{1}{\pi} \langle \phi_{in}(E) | \hat{\varepsilon}_{in}^* \hat{G}^\dagger \hat{\varepsilon}_{out} \hat{G} \hat{\varepsilon}_{in} | \phi_{in}(E) \rangle, \qquad 125.$$

where ϕ_{in} denotes an incoming state in the reactant region and ε_{in} and ε_{out} are the absorbing boundary functions in the reactant (incoming) and product (outgoing) regions, respectively. In the present application, the electron is taken to have two internal states, so that if x is the tunneling direction, $\phi_{in} = \left(e^{ikx}/\sqrt{v}\right)\binom{1}{0}$. The Green's operator is given by $\hat{G} = (E - \hat{H}_0 + i(\hat{\varepsilon}_{in} + \hat{\varepsilon}_{out}))^{-1}$ with $\hat{H}_0$ replaced by

$$\hat{H} = \hat{H}_0 \begin{pmatrix} 1 & 0 \\ 0 & 1 \end{pmatrix} + \lambda \hat{F}(x) \begin{pmatrix} 0 & 1 \\ 1 & 0 \end{pmatrix}, \qquad 126.$$

where λ is a constant and where $F(x) = 1$ in the barrier region and 0 outside it. The approximate scattering wavefunction,

$$|\psi(E)\rangle = i\hat{G}(E)\hat{\varepsilon}_{in} | \phi_{in}(E) \rangle = \begin{pmatrix} \psi_1(E) \\ \psi_2(E) \end{pmatrix}, \qquad 127.$$

is evaluated using iterative inversion methods (148). The transmission probabilities into the $|1\rangle$ and $|2\rangle$ states are obtained from

$$T_i(E) = \langle \psi_i(E) | \hat{\varepsilon}_{out} | \psi_i(E) \rangle; \quad i = 1, 2. \qquad 128.$$

T_i are equivalent to $|c_i|^2$, where c_i ($i = 1, 2$) are defined above in Equation 73. Accordingly,

$$\tau(E) = \lim_{\lambda \to 0} \left(\frac{\hbar}{|\lambda|} \sqrt{\frac{T_2(E)}{T_1(E)}} \right). \qquad 129.$$

Figures 10 and 11 (94) display some results of this calculation. Figure 10 shows calculated traversal times as functions of incident electron energy for an electron transmitted through a layer of three water films between two platinum electrodes (the distance between the electrodes is $d = 18.9$ au. Shown is τ/τ_0 for several configurations of this system, where τ_0 is the tunneling time associated with the bare vacuum barrier (same geometry with no water). The transient nature of the water structures that give rise to the resonance features is seen here. Note that the difference between different configurations practically disappears

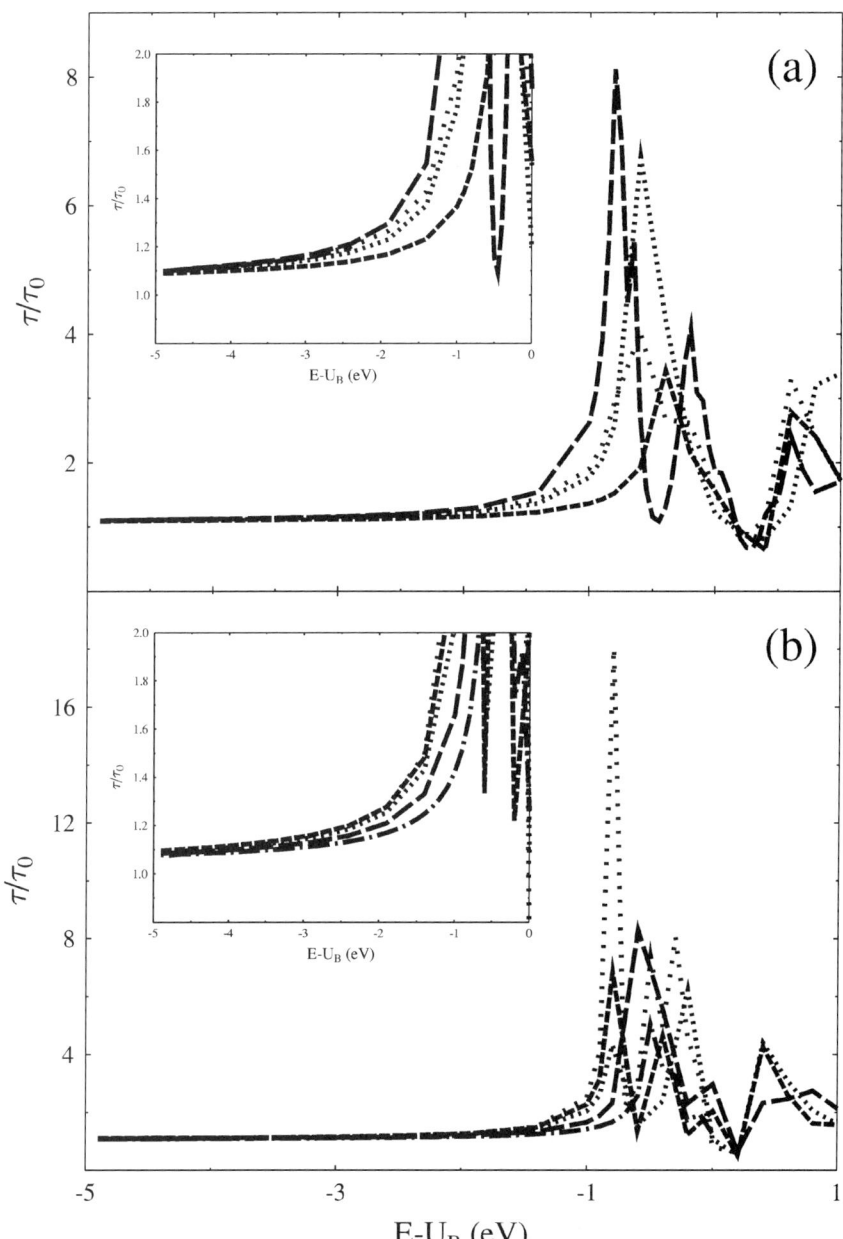

Figure 10 The ratio τ/τ_0 (see text) computed for different static configurations of (a) three and (b) four monolayer water films, displayed against the incident electron energy. The *inset* shows an enlarged vertical scale for the deep tunneling regime. (Reproduced from Reference 94 and used by permission.)

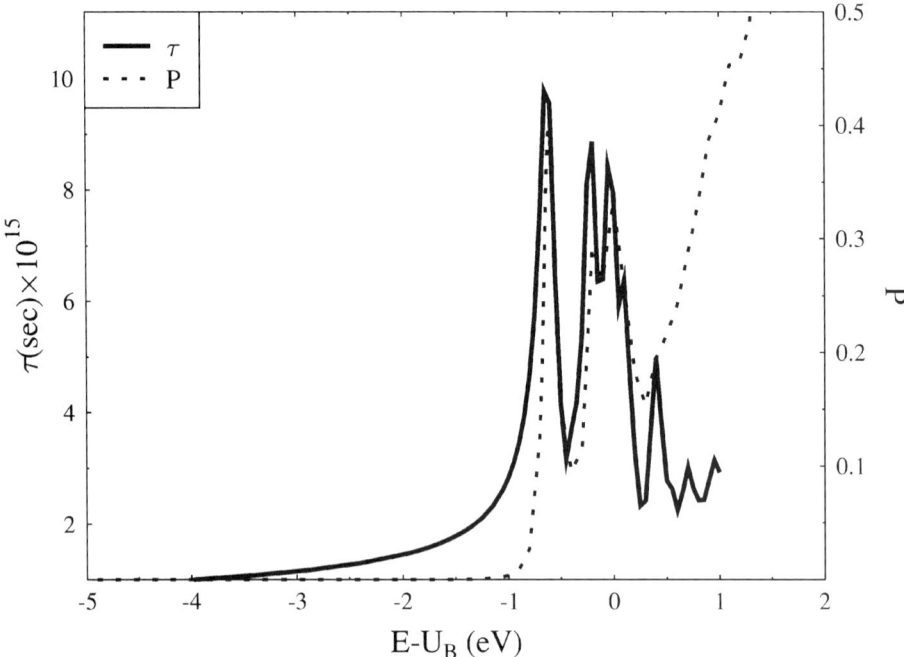

Figure 11 The tunneling traversal time (*full line*; *left vertical scale*) and the transmission probability (*dotted line*; *right vertical scale*) computed as functions of incident electron energy for one static configuration of the 3-monolayer water film. (Reproduced from Reference 94 and used by permission.)

for energies sufficiently below the resonance regime, where the ratio between the time computed in the water system and in the bare barrier is practically constant, approximately 1.1. Figure 11 shows, for one of these configurations, the tunneling time and the transmission probability, both as functions of the incident electron energy. We see that the energy dependence of the tunneling time follows this resonance structure closely. In fact, the times (3–15 fs) obtained from the peaks in Figures 10 are consistent with the resonance lifetimes estimated in Peskin et al.

We conclude this discussion with two more comments. First, in the above analysis, the possibility of transient "contamination" of the tunneling medium by foreign ions has been disregarded. Such ions exist in most systems used in underwater STM studies, and the appearance of even one such ion in the space of 10–20 Å between the electrodes can have a profound effect on the tunneling current behavior. This may add another source of scatter in the experimental results. Second, as discussed, changes in the water structure between the electrodes may appear also as bias-dependent systematic effects. Thus, the asymmetry in the bias dependence of the barrier height observed in Pan et al (128), Hahn et al (22), and Hong et al (131) may be related to the asymmetric transmission properties of orientationally ordered layers.

5. OVERBARRIER TRANSMISSION

Our discussion so far has focused on electron-transmission processes that at zero-temperature can take place only by tunneling. This section provides a brief overview of transmission processes where an electron incident on a molecular barrier carries a positive (above-ground–state vacuum) energy. It should be emphasized that this in itself does not mean that transmission can take place classically. If the incident energy is in the band gap of the molecular spacer, zero-temperature transmission is still a tunneling process. Still, this type of phenomenon is distinct from those discussed in the other parts of this review, for several reasons. First, positive-energy transmission (and reflection), essentially scattering processes, are amenable to initial-state control and to final-state resolution that are not possible in negative-energy processes. Second, a positive-energy electron interacts with a large density of medium states; therefore, the probability for resonance or near resonance transfer is considerably larger, implying also a larger cross-section for dephasing and inelastic energy loss. Third, at this range of energies, conventional quantum-chemistry approaches as well as pseudopotentials derived from low-energy electronic-structure data can be very inaccurate. Finally, at high enough energies, electronic excitations and secondary electron generation become important factors in the transmission mechanism. For the last two reasons the numerical approaches described in Section 2.6–8 are not immediately applicable.

The effect of adsorbates on photoelectrons emitted from surfaces has been studied for almost a century (154, 155). These experiments were partially motivated by their practical ramifications whereby the surface work function was modified by the adsorbate (156). Recently, the development of tunable ultraviolet light sources has enabled studies of energy-resolved photoelectron spectroscopy. This eventually led to studies of photoelectron energy distribution for photoelectrons produced from metal surfaces covered with self-assembled monolayers of organic molecules, or organized organic thin films (147, 157–162). These films are prepared either with the Langmuir-Blodgett technique (163) or by self-assembly from vapor or solution. One of the earlier experiments of this kind was the measurement of transmitted electron energy distribution for photoelectrons produced from a Pt (111, 111a) surface covered with several layers of water (139). The transmission probability decreased exponentially with increasing numbers of water layers; however, this number does not affect the energy distribution of the emitted electrons, indicating that transmission in this system is independent of the electron energy and that inelastic energy loss is small. These results should, however, be regarded with caution in view of LEET data (142) that indicate that energy loss from a transmitted electron to water nuclear motion may be quite efficient. The latter observation is supported by estimates (143) of the distance (20–50 Å) traversed by electrons photoejected into water at subexcitation energies before their capture to form the precursor of solvated electrons.

Unlike water, the electron affinity $A = -V_0$ of hydrorcarbon layers is negative, i.e. their LUMOs, or in the language of solid state physics, the bottom of their electron-condition band is above vacuum energy [$V_0 = 0.8$ eV for bulk hydrocarbons (164)]. Indeed, a threshold for electron photoemission from silver, covered with a monolayer of cadmium stearate [$CH_3(CH_2)_{16}COO^-]_2Cd^{2+}$, or arachdic acid $CH_3(CH_2)_{16}COOH$, is observed (159). Above 0.8 eV, photoemission from these surfaces proceeds with efficiency close to one, turning down again at higher energies. Oscillations in the transmission probability through similar films as a function of the initial electron energy were interpreted in terms of the electronic-band structure of the film (147). This interpretation gains further support from the observation of the large sensitivity of the transmission probability to the film structure in the lateral dimension (161), and from the strong effect of film ordering (161). This does not exclude what is often taken to express a single molecule effect—a strong preference of the phtoemission to be directed along the axis of the molecular adsorbate (158). Finally, using chiral molecular self-assembled monolayers (L or D polyalanine polypeptides) has revealed that electron transmission of spin-polarized electrons depends, with a high degree of selectivity, on the chirality of the layer (162).

Another way to study electron interactions with molecular layers is to send an electron beam from the vacuum side onto a molecular film condensed on a suitable, usually metallic, substrate. In LEET spectroscopy developed by Sanche and coworkers (141), the electron-transmission spectrum is measured by monitoring the current arriving at the metal substrate as a function of the incident electron energy and direction. Similarly, the reflected electron beam can be analyzed with respect to energy and angular distribution, yielding electron-diffraction data, energy-loss spectra, and energy-loss excitation spectra. The same experimental setup can be used to study the effect of electron trapping, electron-stimulated desorption, and electron-induced chemical reactions on the molecular films. [For a recent review of these types of studies and references to earlier work, see Sanche (141).] Here we focus on observations from LEET experiments that are relevant to our present subject. First, the prominence of the elastic and quasielastic component of the transmitted intensity, observed in most experiments of this kind, is in agreement with the photoemission experiments discussed above. Second, a threshold of a few tens of electron volts (relative to the vacuum level) is seen for transmission through alkane and rare gas layers, indicating negative electron affinities of these layers and providing an estimate for the position of the bottom of the layers' conduction bands. Third, conduction peaks below this threshold are attributed to tunneling assisted by local states inside the gap (125). This is the analog of the bridge-assisted tunneling discussed in Section 2, except that the film constitutes a three-dimensional barrier in which the local states are distributed randomly in position and energy. As discussed in Section 3.4, thermal relaxation and dephasing processes manifest themselves in a characteristic thickness dependence of the transmission probability as the processes change from tunneling to hopping dominated with increasing barrier width (see Figure 7). Fourth, the electron transmission spectra closely reflects the band structure of the corresponding layer. This should not be taken as evidence for ballistic transport, in fact this observation holds only for the inelastic components of the emission intensity. Rather, the electron propagation through the molecular environment is viewed as a sequence of scattering events, with cross-sections that are proportional to the density of available states (165). The resulting averaged-mean free path is therefore inversely proportional to the density of states at an energy that (as long as the absolute energy loss is small) may be approximated by the incident energy. Finally, the transmission can be strongly affected by resonances, i.e. negative ion formation. This in turn may greatly increase the probability for inelastic energy loss (141). These processes are observed in the high-resolution electron energy-loss spectroscopy, by monitoring the energy of reflected electrons, but they undoubtedly play an equally important part in the transmission process.

As already mentioned, while the theoretical methods discussed in previous sections of this review are general, their applicability to electron transmission in the positive energy regime needs special work because standard quantum-chemistry calculations usually address negative energy regimes and bound electronic states, and because pseudopotentials are usually derived from fitting results of such ab initio calculations to analytical forms based on physical insight. Model calculations that demonstrate some of the concepts discussed above are shown in Figures 12 and 13 (147). Figure 12 compares the transmission probability ("one to all," with the incident electron perpendicular to the barrier) through a one-dimensional rectangular barrier of height 3 eV and width 1.2 nm as a function of the incident electron energy measured relative to the barrier top with the transmission through a three-dimensional slab of four Ar layers cut out of an Ar crystal in the {100} direction. The latter results are obtained with a spatial-grid technique using the electron-Ar pseudopotential of Space et al. (166). The oscillations of the dotted line in Figure 12 are interference patterns associated with the finite width of the layers. The full line in Figure 12 also shows such oscillations, but in addition, a prominent dip above 4 eV corresponds to a conduction band gap of this thin-ordered layer. The transmission through the disordered layer (dashed line) is considerably less structured (smoother shapes should be obtained with more configurational averaging); in particular, the dip associated with the bandgap has largely disappeared. Figure 13 compares the transmission (one-to-all) versus electron energy for an electron incident in the normal direction on ordered Ar films made of 2, 4, and 6 atomic monolayers (prepared by cutting them off an Ar crystal, as described above). Already at six-layer thickness, the observed transmission dip is very close to its bulk value, indicating that the band structure is already well developed.

These calculations investigate transmission through static nuclear structures and consequently cannot account for thermal relaxation and dephasing effects. In the other extreme limit one uses stochastic models (167) that become accurate when the molecule film is thick enough so that the electron goes through multiple scattering events before being transmitted through or reflected from the film. Such an approach has been used (141, 168) to describe the energy distributions of

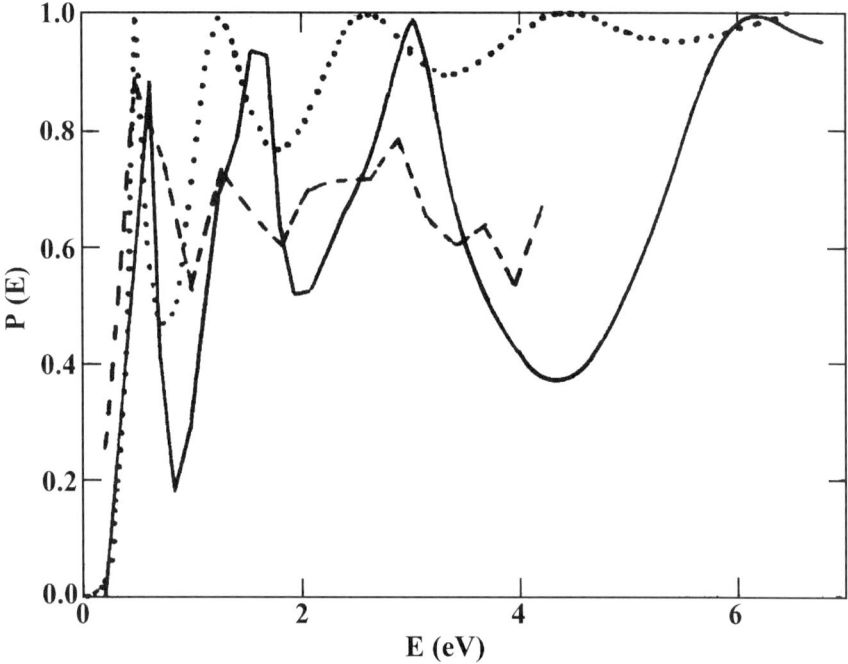

Figure 12 *Dotted line*: Transmission probability through one-dimensional rectangular barrier characterized by height of 3 eV and width of 12 Å, as a function of incident electron energy measured relative to the barrier top. *Full line*: Electron transmission through a slab made of four Ar layers, cut out of an FCC Ar crystal in the {100} direction. *Dashed line*: Same results obtained for a disordered Ar slab, obtained from the crystalline layer by a numerical thermal annealing at 400 K next to an adsorbing wall using molecular dynamics propagation. The results shown are averaged over four such disordered Ar configurations (see 147 for more details). (Reproduced from Reference 147 and used by permission.)

electrons reflected from molecular films and their relation to the density of excess electron states in the film.

6. CONCLUSIONS AND OUTLOOK

This review has described the current status of theoretical approaches to electron transmission and conduction in molecular junctions. In particular, Section 2 constitutes an account of theoretical approaches to this problem for static junctions, whereas Section 3 discusses approaches that focus on dephasing and thermal relaxation effects. It is important to note that even though our methodology follows a stationary, steady state viewpoint of all processes studied, the issue of relative timescales of different processes has played a central role in our analysis.

Current studies of molecular junctions focus on general methodologies on the one hand and on detailed studies of specific systems on the other. We have described

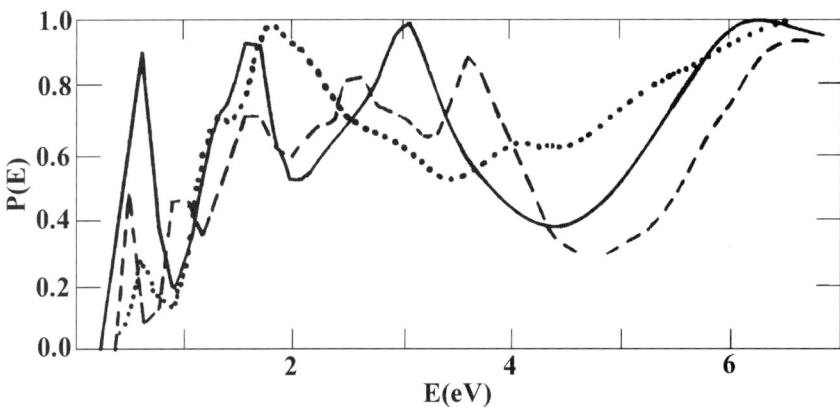

Figure 13 The computed transmission probabilities, vs electron energy, for an electron incident on slabs cut out of an FCC Ar crystal in the {100} direction. Shown are transmission probabilities vs incident electron energy through slabs made of two monolayers (*dotted line*), four monolayers (*full line*), and six monolayers (*dashed line*). (Reproduced from Reference 147 and used by permission.)

in detail recent computations of electron transmission through water layers and have described other studies on prototypes of molecular wires. Two imporant classes of molecular wires have now become subjects of intense research, even development effort. These are DNA wires and carbon nanotubes. Although the general principles discussed in this review apply also to these systems, the scope of recent research on special structure-function properties of these wires merits a separate review.

Returning to theoretical issues, we have outlined some open problems in the methodology of treating these many-body, strongly interacting, nonequilibrium open systems. One additional direction not covered is the possibility of controlling the operation of such junctions using external forces (as opposed to control of function by varying the structure). Several recent studies point out the possibility of controlling transport processes by external fields (169). The specific and selective nature of molecular optical response make molecular junctions strong potential candidates for such applications.

In conclusion, electron transmission and conduction processes in small molecular junctions combine the phenomenology of molecular electron transfer with structural problems associated with design and construction of such junctions, and with the need to understand their macroscopic transport properties. In addition, the potential technological promise suggests that research in this area will intensify.

ACKNOWLEDGMENTS

This work was supported by the US-Israel Binational Science Foundation. I am indebted to many colleagues with whom I have collaborated and/or discussed various aspects of this work: I Benjamin, M Bixon, A Burin, B Davis, D Evans,

M Galperin, P Hänggi, J Jortner, R Kosloff, A Mosyak, V Mujica, R Naaman, U Peskin, E Pollak, M Ratner, D Segal, T Seideman, D Tannor, MR Wasielewski, J Wilkie, and S Yaliraki. I also thank S Datta and L Sanche for allowing me to use figures from their work.

Visit the Annual Reviews home page at www.AnnualReviews.org

LITERATURE CITED

1. Jortner J, Bixon M, eds. 1999. *Advances in Chemical Physics: Electron Transfer—From Isolated Molecules to Biomolecules*, Vol. 106. New York:Wiley
2. Schmickler F. 1996. *Interfacial Electrochemistry*. Oxford, UK: Oxford Univ. Press
3. Joachim C, Roth S, eds. 1997. *Atomic and Molecular Wires*, Vol. 341. Dordrecht: Kluwer
4. Jortner J, Ratner M, eds. 1997. *Molecular Electronics*. Oxford, UK: Blackwell Sci.
5. Marcus RA. 1965. *J. Chem. Phys.* 43:679
6. Gosavi S, Marcus RA. 2000. *J. Phys. Chem.* 104:2067–72
7. Tour JM. 1996. *Chem. Rev.* 96:537
8. Sachs SB, Dudek SP, Sita LR, Newton JF, Feldberg SW, Chidsey CED. 1997. *J. Am. Chem. Soc.* 119:10563
9. Isied SS, Ogawa MY, Wishart JF. 1992. *Chem. Rev.* 92:381
10. Mutz MW, Case MA, Wishart JF, Ghadiri MR, McLendon GL. 1999. *J. Am. Chem. Soc.* 121:858–59
11. Holmlin RE, Dandliker PJ, Barton JK. 1997. *Angew. Chem. Int. Ed. Engl.* 36:2714
12. Lewis FD, Wu TF, Zhang YF, Letsinger RL, Greenfield SR, Wasielewski MR. 1997. *Science* 277:673–76
13. Meggers E, Michel-Beyerle ME, Giese B. 1998. *J. Am. Chem. Soc.* 120:12950–55
14. Ghiggino KP, Clayton AHA, Lawson JM, Paddon-Row MN. 1996. *N. J. Chem.* 20:853
15. Guo HL, Facci JS, Mclendon G. 1995. *J. Phys. Chem.* 99:8458
16. Langlais VJ, Schlittler RR, Tang H, Gourdon A, Joachim C, Gimzewski JK. 1999. *Phys. Rev. Lett.* 83:2809–12
17. Roth S, Burghard M, Fischer CM. 1997. See Ref. 4, pp. 255–80
18. Dhirani A, Lin PH, Guyot-Sionnest P. 1997. *J. Chem. Phys.* 106:5249–53
19. Reed MA, Zhou C, Muller CJ, Burgin TP, Tour JM. 1997. *Science* 278:252–54
20. Datta S, Tian WD, Hong SH, Reifenberger R, Henderson JI, Kubiak CP. 1997. *Phys. Rev. Lett.* 79:2530–33
21. Tian WD, Datta S, Hong SH, Reifenberger R, Henderson JI, Kubiak CP. 1998. *J. Chem. Phys.* 109:2874–82
22. Hahn JR, Hong YA, Kang H. 1998. *Appl. Phys. A* 66:S467–72
23. Zhou C, Muller CJ, Reed MA, Burgin TP, Tour JM. 1997. See Ref. 4, pp. 191–213
24. Porter JD, Zinn AS. 1993. *J. Phys. Chem.* 97:1190–203
25. Slowinski K, Slowinska KU, Majda M. 1999. *J. Phys. Chem.* 103:8544–51
26. McConnel HM. 1961. *J. Chem. Phys.* 35:508–15
27. Onipko A, Klymenko Y. 1998. *J. Phys. Chem. A* 102:4246–55
28. Simmons JG. 1963. *J. Appl. Phys.* 34:1793–803
29. Lang ND. 1988. *Phys. Rev. B* 37:10395
30. Tersoff J, Hamman DR. 1985. *Phys. Rev. B* 5031:805
31. Bardeen J. 1961. *Phys. Rev. Lett.* 6:57
32. Landauer R. 1970. *Philos. Mag.* 21:863–67
33. Imry Y. 1997. *Introduction to Mesoscopic Physics*. Oxford, UK: Oxford Univ. Press
34. Miller WH. 1975. *J. Chem. Phys.* 62:1899–906
35. Miller WH, Schwartz SD, Tromp JW. 1983. *J. Chem. Phys.* 79:4889–98

36. Galperin M, Segal D, Nitzan A. 1999. *J. Chem. Phys.* 111:1569–79
37. Mujica V, Kemp M, Ratner MA. 1994. *J. Chem. Phys.* 101:6849–64
38. Newns DM. 1969. *Phys. Rev.* 178:1123
39. de Andrade PCP, Onuchic JN. 1998. *J. Chem. Phys.* 108:4292–98
40. Nitzan A. 2001. *J. Phys. Chem.* In press
41. Magoga M, Joachim C. 1997. *Phys. Rev. B* 56:4722–9
41a. Magoga M, Joachim C. 1997. *Phys. Rev. B* 57:1820–23
41b. Magoga M, Joachim C. 1997. *Phys. Rev. B* 59:16011–21
42. Yaliraki SN, Kemp M, Ratner MA. 1999. *J. Am. Chem. Soc.* 121:3428–34
43. Biscarini F, Bustamante C, Kenkre VM. 1995. *Phys. Rev. B* 51:11089–101
44. Emberly EG, Kirczenow G. 1999. *Phys. Rev. B* 60:6028–33
45. Hall LE, Reimers JR, Hush NS, Silverbrook K. 2000. *J. Chem. Phys.* 112:1510–21
46. Faglioni F, Claypool CL, Lewis NS, Goddard WA. 1997. *J. Phys. Chem.* 101:5996–6020
47. Larsson S, Klimkans A. 1999. *Theochem. J. Mol. Struct.* 464:59–65
48. Yaliraki SN, Roitberg AE, Gonzalez C, Mujica V, Ratner MA. 1999. *J. Chem. Phys.* 111:6997–7002
49. Newton MD. 1991. *Chem. Rev.* 91:767
50. Emberly EG, Kirczenow G. 1999. *J. Phys. C* 11:6911–26
51. Koopmans T. 1933. *Physica* 1:104
52. Kergueris C, Bourgoin JP, Palacin S, Esteve D, Urbina C, et al. 1999. *Phys. Rev. B* 59:12505–13
53. Miskovic ZI, English RA, Davison SG, Goodman FO. 1997. *J. Phys. Conduct. Matter* 9:10749–60
54. Seideman T, Miller WH. 1992. *J. Chem. Phys.* 96:4412
54a. Seideman T, Miller WH. 1992. *J. Chem. Phys.* 97:2499
55. Liebsch A. 1997. *Electronic Excitations at Metal Surfaces*, pp. 5–48. New York: Plenum
55a. Liebsch A. 1997. See Ref. 55, pp. 283–308
56. Lang ND. 1995. *Phys. Rev. B* 52:5335–42
57. Lang ND, Avouris P. 2000. *Phys. Rev. Lett.* 84:358–61
58. Parr RG, Yang W. 1989. *Density Functional Theory of Atoms and Molecules Density Functional Theory of Atoms and Molecules*. Oxford, UK: Oxford Univ. Press
59. Di Ventra M, Pantelides ST, Lang ND. 2000. *Phys. Rev. Lett.* 84:979–82
60. Hirose K, Tsukada M. 1995. *Phys. Rev. B* 51:5278–90
61. Mujica V, Roitberg AE, Ratner MA. 2000. *J. Chem. Phys.* 112:6834–39
62. Szabo A, Ostlund NS. 1989. *Modern Quantum Chemistry: Introduction to Advanced Electronic Structure Theory*. New York: McGraw-Hill
63. Burke H, Gross EKU. 1998. In *Density Functionals: Theory and Applications*, ed. D Joubert. Berlin: Springer
64. Lamoen D, Ballone P, Parrinello M. 1996. *Phys. Rev. B* 54:5097–105
65. Modinos A. 1984. *Field, Thermionic and Secondary Electron Spectroscopy*. New York: Plenum
66. Harris CB, Ge NH, Lingle RL, McNeill JD, Wong CM. 1997. *Annu. Rev. Phys. Chem.* 48:711
67. Aviram A, Ratner MA. 1974. *Chem. Phys. Lett.* 29:277
68. Waldeck DH, Beratan DN. 1993. *Science* 261:576–77
69. Marcus RA. 1996. *J. Chem. Soc. Faraday Trans.* 92:3905–8
70. Martin AS, Sambles JR, Ashwell GJ. 1993. *Phys. Rev. Lett.* 70:218–21
71. Fischer CM, Burghard M, Roth S, Vonklitzing K. 1994. *Europhys. Lett.* 28:129–34, 375–77
72. Nagesha K, Gamache J, Bass AD, Sanche L. 1997. *Rev. Sci. Instrum.* 68:3883–89

73. Makov G, Nitzan A, Brus LE. 1988. *J. Chem. Phys.* 88:5076–85
74. Ferry DK, Goodnick SM. 1997. *Transport in Nanostructures.* Cambridge, UK: Cambridge Univ. Press
75. Wilkins R, Ben-Jacob E, Jaklevic RC. 1989. *Phys. Rev. Lett.* 63:801–4
76. Andres RP, Bein T, Dorogi M, Feng S, Henderson JI, et al. 1996. *Science* 272:1323–25
77. Kuznetsov AM, Ulstrup J. 1993. *J. Electroanal. Chem.* 362:147–52
78. Fan F-RF, Bard AJ. 1997. *Science* 277:1791–93
79. Malysheva LI, Onipko AI. 1992. *Phys. Rev. B* 46:3906–15
80. Mujica V, Kemp M, Roitberg A, Ratner M. 1996. *J. Chem. Phys.* 104:7296–305
81. Li Y-Q, Gruber C. 1998. *Phys. Rev. Lett.* 80:1034–37
82. Gurvitz SA, Prager YS. 1996. *Phys. Rev. B* 53:15932–43
83. Datta S. 1995. *Electric Transport in Mesoscopic Systems.* Cambridge, UK: Cambridge Univ. Press
84. Wolf EL. 1985. *Principles of Electron Tunneling Spectroscopy.* New York: Oxford Univ. Press
85. Emberly E, Kirczenow G. 2000. *Phys. Rev. B* 61:5740–50
86. Wagner M. 2000. *Phys. Rev. Lett.* 85:174–77
87. Tour JM, Kozaki M, Seminario JM. 1998. *J. Am. Chem. Soc.* 120:8486–93
88. Todorov TN. 1998. *Philos. Mag. B* 77:965–73
89. Büttiker M, Landauer R. 1982. *Phys. Rev. Lett.* 49:1739–42
90. Büttiker M. 1983. *Phys. Rev. B* 27:6178–88
91. Landauer R, Martin T. 1994. *Rev. Mod. Phys.* 66:217–28
92. Hauge EH, Stoveng JA. 1989. *Rev. Mod. Phys.* 61:917–36
93. Nitzan A, Jortner J, Wilkie J, Burin AL, Ratner MA. 2000. *J. Phys. Chem. B* 104:5661–65
94. Galperin M, Nitzan A, Peskin U. Submitted for publication
95. Stipe BC, Rezaei MA, Ho W. 1998. *Science* 280:1732–35
96. Foley ET, Kam AF, Lyding JW, Avouris P. 1998. *Phys. Rev. Lett.* 80:1336–39
97. Gadzuk JW. 1991. *Phys. Rev. B* 44:13466–77
98. Domcke W, Cederbaum LS. 1980. *J. Phys. B* 13:2829–38
99. Wingreen NS, Jacobsen KW, Wilkins JW. 1988. *Phys. Rev. Lett.* 61:1396
99a. Wingreen NS, Jacobsen KW, Wilkins JW. 1989. *Phys. Rev. B* 40:11834
100. Avouris P, Walkup RE. 1989. *Annu. Rev. Phys. Chem.* 40:1989
101. Medvedev ES, Stuchebrukhov AA. 1997. *J. Chem. Phys.* 107:3821–31
102. Xie Q, Archontis G, Skourtis SS. 1999. *Chem. Phys. Lett.* 312:237–46
103. Kuznetsov AM, Vigdorovich MD, Ulstrup J. 1993. *Chem. Phys.* 176:539–54
104. Austin RH, Hong MK, Moser C, Plombon J. 1991. *Chem. Phys.* 158:473–86
105. Sumi H, Kakitani T. 1996. *Chem. Phys. Lett.* 252:85–93
106. Iversen G, Friis EP, Kharkats YI, Kuznetsov AM, Ulstrup J. 1998. *J. Biol. Inorg. Chem.* 3:229–35
107. Schmickler W, Tao N. 1997. *Electrochim. Acta* 42:2809–15
108. Nitzan A. 1979. *Chem. Phys.* 41:163–81
109. Büttiker M. 1988. *IBM J. Res. Dev.* 32:63–75
110. Bixon M, Jortner J. 1997. *J. Chem. Phys.* 107:1470–82, 5154–70
111. Bonča J, Trugman SA. 1995. *Phys. Rev. Lett.* 75:2566–69
111a. Bonča J, Trugman SA. 1995. *Phys. Rev. Lett.* 79:4874–77
112. Yu ZG, Smith DL, Saxena A, Bishop AR. 1999. *Phys. Rev. B* 59:16001–10
113. Heeger AJ, Kivelson S, Schrieffer JR, Su W-P. 1988. *Rev. Mod. Phys.* 60:781–850
114. Ness H, Fisher AJ. 1999. *Phys. Rev. Lett.* 83:452–55
115. Persson BNJ, Baratoff A. 1987. *Phys. Rev. Lett.* 59:339–42

116. Pollard WT, Felts AK, Friesner RA. 1996. *Adv. Chem. Phys.* 93:77–134
117. Okada A, Chernyak V, Mukamel S. 1998. *J. Phys. Chem.* 102:1241–51
118. Segal D, Nitzan A, Davis WB, Wasielewski MR, Ratner MA. 2000. *J. Phys. Chem. B* 104:3817
119. Segal D, Nitzan A, Ratner MA, Davis WB. 2000. *J. Phys. Chem.* 104:2790
120. Redfield AG. 1957. *IBM J. Res. Dev.* 1:19
121. Segal D, Nitzan A. 2001. *Chem. Phys.* In press
122. Bixon M, Jortner J. 1999. See Ref. 1, pp. 35–202
123. Giese B. 2000. *Acc. Chem. Res.* 33:631–36
124. Davis WB, Svec WA, Ratner MA, Wasielewski MR. 1998. *Nature* 396:60–63
125. Caron LG, Perluzzo G, Bader G, Sanche L. 1986. *Phys. Rev. B* 33:3027–38
126. Haiss W, Lackey D, Sass JK, Besocke KH. 1991. *J. Chem. Phys.* 95:2193
127. Meepagala SC. 1994. *Phys. Rev. B* 49:10761
128. Pan J, Jing TW, Lindsay SM. 1994. *Chem. Phys.* 98:4205
129. Vaught A, Jing TW, Lindsay SM. 1995. *Chem. Phys. Lett.* 236:306–10
130. Nagy G. 1996. *J. Electroanal. Chem.* 409:19–23
131. Hong YA, Hahn JR, Kang H. 1998. *J. Chem. Phys.* 108:4367–70
132. Gurevich YY, Pleksov YY, Rotenberg ZA. 1980. *Photo-Electrochemistry*. New York: Consultant Bur.
133. Mosyak A, Nitzan A, Kosloff R. 1996. *J. Chem. Phys.* 104:1549–59
134. Benjamin I, Evans D, Nitzan A. 1997. *J. Chem. Phys.* 106:6647–54
135. Mosyak A, Graf P, Benjamin I, Nitzan A. 1997. *J. Phys. Chem. A* 101:429–33
136. Benjamin I, Evans D, Nitzan A. 1997. *J. Chem. Phys.* 106:1291–93
137. Nitzan A, Benjamin I. 1999. *Acc. Chem. Res.* 32:854–61
138. Peskin U, Edlund A, Bar-On I, Galperin M, Nitzan A. 1999. *J. Chem. Phys.* 111:7558–66
139. Jo SK, White JM. 1991. *J. Chem. Phys.* 94:5761
140. Barnett RN, Landman U, Nitzan A. 1990. *Chem. Phys.* 93:6535–42
141. Sanche L. 1995. *Scanning Microsc.* 9:619
142. Bader G, Chiasson J, Caron LG, Michaud M, Perluzzo G, Sanche L. 1988. *Radiat. Res.* 114:467–79
143. Chernovitz AC, Jonah CD. 1988. *J. Phys. Chem.* 92:5946
144. Rostkier-Edelstein D, Urbakh M, Nitzan A. 1994. *J. Chem. Phys.* 101:8224–37
145. Barnett RN, Landman U, Cleveland CL. 1988. *Chem. Phys.* 88:4420
145a. Rossky PJ, Schnitker J. 1998. *J. Phys. Chem.* 92:4277
146. Tal-Ezer H, Kosslof R. 1984. *J. Chem. Phys.* 81:3967
147. Naaman R, Haran A, Nitzan A, Evans D, Galperin M. 1998. *J. Phys. Chem. B* 102:3658–68
148. Saad Y. 1996. *Iterative Methods for Sparse Linear Systems*. Boston: PWS
149. Staib A, Borgis DJ. 1995. *Chem. Phys.* 103:2642
150. Theil PA, Madey TE. 1987. *Surf. Sci. Rep.* 7:211
151. Halbritter J, Repphun G, Vinzelberg S, Staikov G, Lorentz WJ. 1995. *Electrochim. Acta* 40:1385–94
152. Christoph R, Siegenthaler H, Rohrer H, Wiese W. 1989. *Electrochim. Acta* 34:1011
153. Schmickler W, Henderson DJ. 1990. *Electroanal. Chem.* 290:283
154. Gurney RW. 1935. *Phys. Rev.* 47:479
155. Ueba H. 1991. *Surf. Sci.* 242:266
156. Albano EV. 1982. *Appl. Surf. Sci.* 14:183
157. Ueno N, Azuma Y, Yokota T, Aoki M, Okudaira KK, Harada Y. 1997. *Jpn. J. Appl. Phys.* 36:5731–36
158. Dimitrov DA, Trakhtenberg S, Naaman

R, Smith DJ, Samartzis PC, Kitsopoulos TN. 2000. *Chem. Phys. Lett.* 322:587–91
159. Kadyshevitch A, Naaman R. 1995. *Phys. Rev. Lett.* 74:3443
160. Kadyshevitch A, Naaman R, Cohen R, Cahen D, Libman J, Shanzer A. 1997. *J. Phys. Chem. B* 101:4085–89
161. Kadyshevitch A, Ananthavel SP, Naaman R. 1998. *Thin Solid Films* 329:357–59
162. Ray K, Ananthavel SP, Waldeck DH, Naaman R. 1999. *Science* 283:814–16
163. Blodgett KB, Langmuir I. 1937. *Phys. Rev.* 51:964
164. Sanche L. 1995. *Phys. Rev. Lett.* 75:2904
165. Michaud M, Sanche L. 1984. *Phys. Rev. B* 30:6067
166. Space B, Coker DF, Liu ZH, Berne BJ, Martyna GJ. 1992. *Chem. Phys.* 97:2002
167. Fano U. 1987. *Phys. Rev. A* 36:1929
168. Goulet T, Jung JM, Michaud M, Jay-Gerin JP, Sanche L. 1994. *Phys. Rev. B* 50:5101–9
169. Grifoni M, Hänggi P. 1998. *Phys. Rep.* 304:229–358

EARLY EVENTS IN RNA FOLDING

D Thirumalai[1,2,*], Namkyung Lee[2], Sarah A Woodson[3], and DK Klimov[2]

[1]Department of Chemistry and Biochemistry and [2]Institute for Physical Science and Technology, University of Maryland, College Park, MD 20742;
e-mail: thirum@glue.umd.edu; lee@mpip-mainz.mpg.de
[3]Department of Biophysics, Johns Hopkins University, Baltimore, MD 21218;
e-mail: swoodson@jhu.edu
*corresponding author

Key Words Kinetic partitioning mechanism, RNA folding pathways, *Tetrahymena* ribozyme, specific and nonspecific collapse, counterion-condensation

■ **Abstract** We describe a conceptual framework for understanding the way large RNA molecules fold based on the notion that their free-energy landscape is rugged. A key prediction of our theory is that RNA folding can be described by the kinetic partitioning mechanism (KPM). According to KPM a small fraction of molecules folds rapidly to the native state whereas the remaining fraction is kinetically trapped in a low free-energy non-native state. This model provides a unified description of the way RNA and proteins fold. Single-molecule experiments on *Tetrahymena* ribozyme, which directly validate our theory, are analyzed using KPM. We also describe the earliest events that occur on microsecond time scales in RNA folding. These must involve collapse of RNA molecules that are mediated by counterion-condensation. Estimates of time scales for the initial events in RNA folding are provided for the *Tetrahymena* ribozyme.

INTRODUCTION

The discovery that RNA molecules can function as enzymes (1) has lead to intense efforts to determine their three-dimensional structures and the mechanisms by which they fold. Because RNA molecules are engaged in diverse activities such as ligand binding, protein recognition and catalysis, it is reasonable to suggest that in some ways they are more similar to proteins than to their chemical counterpart, DNA. As in proteins, there are two aspects to RNA folding. The first is the prediction of the folded structures starting from the one-dimensional primary nucleotide sequences. The second is the elucidation of the mechanisms by which they fold. In the past 6 years considerable progress has been made in both aspects of the problem. The determination of the three-dimensional structure of a 160-nucleotide subdomain of *Tetrahymena* group I intron (2, 3), which is the most intensely studied model system for large RNA molecules, has provided considerable insights into the role of discrete metal ion binding sites in directing folding of ribozymes. Similarly, studies from several groups (4–15) have contributed to our understanding of the folding mechanisms of RNA.

A complete understanding of RNA folding and function requires coming to grips with issues that have already been encountered in the study of protein structure and enzymology. To appreciate the common themes that naturally arise in studies of proteins and RNA, it is instructive to compare the properties of their sequence and conformational space. RNA sequences are formed from four nucleotides, whereas proteins are synthesized from 20 naturally occurring amino acids. Despite this difference, it is easy to show that the number of possible sequences for a moderate-sized RNA or a protein is astronomically large. From an evolutionary perspective it is natural to wonder how the dense manifold of sequence space maps onto the sparse structure space. It is believed that the number of distinct folds in proteins is between 1000 and 4000. The tolerance of proteins to multiple mutations and the relatively small number of folds suggest that a given fold may be the native state for a large number of sequences. These arguments have been made precise using simple models for proteins (16). Because estimates of the number of distinct RNA folds are not available, it is not possible to decipher whether a large number of sequences have very similar tertiary folds. For RNA a simple mapping from sequence to structure space may not be straightforward because the restricted chemical diversity of the building blocks (four nucleotides) may lead to a high degree of structural diversity. A recent experiment has shown that a single nucleotide sequence has two functionally different RNA structures (17). Despite the complications in determining RNA structures it is likely that the mapping from the dense sequence space to sparse structure space should be similar in proteins and RNAs.

The conformational space of RNAs and proteins is extremely large because the number of microstates adopted by their building blocks (nucleotide or amino acids) is greater than two. How RNA navigates this large ensemble of unfolded structures in search of the relatively unique native state in finite time may be stated in the form of the Levinthal "paradox" (18). This argument was first made in the context of protein folding. A resolution of this seeming paradox has brought about a deep understanding of the mechanisms by which proteins fold (18). The considerations described above suggest that the folding kinetics of RNA and proteins should have much in common. Using this hypothesis we suggested a conceptual framework for understanding the folding of large RNA molecules (19). The basic prediction of this framework has found experimental support (8, 10, 13).

In this report we present our view of RNA folding kinetics from an energy landscape perspective. Our work reinforces the notion that there are common themes in the folding kinetics of RNA and proteins (5, 19, 20). Despite the expected similarities in the way RNAs and proteins fold there is a fundamental difference, which clearly affects the earliest events in their folding. The driving force for protein folding is the attractive interactions between the hydrophobic residues that cause the initial collapse (21). In RNA each phosphate group is negatively charged,

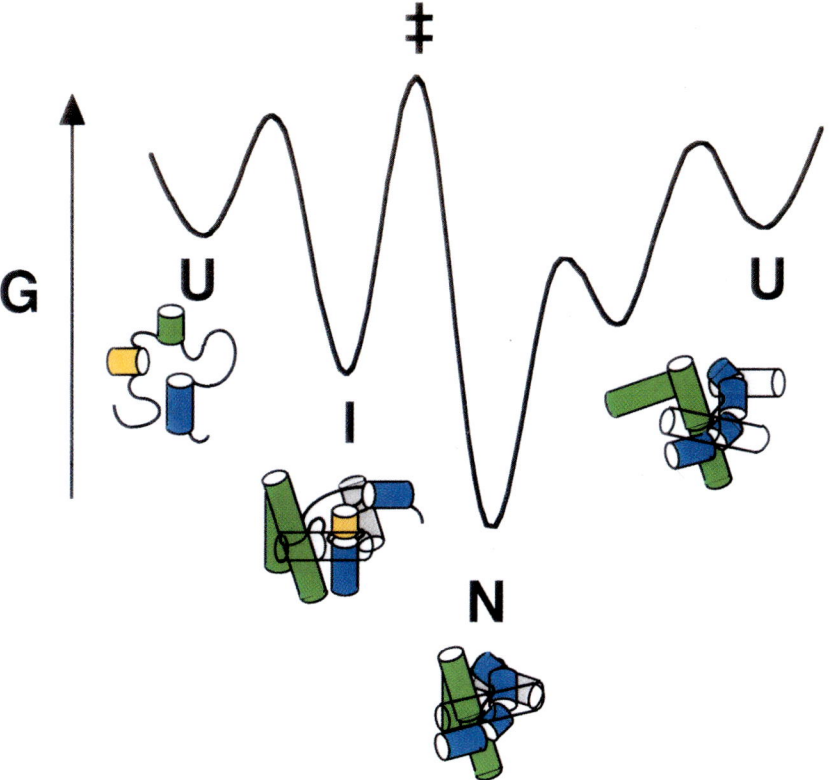

Figure 1 Schematic sketch of the rugged free energy landscape for folding of the *Tetrahymena* ribozyme. The unfolded state is represented as an ensemble of structures in the absence of Mg^{2+}. A small fraction Φ of **U**, whose structures map onto the native state **N**, folds rapidly. The remaining population folds via intermediates (**I**) that are stabilized by native interactions in the P4—P6 domain (*green cylinders*) and misfolded structures in P3 (*yellow*). The slow transition from **I** to **N**, which involves overcoming large activation free energy barrier, requires partial or complete unfolding of the RNA (8).

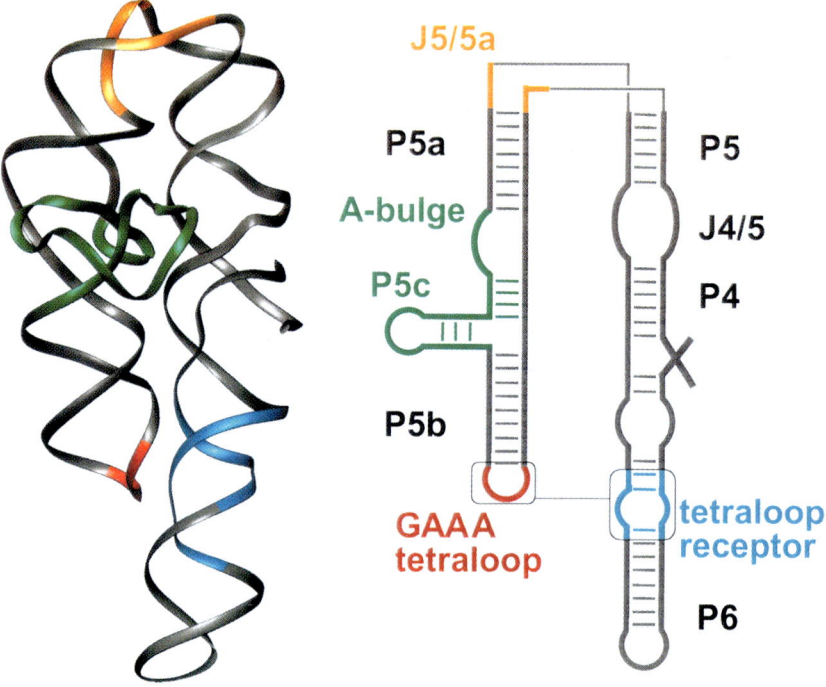

Figure 2 The figure on the right shows a schematic secondary structure of the P4—P6 domain of the *Tetrahymena* ribozyme. The base-paired (P) stems are labeled. On the left a ribbon diagram of the crystal structure of the P4—P6 domain, in the same orientation as the secondary structure, is shown. Comparison of the two gives the locations of the tertiary interactions between the base-paired stems.

so the molecule is a strongly charged polyanion. The initial events must involve substantial neutralization of these charges by counterion condensation. Thus, it is not possible to describe the earliest events in RNA folding without invoking the role of counterions. After describing the overall folding mechanisms we provide a tentative picture of the initial events in RNA folding.

The model system for probing the folding kinetics of large RNA is the *Tetrahymena* ribozyme, an autocatalytic group I intron. This system is an excellent laboratory for probing the folding not only of the intact molecule, but also its subdomains. Throughout this article we use results from this model system (or its variants) to illustrate the fundamental concepts of RNA assembly. Needless to say, our conclusions are general and apply to other large RNA molecules as well.

CONCEPTUAL FRAMEWORK: Kinetic Partitioning Mechanism

We developed a conceptual framework for describing the folding of ribozymes summarized in terms of the kinetic partitioning mechanism (KPM). The underpinnings of our framework are built on the premise that the broad outlines of the way proteins and RNA self-assemble should be similar. The kinetic partitioning mechanism can be understood in terms of the rugged free-energy landscape governing RNA folding. The ruggedness arises for two reasons: (*a*) There are competing interactions in RNA molecules. Attractive interactions (such as base stacking and van der Waals forces) tend to collapse RNA, whereas highly polar and charged moieties are better accommodated by extended structures. Because it is not possible to satisfy all the interactions at every site simultaneously, the molecule is energetically frustrated. (*b*) Due to chain connectivity, there is a high probability of forming local structures to minimize energetic frustration. However, such local structures would be incompatible with the global fold, which requires the formation of specific tertiary contacts. The conflict between local requirements and global considerations, which arises because of the polymeric nature of biomolecules, leads to "topological frustration." If frustration effects are minimized, as is possible for small, well-designed sequences such as tRNA, the free-energy landscape can be relatively smooth. For large RNAs there are many ways of assembling energetically favorable local structures, only one (or few) of which is (are) compatible with the global fold. Some of the incorrectly folded structures can have many aspects in common with the native state. The presence of these low free-energy, native-like misfolded structures could serve as kinetic traps that slow down the folding process.

A schematic sketch of the free-energy surface is shown in Figure 1 (see color insert), from which the basic notions of the KPM can be obtained. Imagine the process, by which an ensemble of unfolded molecules (U) begins to navigate the rugged free-energy landscape in search of the native state. The U state is a collection of structures that rapidly interconvert among each other. However, at the molecular level, the unfolded structures have considerable heterogeneity. As a result, under folding conditions, there is a fraction of molecules, Φ, whose conformations map directly onto the native structure. These molecules fold rapidly without populating any discernible intermediates. The remaining fraction, $1-\Phi$, is trapped by one of the deep local metastable minima (Figure 1, see color insert). Since subsequent rearrangements require activation transitions over free-energy barriers, folding of this set of molecules is slow. Due to the multivalley structure of the free-energy surface (Figure 1, see color insert) the ensemble of initially unfolded molecules partitions into fast folders (Φ being their fraction) and slow folders.

A consequence of KPM is that the time dependence of a fraction of native RNA should be given by

$$f_N(t) = \Phi_{max} - \Phi e^{-k_{fast}t} - \sum_i A_i e^{-k_i t}, \qquad 1.$$

where k_{fast} is the rate for the fast process, k_i is the rate for transitions from the collection of distinct intermediates **I** to **N**, and A_i is the corresponding amplitude satisfying the condition $\sum_i A_i = \Phi_{max} - \Phi$. If the folding reaction goes to completion, as is assumed in the discussions to follow, $\Phi_{max} = 1$.

The partition factor Φ determines whether a high yield of **N** is produced rapidly. The precise value of Φ (which is proportional to the volume of the native basin of attraction (NBA) in the multidimensional free-energy surface) is determined by the structure of the energy landscape. Thus, Φ depends on the sequence and external conditions, such as temperature, ionic strength etc. We expect Φ to depend on mutations as well. This has been found in a recent experiment (7), which shows that a single point mutation (U273A) in *Tetrahymena* ribozyme drastically increases Φ to 80% in the mutant from 2% in the wild type.

Theoretical arguments have shown that the timescale for the fast folding for RNA with M nucleotides $\tau_{fast}(\sim k_{fast}^{-1})$ scales as (23)

$$\tau_{fast} \simeq \tau_0 M^\omega \qquad 2.$$

($\omega \approx 3.8$). The prefactor τ_0, which depends on such factors as viscosity, ionic strength, and monomer diffusion constant, is roughly 10^{-9} s. This formula allows us to estimate τ_{fast} for the fast process within typically an order of magnitude.

KINETICS OF INDEPENDENTLY FOLDING SUBDOMAINS OF *Tetrahymena* RIBOZYME

The framework leading to KPM is also applicable to describe folding of small RNAs (tRNA, for example) as well as independently folding subdomains of the *Tetrahymena* ribozyme. These sequences are expected to fold via two-state kinetics, so that $f_N(t) \simeq 1 - e^{-k_{fast}t}$. Both the P4–P6 domain (Figure 2, see color insert) as well as the three helix P5abc of the *Tetrahymena* group I ribozyme can fold independently. The folding pathways of P5abc depend on the concentration of Mg^{2+}. At low Mg^{2+} concentrations NMR studies showed that the native secondary structure is not stable but becomes so during the tertiary folding at elevated Mg^{2+} concentrations (24). The implication is that at low cation concentrations alternative (misfolded) secondary structures are stabilized. This was noted some time ago in the context of tRNA folding (25). The NMR experiments provide a structural basis for the origin of misfolding at the secondary structure level. At high Mg^{2+} concentrations the native secondary structure is "captured" by tertiary interactions. It has been suggested that the acquisition of the tertiary structure, which forces the formation of native secondary structure in P5abc, occurs by a nucleation-collapse mechanism. The folding nucleus is a substructure, containing discretely bound Mg^{2+} forming the magnesium ion core as well as tertiary interactions involving an A-rich bulge of P5a (26). The 56-nucleotide isolated P5abc is predicted to fold with a time constant of about 20 ms (26), which is consistent with the experimental estimate (24, 27) of 20–35 ms.

The rapid formation of P4–P6 subdomain in the intact *Tetrahymena* ribozyme was established using the time-resolved hydroxyl radical footprinting method. More recently, the same method was used to measure the folding rate of the isolated 160 nucleotides P4–P6 domain in the presence of excess monovalent cations Na^+ (27). The folding time is found to be ~500 ms in the absence of Na^+, and it decreases to 15 ms when 50–100 mM salt is added. This is in a good agreement with time-dependent changes in the fluorescence of a covalently attached pyrene (28). Assuming two-state kinetics our estimate (Equation 2) for folding time for P4–P6 domain (in the absence of Na^+) is $\tau_F \approx \tau_0(l_p)M^\omega \approx 600$ ms. According to this estimate changes in τ_F arise because of the dependence of the prefactor τ_0 on the effective persistence length l_p. For a polyelectrolyte, l_p decreases as C, the concentration of Na^+, increases. Because $\tau_0(l_p) \sim l_p$ we expect τ_F to decrease as C is increased, which is in qualitative accord with experiments (27).

Tetrahymena RIBOZYME FOLDS BY THE KPM

The folding pathways of the *Tetrahymena* ribozyme were first probed by Zarrinkar & Williamson (4) using oligonucleotide hybridization. This ribozyme consists of two major subdomains containing paired (P) regions P4–P6 and P3–P7. As discussed above the stable independently folding subdomain P4–P6 folds in about a second, whereas P3–P7 forms on a much longer timescale (4, 6). The hydroxyl radical footprinting measurements revealed that the peripheral domains (P2, P2.1, and P9.1), which make long-range tertiary interactions, form on an intermediate timescale (6). Thus, under typical experimental conditions (adequate Mg^{2+} and no additional cations), the rate-determining step in the folding of this ribozyme involves proper docking of the P3–P7 domain against the P4–P6 subdomain. Using the kinetics of oligonucleotide hybridization two discrete intermediates were identified. One of them is $\mathbf{I}_1$ (P4–P6 folded) and the other $\mathbf{I}_2$, in which both the major subdomains are nearly formed. According to this hierarchical folding model the **U** state goes through a series of steps (some of which require Mg^{2+}) en route to the native state **N**.

The hierarchical mechanism, resulting in a sequential folding model for RNA, is only valid if Φ is strictly zero. The free-energy landscape perspective predicts that even for a highly frustrated molecule there is a finite probability, however small, that RNA can fold rapidly to the native state. Thus, there must exist a direct pathway from **U** to **N** so that in general, folding of large RNA must involve parallel pathways. Theoretical estimates for the folding timescale for the fast process for $M = 400-650$ is much less than 30 s, making a direct determination of Φ difficult by gel electrophoresis or oligonucleotide hybridization. An estimate of Φ was made by Pan et al (8), who used native polyacrylamide gel containing Mg^{2+} to monitor the time dependence of native state accumulation $f_N(t)$ for $t \geq 1$ min. Accurate double exponential fits of the data for $t > 1$ min showed that f_N ($t = 1$ min) did not coincide with the experimentally measured value. Because of conservation of number of RNA molecules the difference yielded $\Phi \approx 0.08$ for the precursor RNA containing *Tetrahymena* ribozyme (cf Equation 1). Thus, a small fraction (8%) of the molecules reaches the native conformation on timescales less than the earliest time accessible in the gel assay for the native state. This study provided the first experimental demonstration that folding of large RNA can be described in terms of KPM. Subsequent studies have given additional support for our framework (29).

DETERMINATION OF Φ BY SINGLE MOLECULE FLUORESCENCE MEASUREMENTS

The clearest and the most direct evidence that RNA folds by parallel pathways with nonzero Φ comes from the recent single molecule studies of Zhuang et al (29). By attaching fluorescent dyes to the $3'$ and $5'$ ends of the *Tetrahymena* ribozyme, fluorescent resonance energy transfer (FRET) at the single molecule level could be detected. The time-dependent changes in the conformations of the molecule can be probed by measuring the dynamic changes in the FRET signal as a function of the folding conditions. The normalized FRET value in the **U** state is much less than in the **N** state so that time-dependent changes in the FRET signal can be used as a diagnostic for folding. Zhuang et al prepared an initial state, in which RNA is folded, by equilibrating a buffer containing excess Mg^{2+}. At a specified time the solution was diluted and the system was allowed to equilibrate for about 30 s, which is long enough for RNA to fully unfold. This was followed by addition of a buffer containing Mg^{2+}, which subjected the RNA to folding conditions. The method of alternating between folding and unfolding conditions by changing the Mg^{2+} concentration is similar to the NMR H/D exchange method that is used to probe folding kinetics of proteins. Because the mixing time after adding the buffer containing excess Mg^{2+} is much less than τ_{fast}, folding is not due to nonequilibrium effects (such as shear forces, for example). Had τ_{fast} been on the order of milliseconds or less (as is the case in P5abc, for example), then nonequilibrium effects could affect the folding process.

By monitoring the time required for reaching the FRET corresponding to the **N** state the first passage time τ_{1i} for molecule i can be directly measured. Zhuang et al (29) obtained a distribution of first passage times $P_{FP}(t)$ (Figure 3a) by measuring τ_{i1} for several hundred molecules from which the fraction of unfolded molecules, $P_u(t)$, can be obtained using

$$P_u(t) = 1 - \int_0^t P_{FP}(s)ds \qquad 3.$$

On the timescale of the single-molecule experiments the ribozyme (29) folds with two distinct timescales. If the KPM mechanism is valid, we expect $P_u(t)$ to be a sum of at least two exponentials. The time constant for the fastest process can be estimated using Equation 2, which for the 400-nucleotide ribozyme is $\tau_{fast} \approx 7$ s. This is in approximate agreement with the experimental value of 1 s, which is obtained by analyzing the data in (Figure 3a) for $t < 10$ s. From the fits to $P_u(t)$ (Figure 3b) we find that the amplitude of the fastest phase, which is equal to Φ, is the same as that of the slower phase, whose rate constant is $k_1 \simeq 0.016\,\text{s}^{-1}$. Because only 12% of the molecules reach **N** under the conditions of folding, we conclude that $\Phi \approx 0.06$ or 6% of the molecules reach the native state by the fast process. It is noteworthy that the value of Φ is similar for L-21 (29) and for the preRNA (8), which suggests a similarity in the shape of the energy landscape for large RNA molecules.

INITIAL EVENTS IN RNA FOLDING: Overcoming the Electrostatic Repulsion by Counterion-Mediated Collapse

So far we have described events in RNA folding that take place on timescales exceeding at least 1 s. Important folding events that subsequently guide native structure formation occur on subsecond timescales. As in proteins, the earliest events in RNA folding must involve chain compaction. To fold, RNA must overcome the large electrostatic repulsion arising from the Coulomb interactions between the negatively charged phosphate groups. We can estimate the energy due to Coulomb repulsion at temperatures $T > T_m$, the melting temperature. At such temperatures the translational entropy of the counterions is large, and hence they are apparently only weakly attracted to the polyanion. In this case RNA adopts a relatively extended conformation so that the radius of gyration $R_g \sim bM^\nu$ (with $\nu \approx 1$), where b is the Kuhn length. The electrostatic repulsion is $E_R \approx \frac{(Me)^2}{\varepsilon R_g} \approx Mk_BT(\frac{l_B}{b})$, where ε is the dielectric constant and $l_B = \frac{e^2}{\varepsilon k_B T}$ is the Bjerrum length. In water, $l_B \simeq 7$Å. Since b is typically less than l_B, we find that $\frac{E_R}{k_BT} \gg 1$ even for relatively small values of M.

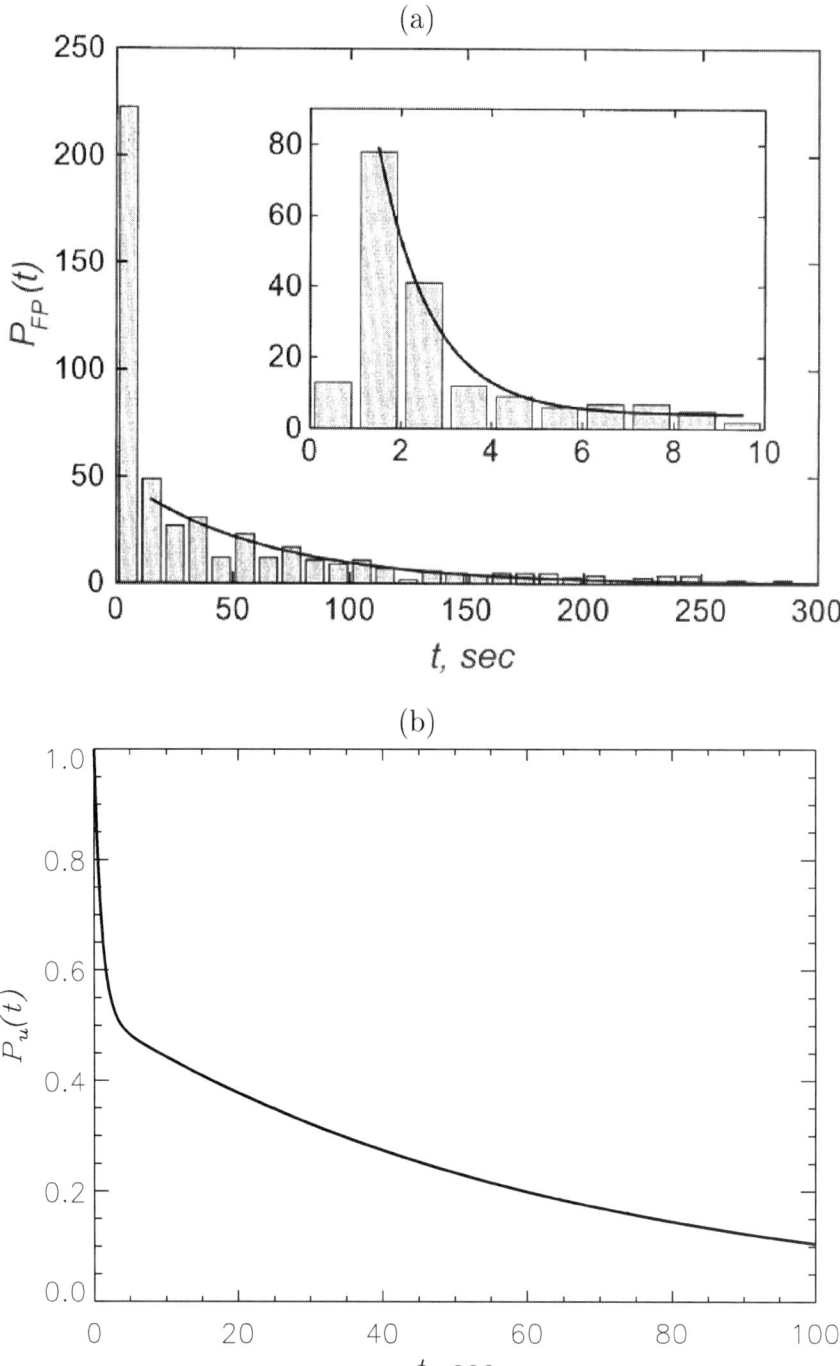

Under folding conditions ($T < T_m$ and a buffer solution containing excess Mg_2^+) the electrostatic interactions are sufficiently softened, because the effective charge on the RNA is drastically reduced from Me. This happens, just as in polyelectrolytes, by counterion (or Manning) condensation (30). At $T < T_m$ the counterions can condense onto the negatively charged phosphate groups of RNA because the gain in the binding energy exceeds TS_{trans}, where S_{trans} is the translational entropy of the counterions. When a sufficiently large fraction of the counterions condenses, the effective charge per phosphate group $q \ll e$. Upon counterion condensation, which typically occurs on a diffusion limited timescale (31), RNA adopts a compact conformation. The extent of collapse is determined by a balance between the renormalized Coulomb repulsion and attractive forces (stacking interactions, van der Waals attraction, etc). The size of the counterion-mediated collapsed state $R_g \approx l_p M^{\frac{1}{3}}$. Small-angle X-ray scattering measurements in RNA seem to suggest that $l_p \approx 7$ Å (32) so that the radius of gyration of the collapsed structure of a 657-nucleotide pre-RNA would be about 60 Å.

The nature of the collapsed states as well as the rates of their formation depend on the pathway. Consider the direct pathway, by which a small fraction Φ of molecules reaches the native conformations. We have proposed that along this pathway the processes of collapse and folding are almost simultaneous (23). Here the counterion-mediated collapse is specific, i.e. in the compact phase $\frac{Q_N}{Q_{NN}} > 1$, where Q_N and Q_{NN} are the number of native and nonnative contacts, respectively. Because the number of all topologically allowed nonnative contacts far exceeds Q_N, it follows that the structures formed by specific collapse are native-like. The probability of forming such native-like collapsed structures becomes smaller as RNA becomes larger, which explains the very small values of Φ observed for the *Tetrahymena* folding. The fast folding pathway occurs by the following steps:

$$U \to I_S \to N, \quad\quad\quad 4.$$

where I_S represents a collection of native-like collapsed structures. The requirement that these structures are native-like implies that the transition from U to I_S may be thermodynamically first order. The rapid conversion of I_S to N also suggests that $\tau_{SC}/\tau_{fast} \leq O(10)$. For the *Tetrahymena* ribozyme we conjecture that the timescale on which I_{SC} forms is on the order of about 0.1 s.

The majority of molecules in the wild-type *Tetrahymena* ribozyme reach the native state through pathways along which one or two discrete intermediates are

---

Figure 3 Distribution of the first passage times $P_{FP}(t)$ is plotted in (*a*) for *Tetrahymena* ribozyme folding studied by single-molecule experiments (29). Each bin gives the number of RNA molecules that reach the native state within a particular time. The inset shows the blowup of $P_{FP}(t)$ distribution on a small timescale. The time dependence of the fraction of unfolded RNAs $P_u(t)$ obtained from $P_{FP}(t)$ (see text for details) is presented in (*b*). The timescales of the fast and slow pathways are approximately 1 s and 63 s, respectively. The amplitudes of both phases are almost equal.

populated (4). Along these pathways the initial step involves a nonspecific collapse of RNA. This step is expected to be very similar to the collapse of strongly charged polyelectrolytes. With this analogy we propose (31) that the timescale for forming the nonspecific collapse structures should be $\tau_{NSC} \approx \tau_0(l_B, z, S)M^\alpha$ (with $\alpha \approx 1$), where z is the valence of counterions. The arguments of τ_0 include the shape S of the counterions. A precise estimate of τ_0 is difficult to make, but we expect it to be between $(10^{-7} - 10^{-6})$s, which implies that $\tau_{NSC} \approx 40$ –$400\,\mu$s for the ribozyme. The slower pathways can be represented as

$$\mathbf{U} \to \mathbf{I}_{NS} \to \{\mathbf{I}_i\} \to \mathbf{N}, \qquad 5.$$

where $\{\mathbf{I}_i\}$ represents a collection of discrete intermediates. It is possible that some of the intermediates convert to **N** only on timescales exceeding hours. Regardless of the timescales in the formation of the intermediates, it is clear that in all pathways the formation of collapsed structures facilitated by Manning condensation, has to be the earliest event in RNA folding.

CONCLUSIONS

By combining the arguments outlined in this article we summarize the folding pathways in large RNA in Figure 4). A hallmark of this scheme is the demonstration (theoretically and experimentally) that a small fraction of the pool of unfolded molecules folds rapidly by a process involving specific collapse. Only the outlines of the predictions sketched in Figure 4 have been validated. Nevertheless, the conceptual framework allows us to formulate precise questions. A few of these are listed below.

(a) For two-state folders (P5abc and P4–P6 for example) with $\Phi \approx 1$ the structures of the transition states connecting **U** and **N** are not known. The importance of polyelectrolyte interactions in RNA has been used to suggest that the transition states in small RNA molecules should be structurally homogeneous (19). A recent experiment supports this prediction (33). However, additional work is required to probe the TS structures in small RNA molecules. The nature of the transition states connecting $\mathbf{I}_i$ and **N** is to a large extent unknown. RNA engineering experiments will be necessary to provide the structure of these elusive states.

(b) The KPM predicts that point mutations can alter the value of Φ. In the mutant U273A about 80% of the molecules fold in fewer than 30 s. This shows that profound changes in Φ can be achieved by a simple point mutation. Further experiments are needed to understand quantitatively the way Φ depends on the sequence and external conditions.

(c) The prediction that the earliest event in RNA folding must involve the collapse, following counterion condensation, suggests that the pathways,

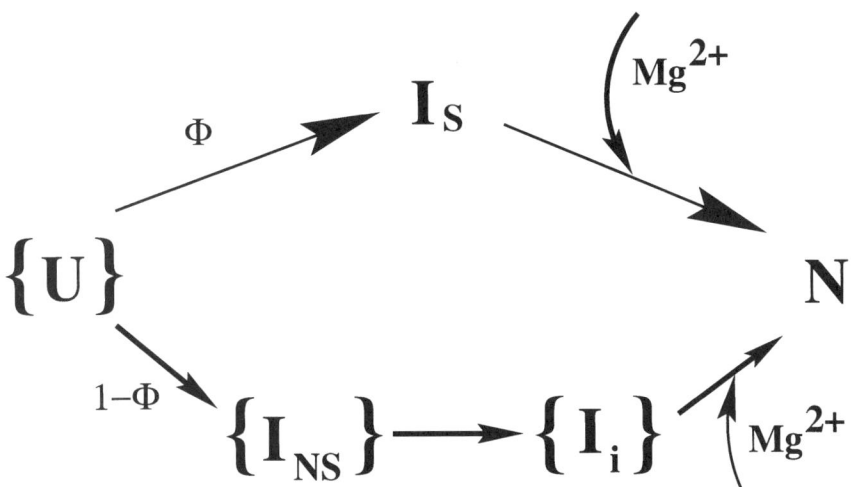

Figure 4 The sketch of the folding pathways for RNA. The fast (upper) folding pathway includes the formation of native-like collapsed states I_S, which rapidly convert into the native state **N**. The fraction of RNA molecules that fold via this pathway is Φ. The lower track (followed by the fraction $1 - \Phi$ of RNAs) represents the slower pathways that involve several stages. First, nonspecific collapse species I_S form that later convert into a collection of discrete intermediates $\{I_i\}$. The transition from $\{I_i\}$ to the native state is slow and represents the rate-limiting step in the slow pathway. The steps requiring binding of magnesium ions Mg^{2+} are also indicated. There could be bifurcations in the pathways leading from $\{I_i\}$ to **N**. The degree of heterogeneity in the folding pathways depends on the sequence and external conditions.

mechanisms, and timescales must depend on the valence and shape of the counterions. Thus, a complete understanding of RNA folding requires a systematic study of RNA folding kinetics by changing the characteristics of counterions. Because the initial counterion-mediated events occur rapidly, the folding pathways of RNA can only be fully mapped out using fast-folding experiments. Finally, our scheme also suggests that the folding pathways must be heterogeneous. A quantitative validation will require experiments at the single molecule level.

Visit the Annual Reviews home page at www.AnnualReviews.org

LITERATURE CITED

1. Cech TR. 1993. In *The RNA World*, ed. RF Gesteland, JF Atkins, pp. 239–69. Plainview, NY: Cold Spring Harbor Lab. Press
2. Cate JH, Gooding AR, Podell E, Zhou K, Golden BL, et al. 1996. *Science* 277:1178–85
3. Batey RT, Rambo RP, Doudna JA. 1999. *Angew Chem. Int. Ed. Eng.* 38:2337–43

4. Zarrinkar PP, Williamson JR. 1994. *Science* 265:918–24
5. Zarrinkar PP, Williamson JR. 1996. *Nat. Struct. Biol.* 3:432–38
6. Scalvi B, Sullivan M, Chance MR, Brenowitz M, Woodson SA. 1998. *Science* 279:1940–43
7. Pan J, Deras ML, Woodson SA. 2000. *J. Mol. Biol.* 296:133–44
8. Pan J, Thirumalai D, Woodson SA. 1997. *J. Mol. Biol.* 273:7–13
9. Treiber DK, Rook MS, Williamson JR. 1998. *Science* 279:1943–46
10. Rook MS, Treiber DK, Williamson JR. 1998. *J. Mol. Biol.* 281:609–20
11. Pan J, Woodson SA. 1998. *J. Mol. Biol.* 280:597–609
12. Sosnick TR, Pan T. 1997. *Nat. Struct. Biol.* 4:931–38
13. Pan T, Fang XW, Sosnick TR. 1999. *J. Mol. Biol.* 286:721–31
14. Treiber DK, Williamson JR. 1999. *Curr. Opin. Struct. Biol.* 9:339–45
15. Thirumalai D, Woodson SA. 2000. *RNA* 6:790–94
16. Thirumalai D, Klimov DK. 2000. In *Stochastic Dynamics and Pattern Formation in Biological and Complex Systems*, ed. S Kim, KJ Lee, W Sung, pp. 95–111. Melville, NY: Am. Inst. Phys.
17. Schultes EA, Bartels DP. 2000. *Science* 289:448–52
18. Dill KA, Chan HS. 1997. *Nat. Struct. Biol.* 4:10–19
19. Thirumalai D, Woodson SA. 1996. *Acc. Chem. Res.* 29:433–39
20. Thirumalai D, Klimov DK, Woodson SA. 1997. *Theor. Chem. Acc.* 96:14–22
21. Dill KA. 1990. *Biochemistry* 29:7133–35
22. Deleted in proof
23. Thirumalai D. 1995. *J. Phys. I* 5:1457–67
24. Wu M, Tinoco I. 1998. *Proc. Natl. Acad. Sci. USA* 95:11555–60
25. Fresco JR, Adams A, Ascione R, Henley D, Linahl T. 1966. *Cold Spring Harbor Symp. Quant. Biol.* 31:527–37
26. Thirumalai D. 1998. *Proc. Natl. Acad. Sci. USA* 95:11506–8
27. Deras ML, Brenowitz M, Ralston M, Chance MR, Woodson SA. 2000. *Biochemistry* 35:10975–85
28. Silverman SK, Deras ML, Woodson SA, Scaringe SA, Cech TR. 2000. *Biochemistry* 39:12465–75
29. Zhuang X, Bartley LE, Babcock AP, Russell R, Ha T, et al. 2000. *Science* 288:2048–51
30. Manning GS. 1977. *Biophys. Chem.* 7:189–92
31. Lee N, Thirumalai D. 2000. *J. Chem. Phys.* 113:5126–29
32. Russell R, Millett IS, Doniach S, Herschlag D. 2000. *Nat. Struct. Biol.* 7:367–70
33. Maglott EJ, Goodwin JJ, Glick GD. 1999. *J. Am. Chem. Soc.* 7461–62

LASER-INDUCED POPULATION TRANSFER BY ADIABATIC PASSAGE TECHNIQUES

Nikolay V Vitanov[1], Thomas Halfmann[2], Bruce W Shore[3], and Klaas Bergmann[2]

[1]*Helsinki Institute of Physics, PL 9, 00014 University of Helsinki, Finland*
e-mail: nikolay.vitanov@helsinki.fi
[2]*Fachbereich Physik, Universität Kaiserslautern, 67653 Kaiserslautern, Germany*
[3]*Lawrence Livermore National Laboratory, Livermore, California 94550*

Key Words stimulated Raman adiabatic passage, level crossing, coherent excitation, Landau-Zener model

■ **Abstract** We review some basic techniques for laser-induced adiabatic population transfer between discrete quantum states in atoms and molecules.

1. INTRODUCTION

Atoms and molecules prepared in well-defined quantum states are instrumental in modern atomic and molecular physics, not only in traditional studies of state-to-state collision dynamics or laser-controlled chemical reactions, but also in various new areas, e.g. atom optics and quantum information. The development of efficient schemes for selective population transfer, such as chirped-pulse excitation and stimulated Raman adiabatic passage (STIRAP), have opened new opportunities for coherent laser control of atomic and molecular processes. This article overviews a variety of techniques, with associated theory, whereby one can accomplish complete population transfer in atoms and molecules by means of adiabatic time evolution induced by properly crafted laser pulses.

The central idea of population transfer is to begin with an atom or molecule in which the internal structure is in a specified discrete quantum state and then, by exposing this system to a controlled pulse of radiation, force the internal structure into a desired target state. Only a radiation field, suitably monochromatized and near-resonant with the atomic Bohr frequency, can provide the selectivity needed to isolate a single final state. A goal of theory is to predict, for a specified set of

---

[1]Also at Department of Physics, Sofia University, James Boucher 5 Blvd., 1126 Sofia, Bulgaria and Institute of Solid State Physics, Bulgarian Academy of Sciences, Tzarigradsko shaussee 72, 1784 Sofia, Bulgaria.

radiation pulses, the probability that atoms will undergo a transition between the initial state and the desired target state. Alternatively, theory should provide a prescription for pulses that will produce a desired population transfer.

The population transfer techniques described in this review have in common the need for coherent laser radiation. To emphasize the qualitative importance of coherence in the radiation field, we first summarize some basic results obtainable with incoherent light, such as that from filtered atomic vapor lamps or from broadband lasers with poor coherence properties.

1.1 Incoherent Population Transfer Schemes

1.1.1 Incoherent Excitation of Two-State Systems   One of the simplest theoretical descriptions of near-resonant excitation of a two-state atom or molecule by incoherent radiation is credited to Einstein, who first postulated that the rate of change in an atomic population within a blackbody cavity is proportional to radiation energy density (1); the Einstein B coefficient quantifies this proportionality. The resulting rate equations (cf (2); Section 2.2) include not only transitions induced by any experimentally controlled radiation field, but also the possibility of spontaneous emission of radiation, as quantified by the Einstein A coefficient ($1/A$ is the spontaneous emission lifetime). If the atoms are initially (at time $t \to -\infty$) unexcited, if the radiation is sufficiently intense that spontaneous emission has a negligible effect on the population (the saturated regime), and if the excited level has the same degeneracy as the initial level, then the excited-state population at time t is

$$P_e(t) = \tfrac{1}{2}[1 - e^{-BF(t)}], \qquad 1.$$

where $F(t) = \int_{-\infty}^{t} I(t')dt'$ is the pulsed radiation fluence (energy per unit area) up to time t, $I(t)$ being the time-varying laser intensity (power per unit area). As this expression shows, when the pulse fluence increases, the excited-state population approaches monotonically the saturation value of 50%, which is the best transfer efficiency one can achieve with incoherent light. Of course, once the radiation ceases, the atom must return to lower-lying levels by spontaneously emitting radiation, and eventually no excitation will remain.

1.1.2 Optical Pumping   Although spontaneous emission hinders direct creation of excitation, it can be used to advantage. Consider the excitation linkage shown in Figure 1a: an initially populated state ψ_1, resonantly excited to state ψ_2, from which spontaneous emission occurs, not only returning to state ψ_1 but possibly also to some third state ψ_3 whose energy lies far from resonance with the initial state or which is prevented, by selection rules based on polarization of the light, from being excited by the intense radiation. Every time the atom is excited from state ψ_1 to state ψ_2 by absorbing a photon there is a chance that spontaneous emission will carry the population to state ψ_3, after which the atom will be immune to further action by the radiation. The resulting population transfer, optical pumping, will eventually place the entire population into state ψ_3.

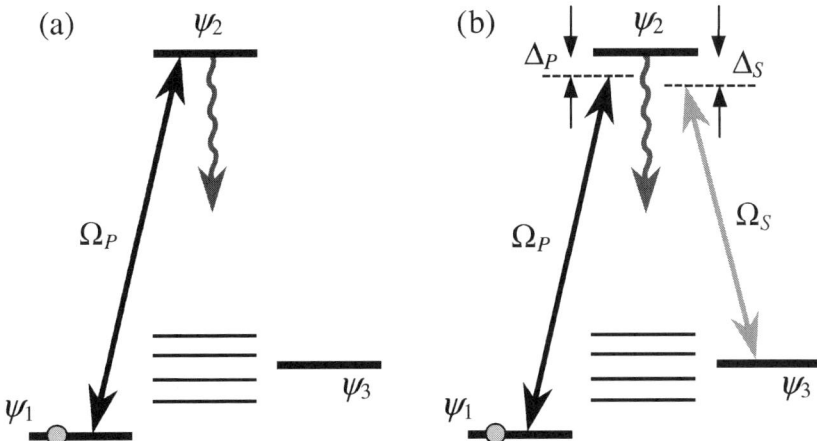

Figure 1 (*a*) Linkage diagram for optical pumping. A pump field (not necessarily a laser) excites state ϕ_2, which spontaneously decays back to state ϕ_1, to state ϕ_3, or possibly to some other states. (*b*) Linkage diagram for stimulated emission pumping. A pump field populates state ϕ_2, and a subsequent Stokes (or dump) field populates state ϕ_3.

The simplicity of optical pumping, which requires only a single light source (not necessarily a laser), has made it a widely used method for preparing atoms or molecules in a well-defined ground or metastable state. Its main limitation is the lack of selectivity: Because the population arrives in the target level by spontaneous emission, it will simultaneously arrive in all levels into which the excited state can decay. Therefore, the pumping procedure will generally populate a statistical mixture of possible final states, not a single state. The distribution of final populations is determined by the relative decay rates that link each final state with the excited state. For vibrational transitions, these rates are proportional to Franck-Condon factors; because the latter rarely exceed 10%, the selectivity is correspondingly low.

1.1.3 Stimulated Emission Pumping Optical pumping uses only a single light source, acting on the pump transition. The overall transfer, however, involves two photons: absorption of a pump photon from the imposed light source and spontaneous emission of a Stokes photon. The names pump and Stokes serve notice that we are dealing with a two-photon Raman process. It is natural to consider a Raman-type process in which both fields are externally supplied (a stimulated Raman process), so that the final step proceeds to a selected final state rather than to a statistical distribution.

One variant of the suggested two-photon process of population transfer, stimulated emission pumping (SEP) (2), uses a pump field to place a population from the initial state ψ_1 into the excited state ψ_2, followed some time later by a Stokes (or dump) field that transfers the population into the desired final state ψ_3 (hence the names pump and dump), as depicted by the Λ-type linkage in Figure 1b. If the intensity of the pump laser is sufficiently strong to saturate the $\psi_1 \leftrightarrow \psi_2$ transition, then, as suggested by the rate equations, the pump step can transfer at most 50% of the population from state ψ_1 to state ψ_2. If the Stokes laser is also sufficiently strong to saturate its transition, then half the population in state ψ_2 will be further transferred to the target state ψ_3. Hence, at most one quarter of the population is transferred to the target state, while one half of the population remains in the initial state. The remaining quarter is distributed statistically according to the branching of spontaneous emission from state ψ_2. The SEP efficiency can be improved slightly if the pump and Stokes pulses are applied simultaneously, rather than successively. If they are sufficiently strong to saturate the transitions, thereby equalizing the populations, then one third of the population can be dispatched to ψ_3.

Because of its simplicity, the SEP technique has enjoyed widespread application in collision dynamics and spectroscopy (3, 4). Its main limitations are low efficiency and low selectivity. Typically, a transfer efficiency of 10% is rarely exceeded, but this is quite adequate for many spectroscopic studies.

1.2 Resonant Coherent Excitation: Rabi Oscillations

The response of a quantum system to coherent (laser) radiation differs in significant and qualitative ways from the response of the same system to light from a lamp, even a very monochromatic lamp. Whereas the sudden application of incoherent radiation to an atom or molecule typically results in a monotonic approach to some equilibrium excitation, the sudden application of steady laser radiation typically produces oscillatory populations. These differences are most clearly seen in the study of two-state systems.

1.2.1 Two-State Systems When coherent radiation is resonant (the carrier frequency equal to the Bohr frequency) and steady, the excitation oscillates sinusoidally (a behavior known as Rabi oscillations),

$$P_e(t) = \tfrac{1}{2}(1 - \cos \Omega t). \qquad 2.$$

The frequency of population oscillation Ω is known as the Rabi frequency; it will be associated below with the strength of the interaction. When the radiation varies in amplitude, the cosine argument is replaced by the so-called pulse area $A(t)$,

$$\Omega t \to \int_{-\infty}^{t} \Omega(t')dt' = A(t). \qquad 3.$$

Unlike the monotonic behavior for incoherent excitation, here the excited-state population oscillates between 0 and 1, depending on the value of $A(t)$. For pulse areas equal to odd multiples of π (odd-π pulses), a complete population transfer to the excited state takes place, whereas for pulse areas equal to even multiples of π (even-π pulses), the system returns to the initial state.

In practice, the sample of atoms or molecules consists of an ensemble, often with a distribution of velocities. Atoms moving in the direction of a traveling wave experience a Doppler shift, so that their excitation is not exactly resonant; their population oscillations are more rapid and have smaller peak values than those of resonant atoms. Atoms moving transversely to a laser beam will experience a pulse area dependent on the duration of their transit time across the beam. These velocity-dependent interactions and the presence of fluctuations in the laser intensity require an averaging over excitation probabilities. The result is an effective excitation probability that has less pronounced oscillations; in extreme cases the averaging can bring the excitation probability to 0.5, the same as with incoherent excitation.

1.2.2 Three-State Systems Rabi cycling is not confined to two-state systems: It can be found in multi-state systems too. One example is a coherently driven three-state system subjected to the same pulse sequence as in SEP: the pump pulse first, followed after its completion (without overlap) by the Stokes pulse. Then the excitation can still be considered as a two-step process, but the probabilities for each step are different from those in SEP. In the case of exact single-photon resonances, the transition probabilities P_{12} from state ψ_1 to state ψ_2 and P_{23} from state ψ_2 to state ψ_3 are

$$P_{12} = \tfrac{1}{2}(1 - \cos A_P), \quad P_{23} = \tfrac{1}{2}(1 - \cos A_S), \qquad 4.$$

where A_P and A_S are the pump and Stokes pulse areas. If the system is initially in state ψ_1, then the population of state ψ_3 after the excitation is the product of the two probabilities,

$$P_{13} = \tfrac{1}{4}(1 - \cos A_P)(1 - \cos A_S). \qquad 5.$$

Hence, for suitably chosen pulse areas (both equal to odd multiples of π) there is complete population transfer from state ψ_1 to state ψ_3. However, the transfer efficiency depends strongly on the pulse areas, and it can even vanish (when A_P or A_S are even multiples of π).

When the pump and Stokes pulses have the same time dependence, the interaction dynamics can no longer be separated into two consecutive, independent two-state transitions. However, an exact solution can still be derived. If the system is initially in state ψ_1, and the two lasers are each resonantly tuned, then the population of state ψ_3 at the end of the excitation is

$$P_{13} = \frac{A_P A_S}{A^2}(1 - \cos \tfrac{1}{2} A), \qquad 6.$$

where $A = \sqrt{A_P^2 + A_S^2}$. Here again the transfer efficiency depends on the pulse areas: Complete population transfer from state ψ_1 to state ψ_3 occurs when $A = 2(2k+1)\pi$ ($k = 0, 1, 2, \ldots$) and $A_P = A_S$, whereas complete population return to the initial state ψ_1 takes place when $A = 4k\pi$.

As for two-state Rabi oscillations, the presence of a distribution of velocities or intensity fluctuations will tend to average out the oscillations and to lower the transfer efficiency. Moreover, because the population passes through the intermediate state ψ_2, inevitable population losses will take place by spontaneous emission unless the excitation time is much shorter than the lifetime of ψ_2.

The Rabi cycling is but one of the ways in which coherent laser pulses can induce population changes. Another class of change, adiabatic evolution, which is the focus of this review, can occur when the Hamiltonian changes slowly. The next section describes the basic principles of adiabatic population transfer by using a level crossing, and the following section describes adiabatic population transfer by sequential laser pulses.

2. ADIABATIC POPULATION TRANSFER VIA A LEVEL CROSSING

2.1 Two-State Systems

2.1.1 Coherent Excitation For coherent laser excitation the correct description of the interaction dynamics is provided by the time-dependent Schrödinger equation (2),

$$i\hbar \frac{d}{dt}\mathbf{C}(t) = \mathsf{H}(t)\mathbf{C}(t). \qquad 7.$$

$\mathbf{C}(t) = [C_1(t), C_2(t)]^T$ is a column vector whose elements are the (complex-valued) probability amplitudes of the two states ψ_1 and ψ_2, and $\mathsf{H}(t)$ is the Hamiltonian of the atom-radiation system. The diagonal elements of $\mathsf{H}(t)$ are the energies of the two states E_1 and E_2, and the off-diagonal elements contain the laser-atom dipole interaction energy. Because the populations $P_n(t) = |C_n(t)|^2$ depend only on the magnitudes of the probability amplitudes, there is some leeway in choosing their phases. It proves particularly convenient to incorporate the state energies as phases. Then the off-diagonal element of the Hamiltonian matrix, which is proportional to the sinusoidally oscillating laser electric field, is multiplied by a phase factor oscillating at the Bohr transition frequency $\omega_0 = (E_2 - E_1)/\hbar$. Because the laser frequency ω is equal or very close to the transition frequency ω_0, the off-diagonal Hamiltonian element can be represented as a sum of two terms: one term oscillating rapidly at nearly twice the transition frequency ($\omega_0 + \omega$) and another term oscillating slowly at frequency $\Delta = \omega_0 - \omega$, the atom-laser detuning. Unless the laser pulse is very short (e.g. a femtosecond pulse) or very intense, the rapidly oscillating term can be neglected; this is known as the rotating-wave approximation (RWA). When this is done, the Hamiltonian reads (2)

$$\mathsf{H}(t) = \hbar \begin{bmatrix} 0 & \frac{1}{2}\Omega(t) \\ \frac{1}{2}\Omega(t) & \Delta(t) \end{bmatrix}. \qquad 8.$$

The off-diagonal element $\Omega(t)$, known as the Rabi frequency, parameterizes the strength of the atom-laser interaction; it is proportional to the atomic transition dipole moment $\mathbf{d}_{12}$ and the laser electric field amplitude $\mathcal{E}(t)$, i.e. $\Omega(t) = \mathbf{d}_{12} \cdot \mathcal{E}(t)/\hbar$. We assume for simplicity and without loss of generality that $\Omega(t)$ is real-valued and positive [its phase can always be attached to one of the probability amplitudes (2)]. The diagonal elements of H(t) are the two RWA energies: The zero element is the energy of state ψ_1 lifted (dressed) by the photon energy $\hbar\omega$ and used as the reference energy level, and Δ is the frequency offset (detuning) of state ψ_2.

2.1.2 Adiabatic States and Adiabatic Following Theoretical discussion of time-evolving quantum systems is greatly facilitated by introducing a special coordinate system in the abstract space in which the state vector $\Psi(t)$ moves. The basis vectors in this moving coordinate system are the instantaneous eigen states $\Phi_k(t)$ of the Hamiltonian H(t)—the adiabatic states. These states are time-dependent superpositions of the unperturbed states ψ_1 and ψ_2 (also known as the diabatic states) (2),

$$\Phi_+(t) = \psi_1 \sin\Theta(t) + \psi_2 \cos\Theta(t), \qquad 9.$$

$$\Phi_-(t) = \psi_1 \cos\Theta(t) - \psi_2 \sin\Theta(t), \qquad 10.$$

where the mixing angle $\Theta(t)$ is defined (modulo π) as $\Theta(t) = \frac{1}{2}\arctan[\Omega(t)/\Delta(t)]$. The energies of the adiabatic states are the two eigenvalues of H(t),

$$\hbar\varepsilon_\pm(t) = \frac{1}{2}\hbar\left[\Delta(t) \pm \sqrt{\Delta^2(t) + \Omega^2(t)}\right]. \qquad 11.$$

When the evolution is adiabatic no transitions between the adiabatic states occur. Mathematically, adiabatic evolution requires that the coupling between the adiabatic states is negligible compared with the difference between their eigen frequencies (2, 5, 6), i.e.,

$$|\langle\dot{\Phi}_+|\Phi_-\rangle| \ll |\varepsilon_+ - \varepsilon_-|, \qquad 12.$$

where the dot denotes a time derivative. Explicitly, the two-state adiabatic condition reads

$$\tfrac{1}{2}\left|\dot{\Omega}\Delta - \Omega\dot{\Delta}\right| \ll (\Omega^2 + \Delta^2)^{3/2}. \qquad 13.$$

Hence, adiabatic evolution requires a smooth pulse, long interaction time, and large Rabi frequency and/or large detuning. When the adiabatic condition holds there are no transitions between the adiabatic states, and their populations are conserved. In particular, if the system state vector $\Psi(t)$ coincides with a single adiabatic state $\Phi(t)$ at some time t, then it will remain in that adiabatic state as long as the evolution is adiabatic: The state vector $\Psi(t)$ will adiabatically follow the adiabatic state $\Phi(t)$. Of course, the relationship of the single adiabatic state $\Phi(t)$ to the diabatic states will change if the mixing angle $\Theta(t)$ changes, and so adiabatic evolution can produce population transfer between those diabatic states.

2.1.3 Rapid Adiabatic Passage There are two distinct types of adiabatic population changes depending on the behavior of the diabatic energies 0 and $\hbar \Delta(t)$ of the Hamiltonians. The no-crossing case is depicted in Figure 2 (*top left frame*) in the particular case of constant detuning; the diabatic energies are parallel to each other. In the absence of interaction, the adiabatic energies coincide with the diabatic ones, but the (pulsed) interaction $\Omega(t)$ pushes them away from each other. As Equation 9 and 10 show, at early and late times each adiabatic state is identified with the same diabatic state: $\Phi_-(t \to \pm\infty) = \psi_1$, $\Phi_+(t \to \pm\infty) = \psi_2$, whereas at intermediate times it is a superposition of diabatic states. Consequently, starting from the ground state ψ_1, the population makes a partial excursion into the excited state ψ_2 at intermediate times and eventually returns to ψ_1 in the end (bottom left frame). Hence, in the no-crossing case adiabatic evolution leads to complete population return.

A rather different situation occurs when the detuning $\Delta(t)$ sweeps slowly from some very large negative value to some very large positive value (or vice versa), as shown in Figure 2 (*top right frame*). That is, the Hamiltonian at the end of

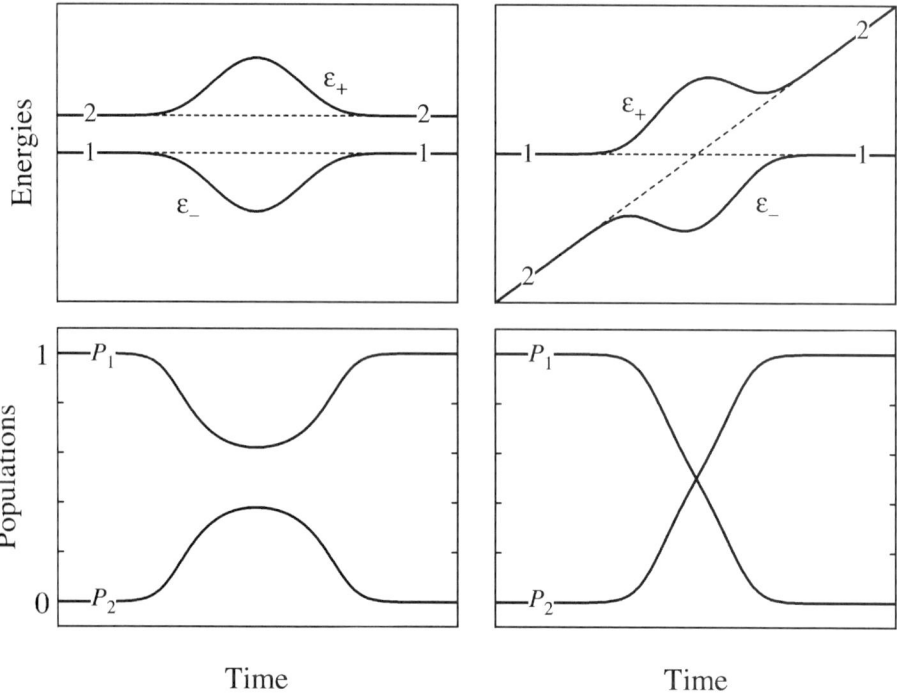

Figure 2 Time evolution of the energies (*upper frames*) and the populations (*lower frames*) in a two-state system. In the upper plots, the dashed lines show the unperturbed (diabatic) energies, and the solid curves show the adiabatic energies. The *left-hand frames* are for the no-crossing case, and the *right-hand frames* are for the level-crossing case.

the pulse differs from the Hamiltonian at the beginning because of the detuning change. Large in this context means much larger than the Rabi frequency $\Omega(t)$. The two diabatic energies 0 and $\hbar\Delta(t)$ intersect at time t_0 when the detuning is zero. The adiabatic energies approach the diabatic energies when $\Delta(t)$ is large (at early and late times), but the presence of interaction prevents their intersection— the adiabatic energies have an avoided crossing. Indeed, as Equation 10 shows, the eigen energy splitting $\hbar\varepsilon_+(t) - \hbar\varepsilon_-(t) = \hbar\sqrt{\Delta^2(t) + \Omega^2(t)}$ is equal to $\hbar\Omega(t_0)$ at the crossing. For constant $\Omega(t)$, this is the minimum value of the splitting, whereas for pulse-shaped $\Omega(t)$ (as in Figure 2), there are two minima near t_0. At extremely early and late times the ratio $\Delta(t)/\Omega(t) \xrightarrow{t\to\pm\infty} \pm\infty$. Hence, during the excitation the mixing angle $\Theta(t)$ rotates clockwise from $\Theta(-\infty) = \pi/2$ to $\Theta(+\infty) = 0$, and the composition of the adiabatic states changes accordingly. Asymptotically, each adiabatic state becomes uniquely identified with a single unperturbed state,

$$\psi_1 \xrightarrow{-\infty \leftarrow t} \Phi_+(t) \xrightarrow{t\to+\infty} \psi_2, \qquad 14.$$

$$-\psi_2 \xleftarrow{-\infty \leftarrow t} \Phi_-(t) \xleftarrow{t\to+\infty} \psi_1. \qquad 15.$$

Consequently, starting from state ψ_1, the system follows adiabatically the adiabatic state $\Phi_+(t)$ and eventually ends up in state ψ_2. The laser pulse, with detuning sweep, has produced complete population transfer, a process known as adiabatic passage or, because it must occur in a time shorter than the radiative lifetime of the excited state, as rapid adiabatic passage. We emphasize that adiabatic passage in a two-state system does not depend on the sign of the detuning slope: It takes place for both $\dot\Delta(t) > 0$ (as was assumed above) and $\dot\Delta(t) < 0$.

Adiabatic passage offers significant advantages over Rabi cycling as a means of producing complete population transfer in an ensemble of atoms. Unlike Rabi cycling, adiabatic passage is robust against small-to-moderate variations in the laser intensity, detuning, and interaction time. Therefore, it can produce uniform excitation for a broad range of Doppler shifts.

2.1.4 Estimating Transition Probabilities

A popular tool for estimating the transition probability between two crossing diabatic states is the Landau-Zener formula (7, 8),

$$P = 1 - p, \quad p = \exp\left[-\frac{\pi\Omega^2(t_0)}{2|\dot\Delta(t_0)|}\right], \qquad 16.$$

where $\dot\Delta(t_0)$ is the rate of change in the detuning evaluated at the crossing time t_0 and $\Omega(t_0)$ is the value of the Rabi frequency at t_0. This formula is exact only for a constant Rabi frequency and a linearly varying detuning over an infinite time interval, so it is only an approximation to any actual adiabatic passage. Nevertheless, it correctly identifies the importance of the ratio of Ω^2 to $\dot\Delta$ as a measure of the likelihood of population transfer. We note here that the probability for (nonadiabatic) transition between the adiabatic states is $p = 1 - P$.

There exist other, more realistic, models of level-crossing excitation. The Allen-Eberly-Hioe model (9, 10) assumes a hyperbolic-secant pulse and a hyperbolic-tangent chirp. The Demkov-Kunike model (11, 12) adds a static detuning to this model.

2.1.5 Experimental Demonstration Population transfer by means of adiabatic passage was first used in nuclear magnetic resonance (13, 14), where the diabatic energies include interactions between nuclear magnetic moments and a magnetic field. The needed diabatic energy crossings are created by slowly changing either the frequency or the direction of an oscillating magnetic field.

Implementation of adiabatic passage in the optical region was suggested in the late 1960s (15). The process was demonstrated in the 1970s, during the early days of the application of laser radiation in atomic and molecular physics, with both one-photon (16–19) and two-photon transitions (20–22). The first demonstration of adiabatic following in laser excitation was achieved by Loy (16) in NH_3 with a fixed-frequency infrared laser; in this work a level crossing was created by Stark-shifting the transition frequency. Hamadani et al (23) observed adiabatic following in NH_3 by sweeping the laser frequency through resonance.

In the near infrared, adiabatic passage was observed by Avrillier et al (24) and Adam et al (25) [see also (26)], whereas Kroon et al (27) and Lorent et al (28) reported adiabatic following with visible light. In these experiments an atomic or molecular beam crossed a focused laser beam at right angles. If the molecules crossed the laser beam near its center, oscillations were observed in the excited-state population as the pulse area changed, as a result of either changing laser intensity (25) or changing time of flight (27). However, if the laser beam was focused off the molecular-beam axis, then the molecules experienced a curved wave front, which appeared to them as a time-varying Doppler shift (24, 25, 27). The detuning from resonance in this case was $\Delta(t) = v^2 \omega t / cr$, v being the longitudinal molecular velocity and r the wave front curvature. This effective chirp enabled adiabatic passage.

Various techniques are currently available for imposing a controlled frequency chirp on the laser field itself. Active electrooptic or acoustooptic modulation techniques provide straightforward means for sweeping the frequency of pulsed radiation with pulse length exceeding a microsecond. It is also straightforward to impose a frequency chirp on picosecond and femtosecond pulses (29–38). The large bandwidth of such ultrashort pulses allows spatial dispersion of the spectrum and subsequent time-delayed spatial recombination of the various frequency components. For example, adiabatic passage has been demonstrated by using picosecond pulses chirped by reflecting them off a suitably arranged pair of parallel gratings, tilted with respect to each other (39–49). The first grating separates the frequency components spatially, and the second one recombines them into a parallel beam. Because the optical path length between the gratings is frequency dependent, a chirp dependent on the path difference is generated. Such a setup can provide an almost-linear frequency chirp. Self-phase modulation (29, 30, 50) is another technique for frequency chirping of picosecond pulses. For femtosecond pulses it suffices to propagate the laser beam through dispersive material, e.g. a thin molecular jet or a prism. For longer pulses a short length of optical fiber can be used (51). The self-phase modulation usually induces a phase shift $\varphi(t)$ proportional to the laser intensity, i.e. to $\Omega^2(t)$ (29, 30, 50). Hence, the detuning, which is proportional to $d\varphi(t)/dt$ (and therefore vanishes at early and late times), crosses resonance at the time when $\Omega(t)$ is maximum.

2.2 Stark-Chirped Rapid Adiabatic Passage

Techniques for producing frequency-swept pulses are not well developed for laser pulses of nanosecond duration. Nanosecond laser systems are used for many applications because they provide a very good combination of sufficiently high intensity (and hence large interaction strength) as well as long interaction time. The successful implementation of adiabatic evolution of laser-matter interaction relies on a combination of both parameters. Laser frequency chirping by active phase modulation, although possible in principle, is difficult for nanosecond laser pulses because it requires driving modulators at GHz frequencies. Alternatively, the spectral bandwidth of nanosecond pulses is too small for successful application of the techniques based on spatial dispersion that are well developed for femtosecond pulses. It appears feasible, however, to use laser-induced Stark shifts to modify the transition frequency.

2.2.1 Theory It has been demonstrated recently (52, 53) that laser-induced Stark shifts can be used as a powerful tool for achieving efficient population transfer in two-state systems. This technique— Stark-chirped rapid adiabatic passage (SCRAP)—utilizes two sequential laser pulses. Figure 3 illustrates these ideas. One of the pulses—the pump pulse—is slightly detuned off resonance with the transition frequency and moderately strong; it serves to drive the population from the ground to the excited state. The other pulse—the Stark pulse—is far off-resonant and strong; it is used merely to modify the atomic transition frequency by inducing Stark shifts in the energies of the two states. Because the Stark shifts $S_g(t)$ and $S_e(t)$ of the ground and excited states are generally different (usually $|S_e(t)| \gg |S_g(t)|$) and each of them is proportional to the intensity of the Stark pulse, the transition frequency will experience a net Stark shift $S(t) = S_e(t) - S_g(t)$. By choosing an appropriate detuning for the pump pulse it is always possible to create two diabatic level crossings in the wings of the Stark pulse: one crossing during the growth and the other during the decline of the Stark pulse. For successful population transfer, the evolution must be adiabatic at one and only one of these crossings. This asymmetry can only occur if the pump and Stark pulses are not simultaneous: The pump pulse must be strong at one and only one of the crossings. It proves appropriate to set the time delay between the two pulses so that the maximum of the pump pulse occurs at one of the crossings in order to optimize the adiabatic passage there. It

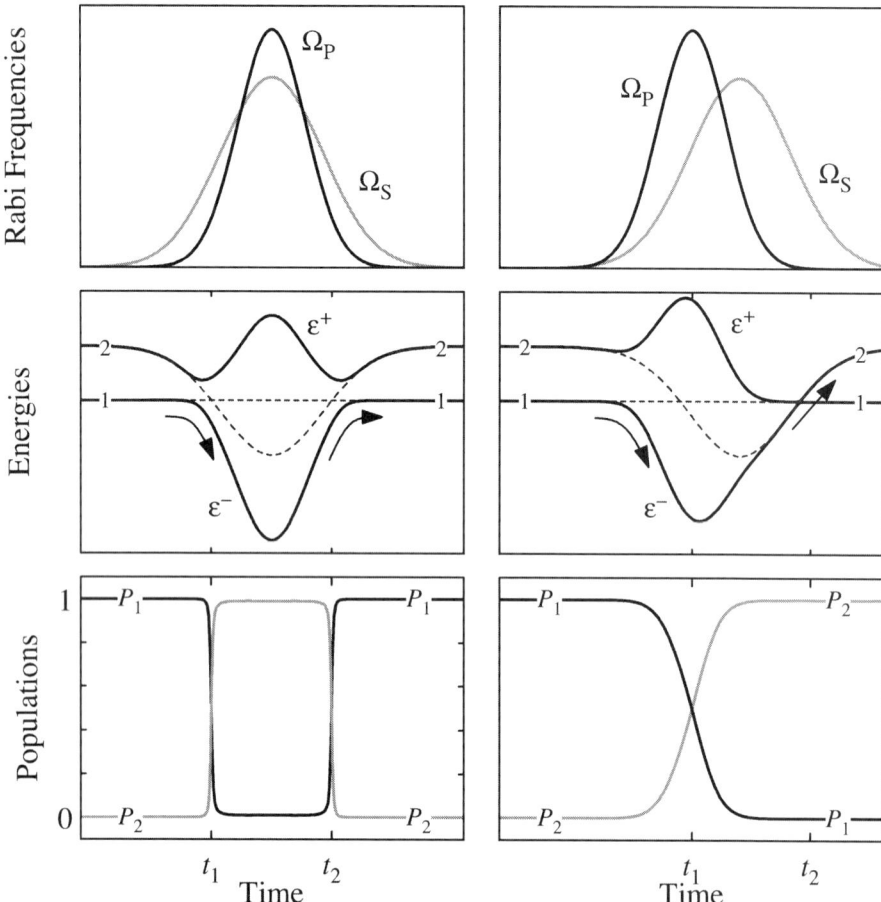

Figure 3 Time evolution of the Rabi frequencies (*top frames*), the level energies (*middle frames*), and the populations (*bottom frames*) in a two-state system driven by a pump pulse Ω_P and a Stark-shifting pulse Ω_S. (*Left*) Simultaneous pump and Stark pulses. (*Right*) Pump pulse before Stark pulse (Stark-chirped rapid adiabatic passage method).

is also appropriate that the pump pulse width be smaller than both the Stark pulse width and the delay between the pulses in order to suppress adiabatic passage at the other crossing. In this adiabatic-diabatic scenario the system will follow the path shown in the *middle right frame* in Figure 3: The state vector will adiabatically follow the lower adiabatic state through the first crossing, whereas during the second crossing it will follow the diabatic state ψ_2 (rather than an adiabatic state) and remain there until the end of the interaction. The net result is complete population transfer from state ψ_1 to state ψ_2. It should be appreciated that the adiabatic and diabatic intervals can occur in either order: The pump pulse may either precede or follow the Stark pulse.

The SCRAP technique resembles the early experiment by Loy (16), who used adiabatic quasistatic pulses of about 5 ms duration to induce Stark shifts. However, he induced two sequential population transfers per pulse—excitation for the leading edge and deexcitation for the trailing edge of each pulse—resulting in no net population transfer. In contrast, the time delay between the pump and Stark pulses in SCRAP ensures that population transfer takes place at just one of the crossings, thus leading to overall population transfer.

It should be obvious from the above description that complete population transfer will only occur within finite ranges of values of the various interaction parameters. For example, in order that there be level crossings the static detuning Δ_0 must be smaller than the maximum Stark shift S_0 and must have the same sign as S_0. Also, the pump pulse should be strong enough to ensure adiabatic passage at one of the crossings, but weak enough to prevent adiabatic passage at the other. For Gaussian pulse shapes, $\Omega(t) = \Omega_0 \exp(-t^2/T_P^2)$ and $S(t) = S_0 \exp[-(t-\tau)^2/T_S^2]$, the latter requirements lead to the conditions (53)

$$1 \ll \frac{(\Omega_0 T_S)^2}{\Delta_0 \tau} \ll \exp\left(8\tau^2/T_P^2\right). \qquad 17.$$

These conditions set upper and lower limits on the peak pump Rabi frequency Ω_0 and the static detuning Δ_0.

The SCRAP technique benefits from the fact that strong fixed-frequency long-wavelength pulsed-laser radiation, suitable for Stark shifting the levels, is often available because it is used to generate (by frequency conversion) the visible or ultraviolet radiation needed for the pump interaction. Moreover, its pulse width is longer than the pump pulse width, which is beneficial for SCRAP.

As with simple adiabatic passage, the SCRAP technique can produce population transfer in an ensemble of atoms having a distribution of Doppler shifts. The peak value of the Stark shift sets the maximum detuning that can be accessed; in turn, this sets the range of Doppler shifts for which population transfer can be produced.

2.2.2 Experimental Demonstration The first experimental demonstration of SCRAP was achieved in metastable helium (53). The initial state $1s2s\ ^3S_1$ was coupled to the target state $1s3s\ ^3S_1$ by a two-photon transition induced by an 855-nm pump laser pulse with a pulse duration of 3 ns (half width at 1/e of intensity), as shown in Figure 4 (*left*). The Stark shift was induced by a 1064-nm laser pulse with a pulse duration of 4.6 ns, delayed by 7 ns with respect to the pump pulse. Both laser pulses were mildly focused into the atomic beam. Nearly complete population transfer was observed with typical intensities of 20–30 MW/cm² for the pump pulse and 200–500 MW/cm² for the Stark pulse.

As an example, Figure 4 (*right*) displays the transfer efficiency plotted versus the static two-photon detuning $\Delta_0 = \omega_0 - 2\omega_P$. Nearly complete population transfer is observed within a certain detuning range, as predicted by analytical estimates. For large positive detuning, the adiabatic condition at the first crossing is violated, and the transfer efficiency decreases. For small positive detunings (near $\Delta_0 = 0$),

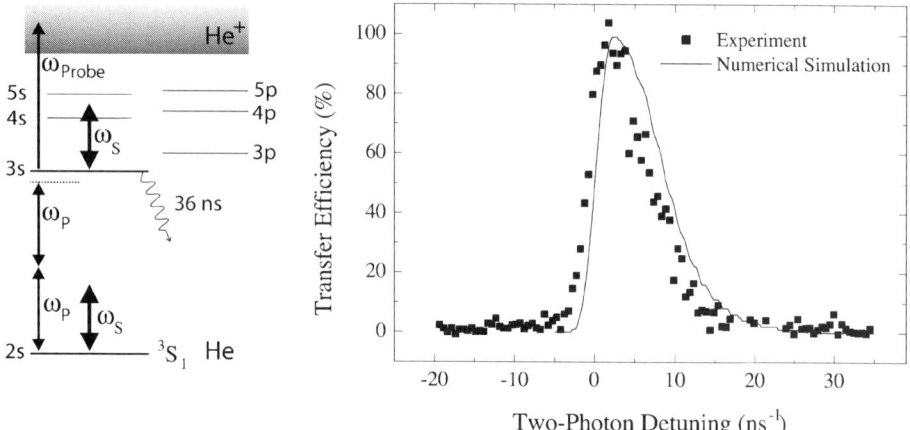

Figure 4 (*Left*) Simplified energy-level diagram of helium atom used in the demonstration of Stark-chirped rapid adiabatic passage. (*Right*) Population transfer efficiency versus the static two-photon detuning Δ_0.

the diabatic condition at the second crossing is violated, and the transfer efficiency decreases. For $\Delta_0 < 0$ and for very large positive Δ_0, there are no level crossings at all, and little population is transferred to the excited state.

An interesting extension of SCRAP—potentially very important for molecules—is the application of two sequential SCRAP processes. For example, the first SCRAP can transfer the population from the electronic ground state to an electronically excited state via a two-photon excitation. The second SCRAP then transfers the population to a target state in the electronic ground state, e.g. a highly vibrationally excited state. The second step can take place via a one-photon process [(2 + 1)-SCRAP] or by a two-photon process [(2 + 2)-SCRAP]. It is easily seen that only one Stark-shifting laser is needed in the (2 + 1)-SCRAP scheme. The (2 + 2)-SCRAP can be realized even without a separate Stark pulse because the Stokes pulse can serve as Stark pulse for the pump transition, and the pump pulse can induce the Stark shift for the Stokes transition (53).

2.3 Multiple Level Crossings

The concept of adiabatic passage via level crossing is not limited to two-state systems. Multi-state systems, however, have multiple eigen energies that often exhibit numerous avoided crossings. Such systems can be quite complicated in general, but there are a number of cases in which one can prescribe simple recipes for successful navigation to the desired target state (54–58).

2.3.1 Electronic Ladder Climbing Using a Single Pulse
Adiabatic passage has been extended to a three-state ladder system by Noordam and co-workers in

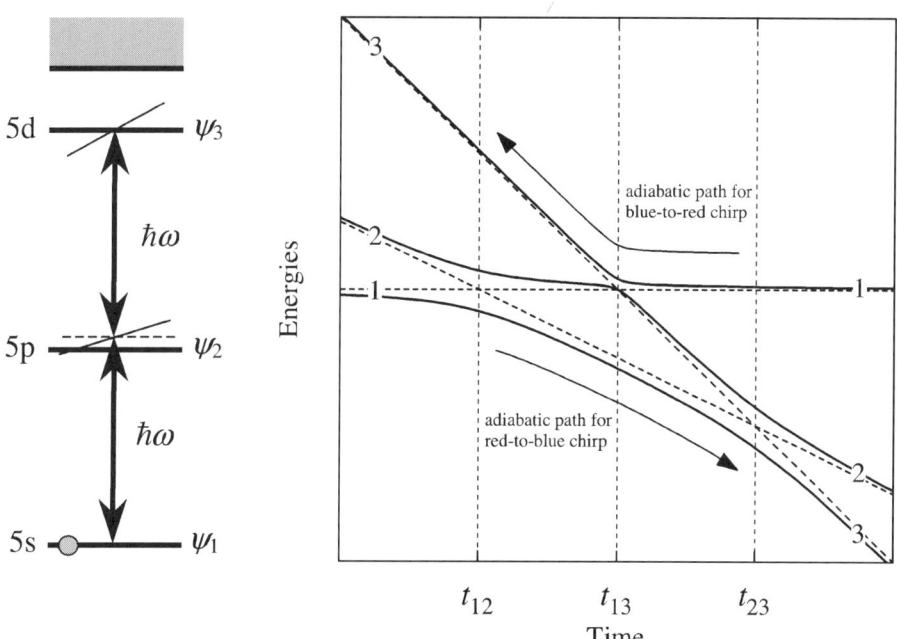

Figure 5 Linkage diagram (left) and energy diagram (right) in chirped-pulse climbing of the three-state electronic ladder 5s-5p-5d in rubidium atom. The dashed lines show the diabatic energies, and the blue solid curves show the adiabatic energies.

several experiments on the 5s-5p-5d transition in rubidium, shown in Figure 5 (40–44). Because the two single-photon transitions have very close Bohr frequencies (780.2 nm for 5s-5p and 775.9 nm for 5p-5d), it is possible to excite both transitions with the same femtosecond pulse. Virtually complete population transfer from the 5s state to the 5d state has been achieved both with red-to-blue and blue-to-red frequency chirp, as shown in Figure 6. However, the transfer mechanisms are different for the two chirps because, unlike the two-state case, multi-state systems are not generally invariant to the detuning signs.

The two transfer mechanisms are readily revealed by examining the adiabatic energies of the Hamiltonian describing the three-state ladder excitation,

$$\mathsf{H}(t) = \hbar \begin{bmatrix} 0 & \frac{1}{2}\Omega_{12}(t) & 0 \\ \frac{1}{2}\Omega_{12}(t) & \Delta_2 & \frac{1}{2}\Omega_{23}(t) \\ 0 & \frac{1}{2}\Omega_{23}(t) & \Delta_3 \end{bmatrix}. \qquad 18.$$

The two detunings are given by differences between Bohr frequencies and multiple photon frequencies: $\hbar\Delta_2(t) = E_2 - E_1 - \hbar\omega(t), \hbar\Delta_3(t) = E_3 - E_1 - 2\hbar\omega(t)$, where we have denoted the 5s, 5p, and 5d states as ψ_1, ψ_2, and ψ_3, respectively. The two Rabi frequencies $\Omega_{12}(t)$ and $\Omega_{23}(t)$ (taken as real-valued without loss of

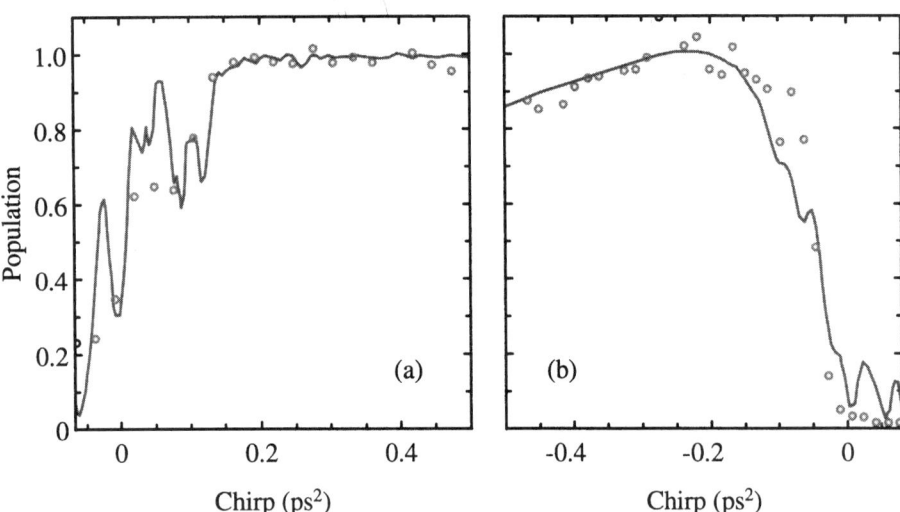

Figure 6 Population of the upper state 5d in the three-state ladder 5s-5p-5d in rubidium atom versus the pulse chirp. (*Left*) Red-to-blue chirp. (*Right*) Blue-to-red chirp. The dots are experimental data and the curves are numeric simulations. Adapted from (40) with permission.

generality) have the same time dependence (matching the laser pulse envelope), but generally different magnitudes owing to different transition dipole moments.

For red-to-blue chirp the transfer occurs via the adiabatic state with the lowest energy in Figure 5 because this state connects state ψ_1 initially with state ψ_3 at the end. While moving along this eigen energy the system encounters two level crossings: first between states ψ_1 and ψ_2 and then between ψ_2 and ψ_3. The red-to-blue chirp is the intuitively correct one because then the laser pulse comes to resonance first with the $\psi_1 \leftrightarrow \psi_2$ transition (which has the smaller transition frequency) and then with the $\psi_2 \leftrightarrow \psi_3$ transition. Consequently, the population flows gradually from state ψ_1 to state ψ_2 via a level crossing and then via another level crossing from state ψ_2 to state ψ_3. If the evolution is not sufficiently adiabatic, some population may leak at the first crossing t_{12} to the adjacent adiabatic state; some of this population will return to the initial adiabatic state at the second crossing t_{23}. This will lead to oscillations in the population of state ψ_3 because of interference caused by the two possible paths from ψ_1 to ψ_3.

For the counterintuitive blue-to-red chirp the laser pulse comes to resonance first with the transition between the initially unpopulated states ψ_2 and ψ_3 and then with the $\psi_1 \leftrightarrow \psi_2$ transition. In this case we can use again the energy plot from Figure 5 because inverting the chirp sign is equivalent to inverting the time direction. Now the evolution progresses from right to left, and the system follows the adiabatic state with the highest energy; this state connects state ψ_1 initially with state ψ_3 at the end. During its evolution along this adiabatic curve the system encounters only one level crossing t_{13} between the energies of the bare states ψ_1 and ψ_3. Because this adiabatic energy is far from the bare energy of the intermediate state ψ_2, the latter will never get appreciable population. In this respect, the counterintuitive chirp excitation resembles stimulated Raman adiabatic passage (STIRAP), which is described below. Moreover, this is the only path linking ψ_1 to ψ_3, which means that there will be no oscillations in the population of ψ_3. Finally, Figure 5 shows that the avoided crossing at t_{13} is narrower than those at t_{12} and t_{23}. Hence, adiabaticity is more difficult to achieve with blue-to-red than with red-to-blue chirp. All these features have been observed experimentally (40–44).

2.3.2 Vibrational Ladder Climbing Using a Single Pulse   Selective excitation of molecular vibrational levels has important potential applications in laser-controlled molecular dissociation. Such excitation requires ultrashort pulses to avoid rotational relaxation that takes place on a picosecond scale. Because the transition frequencies between successive vibrational levels v within a given electronic state differ very little, it is possible to move population along this excitation ladder by using a single chirped laser pulse, provided the chirp is large enough to compensate for the anharmonicity. By using chirped infrared (5 mm) femtosecond laser pulses, Noordam and co-workers have achieved a significant vibrational excitation in the ground electronic state in NO (anharmonicity about 1.5%). First they tuned a broadband pulse from a free-electron laser to the transition frequency of the ladder and found populations in vibrational states up to $v = 5$ (48). Then they observed that the transfer was considerably enhanced when the laser frequency was blue-to-red chirped so as to follow the anharmonicity of the vibrational ladder (49). Compared with the thermal populations, the enhancement was 15 times for $v = 1$ and 700 times for $v = 3$. Furthermore, because there are many paths that lead to a certain rotational state of a high vibrational state, the total population in the ro-vibrational excited states oscillated as a function of the chirp (49). Finally, it has been suggested that vibrational excitation by a single chirped pulse can lead to efficient photodissociation (59–64).

2.3.3 Vibrational Ladder Climbing Using a Pair of Pulses: Raman Chirped Adiabatic Passage   Chelkowski & Gibson (65) and Chelkowski & Bandrauk (66) proposed the technique of Raman-chirped adiabatic passage (RCAP) as a tool for robust and efficient excitation of high vibrational levels in molecules. RCAP employs two laser pulses—pump and Stokes—both of which may be far off their respective single-photon resonances. Figure 7a illustrates the level linkage in RCAP. A successive step-by-step climbing of the vibrational ladder can eventually lead to vibrational dissociation of the molecule. In the original proposal (65), the pump laser was chirped and the Stokes was monochromatic, resulting in frequency sweeping of the two-photon resonance between the initial and final states in each Λ-subsystem in the chain; the feasibility of this scheme has been confirmed numerically.

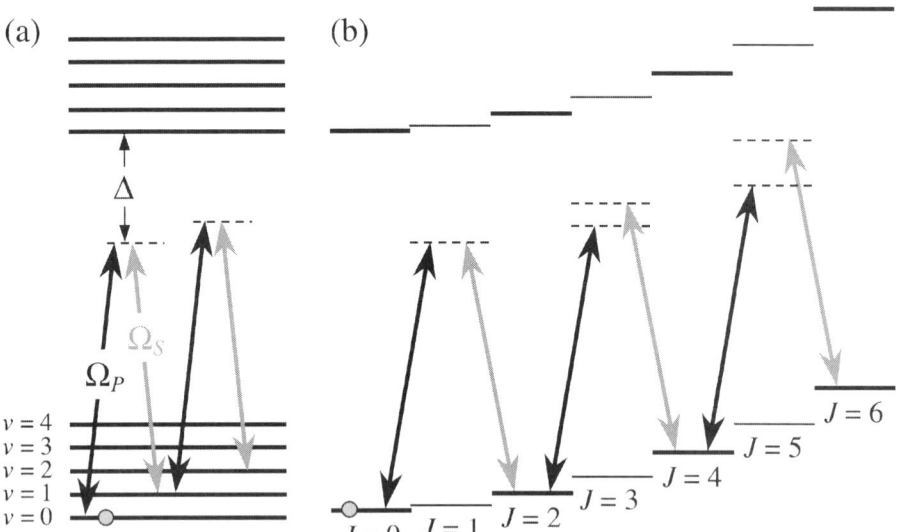

Figure 7 (a) Linkage diagram for vibrational-ladder climbing by a pair of chirped pulses (Raman-chirped adiabatic passage scheme). (b) Linkage diagram for rotational-ladder climbing by a pair of chirped pulses (molecular centrifuge).

Davis & Warren (67) have shown in calculations for oxygen and chlorine that a similar scheme with two chirped pulses provides some additional advantages.

RCAP requires that the frequency difference of the two pulses be tuned across the two-photon transition frequency in each Λ-subsystem, but it provides freedom in choosing each laser frequency, i.e. the location of the "virtual" states shown by dashed lines in Figure 7a. This avoids the necessity for ultraviolet lasers (often needed if coupling to the nearest electronic state is used) and allows the use of near infrared light, where short-pulse laser technology is well developed and can provide sufficient intensities to drive far off-resonance Raman couplings.

2.3.4 Rotational Ladder Climbing Using a Pair of Pulses Recently, Karczmarek et al (68) suggested that a ladder of rotational states can be efficiently excited using a pair of chirped infrared laser pulses with opposite circular polarizations and opposite chirps. Such pulses can induce consecutive level crossings in the chain of serial two-photon Raman transitions between neighboring rotational states with $J = 0, 2, 4, \ldots$, whose energies increase approximately as $J(J+1)$, as shown in Figure 6b. Hence, if the two pulses are sufficiently intense to enable adiabatic evolution, a significant fraction of the population in the low-J states can be driven up the excitation ladder. This scheme for rotational excitation can be viewed as a variant of RCAP. When the rotation becomes sufficiently rapid, above a certain threshold value of J, the molecular bond can no longer withstand the centrifugal force, and the molecule dissociates into two fragments (68).

Very recently, this scheme—called an optical centrifuge for molecules—has been successfully demonstrated experimentally (69) in Cl_2. Laser-induced dissociation has been achieved by inducing excitation up to states with $J = 420$, using chirped near-infrared laser pulses. In this experiment the rate of rotation was accelerated in 50 ps from 0 to 5 THz, at which rate the rotational energy approximately equals the bond energy. Thus, dissociation was achieved in a time scale shorter than the rotational relaxation time (70 ps), at a rate of about one Raman transition $J \rightarrow J + 2$ per 200 fs.

2.3.5 Chirped Excitation of a Manifold of Closely Spaced Levels An interesting situation arises when a chirped laser pulse couples an initially populated state ψ_1 to a manifold of several levels $\psi_2, \psi_3, \ldots, \psi_N$ simultaneously. The energy diagram of such a coupled system, displayed in Figure 8a, shows that adiabatic evolution can produce a very selective excitation even if the Fourier bandwidth of the laser pulse is larger than the level spacing within the manifold. Indeed, for red-to-blue chirp the system will follow from left to right the lowest adiabatic energy, which links the initial state ψ_1 to the lowest unperturbed state ψ_2 of the manifold. In contrast, for blue-to-red chirp the system will follow from right to left the highest adiabatic energy, linking state ψ_1 to the highest unperturbed state ψ_N. Therefore,

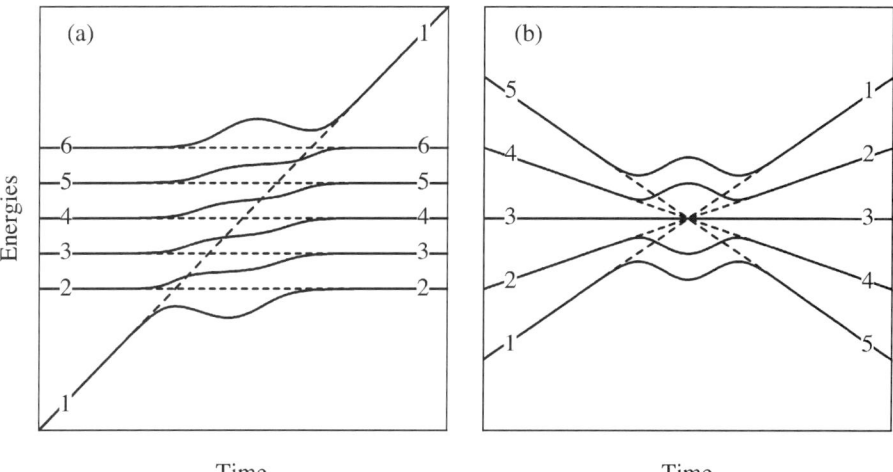

Figure 8 (*a*) Energy diagram for a system in which a ground state is simultaneously coupled by a chirped laser pulse to a manifold of five states. (*b*) Energy diagram in a five-state bowtie model. The dashed lines show the unperturbed (diabatic) energies, and the solid curves show the adiabatic energies.

the chirp sign alone determines if the population will be directed toward the lowest or the highest state of the manifold. Warren and co-workers (32) demonstrated this excitation scheme experimentally on the 3s-3p transition in sodium. Red-to-blue chirped picosecond pulses populated predominantly the lower fine-structure level $3p^2P_{1/2}$, whereas blue-to-red chirped pulses placed the population onto the upper fine-structure level $3p^2P_{3/2}$.

It is interesting to note that this linkage pattern allows an exact analytic solution (not just in the adiabatic limit, but also for any interaction parameters) in the case of constant chirp (i.e. linear detuning) and constant Rabi frequency over infinite time interval; this is the so-called Demkov-Osherov model (70, 71). Somewhat unexpectedly, the transition probabilities in this model are extremely simple and given by products of Landau-Zener probabilities for transition or no transition applied at the relevant crossings. For example, the transition probability from state ψ_1 to state ψ_4 with red-to-blue chirp (left-to-right direction in Figure 8a) is $p_{12}p_{23}(1-p_{34})$, where p_{mn} is the Landau-Zener probability (16) for nonadiabatic transition at the crossing between states ψ_m and ψ_n.

2.3.6 The Bowtie Model Another interesting example of adiabatic passage is the so-called bowtie model (72–77), whose energy diagram is displayed in Figure 7b. It occurs in a sequentially coupled evenly spaced N-state ladder, excited by the same laser pulse. In this case the RWA energy of the nth state is $E_n - E_1 - n\hbar\omega$. Adiabatic sweep of frequency through resonance will transfer all population from the lowest to the highest energy state, for either sign of the chirp. A bowtie-type linkage can also occur in an rf-pulse-controlled Bose-Einstein condensate output coupler (78, 79).

3. ADIABATIC POPULATION TRANSFER USING SEQUENTIAL PULSES

The technique of stimulated Raman adiabatic passage (STIRAP) (56, 80–84) uses the coherence of two pulsed laser fields to achieve a complete population transfer from an initially populated state ψ_1 to a target state ψ_3 via an intermediate state ψ_2 (see Figure 1b). Instead of applying the pulses in the intuitive sequence, where the pump pulse precedes the Stokes pulse (as with SEP), the Stokes pulse precedes the pump pulse (counterintuitive ordering). If there is sufficient overlap of the two pulses and if the pulses are sufficiently strong that the time evolution is adiabatic, then complete population transfer occurs between states ψ_1 and ψ_3.

3.1 Stimulated Raman Adiabatic Passage in Three-State Systems: Theory

3.1.1 Background At first glance, the pulse sequence of STIRAP seems strange (counterintuitive): The first pulse to act on the system (the Stokes pulse) connects two initially unpopulated states. The STIRAP mechanism can be understood when we examine the eigenvalues and the eigen states of the Hamiltonian driving the probability amplitudes $\mathbf{C}(t) = [C_1(t), C_2(t), C_3(t)]^T$ of the Λ-system,

$$\mathsf{H}(t) = \hbar \begin{bmatrix} 0 & \frac{1}{2}\Omega_P(t) & 0 \\ \frac{1}{2}\Omega_P(t) & \Delta_P & \frac{1}{2}\Omega_S(t) \\ 0 & \frac{1}{2}\Omega_S(t) & \Delta_P - \Delta_S \end{bmatrix}. \qquad 19.$$

Here $\Omega_P(t)$ and $\Omega_S(t)$ are the Rabi frequencies of the pump and Stokes pulses, respectively, and Δ_P and Δ_S are the single-photon detunings of the pump and Stokes lasers from their respective transitions, $\hbar\Delta_P = E_2 - E_1 - \hbar\omega_P$, $\hbar\Delta_S = E_2 - E_3 - \hbar\omega_S$. An essential condition for STIRAP is the two-photon resonance between states ψ_1 and ψ_3, $\Delta_P = \Delta_S = \Delta$. Then the three instantaneous eigenstates of $\mathsf{H}(t)$ (the adiabatic states) are given by

$$\Phi_+(t) = \psi_1 \sin\vartheta(t)\sin\varphi(t) + \psi_2 \cos\varphi(t) + \psi_3 \cos\vartheta(t)\sin\varphi(t), \qquad 20.$$

$$\Phi_0(t) = \psi_1 \cos\vartheta(t) - \psi_3 \sin\vartheta(t), \qquad 21.$$

$$\Phi_-(t) = \psi_1 \sin\vartheta(t)\cos\varphi(t) - \psi_2 \sin\varphi(t) + \psi_3 \cos\vartheta(t)\cos\varphi(t), \qquad 22.$$

where the mixing angles $\vartheta(t)$ and $\varphi(t)$ are defined (modulo π) as $\vartheta(t) = \arctan[\Omega_P(t)/\Omega_S(t)]$, $\varphi(t) = \frac{1}{2}\arctan[\sqrt{\Omega_P^2(t) + \Omega_S^2(t)}/\Delta]$. These eigen states have the following eigen frequencies:

$$\varepsilon_0(t) = 0, \quad \varepsilon_\pm(t) = \frac{1}{2}\Delta \pm \frac{1}{2}\sqrt{\Delta^2 + \Omega_P^2(t) + \Omega_S^2(t)}. \qquad 23.$$

STIRAP is based on the zero-eigenvalue adiabatic state $\Phi_0(t)$, which is a coherent superposition of the initial state ψ_1 and the final state ψ_3 only. This adiabatic state has no component of the excited state ψ_2, and hence it has no possibility of radiatively decaying; it is a trapped state (of population) or a radiatively dark state (85–89). For the counterintuitive pulse ordering the relations $\Omega_P(t)/\Omega_S(t) \overset{t\to -\infty}{\longrightarrow} 0$ and $\Omega_P(t)/\Omega_S(t) \overset{t\to +\infty}{\longrightarrow} \infty$ apply; hence, as time progresses from $-\infty$ to $+\infty$, the mixing angle $\vartheta(t)$ rises from 0 to $\pi/2$. Consequently, the adiabatic state $\Phi_0(t)$ evolves from the bare state ψ_1 to a superposition of states ψ_1 and ψ_2 at intermediate times and finally to the target state ψ_3 at the end of the interaction; thus, state $\Phi_0(t)$ links adiabatically the initial state ψ_1 to the target state ψ_3. Because the Hamiltonian is explicitly time dependent, diabatic transitions between the adiabatic states will occur. The goal is to reduce the diabatic transition rates to negligibly small values. When the system can be forced to stay in the dark state at all times, a complete population transfer from ψ_1 to ψ_3 will be achieved, as shown in Figure 9. This can be realized by ensuring adiabatic evolution; then no transitions between the adiabatic states can take place. The adiabatic condition requires

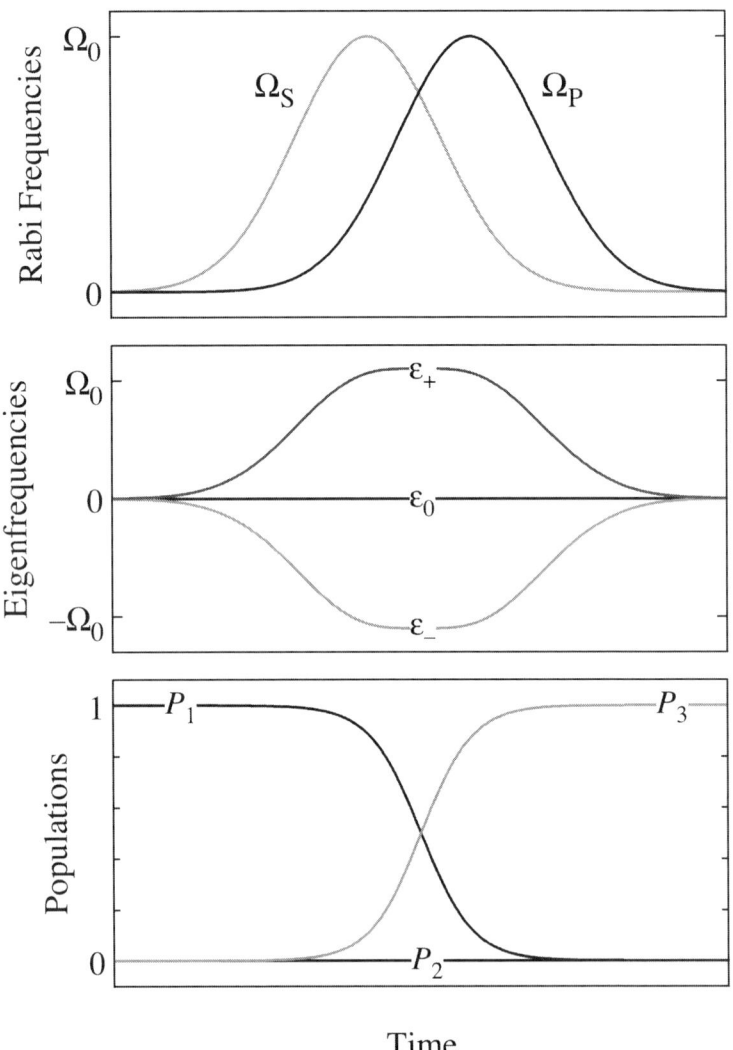

Figure 9 Time dependences of the pump and Stokes Rabi frequencies (*top*), the eigen frequencies (*middle*), and the populations (*bottom*) in three-state stimulated Raman adiabatic passage.

that the coupling between each pair of adiabatic states is negligible compared with the difference between the energies of these states. With respect to the dark state $\Phi_0(t)$, the adiabatic condition reads (5, 6)

$$|\varepsilon_0 - \varepsilon_\pm| \gg |\langle\dot\Phi_0|\Phi_\pm\rangle|. \qquad 24.$$

On one-photon resonance ($\Delta = 0$) the adiabaticity condition simplifies and becomes (81)

$$\sqrt{\Omega_P^2 + \Omega_S^2} \gg |\dot{\vartheta}| \propto T^{-1}, \qquad 25.$$

where T is the pulse width. Assuming that the pump and Stokes pulses have the same peak Rabi frequency Ω_0, this condition can be written as $\Omega_0 T \gg 1$. Hence, adiabaticity demands a large pulse area. In practical applications the pulse area should exceed 10 to provide efficient population transfer. In terms of incoherent excitation, the large pulse area means saturation of the transitions.

Once the conditions for STIRAP are fulfilled (two-photon resonance between states ψ_1 and ψ_3, counterintuitive pulse ordering, and adiabatic evolution), a complete population transfer from ψ_1 to ψ_3 is guaranteed. Moreover, because the dark state does not involve the intermediate state ψ_2, the latter remains unpopulated during the transfer, which means that its properties, such as radiative decay, have little impact on the transfer efficiency. This is a remarkable feature of STIRAP, which allows efficient population transfer on time scales exceeding the intermediate-state lifetimes. For example, such a situation arises in the implementation of STIRAP with continuous lasers when the atomic beam crosses two spatially displaced and partially overlapping continuous-wave (cw) laser beams at right angles (81, 90). The time it takes for the atoms or molecules to cross the laser beams is about two orders of magnitude longer than the lifetime of the excited state, thus prohibiting any population transfer by intermediate storage in the excited state, e.g. by stimulated emission pumping. By contrast, STIRAP still maintains a transfer efficiency of 100%, because the upper state is never populated.

3.1.2 Intuitive versus Counterintuitive Pulse Order When both the pump and Stokes lasers are on resonance with their respective transitions, the two opposite pulse sequences lead to qualitatively different results: The intuitive sequence produces generalized Rabi oscillations, while the counterintuitive sequence induces complete population transfer to state ψ_3 (91–94). When the two lasers are tuned off the respective single-photon resonances, while maintaining the two-photon resonance, both pulse sequences can lead to complete population transfer from state ψ_1 to state ψ_3 (92, 94). This result is easily explained when the single-photon detuning Δ is large; then the intermediate state ψ_2 can be eliminated adiabatically (92, 94). The resulting effective two-state problem involves a coupling $\Omega_{\text{eff}} = \Omega_P \Omega_S / 2\Delta$ and a detuning $\Delta_{\text{eff}} = (\Omega_P^2 - \Omega_S^2)/2\Delta$. Obviously, for delayed pulses the detuning Δ_{eff} passes through resonance at the time t_0 when $\Omega_P(t_0) = \Omega_S(t_0)$; this level crossing leads, in the adiabatic limit, to complete population transfer for both pulse orders because the order reversal is equivalent to the unimportant change of sign in Δ_{eff}. However, there is still a significant difference between the two pulse sequences: The intermediate state ψ_2 receives some transient population for the intuitive sequence, whereas it remains unpopulated for the counterintuitive sequence. Hence, if the lifetime of state ψ_2 is comparable to or shorter than the excitation duration, then efficient population transfer can be achieved only with the counterintuitive pulse sequence.

3.1.3 Dependence on the Interaction Parameters
Various aspects of STIRAP have been studied in detail. Among these are the effects of nonadiabatic transitions (95–101) and in particular, the manner in which the adiabatic limit is approached (95–98). Furthermore, analytical expressions for the single-photon linewidth (102) and the two-photon linewidth (103, 104, 105) have been derived. The results have demonstrated that STIRAP is much less sensitive to single-photon detuning (because the single-photon linewidth grows with the square of the pulse area) than to two-photon detuning (in which the linewidth increases only approximately linearly with the pulse area); Figure 10 shows an example. The influence of spontaneous

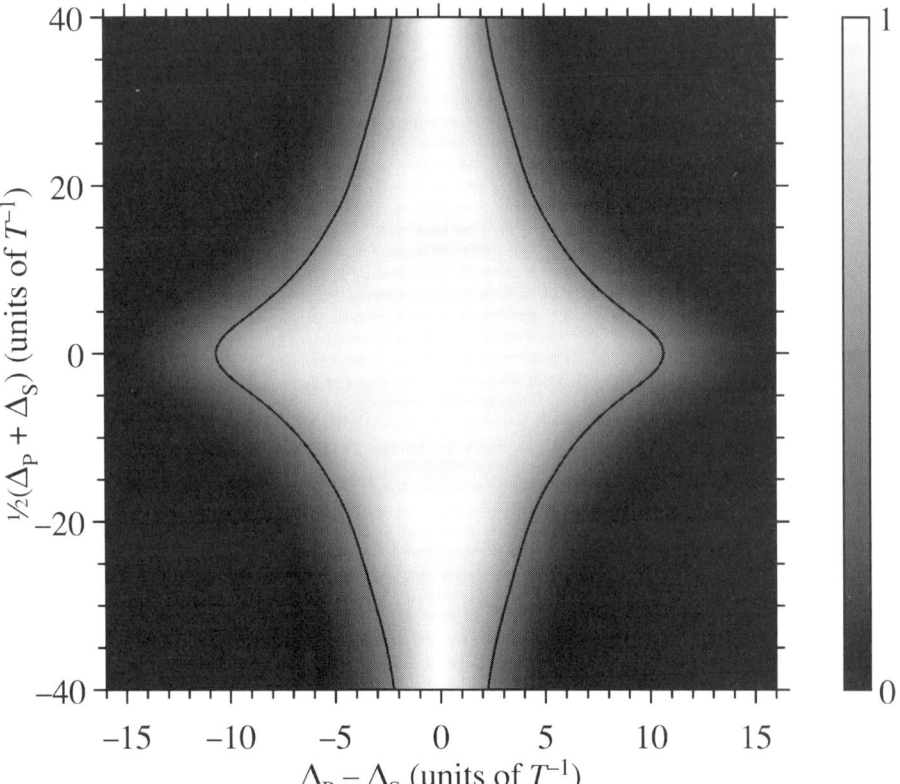

Figure 10 Numerically calculated transfer efficiency in stimulated Raman adiabatic passage plotted versus the sum and the difference of the pump and Stokes detunings (i.e. versus the single-photon and two-photon detunings) for Gaussian pulse shapes, $\Omega_P = \Omega_0 \exp[-(t-\tau)^2/T^2]$, $\Omega_S = \Omega_0 \exp[-(t+\tau)^2/T^2]$, with $\Omega_0 T = 20$, $\tau = 0.5T$. The curves show the $P_3 = 0.5$ value.

emission from the intermediate state ψ_2 has also been studied, including decay both within the Λ-system (back to states ψ_1 and ψ_3) (106) and to other states (107, 108). Finally, the effects of multiple intermediate states (109, 110) and multiple final states (111, 112) have been explored, as has STIRAP beyond the RWA approximation (113).

3.2 STIRAP in Three-State Systems: Experiment

3.2.1 Experimental Demonstration with cw Lasers
After preliminary, yet incomplete, results (114), Bergmann et al achieved the first convincing experimental demonstration of STIRAP in Na_2 (81). A beam of sodium dimers crossed two spatially displaced but partially overlapping cw laser beams. When the sodium dimers interacted first with the Stokes laser (counterintuitive pulse ordering) complete population transfer was observed from the initial level ($v = 0, J = 5$) to the final level ($v = 5, J = 5$) of the Na_2 molecules in their electronic ground state $X^1\Sigma_g^+$ via an intermediate level ($v = 7, J = 6$) of the excited electronic state $A^1\Sigma_u^-$. The duration of the interaction with combined pulses (the time required for a molecule to traverse the two laser beams) was about 0.2 μs. Because the interaction time was much longer than the excited-state lifetime and because the sodium dimers have relatively strong transition moments, only moderate laser intensities were needed to induce large pulse areas. Typical intensities in the range of 100 W/cm^2, provided by cw lasers mildly focused to a few hundred μm into the molecular beam, were sufficient to observe adiabatic passage.

STIRAP has also been observed in metastable neon in a similar crossed-beam geometry (90, 115, 116). In the Theuer & Bergmann experiment (90) the population was transferred from state $2p^53s\,^3P_0$ to state $2p^53s\,^3P_2$ via the intermediate state $2p^53p\,^3P_1$. The intensities used in the experiment, typically a few W/cm^2, were again provided by cw radiation, focused into the atomic beam by cylindrical lenses. Although the typical time required for passage of atoms across the laser beams was more than 20 times longer than the radiative lifetime of the excited intermediate state, no more than 0.5% of the population was detected in this state.

Figure 11 displays a characteristic signature for STIRAP. In this figure the transfer efficiency is plotted versus the spatial displacement between the pump and Stokes laser beams, i.e. versus the pulse delay from the viewpoint of the Ne* atoms. Positive displacement (*left*) corresponds to counterintuitive pulse order (Stokes before pump) and negative displacement (*right*) to intuitive order (pump before Stokes). The population in the target state was detected by a probe laser via laser-induced fluorescence. When the Stokes beam was shifted too far upstream (far left in the figure), it was excluded from the interaction because there was no overlap with the pump pulse; in this case about 25% of the population was optically pumped into the target state because the pump laser excited the atoms to the intermediate state from which they decayed radiatively. As the Stokes beam was shifted toward the pump beam, while still preceding it, the transfer efficiency

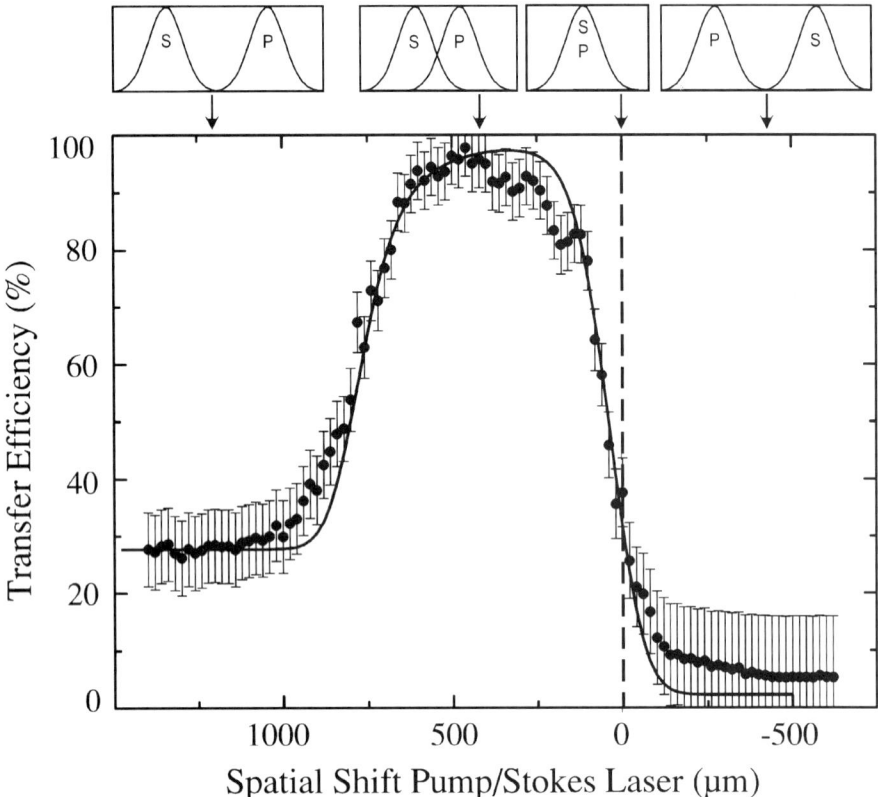

Figure 11 Transfer efficiency versus displacement between pump and Stokes pulses in Ne* experiment. The broad plateau, showing nearly complete population transfer for counterintuitive pulse sequence, is a typical stimulated Raman adiabatic passage signature, as contrasted with the low efficiency for intuitively ordered pulses. The dots are experimental data and the curve shows numeric simulation.

increased dramatically and reached almost unity. When the axes of the two lasers coincided, the transfer efficiency dropped to about 25%. Virtually no transfer was observed when the Stokes beam was moved farther downstream so that the Ne* atoms encountered the pump laser first (intuitive pulse ordering). In this configuration the atoms were initially in a bright adiabatic state and population was transferred to the intermediate state, from which it decayed via a two-step cascade to the ground state of Ne (i.e. out of the laser-coupled three-state system) by spontaneous emission of vacuum ultraviolet radiation. The broad plateau for counterintuitive laser order is a characteristic feature of STIRAP, and it indicates the robustness of the population transfer.

3.2.2 Experimental Demonstration with Pulsed Lasers

A very interesting and important application of STIRAP is the selective excitation of high-lying vibrational levels in molecules. In most molecules the first electronically excited states have energies more than 4 eV above the ground state. Therefore, a Raman-type linkage from the vibrational ground level to a high vibrational level requires ultraviolet lasers. Strong ultraviolet radiation is most readily provided by frequency conversion techniques involving high-intensity pulsed lasers. Furthermore, because the molecular transition dipole moments are usually considerably smaller than for atoms, the adiabaticity condition is difficult to satisfy with cw lasers. Sufficiently strong light intensity, and hence large enough couplings, can be delivered only by pulsed lasers. Pulsed lasers, however, often have inferior coherence properties, e.g. they suffer from phase fluctuations and frequency chirping. A careful analysis of the effect of imperfect laser coherence on the STIRAP efficiency provides the modified adiabaticity condition, $\Omega_0 T \gg [1 + (\Delta\omega/\Delta\omega_{TL})^2]^{1/2}$ (117), where $\Delta\omega_{TL}$ is the transform-limited bandwidth of the pulse and $\Delta\omega$ is its actual bandwidth. The adverse effect of imperfect laser coherence derives from the fact that both phase fluctuations and frequency chirp correspond to time-dependent changes in the laser frequencies. Unless these changes are correlated (e.g. if the pump and Stokes pulses are derived from the same laser), they will result in time-dependent detuning from two-photon resonance. The two-photon detuning induces nonadiabatic couplings between the dark state and the other adiabatic states that reduce the transfer efficiency. These population losses can be reduced by increasing the laser intensity, thereby suppressing nonadiabatic transitions. For conventional nanosecond lasers that are not specially designed to yield nearly transform-limited pulses, the ratio $\Delta\omega/\Delta\omega_{TL}$ is typically bigger than 10. According to the above estimate, the intensity needed for STIRAP has to be increased by a factor of 100 with respect to transform-limited pulses. Although in principle it is possible to obtain higher intensities by focusing the laser beam, this would be detrimental for most applications, in which large volumes of the molecular jet must be excited, e.g. in reactive scattering experiments. Therefore, it is almost impossible to satisfy the adiabaticity criterion for laser pulses whose bandwidth clearly exceeds the bandwidth of transform-limited pulses.

It is also useful to point out that if a pulse energy of 1 mJ were sufficient to ensure adiabatic evolution for a 1-ns laser pulse then the same degree of adiabaticity would require energy of 1 J for a 1-ps laser pulse. Indeed, this conclusion follows readily from the adiabatic condition when it is written as $\Omega_0^2 T \gg 1/T$ (for transform-limited pulses); obviously, the required pulse energy (which is proportional to $\Omega_0^2 T$) grows rapidly when the pulse duration T decreases. Still, this does not mean that long-pulse or cw lasers offer the best possibilities to implement STIRAP. Indeed, the product $\Omega_0 T$, which is proportional to the square root of the pulse energy times the pulse duration T, is typically largest for tunable laser sources of intermediate pulse duration, i.e. nanosecond lasers. Conventional tunable picosecond and femtosecond lasers typically cannot compensate for the shorter pulse duration by an adequate increase in pulse energy. On the other hand, cw lasers may provide longer interaction times, but suffer from weak intensities.

STIRAP has been successfully demonstrated with nanosecond pulses in NO (118, 119) and SO_2 (120) molecules and in rubidium atoms (121). In NO, highly efficient and selective population transfer has been achieved in the electronic ground state from the $X^2\Pi_{1/2}(v = 0, J = \frac{1}{2})$ rovibrational state to the $X^2\Pi_{1/2}(v = 6, J = \frac{1}{2})$ state via the intermediate state $A^2\Sigma(v = 0, J = \frac{1}{2})$. The NO molecule provides an example of complications that may arise owing to hyperfine structure. What seems to be a 3-level system is actually a system of 18 sublevels. Because $^{14}N^{16}O$ has a nuclear spin of $I = 1$, each of the three levels is split into a pair of sublevels with $F = \frac{1}{2}$ and $F = \frac{3}{2}$, which in turn possess magnetic sublevels. For linearly polarized light, and when the pump and Stokes polarizations are parallel, the 18-state system decomposes into 2 independent 3-state systems (one with $F = \frac{3}{2}, m_F = \frac{3}{2}$ and another with $F = \frac{3}{2}, m_F = -\frac{3}{2}$) and 2 6-state systems for $m_F = \frac{1}{2}$ and $m_F = -\frac{1}{2}$. Because the hyperfine splitting of the initial level and the final level is large enough to be resolved experimentally, the complexity of the system can be further reduced. Thus, despite the complications, a nearly complete transfer has been achieved (118, 119).

The population transfer achieved in SO_2 molecules (120) is an example of STIRAP for a polyatomic molecule. The enormously increased density of levels, as compared with atoms or diatomic molecules, results in much smaller transition dipole moments. Nevertheless, efficient population transfer becomes possible with adequate laser power, when the level density in the final state is not too high. Figure 12 shows an example of population transfer from the rotational state 3_{03} of

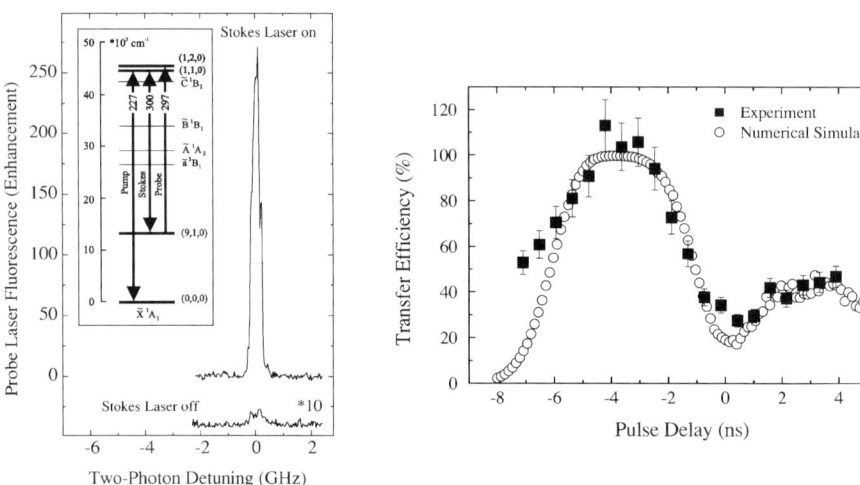

Figure 12 Experimental demonstration of stimulated Raman adiabatic passage in SO_2 molecules versus two-photon detuning (*left*) and pulse delay (*right*). The inset in the left plot shows the level linkage.

the vibrational ground state (0, 0, 0) to the same rotational level of the (9, 1, 0) overtone in the electronic ground state $X\,^1A_1$ via the vibrational level (1, 1, 0) of the excited electronic state $C\,^1B_2$. The wavelengths were 227 nm for the pump and 300 nm for the Stokes laser (120). Pulse durations were 2.7 ns (half width at $1/e$ of maximum electric field) for the pump and 3.1 ns for the Stokes laser pulse. Typical laser intensities were 10 MW/cm^2, yielding Rabi frequencies of about 10^{10} s^{-1}. The population in the target state was probed by laser-induced fluorescence. The lower trace in Figure 12 (*left*) displays signal of the probe-laser-induced fluorescence from the final state (magnified 10 times) in the case in which only the pump pulse was present; then the final state was populated by spontaneous emission from the intermediate state. When the Stokes pulse was turned on and applied before the pump pulse (with an appropriate overlap between them) the final-state population increased by more than two orders of magnitude. When the delay between the pump and Stokes pulses was varied, the typical plateau region of complete population transfer for negative pulse delay was observed (Figure 12 *right*). For positive pulse delay, i.e. the case of SEP with the pump preceding the Stokes pulse, a transfer efficiency of about 25% was observed, as one expects from rate equation calculations.

3.2.3 STIRAP with Degenerate Levels A problem that often arises when implementing STIRAP in real atoms and molecules is the existence of multiple intermediate and final states. These may be present owing to fine and/or hyperfine structure, Zeeman sublevels, or closely spaced rovibrational levels in polyatomic molecules. A detailed numerical, analytical, and experimental investigation of this problem has been presented in a series of papers on STIRAP in metastable neon atoms (122–124). It has been concluded that the presence of closely spaced levels near the intermediate and final state may pose a problem and even be detrimental for STIRAP. In particular, STIRAP can selectively populate a particular target level only if the two-photon linewidth is smaller than the level separation near this level.

A multi-state system has multiple eigen energies that may present a very complicated picture. For instance, narrow avoided crossings between the eigen energies may appear; if such avoided crossings involve the eigen state that provides the adiabatic linkage for STIRAP then the adiabatic path will be blocked and STIRAP may fail. The level scheme in the neon experiment (124) involves states with $J = 0$, 1, and 2, which means that there are $1 + 3 + 5 = 9$ magnetic sublevels in total. Figure 13 (*left*) shows these nine states, with several possible linkage patterns (determined by the dipole selection rules) for different polarizations of the pump and Stokes lasers. Figure 13 (*right*) shows the behavior of the final-state population as a function of the magnetic field, which lifts the Zeeman degeneracy, when the Stokes-laser polarization was parallel to the magnetic field and the pump polarization was perpendicular (124); the ensuing linkage pattern is shown in Figure 13c. The Stokes frequency was tuned to resonance with the Bohr frequency of the degenerate $(B = 0)$ $^3P_2 \leftrightarrow {}^3P_1$ transition, whereas the pump laser frequency was scanned across the resonance. For small magnetic field a single peak in the

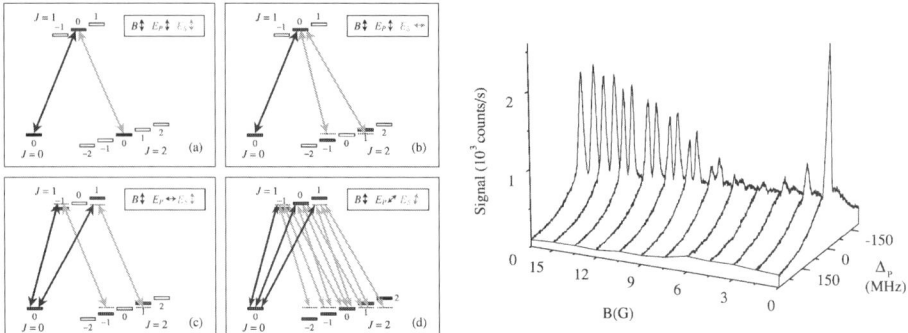

Figure 13 (*Left*) Linkage patterns in Ne* for various choices of pump and Stokes polarizations with respect to the direction of the magnetic field B. (*Right*) Population transfer in Ne* versus magnetic field strength.

target-state population was observed near resonance ($\Delta_P = 0$), because the $M = +1$ and $M = -1$ sublevels were too close to be resolved. For large magnetic field, the Zeeman splitting increased and a symmetric two-peaked structure emerged, indicative of populations of the $M = +1$ and $M = -1$ sublevels. A significant drop in the transfer efficiency was observed at intermediate magnetic field strengths. This drop was identified (124) as due to lack of adiabatic connectivity between the initial and final states, i.e. the adiabatic path between them was blocked because of interference from neighboring adiabatic states.

3.3 Extensions of STIRAP

3.3.1 Multi-State Chains
The success of STIRAP in three-state systems has encouraged its extension to multi-state chainwise-connected systems $\psi_1 \leftrightarrow \psi_2 \leftrightarrow \psi_3 \leftrightarrow \cdots \leftrightarrow \psi_N$ (90, 125–136). It has been discovered that multi-state systems behave differently when they involve odd and even numbers of states.

It has been found that for an odd number of states ($N = 2n + 1$) a STIRAP-like population transfer can take place for delayed and counterintuitively ordered pulses when all lasers are on resonance with the corresponding transitions or when only the even states in the chain are detuned from resonance (125–127). The transfer is carried out via a multi-level dark state that is a time-dependent coherent superposition of the odd states in the chain $\psi_1, \psi_3, \ldots, \psi_{2n+1}$. For example, the dark state in a five-state chain reads (125–127, 137, 138)

$$\Phi_0(t) = \frac{1}{v(t)}[\Omega_{23}(t)\Omega_{45}(t)\psi_1 - \Omega_{12}(t)\Omega_{45}(t)\psi_3 + \Omega_{12}(t)\Omega_{34}(t)\psi_5], \quad 26.$$

where $v(t)$ is a normalization factor. It is easily seen that when the pulse $\Omega_{45}(t)$, driving the last transition, precedes the pulse $\Omega_{12}(t)$, driving the first transition, this state provides adiabatic connection between the initial state ψ_1 and the last state ψ_5 of the chain.

A particularly suitable system for multi-state STIRAP is the chainwise transition formed by the magnetic sublevels of a degenerate two-state system with $J_e = J_g$ or $J_g - 1$, driven by two delayed pulses with opposite circular polarizations. For example, if the system is prepared initially in the $M_g = -J_g$ ground-state sublevel (e.g. by optical pumping), then a STIRAP-like transfer to the $M_g = J_g$ sublevel can be achieved by applying a pulse of σ^- polarization (the Stokes) before a pulse of σ^+ polarization (the pump), i.e. in the counterintuitive order. As in STIRAP, the sublevels of the excited state remain unpopulated throughout the transfer if the interaction is adiabatic; however, the odd states in the chain—the sublevels of the ground state—do acquire some transient populations. In this particular system these transient intermediate-state populations are not a problem because these sublevels do not decay and there are no population losses. Indeed, multi-state STIRAP in such chainwise systems has been demonstrated experimentally by several groups (90, 129–132).

In more general multi-state chains, in which the transiently populated intermediate states can decay radiatively during the transfer, it is desirable to reduce their populations. It has been suggested (133) that these transient populations can be suppressed when the pulses coupling the intermediate transitions are much stronger than the pulses driving the first (the pump) and the last (the Stokes) transitions. A pulse sequence—straddle-STIRAP—has been proposed, in which all intermediate pulses arrive simultaneously with the Stokes pulse and vanish with the pump pulse.

The systems with an even number of states ($N = 2n$) behave very differently in the resonant case because they do not possess a dark state. Consequently, even when such a system is driven adiabatically by counterintuitively ordered resonant pulses, a STIRAP-like population transfer between the initial and final states of the chain cannot occur. Instead, the final-state population exhibits Rabi-like oscillations as the pulse intensities increase (128, 134, 139). These oscillations occur because at early times the initial state is equal to a superposition of adiabatic states (rather than to a single adiabatic state as in STIRAP), and so is the final state at late times; hence, there is interference between the different paths from the initial state to the final state.

When the intermediate states are off resonance while the initial state and final state are still on $(N-1)$-photon resonance, chains with odd and even numbers of states behave quite similarly. In both cases a STIRAP-like transfer may or may not be possible, depending on the laser parameters, particularly on the intermediate detunings (134). For example, in a four-state system, STIRAP-like transfer may take place when the cumulative detunings of the two intermediate states have the same sign, $\Delta_2 \Delta_3 > 0$, whereas it is impossible if $\Delta_2 \Delta_3 \leq 0$. The reason is that in the former case an adiabatic state exists that links the initial and final states, whereas in the latter case, such an adiabatic connection is absent.

A dressed-state approach (135) provides a particularly clear picture of multi-state STIRAP, valid for both odd and even numbers of states. It is most useful when the pulse (or pulses) driving the intermediate transitions arrives before and vanishes

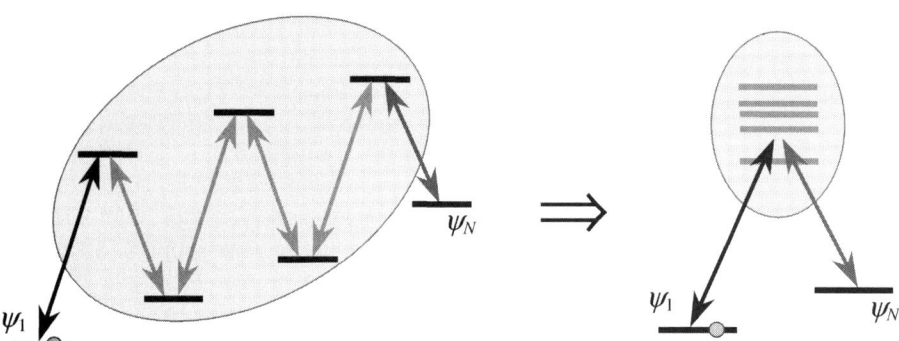

Figure 14 Linkage pattern for chain–stimulated Raman adiabatic passage (serial multi-Λ system) and equivalent parallel multi-Λ system, obtained by diagonalization of the subsystem comprising the intermediate states in the original chainwise system.

after the pulses driving the first (the pump) and the last (the Stokes) transitions, as in (133). Then the $N-2$ intermediate states are coupled into a dressed subsystem (as shown in Figure 14), prior to the arrival of the pump and Stokes pulses. By changing the parameters of the dressing (control) pulse Ω_C (intensity and frequency), one can manipulate the properties of this dressed subsystem and thus, control the population transfer. By tuning the pump and Stokes lasers to one of the dressed eigen states Φ_k, the multi-state dynamics is essentially reduced to a system of three strongly coupled states: $\psi_1 \leftrightarrow \Phi_k \leftrightarrow \psi_N$; this paves the road for an efficient STIRAP-like population transfer from state ψ_1 to state ψ_N. Furthermore, it is beneficial if the dressing pulse(s) is constant (at least when the pump and Stokes pulses are present), because then the couplings between the dressed states vanish. Also, if the dressing pulse(s) is strong the spacing between the dressed energies is large. This makes the multi-Λ system resemble the single-Λ system in STIRAP and therefore place little population in the intermediate states.

The dressed picture also displays the difference between odd and even numbers of states in the on-resonance case (all detunings in the original chain equal to zero). For odd N, one of the dressed states is always on-resonance with the pump and Stokes lasers. In contrast, for even N, the pump and Stokes lasers are tuned in the middle between two adjacent dressed states, and the ensuing interference between different adiabatic paths leads to Rabi-like oscillations. Figure 15 shows the final-state population in a four-state system against the cumulative detunings from the two intermediate states ψ_2 and ψ_3. High transfer efficiency (the white zones) is achieved for sufficiently adiabatic evolution only if $\Delta_2\Delta_3 > 0$, as discussed above. Near the dressed-state resonances (shown by parabolas) the transfer efficiency is high even when the evolution is not very adiabatic elsewhere (left frame). As the pulse areas increase (*right frame*), the two regions where $\Delta_2\Delta_3 > 0$ (first and third quadrants) get filled with white (unity transfer efficiency), whereas no high transfer is possible in the $\Delta_2\Delta_3 < 0$ regions (second and fourth quadrants).

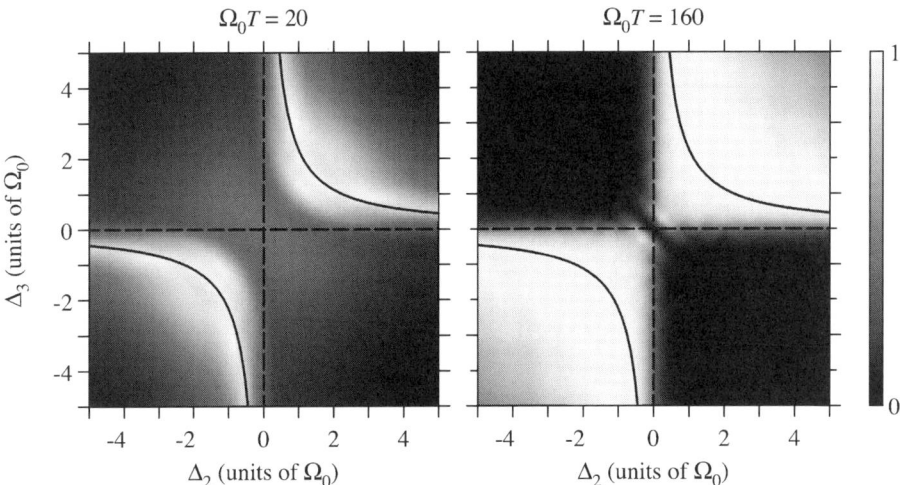

Figure 15 Numerically calculated transfer efficiency for chain-stimulated Raman adiabatic passage in a four-state system versus the two intermediate-state detunings for $\Omega_P = \Omega_0 \exp[-(t-\tau)^2/T^2]$, $\Omega_S = \Omega_0 \exp[-(t+\tau)^2/T^2]$, $\Omega_C = 3\Omega_0$, with $\tau = 0.5T$ and $\Omega_0 T = 20$ (*left*) and $\Omega_0 T = 160$ (*right*). The curves show the two dressed-state resonances.

3.3.2 Population Transfer via Continuum A few years ago it was suggested (140, 141) that a continuum can serve as an intermediary for population transfer between two discrete states in an atom or a molecule by using a sequence of two counterintuitively ordered delayed laser pulses. This intriguing scheme—which has yet to be demonstrated experimentally—can be seen as a variant of STIRAP in which the discrete intermediate state is replaced by a continuum of states. The advantage of this scheme with respect to STIRAP would be that only one tunable laser is needed, because a continuum offers a continuous range of possible combinations to match the pump and Stokes laser frequencies to the two-photon resonance between the initial and target states. The Carroll-Hioe analytic model (140, 141), which involves an infinite quasi-continuum of equidistant discrete states, equally strongly coupled to the two bound states, suggests that complete population transfer is possible, the ionization being suppressed. The physical reason for this unexpected conclusion (because a continuum is traditionally seen as an incoherent medium) is closely related to the laser-induced continuum structure (142–145) created in the structureless, flat continuum by the Stokes laser. Another finding supporting this scheme is that, as in a discrete three-state Λ-system, there exists a dark state—a coherent superposition of the two bound states that is immune to ionization.

Nakajima et al (146) later demonstrated, however, that the completeness of the population transfer in the Carroll-Hioe model derives from the very stringent restrictions of the model that are unlikely to be met in a realistic physical system with a real continuum, in particular with a nonzero Fano parameter and Stark shifts (147).

It has subsequently been recognized that although complete population transfer is unrealistic, significant partial transfer may still be feasible (148–153). It has been shown that, at least in principle, the detrimental effect of the nonzero Fano parameter and the Stark shifts can be compensated by using the Stark shifts induced by a third, nonionizing laser (150) or by using appropriately chirped laser pulses (151, 153). It has been concluded (150, 153) that the main difficulty in achieving efficient population transfer is related to the incoherent ionization channels, of which at least one is always present; these lead to inevitable irreversible population losses. It has been suggested (149, 150) that these losses can be reduced (although not eliminated) by choosing an appropriate region in the continuum where the ionization probability is minimal. It was shown later that incoherent ionization can be suppressed considerably by using a Fano-like resonance induced by an additional laser from a third state tuned near the region in the continuum where the incoherent ionization takes place; this leads to a significant increase in the transfer efficiency (154). Very recently, a similar scheme was proposed using a third laser, but tuned in the same region in the continuum as the pump and Stokes lasers, which provides an enhanced control of the population transfer (155). Finally, it has been suggested that a STIRAP-like process can take place also via an autoionizing state (156, 157).

3.3.3 Hyper-Raman STIRAP   The application of standard STIRAP to molecules, using two single-photon transitions (pump and Stokes), is often impeded by the fact that most molecules require ultraviolet or even vacuum ultraviolet pump photons to reach the first electronically excited states. (For the Stokes pulse, which connects the excited electronic state to a high vibrational level of the ground electronic state, optical wavelengths are usually sufficient.) It is difficult to provide vacuum ultraviolet pulses with adequate power and coherence properties. It is natural to consider achieving the pump excitation (and possibly the Stokes excitation too) by a two-photon transition. The corresponding $(2+1)$ and $(2+2)$ versions of STIRAP have been named hyper-Raman STIRAP (158–161). Although these extensions seem obvious, they turn out to be rather nontrivial.

The main obstacles in hyper-Raman STIRAP are the dynamic Stark shifts induced by the two-photon coupling. These Stark shifts, which are proportional to the laser intensities, modify the Bohr frequencies of the pump and Stokes transitions and destroy the multiphoton resonance between the initial and final states, which is crucial for the existence of the dark state. It has been found, both numerically and analytically, that high transfer efficiency in such a scheme can still be achieved by a suitable choice of static detunings of the carrier frequencies of the two pulses, which suppress the detrimental effect of the Stark shifts; these detuning ranges have been estimated analytically (159). It is interesting to note that, unlike the purely adiabatic evolution in STIRAP, successful population transfer in hyper-Raman STIRAP occurs as a result of a combination of adiabatic and diabatic time evolution, as in SCRAP (Section 2.2). Moreover, unlike STIRAP, the intermediate state does acquire some transient population; again, it can be reduced by suitable static detunings.

It should be pointed out that the Stark shifts are also nonzero in traditional STIRAP, but they are usually negligible compared with the one-photon on resonance couplings (given by the Autler-Townes splittings). In $(2+1)$-STIRAP the fundamental field (ω_P) is very strong, and the related Stark shift is usually not small compared with the two-photon coupling ($2\omega_P$).

3.3.4 Adiabatic Passage by Light-Induced Potentials Recently, Garraway & Suominen (162) (see also 163, 164) have suggested, on the basis of numerical calculations for sodium dimers, that the STIRAP ideas of counterintuitively ordered laser pulses and adiabatic evolution can be applied to the transfer of a wave packet from one molecular potential to the displaced ground vibrational state of another. This process—termed adiabatic passage by light-induced potentials (APLIP)—seemingly violates the Frank-Condon principle because the overlap between the initial and final wavefunctions is very small (the two wavepackets were displaced at a distance seven times larger than their widths). There is, however, no such violation, because the time scale of the process is close to, but longer than, the vibrational time scale. APLIP shares many features with STIRAP, such as high efficiency and insensitivity to pulse parameters. However, in contrast to STIRAP, the two-photon resonance condition in APLIP cannot be satisfied (except at a certain time), and the main mechanism for the transfer of the wave packet is through a "valley" that emerges in the time-dependence of the light-induced potential, as shown in Figure 16 (*left*). Figure 16 (*right*) shows how the wavepacket gradually

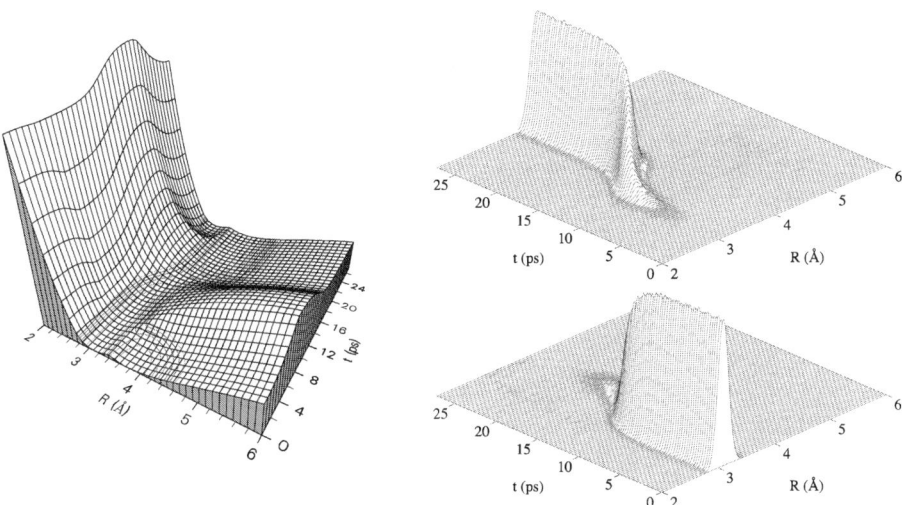

Figure 16 (*Left*) light-induced potential. The population flows through the "valley" in this potential. (*Right*) Time evolution of the ground-state (*lower*) and excited-state (*upper*) populations. Reprinted from (162) with permission.

disappears from the ground-state potential (lower plot) and appears in the excited-state one (upper plot). Although the original proposal assumed transitions between the lowest vibrational states ($v = v' = v'' = 0$), recent calculations (165) extended APLIP to excited vibrational states.

3.3.5 Creation of Coherent Superpositions by Modifications of STIRAP

STIRAP can be used not only to transfer population completely from one state to another, but also to create coherent superpositions of atomic or molecular states. The most obvious approach is to interrupt the time evolution of the dark state (21) at a certain intermediate time, when it is a superposition of the initial state ψ_1 and the final state ψ_3 (126, 166–168). Then only a fraction of the total population is transferred to ψ_3, and the composition of the superposition depends on the ratio between the pump and Stokes Rabi frequencies at the turn-off time. This fractional STIRAP has been demonstrated experimentally (167). A smooth realization of this scheme has also been proposed (169).

A more sophisticated variant of STIRAP has been proposed (170, 171) and demonstrated (172) in which the usual three-state STIRAP is supplied with an additional state ψ_4, coupled to the intermediate state ψ_2 by a third, control laser. Such a scheme (tripod-STIRAP) has two, rather than one, dark states. Because they are degenerate, transitions between them take place even in the adiabatic limit and eventually lead to the creation of a coherent superposition of ψ_1 and ψ_3. The composition of this superposition depends on, and therefore can be controlled by, the time delay between the pump and Stokes pulses and the strength of the control pulse.

3.4 Applications of Stimulated Raman Adiabatic Passage

3.4.1 Control of Chemical Reactions
The remarkable properties of STIRAP have already had applications in many diverse areas. The first implementation of STIRAP has been in a crossed-beam reactive scattering experiment (173). It allowed investigation of the effect of vibrational excitation on the cross section for the chemiluminescent channel in the process $Na_2(v) + Cl \rightarrow NaCl + Na^*$ in crossed particle beams. It was found that the cross sections increased by about 0.75% per vibrational level in the range $3 \leq v \leq 19$.

Another example is the detailed study of the reaction $Na_2(v'', j'') + H \rightarrow NaH(v', j') + Na$, where the angular distribution and the population distribution have been determined for the product molecule NaH for a range of selectively populated levels v'' of the reagent molecule Na_2 (174). STIRAP has also been used to investigate the dependence of the dissociative attachment process $Na_2(v'', j'') + e^- \rightarrow Na + Na^-$ (with electron energies <1 eV) on the vibrational excitation by exciting efficiently and very selectively the Na_2 molecules to a specific vibrationally excited level (175). The vibrational excitation up to $v'' = 12$ has increased the state-dependent dissociative attachment rate by more than three orders of magnitudes.

3.4.2 Atom Optics
Coherent population transfer between atomic states is always accompanied by transfer of photon momenta to the atoms. Momentum transfer is the basis of atom optics, particularly in the design of its key elements—atom mirrors and beam splitters. Because STIRAP enables efficient, robust, and dissipation-free population transfer, it was soon realized that it is a perfect tool for coherent momentum transfer (126). A particularly appropriate system for momentum transfer is the multi-state chain formed from the magnetic sublevels of two degenerate levels, driven by two delayed laser pulses with opposite circular polarizations. An example in the case when $J_g = J_e = 2$, demonstrated in a recent experiment (90), is shown in Figure 17. A beam of metastable neon atoms, prepared by optical pumping into the $M = 2$ magnetic sublevel of the $^3P_2$ metastable level crossed two slightly displaced circularly polarized cw laser beams. The two beams were ordered counterintuitively, so that the atoms encountered the σ^+ beam (the Stokes) first and then the σ^- beam (the pump). In the adiabatic limit, the population was completely transferred to the $M = -2$ sublevel of the $^3P_2$ level, without residing at any time in the $M = -1$ and $M = 1$ sublevels of the decaying excited level $^3D_2$. Because the two laser beams propagated in opposite directions, the atom received a total momentum of $4\hbar k$ in the direction of the σ^- beam during its journey from the $M = 2$ sublevel to the $M = -2$ sublevel: $2\hbar k$ momentum owing to absorption of two photons from the σ^- beam (which transfer their momenta to the atom) and another momentum $2\hbar k$ in the same direction owing to recoil from the stimulated emission of two photons into the σ^+ beam. The experimental results in Figure 17 correspond to a double passage from the $M = 2$ sublevel to the $M = -2$ sublevel and then back to the $M = 2$ sublevel, resulting in the transfer of eight photon momenta.

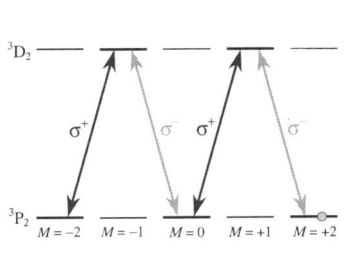
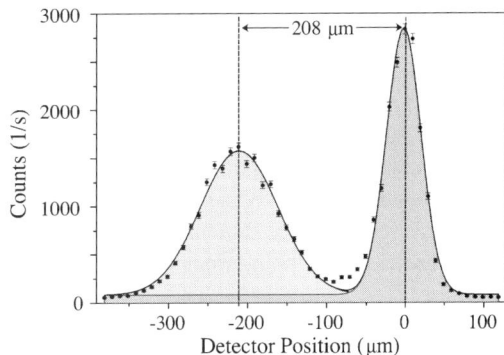

Figure 17 (*Left*) Linkage pattern between the Zeeman sublevels in the $^3P_2$-$^3D_2$ transition in Ne* driven by a pair of counterpropagating and displaced σ^+ and σ^- laser beams. (*Right*) Experimental results showing deflection of a beam of ^{20}Ne* atoms owing to transfer of eight photon momenta after double adiabatic passage from the $M = 2$ sublevel to $M = -2$ and then back to $M = 2$. The narrower, undeflected original distribution is observed owing to the presence of ^{22}Ne isotope atoms that are insensitive to light.

Coherent momentum transfer by adiabatic passage in similar chains of Zeeman sublevels has been demonstrated in a number of other experiments. Pillet et al (129, 130) and Goldner et al (131, 132) have reported momentum transfer of $8\hbar k$ resulting from the single-pass adiabatic passage between the $M_F = -4$ and $M_F = 4$ Zeeman sublevels in the hyperfine transition $F_g = 4 \leftrightarrow F_e = 4$ of the cesium D_2 line with 50% efficiency. Lawall & Prentiss (166) have demonstrated momentum transfer of $4\hbar k$ with 90% efficiency in the $2^3 S_1$-$2^3 P_0$ transition of He^*, after double adiabatic passage ($M = -1 \to M = 1 \to M = -1$) between the ground-state sublevels. Chu et al (167, 168) have built the first atomic interferometer based on STIRAP in the transition between the two cesium hyperfine ground states ($6S_{1/2}, F = 3, M_F = 0$) and ($6S_{1/2}, F = 4, M_F = 0$) via the excited state ($6P_{1/2}, F = 3$ or $4, M_F = 1$). They have achieved multiple-pass coherent transfer of more than 140 photon momenta with 95% efficiency per exchanged photon pair. Figure 18 shows the observed interference fringes. Burnett and coworkers (176–180) have also demonstrated coherent momentum transfer in cesium, both with beams having circular/circular and others having circular/linear polarizations. Finally, a multiple-beam atomic interferometer has been demonstrated wherein a beam of cesium atoms has been split into five spatially distinct beams (separated by two-photon momenta), corresponding to the magnetic sublevels $M = -4, -2, 0, 2, 4$, and then recombined (181).

3.4.3 Laser Cooling The STIRAP technique has been successfully applied in laser-cooling experiments to coherently manipulate the atomic wave packets resulting from subrecoil laser cooling by velocity selective coherent population trapping (182–191). The momentum distribution of atoms cooled by velocity selective coherent population trapping has two peaks at $+\hbar k$ and $-\hbar k$, both with widths smaller than the photon recoil momentum $\hbar k$. Hänsch et al (192) have used adiabatic passage to coherently transfer rubidium atoms cooled by velocity selective coherent population trapping into a single momentum state, still with a subrecoil momentum spread. Kulin et al (193) have demonstrated adiabatic transfer of metastable helium atoms into a single wave packet or into two coherent wave packets, while retaining the subrecoil momentum dispersion of the initial wave packets. They have achieved nearly 100% transfer efficiency in one and two dimensions, and 75% in three dimensions, while being able to choose at will the momentum direction and the internal state of the atoms.

3.4.4 Measurement of Weak Magnetic Fields The potential of STIRAP for atom beam deflection by coherent momentum transfer has been used to create a technique (called Larmor velocity filter) for measuring very small magnetic fields along the axis of the atomic beam (90). The scheme, which was demonstrated with metastable neon atoms, consisted of two STIRAP zones. In the first zone atoms were prepared in the $M = 2$ sublevel of the $^3 P_2$ metastable state and transferred to the $M = -2$ sublevel. They were transferred back to the initial $M = 2$ sublevel in the second zone provided they remained in the $M = -2$ state along the path

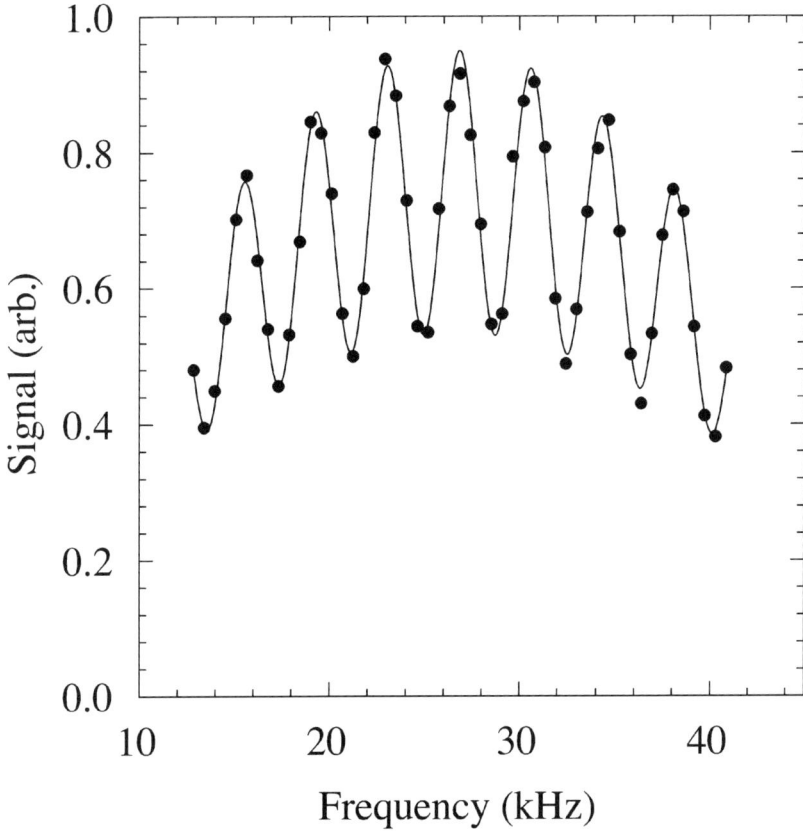

Figure 18 Interference fringes for an atomic interferometer based on adiabatic passage. The dots are experimental data and the curve is a fit by a cosine function with a Gaussian envelope. Reprinted from (167) with permission.

between the two zones (cf Figure 17). Larmor precession due to a magnetic field in the region between the two transfer zones mixed the magnetic sublevels and interfered with the momentum transfer. The resulting narrow-peaked pattern, an example of which is shown in Figure 19, allowed measurement of weak magnetic fields.

3.4.5 Cavity Quantum Electrodynamics The potential of STIRAP in cavity quantum electrodynamics has been pointed out by Kimble et al (194, 195) Parkins & Kimble (196). They proposed to use STIRAP to create coherent superpositions of photon-number states by strongly coupling an atom to a cavity field. Then the atomic ground-state Zeeman coherence (prepared by some preliminary excitation scheme) can be mapped to the cavity mode by adiabatic passage in

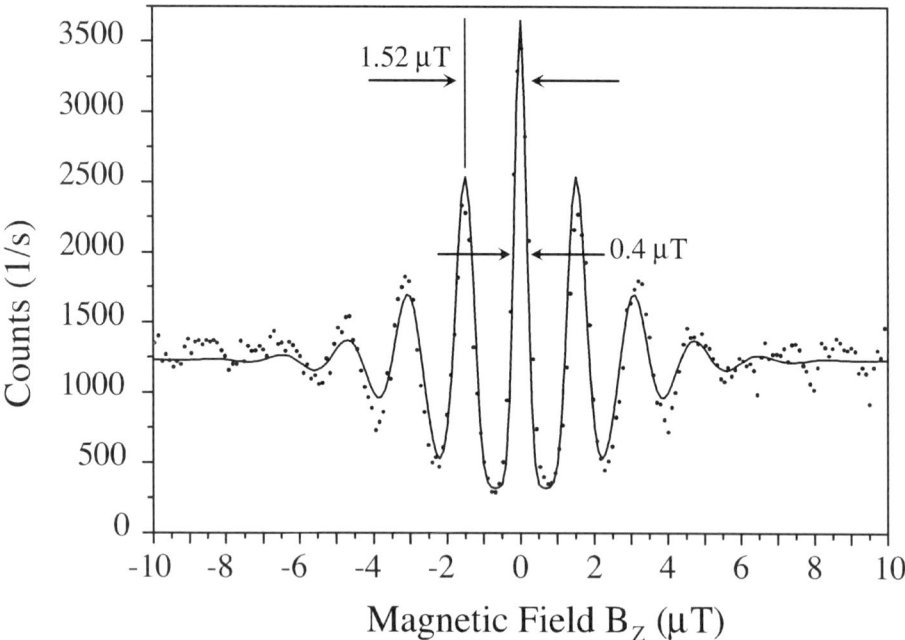

Figure 19 Variation of the flux of deflected Ne* atoms in the Larmor velocity filter with the magnetic field strength.

a decoherence-free fashion by a pair of two delayed laser pulses, one circularly and another linearly polarized. This idea was extended recently to the case of two degenerate cavity modes with orthogonal polarizations (197).

The above scheme is reversible and allows the opposite process, of mapping cavity-mode fields onto atomic ground-state Zeeman coherence. This mapping provides a possibility for measuring cavity fields (195). The atomic angular momentum state can be measured by the Newton-Young method (198) with a finite number of magnetic dipole measurements using Stern-Gerlach analyzers. Alternatively, the parameters of the atomic superposition can be measured by coupling this degenerate atomic state to an excited state by an elliptically polarized laser pulse and measuring the subsequent fluorescence (199, 200).

3.4.6 Bose-Einstein Condensation It has been suggested recently, on the basis of numerical simulations (201–204), that STIRAP can be used to induce coherent two-color photo-association of a Bose-Einstein condensate in free-bound-bound transitions and convert an atomic condensate to a molecular one. It was concluded that Bose stimulation can enhance the atomic free-bound dipole matrix element to the extent enabling photo-associative STIRAP.

4. CONCLUSIONS

Many contemporary fields in atomic, molecular, and optical physics require preparation of samples in which almost all of the population resides in a preselected excited state. Although a variety of methods have been proposed and tried over the years, many of the most successful are based on some application of adiabatic time evolution of the quantum system.

The theoretical description of such time dependence is most easily presented with the aid of adiabatic states: If the evolution is adiabatic then at all times the state vector remains aligned with one of these states. The experimenter has various guides for the applicability of adiabatic passage—the adiabatic conditions. Typically, these require that excitation pulses be strong and smooth and that the interaction act for a "long" time compared with a characteristic atomic time scale. Often this is expressed as the requirement that a time-integrated Rabi frequency be much larger than unity. What typically sets adiabatic techniques apart from other pulsed-excitation techniques is the relative insensitivity of the transfer efficiency to interaction parameters; adiabatic techniques are not sensitive to pulse area, for example.

In simplest form, adiabatic passage can completely invert the population of a two-state system using a level crossing. Numerous extensions have been devised and put to practical use. One very useful extension uses two sequential pulses to stimulate a Raman transition—the STIRAP technique. This too has been extended in a variety of ways from the initial demonstration involving three states. We have mentioned a number of these.

This review is primarily concerned with techniques for producing complete population transfer between two quantum states. There is also much contemporary interest in producing coherent superpositions of quantum states, sometimes using variants of the techniques discussed here to produce partial population transfer.

As laser technology continues to improve and experimenters acquire lasers with ever higher intensity and purer spectral content, one can expect to see imaginative new applications of adiabatic passage to the task of transferring populations between quantum states.

ACKNOWLEDGMENTS

This work has been supported by the European Union Research and Training network COCOMO, contract number HPRN-CT-1999-00129, by NATO grant 1507-826991, and by Deutsche Forschungsgemeinschaft. NVV has received partial support from the Academy of Finland, project 43336. He thanks Prof. Bertrand Girard for hospitality during his visit to Université Paul Sabatier in Toulouse, where a part of his contribution to this work was made. BWS acknowledges the support of Laserzentrum, University of Kaiserslautern. His work in Germany has been supported, in part, by a Research Award from the Alexander von Humboldt

Foundation. His work at Livermore is supported in part under the auspices of the US Department of Energy at Lawrence Livermore National Laboratory under contract W-7405-Eng-48.

Visit the Annual Reviews home page at www.AnnualReviews.org

LITERATURE CITED

1. Einstein A. 1917. *Phys. Z.* 18:121. Translated, reprinted in van der Waerden. 1968. *Sources of Quantum Mechanics.* New York: Dover
2. Shore BW. 1990. *The Theory of Coherent Atomic Excitation.* New York: Wiley
3. Hamilton CE, Kinsey JL, Field RW. 1986. *Annu. Rev. Phys. Chem.* 37:493–524
4. Dai HL, Field RW. 1995. *Molecular Dynamics, Spectroscopy by Stimulated Emission Pumping.* Singapore: World Sci.
5. Messiah A. 1962. *Quantum Mechanics.* New York: North-Holland
6. Crisp MD. 1973. *Phys. Rev. A* 8:2128–35
7. Landau LD. 1932. *Phys. Z. Sowjetunion* 2:46
8. Zener C. 1932. *Proc. R. Soc. London Ser. A* 137:696
9. Allen L, Eberly JH. 1975. *Optical Resonance, Two-Level Atoms.* New York: Dover
10. Hioe FT. 1984. *Phys. Rev. A* 30:2100–3
11. Demkov YN, Kunike M. 1969. *Vestn. Leningr. Univ. Fiz. Khim.* 16:39–45
12. Suominen K-A, Garraway BM. 1992. *Phys. Rev. A* 45:374–86
13. Bloch F. 1946. *Phys. Rev.* 70:460–74
14. Abragam A. 1961. *The Principles of Nuclear Magnetism.* Oxford: Clarendon
15. Treacy EB. 1968. *Phys. Lett. A* 27:421
16. Loy MMT. 1974. *Phys. Rev. Lett.* 32:814–17
17. Grischkowsky D, Loy MMT. 1975. *Phys. Rev. A* 12:1117–20
18. Grischkowsky D, Loy MMT, Liao PF. 1975. *Phys. Rev. A* 12:2514–33
19. Grischkowsky D. 1976. *Phys. Rev. A* 14:802–12
20. Brewer RG, Hahn EL. 1975. *Phys. Rev. A* 11:1641–49
21. Loy MMT. 1976. *Phys. Rev. Lett.* 36:1454–57
22. Loy MMT. 1978. *Phys. Rev. Lett.* 41:473–76
23. Hamadani SM, Mattick AT, Kurnit NA, Javan A. 1975. *Appl. Phys. Lett.* 27:21–24
24. Avrillier S, Raimond J-M, Bordé CJ, Bassi D, Scoles G. 1981. *Opt. Commun.* 30:311
25. Adam AG, Gough TE, Isenor NR, Scoles G. 1985. *Phys. Rev. A* 32:1451–57
26. Leidenbaum C, Stolte S, Reuss J. 1989. *Phys. Rep.* 178:1–24
27. Kroon JPC, Senhorst HAJ, Beijerinck HCW, Verhaar BJ, Verster NF. 1985. *Phys. Rev. A* 31:3724–32
28. Lorent V, Claeys W, Cornet A, Urbain X. 1987. *Opt. Commun.* 64:41–44
29. Diels J-C, Rudolph W. 1996. *Ultrashort Laser Pulse Phenomena: Fundamentals, Techniques, Applications on a Femtosecond Time Scale.* San Diego: Academic
30. Chergui M. 1996. *Femtochemistry: Ultrafast Chemical, Physical Processes in Molecular Systems.* Singapore: World Sci.
31. Melinger JS, Hariharan A, Gandhi SR, Warren WS. 1991. *J. Chem. Phys.* 95:2210–13
32. Melinger JS, Gandhi SR, Hariharan A, Tull JX, Warren WS. 1992. *Phys. Rev. Lett.* 68:2000–3
33. Melinger JS, Gandhi SR, Hariharan A, Goswami D, Warren WS. 1994. *J. Chem. Phys.* 101:6439–54
34. Hillegas CW, Tull JX, Goswami D, Strickland D, Warren WS. 1994. *Opt. Lett.* 19:737–39
35. Weiner AM. 1995. *Prog. Quant. Electr.* 19:161–237

36. Dugan MA, Tull JX, Warren WS. 1997. *J. Opt. Soc. Am. B* 14:2348–58
37. Fetterman MR, Goswami D, Keusters D, Yang W, Rhee J-K, Warren WS. 1998. *Opt. Express* 3:366–75
38. Warren WS, Rabitz H, Dahleh M. 1993. *Science* 259:1581–89
39. Noordam LD, Joosen W, Broers B, ten Wolde A, Lagendijk A, et al. 1991. *Opt. Commun.* 85:331
40. Broers B, van Linden van den Heuvell HB, Noordam LD. 1992. *Phys. Rev. Lett.* 69:2062–65
41. Broers B, Noordam LD, van Linden van den Heuvell HB. 1992. *Phys. Rev. A* 46:2749–56
42. Broers B, van Linden van den Heuvell HB, Noordam LD. 1992. *Opt. Commun.* 91:57–61
43. Balling P, Maas DJ, Noordam LD. 1994. *Phys. Rev. A* 50:4276–85
44. Maas DJ, Rella CW, Antoine P, Toma ES, Noordam LD. 1999. *Phys. Rev. A* 59:1374–81
45. Vrijen RB, Lankhuijzen GM, Maas DJ, Noordam LD. 1996. *Comments At. Mol. Phys.* 33:67–82
46. Vrijen RB, Duncan DI, Noordam LD. 1997. *Phys. Rev. A* 56:2205–12
47. Maas DJ, Duncan DI, van der Meer AFG, van der Zande WJ, Noordam LD. 1997. *Chem. Phys. Lett.* 270:45–49
48. Maas DJ, Duncan DI, Vrijen RB, van der Zande WJ, Noordam LD. 1998. *Chem. Phys. Lett.* 290:75–80
49. Maas DJ, Vrakking MJJ, Noordam LD. 1999. *Phys. Rev. A* 60:1351–62
50. Goswami D, Warren WS. 1994. *Phys. Rev. A* 50:5190–96
51. Herrmann J, Mondry J. 1988. *J. Mod. Opt.* 35:1919–32
52. Yatsenko LP, Shore BW, Halfmann T, Bergmann K, Vardi A. 1999. *Phys. Rev. A* 60:R4237–40
53. Rickes T, Yatsenko LP, Steuerwald S, Halfmann T, Shore BW, et al. 2000. *J. Chem. Phys.* 113:534–46
54. Hioe FT. 1983. *Phys. Lett. A* 99:150
55. Hioe FT, Eberly JH. 1984. *Phys. Rev. A* 29:1164–67
56. Oreg J, Hioe FT, Eberly JH. 1984. *Phys. Rev. A* 29:690–97
57. Oreg J, Hazak G, Eberly JH. 1985. *Phys. Rev. A* 32:2776–83
58. Solá IR, Malinovsky VS, Chang BY, Santamaria J, Bergmann K. 1999. *Phys. Rev. A* 59:4494–501
59. Chelkowski S, Bandrauk AD, Corkum PB. 1990. *Phys. Rev. Lett.* 65:2355–58
60. Chelkowski S, Bandrauk AD. 1991. *Chem. Phys. Lett.* 186:264–69
61. Chelkowski S, Bandrauk AD. 1993. *J. Chem. Phys.* 99:4279–87
62. Melinger JS, McMorrow D, Hillegas C, Warren WS. 1995. *Phys. Rev. A* 51:3366–69
63. Guérin S. 1997. *Phys. Rev. A* 56:1458–62
64. Ghosh S, Sen S, Bhattacharyya SS, Saha S. 1999. *Phys. Rev. A* 59:4475–84
65. Chelkowski S, Gibson GN. 1995. *Phys. Rev. A* 52:R3417–20
66. Chelkowski S, Bandrauk AD. 1997. *J. Raman Spectr.* 28:459–66
67. Davis JC, Warren WS. 1999. *J. Chem. Phys.* 110:4229–42
68. Karczmarek J, Wright J, Corkum P, Ivanov M. 1999. *Phys. Rev. Lett.* 82:3420–23
69. Villeneuve DM, Aseyev SA, Dietrich P, Spanner M, Ivanov MY, Corkum PB. 2000. *Phys. Rev. Lett.* 85:542–45
70. Demkov YN, Osherov VI. 1968. *Sov. Phys. JETP* 26:916–21
71. Demkov YN, Ostrovsky VN. 1995. *J. Phys. B* 28:403–14
72. Carroll CE, Hioe FT. 1985. *J. Opt. Soc. Am. B* 2:1355–60
73. Carroll CE, Hioe FT. 1986. *J. Phys. A* 19:1151–61
74. Carroll CE, Hioe FT. 1986. *J. Phys. A* 19:2061–73
75. Brundobler S, Elser V. 1993. *J. Phys. A* 26:1211–27
76. Harmin DA. 1991. *Phys. Rev. A* 44:433–61

77. Ostrovsky VN, Nakamura H. 1997. *J. Phys. A* 30:6939–50
78. Mewes M-O, rews MR, Kurn DM, Durfee DS, Townsend CG, Ketterle W. 1997. *Phys. Rev. Lett.* 78:582–85
79. Vitanov NV, Suominen K-A. 1997. *Phys. Rev. A* 56:R4377–80
80. Kuklinski JR, Gaubatz U, Hioe FT, Bergmann K. 1989. *Phys. Rev. A* 40:6741–44
81. Gaubatz U, Rudecki P, Schiemann S, Bergmann K. 1990. *J. Chem. Phys.* 92:5363–76
82. Bergmann K, Shore BW. 1995. *Coherent Population Transfer*, See Ref. 4, pp. 315–73
83. Shore BW. 1995. *Contemp. Phys.* 36:15–28
84. Bergmann K, Theuer H, Shore BW. 1998. *Rev. Mod. Phys.* 70:1003–25
85. Alzetta G, Gozzini A, Moi L, Orriols G. 1976. *Nuovo Cim. B* 36:5–20
86. Alzetta G, Moi L, Orriols G. 1979. *Nuovo Cim. B* 52:2333–46
87. Arimondo E, Orriols G. 1976. *Lett. Nuovo Cim.* 17:333–38
88. Arimondo E. 1996. In *Progress in Optics*, Vol. XXXV, ed. E Wolf, pp. 259–356. Amsterdam: North-Holland
89. Gray HR, Whitley RM, Stroud CR. 1978. *Opt. Lett.* 3:218–20
90. Theuer H, Bergmann K. 1998. *Eur. Phys. J. D* 2:279–89
91. He G-Z, Kuhn A, Schiemann S, Bergmann K. 1990. *J. Opt. Soc. Am. B* 7:1960–69
92. Shore BW, Bergmann K, Kuhn A, Schiemann S, Oreg J, Eberly JH. 1992. *Phys. Rev. A* 45:5297–300
93. Shore BW, Bergmann K, Oreg J. 1992. *Z. Phys. D* 23:33–39
94. Vitanov NV, Stenholm S. 1997. *Phys. Rev. A* 55:648–60
95. Laine TA, Stenholm S. 1996. *Phys. Rev. A* 53:2501–12
96. Vitanov NV, Stenholm S. 1996. *Opt. Commun.* 127:215–22
97. Elk M. 1995. *Phys. Rev. A* 52:4017–22
98. Drese K, Holthaus M. 1998. *Eur. Phys. J. D* 3:73–86
99. Fleischhauer M, Manka AS. 1996. *Phys. Rev. A* 54:794–803
100. Fleischhauer M, Unanyan R, Shore BW, Bergmann K. 1999. *Phys. Rev. A* 59:3751–60
101. Kobrak MN, Rice SA. 1998. *Phys. Rev. A* 57:1158–63
102. Vitanov NV, Stenholm S. 1997. *Opt. Commun.* 135:394–405
103. Danileiko MV, Romanenko VI, Yatsenko LP. 1994. *Opt. Commun.* 109:462–66
104. Romanenko VI, Yatsenko LP. 1997. *Opt. Commun.* 140:231–36
105. Fewell MP, Shore BW, Bergmann K. 1997. *Aust. J. Phys.* 50:281–304
106. Band YB, Julienne PS. 1991. *J. Chem. Phys.* 94:5291–98
107. Vitanov NV, Stenholm S. 1997. *Phys. Rev. A* 56:1463–71
108. Glushko B, Kryzhanovsky B. 1992. *Phys. Rev. A* 46:2823–30
109. Coulston GW, Bergmann K. 1992. *J. Chem. Phys.* 96:3467–95
110. Vitanov NV, Stenholm S. 1999. *Phys. Rev. A* 60:3820–32
111. Band YB, Magnes O. 1994. *Phys. Rev. A* 50:584–94
112. Kobrak MN, Rice SA. 1998. *Phys. Rev. A* 58:2885–94
113. Unanyan R, Guérin S, Shore BW, Bergmann K. 2000. *Eur. Phys. J. D* 8:443–49
114. Gaubatz U, Rudecki P, Becker M, Schiemann S, Külz M, Bergmann K. 1988. *Chem. Phys. Lett.* 149:463–68
115. Rubahn H-G, Konz E, Schiemann S, Bergmann K. 1991. *Z. Phys. D* 22:401–6
116. Lindinger A, Verbeek M, Rubahn H-G. 1997. *Z. Phys. D* 39:93–100
117. Kuhn A, Coulston G, He G-Z, Schiemann S, Bergmann K, Warren WS. 1992. *J. Chem. Phys.* 96:4215–23
118. Schiemann S, Kuhn A, Steuerwald S,

Bergmann K. 1993. *Phys. Rev. Lett.* 71:3637–40
119. Kuhn A, Steuerwald S, Bergmann K. 1998. *Eur. Phys. J. D* 1:57–70
120. Halfmann T, Bergmann K. 1996. *J. Chem. Phys.* 104:7068–72
121. Süptitz W, Duncan BC, Gould PL. 1997. *J. Opt. Soc. Am. B* 14:1001–8
122. Shore BW, Martin J, Fewell MP, Bergmann K. 1995. *Phys. Rev. A* 52:566–82
123. Martin J, Shore BW, Bergmann K. 1995. *Phys. Rev. A* 52:583–93
124. Martin J, Shore BW, Bergmann K. 1996. *Phys. Rev. A* 54:1556–69
125. Shore BW, Bergmann K, Oreg J, Rosenwaks S. 1991. *Phys. Rev. A* 44:7442–47
126. Marte P, Zoller P, Hall JL. 1991. *Phys. Rev. A* 44:4118–21
127. Smith AV. 1992. *J. Opt. Soc. Am. B* 9:1543–51
128. Oreg J, Bergmann K, Shore BW, Rosenwaks S. 1992. *Phys. Rev. A* 45:4888–96
129. Pillet P, Valentin C, Yuan R-L, Yu J. 1993. *Phys. Rev. A* 48:845–48
130. Valentin C, Yu J, Pillet P. 1994. *J. Phys. II* 4:1925–37
131. Goldner LS, Gerz C, Spreeuw RJC, Rolston SL, Westbrook CI, et al. 1994. *Phys. Rev. Lett.* 72:997–1000
132. Goldner LS, Gerz C, Spreeuw RJC, Rolston SL, Westbrook CI, et al. 1994. *Quantum Opt.* 6:387–89
133. Malinovsky VS, Tannor DJ. 1997. *Phys. Rev. A* 56:4929–37
134. Vitanov NV. 1998. *Phys. Rev. A* 58:2295–309
135. Vitanov NV, Shore BW, Bergmann K. 1998. *Eur. Phys. J. D* 4:15–29
136. Nakajima T. 1999. *Phys. Rev. A* 59:559–68
137. Morris JR, Shore BW. 1983. *Phys. Rev. A* 27:906–12
138. Hioe FT, Carroll CE. 1988. *Phys. Rev. A* 37:3000–5
139. Band YB, Julienne PS. 1991. *J. Chem. Phys.* 95:5681–85
140. Carroll CE, Hioe FT. 1992. *Phys. Rev. Lett.* 68:3523–26
141. Carroll CE, Hioe FT. 1993. *Phys. Rev. A* 47:571–80
142. Knight PL. 1984. *Comments At. Mol. Phys.* 15:193–214
143. Knight PL, Lauder MA, Dalton BJ. 1990. *Phys. Rep.* 190:1–61
144. Halfmann T, Yatsenko LP, Shapiro M, Shore BW, Bergmann K. 1998. *Phys. Rev. A* 58:R46–49
145. Yatsenko LP, Halfmann T, Shore BW, Bergmann K. 1999. *Phys. Rev. A* 59:2926–47
146. Nakajima T, Elk M, Zhang J, Lambropoulos P. 1994. *Phys. Rev. A* 50:R913–16
147. Fano U. 1961. *Phys. Rev.* 124:1866–78
148. Carroll CE, Hioe FT. 1995. *Phys. Lett. A* 199:145–50
149. Carroll CE, Hioe FT. 1996. *Phys. Rev. A* 54:5147–51
150. Yatsenko LP, Unanyan RG, Bergmann K, Halfmann T, Shore BW. 1997. *Opt. Commun.* 135:406–12
151. Paspalakis E, Protopapas M, Knight PL. 1997. *Opt. Commun.* 142:34–40
152. Paspalakis E, Protopapas M, Knight PL. 1998. *J. Phys. B* 31:775–94
153. Vitanov NV, Stenholm S. 1997. *Phys. Rev. A* 56:741–47
154. Unanyan RG, Vitanov NV, Stenholm S. 1998. *Phys. Rev. A* 57:462–66
155. Unanyan RG, Vitanov NV, Shore BW, Bergmann K. 2000. *Phys. Rev. A* 61:043408
156. Nakajima T, Lambropoulos P. 1996. *Z. Phys. D* 36:17–22
157. Paspalakis E, Knight PL. 1998. *J. Phys. B* 31:2753–67
158. Yatsenko LP, Guérin S, Halfmann T, Boehmer K, Shore BW, Bergmann K. 1998. *Phys. Rev. A* 58:4683–90
159. Guérin S, Yatsenko LP, Halfmann T, Shore BW, Bergmann K. 1998. *Phys. Rev. A* 58:4691–704
160. Guérin S, Jauslin HR. 1998. *Eur. Phys. J. D* 2:99–113

161. Guérin S, Jauslin HR, Unanyan RG, Yatsenko LP. 1999. *Opt. Express* 4:84–90
162. Garraway BM, Suominen K-A. 1998. *Phys. Rev. Lett.* 80:932–35
163. Kallush S, Band YB. 2000. *Phys. Rev. A* 61:041401
164. Solá IR, Santamaría J, Malinovsky VS. 2000. *Phys. Rev. A* 61:043413
165. Rodriguez M, Garraway BM, Suominen K-A. 2000. *Phys. Rev. A* 62:053413
166. Lawall J, Prentiss M. 1994. *Phys. Rev. Lett.* 72:993–96
167. Weitz M, Young BC, Chu S. 1994. *Phys. Rev. Lett.* 73:2563–66
168. Weitz M, Young BC, Chu S. 1994. *Phys. Rev. A* 50:2438–44
169. Vitanov NV, Suominen K-A, Shore BW. 1999. *J. Phys. B* 32:4535–46
170. Unanyan R, Fleischhauer M, Shore BW, Bergmann K. 1998. *Opt. Commun.* 155:144–54
171. Unanyan RG, Shore BW, Bergmann K. 1999. *Phys. Rev. A* 59:2910–19
172. Theuer H, Unanyan RG, Habscheid C, Klein K, Bergmann K. 1999. *Opt. Express* 4:77–83
173. Dittmann P, Pesl FP, Martin J, Coulston GW, He G-Z, Bergmann K. 1992. *J. Chem. Phys.* 97:9472–75
174. Pesl FP, Lutz S, Bergmann K. 2001. In press
175. Külz M, Keil M, Kortyna A, Schellhaass B, Hauck J, Bergmann K, et al. 1996. *Phys. Rev. A* 53:3324–34
176. Featonby PD, Summy GS, Martin JL, Wu H, Zetie KP, et al. 1996. *Phys. Rev. A* 53:373–80
177. Featonby PD, Summy GS, Webb CL, Godun RM, Oberthaler MK, et al. 1998. *Phys. Rev. Lett.* 81:495–99
178. Morigi G, Featonby P, Summy G, Foot C. 1996. *Quant. Semiclass. Opt.* 8:641–53
179. Godun RM, Webb CL, Oberthaler MK, Summy GS, Burnett K. 1999. *Phys. Rev. A* 59:3775–81
180. Webb CL, Godun RM, Summy GS, Oberthaler MK, Featonby PD, et al. 1999. *Phys. Rev. A* 60:R1783–86
181. Weitz M, Heupel T, Hänsch TW. 1996. *Phys. Rev. Lett.* 77:2356–59
182. Aspect A, Armindo E, Kaiser R, Vanteenkiste N, Cohen-Tannoudji C. 1988. *Phys. Rev. Lett.* 61:826–29
183. Aspect A, Armindo E, Kaiser R, Vanteenkiste N, Cohen-Tannoudji C. 1989. *J. Opt. Soc. Am. B* 6:2112–24
184. Chu S. 1998. *Rev. Mod. Phys.* 70:685–706
185. Cohen-Tannoudji CN. 1998. *Rev. Mod. Phys.* 70:707–20
186. Phillips WD. 1998. *Rev. Mod. Phys.* 70:721–42
187. Lawall J, Bardou F, Saubaméa B, Shimizu K, Leduc M, et al. 1994. *Phys. Rev. Lett.* 73:1915–18
188. Lawall J, Kulin S, Saubaméa B, Bigelow N, Leduc M, Cohen-Tannoudji C. 1995. *Phys. Rev. Lett.* 75:4194–97
189. Lawall J, Kulin S, Saubaméa B, Bigelow N, Leduc M, Cohen-Tannoudji C. 1996. *Laser Phys.* 6:153–58
190. Kasevich M, Weiss DS, Riis E, Moler K, Kasapi S, Chu S. 1991. *Phys. Rev. Lett.* 66:2297–300
191. Kasevich M, Chu S. 1992. *Phys. Rev. Lett.* 69:1741–44
192. Esslinger T, Sander F, Weidemüller M, Hemmerich A, Hänsch TW. 1996. *Phys. Rev. Lett.* 76:2432–35
193. Kulin S, Saubaméa B, Peik E, Lawall J, Hijmans TW, et al. 1997. *Phys. Rev. Lett.* 78:4185–88
194. Parkins AS, Marte P, Zoller P, Kimble HJ. 1993. *Phys. Rev. Lett.* 71:3095–98
195. Parkins AS, Marte P, Zoller P, Carnal O, Kimble HJ. 1995. *Phys. Rev. A* 51:1578–96
196. Parkins AS, Kimble HJ. 1999. *J. Opt. B* 1:496–504
197. Lange W, Kimble HJ. 2000. *Phys. Rev. A* 61:063817
198. Newton RG, Young B-L. 1968. *Ann. Phys.* 49:393–402
199. Vitanov NV, Shore BW, Unanyan RG,

Bergmann K. 2000. *Opt. Commun.* 179: 73–83
200. Vitanov NV. 2000. *J. Phys. B* 33:2333–46
201. Javanainen J, Mackie M. 1998. *Phys. Rev. A* 58:R789–92
202. Javanainen J, Mackie M. 1999. *Phys. Rev. A* 59:R3186–89
203. Mackie M, Javanainen J. 1999. *Phys. Rev. A* 60:3174–87
204. Mackie M, Kowalski R, Javanainen J. 2000. *Phys. Rev. Lett.* 84:3803–6

THE DYNAMICS OF "STRETCHED MOLECULES": Experimental Studies of Highly Vibrationally Excited Molecules With Stimulated Emission Pumping

Michelle Silva[1], Rienk Jongma[2], Robert W Field[1], and Alec M Wodtke[3]

[1]*Department of Chemistry, Massachusetts Institute of Technology, 77 Massachusetts Ave., Cambridge, Massachusetts 02139; e-mail: msilva@mit.edu; rwfield@mit.edu*
[2]*Department of Molecular and Laser Physics, University of Nijmegen, Toernooiveld 1, NL-6525 ED Nijmegen and FOM-Institute for Plasma Physics Rijnhuizen, P.O. Box 1207, NL-3430 BE Nieuwegein, The Netherlands; e-mail: jongma@huygens.rijnh.nl*
[3]*Department of Chemistry and Biochemistry, University of California, Santa Barbara, California 93106; e-mail: alec1@silcom.com*

Key Words molecular dynamics, electron transfer, energy transfer, intramolecular vibrational distribution

■ **Abstract** We review stimulated emission pumping as used to study molecular dynamics. The review presents unimolecular as well as scattering studies. Topics include intramolecular vibrational redistribution, unimolecular isomerization and dissociation, van der Waals clusters, rotational energy transfer, vibrational energy transfer, gas-surface interactions, atmospheric effects resulting from nonequilibrium vibrational excitation, and vibrational promotion of electron transfer.

INTRODUCTION

For much of the history of the field of chemistry, scientists have tried to understand the fundamental nature of chemical reactivity by analysis of reactants and products. Many important concepts characterizing reactants (e.g. electrophile, nucleophile) or products (e.g. good or poor leaving group) have been developed, allowing chemical intuition to be refined by systematic study of related families of reactions. Analysis of molecules that reside near stable minima of the reaction hypersurface is the most obvious window into the world of chemical reactions. However, intriguing unanswered questions remain regarding the properties of molecules actually undergoing chemical reactions. That is, how should we think about the chemical properties of transients that reflect the unstable minima of the reaction hypersurface? Is it possible to develop a similar chemical intuition about such fundamentally important but difficult to isolate chemical entities? It is only recently that experimental probes capable of catching molecules in the act of making and breaking bonds have become available (1–5). One such experimental technique that has proven particularly useful in this regard is stimulated emission pumping (SEP) (6). With its ability to prepare vibrationally excited molecules with a near-arbitrary degree of excitation and absolute quantum state selectivity, SEP provides remarkable opportunities to understand the bond-making and bond-breaking processes occurring in chemical reactions in a controlled and quantitative way.

To appreciate this, consider the example shown in Figure 1 (see color insert). Here we show the square of the actual vibrational wave function for two vibrational states of one of the better-studied molecules, nitric oxide (NO). The twenty-fourth excited vibrational level in the ground electronic state, a state easily prepared using SEP, is shown is blue. For comparison, the ground vibrational state is also shown. One aspect of these high vibrational states that becomes immediately obvious from the figure is the enormous amplitude of motion. Furthermore, due to the common fact that the interaction potential possesses a very steep repulsive wall with a gentler attractive potential, there is a substantial bias of the vibrational motion toward the outer turning point. Indeed, 32% of the probability resides in the outer lobe of the probability distribution, and 80% of the probability is found at bond lengths longer than r_e. In light of this, it is useful to think of this excited vibrational state as a "stretched NO molecule," a dynamic system sampling geometries far from the minimum of the potential surface, clearly not a "normal NO reactant," but equally as clearly not atomic products of the dissociation reaction. By preparing molecules in such highly excited states, one approaches many interesting chemical questions from a thoroughly state-resolved, i.e. quantum mechanical, point of view.

This extraordinarily useful experimental method provides answers to questions ranging from intramolecular vibrational redistribution in highly excited molecules to vibrational influences on electron transfer; from quantum-state resolved unimolecular dissociation to direct determinations of hydrogen-bond strengths in clusters; and from nonequilibrium atmospheric ozone chemistry to spectroscopic determinations of chemical reaction hypersurfaces, just to name a few. SEP has been reviewed on several prior occasions (7–15). Most of those articles focused on SEP as a tool in spectroscopy of highly vibrationally excited molecules.

SEP is no longer (if it ever was) simply a spectroscopic tool used to determine vibrational influences on molecular structure. As researchers discovered the increasing complexity of highly vibrationally excited polyatomic molecules, new principles of spectroscopic analysis had to be developed, principles that have revealed the vibrational dynamics present in the spectra of highly vibrationally excited polyatomic molecules (65). New kinds of "simple vibrational motion" distinct from normal and local mode pictures have been discovered and characterized, showing that what once seemed to represent chaotic behavior can, in fact, be understood with deep thinking and intense effort. A second fork in the scientific evolution of SEP exploits a strategy where assigned spectra from SEP spectroscopy act as a road map for preparation of exotic reactants (12). SEP is then used as a means to synthesize novel reactants for use in chemical studies. Once the basic spectroscopy is well characterized, molecules can be prepared in selected states with known vibrational motion (even the wave function itself may be known), and the influence of vibration on chemistry can be explored. With this method, precisely tailored experiments can probe dynamical processes at chemically relevant energies. Consequently, the seemingly static frequency domain spectra offer a wealth of information about unimolecular and bimolecular dynamics.

This article focuses on recent developments, cases in which our knowledge about molecular spectroscopy serves as the starting point for understanding the dynamics of highly vibrationally excited molecules.

First, we briefly describe some of the technical and conceptual issues surrounding the application and implementation of stimulated emission pumping. We then discuss results from recent work that shed light on the nature of molecules in "stretched states." We have broken the paper into two sections: Unimolecular Phenomena (including sections on intramolecular vibrational redistribution and unimolecular isomerization and dissociation, as well as studies of clusters) and Scattering Experiments (including sections on rotational energy transfer, vibrational energy transfer, and interactions at the gas surface interface). We wish to offer apologies for all of the beautiful work that we do not mention in this article. As with any review article, this paper reflects the strengths and weaknesses of its authors. Furthermore, we have intentionally focused only on applications of SEP even though studies using other methods (for example overtone pumping) are thematically similar to those presented here.

SIMULATED EMISSION PUMPING: How the Experiments Are Done

Figure 2 (see color insert) shows the principles of the stimulated emission pumping experiment. SEP is a folded variant of optical-optical double resonance spectroscopy. The sample is excited with a narrow-band pulsed laser operating at the "PUMP" frequency, ω_{pump}, out of thermally populated levels into an excited electronic state. A second pulsed laser, the "DUMP" laser, tuned to ω_{dump}, stimulates emission out of the excited electronic state and transfers population back to a single highly vibrationally excited quantum level in the ground electronic state.

The Franck-Condon principle explains why preparation of highly vibrationally excited molecules can be so straightforward. Generally, the change in the interatomic potential (both equilibrium structure and harmonic frequencies) upon electronic excitation is large; therefore, the "Franck Condon matrix," $Q(v', v'')$, can be and often is highly nondiagonal, thus favoring large changes in the vibrational quantum number for both the PUMP and DUMP transitions. For the example of NO shown in Figure 2, the strongest DUMP transitions result in very highly vibrationally excited states, e.g. $Q(v' = 5, v'' = 19) = 0.2$. Using modern laser technology (in particular, pulsed lasers) that provides light sources of astounding spectral brightness, employing DUMP transitions with $Q(v', v'') < 10^{-6}$ is often possible.

The SEP transition may be detected in many ways, such as ionization dips (16–23) and degenerate four-wave mixing (24–27). However, due to its simplicity, the "fluorescence dip" method (6, 28–31) is used most commonly and has been a powerful tool for spectroscopic investigations of highly vibrationally excited molecules. In this approach the reduction in spontaneous side fluorescence is measured when stimulated emission removes population from the fluorescing state by producing photons along the DUMP laser direction.

One can also use a third laser to probe the sample after the preparation of the DUMP pulse populates a selected highly vibrationally excited state. We refer to this as the "PUMP-DUMP-PROBE" geometry (32–34). When done in a cell, state-to-state kinetics measurements are possible (35, 36). Alternatively, one may use molecular beams (Figure 3, see color insert). In a molecular beam experiment, the molecules prepared by SEP are detected by laser-induced fluorescence or resonant-enhanced multiphoton ionization downstream from and at a specified time delay after the DUMP pulse (37–41). The temporal and spatial separation of the preparation step from the detection step allows detection of very weak transitions and, of course, provides a probe of collisional events.

Recently, a new and as yet unexplored suggestion has appeared in the literature (42). Specifically, it has been proposed that the vibrational and rotational state-specific dipole moment can be used to separate SEP-prepared molecules from a molecular beam and focus them to high densities for detection by electron impact ionization using hexapoles. A hexapole is a simple device, consisting of six identical parallel metal rods in a cylindrically symmetric arrangement with opposing voltages on alternating rods. Along the cylindrical axis of symmetry no field is present, and as one moves in any direction away from the symmetry axis the magnitude of the electric field increases. For molecules with a positive Stark shift, the state's energy increases with distance from the center of the hexapole. The spatial gradient of this energy represents a restoring force, which tends to bring these states back toward the center of the hexapole. States like these are called low-field seeking states and can be focused by a hexapole (43–58). States with a negative Stark shift are called high-field seeking states and are defocused by the hexapole.

Figure 4 (see color insert) shows results of classical trajectory calculations demonstrating that ground state molecules of NO are focused at very different rod voltages than are SEP-populated high vibrational states (43). This is not so much a result of the vibrational dependence of the dipole moment, but rather reflects the fact that one can easily change the Ω quantum number with SEP. Exciting NO from X $^2\Pi_{1/2}$ J = 1/2 to X $^2\Pi_{3/2}$ J = 3/2 nearly doubles the first order Stark effect for the excited molecules because of the $\Omega M/J(J+1)$ dependence of the Stark tuning coefficient.

There are clearly many ways to carry out an SEP experiment, and new ways are still being invented. We now turn to applications of various two- and three-laser incarnations of the SEP/PUMP-DUMP-PROBE experiment and review what has been learned about vibrational dynamics in a host of different molecular problems. The next section describes work designed to answer this intriguing question: What happens to a highly vibrationally excited molecule that cannot suffer collisions?

UNIMOLECULAR DYNAMICS

Major advances have been made in the study of unimolecular dynamics through the use of SEP. SEP has also contributed to the understanding of the structure and dynamics of van der Waals clusters, radicals, and other transient molecules. The relatively high resolution of SEP makes it ideal for observing the ro-vibrational structure in the spectra of small polyatomic molecules. Moreover, the selectivity one gains by preparing a particular rovibronic level in the PUMP step of an SEP experiment profoundly simplifies spectra measured when the DUMP laser is scanned. This selectivity facilitates assignment and converts what might first appear to be a completely chaotic spectrum into a transparently interpretable map of the system under investigation. The following sections summarize some important recent SEP studies that have not been reviewed previously. We attempt to provide an extensive overview of SEP work performed to date, but we specifically focus on selected recent SEP experiments. Our goal is to demonstrate the impact of SEP experiments on our understanding of unimolecular reaction dynamics. We describe advances in our understanding of intramolecular vibrational redistribution (IVR), including a discussion of isomerization as a special type of IVR process. We also review work on quantum state specific unimolecular dissociation dynamics. We briefly mention SEP studies of transient molecules and then move to a discussion of recent advances in van der Waals cluster spectroscopy.

Intramolecular Vibrational Redistribution

One of the cherished dreams of physical chemists has been to understand how the energy of an absorbed photon is initially distributed over the internal motions of a molecule and, more importantly, how this excitation energy redistributes with time, influencing chemistry as it goes. For example, how exactly does a molecule isomerize and what motions facilitate this process? What controls the competition between localized excitation and localization-facilitated chemistry and loss of the local excitation to the rest of the molecule? More importantly, by what mechanism does energy flow from one internal motion to others? van der Waals molecules can be the subject of similar studies, and the possibility exists to characterize the mechanism of energy flow in a model solvent/solute system. To what extent will the weak bonds and weak couplings that characterize van der Waals molecules restrict energy flow in favor of dissociation?

One of the most important contributions of SEP lies in the study of IVR. When vibrational energy is deposited into a polyatomic molecule with a pulsed laser, using light with a bandwidth greater than the inverse density of states, superposition states are created in which the admixture of particular molecular eigenstates is controlled by the relative transition strengths that couple them to the initial state. A good example is overtone pumping of a C–H stretch motion in a large polyatomic molecule. Because of the large anharmonicity of the C–H bond, overtone excitation to eigenstates with a greater component of C–H stretching amplitude are more easily excited than isoenergetic states with less C–H stretch amplitude. Of course, the superposition state evolves from an initially localized excitation that looks like a localized C–H stretching motion to a more complex vibrational motion. This evolution is controlled by couplings between the localized C–H vibration and the rest of the molecule, couplings that are encoded by the potential hypersurface into the eigenstates. Theoretically, the time evolution of such an initial, nonstationary state is very similar to that of a coherently excited Fermi-resonance pair[1], with the difference that the recurrence times may be so long as to be irrelevant.

As just described, IVR would appear to be a topic more suited to time-resolved than frequency-resolved experiments. However, this view is certainly oversimplified. For small molecules, frequency domain experiments can actually provide a more detailed picture of time resolved dynamics than direct measurement in the time domain.

The question boils down to "How does IVR manifest itself in emission spectra?" Typically, at low energies, the rate of IVR is very slow and the spectra are simple. However, at higher energies, IVR rates increase and the spectra become much more complicated. By using low-resolution dispersed fluorescence (DF) to extrapolate from the "long-time simple" regime at low energy to the "short-time simple" regime at high energy, progressions and the most important early-time energy flow pathways can be identified (e.g. Figure 5, see color insert, Figure 7).

SEP offers another advantage. It is often difficult to access highly excited vibrational states with either efficiency or selectivity. By exploiting selected intermediate vibrational levels of an excited electronic state in the PUMP stage, SEP minimizes these difficulties. Thus, one gains access to classes of vibrational levels via the DUMP step that are impossible to examine using conventional absorption techniques because access to those levels from the ground vibrational level is symmetry-forbidden or occurs only through extremely weak high overtone and combination band transitions.

An SEP spectrum encodes, in the frequency domain, the dynamic evolution of one or more individually prepared zero-order bright states (ZOBSs). A ZOBS is not an eigenstate, and if one ZOBS were prepared at $t=0$, the system would evolve in time in such a way that the Fourier transform of the temporal response is the actually recorded SEP eigenstate spectrum.

---

[1] In the now famous words of Prof. Doug MacDonald, "Fermi-resonance and IVR are the *same* thing...".

The power of SEP arises from the ability to initially access, or rather to pluck, a wide variety of a priori perfectly known ZOBSs. Which ZOBSs are accessible in a particular SEP spectrum depends on the choice of the intermediate eigenstate populated by the PUMP laser. More than any other technique for populating highly vibrationally excited states, "brightness" may be defined purposefully by the experimentalist.

The bright states in an SEP spectrum are the "Franck-Condon bright states" (FCBS), and they correspond to a wide range of excitation energies exclusively in the Franck-Condon bright vibrational normal modes. These are the modes that correspond to a change in equilibrium structure between the intermediate and target electronic states. The vibrational selection rule for all of the Franck-Condon dark normal modes is $\Delta v = 0$. Each ZOBS is said to fractionate into finer "feature states" and eventually, at sufficiently high spectral resolution (long time of evolution), into molecular eigenstates.

It is not necessary, and often it is not recommended, to go to the extreme of eigenstate-resolved SEP spectra. SEP spectra encode intramolecular dynamics, and the fastest or earliest time dynamics is often easiest to understand because it involves only the most important anharmonic couplings. As a result, the early-time dynamics, which is encoded in low-resolution SEP spectra, typically explores a lower dimensionality state space than that of the exact molecular Hamiltonian.

Low dimensionality amounts to easier to visualize and easier to understand in terms of harmonic-oscillator matrix-element scaling rules that are built into a spectroscopic effective Hamiltonian, H^{eff}. The dimensionality of the state space actually explored at a time corresponding to the reciprocal spectral resolution is lower than the full density of states because certain approximate quantum numbers, which are not guaranteed by symmetry to be conserved, are conserved on this short time scale.

These approximately conserved quantum numbers are polyad quantum numbers and serve to approximately block-diagonalize the molecular Hamiltonian. Each polyad encodes the early-time dynamics of one or more ZOBSs that are said to illuminate that specific polyad. Each polyad appears in the SEP spectrum as a fractionated pattern resulting from the intrapolyad dynamics of one or more, possibly interfering, ZOBSs. The job of the spectroscopist is to read these patterns and to work backwards from the pattern to the small number of dynamically relevant anharmonic coupling terms that dominate the early-time dynamics.

One problem of restricting one's attention to early-time dynamics is that the corresponding frequency-domain spectra are inherently low-resolution spectra, and dynamically unrelated fractionated patterns often overlap. Statistical pattern-recognition methods have been devised that enable overlapping polyad patterns to be disentangled from each other (59).

One big advantage of restricting one's attention to early-time dynamics is that all matrix elements in one polyad H^{eff} are explicitly scaleable with the vibrational quantum numbers of the interacting states. This means that the information in one polyad H^{eff} is sufficient to generate all members of a large family of related, more highly excited polyads.

The dynamics encoded by the polyad H^{eff} often undergo qualitative changes, either from regular to chaotic or from one kind of regularity (e.g. normal mode) to another (e.g. local mode). These qualitative changes may best be viewed by examining the nodal structure of the polyad eigenstates in coordinate space or by converting the quantum mechanical H^{eff} to a classical Hamiltonian and using various techniques of nonlinear dynamics (especially surfaces of section) to view the classes of quasiperiodic motion present in the classical trajectories.

We now turn to a few specific illustrative examples. Many SEP experiments have been performed to map the energy flow from an initially localized excitation to other modes in the molecules: H_2CO, HCCH, HCN, HCP, HFCO, SO_2, CS_2, CH_3O, $SCCl_2$, and $H_2C_2O_2$ (60–88). Three molecules have provided particularly informative examples for studies of IVR by SEP: HCCH, HFCO, and $SCCl_2$.

Acetylene: Systematic Exploration of Intramolecular Vibrational Redistribution

Acetylene is a prototypical system for studying IVR on an ~1 ps time scale. The first SEP spectra of acetylene revealed dense complex patterns of transitions (65) (e.g. Figure 5, bottom panel). At first it seemed there was no hope of deciphering the dynamics encoded in the spectra at very high vibrational excitation. Nevertheless, it was seen that lines tended to group together. Initially, these groups were called "clumps," and the clumps could be rotationally assigned. The groups of lines observed in the spectra were assigned in terms of the FCBSs. A FCBS illuminates a clump of dark states, each with the same value of the rigorously conserved total rotational angular momentum, J. By tuning the PUMP laser to excite a series of intermediate states with known consecutive values of J, the SEP spectrum reveals clumps illuminated by consecutive rotational levels of the same set of bright vibrational states. As a result, the energies of the clumps follow $BJ(J + 1)$ rotational term curves, where the B value is the rotational constant of the FCBS. Although the dark-state vibrational constituency of each clump changes profoundly from one J to the next, one can follow the energy of a single FCBS as it tunes through the dense manifold of dark states.

One researcher claimed that SEP spectra of acetylene near 27,000 cm^{-1} were the first example of a molecular system exhibiting quantum chaos (65, 71). Statistical examinations of the level spacing and intensity distributions within the clumps of closely spaced lines in the SEP spectrum of the acetylene $\tilde{X}^1\Sigma_g^+$ state revealed a Wigner rather than Poisson distribution, which is characteristic of chaotic behavior (89).

It is now recognized that the bright state fractionates in a predicable and scaleable way. Analysis of the SEP spectra reveals that what was initially thought to be a manifestation of quantum chaos was actually a simple and understandable early-time flow of energy from one ZOBS into dark states. The early-time dynamics is intrapolyad and nonchaotic because the polyad quantum numbers are conserved for ~1 ps. However, the long-time dynamics is not restricted to one polyad and might very well fit the definition of chaotic.

Analysis of SEP spectra allowed the identification of approximately conserved quantum numbers that better describe the short-time motion of acetylene in the high-energy regime. These "polyad quantum numbers" (68, 77, 90–95) are specified as follows:

$$N_{res} = 5v_1 + 3v_2 + 5v_3 + v_4 + v_5,$$
$$N_s = v_1 + v_2 + v_3,$$
$$l = l_4 + l_5,$$
$$g/u \text{ symmetry},$$
$$J \text{ and}$$
$$+/- \text{ parity}.$$

Each fractionated bright state pattern observed in DF and SEP spectra can be assigned to one set of these polyad quantum numbers. Analysis of the polyads reveals the emergence of a new class of vibrational motions that better describe this system at chemically relevant energies, where for example, isomerization becomes feasible. Specifically, it has been shown that local-bender and counter-rotator vibrational modes are ideal for describing the large amplitude bending motions in acetylene when the bending vibrations are highly excited (96).

So far we have not concerned ourselves with how the SEP spectra might be influenced by the isomerization of acetylene to vinylidene. What is the role of IVR in promoting or frustrating near-threshold isomerization? As the excitation energy approaches the adiabatic threshold for isomerization, the SEP spectrum can reveal, for example, how the energetic accessibility of the vinylidene isomer will make itself felt in the spectra illuminated by acetylene bright states. For an initial "pluck" localized in the acetylene region, how (if at all) is the early-time dynamics of the spectrum created by that pluck affected when vinylidene becomes energetically accessible? When isomerization can occur the scalability mandated by the polyad H^{eff} is no longer guaranteed because the frequency of the local bender falls to zero and then becomes imaginary. The fractionation patterns might change in some subtle manner not predicted by polyad scaling or might even exhibit a sudden and total disruption of the scaleable polyad patterns.

Acetylene dynamics is well understood up to about 15,000 cm^{-1} of internal energy, which is just below the energy of the zero-point energy-dressed isomerization barrier. Analysis of the spectra near 15,000 cm^{-1} should yield a wealth of information about acetylene-vinylidene interactions. Thus far, SEP spectra recorded at low internal energy have confirmed the early-time dynamical mechanisms revealed in dispersed fluorescence studies (97). At higher energies the SEP spectra are much more complex, and each feature in the DF spectrum fractionates into many finer ones in the SEP spectrum. Figure 5 (96) illustrates the complementarity of DF and SEP spectra. The spectra in this figure are recorded via the $(2v_3')$ $^{13}C_2H_2$ Ã vibrational level. The top two panels show the broad features observed at low resolution in the DF spectrum, and the lower panel shows a portion of a higher-resolution SEP spectrum (96).

No obvious qualitative changes in DF and SEP spectra of acetylene have been observed near the isomerization barrier, but subtle effects in the spectra of the symmetric acetylene isotopomers ($^{13}C_2H_2, ^{13}C_2D_2$) may provide direct spectroscopic information about isomerization (96). For example, the observation of resolved nuclear permutation splittings in $^{13}C_2H_2$ spectra would be a direct spectral signature of acetylene-vinylidene isomerization. Careful searches for such splittings in $^{13}C_2H_2$ SEP spectra are ongoing.

HFCO: Extreme Motion States Limit Intramolecular Vibrational Redistribution

Statistical theories of unimolecular decay, such as Rice-Ramsperger-Kassel-Marcus (RRKM), are based on the assumption that IVR is complete before reactions occur. Whereas this assumption is usually valid for large polyatomic molecules, it clearly fails for small molecules such as triatomic molecules. It is desirable to find criteria that suggest when the RRKM assumption is likely to be satisfied for a specific case. For example, is McDonald's (98) proposal that above a density of states of $\sim 10^2/cm^{-1}$ RRKM predictions reliable?

SEP spectra have documented many cases of nonstatistical behavior in small polyatomic molecules at high internal energy. It is becoming clear when to expect nonstatistical behavior. A prime example of such a study is the work from Moore's group on HFCO. In this case, IVR is clearly nonstatistical, and mode-specific effects have been shown to profoundly affect the dynamics (60, 61, 63, 99–101). The primary Franck-Condon active modes in jet-cooled SEP spectra of HFCO are the CO stretch, ν_2, and the out-of-plane bend, ν_6. The extent of vibrational mixing of FCBS with background dark states varies dramatically with the zero-order bright state. States at the same total excitation energy with high excitation in ν_2 show much more state mixing than those with ν_6 excited.

Even at energies far above the threshold to dissociation, bright states, which contain a large amount of ν_6 vibration, are stable and exhibit minimal IVR. This is reminiscent of the relative lack of fractionation of highly excited CC-stretch bright states for acetylene at energies well above the HCCH $\leftrightarrow$ CCH$_2$ barrier.

At high energy, where the classical dynamics is typically irregular or chaotic, a few regular or localized states often coexist with the strongly mixed or delocalized states. A particular subset of such states is the "extreme motion" states. In such states, the energy is predominantly localized in a single vibrational mode, regardless of whether that mode is described by traditional normal mode or local mode quantum numbers. These extreme motion states in HFCO are observed to be decoupled from background dark states (101). The extent of the mixing was found to be only weakly dependent on rotational level for states far above the dissociation threshold, in stark contrast to levels in the tunneling region (barrier to dissociation $\sim 17,000$ cm^{-1}), which exhibit a strong dependence on rotational level.

The IVR of DFCO exhibits striking contrasts to IVR for HFCO. SEP spectra of DFCO are much more complicated than spectra of HFCO (61). SEP spectra recorded at different resolutions reveal the existence of two IVR time scales in

DFCO. The first (fastest) stage of IVR results from the near resonant 2:1 interaction between the v_2 (1797 cm^1) and v_6 (857 cm^{-1}) modes (63). The strong 2,66 Fermi resonance leads to a new approximately conserved polyad quantum number:

$$N = v_2 + v_6/2,$$

where the polyad quantum number reflects the approximate energy resonance condition in which it takes two quanta of v_6 to match the energy of one quantum of v_2. Also, v_6, not being a totally symmetric mode, can exchange energy with the totally symmetric mode only in steps of even quanta.

SEP spectra provided a map of 2,66 mixing in DFCO as a function of v_6. Figure 6, from Crane et al (63), illustrates the mixing coefficient $H_{2,66}>/\Delta E$ between zero-order states as a function of v_6. The interaction constant $k_{2,66}$ is found to be on the order of 42–44 cm^{-1}, based on a fit to SEP spectra (61). For DFCO, $\omega_2 - 2\omega_6 \approx 83$ cm^{-1}, whereas in HFCO $\omega_2 - 2\omega_6 \approx -185$ cm^{-1} (61, 63). Figure 6 shows that the mixing coefficient increases as the polyad number, N, increases. Consequently, the quasistability of the HFCO extreme motion states has no counterpart in DFCO. The extreme motion states, which were observed to be quasistable in HFCO, are systematically near degenerate with background dark states in DFCO. The 2,66 Fermi resonance results in a fast initial stage of IVR in DFCO, occurring on a 25 fs time scale. Weaker resonances have also been identified in DFCO: $v_2 \approx 2v_3$, $k_{2,33} = 23$–27 cm^{-1} and $2v_6 \approx v_3 + v_5$, $k_{35,66} = 5$–11 cm^{-1} (61, 63). These weaker resonances destroy the polyad quantum number, N, on a time scale longer than 25 fs. By 500 fs N is no longer conserved because the 2,33 and 35,66 anharmonic resonances couple the $2_n 6_m$ Franck Condon bright states to the background dark states. For many states of DFCO IVR was found to be nearly complete on the time scale of dissociation (63). DFCO does not exhibit the mode-selective dissociation behavior of HFCO, and as a result, statistical theories such as RRKM are well suited to describe its reaction dynamics.

$SCCl_2$: Near the Statistical Limit

In the limit of a dense manifold of nearly equally coupled levels, IVR can be described by Fermi's Golden Rule, in which the exponential decay rate of any initially localized excitation is given by:

$$\Gamma = \frac{2\pi}{\hbar} \rho V_{rms}^2$$

The vibrational density of states is denoted by ρ, and V_{rms}^2 is the average squared off-diagonal coupling matrix element. It should be noted that there is some ambiguity in specifying the appropriate density of states. When IVR is incomplete it is appropriate to use the density of strongly coupled states rather than the total density. Again the central question is the applicability of statistical IVR models to highly vibrationally excited states of small molecules and to not so highly excited states of large molecules. At early times IVR can be described as an exponential process in which the characteristic IVR rate is given by the $1/e$ decay specified by the Golden Rule, but in which the density of states is much smaller than the total

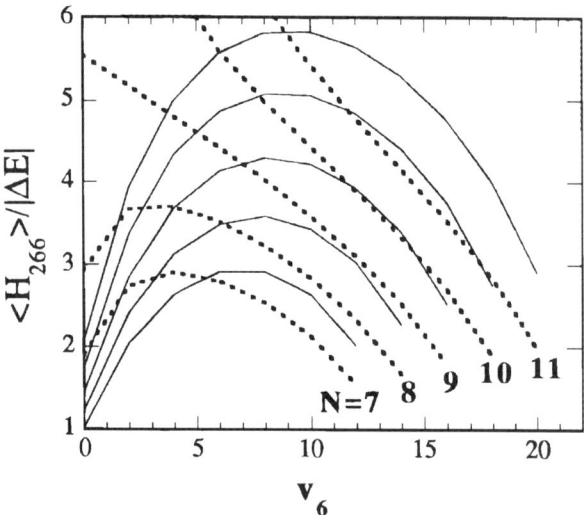

Figure 6 The coupling strength, $<H_{2,66}>/\Delta E$, between the zero-order states, $2_n 6_m$ and $2_{n-1} 6_{m+2}$. This mixing angle is plotted as a function of the v_6 quantum number for the $N = v_2 + v_6/2$ polyads with $N = 7$–11. Solid and dashed lines represent data calculated using two different sets of spectroscopic constants. The coupling strengths increase for higher polyads, and within a polyad the coupling strength is stronger for states near the middle of the polyad. Reprinted with permission from *J. Phys. Chem.* 102, 9433, (1998). Copyright 1998 American Chemical Society.

density of states. However, at longer times the heavy atom low-frequency skeletal modes dominate IVR dynamics instead of the high frequency hydrogenic modes (102), and nonexponential behavior is observed.

This intermediate time regime in which low frequency skeletal modes begin to govern the IVR dynamics bears some discussion. In the case of weak resonances, this region exhibits partial rephasing and nonexponential decays. In the case of a "statistical limit," in which bright-state character is distributed over a larger number of eigenstates, Gruebele and coworkers have developed a power-law description for the survival probability of any localized excitation based on dispersed fluorescence (DF) and stimulated emission pumping (SEP) experiments (102, 103). The power-law description implies less complete vibrational redistribution than that implicit in the Golden Rule and RRKM model.

Experimental modeling and theoretical calculations on thiophosgene, $SCCl_2$, indicate that vibrational energy redistribution among backbone vibrational modes is complete in the sense that it is independent of the initially prepared state. Figure 7 (86) illustrates both the DF and SEP studies of thiophosgene. SEP experiments on the $\tilde{B} \rightarrow \tilde{X}$ transition reveal the presence of weak localized anharmonic resonances at low energy, whereas at high energies ($>v_1 = 8$) highly fractionated progressions, characteristic of resonant IVR, are observed (86). IVR is resonant when a sparse

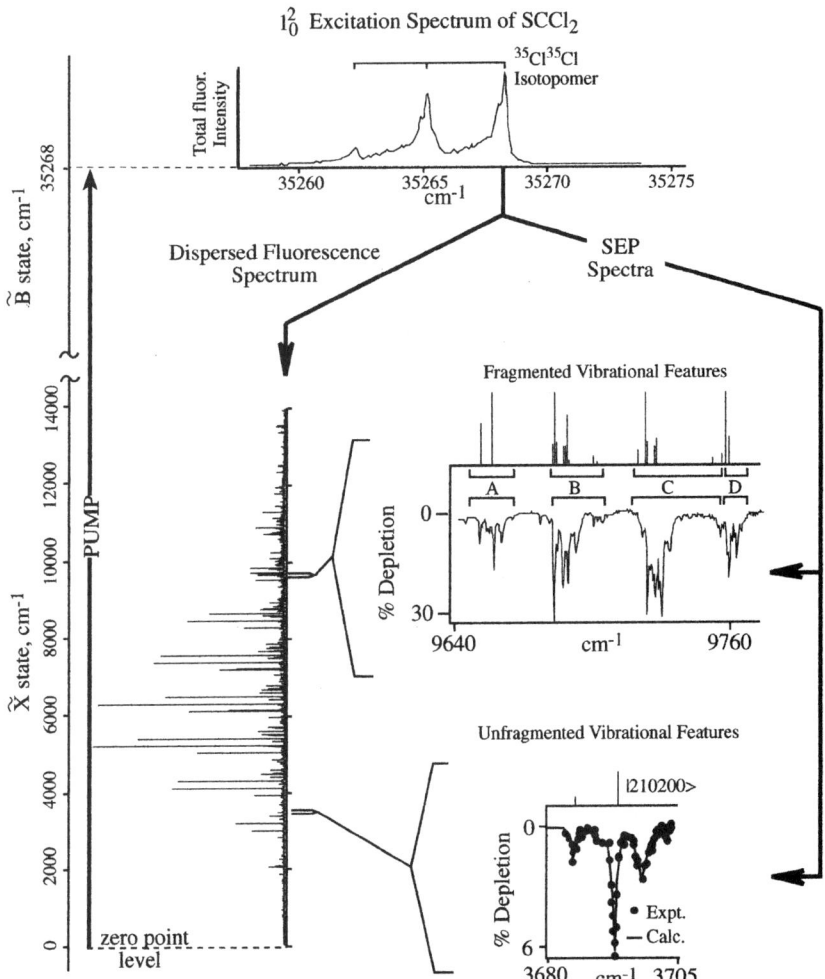

Figure 7 Excitation, dispersed fluorescence, and stimulated emission pumping (SEP) spectra of thiophosgene. The polyad structure in the fluorescence spectrum corresponds to progressions conserving $2\nu_1 + \nu_4$. ($\nu_1 =$ CS stretching mode, $\nu_4 =$ out-of-plane bending mode). The spectra reveal a fragmentation pattern hierarchy that extends over all energy scales from 1000 cm^{-1} down to the 0.1 cm^{-1} scale probed by SEP. (Adapted from 86)

manifold of interacting states are directly coupled (i.e. the coupling matrix element is large relative to the zero order energy mismatch) as opposed to indirectly, where IVR proceeds though a tier structure of intermediate states. The power-law decay observed at intermediate times corresponds to a non-Lorentzian line shape in the frequency domain. Edge states decay more slowly than interior states, for a given total energy (102). Edge states are those with the excitation localized in a few modes and they lie at the energy extremes of the polyads, whereas interior states lie in the middle of a polyad where the initial excitation is divided among several strongly coupled modes.

The work of Gruebele & Bigwood has important ramifications for control experiments. Gruebele & Bigwood have theoretically demonstrated the possibility of "freezing" IVR in $SCCl_2$ by tailoring an optimized excitation pulse. The initial fast IVR decay of $v_1 = 8$ slows by two orders of magnitude when the appropriate pulse shape is used instead of a Fourier-transform-limited Gaussian pulse (102). By controlling the phase and amplitude of the tailored pulse, the coherences between the ground state, the bright state, and a few dark gateway states can be manipulated in order to freeze IVR. Moreover, it is precisely the initial polyad localization of IVR that provides for the possibility of control over IVR.

Isomerization in Triatomics: Test for ab initio Theory

One of our aims is to understand intramolecular dynamics when the internal energy is sufficient to promote chemical reactions such as bond-breaking isomerization and simple bond-breaking dissociation. The simplest bond-breaking isomerization is that of a triatomic monohydride, e.g. HCN ↔ HNC or HCP ↔ HPC. Again, SEP is ideal for examining such processes. SEP allows for the preparation of highly excited vibrational states in the electronic ground state that are inaccessible by conventional direct absorption spectroscopy. In the case of HCP isomerization, HPC corresponds to a saddle-point on the potential surface rather than a local minimum. Transitions observed in SEP spectra of HCP have been assigned to "normal mode" and "isomerization" states, both of which are well described by a polyad model that accounts for the interaction between bending and CP stretching motions (82, 83, 104–108). The delocalized isomerization states in HCP were the first such states in any molecule to be observed, assigned, and fitted.

Because of the role of HCN ↔ HNC as a prototypic isomerization reaction, there has been great interest in carrying out similar experiments on HCN to detect the delocalized states. However, this has thus far proven quite difficult. Past experiments have yielded vibrational-term energies and vibrational state–specific rotation constants for highly vibrationally excited HCN (29, 79, 109–112). This kind of data was used together with expressions based on perturbation theory and the application of simple rotation-vibration to make vibrational assignments. Some states that did not fit into this simple picture of vibrational assignment were observed, but there was no evidence that those states were delocalized isomerizing states of HCN/HNC. In fact, all the states that could not initially be fit within a simple rotation-vibration model were later shown to be simple HCN-localized states accessed via axis-switching transitions (79, 113). This raises the question of how one can identify vibrational states that possess delocalized HCN/HNC character without the necessity of performing a complete rotation-vibration analysis.

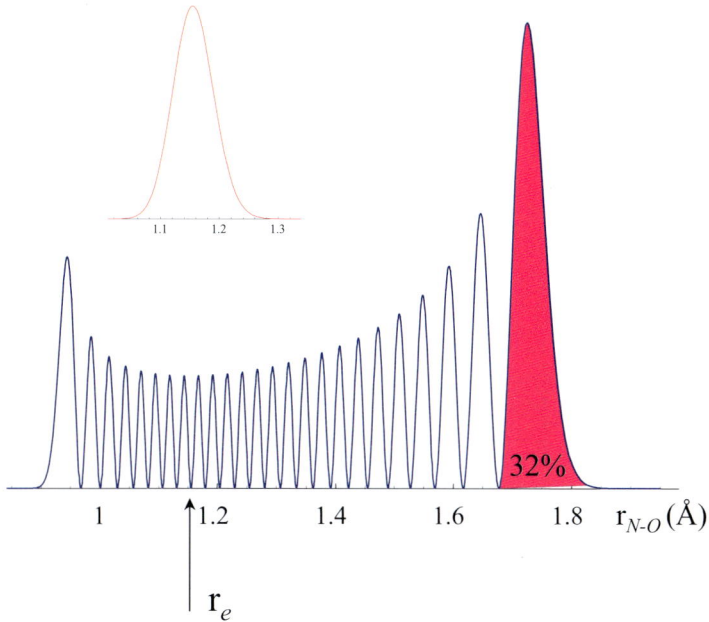

Figure 1 The actual probability distribution for NO in v=24 (*blue*) of its ground electronic state. For comparison, the v=0 (*red*) state is also shown. r_e indicates the N-O separation at the potential minimum. 32% of the probability of the v=24 state is found in the outer probability lobe. The large amplitude of vibrational motion shown here is typical for molecules prepared by stimulated emission pumping.

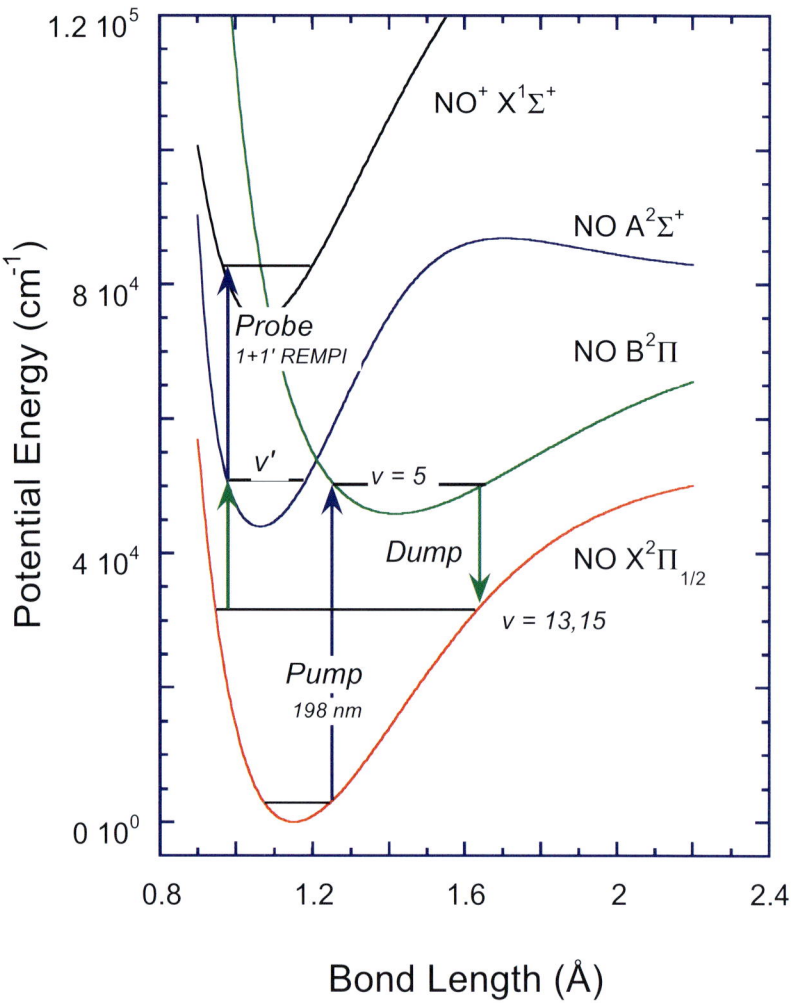

Figure 2 Schematic representation of stimulated emission pumping for the case of NO. The sample is excited with a narrow-band pulsed laser operating at the "PUMP" frequency, ω_{pump}, out of thermally populated states into an excited electronic state. A second pulsed laser, tuned to ω_{dump}, stimulates emission out of the excited electronic state transferring population back to a single highly vibrationally excited quantum level in the ground electronic state. The prepared state may be probed (for example using REMPI).

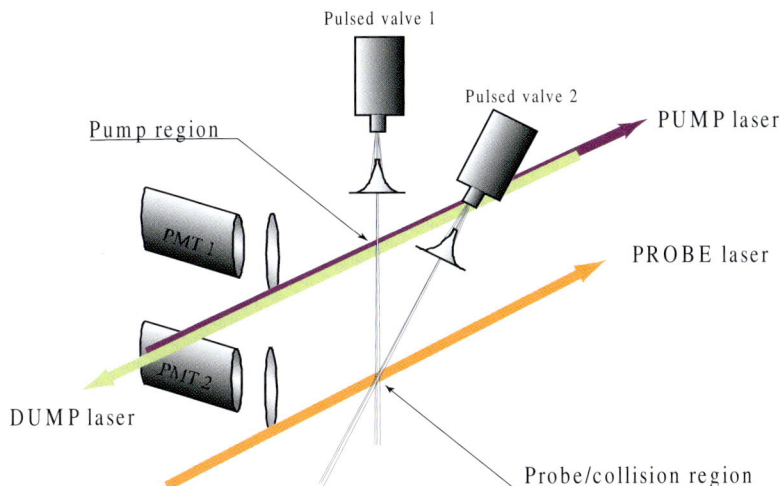

Figure 3 Crossed molecular beams with PUMP-DUMP-PROBE geometry. A molecular beam is excited by SEP preparing single quantum states of highly vibrationally excited molecules at the "PUMP region". PMT-1 views this region allowing fluorescence dip measurements. Several cm downstream the sample is detected by the PROBE-laser. When pulsed valve-2 is turned off, the results of infrared emission occurring during the flight time between preparation and detection can be observed. With pulsed valve-2 turned on, the results of single collisions are detected by the PROBE laser.

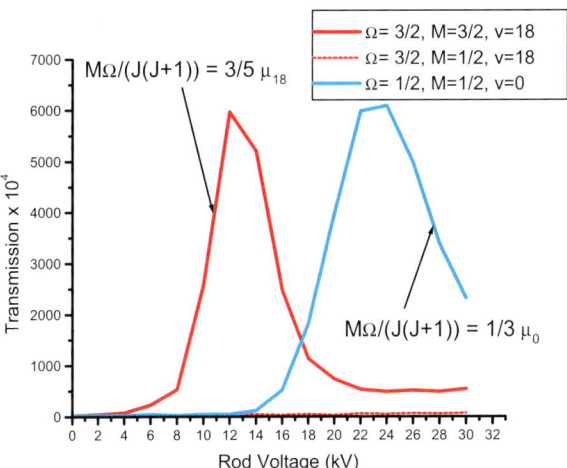

Figure 4 Hexapole focusing of SEP prepared molecules. These classical trajectory calculations are for two states of NO: $v=0\,^2\Pi_{1/2}$ (which dominates the population of the adiabatically cooled molecular beam) and $v=18\,^2\Pi_{3/2}$ (which may be prepared by SEP). Because of the M, Ω and J quantum number dependence of the Stark effect, the SEP prepared molecules focus at a much lower voltage than do the ground state. This provides the approach to production of pure beams of SEP prepared molecules.

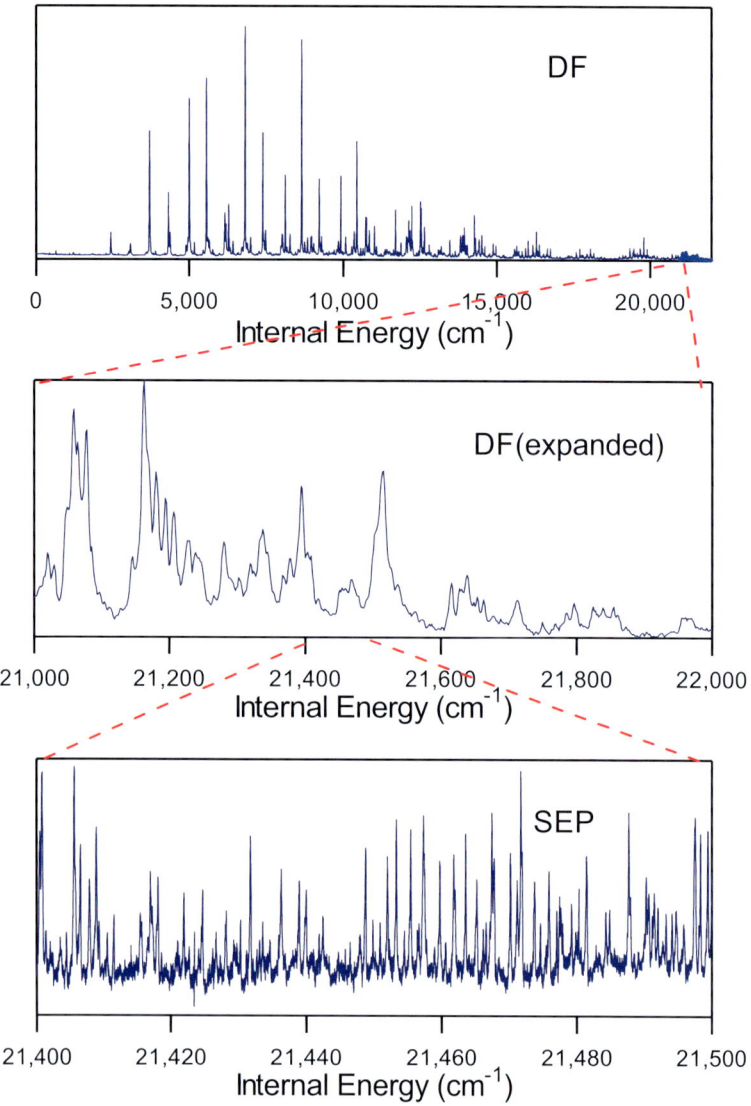

Figure 5 Complementary use of DF and SEP spectroscopy to study acetylene dynamics. The spectra are recorded via the $Q(1)$ line of the $2v_3$' band of $^{13}C_2H_2$. (*Top* and *middle*) DF spectrum (~ 16 cm^{-1} resolution). (*Bottom*) SEP spectrum (~ 0.15 cm^{-1} resolution). At high energy each feature in the DF, assignable to a polyad, corresponds to several SEP transitions. The higher resolution SEP allows for the investigation of acetylene dynamics on a longer time scale (10-100 ps instead of 1ps) where chaos may have a larger role in the dynamics. Reprinted with permission from *J. Phys. Chem. A* 104: 3073. Copyright 2000 American Chemical Society.

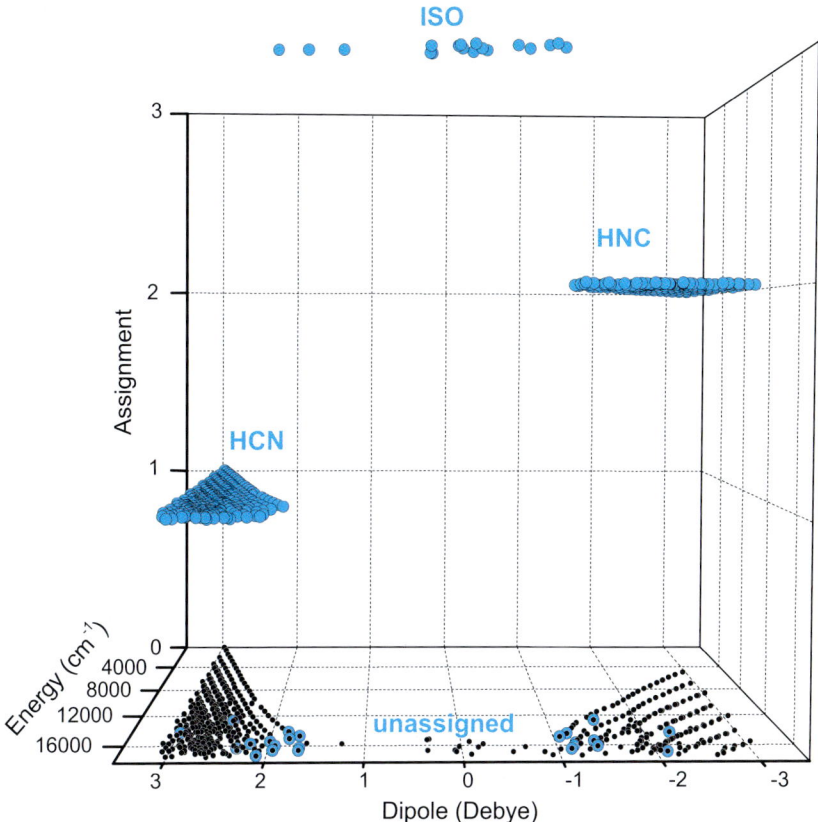

Figure 8 State specific dipole moments are diagnostic of vibrational identity. Blue circles show energy (*X-axis*), signed dipole moment (*Y-axis*) and assignment (*Z-axis*) from ab initio calculations and analysis of vibrational wave functions. The states have been categorized into four groups: HCN-states, HNC-states, delocalized (isomerizing) states and unassigned states. Black circles show the projections of the data onto the Energy/Dipole plane, i.e. ignoring assignments as has been done for the dipole calculations. It is clear that the dipole moment is diagnostic of the vibrational identity and may even allow assignment of some of the unassigned states.

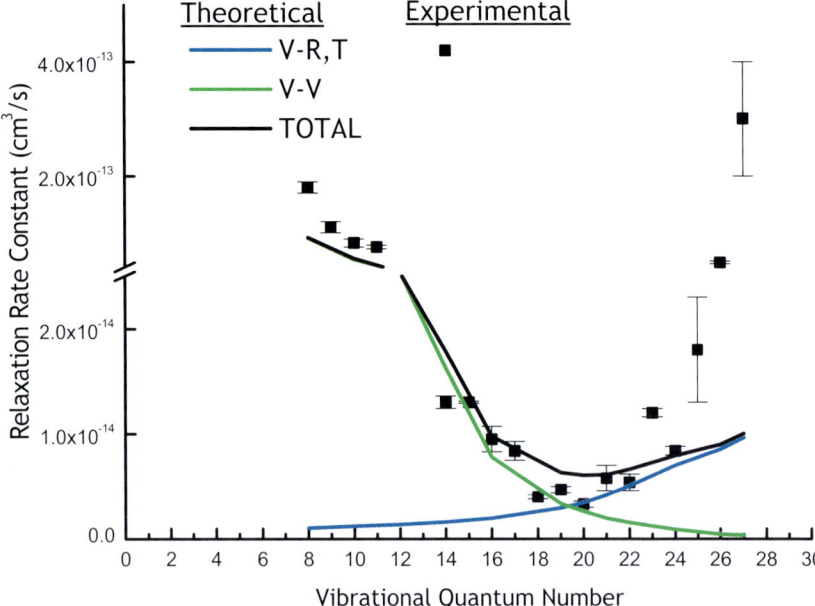

Figure 10 Experiment and theory for vibrational relaxation of $O_2(v) + O_2$ as a function of v. Experimental points come from stimulated emission pumping ($13<v<29$) (177-180) and chemical activation studies ($v<12$) (182, 183). The theoretical results come from ab initio calculations of the interaction potential and molecular motion (188) (*black line*). The contributions from V-R,T (*blue line*) and V-V (*green line*) relaxation are derived from theory.

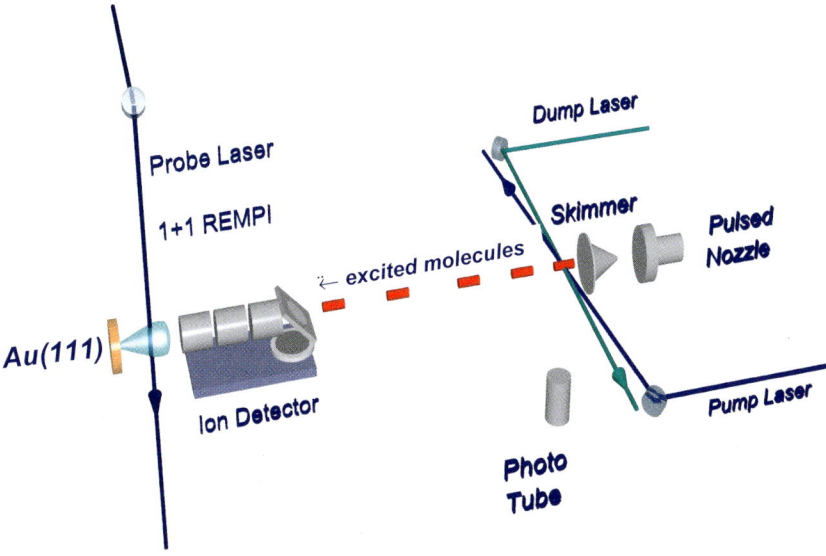

Figure 12 Schematic diagram of an instrument for studying the interactions of highly vibrationally excited molecules with surfaces. PUMP and DUMP lasers prepare highly vibrationally excited NO in the second differential pumping chamber downstream from a pulsed molecular beam of NO. A photomultiplier tube is used for fluorescence dip spectroscopy, which helps control the optical excitation step. Pump laser induced fluorescence was also monitored by a photo-multiplier for signal normalization. The excited molecules travel through one more region of differential pumping and collide with a well characterized surface. Scattered NO is state selectively ionized using 1+1 resonance enhanced multiphoton ionization. The ions are extracted back along the molecular beam direction and deflected to a micro-channel plate detector. By scanning the frequency of the probe laser the vibrational and rotational state distribution of the products can be measured. Translational energy distributions of the products are measured by recording the intensity of the REMPI signal as a function of the time-delay between the excitation and the probe lasers. Angular distributions were recorded by translating the probe laser across the direction of incident molecular beam.

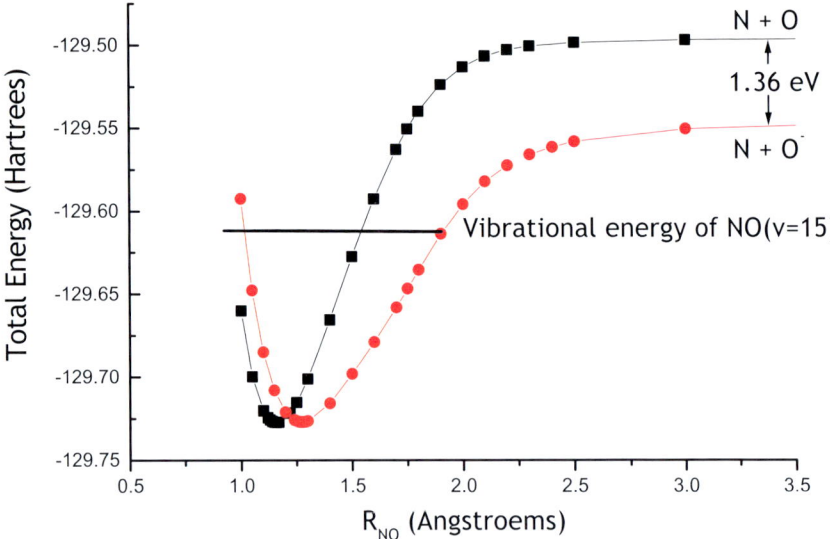

Figure 13 Ab initio calculation of the NO and NO⁻ molecules. The vibrational energy of NO(v=15) is shown. The calculated O atom electron affinity (1.36 eV) is quite close to the experimental value of 1.46 eV. Notice that at the outer turning point of vibration electron capture is strongly exoergic, whereas it is strongly endoergic at the inner turning point of vibration.

Recently, it has been suggested that electric dipole moment measurements would provide a simple way of accomplishing this goal. Using ab initio calculations of dipole moments at a grid of molecular geometries, a global dipole surface was fit for HCN/HNC (42). This was then used with high level calculations of vibrational wave functions (114) to obtain predictions of vibrational state–specific dipole moments.

Figure 8 (see color insert) presents the results of these calculations. An extremely useful observation is that isomerizing states have markedly different dipole moments than their localized counterparts (42). The blue circles show energy (X-axis), signed dipole moment (Y-axis), and assignment (Z-axis). The assignments are based on an analysis of the vibrational wave functions (114). In this context assignment is meant very crudely to indicate 0, unassigned states; 1, HCN-localized states; 2, HNC-localized states; and 3, delocalized (isomerizing) states. The black squares in Figure 8 show the projections of the data onto the energy/dipole plane. One can clearly see that the electric dipole moment is an excellent discriminant of the vibrational identity.

Nonstatistical Unimolecular Dissociation Dynamics

Bond fission is another example of a fundamental chemical reaction that can be influenced by IVR. Statistical rate theories, such as RRKM, provide an adequate description of this process for most systems. However, many small molecules in the gas phase exhibit nonstatistical behavior. SEP experiments permit examination of specific classes of ro-vibrational states, which lie above the dissociation threshold and are therefore in principle predissociated. For small polyatomic molecules the vibrational density of states is sufficiently low that individual predissociated vibrational states can be rotationally resolved in SEP spectra. The dependence of dissociation rates on the vibration-rotation state of the parent molecule can be obtained from careful linewidth measurements (PUMP-DUMP) or lifetime measurements (PUMP-DUMP-PROBE). This state-specific behavior of the dynamics at high energy is not always consistent with the predictions of statistical theories, which assume that vibrational levels at high energy are strongly coupled. SEP has been instrumental in examining state-specific dissociation dynamics in many molecules such as HCO, HFCO, NH_3, CH_3O, ArOH, NeOH, and CS_2 (60–63, 99, 100, 116–144).

State-specific dissociation dynamics in HCO/DCO has been the focus of many theoretical and experimental studies due to the importance of HCO in combustion reactions; it has also been used as a test of theoretical methods (24, 25, 116–126). SEP spectra recorded via the $\tilde{B} \rightarrow \tilde{X}$ transition revealed mode-specific, non-RRKM behavior in a series of metastable ground-state vibrational levels that extended at least 8000 cm^{-1} above the dissociation threshold (24, 25, 116–126). These metastable vibrational levels exhibit high vibrational excitation primarily in the CO stretch coordinate. Analysis of the linewidths observed in series of experiments spanning $E_{vib} = 2,000$–$21,000$ cm^{-1}, using two-color resonant four-wave mixing SEP, demonstrates profound mode specificity in HCO (124). In those experiments pure CO stretch excitations had the narrowest linewidths. As one might expect, the excitations that contained excitation in the CH stretch were the broadest because the CH stretch has a large projection along the dissociation coordinate. Additionally, excitations of combination levels containing CH stretch and bending vibrations were shown to be the most effective at promoting dissociation.

This indicates the nature of the dissociation coordinate. The dissociation coordinate is predominantly a CH stretch combined with a slight angle change. This is illustrated in Figure 9, which is borrowed from Tobiason et al (124). Theoretical investigations, using an ab initio potential surface, reproduced the trends observed in the experimental results. (117)

CO product state distributions have also been measured using SEP that prepared metastable vibrational states on the ground electronic state surface (118, 119). The nascent product state distributions (for low J levels) revealed dependencies on both the rotational state and the number of excited bending quanta in the parent molecule, indicative of nonstatistical dynamics. More recent experiments have focused on DCO (125, 126), for which non-RRKM behavior has also been observed. A 1:1:2 resonance between the DC stretch, CO stretch, and DCO bending vibrations results in strong mixings among all three vibrational modes and results in considerably more complex spectra in DCO than HCO. The DCO dynamics is described as intermediate between mode-specific and statistical limits (126).

HFCO and DFCO also exhibit state-specific dissociation dynamics that is profoundly influenced by the IVR dynamics discussed above. Estimates of the dissociation rates from the linewidths of observed rotationally resolved transitions in SEP spectra provide a wealth of information concerning the unimolecular dissociation dynamics of HFCO (60, 99, 100). Moore and coworkers (60, 99, 100) found that, in the tunneling region near the barrier to dissociation, there exists a strong dependence of the dissociation rate on rotational quantum number. At higher energies the rotational dependence of the dissociation rate is weaker, but vibrational mode-specificity persists.

The rotational-level dependence of the dissociation rates in HFCO also depends on vibrational symmetry. Dissociation rates of states with odd quanta of out-of-plane bending excitation, (v_6 = odd, A'' symmetry), depend much more strongly on rotational level than those with even quanta of v_6 excited (A' symmetry). This was explained in terms of a reduced dissociation rate for vibrational states of A'' symmetry (a node in the plane of the molecule) compared with those of A' symmetry and is consistent with the view that the minimum energy dissociation lies in the plane of the molecule (99).

Another interesting observation is that the dissociation rates increase more rapidly with K_a than with J, implying that A-type Coriolis interactions play a key role in the dissociation dynamics. It is clear that the anharmonic coupling of the out-of-plane bending mode, v_6, to the reaction coordinate is weak and that A-type Coriolis interactions increase this coupling. Classical mechanics predicts that A-type Coriolis interactions convert a large-amplitude out-of-plane bend into

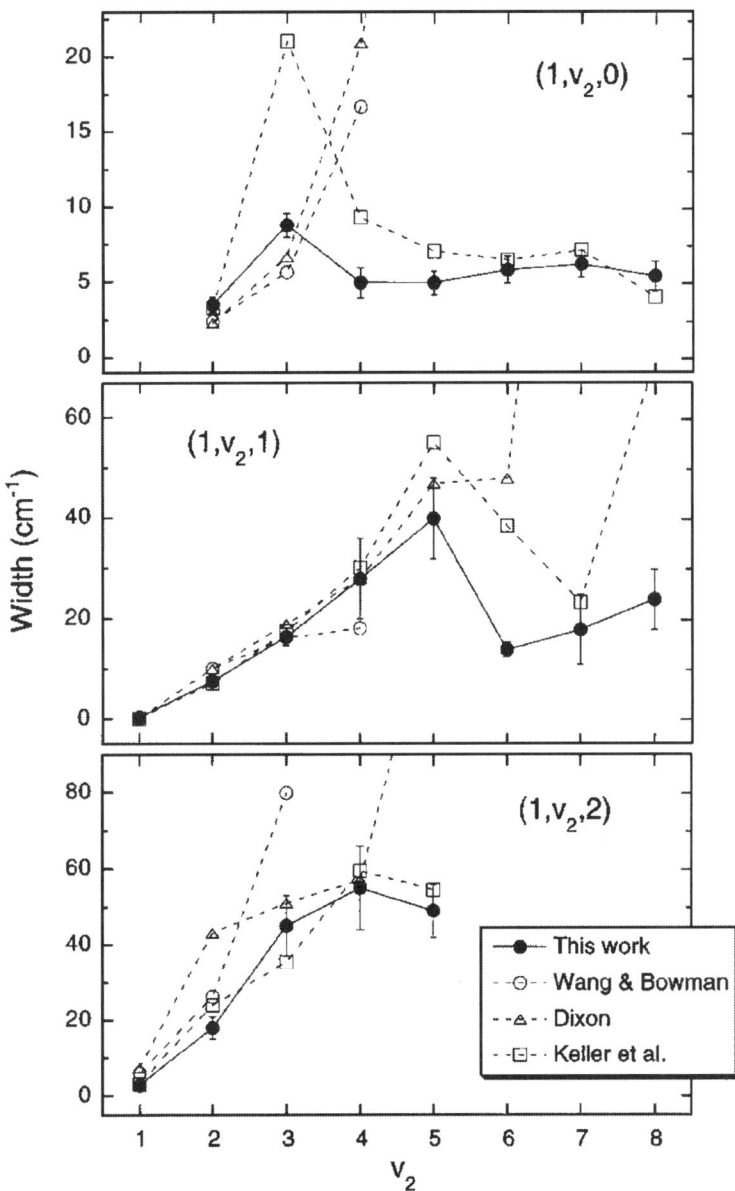

Figure 9 Comparison of measured and calculated widths for the $(1,v_2,0)$, $(1,v_2,1)$ and $(1,v_2,2)$ vibrational levels in HCO from Tobiason and coworkers (124). Clear evidence of vibrational state–specific and highly nonstatistical dissociation rates can be seen. Reprinted with permission from *J. Chem. Phys.* 103, 1448, (1995). Copyright 1995 American Institute of Physics.

an in-plane bend. The anharmonic coupling of v_6 to the other modes (including the reaction coordinate) is weak and of decreasing strength with increasing even quanta in v_6 (because of the decreasing amplitude of in-plane geometry), which accounts for a decrease in the dissociation rate with increasing quanta in v_6. Coriolis coupling of quasistable extreme motion states in the reaction coordinate to v_6 and to background states accounts for the rotational dependence of the dissociation rates in the tunneling region, $E_{vib} \approx 16,500$ cm^{-1} (100). HFCO has provided textbook examples of several distinct aspects of non-RRKM behavior.

SEP Studies on Clusters: Structure, Intramolecular Vibrational Redistribution, and Dissociation

Binary complexes, such as NeOH and ArOH, permit systematic examination of dissociation dynamics for a model solvent/solute system. Here the interactions between the chemically bound atoms and those bound by van der Waals forces are two tiered, similar to expectations for a molecule in solution. In contrast to condensed phase systems, SEP and high level theory provide a near quantitative characterization of the interaction potentials.

Metastable levels, which lie above the dissociation limit, sample the repulsive wall of the intermolecular potential. These open shell complexes are of particular interest because the molecular geometry has a strong effect on the shape and bonding character of the singly occupied orbital. SEP spectra have sampled virtually all of the bound vibrational levels of OH-Ar (145). PUMP-DUMP-PROBE SEP studies of the product-state distributions of the OH photofragments reveal extreme selectivity in the rotational predissociation of OH-Ar (137, 139). The OH fragments were preferentially produced in the highest energetically accessible rotational levels, and in many cases only a single rotational level was populated. The orientation of the singly occupied π orbital of the OH radical [in plane (A$'$) or out of plane (A$''$) with the bent triatomic molecule] results in two Renner-Teller split potential surfaces. These two surfaces are degenerate at linear geometry, $C_{\infty v}$, but when the molecule is bent this degeneracy is removed. There are spin-orbit (rotational parity–independent) and rotation-electronic (rotation-dependent) interactions between the ^{2}A$'$ and ^{2}A$''$ states. As a result, there are spin-rotation-like splittings of the bound rotational levels and strong J, K_a, and K_c dependencies of the predissociation rates. Becuase the spin-orbit coupling constant of the OH X $^2\Pi$ state is well known and the inert gas atom has a negligible effect on the intrinsic spin-orbit coupling strength of the OH π-orbital, the details of the observed spin-splittings and predissociation rate serve as sensitive probes of the shapes of the ^{2}A$'$ and ^{2}A$''$ potentials.

In the case of NeOH, spin-orbit coupling plays a major role in the dissociation dynamics. The bound and predissociated states of NeOH have been investigated using SEP (137, 146). As intermolecular stretching excitation increases, the NeOH linewidths decrease and the spin-orbit predissociation rates decrease. This contradicts the expectation, based on RRKM theory, that the rate of unimolecular reaction increases with vibrational excitation.

SEP studies of intracluster vibrational energy redistribution and cluster vibrational predissociation have been particularly valuable. Dai et al (147, 148) were the first to use SEP to investigate such phenomena, examining the ground-state structure of glyoxal-Ar and glyoxal-Ar_2 complexes. Since then many other clusters have been studied. Larger clusters such as benzene-Ar, phenylacetylene-NH_3, carbazole-R(R = Ne, Ar, Kr, Xe), and 1-naphthol to H_2O, CH_3OH, NH_3, and ND_3 have also been investigated using SEP (21, 149–160).

In addition to dissociation dynamics in van der Waals clusters, SEP is also ideal for studying the intermolecular potentials and interactions of these molecules. Typically, to study clusters one uses ion-dip techniques, in which the DUMP transition causes a resonant decrease in the intensity of a mass selected ionization signal. There are several advantages to using SEP spectroscopy for the study of van der Waals clusters. Low frequency, large amplitude intermolecular vibrations are systematically sampled as the energy difference between the PUMP and DUMP transitions is effortlessly tailored to match them. Additionally, a wide range of inter- and intramolecular vibrational levels on the ground state potential surface is accessible. Moreover, spectral assignment is facilitated because the PUMP transition (plus mass selective ion detection) labels a specific cluster, thereby reducing spectral complexity.

Variants of SEP are also exploited to examine hydrogen bonding. Hydrogen bonded clusters formed in a molecular beam can be manipulated such that solvation of the hydrogen-bonded chromophore can be studied by systematically increasing the number of "solvent" molecules in the cluster. Castleman & Stanley examined phenol•$(H_2O)_n$, phenylacetylene, and phenylacetylene-NH_3 clusters by cluster-ion dip SEP spectroscopy. (21, 155, 156). In this variant of SEP the PUMP pulse prepares the cluster in a known rotation-vibration state and the DUMP photon is chosen so that, in addition to stimulating a downward transition, it has sufficient energy to excite an upward ionizing transition. When the difference between the PUMP and DUMP frequencies is resonant with a transition to a vibrational level in the ground state, stimulated emission causes a dip in the ion current. Castleman & Stanley found that the lifetime of the phenol S_1 state increases dramatically with the addition of only one H_2O molecule. Furthermore, ion dips greater than 50% were observed for some levels. Such large dips imply a dynamical process on the ground electronic state that prevents simple saturation of the DUMP transition. The >50% ion dips are attributed to IVR on the ground state, in which low frequency modes of the water adduct interact with water-phenol intermolecular modes (155). IVR may be viewed as allowing the excitation to move rapidly out of the lower state of the DUMP transition. Alternatively, the fractionation of the SEP DUMP bright state results in a cluster of several unresolved DUMP transitions allowing for >50% ion-dip signals.

Our understanding of the intermolecular forces between large molecules of chemical and biological significance has been enhanced by SEP. Specifically, SEP permits measurement of hydrogen bond dissociation energies for several clusters. Such bond energies are necessary for modeling the structure and dynamics of biological systems (e.g. protein folding). SEP, in conjunction with resonant two-photon ionization, has generated valuable information about these clusters (157–160). Vibrational modes in the ground electronic–states of carbazole-R(R = Ne, Ar, Kr, Xe) complexes were observed up to the dissociation limit, D_0. Accurate binding energies were determined for these complexes. Additionally, transitions in SEP spectra terminating on energy regions above D_0 exhibited efficient vibrational predissociation (157). The dissociation energies for these complexes were found to scale as α_R/R_{cm}^6, where α_R is the polarizability of the rare gas and R_{cm} is the distance from the center of mass of the carbazole to the rare gas. An upper limit of 50 ns was found for the vibrational predissociation lifetime of all these carbazole complexes. Further studies of carbazole-S (S = N_2, CO, CH_4) clusters compare van der Waals binding energies with those of carbazole–rare gas complexes (158). For Carbazole-S the attractive van der Waals interaction is governed by dispersion terms (R^{-6}) in the potential, but for S = N_2, CO additional electrostatic interactions (dipole-dipole, dipole-quadrupole) were shown to play a role.

Additional hydrogen bonding studies have also been performed using SEP. Studies of the binding energies of 1-naphthol to H_2O, CH_3OH, NH_3, ND_3, benzene, and cyclohexane have generated insight into the strength of hydrogen bonds in aromatic π electron systems (159, 160). The hydrogen bond in 1-napthol-H_2O is found to be twice as strong as that in $(H_2O)_2$ because 1-napthol is a stronger hydrogen donor than H_2O.

Detailed studies of benzene clusters have also been completed. Kratzschmar and colleagues have performed mass selective ion-dip experiments on benzene clusters. They accessed vibrational states of benzene-Ar located above the adiabatic dissociation limit of the cluster and saw no trace structure in their spectra attributable to van der Waals vibrational modes (154). However, in benzene trimer structure in the spectra assignable to van der Waals vibrations was observed. Neusser et al used a variant of SEP, coherent ion dip spectroscopy, to measure high overtones of the CH stretch in benzene and benzene-Ar complexes. (149–153) In coherent ion dip spectroscopy one can obtain complete population transfer into a final state in a three-level system through the use of narrow, carefully time-sequenced pulses. In both C_6H_6-Ar and C_6D_6-Ar, overtones of van der Waals vibrations at energies up to 120 cm^{-1} on the ground electronic state were studied by this method so the quality of several benzene-Ar potential surfaces could be tested through comparison with experiment (153). Surprisingly, the van der Waals bend-stretch frequencies of weakly bound benzene-Ar van der Waals complexes exhibited nearly harmonic behavior, and classical harmonic calculations accounted for the observed C_6H_6-Ar and C_6D_6-Ar isotope shifts (153).

Transients

SEP is well suited for the characterization of transient molecules. There are several motivations for studying radicals. Their formation by photolysis or discharge results typically in rotationally and vibrationally hot species. The resultant spectral congestion makes rotation-vibration assignments difficult. Radicals are typically produced in low concentrations; consequently, high sensitivity is essential. Additionally, the relatively short chemical lifetime of many of these species adds another level of difficulty to their spectroscopic characterization. The vibrational-mode selectivity of SEP greatly reduces the congestion of the spectra and facilitates assignment. SEP is based on two electronic transition steps rather than high vibrational overtone transitions. Electronic transitions have much greater oscillator strength than vibrational overtones. Thus, as demonstrated by Dai and colleagues (161), the sensitivity of SEP spectroscopy is better suited to the spectroscopy of radicals. For example, the central problem with CH_2 is the enormous number of perturbations between the $\tilde{X}$ and $\tilde{a}$ states. By using an intermediate level of known rotational and vibrational character, rotational assignments of perturbed bands are greatly simplified.

To date, several transient species such as C_2^+, $BrCN^+$, $C_4H_2^+$, H_3, and CH_2 have been spectroscopically characterized, and their dynamics have been investigated as well. (161–166) The example of H_3 is quite intriguing. This seemingly simple triatomic molecule has proven to have quite complex spectra owing to the unbound nature of the electronic ground state. For the $2s\ ^2A_1'$ Rydberg state the distribution of predissociation products was determined using SEP (167).

Summary of Unimolecular Reaction Dynamics Using SEP

SEP has proven to be a valuable tool for the study of unimolecular reaction dynamics. Information encoded in frequency-domain SEP spectra led to considerable advances in the understanding of intramolecular vibrational redistribution, isomerization, and dissociation—processes that are dynamic in nature. Over the past two decades, SEP and its variants have been routinely applied to the study of stable molecules such as acetylene and to transient species such as benzene clusters. We have briefly touched upon some of the more important recent studies. In the next section we extend our discussion of the role of SEP in bimolecular reaction dynamics.

SCATTERING DYNAMICS

As must now be apparent, SEP can be used to obtain a wealth of information about the unimolecular dynamics of vibrationally excited molecules. A practical advantage of those experiments is their relative simplicity: Only two lasers are needed. For many years it was feared that taking SEP beyond the two-laser fluorescence-dip configuration would prove too unwieldy to yield significant results. Such experiments would require three tunable lasers (PUMP-DUMP-PROBE). More importantly, it was already clear that the spectroscopy of highly vibrationally excited molecules would be richly complex in its own right. If one wished to prepare highly vibrationally excited molecules for scattering experiments, it would be necessary to judiciously choose systems in which one could actually know well what sort of state one was starting with.

The potential for using SEP-prepared molecules in scattering experiments became much clearer, and the entire field drew great optimism from the early work coming out of Alan Knight's laboratory (32, 33, 35, 37, 168–173). That work showed that SEP could be used to prepare vibrationally excited molecules (p-difluorobenzene in $v_3 = 1$) and that laser-induced fluorescence could be combined with a conventional SEP experiment to observe collisional relaxation out of the prepared vibronic state into nearby vibrational levels. The PUMP-DUMP-PROBE experiment was born. Those experiments were soon followed by others showing that rotationally resolved collisional relaxation experiments were possible (174, 175). In other experiments the relaxation of the intermediate electronically excited state was monitored as well (130, 144, 176–177).

More recently, with a few notable exceptions such as CH_3O (130) and CS_2 (144), researchers have focused on PUMP-DUMP-PROBE experiments using diatomic molecules in order to overcome the "spectral complexity problem." This allows one to fully understand the SEP-prepared-state down to the most fundamental level, quantitative knowledge of the vibrational wave function. PUMP-DUMP-PROBE experiments have been carried out in cells, such as the first work of Kable and Knight (33, 37, 170–173), and more recently in crossed molecular beams. In the following sections we review some of the work that has contributed to a better understanding of how highly vibrationally excited molecules behave in collisions. This part of the review includes such topics as rotational energy transfer, vibrational energy transfer, chemical reactivity of highly vibrationally excited molecules, and gas/surface interactions of highly vibrationally excited molecules.

Rotational Energy Transfer

Under many conditions, rotational energy transfer is the most likely outcome of a bimolecular collision. Whether rotational energy transfer follows different rules in highly vibrationally excited states is, therefore, of significant fundamental interest. Most work has shown that observing vibrational influences on rotational energy transfer is not easy. Experiments carried out in gas cells on rotational energy transfer revealed no unusual behavior for highly vibrationally excited I_2 (179–181) or NO (182).

One of the most rigorous studies used crossed molecular beams with SEP to investigate the rotational energy transfer of NO (v = 20) in collisions with He (39). The collision energy could be controlled along with all of the quantum numbers of the excited molecule. Rotational population distributions resulting from scattering of parity-selected NO (v = 20, J = 1/2) were obtained using laser-induced fluorescence under single-collision conditions. A marked oscillation in the scattering cross-section was observed. For collisions that conserved e/f parity, a ΔJ-even propensity was seen. For e/f parity–changing collisions, ΔJ-odd was preferred. This shows that the overall rovibronic parity is approximately conserved. Although this observation is visually striking, similar propensity rules had been observed previously for molecules in low vibrational states: NO (v = 0) (183) and CN (v = 2) (175, 184, 185) (182).

The influence of high vibrational excitation on rotational energy transfer could only be quantified by comparison with high-level calculations. Quantum dynamical calculations carried out by Alexander and co-workers were done on two ab initio potential energy surfaces (39). The first surface was calculated by fixing the bond length at the vibrationally averaged value of the internuclear separation corresponding to NO (v = 0). For the second surface, the NO bond length was held fixed at the vibrationally averaged bond length of NO (v = 20).

Careful comparison of experiment and theory revealed that scattering calculations on the two potential surfaces gave slightly different but experimentally distinguishable results. Not surprisingly, the rotational energy transfer dynamics calculated on the stretched potential energy surface agreed better with observation. Calculations performed with this potential exhibited a larger degree of angular anisotropy, resulting in cross-sections for the larger values of ΔJ more closely matching experiment. This work gives a simple picture of rotational energy transfer of highly vibrationally excited diatomic molecules. Despite the large-amplitude vibrational motion in v = 20, the time scale separation of rotation and vibration simplifies the dynamics. The angular forces exerted on the diatomic by the collision are averaged over the vibrational motion, making a stretched rigid rotor model suitable. Another example shows the creativity with which scientists in the field are employing SEP. Photodissociation of HN_3 above the singlet threshold exclusively produces excited state NH $^1\Delta$. Using intense pulsed lasers, these researchers stimulated emission down to the ground state NH $^3\Sigma^-$, producing single rotational states. This fascinating approach, which stands PUMP and PROBE literally on its head, made rotational relaxation studies of ground-state NH $^3\Sigma^-$ possible (186).

In summary, well-controlled experiments and comparison with high-quality theory are required to observe the subtle differences in rotational energy transfer exhibited by highly vibrationally excited molecules. Observing the unusual vibrational relaxation properties of these species is a much simpler matter.

Vibrational Energy Transfer

Vibrational energy transfer is one of the most important aspects of the entire field of the study of highly vibrationally excited molecules. Vibrational energy transfer represents one way in which vibrationally excited molecules are created in the real world. Furthermore, it is the dominant process that limits the lifetime and number density of vibrationally excited molecules. Although we have a great deal of data and have developed an understanding of the vibrational relaxation of molecules in low vibrational states, it is not obvious that this understanding can be extrapolated to high vibrational levels.

Related to this is the question: What is the expected vibrational quantum number dependence for the total vibrational relaxation rate? There are very few experimental data sets comprehensive enough to reveal the vibrational dependence over a wide range of v. Perhaps the best-studied example is the $O_2(v) + O_2$ collisional relaxation, which has received enormous attention owing to suspicions of its importance to upper stratospheric ozone formation (187).

Experimental measurements of total vibrational relaxation rate constants (shown as points in Figure 10; see color insert) come from PUMP-DUMP-PROBE experiments (188–191) as well as chemical activation studies (192–194). This is one of the few examples of vibrational relaxation of highly vibrationally excited molecules in which substantial theoretical effort has also been applied. Ab initio calculations have provided an accurate picture of the parts of the interaction potential relevant to vibrational relaxation. Quantum (195–199), classical (200, 201), and semiclassical (202, 203) treatments of the O_4 dynamics have all been reported. Figure 10 shows experimental results together with the results of the ab initio quantum dynamics calculations (199). The agreement with experiment below about v = 25 is remarkable.

The theoretical decomposition of the total relaxation rate into its vibration-translation and rotation (V-R,T) and vibration-vibration (V-V) components is also shown. One can see that the relative importance of the two mechanisms changes profoundly from low-to-high v. Furthermore, one expects the temperature dependence of these two mechanisms of energy transfer to be quite different. Therefore, the relative importance of V-R,T versus V-V energy transfer for a given vibrational state will also depend strongly on temperature. One can see that the V-R,T component increases roughly linearly with vibrational quantum number. Linear dependence on v is a prediction of Schwarz-Slawsky-Herzfeld (SSH) theory (204) of V-R,T energy transfer, resulting from linearizing the interaction potential and using harmonic oscillator wave functions. It is fascinating to see that one of the key features of one of the simplest models of vibrational energy transfer is retained in a full ab initio quantum treatment.

PUMP-DUMP-PROBE experiments on NO vibrational self-relaxation showed a similar behavior (36, 205) up to about v = 13, above which vibrational relaxation accelerated faster than linearly with v. This was taken as evidence of the onset of new kinds of vibrational relaxation, peculiar to highly vibrationally excited molecules. Precisely what is going on in self-relaxation of highly vibrationally excited states of NO remains an unanswered and fascinating question. It is interesting to note that NO (v = 13) lies close in energy to the theoretically predicted activation barrier for formation of the lowest of a series of high-energy isomers of the NO dimer (206, 207). Furthermore, the calculated structure at the barrier is asymmetric, with one short (1.184 Å) and one long (1.439 Å) NO bond. The short NO bond resembles ground-state NO. The lengthened NO bond is somewhat shorter than the most probable bond length of NO in v = 13 (r_{mp} = 1.52 Å). We speculate that collisions between one vibrationally excited NO molecule and one in v = 0 could access this isomer. Complex formation would explain more efficient vibrational relaxation above v = 13. We might also expect multiquantum ($\Delta v \geq 1$) relaxation to take place. Experiments carried out on v = 19 indeed confirmed the importance of multiquantum relaxation. In light of these results, it is tempting to speculate that complex formation is responsible for the enhanced vibrational deactivation of NO above v = 13, but the truth behind this conjecture awaits future study.

There are known examples of highly vibrationally excited molecules that do not appear to undergo vibrational relaxation by particularly different dynamics than molecules in low vibrational states. For example, vibrational relaxation from highly vibrationally excited I_2, prepared state-selectively by SEP were reported (179–181). Here individual rotational states in v = 23, 38, and 42 have been prepared, and relaxation by He, Ar, N_2, O_2, Cl_2, I_2, and H_2O was investigated. Vibrational relaxation rate constants were found to depend only weakly on the vibrational quantum number and appeared to be less than linearly proportional to v. In all cases $\Delta v = -1$ relaxation dominated. Owing to the low vibrational frequency of I_2, these experiments are likely probing the V-R,T energy transfer.

Returning to Figure 10, it is evident that V-V vibrational energy transfer can be very important for highly vibrationally excited molecules. If a diatomic molecule formed in a high vibrational state relaxes one vibrational quantum at a time, owing to the $w_e x_e (v + 1/2)^2$ anharmonicity term, the amount of energy transferred increases as the relaxation progresses to lower vibrational levels. For near-resonant V-V energy transfer, the relaxing highly vibrationally excited molecule effectively scans a range of vibrational frequencies in the acceptor molecule. The range of "scanned" acceptor frequencies can be quite large. Take, for example, O_2. If 1-1 and 2-1 resonant processes are considered,

$$O_2(v) + A(0) \rightarrow O_2(v-1) + A(1),$$

$$O_2(v) + A(0) \rightarrow O_2(v-2) + A(1),$$

an O_2 molecule relaxing from v = 27 down to v = 0 scans a range of acceptor molecule frequencies from $\sim$1000 (v = 27$\rightarrow$26) to 3200 (v = 2$\rightarrow$0) cm^{-1}. Clearly, such a molecule experiences manifold opportunities to exchange quanta of vibration with its environment. Extensive investigation of near-resonant V-V energy transfer has been carried out using the PUMP-DUMP-PROBE technique. Again, because of its atmospheric significance, vibrational relaxation rate constants for $O_2(v)$, hereafter referred to as $k_M^{O_2}(v)$, with a number of typical atmospheric collision partners, M = $O_2, O_3, N_2, CO_2,$ and N_2O, have been studied. In every case strong evidence for near-resonant V-V energy transfer was found (188, 192, 208).

An illustrative example of V-V transfer is shown in Figure 11, in which the vibrational relaxation rate constant for the CO_2–O_2 and N_2O–O_2 system, $k_{CO_2}^{O_2}(v)$ and $k_{N_2O}^{O_2}(v)$, are plotted as function of O_2 vibrational quantum number (188). Similar data have been reported using chemical activation methods (192). For relaxation by CO_2, the vibrational relaxation is dramatically enhanced (by about a factor of 100) near $O_2(v = 18)$, where the 2-1 near-resonant energy transfer process,

$$O_2(v) + CO_2(000) \rightarrow O_2(v-2) + CO_2(001),$$

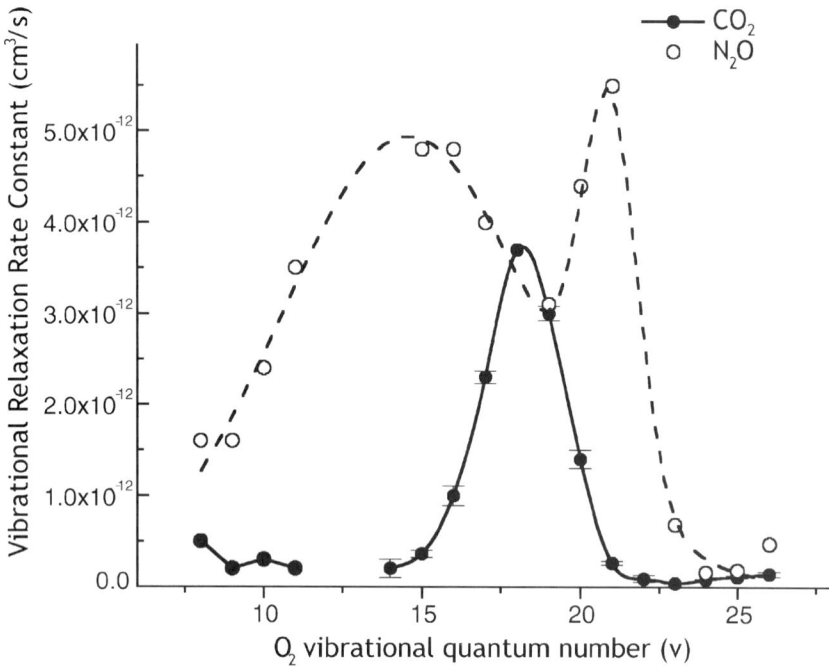

Figure 11 Vibrational relaxation of $O_2(v)$ by CO_2 (solid circles) and N_2O (open circles). Notice the sharp resonance structures. The narrow features at $v = 18$ (CO_2) and $v = 21$ (N_2O) can be assigned to 1-2 resonances with high vibrational states of O_2. The broad feature near $v = 15$ (N_2O) is a 1-1 resonance with O_2 (177). [The data between $v = 8$ and $v = 11$ were graciously provided by Prof. Ian Smith's group (193, 194)].

has a minimum energy defect of only a few cm^{-1} for O_2 ($v = 18$). More direct evidence for the 2-1 resonance came from PUMP-DUMP-PROBE experiments in which O_2 ($v = 17$) was prepared in mixtures with CO_2 and the evolution of the population in $v = 16$ and $v = 15$ was observed. In those experiments population initially in $v = 17$, skips $v = 16$ and appears in $v = 15$. The experimental results for vibrational relaxation of O_2 by N_2O are also shown in Figure 11. Here one sees two resonances, a broad one and a narrow one. These are clearly due to the 2-1 resonance with v_3 and the 1-1 resonance with v_1. Similar 2-1 and 1-1 resonances were found for other systems including $O_2(v) + O_3$ and N_2 (188, 192).

Because near-resonant V-V transfer appears to be so common, we wish to understand its nature better. From a perturbation-theoretical point of view, minimizing the energy defect enhances the transfer probability. However, we do not know which molecular motions effectively compensate the inevitable finite energy defect. Energy defect compensation is the essential physical element that determines the resonance widths, and thereby the overall importance of V-V transfer.

Following the pioneering work of Gentry and colleagues (38), more recent studies have used the PUMP-DUMP-PROBE technique in combination with crossed molecular beams. This provides new insight into the nature of V-V energy transfer (208). Vibrational relaxation was observable for NO (v = 20, 21, and 22) for crossed beam scattering processes:

$$NO(X^2\Pi, v) + N_2O(0, 0, 0) \rightarrow NO(X^2\Pi, v-1) + N_2O(0, 0, 1).$$

Here v_1, v_2, and v_3 indicate the number of quanta excited in the symmetric-stretch, bend, and asymmetric-stretch, respectively. Notice that by varying the initial vibrational state, the energy defect can be systematically tuned: -49 cm^{-1} for v = 20, -18 cm^{-1} for v = 21, and $+14$ cm^{-1} for v = 22. Here a negative value indicates exoergicity. V-V energy transfer was found to be approximately five times more efficient for the v = 21 level than for the other two vibrational levels. For v = 19 and v = 23, which have energy defects of -80 and $+47$ cm^{-1}, respectively, no vibrational energy transfer could be detected. The collision energy in the center-of-mass frame was 125 ± 25 cm^{-1}.

Energy transfer v = 22 → 21 ($\Delta E = +14$ cm^{-1}) was found to be substantially less efficient than v = 21 → 20 ($\Delta E = -18$ cm^{-1}), despite the availability of 125 cm^{-1} of kinetic energy. This is direct evidence that translational energy is not effective in compensating the positive energy defect. The compensation of the energy defect by NO rotation could be seen clearly by comparing the rotational energy distributions for vibrationally elastic and inelastic channels. The rotational distributions were seen to shift with respect to the distribution observed for the vibrationally elastic process. The endoergic channel (v = 22 → 21) was rotationally colder than the elastic channel, whereas the two exoergic channels were rotationally hotter. The observed shifts in the rotational distributions correspond almost exactly to the V-V energy mismatch for these channels. This shows clearly that the energy defect, which is systematically varied in the experiment, is nearly completely compensated for by NO rotational energy.

These results indicate that rotational motion is effective in compensating the energy mismatch in near resonant V-V transfer. Thus, it is clear that the overall importance of V-V energy transfer can be underestimated significantly if the role of rotational excitation is not accounted for adequately.

Returning to Figure 10, we see for vibrational relaxation of O_2 that there is substantial disagreement between theory and experiment above v = 25. The origin of this discrepancy has been the subject of intense experimental and theoretical scrutiny, particularly because of its important atmospheric implications. Initially, it was suspected that the onset of increased relaxation rates above v = 25 could be associated with the chemical reaction, $O_2(X^3\Sigma_g^-, v \gg 0) + O_2 \rightarrow O(^3P) + O_3$ (187, 191). Because the ab initio results of Figure 10 were obtained from a potential surface that did not include chemical reaction (199), the disagreement between experiment and theory could easily be rationalized in this way. Furthermore, the energetic onset of the ozone-forming reaction was close in energy to the onset of enhanced vibrational relaxation shown in Figure 10. Also, observations of vibrational cascading out of initially populated states in the enhanced relaxation regime showed evidence of a dark channel (190). Specifically, molecules prepared in $v = 26$ and higher were not seen to appear in vibrational states one and two quanta lower in energy.

Despite extensive theoretical efforts, evidence for efficient formation of ozone from the reaction of vibrationally excited O_2 has been elusive (195–197, 200). Using ab initio methods, a limited number of points have been calculated in regions of the potential energy surface near the reaction path (196, 197). Reduced dimensionality quantum dynamics calculations on those surfaces have failed to show efficient vibrationally enhanced reactivity.

In one study, enhanced vibrational relaxation was found for vibrational states high enough in energy that the transition state of the reaction could be accessed (196). Although the theoretically derived enhancement was much smaller in magnitude than the experimental one, it appears that some fraction of the enhanced removal above $v = 25$ could be due to the stronger chemical interactions that are possible near a chemical transition state. Such transition state–enhanced vibrational relaxation had been previously hypothesized for vibrational relaxation of highly vibrationally excited NO (205). This suggests that vibrational promotion of chemical reactions, even in a system as small as four atoms can be subject to strong energy dissipation processes.

Others have used semi-empirical potentials and classical trajectory methods to analyze the same O_2 vibrational relaxation dynamics (200). Although the results are qualitatively consistent with the higher-level quantum calculations, there may be problems with the potential energy surface used. The semi-empirical Varandas-Pais potential was originally optimized to model temperature-dependent reaction rates for $O + O_3 \rightarrow 2O_2$. To test the ability of this surface and classical trajectory calculations to reproduce other aspects of the reaction, product state distributions of O_2 produced in $O + O_3 \rightarrow 2O_2$ were measured (209). The results showed a relatively narrow vibrational distribution peaking in $v = 12$–13. O_2 rotational distributions were relatively "cool," and Doppler profiles indicated only minimal release of translational energy. Quasi-classical trajectory calculations carried out by George Schatz on the Varandas-Pais potential (209) showed a broader vibrational distribution, also peaked near $v = 13$, and analysis of the trajectories revealed that the simulated reaction proceeds by a spectator bond mechanism, in which one of the O–O bonds in the ozone reactants is approximately decoupled from the reaction (i.e. quasi-triatomic). However, rotational distributions predicted from simulations were much more energetic than observed experimentally. Furthermore, the simulations predicted a much larger translational energy release than could be rationalized from the experimental Doppler profiles. This analysis strongly suggests that the Varandas-Pais potential, which characterizes the $O + O_3$ reaction as a spectator bond reaction, does not accurately reproduce the dynamics of the reaction.

The complexity of this problem has only recently become apparent. It now appears likely that in order to successfully treat the O_4 system including reaction, several potential energy surfaces must be considered (owing to the high degeneracy of each of the fragment asymptotes). High-resolution SEP spectra of vibrational states of O_2 as high as v = 31 have recently been reported. These experiments provide the first evidence of spectral perturbations in the ground electronic state of O_2 (210). Deperturbation analysis showed that $O_2(X^3\Sigma_g^-)$ interacts strongly with $O_2(b^1\Sigma_g^+)$ by a spin-orbit mechanism, and evidence for a curve crossing was found. This work led to the hypothesis that similar interactions and curve crossings might be important in the O_2/O_2 collision system. Whereas the curve crossing found O_2 was located at too high energy to directly influence vibrational states as low as v = 25, where the enhanced vibrational relaxation begins, it was suggested that the curve crossing could be lowered in energy by as much as 1 eV in O_2/O_2 collisions (210).

The first theoretical attempts to treat the O_4 system with multiple potential surfaces that incorporate spin-orbit coupling are now underway. Using restricted C_{2v} symmetries, several surface crossings have been found in the vicinity of the reaction path, and spin-orbit coupling matrix elements were found to be substantial. However, in this work the crossings are still located at substantially higher energy than v = 25 (211). More recently, calculations exploring a larger range of configuration space have shown that surface crossings may indeed occur at sufficiently low energy to explain enhanced removal (by nonadiabatic relaxation) for O_2 in v as low as 25 (212).

Recently, the dark channel has been reinvestigated experimentally with startling results (213). It was seen that a significant fraction of the initial population of highly vibrationally excited oxygen molecules ($X^3\Sigma_g^-$ v > 24) prepared by SEP, relaxes to much lower vibrational levels ($\Delta v \sim -9$). The time scale of this process is much shorter than the known collisional lifetimes of the intervening vibrational levels, and thus a sequential single-quantum relaxation mechanism could be explicitly ruled out. State-to-state measurements after preparation of v = 28 and v = 30 provided the final-vibrational state population distribution resulting from relaxation of these two states. For v = 28 (v = 30), at least 38% (7.9%) of the initially prepared population underwent multiquantum vibrational relaxation, explaining, at least in part, the previously reported "dark channel" for relaxation of $O_2(v \geq 26)$.

There has been no shortage of surprises in the study of the collision dynamics of highly vibrationally excited O_2. Furthermore, owing to the complementarity of experiment and theory, workers in this area have made significant progress in revealing the unusual properties of this system and elucidating the important chemical issues that are likely to lie at its heart. There remain fascinating unanswered questions, the answers to which are emerging. What is the mechanism of the unexpected multiquantum vibrational relaxation? Could this be evidence for the O_4 complex that has been postulated for years (214, 215) and/or is this the result of complex nonadiabatic interactions? Why does the spectator model of the $O + O_3$ reaction fail so miserably to reproduce the O_2 product energy distribution? Finally, what role does the vibrationally activated ozone reaction play in the atmosphere? These questions are clear targets for future work.

Surface Scattering Dynamics

We have considered unimolecular dynamics and gas-phase bimolecular collisions of highly vibrationally excited molecules. When a vibrationally excited molecule in the gas phase collides with a solid in addition to conversion of vibrational energy to rotation and translation (V → R,T), energy can be transferred to electronic and phononic excitations. For molecules in low-lying vibrational states, several mechanisms for vibrational energy transfer have been identified. These include excitation and de-excitation via trapping (long-lived complexes) (216, 217), nonadiabatic (i.e. vibrational-to-electronic) coupling (218), direct "mechanical coupling" (219), and stretching of bonds in frustrated surface chemical reactions (220, 221).

Which of these or other dynamics are important for vibrational energy transfer involving highly vibrationally excited molecules at surfaces? How will the probability of vibrational energy transfer change for highly excited molecules? Will vibrational relaxation rates scale with the vibrational energy of the initial state as theory predicts (222)? We have almost no direct experimental data to address these important questions.

Figure 12 (see color insert) shows a recently developed experimental apparatus that is providing some of the first information on how highly vibrationally excited molecules interact with solid interfaces (41). The PUMP and DUMP lasers excite a pulsed molecular beam in a conventional molecular beam chamber. A photomultiplier views the volume where the laser beams cross the molecular beam, thus enabling fluorescence dip measurements to characterize the optical excitation efficiency. The highly vibrationally excited molecules then travel through a set of differential pumping apertures and enter an ultra-high vacuum scattering chamber where they collide with a well-characterized solid surface at normal incidence. The scattered molecules can then be ionized by resonant enhanced multiphoton ionization and detected by a two-stage multichannel plate detector.

Scattering from Insulators

Experiments with vibrationally excited molecules scattering from insulators have the potential to probe the vibrational interactions with phonons of the surface. One of the most striking of such measurements for low vibrational states showed that, for low temperature NaCl crystals, the time scale of infrared emission for a monolayer of CO was 4.3 ms (223, 224). Because of the very weak coupling to the surface, energy pooling between CO molecules was directly observed.

There have been few studies of the scattering of vibrationally excited molecules from insulating surfaces. Recently, vibrationally excited acetylene has been prepared by overtone pumping, and scattering from LiF(001) was observed (76, 225). By cooling the surface so that trapping of the vibrationally excited molecules could be ensured, vibrational relaxation could be observed. Comparisons of two isotopomers of acetylene showed that the vibrational relaxation occurred in a stepwise process. The presence of an anharmonic resonance in the light acetylene isotopomer resulted in enhanced rates for the early stages of relaxation. In C_2HD no such Fermi resonance exists, and the relaxation was observed to be less efficient. No vibrational relaxation was observed via a direct mechanism.

A trapping desorption mechanism for vibrational relaxation has also been indicated in studies of vibrational relaxation of CO_2 on LiF (226). HCl (v = 2) was also prepared by overtone excitation, and scattering on MgO has been studied (115, 227). Again, by varying the surface temperature and incidence energy, trapping may be induced and vibrational relaxation effected. Under conditions in which trapping is unimportant, so is vibrational relaxation. Study of the temperature dependence of the vibrational relaxation led these researchers to postulate that vibrational relaxation occurs at defects or steps on the surface.

Very recently, SEP has been used to prepare NO in v = 12, which contains more than 150 kJ/mol of vibrational energy (228), and the vibrational relaxation on LiF has been investigated. This work is one of the first attempts to study the vibrational relaxation dynamics of highly vibrationally excited molecules on insulating surfaces. It showed that survival of NO (v = 12) was close to unity as long as the incidence energy was sufficiently high that trapping was unimportant. Even at low incidence energies, Cos (θ) angular distributions of unrelaxed NO (v = 12) were observed, indicating that translational accommodation was much more efficient than vibrational accommodation.

A small fraction of vibrational relaxation for NO (v = 12) into v = 11 and v = 10 was observed to occur by a direct scattering mechanism, but this represented only a few percent of the total scattering. Because vibrational relaxation on LiF is so sensitive to surface defects, this residual few percent vibrational relaxation cannot be unambiguously attributed to relaxation on the LiF crystal. The angular distribution of the scattered molecules changed dramatically as the incidence energy was lowered, consistent with a transition from the direct to the trapping regime.

Interactions with Metals: Vibrational Promotion of Electron Transfer

Whereas vibrational relaxation on insulating surfaces appears most often to be inefficient, the situation is quite different on metals. This appears to reflect the importance of electronic interactions of the surface with vibrationally excited molecules. Figure 13 (see color insert) provides a basis for understanding this process. The figure shows ab initio electronic structure calculations for NO and NO^- (229). One can see that near their potential minima there is a very small energy difference between the neutral and anion curves, consistent with the low electron affinity of NO. The electron affinity of NO is quite different for vibrationally excited states. Near the outer turning point of v = 15 vibration ($R_{NO} \sim 1.6$ Å) the vertical attachment energy used to form NO^- is exoergic by more than 200 kJ/mole. At the inner turning point electron attachment is endoergic by about the same amount. This simple analysis shows that the ability of NO to accept (or concomitantly, the ability of NO⁻ to donate) electrons depends strongly on internuclear separation.

The ability of "stretched" molecules to exhibit enhanced electron capture chemistry can have a dramatic influence on vibrational relaxation at surfaces. Figure 14 illustrates a simple model for nonadiabatic vibrational interactions of NO with a metal surface (Au) (230). In the lower panel, NO is held at its equilibrium bond length (1.15 Å). The neutral curve (labeled NO) interacts weakly with the gold surface. The anionic interaction (labeled NO⁻) depends strongly on the distance of NO from the surface. At infinite separation the energy required to form NO⁻ is the difference of the Au work function and the NO electron affinity (~5.1 eV). The anion interacts with its image charge and follows (at long range) a Coulomb-like potential. One can see that in the lower panel a large barrier to the curve crossing results and the crossing occurs relatively close to the surface.

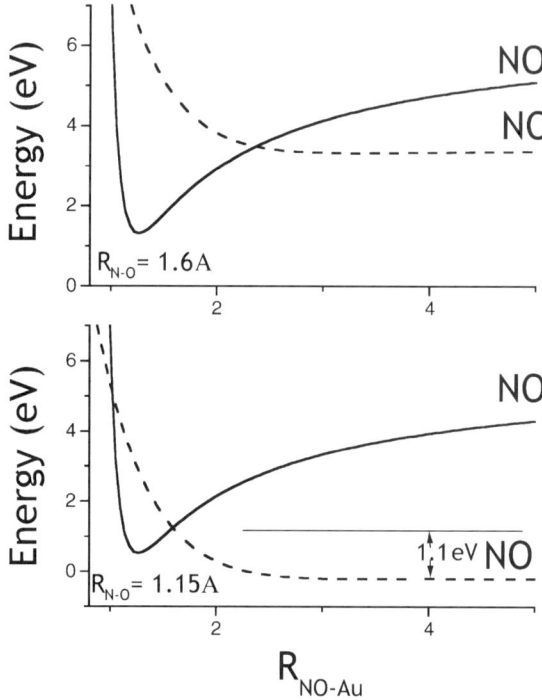

Figure 14 Electron-mediated interactions of NO with a metal (based on analysis of 230). The x-axis indicates the distance from the surface to the O atom. (Bottom) The NO is held at its equilibrium bond length. The asymptotic energy separation is given by the difference between the surface work function and the electron binding strength to the NO. This results in a substantial barrier to electron transfer for NO on Au. (Top) The NO bond length has been extended to 1.6 Å, near the outer classsical turning point of NO (v = 15). The enhanced electron binding sterngth of NO now causes the barrier to electron transfer to disappear.

In the upper panel the NO bond has been stretched to 1.6 Å, increasing the "vertical electronic binding energy" of the projectile in the nonadiabatic model. Now the curve crossing between neutral and anionic potentials lies much farther from the surface and the barrier has disappeared. This argument shows that, near the outer turning point of the NO vibration in v = 15, electron transfer from the Fermi level of the metal to the NO molecule becomes energetically feasible at relatively large distances from the surface.

This argument suggests that one might expect interactions of vibrationally excited NO with a gold surface to result in electron transfer even at very low incident energy. By using SEP to prepare specific quantum states of gas-phase NO, such observations have recently been made (231). The signature of the electron transfer process is a highly efficient multiquantum vibrational relaxation event, in which the NO loses hundreds of kilojoules per mole of energy on a subpicosecond time scale. Figure 15 shows one of the key observations from this work: a comparison of the vibrational relaxation of high vibrational states of NO on Au(111) with

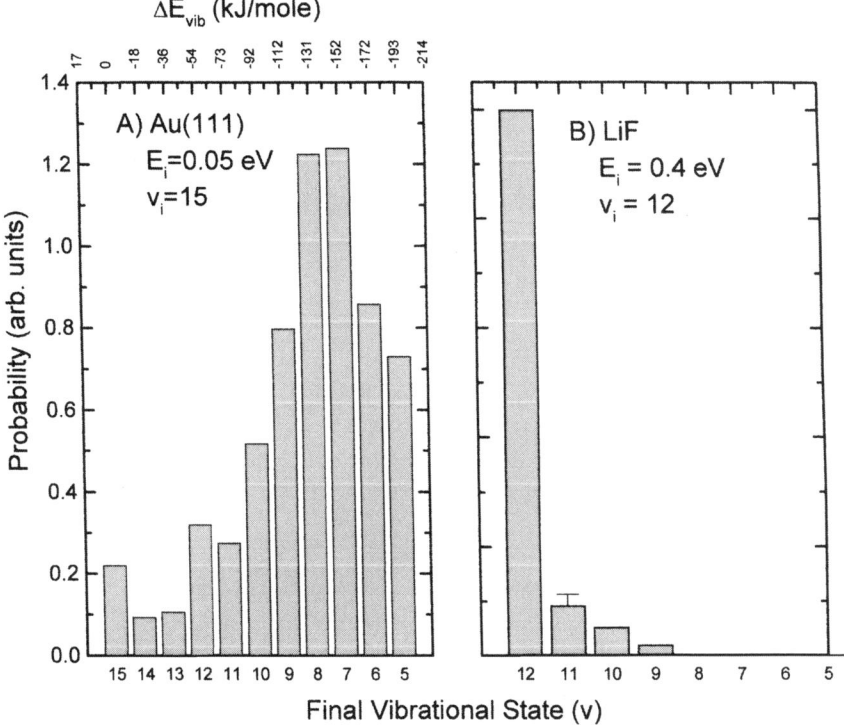

Figure 15 Measured vibrational distribution of NO resulting from scattering of (A) NO (v = 15) from AU(111) at $E_{incidence}$ = 5 kJ/mole and (B) NO (v = 12) from LiF at $E_{incidence}$ = 38 kJ/mole. The multiquantum vibrational relaxation was interpreted as evidence for vibrational promotion of electron transfer from the metal to the NO molecule.

that on LiF single crystals. The contrast could not be more striking. Surprisingly, these results cannot be explained simply on the basis of Franck-Condon factors. The large-amplitude vibrational motion associated with molecules in highly excited vibrational states strongly modulates the energetic driving force of the electron transfer reaction. These results conclusively demonstrate the importance of molecular vibration in promoting electron transfer reactions.

Vibrational Energy Transfer on a Modified Metal

Similar vibrational energy transfer experiments have been carried out on an O atom–covered copper surface, giving some indication of how the vibrational exchange at surfaces is modified by an adsorbate (232, 233). Oxygen adlayers on copper were prepared by exposure of the clean Cu(111) surface to NO_2 followed by elevation of the surface temperature to 800°K for several seconds. Auger analysis revealed a ratio of O adatom to surface Cu atom of $\sim$1:3. These experiments showed substantial vibrational relaxation of NO, including multiquantum vibrational relaxation. Up to five quanta of NO vibration were transferred to an O atom–covered copper surface. Still, $\Delta v = 0$ was the dominant process, and roughly speaking, the results are intermediate between the scattering observation on LiF and Au. The rotational distribution of the scattered molecules was dependent upon the initial translational energy of the molecular beam and far exceeded in rotational excitation that expected from surface desorption. These results indicate that trapping followed by desorption from the surface cannot be responsible for the observed relaxation.

The adsorbate effect on O-covered copper is strong, but it is not obvious how it should be interpreted. As was discussed for scattering on Au, transient negative ion formation could be important in this example as well. Perhaps some property of the O adsorbate diminishes the overall vibrational relaxation compared with Au(111). An alternative explanation is that NO_2 is bound to copper at low temperature. Therefore, highly vibrationally excited NO might transiently form a highly excited NO_2^* adsorbate in collisions with an O-covered copper surface. If the lifetime of adsorbed NO_2^* were short (subpicosecond), it might decompose back to gas phase NO, and the scattered molecules might exhibit the behavior seen in this work. Clearly, there is much to do before these issues can be clarified.

Vibrational Promotion of Dissociative Adsorption

Results presented in the previous section showed that vibrational relaxation at solid surfaces can be remarkably efficient. How effective is vibrational energy in driving surface reactions? Previous studies of this question were restricted to low-lying vibrational states, populated either thermally (234) or by optical excitation (235). The reaction probabilities observed, although strongly influenced by vibration, were low (10^{-2}). Will vibrationally enhanced reaction probabilities greater than this be generally frustrated by efficient vibrational quenching possible at surfaces?

Indirect evidence for the chemical reactivity of highly vibrationally excited NO on clean copper was obtained by monitoring at vibrational relaxation as the copper surface was being oxidized by background NO (v = 0) (41, 236). Regardless of the vibrational state probed, it was found that the strength of the vibrational relaxation signal increased as the copper surface was slowly oxidized to O-covered copper. If the change in scattering signal were exclusively due to changes in the V-V state-to-state vibrational relaxation scattering probabilities, it is unlikely that all of the channels (up to $\Delta v = -5$) would exhibit similar kinetic behavior (first order growth). In fact, because the growth curves for all of the vibrationally inelastic channels are so similar, it appears that they reflect the same kinetic process. This is indirect evidence that highly vibrationally excited NO reacts on a clean copper surface. Presumably, oxidation of the copper surface by ground state NO removes the sites where reaction can take place. Consequently, under surface-oxidized conditions, inelastic channels dominate.

These experiments suggest that the reaction probability for NO (v = 13 and v = 15) on Cu(111) is close to unity. This can be compared with the reaction probability for ground state NO on clean copper ($P_v = 0.0002$), which can be extracted from simple kinetic analysis of O atom build-up from exposure to NO. The vibrational promotion of this reaction is by more than a factor of 5000.

FUTURE OUTLOOK

In any review the focus is on the past; however, it is quite clear that there are still great opportunities for the future applications of SEP in the field of molecular dynamics. For example, there is now conclusive evidence that "stretched molecules" may exhibit much different chemical behavior than the same molecules in their ground vibrational states. Furthermore, it is clear that, at least in the case of electron transfer, this can be understood from a fairly simple point of view. This shows for the first time that chemical properties (unrelated to the large interatomic kinetic energies found in high vibrational states) can be tuned with vibrational excitation. That is, by using vibrational excitation to stretch and/or distort a molecule into an atypical geometry, one can alter the electron cloud around the molecule and influence the chemistry. This reaches one of the fundamental goals of researchers studying highly vibrationally excited molecules, but scratches only the surface. Exploration of this fundamental idea into other kinds of chemical reactions promise great success.

We are also just starting to understand the influence of reactant vibration on surface reactivity and the possible role of electron transfer. Much current thinking regarding vibrational promotion of surface reactivity follows the guiding principles of the Polanyi rules. Can it perhaps be that vibrational promotion of dissociative adsorption at a metal surface is, in fact, fundamentally different? The strong couplings between molecular vibration and surface electrons suggests a different possible mechanism. We know that for many molecules electron transfer forming the anion places electrons in antibonding orbitals and leads to dissociation. Could it be that vibrational excitation promotes electron transfer as the first step in dissociative adsorption?

Another very encouraging sign for the field is the remarkable improvements in our understanding of the nature of vibrationally excited states in polyatomic molecules. We have made profound progress moving from the "age of quantum chaos" to our present understanding that complex spectra can represent simple, albeit unusual, vibrational dynamics. Now, more than ever before, we are in a position to try to understand the chemical nature of highly vibrationally excited polyatomic molecules because we have met a fundamental milestone: We know what we have when we prepare polyatomics in high vibrational states.

Many years of research using lasers as an approach to molecular spectroscopy has led to the exciting state of affairs that we now see. SEP has advanced beyond a spectroscopic tool revealing molecular constants of new states of molecules and is now a means of chemical synthesis. SEP has become one of the most sophisticated preparatory tools, allowing "quantum state resolved reactant distillations." The chemistry of this new class of compounds represents a wide-open field of research.

ACKNOWLEDGMENTS

AMW and RWF would like to thank the AFOSR, NSF, and DOE for support of work described in this article. We would also like to thank Matthew P Jacobson, Martin Gruebele, C Bradley Moore, and Eric A Rohlfing for contributing figures.

Visit the Annual Reviews home page at www.AnnualReviews.org

LITERATURE CITED

1. Polanyi JC, Zewail AH. 1995. *Acc. Chem. Res.* 28:119–32
2. Neumark DM. 1993. *Acc. Chem. Res.* 26:33–39
3. Neumark DM. 1992. *Annu. Rev. Phys. Chem.* 43:153–76
4. Lin SH, Fain B, Hamer N. 1990. *Adv. Chem. Phys.* 79:133–267
5. Johnson BR, Kittrell C, Kelly PB, Kinsey JL. 1996. *J. Phys. Chem.* 100:7743–64
6. Kittrell C, Abramson E, Kinsey JL, McDonald SA, Reisner DE, et al. 1981. *J. Chem. Phys.* 75:2056–59
7. Hamilton CE, Kinsey JL, Field RW. 1986. *Annu. Rev. Phys. Chem.* 37:493–524
8. Northrup FJ, Sears TJ. 1992. *Annu. Rev. Phys. Chem.* 43:127–52
9. Yang XM, Wodtke AM. 1993. *Int. Rev. Phys. Chem.* 12:123–47
10. Dai HL, Field RW, eds. 1995. In *Molecular Dynamics and Spectroscopy by Stimulated Emission Pumping*. River Edge, NJ: World Sci. 1120 pp.
11. Flynn GW, Parmenter CS, Wodtke AM. 1996. *J. Phys. Chem.* 100:12817–38
12. Drabbels M, Wodtke AM. 1999. *J. Phys. Chem. A* 103:7142–54
13. Dai HL. 1998. In *Nonlinear Spectroscopy for Molecular Structure Determination*, ed. RW Field, E Hirota, JP Maier, S Tsuchiya, pp. 55–73. Malden, MA: Blackwell Sci.
14. Dai HL. 1990. *J. Opt. Soc. Am. B* 7:1802–3
15. Dai HL. 1997. *ACS Symp. Ser.* 678:266–74
16. Cooper DE, Klimcak CM, Wessel JE. 1981. *Phys. Rev. Lett.* 46:324–28
17. Jinchun X, Guohe S, Xiaoyuan Z, Cunhao Z. 1986. *Chem. Phys. Lett.* 124:99–104
18. Weber T, Von Bargen A, Riedle E, Neusser HJ. 1990. *J. Chem. Phys.* 92:90–96
19. Ebata T, Furukawa M, Suzuki T, Ito M. 1990. *J. Opt. Soc. Am. B* 7:1890–97

20. Ebata T, Ito M. 1992. *J. Phys. Chem.* 96:3224–31
21. Stanley RJ, Castleman AW Jr. 1990. *J. Chem. Phys.* 92:5770–75
22. Stanley RJ, Castleman AW Jr. 1993. *J. Chem. Phys.* 98:796–99
23. Kratzschmar O, Selzle HL, Schlag EW. 1994. *J. Phys. Chem.* 98:3501–5
24. Williams S, Rohlfing EA, Rahn LA, Zare RN. 1997. *J. Chem. Phys.* 106:3090–102
25. Williams S, Tobiason JD, Dunlop JR, Rohlfing EA. 1995. *J. Chem. Phys.* 102:8342–58
26. Hall G, Whitaker BJ. 1994. *J. Chem. Soc. Faraday Trans.* 90:1–16
27. Zhang Q, Kandel SA, Wasserman TAW, Vaccaro PH. 1992. *J. Chem. Phys.* 96:1640–43
28. Taherian MR, Cosby PC, Slanger TG. 1985. *J. Chem. Phys.* 83:3878–87
29. Yang XM, Rogaski CA, Wodtke AM. 1990. *J. Opt. Soc. Am. B* 7:1835–50
30. Murata Y, Lin CH, Imasaka T. 1993. *Fresenius J. Anal. Chem.* 346:543–44
31. Takayanagi M, Hanazaki I. 1990. *J. Opt. Soc. Am. B* 7:1898–904
32. Lawrance WD, Knight AEW. 1982. *J. Chem. Phys.* 76:5637–39
33. Kable SH, Knight AEW. 1987. *J. Chem. Phys.* 86:4709–11
34. Yang XM, Kim EH, Wodtke AM. 1990. *J. Chem. Phys.* 93:4483–84
35. Lawrance WD, Knight AEW. 1983. *J. Phys. Chem.* 87:389–91
36. Yang XM, Kim EH, Wodtke AM. 1990. *J. Chem. Phys.* 93:4483–84
37. Kable SH, Knight AEW. 1990. *J. Chem. Phys.* 93:3151–59
38. Ma Z, Jons SD, Giese CF, Gentry WR. 1991. *J. Chem. Phys.* 94:8608–10
39. Drabbels M, Wodtke AM, Yang M, Alexander MH. 1997. *J. Phys. Chem. A* 101:6463–74
40. Drabbels M, Wodtke AM. 1997. *J. Chem. Phys.* 106:3024–28
41. Hou H, Huang Y, Gulding SJ, Rettner CT, Auerbach DJ, Wodtke AM. 1999. *Science* 284:1647–50
42. Bowman JM, Irle S, Morokuma K, Wodtke AM. 2001. *J. Chem. Phys.* In press
43. Anderson RW. 1997. *J. Phys. Chem. A* 101:7664–73
44. Blunt DA, Harris SA, Hu WP, Harland PW. 1998. *J. Phys. Chem. A* 102:1482–88
45. Brandt M, Kuhlmann F, Greber T, Bowering N, Heinzmann U. 1999. *Surf. Sci.* 439:49–58
46. Bulthuis J, Vanleuken JJ, Stolte S. 1995. *J. Chem. Soc. Faraday Trans.* 91:205–14
47. Gandhi SR, Bernstein RB. 1987. *J. Chem. Phys.* 87:6457–67
48. Hain TD, Weibel MA, Backstrand KM, Curtiss TJ. 1997. *J. Phys. Chem. A* 101:7674–83
49. Hain TD, Moision RM, Curtiss TJ. 1999. *J. Chem. Phys.* 111:6797–806
50. Harland PW, Hu WP, Vallance C, Brooks PR. 1999. *Phys. Rev. A* 60:3138–43
51. Imura K, Kasai T, Ohoyama H, Takahashi H, Naaman R. 1996. *Chem. Phys. Lett.* 259:356–60
52. Janssen MHM, Mastenbroek JWG, Stolte S. 1997. *J. Phys. Chem. A* 101:7605–13
53. Vallance C, Hu WP, Harland PW. 1999. *J. Phys. Chem. A* 103:665–70
54. Volkmer M, Meier C, Lieschke J, Mihill A, Fink M, Bowering N. 1996. *Phys. Rev. A* 53:1457–68
55. Volkmer M, Meier C, Lieschke J, Dreier R, Fink M, Bowering N. 1997. *Phys. Rev. A* 56:R1690–93
56. Jongma RT, Berden G, Rasing T, Zacharias H, Meijer G. 1997. *J. Chem. Phys.* 107:252–61
57. Jongma RT, Rasing T, Meijer G. 1995. *J. Chem. Phys.* 102:1925–33
58. Jongma RT, Berden G, Rasing T, Zacharias H, Meijer G. 1997. *Chem. Phys. Lett.* 273:147–52
59. Jacobson MP, Coy SL, Field RW. 1997. *J. Chem. Phys.* 107:8349–56

60. Choi YS, Moore CB. 1995. See Ref. 10, pp. 433–57
61. Crane JC, Kawai A, Nam H, Clauberg H, Beal HP, et al. 1997. *J. Mol. Spectrosc.* 183:273–84
62. Crane JC, Hakhyun N, Beal HP, Clauberg H, Choi YS, et al. 1997. *J. Mol. Spectrosc.* 181:56–66
63. Crane JC, Nam H, Clauberg H, Beal HP, Kalinovski IJ, et al. 1998. *J. Phys. Chem. A* 102:9433–44
64. Yamanouchi K, Field RW. 1995. *Laser Chem.* 16:31–41
65. Abramson E, Field RW, Imre D, Innes KK, Kinsey JL. 1984. *J. Chem. Phys.* 80:2298–300
66. Chen YQ, Halle S, Jonas DM, Kinsey JL, Field RW. 1990. *J. Opt. Soc. Am. B* 7:1805–15
67. El Idrissi MI, Lievin J, Campargue A, Herman M. 1999. *J. Chem. Phys.* 110:2074–86
68. Jonas DM, Solina SAB, Rajaram B, Silbey RJ, Field RW, et al. 1993. *J. Chem. Phys.* 99:7350–70
69. Lundberg JK, Field RW, Sherrill CD, Seidl ET, Xie Y, Schaefer HF. 1993. *J. Chem. Phys.* 98:8384–91
70. Moss DB, Duan ZC, Jacobson MP, O'Brien JP, Field RW. 2000. *J. Mol. Spectrosc.* 199:265–74
71. Pique JP, Chen Y, Field RW, Kinsey JL. 1987. *Phys. Rev. Lett.* 58:475–78
72. Smith BC, Winn JS. 1991. *J. Chem. Phys.* 94:4120–30
73. Sundberg RL, Abramson E, Kinsey JL, Field RW. 1985. *J. Chem. Phys.* 83:466–75
74. Wodtke AM, Lee YT. 1985. *J. Phys. Chem.* 89:4744–51
75. Yamanouchi K, Ikeda N, Tsuchiya S, Jonas DM, Lundberg JK, et al. 1991. *J. Chem. Phys.* 95:6330–42
76. Wight AC, Miller RE. 1998. *J. Chem. Phys.* 109:8626–34
77. Holme TA, Levine RD. 1988. *J. Chem. Phys.* 89:3379–81
78. Chen YQ, Jonas DM, Kinsey JL, Field RW. 1989. *J. Chem. Phys.* 91:3976–87
79. Jonas DM, Yang XM, Wodtke AM. 1992. *J. Chem. Phys.* 97:2284–98
80. Xueming Y, Rogaski CA, Wodtke AM. 1990. *J. Chem. Phys.* 92:2111–13
81. Yang XM, Rogaski CA, Wodtke AM. 1990. *J. Opt. Soc. Am. B* 7:1835–50
82. Ishikawa H, Chen YT, Ohshima Y, Wang J, Field RW. 1996. *J. Chem. Phys.* 105:7383–401
83. Ishikawa H, Nagao C, Mikami N, Field RW. 1997. *J. Chem. Phys.* 106:2980–83
84. Ishikawa H, Nagao C, Mikami N. 1999. *Chem. Lett.* 9:941–42
85. Yit-Tsong C, Watt DM, Field RW, Lehmann KK. 1990. *J. Chem. Phys.* 93:2149–51
86. Bigwood R, Milam B, Gruebele M. 1998. *Chem. Phys. Lett.* 287:333–41
87. Yamanouchi K, Yamada H, Tsuchiya S. 1988. *J. Chem. Phys.* 88:4664–70
88. Yamanouchi K, Takeuchi S, Tsuchiya S. 1990. *J. Chem. Phys.* 92:4044–54
89. Pechukas P. 1983. *Phys. Rev. Lett.* 51:943–46
90. Temsamani MA, Herman M. 1995. *J. Chem. Phys.* 102:6371–84
91. Temsamani MA, Herman M. 1996. *J. Chem. Phys.* 105:1355–62
92. Field RW, O'Brien JP, Jacobson MP, Solina SAB, Polik WF, Ishikawa H. 1997. *Adv. Chem. Phys.* 101:463–90
93. Kellman ME. 1990. *J. Chem. Phys.* 93:6630–35
94. Kellman ME, Chen GE. 1991. *J. Chem. Phys.* 95:8671–72
95. Fried LE, Ezra GS. 1987. *J. Chem. Phys.* 86:6270–82
96. Jacobson M, Field RW. 2000. *J. Phys. Chem.* 104:3073–86
97. Moss DB, Duan Z-C, Jacobson MP, O'Brien JP, Field RW. 2000. *J. Mol. Spectrosc.* 199:265–74
98. Stewart GM, McDonald JD. 1983. *J. Chem. Phys.* 78:3907–15

99. Choi YS, Moore CB. 1992. *J. Chem. Phys.* 97:1010–21
100. Choi YS, Moore CB. 1995. *J. Chem. Phys.* 103:9981–88
101. Choi YS, Moore CB. 1991. *J. Chem. Phys.* 94:5414–25
102. Gruebele M, Bigwood R. 1998. *Int. Rev. Phys. Chem.* 17:91–145
103. Wong V, Gruebele M. 1999. *J. Phys. Chem. A* 103:10083–92
104. Joyeux M, Sugny D, Tyng V, Kellman ME, Ishikawa H, et al. 2000. *J. Chem. Phys.* 112:4162–72
105. Ishikawa H, Nagao C, Mikami N, Field RW. 1998. *J. Chem. Phys.* 109:492–503
106. Beck C, Schinke R, Koput J. 2000. *J. Chem. Phys.* 112:8446–57
107. Ishikawa H, Field RW, Farantos SC, Joyeux M, Koput J, et al. 1999. *Annu. Rev. Phys. Chem.* 50:443–84
108. Farantos SC, Keller HM, Schinke R, Yamashita K, Morokuma K. 1996. *J. Chem. Phys.* 104:10055–58
109. Yang X, Rogaski CA, Wodtke AM. 1990. *J. Chem. Phys.* 92:2111–13
110. Quapp W, Klee S, Mellau GC, Albert S, Maki A. 1994. *J. Mol. Spectrosc.* 167:375–82
111. Romanini D, Lehmann KK. 1993. *J. Chem. Phys.* 99:6287–301
112. Smith AM, Coy SL, Klemperer W, Lehmann KK. 1989. *J. Mol. Spectrosc.* 134:134–53
113. Hougen JT, Watson G. 1965. *Can. J. Phys.* 43:298–320
114. Bowman JM, Gazdy B, Bentley JA, Lee TJ, Dateo CE. 1993. *J. Chem. Phys.* 99:308–23
115. Korolik M, Arnold DW, Johnson MJ, Suchan MM, Reisler H, Wittig C. 1998. *Chem Phys. Lett.* 284:164–70
116. Adamson GW, Zhao XS, Field RW. 1993. *J. Mol. Spectrosc.* 160:11–38
117. Keller HM, Floethmann H, Dobbyn AJ, Schinke R, Werner HJ, et al. 1996. *J. Chem. Phys.* 105:4983–5004
118. Neyer DW, Luo X, Burak I, Houston PL. 1995. *J. Chem. Phys.* 102:1645–57
119. Neyer DW, Luo X, Houston PL, Burak I. 1993. *J. Chem. Phys.* 98:5095–98
120. Neyer DW, Kable SH, Loison JC, Houston PL, Burak I, Goldfield EM. 1992. *J. Chem. Phys.* 97:9036–45
121. Loison JC, Kable SH, Houston PL, Burak I. 1991. *J. Chem. Phys.* 94:1796–802
122. Kable SH, Loison JC, Neyer DW, Houston PL, Burak I, Dixon RN. 1991. *J. Phys. Chem.* 95:8013–18
123. Neyer DW, Houston PL. 1995. In *Chemical Dynamics and Kinetics of Small Radicals*, ed. K Lin, A Wagner, pp. 469–503. Singapore: World Sci.
124. Tobiason JD, Dunlop JR, Rohlfing EA. 1995. *J. Chem. Phys.* 103:1448–69
125. Keller HM, Stumpf M, Schroder T, Stock C, Temps F, et al. 1997. *J. Chem. Phys.* 106:5359–78
126. Stock C, Li XN, Keller HM, Schinke R, Temps F. 1997. *J. Chem. Phys.* 106:5333–58
127. Dertinger S, Geers A, Kappert J, Wiebrecht J, Temps F. 1995. *Faraday Discuss. Chem. Soc.* 102:31–52
128. Geers A, Kappert J, Temps F, Wiebrecht JW. 1990. *Ber. Bunsenges Phys. Chem.* 94:1219–24
129. Geers A, Kappert J, Temps F, Wiebrecht JW. 1990. *J. Opt. Soc. Am. B* 7:1935–43
130. Geers A, Kappert J, Temps F, Wiebrecht JW. 1991. *Chem. Phys. Lett.* 178:317–24
131. Geers A, Kappert J, Temps F, Wiebrecht JW. 1994. *J. Chem. Phys.* 101:3634–48
132. Geers A, Kappert J, Temps F, Wiebrecht JW. 1994. *J. Chem. Phys.* 101:3618–33
133. Oguchi T, Miyoshi A, Koshi M, Matsui H. 2000. *Bull. Chem. Soc. Jpn.* 73:53–60
134. Mordaunt DH, Ashfold MNR, Dixon RN. 1996. *J. Chem. Phys.* 104:6460–71
135. Mordaunt DH, Dixon RN, Ashfold MNR. 1996. *J. Chem. Phys.* 104:6472–81
136. Chuang CC, Andrews PM, Lester MI. 1995. *J. Chem. Phys.* 103:3418–29

137. Lester MI, Chuang CC, Andrews PM, Yang MB, Alexander MH. 1995. *Faraday Discus. Chem. Soc.* 102:311–21
138. Yang MB, Alexander MH. 1995. *J. Chem. Phys.* 103:3400–17
139. Lester MI, Choi SE, Giancarlo LC, Randall RW. 1994. *Faraday Discuss. Chem. Soc.* 97:365–78
140. Liou HT, Dan PL, Yang H, Yuh JY. 1991. *Chem. Phys. Lett.* 176:109–17
141. Liou HT, Huang KL, Chen MC. 1997. *Chem. Phys. Lett.* 266:591–96
142. Liou HT, Huang KL. 1999. *Chem. Phys.* 246:391–431
143. Pique JP. 1990. *J. Opt. Soc. Am. B* 7:1816–28
144. Chen XR, Zhao H, Zhang CH, Li L. 1987. *Chem. Phys. Lett.* 136:546–50
145. Lester MI, Green WHJ, Chakravarty C, Clary DC. 1995. See Ref. 10, pp. 659–88
146. Chuang C, Andrews PM, Lester MI. 1995. *J. Chem. Phys.* 103:3418–29
147. Frye D, Lapierre L, Dai HL. 1988. *J. Chem. Phys.* 89:2609–14
148. Frye D, Arias P, Dai HL. 1988. *J. Chem. Phys.* 88:7240–41
149. Sussmann R, Neuhauser R, Neusser HJ. 1994. *J. Chem. Phys.* 100:4784–89
150. Sussmann R, Neuhauser R, Neusser HJ. 1995. *Phys. Rev. Lett.* 74:3141–44
151. Sussmann R, Neuhauser R, Neusser HJ. 1995. *J. Chem. Phys.* 103:3315–24
152. Neuhauser R, Neusser HJ. 1995. *J. Chem. Phys.* 103:5362–65
153. Neuhauser R, Braun J, Neusser HJ, van der Avoird A. 1998. *J. Chem. Phys.* 108:8408–17
154. Kratzschmar O, Selzle HL, Schlag EW. 1994. *J. Phys. Chem.* 98:3501–5
155. Stanley RJ, Castleman AW Jr. 1991. *J. Chem. Phys.* 94:7744–56
156. Stanley RJ, Castleman AW Jr. 1995. *Cluster Ion-Dip Spectroscopy*. Singapore: World Sci.
157. Droz T, Burgi T, Leutwyler S. 1995. *J. Chem. Phys.* 103:4035–45
158. Burgi T, Droz T, Leutwyler S. 1995. *J. Chem. Phys.* 103:7228–39
159. Burgi T, Droz T, Leutwyler S. 1995. *Chem. Phys. Lett.* 246:291–99
160. Wickleder C, Droz T, Burgi T, Leutwyler S. 1997. *Chem. Phys. Lett.* 264:257–64
161. Xie W, Ritter A, Harkin C, Kasturi K, Dai HL. 1988. *J. Chem. Phys.* 89:7033–35
162. Celii FG, Maier JP. 1990. *J. Opt. Soc. Am. B* 7:1944–49
163. Celii FG, Maier JP. 1990. *Chem. Phys. Lett.* 166:517–22
164. Hartland GV, Dai HL. 1995. See Ref. 10, pp. 183–221
165. Xie W, Harkin C, Dai HL. 1990. *J. Chem. Phys.* 93:4615–23
166. Xie W, Harkin C, Dai HL. 1989. *J. Mol. Spectrosc.* 138:596–601
167. Muller U, Cosby PC. 1996. *J. Chem. Phys.* 105:3532–50
168. Lawrance WD, Knight AEW. 1982. *J. Chem. Phys.* 77:570–71
169. Muller DJ, Lawrance WD, Knight AEW. 1983. *J. Phys. Chem.* 87:4952–56
170. Thoman JW Jr, Kable SH, Rock AB, Knight AEW. 1986. *J. Chem. Phys.* 85:6234–35
171. Kable SH, Thoman JW, Beames S, Knight AEW. 1987. *J. Phys. Chem.* 91:1004–6
172. Kable SH, Thoman JW, Knight AEW. 1988. *J. Chem. Phys.* 88:4748–64
173. Kable SH, Knight AEW. 1990. *J. Chem. Phys.* 93:4766–78
174. Temps F, Halle S, Vaccaro PH, Field RW, Kinsey JL. 1987. *J. Chem. Phys.* 87:1895–97
175. Lambert HM, Carrington T, Filseth SV, Sadowski CM. 1990. *Chem. Phys. Lett.* 169:185–90
176. Zhong QH, Wang ZH, Zhu QH, Kong F. 1996. *Prog. Nat. Sci.* 6:436–43
177. Zhong QH, Wang ZH, Sun Y, Zhu QH, Kong FN. 1996. *Chem. Phys. Lett.* 248:277–82
178. Deleted in proof
179. Nowlin ML, Heaven MC. 1993. *J. Chem. Phys.* 99:5654–60

180. Nowlin ML, Heaven MC. 1994. *J. Phys.* 4:729–37
181. Lawrence WG, VanMarter TA, Nowlin ML, Heaven MC. 1997. *J. Chem. Phys.* 106:127–41
182. Yang XM, Wodtke AM. 1992. *J. Chem. Phys.* 96:5123–28
183. Vanleuken JJ, Vanamerom FHW, Bulthuis J, Snijders JG, Stolte S. 1995. *J. Phys. Chem.* 99:15573–79
184. Fei R, Lambert HM, Carrington T, Filseth SV, Sadowski CM, Dugan CH. 1994. *J. Chem. Phys.* 100:1190–201
185. Fei R, Adelman DE, Carrington T, Dugan CH, Filseth SV. 1995. *Chem. Phys. Lett.* 232:547–53
186. Rinnenthal JL, Gericke KH. 1999. *J. Chem. Phys.* 111:9465–68
187. Miller RL, Suits AG, Houston PL, Toumi R, Mack JA, Wodtke AM. 1994. *Science* 265:1831–38
188. Mack JA, Mikulecky K, Wodtke AM. 1996. *J. Chem. Phys.* 105:4105–16
189. Price JM, Mack JA, Rogaski CA, Wodtke AM. 1993. *Chem. Phys.* 175:83–98
190. Rogaski CA, Mack JA, Wodtke AM. 1995. *Faraday Discuss. Chem. Soc.* 100:229–51
191. Rogaski CA, Price JM, Mack JA, Wodtke AM. 1993. *Geophys. Res. Lett.* 20:2885–88
192. Park H, Slanger TG. 1994. *J. Chem. Phys.* 100:287–300
193. Klatt M, Smith IWM, Tuckett RP, Ward GN. 1994. *Chem. Phys. Lett.* 224:253–59
194. Klatt M, Smith IWM, Symonds AC, Tuckett RP, Ward GN. 1996. *J. Chem. Soc. Faraday Trans.* 92:193–99
195. Campos-Martinez J, Carmona-Novillo E, Echave J, Hernandez MI, Hernandez-Lamoneda R, Palma J. 1998. *Eur. Phys. J. D* 4:159–68
196. Campos-Martinez J, Carmona-Novillo E, Echave J, Hernandez MI, Hernandez-Lamoneda R, Palma J. 1998. *Chem. Phys. Lett.* 289:150–55
197. Lauvergnat D, Clary DC. 1998. *J. Chem. Phys.* 108:3566–73
198. Hernandez-Lamoneda R, Hernandez MI, Carmona-Novillo E, Campos-Martinez J, Echave J, Clary DC. 1997. *Chem. Phys. Lett.* 276:152–56
199. Hernandez R, Toumi R, Clary DC. 1995. *J. Chem. Phys.* 102:9544–56
200. Varandas AJC, Wang W. 1997. *Chem. Phys.* 215:167–82
201. Wang W, Varandas AJC. 1998. *Chem. Phys.* 236:181–88
202. Billing GD. 1994. *Chem. Phys.* 179:463–67
203. Balakrishnan N, Dalgarno A, Billing GD. 1998. *Chem. Phys. Lett.* 288:657–62
204. Schwartz R, Slawsky Z, Herzfeld K. 1952. *J. Chem. Phys.* 20:1591–99
205. Yang XM, Kim EH, Wodtke AM. 1992. *J. Chem. Phys.* 96:5111–22
206. East ALL. 1998. *J. Chem. Phys.* 109:2185–93
207. Chaban G, Gordon MS, Nguyen KA. 1997. *J. Phys. Chem. A* 101:4283–89
208. Drabbels M, Wodtke AM. 1998. *J. Chem. Phys.* 109:355–58
209. Mack JA, Huang YH, Wodtke AM, Schatz GC. 1996. *J. Chem. Phys.* 105:7495–503
210. Jongma RT, Shi SM, Wodtke AM. 1999. *J. Chem. Phys.* 111:2588–94
211. Hernandez-Lamoneda R, Ramirez-Solis A. 2000. *Chem. Phys. Lett.* 321:191–96
212. Hernandez-Lamoneda R, Ramirez-Solis A. 2000. *J. Chem. Phys.* 113:4139–45
213. Jongma RT, Wodtke AM. 1999. *J. Chem. Phys.* 111:10957–63
214. Seidl ET, Schaefer HF. 1992. *J. Chem. Phys.* 96:1176–82
215. Peterka DS, Ahmed M, Suits AG, Wilson KJ, Korkin A, et al. 1999. *J. Chem. Phys.* 110:6095–98
216. Asscher M, Guthrie WL, Lin TH, Somorjai GA. 1982. *Phys. Rev. Lett.* 49:76–79
217. Mantell DA, Maa YF, Ryali SB, Haller GL, Fenn JB. 1983. *J. Chem. Phys.* 78:6338–39

218. Rettner CT, Fabre F, Kimman J, Auerbach DJ. 1985. *Phys. Rev. Lett.* 55:1904–7
219. Kay BD, Raymond TD, Coltrin ME. 1987. *Phys. Rev. Lett.* 59:2792–94
220. Hodgson A, Samson P, Wight A, Cottrell C. 1997. *Phys. Rev. Lett.* 78:963–66
221. Rettner CT, Auerbach DJ, Michelsen HA. 1992. *Phys. Rev. Lett.* 68:2547–50
222. Persson BNJ, Gadzuk JW. 1998. *Surf. Sci.* 410:L779–82
223. Ewing GE. 1992. *Acc. Chem. Res.* 25:292–99
224. Huan-Cheng C, Ewing GE. 1990. *Phys. Rev. Lett.* 65:2125–28
225. Wight AC, Penno M, Miller RE. 1999. *J. Chem. Phys.* 111:8622–27
226. Hamers JS, Houston PL, Merrill RP. 1990. *J. Chem. Phys.* 92:5661–76
227. Korolik M, Suchan MM, Johnson MJ, Arnold DW, Reisler H, Wittig C. 2000. *Chem. Phys. Lett.* 326:11–21
228. Huang Y, Rettner CT, Auerbach DJ, Wodtke AM. 2001. In press
229. McCarthy MC, Allington JWR, Griffith KS. 1998. *Chem. Phys. Lett.* 289:156–59
230. Newns DM. 1986. *Surf. Sci.* 171:600–14
231. Huang Y, Rettner C, Auerbach D, Wodtke A. 2000. *Science* 290:111–14
232. Hou H, Rettner CT, Auerbach DJ, Huang Y, Gulding SJ, Wodtke AM. 1999. *Faraday Discuss. Chem. Soc.* 113:181–200
233. Hou H, Huang Y, Gulding SJ, Rettner CT, Auerbach DJ, Wodtke AM. 1999. *J. Chem. Phys.* 110:10660–63
234. Rettner CT, Pfnur HE, Stein H, Auerbach DJ. 1988. *J. Vac. Sci. Technol. A* 6:899–901
235. Gostein M, Parhikhteh H, Sitz GO. 1995. *Phys. Rev. Lett.* 75:342–45
236. Hou H, Gulding SJ, Rettner CT, Wodtke AM, Auerbach DJ. 1999. *J. Electron Spectrosc. Relat. Phenom.* 99:133–38

Author Index

Abe Y, 75
Abella ID, 316, 659
Abkevich V, 500
Abkevich VI, 500, 501
Abragam A, 463, 464, 466, 468, 469, 482, 772
Abramson E, 812, 814, 818
Abrashkevich D, 650
Abstreiter G, 220, 221
Achari A, 512
Achatz U, 424, 425
Achler M, 169, 182
Adachi J, 181
Adachi K, 237
Adam AG, 772
Adams A, 755
Adams H, 43, 49, 50
Adams MWW, 288
Adamson AW, 108
Adamson GW, 825
Adamson JD, 43, 55–57
Adelman DE, 18, 143, 832
Adjari A, 102
Adler F, 220, 221
Adler M, 76
Adrian M (Figure 8, see color insert)
Aebersold R, 84
Agard DA, 234
Agostini P, 669
Ahlström P, 364
Ahmad F, 244
Ahmed M, 141, 155, 839
Ahn J, 641
Ahrenkiel RK, 196, 206, 208, 210
Aisen P, 288
Ajdari A, 102
Akabori K, 297, 301
Akaike T, 80
Akamatsu N, 366, 367

Akasaka K, 578
Aker PM, 27, 32, 35
Akesson E, 644
Akita S, 86
Akutsu H, 580, 582, 587
Alagia M, 53, 152, 155
Alameddin G, 424
Alavi DS, 360, 361
Albano EV, 741
Albert G, 425
Albert S, 824
Albouy PA, 256
Albrecht AC, 670
Albrecht AW, 659
Albrecht M, 219, 220, 223
Albrecht RT (Figure 2, see color insert)
Alconcel LS, 171
Alderman DW, 471, 483
Alexander AJ, 155, 159, 160
Alexander MH, 152, 814, 825, 828, 832
Alexander P, 512
Alexeev O, 424
Alfken J, 75
Alford JM, 424, 425, 436
Ali I, 171
Alivisatos AP, 213, 214, 225, 655, 658
Alla M, 580, 582
Allan BW, 76
Allara DL, 108, 120, 125, 550
Allen HC, 362–64, 372, 373
Allen L, 772
Allen MJ, 78, 80
Allen MP, 501
Allen PJ, 584
Allington JWR, 841
Allison DP, 73, 83
Allison SA, 98
Allison TC, 52

Alm E, 243, 514
Almeida R, 653
Almqvist N, 84
Alonso DOV, 503, 516, 517, 525
Alonso PJ, 294
Als-Nielsen J, 357
Altarelli M, 199, 201, 202
Altenbach C, 299
Althorpe SC, 168, 272
Alvarez JL, 644
Alvarez RA, 43, 52, 58
Alzetta G, 783
Amar JG, 132, 133
Amato I, 72
Ambrose WP, 234, 235
Ambruso DR, 293
Amitay Z, 171
Amoureux JP, 478, 487
Amro NA, 134
Ananthavel SP, 741, 742
Anastasiodis SH, 562, 563
Andersen HC, 176, 318
Anderson CF, 97
Anderson FE, 589
Anderson J, 93
Anderson JB, 143
Anderson JD, 392, 395, 401, 412, 413, 415, 417, 419
Anderson ML, 392, 412, 413, 415
Anderson PA, 542
Anderson PW, 321, 322, 334
Anderson RC, 583
Anderson RW, 814
Anderson SL, 425
Andersson KK, 288
Andersson M, 407, 409, 424
Ando I, 578
Andrae MO, 372
Andres RP, 706

853

Andrew ER, 474, 578, 582
Andrews L, 36
Andrews PM, 825, 828
Andrews S, 327, 328, 340, 351
Aneja R, 298
Anfinrud PA, 316
Anfinsen CB, 500
Angell CA, 324, 332, 334, 343
Anger HO, 171
Angerhofer A, 287, 291, 303
Anglister J, 493, 584, 590
Ansari A, 327
Antipin SA, 641
Antohi O, 583
Antoine P, 772, 777, 779
Antoniadis H, 392, 419
Antonini E, 347
Antzutkin ON, 587, 589, 590, 592, 593, 595, 596
Aoiz FJ, 24, 27, 142, 144, 146, 149, 151, 155
Aoki M, 741
Apkarian VA, 641, 643, 651
Apostolakis J, 501
Appella E, 75
Aquino A, 175, 176
Arakawa Y, 219, 220, 223
Arancibia-Carcamo CV, 80
Arango AC, 225
Arasaki Y, 168, 271, 272
Arata Y, 511
Arbab M, 453
Archontis G, 715
Arduini RM, 316, 330
Arent DJ, 199, 202, 205
Argaman M, 84
Arias P, 828
Arieli D, 288, 304
Arimondo E, 783
Aristov N, 429
ARMENTROUT PB, 423–61; 4, 426–30, 432–34, 441, 444, 445, 447, 449, 453
Armindo E, 800

ARMSTRONG NR, 391–422
Arnett DC, 334, 643, 659
Arno JI, 619
Arnold DW, 149, 172, 176, 177, 841
Arnold FH, 557
Arrivo SM, 641
Arshava B, 583
Asakura T, 585, 587, 590, 597
Aschheim K, 576
Ascione R, 755
Aseyev SA, 667, 781
Ashburn TT, 583, 592, 595
Asher RL, 429, 432
Asher SA, 327
Ashfold MNR, 51, 825
Ashida J, 480
Ashino M, 81
Ashwell GJ, 706
Aspect A, 800
Aspnes DE, 205
Asscher M, 840
Assion A, 261, 262, 266, 270, 640, 641
Astashkin AV, 294, 297
Astinov V, 670
Astumian RD, 557
Atabek O, 650, 667
Atakan B, 48, 58
Athanassenas K, 424
Atkinson DB, 51
Atkinson R, 41
Auer M, 598
Auerbach DJ, 814, 840, 841, 843, 844
Auerbach ID, 72
Auger M, 472, 583, 592, 593, 595
Augspurger JD, 316, 327, 328
Austin RH, 715
Autipin AA, 641
Auzinsh M, 146, 148, 159
Averbukh IS, 640
Aviram A, 705
Avouris P, 702, 712, 743
Avrillier S, 772
Axe JD, 119

Ayers B, 599
Ayers JD, 146, 159
Azinovic D, 650
Aziz H, 413
Azouz IB, 120
Azuma Y, 741

Ba Y, 485, 487
Babcock AP, 756, 757, 759
Babcock GT, 288, 294, 669
Babcock HP, 236, 240, 245
Bačič Z, 614, 619, 21
Bachleitner A, 335
Back JF, 348
Backhaus P, 650, 652, 670
Backstrand KM, 814
Backus S, 265, 669
Bader G, 726, 727, 730, 741, 742
Baer A, 171
Baer M, 17, 141
Bahns JT, 650
Bai C, 77, 84, 234
Bai YS, 316, 317, 321, 322, 334, 511
Bailey DW, 195
Bailly A, 487
Bain CD, 113, 114, 128, 130, 131, 357, 360, 374–76
Baird B, 298
Bajaj KK, 205
Bajema ML, 669
Bak M, 471
Baker D, 243, 245, 510, 513, 514, 518
Baker DB, 484
Balakrishnan N, 834
Balasubramanian S, 239, 240, 293
Balazs AC, 562
Balazs YS, 582, 590
Balbach JJ, 584, 587, 590, 592, 593, 595, 596
Balcoup P, 669
Baldelli S, 363, 373
Baldo MA, 396, 419
Balducci G, 449

AUTHOR INDEX

Baldus M, 480, 489, 490, 492, 580, 582, 587, 589
Baldwin DP, 141
Baldwin RL, 500, 502
Balhorn R, 78
Balint-Kurti GG, 155
Ball JA, 287
Ball JC, 55
Ball RC, 542
Balla RJ, 58
Baller MK, 86, 557
Balling P, 256, 772, 777, 779
Ballone P, 703, 704
Baltisberger JH, 481
Balucani N, 53, 140, 152, 155
Banares L, 142, 144, 146, 149, 151, 155
Band YB, 640, 650, 787, 793, 797
Bandarage G, 627
Bandrauk AD, 641, 667, 669, 779
Banin U, 213, 214, 266, 268
Bankman IN, 73
Banyai L, 212
Bar I, 652
Bar-On I, 730, 731, 735, 737
Barany G, 511
Barat M, 189
Baratoff A, 719
Barbar E, 511
Barbara PF, 392
Barchi JJ Jr, 512
Bard AJ, 114, 391, 394, 396, 398, 401, 404, 407, 410, 417, 706
Bardeen CJ, 214, 641, 647, 649, 655, 658
Bardeen J, 691, 694
Bardou F, 800
Barfield M, 578
Barkema GT, 93, 94, 99
Barlow S, 392, 395, 401, 412, 413, 415, 417
Barnett RN, 731–33
Baro AM, 76
Baronavski AP, 21

Barone SB, 372
Barra AL, 287, 303
Barrat J-L, 98, 245
Barrick D, 316, 330
Bartana A, 266
Bartels C, 501
Bartels DP, 752
Bartels R, 669
Bartlam M, 591
Bartlet JE, 395, 396, 398, 399, 401
Bartley LE, 236, 240, 245, 756, 757, 759
Bartolotti L, 591, 595
Barton AE, 612
Barton JK, 85, 687
Bartz JA, 132
Barynin VV, 294
Basche T, 234
Baselga J, 93
Baselt DR, 71
Bashford D, 501
Baski AA, 119
Baskin JS, 260, 665
Basnar B, 121
Bass AD, 706
Bassi D, 772
Bastard G, 197, 199, 201, 202, 220
Basu P, 294
Basunstein D, 335
Bates FS, 562
Bates RD, 48
Batey RT, 751
Batista VS, 257, 266, 267, 271
Battiste JL, 598
Baughcum SL, 43, 49
Bauknecht A, 220, 221
Baulch DL, 41, 58
Baum G, 644
Baum J, 589
Baumann CG, 560
Baumann P, 79
Baumann T, 480
Baumert T, 261, 262, 264, 266, 270, 640, 641, 653

Bauschlicher CW, 432
Bawendi MG, 211–13, 219, 221, 222, 224, 225
Bax A, 477, 483, 484
Beal HP, 818, 820, 821, 825
Beames S, 831, 832
Bean BD, 146, 159
Bear DG, 75, 76
Beaud P, 663
Beaulieu L, 585, 587, 590, 597
Becerra LR, 303, 599
Becerra R, 55
Beche JF, 189
Bechinger B, 576, 589, 590
Bechtold PS, 424, 438, 445, 454
Beck C, 824
Beck WF, 297, 361
Becker DE, 318, 321, 322
Becker I, 268
Becker JA, 424
Becker JC, 75, 168, 185
Becker M, 787
Becker U, 181, 185
Beckord K, 168, 185
Beece D, 316
Beideman FE, 398
Beijerinck HCW, 772
Beijersbergen JHM, 170
Bein T, 706
Beinert H, 288
Beinstein RB, 139, 140, 142, 154, 158
Beligere GS, 236, 241, 243, 244
Beljonne D, 396, 419
Belkacem A, 171
Bell GR, 357, 360, 374, 375
Bell RC, 424
Bellert D, 429, 432
Bellessa J, 220, 223
Bellew BF, 303, 306
Bellmann E, 413, 415, 417
Beltramini M, 293
Bemish RJ, 610, 614, 625, 634

Bender CJ, 288, 293
Ben-Hamu D, 171
Benisty H, 197, 220
Ben-Jacob E, 706
Benjamin I, 357, 364, 730, 731, 733, 736, 737
Benkovic SJ, 293
Ben-Naim A, 357
Bennati M, 288, 304, 306
Bennett AE, 472, 485, 489, 492, 493, 580, 582, 584, 587, 590
Ben-Nun M, 27, 158
Bensebaa F, 117
Bensky TJ, 669
Bentley JA, 825
Ben-Yoseph G, 83
Benziger JB, 443, 447, 453
Benzinger TLS, 584, 590, 593, 595
Beratan DN, 706
Berden G, 814
Beredzen J, 327
Berendsen HJC, 357, 364, 506
Berg C, 424, 425
Berg M, 316, 317, 320, 322, 324, 327, 328, 334, 336, 346, 347, 351, 352
Berg MA, 322, 337, 341–44, 347
Berge T, 81
Bergeman T, 650
Berger AE, 475, 580, 584, 585, 588
Berger G, 294
BERGMANN K, 763–809; 644, 773, 776, 779, 782, 785–87, 789–93, 795, 796, 798–800
Bergsma-Schutter W, 72
Bergt M, 641
Berkowitz M, 357, 364
Berkowitz ML, 506
Berliner LJ, 296
Berman BL, 77
Bernardo M, 288

Bernards JPC, 578, 585
Bernath PF, 43
Berndt P, 550
Berne BJ, 743
Bernstein ER, 261
Bernstein RB, 21, 23, 255, 643, 650, 669, 814
Bersohn R, 53, 158
Bertolini D, 343, 346
Besocke KH, 730
Beswick JA, 612
Beth AH, 298
Bethardy GA, 44, 64, 65, 608
Beuville E, 189
Bevan JW, 619, 627
Beveridge DL, 95, 502
Beverwyk P, 483, 488
Beyer M, 424, 425
Beyer T, 427
Beyreuther K, 593, 595
Bezanilla M, 76, 77, 81, 84 (Figure 4, see color insert)
Bhagat VR, 295
Bhattacharya P, 220, 223
Bhattacharyya SS, 779
Bian WS, 52–54, 154
Biberian JP, 454
Biehl R, 286
Bielecki A, 473
Bielefeldt H, (Figure 2, see color insert)
Bier M, 557
Bierbaum K, 119
Bieschke J, 234, 247
Bigelow N, 800
Bigelow WC, 108
Bigwood R, 818, 822, 823
Bildsoe H, 490, 580, 582, 584
Billas IML, 424
Billing GD, 834
Bilsel O, 242
Bimberg D, 206, 215, 219–21, 223
Binder K, 567

Binnig G, (Figure 2, see color insert)
Birdi KS, 372
Birke MK, 429, 430, 438
Birshtein TM, 538
Biscarini F, 698
Bise RT, 170
Bishop AR, 718
Bittl R, 287, 291, 293, 294, 299, 301, 302, 304
Biver C, 538, 561
Bixon M, 681, 717, 724
Bjorklund GC, 50
Björling S, 234, 247
Bjorneholm O, 265
Blackmon BW, 44, 45, 46, 152
Blackwell M, 261, 263, 664
Blais NC, 143
Blake CCF, 591, 592
Blake GA, 619
Blake TA, 363
Blanchard GJ, 116, 128, 130, 131
Blanchet V, 168, 264, 272
Blanco FJ, 580, 585
Blandamer MJ, 268
Blank DA, 141, 670
Blank DHA, 120
Bloch F, 772
Block PA, 608, 610, 625, 631
Block SM, 557
Blodgett KB, 741
Bloeß A, 304
Bloembergen N, 358
Bloom CJ, 407, 409
Bloomfield L, 424
Bloomfield VA, 73, 97, 560
Blümich B, 464, 483
Blumler P, 483
Blunt DA, 814
Blye A, 590, 597
Bobst AM, 298
Bock CH, 301
Bockelmann U, 85, 197, 220, 221
Bockmann A, 587

Boczko EM, 500, 504, 506, 508–10, 516, 517, 528, 529
Bodenhausen G, 463, 468, 483, 485, 487, 659
Boehmer K, 796
Boehrer J, 220, 223
Boender GJ, 587
Bohac EJ, 610, 612, 614, 621, 625, 627, 634
Böhm G, 220, 221
Bohmer E, 23
Bohn PW, 117
Bohrer J, 219
Bois C, 293
Bokor J, 256
Bolhuis PG, 500
Bolin KA, 503
Bollinger JM, 303
Bolman PHS, 7
Bolton JR, 4
Bomse DS, 45
Bonačić Koutecky V, 432, 436
Bonagamba TJ, 584
Bonča J, 717, 719, 741
Bond JP, 97
Bondybey VE, 424, 425
Boneberg J, 270
Bonnote P, 224
Booth PJ, 234
Boothroyd AI, 143
Borbat P, 294
Borbat PP, 296
Bordé CJ, 772
Borden WT, 172, 643
Bordui PF, 50
Borgis DJ, 732
Borisov OV, 538, 561
Borkowski J, 142, 144, 145
Bormann BJ, 590
Borovykh IV, 301
Bosch MK, 294, 301
Boszo F, 443, 447
Bottin H, 294
Bottomley LA, 84
Bouchaud JP, 542

Boucher D, 50
Boudreaux DS, 196, 197, 219, 224
Bourgoin JP, 699
Bouwman W, 357
Bowater IC, 7
Bowen KH, 425
Bower JE, 425, 428
Bower PV, 580, 585, 587
Bowering N, 814
Bowers JE, 220, 223
Bowers MT, 424, 434
Bowlby NR, 294
Bowler BE, 298
Bowman CT, 48
Bowman JM, 19, 147, 160, 814, 825
Bowman MK, 298
Bowman RM, 640, 650, 653, 665
Boxer SG, 315, 320, 327, 328, 330, 331, 340, 351
Boyd RW, 640, 655
Bozso F, 455
Brabletz T, 75
Brack M, 437, 454
Bradbury A, 474, 578, 582
Bradbury EM, 75, 76, 78
Bradforth SE, 149, 172
Bradley DDC, 392, 410, 411, 419
Bradley KS, 17, 19, 24
Bragg J, 453
Brand L, 235, 237, 247
Brandt JL, 476
Brandt M, 814
Brasselet S, 236, 240, 248
Bratko D, 541, 544, 548, 549
Bratt PJ, 299, 303
Brault JW, 363
Brauman JI, 176
Braun E, 83
Braun J, 828, 830
Braun M, 168, 188
Braun R, 376
Brauning H, 182
Braunstein DP, 316, 327

Brazda V, 75, 77
Bréchignac Ph, 612
Bredas J-L, 396, 419
Breen JJ, 261
Bremi T, 587
Brems V, 185
Brennan D, 453
Brenot J-C, 169, 171, 180, 181, 189
Brenowitz M, 93, 95, 752, 755
Bresgunov AY, 303
Breslav M, 583
Brewer RG, 318, 772
Brezesinski G, 357
Brickman RO, 424, 436
Bright TB, 125
Brilmyer GH, 404, 407
Brinigar WS, 347
Brinkmann A, 580, 582, 585, 589
Brisson A, 72
Britt RD, 287, 288, 294
Brixner T, 641
Brock PJ, 225
Brockwell DJ, 242
Broers B, 640, 772, 777
Bromley SK, 570
Bronikowski MJ, 19, 157
Brooks BR, 335, 520
BROOKS CL III, 499–535; 52, 242, 243, 500, 501, 503, 504, 506, 508–10, 512, 514–18, 520, 521, 528, 529
Brooks HB, 294
Brooks PR, 650, 814
Brophy JH, 141
Brouard M, 19, 24, 27, 155, 157, 159
Brown AR, 392, 410
Brown EJ, 641, 643–45, 647, 649, 653, 655, 658, 659, 661, 665, 667
Brown FB, 143
Brown G, 562
Brown JM, 7, 589

Brown MG, 362, 364
Brown SP, 478
Brown SS, 51
Brown TG, 287
Brownsword RA, 154
Broyer M, 424
Brucat PJ, 425, 426, 429, 432
Bruccoleri RE, 520
Brucker GA, 23, 261, 650
Brudvig GW, 297
Brügmann O, 303, 306
Brum JA, 199, 201, 202
Brumer P, 640, 651
Brundobler S, 782
Brunel LC, 303
Brunner E, 475, 476
Brunner H, 121, 122, 128
Brunner K, 220, 221
Brunori M, 347
Brus LE, 212, 706
Bryan P, 512
Bryant JA, 288, 303, 304, 306
Bryant Z, 501
Bryngelson JD, 500, 544
Brzoska JB, 120
Bubacco L, 293
Buch V, 363
Buchenau HK, 267
Buck M, 116, 117, 128, 130, 131
Buckner SW, 424
Bucksbaum PH, 256, 641, 669
Budil DE, 298
Budker VG, 80, 298
Buelow S, 21, 23
Buitink J, 298
Buller GS, 220, 223
Bulovic V, 413
Bulthuis J, 814, 832
Buma WJ, 261
Bumgarner RE, 619
Bunker PR, 18, 612, 619
Buntine MA, 141
Buontempo JT, 357
Burak I, 19, 24, 608, 825, 827
Burbage JD, 357

Burgard C, 293
Burghard M, 687, 706
Burghardt I, 651
Burghaus O, 303
Burgi T, 828
Burgin TP, 687, 698, 702, 708
Burie J, 50
Burin AL, 710
Burke H, 703
Burkoth TS, 584, 590, 593, 595
Burnett IJ, 590
Burnett K, 800
Burnett NH, 669
Burns WK, 50
Burroughes JH, 392, 410
Burrows PE, 413
Bursulaya B, 501, 509, 510, 528
Bursulaya BD, 501
Busarow KL, 619
Bush S, 472
Bustamante C, 75–77, 80, 83–85, 246, 557, 560, 698
Bustamante CJ, 75, 76
Butayev BS, 667
Buthelezi T, 429, 432
Butler CS, 288
Butler LJ, 165, 643
Buttenshaw JA, 9
Büttiker M, 710, 711, 717
Buttry DA, 114
Buy C, 294
Byer RL, 51

Cafiso DS, 590
Caflisch A, 501, 503
Cahen D, 741
Cahill AF, 215, 216
Cai W, 206
Calderone A, 117
Caldwell WB, 236, 239
Callaway DJE, 591, 595
Cameron D, 424
Cameron LM, 288
Cammack R, 287, 288, 294

Campagnola PJ, 267
Campargue A, 818
Campbell AA, 583, 584
Campbell CT, 128, 130, 131
Campbell DJ, 373
Campbell GC, 472
Campbell ID, 591
Campbell KA, 287, 288, 294
Campbell ML, 152
Campos VB, 211
Campos-Martinez J, 834, 837, 838
Canet D, 472
Cangshan X, 172, 176, 177
Canters GW, 292, 293, 304
Cao JM, 256
Cao JS, 641, 647, 649
Cao Y, 392, 396, 407, 409, 410
Capasso F, 50
Capp MW, 97
Capson DW, 507
Carlsson N, 219
Carmieli R, 288, 306
Carmona-Novillo E, 834, 837, 838
Carnal O, 801, 802
Caron C, 93, 94, 99
Caron LG, 726, 727, 730, 741, 742
Carpenter IL, 357, 364
Carpenter SD, 264, 641
Carr RT, 293
Carraro C, 120, 121
Carrascosa JL, 75
Carravetta M, 580, 582
CARRINGTON A, 1–13; 2–7, 9, 10
Carrington T, 612, 832
Carrion-Vazquez M, 72
Carroll CE, 782, 792, 795, 796
Cartechini L, 53, 152, 155
Carter RT, 644
Carter SA, 225
Cartland HE, 21
Cartwright E, 56

Cary RB, 75, 76
Casassa MP, 612, 623, 626, 634
Casavecchia P, 53, 140, 152, 155, 165
Case DA, 501–3
Case MA, 687
Casimiro DR, 293
Cassavecchia P, 140
Cassell G, 75
Cassie AB, 124
Casson BD, 375, 376
Castillo JF, 149, 151
Castleman AW, 424, 425, 427, 454, 644
Castleman AW Jr, 814, 828, 829
Castleton KH, 58
Castrillo P, 219
Catalan J, 644
Catanzarite J, 21, 23
Cate JH, 751
Cates ME, 538, 542
Caughey WS, 347
Cavalleri A, 256
Cave JP, 169, 181
Cech TR, 751, 755
Cederbaum LS, 712, 713
Celda B, 511
Celii FG, 831
Cerullo G, 214, 641
Cervantes G, 77
Češpiva L, 432, 436
Chaban G, 834
Chachisvilis M, 644
Chahine J, 242
Chaikin PM, 563
Chakrabarti A, 562, 563
CHAKRABORTY AK, 537–73; 538, 541, 544, 547–50, 552, 557
Chakravarty C, 828
Chamberlain J, 357
Champion PM, 327
Chan HS, 242, 500, 544, 752
Chance MR, 752, 755
Chance RR, 196, 197

Chand A, 87
Chandler DW, 141, 166, 168, 183, 184, 272, 273, 357, 500, 542, 665
Chandross E, 391
Chaney MO, 591
Chang BY, 776, 779
Chang C-H, 108
Chang H, 53
Chang T-M, 364, 368
Chang YC, 195
Chao SD, 151
Chapman MS, 650
Chapman WB, 45, 46, 152
Charlson RJ, 372
Chase MW Jr, 439
Chasovskikh S, 77
Chatelain A, 424
Chatfield DC, 148
Chattopadhyay A, 158
Chattoraj DK, 372
Chaux R, 663
Chawdhury N, 396
Chawla D, 608
Che DC, 64, 141
Che JW, 641, 647, 649
Chelkowski S, 641, 779
CHEMLA DS, 233–53; 235–38, 240, 241, 245
Chen CH, 392, 411, 413
Chen DJ, 75, 76
Chen E, 234
Chen G, 372
Chen GE, 819
Chen H, 563
Chen MC, 825
Chen P, 220
Chen SH, 128, 315
Chen T, 640, 658, 661
Chen W, 50
Chen X, 295, 297
Chen XR, 825, 832
Chen Y, 21, 23, 160, 247, 261, 263, 663, 664, 818
Chen YQ, 818
Chen YT, 818, 824
Cheng GJ, 116, 128

Cheng L-T, 653, 663
Cheng NQ, 598
Cheng SS, 128
Cheong HM, 214, 216, 217, 219
Chergui M, 653, 772, 773
Chernovitz AC, 730, 741
Cherny DI, 74
Chernyak V, 720
Cheshnovsky O, 257, 268, 424
Cheung CL, 86
Chevaleyre J, 424
Chi LF, 119
Chiasson J, 730, 741
Chiba H, 670
Chidsey CED, 125, 687
Chien EYP, 315, 320, 327, 328, 334
Child MS, 151
Chin AH, 256
Chin RP, 375
Chingas GC, 477, 483
Chisholm DA, 287
Chiti F, 591
Chmelka BF, 469, 477
Cho AY, 50
Cho G, 589
Cho MH, 659, 665
Choc MC, 347
Choi G, 590
Choi H, 170
Choi JS, 80
Choi SE, 825, 828
Choi YH, 80
Choi YS, 818, 820, 825, 827
Chopin L, 487
Chottard JC, 293
Chrambach A, 93, 94, 99
Chrisey LA, 85
Christen J, 206, 219
Christensen L, 50, 62, 63
Christoph R, 735, 737
Chu K, 316
Chu S, 93, 94, 236, 240, 245, 798, 800, 801
Chu YC, 363

Chuang CC, 825, 828
Chung C, 21
Ciccotti G, 506
Ciesla CM, 220, 221
Cina JA, 659
Claeys W, 772
Clark A, 591
Clark C, 301
Clark RJH, 665, 667
Clarke JS, 261
Clary DC, 16, 17, 158, 160, 166, 172, 179, 619, 621, 828, 834, 837, 838
Clauberg H, 818, 820, 821, 825
Clausen-Schaumann H, 73, 83, 85, 86
Claypool CL, 698
Clayton AHA, 687
Cleave V, 396
Clementi C, 501, 504, 520
Clements TG, 168, 170, 172, 176, 186, 187
Clemmer DE, 428, 449
Cleveland CL, 731–33
Clifford EP, 50
Cline JF, 288
Cline RA, 650
Clore GM, 316, 512
Cluzel P, 83, 85
Cobos CJ, 41, 58
Cohen FE, 234, 500
Cohen MH, 465, 469, 476
Cohen RC, 608, 619, 741
Cohen Y, 652
Cohen-Tannoudji C, 800
Cohen-Tannoudji CN, 800
Coker DF, 267, 743
Colaneri MJ, 293
Colaneri N, 392, 410
Colby RH, 93, 94, 99
Colchero J, 76
Collings BA, 424
Collins MA, 31
Collinson MM, 398, 399, 401
Colombo MG, 491, 582
Colson SD, 259

Colton MC, 17
Colton RJ, 71, 85
Coltrin ME, 840
Colvin V, 225
Comstock M, 670
Conboy JC, 357, 360, 364, 374, 376, 378, 382
Conceição J, 429, 430, 438, 441, 443, 444
Cong PJ, 641
Connell LL, 665
Conradi MS, 484
Constant E, 669
CONTINETTI RE, 165–92; 166–72, 174–76, 182–84, 186–88, 265, 272, 273
Cook DC, 563
Cookson JL, 48
Coomparat D, 650
Cooper DE, 50, 814
Cooper J, 167
Cooper WF, 48, 49, 58, 59, 61
Copeland K, 141
Copie V, 584
Corchado JC, 25, 31
Corcoran TC, 665
Cordone L, 335
Coremans JWA, 292, 293, 304
Cork C, 189
Corkum PB, 667, 669, 779–81
Cornet A, 772
Coron N, 303
Cosby PC, 170, 814, 831
Cosgrove T, 538
Costa PR, 487, 490, 492, 580, 582, 583, 585, 587, 592, 593, 595
Costen ML, 24, 27
Cottrell C, 840
Coughlin PE, 589
Coulston GW, 787, 789, 798
Coury JE, 84
Cowburn D, 599
Cowen BR, 327, 335

Cowen JA, 288
Cox AJ, 424
Cox DM, 424, 434–36, 440, 447, 452
Cox RA, 41, 58
Coy SL, 817, 824
Cramer WA, 590
Crane JC, 818, 820, 821, 825
Crawford MA, 43
Craycraft MJ, 257, 425, 426
Crccgan K, 425
Creegan KM, 429, 432
Crellin K, 428, 429
Cremer P, 375
Crepeau RH, 296
Creuzet F, 491, 580, 582, 590
Crim FF, 17, 19, 157, 652
Crisp MD, 769, 784
Cronkhite JM, 43, 46
Crooks RM, 114
Cross TA, 576, 578, 589, 590, 597
Crosson E, 315
Crothers DM, 93, 95, 102
Crowder MW, 293
Croxton CA, 357
Crosson E, 315
Crubellier A, 650
Cruchon-Dupeyrat SJN, 118, 128
Cui Y, 83, 85, 246
Cukier RI, 93, 94, 99
Cunhao Z, 814
Cupane A, 335, 347
Curl RF, 43, 49, 50, 52, 55–58, 650
Curtis CJ, 214
Curtiss TJ, 814
Cusack S, 335
Cyr DR, 168, 170, 171, 261
Czajkowsky DM, 72, 81
Czerski L, 590

Dabbagh G, 484, 489, 578, 580, 582, 584, 585
Dabbousi BO, 211, 225
Dagdigian PJ, 26, 27, 61, 152, 157

AUTHOR INDEX 861

Daggett V, 243, 500, 502–4,
    515–18, 525, 528, 529
DAHAN M, 233–53; 235–38,
    240–45
Dahleh M, 772
Dahlquist FW, 578
Dahmann G, 552, 556, 557
Dai HL, 766, 812, 828, 831
Daleman SI, 590
Dalgarno A, 834
Dalickas GA, 316
Dalitz RH, 188
Dalleska NF, 428, 429
Dalnoki-Veress K, 346
Dalton BJ, 795
Dalton LR, 298
Daly NR, 427
Daly TJ, 598
Dan PL, 825
Dandliker PJ, 687
Dane CB, 43
Dang LX, 357, 364, 368
Danileiko MV, 786
Dannenberger O, 116, 128,
    130, 131
DANTUS M, 639–79; 65,
    260, 639–41, 643–45,
    647, 649–53, 655, 658,
    659, 661, 670
Darden TA, 506, 591, 595
Darnton NC, 245
Darwin DC, 52
Dasgupta K, 424
Das Sarma S, 211
Datar R, 557, 561
Date T, 590, 597
Dateo CE, 825
Datta A, 122
Datta S, 687, 698, 700, 704,
    706, 708, 716, 724
Datte P, 189
Daubendiek SL, 83, 84
    (Figure 9, see color
    insert)
Daura X, 501
Davidovits JV, 121
David D, 652

Davidson VL, 294
Davies CA, 439
Davies ER, 284
Davies JA, 168, 182–84, 265,
    272, 273
Davis AV, 266, 268, 269, 658
Davis D, 372
Davis DD, 372
Davis DM, 346, 372
Davis HF, 141, 160
Davis IH, 299
Davis JC, 779
Davis JH, 590
Davis JL, 291
Davis MM, 570
Davis PB, 80
Davis WB, 720, 721, 725, 726
Davison SG, 699
Davoust CE, 288
Dawson DV, 80
Dayton DC, 608, 610, 612,
    631
D'Costa NP, 73
Deacon DAG, 51
de Andrade PCP, 696
deAzevedo ER, 584
Debad JD, 394
de Beer E, 172
De Beer R, 294
de Boeij WP, 653, 655, 658,
    659
DeBono RF, 114, 128
DeBruijn DP, 170
Decatur SM, 330, 331
deDios AC, 578
DedonderLardeux C, 264,
    265
Defontaines AD, 93, 94, 99
de Gennes PG, 93, 94, 99
DeGrado WF, 236, 243, 244
De Groot A, 294
de Groot HJM, 584, 587
de Grooth BG, 76, 87
deHeer WA, 424, 437, 454
Dei A, 303
Deinum G, 669
Dejardin P, 93, 94

Dejong AF, 585
Delain E, 74
Delaney N, 267
Delange CA, 261
Delbrück M, 298
Deligiannakis Y, 291, 294,
    297
Dellago C, 500
Dellamanna D, 114, 128
Deller MC, 234
Deluca MJ, 173
del Valle JC, 644
Demchuk E, 501
Demco DE, 476
deMello JC, 407, 410
D'Mello MJ, 142, 144
Demkov YN, 772, 782
DeMore WB, 41
Demura M, 585, 587, 590,
    597
den Blaauwen T, 304
Deng HT, 454
DENIZ AA, 233–53; 235–45
Denny J, 590
de Pablo JJ, 562, 563, 567,
    569
DePillis GD, 330, 331
DePristo AE, 436, 438
Deras ML, 752, 754, 755
Derome AE, 463, 469
DeRose VJ, 288, 294
Dertinger S, 825
Desai PR, 293
DeSain JD, 43, 50
de Silvestri S, 647
Dessent CEH, 268
Deutsch JM, 93, 94, 99
Devlin JP, 363
de Vos AM, 293
Dewar MJS, 643
Dewey G, 540
DeWitt MJ, 264
Dexter AF, 288
Deyerl H-J, 171, 172, 176,
    261
Dhar L, 663
Dharmasena G, 141

Dhirani A, 687, 706
DiCapua FM, 502
Dicker AIM, 320
Diederich F, 294, 424
Diels J-C, 772, 773
Dietrich G, 424, 429, 430, 432, 433, 437, 439
Dietrich P, 667, 669, 781
Dietz TG, 426
Dikanov SA, 288, 291, 293, 294
Dill D, 167
Dill KA, 242–44, 500, 540, 544, 752
diMarzio EA, 545
DiMauro LF, 669
Dimitrov DA, 741, 742
DiNatale JA, 589
Dindot JL, 583, 584
Diner BA, 287, 294, 297
Dingle R, 197, 199, 202, 205
Dinner AR, 525
Dinse KP, 286
Diol SJ, 220, 223
Dion CM, 667
Disselhorst JAJM, 303, 304
Dittmann P, 798
DiValentin M, 287
Dixon RN, 608, 610, 825
Dixon WT, 485
Djaoui A, 669
Dlakic M, 81 (Figure 8, see color insert)
Dlott DD, 20, 315, 317, 320, 322, 323, 325, 327, 328, 330, 331, 334, 336, 337, 341, 345, 346, 349, 350
Dluhy RA, 357
Do KG, 590, 597
Doan PE, 288
Doany FE, 639
Dobber MR, 261
Dobbyn AJ, 155, 825, 827
Doblhofer K, 114
Dobson CM, 242, 500, 503, 591
Doerner R, 169, 182

Doetschman DC, 299
Doi K, 288
Doi M, 93, 94, 99
Doktycz MJ, 73
Domcke W, 270, 712, 713
Domen K, 366, 367
Domingo C, 56
Dong F, 149, 150
Dong X-F, 80
Doniach S, 245, 759
Donovan RJ, 152
Dooley DM, 293
Dorlet P, 294, 303
Dorner R, 171
Dorogi M, 706
Doster W, 335
Doudevski I, 122, 123, 125, 128, 132, 133
Doudna JA, 751
Douglass DC, 424
Douhal A, 265, 644
Downey JR Jr, 439
Downie P, 169, 181, 182
Dows DA, 652
Drabbels M, 812–14, 832, 835, 836
Drake B, 72, 81
Draunstein D, 335
Dravnieks F, 3
Dreier R, 814
Drese K, 786
Dresner J, 391
Drewello T, 170
Drexhage KH, 235, 248
Dritschilo A, 75–77
Drmanac R, 93
Drobny GP, 490, 578, 580, 583–85, 587
Droz T, 828
Drukker K, 155
Dryden DT, 81
Du JL, 298
Du M, 665
Du Q, 357, 362–64, 374, 375
Duan L, 501, 528

Duan Z-C, 818, 819
Dubach J, 293
Dubinskii AA, 295, 303
Dubov VS, 650
Dubovoy T, 419
Dudek SP, 687
Duer MJ, 478
Duffy DC, 374, 375
Dufrene YF, 71
Dugan CH, 832
Dugan MA, 772
Dugourd P, 424
Duke TA, 93, 94, 99
Dulieu O, 650
Dulligan M, 609
Dunau R, 335
Duncan BC, 790
Duncan DI, 772
Duncan MA, 426
Dunet H, 189
Dunlap DD, 78
Dunlavy DJ, 208, 210
Dunlop JR, 814, 825, 826
Dunning TH Jr, 16, 17
Duppen K, 670
Durant JL, 58
Durbin MK, 122
Durfee CG, 265
Durfee DS, 782
Durup J, 154
Durup-Ferguson M, 171
Duschek F, 644
Dusold S, 485, 489
Dutcher JR, 346
Dutta P, 112, 122
Duxon SP, 24, 27, 272
Dybowski C, 463, 468, 469, 473, 475
Dyer PN, 5
Dyer RB, 234, 513, 515
Dyke TR, 612
Dykstra CE, 316, 327, 328, 612
Dynan WS, 75, 76
Dyne PJ, 644
Dyson HJ, 502, 511
Dzuba SA, 297, 298, 301

AUTHOR INDEX

Eades RG, 474, 578, 582
Eap CB, 79
Earl WL, 580, 582, 584
Earle KA, 303
Earnest T, 189
East ALL, 834
Eastman MA, 483
Eaton GR, 293, 298
Eaton SS, 293, 298
Eaton WA, 242, 244, 316
Ebata T, 814
Eberhardt A, 113
Eberhardt W, 268, 424
Eberly JH, 772, 776, 782, 785
Echave J, 834, 837, 838
Eckbreth AC, 643
Eckert HJ, 301
Eckert T, 168, 188
Eckhoff WC, 50
Edelstein DC, 206
Eden JG, 650
Eden M, 471, 472, 491, 580, 582, 585, 589, 590
Edlund A, 730, 731, 735, 737
Edman L, 248
Edwards DP, 47
Edwards SF, 93, 94, 99, 542
Edwardson JM, 81
Edzes HT, 578, 585
Efros AL, 211–13, 219, 220
Egeberg KD, 316
Egeler T, 220
Eggeling C, 235, 237, 247
Egorov VG, 650
Eichen Y, 83
Eigen M, 234, 235, 247
Eilers M, 599
Einstein A, 764
Eisels FL, 372
Eisenberg D, 361, 362, 508
Eisenberger P, 113
Eisenstein L, 316
Eisenthal KB, 357, 372, 375
Eisert F, 116, 117, 128
Ekberg CA, 306
Eker AP, 76
Ekimov AI, 212

Ekstrom CR, 650
Eland JHD, 168, 169, 180, 181, 185, 186
Elber R, 316, 346, 501
Elgersma H, 16, 17
Elhanine M, 266, 268, 658
El Idrissi MI, 818
Eliel ER, 361
Elings VB, (Figure 8, see color insert)
Elioff MS, 62
Elk M, 786, 795
Elkind JL, 425, 436, 449
Ellingson RJ, 195, 196
Elliott CM, 407, 409
Ellis DJ, 81
Ellis TH, 117
Ellman JA, 599
Elrod MJ, 619
El-Sayed MA, 424
Elschner A, 316, 321, 322, 334
Elser V, 782
Elson E, 247
Emberly EG, 698, 708, 719
Empedocles SA, 219
Emshwiller M, 485
Emsley L, 471, 493
Enderle T, 236
Enemark J, 294
Enemark JH, 294
Engel A, 71, 72
Engel V, 263, 270, 271, 653, 658, 661
Engel YM, 21, 23
Engelhard M, 590
Engholm JR, 315, 317, 320, 322, 325, 327, 328, 330, 331, 334, 336, 337, 341, 345, 346, 349, 350
English RA, 699
Engström S, 364
Enriquez PA, 24, 27
Epa VC, 619
Epel B, 304
Eppink ATJB, 142
Erickson BW, 503

Eriksson L, 585, 589
Ermolaev K, 475
Ernst M, 472, 492, 584
Ernst P, 220, 221
Ernst RR, 463, 468, 471, 480, 483, 485, 490, 491, 578, 580, 582, 585, 587, 659
Erskine DJ, 639
Ertl G, 256, 443, 447, 455
Ervin J, 242
Ervin KM, 424, 427–30, 435–37, 439, 454
Espe MP, 294
Espinosa-Garcia J, 25, 31
Esquerra RM, 234
Essevaz-Roulet B, 85
Essig NZ, 512
Esslinger T, 800
Essmann U, 357, 364, 506
Esteve D, 699
Evall J, 113, 114, 128, 130, 131
Evans DF, 97, 730, 731, 736, 737, 741
Evans JW, 132
Evans MCW, 303
Evelo RG, 297
Ewing GE, 151, 608, 614, 840
Eychmüller A, 213, 424
Eyring H, 639
Ezra GS, 819

Fabre F, 840
Fabris AR, 7
Facci JS, 687
Facelli JC, 578
Faeder J, 267
Faeder SMG, 659
Fafard S, 220, 223
Faglioni F, 698
Fahr A, 56
Faibis A, 171
Fain B, 812
Faist J, 50
Fakhr A, 48
Falk M, 363
Faller P, 287

Family F, 132, 133
Fan C, 288
Fan F-RF, 410, 706
Fang X, 752
Fang Y, 72, 73, 77, 78
Fann YCh, 288
Fano U, 743, 795
Fantucci P, 432, 436
Farach HA, 282
Farantos SC, 824
Farhat SK, 55–58
Farid S, 396, 398
Farmanara P, 264, 270, 644
Farrar CT, 288, 293, 304, 306
Farrar TC, 318, 321, 322
Farrell JT, 50, 62
Farrell JT Jr, 612, 616
Farver O, 293
Faubel M, 141, 274
Faucher O, 663
Faulhaber AE, 235, 241, 242, 245
Faulkner LR, 391, 394, 396, 398, 401, 407
Fauster T, 256
Fauth JM, 293
Fay N, 247
FAYER MD, 315–56; 20, 315–17, 320–22, 325, 327, 328, 330, 331, 334, 336, 337, 341, 345, 346, 348–50
Fayeton JA, 189
Featonby PD, 800
Feher G, 284, 287
Feher VA, 502
Fei R, 146, 159, 832
Feix JB, 299
Fejer MM, 50
Feldberg SW, 391, 396, 687
Felder P, 643, 644
Feldman EB, 589
Feldman H, 171
Feldman LC, 133
Feldmann J, 220, 223
Felker PM, 608, 665, 667
Feller SE, 357, 364

Felts AK, 720, 721
Feng L, 424
Feng S, 706
Feng W, 515
Feng X, 77, 585, 590
Feng XL, 590
Fenn JB, 840
Fenter P, 113
Ferguson-Miller S, 294
Ferkol T, 80
Fernandes PB, 234
Fernandez C, 478, 487
Fernandez JM, 72, 234
Fernandez-Aloso F, 146, 159
Ferrante C, 315, 348
Ferrara P, 501
Ferrenberg AM, 509, 520
Ferry DK, 211, 706
Fersht AR, 243, 244, 504, 515, 516, 518, 525, 528, 529
Fesik SW, 491, 580, 582, 584, 587
Fetterman MR, 772
Fewell MP, 786, 791
Feynman RP, 542
Fiebig T, 644
Fiehrer KM, 357
Field KA, 298
FIELD RW, 811–52; 160, 607, 766, 812, 817–20, 824, 825, 832
Field TA, 168, 272
Fields GB, 550
Filip C, 476
Filpula DR, 512
Filseth SV, 18, 832
Finch JN, 361
Finer JT, 557, 561
Fink M, 814
Fink WH, 59, 61
Finkelstein AV, 243
Fioretti A, 650
Fiori WR, 503
Fischer CM, 687, 706
Fischer I, 261
Fischer J, 116, 128

Fischer KH, 542, 544
Fischer P, 374, 578, 585
Fishbein KW, 476
Fisher A, 205
Fisher AJ, 719, 727
Fisher ER, 449
Fisher TE, 72, 234
Fiss JA, 640
Fitzcharles MS, 23
Flanagan HL, 290
Fleischhauer M, 786, 798
Fleming GR, 659, 665, 670
Floethmann H, 825, 827
Flores A, 590, 597
Florian P, 478
Florin E-L, 72
Fluegel B, 212, 220, 223
Fluegel BD, 220, 223
Flynn GW, 16, 26, 27, 62, 812
Flytzanis C, 212
Fodor SP, 73
Fogg DE, 225
Foldes-Papp Z, 248
Foley ET, 712
Folmer DE, 644
Fominykh NG, 181
Foot C, 800
Forbes RA, 425
Force DA, 287, 294
Forchel A, 219, 220, 223
Fork RL, 641, 653
Forrest JA, 346
Forrest SR, 396, 413, 419
Forster H, 587
Foster S, 245
Fotiadis D, 72
Fourcade A, 74
Fourkas JT, 322
Fournier MJ, 556
Fox MF, 268
Fraelich M, 62
Francis RS, 315, 317, 348
Francisco JS, 43
Franck J, 268
Frank CW, 73, 128
Frank MK, 512
Frank P, 41, 58

Franks F, 361, 362
Franses EI, 108
Frasch WD, 294
Fraser GT, 608, 612, 616, 623
Fraser PE, 591, 593
Frauenfelder H, 234, 316, 327, 335
Frechet JMJ, 550
Fredkin DR, 367
Fredrickson GH, 324, 332, 334, 343, 562, 563
Freed JH, 296, 298, 303
Freed KF, 562
Freeman RR, 256
Freiberg H, 570
Freiser BS, 424, 425, 429
Fresco JR, 755
Frese KW, 443, 447
Freude D, 469, 475, 476
Freudenberg T, 264, 265
Frey H, 256
Frey HM, 663
Freysz E, 357, 362–64
Fridovich B, 42
Fried A, 43, 44, 46, 52, 62
Fried LE, 819
Fried M, 93, 95
Friedbacher G, 122
Friedman RS, 148
Friedrich B, 610, 667
Friedrich J, 334, 335
Friend RH, 392, 396, 407, 410, 411
Fries JR, 235, 237, 247
Friesner RA, 720, 721
Friis EP, 715
Frischkorn C, 168, 266, 268, 269, 658
Frishman E, 650
Fritsch K, 334
Fritz J, 86, 557
Frohnmeyer T, 264
Froitzheim H, 454
Fromme P, 301, 304
Froncisz W, 299
Fronticelli C, 347

Frost MJ, 44, 64
Fruböse C, 114
Frurip DJ, 439
FRYDMAN L, 463–98; 477, 478, 483, 488
Frydman V, 488
Frye D, 828
Fu H, 213, 214, 216, 217, 219
Fu R, 487
Fu ZW, 424, 429, 432, 436
Fuchs H, 119
Fuchs M, 303
Fuentes JH, 93
Fujito T, 490, 580, 582
Fujiwara T, 490, 580, 582, 587
Fulara J, 270
Funatsu T, 236, 243
Fung BM, 472, 475
Furue H, 31
Furukawa M, 814
Furuno T, 80
Fyfe CA, 463, 468, 469, 475, 477, 483, 487

Gadzuk JW, 712, 840
Gaffney D, 237
Gaffney KJ, 256
Gai F, 234, 513, 515
Gall AA, 84
Gall K, 247
Gallagher T, 512
Gallot Y, 563
Galperin M, 694, 712, 721, 730, 731, 735, 737, 738, 741
Galzitskaya OV, 243
Gamache J, 706
Gammie DI, 10
Gammon D, 219
Gan Z, 478, 483
Gan ZH, 491
Ganapathy S, 578
Gandhi SR, 640, 772, 782, 814
Gann SL, 481
Gantefor G, 268, 270, 424

Gao J, 357, 364, 392, 407, 409
Gao S, 76
Garbow JR, 486, 583
García AE, 242, 357
Garcia R, 75, 76
Garcia RA, 75, 76, 84
Garde S, 357
Gardiner AT, 287, 294, 301, 304
Garner MC, 170, 172, 174–76, 187, 188
Garno JC, 118, 128
Garraway BM, 772, 797, 798
Garrett BC, 357
Garroway AN, 589
Gaspard P, 640, 651
Gast AP, 108
Gast P, 294, 299, 301, 304
Gatteschi D, 303
Gaub HE, 72, 73, 83, 85–87
Gaubatz U, 782, 785, 787
Gausa M, 424
Gay E, 288
Gazdy B, 825
Ge M, 298
Ge NH, 256, 704
Gebauer C, 141
Gebeshuber IC, 87
Gebhard R, 590
Geers A, 825, 832
Geerts Y, 394, 417
Geiger M, 220, 221
Geindre JP, 256
Geisler M, 261, 262, 266, 270
Gelfi C, 97–99
Gennis RB, 287
Gensemer SD, 650
Gentry WR, 814, 836
George AR, 591
Georges A, 542
Georgiadis R, 114, 115, 123, 128, 130, 131, 449
Gerber C (Figure 2, see color insert)
Gerber G, 261, 640, 653

Gerber T, 663
Gerfen GJ, 303, 306, 599
Gericke KH, 157, 168, 185, 189, 833
Gerken T, 80
Gerlich D, 427
Gerlsma SY, 348
Germann GJ, 32
Gershfeld NL, 381
Gershikov AG, 667
Gerstein BC, 463, 468, 469, 473, 475, 476
Gerz C, 792, 793, 800
Gesell JJ, 590
Gessner O, 181, 185
Geurts DG, 480, 489, 490, 492, 580, 582
Geusic ME, 424, 434, 435, 440
Gewirth AA (Figure 2, see color insert)
Geyer M, 303
Gfroerer TH, 220, 221
Ghadiri MR, 687
Ghiggino KP, 687
Ghosh S, 779
Giacometti G, 287
Giancarlo LC, 825, 828
Gibby MG, 472, 479
Gibson GN, 779
Giebeler C, 419
Gierasch LM, 488, 590
Giese A, 234, 247
Giese B, 687, 726
Giese CF, 814, 836
Giessen H, 220, 223
Giessibl FJ (Figure 2, see color insert)
Gigli G, 449
Giguere PA, 363
Gilbert RG, 31, 427, 428
Gilbert T, 261
Gilchrist ML, 287, 294
Gillilan RE, 640
Gilliland GL, 512
Gilliot P, 212
Gilroy KS, 335

Gimzewski JK, 687
Gindele F, 220, 223
Gippert GP, 501
Girard B, 141, 157
Girault G, 294
Girvin ME, 557
Glass GP, 43, 49, 52, 55–58
Glass R, 75
Glaubitz C, 590
Gleason KK, 589
Gleichmann MM, 650
Glick GD, 760
Glover TE, 256
Glushko B, 787
Gmachl C, 50, 51
Go JS, 608
Gō N, 500, 519
Goates SR, 375
Goddard WA, 698
Godt A, 294–97
Godun RM, 800
Goetz JM, 580, 585
Golan R, 72, 77–80, 83, 84 (Figures 3, 9, see color insert)
Goldbeck RA, 234
Goldberg L, 49, 50
Goldbourt A, 478
Goldburg WI, 472
Golden BL, 751
Golden DM, 41
Goldfarb D, 288, 293, 304, 306
Goldfield EM, 155, 825
Goldman A, 47
Goldman M, 463, 469–71
Goldmann M, 121
Goldner LS, 792, 793, 800
Goldstein AN, 214
Golovin AV, 181, 185
GOLUMBFSKIE AJ, 537–73; 541, 549, 550, 557
Gomez PC, 612
Gomez-Moreno C, 287, 294
Gontijo I, 220, 223
Gonzalez AI, 155

Gonzalez C, 698, 699, 703
Good D, 316
Goodin DB, 287
Gooding AR, 751
Goodman FO, 699
Goodman MF, 652
Goodnick SM, 211, 706
Goodwin JJ, 760
Goodwin PM, 234, 235, 246, 248
Gord JR, 424
Gordon EB, 650
Gordon JP, 641, 653
Gordon MS, 31, 52, 834
Gordon RG, 317
Gordon RJ, 640, 651
Gorzelle BM, 590
Gosavi S, 686, 692
Goss LM, 363
Gossard AC, 202, 205
Gostein M, 844
Gostev FE, 641
Goswami D, 772, 773
Götzinger T, 303
Gouda H, 511
Gouedard G, 141
Gough TE, 772
Gould IR, 396, 398
Gould PL, 650, 790
Gould SA, 72
Goulet T, 743
Gourdon A, 687
Gourdon C, 220, 223
Goy P, 303
Gozzini A, 783
Grabow MH, 133
Graener H, 357
Graessley WW, 93, 94, 99
Graf P, 730–35
Gragson DE, 360, 361, 364, 365, 368, 369, 372, 373, 381
Graham MD, 563, 569
Graham SC, 407, 410
Grakoui A, 570
Gramlich V, 294

Grandinetti PJ, 471, 478, 481, 483
Grant CV, 294
Grant DM, 464, 471, 483, 491
Grant JL, 17
Grantcharova VP, 514
Grasberger B, 512
Grasserbauer M, 122
Gratkowski H, 590
Gratton E, 247
Grätzel M, 224
Graumann PL, 515, 519
Gray HR, 783
Gray SK, 155
Grayce BB, 149
Greber T, 814
Green F, 24, 27
Green JBD, 71
Green MA, 196
Green WH, 48, 148
Green WHJ, 828
Greenblatt BJ, 257, 261, 266–69, 271
Greene BI, 639, 641, 653
Greenfield SR, 315, 687
Greenham NC, 225, 392, 396, 410, 411
Gregg BA, 225
Gregoret LM, 234, 513, 515, 519
Gregory DM, 490, 580, 582, 584, 585, 587, 590, 593, 595
Gregory JK, 172, 179
Gremlich HU, 598
Greschik S, 263
Greve J, 76, 87
Grey CP, 487
Grief D, 31
Griffin JB, 427, 428, 444, 447, 449, 453
Griffin JC, 610
Griffin RG, 303, 306, 471, 472, 475, 476, 478, 484, 485, 487, 489–92, 576, 578, 580, 582, 587
Griffith KS, 841

Griffiths JM, 485, 489, 491, 580, 582–84, 587, 590, 592, 593, 595
Grifoni M, 745
Grigoriev M, 74, 75
Grigoryan B, 212, 213
Grimaldi S, 294, 295
Grimberg BI, 643, 653, 655, 658, 659, 661
Grinberg OY, 303
Grischkowsky D, 772
Grishin Y, 304
Grobner G, 590
GrobPisano C, 578, 585
Groenen EJJ, 292, 293, 304
Groesbeek M, 576
Gromov I, 288, 304
Gronbeck H, 432, 438
Grondey H, 487
Gronenborn AM, 316, 512
Grosberg AY, 242, 500, 540
Gross EKU, 703
GROSS EM, 391–422; 392, 395, 401, 412, 413, 415
Gross JD, 487, 580, 582, 585, 587, 589
Gross P, 650, 670
Grossberg AY, 544, 545, 548
Grosse S, 220, 223
Grosser J, 160
Grotkopp E, 76
Grousson R, 220, 223
Grubbs RH, 413, 415, 417
Gruber C, 707
Gruber D, 650
Gruebele M, 242, 650, 818, 822, 823
Gruetzmacher JA, 364, 383
Grundmann M, 215, 219, 220, 223
Gruner G, 424
Gruner SM, 245
Grunwell JR, 235, 241, 242, 245
Grunze M, 116, 117, 119, 128, 130, 131, 443, 447, 455

Grupp A, 284, 290, 304
Grushow A, 429, 436, 437, 454
Grynberg M, 303
Gu ZT, 578, 590
Gudzenko LI, 650
Guérin S, 779, 787, 796
Guido M, 449
Guijarro JI, 591
Guillot G, 293
Gulding SJ, 814, 840, 844
Gullion T, 484–87, 492, 580, 582–84
Gunion RF, 172
Gunning AP, 556
Guntherodt HJ, 73, 85
Guo BC, 454
Guo H, 22
Guo HL, 687
Guo K, 316, 327, 328
Guo T, 256
Guo Y, 43, 58
Guo Z, 500, 504, 506, 508, 510, 514, 516, 517, 521, 528, 529
Guohe S, 814
Gupta SKD, 580, 582
Gurbiel RJ, 288
Gurevich YY, 730, 733, 736
Gurney RW, 741
Gurvich LV, 650
Gurvitz SA, 707, 708
Gustafson TK, 665, 667
Gustafsson A, 219
Gustafsson G, 392, 410
Gustev GL, 36
Guthold M, 75–77, 84, 85
Guthrie WL, 840
Gutin AM, 500, 501, 544
Gutmann M, 261
Gutow J, 375
Güvenç ZB, 450
Guyon PM, 169, 180, 181
Guyot-Sionnest P, 220, 360, 374, 687, 706
Guzelian AA, 213, 214

Gygax R, 176
Gymer RW, 392

Ha T, 235–38, 240–42, 245, 756, 757, 759
Ha TJ, 236, 239
Haarer D, 316
Haas E, 241, 249
Haase D, 220, 221
Haase J, 469, 476
Haber E, 500
Habscheid C, 798
Hache F, 212
Hackett PA, 424
Hadjiliadis N, 291
Haeberlen U, 463, 464, 467, 468, 472, 473, 578
Haeffler G, 669
Hafner JH, 86
Hafner S, 476, 493
Hagen SJ, 242, 244
Hager LP, 288
Hagerman PJ, 246
Hagfeldt A, 224
Haggerty L, 93
Hagstrom JE, 80
Hahn EL, 315, 318, 479, 485, 659, 772
Hahn JR, 687, 730, 731, 735, 737, 740
Hahne S, 591, 595
Hähner G, 116
Haight R, 256
Hain TD, 814
Haire RN, 348
Haiss W, 730
Hakhyun N, 818, 825
Halberstadt N, 612, 616
Halbritter J, 735
Hale JM, 391
Hales B, 288
Hales DA, 426, 428–30, 432, 433, 438, 441
HALFMANN T, 763–809; 773, 775, 776, 790, 795, 796
Halkides CJ, 293, 303

Hall DA, 303, 599
Hall GE, 24, 27, 50, 146, 159, 160, 608, 609, 644, 814
Hall JL, 43, 49, 50, 58, 792, 798, 799
Hall LE, 698
Hall R, 181
Hall RJ, 643
Halle S, 818, 832
Haller GL, 840
Halperin BI, 321, 334
Halverson KJ, 583, 595
Hamad K, 214
Hamadani SM, 772
Hamer N, 812
Hamers JS, 840
Hamilton CE, 766, 812
Hamley IW, 562
Hamman DR, 691, 692
Hammarström P, 297
Hammerich AD, 647
Hammond GS, 525
Hammond TD, 650
Hampson RF Jr, 41
Hamrick Y, 424, 436
Hamrick YM, 424, 436
Hamza AV, 450
Han KT, 235, 248
Han W, 81 (Figure 8, see color insert)
Hanazaki I, 17, 18, 814
Hancock G, 24, 27, 48
Hancock GC, 612
Hancock JK, 48
Hänggi P, 745
Hanley J, 294
Hanley L, 425
Hann BC, 76
Hanna MC, 196, 206, 215, 216, 220, 223
Hannon TE, 381
Hanold KA, 168, 170, 172, 174–76, 186–88
Hänsch TW, 800
Hansen AP, 288
Hansen GP, 434, 435
Hansen K, 557, 561

HANSMA HG, 71–92; 71, 72, 74, 76–81, 83, 84 (Figures 3, 4, 8, 9, see color insert)
Hansma PK, 72, 76, 77, 81, 87 (Figures 2, 4, 8, see color insert)
Hansmann UHE, 501
Hao M-H, 501
Hara H, 297, 298, 301
Hara M, 119
Harada Y, 237, 741
Haraharan R, 538
Haran A, 731, 741
Harb CC, 51
Harbison GS, 484, 578
Harbridge JR, 297
Harding LB, 16, 17, 23, 45, 50, 64, 155
Hare DE, 363
Harich SA, 155
Hariharan A, 640, 772, 782
Hariharan R, 561
Harkin C, 831
Harland PW, 814
Harmin DA, 782
Harms AC, 424
Harper WW, 44, 152
Harrington JE, 434
Harrington RE, 75, 81, 84 (Figure 8, see color insert)
Harris CB, 256, 644, 704
Harris RK, 464
Harris SA, 814
Harrison C, 563
Harrison G, 346
Harrison JA, 43, 52
Hart SL, 80
Hartland GV, 831
Hartmann SR, 316, 322, 479, 659
Hartz CL, 619
Hartzell CJ, 578
Harvey S, 501
Harwood JS, 477
Harzer U, 576

Hasegawa N, 669
Hashimoto J, 670
Hashimoto S, 51
Hassan AK, 303
Hatcher ME, 491, 580, 582, 584, 587
Hauck J, 798
Hauge EH, 710
Haugland RP, 236, 242
Haupts U, 235
Hausinger RP, 293
Hawe WP, 288
Hawker CJ, 563
Hawn DD, 84
Hayden CC, 141, 149, 151, 168, 182–84, 261, 265, 272, 273, 665
Hayes E, 237
Hayes G, 220, 223
Hayes WA, 122, 123, 128, 132, 133, 550
Hayman G, 41, 58
Hayter JB, 97
Hayward DO, 453
Hazak G, 776
Hazzard JH, 347
He G, 43–45, 50, 64, 65
He G-Z, 785, 789, 798
Heath JR, 424
Heaven MC, 160, 832, 835
Heben MJ, 215, 216
Heber O, 171
Heck AJR, 141, 166
Heckl WM, 80
Hediger S, 480
Hedlund BE, 348
Heeger AJ, 392, 396, 407–10, 718
Hehre WJ, 357, 364
Heid M, 640, 658, 661
Heidecke SA, 429, 430, 438
Heikal A, 261
Heilweil EJ, 641
Heim G, 73
Heinemann U, 515
Heinzen DJ, 650
Heinzmann U, 814

Heiser F, 181, 185
Heitz R, 220, 221, 223
Heiz U, 424
Helbing J, 261, 262, 266, 270, 640
Helfand E, 93, 94, 99
Heller C, 83, 85
Helm H, 168, 170, 188
Helmle M, 491, 576, 580, 582
Hembacher S (Figure 2, see color insert)
Hemingway RE, 398
Hemmerich A, 800
Hemmi N, 141
Henderson DJ, 736
Henderson E, 73, 74, 80
Henderson JI, 687, 698, 700, 704, 706
Henderson RM, 81
Henley D, 755
Henneberger F, 220, 223
Henrichs PM, 578, 585
Henry CH, 197, 205
Hentges R, 181, 185
Herbert A, 75
Hercules DM, 398
Herek JL, 256, 643, 651
Heritage JP, 665, 667
Herman MF, 93, 818, 819
Hermans J, 503
Hermine P, 142, 155
Hernandez MI, 834, 837, 838
Hernandez R, 834, 837
Hernandez-Lamoneda R, 834, 837–39
Herne C, 265
Herrero P, 76
Herrero VJ, 142, 144, 149, 151, 155
Herriott D, 51
Herrler M, 515, 519
Herrmann A, 394, 417
Herrmann C, 303
Herrmann J, 773
Herschbach DR, 610, 650, 667, 669
Herschlag D, 759

HERSHBERGER JF, 41–70; 26, 27, 48, 49, 56, 58, 59, 61
Hertel IV, 169, 264, 265
Hertel T, 256
Hertlein MP, 669
Hertz JA, 542, 544
Herzberg G, 18, 441, 665
Herzfeld J, 475, 476, 578, 580, 582, 589
Herzfeld K, 834
Heslot F, 85
Hess BA, 650
Hess K, 195
Hessler JP, 58
Hessman D, 219
Hettich RL, 429
Hetzler JR, 626, 634
Heupel T, 800
Hiebl M, 335
Hijirida DH, 590, 597
Hijmans TW, 800
Hilbich C, 593, 595
Hild U, 429, 430, 432
Hill JR, 315, 317, 320, 322, 323, 325, 327, 328, 330, 331, 334, 336, 337, 341, 345, 346, 349, 350
Hille R, 294
Hillegas CW, 772, 779
Hillier BJ, 519
Himmelhaus M, 117, 128
Hineman MF, 261
Hines M, 220–22
Hing AW, 487, 590
Hinkley ED, 42
Hino T, 93
Hinrichs RZ, 141
Hintsa EJ, 643
Hintz PA, 424, 427
Hioe FT, 772, 776, 782, 792, 795, 796
Hioka K, 490, 580, 582
Hiraishi Y, 590, 597
Hirao K, 31, 585, 587
Hirose C, 366, 367
Hirose K, 703, 704

Hirota E, 43, 64
Hirschfelder JO, 639
Hirsh DJ, 297
Hirst JD, 502
Hirst M, 84
Hjelte I, 265
Ho J, 424, 435, 441
Ho JH, 77 (Figure 4, see color insert)
Ho PKH, 396
Ho TS, 155
Ho W, 712
Hochstrasser RM, 236, 243, 244, 316, 639
Hodgkinson P, 471, 493
Hodgson A, 840
Hoekstra FA, 298
Hoelger C, 492
Hoeve CAJ, 545
Hofbauer W, 293
Höfer P, 281, 290, 303, 304
Hofer U, 256
Hoff AJ, 294, 297, 301
Hoffman BM, 287, 288
Hoffman WF, 424, 438, 445
Hoffmann A, 220, 221, 223
Hoffmann G, 21, 23
Hoffmann H, 121, 128
Hoffmann O, 160
Hoffmann R, 643
Hofmann H, 264
Hogan DA, 293
Hoganson CW, 669
Hoh J, 71, 76, 77
Hoh JH, 73, 77, 78
Hohener A, 485
Hohmura KI, 80, 86
Hohwy M, 473, 491, 580, 582, 584, 585, 589
Holbrook KA, 428, 653
Holczer K, 281, 303, 424
Holden PA, 72
Hole UH, 288
Holland M, 220
Hollberg L, 50
Hollebeck T, 155
Hollingworth AR, 220, 221

Holme TA, 818, 819
Holmes AB, 392, 410, 411
Holmes DL, 93
Holmgren L, 424
Holmlin RE, 687
Holowka D, 298
Holthaus M, 786
Holzer L, 407
Homans SW, 589
Homer ML, 424
Honda K, 17, 18
Honea EC, 424
Honeycutt JN, 500, 514
Hong J, 484
Hong MK, 335, 487, 488, 585, 587, 589, 590, 715
Hong S, 134
Hong SH, 134, 687, 698, 700, 704, 706
Hong YA, 687, 730, 731, 735, 737, 740
Honma K, 428, 429
Hoogstraten CG, 294
Hopkins JB, 260, 426
Hopkins MA, 303
Hopkins TA, 413, 415
Hopkinson AC, 36
Hor A-M, 413
Horchstrasser RM, 639
Hore PJ, 301
Hornung T, 658
Horowitz CJ, 143
Horton TE, 294
Hosler JP, 294
Hossenlopp JM, 26, 27
Hou H, 814, 840, 844
Hougen JT, 824
Houghten RA, 502
Houk KN, 643
House PG, 43
Houseman ALP, 294
Houston PL, 160, 166, 608, 609, 643, 825, 827, 833, 837, 840
Hovda N, 58
Howard BJ, 7, 612
Howlett DR, 591

Hoy AR, 18
Hoyer P, 225
Hrovat DA, 172
Hruby VJ, 578
Hsiao MC, 17, 19, 157
Hsieh P, 74, 75
Hsieh S, 168, 185, 186
Hsieh W, 76–80, 83, 84 (Figures 3, 9, see color insert)
Hsu MT, 424
Hsu SL, 357
Hsu YT, 142, 155, 159
Hu JZ, 483
Hu N-X, 413
Hu P, 322
Hu R, 315, 320, 327, 328, 330, 331
Hu W, 578, 585, 587, 589
Hu WP, 814
Hu YZ, 212
Hu ZB, 93
Huan-Cheng C, 840
Huang C, 384
Huang E, 563
Huang J, 35
Huang JL, 86
Huang KL, 825
Huang X, 53
Huang Y, 487, 814, 840, 841, 843, 844
Huang YH, 838
Huang ZS, 608, 612, 625
Hubbell WL, 299
Hubble HW, 341, 342
Huber JR, 643, 644
Huber KP, 18, 441
Hubrich M, 294
Hudgens JW III, 36, 51
Hudson JB, 453
Hudson JH, 93
Huenenberger PH, 503, 507
Hueschen M, 392, 419
Hughes DW, 24
Huh YD, 32
Humbert C, 117
Hummer G, 357

Hunenberger PH, 507
Hunt JH, 374
Hunt SA, 589
Hunter J, 424
Hunziker L, 664
Huo S, 590
Hurley MD, 50, 62, 63
Huron L, 117
Husain D, 152
Hush NS, 698
Hutson JM, 10, 149, 608
Hüttermann J, 288, 293
Huynh WU, 225
Hybl JD, 659
Hyde JS, 298, 299

Iakovlenko SI, 650
Iben IET, 335
Ideno H, 75
Ihee H, 256
Ikeda N, 818
Illies AJ, 434
Imasaka T, 814
Imre D, 812, 818
Imry Y, 692, 693, 716, 717
Imura K, 814
Inati SJ, 288, 304, 306, 599
Ingram DJE, 2
Innes JB, 297
Innes KK, 812, 818
Inoshita T, 220
Inoue G, 650
Inouye H, 593
Inouye M, 515
Insepov ZA, 631
Ionov S, 23
Ionov SI, 23, 650
Iqbal M, 578
Irback A, 540
Irion MP, 424
Irle S, 814, 825
Isaacson AD, 16, 17
Isbester PK, 476
Isenor NR, 772

Ishii Y, 236, 243, 480, 487, 580, 582, 584, 585, 587, 590, 592, 593, 595, 596, 599
Ishikawa H, 818, 819, 824
Isied SS, 687
Israelachvili J, 119
Israelachvili JN, 357
Ito F, 64
Ito K, 181
Ito M, 814
Ito N, 599
Ito T, 590, 597
Itoh H, 237
Itoh S, 301
Itzhaki LS, 243, 504, 515, 516, 518, 525, 528, 529
Ivancich A, 294
Ivanecky JE, 670
Ivanov D, 327
Ivanov MG, 212, 669, 780, 781
Ivanov MY, 667, 669, 781
Iversen G, 715
Iwaki M, 301

Jabbour GE, 396, 417, 419
Jackson NM, 85
Jackson SE, 243
Jackson TA, 316
Jackson WM, 59, 61
Jacobs A, 16, 21
Jacobsen KW, 712, 713
Jacobson MP, 607, 817–20
Jaecks DH, 188
Jagutzki O, 169, 171, 180–82
Jakes K, 587, 589, 590
Jaklevic RC, 706
Jakobsen HJ, 490, 491, 580, 582, 584
James MNG, 243
James TL, 97
Janda KC, 616
Jane-Dyson H, 293
Janes SM, 316
Jannsen R, 481
Jansen J, 483

Jansen M, 475
Janssen MHM, 814
Jaques C, 23, 650
Jaroniec CP, 580, 582, 584, 585, 589
Jaros M, 197, 220
Jarrold MF, 425, 428, 429, 432, 434
Jas GS, 242
Jasinski JM, 55
Jaun B, 501
Jauslin HR, 796
Javan A, 772
Javanainen J, 802
Jayaram B, 95
Jayasinghe L, 146, 148, 159
Jay-Gerin JP, 743
Jefferson A, 372
Jelinski LW, 578, 590, 597
Jelinsky P, 171
Jellinek J, 450
Jena P, 424
Jendeberg L, 511
Jennings GK, 118, 128
Jennings RT, 51
Jensen P, 612, 619
Jeschke G, 286, 291, 293, 294, 301, 475
Jett JH, 234, 235
Jeuken LJC, 293
Ji H, 557, 561
Jia YW, 236, 243, 244
Jiang H, 220, 221, 223
Jiang J, 320, 322, 324, 327, 328, 336, 346, 347, 351, 352
Jiang W, 515
Jiao D, 578
Jinchun X, 814
Jing TW, 730, 735, 737, 740
Jo SK, 730, 741
Joachim C, 682, 687, 698, 699
Joanny J-F, 98, 245
Jodkowski JT, 43
Joers JM, 583

Johannessen OG, 491, 580, 582
Johns JWC, 619
Johnson BR, 147, 812
Johnson HS, 152
Johnson KH, 424, 434, 435
Johnson MA, 173, 268
Johnson ME, 298
Johnson MJ, 841
Johnson RD III, 36, 557
Johnson TJ, 215
Johnston GW, 54
Johnston HS, 52
Joireman PW, 665
Joly AG, 653
Jonah CD, 730, 741
Jonas DM, 659, 818, 819, 824
Jones EY, 234
Jones JA, 591
Jones KM, 214
Jones RB, 650
Jones RR, 669
JONGMA R, 811–52
Jongma RT, 814, 838, 839
Jons SD, 814, 836
Jonsson B, 364
Joos S, 424, 425
Joosen W, 772
Jordan MJT, 31, 172, 179
Jorgensen WL, 357, 364, 502–4, 528, 529
Jortner J, 268, 652, 681, 682, 710, 717, 724
Joshi RP, 211
Jouvet C, 264, 265
Jovin TM, 73
Joyeux M, 824
Jucks KW, 608, 610, 612, 621, 625
Judson RS, 640
Julienne PS, 640, 650, 787, 793
Jung H, 205
Jung JM, 743
Jung LS, 128, 130, 131
Jung M, 75, 76

Jungen M, 270
Juvvadi P, 590

Käß H, 294, 303
Kable SH, 814, 825, 831, 832
Kadavanich AV, 213, 214
Kadyshevitch A, 741, 742
Kaess H, 294
Kafali SA, 36
Kagan CR, 212, 224
Kaganer VM, 112
Kainosho M, 585, 587, 589
Kaiser EW, 43, 55
Kaiser R, 800
Kaiser W, 639
Kakitani T, 715
Kalb AJ, 288, 306
Kalbitzer HR, 303
Kalburge A, 220, 223
Kaldor A, 424, 434–36, 440, 447, 452
Kalinovski IJ, 818, 820, 821
Kallenbach NR, 83, 84
Kallush S, 797
Kalogerakis KS, 24, 155, 159
Kalyanasundaram K, 224
Kam AF, 712
Kamal-Saadi M, 220, 221
Kamat PV, 214, 225
Kamath K, 220, 221, 223
Kamath MV, 507
Kameda T, 590, 597
Kamo S, 86
Kandel SA, 26, 28, 31, 157, 158, 814
Kandrashkin YE, 301
Kane DJ, 641
Kang CH, 287
Kang H, 687, 730, 731, 735, 737, 740
Kanter EP, 171
Kao HM, 487
Kapiloff E, 427
Kaplan DE, 288, 485
Kaplan IG, 436
Kappel C, 154
Kappert J, 825, 832

Kappes MM, 424
Kapteyn HC, 265
Kar L, 298
Karataev EM, 631
Karavitis M, 347
Karczmarek J, 780, 781
Karimi A, 502, 503, 511
Karlsson T, 492
Karpen ME, 502
Karpfen A, 612, 619
Karplus M, 242, 243, 316, 346, 500, 501, 503, 511, 516, 519, 520, 525, 544
Karpovich DS, 116, 128, 130, 131
Kasai T, 814
Kasapi S, 800
Kasas S, 76
Kasevich M, 800
Kash K, 220, 221
Kasha M, 644
Kashlev M, 81, 84
Kasibhatla P, 372
Kask P, 247
Kasper JVV, 43, 49, 50, 58
Kastrau DHW, 288
Kasturi K, 831
Katari JEB, 214
Katchalski-Katzir E, 241, 249
Kato K, 511
Kato T, 154
Kattner J, 121, 128
Katz B, 53, 54
Katzer DS, 219
Kauffman SA, 552
Kaufman LJ, 670
Kauppi E, 172, 179
Kauzmann W, 361, 362
Kawabe Y, 396
Kawaguchi K, 43
Kawaguchi M, 538
Kawai A, 818, 820, 821, 825
Kawamori A, 287, 297, 298, 301
Kawana S, 669
Kawanishi S, 347
Kawasaki K, 567

AUTHOR INDEX

Kawasaki M, 51, 116, 644
Kawaura C, 80
Kawi S, 424
Kay BD, 840
Kay CWM, 287
Kaya K, 424
Kayanuma Y, 212
Kazaryan EM, 212, 213
Ke SC, 294
Kebabian PL, 51
Kebler K, 146
Kegley-Owen CS, 43, 44, 46, 52, 62
Keil M, 141, 798
Keim M, 327
Keller A, 667
Keller D, 246, 557
Keller HM, 824, 825, 827
Keller RA, 234, 235, 246, 248
Kelley DF, 261
Kelley SO, 85
Kellman ME, 819, 824
Kellogg GJ, 563
Kelly D, 450
Kelly PB, 812
Kelz T, 50
Kemp M, 695, 696, 698, 699, 707
Kemper PR, 424
Kendrick BK, 144, 146, 148, 149, 159
Kenkre VM, 698
Kennedy MC, 288
Kennedy RA, 9, 10
Kentgens APM, 478, 492, 584, 585
Kentsis A, 234
Keogh GP, 670
Keogh WJ, 143
Kergueris C, 699
Kerns KP, 454
Kerper PS, 73
Kerr JA, 41
Kessler K, 16, 17
Kestner TA, 476
Keszthelyi CP, 404
Ketchem RR, 589, 590

Ketterle W, 782
Kettling U, 235, 247
Keusters D, 659, 772
Kevan L, 294
Keyes BM, 208, 210
Keyes RS, 298
Khan FA, 428
Khandelwal P, 580, 582
Khanna SN, 424
Kharchenko VA, 220
Kharkats YI, 715
Khas Z, 220, 223
Khayyat K, 169, 182
Khil PP, 74, 75
Khitrin A, 472
Khokhlov AK, 545, 548
Kholer B, 653
Khorana HG, 599
Khundkar LR, 23, 255, 650
Kickel BL, 432, 433, 439
Kiefer B, 641
Kiefer W, 640, 644, 658, 661
Kietzmann H, 424
Kihlgren T, 265
Kiihne S, 490, 580, 582, 593
Kikuchi M, 567
Kilic HS, 183
Kim BI, 80, 260
Kim EH, 172, 814, 834, 838
Kim HD, 236, 240, 245
Kim HL, 24, 27
Kim IK, 45, 46
Kim JH, 294, 396
Kim JS, 80
Kim K, 74, 84
Kim KJ, 84
Kim P, 86
Kim PS, 500, 502
Kim SH, 293, 590
Kim SK, 261, 265, 643
Kim SS, 299
Kim SW, 80
Kim TK, 563
Kim W, 667
Kim YD, 423
Kim Y-T, 114, 121, 590
Kimble HJ, 801, 802

Kimman J, 840
Kinder R, 576
King DS, 612, 623, 626, 634
King JC, 644
Kinosita K Jr, 237
Kinsey JL, 141, 766, 812, 814, 818, 832
Kinzer B, 306
Kippelen B, 415, 417, 419
Kirby AR, 556
Kirczenow G, 698, 708, 719
Kirkwood JC, 670
Kishi T, 424
Kistemaker PG, 170
Kisterswoike B, 593, 595
Kitchen DB, 503
Kitsopoulos TN, 141, 149, 741, 742
Kittrell C, 812, 814
Kivelson S, 718
Kjaer K, 357
Klatt M, 833, 836
Klauder JR, 322
Klavetter F, 392, 410
Klee S, 824
Kleiman V, 640
Kleiman VD, 641
Klein FS, 52, 53
Klein K, 798
Klein MP, 288, 294
Klein O, 424
Kleinermanns K, 16, 17, 21, 146
Klemperer W, 612, 824
Klenerman D, 239, 240
Kliger DS, 234
Klimcak CM, 814
Klimkans A, 698
KLIMOV DK, 751–62; 243, 500, 752
Klimov VI, 220–23
Kliner DAV, 143
Klinov DV, 74, 75
Klopper W, 614
Klots TD, 424, 434, 435, 454
Klotzkin D, 220, 221
Klug CA, 490

Klug DD, 363
Klymenko Y, 689, 723
Kmetko J, 122
Knapp EW, 335
Knee JL, 260, 639
Knickelbein MB, 424, 425, 434–36, 438, 440, 447, 454
Knight AEW, 814, 831, 832
Knight JB, 245
Knight PL, 795, 796
Knobler CM, 357
Knock MM, 375
Knoesel E, 256
Knoll W, 119
Knopp G, 658, 659, 661
Knowles PJ, 10, 155, 619, 621
Kobayashi GS, 590
Kobayashi T, 212
Kobrak MN, 786, 787
Koch SW, 212, 220, 223
Koch T, 219
Kocha SS, 220, 223
Kocisko D, 580, 582
Kodera Y, 297, 298
Koenig M, 143
Koenig W, 171
Kofman V, 293
Kofranek M, 612
Kogelnik H, 51
Kohguchi H, 141, 168, 265, 272
Kohler A, 396
Kohler B, 641
Köhler R, 50
Kohler W, 335
Kolbert AC, 472, 473
Kolling E, 639
Kollman PA, 501, 528
Kollner M, 235, 237, 247
Kolodinski S, 196, 224
Koltermann A, 235, 247
Kominos D, 591, 595
Komives EA, 590
Komornicki A, 367
Kompfner R, 51

Könenkamp R, 225
Kong CY, 556
Kong F, 832
Kong FN, 832, 836
Konz E, 787
Kool ET, 83, 84 (Figure 9, see color insert)
Koopmans T, 699
Koot W, 170
Kopelman R, 119
Koplitz B, 26
Koplow J, 49
Koppe S, 16–19
Koput J, 824
Korkin A, 839
Kornweitz H, 54
Korolik M, 841
Kortyna A, 798
Kosfeld R, 464
Koshi M, 51, 55, 825
Kosisko D Jr, 490
Kosloff R, 266, 640, 647, 653, 730, 731, 733, 736
Kosmidis C, 183
Kosslof R, 731
Kosterev A, 50
Kothe G, 301
Kotitschke RT, 220, 221
Koulougliotis D, 297
Kouteck J, 432, 436
Kovacs FA, 590
Kovacs H, 503, 507
Kowalczykowski SC, 80
Kowalewski T, 75, 76, 80
Kowalski R, 802
Koyano I, 154
Kozaki M, 709
Krabben L, 287
Kraka E, 16, 17
Kral K, 220, 223
Kratzschmar O, 814, 828, 830
Kraus JS, 425
Kraus S, 268
Krause JL, 640, 641, 651
Kreher C, 157
Kreisle D, 424
Kreller F, 220, 223

Krimov VN, 303
Kroneck PMH, 288
Kronfeldt HD, 50
Kroon JPC, 772
Kroon P, 556
Kropman BL, 120
Krückeberg S, 429, 430, 432, 433, 437, 439
Kruit P, 257
Krull UJ, 114, 128
Kruus E, 117
Krymov V, 288, 304
Kryzhanovsky B, 787
Krzystek J, 303
Ku JK, 650
Kubarych KJ, 670
Kubiak CP, 687, 698, 700, 704, 706
Kubo A, 483, 582
Kubo R, 321
Kucheida D, 335
Kudla K, 17, 23
Küglz M, 787, 798
Kuhlmann F, 814
Kuhn A, 785, 789, 790
Kuhne T, 268
Kuimelis RG, 102
Kuklinski JR, 782
Kulcke A, 45, 46
Kulikov AV, 297
Kulin S, 800
Kumar A, 480
Kumaran SS, 52, 53, 152
Kumashiro KK, 590
Kummerlen J, 590, 597
Kundla EI, 580, 582
Kunike M, 772
Kuno M, 219
Kuntz PJ, 155
Kuo SC, 557
Kuo YM, 591
Kupfer A, 570
Kupfer H, 570
Kuppermann A, 143, 144, 146, 147
Kurn DM, 782
Kurnit NA, 316, 659, 772

Kurochkin VI, 303
Kurosawa T, 670
Kurreck J, 301
Kurshev VV, 294, 295
Kurur ND, 491, 580, 582
Kurylo MJ, 41
Kustanovich I, 480
Kuster R, 268
Kustu S, 75, 76
Kuszewski J, 512
Kutter C, 303
Kutyavin I, 75
Kuznetsov AM, 706, 715
Kuznetsov S, 424
Kwoh D, 77, 78, 80
Kwok AS, 315, 317, 320, 323, 327, 328, 334, 348
Kwok WM, 644
Kwon O, 558
Kwong RC, 419
Kyogoku Y, 599

Laaksonen A, 357, 364
Laaksonen RT, 425, 436
Labastie P, 424
Lacelle S, 589
Lackey D, 730
Ladizhansky V, 476
Ladurner AG, 243, 518
Lafferty WJ, 612
Lafosse A, 169, 180, 181
Lagally MG, 110, 132
Lagendijk A, 772
Laguna GA, 43, 49
Lagutina IV, 74, 75
Lai LH, 53, 54, 64, 154
Laibinis PE, 134
Laine TA, 786
Lakin ND, 76
Lakner FJ, 288
Lakowicz JR, 236, 237, 241, 249
Lakshmi KV, 472, 590
Lakshminarayan C, 260
Lam PM, 132, 133
Lamb DC, 248
Lambert HM, 24, 832

Lambert O, 72
Lambooy P, 563
Lambrecht RK, 58
Lambropoulos P, 795, 796
Lammers A, 424, 425
Lamoen D, 703, 704
Lampton M, 171
Lancaster DG, 49
Landau LD, 771
Landauer R, 692, 710, 711
Lander DR, 43, 52, 58
Landis GC, 589, 590
Landman U, 731–33
Landsberg PT, 196, 224
Laney DE, 74, 84, (Figure 8, see color insert)
Lang DP, 487
Lang HP, 86, 557
Lang MJ, 670
Lang ND, 691, 702, 735
Lange W, 802
Langlais VJ, 687
Langmuir I, 741
Langowski J, 73, 77
Langridge-Smith PRR, 426, 434, 435
Lankhuijzen GM, 772
Lansbury PT, 490, 580, 582–84, 587, 592, 593, 595
Lanzisera DV, 36
Lapidus LJ, 242
Lapierre L, 828
Laporta P, 647
Larsen BS, 294
Larsen DS, 670
Larsen FH, 478
Larsen JJ, 272
Larsen RG, 293, 294
Larsson M, 267
Larsson S, 698
Lartius R, 73
Lascola R, 619
Lasell RA, 141
Laser D, 404
Lau KF, 540
Lau WL, 236, 243, 244

Laubereau A, 357, 376
Lauder MA, 795
Laufer AH, 56
Laughlin KB, 619
Launay JM, 151
LAURENCE TA, 233–53; 236, 237, 241, 243–45
Laurent T, 16–19
Lauvergnat D, 834, 837, 838
Lauzon AM, 237
Lavender HB, 179
LaVerde G, 590, 597
Lavery R, 83, 85
Lavollée M, 171, 185, 186
Lavorel B, 663
Lavrich DJ, 113
Law MM, 10
Lawall J, 798, 800
Lawrance WD, 814, 831
Lawrence WG, 832, 835
Lawson JM, 687
Lax M, 206
Lazaridis T, 242, 243, 501, 519, 525
Leach CA, 10
Leahy DJ, 170, 171, 272
Leapman RD, 587, 590, 592, 593, 595, 596
Leatherdale CA, 221, 222
Le Barny P, 396
Lebeault-Dorget MA, 424
Lebech M, 169, 180, 181
Lebedev YS, 303
Lebrun A, 83, 85
Lech LM, 425
LeClaire JE, 168, 182, 183, 265
Ledentsov NN, 215, 219–21, 223
Ledingham KWD, 183
LeDourneuf M, 151
Leduc MR, 550, 800
Lee CH, 358
Lee GU, 71, 73, 85
Lee H, 215, 506
Lee HC, 295, 297
Lee HI, 288, 294

Lee JC, 213, 392, 407
Lee J-K, 392, 407, 409
Lee KC, 590
Lee M, 472
LEE N, 751–62; 759, 760
Lee PA, 392, 412, 419
Lee S-h, 53, 54, 298
Lee SH, 142, 149, 150, 152, 159
Lee SK, 394, 417
Lee TH, 429, 430, 436, 437, 439, 454
Lee TJ, 825
Lee Y, 26, 27
Lee YK, 471, 483, 491, 580, 582, 585, 590
Lee YT, 140, 141, 149, 151, 619, 643, 650, 669, 818
Leemans WP, 256
Leeson DT, 234, 320, 334, 513, 515
Lefebre R, 650
Legare F, 641
Lehmann KK, 818, 824
Lehr L, 168, 269
Leibler L, 562
Leidenbaum C, 772
Leiderer P, 270
Leigh JS Jr, 293
Leiserowitz L, 357
Leising G, 407
Leitner T, 121
Lemire GW, 424, 436
Lendvay G, 19
Lendzian F, 287, 294
Lenhert S, 76
Lenth W, 50
Leonard D, 220, 223
Leonard W, 392, 419
Leone M, 335, 347
Leone SR, 43, 641
Leopold DG, 424
Leopold PE, 500
Lerner RA, 502
Lessen DE, 429
Lester MI, 160, 825, 828
Lesurf JCG, 303

Letokhov VS, 652
Letsinger RL, 687
Lett PD, 650
Leuba SH, 80
Leuchtner RE, 424
Leutwyler S, 828
Levante TO, 471
Levene SD, 93, 94
Levenson MD, 50, 51, 321, 322
Levi DH, 196, 206
Levin J, 171
Levine RD, 21, 27, 54, 139, 140, 142, 147, 154, 158, 818, 819
Levine YK, 299
Levinger NE, 357
Levinthal C, 500
Levis RJ, 264
Levitt MH, 471–73, 487, 490–92, 500, 502, 514–16, 518, 525, 528, 529, 580, 582, 584, 589, 590
Lev-On T, 26, 28, 31, 157
Levy DH, 5, 6, 589
Levy RM, 503
Lewenstein M, 669
Lewis FD, 687
Lewis J, 396
Lewis NS, 698
Lewis RS, 26, 590, 597
L'Hermite JM, 141
Lhuillier A, 669
Li A, 504, 515, 516, 518, 525, 528, 529
Li CF, 93
Li J, 77, 84
Li L, 500, 825, 832
Li LP, 591, 595
Li Q, 43
Li R, 511
Li XJ, 83
Li XN, 640
Li X-Q, 220, 223, 650
Li Y, 93
Li YF, 407

Li Y-Q, 707
Li ZX, 357, 374
Lian L, 426, 428–30, 432, 433, 438, 441, 443–45
Liang J, 236, 239
Liang Z, 298
Liao PF, 650, 772
Libman J, 741
LiCata VJ, 511
Lieber CM, 86
Lieber MR, 75, 76
Liebsch A, 702
Lieschke J, 814
Lievin J, 818
Light JC, 17, 19
Liivak O, 590, 597
Likhtenstein GI, 297
Lilliehook C, 591, 595
Lim KP, 52, 53, 152
Lim M, 316
Limbach HH, 492
Limberger HG, 424
Lin C, 160
Lin CH, 665, 667, 814
Lin J, 84
Lin JJ, 155
Lin MC, 51
Lin PH, 687, 706
Lin SH, 812
Lin TH, 840
Lin Z, 62, 77, 84
Linahl T, 755
Lindberg M, 212
Linder M, 485, 578, 585
Lindholm N, 58
Lindinger A, 787
Lindinger M, 429, 430, 432
Lindsay SM, 75, 81, 83, 84 (Figure 8, see color insert), 730, 735, 737, 740
Lineberger WC, 172, 267, 424, 435
Lingle RL, 256, 704
Linnebach E, 21
Linse P, 357, 364
Liorente MA, 93

AUTHOR INDEX 877

Liou HT, 825
Lippincott ER, 361
Lippmaa E, 477
Lippmaa ET, 580, 582
Lischka H, 612
Lising LJ, 424
Litosh V, 50
Litovitz TA, 343, 346
Littau KA, 316, 317, 321, 322, 334
Litton CW, 205
Liu B, 143
Liu D, 83, 225
Liu F, 80, 83
Liu GY, 118, 128, 134
LIU K, 139–64; 53, 54, 64, 141, 142, 149, 150, 152, 154, 155, 159, 424, 441, 443, 444, 447
Liu M, 77
Liu Q, 651
Liu SB, 589
Liu X, 155
Liu Y, 567
Liu ZF, 116, 128
Liu ZH, 743
Liyanage R, 441, 444
Llor A, 471, 477
Loa I, 50
LoBrutto R, 293, 294
Lochbrunner S, 272
Lodge TP, 93, 94
Loew GH, 36
Loh SK, 55, 426, 429, 433, 441, 443, 444, 449
Lohr D, 80
Loison JC, 825
Lollo CP, 77, 78, 80
Loncharich RJ, 335, 643
Long D, 102
Long HW, 597
Long JR, 580, 583–85, 587
Long P, 264
Longuet-Higgins HC, 4, 5, 143
Lonzer WL, 590
Loomis A, 80

Loomis RA, 160
Lopez-Martens R, 264
Lopez-Martens RB, 667
Lorent V, 772
Lorentz WJ, 735
Lorenz KT, 141
Lorigan GA, 294
Loring RF, 317
Los J, 170
Lostao A, 287, 294
Lott KAK, 2
Loucks GD, 114, 128
Louderback JG, 424
Loudoudi M, 291
Lounis B, 234
Lovejoy CM, 608, 619, 621
Lovelock JE, 372
Lowe IJ, 474, 578
Lowenhaupt K, 75
Lowe-Webb R, 215
Lowisch M, 220, 223
Loy MMT, 772, 775, 782, 798
Lozovoy VV, 641, 643, 653, 655, 658, 659, 661, 670
Lu H, 234
Lu HP, 248
Lu HSM, 236, 243, 244
Lu J, 293
Lu JR, 357
Lu SP, 640
Lu W, 424
Lu Z, 214, 216
Lu ZH, 215, 216
Lubitz W, 287, 294, 301, 303, 304
Lucas NJD, 7
Lucchese RR, 619, 627
Lucia J, 260
Luckhurst GR, 5
Ludowise P, 261, 263, 663, 664
Ludwig B, 294, 295, 303
Lugli P, 211
Lugtenburg J, 590
Luke BT, 36
Lum K, 357
Lumpkin OJ, 93, 94

Lundberg JK, 818
Lundgren E, 423
Luo X, 825, 827
Luong AK, 168, 170, 172, 176, 186, 187
Luscher E, 335
Luthey-Schulten Z, 242, 243, 500, 511, 513
Luthman H, 585, 589
Lutz HO, 168, 185, 424
Lutz S, 798
Lützenkirchen K, 424, 429, 430, 432, 433, 437, 439
Luz Z, 585
Lycett GJ, 7
Lycett S, 220, 221
Lyding JW, 712
Lynch GC, 52
Lynch JB, 297
Lynch K, 298
Lynch V, 394
Lynn DG, 584, 590, 593, 595
Lyons BA, 511
Lyubchenko YL, 74, 75, 80, 84

Ma C, 590
Ma J, 322
Ma S, 515, 519
Ma Z, 814, 836
Maa YF, 840
Maali A, 234
Maan JC, 303
Maas DJ, 772, 777, 779
Maas W, 582
Maberry JP, 429, 430, 436, 437, 439
Maboudian R, 120, 121
Mac M, 294
Macdonald RG, 43–45, 50, 64, 65
Machara NP, 246, 248
Machol JL, 214
Maciel A, 206
Maciel GE, 477, 480, 483, 484
Mack JA, 833, 835–38

Mackay K, 392, 410
MacKerell AD Jr, 73, 85
Mackie AR, 556
Mackie M, 802
Macklin JJ, 211
MacLeod MC, 77
MACMILLAN F, 279–313; 287, 294, 295, 303, 304
Mader ML, 294
Madey TE, 733
Madhu PK, 478
Madix RJ, 450
Madou MJ, 558
Maegley K, 76
Magde D, 247
Maggi A, 78
Magliozzo RS, 293
Maglott EJ, 760
Magnes O, 787
Magnus P, 394
Magoga M, 698, 699
Maguire TC, 650
Mahadevan L, 557
Maier JP, 270, 831
Mailhos C, 80
Maisel H, 585
Maiti S, 235
Majda M, 687
Majumdar A, 558
Makarov AA, 652
Maki A, 824
Makov G, 706
Malik P, 489
Malina RF, 171
Malinovsky VS, 776, 779, 792, 797
Malvy C, 74
Maly T, 287
Malysheva LI, 707
Mammen M, 552, 556, 557
Maness KM, 392, 395, 396, 398, 399, 401, 407, 409
Manger M, 644
Manikandan P, 288, 304, 306
Manka AS, 786
Manne S, 84 (Figure 2, see color insert)

Mannhart J (Figure 2, see color insert)
Manning GS, 95, 99, 102, 759
Manolopoulos DE, 53, 54, 149, 151, 152, 154, 172
Mansfield P, 489
Mansky P, 563, 567
Manson EJ, 476
Mantell DA, 840
Manthe U, 154
Mantsch HH, 316
Mantz AW, 49
Manz J, 653
Manzanares JA, 407, 409
Mao HY, 598
Mao M, 558
Maoz R, 108
Marahiel MA, 515, 519
Marassi FM, 576, 590
Marchetti MC, 206
Marcus RA, 391, 395, 396, 607, 608, 653, 686, 692, 706
Marden MC, 316
Maresch GG, 281, 295, 303
Margoliash E, 287
Margulis CJ, 267
Maricq MM, 43, 55, 469, 474, 475
Marijnissen A, 424
Marinelli L, 478
Mark AE, 501, 503, 504, 507, 528, 529
Markillie GAJ, 19
Marko JF, 93, 94, 99
Markovich G, 268
Marks D, 480
Marks RN, 392, 410
Marr AJ, 10
Marrink S-J, 357, 364
Marsh D, 298
Marsh TC, 74
Marshall E, 80
Marshall MD, 610, 612, 614, 634
Marszalek PE, 72, 234

Marte P, 792, 798, 799, 801, 802
Mårtensson LG, 297
Martin AB, 236, 241, 243, 244
Martin AS, 706
Martin C, 171
Martin JDD, 669, 791, 792, 798
Martin JL, 800
Martin PG, 143
Martin RE, 294
Martin TP, 424, 710
Martinez G, 303
Martinez GV, 503
Martinez JI, 294
Martinez-Haya B, 141
Martyna GJ, 743
Marui S, 51
Maruyama A, 80
Marvet U, 644, 645, 650, 651, 670
Maryasov AG, 298
Maslen PE, 267
Masnou-Seeuws F, 650
Mason AJ, 590
Mason JR, 288
Mason JT, 384
Mason TL, 556
Massa JS, 220, 223
Massie J (Figure 2, see color insert)
Massie ST, 47
Massiot D, 478
Mastenbroek JWG, 814
Masters CL, 593, 595
Masui H, 392, 407, 409
Materny A, 644, 658, 661
Matousek P, 168, 272
Matranga C, 220–22
Matsen MW, 562, 567
Matsudaira P, 557
Matsui H, 55, 825
Matsumoto K, 55
Matsumura C, 64
Matthews CR, 242
Mattick AT, 772

Mattoussi H, 225
Maul C, 168, 185, 189
Maus DC, 599
Mayer E, 335
Mayer PM, 166
Mayer U, 121, 128
Mayes AM, 563
Mays J, 93, 94, 561, 563
Mazzola LT, 73
McBranch DW, 220–23
McBrierty VJ, 463, 464, 469
McCammon JA, 501
McCanny T, 183
McCarley RL, 114
McCarthy MC, 841
McCarty BM, 360, 368, 369
McCarty DM, 413
McClellan AL, 362
McConnel HM, 687
McCormick JM, 288
McCoy AB, 179
McCracken J, 287, 288, 293, 294
McCracken JL, 294
McDermott AE, 294, 303, 578, 583, 585, 587, 590, 595, 599
McDonald EM, 392, 412
McDonald JD, 820
McDonald RA, 439
McDonald SA, 812, 814
McDowell CA, 578, 582
McDowell LM, 576
McElhannon RW, 50
McFail-Isom L, 84
McGee JD, 288
McGrath K, 19
McGuire JA, 361
McIlroy A, 612
McIntosh AL, 619
McKay RA, 485, 486
McKellar ARW, 619
McKendrick KG, 48
McKoy V, 168, 271, 272
MCLAUGHLIN LW, 93–106; 93, 95, 99, 101, 102, 104
McLean AD, 36

McLelland H, 220
McLendon G, 197, 201, 202, 205, 206, 220, 223, 687
McLendon GL, 687
McMahon TJ, 578
McMahon WE, 199, 202
McManus JB, 51
McMillin DR, 293
McMorrow D, 779
McNab IR, 9, 10
McNeill JD, 256, 704
McPeters HL, 176
McPhalen CA, 243
McPherson LD, 85
McSherry D, 669
Mead CA, 143, 146
Meads RF, 619
Measures RM, 391
Medek A, 477, 478, 483
Medina M, 287, 294
Medvedev ES, 714, 715
Meepagala SC, 730
Meerholz K, 413, 415
Meggers E, 687, 726
Mehl W, 391
Mehring M, 284, 290, 304, 463, 464, 468, 469, 473, 578, 580
Mehta MA, 490, 580, 585, 587
Meier A, 220, 223
Meier BH, 471, 480, 489–93, 578, 580, 582, 584, 585, 587, 590, 597
Meier C, 263, 270, 271, 814
Meier R, 658
Meijer G, 814
Meiwes-Broer KH, 424
Melinger JS, 640, 641, 772, 779, 782
Melius CF, 58
Mellau GC, 824
Meller J, 501
Memory JD, 463, 464, 472
Mendel D, 599
Mendelman LV, 102
Menzies RT, 42

Merchant K, 317, 320, 323, 348
Mercouris T, 650
Mergel V, 169, 171, 182
Merrifield RB, 590
Merrill RP, 840
Merritt ME, 490
Mertz E, 502, 503
Merz JL, 220, 223
Messerschmidt A, 292, 293, 304
Messiah A, 769, 784
Messina M, 641
Messmer MC, 357, 360, 364, 374, 376, 378
Mestdagh JM, 141
Metiu H, 653
Mets U, 247
Metz RB, 19, 157, 172
Meunier H, 43, 58
Meunier V, 117
Meuse CW, 357
Mewes M-O, 782
Meyer S, 661
Meyer TJ, 392, 407, 409
Meyer-Almes FJ, 235
Mezard M, 544
Michael JV, 52, 53, 152
Michaels CA, 62
Michal C, 590, 597
Michal CA, 590, 597, 599
Michaud M, 730, 741, 743
Michel H, 294, 295
Michel M, 219
Michel-Beyerle ME, 687, 726
Michelis T, 661
Michelsen HA, 840
Mićić OI, 214, 216, 217, 219, 225
Middleton DA, 590
Miedema AR, 439
Mihill A, 814
Miick SM, 503
Mikami N, 818, 824
Mikhailovsky AA, 221, 222
Mikulecky K, 833, 835, 836
Milam B, 818, 822, 823

Milburn MV, 293
Mildvan AS, 293
Miles M, 72, 80
Militello V, 347
Millard RR, 639
Miller JA, 48, 392, 419
Miller JD, 650
Miller RDJ, 197, 201, 202, 205, 206, 220, 223
MILLER RE, 607–37; 608, 610, 612, 614, 616, 621, 623, 625, 627, 629, 631, 632, 634, 818
Miller RJD, 220, 223, 670
Miller RL, 833, 837
Miller S, 134
Miller TA, 5, 6
Miller WH, 172, 257, 266, 271, 693, 700
Miller-Auer H, 584, 590, 593, 595
Millett IS, 245, 759
Millhauser GL, 298, 503
Millie P, 264
Milligan RA, 557
Millikan RC, 48
Mills JB, 246
Mills PA, 97
Milne CJ, 670
Milner ST, 538
Milosevic S, 650
Milov AD, 294
Milverton DRJ, 9
Mims WB, 284, 288, 290–93
Minami M, 590, 597
Minemoto S, 270
Mino H, 287
Minoghchi S, 236
Miranda PB, 374
Mirau PA, 484, 580, 582
Mirkin CA, 134
Mirny LA, 500
Mischler B, 663
Miskovic ZI, 699
Misoguti L, 669
Misra M, 598
Mitani M, 396

Mitchell DJ, 97, 490, 580, 582, 593
Mitchell RE, 171
Mitchell RH, 303
Mitri R, 297
Mittleman DM, 655, 658
Miura H, 485
Miyashita N, 590, 597
Miyawaki A, 236, 240, 248
Miyoshi A, 825
Mo Y, 590
Moazzen-Ahmadi N, 619
Möbius K, 286, 287, 299, 301, 303, 304, 306
Mochrie SGJ, 567
Modinos A, 704
Modrich P, 73
Moerdijk AJ, 650
Moerner WE, 234, 236, 240, 248
Mohammad F, 59, 61
Mohanty JF, 99, 102
MOHANTY U, 93–106; 93, 95, 97–99, 101, 102, 104
Mohwald H, 112, 357
Moi L, 783
Moision R, 75
Moision RM, 814
Mok MH, 17
Molenkamp LW, 316
Moler K, 800
Moll HP, 303
Momiji H, 212
Monaco L, 78
Mondry J, 773
Monks CR, 570
Montal M, 500
Montelione GT, 511, 515
Montgomerie CA, 10
Moore CB, 43, 48–50, 52, 58, 148, 818, 820, 825, 827
Moore DT, 629, 634
Moore JW, 54
Moran LB, 234
Moratti SC, 392, 410, 411
Mordaunt DH, 825
More KM, 293

More MB, 428
Moreno F, 76
Moreno V, 77
Moreno-Herrero F, 76
Morenzin J, 424
Morgan L, 294
Morgan RA, 212
Morgen M, 663, 664
Morigi G, 800
Morin P, 181, 185
Morioka Y, 166
Morokuma K, 814, 824, 825
Morozov VN, 84
Morozova TY, 84
Morrell MM, 396, 419
Morris JC, 394
Morris JR, 792
Morris VJ, 556
Morris VR, 59, 61
Morrissey SR, 294
Morse MD, 424–26, 429, 430, 432, 434–36, 438, 440
Morse PM, 373
Mort J, 76
Morter CL, 43, 55–58
Moser C, 715
Moser J, 224
Moser S, 146, 148, 159
Mosher C, 73
Moss DB, 818, 819
Mosyak A, 730, 731, 733, 736
Motoki SES, 181
Motzkus M, 658
Mou J, 72, 81
Mountain RD, 343
Mourant JR, 316
Moy VT, 72
Mrozowski S, 650
Mucke N, 77
Mueller KT, 477, 486, 487
Mueller LJ, 491, 580, 582, 584, 587
Mueller U, 515
Mueller WE, 79
Muhandiram DR, 598
Muir TW, 599

Mujica V, 695, 696, 698, 699, 703, 704, 707, 708
Mukai K, 220, 221
Mukamel S, 317, 322, 640, 652, 653, 655, 661, 663, 670, 720
Mullen K, 394, 417
Mullen-Ley P, 77, 78, 80
Muller CJ, 687, 698, 702, 708
Müller DJ, 72, 831
Müller F, 303
Muller HG, 669
Muller J, 270
Müller JD, 247
Müller L, 480
Muller S, 80
Müller U, 168, 188, 831
Muller-Dethlefs K, 258
Mulliez E, 293
Mullin AS, 62
Munowitz MG, 463, 469, 485, 580, 589
Munoz V, 242
Munzer HJ, 270
Murata Y, 814
Murdin BN, 220, 221
Murnane MM, 265
Murphy EJ, 141
Murphy OJ, 590
Murray CB, 212, 213, 224
Murray KK, 52
Murray R, 220, 221
Murray RW, 392, 407, 409
Muthukumar M, 556, 563, 566
Mutz MW, 687
Muyskens MA, 62
Myers DJ, 315, 320, 327, 328, 330, 331

Naaman R, 631, 731, 741, 742, 814
Nadal ME, 267
Nagamune Y, 219
Nagano H, 256
Nagao C, 818, 824
Nagayama K, 490, 580, 582

Nagesha K, 706
Nagy G, 730, 736
Nagy JK, 590
Naider F, 583
Naik PD, 16–19
Naito A, 578
Naitoh Y, 670
Nakajima A, 424
Nakajima T, 792, 795, 796
Nakamura H, 782
Nakanaga T, 64
Nakanishi M, 80
Nakata Y, 224
Nakayama H, 220, 223
Nakayama Y, 86
Nalivaiko SE, 650
Nam H, 818, 820, 821, 825
Nandi S, 267
Napari I, 357, 364
Napolitano J, 650
Nar H, 292, 293, 304
Narasimhan LR, 316, 317, 321, 322, 334
Nardi E, 669
Nash P, 234
Nassau K, 288
Nath SK, 563, 569
Nathanson GM, 357
Nazarova O, 80
Nealey PF, 562, 563, 567, 569
Neese F, 288
Nelson DD, 43
Nelson DD Jr, 619, 621
Nelson HC, 76
Nelson KA, 653, 663
Neretina T, 74, 75
Nesbitt DJ, 43–46, 49, 50, 152, 160, 607, 608, 612, 616, 619, 621
Ness H, 719, 727
Nettikadan S, 75
Neuhauser R, 828, 830
NEUMARK DM, 255–77; 141, 149, 151, 165, 168, 170–72, 176, 177, 257, 261, 266–69, 271, 658, 812

Neusser HJ, 814, 828, 830
Newman SM, 51
Newns DM, 696, 841
Newton JF, 687
Newton MD, 698
Newton RG, 802
Neyer DW, 825, 827
Nguyen JT, 593
Nguyen KA, 834
Ni H, 610, 619, 634
Nibbering NMM, 170
Nicholas JE, 31
Nicholson LK, 590, 597
Nickerson DA, 84
Nickolaisen SL, 21
Nicolaides CA, 650
Nicovich JM, 43, 46
Nie S, 234
Niederjohann B, 142, 144
Niedner-Schatteburg G, 424, 425
Nielsen NC, 471, 473, 478, 490, 491, 580, 582, 584
Nielsen PE, 74
Nieman GC, 424, 438, 445, 447
Niemeyer CM, 76
Nienhaus GU, 248, 316
Nijman M, 492
Nikroo A, 75
Nilsson B, 511
Nilsson L, 503
Ninham BW, 97
Ninomiya Y, 51
Nirmal M, 211, 212, 219
Nishijima H, 86
Nishioka M, 219
Nishiya T, 17, 18
NITZAN A, 681–750; 694, 696, 697, 706, 710, 712, 716, 720, 721, 723, 725, 726, 730, 731, 733, 735–38, 741
Nizkorodov SA, 44, 152
Nochomovitz YD, 500, 504, 514, 521
Noguchi A, 80

Noji H, 237
Nomura K, 582, 587, 589
Nomura S, 212
Nonose S, 424
Noordam LD, 640, 772, 777, 779
Norman AG, 215
Norris DJ, 213, 219
Norris JR, 301, 303
Norris TB, 220, 223
North AK, 75, 76
North S, 50
Northrup FJ, 44, 64, 65, 812
Notarnicola SM, 102
Novaro O, 436
Novicki SW, 610
Nowlin ML, 832, 835
NOZIK AJ, 193–231; 2, 195–97, 199, 201, 202, 205, 206, 208, 210, 211, 214, 216, 219, 220, 223–25
Nudler E, 76, 77, 81 (Figure 4, see color insert)
Nuss MC, 639
Nussbaumer H, 196, 224
Nuttgens S, 171
Nuzzo RG, 108, 113, 114, 128, 130, 131
Nyman G, 17, 27, 158
Nymeyer H, 242, 501, 504, 520

Oakenfull D, 348
Oas TG, 487, 578, 580, 582
Oba D, 650
Oba M, 585, 587
Oberhauser AF, 72
Oberthaler MK, 800
O'Brien DF, 396, 419
O'Brien JP, 818, 819
O'Brien SC, 424
Ochoa de Aspuru G, 16, 17
Ocko BM, 119
O'Connell T, 503
O'Connor EJ, 585, 587
Oden PI, 84

Odian GG, 549
Odijk T, 97
Odom TW, 86
Oesterhelt D, 639
Ogawa K, 590, 597
Ogawa MY, 687
Ogawa S, 256
Ogilby PR, 43, 49, 50
Ogletree DF, 236
Ogstron AG, 93, 94, 99
Oguchi T, 825
Oh D, 21, 23
Oh DB, 50, 51
Ohmes E, 301
Ohmori K, 670
Ohnesorge B, 220, 223
O'Neil KT, 512
O'Neill TE, 80
Ohnishi T, 84
Ohoyama H, 17, 18, 814
Ohrwall G, 265
Ohshima Y, 818, 824
Ohta T, 75, 567
Ohtsuka N, 220, 221
Ok JH, 492, 580, 582, 587
Okabe H, 644
Okada A, 720
Okamoto Y, 501
Okawa K, 55
Okazaki S, 396
O'Keefe A, 51
Okudaira KK, 741
Okudaira M, 590, 597
Okunishi M, 670
Olafson BD, 520
Olbright GR, 220, 223
Oldershaw GA, 31, 36
Oldfield E, 316, 327, 328, 578
Olejniczak ET, 471, 475, 476, 484
Olender R, 501
Oleniczak Z, 471
Oling F, 72
Oliver EJ, 589
Olmsted MC, 97
Olshavsky MA, 214
Olson JS, 316, 330

Olson WK, 245
Ong S, 372
Onipko A, 689, 723
Onipko AI, 707
Onitsuka O, 225
Ono A, 587
Ono MY, 71, 72, 79
Ono TA, 287
Onoa GB, 77
Onuchic JN, 242, 243, 500, 501, 504, 510–13, 518, 520, 528, 529, 544
Onuchik JN, 696
Onushenko AA, 212
Opansky BJ, 43
Opella SJ, 488, 489, 576, 578, 589, 590
Oppenheim I, 97
Oppermann W, 93
Oranskii LG, 303
Oreg J, 776, 782, 785, 792, 793
Orlando JJ, 43, 44, 46, 52, 62
Orme-Johnson WH, 293
Ormos P, 316
Oroszlan K, 73, 85
Orozco M, 503
Orr JW, 236, 240, 245
Orr-Ewing AJ, 24, 27, 28, 31, 51, 157–59
Orriols G, 783
Orrit M, 234
Orsini F, 76
Ortiz C, 50
Osborn DL, 170, 171
Osborne JP, 287
Osborne MC, 43
Oschkinat H, 587
Osherov VI, 782
Oshinowo J, 220, 223
Ostermann T, 294, 295
Ostlund NS, 703
Ostrovsky VN, 782
Otomo T, 599
Ottolenghi M, 268
Otzen DE, 243, 504, 515, 516, 518, 525, 528, 529

OUDEJANS L, 607–37; 616, 623, 625, 627, 629, 631, 632, 634
Ouellette KL, 590
Oura M, 181
Oussatcheva EA, 75
Over H, 423
Overbeek JTG, 213
Oxenoid K, 590
Oxtoby DW, 317, 357, 364
Oyama M, 396
Oyler N, 580, 585, 587
Ozaki T, 578

Pacchioni G, 426
Pacey PD, 31
Pacheco A, 294
Pack DW, 316, 321, 322, 334
Pack RT, 145
Packer KJ, 463, 464, 469
Paddon-Row MN, 687
Page JC, 10
Pagsberg P, 43, 58
Pake GE, 578
Palacin S, 699
Paldus BA, 51
Palecek E, 75, 77
Pallix JB, 259
Palma J, 166, 834, 837, 838
Palmas P, 472
Palo K, 247
Paloczi GT, 87
Pan J, 730, 735, 737, 740, 752, 754, 756, 757
Pan T, 752
Pande VS, 242, 243, 500, 501, 540, 541, 544, 549, 550
Pang D, 75–77
Pankove JI, 194, 196
Pannier M, 294
Pantelides ST, 702
Papadopoulos NJ, 293
Papaioannou A, 155
Papanikolas JM, 267, 641
Parak F, 335
Pardi LA, 303

Parent DC, 425
Pargas RE, 578
Parhikhteh H, 844
Parikh AN, 120, 550
Parisi G, 544
Park E, 327, 328, 340, 351
Park H, 167, 833, 835, 836
Park J, 26, 27, 48, 58, 59, 61, 158
Park JS, 80
Park SH, 212, 512
Park S-M, 396
Parker DH, 142, 166
Parker GA, 141, 145
Parker ID, 396
Parkins AS, 801, 802
Parkinson CJ, 166
Parks EK, 424, 434, 435, 438, 441, 443–45, 447, 454
Parmenter CS, 608, 653, 812
Parr RG, 702
Parrinello M, 703, 704
Parslow TG, 598
Parson R, 267
Parsons CA, 199, 202, 205, 208, 210
Paspalakis E, 796
Pastirk I, 641, 643, 644, 647, 649, 653, 655, 658, 659, 661, 670
Pastor RW, 357, 364
Pastore P, 398, 399, 401
Patel CKN, 659
Patel SS, 538
Patel-Misra D, 61
Paul JB, 51
Paulaitis ME, 357
Pauli J, 587
Pauling L, 669
Pauls S, 260
Paulsen EK, 587
Pausch R, 640
Pavlenko VS, 650
Pavylchev AA, 181
Pawson T, 234
Payner SP, 24
Pearson DS, 93, 94, 99

Pearson RG, 54
Pecht I, 293
Pechukas P, 818
Pedersen JOP, 43
Pedersen LG, 506, 591, 595
Pedersen S, 256, 643, 651
Pederson LA, 142, 155, 159
Peersen OB, 480
Pei Q, 392, 407, 408
Peik E, 800
Peisach J, 288, 291–93, 295, 297
Peloquin JM, 287, 288, 294
Pelouch WS, 195, 196
Pelupessy TPH, 299
Pemberton JE, 357
Penfold J, 357
Peng LW, 261
Peng X, 214
Peng XG, 225
Pennington CH, 485, 580, 582, 584
Penno M, 840
Peoples R, 288
Pepys MB, 591
Perazzo T, 558
Pereira GG, 563, 567, 568
Peremans A, 360
Perera L, 506
Perham RN, 489
Perkins TT, 93, 94
Perl D, 515, 519
Perlmutter DH, 80
Perluzzo G, 726, 727, 730, 741, 742
Perry DS, 608
Perry RA, 48
Pershan PS, 119
Persky A, 52–54
Personov RI, 316
Persson BNJ, 719, 840
Persson JL, 424
Persson M, 297
Peskin U, 712, 730, 731, 735, 737, 738
Pesl FP, 798
Petek H, 43, 49, 50, 256

Petera D, 563, 566
Peterka DS, 141, 155, 839
Peterlinz KA, 114, 115, 123, 128, 130, 131
Peterman EJG, 236, 240, 248
Peters RD, 563
Peterson C, 540
Peterson KA, 331
Peterson MR, 143
Peterson MW, 199, 202, 205
Peterson SR, 75, 76
Petkova AT, 582, 589
Petralli-Mallo TP, 361
Petroff PM, 220, 223
Petrongolo C, 155
Petrov KP, 50
Petrov YG, 303
Petry W, 335
Petsalakis ID, 650
Petsko GA, 316, 346
Pettiette CL, 257, 425, 426, 429
Pettiette-Hall CL, 424
Pettitt BM, 501
Petty JT, 43, 52
Peyghambarian N, 212, 220, 223, 415, 417, 419
Peyser P, 545
Pfannschmidt C, 73
Pfeifer H, 475, 476
Pfeiffer JM, 19, 157
Pfeiffer M, 299, 306
Pflumio V, 373
Pfnur HE, 844
Phillips DL, 644
Phillips GN Jr, 316, 330
Phillips J, 220, 221
Phillips LF, 55, 61
Phillips TR, 141
Phillips WA, 321, 334, 335
Phillips WD, 650, 800
Picard YJ, 189
Piccirelli R, 343
Picconatto CA, 32–36
Pichler G, 650
Pichot F, 407, 409
Pickett DL, 108

Pickett GT, 562
Piel J, 643, 658
Pierola IF, 93
Pietrasanta L, 71, 76
Pietrasanta LI, 72, 76–79, 87 (Figure 3, see color insert)
Pignataro B, 76
Pilgrim JS, 51, 62
Pillet P, 650, 792, 793, 800
Pilling MJ, 45, 52
Pimental GC, 362
Pincus P, 561
Pine AS, 50, 608, 612, 616, 623
Piner RD, 134
Pines A, 143, 466, 472, 477, 479, 481, 589
Pinheiro TJT, 299
Pinkas I, 658, 659
Pique JP, 818, 825
Pispas S, 93, 94
Pitsikalis M, 563
Pittner J, 436
Place C, 293
Plato M, 299, 303, 304, 306
Plaut AS, 220, 221
Plaxco KW, 245, 510, 513, 514
Pleksov YY, 730, 733, 736
Plombon J, 715
Ploog K, 205
Pobo LG, 424, 441, 443, 444, 447, 454
Podell E, 751
Pohorille A, 357
Poirier GE, 108, 113
Polanyi JC, 17, 255, 669, 812
Polenova T, 587, 599
Poles E, 220, 223
Polik WF, 148, 819
Pollack L, 245
Pollak E, 151
Pollard WT, 720, 721
Pollock EL, 212
Pollock RC, 288

Poluektov OG, 292, 293, 303, 304
Pomerantz AE, 146, 159
Ponomarev AB, 294
Ponti A, 288
Poole CP, 282
Poolman B, 590
Pope M, 391, 396
Popescu G, 77
Popovic ZD, 413
Porter G, 42, 639, 669
Porter JD, 687, 730
Porter MD, 125
Porter MR, 373
Poser S, 234, 247
Posey LA, 173
Potaman VN, 74, 75
Potanin AA, 538
Pötsch S, 287
Potter AB, 85
Potter ED, 650, 651
Potter WT, 347
Potthast F, 540
Powell JR, 327
Powers DE, 260, 426
Powers PE, 195, 196
Powis I, 169, 181, 182
Powles JG, 489
Poyner R, 288
Prager S, 93, 94
Prager YS, 707, 708
Pramanik A, 234, 247
Prater CB, 72
Pratt LR, 357
Prausnitz JM, 93
Prentiss M, 798, 800
Prestegard JH, 589
Prevo LJ, 303
Price JM, 833, 837
Price W, 663, 664
Prieto MJ, 77
Prior Y, 643, 658, 659
PRISNER T, 279–313
Prisner TF, 287, 294, 295, 299, 301, 303, 304, 306
Prokhorov VV, 74, 75
Proksch R, 83

Proskuryakov II, 301
Prosser RS, 589, 590
Protopapas M, 796
Prud'homme RK, 93
Prusiner SB, 234
Pruski M, 487
Prytulla S, 587
Pshenichnikov MS, 653, 655, 658, 659
Ptitsyn O, 500
Pugh LA, 42
Pugmire RJ, 483
Pulver S, 288
Purnell SK, 424
Putnam RS, 50
Pylant ED, 113
Pyne CH, 10

Qi J, 83
Qian H-b, 45, 52
Qiu Y, 619
Quack M, 612, 614, 632, 652
Quandt RW, 56, 58
Quapp W, 824
Quate CF (Figure 2, see color insert)
Quayle CJK, 181, 185
Que L, 297
Queisser HJ, 196, 224
Quillin ML, 316, 330
Quiñones A, 627

Raap J, 587, 590
Rabani E, 75
Rabanos VS, 149, 151
Rabe M, 220, 223
Rabinovitch BS, 427
Rabitz H, 155, 640, 772
Rader SD, 234
Radford SE, 242
Radhakrishnan G, 21, 23
Radi PP, 424, 663
Radloff W, 168, 169, 183, 264, 265, 270, 644
Radom L, 166
Raduge C, 373
Radzilowski LH, 225

Raghavan K, 438
Raghunathan P, 578
Rahmat G, 141
Rahn LA, 814, 825
Raimond J-M, 772
Raitsimring AM, 294–97
Rajaram B, 818, 819
Rakete C, 160
Rakhmatullin R, 291
Rakitzis TP, 26, 28, 31, 157, 158
Raleigh DP, 491, 580, 582
Ralston M, 755
Ramamoorthy A, 488, 490, 580, 582, 590
Ramanathan G, 97
Rambo RP, 751
Ramirez-Solis A, 839
Ramshaw JD, 473
Rance M, 502
Randall DW, 287, 294
Randall RW, 825, 828
Rangan SK, 295
Rao KN, 42
Rao NS, 598
Rao S, 76 (Figure 3, see color insert)
Rasch P, 80
Rashid A, 97
Rasing T, 814
Rastogi VK, 557
Ratajczak E, 43
Ratcliffe CI, 485
Ratner MA, 647, 682, 695, 696, 698, 699, 703–5, 707, 708, 710, 720, 721, 725, 726
Rautter J, 294
Ravishankara AR, 51, 372
Ray AK, 432, 438
Ray D, 428
Ray E, 590, 597
Ray K, 741, 742
Ray S, 476
Rayane D, 424
Raymond EA, 362–64, 372, 373

Raymond TD, 840
Rayner DM, 424
Read FH, 257
Record MT Jr, 97
Rector KD, 315, 317, 320, 322, 323, 325, 327, 328, 330, 331, 334, 336, 337, 341, 345, 346, 348–50
Reddy GA, 234
Redfield AG, 293, 721
Reed GH, 288, 293
Reed J, 590, 593, 595
Reed KJ, 176
Reed LH, 503
Reed MA, 687, 698, 702, 708
Reeves PJ, 599
Reeves RH, 73
Register RA, 563
Rehage G, 93
Rehle D, 50
Reich NO, 75, 76
Reichmann KC, 424, 434, 435
Reid KL, 168, 272, 519
Reid SA, 147
Reif B, 585, 589
Reif F, 465, 469, 476
Reifenberger R, 687, 698, 700, 704, 706
Reilly JP, 259, 260
Reimer JA, 589, 599
Reimers JR, 698
Reinecke TL, 219
Reinisch L, 316
Reisfeld RA, 75
Reisler H, 147, 841
Reisner DE, 812, 814
Reiss H, 407, 409
Rekesh D, 75
Rella CW, 315, 317, 320, 322, 323, 325, 327, 328, 331, 334, 336, 337, 341, 345, 346, 348–50, 772, 777, 779
Remsen EE, 80
Ren Y, 357
Renger G, 301

Repphun G, 735
Resat MS, 175, 187, 188
Resch R, 122
Rettner CT, 450, 814, 840, 841, 843, 844
Reuss C, 256
Reuss J, 772
Revenko I, 74, 83, 84
Revet B, 74
Reviakine I, 72
Rews MR, 782
Rey A, 500
Reynolds AH, 316
Reynolds DC, 205
Rezaei MA, 712
Rhee J-K, 772
Ricard D, 212
Rice J, 21, 23
Rice JK, 21
Rice SA, 357, 640, 651, 653, 786, 787
Richard FM, 500
Richardson CC, 102
Richardson DM, 542
Richman DC, 48
RICHMOND GL, 357–89; 357, 360–65, 368, 369, 372–74, 376, 378, 379, 381–83
Richmond TJ, 500
Richter AG, 122
Richter D, 49
Richter LJ, 361
Richter MM, 410
Richtsmeier SC, 424, 441, 443, 444, 447
Rickes T, 773, 775, 776
Riddle DS, 514
Rider LS, 407, 409
Riedi PC, 303
Riedle E, 643, 658, 814
Rief M, 73, 83, 85–87
Rieley H, 610
Rienstra CM, 472, 491, 580, 582, 584, 585, 587, 589
Righetti GP, 97–99
Rigler R, 234, 235, 247, 248

Riis E, 800
Riley D, 669
Riley SJ, 424, 434, 435, 438, 441, 443, 444, 447, 454
Rim KT, 58–61
Rimai L, 43
Ringe D, 316, 346
Rinnenthal JL, 157, 833
Rinsland CP, 42, 47
Ripmeester JA, 485
Rippe K, 75–77, 235, 247
Rippon DM, 665, 667
Rischel C, 256
Ritchie EL, 392, 399, 401, 412
Riter RE, 357
Ritter A, 831
Rivetti C, 76, 77, 84, 85
Rizzo NW, 587, 590, 592, 593, 595, 596
Roberts C, 220, 221
Roberts G, 650, 667
Roberts JE, 287
Roberts VA, 293
Robins R, 590
Robinson BH, 298
Robinson EA, 31, 36
Robinson GN, 141, 149, 151
Robinson MS, 43
Robinson PJ, 428, 653
Robyr P, 578, 585
Rock AB, 831, 832
Rocker C, 248
Rockford L, 567
Rodbard D, 93, 94, 99
Roder H, 245, 512
Rodgers JE, 77
Rodgers MT, 428, 444, 447
Roditchev D, 220, 223
Rodriguez CF, 36
Rodriguez G, 650
Rodriguez HM, 234, 513, 515, 519
Rodriguez M, 798
Rogalla H, 120
Rogaski CA, 814, 818, 824, 833, 837

Rogers JA, 663
Roher AE, 591
Rohlfing EA, 424, 434, 435, 814, 825, 826
Rohrer H, 735, 737
ROHRER M, 279–313; 287, 299, 303, 304, 306
Roitberg AE, 698, 699, 703, 704, 707, 708
Rokhsar DS, 242, 243, 500, 501
Rolston SL, 792, 793, 800
Romanenko VI, 786
Romanini D, 824
Rombel I, 75, 76
Romijn JC, 301
Roncero O, 616
Rondelez F, 120
Rook MS, 752, 759
Roos BO, 166
Rosch N, 426
Rose ME, 467
Rose S, 93
Rose TS, 653, 663
Rosen A, 424, 432, 438
Rosen M, 211, 213, 219, 220
Rosenwaks S, 652, 792, 793
Rosenwaks Y, 196, 206, 220, 223
Rosker MJ, 206, 639, 643, 653
Ross RS, 561
Ross RT, 196, 224
Rossky PJ, 731
Rost JM, 610
Rostkier-Edelstein D, 730
Rotenberg ZA, 730, 733, 736
Roth S, 682, 687, 706
Rothman LS, 47
Rothuizen H, 86, 557
Rotsein NA, 93, 94
Rousse A, 256
Roux B, 590
Rouzina I, 73, 97
Rovnyak D, 478
Rowntree PA, 116
Ruatta SA, 425

Rubahn H-G, 787
Ruben DJ, 578
Rubenstein RA, 650
Rubinstein M, 93, 94, 99
Rubner MF, 225, 392, 407, 409
Ruczinski I, 514
Rudecki P, 782, 785, 787
Rudich Y, 631
Rudolph W, 772, 773
Ruhman S, 266, 268, 640, 653
Rumbles G, 225
Ruminov S, 215
Rundquist AR, 265
Rusin LY, 141
Russel WB, 538, 561
Russell LA, 43, 49, 50
Russell R, 236, 240, 245, 756, 757, 759
Russell TP, 562, 563, 567
Russon LM, 429, 430, 432, 438
Rutherford AW, 287, 294, 297, 303
Rutz S, 263
Ryali SB, 840
Ryan JF, 206
Ryan K, 83
Ryckaert JP, 506
Rzhevskii OS, 650

Saad Y, 732, 738
Saari EA, 293
Sabelko J, 242
Sachleben JR, 471, 483, 488
Sachs SB, 687
Sadowski CM, 832
Saffman PG, 298
Safran SA, 567
Sage JT, 327
Sagiv J, 108
Saha S, 779
Saika A, 485
Saito A, 511
Saito H, 578
Sakaki H, 220
Sako Y, 236

Sakurai JJ, 466, 467
Salh JS, 64
Sali A, 500, 516
Salikhov AK, 301
Salikhov KM, 301
Salvato B, 293
Samartzis PC, 741, 742
Sambles JR, 706
Samoson A, 477
Samson P, 840
Samuelson L, 219
Sanche L, 706, 726, 727, 730, 741–43
Sanchez A, 424
Sancho J, 287, 294
Sander F, 800
Sander SP, 26, 41
Sanders CR, 589, 590
Sanders DE, 132
Sandmann JHH, 220, 223
Sandorfy C, 361
Sandström D, 487, 585, 590
Sansom RL, 62
Santamaría J, 776, 779, 797
Santamaria R, 436
Santiago JV, 514
Sappey AD, 434
Sardashti M, 480
Sarikaya M, 84
Sarkisov OM, 641
Sarre PJ, 9
Sasabe H, 119
Sasaki K, 84
Sass JK, 730
Sastry VSS, 295
Satija SK, 93, 94, 563
Sato MH, 80, 511
Sato T, 116
Sato Y, 114, 670
Satyapal S, 158
Saubaméa B, 800
Sauder DG, 61
Sauer K, 294
Sauer M, 235, 248
Sauter B, 331
Savitsky A, 299, 306
Saxena A, 718

Saxena S, 296
Saykally RJ, 51, 608, 619
Sayos R, 24
Scalfi-Happ C, 248
Scalvi B, 752, 755
Scaringe SA, 755
Scatena LF, 362, 364
Schacht EH, 80
Schade M, 75
Schaefer HF, 818, 839
Schaefer J, 472, 475, 485–87, 490, 576, 580, 582, 584, 585, 590
Schaefer M, 501
Schaeffer TE, 87
Schafer H, 585, 587, 590, 597
Schafer KJ, 669
Schafer R, 73, 85
Schaffer TE, 87
Schanne-Klein MC, 212
Schaper A, 73
Schatz GC, 16, 17, 19, 22–24, 62, 64, 142, 147, 155, 159, 160, 838
Scheafer J, 480
Schechter I, 54
Scheer H, 294
Scheerschmidt K, 215
Scheibe G, 268
Schellhaass B, 798
Schelvis JPM, 669
Schendel SL, 590
Schenk A, 248
Scheraga HA, 501
Scherer JJ, 51
Scherer JR, 361, 363
Scherer NF, 23, 255, 334, 639, 643, 650, 659, 665
Scherson DA, 128
Schessler HM, 116, 128, 131
Schick CP, 260, 264
Schiemann S, 782, 785, 787, 789, 790
Schiffer CA, 503, 507
Schiffman A, 43
Schindelin H, 515

Schindler T, 424, 425, 515, 519
Schinke R, 824, 825, 827
Schinzel S, 576
Schlag EW, 258, 814, 828, 830
Schlamp M, 225
Schlamp MC, 225
Schlittler RR, 687
Schloss JH, 650
Schmalbein D, 281, 303
Schmatjko KJ, 59
Schmickler F, 681, 682, 686
Schmickler W, 197, 201, 202, 205, 206, 220, 223, 715, 736
Schmid FX, 515, 519
Schmidt A, 413, 415, 486
Schmidt B, 650, 652, 670
Schmidt CF, 557
Schmidt HM, 212
Schmidt J, 303, 304
Schmidt M, 245
Schmidt PP, 287
Schmidt T, 236
Schmidt TW, 667
Schmidt-Rohr K, 316, 463, 464, 469, 474, 479, 483, 487, 578, 584, 585, 587, 590, 597
Schmiechen P, 154
Schmiedmayer J, 650
Schmitt L, 84
Schmitt M, 272, 644, 658, 661
Schnabel P, 424
Schnapp BJ, 557
Schneider JP, 234
Schneider TW, 114
Schneider WD, 424
Schnell I, 476
Schnieder L, 142, 144–46, 155, 609
Schnitker J, 731
Schnitzer C, 363, 373
Schoenlein RW, 256, 655, 658
Schofield DA, 375

Scholes CP, 287, 288
Scholz F, 220, 221
Schonland DS, 2, 3
Schreiber E, 263, 644
Schreiber F, 113
Schrepp W, 119
Schrieffer JR, 718
Schrock RR, 225
Schroder T, 825, 827
Schroeder R, 361
Schuder MD, 619
Schultes EA, 752
SCHULTZ PG, 233–53; 235, 237, 238, 240, 241, 599
Schultz RH, 428, 429
Schultz T, 261
Schulz A, 77
Schulz CP, 169, 265
Schulze M, 454
Schulz-Schaeffer W, 234, 247
Schumacher DW, 669
Schumacher W, 288
Schuster P, 361
Schütz GJ, 236
Schwartz BJ, 644
Schwartz DC, 84
SCHWARTZ DK, 107–37; 119, 122, 123, 125, 128, 132, 133
Schwartz P, 113
Schwartz R, 834
Schwartz SD, 693
Schwartz T, 75
Schweiger A, 284, 286, 291, 293
Schweighofer K, 357, 364
Schweighofer KJ, 357, 364
Schweikhard L, 424, 429, 432, 433, 437, 439
Schweins T, 303
Schweizer EK, 450
Schweizer H, 220, 221
Schwenke DW, 148
Schwenter N, 641
Schwettman HA, 315, 320, 323, 327, 334
Schwille P, 235, 239, 247

Schwoerer H, 640
Schwonek JP, 589
Sciaky N, 570
Scoles G, 113, 772
Scott G, 50
Scott JL, 652
Scruggs BE, 589
Seakins PW, 43, 45, 52
Searls T, 99, 101, 102, 104
Sears TJ, 812
Sebald A, 485, 489
Seebach D, 501
Seekamp-Rahn K, 142, 144, 145, 155, 609
Seel M, 270
Seeman NC, 83
Segal D, 694, 720, 721, 723, 725, 726
Segall A, 75
Segall J, 609
Segel DJ, 245
Seidel CAM, 235, 237, 247
Seideman T, 168, 264, 272, 640, 670, 700
Seidl ET, 818, 839
Seifert JM, 598
Seiter M, 298
Seitsonen AP, 423
Sekharudu YC, 503
Sekreta E, 259
Sela M, 500
Selinger A, 424
Selkoe DJ, 593
Selmarten DC, 220, 223, 225
Selvin PR, 236
Selzle HL, 814, 828, 830
Semenov AN, 93, 94, 99, 567
Semenza G, 72
Seminario JM, 709
Sen S, 779
Senear DF, 93, 95
Senhorst HAJ, 772
Serafin JM, 610, 619, 634
Sercel PC, 212, 213, 215, 220, 223
Serna S, 616
Serpell L, 591, 592

AUTHOR INDEX

Serpell LC, 591
Serpersu EH, 293
Serxner D, 268
Setser DW, 650
Setzler JV, 22
Severns JC, 293
Sevy ET, 62
Seyfried V, 261, 262, 266, 270, 640
Seymour CK, 73
Seymour LW, 80
Shafer NE, 24, 27
Shafer-Ray NE, 24, 27, 146, 148, 159
Shaffer JP, 272
Shaheen SE, 396, 413, 415, 417, 419
Shahidzadeh N, 120
Shakhnovich EI, 500, 501, 541, 544, 547–49, 552
Shane SF, 643
Shane T, 288, 306
Shank CV, 214, 256, 639, 641, 647, 653, 655, 658
Shanzer A, 741
Shao HB, 116, 128
Shao Z, 72, 81
Shapiro M, 583, 612, 640, 650, 651, 795
Sharkey P, 44, 64
Sharpe SW, 363
Shastry MC, 234, 245
Shaw AM, 10
Shaw AS, 570
Shaw WJ, 583, 584
SHEA J-E, 499–535; 500, 501, 504, 510, 512, 514, 518, 520, 521, 528, 529
Sheats JR, 392, 419
Sheehy B, 669
Sheetz MP, 557
Sheinerman FB, 503, 506, 508, 510, 512, 513, 528, 529
Shen YR, 357, 358, 361–64, 367, 373–75, 640, 655, 665

Shen Z, 413
Shergill JK, 288
Sherrill CD, 818
Sherwood CR, 170, 172, 175
Shi J, 43
Shi JM, 415
Shi S, 640
Shi SM, 838, 839
Shi Y, 429, 430, 436, 437
Shiang JJ, 655, 658
Shiels JC, 490, 580, 582, 585, 587, 593
Shigemasa E, 181
Shigemori K, 297
Shim M, 220–22
Shimada H, 347
Shimada I, 511
Shimazu K, 114
Shimizu K, 800
Shimoyama Y, 298
Shiraishi M, 511
Shirley JA, 643
Shirley JH, 471
Shirota Y, 419
Shlyakhtenko LS, 74, 75, 80, 84
Shobatake K, 141, 149, 151
Shoji A, 578
Shoji H, 220, 221
Shokhirev KN, 141
SHORE BW, 763–809; 764, 765, 768, 769, 773, 775, 776, 782, 785, 786, 787, 791–93, 795, 796, 798, 802
Shorter JH, 612, 623, 634
Shortle D, 244
Shoustikov A, 396, 419
Shuai Z, 396, 419
Shubin AA, 288
Shui JC, 424, 436
Shull KR, 562, 563
Shultz MJ, 363, 373
Shumay IL, 256
Shumilov SK, 212
Shustorovich E, 455
Shuvalov VF, 303

Sibley S, 396, 419
Siders CW, 256
Siebers DL, 48
Siebert F, 590
Siebert T, 658, 661
Siegbahn P, 143, 443
Siegenthaler H, 735, 737
Siemoneit K, 220, 223
Sienkiewicz A, 303
Sigal G, 552, 556, 557
Sikes HD, 122, 123, 125
Silberzan P, 121
Silbey RJ, 396, 419, 818, 819
Sillesen A, 43, 58
SILVA M, 811–52
Silverbrook K, 698
Silverman RB, 288
Silverman SK, 755
Simard B, 424
Simmon HJ, 358
Simmons A, 590, 597
Simmons JG, 690, 691, 704
Simmons RB, 557, 561
Simon A, 72
Simon EJ, 583, 592, 593, 595
Simon M, 181, 185
Simon U, 50
Simonelli DM, 363
Simons JP, 24, 27, 155, 157, 159
Simons KT, 510, 513, 514
Simpson WR, 19, 24, 26–28, 31, 157, 158
Sims IR, 58, 60, 650
Sinden RR, 74, 75
Singel D, 290, 303
Singel DJ, 293, 294, 306, 599
Singh J, 220, 223
Singh Y, 542
Singhal RP, 183
Sinha A, 17, 19, 157
Sinsheimer RL (Figure 8, see color insert)
Sipes C, 23
Sirois S, 36
Sirtori C, 50
Sita LR, 687

Sitko JC, 78 (Figure 3, see color insert)
Sitz GO, 844
Sivakumar N, 608
Sivan U, 83
Sivaraja M, 287
Sivco DL, 50
Skinner JL, 318
Skodje RT, 149, 151
Skolnick J, 500, 511
Skourtis SS, 715
Skouteris D, 53, 54, 149, 154
Slanger TG, 814, 833, 835, 836
Slaterbeck AF, 392, 395, 401, 412, 413, 415
Slawsky Z, 834
Slichter CP, 321, 322, 463, 468, 469, 485, 489
Sligar SG, 234, 315, 316, 320, 327, 328, 334, 335
Sloan JJ, 62
Slowinska KU, 687
Slowinski K, 687
Slusher RE, 659
Smalley RE, 257, 260, 424–26, 429, 436, 440, 608
Smiley BE, 364, 381, 383, 384
Smith AM, 824
Smith AV, 792
Smith BB, 220, 223
Smith BC, 818
Smith BL, 76, 77, 84, 87
Smith CS, 590
Smith DA, 242
Smith DD, 639
Smith DE, 93, 94
Smith DJ, 741, 742
Smith DL, 324, 332, 334, 343, 718
Smith DM, 166
Smith GC, 76
Smith GM, 303
Smith ICP, 5

Smith IWM, 43, 44, 58, 60, 61, 64, 833, 836
Smith JM, 260, 591
Smith LJ, 503
Smith MAH, 42, 287
Smith MD, 348
Smith SA, 471, 487
Smith SB, 83, 85, 246, 560
Smith SC, 58, 427, 428
Smith SO, 480, 489, 576, 580, 582, 584, 590, 599
Smith SR, 293
Smith TI, 315
Smithers GW, 293
Sneddon SF, 502, 503, 505
Snijders JG, 832
Snow ES, 219
Snyder SW, 301
So PT, 247
Sobolewski A, 270
Socci ND, 242, 500, 501, 511, 513, 520, 544
Soejima K, 181
Soep B, 264, 266, 650
Sohn BH, 563
Sokhan VP, 357
Sokolowski-Tinten K, 256
Solá IR, 776, 779, 797
Solgadi D, 264, 265
Solina SAB, 818, 819
Solum MS, 471
Soman KV, 502, 503
Somorjai GA, 375, 423, 443, 447, 454, 840
Sone Y, 424
Song L, 424
Song XB, 260
Sorensen SL, 265
Sorenson C, 72
Sorenson CM, 363
Soria MR, 78
Sosnick TR, 234, 752
Sosnowski T, 220, 221
Sosnowski TS, 220, 223
Sotomayor-Torres CM, 197, 220
Space B, 743

Spain EM, 71, 72, 79, 434, 435
Spain JA, 73
Spaltenstein A, 552, 556, 557
Spaniol T, 483
Spanner M, 667, 781
Spasov VA, 429, 430, 436, 437, 439, 454
Spence TG, 51
Spencer RGS, 580, 582, 583, 587, 595
Spielberger L, 169, 182
Spiers ID, 590
Spiess HW, 294, 295, 316, 463, 464, 468, 469, 474, 476, 479, 483, 493, 585
Spijkerassink MB, 590
Spiridonov VP, 667
Spisz TS, 73, 78
Spooner PJR, 590
Spörner M, 303
Spoyalov AP, 294
Sprague JR, 214, 216, 217, 219
Spreeuw RJC, 792, 793, 800
Springer BA, 316
Spuldich JA, 557, 561
Srebnik S, 541, 547, 548, 552
Srivastava A, 32–36
Stahl SJ, 598
Staib A, 732
Staikov G, 735
Stankova V, 75, 77
Stanley CR, 220
Stanley RJ, 814, 828, 829
Stanners CD, 375
Stanton CJ, 195
St. Pierre RJ, 424
Stark H, 51
Stark J, 256
Stark K, 149, 151, 172
Stark RE, 578, 583
Stark RW, 80
States DJ, 520
Stauffer HU, 141
Stave MS, 436, 438
Stayton PS, 580, 583–85, 587

AUTHOR INDEX

Steadman J, 257
Stefani M, 591
Steffen R, 219
Steffen T, 670
Stehlik D, 299, 301, 304
Stein G, 268
Stein H, 450, 844
Stein SE, 427
Stein VM, 97
Steinbach PJ, 493, 584, 590
Steinberg IZ, 241, 249
Steinberg S, 119
Steiner B, 274
Steinhoff HJ, 299, 306
Steinruck HP, 450
Stejskal EO, 463, 464, 472, 475, 480, 485
Stellwagen J, 93, 95
Stellwagen NC, 93, 95, 97, 98
Stemmann O, 76 (Figure 3, see color insert)
Stenholm S, 785, 786, 796
Stephens JW, 49, 52, 58
Stephenson JC, 48, 361, 626
Stert V, 169, 264, 265, 270, 644
Steuernagel S, 478
Steuerwald S, 773, 775, 776, 790
Stewart GM, 820
Stewart JD, 293
Stickel RE, 43, 46, 58
Stidham HD, 357
Stier O, 220, 223
Stigter D, 98
Stillinger FH, 357
Stipe BC, 712
Stock C, 825, 827
Stockmayer WH, 550
Stohner J, 632
Stokstad E, 72
Stoll E (Figure 2, see color insert)
Stoll ME, 483
Stoll S, 293
Stolow A, 168, 261, 264, 640, 669

Stolte S, 772, 814, 832
Stone J, 652
Storz RH, 256
Stourton C, 478
Stoveng JA, 710
Straccia AM, 55
Stranges D, 155
Stranick SJ, 550
Straus SK, 587, 590
Strauss CEM, 643
Strazisar BR, 160
Strebel K, 590
Strehle M, 264
Strickland D, 772
Stringer JA, 490, 580, 582, 593
Strohmaier KG, 304
Strong KM, 172, 175
Stroscio MA, 211
Stroud CR, 783
Strunz T, 73, 85
Struppe J, 590
Struth B, 357
Stryer L, 236, 242, 316, 348
Stuart MAC, 538
Stubbe JA, 303
Stuchebrukhov AA, 608, 714, 715
Stumpf M, 825, 827
Sturge MD, 220, 221
Stuur ER, 348
Stwaley WC, 650
Style DWG, 644
Styring S, 297
Su CX, 428–30, 432, 433, 445
Su W-P, 718
Suchan MM, 841
Sugarman JH, 93
Sugase K, 587
Sugawara KI, 64
Sugawara M, 196, 215, 220, 221, 224
Sugawara Y, 81
Sugita Y, 501
Sugiyama Y, 224
Sugny D, 824
Suhm MA, 612, 614, 616, 632

Suits AG, 141, 155, 166, 833, 837, 839
Sukenik CN, 128
Sullivan M, 752, 755
Sumen C, 570
Sumi H, 715
Sumiyoshi Y, 56
Summerfield D, 24, 27
Summy GS, 800
Sumpf B, 50
Sun BQ, 477, 490–92, 580, 582, 584, 587
Sun F, 50
Sun H, 612
Sun L, 114
Sun Y, 832, 836
Sundaralingam M, 503
Sundberg RL, 818
Sunde M, 591
Sundström V, 644, 653
Sung MM, 120, 121
Sunwoo J, 84
Suominen K-A, 772, 782, 797, 798
Superfine R, 362, 364
Supplee JM, 50
Süptitz W, 790
Surewicz WK, 316, 593
Sussmann R, 828, 830
Suter D, 589
Suter UW, 578, 585
Sutin N, 395
Suto K, 640
Suzuki T, 64, 141, 168, 265, 272, 814
Svec WA, 726
Svetlo O, 647
Svinarchuk F, 74
Svoboda K, 557
Swaminathan DJ, 520
Swaminathan S, 502
Swenberg CE, 391, 396
Swendsen RH, 509, 520
Swinehart DF, 427
Syage JA, 257, 260, 264
Symonds AC, 833, 836
Symons MCR, 2, 3

Syverud AN, 439
Szabo A, 703
Sze S, 194, 196
Szente JJ, 43, 55
Szeverenyi NM, 477, 483, 484
Szichman H, 17
Szmyd DM, 195, 196, 199, 202, 206
Szuromi P, 86

TAATJES CA, 41–70; 44, 50–53, 62, 63
Tabeta R, 578
Tachikawa H, 398
Tacke M, 49
Taddei N, 591
Tadjeddine A, 360
Taherian MR, 814
Takagahara T, 212
Takahashi A, 538
Takahashi H, 511, 814
Takahashi M, 169, 181
Takao K, 116
Takata T, 31
Takatsuka K, 168, 271, 272
Takayanagi M, 17, 18, 814
Takayanagi T, 64
Takegoshi K, 582, 587, 589
Takeo H, 64
Taketsugu T, 31
Takeuchi S, 818
Takura K, 297
Talaga DS, 236, 243, 244
Talanquer V, 357, 364
Tal-Ezer H, 731
Tamada K, 119
Tamarat P, 234
Tamayo J, 75, 80
Tan HS, 659
Tan RY, 598
Tan SL, 293
Tan WM, 489, 590
Tan X, 288
Tanaka K, 56, 154
Tanaka T, 56, 116, 242, 500, 540, 544

Tanaka Y, 644
Tanford C, 357, 361
Tang CL, 195, 196, 206, 639, 653
Tang CW, 391, 392, 411, 413, 415
Tang H, 562, 687
Tang J, 301
Tang LH, 132, 133
Tang XS, 287, 294, 297
Tani A, 343, 346
Tanigawa M, 81
Tanimura Y, 670
Tanner D, 372
Tanner DJ, 372
Tannor DJ, 640, 653, 792
Tanzer TA, 117
Tao N, 715
Tao Y, 113, 114, 128, 130, 131
Tao YT, 550
Tapalian HC, 62
Tappe U, 141
Taranukhin VD, 669
Tarnovsky AN, 644
Tasaki S, 158
Tasch S, 407
Tashiro M, 511
Tate MW, 245
Taylor AJ, 639
Taylor KJ, 424
Taylor PR, 175, 176
Taylor RA, 206
Taylor RS, 357
Taylor SM, 10, 424, 436
Taylor SS, 590
Taylor TR, 266
Tejero R, 511, 515
Tekely P, 472
Teloy E, 427
Temkin RJ, 599
Temps F, 825, 827, 832
Temsamani MA, 819
Teng Q, 578
Tennant LL, 502
Ten Wolde A, 772
Teo SH, 76

Terao T, 477, 480, 483, 485, 487, 580, 582, 585, 587, 589
Terauchi T, 585, 587
Ter Horst M, 64, 142
Ter Horst MA, 64
Ter Meulen JJ, 424
Terrill RH, 117, 392, 407, 409
Terry TD, 489
Tersigni S, 640
Tersoff J, 691, 692
Tessler N, 396, 407, 410
Teutloff C, 287
Tevelrakh E, 293
Thacker BR, 199, 202
Thaden B, 188
Thalhammer S, 80
Thalweiser R, 261, 640
Thantu N, 260
Thayer AM, 466
Thayumanavan S, 392, 395, 401, 412, 413, 415, 417
Theil PA, 733
Theinl R, 157
Theuer H, 782, 785, 787, 792, 793, 798–800
Thijssen HPH, 320, 334
THIRUMALAI D, 751–62; 243, 500, 514, 752, 754–57, 759, 760
Thoemke JD, 19, 157
Thoma M, 357
Thoman JW Jr, 831, 832
Thomann H, 288, 304
Thomann U, 256
Thomas EL, 225
Thomas RK, 357
Thomas YG, 234
Thompson DE, 317, 320, 323, 348
Thompson JB, 87
Thompson JD, 424
Thompson KC, 31
Thompson LK, 582, 590
Thompson ME, 396, 419
Thompson WH, 172, 179
Thoms S, 220

Thomson NH, 76, 84
Thorsheim HR, 650
Thrower D, 76 (Figure 3, see color insert)
Thrush BA, 43, 49, 55
Thundat T, 73, 83, 557, 561
Thuranuer MC, 301
Thurman G, 293
Thurmond KB II, 80
Thurnauer MC, 293, 301, 303
Thyberg J, 234, 247
Thyberg P, 234, 247
Tian C, 590
Tian WD, 687, 698, 700, 704, 706
Tianwei J, 83
Tidesley DJ, 501
Tidswell IM, 119
Tiede DM, 293
Tildesley DJ, 357
Tillement J-P, 79
Ting AY, 236, 239
Tinoco I, 755
Tipikin DS, 303
Tirado-Rives J, 502–4, 528, 529
Tirrell DA, 556
Tirrell M, 538, 550
Titman JJ, 585
Titov AA, 641
Tittel FK, 43, 49, 50
Tjernberg A, 591, 595
Tjernberg LO, 234, 247, 591, 595
Tobias DJ, 502, 503, 505
Tobiason JD, 814, 825, 826
Todd AP, 503
Todd PF, 5
Todd RJ, 557
Todorov TN, 709
Toennies JP, 141, 274
Tokei-Takvoryan NE, 398
Tokmakoff A, 315, 317, 323, 331, 348, 670
Tokue I, 43–45, 50, 64, 65
Tokumasu F, 75, 80

Tolksdorf C, 73, 85
Toma ES, 772, 777, 779
Tomaselli M, 490, 578, 580, 582, 585
Tominaga K, 670
Tomita Y, 585, 587
Tomkins FC, 453
Tommos C, 669
Tomo T, 301
Tonaka M, 301
Toncheva V, 80
Tong J, 93
Tong L, 293
Tonokura K, 51
Topley B, 639
Torchia DA, 578
Torigoe H, 511
Toth C, 256
Toukmaji A, 506
Toumi R, 833, 834, 837
Tour JM, 687, 698, 702, 708, 709
Tovbin DG, 641
Townsend CG, 782
Townsend RM, 357
Towrie M, 168, 272
Toyoshima I, 443, 447
Trabesinger W, 236
Tradnell BMW, 453
Trager F, 116, 128
Trakhtenberg S, 741, 742
Trask BJ, 84
Trautman JK, 211, 234
Treacy EB, 647, 772
Treacy GM, 392, 410
Trebino R, 641
Tredicucci A, 50
Treiber DK, 752, 759
Trentelman K, 640
Trevor DJ, 424, 434, 435, 440, 447, 452
Tromp JW, 693
Tropsha A, 503
Troughton EB, 113, 114, 128, 130, 131
Trubetskoy VS, 80
Trugman SA, 717, 719, 741

Truhlar DG, 25, 52, 143, 148, 612
Truong KD, 116
Tsai BP, 36
Tsai J, 514, 518
Tsuchiya S, 818
Tsuda S, 316
Tsui OKC, 563
Tsukada M, 703, 704
Tsukiyama K, 53
Tsvetkov YD, 288, 291, 294, 298
Tsvetkov YuD, 294, 295
Tuberfield AJ, 206
Tucker MP, 347
Tuckett RP, 61, 833, 836
Tugarinov V, 590
Tulej M, 270
Tull JX, 772, 782
Tully FP, 58
Turner JA, 199, 202, 205
Turner MS, 567
Turnipseed AA, 372
Turton D, 45, 52
Tutunjian PN, 484, 580, 582
TYCKO R, 575–606; 484, 489, 493, 578, 580, 582, 584, 585, 587–90, 599
Tyndall GS, 43, 44, 46, 49, 52, 55, 62
Tyng V, 824
Tzannis AP, 663

Uchida H, 214
Uchida K, 587, 589
Uchihashi T, 81
Udgaonkar JB, 234
Ueba H, 741
Ueda K, 669, 670
Ueno N, 741
Uhlen M, 511
Ulbrich K, 80
Ullmann D, 247
Ullrich J, 182
Ullrich V, 288
Ulman A, 108, 122
Ulness DJ, 670

Ulrich AS, 590
Ulstrup J, 706, 715
Umanskii SY, 641
Un S, 303
Unanyan R, 786, 787, 798
Unanyan RG, 796, 798, 802
Underwood JG, 168, 272
Unfried KG, 52, 58
Uosaki K, 114, 117
Upadhyaya HP, 154
Ura K, 80
Urbain X, 772
Urbakh M, 730
Urbina C, 699
Urdahl RS, 315, 317, 348
Uschmann I, 256
Ushiki T, 84
Utschig LM, 293
Utz M, 597
Uyama A, 590, 597
Uzer T, 608

Vacano E, 246
Vaccaro PH, 814, 832
Vaghjiani G, 31
Vahala KJ, 212, 213
Vaida V, 363
Vainer YG, 316
Valachovic L, 23, 650
Valdmanis JA, 641, 653
Vale RD, 557
Valente AP, 590
Valentin C, 792, 793, 800
Valentin MD, 294
Valentine K, 590
VALENTINI JJ, 15–39; 27, 32, 35, 36, 610, 619, 634, 652
Vallance C, 19, 814
Vallant T, 121, 122, 128
Valle M, 75
Valpuesta JM, 75
Vanamerom FHW, 832
van Beek JD, 585, 587, 590, 597
vanBuuren AR, 357, 364
van Dam PJ, 287

van den Berg R, 334
vanden Bout DA, 322
van den Engh G, 84
van den Heuvell HBV, 640
van der Avoird A, 828, 830
van der Est A, 301, 304
van der Ham EWM, 361
vanderHart DL, 472
Vanderhoef K, 590
van der Meer AFG, 772
van der Meer H, 303, 304
van der Struijf C, 299
Vander Wal RL, 652
van der Werf KO, 76, 87
Vanderwielen CM, 590
van der Zande WJ, 170, 772
van Faassen EE, 299
van Gastel M, 293, 304
van Gunsteren WF, 501, 503, 504, 507, 528, 529
van Holde K, 80
Van Hove MA, 454
van Kleef EH, 53, 152
van Leeuwen R, 669
Vanleuken JJ, 814, 832
VanLindenvandenHeuvel HB, 772, 777
VanMarter TA, 832, 835
Vann WD, 424, 427
van Noort J, 76
van Noort SJT, 76, 87
Vanolli F, 424
Van Orden A, 234, 235, 246, 248
vanOs JWM, 492
van Rossum B, 587
Van Slyke SA, 391, 392, 411, 413
Vanteenkiste N, 800
van Tol J, 303
Varandas AJC, 143, 834, 837, 838
Vardi A, 650, 773
Varley DF, 26, 27, 157
Varma CM, 321, 334
Varotsis C, 669
Vasquez GB, 347

Vasudev R, 608, 610
Vaughan DEW, 304
Vaughan RW, 483
Vaught A, 730
Vdovin G, 669
Veeman WS, 481, 487, 582, 585
Veenhoff LM, 590
Vega AJ, 481, 483, 487
Vega S, 471, 475, 476, 478, 480, 484, 485, 487, 489, 492, 580, 582
Veit M, 220, 221, 223
Veit S, 294–97
Venkataraman B, 295
Verbeek M, 787
Verdegem PJE, 590
Verdier PH, 550
Verel R, 492
Verhaar BJ, 650, 772
Verhagen R, 478
Vernon JL Jr, 542
Verster NF, 772
Veselov AV, 287
Vesenka J, 74
Vesenka JC (Figure 4, see color insert)
Vetchinkin AS, 641
Vetter R, 141
Vettiger P, 86, 557
Vezin B, 424
Viani M, 78 (Figure 3, see color insert)
Viani MB, 87
Vidic B, 77
Viefhaus J, 181, 185
Vierheilig A, 658
Vigdorovich MD, 715
Vigneron JP, 117
Vigue J, 141
Villafranca JJ, 293
Villeneuve DM, 261, 640, 667, 669, 781
Villhauer EB, 583
Vincent B, 538
Vinckier A, 72
Vinter B, 197, 202, 204

Vinzelberg S, 735
Viovy J-L, 83, 85, 93, 94, 99, 102
Virlet J, 477
Visco RE, 391
Visticot JP, 141, 650
VITANOV NV, 763–809; 782, 785, 792, 793, 796, 798, 802
Vitkup D, 316, 346
Vitrano E, 335
Vlk D, 75, 77
Vogel R, 225
Vogt P, 644
Vogt TCB, 576
Vohringer P, 268, 334, 643, 659
Voicu R, 117
Vojtesek B, 75, 77
Vold RR, 589, 590
Voliotis V, 220, 223
Vjlker S, 320, 334
Volkman BF, 599
Volkmer M, 814
Vollrath F, 590, 597
Volpi GG, 53, 140, 155
Volpp HR, 16, 17, 21, 154
Von Bargen A, 814
von Dirke M, 614
von Helden G, 424
von Hippel PH, 75, 76
Vonklitzing K, 706
von Plessen G, 220, 223
von Puttkamer K, 612
Vorsa V, 267
Vosegaard T, 478
Vrakking MJJ, 261, 640, 772
Vrehen QHR, 361
Vrijen RB, 772
Vurgaftman I, 220

Wäckerle G, 663
Wada R, 640
Wagner AF, 64
Wagner M, 709
Wagner RL, 424, 427
Wagner S, 26
Wahl M, 21
Wahlgren U, 443
Walch SP, 16, 17, 175, 176, 432
Waldeck DH, 706, 741, 742
Walker AC, 220, 223
Walker C, 84, 85
Walker LR, 322
Walker RA, 364, 368, 381–83
Walker RB, 27, 364, 376
Walkup RE, 712, 743
Wallace MI, 239, 240
Wallauer W, 256
Waller IM, 149
Walling AE, 578
Wallington TJ, 43, 50, 51, 62, 63
Wallqvist A, 364
Walrafen GE, 363
Walsh CA, 316, 317, 322, 334
Walsh R, 55
Walter CW, 170
Walther C, 429, 432, 433, 437, 439
Walther M, 220, 221
Waltho JP, 502
Waltman S, 50
Walton DG, 563
Wang C, 77, 84, 234
Wang CH, 473
Wang DL, 407
Wang G, 220, 223
Wang H, 650
Wang JF, 396, 415
Wang J-H, 64, 75, 76, 142, 155, 590, 818, 824
Wang JX, 590
Wang K, 168, 271, 272
Wang L, 168, 265, 272, 501, 503, 528
Wang L-W, 213
Wang PD, 220
Wang Q, 562, 563, 567, 569, 641, 647
Wang S, 50
Wang TH, 485
Wang W, 84, 834, 837, 838
Wang X, 27, 158
Wang XT, 650
Wang Z, 26
Wang Z-G, 557, 567
Wang ZH, 832, 836
Wannenmacher EAJ, 643, 644
Ward B, 639, 669
Ward GN, 833, 836
Ward JC, 644
Ward JH Jr, 507
Ward RN, 357, 360, 375
Wardlaw DM, 607
Warmack RJ, 83
Warncke K, 287, 294
Warren SG, 372
Warren WS, 589, 640, 641, 659, 772, 773, 779, 782, 789
Warshaw DM, 237
Wasielewski MR, 687, 720, 721, 725, 726
Wasserman SR, 119
Wasserman TAW, 814
Watabe H, 219
Watanabe N, 181
Watari H, 298
Watjen JP, 50
Watry M, 378, 379, 381
Watson G, 824
Watson RT, 26
Watt DM, 818
Watts A, 299, 576, 590
Watts NR, 598
Watts RO, 612
Waugh JS, 468, 469, 472–75, 479, 480, 483, 484, 580, 582
Wazawa T, 236, 243
Weaver A, 149
Weaver DL, 500
Weaver I, 669
Webb CL, 800
Webb WW, 235, 247
Weber G, 347
Weber HP, 583
Weber M, 295

Weber PM, 260, 264, 641
Weber RT, 303
Weber S, 301
Weber T, 814
Webster CR, 42
Webster P, 591
Webster SD, 591
Wedler G, 453
Weeks JD, 357
Wegener C, 299, 306
Wei X, 361
Wei YF, 578
Weibel MA, 814
Weida MJ, 256
Weidemüller M, 800
Weil KG, 424
Weiller BH, 424, 438, 445, 454
Weimer JJ, 84
Weinacht TC, 641
Weinbach SP, 357
Weinberg WH, 450
Weiner AM, 772
Weiner J, 650
Weinkauf R, 168, 269
Weintraub O, 492
Weis V, 288, 304, 306
Weisbuch C, 197, 202, 204, 220
Weisenhorn AL, 72
Weisman SI, 299
Weiss DS, 800
Weiss FD, 425, 436
Weiss M, 443, 447, 455
Weiss PS, 550
WEISS S, 233–53; 234–38, 240, 241
Weiss V, 640
Weisshaar JC, 434
Weitekamp DP, 589
Weitz E, 43
Weitz M, 798, 800, 801
Weitzel K-M, 166
Welge KH, 142, 144–46, 155, 609
Weliky DP, 493, 578, 580, 584, 585, 587, 590

Weller A, 396
Weller H, 212, 213, 225
Weller R, 21
Welp KA, 93, 94
Wemmer DE, 599
Wen Y, 23, 609
Wenckebach WT, 303
Wendel R, 424
Wendt S, 423
Wenhai H, 83
Wennmalm S, 248
Wenthold P, 172
Wenzl FP, 407
Wenzler LA, 83
Werle P, 49
Werner HJ, 52–54, 149, 151, 152, 154, 172, 825, 827
Werner JH, 196, 224, 234, 235
Werner U, 168, 185
Wessel JE, 814
Wesson L, 508
West MA, 42
West YD, 9
Westbrook CI, 792, 793, 800
Westley MS, 141
Weston RE Jr, 16, 62
Wetterer SM, 113
Wexler S, 447
Whalley E, 363
Wheeler MD, 51
Whetten RL, 424, 434, 435, 440, 447, 452
Whitaker BJ, 814
White AC, 288
White FH, 500
White JL, 641
White JM, 730, 741
White JU, 51
White SR, 367
Whitesides GM, 113, 114, 119, 128, 130, 131, 552, 556, 557
Whitfield M, 578
Whitham CJ, 61, 650
Whitley RM, 783
Whitlow M, 512

Whitnell RM, 640
Whittaker EA, 50
Whittaker JW, 306
Whitten JL, 425
Wickelgren I, 93
Wickleder C, 828–30
Wickramaaratchi MA, 24, 27
Widengren J, 235, 239, 247
Widom B, 93, 94, 99
Wiebrecht J, 825
Wiebrecht JW, 825, 832
Wiedemann U, 80
Wiegmann W, 197, 202, 205
Wieliczek K, 181, 185
Wienberg J, 80
Wiersma DA, 316, 320, 334, 653, 655, 658, 659
Wiese LM, 188
Wiese W, 735, 737
Wiesenfeld JM, 639
Wight AC, 818, 840
WIGHTMAN RM, 391–422; 392, 395, 396, 398, 399, 401, 407, 409
Wilchek M, 241, 249
Wild UP, 234
Wilhelm T, 643, 658
Wilkerson CW, 260
Wilkie J, 710
Wilkins JW, 712, 713
Wilkins R, 706
Willard DM, 357
Willberg DM, 261
Willeke G, 196, 224
Willems JP, 287
Willer M, 293
Williams CJ, 650
Williams DRM, 563, 567, 568
Williams F, 196, 197, 219, 224
Williams FE, 196, 197
Williams LD, 84
Williams LR, 653
Williams PE, 424
Williams RM, 641
Williams S, 814, 825
Williams VS, 220, 223

Williamson G, 556
Williamson JC, 256
Williamson JR, 236, 240, 245, 587, 752, 755, 759
Williamson MP, 590, 597
Willig F, 197, 201, 202, 205, 206, 220, 223
Willis PA, 141
Willke B, 51
Wilson KJ, 839
Wilson KR, 367, 640, 641, 647, 649
Wilson MA, 357
Wilson WL, 663
Wiltshire T, 73
Wimperis S, 478
Wind O, 220, 223
Wind RA, 483
Windholz L, 650
Wine PH, 43, 46, 58
Winfree E, 83
Wingfield PT, 598
Wingreen NS, 712, 713
Winkel C, 584
Winkler B, 407
Winn JS, 818
Winter BJ, 454
Wirt MD, 293
Wirth MJ, 357
Wise FW, 206, 639, 653
Wishart JF, 687
Wisniewski ES, 644
Witt HT, 301
Wittchen T, 196, 224
Witte K, 511
Witten TA, 538
Wittig C, 21, 23, 609, 841
Witting C, 650
Wittinghofer A, 303
Wiza JL, 169
WODTKE AM, 811–52; 141, 149, 151, 812–14, 818, 824, 825, 832–38, 840, 844
Woggon U, 213, 220, 223
Wokaun A, 463, 468, 483, 659

Wolf EL, 707, 712
Wolf M, 256
Wolfert MA, 80
Wolff JA, 80
Wolff JJ, 128
Wolfrum J, 16, 17, 21, 48, 58, 59, 235, 248
Wolfrum K, 357, 376
Woll C, 116
Wolynes PG, 234, 242, 243, 335, 500, 511, 513, 544
Wong C-H, 511
Wong CM, 256, 704
Wong D, 76
Wong SS, 86, 590, 597
Wong V, 822
Wong WH, 17
Wong WY, 396
Wong-Mong KC, 487
Woo JCG, 288
Woodbury CP Jr, 97
Woodbury N, 80
Woods E III, 157
WOODSON SA, 751–62; 752, 754–57, 760
Woodward C, 511
Woodward JT, 122, 123, 125
Woodward RB, 643
Woody A, 640
Wool RP, 93, 94
Wooley KL, 80
Woolley AT, 86
Woolsey NC, 669
Wooten EW, 477
Worlock JM, 206
Wöste L, 263, 653
Wrede E, 142, 144–46, 155, 609
Wright JC, 670, 780, 781
Wright PE, 502, 511
Wu AP, 392, 407, 409
Wu CH, 488, 590
Wu G, 478, 557, 561
Wu HS, 102, 800
Wu JR, 237
Wu M, 610, 614, 625, 634, 755

Wu Q, 424
Wu SF, 147
Wu TF, 687
Wu X, 432, 438, 480
Wu YS, 143, 144, 146
Wuite GJ, 246
Wyder P, 303
Wyman C, 75, 76
Wynn VT, 582

Xia J-B, 213
Xiao FS, 424
Xiaonong L, 825, 827
Xiaoyuan Z, 814
Xie Q, 220, 223, 715
Xie W, 831
Xie XS, 234, 248
Xie Y, 818
Xing G, 53
Xu F, 134
Xu G, 413
Xu H, 24, 27
Xu J, 444, 447
Xu R, 599
Xu S, 118, 128, 134, 507
Xu ZY, 206, 424
Xue B, 26
Xun L, 248

Yaakov YB, 52, 53
Yachandra VK, 294
Yadav S, 244
Yagi I, 114
Yagishita A, 181
Yahioglu G, 396
Yakovlev VV, 641, 647, 649
Yaliraki SN, 698, 699, 703
Yamada H, 818
Yamada R, 117
Yamanobe T, 578
Yamanouchi K, 818
Yamashita K, 824
Yamazaki T, 599
Yamazaki Y, 590, 597
Yan Q, 562, 563, 567
Yan YJ, 322, 503, 640
Yan Y-X, 653, 663

Yan ZC, 670
Yanagida T, 236, 243
Yaneva M, 75, 76
Yang GL, 76, 77, 84 (Figure 4, see color insert)
Yang H, 825
Yang J, 72, 590
Yang M, 814, 832
Yang MB, 825, 828
Yang SH, 257, 424–26, 429, 434–36
Yang TS, 334, 643, 659
Yang W, 215, 702, 772
Yang WH, 363
Yang XM, 563, 812, 814, 818, 824, 832, 834, 838
Yang XP, 83, 155
Yang Y, 392, 407, 408
Yarim-Agaev Y, 484, 580, 582
Yarkony DR, 143
Yartsev AP, 644
Yasuda R, 237
Yater JA, 220, 221
Yatsenko LP, 773, 775, 776, 786, 795, 796
Yau P, 80
Yauw OW, 120, 121
Yazeva TV, 212, 213
Yen RT, 641, 653
Yen YF, 26
Yen YS, 589
Yenen O, 188
Yesinowski JP, 589
Yevich R, 245
Yi W, 158
Yin JJ, 299
Ying LM, 239, 240
Ying WW, 599
Yit-Tsong C, 818
Yodh JG, 80
Yoffe AD, 197
Yokota H, 84
Yokota T, 741
Yokoyama K, 81
Yoneda H, 669
Yonekura N, 141

Yoo D, 392, 407, 409
Yoo S, 75, 76
Yoon M, 567
York DM, 506
Yoshida K, 84
Yoshida T, 236, 243
Yoshihara K, 670
Yoshii T, 301
Yoshimoto M, 84
Yoshimura SH, 80
You Y, 396, 419
Young AT, 52
Young BC, 798, 800, 801
Young B-L, 802
Young JH, 141
Young MA, 95, 246
Young RD, 316
Young WS, 503
Yu C-J, 122
Yu G, 392, 396, 407–9
Yu HG, 27, 220, 221
Yu HT, 669
Yu HZ, 116, 128
Yu J, 792, 793, 800
Yu T, 51
Yu ZG, 718
Yuan R-L, 792, 793, 800
Yuh JY, 825

Zaban A, 225
Zacharias H, 814
Zadoyan R, 641, 643
Zahniser MS, 51
Zaitsev SV, 220, 223
Zajfman D, 171
Zakin MR, 424, 434–36
Zaloj V, 501
Zambrano R, 578
Zander C, 235, 248
Zanni MT, 168, 257, 261, 266–69, 271, 658
Zare RN, 18, 19, 24, 26–28, 31, 51, 141, 143, 146, 157–60, 167, 168, 234, 272, 608, 610
Zarrinkar PP, 752, 755
Zasadzinski JAN, 119

Zasloff M, 590
Zax D, 590, 597
Zech SG, 287, 294, 299, 301, 302
Zeek E, 669
Zeigler RC, 483
Zeitz D, 49, 58
Zelink R, 489
Zemski KA, 424
Zener C, 771
Zengin V, 175
Zenhausern F (Figure 8, see color insert)
Zeri AC, 590
Zerr I, 234, 247
Zetie KP, 800
Zewail AH, 23, 139, 255, 256, 260, 261, 265, 271, 608, 639, 640, 643, 644, 650, 651, 653, 665, 812
Zgierski M, 168
Zgierski MZ, 264
Zhang C, 392, 407, 408
Zhang CH, 825, 832
Zhang DH, 17, 19, 375, 612, 614, 615, 621
Zhang HL, 116, 128
Zhang J, 609, 795
Zhang JZH, 612, 614, 615, 619, 621, 644
Zhang LS, 425, 426, 429, 503
Zhang Q, 641, 644, 645, 647, 649, 650, 665, 667, 814
Zhang X, 260
Zhang Y, 72, 81, 357, 364
Zhang YF, 687
Zhang ZY, 110, 132
Zhao H, 825, 832
Zhao XL, 119, 372, 580, 582
Zhao XS, 825
Zhao Z, 58
Zhen WG, 580, 582, 584, 587
Zheng LS, 434, 435, 476
Zheng XN, 24, 27

Zheng XS, 146, 159
Zhong QH, 832, 836
Zhou C, 687, 698, 702, 708
Zhou K, 751
Zhou Y, 298, 511
Zhu C, 84
Zhu J, 134
Zhu L, 441
Zhu LC, 640
Zhu QH, 832, 836
Zhu W, 490
Zhu XS, 76, 77, 84
Zhu YJ, 81 (Figure 8, see color insert)
Zhuang X, 236, 240, 245, 756, 757, 759
Zhulina EB, 538, 561

Ziady AG, 80
Ziegler J, 424, 429, 432, 433, 437, 439
Zilker S, 316
Zillman MA, 83
Zilm KW, 480, 587
Zimdars D, 315, 317
Zimdars DA, 315, 331
Zimm BH, 93, 94
Zimmerman AH, 176
Zimmerman DE, 511, 515
Zimmermann JL, 293, 294
Zinke-Allmang M, 133
Zinkel SS, 102
Zinn AS, 687, 730
Zinth W, 639
Zisman WA, 108
Zlatanova J, 80

Zoller P, 792, 798, 799, 801, 802
Zollfrank J, 335
Zu YB, 394, 417
Zuckermann H, 303
Zumbulyadis N, 578
Zumft WG, 288
Zundel G, 361, 363
Zunger A, 213, 214, 216, 217, 219
Zuo T, 669
Zvi A, 493, 584, 590
Zwanenburg G, 301
Zwanzig R, 343
Zwanziger JW, 143, 469, 471, 477
Zweier JJ, 291, 293
Zysmilich MG, 599

Subject Index

A
A-type Coriolis interactions, 827
Absolute rate coefficients, 52
Absorption
 determining integrated, 45–46
 optical, 203–5
Absorption spectroscopy
 time-resolved, 42
Absorption techniques
 infrared, for elementary gas-phase reaction kinetics, 41–70
Abstraction
 direct, chemical reaction dynamics of reactions proceeding by, 21–25
 of a hydrogen atom from an alkane, chemical reaction dynamics involving, 31–36
Acetylene-HF mode specific effects, 627–31
Acousto-optic modulator (AOM), 323
Adiabatic passage techniques
 laser-induced population transfer by, 763–809
 by light-induced potentials, 797–98
Adiabatic population transfer
 applications of stimulated Raman adiabatic passage, 798–802
 extensions of STIRAP, 792–98
 multiple level crossings, 776–82
 Stark-chirped rapid passage, 773–76
 stimulated Raman adiabatic passage in three-state systems, 782–87
 STIRAP in three-state systems, 787–92
 two-state systems for, 768–73
 using sequential pulses, 782–802
 via level crossing, 768–82
Adiabatic states, 482
 and adiabatic following, 769
 testing, 17–18
 See also Vibrationally adiabatic barrier heights
Adsorption
 nanomechanical motion due to biopolymer, 557–62
 surfactant, 368–72
Adsorption energetics
 in growth of self-assembled monolayers, 108, 110, 131
AFM
 See Atomic force microscopy
Air-water interface
 surfactant studies at, 374–76, 378–80
Alkanes
 chemical reaction dynamics involving abstraction of a hydrogen atom from, 31–36
 selectively deuterated, 26
All-atom simulations
 shape and folding, 510–15
Amphiphilic polymers, 538
Amphiphilic surfactant molecules, 107–8
Amyloid fibrils, 591–96
 peptide conformations, 595–96
 structural issues, 591–92
 supramolecular organization of -sheets, 592–95
Analysis
 advanced, 246–49
 burst and fluctuation, 247–48
 via histograms, 237–40
Angles
 of DNA bending, 76
 specification of atom-atom, 16
Angular correlations
 O_{4^-}, 171–76
Angular distributions
 molecular-frame photoelectron, 167–68
 photoelectron-photoion, 182–85
 time-resolved photoelectron, 271–73
Angular measurements
 in biomolecular solid state NMR, 585–87
Angular resolution
 limited, 159
Anharmonic mean square displacements, 335
Anharmonic resonance, 625
Anion TRPES experiments, 268
Anionic transition metal clusters
 bond energies of, 436–37
Anisotropic MF-PADs, 175

901

Anisotropies
  dipole-dipole, 484
  first- and second-order, in solids NMR, 467–69
  polarization, 236
  selective reintroduction of spin, 483–93
Anomalous migration
  in oligomeric DNA, 99–105
Antiphase coherences
  single-quantum, 470
AOM
  See Acousto-optic modulator
APDs
  See Avalanche photodiode detectors
APLIP
  See Adiabatic passage by light-induced potentials
Aqueous-phase-air interfaces
  environmentally relevant, 372–73
Aqueous surfaces
  biologically relevant systems, 380–85
  environmentally relevant aqueous-phase-air interfaces, 372–73
  hydrogen bonding of interfacial water at, 361–68
  structure and bonding of molecules at, 357–89
  surfactant adsorption at, 368–72
  surfactant studies at liquid surfaces, 373–80
  vibrational sum frequency spectroscopy of, 358–61
Arrhenius-like behavior, 131
Astigmatic Herriott cells, 51
Atom-atom distances and angles
  reactions requiring specification of, 16

Atom optics, 799–800
Atomic force microscopy (AFM)
  time-resolved, 556
Atomic force microscopy (AFM) of DNA, 71–92
  biotechnological and other novel applications for, 83–84
  in cancer research, 77
  condensation and enzymology, 81–83
  condensed, 77–80
  conformations, binding sites, and stoichiometries of DNA-protein interactions, 75–77
  for determining Michaelis-Menten kinetics, 83–84
  fast, 87
  fine-resolution, 81–83
  in fluid, 81
  mapping, sizing, and sequence recognition, 73–75
  new technology for, 86–87
  nonimaging, 85–86
  for polymerase assays, 83
  ways of preparing samples for, 84–85
Atomic force microscopy (AFM) of gold, 118–19
Atomic force microscopy (AFM) of silane, 119–22
Atoms
  reactions with diatoms, 15–16
Auger mechanism, 211, 222, 225
  inverse, 196
Autocorrelation analysis
  fluorescence, 247
Autodetachment dynamics, 266
Automated DNA sizing, 73
Avalanche photodiode detectors (APDs), 238

B
Bardeen's transmission formula, 691, 708
Barrier heights of reactions, 16, 25, 31, 60
  vibrationally adiabatic, 25
BDEs
  See Bond dissociation energies
Beer's law, 47
Biased sampling
  generation of initial conditions for, 507–8
  of the initial conditions, 508–9
Bimolecular reactions
  free-bound, 649–52
  rate constants for, 54
Binary complexes
  dissociation energies for, 632–34
  HF-HCl and HCl-HF, 621–23
Binding
  nanomechanical motion due to biopolymer, 557–62
Binding sites
  of DNA-protein interactions, 75–77
  See also Unbinding forces
Biological single-molecule spectroscopy, 234
Biologically relevant systems at aqueous surfaces, 380–85
Biomimetic recognition between polymers and surfaces, 539–57
Biomolecular solid state NMR
  angular measurements, 585–87
  applications to peptide and protein fibrils, 575–606

SUBJECT INDEX 903

distance measurements, 583–84
multiple quantum spectroscopy, 589
multiplying labeled samples, 587–89
nuclear spin interactions, 577–78
powder patterns, magic angle spinning, and recoupling, 578–83
spectroscopic background, 577–90
spectroscopy of oriented samples, 589–90
Biomolecules
freely diffusing, 233–53
Biopolymer binding and adsorption
nanomechanical motion due to, 557–62
Biopolymers, 555
Biotechnological applications for AFM, 83–84
Birefringence method
picosecond transient, 665
Bloch functions, 204
Blue shifting, 267
Boltzmann distributions, 48
Bond dissociation energies (BDEs), 430, 432, 434, 436, 439–40, 443, 445, 449–50
Bond energies
of anionic transition metal clusters, 436–37
of cationic transition metal clusters, 430–34
of cluster-deuterides, 443–45
of cluster-oxides, 449–50, 452–54
of neutral clusters, 434–36
Bond-selective reactions
in polyatomic molecules, 19
Bonds

of molecules at aqueous surfaces, 357–89
numbers of among reactants, 16
See also Weakly bound complexes
Born-Oppenheimer PES, 149
Born solvation energy, 395
Bose-Einstein condensation, 802
Bound-bound molecular transitions
vibrational dynamics and coherence of, 652–61
Bound-free transitions
concerted-elimination chemical reactions, 643–49
Bowtie model
for multiple level crossings, 782
Bragg diffracts, 663
Bridge Hamiltonian, 715
Bulk phase thermodynamics
of transition metal clusters, 439–40
Bulk transport
in growth of self-assembled monolayers, 108
Burst analysis, 247–48
Büttiker-Landauer approach, 710–11

C

Cahn-Hilliard formalism, 563
Calibration
of IR absorption signals, 46–49
Cancer research
atomic force microscopy in, 77
Cantilever deflections
induced by intermolecular forces, 86
Carrier-carrier interactions
in molecular conduction, 706–7

Carrier cooling, 195–96, 221
Cartesian tensors, 467
Cationic transition metal clusters
bond energies of, 430–34
Cavity quantum electrodynamics, 801–2
Cavity ringdown (CRD) spectroscopy, 51
CCD
See Charge-coupled-device-based detection
Center-of-mass (CM) frame, 169–70
CH stretching, 382, 816
Chain length effects
on SAM growth mechanisms, 130–31
Chain mismatch, 384
Charge-coupled-device (CCD)-based detection, 166, 171, 189, 361
Charge-transfer-to-solvent (CTTS) bands, 268
CHARMM molecular dynamics program, 520
Chemical force microscopy, 73
Chemical reaction dynamics, 139
of endoergic reactions, 16–20
involving abstraction of a hydrogen atom from an alkane, 31–36
of polyatomic reactions, 25–31
of reactions proceeding by direct abstraction, 21–25
state-to-state, in polyatomic systems, 15–39
STIRAP control of, 798
Chemical shift
correlations with

heteronuclear coupling, 484–88
correlations with homonuclear coupling, 488–93
Chemical shift anisotropy (CSA) tensor, 577
Chemisorption
  Newns-Anderson model of, 696
Chirped excitation of a manifold of closely spaced levels for multiple level crossings, 781–82
Chromosomes
  AFM analysis of, 80
Chymotrypsin inhibitor 2, 518
CID
  See Collision-induced dissociation processes
Cl + $CH_4(\nu_i)$
  mode-specific reactivity, 157–58
Cl + RH → HCl + R
  chemical reaction dynamics of, 25–31
Cl(^2P) + H_2
  spin-orbit reactivity, 152–54
Clusters, 828–30
  electron solvation dynamics in, 266
  small transition metal ions, 423–61
  time-resolved photoelectron spectroscopy of, 255–77
  TRPES studies of neutral, 259–66
Clusters with CO_2
  bond energies of cluster-oxides, 452–54
  reaction pathways, 450–52
  reactions of, 450–54
Clusters with D_2
  bond energies of cluster-deuterides, 443–45
  probes of electronic structure, 445–46
  reaction pathways, 441–43
  reactions of, 440–46
Clusters with O_2
  bond energies of cluster-oxides, 449–50
  reaction pathways, 447–49
  reactions of, 447–50
CM
  See Center-of-mass frame
CN + O_2
  product branching measurements using infrared absorption, 57–62
CO dephasing
  protein structural dynamics and, 325–31
CO_2-HF binary complex, 627
Coherence
  double quantum, 585
Coherent excitation
  in two-state systems, 768–69
Coherent IR pulse sequences, 317
Coherent momentum transfer, 800
Coherent nonlinear spectroscopy, 639–79
Coherent superpositions
  creation of by modifications of STIRAP, 798
Coincidence spectroscopy, 165–92
Coincidence studies of dissociative photodetachment, 171–79
  energy and angular correlations, 171–76
  transition-state dynamics, 176–79
Coincidence studies of dissociative photoionization, 180–82
Cold shock protein A, 515, 519
Cold target recoil ion momentum spectroscopy (COLTRIMS), 182
Collision energies
  accessing above the reaction threshold, 16
Collision-induced dissociation (CID) processes, 428–29, 432, 434, 436, 441, 449
Collisional vibrational relaxation, 44
Collisions
  large impact parameter grazing, 27
  See also Local collision model
Colloidal gold
  initial synthesis of, 213
Colloidal quantum dots
  synthesis of, 213–14
COLTRIMS
  See Cold target recoil ion momentum spectroscopy
Comparison studies of protein folding/unfolding, 523–27
  with detailed model unfolding trajectories and folding potentials of mean force, 526–27
  with potentials of mean force generated from biased sampling, 525–26
  using unfolding simulations to study folding, 523–25
Complex conformational-energy landscape, 316

Concentration dependence
of ECL response, 403, 406
Concerted-elimination
chemical reactions
bound-free transitions,
643–49
Condensed DNA
compact, 80
for gene therapy, 79–80
imaging by AFM, 77–83
in solution, 77–79
surface-directed, 78–79
Conducting leads
transmission between,
689–95
Conduction
highest unoccupied
molecular orbital
(HUMO)-dominated, 699,
703–7
lowest unoccupied
molecular orbital
(LUMO)-dominated, 699,
703–7
Conformations
of DNA-protein
interactions, 75–77
Constant-time
double-quantum-filtered
dipolar evolution
(CTDQFD), 584, 595
Contact radical-ion pairing
(CRIP), 396
Continuum
population transfer via,
795–96
Copolymers
adsorption of diblock, 538
molten diblock, 538
"Core extraction," 28
Correlated radical pairs
transient EPR, 299–301
Correlation spectroscopy
(COSY)
free induction decay-based,
298
COSY

See Correlation
spectroscopy
Coulomb blockade
phenomena, 707
Coulomb interactions, 212
Coulomb repulsion
renormalized, 759
Counterintuitive pulse
ordering, 787
Counterion-mediated collapse
overcoming electrostatic
repulsion by, 757–60
Coupling Hamiltonians, 466,
468
Coverage kinetics
functional form of SAM
growth mechanisms,
128–29
CP
See Cross polarization
CRD
See Cavity ringdown
spectroscopy
CRIP
See Contact radical-ion
pairing
Cross-linking, 121
Cross polarization (CP)
double-resonance, 478–79
to half-integer quadrupoles,
480–82
pulsed, 479
in solid state NMR, 478–82
transfers between spin-$\frac{1}{2}$,
478–80
Cross-reactions, 401
Cross sections
state-specific differential,
139–64
Crossed beam scattering
processes, 836
Crossed-molecular-beam
spectroscopy
studies of neutral reactions,
139–64
Cruciforms
in DNA plasmids, 74–75

Cryo-electron microscopy,
591
CSA
See Chemical shift
anisotropy tensor
CTDQFD
See Constant-time
double-quantum-filtered
dipolar evolution
CTTS
See Charge-transfer-to-
solvent
bands
Current/voltage behavior
light-emitting
electrochemical cells with
and without rectifying,
404–10
Cw-ENDOR, 284
Cw lasers
experimental
demonstration of STIRAP
with, 787–88
transient digitization of,
43
Cyclotron resonance
techniques, 425

D

D_2- HF intermolecular energy
transfer, 619–21
DAC
See Dodecylammonium
chloride studies
Dark channel
investigations of, 839
Data acquisition and analysis
advanced, 246–49
ratiometric, 237–40
DBA
See Donor-bridge-acceptor
system
DBS
See Sodium
dodecylbenzenesulfonate
studies
DCS

See Differential cross
 sections
DDS
 See Sodium dodecyl
  sulfonate studies
Debye-Huckel theory
 for electrophoretic mobility
  of oligomeric DNA,
  95–97
Decay
 free-induction, 321
 vibrational echo, 318
Decay-based correlation
  spectroscopy
 free induction, 298
Decoupling
 heteronuclear, 472
Degenerate levels
 STIRAP with, 791–92
Delocalization
 electron, 268
 transition state, 511
Density functional
  calculations
 in molecular conduction,
  702–3
Density matrix theory of
  relaxation, 5
Density of states
 in quantum dots, 212–13
Dephasing
 fluctuating-electric-field,
  330
 homogeneous, 321
 maximum, 484
 optical, 334
 vibrational, 334
Dephasing and relaxation
  effects, 709–30
 nuclear relaxation during
  electron transmission,
  712–16
 reduced density matrix
  approaches, 719–30
 thermal interactions in
  molecular conduction,
  716–19

tunneling traversal times,
 709–12
Dephasing time
 pure, 321
 rates, 336
Designer oligonucleotides, 83
Desorption mechanism
 trapping, 840
Deuterated alkanes, 26
DFG
 See Difference-frequency
  generation
DHPs
 See Disordered
  heteropolymers
Diatoms
 reactions with atoms,
  15–16
Diblock copolymers
 adsorption of, 538
 molten, 538
Diblock melts
 near patterned surfaces,
  ordering of, 562–68
Difference-frequency
  generation (DFG), 50
 molecular-frame, 168
Differential cross sections
  (DCSs)
 bulb studies, 158–59
 doubly, 152
 experimental
  methodologies, 140–58
 mode-specific, 157
 molecular-frame, 187–89
 ro-vibrational
  state-resolved, 155
 state-specific, 139–64
Diffusing biomolecules,
 233–53
Diffusion FRET
 methodology, 236
Diffusion methods
 radiometric, 235–36
Dilauroyl-PC (DLPC)
 studies, 382
Dimethyl sulfoxide (DMSO)

 studies, 372–73
Dimyristoyl-PC (DMPC),
 382
Diode laser detection, 49
 See also Avalanche
  photodiode detectors
Diodes
 organic light-emitting,
  391–94, 411–19
Dip-pen nanolithography,
 134
Dipalmitoyl-PC (DPPC), 382
Dipolar recovery at the magic
  angle (DRAMA)
  sequence, 489–92, 580,
  582
Dipolar recovery with a
  windowless sequence
  (DRAWS), 490, 595
Dipole-dipole anisotropies,
 484
Dirac delta function, 545
Direct abstraction
 chemical reaction
  dynamics of reactions
  proceeding by, 21–25
Disease conditions, 591–92
Disordered heteropolymers
 (DHPs), 540–44,
 546–50, 552–54, 557
Disposal
 See Energy disposal
Dissociation
 electric field, for
  state-selection, 9
 vibrationally induced, 631
Dissociation asymptote
 ion beam studies of energy
  levels near, 9
Dissociation dynamics
 nonstatistical unimolecular,
  825–27
 three-body, 168, 185–89
Dissociation energies for
  binary complexes,
  632–34
Dissociation pathways

in transition metal clusters, 429–30
Dissociation processes
 collision-induced, 428–29, 432, 434, 436
Dissociative adsorption
 vibrational promotion of, 844–45
Dissociative photodetachment (DPD), 167, 170
 coincidence studies of, 171–79
Dissociative photoionization (DPI), 167
 coincidence studies of, 180–82
Distance dependence
 Förster, and subpopulations, 241–42
Distance measurements
 in biomolecular solid state NMR, 583–84
Distances
 reactions requiring specification of atom-atom, 16
Distearoyl-PC (DSPC), 382
DLPC
 See Dilauroyl-PC studies
DMBE
 See Double many-body expansion
DMPC
 See Dimyristoyl-PC
DMSO
 See Dimethyl sulfoxide studies
DNA
 automated sizing, 73
 bending and bend angles of, 76
 condensates of, 72, 78–79
 migration of oligomeric, 93–106
 preparing surfaces for binding, 84
 sequence-specific

structures, 73–74
 stretched, 83
 surface biology of, 71–92
 targeted conformational changes, 245
DNA constructs, 83
DNA crystals
 two-dimensional, 83
DNA imaging by AFM, 73–85
 biotechnological and other novel applications for, 83–84
 condensation and enzymology, 81–83
 condensed, 77–80
 conformations, binding sites, and stoichiometries of DNA-protein interactions, 75–77
 fine-resolution, 81–83
 in fluid, 81
 mapping, sizing, and sequence recognition, 73–75
 preparing surfaces for DNA binding, 84
 ways of preparing samples, 84–85
DNA molecules
 equilibrium vs kinetic trapping on surfaces, 84–85
 orienting and elongating on surfaces, 84
 stretching single, 85–86
DNA plasmids
 cruciforms in, 74–75
DNA-protein interactions
 conformations, binding sites, and stoichiometries of, 75–77
Dodecylammonium chloride (DAC) studies, 376
Dodecyltrimethylammonium chloride (DTAC) studies, 376

Donor-bridge-acceptor (DBA) system, 696, 714
Dopants
 molecular, for multilayer OLEDs, 415–17
Doppler broadening, 47, 660
Doppler-resolved
 laser-induced fluorescence, 146, 158
Dots
 See Quantum wells and quantum dots
Double many-body expansion (DMBE), 143
Double quantum (DQ)
 coherence, 585
 filtering, 584
DPD
 See Dissociative photodetachment
DPI
 See Dissociative photoionization
DPPC
 See Dipalmitoyl-PC
DQ
 See Double quantum
DRAMA
 See Dipolar recovery at the magic angle sequence
DRAWS
 See Dipolar recovery with a windowless sequence
ds DNA, 560
 gel electrophoresis on, 102, 105
DSPC
 See Distearoyl-PC
DTAC
 See Dodecyltrimethylammonium chloride studies
DUMP transitions
 See PUMP-DUMP-PROBE geometry
Durosemiquinone cation
 hyperfine line-width

alternation in, 4–5
Dye molecules
   photoexcitation of, 224–25, 238
Dynamic isomerization
   in the durosemiquinone cation, 5
Dynamical resonance
   quantum, in $F + H_2$, 147–52
   See Chemical reaction dynamics

E

Echo experiments
   infrared vibrational, 315–56
ECL
   See Electrogenerated chemiluminescence
Effective mass approximation (EMA) model, 212–13
EL
   See Electrogenerated luminescence
Elastic Rayleigh scattering, 715
Electric couplings, 465
Electric field dissociation
   for state-selection, 9
Electrochemical cells
   light-emitting, 392–410
Electrochemical characterization
   of hole- and electron-transport materials of multilayer OLEDs, 411–15
   of molecular dopants for multilayer OLEDs, 415–17
   of single-layer light-emitting polymers, 410–11
Electrochemical processes
   light-emitting, 391–422
Electroendosmosis flows

(EOFs) causes of, 97
Electrogenerated chemiluminescence (ECL), 394–404
   concentration dependence of, 403, 406
   in model systems, 398–404
   solution-based, 394–98
Electrogenerated luminescence (EL), 391–92
Electron-conducting polymers, 225
Electron delocalization, 268
Electron nuclear double resonance (ENDOR) experiments, 280
Electron paramagnetic resonance (EPR) spectroscopy
   pulsed, 279–313
Electron solvation dynamics
   in clusters, 266
Electron spin echo envelope modulation (ESEEM), 280, 288–94
   multifrequency, 294
Electron spin resonance (ESR) spectrometry
   liquid phase, 5
   pioneering work in, 2–4
Electron-transfer rates
   in molecular conduction, 696–97
Electron-transfer theory
   in molecular conduction, 684–89
Electron transmission
   through molecules and molecular interfaces, 681–750
Electron-transport layers (ETLs), 411, 413
Electron-transport materials
   of multilayer OLEDs, 411–15

Electron tunneling through water, 730–41
Electronic Hamiltonians, 713
Electronic ladder climbing for multiple level crossings using a single pulse, 776–79
Electronic nonadiabatic transition, 152
Electronic states
   in quantum dots, 213
Electronic structure
   probes of clusters with D_2, 445–46
   of transition metal clusters, 437–38
Electrophoretic mobility of oligomeric DNA, 95–105
   anomalous migration, 99–105
   extended Debye-Huckel theory for, 95–97
   in free-solutions, 97–99
   plotting, 99
Electrostatic repulsion
   overcoming by counterion-mediated collapse, 757–60
Elementary gas-phase reaction kinetics
   infrared absorption techniques for, 41–70
Elongating DNA molecules
   on surfaces, 84
EMA
   See Effective mass approximation
Emission pumping
   stimulated, 765–66
Endoergic reactions
   chemical reaction dynamics of, 16–20
ENDOR
   See Electron nuclear double resonance experiments

Energetic correlations,
  166–67
O_{4^-}, 171–76
Energy disposal
  rovibrational, 36
  vibrational, 62–65
Energy distributions
  photoelectron-photoion,
    182–85
Energy levels
  near a dissociation
    asymptote, 9
  in quantum dots, 212–13
Energy splittings
  Zeeman, 284, 291, 303–4
Energy transfer
  vibrational, 48
Environmental relevance
  of aqueous-phase-air
    interfaces, 372–73
Enzymology
  DNA imaging of by AFM,
    81–83
EOFs
  See Electroendosmosis
    flows
Epitaxial growth
  Stranski-Krastinow,
    214–16
EPR
  See Electron paramagnetic
    resonance spectroscopy
ESEEM
  See Electron spin echo
    envelope modulation
ESR
  See Electron spin
    resonance spectrometry
ETLs
  See Electron-transport
    layers
Excitation, 155, 157
  See also Photoexcitation;
    Pulsed excitation;
    Resonant coherent
    excitation; Rotational
    excitation; Stretched

molecules; UV excitation;
  Vibrational excitation
Experimental DCS
  methodologies, 140–58
  $Cl + CH_4(v_i)$, 157–58
  $Cl(^2P) + H_2$, 152–54
  $F + H_2$, 147–52
  $H + H_2$, 143–47
  $O(^1D) + H_2$, 154–57
Experimental demonstration
  of STIRAP
  with cw lasers, 787–88
  with pulsed lasers, 789–91
Experimental probes
  manipulating growth
  mechanisms with, 133–34
Extended Debye-Huckel
  theory
  for electrophoretic mobility
  of oligomeric DNA,
    95–97

F
$F + H_2$
  quantum dynamical
    resonance, 147–52
Fabrication
  of quantum wells and
    superlattices, 202–3
Far-infrared laser magnetic
  resonance, 8
Fast atomic force microscopy
  imaging, 87
Fast protein dynamics
  probed with infrared
    vibrational echo
    experiments, 315–56
FC
  See Franck-Condon
    overlap
FEL
  See Free-electron laser
Femtosecond dynamics, 639
Femtosecond resolution,
  255–56, 262, 264,
  267–68
Femtosecond stimulated

emission pumping, 266,
  361
Femtosecond transition state
  spectroscopy, 643
Fermi distribution functions,
  708
Fermi's Golden Rule, 821–22
Feshbach resonance, 149
Fibrils
  amyloid, 591–96
  other protein, 597–98
Films
  defect structure in, 113
Filtering
  double quantum, 584
Final state distributions
  inter-fragment correlated,
    607–37
Fine-resolution AFM
  of helix turns on DNA,
    81–83
First-order anisotropies
  in solids NMR, 467–69
  spatial, 474
First-order couplings
  averaging via spatial space
    manipulations, 474–76
First-order Hamiltonians, 468
Flash kinetic spectroscopy,
  42–43
FLN
  See Fluorescence line
    narrowing
Floquet theory, 471
Flory interaction, 563
Fluctuating-electric-field
  dephasing mechanism,
  330
Fluctuation analysis, 247–48
Fluorescence autocorrelation
  analysis, 247
Fluorescence detection
  laser-induced, 24, 255
  observables in, 236–37
  polarization of, 237
  single-molecule, 234–35
Fluorescence line narrowing

(FLN), 217, 219
Fluorescence resonance energy transfer (FRET) measurements, 236, 238–43, 247, 249, 756–57
histograms of, 249
Folding
under native conditions, 523
RNA, early events in, 751–62
simulation studies of protein, 499–535
unfolding simulations used to study, 523–25
Folding landscapes
characterizing, 519–27
Folding potentials of mean force
unfolding trajectories and, 526–27
Folding transition temperature
free energy profiles relative to, 522–23
Förster distance dependence and subpopulations, 241–42
Förster radius, 241
Förster resonance energy transfer, 236
Four-wave mixing (FWM), 641, 643, 653–57, 661
Fourier transform (FT) spectroscopy, 280, 361
Fourier-transform-limited Gaussian pulse, 824
Fragment B of staphylococcal protein A, 510–12, 526–27
Franck-Condon bright states (FCBS), 817–18, 821
Franck-Condon (FC)
active modes, 820
factors, 729, 765, 797, 813, 843

matrix, 813
overlaps, 166–67, 175, 257, 651–52, 717
weighted density, 697
Free-bound bimolecular reactions
photoassociation, 649–52
Free-electron laser (FEL), 323, 348
Free energy profiles
relative to folding transition temperature, 522–23
Free energy surfaces for protein folding, 505–15
potentials of mean force generated from, 505–9
Free-induction decay, 321
Free-induction decay-based correlation spectroscopy, 298
Free radicals, 11–12
microwave spectra of, 7
organic, 3
partial alignment in a nematic liquid crystal, 5
pioneering work on, 1–13
Freely diffusing biomolecules
ratiometric single-molecule studies of, 233–53
Frequency-modulation (FM) spectroscopy, 50–51
Fresnel coefficients, 363
FRET
See Fluorescence resonance energy transfer
Frumkin isotherm approach, 126
FT
See Fourier transform spectroscopy
FWM
See Four-wave mixing

G

Gas phase molecular ion studies, 8

Gas-phase radical reactions
measurements of rate coefficients for, 51–57
measuring second order rate constants of, 45
radical-molecule reactions, 51–54
radical-radical kinetics, 54–57
time-resolved infrared laser absorption spectroscopy for, 43–51
vibrational energy disposal in thermal, 62–65
Gas-phase reaction kinetics
infrared absorption techniques for elementary, 41–70
Gaussian pulse
Fourier-transform-limited, 824
Gel electrophoresis
on end-labeled ss DNA, 102
Ogstron model for polyacrylamide, 93–94
Gene therapy
DNA condensation for, 79–80
Geometric phase (GP) effects in $H + H_2$, 143–47
Geometric structures
of transition metal clusters, 437–38
Glass transitions, 331–37
"slaved," 335
Global emission peaks, 217
Global photoluminescence
optical absorption and, 216–17
Gold
initial synthesis of colloidal, 213
Gold surfaces
short DNA on, 85
thiol on, in growth of

self-assembled
  monolayers, 113–19
Gouy-Chapman model, 372
GP
  See Geometric phase
  effects
Green's function, 695–96,
  698–702, 731
Ground state dynamics
  off-resonance probing of,
  661–69
Ground state PESs, 155
Growth mechanisms
  manipulating with
    experimental probes,
    133–34
  quantitative aspects of,
    126–33
  of self-assembled
    monolayers, 108–26
Guided-ion-beam mass
  spectrometry, 425, 447,
  455

H
$H + CO_2 \rightarrow OH + CO$
  chemical reaction
    dynamics of, 21–25
$H + H_2$, 639
  geometric phase effects,
  143–47
$H + H_2O \rightarrow H_2 + OH$
  chemical reaction
    dynamics of, 16–20
  rotational manifolds of, 18
$H + RH \rightarrow H_2 + R$
  chemical reaction
    dynamics of, 31–36
H_2-HF intermolecular energy
  transfer, 619–21
Hahn spin echo, 659
Half-integer quadrupoles
  cross polarization to,
    480–82
Hamiltonians, 695
  average, 582
  bridge, 715

coupling, 466, 468
  effective, 542, 817
  electronic, 713
  first-order, 468
  heteronuclear dipole, 479
  imaginary
    boundary-condition, 738
  interaction-frame, 473
  Ising-like, 550
  matrix, 768
  microscopic, 567
  NMR, 471
  polaron-type, 728
  prototypic, 713
  rotating frame, 469
  single-spin, 470
  spin, 282–84
Hartree Fock (HF)
  calculations, 698–99
HCl dimer
  vs HF dimer, 612–19
HCN-HF mode specific
  effects, 627–31
Hemoglobin-CO
  vibrational echo
    experiments on, 347–51
Herriott design, 51
Heteronuclear coupling
  correlations with chemical
    shift, 484–88
Heteronuclear decoupling,
  472
Heteronuclear dipole
  Hamiltonians, 479
Heteropolymers
  disordered, 540–44
HF-HCl and HCl-HF binary
  complexes, 621–23
HF stretching, 614, 632
  See also Hartree Fock
  calculations
High-field/high-frequency
  EPR, 301–6
High resolution in solids
  NMR, 471–78
  averaging first-order
    couplings via spatial

space manipulations,
  474–76
averaging via spin-space
  manipulations, 471–74
avoiding penalties of,
  483–84
manipulating second-order
  quadrupolar interactions,
  476–78
High temperatures
  unfolding at, 524–25
Highest unoccupied
  molecular orbital
  (HUMO)-dominated
  conduction, 699, 703–7
Highly vibrationally excited
  molecules
  with stimulated emission
    pumping, 811–52
Histograms
  analysis via, 237–40
  FRET, 249
  moments of, 240–41
HITRAN database, 47
Hole-conducting polymers,
  225
Hole-transport layers (HTLs)
  vacuum-deposited, 411,
  413
Hole-transport materials
  of multilayer OLEDs,
  411–15
Homogeneous dephasing, 321
Homonuclear coupling
  correlations with chemical
    shift, 488–93
Homonuclear rotary
  resonance (HORROR)
  variant, 490–92
Homopolymers
  adsorbed, 538, 548
HORROR
  See Homonuclear rotary
    resonance variant
Hot-electron cooling
  in quantum wells and
    superlattices, 206–11

slowed, in quantum dots, 219–20
Hot-electron relaxation dynamics, 197–211
in semiconductor quantum wells and quantum dots, 193–231
HTLs
See Hole-transport layers (HTLs)
HUMO
See Highest unoccupied molecular orbital-dominated conduction
Hydrogen bonding of interfacial water, 361–68
"icelike," 363
VSFS of interfacial H_2O, 366–68
VSFS of the vapor-water interface, 361–63
water hydrogen bonding at the organic-liquid/water interface, 363–66
Hydrogen fluoride (HF) dimer vs HCl dimer, 612–19
Hydrogen Rydberg atom photofragment translational spectroscopy, 609
Hyper-Raman STIRAP, 796–97
Hyperfine line-width alternation
in the durosemiquinone cation, 4–5
Hyperfine structure
proton, in organic free radicals, 3
Hyperfine sublevel correlation experiment (HYSCORE), 290, 294
Hyperfine transitions
radiofrequency quadrupole trap for recording, 8
HYSCORE

See Hyperfine sublevel correlation experiment

I

"Icelike" hydrogen bonding, 363
ICS
See Integral cross section measurements
Impact ionization, 196
Impulsive stimulated Raman scattering
resonant, 266–67
Incoherent excitation
of two-state systems, 764
Incoherent population transfer schemes, 764–66
incoherent excitation of two-state systems, 764
optical pumping, 764–65
stimulated emission pumping, 765–66
Independently folding subdomains of *Tetrahymena* ribozyme kinetics of, 754–55
Inelastic Raman scattering, 716
Infrared absorption spectroscopy
of HCl, 52
product branching measurements using, 57–62
Infrared absorption techniques
for elementary gas-phase reaction kinetics, 41–70
Infrared laser flash kinetic spectroscopy, 43
experimental techniques, 49–51
increasing sensitivity of, 50
Infrared spectroscopy
transient, 58
Infrared vibrational echo experiments

probing fast protein dynamics with, 315–56
Inhomogeneous spectrum broadening, 320–21
Insertion reactions
prototypical, in $O(^1D) + H_2$, 154–57
Insulators
scattering dynamics of SEP from, 840–41
Integral cross section (ICS) measurements, 146, 151
Integrated absorption
determining, 45–46
Inter-fragment correlated final state distributions, 607–37
Interaction-frame Hamiltonian, 473
Interfaces
See Air-water interface; Aqueous-phase-air interfaces; Liquid-liquid interfaces; Organic-liquid/ water interface; Vapor-water interface
Interfacial water
hydrogen bonding of, 361–68
VSFS of, 366–68
Intermolecular energy transfer
D_2–HF and H_2–HF, 619–21
Intermolecular forces
inducing cantilever deflections, 86
Intramolecular motion
accounting for hyperfine line-width alternation, 4–5
Intramolecular vibrational redistribution (IVR), 815–24
of acetylene, 818–20
of HFCO, 820–21
of $SCCl_2$, 821–24

Intramolecular vibrational
    relaxation (IVR), 260,
    646, 652–53
Intramolecular vibrations,
    614, 616
Intuitive vs counterintuitive
    pulse order, 785–86
Ion beam studies, 8–10
  of energy levels near a
    dissociation asymptote,
    9
Ion cyclotron resonance
    techniques, 425
Ion studies
  vs neutral, 425–26
Ion-TOF measurement,
    157
Ionic oligomers
  extended Debye-Huckel
    theory for, 95–97
Ionization
  impact, 196
IR absorption signals
  calibration of, 46–49
Isaacson 5 potential, 17
Ising-like Hamiltonian,
    550
Islands
  dilute phase surrounding,
    126
  growth kinetics of, 119,
    121, 133
  number density of, 132–33
Isomerization
  in triatomics, 824–25
IVR
  See Intramolecular
    vibrational redistribution;
    Intramolecular vibrational
    relaxation

J
Junctions
  metal-insulator-metal,
    690
  metal-molecule-metal, 689,
    704–6, 709

K
Kinetic partitioning
    mechanism (KPM),
    753–54
Kinetics of independently
    folding subdomains of
    Tetrahymena ribozyme,
    754–55
Koopmans theorem, 699
KPM
  See Kinetic partitioning
    mechanism
Krylov space-based iterative
    methods, 732
KTP
  See Potassium titanyl
    phosphate (KTP)
    assembly

L
Labeled samples
  multiplying in
    biomolecular solid state
    NMR, 587–89
Ladder climbing for multiple
    level crossings
  electronic, using a single
    pulse, 776–79
  rotational, using a pair of
    pulses, 780–81
  vibrational, using a pair of
    pulses, 779–80
  vibrational, using a single
    pulse, 779
Lamellae
  perpendicular vs parallel,
    562–63
Landau-Ginzburg free energy
    functional, 563
Landau-Zener formula, 771,
    782
Landauer formula, 696, 698,
    716
  transmission between
    conducting leads, 692–95
Langmuir-Blodgett films,
    112, 687

Langmuir monolayers, 112,
    128–29
Large impact parameter
    grazing collisions, 27
Larmor frequency, 301, 304
Larmor precession, 801
Larmor velocity filter, 800
Laser cooling, 800
Laser flash kinetic
    spectroscopy, 43
Laser-induced fluorescence,
    255
  polarized Doppler-resolved
    detection using, 24
Laser-induced population
    transfer
  by adiabatic passage
    techniques, 763–809
  adiabatic population
    transfer using sequential
    pulses, 782–802
  adiabatic population
    transfer via level crossing,
    768–82
  incoherent population
    transfer schemes, 764–66
  resonant coherent
    excitation, 766–68
Lattice temperature, 195
LECs
  See Light-emitting
    electrochemical cells
LEET
  See Low-energy electron
    transmission
LEPs
  See Light-emitting
    polymers
Ligation assays, 75
Light-emitting diodes
  organic, 391–94, 411–19
Light-emitting
    electrochemical cells
    (LECs), 392–410
  electrogenerated
    chemiluminescence,
    394–404

with rectifying
current/voltage behavior,
407–10
thin-layer, 404–10
without rectifying
current/voltage behavior,
404–7
Light-emitting
electrochemical
processes, 391–422
Light-emitting polymers
(LEPs), 392
electrochemical
characterization of
single-layer, 410–11
Light-induced potentials
adiabatic passage by,
797–98
Line-narrowing
methodologies, 481
Liouville equation, 721–22
stochastic, 282
Liouville pathways, 657, 670
Liquid crystals
free radical partial
alignment in nematic, 5
Liquid-liquid interfaces
surfactant studies at,
376–80
Liquid state NMR, 576
Liquid surfaces
surfactant studies at,
373–80
Lithography
scanning probe, 133–34
LO
See Longitudinal optical
phonons
Local collision model, 35–36
Longitudinal optical (LO)
phonons, 195, 206, 211,
220
Low-energy electron
transmission (LEET),
682, 726, 741–42
Lowest unoccupied molecular
orbital (LUMO)-

dominated conduction,
699, 703–7, 742
LUMO
See Lowest unoccupied
molecular
orbital-dominated
conduction

M

Macromolecules
action of random forces on,
98
Magic-angle spinning (MAS)
in biomolecular solid state
NMR, 578–83
Magnetic bottle analyzers,
256
Manning counterion
condensation
framework, 99
Marcus theory for electron
transfer, 395
MAS
See Magic-angle spinning
Mass spectrometry
guided-ion-beam, 425,
447, 455
Mb-CO
vibrational echo decay of,
325–26
MBE
See Molecular beam
epitaxy
Mean square displacements
anharmonic, 335
Metal-insulator-metal (MIM)
junctions, 690
Metal-molecule-metal
(MMM) junctions, 689,
704–6, 709
Metallo-organic chemical
vapor deposition
(MOCVD), 197, 202–3,
214
Metals
small transition ion
clusters, 423–61

vibrational promotion of
electron transfer in
interactions with, 841–43
Methane sulfonic acid (MSA)
studies, 372–73
MF-DCSs
See Molecular-frame
differential cross sections
MF-DFG
See Molecular-frame
difference-frequency
generation
MF-PADs
See Molecular-frame
photoelectron angular
distributions
Michaelis-Menten kinetics
determining with atomic
force microscopy, 83–84
Microcantilever bending,
560–61
Microscopy
See Atomic force
microscopy; Chemical
force microscopy;
Cryo-electron
microscopy; Scanning-
probe-microscopy;
Scanning tunneling
microscopy
Microwave cavity
early developments using,
5–6
Microwave magnetic
resonance methods, 8
Microwave spectra of free
radicals
early studies of, 7
Migration of oligomeric DNA
in polyacrylamide gels and
in free solution, 93–106
MIM
See Metal-insulator-metal
junctions
Minibands, 199
MMM
See Metal-molecule-metal

SUBJECT INDEX 915

junctions
MOCVD
  See Metallo-organic
    chemical vapor deposition
Mode-selective reactions
  in polyatomic molecules,
  19
Mode-specific reactivity
  in Cl + CH$_4$(v_i), 157–58
Mode suppression
  stimulated photon echo
  and, 657
Modified metal
  vibrational energy transfer
  on, 844
Molecular beam epitaxy
  (MBE), 197, 202–3, 214
Molecular conduction
  carrier-carrier interactions,
  706–7
  density functional
    calculations, 702–3
  notations used, 684–85
  potential profiles, 703–5
  quantum chemical
    calculations, 698–99
  rectification, 705–6
  relation to electron-transfer
    rates, 696–97
  spatial-grid based
    pseudopotential
    approaches, 699–702
  standard electron-transfer
    theory, 684–89
  theoretical approaches to,
    684–709
  thermal interactions in,
    716–19
  transmission between
    conducting leads, 689–95
Molecular dopants
  for multilayer OLEDs,
    415–17
Molecular-frame
  difference-frequency
    generation (MF-DFG),
    168

Molecular-frame differential
  cross sections
  (MF-DCSs), 187–89
Molecular-frame
  photoelectron angular
  distributions
  (MF-PADs), 167–68,
  180–81, 183, 186
  anisotropic, 175
Molecular motors, 557
Molecular switches, 74–75
Molecular transitions
  bound-bound, 652–61
Molecules
  electron transmission
    through, 681–750
  highly vibrationally
    excited, 811–52
  optical centrifuge for, 781
  time-resolved
    photoelectron
    spectroscopy of, 255–77
  TRPES studies of neutral,
    259–66
Monolayers
  formation mechanisms and
    kinetics of
    self-assembled, 107–37
  Langmuir, 112
  phosphatidylcholine,
    380–85
MQ
  See Multiple quantum
MQW
  See Multiple quantum well
    structures
MSA
  See Methane sulfonic acid
    studies
Multi-state chains
  STIRAP in three-state
    systems, 792–94
Multifrequency ESEEM, 294
Multilayer organic
  light-emitting diodes
  (OLEDs), 411–17
  electrochemical

characterization of hole-
  and electron-transport
  materials of, 411–15
electrochemical
  characterization of
  molecular dopants for,
  415–17
Multipass cells, 51
Multiple level crossings
  for adiabatic population
    transfer, 776–82
  bowtie model, 782
  chirped excitation of a
    manifold of closely
    spaced levels, 781–82
  electronic ladder climbing
    using a single pulse,
    776–79
  rotational ladder climbing
    using a pair of pulses,
    780–81
  vibrational ladder climbing
    using a pair of pulses,
    779–80
  vibrational ladder climbing
    using a single pulse, 779
Multiple quantum (MQ)
  coherences, 589
  correlating numbers of
    products, 16
  transition, 477
Multiple quantum
  spectroscopy, 589
Multiple quantum well
  (MQW) structures, 199,
  206, 208
Multiplying labeled samples
  in biomolecular solid state
  NMR, 587–89
Mutant studies
  protein structural dynamics
    and CO dephasing,
    325–31

N

N$_2$- HF energy transfer,
  623–26

Nanomechanical motion
  due to biopolymer binding
  and adsorption, 557–62
Nanoshaving/nanografting,
  134
Nanostructures
  polymer adsorption driven
  self-assembly of, 537–73
Nanotube tips
  single-walled, 86–87
Near-field optical
  microscopy, 236
Negative ions
  time-resolved
  photoelectron
  spectroscopy of, 266–70
Negative time
  reverse transient grating
  measurements, 660–62
Nematic liquid crystals
  free radical partial
  alignment in, 5
Neutral clusters
  bond energies of, 434–36
  TRPES studies of, 259–66
Neutral reactions
  $Cl + CH_4(v_i)$, 157–58
  $Cl(^2P) + H_2$, 152–54
  crossed-beam studies of,
  139–64
  $F + H_2$, 147–52
  $H + H_2$, 143–47
  $O(^1D) + H_2$, 154–57
Neutral studies
  vs ion, 425–26
Newns-Anderson model of
  chemisorption, 696
NMR Hamiltonians, 471
NMR interactions
  as scalar products between
  spin and spatial tensors,
  464–67
  See also Nuclear magnetic
  resonance spectrometry
NMR spectra
  shift-based, 473
  spin evolution and the
  calculation of, 469–71
NO_2
  photoelectron-photoion
  energy and angular
  distributions, 182–85
NO dimer, 626
NO-HF-V-V energy transfer,
  623–26
NO stretching, 812
Non-Markovian dynamics,
  248
Non-RRKM behavior, 825,
  827
Nonadiabatic processes, 626
Nonadiabatic transition
  electronic, 152
Nonequilibrium vibrational
  distributions, 47
Nonstatistical unimolecular
  dissociation dynamics,
  825–27
Notations used
  in molecular conduction,
  684–85
Nuclear magnetic resonance
  (NMR) spectrometry,
  279–80, 296, 755
  early work on, 2
  quadrupolar, 480
  spin-$1/2$ in solid state,
  463–98
  spin echo experiments
  using, 315–16
  See also Biomolecular
  solid state NMR;
  Quadrupolar NMR; Solid
  state NMR spectroscopy;
  Solids NMR
Nuclear Overhauser effect
  spectroscopy, 659–60
Nuclear relaxation during
  electron transmission,
  712–16
Nuclear spin interactions
  in biomolecular solid state
  NMR, 577–78
Nucleation centers, 213
Nucleosomes
  AFM analysis of, 80

O

O_4^-
  energy and angular
  correlations of, 171–76
OC-HF energy transfer,
  623–26
Octadecylphosphonic acid
  (OPA) growth, 122–25
Octadecyltrimethylammonium
  bromide (OTAB)
  growth, 123–26
$O(^1D) + H_2$
  prototypical insertion
  reaction, 154–57
Off-resonance probing
  of ground state dynamics,
  661–69
Off-resonance TG
  measurements
  time-resolved, 665
Ogstron model
  for polyacrylamide gel
  electrophoresis, 93–94
$OH + H_2O \rightarrow H_2O + OH$
  transition-state dynamics
  of, 176–79
OH adiabaticity, 18
OH stretching, 362–63, 368,
  371
OLEDs
  See Organic light-emitting
  diodes
Oligomeric DNA
  electrophoretic mobility of,
  95–105
  migration in
  polyacrylamide gels and
  in free solution, 93–106
Oligomers
  extended Debye-Huckel
  theory for ionic, 95–97
  mobility in tris-acetate-
  EDTA, 97
Oligonucleotides

designer, 83
OPA
 See Octadecylphosphonic acid growth; Optical-parametric-amplifier system
OPG
 See Optical parametric generation
OPO
 See Optical parametric oscillation
Optical absorption and global photoluminescence, 216–17
Optical centrifuge for molecules, 781
Optical dephasing, 334
Optical microscopy
 near-field, 236
Optical parametric amplifier (OPA) system, 323, 348, 360
Optical parametric generation (OPG), 360
Optical parametric oscillation (OPO), 360
Optical properties of quantum dots, 216–19
 optical absorption and global photoluminescence, 216–17
 origin of resonant red shift, 217–19
 size-selected photoluminescence, 217
Optical pumping, 764–65
Optical spectroscopy
 optical absorption, 203–5
 photoreflectance spectroscopy, 205–6
 of quantum wells and superlattices, 203–6
Organic acid and ion systems
 in growth of self-assembled monolayers, 122–26
Organic free radicals
 proton hyperfine structure in, 3
Organic light-emitting diodes (OLEDs), 391–94
 multilayer, 411–17
 two-layer small-molecule, 417–19
Organic-liquid/water interface
 water hydrogen bonding at, 363–66
Organometallic compounds, 398–99
Oriented samples
 in biomolecular solid state NMR, 589–90
Orienting DNA molecules
 on surfaces, 84
Ostwald ripening, 133, 213
OTAB
 See Octadecyltrimethylammonium bromide growth
Overbarrier transmission, 741–44
Overhauser effect, 659–60

P

PADs
 See Photoelectron angular distributions
Pairs of pulses
 rotational ladder climbing using, 780–81
 vibrational ladder climbing using, 779–80
Paramagnetic resonance spectra
 of crystals, 2
Partition factor (Φ), 754
 determination by single molecule fluorescence measurements, 756–57
Partitioning mechanism
 kinetic, 753–54

Passage techniques
 adiabatic, 763–809
PC
 See Phosphatidylcholine monolayers
PCRs
 See Polymerase chain reactions
PD
 See Photodissociation methods
PE
 See Photoemission
Peierls distortion, 718
PELDOR
 See Pulsed electron double resonance experiments
PEO
 See Polyethylene oxide
PEPICO
 See Photoelectron-photoion coincidence experiments
Peptide fibrils
 applications of biomolecular solid state NMR to, 575–606
 structural studies of, 590–98
PESs
 See Potential energy surfaces
Phase effects
 geometric, in H + H_2, 143–47
Φ
 See Partition factor
PHOFAD
 See Photofragment angular distribution experiments
Phonon bottleneck
 in quantum dots, 219–23
Phonons
 longitudinal optical, 195
Phosphatidylcholine (PC) monolayers, 380–85

Phospholipid assembly, 380–85
Photoassociation
  free-bound bimolecular reactions, 649–52
Photoassociative STIRAP, 802
Photodetachment spectroscopy, 149
Photodissociation (PD), 266, 429, 432, 434
  two-photon infrared method dependent on, 10
Photodissociation (PD) of cyclic (HF)n, 631–32
Photoelectron angular distributions (PADs)
  molecular-frame, 167–68, 185
Photoelectron (PE) spectroscopy
  time-resolved, 255–77
Photoelectron-photoion coincidence (PEPICO) experiments, 166, 169, 180–83
  threshold, 180
Photoelectron-photoion energy and angular distributions
  time-resolved dynamics of, 182–85
Photoemission (PE), 682
Photoexcitation
  of dye molecules, 224–25
Photofragment angular distribution (PHOFAD) experiments, 610–17, 621, 623, 625, 627–33
Photofragment translational spectroscopy
  hydrogen Rydberg atom, 609
  of weakly bound complexes, 607–37
Photogenerated carriers, 194–95

Photoluminescence
  global, 216–17
  size-selected, 217
Photolysis
  methodology using, 43–44
  pulsed, 42
Photon echo (PE) experiments, 316, 659–61
  dephasing measurements, 335
Photons
  Stokes, 765
Photoreflectance (PR) spectroscopy, 205–6
Physical chemistry
  contributions from biological AFM, 72–73
Picosecond resolution, 260–61, 363
Picosecond transient birefringence method, 665
PISEMA
  See Polarization-inversion with spin-exchange at the magic-angle sequence
PMF
  See Potentials of mean force
Polanyi rules, 845
Polarization
  of emitted fluorescence, 237
Polarization anisotropy, 236
Polarization-inversion with spin-exchange at the magic-angle (PISEMA) sequence, 487–88
Polarized Doppler-resolved detection, 19, 158
  using laser-induced fluorescence, 24
Polaron-type Hamiltonian, 728
Polyacrylamide gel electrophoresis

Ogstron model for, 93–94
Polyacrylamide gels and free solutions
  migration of oligomeric DNA in, 93–106
Polyatomic molecules
  chemical reaction dynamics of, 25–31
  mode-selective *versus* bond-selective reactions, 19
Polyatomic systems
  state-to-state chemical reaction dynamics in, 15–39
Polyethylene oxide (PEO), 408–9
Polymer adsorption driven self-assembly
  of nanostructures, 537–73
Polymer physics, 245–46
Polymerase assays
  using atomic force microscopy, 83
Polymerase chain reactions (PCRs), 73
Polymers
  amphiphilic, 538
  biomimetic recognition with surfaces, 539–57
  electron- and hole-conducting, 225
  light-emitting, 392, 410–11
Poly(*p*-phenylenes) (PPPs), 407
Poly(*p*-phenylenevinylene) (PPV) devices, 392, 407–8, 410–11
Population transfer
  laser-induced, 763–809
  via continuum, 795–96
Position sensitive translational spectroscopy (POSTS), 610
Positive time
  transient grating (TG)

method, 661
POSTS
　See Position sensitive translational spectroscopy
Potassium permanganate
　color studies on, 2
Potassium titanyl phosphate (KTP) assembly, 360–61
Potential energy surfaces (PESs), 140, 143–44
　Born-Oppenheimer, 149
　empirical, 147
　excited, 155
　ground state, 155
　singlet, 154
　Stark-Werner, 149
Potential profiles
　in molecular conduction, 703–5
Potentials of mean force (PMF)
　generated from biased sampling, 525–26
　generated from free energy surfaces for protein folding, 505–9
　shape and folding, 510–15
　unfolding trajectories and folding, 526–27
Powder patterns
　in biomolecular solid state NMR, 578–83
Power law temperature dependence, 334–35
PPC spectroscopy, 172, 175–77, 186
PPPs
　See Poly(p-phenylenes)
PPV
　See Poly (p-phenylenevinylene) devices
PR
　See Photoreflectance spectroscopy
Predissociation
　vibrational, 609–12

Pressure broadening, 47
PROBE transitions
　See PUMP-DUMP-PROBE geometry
Product angular distributions quantum-state-resolved, 27
Product branching measurements using infrared absorption, 57–62
Propargyl radical recombination, 56–57
Protein A
　fragment B of staphylococcal, 510–512, 526–527
Protein banding to DNA, 76
Protein dynamics
　fast, 315–56
　solvent viscosity and, 336–47
　structural, and CO dephasing, 325–31
Protein fibrils
　applications of biomolecular solid state NMR to, 575–606
　structural studies of, 590–98
Protein folding/unfolding, 242–45, 540
　comparison studies of, 523–27
　free energy surfaces for, 505–15
　historical perspectives, 502–5
　mean force and temperature induced, 519–27
　probing through unfolding simulations, 515–19
　simulation studies of, 499–535
Protein-glass transition temperature dependence, 331–36

Proton hyperfine structure
　in organic free radicals, 3–4
Prototypic Hamiltonians, 712
Prototypic insertion reactions
　in $O(^1D) + H_2$, 154–57
Pseudopotential approaches
　spatial-grid based, 699–702
Pulse
　Stokes, 766–67
Pulse order
　intuitive vs counterintuitive, 785–86
Pulsed cross polarization, 479
Pulsed electron double resonance (PELDOR) experiments, 280, 294–97
Pulsed EPR spectroscopy
　biological applications, 279–313
　electron spin echo envelope modulation, 288–94
　high-field/high-frequency EPR, 301–6
　pulsed electron double resonance, 294–97
　pulsed ENDOR, 284–88
　pulsed EPR/relaxation measurements, 297–99
　spin Hamilton operator, 282–84
　technical aspects, 280–82
　transient EPR/correlated radical pairs, 299–301
Pulsed excitation, 195
Pulsed lasers
　experimental demonstration of STIRAP with, 789–91
　photolytic generation of translationally fast H atoms, 16
Pulsed photolytic methods, 42
PUMP-DUMP-PROBE
　geometry, 814–15, 825, 828–29, 831–36
Pure-dephasing temperature

dependence, 327–28, 331
Pure dephasing time, 321, 337

Q

QCM
  See Quartz crystal microbalance measurements
QCT
  See Quasi-classical trajectory calculations
QDs
  See Quantum dots
QM
  See Quantum mechanical calculations
QPM
  See Quasi phase-matched materials
Quadrupolar interactions
  manipulating second-order, 476–78
Quadrupolar NMR, 480
Quadrupole trap
  radiofrequency, 8
Quadrupoles
  cross polarization to half-integer, 480–82
Quantitative aspects of SAM growth mechanisms, 126–33
  adsorption energetics of, 131
  chain length effects, 130–31
  functional form of coverage kinetics, 128–29
  rate constants and time scales, 126–28
  solvent effects, 129–30
  submonolayer island nucleation, growth, and size distributions, 132–33
Quantum calculations, 17

chemical, in molecular conduction, 698–99
Quantum dots (QDs)
  electronic states in, 213
  energy levels and density of states in, 212–13
  experimental determination of relaxation/cooling dynamics and phonon bottleneck in, 220–23
  hot electron relaxation dynamics in semiconductor, 193–231
  optical properties of, 216–19
  phonon bottleneck and slowed hot-electron cooling in, 219–20
  relaxation dynamics of hot electrons in, 211–26
  synthesis of, 213–16
  3-D, 224
Quantum dots solar cells, 223–26
  photoelectrodes composed of quantum dot arrays, 224
  quantum dot-sensitized nanocrystalline TiO_2 solar cells, 224–25
  quantum dots dispersed in organic semiconductor polymer matrices, 225–26
Quantum dynamical resonance
  in $F + H_2$, 147–52
Quantum electrodynamics
  cavity, 801–2
Quantum mechanical (QM) calculations, 140, 143–44
  multisurface, 152
  wave packet, 155
Quantum numbers of products
  correlating multiple, 16
Quantum-state-resolved

product angular distributions, 27
Quantum wells (QWs), 197–202
  fabrication of, 202–3
  hot electron cooling dynamics in, 206–11
  hot electron relaxation dynamics in semiconductor, 193–231
  multiple, 199
  optical spectroscopy of, 203–6
  relaxation dynamics of hot electrons in, 197–211
Quartz crystal microbalance (QCM) measurements, 114, 116
Quasi-classical trajectory (QCT) calculations, 17, 19, 27–28, 140, 143–44, 151, 155
Quasi phase-matched (QPM) materials, 50
Quenched disorder, 542
QWs
  See Quantum wells

R

Rabi cycling, 771
Rabi frequencies, 768, 777, 783, 785
Rabi oscillations, 766–68
Radiationless transitions, 259
Radical-molecule reactions
  gas-phase, 51–54
  See also Free radicals
Radical pairs
  correlated, 299–301
Radical-radical reactions
  kinetics of gas-phase, 54–57
  minimizing competition of, 50
Radical reaction kinetics, 42
Radical recombination
  propargyl, 56–57

SUBJECT INDEX 921

Radio-frequency-driven recoupling (RFDR), 582
Radiofrequency quadrupole trap
 recording hyperfine transitions, 8
Radiometric diffusion methods, 235–36
Raman-active mode, 665–66
Raman-chirped adiabatic passage (RCAP), 779–80
Raman-induced polarization spectroscopy, 664
Raman scattering, 362–63, 367, 641, 644
 inelastic, 716
 resonant impulsive stimulated, 266–67
Random energy model (REM), 544, 547
Rapid adiabatic passage, 770–71
Rate coefficients
 absolute, 52
 measurements of, for gas-phase radical reactions, 51–57
 thermal, 53
Rate constants
 bimolecular, 54
 of SAM growth mechanisms, 126–28
Ratiometric data acquisition, 237–40
Ratiometric single-molecule studies
 of freely diffusing biomolecules, 233–53
Rayleigh scattering
 elastic, 715
RCAP
 See Raman-chirped adiabatic passage
Reaction kinetics, 165
 adiabatic vs diabatic, 159
 infrared absorption techniques for elementary gas-phase, 41–70
 radical, 42
Reactions
 barrier heights for, 16, 25, 31, 60
 of clusters with CO_2, 450–54
 of clusters with D_2, 440–46
 of clusters with O_2, 447–50
 of small transition metal cluster ions, 423–61
 in van der Waals complexes, 23
 vibrationally enhanced, 28
 See also Concerted-elimination chemical reactions; Free-bound bimolecular reactions; Gas-phase radical reactions; Insertion reactions; Mode-selective reactions; Neutral reactions; Radical-molecule reactions; Radical-radical reactions; Thermal gas-phase radical reactions
Reactivity
 mode-specific, in Cl + $CH_4(v_i)$, 157–58
 spin-orbit, in $Cl(^2P) + H_2$, 152–54
REAPDOR
 See Rotational echo by adiabatic passage double resonance sequence
Recombination
 propargyl radical, 56–57
Recoupling
 in biomolecular solid state NMR, 578–83
Rectification
 in molecular conduction, 705–6
Rectifying current/voltage behavior
 light-emitting electrochemical cells with, 407–10
 light-emitting electrochemical cells without, 404–7
Red shift
 origin of resonant, 217–19
Redfield approximation, 721
REDOR
 See Rotational-echo double-resonance protocol
Reduced density matrix approaches, 719–30
 strong thermal coupling, 726–30
 weak thermal coupling, 719–26
Reduced dimensionality quantum calculations, 17
Relaxation
 collisional vibrational, 44
 density matrix theory of, 5
 intramolecular vibrational, 260
 See also Dephasing and relaxation effects
Relaxation/cooling dynamics
 experimental determination of in quantum dots, 220–23
Relaxation dynamics of hot electrons
 energy levels and density of states, 199–202
 in fabrication, 202–3
 hot electron cooling dynamics, 206–11
 optical spectroscopy, 203–6
 in quantum dots, 211–26
 in quantum wells and superlattices, 197–211
Relaxation measurements, 48
 pulsed EPR, 297–99

REM
  See Random energy model
REMPI
  See Resonant enhanced multi photon ionization detection
Renner-Teller coupling, 7
Renormalized Coulomb repulsion, 759
Resolution
  of AFM for imaging helix turns on DNA, 81–83
  avoiding penalties of high, 483–84
  femtosecond, 255–56, 262, 264, 267–68
  limited angular, 159
  picosecond, 260–61
  single-molecule, 236–37
  See also High resolution in solids NMR
Resonances
  Feshbach, 149
  quantum dynamical, in F + H_2, 147–52
  trapped state, 149
  vibrational threshold, 147–49
Resonant coherent excitation, 766–68
  three-state systems, 767–68
  two-state systems, 766–67
  in two-state systems, 766–67
Resonant enhanced multi photon ionization (REMPI) detection, 28, 612
Resonant impulsive stimulated Raman scattering (RISRS), 266–67
Resonant multiphoton ionization, 255
Resonant red shift
  origin of, 217–19
Reverse transient grating (RTG) measurements
  negative time, 660–62
Rf couplings, 464
RFDR See Radio-frequency-driven recoupling
Rice-Ramsperger-Kassel-Marcus (RRKM) theory, 428, 607, 820–22, 828
  calculations based on, 23
  non-RRKM behavior, 825
RISRS
  See Resonant impulsive stimulated Raman scattering
RNA folding
  early events in, 751–62
  initial events in, 757–60
RNA polymerase
  imaging, 76–77
  orienting, 84
RNA-targeted conformational changes, 245
Ro-vibrational state-resolved DCSs, 155
Room temperature effects, 5
Rotating frame Hamiltonian, 469
Rotation-sequenced pulse sequences, 580
Rotational echo by adiabatic passage double resonance (REAPDOR) sequence, 487
Rotational-echo double-resonance (REDOR) protocol, 485–87, 489–90, 584, 597
Rotational energy transfer of SEP, 832–33
Rotational excitation
  correlated with vibrational excitation, 32, 35–36
Rotational ladder climbing for multiple level crossings using a pair of pulses, 780–81
Rotational spectroscopy
  of diatomic molecules, 11
Rovibrational energy disposal, 36
RRKM
  See Rice-Ramsperger-Kassel-Marcus theory
RTG
  See Reverse transient grating measurements
Rupture forces
  adiabatic, 72–73
Rydberg atom photofragment translational spectroscopy
  hydrogen, 609
Rydberg H atom TOF spectroscopy, 144, 156
Rydberg states, 188, 831

S
SAMs
  See Self-assembled monolayers
Scalar products
  NMR interactions as, 464–67
Scanning-probe lithography, 133–34
Scanning-probe microscopy, 117
Scanning tunneling microscopy (STM), 108, 682, 687, 691, 730, 735–36
Scattering
  crossed beam, 836
  elastic Rayleigh, 715
  inelastic Raman, 716
  Raman and Rayleigh, 238
  resonant impulsive stimulated Raman, 266–67

Scattering dynamics of SEP, 831–45
 from insulators, 840–41
 rotational energy transfer, 832–33
 surface, 839–40
 vibrational energy transfer, 833–39
 vibrational energy transfer on a modified metal, 844
 vibrational promotion of dissociative adsorption, 844–45
 vibrational promotion of electron transfer, 841–43
Schrödinger equation, 199, 469, 708, 733, 768
Schwarz-Slawsky-Herzfeld (SSH) theory, 834
Science
 internationality of, 12
SCRAP
 See Stark-chirped rapid adiabatic passage
SDS
 See Sodium dodecyl sulfate studies
Second-order anisotropies
 in solids NMR, 467–69
Second-order quadrupolar interactions
 manipulating, 476–78
Second-order rate constants
 measuring, of gas-phase radical kinetics, 45
SECSY
 See Spin echo correlation spectroscopy
SEDOR
 See Spin-echo double-resonance experiment
SEDRA
 See Simple excitation for the dephasing of rotational-echo amplitudes protocol

Segment B1
 of streptococcal protein G, 512–13
Selectively deuterated alkanes, 26
Self-assembled monolayers (SAMs)
 bulk transport and adsorption of, 108
 general growth mechanisms of, 108–26
 mechanisms and kinetics of formation of, 107–37
 organic acid and ion systems, 122–26
 self-organization of on the surface, 108–12
 silane, 119–22
 thiol on gold, 113–19
 vapor phase-deposited thiol films, 112–13
Self-assembly of nanostructures
 polymer adsorption driven, 537–73
Self-organization
 in growth of self-assembled monolayers, 108–12
Semiconductor quantum wells and quantum dots
 hot electron relaxation dynamics in, 193–231
SEP
 See Stimulated emission pumping
Sequence-dependent unbinding forces, 85–86
Sequence-specific DNA structures, 73–74
SF
 See Sum frequency signal intensity
SFAM
 See Simultaneous frequency and amplitude modulations
SHAKE algorithm, 506

Shaving, 134
Shift-based NMR spectra, 473
Short DNA
 on gold surfaces, 85
Signal-enhancement techniques, 482
Signal-to-noise (S/N) ratios, 478
 increasing by signal averaging, 42
$SiH_3 + SiH_3$ reaction, 55–56
Silane surfaces
 in growth of self-assembled monolayers, 119–22
Silks, 597
Simple excitation for the dephasing of rotational-echo amplitudes (SEDRA) protocol, 492
Simulation studies
 of protein folding and unfolding, 499–535
Simultaneous frequency and amplitude modulations (SFAM), 487
Single-layer light-emitting polymers
 electrochemical characterization of, 410–11
Single-molecule fluorescence detection
 and biological applications, 234–35
 determination of by, 756–57
Single-molecule studies
 ratiometric, of freely diffusing biomolecules, 233–53
Single-photon counting time-correlated, 248–49
Single pulse
 electronic ladder climbing using, 776–79

vibrational ladder climbing using, 779
Single-quantum antiphase coherences, 470
Single-spin Hamiltonians, 470
Single-walled nanotube (SWNT) tips, 86–87
Singlet PESs, 154
Size-selected photoluminescence, 217
SK
 See Stranski-Krastinow epitaxial growth
Slowed hot-electron cooling in quantum dots, 219–20
SLs
 See Superlattices
Small-molecule organic light-emitting diodes optimization of two-layer, 417–19
Small transition metal cluster ions
 reactions and thermochemistry of, 423–61
Sodium dodecyl sulfate (SDS) studies, 375–76
Sodium dodecyl sulfonate (DDS) studies, 376–79
Sodium dodecylbenzenesulfonate (DBS) studies, 378–80
Solid state NMR spectroscopy
 cross polarization in, 478–82
 selective reintroduction of spin anisotropies, 483–93
 spin-$1/2$ in, 463–98
Solids NMR, 576
 first- and second-order anisotropies, 467–69
 high resolution in, 471–78
 interactions and spectra, 464–71
 interactions as scalar products between spin and spatial tensors, 464–67
 spin evolution and the calculation of NMR spectra, 469–71
Solution electrogenerated chemiluminescence, 394–98
Solution phase transport in SAM growth, 108
Solvation dynamics
 electron, 266
Solvation energy
 Born, 395
Solvent effects
 on SAM growth mechanisms, 129–30
Solvent-separated radical-ion pairs (SSRIP), 396
Solvent viscosity
 and protein dynamics, 336–47
Spatial anisotropies
 first-order, 474
Spatial-grid based pseudopotential approaches
 in molecular conduction, 699–702
Spatial space manipulations
 averaging first-order couplings via, 474–76
Spatial tensors, 464–67
Spectra
 inhomogeneous broadening of, 320–21
 of linear and nonlinear triatomics, 7
 paramagnetic resonance, of crystals, 2
 theoretic analysis of, 4–5
 vibration-rotation, 9
Spectral diffusion, 322–23
Spectral peak position, 236
Spectrometry
 See Electron spin resonance spectrometry; Guided-ion-beam mass spectrometry
Spectroscopy
 and hot electron relaxation dynamics, 193–231
 See also Biological single-molecule spectroscopy; Cavity ringdown spectroscopy; Coherent nonlinear spectroscopy; Coincidence spectroscopy; Cold target recoil ion momentum spectroscopy; Correlation spectroscopy; Crossed-molecular-beam spectroscopy; Electron paramagnetic resonance spectroscopy; Femtosecond transition state spectroscopy; Flash kinetic spectroscopy; Fourier transform spectroscopy; Frequency-modulation spectroscopy; Infrared absorption spectroscopy; Nuclear magnetic resonance spectroscopy; Nuclear Overhauser effect spectroscopy; Optical spectroscopy; Photodetachment spectroscopy; Photofragment translational spectroscopy; Photoreflectance spectroscopy; Position sensitive translational spectroscopy; PPC spectroscopy; Ramen-induced polarization spectroscopy; Rotational spectroscopy;

Rydberg H atom TOF
  spectroscopy; Spin echo
  correlation spectroscopy;
  Sum frequency generation
  spectroscopy;
  Time-resolved infrared
  laser absorption
  spectroscopy;
  Time-resolved
  photoelectron
  spectroscopy; Transient
  infrared spectroscopy;
  Translational
  spectroscopy; Vibrational
  sum frequency
  spectroscopy;
  Wavelength-modulation
  spectroscopy; Zero-
  electron-kinetic-energy
  spectroscopy
spFRET measurements, 236,
  241–43, 245–47
Spherical drop model, 439
Spherical Herriott design,
  51
Spherical tensor methods
  irreducible, 7
Spin-$^1/_2$
  cross polarization transfers
    between, 478–80
  in solid state NMR
    spectroscopy, 463–98
Spin and spatial tensors
  NMR interactions as scalar
    products between, 464–67
Spin anisotropies
  avoiding penalties of high
    resolution, 483–84
  chemical shift/
    heteronuclear coupling
    correlations, 484–88
  chemical shift/
    homonuclear coupling
    correlations, 488–93
  selective reintroduction of,
    483–93
Spin-echo correlation
  spectroscopy (SECSY),
    298
Spin-echo double-resonance
  (SEDOR) experiment,
    485, 488–89
Spin-echo experiments
  using NMR, 315–16
Spin evolution
  and the calculation of
    NMR spectra, 469–71
Spin glass physics, 547
Spin Hamilton operator
  in pulsed EPR
    spectroscopy, 282–84
Spin-interaction parameters
  determining, 580
Spin-orbit reactivity
  in Cl(^2P) + H$_2$, 152–54
Spin-space manipulations
  averaging via, 471–74
Spinning
  magic angle, in
    biomolecular solid state
    NMR, 578–83
Splittings
  Zeeman energy, 284, 291,
    303–4
SPR
  See Surface plasmon
    resonance experiments
Squeezed atom effect, 23
src-SH3 domain, 513–15,
  518–19, 527
ss DNA, 558, 561
  gel electrophoresis on, 102
SSH
  See Schwarz-Slawsky-
    Herzfeld theory
SSRIP
  See Solvent-separated
    radical-ion pairs
Staphylococcal protein A
  fragment B of, 526–27
Stark-chirped rapid adiabatic
  passage (SCRAP),
    773–76
  experimental
    demonstration, 775–76
  theory of, 773–75
Stark effects, 5, 264, 327,
  330, 351
Stark pulse, 773–76
Stark shifting, 772–73, 775,
  795–97
Stark-Werner (SW) PES,
  149–51
State-resolved DCSs
  ro-vibrational, 155
State-selection
  electric field dissociation
    for, 9
State-specific differential
  cross sections, 139–64
State-to-state chemical
  reaction dynamics
  in polyatomic systems,
    15–39
Sticky ends
  analyzing, 74
Stimulated emission pumping
  (SEP), 765–66
  experiments with, 813–15
  femtosecond, 266
  highly vibrationally excited
    molecules with, 811–52
  scattering dynamics,
    831–45
  unimolecular dynamics,
    815–31
Stimulated photon echo
  and mode suppression,
    657
Stimulated Raman adiabatic
  passage (STIRAP)
  applications of, 798–802
  atom optics, 799–800
  background, 782–85
  Bose-Einstein
    condensation, 802
  cavity quantum
    electrodynamics, 801–2
  control of chemical
    reactions, 798
  dependence on the

interaction parameters,
786–87
intuitive vs counterintuitive
pulse order, 785–86
laser cooling, 800
measurement of weak
magnetic fields, 800–1
photoassociative, 802
in three-state systems,
782–87
Stimulated Raman adiabatic
passage (STIRAP) in
three-state systems,
787–92
adiabatic passage by
light-induced potentials,
797–98
creation of coherent
superpositions by
modifications of,
798
with degenerate levels,
791–92
experimental
demonstration with cw
lasers, 787–88
experimental
demonstration with
pulsed lasers, 789–91
extensions of, 792–98
hyper-Raman, 796–97
multi-state chains, 792–94
population transfer via
continuum, 795–96
Stimulated Raman scattering
resonant impulsive, 266–67
STIRAP
See Stimulated Raman
adiabatic passage
STM
See Scanning tunneling
microscopy
Stochastic Liouville equation,
282
Stoichiometries
of DNA-protein
interactions, 75–77

Stokes laser, 787–88, 791,
794–95
Stokes photon, 765
Stokes pulse, 766–67,
782–83
Stokes transition, 776
Stranski-Krastinow (SK)
epitaxial growth
synthesis of quantum dots
by, 214–16, 221
Streptococcal protein G
segment B1 of, 512–13
Stretched molecules, 64,
811–52
CH, 382, 816
DNA, 83, 85–86
HF, 614, 632
NO, 812
OH, 362–63, 368,
371
Strong thermal coupling,
726–30
Submonolayer island
nucleation, growth, and
size distributions
in SAM growth
mechanisms, 132–33
Subpopulations
Förster distance
dependence and, 241–42
Sum frequency (SF)
generation spectroscopy,
117
Sum frequency (SF) signal
intensity, 358–59, 366
Sum rejection criterion,
239–40
Superlattices (SLs), 199
fabrication of, 202–3
hot electron cooling
dynamics in, 206–11
optical spectroscopy of,
203–6
relaxation dynamics of hot
electrons in, 197–211
Supermolecules, 698–99
Surface biology, 72

of DNA, studying by
atomic force microscopy,
71–92
Surface plasmon resonance
(SPR) experiments, 114
Surface scattering dynamics
of SEP, 839–40
Surfaces
biomimetic recognition
between polymers and,
539–57
DNA equilibrium vs
kinetic trapping on,
84–85
orienting and elongating
DNA molecules on, 84
preparing for DNA
binding, 84
self-organization of
self-assembled
monolayers on, 108–12
short DNA on gold, 85
susceptibility of, 359
Surfactant adsorption, 368–72
Surfactant molecules
amphiphilic, 107–8
Surfactant studies
at the air-water interface,
374–76, 378–80
comparative, at both
liquid-liquid and
liquid-air interfaces,
378–80
at liquid-liquid interfaces,
376–80
at liquid surfaces, 373–80
Surprisal analysis, 21, 23
SW
See Stark-Werner PES
SWNT
See Single-walled
nanotube tips
Synthesis of quantum dots,
213–16
colloidal, 213–14
by Stranski-Krastinow
epitaxial growth, 214–16

SUBJECT INDEX 927

T

TAE
  See Tris-acetate-EDTA
Tannor-Rice pump-dump
    scheme
  optimization of, 640
TBE
  See Tris-borate-EDTA
TDSCF
  See Time-dependent,
    self-consistent field
    approximation
TEDOR
  See Transferred-echo
    double-resonance
    protocol
Temperature dependence, 48,
    60–62
  of dynamic isomerization,
    5
  power law, 334–35
  and the protein-glass
    transition, 331–36
  pure-dephasing, 327–28,
    331
Tensors
  NMR interactions as scalar
    products between spin
    and spatial, 464–67
Tetrahymena ribozyme
  folds by the KPM, 755–56
  kinetics of independently
    folding subdomains of,
    754–55
Tetramethylrhodamine
    (TMR) dye, 238
TG
  See Transient grating
    method
Theories
  See Debye-Huckel theory;
    Density matrix theory of
    relaxation; Marcus theory
    for electron transfer;
    Rice-Ramsperger-Kassel-
    Marcus theory;
    Schwarz-Slawsky-

Herzfeld theory;
  Time-independent
    perturbation theory;
  Viscoelastic-continuum
    theory of solvent
    dynamics
Thermal coupling
  strong, 726–30
  weak, 719–26
Thermal gas-phase radical
    reactions
  vibrational energy disposal
    in, 62–65
Thermal interactions
  in molecular conduction,
    716–19
Thermal kinetics
    measurements
  comparing, 54
Thermal rate coefficients, 53
  relative, 54
Thermal-relaxation process,
    726
Thermochemistry
  of small transition metal
    cluster ions, 423–61
Thermodynamic simulations,
    520–23
  free energy profiles relative
    to folding transition
    temperature, 522–23
  two-state thermodynamics
    of folding, 521–22
Thermodynamics
  bulk phase, of transition
    metal clusters, 439–40
Thiol films
  vapor phase-deposited,
    112–13
Thiol on gold
  in growth of self-assembled
    monolayers, 113–19
Three-body dissociation
    dynamics, 168, 185–89
3-D arrays
  quantum dot, 224
Three-state systems

resonant coherent
  excitation in, 767–68
Threshold
  photoelectron-photoion
    coincidence (TPEPICO)
    experiments, 180
Threshold resonances
  vibrational, 147–49
Time-correlated
  single-photon counting,
    248–49
Time-dependent,
  self-consistent field
    (TDSCF)
    approximation, 718
Time-dependent signals, 464
Time-independent
  perturbation theory, 468
Time-of-flight (TOF)
  measurements, 140–42,
    609
Time-resolved, off-resonance
  TG measurements, 665
Time-resolved dynamics, 640
  photoelectron-photoion
    energy and angular
    distributions, 182–85
Time-resolved infrared laser
  absorption spectroscopy,
    42
  experimental techniques,
    49–51
  for gas-phase radical
    kinetics, 43–51
Time-resolved PD (TRPD)
  experiments, 432, 436
Time-resolved photoelectron
  angular distributions,
    271–73
Time-resolved photoelectron
  spectroscopy (TRPES)
  of molecules and clusters,
    255–77
  of negative ions, 266–70
  theoretical treatment of,
    270–71
Time scales

of SAM growth
    mechanisms, 126–28
TIR
    See Total internal reflection
    mechanisms
TMA
    See Trimethyl aluminum
TMG
    See Trimethyl gallium
TMI
    See Trimethyl indium
TMR
    See Tetramethylrhodamine
    dye
TOF
    See Time-of-flight
    measurements
Torsional transitions, 335
Total internal reflection (TIR)
    mechanisms, 360, 364
TPEPICO
    See Threshold
    photoelectron-photoion
    coincidence experiments
TPPM
    See Two-pulse phase
    modulated scheme
Transfer
    See Energy transfer
Transferred-echo
    double-resonance
    (TEDOR) protocol, 487
Transient birefringence
    method
    picosecond, 665
Transient digitization
    of CW laser signal, 43
Transient EPR
    correlated radical pairs,
    299–301
Transient grating (TG)
    method
    positive time, 661
Transient infrared
    spectroscopy, 58
Transients, 830–31
Transition metal cluster ions

data analysis, 427–29
experimental methods, 427
instrumentation, 427
reactions and
    thermochemistry of small,
    423–61
sources of, 426–27
Transition metal clusters
    approach to bulk phase
    thermodynamics of,
    439–40
bond energies of anionic,
    436–37
bond energies of cationic,
    430–34
dissociation pathways in,
    429–30
electronic and geometric
    structures of, 437–38
stabilities and structures of,
    429–40
Transition metal oxyanions
    early research on, 2
Transition probabilities
    estimating, 771–72
Transition-state barrier
    for folding, 511
Transition-state
    delocalization, 511
Transition-state dynamics
    $OH + H_2O \rightarrow H_2O + OH$,
    176–79
Transition temperature
    folding and unfolding at,
    523–24
Transitions
    radiationless, 259
Translational spectroscopy
    dissociation energies
    determined from, 634
    as a probe of vibrational
    predissociation, 609–12
Trapped state resonance, 149
Trapping desorption
    mechanism, 840
Triatomics
    isomerization in, 824–25

Trimethyl aluminum (TMA),
    203
Trimethyl gallium (TMG),
    203
Trimethyl indium (TMI),
    203
Tris-acetate-EDTA (TAE)
    oligomeric mobility in, 97
Tris-borate-EDTA (TBE)
    buffers, 94, 97–98
TRPD
    See Time-resolved PD
    (TRPD) experiments
TRPES
    See Time-resolved
    photoelectron
    spectroscopy
Tunneling traversal times,
    709–12
Two-dimensional DNA
    crystals, 83
Two-layer small-molecule
    organic light-emitting
    diodes
    optimization of, 417–19
Two-photon infrared method
    dependent on
    photodissociation, 10
Two-pulse phase modulated
    (TPPM) scheme, 472
Two-state systems, 768–73
    adiabatic states and
    adiabatic following, 769
    coherent excitation in,
    768–69
    estimating transition
    probabilities, 771–72
    experimental
    demonstration, 772–73
    incoherent excitation of,
    764
    rapid adiabatic passage,
    770–71
    resonant coherent
    excitation in, 766–67
    thermodynamics of
    folding, 521–22

SUBJECT INDEX 929

U

Unbinding forces
  sequence-dependent,
  85–86
Unfolding simulations
  folding and unfolding at
    the transition temperature,
    523–24
  folding under native
    conditions, 523
  probing protein folding
    through, 515–19
  studies of protein, 499–535
  unfolding at high
    temperatures, 524–25
  used to study folding,
    523–25
Unfolding trajectories
  and folding potentials of
    mean force, 526–27
Unimolecular dissociation
  dynamics
  nonstatistical, 825–27
Unimolecular dynamics of
  SEP, 815–31
  intramolecular vibrational
    redistribution, 815–24
  isomerization in triatomics,
    824–25
  nonstatistical unimolecular
    dissociation dynamics,
    825–27
  studies on clusters, 828–30
  transients, 830–31
UV excitation, 609

V

V-R,T
  See Vibration-translation
    and rotation components
V-V
  See Vibration-vibration
    components
V-V energy transfer, 621
van der Waals attraction,
  759
van der Waals complexes

reactions in, 23, 650–51,
  815
van der Waals vibrational
  modes, 830
Vapor phase-deposited thiol
  films
  in growth of self-assembled
    monolayers, 112–13
Vapor-water interface
  vibrational sum frequency
    spectroscopy of, 361–63
Varandas-Pais potential,
  838
Velocity-dependent
  interactions, 767
Verdier-Stockmayer
  algorithm, 550
Vertical electronic binding
  energy, 842
Vibration-rotation spectra
  studies of, 9
Vibration/rotation/tunneling
  (VRT) spectroscopy,
  608
Vibration-translation and
  rotation (V-R,T)
  components, 834–35
Vibration-vibration (V-V)
  components, 834–37
Vibrational density of states,
  821
Vibrational dephasing,
  334
Vibrational dynamics and
  coherence
  bound-bound molecular
    transitions, 652–61
Vibrational echo decay, 318,
  346
  of Mb-CO, 325–26
Vibrational echo experiments,
  316–52
  on hemoglobin-CO,
    347–51
  infrared, 315–56
  procedures, 323–25
Vibrational energy disposal

in thermal gas-phase
  radical reactions, 62–65
Vibrational energy transfer,
  48
  on modified metal, 844
  of SEP, 833–39
Vibrational excitation, 28
  correlated with rotational
    excitation, 32, 35–36
  nonequilibrium, 47
Vibrational-frequency
  transition, 320
Vibrational ladder climbing
  for multiple level
    crossings
  using a pair of pulses,
    779–80
  using a single pulse,
    779
Vibrational predissociation,
  608, 627
  translational spectroscopy
    as a probe of, 609–12
Vibrational promotion of
  dissociative adsorption,
  844–45
Vibrational promotion of
  electron transfer
  interactions with metals,
    841–43
Vibrational relaxation
  enhanced, 837–38
  intramolecular, 260
Vibrational sum frequency
  spectroscopy (VSFS),
  358–61
  experimental
    considerations, 360–61
  of interfacial H_2O, 366–68
  of surfactants, 374
  of the vapor-water
    interface, 361–63
Vibrational threshold
  resonances, 147–49
Vibrationally adiabatic barrier
  heights, 25
Vibrationally excited

molecules highly, with stimulated emission pumping, 811–52
Vibrationally induced dissociation, 631
Viral inhibition experiments, 556
Viscoelastic-continuum theory of solvent dynamics, 336–37, 341–47
Viscosity dependence, 336–77
Vogel-Tammann-Fulcher (VTF) equation modeling, 324
Voigt profiles, 367
Voltammograms, 399–400, 414
VRT
   See Vibration/rotation/tunneling spectroscopy
VSFS
   See Vibrational sum frequency spectroscopy

W

Walch-Dunning-Schatz-Elgersma potential, 17
Water hydrogen bonding at the organic-liquid/water interface, 363–66
   See also Interfacial water
Wave packet dynamics, 263
Wave packet QM calculations, 155
Wavelength-modulation (WM) spectroscopy, 50
Weak magnetic fields measurement of, 800–1
Weak thermal coupling, 719–26
Weakly bound complexes photofragment translational spectroscopy of, 607–37
Weighted histogram analysis method (WHAM), 509
Wells
   See Quantum wells and quantum dots
WHAM
   See Weighted histogram analysis method
White cell overlap, 51
Wigner D-functions, 145
Wigner rotation matrix, 467

X

X-ray crystallography, 330

Z

Zeeman coherence, 801–2
Zeeman couplings, 464, 468
Zeeman degeneracy, 791
Zeeman effect, 7, 10, 470
Zeeman energy splittings, 284, 291, 303–4
ZEKE
   See Zero-electron-kinetic-energy spectroscopy
Zero-electron-kinetic-energy (ZEKE) spectroscopy, 258–61
Zero-order bright states (ZOBSs), 816–18
ZOBSs
   See Zero-order bright states

CUMULATIVE INDEXES

CONTRIBUTING AUTHORS, VOLUMES 48–52

Abramson EH, 50:279–313
Aksay IA, 51:601–22
Anderson JB, 51:501–26
Andreoni W, 49:405–39
Armentrout PB, 52:423–61
Armstrong NR, 52:391–422

Balucani N, 50:347–76
Bartell LS, 49:43–72
Bashford D, 51:129–52
Beck C, 50:443–84
Benjamin I, 48:407–51
Bergmann K, 52:763–809
Blomberg MRA, 50:221–49
Boato G, 50:23–50
Boxer SG, 48:213–42
Brockman JM, 51:41–63
Brooks CL III, 52:499–535
Brown JM, 50:279–313
Bublitz GU, 48:213–42
Buckingham AD, 49:xiii–xxxv
Bürgi HB, 51:275–96
Butler LJ, 49:125–71

Callender RH, 49:173–202
Callis PR, 48:271–97
Campbell EEB, 51:65–98
Carrington A, 52:1–13
Carter EA, 48:243–70
Casavecchia P, 50:347–76
Case DA, 51:129–52
Chakraborty AK, 52:537–73
Chapovsky PL, 50:315–45
Cheatham TE III, 51:435–71
Chemla DS, 52:233–53
Collier CP, 49:371–404
Continetti RE, 52:165–92
Corkum PB, 48:387–406

Corn RM, 51:41–63
Cukier RI, 49:337–69

Dabbs DM, 51:601–22
Dagdigian PJ, 48:95–123
Dahan M, 52:233–53
Dantus M, 52:639–79
De Boeij WP, 49:99–123
Deniz AA, 52:233–53
De Pablo JJ, 50:377–411
Dlott DD, 50:251–78
Donley EA, 48:181–212
Duncan MA, 48:69–93
Dyer RB, 49:173–202

Eachus RS, 50:117–44
Ediger MD, 51:99–128
Emmett MR, 50:517–36
Escobedo FA, 50:377–411

Farantos SC, 50:443–84
Fayer MD, 52:315–56
Field RW, 50:443–84; 52:811–52
Flynn GW, 49:297–336
Freed JH, 51:655–89
Frydman L, 52:463–98

Gallagher SC, 51:355–80
Gallicchio E, 49:531–67
Garashchuk S, 51:553–600
Ge N-H, 48:711–44
Giancarlo LC, 49:297–336
Gilmanshin R, 49:173–202
Goldbeck RA, 48:453–79
Golumbfskie AJ, 52:537–73
Goodman DW, 48:43–68
Gordon MS, 49:233–66
Gordon RJ, 48:601–41

Gross EM, 52:391–422
Gruebele M, 50:485–516
Gutman M, 48:329–56

Halberstadt N, 51:405–33
Halfmann T, 52:763–809
Hall GE, 51:243–74
Hansen J-P, 51:209–42
Hansma HG, 52:71–92
Harata A, 50:193–219
Harris CB, 48:711–44
Heath JR, 49:371–404
Hemley RJ, 51:763–800
Hendrickson CL, 50:517–36
Hermans LJF, 50:315–45
Herschbach D, 51:1–39
Hershberger JF, 52:41–70
Ho T-S, 50:537–70
Hollebeek T, 50:537–70
Hudson PK, 51:473–99

Ishikawa H, 50:443–84
Ivanov MYu, 48:387–406

Janda KC, 51:405–33
Jarrold MF, 51:179–207
Johnson MR, 51:297–321
Jongma R, 52:811–52
Jónsson H, 51:623–53
Joyeux M, 50:443–84

Kamins TI, 51:527–51
Kearley GJ, 51:297–321
Keske JC, 51:323–53
Kim-Shapiro DB, 48:453–79
Kliger DS, 48:453–79
Klimov DK, 52:751–62
Knickelbein MB, 50:79–115
Knoll W, 49:569–638

931

Kollman PA, 51:435–71
Kondow T, 51:731–61
Koput J, 50:443–84

Larsson M, 48:151–79
Laurence TA, 52:233–53
Leach S, 48:1–41
Lee N, 52:751–62
Lester MI, 48:643–73
Levine RD, 51:65–98
Levy RM, 49:531–67
Lingle RL Jr, 48:711–44
Liu K, 52:139–64
Loomis RA, 48:643–73
Löwen H, 51:209–42
Lüchow A, 51:501–26
Luthey-Schulten Z, 48:545–600

MacMillan F, 52:279–313
Mafuné F, 51:731–61
Makri N, 50:167–91
Marchetti AP, 50:117–44
McLaughlin L, 52:93–106
McNeill JD, 48:711–44
Medeiros-Ribeiro G, 51:527–51
Merkt F, 48:675–709
Miller RE, 52:607–37
Möbius K, 48:745–84
Mohanty U, 52:93–106
Muenter AA, 50:117–44
Mukamel S, 51:691–729
Murad E, 49:73–98
Myers AB, 49:267–95

Nachliel E, 48:329–56
Nafie LA, 48:357–86
Nakamura H, 48:299–328
Nelson BP, 51:41–63
Neumark DM, 52:255–77
Nielson JB, 48:511–44
Nikitin EE, 50:1–22

Nitzan A, 52:681–750
Nocera DG, 49:337–69
North SW, 51:243–74
Nozik AJ, 52:193–231

Ohlberg DAA, 51:527–51
Onuchic JN, 48:545–600
Oudejans L, 52:607–37

Pate BH, 51:323–53
Perrin CL, 48:511–44
Peter T, 48:785–822
Plakhotnik T, 48:181–212
Pratt DW, 49:481–530
Prenni AJ, 51:473–99
Price DL, 50:571–601
Prisner T, 52:279–313
Pshenichnikov MS, 49:99–123

Rabitz H, 50:537–70
Radeke MR, 48:243–70
Rice SA, 48:601–41
Richmond GL, 52:357–89
Rohrbacher A, 51:405–33
Rohrer M, 52:279–313
Ross J, 50:51–78

Sawada T, 50:193–219
Schinke R, 50:443–84
Schmidt MW, 49:233–66
Schultz PG, 52:233–53
Schwartz DK, 52:107–37
Shea J-E, 52:499–535
Shen Q, 50:193–219
Shore BW, 52:763–809
Siegbahn PEM, 50:221–49
Silva M, 52:811–52
Slutsky LJ, 50:279–313
Spangler LH, 48:481–510
Stehlik D, 48:745–84
Steinfeld JI, 49:203–32
Street SC, 48:43–68

Taatjes CA, 52:41–70
Tannor DJ, 51:553–600
Thirumalai D, 52:751–62
Toennies JP, 49:1–41
Tolbert MA, 51:473–99
Trautman JK, 49:441–80
Trenary M, 51:381–403
Trewhella J, 51:355–80
Trouw FR, 50:571–601
Tully JC, 51:153–78
Tycko R, 52:575–606

Valentini JJ, 52:15–39
Vilesov AF, 49:1–41
Vitanov NV, 52:763–809
Vlachy V, 50:145–65
Vlad MO, 50:51–78
Volpi GG, 50:23–50, 347–76
Vossmeyer T, 49:371–404
Vyazovkin S, 48:125–49

Wall ME, 51:355–80
Weiss S, 52:233–53
Wiersma DA, 49:99–123
Wight CA, 48:125–49
Wightman RM, 52:391–422
Wild UP, 48:181–212
Williams RS, 51:527–51
Wodtke AM, 52:811–52
Wolkow RA, 50:413–41
Wolynes PG, 48:545–600
Wong CM, 48:711–44
Woodruff WH, 49:173–202
Woodson SA, 52:751–62
Wormhoudt J, 49:203–32
Wright JS, 48:387–406

Xie XS, 49:441–80
Xu C, 48:43–68

Yan Q, 50:377–411

Zondlo MA, 51:473–99

CHAPTER TITLES, VOLUMES 48–52

Biophysical Chemistry

Stark Spectroscopy: Applications in Chemistry, Biology, and Materials Science	GU Bublitz, SG Boxer	48:213–42
Two-Photon-Induced Fluorescence	PR Callis	48:271–97
Time-Resolved Dynamics of Proton Transfer in Proteinous Systems	M Gutman, E Nachliel	48:329–56
"Strong" Hydrogen Bonds in Chemistry and Biology	CL Perrin, JB Nielson	48:511–44
Theory of Protein Folding: The Energy Landscape Perspective	JN Onuchic, Z Luthey-Schulten, PG Wolynes	48:545–600
Fast Events in Protein Folding: The Time Evolution of Primary Processes	RH Callender, RB Dyer, R Gilmanshin, WH Woodruff	49:173–202
Molecular Electronic Spectral Broadening in Liquids and Glasses	AB Myers	49:267–95
Computer Simulations with Explicit Solvent: Recent Progress in the Thermodynamic Decomposition of Free Energies, and in Modeling Electrostatic Effects	RM Levy, E Gallicchio	49:531–67
Density Functional Theory of Biologically Relevant Metal Centers	PEM Siegbahn, MRA Blomberg	50:221–49
The Fast Protein Folding Problem	M Gruebele	50:485–516
Peptides and Proteins in the Vapor Phase	MF Jarrold	51:179–207
Large-Scale Shape Changes in Proteins and Macromolecular Complexes	ME Wall, SC Gallagher, J Trewhella	51:355–80
Molecular Dynamics Simulation of Nucleic Acids	TE Cheatham III, PA Kollman	51:435–71
On the Characteristics of Migration of Oligomeric DNA in Polyacrylamide Gels and in Free Solution	U Mohanty, L McLaughlin	52:93–106

From Folding Theories to Folding Proteins: A Review and Assessment of Simulation Studies of Protein Folding and Unfolding	J-E Shea, CL Brooks III	52:499–535
Biomolecular Solid State NMR: Advances in Structural Methodology and Applications to Peptide and Protein Fibrils	R Tycko	52:575–606
Early Events in RNA Folding	D Thirumalai, N Lee, SA Woodson, DK Klimov	52:751–62

Chemical Kinetics–Reactions

Kinetics in Solids	S Vyazovkin, CA Wight	48:125–49
Chemical Reaction Dynamics Beyond the Born-Oppenheimer Approximation	LJ Butler	49:125–71
Nonlinear Kinetics and New Approaches to Complex Reaction Mechanisms	J Ross, MO Vlad	50:51–78
Reactions of Transition Metal Clusters with Small Molecules	MB Knickelbein	50:79–115
Constructing Multi-Dimensional Molecular Potential Energy Surfaces from Ab Initio Data	T Hollebeek, T-S Ho, H Rabitz	50:537–70
Delayed Ionization and Fragmentation En Route to Thermionic Emission: Statistics and Dynamics	EEB Campbell, RD Levine	51:65–98
The Dynamics of Noble Gas-Halogen Molecules and Clusters	A Rohrbacher, N Halberstadt, KC Janda	51:405–33
Semiclassical Calculation of Chemical Reaction Dynamics via Wavepacket Correlation Functions	DJ Tannor, S Garashchuk	51:553–600

Chemical Kinetics–State-to-State

State-Resolved Collision-Induced Electronic Transitions	PJ Dagdigian	48:95–123
Dissociative Recombination with Ion Storage Rings	M Larsson	48:151–79
Active Control of the Dynamics of Atoms and Molecules	RJ Gordon, SA Rice	48:601–41
Proton-Coupled Electron Transfer	RI Cukier, DG Nocera	49:337–69
Crossed-Beam Studies of Reaction Dynamics	P Casavecchia, N Balucani, GG Volpi	50:347–76

HCP ↔ CPH Isomerization: Caught in the Act	H Ishikawa, RW Field, SC Farantos, M Joyeux, J Koput, C Beck, R Schinke	50:443–84
State-to-State Chemical Reaction Dynamics in Polyatomic Systems: Case Studies	JJ Valentini	52:15–39
Recent Progress in Infrared Absorption Techniques for Elementary Gas-Phase Reaction Kinetics	CA Taatjes, JF Hershberger	52:41–70
Crossed-Beam Studies of Neutral Reactions: State-Specific Differential Cross Sections	K Liu	52:139–64
Photofragment Translational Spectroscopy of Weakly Bound Complexes: Probing the Interfragment Correlated Final State Distributions	L Oudejans, RE Miller	52:607–37
The Dynamics of "Stretched Molecules": Experimental Studies of Highly Vibrationally Excited Molecules with Stimulated Emission Pumping	M Silva, R Jongma, RW Field, AM Wodtke	52:811–52

Colloids

Effective Interactions Between Electric Double Layers	J-P Hansen, H Löwen	51:209–42

Electrochemistry

Light-Emitting Electrochemical Processes	NR Armstrong, RM Wightman, EM Gross	52:391–422

Geochemistry and Cosmochemistry

Microphysics and Heterogeneous Chemistry of Polar Stratospheric Clouds	T Peter	48:785–822
The Shuttle Glow Phenomenon	E Murad	49:73–98
Chemistry and Microphysics of Polar Stratospheric Clouds and Cirrus Clouds	MA Zondlo, PK Hudson, AJ Prenni, MA Tolbert	51:473–99

Laser Chemistry

Subfemtosecond Processes in Strong Laser Fields	PB Corkum, MYu Ivanov, JS Wright	48:387–406
Photothermal Applications of Lasers: Study of Fast and Ultrafast Photothermal Phenomena at Metal-Liquid Interfaces	A Harata, Q Shen, T Sawada	50:193–219
Ultrafast Spectroscopy of Shock Waves in Molecular Materials	DD Dlott	50:251–78
Transient Lasar Frequency Modulation Spectroscopy	GE Hall, SW North	51:243–74

Liquid State

Simulation of Phase Transitions in Fluids	JJ de Pablo, Q Yan, FA Escobedo	50:377–411
Generalized Born Models of Macromolecular Solvation Effects	D Bashford, DA Case	51:129–52
Structures and Dynamics of Molecules on Liquid Beam Surfaces	T Kondow, F Mafuné	51:731–61
Coherent Nonlinear Spectroscopy: From Femtosecond Dynamics to Control	M Dantus	52:639–79

Magnetic Resonance

New EPR Methods for Investigating Photoprocesses with Paramagnetic Intermediates	D Stehlik, K Möbius	48:745–84
Nuclear Spin Conversion in Polyatomic Molecules	PL Chapovsky, LJF Hermans	50:315–45
New Technologies in Electron Spin Resonance	JH Freed	51:655–89
Pulsed EPR Spectroscopy: Biological Applications	T Prisner, M Rohrer, F MacMillan	52:279–313
Spin-1/2 and Beyond: A Perspective in Solid State NMR Spectroscopy	L Frydman	52:463–98

Miscellaneous

Explosives Detection: A Challenge for Physical Chemistry	JI Steinfeld, J Wormhoudt	49:203–32
Electrospray Ionization Fourier Transform Ion Cyclotron Resonance Mass Spectrometry	CL Hendrickson, MR Emmett	50:517–36
Effects of High Pressure on Molecules	RJ Hemley	51:763–800

Physical Organic

"Strong" Hydrogen Bonds in Chemistry and Biology	CL Perrin, JB Nielson	48:511–44

Polymers and Macromolecules

Interfaces and Thin Films as Seen by Bound Electromagnetic Waves	W Knoll	49:569–638
Spatially Heterogeneous Dynamics in Supercooled Liquids	MD Ediger	51:99–128
Generalized Born Models of Macromolecular Solvation Effects	D Bashford, DA Case	51:129–52
Self-Assembled Ceramics Produced by Complex-Fluid Templation	DM Dabbs, IA Aksay	51:601–22
Polymer Adsorption-Driven Self-Assembly of Nanostructures	AK Chakraborty, AJ Golumbfskie	52:537–73

Prefatory Chapters

In My Time: Scenes of Scientific Life	S Leach	48:1–41
Molecules in Optical, Electric, and Magnetic Fields: A Personal Perspective	AD Buckingham	49:xiii–xxxv
Nonadiabatic Transitions: What We Learned from Old Masters and How Much We Owe Them	EE Nikitin	50:1–21
Experiments on the Dynamics of Molecular Processes: A Chronicle of Fifty Years	G Boato, GG Volpi	50:23–50
Fifty Years in Physical Chemistry: Homage to Mentors, Methods, and Molecules	D Herschbach	51:1–39
A Free Radical	A Carrington	52:1–13

Quantum Chemistry

Ab Initio Dynamics of Surface Chemistry	MR Radeke, EA Carter	48:243–70
Theoretical Studies of Chemical Dynamics: Overview of Some Fundamental Mechanisms	H Nakamura	48:299–328
The Construction and Interpretation of MCSCF Wavefunctions	MW Schmidt, MS Gordon	49:233–66
Computational Approach to the Physical Chemistry of Fullerenes and Their Derivatives	W Andreoni	49:405–35
Time-Dependent Quantum Methods for Large Systems	N Makri	50:167–91

Chemical Dynamics at Metal Surfaces	JC Tully	51:153–78
Monte Carlo Methods in Electronic Structures for Large Systems	A Lüchow, JB Anderson	51:501–26
Semiclassical Calculation of Chemical Reaction Dynamics via Wavepacket Correlation Functions	DJ Tannor, S Garashchuk	51:553–600
Laser-Induced Population Transfer by Adiabatic Passage Techniques	NV Vitanov, T Halfmann, BW Shore, K Bergmann	52:763–809

Scattering–Elastic and Inelastic

Structure and Transformation: Large Molecular Clusters as Models of Condensed Matter	LS Bartell	49:43–72
Chemical Applications of Neutron Scattering	FR Trouw, DL Price	50:571–601
Motion and Disorder in Crystal Structure Analysis: Measuring and Distinguishing Them	HB Bürgi	51:275–96
Quantitative Atom-Atom Potentials from Rotational Tunneling: Their Extraction and Their Use	MR Johnson, GJ Kearley	51:297–321

Solids and Ordered Arrays

Nanocrystal Superlattices	CP Collier, T Vossmeyer, JR Heath	49:371–404
The Photophysics of Silver Halide Imaging Materials	RS Eachus, AP Marchetti, AA Muenter	50:117–44
Applications of Impulsive Stimulated Scattering in the Earth and Planetary Sciences	EH Abramson, JM Brown, LJ Slutsky	50:279–313
Surface Plasmon Resonance Imaging Measurements of Ultrathin Organic Films	JM Brockman, BP Nelson, RM Corn	51:41–63
Thermodynamics of the Size and Shape of Nanocrystals: Epitaxial Ge on Si(001)	RS Williams, G Medeiros-Ribeiro, TI Kamins, DAA Ohlberg	51:527–51

Self-Assembled Ceramics Produced by Complex-Fluid Templation	DM Dabbs, IA Aksay	51:601–22
Theoretical Studies of Atomic-Scale Processes Relevant to Crystal Growth	H Jónsson	51:623–53
Spectroscopy and Hot Electron Relaxation Dynamics in Semiconductor Quantum Wells and Quantum Dots	AJ Nozik	52:193–231

Spectroscopy–Infrared, Raman, and Electronic

Spectroscopy of Metal Ion Complexes: Gas-Phase Models for Solvation	MA Duncan	48:69–93
Single-Molecule Spectroscopy	T Plakhotnik, EA Donley, UP Wild	48:181–212
Stark Spectroscopy: Applications in Chemistry, Biology, and Materials Science	GU Bublitz, SG Boxer	48:213–42
Infrared and Raman Vibrational Optical Activity: Theoretical and Experimental Aspects	LA Nafie	48:357–86
Fast Natural and Magnetic Circular Dichroism Spectroscopy	RA Goldbeck, DB Kim-Shapiro, DS Kliger	48:453–79
Structural Information from Methyl Internal Rotation Spectroscopy	LH Spangler	48:481–510
$OH-H_2$ Entrance Channel Complexes	RA Loomis, MI Lester	48:643–73
Molecules in High Rydberg States	F Merkt	48:675–709
Spectroscopy of Atoms and Molecules in Liquid Helium	JP Toennies, AF Vilesov	49:1–41
Ultrafast Solvation Dynamics Explored by Femtosecond Photon Echo Spectroscopies	WP de Boeij, MS Pshenichnikov, DA Wiersma	49:99–123
Explosives Detection: A Challenge for Physical Chemistry	JI Steinfeld, J Wormhoudt	49:203–32
Molecular Electronic Spectral Broadening in Liquids and Glasses	AB Myers	49:267–95
Optical Studies of Single Molecules at Room Temperature	XS Xie, JK Trautman	49:441–80
High Resolution Spectroscopy in the Gas Phase: Even Large Molecules Have Well-Defined Shapes	DW Pratt	49:481–530
Transient Laser Frequency Modulation Spectroscopy	GE Hall, SW North	51:243–74
Quantitative Atom-Atom Potentials from Rotational Tunneling: Their Extraction and Their Use	MR Johnson, GJ Kearley	51:297–321

Decoding the Dynamical Information Embedded in Highly Mixed Quantum States	JC Keske, BH Pate	51:323–53
Reflection Absorption Infrared Spectroscopy and the Structure of Molecular Adsorbates on Metal Surfaces	M Trenary	51:381–403
Multidimensional Femtosecond Correlation Spectroscopies of Electronic and Vibrational Excitations	S Mukamel	51:691–729
Coincidence Spectroscopy	RE Continetti	52:165–92
Ratiometric Single-Molecule Studies of Freely Diffusing Biomolecules	AA Deniz, TA Laurence, M Dahan, DS Chemla, PG Schultz, S Weiss	52:233–53
Time-Resolved Photoelectron Spectroscopy of Molecules and Clusters	DM Neumark	52:255–77
Fast Protein Dynamics Probed with Infrared Vibrational Echo Experiments	MD Fayer	52:315–56

Statistical Mechanics

Structure and Transformation: Large Molecular Clusters as Models of Condensed Matter	LS Bartell	49:43–72
Ionic Effects Beyond Poisson-Boltzmann Theory	V Vlachy	50:145–65

Surfaces

The Physical and Chemical Properties of Ultrathin Oxide Films	SC Street, C Xu, DW Goodman	48:43–68
Molecular Structure and Dynamics at Liquid-Liquid Interfaces	I Benjamin	48:407–51
Femtosecond Dynamics of Electrons on Surfaces and at Interfaces	CB Harris, N-H Ge, RL Lingle Jr, JD McNeill, CM Wong	48:711–44
Scanning Tunneling and Atomic Force Microscopy Probes of Self-Assembled, Physisorbed Monolayers: Peeking at the Peaks	LC Giancarlo, GW Flynn	49:297–336
Interfaces and Thin Films as Seen by Bound Electromagnetic Waves	W Knoll	49:569–638
Controlled Molecular Adsorption on Silicon: Laying a Foundation for Molecular Devices	RA Wolkow	50:413–41

Surface Plasmon Resonance Imaging Measurements of Ultrathin Organic Films	JM Brockman, BP Nelson, RM Corn	51:41–63
Chemical Dynamics at Metal Surfaces	JC Tully	51:153–78
Reflection Absorption Infrared Spectroscopy and the Structure of Molecular Adsorbates on Metal Surfaces	M Trenary	51:381–403
Structures and Dynamics of Molecules on Liquid Beam Surfaces	T Kondow, F Mafuné	51:731–61
Surface Biology of DNA by Atomic Force Microscopy	HG Hansma	52:71–92
Mechanisms and Kinetics of Self-Assembled Monolayer Formation	DK Schwartz	52:107–37
Structure and Bonding of Molecules at Aqueous Surfaces	GL Richmond	52:357–89
Electron Transmission through Molecules and Molecular Interfaces	A Nitzan	52:681–750

Thermochemistry and Thermodynamics

Computer Simulations with Explicit Solvent: Recent Progress in the Thermodynamic Decomposition of Free Energies and in Modeling Electrostatic Effects	RM Levy, E Gallicchio	49:531–67
Reactions and Thermochemistry of Small Transition Metal Cluster Ions	PB Armentrout	52:423–61

William F. Maag Library
Youngstown State University